CURRENT TOPICS IN IONIZING RADIATION RESEARCH

Edited by **Mitsuru Nenoi**

Current Topics in Ionizing Radiation Research
Edited by Mitsuru Nenoi

Published by InTech
Janeza Trdine 9, 51000 Rijeka, Croatia

Publishing Process Manager Sandra Bakic

Technical Editor Teodora Smiljanic

Cover Designer InTech Design Team

Additional hard copies can be obtained from orders@intechopen.com

Current Topics in Ionizing Radiation Research, Edited by Mitsuru Nenoi
p. cm.
ISBN 978-953-51-0196-3

Meet the editor

Dr Mitsuru Nenoi graduated from Kyoto University, Graduate School of Sciences, Japan in 1983, and started his career as a scientist at National Institute of Radiological Sciences (NIRS), Japan. He received a Ph.D from Kyoto University in 1992 for the study on induced accumulation of polyubiquitin gene transcripts after exposure to ultraviolet light and treatment with 12-O-Tetradecanoylphorbol 13-acetate. In 1991, he stayed at University of Cincinnati College of Medicine, USA as a visiting scientist, and was involved in the study of mechanisms for transcriptional regulation of small heat shock genes of Drosophila. Dr. Nenoi is now a Director, Radiation Risk Reduction Research Program, Research Center for Radiation Protection of NIRS. His research interest is radiation biology, especially on the mechanism for gene regulation after exposure to ionizing radiation.

Contents

Preface

Since the discovery of X rays by Roentgen in 1895, the ionizing radiation has been extensively utilized in a variety of medical and industrial applications And it has also played crucial roles in development of modern sciences and technologies, as witnessed by more than 60 Nobel Prize winners awarded for achievements in atomic sciences and ionizing radiation-related researches. However people have shortly recognized the harmful aspects of ionizing radiation through inadvertent uses of the medical radiation and intake/inhalation of radioactive materials in industries. Subsequently people experienced the atomic bombs in Hiroshima and Nagasaki of Japan as well as nuclear power plant accidents in Chernobyl and Fukushima, which taught us that the risk of ionizing radiation is closely and seriously involved in the modern society.

In these circumstances, it becomes increasingly important that more scientists, engineers and students get familiar with ionizing radiation research regardless of the research field they are working. Based on this idea, the book "Current Topics in Ionizing Radiation Research" was designed to overview the recent achievements in ionizing radiation research including biological effects, medical uses and principles of radiation measurement. The book is divided into following sections: "Molecular Process", "Biological Process", "Health Effects", "Medical Uses", "Detection and Measurement" and "Effects on Materials", which contain advanced and challenging developments as well as basic reviews of the research field.

In the section "Molecular Process", radiation-induced DNA damage and its biological effects are *discussed*. And recent development in biological dosimetry including a method for fast dose estimation by use of γH2AX is reviewed. In addition, new findings in DNA repair and the effects of higher order of chromatin structures on DNA repair capability are discussed. As a topic for discussion, the *response of mealybug to high doses of ionizing radiation is presented.*

In the section "Biological Process", biological responses to ionizing radiation on cellular and organ levels are reviewed. They include the cancer cell responses to ionizing radiation depending on the mutation type of p53, modulation of gene expression in organs of irradiated animals, expression of microRNA relating to ionizing radiation, ubiquitin-related cell cycle regulation, as well as *in vivo* radioadaptive responses. In addition, novel and challenging researches, such as toxico-dynamic model representing interactions between cell component-derived radiation toxins and innocent bystander cells leading to systemic toxicity, the nonlinear dynamical aspects of radiation-induced apoptosis, as well as *numerical simulation for tumorigenesis after irradiation* are presented.

In the section "Health Effects", the epidemiological studies concerning radiation carcinogenesis is reviewed, followed by the topic of structural changes of teeth taken from people who were exposed to high doses of external ionizing radiation at the Chernobyl accident. Issues of radiation exposure from medical imaging such as CT diagnosis and industrial activities are also reviewed.

In the section "Medical Uses", effects of antioxidants on radiation-induced chromosomal damage under radiation therapy conditions are reviewed, followed by the study of antioxidant features of melatonin as a potentially efficient pharmacological radioprotector. In addition, the effects of hyperthermia and chemotherapeutic agents on cell survival curves after irradiation, as well as the phage display as a promising technology in discovery of diagnostic biomarker or therapeutic target are also reviewed. A historical background of the different imaging techniques used in orthodontics were reviewed, as well as the study of biological effects comparison between laser radiation, intense pulsed light and infrared and ultraviolet lamps.

In the section "Detection and Measurement", the theory and experimental methods for the detection, spectroscopy, and imaging of ionizing radiation is overviewed. Then application of glasses as radiation sensors is reviewed, followed by discussion on the effect of the irradiation dose on the formation and the decay of the induced color centers in irradiated silicate glass. Modern storage phosphors are also discussed from the viewpoint of their applications in dosimetry and computed radiography. In addition, studies of radiation-induced radicals in relation to materials sciences are reviewed. Detection of the atmospheric ionizing radiation from galactic and solar cosmic rays is also described.

In the section "Effects on Materials", the aspects of total ionizing dose and dose rate effects on semiconducting devices are reviewed, followed by the study of effects of ionizing radiation and hot carrier injection on the characteristics of semiconductor transistors. Effects of aromaticity on radiostability are also discussed.

The editor wishes that publication of "Current Topics in Ionizing Radiation Research" could extensively drive research activities in the related areas, which results in deeper understanding of ionizing radiation research by people, especially young researchers and engineers.

Dr. Mitsuru Nenoi
Research Center for Radiation Protection
National Institute of Radiological Sciences, Chiba
Japan

Part 1

Molecular Process

1

Measurement of H2AX Phosphorylation as a Marker of Ionizing Radiation Induced Cell Damage

Aida Muslimovic, Pegah Johansson and Ola Hammarsten*
Department of Clinical Chemistry and Transfusion Medicine, The Institute of Biomedicine, The Sahlgrenska Academy at University of Gothenburg, Sweden

1. Introduction

Cancer therapy often involves drugs or radiation, which induce DNA double-strand breaks (DSBs) and cause death of tumor cells. A dilemma with radiation and chemotherapy is the narrow therapeutic window and the fact that the sensitivity to DNA damage varies greatly in the human population, resulting in severe side effects and therapy-related deaths. Radiation and chemotherapy are usually planned during 1-3 weeks and delivered over a period of weeks to months. DNA damage testing of the patient could therefore be performed as part of the therapy planning. The level of DNA damage and DNA repair capacity could be measured after the first dose of chemotherapy or radiation, to allow personalized dosing during the following weeks of treatment. Induction of DSBs is followed by rapid formation of gamma-H2AX foci, with thousands of H2AX molecules being phosphorylated in the chromatin flanking the DSB site. Gamma-H2AX foci, each representing one radiation-induced DSB, can be measured by flow cytometry or counted in cell nuclei by fluorescence microscopy. In this chapter we will focus on radiation-induced DNA damage and its biological effects. We will also describe available methods for gamma-H2AX detection and discuss its possible use in clinical practice.

2. Radiation-induced DNA damage

The two DNA strands in the double-stranded DNA helix are over 20Å apart. Therefore, DNA-damaging chemicals like hydroxyl radicals or methylating agents for the most part generate single-stranded DNA damage, leaving the opposite strand intact. To take care of this type of DNA damage, human cells have several template-directed DNA repair systems, like nucleotide excision repair and base excision repair that require at least 10-30 nucleotides of an undamaged template strand to direct the synthesis of new DNA after removal of the DNA damage. These DNA repair systems are highly effective and accurate. Therefore, most forms of single-stranded DNA damage are removed within minutes without mutations. Single-stranded DNA damage is therefore not very toxic. As an example, over 100 000

* Corresponding Author

hydrogen peroxide-induced single-stranded DNA damage is required to affect the viability of a human cell (Muslimovic, Nystrom et al. 2009). However, template-directed repair is not possible when both DNA strands within 10-20 bp are damaged, called clustered DNA damage (Fig. 1).

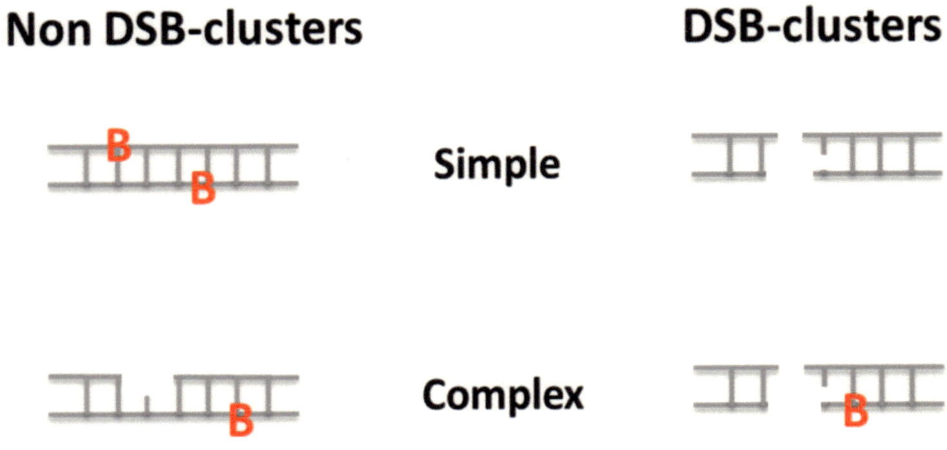

Fig. 1. Different types of radiation induced clustered DNA damage.

In this instance the cell has to repair the broken DNA by direct end joining (NHEJ) or homology-directed DNA repair (HR) (see next section) that are error prone and slow. Ionizing radiation is an efficient inducer of clustered DNA damage and is therefore highly toxic and mutagenic.

Ionizing photons or particles (ionizing radiation (IR)) damage DNA by direct energy deposition, leading to ionization events within the DNA molecule as well as ionization of water molecules and the formation of highly reactive hydroxyl radicals that damage the DNA indirectly. The passing photon or particle induces multiple ionizations as it passes through biological tissue and the number of such ionization events per unit of distance depends on the quality of the IR. High linear energy transfer (LET) IR induces denser ionization energy tracks than low LET IR. It is this ability to produce multiple ionization events within a short distance that is the reason behind the unique capacity of IR to generate damage to both DNA strands in close proximity (Hall 2000).

The clustered DNA damage induced by IR may include base damage to both strands, one-strand break opposed by base damage or the simple DNA double-strand break (DSB) with two strand beaks on opposing strands within 10-20 bp apart. In addition, the DSB can contain additional base damage close to the broken strands called complex clustered DSBs (Fig. 1). The complexity of the lesions increases with increasing LET of the radiation. For low LET, 30-40% of the DNA damage consists of complex clustered DSBs, as opposed to 90% for high LET (Nikjoo, O'Neill et al. 1999). Therefore, high LET IR is more toxic than low LET IR at the same absorbed dose.

IR induces ionization and molecular damage in all cellular components but it is its ability to damage DNA that is biologically important. For instance, it is possible to irradiate only parts of a cell using microscopic needle tips coated with alpha-emitting polonium, due to the short range of this radiation. In this setting, the mean lethal dose for a cell is 1.5 Gy if the irradiated area includes the nucleus, whereas the viability of the cell remains unaffected

even after 250 Gy if only the cytoplasm is irradiated (Hall 2000). This clearly demonstrates that the biological effects of irradiation are due to its interaction with the nucleus. Furthermore, studies using selective irradiation of DNA by incorporation of radioactive nucleotides in cellular DNA have shown that DNA is the relevant biological target of IR.
Although 95% of the DNA damage induced by IR is single-stranded DNA lesions, it is the few DSBs that are responsible for the toxic effects. Microinjection of more than 300 molecules of circular DNA with different forms of single-stranded DNA damage into the nucleus of human cells failed to induce significant cell cycle arrest, compared with injection of undamaged DNA. In contrast, nuclear injection of only 1-10 molecules of DNA containing one DSB per molecule were able to induce permanent cell cycle arrest (Huang, Clarkin et al. 1996), demonstrating that DSBs are particularly capable of inducing clonogenic cell death, the type of cell death typically induced by IR. Therefore, if we can measure the levels of DSBs induced after a dose of irradiation we will have a good measure of the biological effects of this dose.

3. DSB repair

In mammalian cells, there are two main pathways of DSB repair, homologous recombination (HR) and non-homologous end joining (NHEJ) (Mladenov and Iliakis; Mahaney, Meek et al. 2009). HR is the pathway of choice during the S and G2 phases when an identical sister chromatid, produced during the S phase, can be used as a template for accurate DSB repair (Rothkamm, Kruger et al. 2003). Therefore, HR is considered to be an error-free repair pathway. However, since over 99% of the cells in a human body are in the G1 or G0 stage of the cell cycle, where HR is supressed, NHEJ will repair most of the DSBs induced by IR in humans. Most of the DSBs produced by IR can not be directly ligated, as they contain base damage and damage to the sugar phosphate backbone that must be processed prior to repair. Therefore the broken DNA ends are first trimmed and then ligated. As a consequence of the processing, nucleotides may be lost, generating deletion mutations at the breakpoint. Therefore, NHEJ is an error-prone DNA repair pathway, explaining the highly mutagenic nature of IR.
In canonical NHEJ, DSBs are repaired by a process involving DNA-dependent protein kinase, DNA-PK. Once a DSB is induced, Ku, the DNA-binding subunit of DNA-PK, binds specifically to the DNA ends, due to a preformed channel in its structure that allows Ku to slide onto the end like a nut on a screw (Mimori and Hardin 1986; de Vries, van Driel et al. 1989; Walker, Corpina et al. 2001). Once bound to DNA ends, Ku will translocate inwards onto the DNA, allowing two DNA-PKcs molecules to bind to DNA ends (Hammarsten, DeFazio et al. 2000; DeFazio, Stansel et al. 2002) and interact in a synaptic complex across the DSB (DeFazio, Stansel et al. 2002), leading to activation of the DNA-PK holoenzyme (Meek, Gupta et al. 2004; Meek 2009) and subsequent autophosphorylation (Mahaney, Meek et al. 2009). DNA-PKcs binds to and activates the nuclease Artemis which is believed to be involved in the end processing of more complex DSBs (Ma, Pannicke et al. 2002; Wang, Pluth et al. 2005; Drouet, Frit et al. 2006; Goodarzi, Yu et al. 2006). Autophosphorylation is also thought to be required for the conformational change between opened and closed holoenzyme conformations, which allows end processing by polynucleotide kinase (PNK), polymerization by the DNA polymerases micro or gamma and finally ligation by LigIV/XRCC4/XLF complex. This DNA-PK dependent NHEJ pathway, dependent on DNA-PK is referred to as D-NHEJ, where D is short for DNA-PK or C-NHEJ, short for canonical NHEJ.

Lately, a back-up NHEJ pathway (B-NHEJ) has been discussed in the literature. Studies where cells defective in HR and D-NHEJ were treated with DSB-inducing agents have revealed an alternative DSB-repair pathway with yet slower kinetics and with an increased number of chromosomal abberations, see review (Mladenov and Iliakis). In this pathway, PARP1 binds to DNA ends instead of to Ku and the LigIII/XRCC1 complex performs the ligation.

DSBs induced by IR are repaired by different kinetics. After high doses of IR, repair of 85% of all DSBs occur during the first 10-30 minutes, whereas the remaining 15% of the breaks can take up to 24 hours to be fully removed (Goodarzi, Jeggo et al.). This dual kinetic profile probably reflects a different chemical complexity at the DSB ends and the local chromatin compaction. In agreement with this possibility, complex DSBs induced by high LET and DSBs induced in heterochromatin are repaired more slowly, whereas a larger portion of low LET-induced DSBs and DSBs in low-complexity euchromatic regions are repaired with fast kinetics (Goodarzi, Jeggo et al.). In G1 cells, all components of the NHEJ are required for both slow and fast DSB-repair. An exception is Artemis, which probably is only required for the slow repair of more complex lesions. The slow repair of DSBs in high-complexity heterochromatin regions also involve ATM/p-KAP1, the MRN complex, gamma-H2AX, MDC1, RNF8, RNF168 and 53BP1 (Goodarzi, Jeggo et al.). It is likely that DSBs with different chemical modifications require distinct repair complexes involving polymerase, polynucleotide kinase and nuclease activities. As we still lack a reliable *in vitro* system for DSB repair and methods to induce DSBs with a defined chemical structure in cells, our knowledge of how different repair complexes assemble in response to different forms of DSBs is still incomplete.

4. DSB response and H2AX phosphorylation

The DNA of eukaryotic cells is organized in DNA protein complexes called chromatin. Chromatin proteins mostly consist of histones, small arginine-lysine-rich proteins with a high binding affinity for the negatively charged DNA. Chromatin is organized in smaller structural subunits called nucleosomes. The bead particles called nucleosome core particles contain about 150bp DNA coiled 1.7 times around the core consisting of eight histones, two from each of the H2A, H2B, H3 and H4 types, and, additional 50bp DNA for the linker-histone H1 connecting two nucleosomes. The nucleosome structure results in a packaging of DNA into 10nm fibers, which can be further compacted into 30nm fibers and chromatin loops.

One of the first cellular responses to DSBs is the rapid phosphorylation of histone H2AX (Rogakou, Pilch et al. 1998). H2AX is one of the most evolutionary conserved H2A variants and constitutes 2-25% of all nucleosomes, depending on the cell type (Kinner, Wu et al. 2008). H2AX phosphorylation occurs on serine- 139 in the SQ motif that is positioned at the carboxy terminus of H2AX (Rogakou, Pilch et al. 1998; Rogakou, Boon et al. 1999). H2AX phosphorylation occurs minutes after DSB induction, leading to the formation of distinct gamma-H2AX foci (Rogakou, Boon et al. 1999). Several reports indicate that each IR-induced gamma-H2AX focus represents a single DSB, at least in G1 cells (Sedelnikova, Rogakou et al. 2002; Pilch, Sedelnikova et al. 2003; Rothkamm and Lobrich 2003). The level of gamma-H2AX induced by different drugs and radiation is closely correlated to the level of cell death (Banath and Olive 2003; Muslimovic, Nystrom et al. 2009). This indicates that measurement of gamma-H2AX in patient cells will be a biologically important marker for the induction of DSBs and toxicity following chemotherapy and radiotherapy.

Ataxia-telangiectasia-mutated (ATM), DNA-dependent protein kinase (DNA-PK) and Rad-3-related (ATR) kinases that belong to the phosphatidylinositol-OH-kinase-like family of protein kinases (PIKKs), can phosphorylate H2AX (Paull, Rogakou et al. 2000; Ward and Chen 2001; Burma, Chen et al. 2002; Furuta, Takemura et al. 2003; Stiff, O'Driscoll et al. 2004; Ward, Minn et al. 2004). ATM and DNA-PK phosphorylate H2AX redundantly upon DSB induction in the G1 and G2 phases of the cell cycle (Stiff, O'Driscoll et al. 2004), while ATR is responsible for replication stress and UV damage related H2AX phosphorylation during the S phase (Ward and Chen 2001; Ward, Minn et al. 2004).

The first event in the recognition of a newly formed DSB is the binding of a complex of three proteins, Mre11, Rad50 and Nbs1 (MRN) to the DNA ends. The MRN complex is a multifunctional DNA-binding protein complex with a weak preference for DNA ends, a nuclease function (Mre11) and the ability to recruit ATM dimers to the DNA via the C terminus of Nbs1. In this setting, ATM dimers are activated for their kinase activity. Active ATM monomers then phosphorylate H2AX close to the end, which leads to recruitment of mediator of DNA damage checkpoint protein 1 (MDC1), a protein that binds specifically to gamma-H2AX. As MDC1 also binds to the MRN complex, this leads to a positive feed-back loop where additional MRN complexes and ATM are recruited to the DSB site, resulting in extended H2AX phosphorylation by ATM (see below).

Initially, gamma-H2AX foci are formed as small, distinct foci at the site of the DSB, visible as soon as 1-3 minutes after irradiation. With time, these gamma-H2AX foci grow and spread along the chromatin, further away from the DSB site, and reach a plateau after 10-30 minutes (Rogakou, Pilch et al. 1998; Rogakou, Boon et al. 1999; Sedelnikova, Rogakou et al. 2002). Depending on the celline, 0.03-0.06% of H2AX molecules per DSB become phosphorylated and spread over 2 Mbp regions of chromatin that covers ~2000 gamma-H2AX molecules (Rogakou, Pilch et al. 1998). However, the phosphorylated gamma-H2AX regions can reach up to 30Mbp from the break (Rogakou, Boon et al. 1999; Pilch, Sedelnikova et al. 2003). In this way, the DSB signal is locally amplified by the spreading of the H2AX phosphorylation to large chromatin regions. Several reports have described large residual gamma-H2AX foci that form several hours after irradiation (Bocker and Iliakis 2006; Markova, Schultz et al. 2007; Yamauchi, Oka et al. 2008; Yoshikawa, Kashino et al. 2009) and these foci can persist for months (Rodier, Coppe et al. 2009).

When DSBs are repaired, H2AX phosphorylation declines with time. Dephosphorylation of H2AX has been proposed as the possible mechanism. Four serine/threonine phosphatases have been reported to dephosphorylate H2AX in human cells: PP2A, PP4, PP6 and PP2C. PP2A dephosphorylates H2AX at the site of the DSB and siRNA knockdown of the catalytic subunit of PP2A leads to persistence of foci and inefficient DSB repair (Chowdhury, Keogh et al. 2005). PP4 associates with chromatin and dephosphorylates chromatin-bound H2AX molecules phosphorylated by ATR during replication (Douglas, Zhong et al. ; Chowdhury, Xu et al. 2008; Nakada, Chen et al. 2008). PP6 also dephosphorylates H2AX (Douglas, Zhong et al.) and knockdown of its catalytic subunit results in persistent gamma-H2AX and 53BP1 foci. PP2A and PP6 seem to be involved in the recovery from the G2/M checkpoint. PPC2 has been co-purified with H2A-H2B histone dimer and may be involved in the dephosphorylation of dissociated H2A-H2B and the incorporation of the unphosphorylated H2A-H2B into the nucleosome (Kimura, Takizawa et al. 2006). Recently, WIP1 (wild-type-53-induced phosphatase) WIP1 has been reported to dephosphorylate H2AX, resulting in suppression of the DSB repair (Macurek, Lindqvist et al.; Moon, Nguyen et al.).

5. Gamma-H2AX detection and its usage

Measurements of gamma-H2AX phosphorylation are currently the most sensitive way to detect IR-induced DNA damage and to measure the extent of DSB repair. Measurements of gamma-H2AX foci could therefore be a possible approach to monitor DSBs or DSB repair in cancer patients during treatment with radiotherapy or chemotherapy as a way to personalize the dosing. It is also possible that gamma-H2AX measurements could be used for radiation biodosimetry for nuclear plant workers or in the general public after nuclear accidents as well as exposure to potentially DNA damaging environmental toxins. In the following sections we will summarize the methods that have been developed for gamma-H2AX measurement in humans and its possible future use in cancer treatment and public dosimetric testing.

6. Methods for measurement of gamma-H2AX in patient cells

6.1 General considerations

One major consideration in gamma-H2AX foci biodosimetry is the source of patient cells to be assayed after exposure to DNA-damaging agents.

Peripheral blood mononuclear cells prepared by density gradient centrifugation are perhaps the optimal source of cells for this purpose, for several reasons. Firstly, over a million cells can routinely be obtained from each milliliter of blood and prepared within 30 minutes by a simple centrifugation step. Secondly, the acquisition is not as invasive as obtaining biopsies and poses little risk to patients. Finally, the possibility of quick sampling post-radiation is important for the gamma-H2AX assay, due to the fast repair kinetics of foci. A potential problem with blood mononuclear cells is that at least three different cell types are present in these preparations. The most prominent cell type is lymphocytes (80-95% T-lymphocytes and 5-20 % B-lymphocytes) that constitute 80-90% of the cells. In addition, the mononuclear cell preparation contains monocytes (5-10%) and a variable amount of contaminating neutrophiles. All these cells are able to induce gamma-H2AX in response to IR, although the level of the gamma-H2AX response is lower in neutrophiles (Ismail, Wadhra et al. 2007) and in some subsets of lymphocytes (Andrievski and Wilkins 2009).

Another viable option that has been tested is hair bulb cells from plucked hair (Redon, Nakamura et al.). The sheath around the base of a plucked hair contains proliferating cells that are easily accessible. The hairs are collected by gripping them close to the skin using blunt-nosed forceps and plucking them with a single, sharp action, checking them for the presence of sheath and immersing them directly in a fixation medium. In humans, scalp hair bulbs have been shown to contain more viable cells than hair bulbs from the eyebrow. In a study using macaque monkeys receiving total body irradiation, hair bulbs from the fur did not contain enough cells for analysis, indicating that body hair from humans might not be useful. This study also showed that mononuclear cells yielded more reproducible results for the gamma-H2AX analysis than cells from hair bulbs. A problem with hair bulb cells is that this cell type can only be used for gamma-H2AX measurements following systemic treatment, unless the scalp or eyebrows have been part of the irradiated area. On the other hand, active hair bulbs contain proliferating cells as opposed to mononuclear cell preparations. This opens up the possibility of using cells from hair bulbs to monitor DNA damage induced by nucleoside analogues and other drugs that induce DNA damage only in proliferating cells.

6.2 Gamma-H2AX detection in patients with immunofluorescence microscopy

The most sensitive method for gamma-H2AX detection is immunofluorescence microscopy, which enables quantification of single foci in individual cells. In this method, fixed cells or tissue are attached to glass slides and incubated with gamma-H2AX antibodies, usually overnight. After washing, the slides are stained with secondary antibodies and DNA dyes. After a final wash, the slides are dried and mounted with cover glasses before analysis by fluorescence microscopy (Fig. 2).

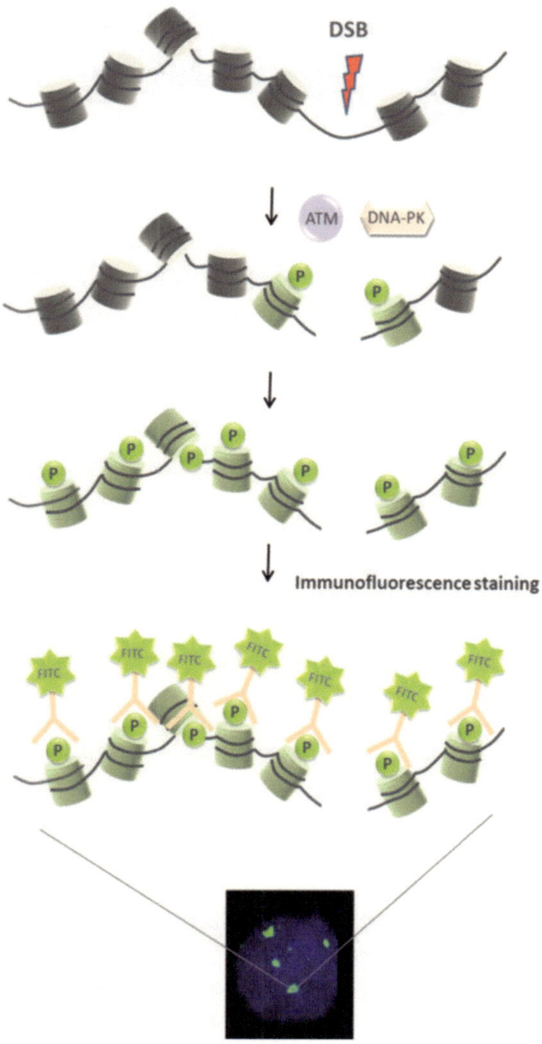

Fig. 2. Immunofluorescence analysis of H2AX phosphorylation.

Gamma-H2AX are then quantified either by manual counting in fluorescence microscope or by foci counting software from digital images. Both methods of counting are error-prone, especially at high IR doses. A general problem is that gamma-H2AX foci are heterogeneous with large and small foci coexisting side by side in the same nucleus (Fig. 3, white arrows).

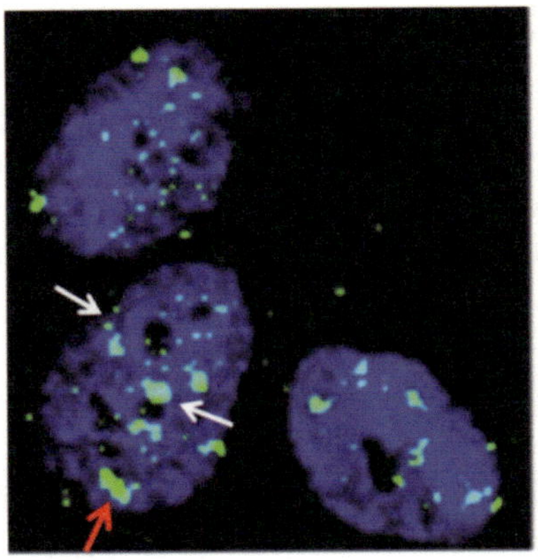

Fig. 3. Heterogeneous sizes of gamma-H2AX foci.

In addition, foci that are close together can be incorrectly scored as a single large focus (Fig. 3, red arrow). This problem is especially prominent following high-LET irradiation, where multiple closely spaced DSBs are generated along the high-LET particle track. The problem of superimposed foci increases as the IR dose increases and limits foci counting to doses below 2-3 Gy where 20-40 foci are found in many nuclei. In addition, the number of foci per cell is often highly variable due to the distributive nature of DSB induction also when cells have been irradiated *in vitro*. After local irradiation of patients blood cells from the irradiated area are mixed with unirradiated cells, further adding to the heterogeneity. Therefore, over 50 cells must be counted to allow reliable measurements of gamma-H2AX induction in an *in vitro*-irradiated sample. In samples of mononuclear cells from patients receiving local irradiation, even more cells must be scored for reliable results. The requirement to count foci in a large number of nuclei in each sample makes the immunofluorescence microscopy method labor-intensive when the counting is done manually. On the other hand, spot-finding software have difficulties with the quantification settings, so that foci of different sizes and brightness are sometimes counted incorrectly. In addition, digital images are usually taken in a single plane where only a subset of the nuclei is in perfect focus, affecting the focus brightness and size. Lastly, background staining from debris in the sample may affect the counting when digital systems are used. However, microscopic systems dedicated to foci counting are now able to overcome many of these problems (Lobrich, Rief et al. 2005). These microscopic systems are currently expensive and rarely available in routine clinical laboratories.

Another problem with immunofluorescence microscopy is the handling of microscopic slides. Lymphocytes are normally attached to the microscopic slides via centrifugation (cytospin) that only allows one patient sample per slide. In addition, the staining of microscopic slides is time-consuming and requires large volumes of antibody solutions, adding to the analysis time and cost. We have therefore developed a modification of the gamma-H2AX immunofluorescence method, where fixed cells are stained with the fluorescent gamma-H2AX antibody in solution before being attached to microscopic slides by drying a droplet of the cell sample in a salt-free protein solution. This allows arraying of up to 16 patient samples per slide ready for foci counting in two hours (Johansson, Muslimovic et al. 2011). Together with automatic microscopic foci counting systems that are available (Lobrich, Rief et al. 2005), this new staining method holds great promise in future implementation of the gamma-H2AX foci analysis in clinical routine.

Immunofluorescence microscopy and foci counting has been used both *in vitro* and in patients to measure DNA-damage following radiotherapy and chemotherapy. Initial studies using low doses of irradiation of human fibroblast cultures showed that there was a correlation between the IR dose, the numbers of gamma-H2AX foci, and amount of DSBs. This study also established that increased levels of gamma-H2AX foci are detectable at doses as low as 1mGy (Rothkamm and Lobrich 2003). The ability to detect very low levels of radiation has inspired the use of gamma-H2AX foci in mononuclear cells from patients exposed to X-rays in computerized tomography (CT) (Lobrich, Rief et al. 2005) and other X-ray examinations. Gamma-H2AX foci analysis was used as a biomarker for individual radiation damage in pediatric patients undergoing cardiac catherization showing that previous estimates according to linear-no-treshold hypothesis have been underestimating radiation damage. Therefore, gamma-H2AX foci analysis could be used as a tool for estimations of DNA-damage effects after cardiac catherization (Beels, Bacher et al. 2009). Another study showed that the amount of gamma-H2AX foci in lymphocytes after coronary CT angiography correlates to dose-length product (DLP) which is a physical parameter used to calculate delivered dose after CT. Therefore, gamma-H2AX foci analysis could be used to measure individual dose-related effects of X-rays (Kuefner, Grudzenski et al.). Gamma-H2AX foci analysis has also been used for estimations of individual radiation doses by comparisons of *in vivo* and *in vitro* exposed lymphocytes to fractionated irradiation or angiography indicating that gamma-H2AX foci analysis could be used for dose estimations (Kuefner, Grudzenski et al. 2009). In addition, several reports now show that gamma-H2AX foci analysis can be used to measure DNA-damage induced by local radiotherapy. Analysis of gamma-H2AX foci in prostate cancer biopsies after *in vivo* radiation showed reproducible quantifications of foci-numbers in the prostata specimens in the dose region 0-1Gy (Qvarnstrom, Simonsson et al. 2004). Gamma-H2AX foci has also been be used to follow induction of DNA-damage in lymphocytes of cancer patients with different cancer types and exposed to local radiotherapy treatments to different sites of the body. There was a correlation between mean number of gamma-H2AX foci per lymphocyte and integrated total body radiation dose indicating that measurements of gamma-H2AX foci in lymphocytes can be used for *in vivo* dosimetry (Sak, Grehl et al. 2007). Immunofluorescence microscopy has also been successful in the detection of a DSB response after treatment with cancer drugs, like the Topoisomerase-I inhibitor camptothecin (Kinders, Hollingshead et al.), in human biopsies (Qvarnstrom, Simonsson et al. 2004), circulating tumor cells (Wang, Pfister et al.), lymphocytes and plucked hair bulbs (Redon, Nakamura et al.). A fully automated high-throughput system, the RABIT (Rapid Automated Biodosimetry Tool), has been developed for gamma-H2AX analysis on human lymphocytes able to process several thousands of blood samples per day that might be useful in larger nuclear accidents (Garty, Chen et al.). Finally, it was reported that radioimmunoconjugates with a peptide that allows cellular uptake of labeled gamma-H2AX antibody complexes could be used to measure gamma-H2AX foci in living cells and in whole mice. The method was less sensitive than conventional gamma-H2AX foci staining of fixed and permeabilized cells, but this technique opens up the interesting possibility of monitoring DSB induction in patient tumors by imaging techniques, such as gamma camera or even MRT, in the future (Cornelissen, Kersemans et al.).

While quantification of the gamma-H2AX foci is clearly sensitive enough to allow reliable measurement of DSB induction in patients, the method is still too labor-intensive, time-consuming and operator-dependent. In addition, automatic fluorescence microscopes equipped with software for automatic counting of foci are rarely available in routine clinical laboratories. Therefore, the gamma-H2AX foci method is currently difficult to implement in clinical routine.

6.3 Gamma-H2AX detection in patients using flow cytometry

Flow cytometry has also been used to measure gamma-H2AX phosphorylation in patient cells. In this method, the possibility to detect single gamma-H2AX foci and DSBs is lost. Instead, the gamma-H2AX phosphorylation signal from the whole cell is measured in a large number of cells. The flow cytometry method is less labor-intensive but, also less sensitive compared to counting gamma-H2AX foci by immunofluorescence microscopy.

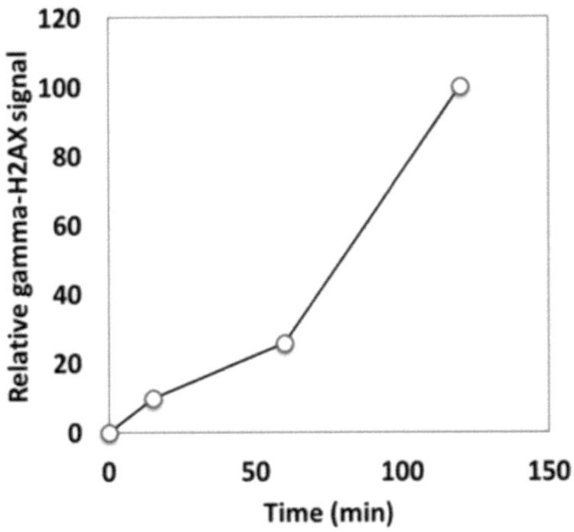

Fig. 4. Staining-time dependent gamma-H2AX signal analysed by flow cytometry.

In the flow cytometry method, cells are stained and permeabilized in a staining buffer containing fluorescently labeled gamma-H2AX antibodies and DNA dyes. After staining for 15 minutes up to a few hours (Fig. 4), the samples can be analyzed directly by flow cytometry without further processing. In flow cytometry, up to 100 000 cells can be analyzed in each sample within minutes, allowing detection of even small subsets of gamma-H2AX-positive cells.

Due to its simplicity, this method can be implemented in its current form in clinical laboratories where flow cytometers often are available. An important shortcoming of this method is its low analytical sensitivity compared to counting of separate gamma-H2AX foci (Fig. 5).

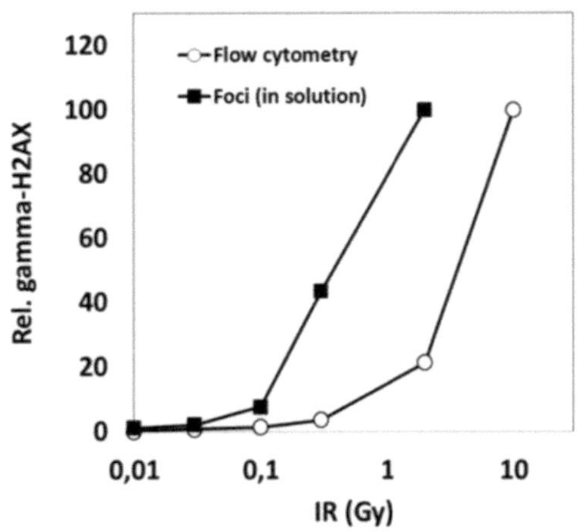

Fig. 5. Sensitivity of the gamma-H2AX detection by foci method and flow cytometry method.

A contributing factor to the low sensitivity is that undamaged cells have a diffuse background signal from the gamma-H2AX antibodies. The nature of this background is unclear but results in a requirement of at least 2-5 gamma-H2AX foci per cell to separate the signal from undamaged cells. At low IR doses (<0.1Gy) most cells contain 0-1 foci per nucleus. The sensitivity of the foci method relies on its ability to score cells with one focus among cells without foci. Radiation doses generating an increase from one focus in every 20 cells (0.05 foci/cell) to one focus in every 10 cells (0.1 foci/cell) is a significant increase that can be scored with the foci method. In contrast, this increase remains below the detection limit in the flow cytometer, as the gamma-H2AX signal from cells with 1-2 foci is not separated from undamaged cells. For that reason, flow cytometry analysis is only able to measure gamma-H2AX signal at IR doses of over 0.1-0.3 Gy, whereas the foci method is able to measure in the mGy range (Rothkamm and Lobrich 2003) and in our hands down to 0.01Gy (Fig. 6).

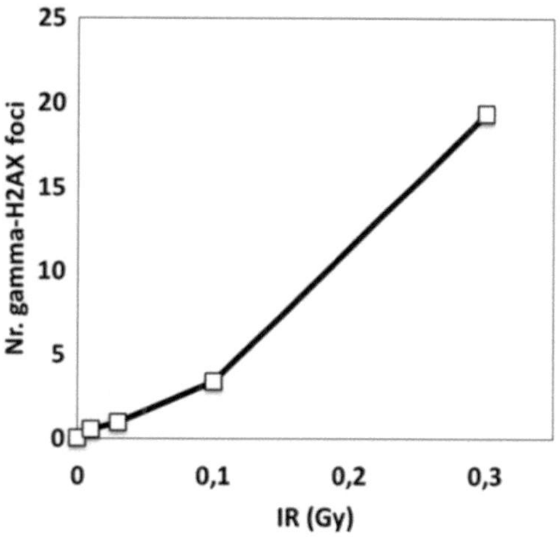

Fig. 6. Dose-dependence of gamma-H2AX foci formation in human lymphocytes after *in vitro* irradiation as measured by immunofluorescence microscopy with "in-solution" staining.

On the other hand, most radiotherapy regimens include fractionated doses of 1.5-5 Gy. For the subset of mononuclear cells residing in the irradiated area, this dose is well within the measurable range for flow cytometry analysis. In addition, flow cytometry allows analysis of small subsets of gamma-H2AX-positive cells, as up to 100 000 cells are analyzed in each sample. This is particularly helpful when mononuclear cells are analyzed from patients following local irradiation, where blood cells from the irradiated area are mixed with unirradiated cells. In one study, blood samples were collected from three rectal cancer patients before and at different times after 5Gy pelvic irradiation (Muslimovic, Ismail et al. 2008). The gamma-H2AX signal was analyzed by flow cytometry and compared to *in vitro*-irradiated mononuclear cells from the same patients. This analysis revealed that approximately 5% of the mononuclear cells retrieved one hour after the pelvic irradiation had the same gamma-H2AX signal as cells irradiated with 5 Gy *in vitro*. Apparently, these "sentinel cells" were trapped in the microcirculation in the irradiated area and therefore received the full 5 Gy dose. This pilot study indicated that flow cytometry may be particularly well suited for the measurement of gamma-H2AX induction in blood cells following local radiotherapy. The same flow cytometry method was tested on *in vitro*-irradiated blood samples and found to be robust, reproducible and able to detect radiation doses as low as 0.1Gy with a linear increase of the gamma-H2AX signal at doses over 0.3 Gy. The signal was specific for H2AX and was absent in mouse embryonic cells from H2AX-/- mice. All nucleated blood cells displayed the gamma-H2AX signal following DSB induction. However, the signal was close to 50% lower from neutrophiles compared to monocytes and lymphocytes. In addition, the study revealed a marked inter-individual difference of the gamma-H2AX signal intensity following DSB-induction *in vitro*, with a twofold difference in gamma-H2AX signal intensity between the highest and lowest individual tested. This finding has been confirmed in other studies using similar methods (Bourton, Plowman et al.).

Flow cytometry measurement of gamma-H2AX has been used to diagnose Ataxia telangiectasia, a disease where patients are radiosensitive and show a delayed clearance of H2AX foci after *in vitro* irradiation of lymphocytes (Porcedda, Turinetto et al. 2008). Similar methods have been used to find patients with severe normal tissue toxicity following radiotherapy (Bourton, Plowman et al.) and to study the gamma-H2AX response in several subsets of nucleated blood cells (Andrievski and Wilkins 2009). However, apart from the pilot study of rectal cancer patients receiving pelvic irradation, no clinical study has used flow cytometry to measure gamma-H2AX levels in patient samples after radiotherapy. Future studies will reveal the usefulness of flow cytometry methods for gamma-H2AX measurement in the clinical setting.

6.4 Other methods for gamma-H2AX measurement

Gamma-H2AX analysis has been used in genotoxicity measurements with ELISA-based assays (Matsuzaki, Harada et al.) and high content screening assays (Yip, Cuddy et al.).

7. Internal controls and standards

Like all clinical laboratory methods, measurements of gamma-H2AX by flow cytometry or by immunofluorescence microscopy are subject to day-to-day variations due to many factors including different quality among antibody batches and other reagents, fluctuations in detector performance and staining efficiency. In addition, inter-laboratory variations in foci detection software and flow cytometry detector sensitivity must also be calibrated to allow data comparison between clinical laboratories. Therefore, in order to implement H2AX methods in clinical practice there will be a need for calibrators and internal controls to monitor method performance. For that purpose, we have developed two types of controls based on phosphopeptide coated plastic beads and fixed cells. Control beads are generated by binding biotinylated phosphopetides derived from H2AX to streptavidin coated plastic beads, dried and stored at room temperature. Control cells are generated from primary human fibroblasts treated with the DSB inducing drug calicheamicin-γ1 for 30 minutes followed by fixation and storage in 90% EtOH at -70 °C. Both controls are stable over 40 days and are equally capable of correcting flow cytometry based gamma-H2AX signal from two different antibody batches (Muslimovic A., manuscript in preparation). These two types of controls have their own merits. The phosphopeptide coated beads are easy to manufacture in large quantities and can be shipped at ambient temperature facilitating inter-laboratory comparisons. A potential problem with the beads is that they do not present the antigen in a way that resembles the cellular structure in fixed cells. This could generate problems if phosphopeptide coated beads are used to harmonize phospho-H2AX methods using different antibodies with variable antigen specificity. We therefore also established conditions that allowed fixed cells to serve as controls. Control cells are less amenable to large scale production and must be shipped on dry-ice that could make inter-laboratory harmonization impractical. However, these control cells can also be used as controls for gamma-H2AX foci staining as the foci morphology and brightness are preserved during storage (Johansson 2011) (Fig. 7). Thus, control cells can be used in the foci method and added to each microscopic slide with patient samples as an internal control as well as inter-laboratory comparisons of the gamma-H2AX foci method.

A B

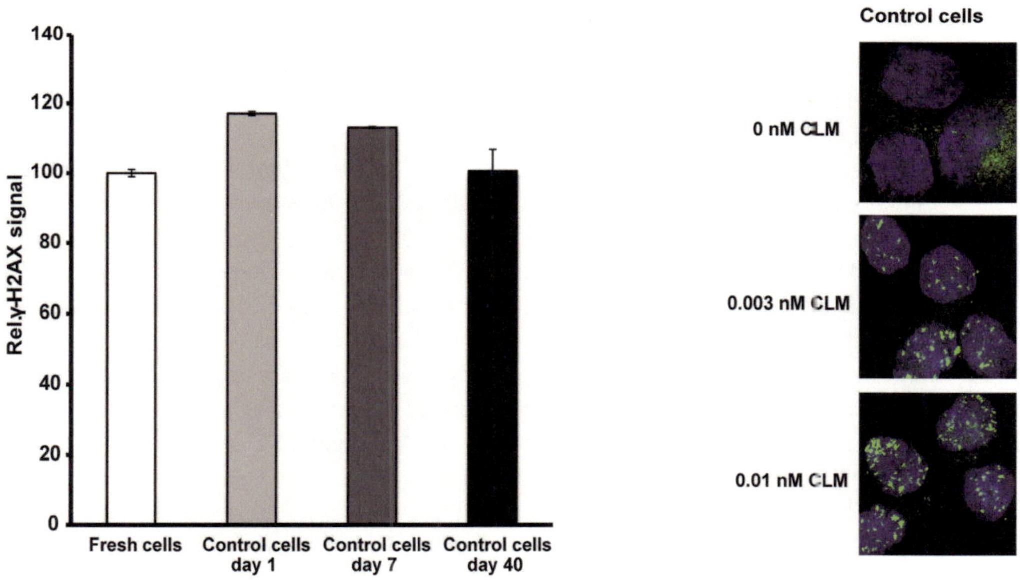

Fig. 7. A: Control cells in ethanol solution are stable up to 40 days and -70^{0}C as shown by flow cytometry. **B:** Control cells treated with different levels of DSB-inducing drug CLM preserve intact gamma-H2AX foci after storage in -70^{0}C and ethanol.

8. Conclusion

Gamma-H2AX analysis has enhanced our ability to detect DNA damage in individual patients and opens for the first time the possibility to measure DNA-damage during cancer treatment and in bio-monitoring purposes in clinical laboratories. In addition, a wide range of genotoxic events can now be monitored in individuals with the gamma-H2AX methods due to its unique sensitivity and tight correlation to the toxicity of the DNA-damage. In some cases there is substantial induction of gamma-H2AX foci with exposure to high levels of ionizing radiation and chemotherapeutic drugs that can be monitored with simple flow cytometry methods. In other instances, the DSB induction is of smaller magnitude, such as with diagnostic CT scans or exposure to environmental toxins, that can be measured using the more sensitive gamma-H2AX foci method. In addition, gamma-H2AX measurements can also be used following nuclear accidents, allowing rapid identification of individuals exposed to critically high radiation doses for initial triage and medical treatment decisions. There is, however, a need to further elucidate the molecular steps leading to gamma-H2AX formation to allow safe use of the technique. In addition, the technical limitations of the foci method need to be solved so that this method could be implemented in routine clinical labs. Alternatively, the sensitivity of the flow cytometry methods should be enhanced to allow detection of cells with single foci. When these limitations are solved the gamma-H2AX methods holds great promise as an important asset to allow reliable measurement of DNA-damage in clinical routine.

9. References

Andrievski, A. and R. C. Wilkins (2009). "The response of gamma-H2AX in human lymphocytes and lymphocytes subsets measured in whole blood cultures." *Int J Radiat Biol* 85(4): 369-76.

Banath, J. P. and P. L. Olive (2003). "Expression of phosphorylated histone H2AX as a surrogate of cell killing by drugs that create DNA double-strand breaks." *Cancer Res* 63(15): 4347-50.

Beels, L., K. Bacher, et al. (2009). "gamma-H2AX foci as a biomarker for patient X-ray exposure in pediatric cardiac catheterization: are we underestimating radiation risks?" *Circulation* 120(19): 1903-9.

Bocker, W. and G. Iliakis (2006). "Computational Methods for analysis of foci: validation for radiation-induced gamma-H2AX foci in human cells." *Radiat Res* 165(1): 113-24.

Bourton, E. C., P. N. Plowman, et al. "Prolonged expression of the gamma-H2AX DNA repair biomarker correlates with excess acute and chronic toxicity from radiotherapy treatment." *Int J Cancer*.

Burma, S., B. P. Chen, et al. (2002). "ATM Phosphorylates Histone H2AX in Response to DNA Double-strand Breaks*." *J Biol Chem* 276(Nov 9): 42462-7.

Chowdhury, D., M. C. Keogh, et al. (2005). "gamma-H2AX dephosphorylation by protein phosphatase 2A facilitates DNA double-strand break repair." *Mol Cell* 20(5): 801-9.

Chowdhury, D., X. Xu, et al. (2008). "A PP4-phosphatase complex dephosphorylates gamma-H2AX generated during DNA replication." *Mol Cell* 31(1): 33-46.

Cornelissen, B., V. Kersemans, et al. "Imaging DNA damage in vivo using gammaH2AX-targeted immunoconjugates." *Cancer Res* 71(13): 4539-49.

de Vries, E., W. van Driel, et al. (1989). "HeLa nuclear protein recognizing DNA termini and translocating on DNA forming a regular DNA-multimeric protein complex." *J Mol Biol* 208(1): 65-78.

DeFazio, L. G., R. M. Stansel, et al. (2002). "Synapsis of DNA ends by DNA-dependent protein kinase." *EMBO J* 21(12): 3192-200.

Douglas, P., J. Zhong, et al. "Protein phosphatase 6 interacts with the DNA-dependent protein kinase catalytic subunit and dephosphorylates gamma-H2AX." *Mol Cell Biol* 30(6): 1368-81.

Drouet, J., P. Frit, et al. (2006). "Interplay between Ku, Artemis, and the DNA-dependent protein kinase catalytic subunit at DNA ends." *J Biol Chem* 281(38): 27784-93.

Furuta, T., H. Takemura, et al. (2003). "Phosphorylation of histone H2AX and activation of Mre11, Rad50, and Nbs1 in response to replication-dependent DNA double-strand breaks induced by mammalian DNA topoisomerase I cleavage complexes." *J Biol Chem* 278(22): 20303-12.

Garty, G., Y. Chen, et al. "The RABIT: a rapid automated biodosimetry tool for radiological triage." *Health Phys* 98(2): 209-17.

Goodarzi, A. A., P. Jeggo, et al. "The influence of heterochromatin on DNA double strand break repair: Getting the strong, silent type to relax." *DNA Repair (Amst)* 9(12): 1273-82.

Goodarzi, A. A., Y. Yu, et al. (2006). "DNA-PK autophosphorylation facilitates Artemis endonuclease activity." *EMBO J* 25(16): 3880-9.

Hall, Ë. J. (2000). *Radiobiology for the Radiologist*. Philadelphia, PA, Lippincott Williams & Wilkins.

Hammarsten, O., L. G. DeFazio, et al. (2000). "Activation of DNA-dependent protein kinase by single-stranded DNA ends." *J Biol Chem* 275(3): 1541-50.

Huang, L. C., K. C. Clarkin, et al. (1996). "Sensitivity and selectivity of the DNA damage sensor responsible for activating p53-dependent G1 arrest." *Proc Natl Acad Sci U S A* 93(10): 4827-32.

Ismail, I. H., T. I. Wadhra, et al. (2007). "An optimized method for detecting gamma-H2AX in blood cells reveals a significant interindividual variation in the gamma-H2AX response among humans." *Nucleic Acids Res* 35(5): e36.

Johansson, P., A. Muslimovic, et al. "In-solution staining and arraying method for the immunofluorescence detection of gammaH2AX foci optimized for clinical applications." *Biotechniques* 51(3): 185-9.

Kimura, H., N. Takizawa, et al. (2006). "A novel histone exchange factor, protein phosphatase 2Cgamma, mediates the exchange and dephosphorylation of H2A-H2B." *J Cell Biol* 175(3): 389-400.

Kinders, R. J., M. Hollingshead, et al. "Development of a validated immunofluorescence assay for gammaH2AX as a pharmacodynamic marker of topoisomerase I inhibitor activity." *Clin Cancer Res* 16(22): 5447-57.

Kinner, A., W. Wu, et al. (2008). "Gamma-H2AX in recognition and signaling of DNA double-strand breaks in the context of chromatin." *Nucleic Acids Res* 36(17): 5678-94.

Kuefner, M. A., S. Grudzenski, et al. "Effect of CT scan protocols on x-ray-induced DNA double-strand breaks in blood lymphocytes of patients undergoing coronary CT angiography." *Eur Radiol* 20(12): 2917-24.

Kuefner, M. A., S. Grudzenski, et al. (2009). "DNA double-strand breaks and their repair in blood lymphocytes of patients undergoing angiographic procedures." *Invest Radiol* 44(8): 440-6.

Lobrich, M., N. Rief, et al. (2005). "In vivo formation and repair of DNA double-strand breaks after computed tomography examinations." *Proc Natl Acad Sci U S A* 102(25): 8984-9.

Ma, Y., U. Pannicke, et al. (2002). "Hairpin opening and overhang processing by an Artemis/DNA-dependent protein kinase complex in nonhomologous end joining and V(D)J recombination." *Cell* 108(6): 781-94.

Macurek, L., A. Lindqvist, et al. "Wip1 phosphatase is associated with chromatin and dephosphorylates gammaH2AX to promote checkpoint inhibition." *Oncogene* 29(15): 2281-91.

Mahaney, B. L., K. Meek, et al. (2009). "Repair of ionizing radiation-induced DNA double-strand breaks by non-homologous end-joining." *Biochem J* 417(3): 639-50.

Markova, E., N. Schultz, et al. (2007). "Kinetics and dose-response of residual 53BP1/gamma-H2AX foci: co-localization, relationship with DSB repair and clonogenic survival." *Int J Radiat Biol* 83(5): 319-29.

Matsuzaki, K., A. Harada, et al. "Whole cell-ELISA to measure the gammaH2AX response of six aneugens and eight DNA-damaging chemicals." *Mutat Res* 700(1-2): 71-9.

Meek, K. (2009). "New targets to translate DNA-PK signals." *Cell Cycle* 8(23): 3809.

Meek, K., S. Gupta, et al. (2004). "The DNA-dependent protein kinase: the director at the end." *Immunol Rev* 200: 132-41.

Mimori, T. and J. A. Hardin (1986). "Mechanism of interaction between Ku protein and DNA." *J Biol Chem* 261(22): 10375-9.

Mladenov, E. and G. Iliakis "Induction and repair of DNA double strand breaks: the increasing spectrum of non-homologous end joining pathways." *Mutat Res* 711(1-2): 61-72.

Moon, S. H., T. A. Nguyen, et al. "Dephosphorylation of gamma-H2AX by WIP1: an important homeostatic regulatory event in DNA repair and cell cycle control." *Cell Cycle* 9(11): 2092-6.

Muslimovic, A., I. H. Ismail, et al. (2008). "An optimized method for measurement of gamma-H2AX in blood mononuclear and cultured cells." *Nat Protoc* 3(7): 1187-93.

Muslimovic, A., S. Nystrom, et al. (2009). "Numerical analysis of etoposide induced DNA breaks." *PLoS One* 4(6): e5859.

Nakada, S., G. I. Chen, et al. (2008). "PP4 is a gamma H2AX phosphatase required for recovery from the DNA damage checkpoint." *EMBO Rep* 9(10): 1019-26.

Nikjoo, H., P. O'Neill, et al. (1999). "Quantitative modelling of DNA damage using Monte Carlo track structure method." *Radiat Environ Biophys* 38(1): 31-8.

Paull, T. T., E. P. Rogakou, et al. (2000). "A critical role for histone H2AX in recruitment of repair factors to nuclear foci after DNA damage." *Curr Biol* 10(15): 886-95.

Pilch, D. R., O. A. Sedelnikova, et al. (2003). "Characteristics of gamma-H2AX foci at DNA double-strand breaks sites." *Biochem Cell Biol* 81(3): 123-9.

Porcedda, P., V. Turinetto, et al. (2008). "A rapid flow cytometry test based on histone H2AX phosphorylation for the sensitive and specific diagnosis of ataxia telangiectasia." *Cytometry A* 73(6): 508-16.

Qvarnstrom, O. F., M. Simonsson, et al. (2004). "DNA double strand break quantification in skin biopsies." *Radiother Oncol* 72(3): 311-7.

Redon, C. E., A. J. Nakamura, et al. "The use of gamma-H2AX as a biodosimeter for total-body radiation exposure in non-human primates." *PLoS One* 5(11): e15544.

Rodier, F., J. P. Coppe, et al. (2009). "Persistent DNA damage signalling triggers senescence-associated inflammatory cytokine secretion." *Nat Cell Biol* 11(8): 973-9.

Rogakou, E. P., C. Boon, et al. (1999). "Megabase chromatin domains involved in DNA double-strand breaks in vivo." *J Cell Biol* 146(5): 905-16.

Rogakou, E. P., D. R. Pilch, et al. (1998). "DNA double-stranded breaks induce histone H2AX phosphorylation on serine 139." *J Biol Chem* 273(10): 5858-68.

Rothkamm, K., I. Kruger, et al. (2003). "Pathways of DNA double-strand break repair during the mammalian cell cycle." *Mol Cell Biol* 23(16): 5706-15.

Rothkamm, K. and M. Lobrich (2003). "Evidence for a lack of DNA double-strand break repair in human cells exposed to very low x-ray doses." *Proc Natl Acad Sci U S A* 100(9): 5057-62.

Sak, A., S. Grehl, et al. (2007). "gamma-H2AX foci formation in peripheral blood lymphocytes of tumor patients after local radiotherapy to different sites of the body: dependence on the dose-distribution, irradiated site and time from start of treatment." *Int J Radiat Biol* 83(10): 639-52.

Sedelnikova, O. A., E. P. Rogakou, et al. (2002). "Quantitative detection of (125)IdU-induced DNA double-strand breaks with gamma-H2AX antibody." *Radiat Res* 158(4): 486-92.

Stiff, T., M. O'Driscoll, et al. (2004). "ATM and DNA-PK function redundantly to phosphorylate H2AX after exposure to ionizing radiation." *Cancer Res* 64(7): 2390-6.

Walker, J. R., R. A. Corpina, et al. (2001). "Structure of the Ku heterodimer bound to DNA and its implications for double-strand break repair." *Nature* 412(6847): 607-14.

Wang, J., J. M. Pluth, et al. (2005). "Artemis deficiency confers a DNA double-strand break repair defect and Artemis phosphorylation status is altered by DNA damage and cell cycle progression." *DNA Repair (Amst)* 4(5): 556-70.

Wang, L. H., T. D. Pfister, et al. "Monitoring drug-induced gammaH2AX as a pharmacodynamic biomarker in individual circulating tumor cells." *Clin Cancer Res* 16(3): 1073-84.

Ward, I. M. and J. Chen (2001). "Histone H2AX is phosphorylated in an ATR-dependent manner in response to replicational stress." *J Biol Chem* 276(51): 47759-62.

Ward, I. M., K. Minn, et al. (2004). "UV-induced ataxia-telangiectasia-mutated and Rad3-related (ATR) activation requires replication stress." *J Biol Chem* 279(11): 9677-80.

Yamauchi, M., Y. Oka, et al. (2008). "Growth of persistent foci of DNA damage checkpoint factors is essential for amplification of G1 checkpoint signaling." *DNA Repair (Amst)* 7(3): 405-17.

Yip, K. W., M. Cuddy, et al. "A high-content screening (HCS) assay for the identification of chemical inducers of PML oncogenic domains (PODs)." *J Biomol Screen* 16(2): 251-8.

Yoshikawa, T., G. Kashino, et al. (2009). "Phosphorylated H2AX foci in tumor cells have no correlation with their radiation sensitivities." *J Radiat Res (Tokyo)* 50(2): 151-60.

2

Suitability of the γ-H2AX Assay for Human Radiation Biodosimetry

Sandrine Roch-Lefèvre, Marco Valente,
Philippe Voisin and Joan-Francesc Barquinero
The Institute of Radiation Protection and Nuclear Safety, IRSN/DRPH/SRBE/LDB, France

1. Introduction

For biodosimetry purposes, there are only a few minimally invasive procedures. Two of them, fast and painless are particularly suited for large scale sample collection. One is to collect buccal cells, either by scraping the patient's inner cheek or using mouthwash. The other is to collect venous blood samples that are commonly used to obtain lymphocytes for chromosome analysis. Chromosome aberrations, such as dicentrics or translocations, have long been used as a dose indicator in biological dosimetry in cases of exposure or suspected overexposure to ionizing radiation (IAEA, 2001). The dicentric assay in peripheral blood lymphocytes is considered to be the reference method for biodosimetry: it is specific to ionizing radiation and stable enough to be used for dose estimation several months after exposure. However, it has several limitations; it is time consuming, it requires skilled personnel and needs a non-reducible period of 48-hour culture to obtain lymphocyte metaphases before chromosome scoring. One major application of biodosimetry is the identification, in the event of a large-scale radiation emergency, of the most severely exposed individuals. In this situation, a reliable bioassay is needed for population triage during the first few hours. Then, for faster dose estimation, current protocols for triage assessment call for the analysis of only 50 metaphases. This reduces the scoring delay somewhat but substantially augments the confidence interval and consequently decreases sensitivity, to 1 Gy (Miller, 2007; Voisin, 2004). The development of efficient new biodosimeters could overcome the limitations of dose sensitivity and avoid the lymphocyte culture step inherent to the dicentric assay.

Potential candidates include several proteins that are involved in the early steps of cellular response to ionizing radiation and specifically to DNA damage (Marchetti, 2006). One of the primary cellular effects of ionizing radiation is the induction of DSB (double-strand breaks). Following DSB induction, hundreds of histone H2AX molecules are phosphorylated in the chromatin flanking the DSB site and generates so-called γ-H2AX (Rogakou, 1998). The production of fluorescent antibodies specific for γ-H2AX coupled with fluorescence microscopy led to the development of sensitive assays that make it possible to visualize discrete nuclear foci at DSB sites (Rogakou, 1999). Foci induction and disappearance over time follows DSB rejoining in repair-competent cells, suggesting a correlation between initial as well as residual radiation-induced DSB and γ-H2AX foci (Rothkamm & Horn,

2009). The scoring of γ-H2AX foci is now widely used for quantitative evaluation of DSB formation and repair (Rothkamm, 2003) (Olive & Banath, 2004). Recent immunofluorescence studies show that the yield of these foci induced by ionizing radiation in humans increases linearly with the radiation dose after both *in vitro* and *in vivo* exposure (Leatherbarrow, 2006; Lobrich, 2005; Rothkamm, 2007; Sak, 2007). It has been shown that the scoring of γ-H2AX foci in human lymphocytes can be used to estimate very low doses (down to few cGy) at 30 minutes after *in vivo* radiation exposure (Lobrich, 2005; Rothkamm, 2007). However, since the level of γ-H2AX foci varies with time after irradiation, the sensitivity of its detection is necessarily time-dependent: very high sensitivity after few minutes but fair sensitivity at several hours post-exposure. Whilst microscopic imaging and scoring of γ-H2AX foci offers the highest sensitivity, intensity-based assays for γ-H2AX are widely used in experimental research and may offer some advantages in terms of throughput and automation (Rothkamm & Horn, 2009). In the contexts where the user is particularly interested in quickly quantifying the γ-H2AX signalling response, most attempts of developing a fast H2AX assay used flow cytometry (Ismail, 2007). This technique measures a relative intensity of the γ-H2AX staining instead of scoring the actual number of foci and is known to have a large level of inter-individual variation (Andrievski & Wilkins, 2009; Hamasaki, 2007; Ismail, 2007). Total γ-H2AX intensity levels are dose dependent and approximately linear up to a supralethal dose of 100 Gy (Ismail, 2007). Recently, a dual method has been developed to determine fluorescence yield using high-speed microscope imaging analysis. This workstation has been designed to fully automate the γ-H2AX immunocytochemical protocol, from the isolation of human blood lymphocytes to the image acquisition step (Turner, 2011).

The translation of γ-H2AX analysis into a reliable dosimetry device nonetheless requires both further validation and better automated methods in microscopy-based scoring. Despite the wide application of this appraoch, γ-H2AX counting is frequently carried out manually by eye and may be prone to investigator-related biases. Currently, manual counting of γ-H2AX foci on microscopy images is a time-consuming, tedious process, especially for doses higher than 0.1 Gy, where there are many foci throughout the nucleus. In the laboratories that manually score γ-H2AX foci today, there are a few numbers of trained investigators, to minimize scoring artefacts. Indeed, focus scoring uses measurement features like the focus size or brightness which are very difficult to evaluate objectively by eye. Therefore, automatic systems for γ-H2AX analysis may be used to allow a consistent and a fast quantification of γ-H2AX in a wide dose range that would be compatible with biodosimetry. Here, different methods that use either software already available or home-made developed for the automatic analysis of γ-H2AX are reviewed. A comparison of the results obtained for γ-H2AX responses in relation to dose, time since exposure, lower and higher limit of detection is made in this chapter.

2. γ-H2AX assay in peripheral lymphocytes

Peripheral blood mononuclear cells (PBMCs) are useful to evaluate the effects of ionizing radiation exposure, as they can be obtained with minimal invasiveness and under standard conditions. For the evaluation of γ-H2AX foci formation, PBMCs are very attractive because considerable amounts of cells can be easily obtained within a short time. Because monocytes and granulocytes can be excluded from the γ-H2AX analysis, the data obtained in PBMCs mainly represent the result of a mixture of the different lymphocyte subpopulations. This cell selection can be done during the analysis either visually by the scorer according to morphological features or by a software using nucleus staining intensity criteria.

2.1 Microscopy-based γ-H2AX analysis

To count the foci present in the nuclei is currently the most sensitive method for γ-H2AX analysis. γ -H2AX foci become microscopically visible within minutes after irradiation, indicating the rapid phosphorylation of thousands γ-H2AX molecules. Thanks to this large scale formation of γ-H2AX, foci can be easily distinguished from a relatively homogeneous background signal so that one individual DSB can be detected (Fig. 1). In case of manual scoring by eye directly at the microscope, the speed can be relatively fast for a well-trained scorer, but this approach quickly becomes tiresome if many samples need to be analysed. This method of γ-H2AX-based visualization and quantification in PBMCs has been used to estimate the radiation dose received by adult patients who had undergone either multidetector computed tomography (CT) or radiotherapy treatment (Rothkamm, 2007; Sak, 2007). In these studies, blood samples obtained from patients before the exposure were

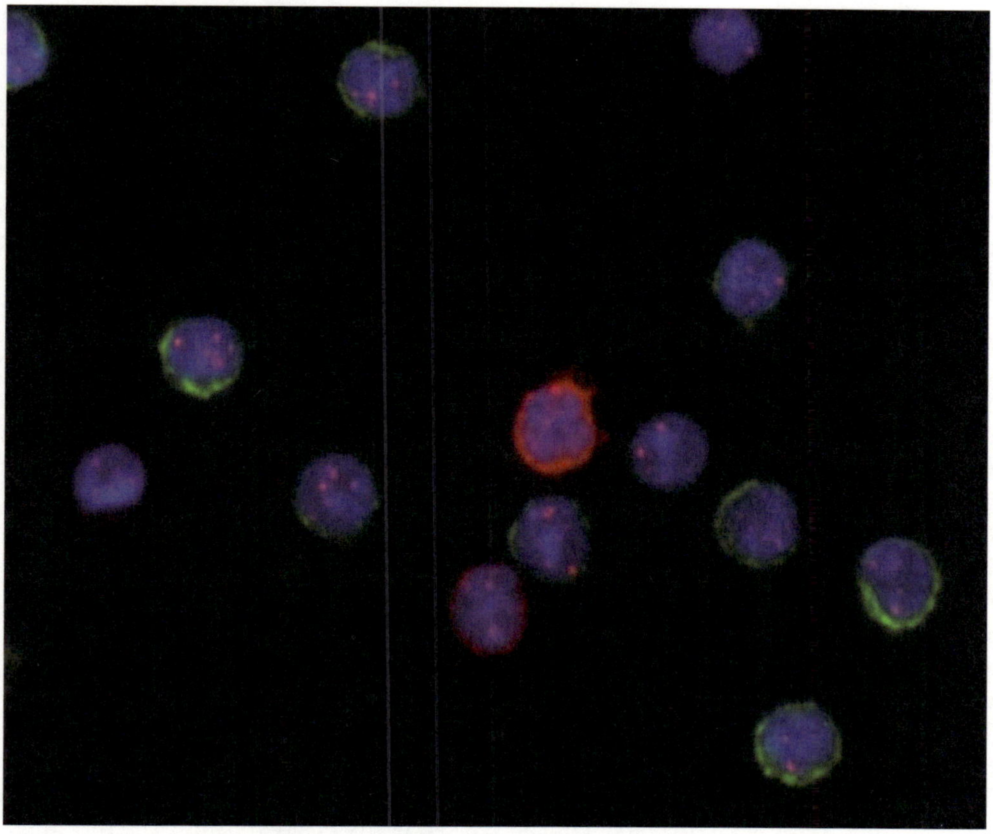

Fig. 1. Gamma-H2AX foci in human lymphocytes (CD4 and CD8 sub-types) after 0.5 Gy of γ-rays, 30 minutes post-exposure. The blue channel corresponds to DAPI staining; the magenta channel corresponds to γH2AX staining; the green and red channels correspond to CD4+ and CD8+ membrane staining, respectively.

irradiated *in vitro* at doses from 0 to 1 Gy. The results obtained *in vitro* from these studies show a good linear dose-effect relationship in PBMCs. However, although mean background of γ-H2AX were very similar in both studies (0.06 and 0.057 focus per cell, respectively) the induced foci yields 30 minutes after *in vitro* irradiation were quite different: 3 times higher in the CT study than in the radiotherapy study. This discrepancy in radiation-induced response of γ-H2AX could have been attributed to the difference in radiation quality types: X-rays 150 kV vs. γ-rays of Cobalt 60, respectively. Concerning the results obtained *in vivo*, the CT study shows that the doses obtained from the γ-H2AX focus yields after whole-body CT (16.4 mGy) or chest CT (6.3 mGy) were similar to phantom dosimetry-calculated doses (13.85 mGy and 5.16 mGy, respectively). In the radiotherapy study, it has been shown that the *in vivo* formation of γ -H2AX foci in dependence of the applied integral dose in cancer patients irradiated at different sites of the body follows a linear relationship. In another publication where γ-H2AX foci yields were also evaluated visually in PBMCs irradiated *in vitro* either with X-rays 100 kV or γ-rays of Cobalt 60; the slopes of the dose-effect relations were quite similar (about 10 foci per cell Gy^{-1}) in a dose range of 0.01 – 0.5 Gy (Beels, 2010). These results mean that the discrepancy between γ-H2AX foci yields observed in studies from different laboratories is rather due to the differences in scoring criteria than to true differences in radiation-induced responses. Furthermore, it is important to consider that these estimates are only true for blood samples that are processed within less than one hour after irradiation.

Although γ-H2AX foci can be detected just few minutes after irradiation, in a radiation emergency scene, it is however quite unrealistic to expect that blood samples can be obtained within an hour of exposure. Since the method will only be practical if it can be applied to samples taken at longer times post exposure, dose-response curves are needed for later time points. The response of lymphocytes to doses was examined visually in blood samples from individuals exposed *in vitro* to increasing doses (0.02–5 Gy) up to 2 days after irradiation (Redon, 2009). Thirty minutes post-exposure, doses as low as 0.02 Gy were detectable, and, for doses up to 2 Gy, the data followed a linear relationship between dose and γ-H2AX signal. For doses greater than 1 Gy, a substantial response was detected in the lymphocytes for up to 2 days after exposure (Redon, 2009). These results - obtained visually– have been confirmed *in vivo* in non-human PBMCs after irradiation of macaques receiving total body γ-ray doses ranging from 1 Gy to 8.5 Gy, for times up to 4 days after exposure (Redon, 2010).

Automatic systems may be used to avoid focus scoring bias that may prevent inter-laboratory comparisons. It allows fast scoring of γ-H2AX foci for biodosimetry purposes. Several groups have developed image analysis solutions for automated foci scoring in human cellular models (Bocker & Iliakis, 2006; Costes, 2006; Leatherbarrow, 2006; Mistrik, 2009). A semi-automatic approach, commercially available, has been used to score γ-H2AX foci in PBMCs for radiation doses assessment up to 16h after radiation exposure. Dose response curves were designed at different times post-exposure and the threshold of detection of the technique was evaluated at each time point (Roch-Lefèvre, 2010). The slopes of the dose-effect relation decreased with time post-exposure, from 10.7 γ-H2AX foci per cell Gy^{-1} at 30 minutes down to 0.50 foci per cell Gy^{-1} 16 hours post-exposure. The threshold of detection - defined as the lowest dose (D_{LD}) that could be detected was calculated for each time post-exposure. The calculated D_{LD} values were 0.05 Gy at 30 minutes, 0.3 Gy at 8 hours and 0.6 Gy, 16 hours post-exposure. It should be noticed that the disappearance of γ-H2AX can be inhibited by incubating the whole blood on ice at 0°C which is important in biological dosimetry where blood samples are analyzed several hours after they have been withdrawn (for blood sample transport to the expert laboratory, for example) (Roch-Lefevre, 2010). It has been shown that blood samples could be stored on ice for extended periods of time without loss of the gamma-H2AX signal (Ismail, 2007).

2.1.1 Automation of image acquisition

Automation of image acquisition is a major issue for high throughout microscopy analysis. Resources commonly found in laboratories for automatic acquisition coupled with export of gray-scale TIFF files can be used to acquire images of lymphocyte nuclei and γ-H2AX foci. Actually, automated multi-colour image acquisition is becoming a standard function in many systems of computer-controlled fluorescent microscopes. Numerous possibilities exist, both commercial and open source, like micro-manager (http://micro-manager.org/) (Valente, 2011). Also in the interest of shortening both acquisition and analysis time, several technical parameters, like the lymphocyte density on slide and the minimum number of cells to score has to be optimized.

2.1.2 Automation of image analysis: Image cytometry

While fluorescence microscope in combination with a digital camera is standard equipment in many laboratories, the availability of specialized software, especially freeware, for the analysis of γ-H2AX foci is increasing (Carpenter, 2006) (Jucha, 2010) (Ivashkevich, 2011). Image analysis implies two distinct steps: selection of the cells and detection of γ-H2AX foci; therefore it needs to load two colour-types of grey scale images: one with counterstained nuclei for cell selection and the other one with fluorescent-coupled anti-γ-H2AX for focus detection. These two steps of analysis can be done manually, either fully automatically or semi-automatically, which all lead to advantages and limitations as described in Fig. 2. In this chapter, are compared two applications: Histolab™, a semi-automatic software and CellProfiler, a freeware that allows full-automated analysis (Roch-Lefevre, 2010) (Valente, 2011). Histolab™ is considered as semi-automatic since it still requires that the images be loaded separately by the user during analysis. It detects both nuclei and foci by applying a fixed threshold and a"top hat" threshold to the greyscale images of the nucleus and γ - H2AX staining, respectively. The major advantage of this software is that user interface is accessible, rendering parameters setting for foci detection very easy. Furthermore, in the current version of HistoLab™, the operator has the possibility of removing manually the cells with aberrant staining or morphology. But, with other software approaches where there is no operator intervention (as with CellProfiler for example); the cell type selection has to be done graphically using measurements of nucleus staining and morphology (Valente, 2011). For the nucleus analysis with CellProfiler an adaptive algorithm is applied on the counterstained-nucleus pictures to detect nuclei with a range of diameters. Inside these nuclei, another module can use a various choice of algorithms on the H2AX pictures (pre-subjected or not to a treatment) to detect foci with possibly minimum and maximum diameters. The last modules measure the area and intensity of both nuclei and foci. CellProfiler is a program that offers a great number of measurements and detection options. The main drawback of this freeware is a less user-friendly graphical interface that may render the vast number of detection algorithms and measurements overwhelming for a beginner. When CD-specific membrane staining takes place (as seen in Fig. 1.) CellProfiler is also able to measure the mean intensity of the associated colour channel inside each cell. The optimization of this approach has been described for PBMCs isolated from blood exposed *ex vivo* to different doses of radiation (Valente, 2011).

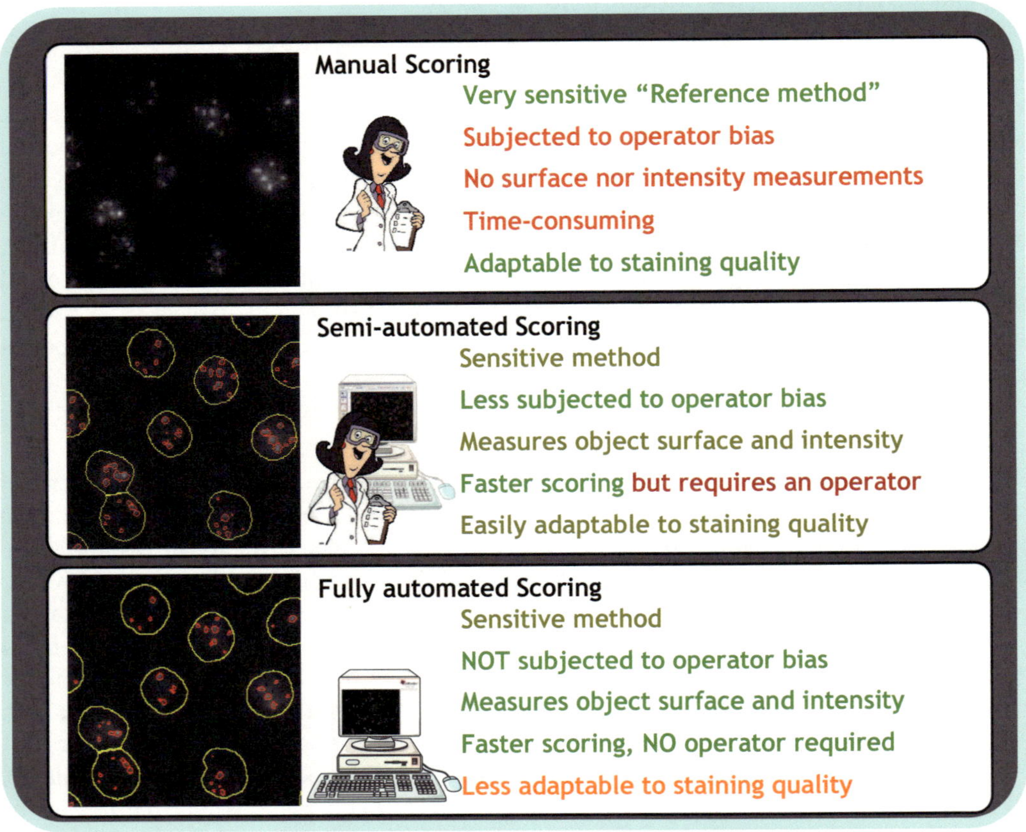

Fig. 2. Advantages and limitations for 3 methods of γ-H2AX focus scoring

All free (and paid) automated scoring applications seem to present good correlations to manual scoring once the right parameters are inputted. The free alternative that resembles most HistoLab™ is FociCounter (Jucha, 2010): it was developed specifically to score foci, with a simple interface and few parameters to change (faster to set up). These programs are ideal for simple scoring of a reduced number of cells/conditions, where the quality of the images/slides frequently requires operator intervention (to eliminate aberrant objects, for example). Since these programs require user intervention to select nuclei, the time of the analysis is their main disadvantage. Using CellProfiler, once the pipeline is set up, it requires very little operator intervention for analysis. A free equivalent is difficult to find. We do not compare it to imageJ as a CellProfiler module has been recently created to allow the user to run ImageJ macros and plugins as part of a CellProfiler image processing pipeline (http://cellprofiler.org/CPmanual/RunImageJ.html). In a recent publication, a new image cytometry program that will become freely available was presented (Ivashkevich, 2011). The details on the measurements that will be possible to obtain with this software are not yet disclosed, so we cannot fairly compare it with CellProfiler.

However, the computational approach they present is a good reference to help researchers select the parameters of other programs of this type (Ivashkevich, 2011).

2.2 Intensity-based γ-H2AX analysis

While fluorescence microscopy enables individual γ-H2AX foci and their characteristics to be imaged and analysed, flow cytometry provides a more rapid and straightforward method of γ-H2AX quantification, that is by measuring the total fluorescence intensity for each cell.

2.2.1 Flow cytometry γ-H2AX analysis

Using a rapid flow cytometry method, the dose response of γ-H2AX induction in lymphocytes and lymphocyte subsets was found to be linear in the dose range examined (0–10 Gy) (Andrievski & Wilkins, 2009). Although this linear dose relationship and speed of this assay are advantageous for using γ-H2AX as a biological dosimeter, it is limited by the high variability between experiments with standard errors of the mean often reaching more than 20% of mean values, indicating high intra-donor variability. This level of inter and intra-individual variation is consistent with previously reported data in human lymphocytes (Hamasaki, 2007). After an accidental exposure, the dosimetry will be complicated by the lack of control samples for each individual that could be used as a comparison to the post-exposure levels of expression.

2.2.2 Other intensity-based γ-H2AX analysis

γ-H2AX total fluorescence measurements were determined in lymphocyte nuclei from blood samples collected from four healthy donors, irradiated (*ex vivo*) with a range of γ-ray doses between 0 and 8 Gy (Turner, 2011). Dose response for γ-H2AX was analyzed 30 min after irradiation. Curve-fitting analysis showed that the induction of total γ-H2AX fluorescence was linear with increasing γ-ray doses up to 8 Gy. Statistical analysis of the individual data points at 2 Gy showed that there was a significant induction of total γ-H2AX fluorescence. However, with this method, the data for one of the four donors show that there was no significant difference in the γ-H2AX levels at 1 Gy, 30 min postirradiation, suggesting that the sensitivity of this method is above 1 Gy whereas sensitivity of γ-H2AX foci scoring methods are above 0.05 Gy at 30 minutes post-exposure (Roch-Lefèvre, 2010).

3. γ-H2AX assay in epithelial cells

3.1 γ-H2AX assay in exfoliated buccal cells

Buccal cells are expected to accumulate DNA damage after exposure to DNA strand-breaking agents such as ionizing radiation; making the system potentially useful to estimate a dose received by the very upper part of the body. To test the γ-H2AX assay for the detection of ionizing radiation-induced DNA damage in buccal exfoliated cells, an *in vitro* study was carried out with these cells from healthy donors exposed to γ-rays in a dose range of 0 – 4Gy (Gonzalez, 2010). The γ-H2AX foci rate of 0.08 foci per cell observed in non-irradiated buccal cells was similar to the one usually obtained in sham-irradiated lymphocytes. The number of γ-H2AX foci increased linearly with ionizing radiation dose in the interval from 0 to 4 Gy, and reached a rate of 0.82 foci per cell at 4 Gy which is much lower than in lymphocytes. Incubation experiments after *in vitro* irradiation revealed that the number of γ -H2AX foci did not show a significant decrease 5 hours post-exposure (Gonzalez, 2010). It was concluded that it is possible to apply the γ -H2AX foci assay for the detection of ionising radiation-induced DNA damage in buccal exfoliated cells. The low removal of ionising radiation induced γ-H2AX foci in buccal cells may be a potential advantage for a biological dosimetry application.

3.2 γ-H2AX assay in plucked-hair cells

Plucking hairs is non-invasive and fairly painless compared to other types of sample collection. As buccal cells, it could be withdrawn by non-clinical personnel and then, it may be suited as well for large scale sample collection. It has been shown in macaques that the detection of γ-H2AX in plucked hairs can be used to monitor total body irradiation for several days after exposure (Redon, 2010). Moreover, the γ-H2AX signal decreased little or not at all between 1 and 2 days post-irradiation. It was concluded, as for buccal cells, that this slower rate of γ-H2AX foci loss compared to lymphocytes may be due to a lack of repair in hair differentiating hair cells and may be a potential advantage for biological dosimetry purposes (Redon, 2010) (Gonzalez, 2010).

4. Main issues to address for biodosimetry purposes

In contrast to chromosome-based dosimetry methods, nowadays well-established, a wide range of issues have to be addressed for the γ-H2AX assay before it will become completely relevant to its application in biological dosimetry. There are two main limitations that have to be addressed: inter- or intra-individual variability and partial-body exposure. A high individual variability leads to a decreased sensitivity of the γ-H2AX assay. This individual variability in radiation induced γ-H2AX focus yields was measured *in vitro* in the lymphocytes (from up to 27 healthy donors) before exposure and at different times after exposure (Roch-Lèfevre, 2010). In non-exposed lymphocytes, the inter-individual variation of the γ-H2AX yield was two to three times higher than the intra-individual variation, with a coefficient of variation of 31%. On the other hand, after irradiation, the inter-individual variation of the γ-H2AX yield was similar to the intra-individual one. Furthermore, radiation accidents are more commonly characterized by acute partial-body or inhomogeneous rather than homogeneous total-body radiation exposures. It has been shown that the yield of γ-H2AX foci formation can allow the estimation of the applied integral body dose after local radiotherapy to different sites of the body (Sak, 2007). However, further work is required to determine the effect of partial body exposure on γ-H2AX levels. In vivo exposures involving radiotherapy patients may help to determine the effects of lymphocyte circulation through the various body organs on γ-H2AX levels in blood samples taken at different time points after treatment. Also, changes in the distribution of γ-H2AX foci over time would have to be determined.

5. Conclusion

The gamma-H2AX assay will require additional *in vivo* experiments for validation. In particular, all usual limitations for human biodosimetry, particularly in regards to the different accident scenarios (delayed, fractioned, inhomogeneous exposures…) should be addressed. The γ-H2AX signal is also fairly short lived such that samples from potentially exposed individuals would need to be collected and processed quickly after exposure. These issues make this assay have limited applications, for rapid screening of individuals when samples can be collected and processed within 24 hours.

6. References

Andrievski, A. & Wilkins, R. C. (2009). The response of gamma-H2AX in human lymphocytes and lymphocytes subsets measured in whole blood cultures. *Int J Radiat Biol.* 85(4): 369-376.

Beels, L.,Werbrouck, J., et al. (2010). Dose response and repair kinetics of gamma-H2AX foci induced by in vitro irradiation of whole blood and T-lymphocytes with X- and gamma-radiation. *Int J Radiat Biol.* 86(9): 760-768.

Bocker, W. & Iliakis, G. (2006). Computational Methods for analysis of foci: validation for radiation-induced gamma-H2AX foci in human cells. *Radiat Res.* 165(1): 113-124.

Carpenter, A. E.,Jones, T. R., et al. (2006). CellProfiler: image analysis software for identifying and quantifying cell phenotypes. *Genome Biol.* 7(10): R100.

Costes, S. V.,Boissiere, A., et al. (2006). Imaging features that discriminate between foci induced by high- and low-LET radiation in human fibroblasts. *Radiat Res.* 165(5): 505-515.

Gonzalez, J. E.,Roch-Lefevre, S. H., et al. (2010). Induction of gamma-H2AX foci in human exfoliated buccal cells after in vitro exposure to ionising radiation. *Int J Radiat Biol.* 86(9): 752-759.

Hamasaki, K.,Imai, K., et al. (2007). Short-term culture and gammaH2AX flow cytometry determine differences in individual radiosensitivity in human peripheral T lymphocytes. *Environ Mol Mutagen.* 48(1): 38-47.

IAEA (2001). Cytogenetic Analysis for Radiation Dose Assessment A Manual. Vienna, Austria, International Atomic Energy Agency. Technical reports series No. 405.

Ismail, I. H.,Wadhra, T. I., et al. (2007). An optimized method for detecting gamma-H2AX in blood cells reveals a significant interindividual variation in the gamma-H2AX response among humans. *Nucleic Acids Res.* 35(5): e36.

Ivashkevich, A. N.,Martin, O. A., et al. (2011). gammaH2AX foci as a measure of DNA damage: A computational approach to automatic analysis. *Mutat Res.*

Jucha, A.,Wegierek-Ciuk, A., et al. (2010). FociCounter: a freely available PC programme for quantitative and qualitative analysis of gamma-H2AX foci. *Mutat Res.* 696(1): 16-20.

Leatherbarrow, E. L.,Harper, J. V., et al. (2006). Induction and quantification of gamma-H2AX foci following low and high LET-irradiation. *Int J Radiat Biol.* 82(2): 111-118.

Lobrich, M.,Rief, N., et al. (2005). In vivo formation and repair of DNA double-strand breaks after computed tomography examinations. *Proc Natl Acad Sci U S A.* 102(25): 8984-8989.

Marchetti, F.,Coleman, M. A., et al. (2006). Candidate protein biodosimeters of human exposure to ionizing radiation. *Int J Radiat Biol.* 82(9): 605-639.

Miller, S. M.,Ferrarotto, C. L., et al. (2007). Canadian Cytogenetic Emergency network (CEN) for biological dosimetry following radiological/nuclear accidents. *Int J Radiat Biol.* 83(7): 471-477.

Mistrik, M.,Oplustilova, L., et al. (2009). Low-dose DNA damage and replication stress responses quantified by optimized automated single-cell image analysis. *Cell Cycle.* 8(16): 2592-2599.

Olive, P. L. & Banath, J. P. (2004). Phosphorylation of histone H2AX as a measure of radiosensitivity. *Int J Radiat Oncol Biol Phys.* 58(2): 331-335.

Redon, C. E.,Dickey, J. S., et al. (2009). gamma-H2AX as a biomarker of DNA damage induced by ionizing radiation in human peripheral blood lymphocytes and artificial skin. *Adv Space Res.* 43(8): 1171-1178.

Redon, C. E.,Nakamura, A. J., et al. (2010). The use of gamma-H2AX as a biodosimeter for total-body radiation exposure in non-human primates. *PLoS One.* 5(11): e15544.

Roch-Lefèvre, S.,Mandina, T., et al. (2010). Quantification of gamma-H2AX foci in human lymphocytes: a method for biological dosimetry after ionizing radiation exposure. *Radiat Res.* 174(2): 185-194.

Rogakou, E. P.,Boon, C., et al. (1999). Megabase chromatin domains involved in DNA double-strand breaks in vivo. *J Cell Biol.* 146(5): 905-916.

Rogakou, E. P.,Pilch, D. R., et al. (1998). DNA double-stranded breaks induce histone H2AX phosphorylation on serine 139. *J Biol Chem.* 273(10): 5858-5868.

Rothkamm, K.,Balroop, S., et al. (2007). Leukocyte DNA damage after multi-detector row CT: a quantitative biomarker of low-level radiation exposure. *Radiology.* 242(1): 244-251.

Rothkamm, K. & Horn, S. (2009). gamma-H2AX as protein biomarker for radiation exposure. *Ann Ist Super Sanita.* 45(3): 265-271.

Rothkamm, K.,Kruger, I., et al. (2003). Pathways of DNA double-strand break repair during the mammalian cell cycle. *Mol Cell Biol.* 23(16): 5706-5715.

Sak, A.,Grehl, S., et al. (2007). gamma-H2AX foci formation in peripheral blood lymphocytes of tumor patients after local radiotherapy to different sites of the body: dependence on the dose-distribution, irradiated site and time from start of treatment. *Int J Radiat Biol.* 83(10): 639-652.

Turner, H. C.,Brenner, D. J., et al. (2011). Adapting the gamma-H2AX assay for automated processing in human lymphocytes. 1. Technological aspects. *Radiat Res.* 175(3): 282-290.

Valente, M.,Voisin, P., et al. (2011). Automated gamma-H2AX focus scoring method for human lymphocytes after ionizing radiation exposure. *Radiat Measurements.* In press.

Voisin, P.,Roy, L., et al. (2004). Criticality accident dosimetry by chromosomal analysis. *Radiat Prot Dosimetry.* 110(1-4): 443-447.

3

Biological Dosimetry of Ionizing Radiation

Aurélie Vaurijoux, Gaëtan Gruel,
Sandrine Roch-Lefèvre and Philippe Voisin
*Laboratory of Biological Dosimetry,
Institute of Radioprotection and Nuclear Safety,
France*

1. Introduction

A worsening of the accidental hazards linked to the use of ionizing radiation is currently being observed for four reasons. First, the increasing need for radiation sources in numerous industrial applications (food sterilization, construction, engineering...) leads to an increasing probability of loss of the sources or abnormal/unsuitable use and storage. Second, advances in medicine generate new protocols and tools that are more efficient but also much more complex to carry out, increasing the risk of accidental overexposure. Third, the possibility of a terrorist attack using radiological or nuclear devices has to be taken into account. Finally, recent events in Fukushima (Japan) highlight the risks of exposure in the case of nuclear power plant accidents. All these issues could lead to the accidental exposure of one to several thousand individuals not wearing dosimeters. Thus, it is essential to be able to estimate the exposure level of victims. Nowadays, this evaluation is based on clinical diagnosis (mainly irradiation symptoms and hematological variations) supplemented with biological dosimetry and physical dose reconstruction. Biological dosimetry is especially important when the personal dosimeter is lacking or when the accidental context is unclear. All this information should help the medical staff to deliver appropriate medical care and to manage the long-term medical follow-up, if required.

It has been known since the last century that ionizing radiation causes DNA damage and that DNA misrepair can induce chromosome aberrations: stable (translocations, deletions, insertions) or unstable (dicentrics, centric rings, acentric fragments). These aberrations are observed in metaphase cells. A misrepair can be also observed after anaphase in the form of micronuclei. The applicability of the available assays of biodosimetry is based on the analysis of the chromosome damage present in peripheral blood lymphocytes, which is a convenient because its collection is non-invasive and it is easy to obtain.

Dicentric assay is currently the gold standard method for classic biodosimetry in cases of recent accidental exposure. The scoring of dicentrics allows assessment of the whole-body dose received by the individual. Moreover, in numerous accident contexts the exposure of victims is heterogeneous. In these cases, the fraction of the body irradiated and the dose received by this fraction can be estimated by dicentric scoring. In the case of large-scale accidents the dicentric aberration is always used as well as the micronuclei because its analysis is faster and easier. However, dicentrics and micronuclei are unstable aberrations and their rate decreases with time. Consequently, for past accidental exposure, the analysis of stable chromosome aberrations like translocations is required (Figure 1).

Nevertheless, these bioindicators have many limitations. Currently, a lot of research is performed to find new indicators of exposure (such as γH2AX and gene expression) (Figure 1). It is important to note that there are several essential requirements for biological parameters as meaningful dosimeters: low background level, clear dose-effect relationship for different radiation qualities and dose rates, specificity to ionizing radiation, noninvasive, fast availability of dose estimate, good reproducibility and comparability of in vitro and in vivo results (Romm, 2009).

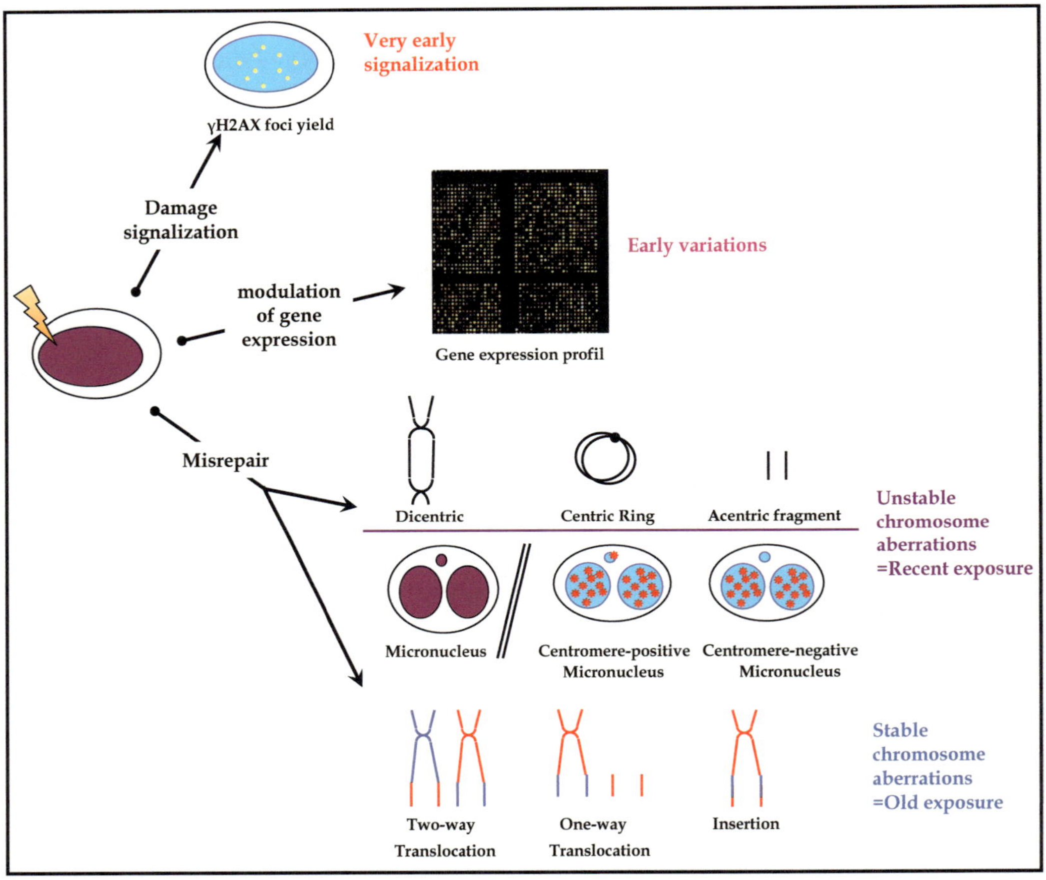

Fig. 1. After irradiation of cells, the induced DNA damage was observed i) very early with γH2AX foci yield, ii) early with variations of gene expression profile, and iii) later with chromosome aberrations.

2. Recent accidental exposure

2.1 Individual biodosimetry

Due to their instability and low background level, the dicentric rate in lymphocytes is an indicator of recent exposure. For dose assessment using dicentric assay, it is necessary to perform a dose-effect curve range to 0.1-5 Gy representing the relationship between the yield of dicentrics and the dose. The yield of dicentrics is obtained by manual scoring of dicentrics, rings and acentric fragments among 500 or more metaphase cells obtained from peripheral blood lymphocytes after Giemsa staining (Figure 2). This technique requires a lymphocyte culture time from 48 to 50 hours. Additionally, the manual scoring of unstable aberrations on 500 metaphase images requires about 10 hours for a trained operator. This technique has a minimum detection threshold ranging from 0.1 to 0.2 Gy depending on the number of metaphase cells scored, background level and the radiation used (quality and dose-rate) (Agency, 2001).

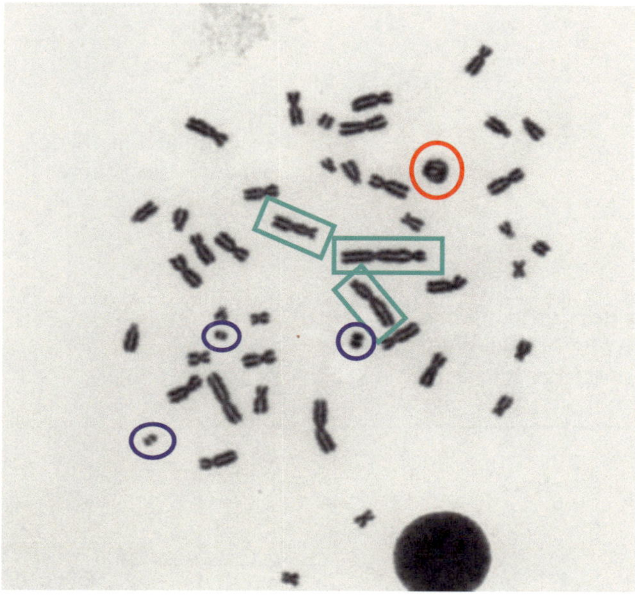

Fig. 2. Metaphase of peripheral blood lymphocytes. Unstable chromosomal aberrations are analyzed with Giemsa staining. Dicentrics are framed in green, rings in red and acentric fragments in blue.

2.1.1 Dose-effect relationship

Ionizations induce damages in the DNA (base damage, double-strand break, single-strand break…), their number in cells increases with the ionizing capacity of the radiation. Besides, the damage distribution between cells will differ according to the type of radiation because the quantity of ionization induced per track varies. Indeed, a track is formed by one more or less ionizing particle. This can be defined by the linear energy transfer (LET) value which describes the energy deposition per micrometer of matter.

Dicentric damage requires at least two double-strand breaks to be formed. The frequency of production of dicentrics induced by one track is proportional to the dose. By contrast, the frequency of production of dicentrics induced by two tracks is proportional to the square of the dose.

For low LET radiation (X or gamma rays) produces many tracks containing few primary events. The distribution of tracks is more randomized. Therefore, the distribution of damage between cells will be more uniform. There is a greater probability that two tracks induce one dicentric in the same cell. The dose-effect relationship is then linear in the low dose range (dicentrics induced exclusively by two tracks) and becomes quadratic at high doses (Figure 3). The dose-effect curve then fits the following equation:

$$Y = \alpha D + \beta D^2 + c$$

where β is the dose squared coefficient and the constant is the background frequency (Edwards, 1979). In this case, the distribution of dicentrics per cell follows a Poisson law. The coefficients α, β and the constant, with their standard errors, are calculated using a maximum likelihood method (Papworth, 1975).

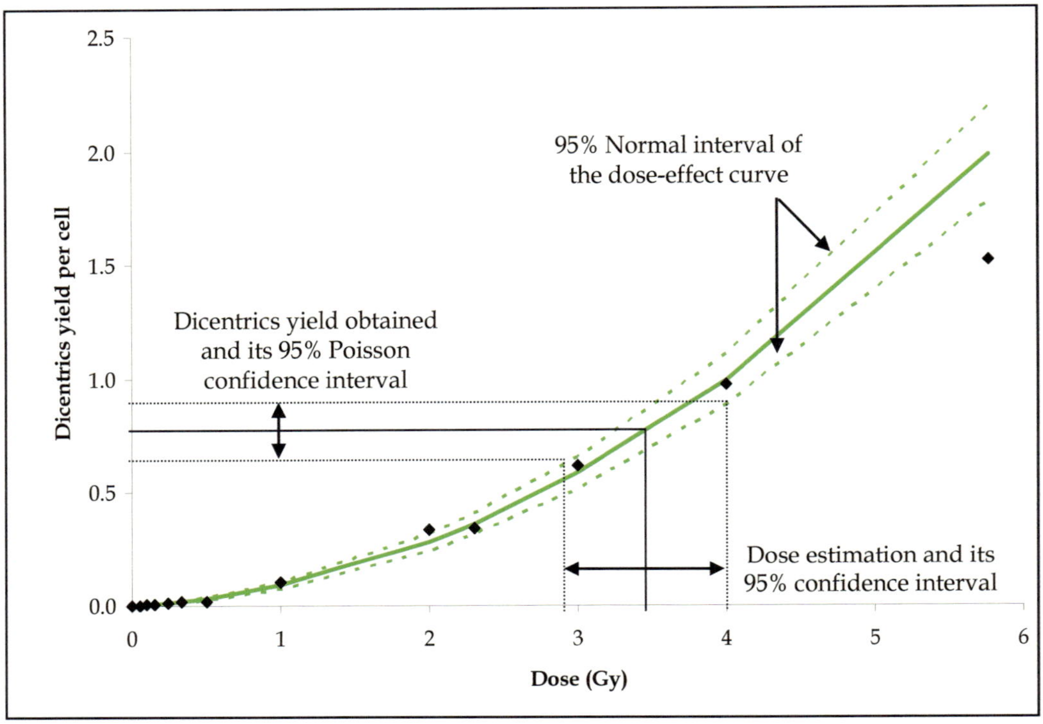

Fig. 3. Dose-effect relationship fitted by scoring of dicentrics induced in peripheral blood lymphocytes after in vitro irradiation by 137Cs gamma rays (dose rate of 0.5 Gy/min). The equation of the curve is $Y=0.0338D+0.0536D^2+0.0010$.

From the dicentric yield obtained for an individual a 95% Poisson confidence interval is calculated using the Poisson table. The dose is estimated by correspondence of the dicentric yield to the dose-effect curves (Figure 3). The confidence interval of the dose is estimated by correspondence of the lower and upper confidence interval of the dicentric yield on the upper and lower curves, respectively (Papworth, 1975).

High LET radiation (neutrons or alpha particles) produces few tracks with many primary events (ionizations, excitations) very close together. There is a lot of damage at the same point and the misrepair of these damages induces multi-aberrant cells. With high LET radiation exposure, there is a high probability of one track inducing one dicentric. In this case, the dose-effect relationship is linear (Figure 4) and the dose-effect curve fits the following equation:

$$Y = \alpha D + c$$

where Y is the dicentric yield, α is the linear coefficient and D is the dose (Lloyd, 1976).

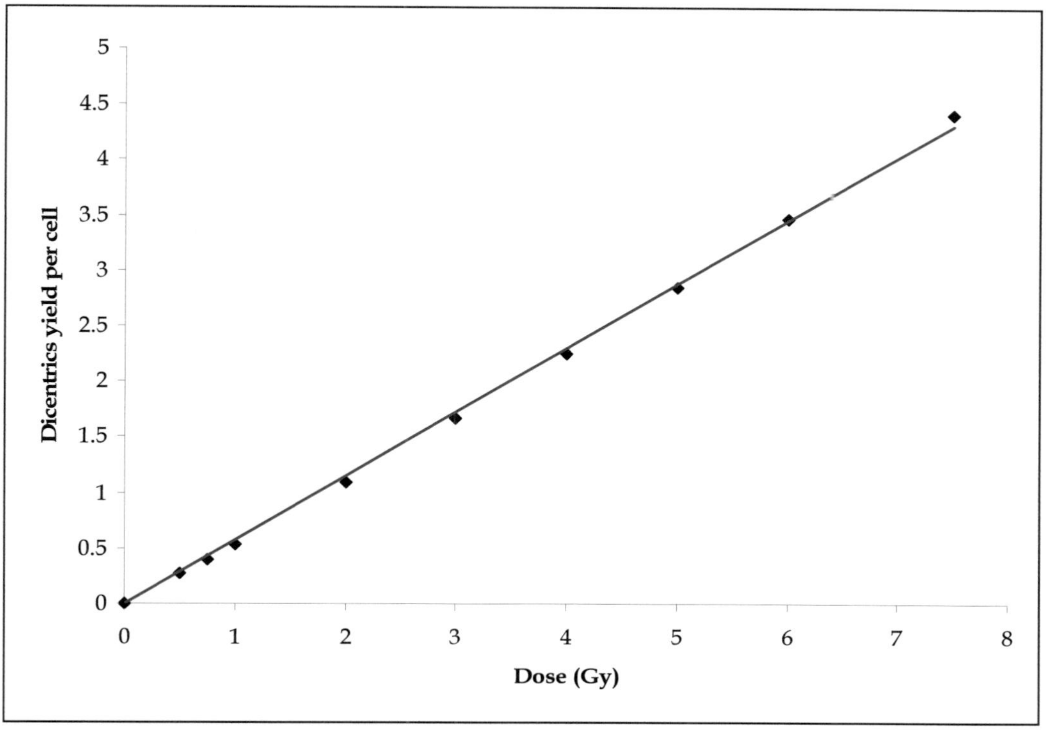

Fig. 4. Dose-effect relationship fitted by scoring of dicentrics induced in peripheral blood lymphocytes after in vitro irradiation with fission neutrons (1 MeV). The equation of the curve is $Y=0.5356D+3x10^{-16}$

2.1.2 Case of partial-body exposure

During exposure to low LET radiation, a dicentric is considered a rare event. It is mostly accepted that its distribution among the analyzed metaphases follows Poisson's law. In biodosimetry, we check that the dicentric distribution obtained is in agreement with Poisson distribution by u-test. In the case of partial-body exposure, mathematical models have been developed to estimate the fraction of the body irradiated and the dose received by this fraction, the contaminated Poisson's method (Dolphin, 1969) and the Qdr method (Sasaki&Miyata, 1968).

If the exposure is partial, this distribution does not follow Poisson's law. Indeed, unirradiated cells are now scored within the population of analyzed metaphases and increase the number of undamaged cells. This contamination produces an overdispersion of the distribution and this is tested by u-test on each distribution obtained. The deviation of the variance was calculated and used to calculate the u value which approximates to a unit normal deviate. Thus, the values of u were compared with the theoretical value of 1.96, and this was used to identify a significant under- or overdispersion of the experimental distributions (u >1.96 then p-value <0.05) (Edwards, Lloyd, 1979).

If the distribution of dicentrics per cell is overdispersed due to nonuniform exposure (u >1.96), two mathematical models can be used to calculate the yield of dicentrics of the fraction irradiated: contaminated Poisson's (Dolphin, 1969) and Qdr (Sasaki&Miyata, 1968) allowing the calculation of the dose received by the fraction.

The calculation of the dicentric yield of the fraction irradiated, with the contaminated Poisson method, uses the following equation:

$$\frac{y}{\left(1-e^{-y}\right)}=\frac{x}{\left(N-n_0\right)}$$

where y is the mean yield of dicentrics of the irradiated fraction, e^{-y} represents the Poisson probability of cells without dicentric in the irradiated fraction, x is the number of dicentrics observed, N is the number of cells scored and n0 is the number of cells free of dicentrics.
The 95% confidence intervals of y are calculated with the following equation:

$$y_{min/max}=y*\pm 1.96\left[Var\left(y*\right)^{0.5}\right]$$

where y* is the yield of dicentrics per unstable cell ($x/(N - n_0)$) and Var(y*) is its variance ($x(1 + y - y^*) / (N - n_0)^2$) (Barquinero, 1997).
The dose of the irradiated fraction and its 95% confidence interval are estimated by correspondence of the yield of dicentrics per unstable cell and its 95% confidence interval to dose-effect curve.
Furthermore, the fraction of cells exposed (F) and its 95% confidence interval are estimated using the following equation:

$$F=\frac{f}{p}\left(1-f+\frac{f}{p}\right)$$

where f is the fraction of cells scored that were irradiated (x/(yN)) and p is the fraction of surviving cells, taking into account the selective loss of irradiated cells due to interphase death and mitotic delay. For each condition, p is estimated using the following equation:

$$p=e^{-\frac{D}{D_0}}$$

where D is the estimated dose for the irradiated fraction (f) and D_0 is the dose for which 37% of irradiated cells survived. Various values have been reported for D_0 depending on the studies and the irradiation conditions: 2.7 Gy (Lloyd, 1973) and 3.8 Gy (Barquinero, Barrios, 1997) for X-rays and 3.5 Gy for 60Co γ-rays (Matsubara, 1974).
The yield of dicentrics of the irradiated fraction can also be calculated with a second method, Qdr (Sasaki&Miyata, 1968). This method uses the same parameters as the contaminated Poisson method and in addition the excess acentric fragments yield. Acentric fragments are eliminated rapidly during mitosis, so their presence indicates that aberrant cells counted are due to exposure. It is necessary to plot a dose-effect curve based on the excess acentric fragments yield observed per exposure dose. The Qdr method uses the following equation:

$$\frac{x}{\left(N-n_0\right)}=\frac{y*}{\left(1-e^{-\left(y*-y_{ace}\right)}\right)}$$

where y_{ace} is the yield of excess acentric fragments per unstable cell (Sasaki&Miyata, 1968).

The 95% confidence intervals of y are calculated with a similar equation to the contaminated Poisson method. The dose of the irradiated fraction and its 95% confidence intervals are estimated by correspondence of the yield obtained with the Qdr method and its 95% confidence intervals to the dose-effect curve.

2.1.3 Case of protracted exposure

Protracted exposures have an effect on the yield of dicentrics. Indeed the time to repair the induced damage is lower than the exposure time. Some double-strand break can be repaired before and to coexist with new double-strand break produced later. Despite that, studies have shown that there is also a good relationship between the dose received and the dicentric yield (Bauchinger, 1979). For low LET radiation, protracted exposures modify the number of dicentrics produce by two tracks. For this reason and to correct the β coefficient obtained at the acute irradiation the G-function is need. Several studies have shown that the mean time to repair a lesion is about 2 hours (Schmid, 1976), (Purrott&Reeder, 1976), (Virsik&Harder, 1980). The equation representing the dose-effect relationship can be written as:

$$Y = \alpha D + \beta G(t / t_0)D^2 + c$$

where the G-function follows this equation:

$$G(t / t_0) = \frac{2}{(t / t_0)}\left[(t / t_0) - 1 + e^{-(t/t_0)}\right]$$

where t is the time over which irradiation occurs and t_0 represents the mean lesion repair time estimated (2 hours). If the time of exposure is long, the G-function may be reduced to zero (Lloyd, 1984). Therefore, in case of fractioned irradiations if the time between exposures is greater than 6 hours, the different exposures will be considered as separate and the effect as additive. For high LET radiation, no effect on the yield of dicentric produced will be expected.

2.1.4 Case of high dose exposure

The dose range detectable with the dicentric assay is 0.1-5 Gy. In the case of exposure to a highest dose, there is a strong influence of cell death and mitotic delay. It is then necessary to observe chromosome aberrations induced by ionizing radiation in interphase cells. In this sense, two different techniques based on premature chromosome condensation (PCC) have been proposed. The first method used the fusion of interphase lymphocytes with mitotic cells, it does not need to stimulate the cell division. With this methodology excess fragments can be detected (Pantelias&Maillie, 1984) and results can be obtained in a few hours after sample reception. The second method is able to condensate chromatide before metaphase, it is used chemical agents (calyculin A or okadaic acid). In this second method lymphocytes need to be stimulated and aberrations can be detected in G1 and G2 phases of the cell cycle (Kanda, 1999). One of chromosome aberrations observed are ring chromosomes (Figure 5). For both chromosome aberrations (excess fragments, rings), the dose-effect curves are fitted to a linear model. The methodology dose up to 25 Gy is estimated (Lamadrid, 2007). The PCC assay, using chemical agent, was successfully used to estimate dose in the Tokai-Mura radiation accident (1999, Japan) (Hayata, 2001).

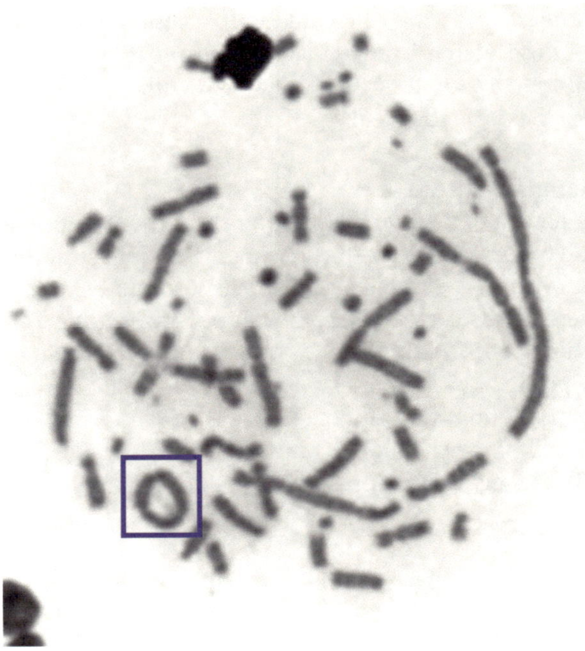

Fig. 5. PCC of peripheral blood lymphocytes in G2. Unstable chromosomal aberrations are analyzed with Giemsa staining. The ring is framed in blue.

2.2 Biodosimetry for population triage purposes

The standard biological dosimetry technique, based on dicentric, is labor-intensive and time-consuming. In the case of a large-scale radiological event, the standard dicentric assay cannot be used to perform rapid triage of the numerous potential victims and triage of badly injured victims would have to be done rapidly. Besides, following this triage step it will be necessary to estimate the dose received as accurately as possible in order to manage the long-term medical follow-up of the victims.

2.2.1 Application of the dicentric assay

Currently, the strategy for triage is to use the dicentric assay by reducing the number of metaphases scored. The IAEA advises to score unstable chromosomal aberrations among only 50 metaphases. This increases the minimum detection threshold from 0.1-0.2 Gy to 0.5-0.6 Gy depending on the calibration curve (Agency, 2001). In vitro studies have shown good dose assessment for whole-body exposure (Lloyd, 2000), (Romm, 2011). The triage mode was applied after accident in Georgia (1998) where 85 people were potentially exposed. The results showed an under-estimation of the dose for 82% of individuals when the scoring of 50 metaphases was compared with the scoring of 250 metaphases (Voisin, 2001). This trend seemed to be correlated with another accident in Senegal (2006) where 63 people were potentially exposed.. Indeed, under-estimation was observed for 50% of individuals with the scoring of 50 metaphases (Vaurijoux, 2009). These results indicate that the good agreement observed in in vitro experiments is not left with real cases of accident.

Additionally, it should be expected that triage mode estimation of partial-body exposure will become much less accurate, because there is a low probability to observe an overdispersion with few metaphases scored. The results of in vitro simulation of partial-body exposure suggested that the estimation of the irradiated fraction was good in 25% of cases for doses below 3 Gy and in 75% of cases for doses above 3 Gy (Lloyd, Edwards, 2000). There are no published results on the detection of partial-body exposure with scoring of 50 metaphases in comparison with scoring of 500 metaphases in the case of real accidents.

2.2.2 Application of the micronucleus assay

The other alternative is the in vitro cytokinesis-block micronucleus (CBMN) assay performed on peripheral blood lymphocytes (Fenech&Morley, 1985). The micronucleus is a small spherical object of same appearance as that nuclei after Giemsa staining. It is composed of an acentric fragment or a whole chromosome that is not included in the daughter nuclei during cell division. It is observed in binucleated cells, though if the micronucleus contains an acentric fragment (Figure 6) it will be eliminated after anaphase.

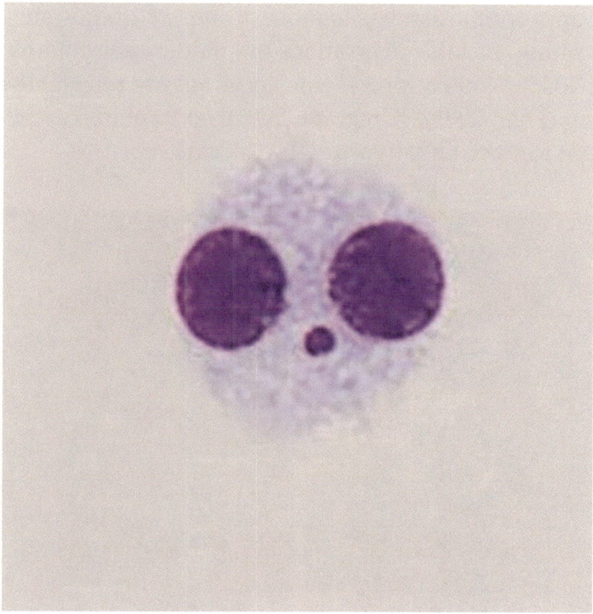

Fig. 6. Binucleated cell of peripheral blood lymphocytes containing a micronucleus (Giemsa straining).

Scoring micronuclei is easier than scoring dicentrics and therefore allows a faster analysis. Moreover, the precision of the micronucleus assay is better than that of the dicentric assay on 50 metaphases (Roy, 2007). The dose-effect response fits the linear quadratic model for low LET radiation and the linear model for high LET radiation (Vral, 2011). Dose is estimated using the correspondence between micronucleus yield and the dose-effect curve. However, the CBMN assay does not allow the detection of partial-body exposure. Furthermore, the micronucleus is not radiation-specific and can be induced by many genotoxic agents. There is also marked inter-individual variability due to age and gender (Fenech, 1993).

It is well known that most micronuclei induced by ionizing radiation are formed of acentric fragments because they are the result of chromosome breakage. However, a minority of micronuclei contain a whole chromosome because of time lag during anaphase caused by some defect in the spindle or the kinetochore protein (Vral, 1997). The use of fluorescence in situ hybridization (FISH) to highlight chromosome centromere indicates whether the micronucleus contains an acentrics fragment (MNCM$^-$) or if it contains one or more chromosome (MNCM$^+$). The scoring of centromere-negative micronuclei (MNCM$^-$) reduces the detection threshold to 0.1 Gy for 2000 binucleated cells scored (Vral, Thierens, 1997). However to apply FISH technique increase the time need to obtain the result and more expensive and for these reasons in large-scale radiation seems to be less suitable.

3. Past exposure to ionizing radiation

Past exposures to ionizing radiation are evaluated differently from recent exposures. Dicentric is an unstable aberration and its yield decreases with time, so an accurate dose estimate can only be obtained up to 1 year after exposure (Lloyd, 1998). For past dose assessment the most appropriate assay is to score of stable chromosome aberrations in stable cells, since cells containing unstable aberrations or complex aberrations decrease in number over time (Edwards, 2005). Non-reciprocal translocations are not stable because there is loss of genetic information (Pala, 2001), (Gregoire, 2006), so to obtain an accurate estimation of the dose it is important to score only two-way translocations.

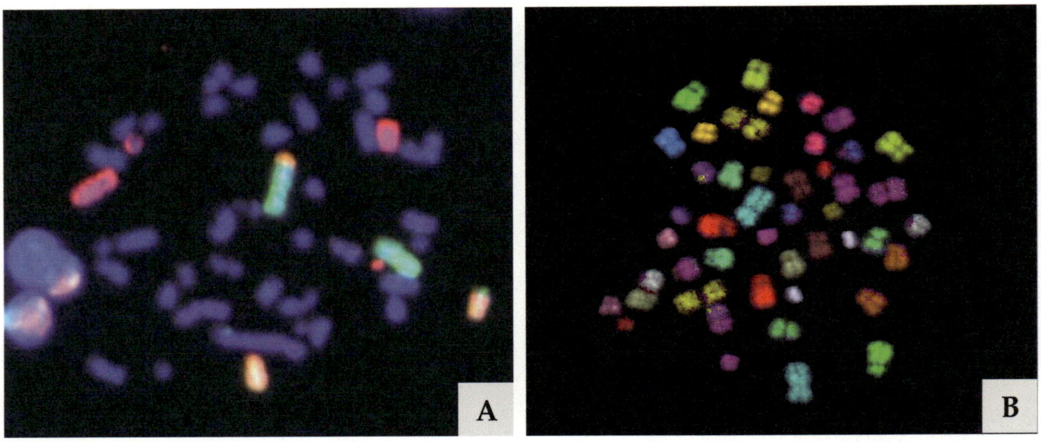

Fig. 7. FISH staining of peripheral blood lymphocytes in metaphase. A) Three chromosomes are painted: 2 (green), 4 (red) and 13 (orange) and B) All chromosomes are painted with a combination of fluorochromes (multicolor-FISH).

Translocations are analyzed by FISH, which paints one or more chromosome with the aid of DNA probes associate with a fluorochrome (Figure 7). In case of a part of genome is visualized, the yield of translocations obtained represents is not directly applicable to full genome. Lucas et al have developed mathematical models to reconstitute the yield of the full genome (Lucas, 1997):

$$F_G = \frac{F_P}{2.05 f_P (1 - f_P)}$$

where F_G is the yield of the full genome, F_P is the yield of translocations observed by FISH, and f_P is the fraction of the genome hybridized. This equation is general. In the case where more than two colors are used, the following equation is used (Lucas, 1997):

$$F_G = \frac{F_P}{2.05\left[\sum_i f_i(1-fi)-\sum_{i \triangleleft j} f_i f_j\right]}$$

However, stable chromosome aberrations are less specific to ionizing radiation exposure than dicentrics. Furthermore, translocations persist for many years in circulating lymphocytes and their frequency increases significantly with age and the lifetime conditions (alcohol, tobacco, pesticide exposure...) (Whitehouse, 2005), (Sigurdson, 2008), (Grégoire, 2010). This increase has to be taken into consideration in evaluation of exposure and corrected for as a function of the age of the individual. The relationship between translocation yield and age is based on a linear-exponential model (Sigurdson, Ha, 2008):

$$F_T = \alpha + \beta_{age} e^{(\gamma_{age})}$$

where F_T is translocation yield, α represents the translocation yield at age 0, β_{age} represents a linear slope indicating the increase in translocation yield with age and γ_{age} represents a loglinear curvature parameter.

In a study of blood from nuclear test veterans, 50 years after exposure to ionizing radiation (Wahab, 2008), the frequency of total chromosome translocations was three times higher than the one observed in the control group. Analysis of potential confounding factors suggested that this high frequency was most likely attributable to radiation exposure. It is hard to estimate a dose accurately 50 years after the fact, but exposure may be assumed if the frequency of translocation is above the background level in the population (Lindholm&Edwards, 2004).

4. Improving biological dosimetry

Current research in biological dosimetry is seeking shorter analysis time, lower threshold detection and accurate localization and dose estimation in the case of partial-body exposure. New methods of scoring dicentrics and micronuclei, as well as new biomarkers such as foci γH2AX, and gene expression are investigated. Advantages and limits have been observed in all cases, and further research is needed remains to be done in the development of ionizing radiation biodosimetry.

4.1 Reducing analysis time

4.1.1 Automation of dicentric scoring

Dicentrics assay is the gold standard method in biodosimetry. One improvement has been automation of scoring using image analysis software (DCScore software; MetaSystems). Briefly, the software identifies as chromosomes all objects corresponding to the shape and size of benchmarks and detects dicentrics among them (Schunck, 2004) (Figure 8). Putative dicentrics are validated by an operator.

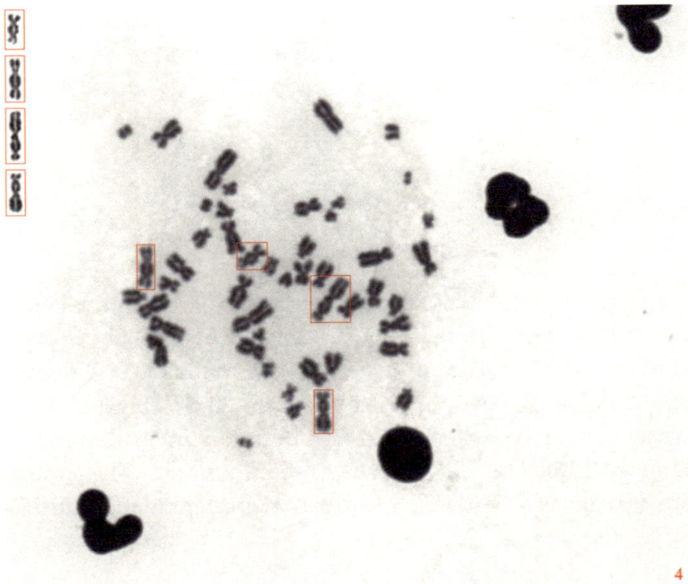

Fig. 8. Image of metaphase obtained after analysis by DCScore software (MetaSystems). Chromosomes detected as dicentrics are framed in red.

One thousand metaphases can be analyzed for triage in 1 hour and 3000 for individual dosimetry in 3 hours, with a 3-fold reduction in analysis time. For triage the threshold obtained with automatic scoring is better than that with manual scoring of 50 metaphases. For triage the automatic detection of dicentrics has been validated of the accident in Senegal (2006). We have show that use of this methodology for a large population dose estimation as it can replace the usual manual scoring of both the 50MS and 500MS methods (Vaurijoux, 2008).

4.1.2 Automation of micronucleus scoring

The scoring of micronuclei poses a problem of inter- and intra-laboratory variability (Fenech, 2003), and in response to this and to speed up scoring use is made of image analysis software (Metafer MicroNuclei; MetaSystems). Briefly, the software first detects binucleated cells according to morphometric criteria: size, ratio of the longest to the shortest diameter, relative concavity depth, and distance between objects. Then, it detects the presence of a micronucleus with the same criteria (Varga, 2004). It is interesting to note that the correlation is good between manual and automatic scoring. When 1000 binucleated cells are scored, the detection threshold is 0.4 Gy (Willems, 2010). When 5000 binucleated cells are scored, the detection threshold is 0.2 Gy. This scoring requires 40 min (Baeyens, 2011). Inter-laboratory variability appears to be limited by the use of this software (Willems, August, 2010).

It is describe that the scoring of centromere-negative micronuclei ($MNCM^-$) improves the accuracy of the dose estimation. Semi-automation of $MNCM^-$ scoring (automation of micronucleus detection and manual analysis of $MNCM^-$) enables analysis of 5000 binucleated cells in approximately 2 hours. This is longer than automation of all micronuclei seems more accurate in the low dose range (Baeyens, Swanson, 2011).

4.2 Application of early biomarkers

Dicentric assay and micronucleus assay have two main limitations. First, the lymphocyte culture step requires 48 or 64 hours, respectively. Second, both methods are not able to estimate a dose below 0.1 Gy. Current research is investigating the relevance of early biomarkers after radiation exposure for biodosimetry purposes.

4.2.1 H2AX protein

H2AX protein is a histone whose serine 139 is phosphorylated in response to a DNA double-strand break. This phosphorylated form is called γH2AX and is visualized in the nucleus after immunofluorescence staining with specific fluorescent antibodies (Rogakou, 1998) (Figure 9).

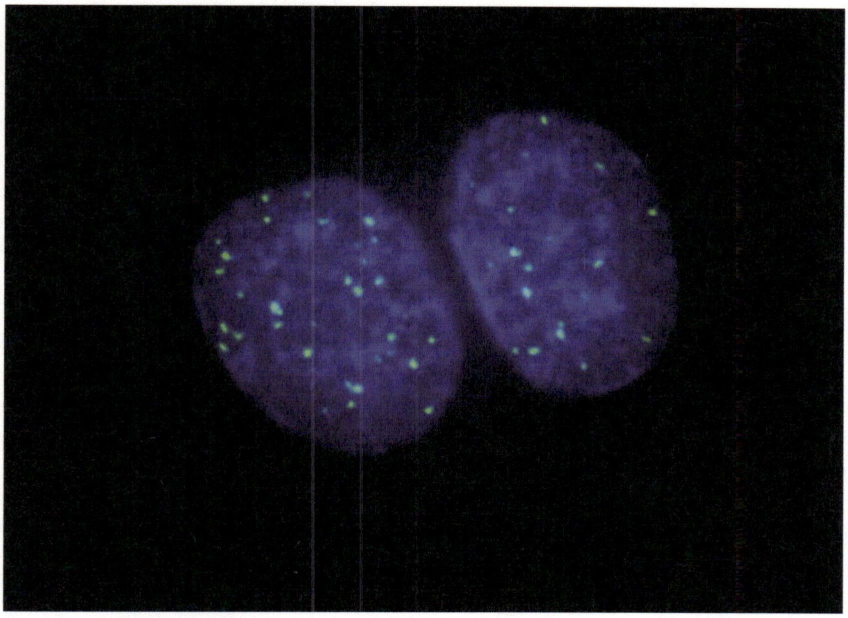

Fig. 9. Nuclei of peripheral blood lymphocytes with γH2AX foci visualized by fluorescence.

The number of γH2AX foci and their relative fluorescence show a linear relationship with the dose received (Leatherbarrow, 2006). Doses as low as 0.05 Gy can be detected with this marker after in vivo exposure (Rothkamm, 2007). The linear relationship is maintained for 30 min to 16 hours in human peripheral lymphocytes after irradiation with γ rays for doses in the range 0.05 to 2 Gy (Roch-Lefevre, 2010).

The main limitations of γH2AX foci quantification are the marked inter-individual variation in baseline values and the decrease of γH2AX foci yield with time (Roch-Lefevre, Mandina, 2010).

4.2.2 Gene expression

Study of gene expression profiles in the response to radiation exposure is an alternative approach to biodosimetry (Amundson, 2000), (Amundson, 2001). Cellular damages usually induce cellular stress which leads to a response through activation of several cellular pathways that result in modulations of gene expression. Microarray technology is used to

study these modulations with the aim of identifying the corresponding gene or group of genes whose profile shows a dose-effect relationship (Paul, 2011). Amundson et al in 2000 were among the first to use gene expression as biological dosimetry in peripheral blood lymphocytes. They showed that modulation of several target genes of p53 protein seems to be correlated with the dose in the range 0.2 to 2 Gy. The maximal response for doses of 0.5 Gy or less could be observed early (about 4 hours after exposure), but this ex-vivo model seems to be limited for times longer than 48 h post-exposure (Amundson, Do, 2000).
Studies of doses below 0.1 Gy, especially on human blood cells, are still rare (Gruel, 2008), (Morandi, 2009). Globally, the level of induction of known target genes of p53 protein appears to decrease with dose and even becomes undetectable at very low doses (>0.025 Gy). This suggests loss of the characteristics of a typical stress response at very low doses and that the response is more diverse and less specialized. It is interesting to note that many genes modulated at this level of dose are known to play a role in mechanisms such as cytoskeleton metabolism, cell-cell signaling, chromatin modeling, RNA and protein processing, proliferation, etc (Gruel, Voisin, 2008), (Morandi, Severini, 2009).
There are two main limitations on immediate implementation of these results in operational dosimetry. First, given the time limit of current studies (48-64 hours), it is hard to keep peripheral blood lymphocytes in culture without creating bias in the analysis. Second, RNA is not stable in blood collection tubes with anticoagulant which poses a problem for the storage and handling of samples during shipment from the sampling site to the specialized laboratory.

5. Standardisation of biodosimetry

Biological dosimetry, based on the study of the radio-induced chromosomal aberrations, mainly the dicentric assay, has become a routine component of the accidental dose assessment. Experience of its application in hundreds of cases of suspected or proven overexposures has proved the value of the method and also defined its limitations.
Biological dosimetry is incorporated into radiation protection programmes of several countries, to confirm or discount a suspected radiation exposure. By contrast, the absence of real concurrence (only one or two lab per country) underlines the needs of homogeneously and largely established bases to assure its credibility. Therefore, an ISO standard was developed to address the critical aspects of the use of the dicentric assay as a biodosimeter. The first publication of 19238 ISO standard in 2004 provides for expertise, minimum requirement for experimental processes, quality assurance and quality control programmes, and evaluation of performance. Another 21243 ISO standard published in 2008 was intended to define performance criteria for cytogenetic triage. The primary purpose of this standard is to provide a guideline in order to perform the dicentric assay for dose categorization in triage mode using documented and validated procedures. The described approaches included pre-planning, reagent stockpiling, simplified sample processing, automation, networking, and modification of some of the ISO 19238 scoring criteria.
The standards are written in the form of procedures to be adopted for dicentric expertise/cytogenetic triage depending upon the application of the results: medical management when appropriate, radiation protection management, record keeping and medical/legal requirements. Whatever the laboratory in any country, the application of such standards ensures quality of practice which is very important for credibility. Second, it helps to compare the results obtained in one laboratory to another one, particularly in case of an

international collaboration or intercomparison. Finally, each new laboratory should get from this standard the most useful information to perform dicentrics assay in the best experimental and reproducible conditions.
To be qualified, the biological dosimetry laboratory must establish technical validation of the bioassay used, besides the precise description of the dose assessment process (relationship with customer, confidentiality of information, capability of laboratory staff, QA&QC program...) For instance, in order to test the influence of protocol variations the mitotic index and dicentrics rates were measured under different experimental conditions (L. Roy et al, 2011, Radiation Protection Dosimetry, submitted). The effect of seven parameters was tested: BrdU, PHA and colcemid concentrations, blood and medium volumes, culture duration and incubation temperature. The results show that mitotic index was influenced by the concentration of BrdU, medium and blood, the culture duration and the temperature. However none of the factors has a significant impact on the yield of dicentrics. We can conclude that the dicentric assay is robust against reagents variations within the range tested. These results could be used by relevant laboratories as element of the quality of their dose assessment and their procedures robustness in any event requiring such demonstration.
There is also some limitation to systematically introduce specific QA&QM programs in the normal activity of the biological dosimetry laboratory. While the quality system is a natural way for any R&D activity, the application of such standard is time consuming because all the process must be checked for deviation and this checking is required regularly. It is especially true when a specific ISO standard is required for supplementing classical accreditation process following more general ISO 17025:2000. For instance the 19238 ISO standard was heavily updated by more detailed description of the experimental and statistical steps for satisfying the accreditation requirements and this implementation come into force probably next year.
A new ISO standard on the use of micronuclei assay in individual biodosimetry and population triage is in preparation and it is expected a future ISO standard on automation in cytogenetic dosimetry.

6. Conclusion

Currently, the dicentric assay seems to be the best bioindicator of recent radiation exposure and their assay is the only one to offer all of the following advantages: low background level, clear dose-effect relationship for different radiation qualities and dose rates, specific to ionizing radiation, non-invasive, good reproducibility and comparability of in vitro and in vivo results. The only drawback is the time required to obtain a dose estimation, especially in the case of large-scale accidental exposure. However, new advances in the automation of dicentric scoring enormously reduce the time needed to estimate the dose and this method remains the most promising.
Nevertheless, the detection of accidental exposure below 0.1 Gy remains difficult. Markers as γH2AX foci and gene expression lead to reduce this detection threshold but some limitations shill exists to be used them at the real accidental biodosimetry.
Translocation is the only biomarker used in biodosimetry for past exposure. However, it is difficult to estimate an accurate dose because the background level depends on age and life conditions of the individual.

7. Acknowledgment

The authors gratefully acknowledge Joan Francesc Barquinero for his rereading and relevance of these comments.

8. References

Agency, I. A. E. (2001) Cytogenetic analysis for radiation dose assessment, Vienna.

Amundson, S. A., Bittner, M., Meltzer, P., Trent, J. & Fornace, A. J., Jr. (2001). Induction of gene expression as a monitor of exposure to ionizing radiation. *Radiat Res*, Vol.156, No.5 Pt 2, pp. 657-661.

Amundson, S. A., Do, K. T., Shahab, S., Bittner, M., Meltzer, P., Trent, J. & Fornace, A. J., Jr. (2000). Identification of potential mRNA biomarkers in peripheral blood lymphocytes for human exposure to ionizing radiation. *Radiat Res*, Vol.154, No.3, pp. 342-346.

Baeyens, A., Swanson, R., Herd, O., Ainsbury, E., Mabhengu, T., Willem, P., Thierens, H., Slabbert, J. P. & Vral, A. (2011). A semi-automated micronucleus-centromere assay to assess low-dose radiation exposure in human lymphocytes. *Int J Radiat Biol*, 1-11.

Barquinero, J. F., Barrios, L., Caballin, M. R., Miro, R., Ribas, M. & Egozcue, J. (1997). Biological dosimetry in simulated in vitro partial irradiations. *Int J Radiat Biol*, Vol.71, No.4, pp. 435-440.

Bauchinger, M., Schmid, E. & Dresp, J. (1979). Calculation of the dose-rate dependence of the decentric yield after Co gamma-irradiation of human lymphocytes. *Int J Radiat Biol Relat Stud Phys Chem Med*, Vol.35, No.3, pp. 229-233.

Dolphin, G. W. (1969). Biological dosimetry with particular reference to chromosome aberration analysis. A review of methods. *IAEA, Vienna*, Vol.Handling of Radiation Accidents, 215-224.

Edwards, A. A., Lindholm, C., Darroudi, F., Stephan, G., Romm, H., Barquinero, J., Barrios, L., Caballin, M. R., Roy, L., Whitehouse, C. A., Tawn, E. J., Moquet, J., Lloyd, D. C. & Voisin, P. (2005). Review of translocations detected by FISH for retrospective biological dosimetry applications. *Radiat Prot Dosimetry*, Vol.113, No.4, pp. 396-402.

Edwards, A. A., Lloyd, D. C. & Purrott, R. J. (1979). Radiation induced chromosome aberrations and the Poisson distribution. *Radiat Environ Biophys*, Vol.16, No.2, pp. 89-100.

Fenech, M. (1993). The cytokinesis-block micronucleus technique: a detailed description of the method and its application to genotoxicity studies in human populations. *Mutat Res*, Vol.285, No.1, pp. 35-44.

Fenech, M., Bonassi, S., Turner, J., Lando, C., Ceppi, M., Chang, W. P., Holland, N., Kirsch-Volders, M., Zeiger, E., Bigatti, M. P., Bolognesi, C., Cao, J., De Luca, G., Di Giorgio, M., Ferguson, L. R., Fucic, A., Lima, O. G., Hadjidekova, V. V., Hrelia, P., Jaworska, A., Joksic, G., Krishnaja, A. P., Lee, T. K., Martelli, A., McKay, M. J., Migliore, L., Mirkova, E., Muller, W. U., Odagiri, Y., Orsiere, T., Scarfi, M. R., Silva, M. J., Sofuni, T., Surralles, J., Trenta, G., Vorobtsova, I., Vral, A. & Zijno, A. (2003). Intra- and inter-laboratory variation in the scoring of micronuclei and nucleoplasmic bridges in binucleated human lymphocytes. Results of an international slide-scoring exercise by the HUMN project. *Mutat Res*, Vol.534, No.1-2, pp. 45-64.

Fenech, M. & Morley, A. A. (1985). Measurement of micronuclei in lymphocytes. *Mutat Res*, Vol.147, No.1-2, pp. 29-36.

Grégoire, E., Gruel, G., Martin, C., Roch-Lefevre, S., Vaurijoux, A., Voisin, P. & Roy, L. (2010). Impact des facteurs individuels et environnementaux sur le taux d'aberrations chromosomiques de type translocations Partie 1 : âge, sexe, tabac, alcool *Radioprotection*, Vol.45, No.2, pp. 153-169.

Gregoire, E., Sorokine-Durm, I., Bertho, J. M., Jacquet, N., Delbos, M., Demarquay, C., Voisin, P. & Roy, L. (2006). Follow-up of stable chromosomal aberrations in gamma-ray irradiated non-human primates. *Int J Radiat Biol*, Vol.82, No.7, pp. 493-502.

Gruel, G., Voisin, P., Vaurijoux, A., Roch-Lefevre, S., Gregoire, E., Maltere, P., Petat, C., Gidrol, X., Voisin, P. & Roy, L. (2008). Broad modulation of gene expression in CD4+ lymphocyte subpopulations in response to low doses of ionizing radiation. *Radiat Res*, Vol.170, No.3, pp. 335-344.

Hayata, I., Kanda, R., Minamihisamatsu, M., Furukawa, M. & Sasaki, M. S. (2001). Cytogenetical dose estimation for 3 severely exposed patients in the JCO criticality accident in Tokai-mura. *J Radiat Res (Tokyo)*, Vol.42 Suppl, S149-155.

Kanda, R., Hayata, I. & Lloyd, D. C. (1999). Easy biodosimetry for high-dose radiation exposures using drug-induced, prematurely condensed chromosomes. *Int J Radiat Biol*, Vol.75, No.4, pp. 441-446.

Lamadrid, A. I., Garcia, O., Delbos, M., Voisin, P. & Roy, L. (2007). PCC-ring induction in human lymphocytes exposed to gamma and neutron irradiation. *J Radiat Res (Tokyo)*, Vol.48, No.1, pp. 1-6.

Leatherbarrow, E. L., Harper, J. V., Cucinotta, F. A. & O'Neill, P. (2006). Induction and quantification of gamma-H2AX foci following low and high LET-irradiation. *Int J Radiat Biol*, Vol.82, No.2, pp. 111-118.

Lindholm, C. & Edwards, A. (2004). Long-term persistence of translocations in stable lymphocytes from victims of a radiological accident. *Int J Radiat Biol*, Vol.80, No.8, pp. 559-566.

Lloyd, D. C., Edwards, A. A., Moquet, J. E. & Guerrero-Carbajal, Y. C. (2000). The role of cytogenetics in early triage of radiation casualties. *Appl Radiat Isot*, Vol.52, No.5, pp. 1107-1112.

Lloyd, D. C., Edwards, A. A., Prosser, J. S. & Corp, M. J. (1984). The dose response relationship obtained at constant irradiation times for the induction of chromosome aberrations in human lymphocytes by cobalt-60 gamma rays. *Radiat Environ Biophys*, Vol.23, No.3, pp. 179-189.

Lloyd, D. C., Moquet, J. E., Oram, S., Edwards, A. A. & Lucas, J. N. (1998). Accidental intake of tritiated water: a cytogenetic follow-up case on translocation stability and dose reconstruction. *Int J Radiat Biol*, Vol.73, No.5, pp. 543-547.

Lloyd, D. C., Purrott, R. J. & Dolphin, G. W. (1973). Chromosome aberration dosimetry using human lymphocytes in simulated partial body irradiation. *Phys Med Biol*, Vol.18, No.3, pp. 421-431.

Lloyd, D. C., Purrott, R. J., Dolphin, G. W. & Edwards, A. A. (1976). Chromosome aberrations induced in human lymphocytes by neutron irradiation. *Int J Radiat Biol Relat Stud Phys Chem Med*, Vol.29, No.2, pp. 169-182.

Lucas, J. N. (1997). Dose reconstruction for individuals exposed to ionizing radiation using chromosome painting. *Radiat Res*, Vol.148, No.5 Suppl, pp. S33-38.

Matsubara, S., Sasaki, M. S. & Adachi, T. (1974). Dose-response relationship of lymphocyte chromosome aberrations in locally irradiated persons. *J Radiat Res (Tokyo)*, Vol.15, No.4, pp. 189-196.

Morandi, E., Severini, C., Quercioli, D., Perdichizzi, S., Mascolo, M. G., Horn, W., Vaccari, M., Nucci, M. C., Lodi, V., Violante, F. S., Bolognesi, C., Grilli, S., Silingardi, P. & Colacci, A. (2009). Gene expression changes in medical workers exposed to radiation. *Radiat Res*, Vol.172, No.4, pp. 500-508.

Pala, F. S., Moquet, J. E., Edwards, A. A. & Lloyd, D. C. (2001). In vitro transmission of chromosomal aberrations through mitosis in human lymphocytes. *Mutat Res*, Vol.474, No.1-2, pp. 139-146.

Pantelias, G. E. & Maillie, H. D. (1984). The use of peripheral blood mononuclear cell prematurely condensed chromosomes for biological dosimetry. *Radiat Res*, Vol.99, No.1, pp. 140-150.

Papworth, D. (1975). Curve fitting by maximum likelihood. Appendix to paper by SAVAGE JRK: Radiation-induced chromosomal aberrations in plant Tradescantia: Dose response curves. *Radiation Botanics*, Vol.15, 127-131.

Paul, S., Barker, C. A., Turner, H. C., McLane, A., Wolden, S. L. & Amundson, S. A. (2011). Prediction of in vivo radiation dose status in radiotherapy patients using ex vivo and in vivo gene expression signatures. *Radiat Res*, Vol.175, No.3, pp. 257-265.

Purrott, R. J. & Reeder, E. (1976). Chromosome aberration yields in human lymphocytes induced by fractionated doses of x-radiation. *Mutat Res*, Vol.34, No.3, pp. 437-445.

Roch-Lefevre, S., Mandina, T., Voisin, P., Gaetan, G., Mesa, J. E., Valente, M., Bonnesoeur, P., Garcia, O., Voisin, P. & Roy, L. (2010). Quantification of gamma-H2AX Foci in Human Lymphocytes: A Method for Biological Dosimetry after Ionizing Radiation Exposure. *Radiat Res*, Vol.174, No.2, pp. 185-194.

Rogakou, E. P., Pilch, D. R., Orr, A. H., Ivanova, V. S. & Bonner, W. M. (1998). DNA double-stranded breaks induce histone H2AX phosphorylation on serine 139. *J Biol Chem*, Vol.273, No.10, pp. 5858-5868.

Romm, H., Oestreicher, U. & Kulka, U. (2009). Cytogenetic damage analysed by the dicentric assay. *Ann Ist Super Sanita*, Vol.45, No.3, pp. 251-259.

Romm, H., Wilkins, R. C., Coleman, C. N., Lillis-Hearne, P. K., Pellmar, T. C., Livingston, G. K., Awa, A. A., Jenkins, M. S., Yoshida, M. A., Oestreicher, U. & Prasanna, P. G. (2011). Biological dosimetry by the triage dicentric chromosome assay: potential implications for treatment of acute radiation syndrome in radiological mass casualties. *Radiat Res*, Vol.175, No.3, pp. 397-404.

Rothkamm, K., Balroop, S., Shekhdar, J., Fernie, P. & Goh, V. (2007). Leukocyte DNA damage after multi-detector row CT: a quantitative biomarker of low-level radiation exposure. *Radiology*, Vol.242, No.1, pp. 244-251.

Roy, L., Roch-Lefevre, S., Vaurijoux, A., Voisin, P., Martin, C., Gregoire, E. & Voisin, P. (2007). Optimization of cytogenetic procedures for population triage in case of radiological emergency. *Radiation Measurements*, Vol.42, No.6-7, pp. 1143-1146.

Sasaki, M. S. & Miyata, H. (1968). Biological dosimetry in atomic bomb survivors. *Nature*, Vol.220, No.5173, pp. 1189-1193.

Schmid, E., Bauchinger, M. & Mergenthaler, W. (1976). Analysis of the time relationship for the interaction of X-ray induced primary breaks in the formation of dicentric chromosomes. *Int J Radiat Biol Relat Stud Phys Chem Med*, Vol.30, No.4, pp. 339-346.

Schunck, C., Johannes, T., Varga, D., Lorch, T. & Plesch, A. (2004). New developments in automated cytogenetic imaging: unattended scoring of dicentric chromosomes, micronuclei, single cell gel electrophoresis, and fluorescence signals. *Cytogenet Genome Res*, Vol.104, No.1-4, pp. 383-389.

Sigurdson, A. J., Ha, M., Hauptmann, M., Bhatti, P., Sram, R. J., Beskid, O., Tawn, E. J., Whitehouse, C. A., Lindholm, C., Nakano, M., Kodama, Y, Nakamura, N., Vorobtsova, I., Oestreicher, U., Stephan, G., Yong, L. C., Bauchinger, M., Schmid, E., Chung, H. W., Darroudi, F., Roy, L., Voisin, P., Barquinero, J. F., Livingston, G., Blakey, D., Hayata, I., Zhang, W., Wang, C., Bennett, L. M., Littlefield, L. G., Edwards, A. A., Kleinerman, R. A. & Tucker, J. D. (2008). International study of factors affecting human chromosome translocations. *Mutat Res*, Vol.652, No.2, pp. 112-121.

Varga, D., Johannes, T., Jainta, S., Schuster, S., Schwarz-Boeger, U., Kiechle, M., Patino Garcia, B. & Vogel, W. (2004). An automated scoring procedure for the micronucleus test by image analysis. *Mutagenesis*, Vol.19, No.5, pp. 391-397.

Vaurijoux, A., Gruel, G., Pouzoulet, F., Gregoire, E., Martin, C., Roch-Lefevre, S., Voisin, P., Voisin, P. & Roy, L. (2008). Strategy for Population Triage Based on Chromosome Analysis. *European Radiation Research Congress, Tours, France.*

Vaurijoux, A., Gruel, G., Pouzoulet, F., Gregoire, E., Martin, C., Roch-Lefevre, S., Voisin, P., Voisin, P. & Roy, L. (2009). Strategy for population triage based on dicentric analysis. *Radiat Res*, Vol.171, No.5, pp. 541-548.

Virsik, R. P. & Harder, D. (1980). Recovery kinetics of radiation-induced chromosome aberrations in human G0 lymphocytes. *Radiat Environ Biophys*, Vol.18, No.3, pp. 221-238.

Voisin, P., Benderitter, M., Claraz, M., Chambrette, V., Sorokine-Durm, I., Delbos, M., Durand, V., Leroy, A. & Paillole, N. (2001). The cytogenetic dosimetry of recent accidental overexposure. *Cell Mol Biol (Noisy-le-grand)*, Vol.47, No.3, pp. 557-564.

Vral, A., Fenech, M. & Thierens, H. (2011). The micronucleus assay as a biological dosimeter of in vivo ionising radiation exposure. *Mutagenesis*, Vol.26, No.1, pp. 11-17.

Vral, A., Thierens, H. & De Ridder, L. (1997). In vitro micronucleus-centromere assay to detect radiation-damage induced by low doses in human lymphocytes. *Int J Radiat Biol*, Vol.71, No.1, pp. 61-68.

Wahab, M. A., Nickless, E. M., Najar-M'kacher, R., Parmentier, C., Podd, J. V. & Rowland, R. E. (2008). Elevated chromosome translocation frequencies in New Zealand nuclear test veterans. *Cytogenet Genome Res*, Vol.121, No.2, pp. 79-87.

Whitehouse, C. A., Edwards, A. A., Tawn, E. J., Stephan, G., Oestreicher, U., Moquet, J. E., Lloyd, D. C., Roy, L., Voisin, P., Lindholm, C., Barquinero, J., Barrios, L., Caballin, M. R., Darroudi, F. & Fomina, J. (2005). Translocation yields in peripheral blood lymphocytes from control populations. *Int J Radiat Biol*, Vol.81, No.2, pp. 139-145.

Willems, P., August, L., Slabbert, J., Romm, H., Oestreicher, U., Thierens, H. & Vral, A. (2010). Automated micronucleus (MN) scoring for population triage in case of large scale radiation events. *Int J Radiat Biol*, Vol.86, No.1, pp. 2-11.

4

Limited Repair of Critical DNA Damage in Cells Exposed to Low Dose Radiation

Azhub Gaziev[1,*] and Gadzhi Shaikhaev[2]
[1]*Institute of Theoretical and Experimental Biophysics, Russian Academy of Sciences, Pushchino,*
[2]*Vavilov Institute of General Genetics, Russian Academy of Sciences, Moscow, Russia*

1. Introduction

Studies on the effects of low doses of ionizing radiation (IR) on living organisms in the last decade have received considerable development. There are numerous investigations that contribute to our knowledge of how post-radiation processes develop, and after small and low-intensity doses of IR in particular.

Nevertheless discussions about the quantitative threshold of radiation dose have been continued. A linear non-threshold (LNT) model has been accepted as a basic standard for radiation protection and risk estimates over many years. The LNT model postulates a linear dependence of biological effects on IR dose. Also, this model suggests that low-dose induced damage does not markedly contribute to the risk because a great deal of endogenous (spontaneous) lesions to the genome continuously arise and are repaired in the cell. However there is considerable experimental evidence that allows the suggestion that IR-induced DNA lesions increase linearly with dose, but the reaction of cells to these lesions and the efficiency of repair of most critical injuries may be nonlinear (Tubiana et al., 2006).

Low radiation doses may increase the risk of cancer development (NRC, 2006, Brenner et al., 2003) and, probably, of non-cancer pathologies (Hildebrandt, 2010).

Though the notion of "low doses" is not formally determined, this dose interval is defined in many reports as ≤ 100-200 mSv. In the BEIR VII report of the National Academy of Sciences of USA, the low dose is determined as a dose not exceeding 100 mSv (NRC, 2006). Most likely, such doses (not more than 100 mSv) may be considered as "low doses". To the point, 100 mSv exceeds nearly 100 times the average yearly natural background (~1.0 mGy) of low linear energy transfer (LET).

This paper is a review of studies of cellular response to the critical DNA damage induced by ionizing radiation. Mechanisms are elucidated that limit the efficiency of repair of such lesions arising in cells irradiated with low doses of ionizing radiation. However, the analysis requires clarification of specific differences between the DNA damage induced by IR and endogenous lesions.

2. Endogenous and radiation-induced DNA damage

Endogenous DNA lesions constantly arise in cells. So, according to Lindahl (1993), in the DNA of one cell during one day there may be around 50 000 single-strand breaks (SSB), 10 000 apurinic/apyrimidinic sites (AP sites), 2000 oxidative lesions, 5000 alkylating lesions, 10 double-strand breaks (DSB) and 10 interstrand cross-linkings (ICL). These lesions to DNA result from the hydrolytic reactions, the action of reactive oxygen species (ROS) and other reactive metabolites (Lindahl, 1993; Bont, and van Larebeke, 2004). ROS induce modifications in DNA bases, SSB, AP sites, and DSB as a result of a close match of SSB (Bont, and van Larebeke, 2004).

ROS cause damage not only to DNA but also to lipids and other intracellular molecules. Therewith, the products of lipid: malondialdehyde (MDA), crotonaldehyde, acrolein, 4-hydroxynonenal can interact with DNA to form injuries. In particular, MDA is able to form in DNA ICL (Marnett, 2002). Apart from ROS, the cells contain other reactive molecules that may damage DNA. The most significant of them is S-adenosylmethionine (SAM). This is a donor of the methyl group of the enzymatic DNA methylation, which plays a role in the regulation of gene expression. However, SAM can also exercise the non-enzymatic methylation of DNA to yield mutagenic base adducts, such as 7-methylguanine, 3-methyladenine and O^6-methylguanine in mammalian cells (Barnes and Lindahl, 2004). Other endogenous non-enzymatic methylating agents are betaine and choline.

Endogenous hydrolytic DNA lesions arise due to instability of the glycoside bond between the bases and deoxyribose. This connection is broken upon temperature increase, base alkylation or as a result of action of DNA-N- glycosylases. Break of the N-glycoside bond leads to arising of AP sites in DNA (Lindahl, 1993; Barnes and Lindahl, 2004). DNA bases are also subject to hydrolysis and deamination. These events more intensively occur in the single-stranded DNA than in double-stranded. Cytosine and its homologue 5-methylcytosine are largely susceptible to deamination. Every day about 100-500 cytosines within DNA of one cell are subject to deamination and convert into uracils (Barnes and Lindahl, 2004).

Endogenous DNA lesions such as DSB and ICL, although arising in small proportion, approximately 10 per day on cell genome, are most critical for the fate of cells compared to a multiplicity of other injuries (Lieber, 2010; Dextraze et al., 2010). Endogenous DSB may arise in DNA not only as a result of ROS action, but also by programmed genome rearrangements such as V(D)J recombination (in lymphoid cells of the immune system) or due to meiotic recombination, shortening of telomeres, in the processes of transcription, replication, and as a result of random influence of DNA metabolic enzymes (Lieber, 2010).

Normally, the endogenous DNA lesions are not accumulated because in the cell there are constantly functioning repair mechanisms targeted at removing various types of damage. Most of these injuries are repaired by the mechanisms of base excision repair (BER). Adducts of non-enzymatic methylation may also be removed by BER enzymes, except for O^6-methylguanine. The latter is removed from DNA by excision of the methyl group from guanine by O^6-methylguanine-DNA methyltransferase (Barnes and Lindahl, 2004).

Thus, to maintain genome stability and long-term viability, normal cells possess a repair system for eliminating constantly arising endogenous lesions at maximum accuracy and efficiency.

This conclusion entails two questions. First: what is the difference between an endogenous DNA damage and a DNA damage induced by IR? Second: if in the cells significant amounts of endogenous DNA lesions are constantly repaired, then whether or not the additional damage induced by low doses of IR may "overload" the repair systems of the cell? Let us consider both of these versions.

IR induces multiple injuries to DNA of various chemical natures either as a result of direct ionization of DNA molecules, or indirectly through free radicals - the products of radiolysis of water and other intracellular compounds. In the cell there arise a variety of single DNA lesions: modified bases, SSB, AP sites, DNA-protein crosslinks, as well as complex injuries, such as DSB, ICL, replication fork destruction and the sites with long single-strand gaps (ssDNA) (Ward, 1988; Gaziev, 1999).

The main distinction of IR-induced DNA lesions from endogenous ones is their complex chemical nature and clustering of the former. Furthermore, the percentage of complex DNA lesions, critical for the fate of the cell, is much higher in the case of IR influence. As noted above, among endogenous DNA lesions arising in one cell per day, SSB and DSB are in the ratio 5000:1 (Lindahl, 1993; Barnes and Lindahl, 2004). Whereas for IR-induced lesions this ratio is: 20:1 (Ward, 1988).

The Biological efficiency of some amount of IR-induced DNA damage is much higher than that of the same amount of endogenous DNA damage, or that induced by chemical oxidants like H_2O_2 (Dextraze et al., 2010; Shikazono et al., 2009). Thus, much of the DNA lesions induced by IR, in contrast to endogenous ones, may be accompanied by loss of encoded information; they contain more complex groups at break ends.

Quantitative analysis of DNA damage in a single mammalian cell, immediately after its exposure to IR with low-LET at a dose of 1 Gy shows the formation of about 40 DSB and 40 ICL, about 150 DNA-protein crosslinks (DPC), some 2000 base modifications, about 3000 AP sites, damaged deoxyribose residues, SSB and alkali-labile sites (Ward, 1988; Gaziev, 1999). Many of IR-induced lesions are not randomly distributed but form, to a large extent, clusters, that is, they can be located in close proximity to each other. Typically, cluster DNA damage results from local matches of two or more single lesions within 1-2 turns of the DNA helix (Hada and Georgakilas, 2008; Sage and Harrison, 2011). A clustered DNA damage may represent damaged bases, AP sites located next to DSB or SSB. Clustered injuries are generated both at high and at low doses of IR (Sutherland et al., 2000). Clustering of DNA damage occurs on a large scale and depends on DNA packing and the structural state of chromatin in the cell nucleus during exposure of cell to IR. Most pronounced clustering occurs when the ionization tracks run along the chromatin fiber. In this case the clustered damage may cover DNA regions of about 2 thousand bp (Radulescu et al., 2004; Regulus et al., 2007). It was shown that clustered DNA damage in the form of modified bases, AP sites and SSB, localized close to each other, is 4-8 times more than the amount of DSB in DNA of mammalian cells after radiation exposure (Radulescu et al., 2004; Hada and Georgakilas, 2008). Such clustered lesions are turned to DSB in the initial stage of BER functioning (Asaithamby et al., 2011). DSB can be considered as a kind of clustered DNA lesions, in which the SSB or gaps are formed on opposite chains in the immediate vicinity. DSB may be formed not only directly but also indirectly (via production of ROS) as a result of closely spaced two or more SSB on opposite strands within approximately 10-20 bp. In addition, IR induced DSB may contain long single-stranded overhangs on the ends (Eccles et al., 2010). The significant amount of IR-induced ICL in cellular DNA can also be viewed as the result of clustered or multiple lesions (Gantchev et al., 2009). Formation of long gaps can also occur as a result of the clustered arrangement of SSB and the damaged DNA bases in chromatin structure (Gulston et al., 2004).

It is well established that the level of clustering of DNA damage and their complexity increase with increasing LET of radiation (Terato and Ide, 2004; Sage and Harrison, 2011). For low-LET IR, this clustering occurs mainly on small sites of DNA and nucleosomes in the chromatin. High-LET radiation can produce in DNA several DSB close to each other (Radulescu et al., 2004). The relationship between various DNA lesions in cells exposed to radiation with high-LET is changing in favor of more complex DNA damage (Sutherland et al., 2002; Sage and Harrison, 2011). Low-LET radiation, γ-rays in particular, causes in DNA DSB, 30% of which represent cluster formations. High-LET radiation, such as α-rays, yields more than 90% of DSB with ends carrying additional single lesions (Terato and Ide, 2004; Hada and Georgakilas, 2008). It was shown that in the case of irradiation of cells with protons (LET 9.2 keV/m), 45% of the resulting SSB in the DNA are accompanied by additional damage of bases in close proximity to these breaks, and this value increases to 78% after exposure of these cells to α-particles (LET 129 keV/m) (Terato and Ide, 2004). Other studies have also shown the formation of more complex DNA lesions after exposure of cells to high-LET radiation (Semenenko and Stewart, 2004). Clustered DNA damage occurs after irradiated of cells by low doses. So it was shown the linear dependence of the induction of clustered DNA damage in human primary fibroblast on the dose (10-100 cGy) X-rays (Das et al., 2011). Note that the probability of endogenous clustered DNA damage is extremely low (Bennett et al., 2004).

It is well known that the choice of repair pathways, as well as the efficiency and accuracy of this process in the cell, depend on the type of injuries and their complexity. Many DNA lesions, such as SSB, AP sites, base modifications, induced by IR or emerging endogenously in the cell, are mainly repaired by the BER pathway. The repair of DNA-protein crosslinks in mammalian cells occurs mostly by homologous recombination (Ide et al., 2011). The most complex pathways of DNA repair are responsible for correction of DSB and ICL induced by IR (Lieber, 2010; Dextraze et al., 2010).

Repair of a single lesion within a cluster may lead to formation of additional DSB or cause errors, which are critical for cell survival, mutagenesis, and malignization. There is literature evidence that DSB repair in cells irradiated with high-LET IR is less efficient than in the case of low-LET radiation. This is due, exclusively, to the complexity of these breaks (Heilmann et al., 1993; Dianov et al., 2001). Tandems of different single lesions with DSB may further complicate the repair of both types of injury and increase their biological significance (Sutherland et al., 2000; Sage and Harrison, 2011).

Thus, it is believed that the complexity and clustering are the most important specific characteristics of radiation-induced DNA lesions, which distinguishes them from the endogenous damage. Now also clear that a genetic effects and lethality of IR is not just related to the types of DNA lesions introduced, but also to how the damage is distributed in the DNA (Sage & Harrison, 2011). The most critical DNA injuries induced by IR may be thought to be DSB, ICL, damage due to destruction of the replication fork and formation of long single-stranded gaps. A complex radiation-induced DNA lesion is a serious burden for constitutive cellular repair systems. To repair such damage requires additional activation of many processes. In contrast to endogenous damage, repair of radiation-induced damage is accompanied by significant errors.

3. Mechanisms of cell response to DNA damage

Since DNA is constantly damaged by the action of endogenous and exogenous factors, living organisms in their evolution developed complex systems that monitor the structural integrity and stable operation of the genome. The components of these systems are able to recognize the most critical DNA lesions and to trigger a cascade of multiple molecular and biochemical reactions for the safety of genetic information. The extensive network of these reactions is activated as critical DNA lesions appear and is included in cell cycle regulation, DNA repair, reconstruction of chromatin, induction of transcription and cell death. The total sum of these reactions in the cell is called DNA damage response (DDR) (Kastan and Bartek, 2004).

Recent results of many studies on allow us to understand the mechanisms of DDR development in the cells after its exposure to IR, although much is still not clear. The development of DDR depends on many of proteins with different functions. These proteins have been classified as sensors, mediators, transducers, and effectors (Ding et al., 2005; Harper and Elledge, 2007). Although this classification is an arbitrary one, and the differences between protein groups are sometimes blurred, it facilitates our understanding of the mechanisms of process development.

The activation of cell cycle checkpoints is of vital importance for an effective DDR. The induction of gene expression is also associated with the development of the whole complex of DDR events. And the expression level of many genes may change depending on IR dose and the time after irradiation of cells (Ding et al., 2005; Taki et al., 2009; Ghandhi et al., 2011). These processes are influenced by microRNAs, the endogenous gene regulators (Hu and Gatti, 2011). DDR is also accompanied by reconstruction of chromatin by modifying proteins, which contributes to the recognition of DNA damage and repair (Pandita and Richardson, 2009; Misteli and Soutoglou, 2009).

DDR is regarded as an anti-carcinogenic mechanism (Costes et al., 2010), it is associated with a wide range of essential cellular processes and contributes to prevention of multiple human diseases (Kastan and Bartek, 2004; Bartek et al., 2007). Cells, deficient in different components of DDR development, have an increased sensitivity to IR and other DNA-damaging agents, and many of these defects are causative factors of human diseases (Jackson and Bartek, 2009; Jeggo, 2010).

The development of DSB-initiated DDR has been studied most of all. However, the development of DDR may also occur with other critical DNA lesions involving the same protein hierarchy as in the case of DSB (Fig. 1). At least, phosphorylated proteins, the characteristic markers for the development of DDR are also registered in case of ICL, destruction of replication fork and other complex DNA lesions of cells exposed to IR, UV light, chemical genotoxicants, as well as heat or adenovirus infection (Nagy and Soutoglou, 2009; Olive and Banath, 2009; Nichols et al., 2009; Callegari et al., 2010; Takahashi et al., 2010). Although it is possible that DDR markers detected on such complex injuries are caused by occurrence of DSB as an intermediate stage in the process of their repair.

DDR develops rapidly by migration of the protein complex MRN, which consists of three subunits (Mre11-Rad50-Nbs1), and the 9-1-1 complex, consisting also of three proteins (Rad9-Rad1-Hus1), which are the primary sensors of DSB and other critical DNA lesions (Parrilla-Castellar et al., 2004; Rupnik et al., 2010). Along with the complexes MRN and 9-1-1, the protein CtIP (tumor suppressor) also binds to damaged DNA sites (You and Bailis, 2010). The protein CtIP, by interacting with the subunit Mre11 (endonuclease), stimulates the activity of the complex MRN (Rupnik et al., 2010; You and Bailis, 2010). The complex MRN and the protein CtIP, when interacting with a DNA sequence containing DSB may unwind and resects the terminal sequences with removal of adducts and formation of short

single-stranded fragments at the 3' ends, which can bind proteins of the 9-1-1 complex (Kastan and Bartek, 2004; You and Bailis, 2010).
The signals of occurrence of critical injuries, DSB in particular, and of the interaction of sensory proteins with damaged DNA regions are received by other transducer proteins: ATM/ATR-kinases from the family of phosphatidylinositol- 3-kinase-like protein kinases. In the interaction of the MRN complex with a broken site of DNA, a rapid (within tens of seconds or one minute) recruitment of ATM (ATM, ataxia-telangiectasia mutated) to the same site takes place. ATM is a major kinase of the early response to a signal from the IR-induced DSB of DNA (Kastan and Bartek, 2004; Borde and Cobb, 2009). In this case, ATM, which in intact cells is in the form of a dimer, undergoes activation by autophosphorylation and splits into monomers. The monomer ATM molecules are further activated as a result of interaction with the C-terminal of the Nbs1-subunit (a protein mutated in the Nijmegen syndrome) of the MRN complex (Borde and Cobb, 2009; Huertas, 2010).

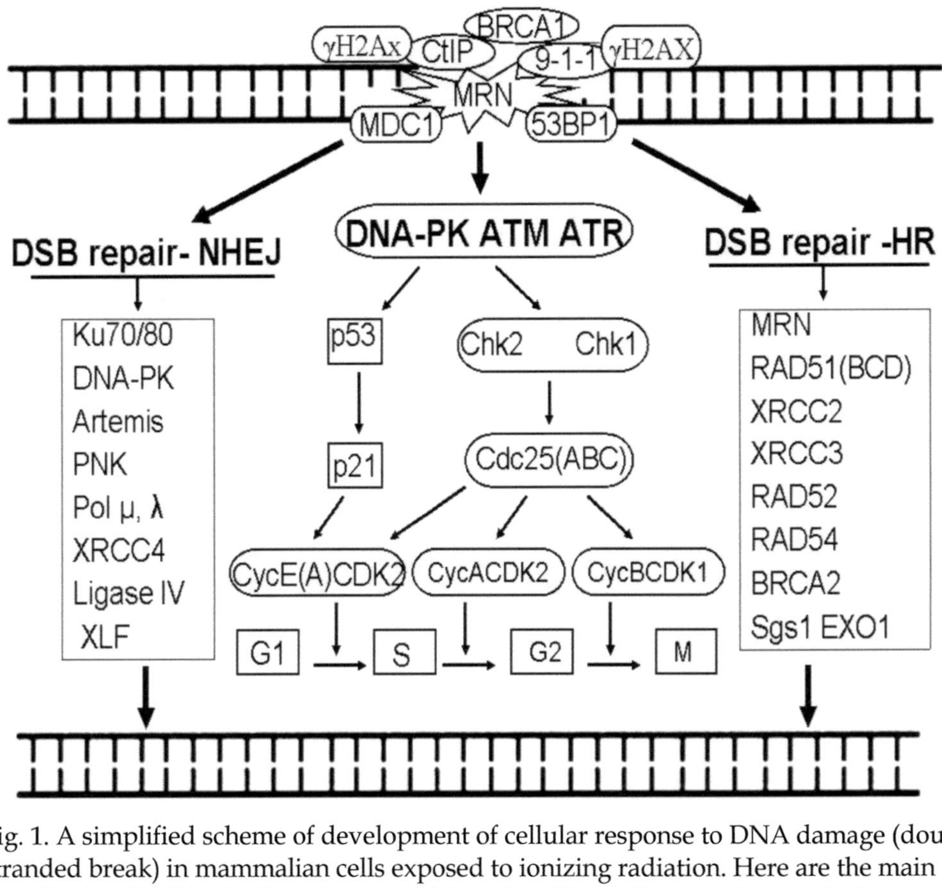

Fig. 1. A simplified scheme of development of cellular response to DNA damage (double-stranded break) in mammalian cells exposed to ionizing radiation. Here are the main proteins involved in the damage recognition, signaling and reparation. At the heart of the scheme - the direction of activation of cell cycle checkpoint that arrest cell transition in phases until the completion of DNA repair. Left and right of the scheme shows the basic components of DSB DNA repair by two principal pathways: non-homologous end-joining (NHEJ) and homologous recombination (HR). Explanations are given in the text.

Activated ATM stimulates ATP-dependent unwinding of terminal regions to form single-strand DNA. It has been also shown that, due to resection of DSB ends by MRN and CtIP, small overhanging ends of single strand DNA are formed (Huertas, 2010). Subsequent action of proteins ExoI and Dna2 (nucleases) alone or in combination with Sgs1 (helicase), create more extensive sequences of single-strand DNA (Lee and Paull, 2007; Lavin and Kozlov, 2007).

The formation of single-strand DNA is a signal that is perceived by ATR-kinase (ATM-related Rad3), which is also activated by autophosphorylation and then phosphorylates itself a set of proteins (on serine and threonine) participating in DDR (Mimitou and Symington, 2008; Cimprich and Cortez, 2008). Actually, after the DSB end resection has occurred and single-stranded oligonucleotides at these ends has appeared, the activity of ATM kinase is suppressed and ATR-kinase is activated through its interaction with single-strand DNA. In the process of replacing of ATM by ATR, the sensory complexes MRN, 9-1-1, RAD17-RFC and TopBP1 are involved (Cimprich and Cortez, 2008). The single-strand DNA regions are loaded by molecules of the protein RPA, participating in DNA replication and these proteins provide protection of single-strand DNA from nuclease attacks. ATR-kinase, in conjunction with the protein ATRIP (ATR binding protein), move to the regions with which RPA molecules are associated and covers single-strand DNA (Zou & Elledge, 2003; Smits et al., 2010).

Thus, ATR, like ATM, is a transducer of the signal received from the sensors of DNA damage. ATR-kinase can be activated in cells irradiated in the S-phase (when the damage blocks DNA replication) (Cimprich and Cortez, 2008; Smits et al., 2010). Although ATM is crucial in the primary response of the cell to DNA damage and in the subsequent activation of ATR, the latter has been proved to be more important for the survival of mammals. Deficient ATR in mouse or human cells leads to cell cycle arrest or to cell death, even in the absence of exogenous DNA damage (Falck et al., 2005).

In response to critical DNA damage in mammalian cells, not only ATM and ATR are activated, but also the catalytic subunit (cs) of DNA-dependent protein kinase (DNA-PKcs). These three kinases phosphorylate more than 700 proteins (Falck et al., 2005; Matsuoka et al., 2007). Among them are histone H2AX molecules, which after phosphorylation (γH2AX) contribute to recruitment of other proteins to the damaged region of DNA. The major proteins recruited for the damaged region of DNA are mediator proteins of DDR, such as Nbs1, BRCA1, Artemis (nuclease), 53BP1 (p53 binding protein), RAD51, MDC1, MCPH1/BRIT1, ubiquitin ligase RNF8, the complex RNF168/RIDDLIN, RAD17, RPA, ATRIP, the complex Rad9-Rad1-Hus1, TopBP1, Claspin and other target factors able to be phosphorylated. Many mediator proteins, such as γH2AX, BRCA1, MDC1, Claspin and 53BP1 usually function in coordination with other DDR factors, contribute to their activation (Matsuoka et al., 2007; Cimprich and Cortez, 2008; Smits et al., 2010).

Thus, the protein MDC1 binds to γH2AX, and the complex MRN recruits ubiquitin ligase RNF8 for the damage region to form a structure that holds the proteins RNF168, BRCA1 and 53BP1 in the DSB region of DNA. In this system, the protein MDC1 (checkpoint mediator) plays a major role in forming the molecular scaffold, which stabilizes the binding of the MRN complex to the damaged region of DNA and the activated ATM (Matsuoka et al., 2007; Kolas et al., 2007). The protein MDC1, along with γH2AX, also promotes the assembly of repair and checkpoint proteins in the chromatin portion with damaged DNA and contributes to its transition to a more relaxed conformation (Goldberg et al., 2003; Stewart et al., 2003; Kolas et al., 2007).

Numerous proteins localized, along with histone γH2AX, in DNA regions with DSB form large local structures (foci). These nuclear structures formed on DNA sites with DSB, or, as they are often referred to, radiation-induced foci (RIF), may cover a nucleus area of 1.7-2 μm^2 or DNA up to $15x10^6$ bp (Rogakou et al., 1999; Celeste et al., 2002; Wang and Elledge, 2007). The number of RIF in the nuclei can be determined by immunofluorescence, with matching specific antibodies against proteins that make up these foci, such as histone γH2AX or 53BP1. It has been well established that the number of RIF determined by means of γH2AX or other proteins, characterizes both the number of DSB in DNA and the state of the kinetics of their repair (Celeste et al., 2002; Wang and Elledge, 2007; Löbrich et al., 2010). These foci are eliminated from the chromatin as a result of dephosphorylation of γH2AX after DSB repair (Svetlova et al., 2010).

4. Activation of cell cycle checkpoints

It is well known that cell cycle progression is controlled by a series of checkpoints. If this system is active enough, the entry of cells with damaged DNA into the next phase will be arrested until the completion of its repair. However, if the checkpoint system is not activated, the entry transition of cells with critical DNA damage into another phase of the cell cycle can be completed either by fixation of chromosomal disorders or by apoptosis (Costes et al., 2006; Harrison and Haber, 2006; Yamauchi et al., 2008). Therefore, the arrival of a signal to activate the cell cycle checkpoint is a critical event in the development of the complex network of DDR. The signal to activate the cell cycle checkpoint is transmitted by ATM/ATR inverters, which themselves take this signal from the sensors of DNA damage.

Activation of the checkpoint occurs by the following main mechanisms (Fig. 1). In the scheme involves the activation of cell cycle checkpoint due to DSB. ATM and ATR take signals from sensor proteins capable of recognizing the critical DNA damage, and then phosphorylate the effector kinases Chk1 and Chk2 (checkpoint-kinases 1, 2) (Costes et al., 2006; Yamauchi et al., 2008). Activation of Chk2 occurs through its phosphorylation by ATM kinase at the amino acid T68, followed by oligomerization and autophosphorylation at T383 and T387 (Harrison and Haber, 2006).

Activation of Chk1 occurs through its phosphorylation by ATR-kinase at amino acids S317 and S345 (Cook, 2009). ATP-mediated activation of the checkpoint occurs also due to the destruction of the replication forks (Warmerdam et al., 2010). In this case, as shown by biochemical data ATR in complex with ATRIP, binds to RPA-coated single-strand DNA (Zou & Elledge, 2003). This obviously protects the single-strand DNA against nuclease attack.

Activated kinases Chk1 and Chk2 in turn phosphorylate other proteins–phosphatases (Cdc25A, Cdc25V, Cdc25S), which control the activity of cyclin-dependent kinases (CDKs) by splitting out their phosphate groups. However, the phase-specific CDKs are active only in association with cellular cyclins. Thus, in the cell transition G1/S a cyclin E (A)-CDK2 complex is functioning, within the S phase the complex cyclin A-CDK2 and in the G2/M transition – the cyclin B-CDK1 complex (Harrison and Haber, 2006; Bartek and Lukas, 2007). After phosphorylation, the phosphatases Cdc25 (ABC) lose their activity and undergo ubiquitin-mediated proteosomal degradation. Under conditions of suppressed phosphatase activity, the CDK dephosphorylation stops and there occurs a complete phosphorylation of

the phase-specific cyclin-CDK complexes, which block the G1/S and G2/M phase transitions of the cells. In this case, the initiation of replication of new DNA replicons in the S-phase is also blocked (Bartek and Lukas, 2007; Stracker et al., 2009).
There is also another p53-dependent pathway to activate the checkpoint. Thus, the chain of reactions ATM → CHK2 (ATR → CHK1) → p53 → MDM2 → p21 → cyclin-CDK is also one of the main pathways to arrest G1/S and G2/M transitions in the presence of DNA lesions. In this chain of reactions (with stabilization of p53 by phosphorylation) there occurs the activation of *p21* gene expression (Chung and Bunz, 2010). Accumulation of protein p21 leads to inhibition of the activity of cyclin E (A)-CDK2 and cyclin B-CDK1complexes. It should be noted that the p53-independent (CHK1/CHK2 → Cdc25 (ABC) pathway of checkpoint delays the G1/S transition for no more than a few hours, while the p53-dependent pathway extends significantly the block of the G1phase (Niida et al., 2010; Freeman and Monteiro, 2010; Shibata et al., 2010).
Thus, the signaling of DNA damage perceived by ATM/ATR, is transmitted to the key effectors of checkpoint activation - Chk1 and Chk2 kinases, and the subsequent reaction chain provides an arrest of cell cycle progression prior to the completion of DNA repair. After the repair of critical DNA damage has been completed, the checkpoints can be switched off and the cell cycle progression resumed in the cell.

5. Repair of double strand breaks and interstrand cross-links in DNA

The ATM/ATR signaling, simultaneously with the activation of cell cycle checkpoints, activates the repair process by induction of repair protein transcription, mobilization of other components involved in the repair and by modification of chromatin proteins (Costes et al., 2010; Warmerdam and Kanaar, 2010). This can be seen during destruction of the replication forks, induction of DSB, ICL and other complex lesions. So, DNA replication forks are stalled by many DNA lesions, activating checkpoint proteins that stabilize stalled forks (Allen et al., 2011). The activation of the checkpoint and the repair of critical lesions are interrelated. The efficiency of repair of such DNA lesions depends on their complexity.
Let us consider pathways to repair the two types of critical DNA damage: DSB and ICL. The repair of DSB and ICL are important for maintaining the physical and genetic integrity of the genome. Moreover, in humans it is associated with the prevention of diseases such as cancer.

5.1 Repair of DNA double strand breaks

DNA DSBs induce breakage of chromosomes. In the course of cell division, chromosome fragments can unevenly be redistributed among daughter cells, so that they can generate translocations, i.e. appear in other parts of the genome, where they are not supposed to be. Deletions or translocations of chromosome fragments lead to inactivation of tumor suppressor genes or activation of oncogenes. Both events may initiate carcinogenesis in the surviving cells (Hoeijmakers, 2001; Halazonetis, et al., 2008; Natarajan & Palitti, 2008).
In mammalian cells, the repair of DSBs in DNA proceeds by two main pathways: by non-homologous end-joining (NHEJ) and by homologous recombination (HR) (Fig. 1) (San Filippo et al., 2008; Lieber, 2010). The repair of DSBs through NHEJ occurs in the G0, G1, and G2 phases of the cell cycle whereas the mechanism of HR may function in the S and G2 phases, because in these phases two sister chromatids are present. However, if in the course

of S and G2 phases there is no homologous variant alongside the damaged DNA, then the mechanism of HR cannot operate. In haploid cells in the stationary phase, in the absence of a homology of damaged DNA, the DSB repair by HR pathway is likewise impossible (Heyer et al., 2010). Also, in the case of non-dividing diploid cells, the occurrence of the damaged DNA region in the vicinity of a homologous donor is quite problematic. Moreover, the repair of IR-induced DSB by HR mechanisms can not always be successful since IR induces cluster lesions which produce complex adducts at DNA ends (Shikazono et al., 2009). Probably, the reunion of most DSB in DNA regions not involved in replication during the S-phase is accomplished by NHEJ. Therefore, the repair of DSBs by an error-prone NHEJ mechanism in the cells of IR-exposed mammals may be prevailing throughout the cell cycle (Mahaney et al., 2009; Lieber, 2010).

The repair of DSBs by HR mechanisms is more accurate because homologous regions of sister chromatides are used as a template to restore the broken DNA strand (San Filippo et al., 2008; Heyer et al., 2010). This process is initiated by binding of the complex MRN (Mre11/Rad50/Nbs1) to the ends of DSBs. Together with the protein CtIP, the complex MRN provides the initial resection of broken DNA ends while forming small 3' overhanging of single-strand DNA (Sartori et al., 2007). Nucleases ExoI or Dna2 in combination with Sgs1- helicase produce, by further resection of the 5' end of broken DNA strand, longer 3' overhangs - single-strand regions covered with RPA molecules. Then the protein Rad51, with the participation of BRCA2, replaces RPA on single-strand DNA to form a nucleoprotein filament. Thus, recombinant 3' single-strand tails are formed at broken DNA ends. Hereafter, these single-strand regions invade, with the help of their associated protein Rad51, the homologous double-strand DNA-template to generate homologous DNA duplex (Modesti et al., 2007; Cejka et al., 2010). Then the branch migrates with the formation of so called Holliday structures, which favors error-free DNA repair (San Filippo et al., 2008; Cejka et al., 2010).

In mammalian cells, this process is mainly provided by multiple proteins encoded by a group of the gene family *Rad* (RAD51, RAD52, RAD54, RAD55, RAD57, RAD59, and RDH54). In addition to the products of these genes, involved are also nucleases (ExoI, Sae2, Rad1-Rad10) helicases (Sgs1, Srs2), topoisomerase (Top3), polymerases (Pol32) as well as ligases, which terminate the repair (San Filippo et al., 2008; Cejka et al., 2010).

Unlike DSB repair by HR, the repair mechanisms of NHEJ remain in many respects unclear. The repair of DSBs by NHEJ is rather complex and dynamic and requires coordinated interactions of many active proteins throughout the process. Repair of a certain number of DSBs through NHEJ can occur independently of ATM activation, whereas the repair of the other part of DSB is extremely ATM-dependent (Jeggo, 2009; Warmerdam and Kanaar, 2010).

On the other hand the DSB repair occurs in DNA not only in the major, "classic" pathway - NHEJ, but also in an "alternative" pathway - by binding the ends of terminal microhomologies (Haber, 2008; McVey and Lee, 2008). In this case, microhomologies of about 4 bp can be formed at broken DNA ends. This kind of DSB repair is likely to occur in DNA regions that contain repetitive sequences. This process does not depend on Ku-proteins, the complex of ligase IV-XRCC4 (the most important components of NHEJ), but it depends on Nbs1 and Fen-1. In addition, it involves proteins of the system for mismatch repair: MLH1 and MSH2 (McVey and Lee, 2008). However, repair by binding the ends of terminal microhomologies, perhaps, does not play any significant role in stabilizing the genome of mammalian cells (Lieber, 2010).

The main pathway to repair DSBs by NHEJ is initiated by binding heterodimer Ku70/80 protein complexes to both ends of broken DNA molecules (Pardo et al., 2009). The complex Ku70/80 protects the ends of DSB from early degradation and ligation, and also contributes to recruitment of DNA-PK_{CS} to broken DNA region. DNA-PK_{CS} when associated with both ends of DSB and with the dimer Ku70/80 is activated by autophosphorylation (Costantini et al., 2007). DNA-PKcs and Ku heterodimer together form the complex that plays key roles in the repair DSB through the NHEJ pathway.

DNA-PK_{CS} and ATM phosphorylate a number of proteins involved in the process of NHEJ of DNA. When captured by the complex Ku70/80/DNK-RK_{CS}, both ends of DNA are processed by various enzymes for removal of terminal groups or adding nucleotides to these ends. This stage of processing in the repair by NHEJ is believed to be responsible for errors resulting from losses and additions of nucleotides at the sites of DNA DSB rejoining (Mahaney et al., 2009; Lieber, 2010). In the processing of broken DNA ends, the following enzymes are involved: the nuclease Artemis (Yannone et al., 2008), the DNA polymerases of the family X (Pol μ and Pol λ) (Moon et al., 2007), and the terminal deoxynucleotidyl transferase (Mahaney et al., 2009), as well as polynucleotide kinase having both the kinase and phosphatase activities (Bernstein et al., 2008). Also, some other substances participate in the process, such as aprataxin (APTX) which catalyzes the removal of AMP groups at the 5 'ends of the DSB (Rass et al., 2008), the PALF-factor, and the protein cernunnos (XLF), which interact with XRCC4 in the activation of the DNA-ligase complex XRCC4/DNK-ligase IV/XLF (Tsai et al., 2007; Yano et al., 2009).

The final stage of DSBs repair by NHEJ is binding the broken DNA ends by the ligase complexes (Mahaney et al., 2009; Heyer et al., 2010). In the presence of Ku proteins, the full ligase complex (XRCC4/DNK-ligase IV/XLF) can couple incompatible DNA ends and, more effectively, blunt ends of breaks (Gu et al., 2007).

Thus, we can assume that the main mechanism for repairing DNA DSB induced by low doses of IR is the NHEJ pathway.

5.2 Repair of interstrand cross-links of DNA

The interstrand cross-links (ICLs) are highly toxic DNA lesions that block replication and transcription by inhibiting DNA strand separation, can also retard the repair of monoadducts in DNA. ICL may provoke genomic instability, chromosomal aberrations, and cell death (Noll et al., 2006; Dextraze et al., 2010). Many studies on the formation and repair of DNA ICL in the cells are related with the influence of chemical agents. As for the data on the repair of IR-induced ICL, they are rather limited. This is most likely caused by methodological difficulties of their acquisition against a variety of other IR-induced DNA lesions and, in particular, because of their association with clustered lesions (Ding and Greenberg, 2007; Hong et al., 2007; Gantchev et al., 2009). It should be noted that the cells of patients with the syndrome of Fanconi anemia (FA) are characterized by genome instability, and predisposition to malignancy, and by increased sensitivity to IR due to their inability to repair ICL of DNA (Alter, 2002; Kuhnert et al., 2009).

ICL initiate the development of DDR with the activation of cell cycle checkpoints, with transmission of the signal along the chain ATM/ATR → Chk1/Chk2 → Cdc25 (A, B, C) → cyclin-CDK (Wu and Vasquez, 2008; Ben-Yehoyada et al., 2009; Vasquez, 2010). It has been also shown that the induction of ICL in DNA is accompanied by formation of γH2AX-foci (Huang et al., 2004; Mogi and Oh, 2006; Mogi et al., 2008). Moreover, γH2AX-foci occur in the nuclei of cells after their treatment with compounds that cause ICL, but not directly induce DSB in DNA (Huang et al., 2004; Mogi and Oh, 2006; Olive and Banath, 2009). However, this does not exclude that γH2AX-foci in the nuclei of cells exposed to agents that induce ICL are caused by DSB, since ICL repair processes are accompanied by DSB generation (Olive and Banath, 2009; Sczepanski et al., 2009; Muniandy et al., 2010).

Removal of ICL from replicating and non-replicating DNA of mammalian cells can occur in two ways (depending on cell cycle phase) with the participation of enzymes produced by a number of repair systems, as well as proteins produced by *Fanc* genes (mutated genes in patients with the syndrome FA) (Moldovan and D'Andrea, 2009; Thompson and Hinz, 2009; Knipscheer et al., 2009). Thus, in the phases G0, G1 of the cell cycle, the removal of ICL is mostly provided by the system of nucleotide excision repair (NER), which ensures the global genome repair and the transcription-related repair (Sczepanski et al., 2009; Wood, 2010). Through this pathway, the repair of ICL begins with recognition of the injury with the complex XPC-HR23B. However, the endonuclease complex ERCC1-XPF is a key factor in the repair of ICL on non-replicating DNA (Cole et al., 2010). This complex, while inducing breaks with 5 'and 3' ends on both sides of ICL on one strand, creates a gap that can serve as a site for DNA synthesis bypassing the damage (Cole et al., 2010; Liu et al., 2010), after processing DNA break ends by the nuclease DCLRE1A/SNM1A. The duplex formed as a result of the DNA synthesis bypassing the damage and bearing cross-link remains on one strand becomes a substrate for the classical system of NER, which provides the final stage of removal of ICL from non-replicating DNA. This process is accompanied by a loss of 11-14 bp, and is concluded with formation of DSBs which can be restored by HR (Peng et al., 2010; Legerski, 2010; Muniandy et al., 2010).

The removal of ICL in the S phase of cells proceeds by a different pathway. Collision of the replication fork with ICL triggers a cascade of reactions, which promotes removal of the cross-link and restoration of DNA replication (Moldovan and D'Andrea, 2009; Thompson and Hinz, 2009; Knipscheer et al., 2009). Recent studies have shown that the main function of FA-proteins is associated with activation of signaling from DNA damage and repair of ICL in the S-phase of the cell cycle (Knipscheer et al., 2009; Liu et al., 2010). The removal of ICL from replicating DNA is accompanied by activation of cell cycle checkpoints (Noll et al., 2006; Legerski, 2010). ATR-dependent activation of the major FA- complex (A-C, E-G, L, and FAAP100) is the central event of repair of ICL that blocks the replication fork. The occurrence of ICL in the replication fork is recognized by several FA-protein complexes, including FANCM and FAAP24, which provide access of other proteins to the damage to repair it (Grillari et al., 2007; Dextraze et al., 2010).

An important role in these events is played by the complex FANCI/FANCD2, which binds to DNA sites with ICL (Smogorzewska et al., 2010). This complex, in collaboration with the Tip60 protein, also contributes to remodeling of chromatin sites with damaged DNA (Hejna et al., 2008). When this complex encounters a restructured DNA replication fork, due to localization of ICL, it is phosphorylated by ATR/ATRIP kinases localized in ssDNA, together with the protein RPA. The phosphorylated complex FANCI/FANCD2 is fully activated after its monoubiquitination by the multi-subunit ligase E3. The activated complex FANCI/FANCD2 moves to DNA sites with ICL. This activated complex, together with other FA-protein complexes FANCD1/BRCA2 and FANCN/PALB2 is involved also in a subsequent stage of recovery of broken replication fork via HR (Knipscheer et al., 2009; Liu et al., 2010; Smogorzewska et al., 2010). Protein complexes, such as MRN, Snm1B and MUS81-EME1/2, also bind to the DNA sites of the replication fork blocked by ICL. The last-named complex has a nuclease activity, it makes an incision on one side flanking the ICL, and the endonuclease ERCC1-XPF makes the second incision, on the other side of the crosslink (Knipscheer et al., 2009; Liu et al., 2010; Cole et al., 2010). As a result of these endonuclease attacks, one strand gets free of the cross-link and the ends of the resulting gap undergo processing. Then the gap is filled due to DNA synthesis (bypassing lesions on the template strand bearing cross-link remains) with the participation of DNA polymerases: polζ, polκ, polι, polN, polη and Rev1 (Ho and Schärer, 2010). After filling in the gap, the cross-link remains bound to the second strand undergo excision by NER enzymes. The repair of DSB formed during removal of ICL from the replicating DNA with the participation of FA-proteins, is also provided by the mechanism of HR (Hinz, 2010).

6. Low efficiency of repair of critical DNA damage induced by low dose radiation

Over the last 10-15 years evidence has been accumulated pointing to that the responses of mammalian cells to small and large doses of ionizing radiation may be qualitatively different in a number of characteristics, including the development of DDR. First of all, these are the reactions of cells to critical DNA damage, such as signals to activate checkpoints, DNA repair, chromatin reconstruction and postradiation changes in the expression of many genes. To this must be added epigenetic effects, induced genome instability and the bystander effect (Mullenders et al., 2009). The factual data on these characteristics obtained for low-dose-irradiated organisms do not fit into the model of the linear dependence of biological effects on IR dose (Hayes, 2008; Averbeck, 2009; Ulsh, 2010).
In this section we analyze the results of a number of independent studies on DDR development, on the signaling to activate cell cycle checkpoints and on the repair efficiency of critical DNA lesions induced by low-dose or low-intensity radiation in mammalian cells.

6.1 Inadequate activation of cell cycle checkpoint at low-dose radiation

For the most part, the data on this subject relate to DNA DSB. Much of the evidence presented earlier (before the advent of more sensitive methods) on assessment of IR-induced DNA DSB and their repair, were obtained in studies using high radiation doses. These methods, although quite adequate, appeared not sensitive enough to detect these lesions and to follow the kinetics of their repair in cells exposed to low-dose radiation. Actually, sensitive methods to assess DSB that arise under the influence of low dose radiation have been developed in studies of DDR. An important step in this direction was the development of methods using fluorescent antibodies to phosphorylated proteins, specifically localized in discrete nuclear radiation-induced foci (RIF) at DNA sites containing DSB. RIF can be visualized and counted under a microscope. These methods are most commonly based on the use of fluorescent antibodies specific for phosphorylated proteins involved in RIF formation, such as histone H2AX (γH2AX), ATM-kinase (pATM), 53BP1 (p53BP1) and others. These RIF localized in the nuclei of irradiated cells, enable counting of DSB, and, from reduction of their number, judging the kinetics of their repair, as well as analyzing other critical injuries that can be realized in the form of DSB. Thus, the determination of DNA DSB induced by low-dose radiation (1-100 mGy), has become possible by analysis of histone γH2AX immunofluorescence in RIF formed by accumulation of their molecules at DSB sites of DNA within the chromatin (Celeste et al., 2002; Wang and Elledge, 2007; Nagy and Soutoglou, 2009; Löbrich et al., 2010; Redon et al., 2011). This method also allows determination of the endogenous level of DSB in cellular DNA. Currently, RIFs are regarded as a highly sensitive biomarker to assess the damage to cell genome due to radiation or other genotoxic effects, the method can be applied in the clinics (Sedelnikova et al., 2008; Lassmann et al., 2010).

By the method of identifying RIF by γH2AX it was revealed that DSB induced in DNA of non-dividing primary human fibroblasts, after their exposure to low doses of radiation (~ 1 mGy), remained non-repaired for many days. On the other hand, the number of DSB in cells irradiated with low doses, decreased in the postradiation period up to the control level if these cells were allowed to proliferate after irradiation. The authors suggested that this effect might be due to elimination of cells with non-repairable DSB from the total cell population by apoptosis (Rothkamm and Löbrich, 2003). In a recent study, by analysis of RIF marked at phosphorylated proteins γH2AX, ATM (pATM) and 53BP1, DNA DSB repair was monitored in human fibroblasts exposed to IR at doses between 2.5 to 200 mGy (Grudzenskia et al., 2010). The kinetics of disappearance of RIF recorded in these cells during post-radiation time depended on irradiation dose. The repair of DSB in low-dose irradiated cells was significantly lower than in cells irradiated with higher doses. In the case of exposure of cells to 2.5 mGy, the loss of RIF was not observed for a long time after irradiation. The results indicate that the smaller the number of radiation induced DSB, the lower the efficiency of their repair. This study was continued on mice by analyzing RIF (γH2AX and 53BP1 foci) in various tissues irradiated at doses of 10, 100 mGy and 1 Gy. In the cells of all analyzed tissues of mice exposed to a low dose (10 mGy), but not to a moderate (100 mGy) or a high dose (1 Gy), RIF persisted (γH2AX and 53BP1) during 24 hours (Grudzenskia et al., 2010).

It should be noted that at a dose of 10 mGy, DSB were induced not in all cells of the tissue. It is known, that only after whole body irradiation with 10 mGy with low LET (photons or electrons) there may arise about 20 injuries in each cell, on the average, including 0.4-0.5 DSB (NRC, 2006). In fact, at a dose of 10 mGy the DNA of only every second or third cell induces a single DSB (NRC, 2006). On this basis, the authors suggest that the absence of efficient DSB repair after exposure of cells or animals to 10 mGy, may be associated with the fact that DSB DNA occur not in all cells being in contact with each other, and the activation of repair occurs not in all cells. This hypothesis assumes that the efficiency of DNA DSB repair in a single cell may be influenced be active repair processes in adjacent cells. That is, the effective repair of DNA DSB in cells is only possible, when all contacting cells contain at least 1 DSB per cell genome, which can occur at doses of 40-80 mGy (Grudzenskia et al., 2010). Such an explanation suggests that the intercellular communication mechanisms or the "bystander effect" play a role in the repair of DNA DSB (Rothkamm and Löbrich, 2003).

In the context of these explanations on the possible role of the "bystander effect", of interest are several publications related to this topic. Thus, it has been shown that irradiation of individual human astrocytes and glioma cells by a microbeam of helium ions generates signals that induce γH2AX foci in unirradiated neighboring cells, that persist for 48 hours. This bystander effect is not observed in cells mutant in ATR-kinase, but retains on inhibition of their ATM and DNA-PK. In addition, the ATR-dependent induction of the bystander effect, on γH2AX foci formation, was limited in the S-phase (Belyakov et al., 2005). The recent paper by Ojima et al., (2011) deals with the kinetics of disappearance of RIF (as determined from phosphorylated ATM) in the course of DNA DSB repair in X-ray-irradiated human fibroblasts at doses of 1.2, 20, 200 and 1000 mGy. It has been shown that the numbers of DSB determined by ATM foci formation 24 h after irradiation were: 100%

(1.2 mGy), 58% (20 mGy), 12% (200 mGy) and 8.5% (1000 mGy) of the original amounts of ATM foci (Ojima et al., 2011). These data are to some extent consistent with the results of studies mentioned above (Rothkamm and Löbrich, 2003; Grudzenskia et al., 2010). However, in studies using the inhibitor of the "bystander effect" (lindane), the authors concluded that the ATM foci detected in the nuclei of cells exposed to low doses (1.2-5 mGy) and persisting in them for a long time after exposure, are mostly related to the bystander effect (Burdak-Rothkamm et al., 2007).

Previously evidence was obtained for an extremely high sensitivity of cells defective in mechanisms for repair of DSB by NHEJ to low doses of radiation (Tomita et al., 2008). The authors have concluded that the process of DNA DSB repair occurring upon low-dose irradiation of cells, is primarily controlled by NHEJ mechanisms, whereas for repairing DSB caused by higher radiation doses, HR is more appropriate (Tomita et al., 2008). Probably, with a small number of DSB the chances that the damaged DNA site and a homologous intact DNA portion in the interphase nucleus will be in close proximity to each other are rather small for a fast start of DSB repair by the HR mechanism.

The data by different authors presented here, though contradictory to some extent, confirm that DDR induced in cells exposed to low-dose radiation differs significantly from that observed under the action of high doses. In the case of low doses or low dose rates, the level of signaling from DNA lesions perceived by transducers is of primary importance for subsequent DDR development, and, above all, for cell cycle checkpoints. Based on studies of the DSB-induced DDR development in yeast, it was assumed that the activation of cell cycle checkpoints occurs as a result of induction of one DNA DSB per cell, and the arrest of the cell cycle will continue until the completion of repair of these injuries. However, studies on mammalian cells have shown that IR doses that cause less than 10-20 DSB/per cell do not contribute to arresting the phase transition G2/M by checkpoints, and this greatly reduces the efficiency of DSB repair (Deckbar et al., 2007; Grudzenskia et al., 2010). On higher doses of ionizing radiation, which induce more than 20 DNA DSB per cell, the checkpoints are activated and arrest the cell cycle phase transition G2/M. But this arrest may not persist until completion of DNA DSB repair (Deckbar et al., 2007; Jeggo, 2009). These results indicate that in the absence of cell cycle arrest, mammalian cells are unable to provide complete repair of DSB induced even at low IR doses. The lack of G2/M transition arrest by the checkpoint until completion of DNA DSB repair leads to rearrangements of chromosomes and their unequal distribution between the two daughter cells.

Obviously, checkpoints must operate efficiently enough to stop the cell cycle progression, and to prevent the transition of cells with DNA DSB to mitosis (Shibata et al., 2010; Warmerdam and Kanaar, 2010). Thus, when DSB in DNA are small in number, the cell cycle checkpoint fails, and mammalian cells in the G2 phase enter mitosis with DNA lesions (Deckbar et al., 2007). Note that similar results were obtained after UV irradiation of yeast cells. The cells did not stop at the G2/M boundary after exposure to low UV light doses, whereas high doses might cause a delay in the cell cycle at the G2/M transition (Callegari and Kelly, 2007).

With a small amount of DSB in DNA, transducer proteins do not sense the signals of damage, so the cell cycle checkpoint is not activated. This situation is confirmed by studies of Collis and colleagues (Collis et al., 2004), which showed that if the alarm signal is weak, that is the amount of critical DNA lesions is small, the transducer proteins did not perceive it, and are not activated by phosphorylation. In the study of these authors several lines of normal and tumor human cells were irradiated by gamma rays at two dose rates. In the first case, the dose rate was 9.4 cGy/h (~ 4-5 DSB/ h), the second dose rate was ~ 4500 cGy / h (~ 1800 DSB/ h). The activation of ATM kinase in response to DNA damage and the formation of γH2AX foci in cells exposed to low dose rate were shown to be significantly lower than those in cells irradiated with an equivalent dose, but at a high dose rate. This lack of signaling of DNA damage was associated with an increase in cell death after exposure to low intensity radiation. Thus, after exposure to low intensity radiation cells died due to lack of a developed DDR, because at that point the level of damage was not sufficient for its full development. The "molecular radar", (ATM/ATR), able to recognize the DNA DSB did not catch a weak signal from sensor proteins, because the damage was small (~ 4-5 DSB DNA/h) during prolonged exposure (Collis et al., 2004). These results, to some extent, are in agreement with studies by Ishizaki et al.(2004) who showed that after exposure to IR in an acute dose, human cells accumulated a significant amount of γH2AX foci. However, after chronic exposure of the same line cells to an equivalent dose of radiation there was only a small amount of γH2AX foci, indicating a low level of DDR in these cells (Ishizaki et al., 2004). The lack of signals of damage and hence of activation of cell cycle checkpoints reduces the efficiency of DSB repair, which contributes to cell death. This event seems to be of vital importance for multicellular organisms because it prevents genetic and carcinogenic consequences.

Previously, evidence was obtained suggesting that the increase in cell death at low DNA damage and low repair efficiency, is a pathway by which cells prevent the transmission of potentially promutagenous DNA damage to the offspring (Joiner et al., 2001). Analysis of cell survival reveals an increased radiosensitivity (to judge from the decay of the initial curve segment) of a particular fraction of cell population at doses of 0.02-0.3 Gy (Joiner et al., 2001; Marples et al., 2004). This phenomenon has come to be known as hyper-radiosensitivity of cells. Now it has been determined that hyper-radiosensitivity is associated with failures in the ATM-dependent stage of DDR development and is due to the low efficiency of repair of DSB induced in DNA by low IR doses of both low- and high LET (Wykes et al., 2006; Xue et al., 2009). It was established a direct relation between low-dose hyper-radiosensitivity and the lack of ATM-dependent "early" activation of cell cycle checkpoint at the G2/M transition (Marples et al., 2004). This "early" activation is necessary to prevent the transition of cells irradiated in the G2 phase to mitosis without completion of DNA repair (Marples et al., 2004). It has been shown that an increase in the share of cells at the G2 phase in the overall population, leads to an increase in hyper-radiosensitivity, as most of the damaged cells escape the early delay of G2/M transition thus entering mitosis with non-repaired DSB. In addition, the absence of cell cycle arrest at G2/M transitions, as a result of inhibition of kinases Chk1, Chk2, also leads to an increase in hyper-radiosensitivity (Krueger et al., 2010).

In a recent study it has been found that after exposure of cells to IR, there occurs a dose-dependent induction of molecular arrest at G2/M transition (Fernet et al., 2010). Cells irradiated with small doses (about 0.2 Gy) at G1 or S phase further progress through the cell cycle, but they are less accumulated in the G2 phase. A block of the G2/M transition for a large number of cells was detected only after their irradiation at higher doses (above 0.5 Gy). And the absence of early cell cycle arrest at the G2/M transition by checkpoint control mechanisms correlates with hyper-radiosensitivity of these cells (Fernet et al., 2010). Thus, the increased sensitivity of cells to low IR doses is associated with low levels of DDR due to a small number of critical DNA lesions necessary for the activation of G2 checkpoints, i.e. the cell cycle transitions continue in the presence of critical DNA damage (Marples et al.,

2004; Krueger et al., 2007). As noted above, a possible threshold for full induction of DDR (with activation the checkpoint arrest of cell cycle) in human cells can be 10 to 20 DSB DNA/per cell (Rothkamm and Löbrich, 2003).

6.2 Limited repair of complex DNA lesions in compacted chromatin

In eukaryotic cells, the whole cascade of events of DDR development, including recognition of DNA damage, signaling, transduction with activation of the checkpoints, and the repair occur in the chromatin. It is well known that chromatin is a highly compact and dynamic structure in which DNA is packaged with histones and nonhistone proteins. The rigid compaction of chromatin creates a barrier to restrict access to damaged DNA sites for various enzymes and factors (Tang and Greenberg, 2010; Costes et al., 2010). In the case of complex DNA lesions, however, local changes in chromatin structure occur during their repair. The development of DDR is associated with active movement of many proteins to damaged DNA sites within chromatin (Pandita and Richardson, 2009). It is probably because of this DDR-related processes are accompanied with structural reorganization of chromatin and with relaxation of its individual parts. The mechanisms of alterations of local chromatin areas related to appearance of DSB or ICL in DNA are still not clearly understood.

Many studies have shown that modification of chromatin proteins by phosphorylation, methylation, acetylation, ubiquitination, sumoylation and poly ADP-ribosylation plays a critical role not only in the regulation of many genetic processes, but also in DDR development (Pandita and Richardson, 2009; Huertas et al., 2009). Decondensation of chromatin near the region of DNA DSB is considered an important trigger for ATM-dimer dissociation and subsequent autophosphorylation of ATM kinase (Goodarzi et al., 2009).

The results of studies suggest that chromatin at damage sites may be quite structurally dynamic during the repair process, with transient intervals of 'closed' configurations before a more 'open' arrangement that allows the repair machinery to access damaged DNA (Ball & Yokomori, 2011). DNA damage, in particular DSBs promote to the formation of open, relaxed chromatin regions. These relaxed chromatin structures are created through the coupled action of the p400 SWI/SNF ATPase and histone acetylation by the Tip60 acetyltransferase. The resulting destabilization of nucleosomes at the DSB by Tip60 and p400 is required for ubiquitination of the chromatin by the RNF8 ubiquitin ligase, and for the subsequent recruitment of the BRCA1 and others proteins (Xu & Price, 2011).

Chromatin dynamics at DSBs can therefore exert a powerful influence on the process of DSB repair. Further, there is emerging evidence that the different chromatin structures in the cell, such as heterochromatin and euchromatin, utilize distinct remodeling complexes and pathways to facilitate DSB. The processing and repair of DSB is therefore critically influenced by the nuclear architecture in which the lesion arises (Xu & Price, 2011; Ball & Yokomori, 2011).

Phosphorylation of histone H2AX and other proteins by ATM, ATR and DNA-PKcs near DSB also modifies chromatin, which is crucial in the interactions between many proteins after exposure to IR and, above all, for the process of DNA repair. Thus, chromatin protein modifications contribute to active regulation of multiple reactions associated with the development of whole DDR (Goodarzi et al., 2009; Huertas et al., 2009; Costes et al., 2010).

As a rule, RIF related to chromatin sites with damaged DNA in the cell nuclei are formed not immediately after irradiation. The RIF frequency reaches its maximum (40 RIF/nucleus/Gy) about 15-30 minutes after the action of IR with low LET. Obviously, this is the time required for recruitment of proteins that form RIF to the portion of damaged DNA in the chromatin.

Costes and colleagues believe that the long-term persistence RIF (recorded by proteins γH2AX, ATM, 53BP1) in the nuclei of irradiated cells is associated with constant rearrangements of chromatin architecture (Costes et al., 2010). On the other hand, the long-term persistent of RIF in compacted sites of chromatin, following exposure of cells to IR suggests that the kinetics of repair of complex DNA lesions is affected by the slow reconstruction of the chromatin, as a result of a "delayed" signal from DNA damage to DDR proteins (Aten et al., 2004; Costes et al., 2010). In addition, the time of formation of RIF and their long-term persistence may depend not only on the original hard compaction of chromatin and cell cycle phase (Goodarzi et al., 2009), but also on the amount of lesions, such as DSB or ICL, as well as single lesions contributing to the relaxation of chromatin (Ploskonosova et al., 1999; Duan and Smerdon, 2010). Thus, a comparison of times of DNA damage persistence in peripheral blood cells of mice irradiated with X-rays at doses of 0.1 and 1 Gy, has shown that in mice irradiated at a dose of 0.1 Gy, the DNA lesions remained for a long time, whereas similar DNA lesions disappeared much faster in mice irradiated at a dose of 1 Gy (Giovanetti et al., 2008).

This difference is to some extent depending on the degree of chromatin relaxation. Another example the results obtained in studying (by RIF's analysis) the repair of DNA DSB induced by low-dose radiation (Grudzenskia et al., 2010). It was shown that primary human fibroblasts are unable for a long time to repair DNA DSB induced by a dose of 10 mGy. However, they were able to effectively restore these breaks, if the cells were preliminarily treated with 10 μM H_2O_2 prior to irradiation. At this H_2O_2 concentration, no DSB arise in DNA, and only SSB and base modifications are formed (Grudzenskia et al., 2010). It is believed that additional single DNA lesions cause relaxation areas of chromatin and thus facilitate the process of DNA DSB repair in cells irradiated at a dose of 10 mGy.

DNA in heterochromatin and euchromatin may differ both in sensitivity to damaging agents and in the kinetics of damage repair. In compacted heterochromatin DNA is more protected from damaging agents. However, when there is damage, the compaction of heterochromatin restricts the access of repair enzymes to DNA damage (Cowell et al., 2007; Karagiannis et al., 2007; Goodarzi et al., 2008; 2009; 2010). Therefore, the kinetics of RIF appearance and disappearance are heterogeneous in the regions of nuclei of irradiated cells and are significantly different in the heterochromatin and euchromatin. It was found that after irradiation of cells with X-rays, preferential phosphorylation of H2AX histones occurred in the euchromatic areas of nuclei (Goodarzi et al., 2009; 2010). In the heterochromatin, RIF are formed slowly and persist much longer. This characterizes retardation of recognition and repair of DNA damage associated with RIF formation. Recent studies have shown that DNA segments with DSB may move to the periphery of heterochromatin in irradiated human and mouse cells (Jakob et al., 2011). Thus, it is believed that retention of RIF in the nucleus of irradiated cells for a long time is due to a delayed development of DDR and the presence of restrictions on the repair of complex DNA damage (Muniandy et al., 2010; Belyaev, 2010; Goodarzi et al., 2010).

ATM, as noted above, is a key unit of signaling in the development of DDR, and contributes to changes in the structure of chromatin in the vicinity of DSB and to the activation of repair of this damage. It is well known that the majority of DSB (85%) can be repaired with faster kinetics, mainly by an ATM-independent pathway. The repair of the other part of IR-induced DSB (15%) occurs by the slow kinetics, and this process requires involvement of

ATM and mediator proteins such as MDC1 and 53BP1, which are accumulated in RIF. Slowly repaired DSB are predominantly localized in DNA within heterochromatin (Cowell et al., 2007;). Further investigations showed that the heterochromatic superstructure creates a barrier to the functioning of ATM in the development of DDR and DNA DSB repair (Goodarzi et al., 2009; Jakob et al., 2011). However, although slowly, the DSB repair of DNA within heterochromatin does occur. The repair of DNA in heterochromatin requires relaxation of its structure. This relaxation is achieved by various modifications of chromatin proteins, in particular, by ATM-kinase phosphorylation of heterochromatin proteins (Noon et al., 2010; Cann and Dellaire 2011). An important role in this process belongs to the protein KAP-1 (a protein known as TIF1beta, TRIM28), which undergoes phosphorylation by ATM. Protein KAP-1 is a critical factor that contributes to repair of DSB DNA in the heterochromatin package. Phosphorylated KAP-1 (pKAP-1) is found in RIF at heterochromatin sites and promotes DSB repair and the mitotic progression (Noon et al., 2010).

Thus, the critical lesions of DNA in the heterochromatin packing persist for a long post-irradiation time, due to the necessity of its "opening" to make damaged DNA sites easily accessible for repair system components. At low IR doses, heterochromatin relaxes to a less extent, so that the access of repair enzymes to DNA damage is limited.

It is precisely this situation that arises in DNA damage in post meiotic male germ cells - the spermatozoa. This is because male germ cells gradually, from stage to stage, lose their ability to effectively repair complex DNA lesions during spermatogenesis. The causative factor in the low activity of DNA repair systems in spermatozoa is the rigid package of the haploid genome, created by protamines in these cells. Haploid male germ cells pack their DNA in a volume of less than 10% of a somatic cell nucleus. To achieve this level of DNA compaction, sperm cells have replaced most of their histones by protamines (Oliva, 2006; Marchetti et al., 2007).

So in the male germ cells DDR development is very slow, and formed RIF (γH2AX) are stored for longer periods of time than those in somatic cells (Paris et al., 2011). Studies in mice show that many radiation-induced DNA lesions in mature spermatozoa are not subject to repair and persist for at least 7 days (Marchetti and Wyrobek, 2008; Ahmed et al., 2010). However, male gametes with DNA damage are able to fertilize oocytes. DNA repair in the spermatozoon occurs, to a large extent, after their penetration into the ovum, the repair systems of the latter being involved in the process. Prior to DNA repair, reconstruction of chromatin is, obviously, required, accompanied by substitution of oocyte histones for protamines (Marchetti and Wyrobek, 2008; Ahmed et al., 2010; Paris et al., 2011).

7. Conclusion

In recent years our understandings of the effects of low doses of ionizing radiation on organisms, as well as features of the formation of DNA damage and repair have increased significantly. Based on data from many studies, we conclude that the majority of lesions occurring in the DNA of IR-exposed cells differ significantly in their chemical nature from the endogenous damage. The most important characteristics of radiation-induced DNA damage are their complexity and clustering. DSB, ICL or destruction of the replication fork and formation of long single-stranded gaps in DNA are considered as critical for the fate of damaged cells. The occurrence of such lesions in DNA may be a key event in both the etiology and the treatment of cancer. A complex radiation-induced DNA lesion is a serious burden for constitutive cellular repair systems.

In contrast to endogenous damage, repair of radiation-induced damage is accompanied by significant errors. With appearance of critical lesions in DNA, the cells rapidly develop a complex branched network of molecular and biochemical reactions, which are known as DNA damage response (DDR). Induction of DDR is associated with activation of cell cycle checkpoints, DNA repair, changes in the expression of many genes, chromatin reconstruction or apoptosis. However, the efficiency of repair of complex DNA lesions in cells exposed to low-dose radiation remains at low levels. The development of cell response to DNA damage at low radiation doses does not reach of the desired result, due to a small amount of damage, so that the progression of the cell cycle through the phases is ahead of DNA repair processes. This is primarily due to the failure of signaling from critical DNA lesions (low damage at low doses of radiation) to activation checkpoint to the cell cycle arrest. In the absence of cell cycle arrest, especially in the G2/M transition, the repair systems do not have time to complete DNA recovery, so that the cells enter mitosis with damaged DNA.

On the other hand, since eukaryotic chromosomal DNA is packaged into chromatin, there is evidence that the modulation of chromatin structure is crucial for the development of the DDR checkpoint signaling and repair. The compaction of DNA within chromatin, and most of all in the heterochromatin, poses a unique hindrance with regards to the accessibility of the DNA to enzymes involved in repair.

Therefore, another cause of the low efficiency of DNA repair in cells exposed to low radiation doses may be a slow reconstruction of chromatin and limited access of repair system components to complex lesions of DNA packed in compacted chromatin.

Now become clear that the responses of higher eukaryotes cells to low doses of radiation are different from those induced by high doses, not only quantitatively but qualitatively. Further research is needed on many aspects to this problem.

8. Acknowledgments

The author would like to thank Margaret Ilina for her excellent help in the preparation of this manuscript. This work was supported by the Basic Sciences to Medicine program of the Presidium of the Russian Academy of Sciences and the Russian Foundation for Basic Research (Grant 08-04-00163).

9. References

Ahmed, E.A., de Boer, P., Philippens. M.E., Kal, H.B. & de Rooij, D.G. (2010). Parp1-XRCC1 and the repair of DNA double strand breaks in mouse round spermatids. *Mutat. Res.*, 683 (1-2), 84-90.

Allen, C., Ashley, A.K., Hromas, R. & Nickoloff J.A. (2011). More forks on the road to replication stress recovery. J. Mol.Cell Biol., 3, (1) 4–12, doi:10.1093/jmcb/mjq049

Alter, B.P. (2002). Radiosensitivity in Fanconi's anemia patients. *Radiother. Oncol.*, 62 (3), 345-347.

Asaithamby, A., Hu, B., & Chen, D.J. (2011). Unrepaired clustered DNA lesions induce chromosome breakage in humancells. *Proc. Natl. Acad. Sci. USA*, 108 (20), 8293-8298.

Aten, J.A., Stap, J., Krawczyk, P.M., van Oven, C.H., Hoebe, R.A., Essers, J. & Kanaar, R., (2004). Dynamics of DNA double-strand breaks revealed by clustering of damaged chromosome domains. *Science*, 303 (5654), 92-95.

Averbeck, D. (2009). Does scientific evidence support a change from the LNT model for low-dose radiation risk extrapolation? *Health Phys.*, 97 (5), 493-504.

Ball, A.R. Jr. & Yokomori, K. (2011). Damage site chromatin: open or closed? *Curr. Opin. Cell Biol.*, 23 (3), 277-283.

Barnes, D.E., & Lindahl, T. (2004) Repair and genetic consequences of endogenous DNA base damage in mammalian cells. *Annu. Rev. Genet.*, 38, 445-476.

Bartek, J. & Lukas,J. (2007). DNA damage checkpoints: from initiation to recovery or adaptation. *Curr. Opinion Cell Biol.*, 19 (2), 238-245.

Bartek, J., Bartkova, J., & Lukas, J. (2007). DNA damage signalling guards against activated oncogenes and tumour progression. *Oncogene*, 26 (56), 7773-7779.

Belyaev, I.Y. (2010). Radiation-induced DNA repair foci: Spatio-temporal aspects of formation,application for assessment of radiosensitivity and biological dosimetry. *Mutat. Res.*, 704 (1-3), 132-141.

Belyakov, O.V., Mitchell, S.A., Parikh, D., Randers-Pehrson, G., Marino, S., Amundson, S.A., Geard, C.R. & Brenner, D.J. (2005). Biological effects in unirradiated human tissue induced by radiation damage up to 1 mm away. *Proc. Natl. Acad. Sci. USA*, 102 (40), 14203-14208.

Bennett, P.V., Cintron, N.S., Gros, L., Laval, J. & Sutherland, B.M. (2004) Are endogenous clustered DNA damages induced in human cells? *Free Radic. Biol. Med.*, 37 (4), 488–499.

Ben-Yehoyada, M., Wang, L.C., Kozekov, I.D., Rizzo, C.J., Gottesman, M.E. & Gautier, J. (2009). Checkpoint signaling from a single DNA interstrand crosslink. *Mol. Cell.*, 35 (5), 704-715.

Bernstein, N.K., Karimi-Busheri, F., Rasouli-Nia, A., Mani, R., Dianov, G., Glover, J.N. & Weinfeld, M. (2008). Polynucleotide kinase as a potential target for enhancing cytotoxicity by ionizing radiation and topoisomerase I inhibitors. *Anticancer Agents Med. Chem.*, 8 (4), 358-367

Bont, R.D. & van Larebeke, N. (2004). Endogenous DNA damage in humans: a review of quantitative data. *Mutagenesis*, 19 (3), 169-185.

Borde, V, & Cobb, J. (2009). Double functions for the Mre11 complex during DNA double-strand break repair and replication. *Int. J. Biochem. Cell Biol.*, 41 (6), 1249-1253.

Brenner, D.J., Doll, R., Goodhead, D.T., Hall, E.J., Land, C.E., Little, J.B., Lubin, J.H., Preston, D.L., Preston, R.J., Puskin, J.S., Ron, E., Sachs, R.K., Samet, J.M., Setlow, R.B., & Zaider, M. (2003). Cancer risk attributable to low doses of ionizing radiation: assessing what we really know. *Proc Natl Acad Sci USA*, 100 (24), 13761-13766.

Burdak-Rothkamm, S., Short, S., Folkard, M., Rothkamm, K. & Prise, K.M. (2007). ATR-dependent radiation-induced gamma H2AX foci in bystander primary human astrocytes and glioma cells. *Oncogene*, 26 (7), 993-1002.

Callegari, A.J. & Kelly T.J. (2007). Shedding light on the DNA damage checkpoint. *Cell Cycle*, 6 (6), 660-666.

Callegari, A.J., Clark, E., Pneuman, A., & Kelly T.J. (2010). Postreplication gaps at UV lesions are signals for checkpoint activation. *Proc. Natl. Acad. Sci. U S A*, 107 (18), 8219-8224.

Cann, K.L. & Dellaire, G. (2011). Heterochromatin and the DNA damage response: the need to relax. *Biochem. Cell Biol.*, 89 (1), 45-60.

Cejka, P., Cannavo, E., Polaczek, P., Masuda-Sasa, T., Pokharel, S., Campbell, J. & Kowalczykowski, S. C. (2010). DNA end resection by Dna2-Sgs1-RPA and its stimulation by Top3-Rmi1 and Mre11-Rad50-Xrs2. *Nature*, 467 (7311), 112-116.

Celeste, A., Petersen, S., Romanienko, P.J., Fernandez-Capetillo, O., Chen, H.T., Sedelnikova, O.A., Reina-San-Martin, B.,

Coppola, V., Meffre, E., Difilippantonio, M.J., Redon, C., Pilch, D.R., Olaru, A., Eckhaus, M., Camerini-Otero, R.D., Tessarollo, L., Livak, F., Manova, K., Bonner, W.M., Nussenzweig, M.C. & Nussenzweig, A. (2002). Genomic instability in mice lacking histone H2AX. *Science*, 296 (5569), 922-927.

Chung, J. & Bunz, F. (2010). Cdk2 is required for p53 independent G2/M checkpoint control. *PLoS Genet.*, 6 (2), e1000863.

Cimprich, K. & Cortez, D. (2008). ATR: an essential regulator of genome integrity, *Nat. Rev. Mol. Cell Biol.*, 9 (8), 616-627.

Cole, A.R., Lewis, L.P. & Walden, H. (2010). The structure of the catalytic subunit FANCL of the Fanconi anemia core complex. *Nat. Struct. Mol. Biol.*, 17 (3), 294-298.

Collis, S.J., Schwaninger, J., Ntambi, A., Keller, T.W., Nelson, W.G., Dillehay, L.E. & DeWeese, T.L. (2004). Evasion of

Early Cellular Response Mechanisms following Low Level Radiation-induced DNA Damage. *J. Biol. Chem.*, 279 (48), 49624-49632.

Cook, J.G. (2009). Replication licensing and the DNA damage checkpoint. *Front Biosci.*, 14, 5013-5030.

Costantini, S.,Woodbine, L., Andreoli, L., Jeggo, P.A. & Vindigni, A. (2007). Interaction of the Ku heterodimer with the DNA ligase IV/Xrcc4 complex and its regulation by DNA-PK. *DNA Repair(Amst)*, 6 (6), 712-722

Costes, S.V., Boissiere, A., Ravani, S., Romano, R., Parvin, B. & Barcellos-Hoff, M.H. (2006). Imaging features that discriminate between foci induced by high- and low-LET radiation in human fibroblasts. *Radiat. Res.*, 165 (5), 505-515.

Costes, S.V., Chiolo, I., Pluth, J.M., Barcellos-Hoff, M.H. & Jakob, B. (2010). Spatiotemporal characterization of ionizing radiation induced DNA damage foci and their relation to chromatin organization. *Mutat. Res.*, 704 (1-3), 78-87.

Cowell, I.G., Sunter, N.J., Singh, P.B., Austin, C.A., Durkacz, B.W., Tilby, M.J., (2007). Gamma- H2AX foci form preferentially in euchromatin after ionising-radiation, *PLoS One*, 2 (10), e1057. doi: 10.1371.

Das, P., Bennett, P. & Sutherland, B.M. (2011). Processing of abasic DNA clusters in hApeI-silenced primary fibroblasts exposed to low doses of X-irradiation. J. Biosci., 36 (1), 105-116.

Deckbar, D., Birraux, J., Krempler, A., Tchouandong, L., Beucher, A., Walker, S., Stiff, T., Jeggo, P.A. & Löbrich, M. (2007). Chromosome breakage after G2 checkpoint release. *J. Cell Biol.*, 176 (6), 748-755.

Dextraze, M.E., Gantchev, T., Girouard, S. & Hunting, D. (2010). DNA interstrand cross-links induced by ionizing radiation: An unsung lesion. *Mutat. Res.*, 704 (1-3), 101-107.

Dianov, G.L., O'Neill, P. & Goodhead, D.T. (2001). Securing genome stability by orchestrating DNA repair: removal of radiation-induced clustered lesions in DNA. *Bioessays*, 23 (8), 745-749

Ding, H. & Greenberg, M.M., (2007). Gamma-radiolysis and hydroxyl radical produce interstrand cross-links in DNA involving thymidine. *Chem. Res. Toxicol.*, 20 (11), 1623-1628.

Ding, L.H., Shingyoji, M., Chen, F., Hwang, J.J., Burma, S., Lee, C., Cheng, J.F. & Chen, D.J. (2005). Gene expression profiles of normal human fibroblasts after exposure to ionizing radiation: a comparative study of low and high doses. *Radiat. Res.*, 164 (1), 17-26.

Duan, M. & Smerdon, M.J. (2010). UV damage in DNA promotes nucleosome unwrapping. *J. Biol. Chem.*, 285 (34), 26295-26303.

Eccles, L.J., Lomax, M.E., O'Neill, P. (2010). Hierarchy of lesion processing governs the repair, double-strand break formation and mutability of three-lesion clustered DNA damage. *Nucleic Acids Res.*, 38 (4), 1123-1134

Falck, J., Coates, J. & Jackson S.P. (2005). Conserved modes of recruitment of ATM, ATR and DNA-PKcs to sites of DNA damage. *Nature*, 434 (7033), 605-611.

Fernet, M., Mégnin-Chanet, F., Hall, J. & Favaudon,V. (2010). Control of the G2/M checkpoints after exposure to low doses of ionising radiation: implications for hyper-radiosensitivity. *DNA Repair (Amst)*, 9 (1), 48-57.

Freeman, A.K. & (Monteiro, A.N. (2010). Phosphatases in the cellular response to DNA Damage. *Cell Commun. Signal.*, 8 (27), doi: 10.1186-8-27

Gantchev, T.G., Girouard, S., Dodd, D.W., Wojciechowski, F., Hudson, R.H. & Hunting, D.J. (2009). Gamma-radiation induced interstrand cross-links in PNA:DNA heteroduplexes. *Biochemistry*, 48 (29), 7032-7044.

Gaziev, A.I. (1999). DNA damage in cells exposed to ionizing radiation. *Radiat. Biol. Radioecol.*, 39 (5), 630-639.

Ghandhi, S., Sinha, A., Markatou, M. & Amundson, S. (2011). Time-series clustering of gene expression in irradiated and Bystander fibroblasts: an application of FBPA clustering. *BMC Genomics*, 12 (2), 121-133.

Giovanetti, A., Deshpande, T. & Basso, E. (2008). Persistence of genetic damage in mice exposed to low dose of X rays. *Int. J. Radiat. Biol.*, 84 (3), 227-235.

Goldberg, M., Stucki, M., Falck, J., D'Amours, D., Rahman, D., Pappin, D.,Bartek J. & Jackson, S.P. (2003). MDC1 is = required for the intra-S-phase DNA damage checkpoint. *Nature*, 421 (6926), 952-956.

Goodarzi, A., Noon, A., Deckbar, D., Ziv, Y., Shiloh, Y., Löbrich, M. & Jeggo, P.A. (2008) ATM signaling facilitates repair of DNA double-strand breaks associated with heterochromatin. *Mol. Cell.*, 31 (2), 167-177.

Goodarzi, A.A., Noon, A.T. & Jeggo, P.A. (2009). The impact of heterochromatin on DSB repair. *Biochem. Soc. Trans.* 37 (Pt.3), 569-576.

Goodarzi, A.A., Jeggo, P. & Löbrich, M. (2010). The influence of heterochromatin on DNA double strand break repair:

Getting the strong, silent type to relax. *DNA Repair (Amst)*, 9 (12), 1273-1282.

Grillari, J., Katinger, H. & Voglaue, R. (2007). Contributions of DNA interstrand cross-links to aging of cells and organisms. *Nucleic Acids Res.*, 35 (22), 7566-7576

Grudzenskia, S., Rathsa, A., Conrada, S., Rubeb, C.E. & Löbrich, M. (2010). Inducible response required for repair of low-dose radiation damage in human fibroblasts. *Proc. Natl. Acad. Sci. USA*, 107 (32), 14205-14210.

Gu, J., Lu, H.,Tsai, A.G., Schwarz, K. & Lieber, M.R. (2007). Single-stranded DNA ligation and XLF-stimulated incompatible DNA end ligation by the XRCC4-DNA ligase IV complex: influence of terminal DNA sequence. *Nucleic Acids Res.*, 35 (17), 5755-5762.

Gulston, M., de Lara, C., Jenner, T., Davis, E. & O'Neill, P. (2004). Processing of clustered DNA damage generates additional double-strand breaks in mammalian cells post-irradiation. *Nucleic Acids Res.*, 32 (41), 602-1609.

Haber, J.E. (2008). Alternative endings. *Proc. Natl. Acad. Sci. USA*, 105 (2), 405-406.

Hada M. & Georgakilas A.G. (2008). Formation o cluster DNA damage after high LET irradiation: A review. *J. Radiat. Res.*, 49 (3), 203-210.

Halazonetis, T.D., Gorgoulis, V.G. & Bartek, J. (2008). An oncogene-induced DNA damage model for cancer development. *Science*, 319 (5868), 1352-1355.

Harper, J.W. & Elledge, S.J. (2007). The DNA damage response: ten years after. *Mol. Cell.*, 28 (5), 739-745.

Harrison, J.C. & Haber, J.E. (2006). Surviving the breakup: The DNA damage checkpoint. *Annu. Rev. Genet.*, 40, 209-235.

Hayes D.P. (2008). Non-problematic risks from low-dose radiation-induced DNA damage clusters. *Dose-Response*, 6 (1), 30-52.

Heyer, W.D., Ehmsen, K.T. & Liu, J. (2010). Regulation of homologous recombination in eukaryotes. *Annu. Rev. Genet.*, 44, 113-139.

Heilmann, J., Rink, H., Taucher-Scholz, G. & Kraft, G. (1993). DNA strand break induction and rejoining and cellular recovery in mammalian cells after heavy-ion irradiation. *Radiat. Res.*, 135 (1), 46-55.

Hejna, J., Holtorf, M., Hines, J., Mathewson, L., Hemphill, A., Olson, S.B. & Moses, R.E. (2008). Tip60 is required for DNA interstrand cross-link repair in the Fanconi anemia pathway. *J. Biol. Chem.*, 283 (15), 9844-9851.

Heyer, W.D., Ehmsen, K.T. & Liu, J. (2010). Regulation of homologous recombination in eukaryotes. *Annu. Rev. Genet.*, 44, 113-139.

Hildebrandt, G. (2010). Non-cancer diseases and non-targeted effects. *Mutat. Res.*, 687 (1-2), 73-77.

Hinz, J.M. (2010). Role of homologous recombination in DNA interstrand crosslink repair. *Environ. Mol. Mutagen.*, 51 (6), 582-603.

Ho, T.V. & Schärer, O.D. (2010). Translesion DNA synthesis polymerases in DNA interstrand crosslink repair. *Environ. Mol. Mutagen.*, 51 (6), 552-566.

Hoeijmakers, J.H. (2001). Genome maintenance mechanisms for preventing cancer. *Nature*, 411 (6835), 366-374.

Hong, H., Cao, H. & Wang, Y. (2007). Formation and genotoxicity of a guanine-cytosine intrastrand cross-link lesion *in vivo*. *Nucleic Acids Res.*, 35 (21), 7118-7127.

Hu, H. & Gatti R.A. (2011). MicroRNAs: new players in the DNA damage response. *J. Mol. Cell Biol.*, 3 (3), 151-158.

Huang, X., Okafuji, M., Traganos, F., Luther, E., Holden, E. & Darzynkiewicz, Z. (2004). Assessment of histone H2AX phosphorylation induced by DNA topoisomerase I and II inhibitors topotecan and mitoxantrone and by the DNA cross-linking agent cisplatin. *Cytometry A*, 58 (2), 99-110.

Huertas, P. (2010). DNA resection in eukaryotes: deciding how to fix the break. *Nat. Struct. Mol. Biol.*, 17 (1), 11-16.

Huertas, P., Sendra, R. & Muñoz, P. (2009). Chromatin dynamics coupled to DNA repair. *Epigenetics*, 4 (1), 31-42.

Ide, H, Shoulkamy, M.I., Nakano, T., Miyamoto-Matsubara, M. & Salem, A.M. (2011). Repair and biochemical effects of DNA-protein crosslinks. *Mutat. Res.*, 711 (1-2), 113-122.

Ishizaki, K., Hayashi, Y., Nakamura, H., Yasui, Y., Komatsu, K. & Tachibana, A. (2004). No induction of p53 phosphorylation and few focus formation of phosphorylated H2AX suggest efficient repair of DNA damage during chronic low-dose-irradiation in human cells. *J. Radiat. Res.*, 45 (4), 521-525.

Jackson, S. & Bartek, J. (2009). The DNA-damage response in human biology and disease. *Nature,* 461 (7267), 1071-1078.

Jakob, B., Splinter, J., Conrad, S., Voss, K., Zink, D., Durante, M., Löbrich, M. & Taucher-Scholz, G. (2011). DNA double- strand breaks in heterochromatin elicit fast repair protein recruitment, histone H2AX phosphorylation and relocation to euchromatin. *Nucleic Acids Res.*, doi:10.1093/nar/gkr230.

Jeggo, P.A. (2009). Risks from low dose/dose rate radiation: what an understanding of DNA damage response mechanisms can tell us. *Health Phys.*, 97 (5), 1-10.

Jeggo, P. (2010). The role of the DNA damage response mechanisms after low-dose radiation exposure and a consideration of potentially sensitive individuals. *Radiat. Res.*, 174 (6), 825-832.

Joiner, M.C., Marples, B., Lambin, P., Short, S.C. & Turesson, I., (2001). Low-dose hypersensitivity: Current status and possible mechanisms. *Int. J. Radiat. Oncol. Biol. Phys.*, 49 (2), 379-389.

Karagiannis, T.C., Harikrishnan, K.N. & El-Osta, A. (2007). Disparity of histone deacetylase inhibition on repair of radiation-induced DNA damage on euchromatin and constitutive heterochromatin compartments. *Oncogene,* 26 (27), 3963-3971.

Kastan, M.B. & Bartek, J. (2004). Cell-cycle checkpoints and cancer. *Nature,* 432 (7015), 316-23.

Knipscheer, P., Räschle, M., Smogorzewska, A., Enoiu, M., Ho, T.V., Schärer, O.D., Elledge, S.J. & Walter, J.C. (2009). The Fanconi anemia pathway promotes replication-dependent DNA interstrand crosslink repair. *Science,* 326 (5960), 1698-1701.

Kolas, N.K., Chapman, J.R., Nakada, S., Ylanko, J., Chahwan, R., Sweeney, F.D., Panier, S., Mendez, M., Wildenhain, J.,

Thomson, T.M., Pelletier, L., Jackson, S.P. & Durocher, D. (2007). Orchestration of the DNA-damage response by the RNF8 ubiquitin ligase. *Science,* 318 (5856), 1637-1640.

Krueger, S.A., Collis, S.J., Joiner, M.C., Wilson, G.D. & Marples, B. (2007). Transition in survival from low dose hyper-radiosensitivity to increased radioresistance is independent of activation of ATM Ser1981 activity. *Int. J. Radiat. Oncol. Biol. Phys.,* 69 (4), 1262-1271.

Krueger, S.A., Wilson, G.D., Piasentin, E., Joiner, M.C. & Marples, B. (2010). The effects of G2-phase enrichment and checkpoint abrogation on low-dose hyper-radiosensitivity. *Int. J. Radiat. Oncol. Biol. Phys.,* 77 (5), 1509-1517.

Kuhnert, V.M., Kachnic, L.A., Li, L., Purschke, M., Gheorghiu, L., Lee, R., Held, K.D. & Willers, H. (2009). FANCD2-deficient human fibroblasts are hypersensitive to ionizing radiation at oxygen concentrations of 0% and 3% but not under normoxic conditions. *Int. J. Radiat. Biol.,* 85 (6), 523-531.

Lassmann, M., Hänscheid, H., Gassen, D., Biko, J., Meineke, V., Reiners, C. & Scherthan, H. (2010). *In vivo* formation of gamma-H2AX and 53BP1 DNA repair foci in blood cells after radioiodine therapy of differentiated thyroid cancer. *J. Nucl. Med.,* 51 (8), 1318-1325.

Lavin, M.F., Kozlov, S. (2007). ATM Activation and DNA Damage Response. *Cell Cycle,* 6 (8), 931-942.

Lee, J.H. & Paull, T.T. (2007). Activation and regulation of ATM kinase activity in response to DNA double-strand breaks. *Oncogene,* 26 (56), 7741-7748

Legerski, R.J. (2010). Repair of DNA interstrand cross-links during S phase of the mammalian cell cycle. *Environ. Mol. Mutagen.,* 51 (6), 540-551.

Lieber, M.R. (2010). The mechanism of double-strand DNA break repair by the nonhomologous DNA end-joining pathway. *Annu. Rev. Biochem.*, 79, 181-211.

Lindahl, T. (1993). Instability and decay of the primary structure of DNA. *Nature*, 362 (6422), 709-715.

Liu, T., Ghosal, G., Yuan, J., Chen, J. & Huang, J. (2010). FAN1 acts with FANCI-FANCD2 to promote DNA interstrand cross-link repair. *Science*, 329 (5992), 693-696

Löbrich, M., Shibata, A., Beucher, A., Fisher, A., Ensminger, M., Goodarzi, A.A., Barton, O. & Jeggo, P.A. (2010). γ-H2AX foci analysis for monitoring DNA double-strand break repair. Strengths, limitations and optimization. *Cell Cycle*, 9 (4), 662-669.

Mahaney, B.L., Meek, K. & Lees-Miller, S.P. (2009). Repair of ionizing radiation-induced DNA double-strand breaks by non-homologous end-joining. *Biochem. J.*, 417 (3), 639-650.

Marchetti, F., Essers, J., Kanaar, R. & Wyrobek, A.J. (2007). Disruption of maternal DNA repair increases sperm-derived chromosomal aberrations. *Proc. Natl. Acad. Sci. USA*, 107 (45), 17725-17729.

Marchetti, F. & Wyrobek, A.J. (2008). DNA repair decline during mouse spermiogenesis results in the accumulation of heritable DNA damage. *DNA Repair (Amst)*, 7 (4), 572-581.

Marples, B., Wouters, B., Collis, S., Chalmers, A. & Joiner, M. (2004). Low-dose hyper-radiosensitivity: a consequence of ineffective cell cycle arrest of radiation-damaged G2-phase cells. *Radiat. Res.*, 161 (3), 247-255.

Matsuoka, S., Ballif, B., Smogorzewska, A., McDonald, E., Hurov, K.E., Luo, J., Bakalarski, C.E., Zhao, Z., Solimini, N., Lerenthal, Y., Shiloh, Y., Gygi, S.P., & Elledge, S.J. (2007). ATM and ATR substrate analysis reveals extensive protein networks responsive to DNA damage. *Science*, 316 (5828), 1160-1166.

McVey, M. & Lee, S. (2008). MMEJ repair of double-strand breaks (director's cut): deleted sequences and alternative endings. *Trends Genet.*, 24 (11), 529-538.

Mimitou, E.P. & Symington, L.S. (2008). Sae2, Exo1 and Sgs1 collaborate in DNA double strand break processing. *Nature*, 455 (7214), 770-774.

Misteli, T. & Soutoglou, E. (2009). The emerging role of nuclear architecture in DNA repair and genome maintenance. *Nat. Rev. Mol. Cell Biol.*, 10 (4), 243-254

Modesti, M., Budzowska, M., Baldeyron, C., Demmers, J.A., Ghirlando, R., & Kanaar, R., (2007). RAD51AP1 is a structure-specific DNA binding protein that stimulates joint molecule formation during RAD51-mediated homologous recombination. *Mol. Cell*, 28 (3), 468-481.

Mogi, S. & Oh, D.H. (2006). γ-H2AX formation in response to interstrand crosslinks requires XPF in human cells. *DNA Repair (Amst)*, 5 (6), 731-740.

Mogi, S., Butcher, C.E. & Oh, D.H. (2008). DNA polymerase η reduces the γ-H2AX response to psoralen interstrand crosslinks in human cells. *Exp. Cell Res.*, 314 (4), 887-895.

Moldovan, G.L. & D'Andrea, A.D. (2009). How the Fanconi anemia pathway guards the genome. *Annu. Rev. Genet.*, 43, 223-249.

Moon, A.F., Garcia-Diaz, M., Batra, V.K., Beard, W.A., Bebenek, K., Kunkel, T.A., Wilson, S.H. & Pedersen, L.C. (2007). The X family portrait: structural insights into biological functions of X family polymerases. *DNA Repair (Amst)*, 6 (12), 1709-1725.

Mullenders, L., Atkinson, M., Paretzke, H., Sabatier, L. & Bouffler, S. (2009). Assessing cancer risks of low-dose radiation. *Nat. Rev. Cancer*, 9 (8), 596-604.

Muniandy, P., Liu. J., Majumdar, A., Liu, S. & Seidman M.M. (2010). DNA interstrand cross links repair in mammalian cells: step by step. *Crit. Rev. Biochem. Mol. Biol.*, 45 (1), 23-49.

Nagy, Z. & Soutoglou, E. (2009). DNA repair: easy to visualize, difficult to elucidate. *Trends Cell Biol.*, 19 (11), 617-629.

Natarajan, A.T. & Palitti, F. (2008). DNA repair and chromosomal alterations. *Mutat. Res.* 657 (1), 3-7.

Nichols, G. J., Schaack, J. & Ornelles, D. A. (2009). Widespread phosphorylation of histone H2AX by species C adenovirus infection requires viral DNA replication. *J. Virol.*, 83 (12), 5987–5998.

Niedernhofer, L.J.,Daniels, J.S., Rouzer, C.A., Greene, R.E., Marnett, L.J. (2003). Malondialdehyde, a product of lipid peroxidation, is mutagenic in human cells. *J. Biol. Chem.*, 278 (33), 31426-31433.

Niida, H., Murata, K., Shimada, M., Ogawa, K., Ohta, K., Suzuki, K., Fujigaki, H., Khaw, A.K., Banerjee, B., Hande, M.P., Miyamoto, T., Miyoshi, I., Shirai, T., Motoyama, N., Delhase, M., Appella, E. & Nakanishi, M. (2010). Cooperative functions of Chk1 and Chk2 reduce tumour susceptibility *in vivo*. *EMBO J.*, 29 (20), 3558-3570.

Noll, D., Mason, T. & Miller, P.S. (2006). Formation and repair of interstrand cross-links in DNA. *Chem. Rev.*, 106 (2), 277-301.

Noon, A.T., Shibata, A., Rief, N., Löbrich, M., Stewart, G.S., Jeggo, P.A. & Goodarzi, A.A. (2010). 53BP1-dependent robust localized KAP-1 phosphorylation is essential for heterochromatic DNA double-strand break repair. *Nat. Cell. Biol.*, 12 (2), 177-184.

NRC (National Research Council of the NAS). (2006). *Health Risks from Exposure to Low Levels of Ionizing Radiation (BEIR VII, Phase 2)*. National Academies Press, Washington, DC. pp 30-143.

Ojima, M., Furutani, A., Ban, N. & Kai, M. (2011). Persistence of DNA double-strand breaks in normal human cells induced by radiation-induced bystander effect. *Radiat. Res.* 175 (1), 90-96.

Oliva, R. (2006). Protamines and male infertility. *Hum. Reproduc. Update*, 12 (4), 417-435.

Olive, P.L. & Banath, J.P. (2009). Kinetics of H2AX Phosphorylation After Exposure to Cisplatin. *Cytometry B. Clinic. Cytom.* 76 (2), 79-90.

Pandita, T.K. & Richardson, C. (2009). Chromatin remodeling finds its place in the DNA double-strand break response. *Nucleic Acids Res.*, 37 (5), 1363-1377.

Pardo, B., Gomez-Gonzalez, B. & Aguilera, A. (2009). DNA double-strand break repair: how to fix a broken relationship. *Cell. Mol. Life Sci.*, 66 (6), 1039-1056.

Paris, L., Cordelli, E., Eleuteri, P., Grollino, M.G., Pasquali, E., Ranaldi, R., Meschini, R. & Pacchierotti, F. (2011). Kinetics of g-H2AX induction and removal in bone marrow and testicular cells of mice after X-ray irradiation. *Mutagenesis*, 26 (4), 563-572.

Parrilla-Castellar, E.R., Arlander, S.J. & Karnitz, L. (2004). Dial 9-1-1 for DNA damage: the Rad9-Hus1-Rad1 (9-1-1) clamp complex. *DNA Repair (Amst)*, 3 (8-9), 1009-1014.

Ploskonosova, I.I., Baranov, V.I. & Gaziev, A.I. (1999). PCR assay of DNA damage and repair at the gene level in brain and spleen of gamma-irradiated young and old rats. *Mutat. Res.*, 434 (2), 109-117.

Pebg, X., Ghosh, A.K., Van Houten, B., & Greenberg, M.M. (2010). Nucleotide excision repair of a DNA interstrand cross-link produces single- and double-strand breaks. *Biochemistry*, 49 (1), 11-19.

Radulescu, I., Elmroth, K. & Stenerlöw, B. (2004). Chromatin organization contributes to non-randomly distributed double-strand breaks after exposure to high-LET radiation. *Radiat. Res.*, 161 (1), 1-8.

Rass, U., Ahel, I. & West, S.C. (2008). Molecular mechanism of DNA deadenylation by the neurological disease protein aprataxin. *J. Biol. Chem.*, 283 (49), 3394-4001.

Redon, C.E., Nakamura, A.J., Sordet, O., Dickey, J.S., Gouliaeva, K., Tabb, B., Lawrence, S., Kinders, R.J., Bonner, W.M.,

Sedelnikova, O.A. (2011). γ-H2AX Detection in peripheral blood lymphocytes, splenocytes, bone marrow, xenografts, and skin. *Methods Mol. Biol.*, 682, 249-270.

Regulus, P., Duroux, B., Bayle, P. A., Favier, A., Cadet, J. & Ravanat, J. L. (2007). Oxidation of the sugar moiety of DNA by ionizing radiation or bleomycin could induce the formation of a cluster DNA lesion. *Proc. Natl. Acad. Sci. USA*, 104 (35), 14032-14037.

Rogakou, E.P., Boon, C., Redon, C. & Bonner, W.M. (1999). Megabase chromatin domains involved in DNA double-strand breaks *in vivo*. *J. Cell Biol.*, 146 (5), 905-916.

Rothkamm, K. & Löbrich, M. (2003). Evidence for a lack of DNA double-strand break repair in human cells exposed to very low x-ray doses. *Proc. Natl. Acad. Sci. USA*, 100 (9), 5057-5062.

Rupnik, A., Lowndes, N.F. & Grenon, M. (2010). MRN and the race to the break. *Chromosoma*, 119 (2), 115-135.

Sage, E. & Harrison, L. (2011). Clustered DNA lesion repair in eukaryotes: Relevance to mutagenesis and cell survival. *Mutat. Res.*, 711 (1-2), 123-133.

San Filippo, J., Sung, P. & Klein, H. (2008). Mechanism of eukaryotic homologous recombination. *Annu. Rev. Biochem.*, 77, 229-257.

Sartori, A.A., Lukas, C., Coates, J., Mistrik, M., Fu, S., Bartek, J., Baer, R., Lukas, J. & Jackson, S.P. (2007). Human CtIP promotes DNA end resection. *Nature*, 450 (7169), 509-514.

Sczepanski, J., Jacobs, A.C., Van Houten, B. & Greenberg M. (2009). Double-strand break formation during nucleotide excision repair of a DNA interstrand cross-link. *Biochemistry*, 48 (32), 7565-7567.

Sedelnikova, O.A., Horikawa, I., Redon, C., Nakamura, A., Zimonjic, D., Popescu, N. & Bonner, W.M. (2008). Delayed kinetics of DNA double-strand break processing in normal and pathological aging. *Aging Cell*, 7 (1), 89-100.

Semenenko, V.A. & Stewart, R.D. (2004). A fast Monte Carlo algorithm to simulate the spectrum of DNA damages formed by ionizing radiation. *Radiat. Res.*, 161 (4), 451-457.

Shibata, A., Barton, O., Noon, A.T., Dahm, K., Deckbar, D., Goodarzi, A.A., Lőbrich, M. & Jeggo, P.A. (2010). Role of ATM and the damage response mediator proteins 53BP1 and MDC1 in the maintenance of G(2)/M checkpoint arrest. *Mol. Cell Biol.*, 30 (13), 3371-3383.

Shikazono, N., Noguchi, M., Fujii, K., Urushibara, A. & Yokoya, A. (2009). The yield, processing, and biological consequences of clustered DNA damage induced by ionizing radiation. *J. Radiat. Res.*, 50 (1), 27-36.

Smits, V.A., Warmerdam, D.O., Martin, Y. & Freire, R. (2010). Mechanisms of ATR-mediated checkpoint signalling. *Front Biosci.*, 15, 840-853.

Smogorzewska, A., Desetty, R., Saito, T.T., Schlabach, M., Lach, F.P., Sowa, M. E., Clark, A.B., Kunkel T.A., Harper J.W.,

Colaiacovo M. & Elledge, S.J. (2010). A genetic screen identifies FAN1, a Fanconi anemia-associated nuclease necessary for DNA interstrand crosslink repair. *Mol. Cell*, 39 (1), 36-47.

Stewart, G.S., Wang, B., Bignell, C., Taylor, A.M. & Elledge, S.J. (2003). MDC1 is a mediator of the mammalian DNA damage checkpoint. *Nature*, 421 (6926), 961-966.

Stracker, T.H., Usui, T. & Petrini, J.H. (2009). Taking the time to make important decisions: the checkpoint effector kinases Chk1 and Chk2 and the DNA damage response. *DNA Repair (Amst)*, 8 (9), 1047-1054.

Sutherland, B.M., Bennett, P.V., Sidorkina, O., & Laval, J. (2000). Clustered DNA damages induced in isolated DNA and in human cells by low doses of ionizing radiation. *Proc. Natl. Acad. Sci. USA*, 97 (1), 103-108.

Sutherland, B.M., Bennett, P.V., Sutherland, J.C. & Laval, J. (2002). Clustered DNA damages induced by x rays in human cells. *Radiat. Res.*, 157 (6), 611–616.

Svetlova, M.P., Solovjeva, L.V. & Tomilin, N.V. (2010). Mechanism of elimination of phosphorylated histone from chromatin after repair of DNA double-strand breaks. *Mutat. Res.*, 685 (1-2), 54-60.

Takahashi, A., Mori, E. & Ohnishi T. (2010). The foci of DNA double strand break-recognition proteins localize with γH2AX after heat treatment. *J. Radiat. Res.*, 51 (1), 91-95.

Taki, K., Wang, B., Nakajima, T., Wu, J., Ono, T., Uenara, Y., Matsumoto, T., Oghiso, Y., Tanaka, K., Ichinohe, K., Nakamura, S., Tanaka, S., Magae, J., Kakimoto, A. & Nenoi, M. (2009). Microarray analysis of differentially expressed genes in the kidneys and testes of mice after long-term irradiation with low-dose-rate γ-rays. *J. Radiat. Res.*, 50 (3), 241–252

Tang, J., & Greenberg, R.A. (2010). Connecting the dots: Interplay between ubiquitylation and sumoylation at DNA double strand breaks. *Gen. Cancer*, 1 (7), 787-796.

Terato, H. & Ide, H. (2004). Clustered DNA damage induced heavy ion particles. *Biol. Sci. Space*, 18 (4), 206-212.

Thompson, L.H. & Hinz, J.M. (2009). Cellular and molecular consequences of defective Fanconi anemia proteins in replication-coupled DNA repair: mechanistic insights. *Mutat. Res.*, 668 (1-2), 54-72.

Tomita, M., Morohoshi, F., Matsumoto, Y., Otsuka, K. & Sakai, K. (2008). Role DNA double-strand breaks repair genes in cell proliferation under low-dose rate irradiation condition. *J. Radiat. Res.*, 49 (5), 557-564.

Tsai, C.J., Kim, S.A. & Chu, G. (2007). Cernunnos/XLF promotes the ligation of mismatched and non cohesive DNA ends. *Proc. Natl. Acad. Sci. USA*, 104 (19), 7851-7856.

Tubiana M ., Aurengo, A., Averbeck, D., & Masse, R. (2006). The debate on the use of linear no threshold for assessing the effects of low doses. J. Radiol. Prot. 26 (3), 317-324.

Ulsh, B.A. (2010). Checking the foundation: recent radiobiology and the linear no-threshold theory. *Health Phys.*, 99 (6), 747-758.

Vasquez, K.M. (2010). Targeting and processing of site-specific DNA interstrand crosslinks. *Environ. Mol. Mutagen.*, 51 (6), 527-539.

Wang, B. & Elledge, S.J. (2007). Ubc13/Rnf8 ubiquitin ligases control foci formation of the Rap80/Abraxas/Brca1/Brcc36 complex in response to DNA damage. *Proc. Natl. Acad. Sci. USA*, 104 (52), 20759-20763.

Ward, J.F. (1988) DNA damage produced by ionizing radiation in mammalian cells: identities, mechanisms of formation, and reparability. *Prog. Nucleic Acid Res. Mol. Biol.*, 35, 95–125.

Warmerdam, D.O. & Kanaar, R. (2010). Dealing with DNA damage: Relationships between checkpoint and repair pathways. *Mutat. Res.*, 704 (1-3), 2-11.

Warmerdam, D.O., Kanaar, R. &. Smits, V.J. (2010). Differential Dynamics of ATR-Mediated Checkpoint Regulators. J. Nucleic Acids, aug. 17, Pii: 319142, doi: 10.4061/2010/319142

Wood, R.D. (2010). Mammalian nucleotide excision repair proteins and interstrand crosslink repair. *Environ. Mol.Mutagen.*, 51 (6), 520-526.

Wu, Q. & Vasquez, K.M. (2008). Human MLH1 protein participates in genomic damage checkpoint signaling in response to DNA interstrand crosslinks, while MSH2 functions in DNA repair. *PLoS Genet.* 4 (9), e1000189.

Wykes, S., Piasentin, E., Joiner, M.C., Wilson, G.D. & Marples, B. (2006). Low-dose hyper-radiosensitivity is not caused by a failure to recognize DNA double-strand breaks. *Radiat. Res.*, 165 (5), 516-524.

Xu, Y., & Price, B.D. (2011). Chromatin dynamics and the repair of DNA double strand breaks. Cell Cycle, 10 (2), 261-267. Xue, L., Yu, D., Furusawa, Y., Cao, J., Okayasu, R. & Fan, S. (2009). ATM-dependent hyper-radiosensitivity inmammalian cells irradiated by heavy ions. *Int. J. Radiat. Oncol. Biol. Phys.*, 75 (1), 235-243.

Yamauchi, M., Oka, Y., Yamamoto, M., Niimura, K., Uchida, M., Kodama, S., Watanabe, M., Sekine, I., Yamashita, S. & Suzuki, K. (2008). Growth of persistent foci of DNA damage checkpoint factors is essential for amplification of G1 checkpoint signaling. *DNA Repair (Amst.)*, 7 (3), 405-417.

Yannone, S.M., Khan, I.S., Zhou, R.Z., Zhou, T., Valerie, K. & Povirk, L.F. (2008). Coordinate 5' and 3' endonucleolytic trimming of terminally blocked blunt DNA double-strand break ends by Artemis nuclease and DNA-dependent protein kinase. *Nucleic Acids Res.*, 36 (10), 3354-3365.

Yano, K., Moratomi-Yano, K., Adachi, N. & Aiyama, H. (2009). Molecular mechanism of protein assembly on DNA double-strand breaks in the non-homologous end-joining pathway. *J. Radiat. Res.*, 50 (1), 97-108

You, Z. & Bailis, J.M. (2010). DNA damage and decisions: CtIP coordinates DNA repair and cell cycle checkpoints.*Trends Cell Biol.*, 20 (7), 402-409.

Zou, L. & Elledge, S.J. (2003). Sensing DNA damage through ATRIP recognition of RPA-ssDNA complexes. Science, 300 (5625), 1542–1548.

5

Genome Integrity and Organization in the Context of Radiobiology

Vladan Ondřej
Laboratory of Genome Integrity, Faculty of Science,
Palacký University in Olomouc,
Czech Republic

1. Introduction

The genomes of all organisms on Earth are exposed to various stimuli that induce DNA damage in the form of single- or double-stranded DNA breaks. One of the stimuli that impair genome stability is the irradiation of cells by ionizing radiation. Organisms have mechanisms to prevent the disruption of genome integrity by DNA breaks. The repair of double-stranded DNA breaks (DSBs) is mediated by two major pathways: nonhomologous recombination (NHEJ), and homologous recombination repair (HRR). However, a lot of the studies are focused on the repair pathways, the DNA repair must also be studied in the context of genome organization. Genomes are organized into complex higher-order structures by folding of the DNA into chromatin fibres, loops, domains, and chromosome territories. This functional organization is important, not only for gene regulation and control of gene expression patterns; but further, they also provide an environment for DNA repair processes. Defects in higher-order chromatin structures affects DNA repair, making cells more sensitive to ionizing radiation, and they are relevant in many physiological and pathological processes. An understanding of the global higher-order chromatin structures also shines a light into those processes causing chromosomal abnormalities, such as translocations, which are a common hallmark of cancer cells.

2. Higher-order chromatin structure and DSB repair

Most of the genome of Eukaryotes is deposed within the cell nucleus, with the remainder in the mitochondria (in the case of plants, in the chloroplasts). The cell nucleus provides not only the space and environment for the genome, but also the organization and machinery required for genomic function and maintenance. Nuclear architecture is not a rigid structure just for DNA deposition, as was thought for many decades prior. Rather, it is a dynamic and highly organized organelle, reflecting all of the requirements for managing the genetic data such as replication and transcription; additionally, this includes DNA repair.

2.1 Chromatin structure

In their native state, genomes are folded into complex higher-order structures that interact with nuclear proteins, as well as protein complexes at the nuclear periphery, and they communicate with the cytoplasm through the nuclear pores. At the lowest level, the DNA is wrapped around an octamer of core histone proteins (which are the primary structural elements of the chromatin fibre) in order to form a nucleosome. Multiple nucleosomes are linked by stretches of DNA, often occupied by a linker histone, into a 'beads-on-a string' fibre of 10 nm in diameter. These strings of linked nucleosomes are helically twisted into a 10 nm fibre, which in turn is folded into a 30 nm fibre. There are indications that this fibre is folded into yet higher order chromatin loops, although the topological organization of these higher order structures *in vivo* is unclear (reviewed in Misteli, 2010). In some models, the compaction of chromatin is linked to the size of the loops, and distinguishes the chromatin from the compact heterochromatin and decondensed euchromatin (Goetze et al., 2007).

2.2 Epigenetic modifications of the chromatin

Chromatin, the physiological template of all eukaryotic genetic information, is subject to diverse epigenetic modifications of the DNA and histones. Large arrays of posttranslational modifications of the histones, which largely impinge on the histone amino termini, regulate access to the underlying DNA. Distinct histone amino-terminal modifications can generate synergistic or antagonistic interaction affinities for chromatin-associated proteins; in turn, these dictate the dynamic transitions between transcriptionally active or transcriptionally silent chromatin states. Thus, the combinatorial nature of histone amino-terminal modifications reveals a "histone code" that considerably extends the information potential of the genetic code (Jenuwain & Allis, 2001). The histone code influences higher-order chromatin structures by affecting contacts between different histones, and between the histones and the DNA. Specific histone modifications are responsible for the compartmentalization of the genome into distinct domains, such as transcriptionally silent heterochromatin and transcriptionally active euchromatin (summarized by Martin & Zhang, 2005). The recent term 'heterochromatin' is more loosely applied; it is often extended to include transcriptionally silent regions of the chromatin, regardless of their staining properties. An important distinction is made between the constitutive and facultative heterochromatin. Constitutive heterochromatin is always compact, and tends to be enriched in repetitive, gene-poor, and late replicating DNA sequences; whereas, facultative heterochromatin can undergo reversible transitions from a compact, transcriptionally inactive state to become more open and transcriptionally competent. In a review, Trojer and Reinberg (2007) suggested that facultative heterochromatin should be molecularly defined as condensed, transcriptionally silent chromatin regions that decondense, and allow for transcription within temporal, spatial, or parental/heritable contexts. For example, during embryogenesis, the amount of facultative heterochromatin increases, as unwanted sets of genes are progressively shut down until maturity, when a cell expresses only the genes appropriate for that tissue. The reverse occurs when, for example, differentiated cells are reprogrammed to become stem cells. These events are typically accompanied by profound changes in histone variants, histone modifications, and the presence of chromatin attaching proteins (reviewed in Woodcock & Ghosh, 2010). The ability of the histone code to dictate the chromatin environment allows it to regulate nuclear processes, such as replication, transcription, DNA repair, and chromosome condensation (Kouzarides, 2007).

Next to DNA methylation, histone acetylation and histone methylation are the best-characterized epigenetic markers. Trimethylation at H3K4, H3K36, or H3K79 results in an open chromatin configuration, and therefore is characteristic of euchromatin. Euchromatin is also characterized by a high level of histone acetylation, which is mediated by histone acetyl transferases (HATs). Conversely, histone deacetylases (HDACs) have the ability to remove this epigenetic mark, which leads to transcriptional repression. Condensed heterochromatin is enriched in the trimethylation of H3K9, K3K27, and H4K20 (Kouzarides, 2007), which along with the silencing of euchromatin loci, caused by histone deacetylation, involves the recruitment of specific K9 histone methyltransferases (HMTs). Methylated H3K9 provides a binding site for the chromodomain-containing heterochromatin protein 1 (HP1), which induces transcriptional repression and heterochromatinization. At euchromatic loci, this process is mediated by co-repressors, such as retinoblastoma protein pRb or KAP1 (summarized by Kouzarides, 2007). Histone demethylases, first described by Shi et al. (2004), have the opposite effect on transcription. The histone demethylase LSD1 is responsible for H3K4 demethylation, which leads to transcriptional inactivation. (reviewed in Bartova et al., 2008).

In contrast to the large amounts of available information on the functions of HATs, HDACs, and enzymes that mediate histone methylation (listed in Kouzarides, 2007), enzymes responsible for other types of histone modifications, such as phosphorylation and ubiquitination, are not as intensively studied (Grant, 2001, as cited in Bartova et al., 2008). This is despite the fact that these modifications have important roles in transcription, DNA repair, the induction of apoptosis, and chromosome condensation (Cheung et al., 2000, as cited in Bartova et al., 2008). For example, the phosphorylation of serine 10 in histone H3 is associated with transcriptional activation in mammalian cells (Thomson *et al.*, 1999, as cited in Bartova et al., 2008); and H2A phosphorylation is responsible for chromosome condensation (Grant, 2001, as cited in Bartova et al., 2008). These observations underscore the importance and complexity of histone modifications in the regulation of nuclear, and subsequently, cellular processes (Bartova et al., 2008).

2.3 DNA repair in heterochromatin and euchromatin

Phosphorylated histone H2AX (γH2AX) are widely used as a specific marker of DSBs, and the immunostaining of γH2AX foci allows for the most sensitive monitoring of DSB induction (with ionizing radiation doses in the order of cGy or even mGy), with the cell nucleus and also of DSB repair. One striking observation about γH2AX is that foci are more rarely detected at heterochromatic sites than in the euchromatin. γH2AX foci distribution within irradiated cells is uneven, as foci can only be detected at the periphery of heterochromatic regions, rather than within them; the boundaries of which are maintained by methylation of lysine at position 9 on histone H3 (H3K9), an important epigenomic imprint of heterochromatic regions. However, there is still debate as to the significance of this resistance of heterochromatin to gH2AX foci formation. The H2AX in the heterochromatin may simply be inaccessible to phosphorylation because of the compact nature of the chromatin; or in fact, the heterochromatin might be more resistant to DNA damage (Cowell et al., 2007; Falk et al., 2010; reviewed in Cann & Dellaire, 2011). Falk et al. (2008) used an immunoFISH method (immunocytochemistry combined with fluorescence *in situ* hybridization) to analyse DSB formation in chromosomes as well as in chromatin regions with known chromatin compaction. They found that condensed regions of chromatin (with a low density of expressed genes) known as anti-Ridges, and also interphase domains of gene-poor and transcriptionally less active chromosomes were less susceptible to DSB induction, compared with decondensed chromosomes and decondensed gene-rich chromatin; with the caveat that gH2AX was still being used as the DSB marker (Falk et al., 2008). They speculated that the high amount of protein bound to the heterochromatin, including HP1, acts as a protective layer that prevents access to the DNA by ROS (Falk et al., 2008). Indeed, approximately 56% of g-irradiation-induced DSBs are indirectly caused by hydroxyl radicals, rather than directly by the g-irradiation itself. As such, the decondensed and hydrated euchromatin would be much more susceptible to this secondary damage (reviewed in Falk et al., 2010). It is easy to see how this model of heterochromatin protein-based protection could be extrapolated to other DNA-damaging agents, such as UV irradiation and mutagenic chemicals. Furthermore, increasing the accessibility of chromatin to DNA-damaging agents is one of the proposed mechanisms of action of histone deacetylase (HDAC) inhibitors, with respect to their ability to sensitize cells to chemotherapeutic agents (Kim et al., 2003, as cited in Cann & Dellaire, 2011).

An additional confounding observation is that following DSBs, chromatin rapidly decondenses in the vicinity of the break (see below Kruhlak et al., 2006; Falk et al., 2007; Dellaire et al., 2009) and undergoes histone acetylation (Falk et al., 2007). Goodarzi et al. (2008) proposed a model for heterochromatic DSBs, which are generally repaired more slowly than are euchromatic ones. This model involves ATM signalling, required for DSB repair in heterochromatin, which affect <25% of the DSBs. The ATM protein belongs to a growing family of PIKK kinases, which is observed at the sites of DNA damage, and where it is autophosphorylated and dissociated from its nonactive dimeric form into the active monomeric form. Direct connections between chromatin alteration and ATM were made when the transcriptional corepressor Krűppel-associated box (KRAB)-associated protein (KAP)-1 was identified as an ATM substrate, being robustly phosphorylated at S824; causing transient chromatin relaxation (Ziv et al., 2006, as cited in Cann & Dellaire, 2011). KAP-1 (also called TIF1b, TRIM28, or KRIP-1) is an abundant nuclear protein that binds to KRAB domains within sequence-specific transcriptional repressors to trigger heterochromatin formation via interactions with proteins, such as heterochromatin protein 1 (HP1), HDACs, SET-domain histone methyltransferases, and ATP-dependent chromatin remodelers (Craig, 2005, as cited in Goodarzi et al., 2008). Nevertheless, exactly how KAP1 mediates this decondensation is still unknown, as its mobility, its association with chromatin, and its interaction with its heterochromatic mediators are relatively unchanged following phosphorylation (reviewed by Cann & Dellaire, 2011).

In the absence of ATM signalling, heterochromatic repair stalls and the lesion persists. Significantly, the knockdown of the transcriptional repressor KAP-1, an ATM substrate, or of the heterochromatin-building factors HP1 or HDAC1/2, alleviates the requirement for ATM in DSB repair. Moreover, cells exhibiting heterochromatin disorganization caused by mutations in the DNA or histone methyltransferases do not require ATM for heterochromatic DSB repair. The requirement of ATM in DNA repair is also affected by lamina disruption, as the result of LMNA gene mutations. Lamins play a significant role in heterochromatin; moreover, in the interphase chromosome organization at the nuclear periphery (Ondrej et al., 2008). Interestingly, transient induction of chromatin hypercondensation by hyperosmotic treatment of the cells led to the formation of a small dot of γH2AX, dispersion of ATM, and its additional partners in the signalling pathways (53BP, NBS1) through the cell nucleus. The signals of γH2AX and ATM or 53BP did not overlap. Cell transfer into isotonic medium for the subsequent 10 minutes led to the complete restoration of chromatin structure, as well as the formation of γH2AX foci and the assembly of ATM and 53BP at the sites of DNA damage (Falk et al., 2008).

In euchromatin, ATM signalling perturbs the local chromatin architecture which, while important for optimal signalling, is dispensable for repair, since NHEJ factors can freely access or manipulate the DSB. Hence, >75% of DSBs in G0/G1 are repaired in an ATM-independent manner. Interestingly, lower eukaryotes, which have minimal heterochromatin, have a diminished role for ATM signalling in their DSB response (Morrow et al., 1995, as cited in Goodarzi et al., 2008). By contrast, ATM signalling is of major importance to the higher eukaryotic DSB response (Lavin et al., 2005, as cited in Goodarzi et al., 2008), where a larger genome and complex developmental program resulted in more constitutive and facultative heterochromatin. It suggests that the increased prominence of ATM signalling in the DSB response (compared with TEL1 signalling in yeast) correlates with the increasing complexity of chromatin architecture observed throughout eukaryotic evolution (Goodarzi et al., 2008).

As introduced above, HP1 is important in heterochromatin formation. It can bind H3K9me3, as well as the DNA methyltransferases DNMT1 and DNMT3a, which are involved in CpG island methylation, and the histone methyltransferase Suv39H1 (reviewed in Cann & Dellaire, 2011). Given that Suv39H1 can trimethylate H3K9, and that HP1 can interact with both H3K9 and Suv39H1, HP1 function has been proposed as a mechanism for heterochromatin expansion (Maison & Almouzni, 2004, as cited in Cann & Dellaire, 2011). Mammalian cells have 3 HP1 isoforms: HP1a, HP1b, and HP1g. These proteins can form homo- and heterodimers through their C-terminal chromoshadow domains; with this self-association potentially helping to compact the chromatin (Maison & Almouzni, 2004, as cited in Cann & Dellaire, 2011). The dimerization of 2 chromoshadow domains also creates a hydrophobic surface that can bind proteins such as KAP1 (Lechner et al., 2005, as cited in Cann & Dellaire, 2011). While the HP1 proteins localize to, and are involved in, the assembly and maintenance of heterochromatin at the centromeres and telomeres, they are also involved in the regulation of transcription in both euchromatin and heterochromatin (Dinant & Luijsterburg, 2009). Goodarzi et al. (2008) demonstrated that depletion of all three HP1 isoforms removed the requirement for ATM in the repair of heterochromatic DSBs, suggesting that HP1-mediated chromatin compaction is inhibitory to DNA repair. Indeed, HP1b exhibits a transient release from damaged sites, with casein kinase 2-mediated phosphorylation of Thr51 being the proposed initiator of release (Ayoub et al. 2009b). This residue is within the chromodomain of HP1, which is responsible for its ability to bind H3K9me3 (Ayoub et al., 2009b). However, all three HP1 isoforms have also been shown to be recruited to sites of DNA damage, including UV-induced cyclobutane pyrimidine dimmers and pyrimidine(6-4)pyrimidone photoproducts (Luijsterburg et al., 2009), oxidative lesions (Zarebski et al., 2009), and DNA DSBs (Ayoub et al., 2008, 2009a; Luijsterburg et al., 2009, reviewed in Cann & Dellaire, 2011). The recruitment of HP1 to sites of DSBs was found to be dependent on the chromoshadow domain and independent of the chromodomain and H3K9me3 (Luijsterburg et al., 2009). In an attempt to rationalize these conflicting results, a bimodal model for HP1 dynamics following DSBs has been proposed, with an initial phosphorylation-dependent release of HP1, and a subsequent accumulation of HP1 at sites of DNA damage (Ayoub et al., 2009b; Dinant & Luijsterburg, 2009; Zarebski et al., 2009). In this model (reviewed in Cann & Dellaire, 2011), the initial release of HP1 from H3K9me3 would help effect chromatin decondensation. The subsequent chromoshadow dependent accumulation of HP1 might then help mediate and enhance the DNA damage signalling. Indeed, HP1 has also been shown to be recruited to sites of heterochromatic perturbation (Zhang et al. 2007), suggesting that it could help stabilize damaged chromatin. Furthermore, phosphorylated HP1 might provide an epitope for mediators of the DNA damage response pathway, many of which bind phosphorylated proteins (Dinant & Luijsterburg, 2009).

At present, a lot is known about the biochemical aspects of DSB repair, but very little about how chromatin structure influences this process and the sensitivity of DNA to DSB induction. We lack the answers even to the very basic questions formulated by Falk et al. (2010) such as: Are DSBs induced equally in functionally and structurally distinct chromatin domains? Where and how are DSBs actually repaired: individually at their sites of origin, or together in specialized, repair-competent nuclear sub-compartments. Are DSBs spatially stable or do they migrate into "repair factories". Further, what are the DSB clusters observed by several authors: (Kruhlak et al., 2006; Falk et al., 2007) sites of multiple DNA damage, repair factories, or by-products of DSB repair that potentially increase the risk of chromatin mis-rejoining? Consequently, which is the more important factor that underlies chromosomal translocations - the predetermined nuclear proximity of DSBs ("position first" hypothesis) in the range of Brownian movement of chromatin (about 0.5 mm), or in their more extensive, potentially directed movement during the repair process that brings them close to one another (the "breakage first" hypothesis)?

2.4 Chromatin loops – Structure, organization and mobility

Chromatin loops are an ubiquitous structural element of chromatin, which is important in many scenarios of chromatin function and dynamics (van Driel et al., 2003; Fraser, 2006). They are attractive organizational and regulatory features, because they provide structural support to the chromatin fibre; at the same time, they bring distantly located sequence elements into spatial proximity, allowing for regulatory communication between these sites. *Vice-versa*, loops can spatially segregate genomic regions from one another, and ensure their independent function. Loops have been implicated at virtually all levels of chromatin organization, and they function ranging from kilobase-sized loops involved in the interaction of upstream elements with promoters, to giant loops of hundreds of kilobases which might contribute to gene placement away from the chromosome body and into distinct nuclear environments (van Driel et al., 2003; Cremer et al., 2006; Misteli, 2007). The existence and physiological relevance of the various types of loops at times is difficult to ascertain, as they often cannot be detected under native conditions and are generally refractory to visualization *in situ*. Regardless, the relevance of loops in several gene-regulatory events has recently been reinforced (Fraser, 2006; reviewed in Misteli, 2007). In the past, the concept of chromatin loops involved an insoluble proteinaceous nuclear matrix or nucleoskeleton (to which chromatin loops are anchored by DNA sequences referred to as matrix attachment regions (MARs)). However, because it has not been possible to define its composition and structure, the matrix remains a useful working concept, rather than a well-accepted structure in the same sense as the cytoskeleton (Pederson, 2000). Nonetheless, recent findings concerning lamins and nuclear actin could revive the idea of a nucleoskeleton; however, not as the main foundation of chromatin organization.

To address the topology and organization of the loops, several authors have measured the two-dimensional spatial distance between the genomic markers as a function of their genomic separation (distance). Two different models have been presented. The first is called the Random-Walk/Giant-Loop (RW/GL) Interphase Model. In this model, below ~2Mbp, the chromatin fibre seems to rapidly extend with the genomic separation; and above ~2Mbp a random walk behaviour was found. Therefore, Sachs et al. (1995) assumed loops of 1 to 5

Mbp, which are attached to a non-DNA (and presumably protein) backbone, which is in agreement with the nuclear matrix hypothesis (although not yet supported). Fitting the experimental distance measurements (Yokota et al., 1995) with this extended analytical model revealed a loop size of ~3Mbp, and separations of the attachment points of the giant loops of 620 nm. More accurate spatial distance measurements between the genetic markers, as a function of their genomic separation using a more structure preserving protocol, also showed biphasic behaviours. However, the increase in the distance was slower with growing separations below ~2Mbp; implying much smaller loops than in the RW/GL model. Thus, the smaller loops need to be aggregated into connected clusters, which for larger genetic separations are arranged like a random walk. A detailed interpretation of these findings resulted in the proposal of a Multi-Loop-Subcompartment (MLS) interphase–metaphase model (Műnkel et al., 1998). In this model, small loops of 60 - 120 kbp from rosettes in the interphase, are linked by chromatin linkers (again of ~120kbp). This is also in agreement with the general dynamics of chromatin, showing a short-range motion that is locally constrained within random directions. Nevertheless, as mentioned above, Goetze et al. (2006) showed that the compaction of chromatin is linked to the size of the loops, and distinguishes chromatin from compact heterochromatin and decondensed euchromatin. Except for short-range motion, the long-range motion of the chromatin has also been observed within the nucleus. Some chromatin regions occasionally showed large (~0.5 µm) directional movements, over time-periods as short as 10 seconds, which were inhibited in ATP-depleted cells. The large-range motion of the loops (giant loops) is connected with the transcription of the genes, located in that chromatin region. Dundr et al. (2007) showed that the movement of synthetic inducible U2 snRNA gene arrays over long distances inside the nucleus (2-3 µm) became associated with the Cajal bodies. Interestingly, the expression of these arrays and their movement was dependent on the ability of nuclear actin to polymerize. These findings showed that some parts of the chromatin could undergo changes via the RW/GL model.

As described above, chromatin has the potential for large-range motion in some cases; however, we do not know if DSBs are similarly able to move. That DSBs can form clusters has been described. For example, Markova et al. (2007) correspondingly quantified the area of gH2AX/53BP1 foci induced by g-radiation during the post-irradiation period, and surprisingly found that a progressive decrease in the number of gH2AX/53BP1 foci was not followed by an equivalent decrease in the overall area of foci; instead, it remained unchanged, up to 4 h post irradiation. However, if these clusters are the results of long-range motion of the chromatin is still not clear.

2.5 DSB mobility and clustering

Chromatin decondensation at the break sites is probably connected with frequent DSB protruding into "chromatin holes"; an area of low chromatin density from boundary regions, with areas of high chromatin density (Falk et al., 2007). It has been demonstrated that most gH2AX foci induced by g-radiation are spatially stable; however, some "breaks" show a significantly higher mobility. When the mobility was quantified as the mean change of the distances between all possible DSB pairs over time (53BP1 or NBS1 foci), the mobility of the damaged chromatin was equivalent to that of the undamaged. However, when tracking individual foci, a subgroup with significantly increased mobility was identified. To explain this, Falk et al. (2007) double-transfected MCF7 cells with the 53BP1 protein, and H2B histone fused with GFP and RFP, respectively (so that one could follow the movement of 53BP1 foci relative to chromatin density). Interestingly, the foci showing increased (but restricted) movement frequently corresponded with heterochromatic DSBs. This movement, although seemingly random when studied at the pan-nuclear level, was from high-density chromatin at the sites of DSB origin, to immediately adjacent domains of sparse chromatin.

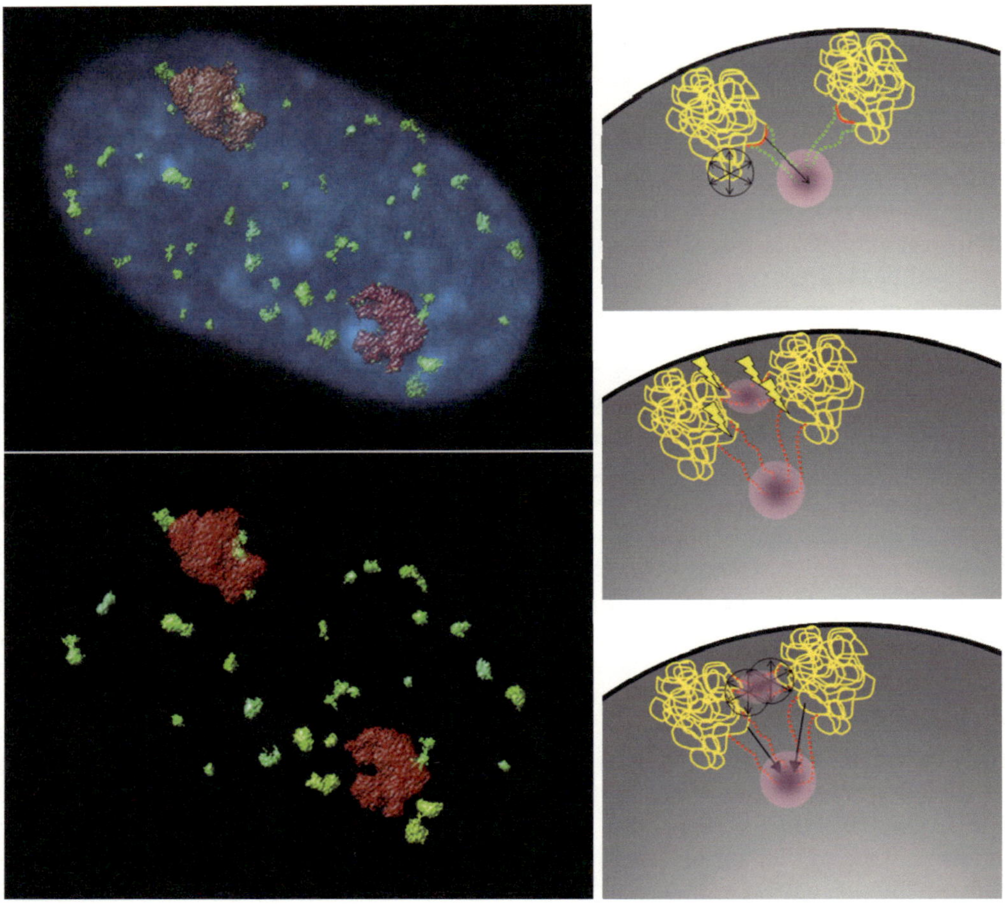

Fig. 1. A deconvolved image of the human fibroblast nucleus, irradiated with a dose of 2 Gy. γH2AX foci represented DSBs are dispersed within the nucleus, mostly out of heterochromatin (chromatin is stained by DAPI – blue). Some of γH2AX foci are in close proximity to form signals such as double-dots. CT of HSA1 (red) is visualized by the FISH technique. γH2AX foci are located on the surface of the chromatin mass of the CTs, not inside. A schematic drawing of chromatin mobility. The upper subset figure shows the possibilities of chromatin motion - constrained random short-range motion (black circle with arrows) and directional long-range motion (arrows) during transcriptional activation of a given region (red – inactive changes to active – green interrupted line). The active genes are shifted to the transcription factory (pink circle). The middle figure of the subset demonstrates clustering of the damaged chromatin loci (red interrupted lines). The pink circles represent only the location of random clustering or hypothetical repair factories. The lower figure of the subset shows the hypothetical possibility of DSBs' motion to form the clusters or repair factories (pink circles) - constrained random short-range motion (black circle with arrows) and active directional long-range motion (arrows)

As reviewed by Falk et al. (2010), DSB clusters are formed at sites that differ from the original positions of DSB lesions. At first, it was thought that the presence of these clusters confirmed the existence of the "repair factories" described above. However, contrary to the frequent clustering of foci in cells irradiated with high-LET radiation, only a few clusters per cell (each usually containing only 2 or 3 gH2AX foci) appear after a dose of 1 Gy (1 Gy/min) of g-rays (Falk et al., 2007). This means that no more than 5 - 8 gH2AX foci out of about 35 DSBs induced per cell (on average) after this dosage usually form clusters; and most foci have not entered clusters, even several hours after g-irradiation. These results do not support the hypothesis of "repair factories", and the clustering of gH2AX foci after α-irradiation apparently reflects only the random "movement" of chromatin. Moreover, soon after irradiation (5 min. post irradiation), the frequency of clustered foci was low, but increased with time post irradiation, and DSB clustering occurs preferentially in chromatin holes, regions with very low chromatin concentration (Falk et al., 2007). Falk and co-workers (2010) also proposed that clusters represent unrepaired DSBs, since they still co-localized with other repair proteins (MRE11, NBS1, and 53BP1). It indicates that DSBs in clusters could only be repaired with difficulty; it is therefore tempting to speculate that they represent sites with an increased risk of chromosomal aberrations. Indeed, Scherthan et al. (2008, as cited in Falk et al., 2010) recently reported results that strongly support this hypothesis. Comparing the numbers of gH2AX foci induced by different doses of X-rays with corresponding frequencies of chromosomal aberrations detected by spectral karyotyping (SKY), they showed that more DSBs accumulate in a single gH2AX focus after a higher dose of IR; this correlates with an increased risk of DNA mis-rejoining, and the formation of complex aberrations. In addition, it seems that clustered foci also contribute to the process of cell adaptation to ionizing radiation, as reviewed in Falk et al., 2010. This process enhances the risk of fixing chromosomal aberrations, because it allows the cell to enter mitosis with unrepaired DSB damage (reviewed in Falk et al., 2010).

3. Chromosome territories in theories of chromosomal translocation origin

To answer questions concerning chromosomal aberrations and the hypothesis of their origin (the "position first" and "breakage first" hypotheses), it is necessary to clearly understand the spatial organization of the genome. The most global level of cellular genome organization is the arrangement of the genome regions within the 3D space of the cell nucleus (Cremer et al., 2006; Meaburn & Misteli, 2007; Misteli, 2007). The nonrandom nature of spatial genome organization is indicated by the age-old observation of the segregation of transcriptionally active and inactive regions into physically separate domains of euchromatin and heterochromatin, respectively. Recently, more-detailed mapping studies of smaller genome regions have significantly extended this concept and have made it clear that chromosomes, genome regions, and single genes are nonrandomly arranged within the nucleus (Cremer et al., 2006). Changes in positioning patterns occur during differentiation and development; strongly suggesting a link between the positioning and genome function (Parada et al., 2004; Cremer et al., 2006; reviewed in Misteli, 2007).

3.1 CT theory, structure, organizations, and models

In the interphase nuclei, decondensed mitotic chromosomes exist as chromosome territories. They are defined as the nuclear space taken up by the DNA of a given chromosome; the term "territory" refers to the fact that the occupied nuclear space is compact, typically roughly in shape, and with a volume of about 2 - 3 μm in diameter (reviewed in Cremer & Cremer, 2001; Misteli, 2007; Misteli, 2010). Chromosome territories (CTs) represent the highest order of chromatin structure and the largest unit of organization of the eukaryotic genome. Speculation about a chromosome territory concept has developed since the late 19th century. The territorial organization of interphase chromosomes was first suggested for animal cell nuclei by Carl Rabl (Rabl, 1885, as cited in Cremer & Cremer, 2010), but it was Theodor Boveri who introduced the term chromosome territory in his seminal studies of blastomere stages of the horse roundworm Parascaris equorum or Ascaris megalocephala (Cremer & Cremer, 2010). However, the direct visualization of individual CTs was made possible by *in situ* hybridization techniques developed during the mid-1980s. The achievement of chromosome sorting by flow cytometry of fluorescently-labelled mitotic chromosomes (Cremer et al., 1984a; Gray et al., 1987; Fawcett et al., 1994, all as cited in Cremer & Cremer, 2010) enabled the generation of chromosome specific painting probes for a large number of species. Subsequent amplification of DNA by cloning in bacterial vectors; or by universal PCR (Telenius et al., 1992, as cited in Cremer & Cremer, 2010), as well as novel techniques for the suppression of ubiquitous repetitive sequences by COT-1 DNA (Cremer et al., 1988; Pinkel et al., 1988; Lichter et al., 1988a; Lichter et al., 1988b, all as cited in Cremer & Cremer, 2010); likewise, the depletion of these sequences from the respective probes (Bolzer et al., 1999, as cited in Cremer & Cremer, 2010), made it possible to delineate individual chromosomes in metaphase plates and their territories in the interphase nucleus (reviewed in Cremer & Cremer, 2010).

In the conventional model of mammalian nuclear architecture, the arrangement of the chromosome territories within the interphase nucleus is nonrandom in two ways – radial distribution and nonrandom neighborhood arrangements. Radial distribution is characterized by gene-rich chromosome territories' occupation of the interior regions of the interphase nucleus; whereas, gene-poor domains are localized to the nuclear periphery. This general arrangement was confirmed by analyses comprising all human chromosomes (Boyle et al., 2001). An evolutionary comparison of lymphoblastoid cells from various primate species showed that this nonrandom radial nuclear distribution has been evolutionarily conserved despite major evolutionary chromosome rearrangements (reviewed in Cremer & Cremer, 2010). Nonrandom radial nuclear arrangements of CTs, depending on their gene density, were also found in other mammals like rodents (Mayer et al., 2005; Neusser et al., 2007, all as cited in Cremer & Cremer, 2010) and cattle (Koehler et al., 2009, as cited in Cremer & Cremer, 2010), additionally in birds (Habermann et al., 2001, as cited in Cremer & Cremer, 2010). Recently, it was shown in bovine pre-implantation embryos that this difference was not yet present in the nuclei of early blastomere stages. Its first appearance in correlation with major genome activation, and was fully established in blastocysts (Koehler et al., 2009; as cited in Cremer & Cremer, 2010).

CTs have preferred positions, not only with respect to the centre or periphery of the nucleus, but also with respect to one another (Parada & Misteli, 2002). This circumstance of the nonrandom arrangement is referred to as proximity patterns. However, these proximity patterns were of a rather probabilistic nature, i.e., their presence was shown by an excess of certain heterologous (and occasionally also homologous) CTs in a population of nuclei, not as an event consistently observed in each nucleus (Cremer & Cremer, 2010). Differences of statistically preferred proximity patterns varies between cell types, and it has consequences for a chromosome's ability to interact in trans with other parts of the genome, as revealed by the frequency of specific chromosome translocations (reviewed in Fraser & Bickmore, 2007). The nonrandom organization of the genome in 3D space of the nucleus is explained by self-organization of the chromatin (reviewed in Misteli, 2010). Although the nuclear model with protein matrix was rejected, recent studies increased the importance of a peripheral protein network based on lamins - the nuclear lamina (NL). Microscopy studies have indicated that specific regions of the chromosomes are located in close proximity to the NL, and that the chromatin is specifically attached to the NL. It has also been demonstrated that the human genome contains more than 1,300 sharply defined large domains 0.1 - 10 megabases in size, characterized by genome–lamina interactions. These lamina-associated domains (LADs) are typified by low gene-expression levels, indicating that LADs represent a repressive chromatin environment. Taken together, it demonstrates that the human genome is divided into large, discrete domains that are units of chromosome organization within the nucleus (Guelen et al., 2008).

CTs can be further classified into sub-chromosomal domains such as chromosome arms, centromeres, or telomeres. It has not only been demonstrated that CTs as a whole, but also sub-domains like the chromosome arms occupy discrete areas in the nuclear volume (Dietzel et al., 1998). Centromeres and telomeres, whose main functional role is to maintain the chromosome integrity, also do locate in distinct sub-volumes shown in various FISH experiments with specific probes (Molenaar et al., 2003; Weierich et al., 2003). In addition, sub-chromosomal domains undergo peripheral genome positioning, depending on gene-density and transcriptional activity of a given domain. For instance, R bands of chromosomes are generally found toward the centre of the nucleus, whereas late-replicating, inactive G bands are often located toward the periphery (Ferreira et al., 1997; Sadoni et al., 1999; Gilbert et al., 2004). Goetze et al. (2007) found that ridges (gene rich domains containing both housekeeping genes and highly expressed genes) and anti-ridges (relative gene poor domains with low level of gene expression) of chromosomes 1 and 11 occupy different radial positions. The two ridges are predominantly located more towards the nuclear interior than the three anti-ridges, which on average are closer to the nuclear envelope.

Several groups have documented the relationship between gene topology within the CT and its transcriptional activity (Dietzel et al., 1999; Bartova et al., 2002). Active gene loci were found on the surface of the chromatin mass of CT, and the same inactive genes were positioned towards the territory's interior (reviewed in Cremer & Cremer, 2001). Moreover, a few chromatin regions with particularly high gene density and/or transcriptional activity, such as the 11p15.5 segment, the MHC and EDC loci, or the HOX gene cluster have been consistently found looping out as protrusions from the core territory (Volpi et al., 2000; Williams et al., 2002; Chambeyron et al., 2005; Kupper et al., 2007).

All of the above-mentioned findings led to the creation of models of the functional architecture of CTs and the nucleus. The chromosome territory-interchromatin compartment (CT-IC) model proposes that nuclei are built up from two principal components: chromosome territories, and the interchromatin compartment (IC). The IC represents a The DNA-free (or at least largely free) contiguous space of the channels, starts at the nuclear pores and expands as larger channels and lacunas between the higher-order chromatin network, described above. The IC harbours splicing speckles and a variety of nonchromatin nuclear bodies (Verschure et al., 1999; Visser et al., 2000; Albiez et al., 2006). The IC concept

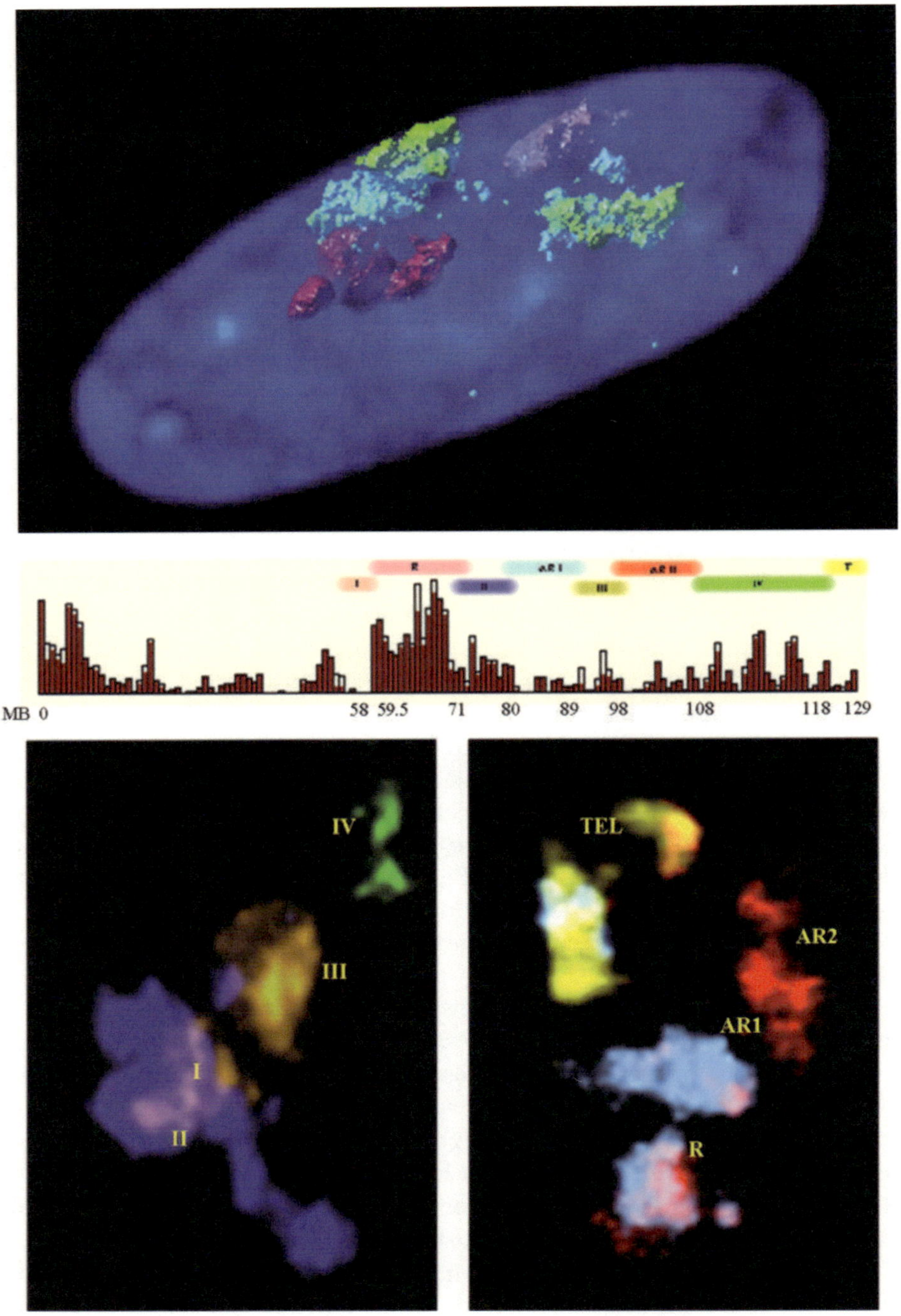

Fig. 2. A deconvolved image of chromosome territories of HAS 1 (green) and 2 (red) within the nucleus of a human fibroblast. Multicolor FISH of specific subdomains of chromosome territory HAS 11, showing radial arrangements of the subdomains in relationship to gene expression activity. Highly transcriptionally active domains (R – ridges, and domains marked I and II) are oriented to the nuclear centre, less active domains (AR1, AR2 – anti-ridges, domains marked as III, IV, and the telomeric region) are located more peripherally.

evolved from the interchromosomal domain (ICD) concept originally proposed by Zirbel et al. (1993, as cited in Cremer & Cremer, 2010), who defined the ICD as a network-like space mainly expanding around CTs, with little penetration into the CT interior (Cremer et al., 1993, as cited in Cremer & Cremer, 2010). Supposedly, genes were preferentially transcribed in a region of decondensed chromatin at the CT periphery, and RNA transcripts would be directly released into the ICD compartment (reviewed in Cremer & Cremer, 2010). However, a physical separation between CTs is not supported by the data on translocation frequencies and chromatin dynamics. Simulations of chromosome translocations, based on models of chromosome organization, have suggested the existence of a significant degree of intermingling between CTs (Hlatky et al., 2002; Holley et al., 2002; Sachs et al., 2000, all as cited in Branco & Pombo, 2006).

Using the novel FISH procedure for ultrathin cryosections (approximately 150 nm thick; cryo-FISH) of well-fixed, sucrose-embedded cells, which maximizes chromosome-painting efficiency, provides high resolution; and it simultaneously preserves chromatin nanostructure (Branco & Pombo, 2006). It has been shown that chromosomes intermingle significantly in the interphase nuclei of human cells, and depending on CT compaction, as well; this argues against the presence of an interchromosomal domain that separates CTs. The extent with which particular pairs of CTs intermingle correlates with the frequency of chromosome translocations in the same cell type (Branco & Pombo, 2006). According to these results, the interchromatin network (ICN) model was proposed (Branco & Pombo, 2006). This model predicts that chromatin fibres and loops intermingle in a rather uniform fashion, both in the interior of individual CTs and between differentially labelled neighbouring CTs; making any distinction between the interior or periphery of distinct chromatin domains functionally meaningless (reviewed in Cremer & Cremer, 2010). Albiez et al. (2006) undertook to resolve these conflicting views. Using experimental manipulations of the nuclear architecture, these authors obtained results supporting the CT-IC model, but they did not confirm the conclusions of the ICN model. On the other hand, a certain degree of intermingling between neighbouring territories and their physical distance may be important for the occurrence of chromosome translocations. Thus, IC represents space where chromatin loops of a different CT, with genes and regulatory elements, could meet, co-localize, and regulate themselves; furthermore, they could also create a transcription factory (Fraser & Bickmore, 2007).

3.2 Spatial CT organization in cancer

The organization of chromosome territories has functional consequences for genome organization and its maintenance; but also for the regulation of gene expression (reviewed in Cremer & Cremer, 2010). It is therefore not surprising that defects in higher-order chromatin structures and chromosome organization, such as translocations, cause diseases. Chromosomal abnormalities in the form of translocations are a general hallmark of cancer cells. The translocations occur during interphase by illegitimate joining of two or multiple chromosomes containing persistently damaged DNA such as DSBs requiring physical interaction of them. Thus, it is not surprising that the 3D proximity, CT volume, and the intermingling volume of CTs increases the translocation frequency between them (Branco & Pombo, 2006). This is a particularly intriguing and important finding in light of the nonrandomness in spatial genome organization, enabling determination of potential translocation partners (reviewed in Meaburn et al., 2007; Misteli, 2010). The contribution of spatial proximity to translocation formation is illustrated by the frequent formation of Robertsonian translocations, arising by the fusion of two acrocentric chromosomes, which in humans always contains nucleolar organizing regions (NORs), which cluster together in the nucleoli. Furthermore, translocations amongst peripheral chromosomes such as HSA4, 9, 13, and 18 occur at higher frequencies than with internally localized chromosomes, presumably due to their closer physical localization to each other (reviewed in Maeburn et al., 2007). Furthermore, when cells which contain nonrandom pairs of closely positioned chromosome 2 and 5 were irradiated to induce DSBs, these two chromosomes formed translocations with a significantly higher frequency than in cells where they were distant from one another; suggesting that their proximity facilitated their translocation (Mathas et al., 2009).

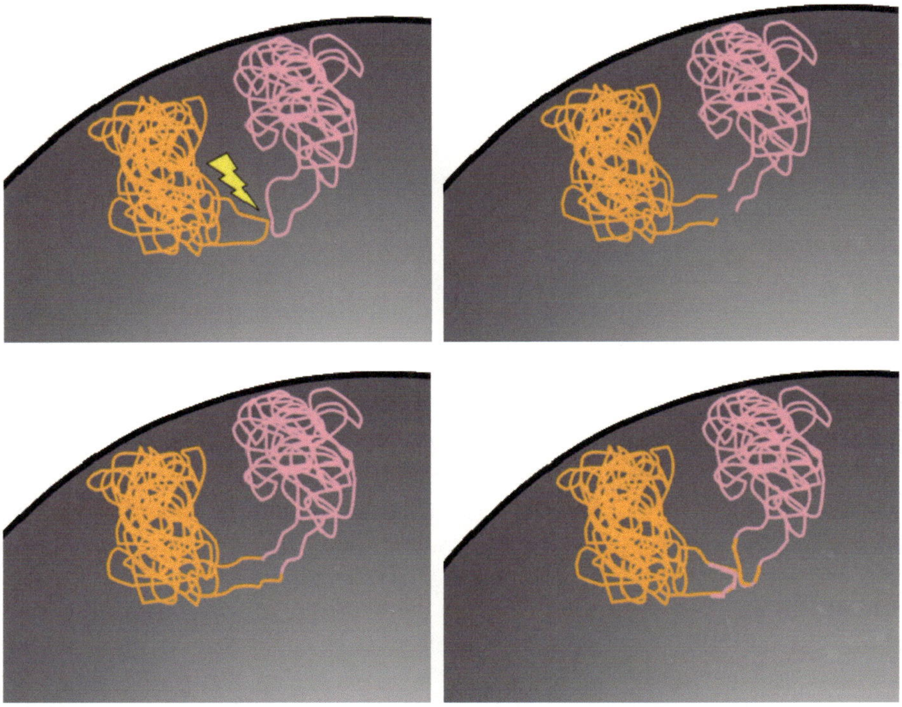

Fig. 3. Spatial organization of chromosomes in the formation of translocations after irradiation. Neighbouring CTs (yellow and red) are impaired by radiation, and closely juxtaposed DSBs created free chromosome ends. These ends could improperly join and formed a chromosome translocation by recombination.

Interestingly, the positioning of the specific regions or gene loci of the two chromosomes is involved in the translocation event, which is another important fact in cancerous transformation. The t(9;22)(q34;q11) translocation in chronic myeloid leukaemia results in the fusion of BCR and ABL, and these genes are in closer spatial proximity in normal heamatopoetic cells than other, non-translocating gene pairs (Lukasova et al., 1997; Kozubek et al., 1997; Neves et al., 1997; as cited in Meaburn et al., 2007). Significantly, exposure to radiation, a known inducer of leukaemia, results in BCR and ABL locating nearer to each other, in a more central part of the nucleus (Lukasova et al., 1997; Kozubek et al., 1997, as cited in Meaburn et al., 2007). A translocation involving HSA15 and 17, and the consequential fusion of the PML and RARα genes, is common in promyeloctic leukemia. PML and RARα are recurrently found in close proximity in normal B cells at a higher rate than the control gene pairs (Neves et al., 1997). An equivalent correlation between the proximity of genes and translocation frequency is observed in Burkitt's lymphoma. Notably, in B-cells the spatial proximity between MYC and any one of its multiple translocation partners (IGH, IGL, or IGK) decreases as the clinically observed incidence of translocation for the given pair increases (reviewed in Meaburn et al., 2007). Proximity effects on translocation formation are not limited to neighbour chromosome territories nor gene loci located on different chromosomes, but also apply to events on the same chromosome. The RET and the H4 genes are frequent translocation partners in thyroid tumours, and they are both located on chromosome 10 about 30 MB apart, but are brought together in the 3D space of the cell nucleus by the looping of the chromosome fibre; thus juxtapositioning and predisposing them to undergo a mutual translocation, resulting in an internal chromosome reversion (Nikiforova et al., 2000). Interestingly, translocation events could lead to heterochromatin-mediated gene silencing. Bártová et al. (2002) demonstrated that t(X;13) contributes to RB1 gene (13q14) silencing, also involving DNA methylation in a gene copy close to the translocated chromatin mass of HAS X. The changes in RB1 gene expression play an important role in the progression of the retinoblastoma tumour.

4. Conclusions

Higher-order genomic organization, involving the structure of the chromatin fibre, epigenetic modifications of chromatin, chromatin loops, chromatin compaction, and chromosome territories is a fundamental element, but not for storage of genetic information and it's processing alone. As was described above, it is also the basis for the maintenance of genome stability and its integrity impaired by external and internal damaging agents, including ionizing radiation. Here, it was shown how the higher-order chromatin structure influences DSB induction, mobility, repair, and mis-repair. In addition, chromatin structural defects lead to a list of specific diseases. The function of global higher-order chromatin organization, which maintains genome integrity, is emerging as an integral mechanism in the pathological processes, such as the formation of cancer translocations. Elucidation of the precise role and molecular mechanisms involved in maintaining chromatin structure and its role will be essential for our understanding of these all-important events. The understanding of genome functional organization; also in the context of the nuclear architecture will, in the future, provide promising developments of novel diagnostics and therapeutic strategies.

5. Acknowledgment

This work was supported by the grant of Ministry of Interior: VG20102014001.

6. References

Albiez, H.; Cremer, M.; Tiberi, C.; Vecchio, L.; Schermelleh, L.; Dittrich, S ; Kupper, K.; Joffe, B.; Thormeyer, T.; von Hase, J.; Yang, S.; Rohr, K.; Leonhardt, H.; Solovei, I.; Cremer, C.; Fakan, S. & Cremer, T. (2006). Chromatin domains and the interchromatin compartment form structurally defined and functionally interacting nuclear networks. *Chromosome Research*, Vol. 14, pp. 707-733, ISSN 0967-3849.

Ayoub, N.; Jeyasekharan, A.D.; Bernal, J.A. & Venkitaraman, A.R. (2008). HP1-β mobilization promotes chromatin changes that initiate the DNA damage response. *Nature*, Vol. 453, No. 7195, pp. 682–686, ISSN 0028-0836.

Ayoub, N.; Jeyasekharan, A.D.; Bernal, J.A. & Venkitaraman, A.R. (2009a). Paving the way for H2AX phosphorylation: chromatin changes in the DNA damage response. *Cell Cycle*, Vol. 8, No. 10, pp. 1494–1500, ISSN 1538-4101.

Ayoub, N.; Jeyasekharan, A.D.; & Venkitaraman, A.R. (2009b). Mobilization and recruitment of HP1: a bimodal response to DNA breakage. *Cell Cycle*, Vol. 8, No. 18, pp. 2945–2950, ISSN 1538-4101.

Bártová, E.; Kozubek, S.; Jirsová, P.; Kozubek, M.; Gajová, H.; Lukášová, E.; Skalníková, M.; Gaňová, A.; Koutná, I. & Hausmann, M. (2002). Nuclear structure and gene activity in human differentiated cells. *Journal of Structural Biology*, Vol. 139, pp. 76-89, ISSN 1047-8477.

Bártová, E.; Krejčí, J.; Harničarová, A.; Galiová, G. & Kozubek, S. (2008). Histone modifications and nuclear architecture: A review. *Journal of Histochemistry & Cytochemistry*, Vol. 56, No.8, pp. 711-721, ISSN 0022-1554.

Boyle, S.; Gilchrist, S.; Bridger, J.M.; Mahy, N.L.; Ellis, J.A. & Bickmore, W.A. (2001). The spatial organization of human chromosomes within the nuclei of normal and emerinmutant cells. *Human Molecular Genetics*, Vol. 10, pp. 211–219, ISSN 0964-6906.

Branco, M.R. & Pombo, A. (2006). Intermingling of chromosome territories in interphase suggests role in translocations and transcription-dependent associations. *PLoS Biology*, Vol. 4, e138, ISSN 1544-9173.

Cann, K.L. & Dellaire, G. (2011). Heterochromatin and the DNA damage response: the need to relax. *Biochemistry and Cell Biology*, Vol.89, pp. 45-60, ISSN 0829-8211.

Chambeyron, S.; Da Silva, N.R.; Lawson, K.A. & Bickmore, W.A. (2005). Nuclear re-organisation of the Hoxb complex during mouse embryonic development. *Development*, Vol. 132, pp. 2215–2223, ISSN 0950-1991.

Cowell, I.G.; Sunter, N.J.; Singh, P.B.; Austin, C.A.; Durkacz, B.W.; Tilby, M.J. & Sullivan, B. (2007). γH2AX foci form preferentially in euchromatin after ionizing-radiation. *PLoS One*, Vol. 2, No. 10, e1057, ISSN 1932-6203.

Cremer, T. & Cremer, C. (2001). Chromosome territories, nuclear architecture and gene regulation in mammalian cells. *Nature Reviews Genetics*, Vol. 2, pp. 292–301, ISSN 1471-0056.

Cremer, T.; Cremer, M.; Dietzel, S.; Muller, S.; Solovei, I. & Fakan, S. (2006). Chromosome territories - a functional nuclear landscape. *Current Opinion in Cell Biology*, Vol. 18, pp. 307-316, ISSN 0955-0674.

Cremer, T. & Cremer, M. (2010). Chromosome territories. *Cold Spring Harbour Perspectives in Biology*, Vol. 2, a003889, ISSN 1943-0264.

Dinant, C. & Luijsterburg, M.S. (2009). The emerging role of HP1 in the DNA damage response. *Molecular and Cellular Biology*, Vol. 29, No. 24, pp. 6335-6340, ISSN 0270-7306.

Dietzel, S.; Jauch, A.; Kienle, D.; Qu, G.; Holtgreve-Grez, H.; Eils, R.; Munkel, C.; Bittner, M.; Meltzer, P.S.; Trent, J.M. & Cremer, T. (1998). Separate and variably shaped chromosome arm domains are disclosed by chromosome arm painting in human cell nuclei. *Chromosome Research*, Vol. 6, No. 1, pp. 25–33, ISSN 0967-3849.

Dietzel, S.; Schiebel, K.; Little, G.; Edelmann, P.; Rappold, G.A.; Eils, R.; Cremer, C. & Cremer, T. (1999). The 3D positioning of ANT2 and ANT3 genes within female X chromosome territories correlates with gene activity. *Experimental Cell Research*, Vol. 252, pp. 363-375, ISSN 0014-4827.

Dellaire, G.; Kepkay, R. & Bazett-Jones, D.P. (2009). High resolution imaging of changes in the structure and spatial organization of chromatin, γ-H2AX and the MRN complex within etoposide-induced DNA repair foci. *Cell Cycle*, Vol. 8, No. 22, pp. 3750-3769, ISSN 1538-4101.

Dundr, M.; Ospina, J.K.; Sung, M.H.; John, S.; Upender, M.; Ried, T.; Hager, G.L. & Matera, G. (2007). Actin-dependent intranuclear repositioning of an active gene locus in vivo. *Journal of Cell Biology*, Vol. 179, No. 6, pp. 1095-1103, ISSN 0021-9525.

Falk, M. ; Lukasova, E. ; Gabrielova, B. ; Ondrej, V. & Kozubek, S. (2007). Chromatin dynamics during DSB repair. *BBA Molecular Cell Research*, Vol. 1773, pp. 1534-1545, ISSN 0167-4889.

Falk, M.; Lukasova, E. & Kozubek, S. (2008). Chromatin structure influences the sensitivity of DNA to γ-radiation. *BBA Molecular Cell Research*, Vol. 1783, pp. 2398-2414, ISSN 0167-4889.

Falk, M.; Lukasova, E. & Kozubek, S. (2010). Higher order chromatin structure in DSB induction, repair and misrepair. *Mutation Research*, Vol. 704, pp. 88-100, ISSN 1383-5742.

Ferreira, J.; Paolella, G.; Ramos, C. & Lamond, A.I. (1997). Spatial organization of large-scale chromatin domains in the nucleus: A magnified view of single chromosome territories. *Journal of Cell Biology*, Vol. 139, pp. 1597–1610, ISSN 0021-9525.

Fraser, P. (2006). Transcriptional control thrown for a loop. *Current Opinion in Genetic Development*, Vol. 16, pp. 490-495, ISSN 0950-1991.

Fraser, P. & Bickmore, W. (2007). Nuclear organization of the genome and the potential for gene regulation. *Nature*, Vol. 447, pp. 413–417, ISSN 0028-0836.

Gilbert, N.; Boyle, S.; Fiegler, H.; Woodfine, K.; Carter, N.P. & Bickmore, W.A. (2004). Chromatin architecture of the human genome: gene-rich domains are enriched in open chromatin fibers. *Cell*, Vol. 118, pp. 555–566, ISSN 0092-8674.

Goetze, S.; Mateos-Langerak, J.; Gierman, H.J.; de Leeuw, W.; Giromus, O.; Indemans, M.H.G.; Koster, J.; Ondrej, V.; Versteeg, R. & van Driel, R. (2007). The 3D structure of human interphase chromosome is related to the transcriptome map. *Molecular and Cellular Biology*, Vol.27, pp. 4475-4487, ISSN 0270-7306.

Goodarzi, A.A. ; Noon, A.T. ; Deckbar, D. ; Ziv, Y. ; Shiloh, Y. ; Lobrich, M. & Jeggo, P.A. (2008). ATM signaling facilitates repair of DNA double-strand breaks associated with heterochromatin. *Molecular Cell*, Vol. 31, pp. 167–177, ISSN 1097-2765.

Guelen, L.; Pagie, L.; Brasset, E.; Meuleman, W.; Faza, M.B.; Talhout, W.; Eussen, B.H.; de Klein, A.; Wessels, L.; de Laat, W. & van Steelsen, B. (2008). Domain organization of human chromosomes revealed by mapping of nuclear lamina interactions. *Nature*, Vol. 453, pp. 948-951, ISSN 0028-0836.

Jenuwein, T. & Allis, C.D. (2001). Translating the histone code. *Science*, Vol.293, pp. 1074-1080, ISSN 0036-8075.

Kouzarides, T. (2007). Chromatin modifications and their function. *Cell*, Vol.128, pp. 693-705, ISSN 0092-8674.

Kruhlak, M.J.; Celeste, A.; Dellaire, G.; Fernandez-Capetillo, O.; Müller, W.G.; McNally, J.G.; Bazett-Jones, D.P. & Nussenzweig, A. (2006). Changes in chromatin structure and mobility in living cells at sites of DNA double-strand breaks. *Journal of Cell Biology*, Vol. 172, No. 6, pp. 823-834, ISSN 0021-9525.

Kűpper, K.; Kőlbl, A.; Biener, D.; Dittrich, S.; von Hase, J.; Thormeyer, T.; Fiegler, H.; Carter, N.P.; Speicher, M.R.; Cremer, T. & Cremer, M. (2007). Radial chromatin positioning is shaped by local gene density, not by gene expression. *Chromosoma*, Vol. 116, pp. 285–306, ISSN 0009-5915.

Luijsterburg, M.C.; Dinant, C.; Lans, H.; Stap, J.; Wienrnasz, E., Lagerwerf, S.; Warmerdam, D.O.; Lindh, M.; Brink, M.C.; Dobrucki, J.W.; Aten, J.A.; Fousteri, M.I.; Jansen, G.; Dantuma, N.P.; Vermeulen, W.; Mullenders, L.H.F.; Houtsmuller, A.B.; Verschure, P.J. & van Driel, R. (2009). Heterochromatin protein 1 is recruited to various types of DNA damage. *Journal of Cell Biology*, Vol. 185, No. 4, pp. 577-586, ISSN 0021-9525.

Markova, E.; Schultz, N. & Belyaev, I.Y. (2007). Kinetics and dose response of residual 53BP1/gamma-H2AX foci: co-localization, relationship with DSB repair and clonogenic survival. *International Journal of Radiation Biology*, Vol. 83, No. 5, pp. 319-329, ISSN 0955-3002.

Mathas, S.; Kreher, S.; Meaburn, K.J.; Johrens, K.; Lamprecht, B.; Assaf, C.; Sterry, W. ; Kadin, M.E.; Daibat, M.; Joos, S.; Hummel, M.; Stein, H.; Janz, M.; Anagnostopoulos, I.; Schrock, E.; Misteli, T. & Dörken, B. (2009). Gene deregulation and spatial genome reorganization near breakpoints prior to formation of translocations in anaplastic large cell lymphoma. P*roceedings of the National Academyof Sciences of the USA*, Vol. 106, No. 14, pp. 5831–5836, ISSN 0027-8424.

Meaburn, K.J. & Misteli, T. (2007). Cell biology: Chromosome territories. *Nature*, Vol. 445, pp. 379-381, ISSN 0028-0836.

Meaburn, K.J.; Misteli, T. & Soutoglou, E. (2007). Spatial genome organization in the formation of chromosomal translocations. *Seminars in Cancer Biology*, Vol. 17, No. 1, pp. 80-90, ISSN 1044-579X.

Misteli, T. (2007). Beyond the sequence: cellular organization of genome function. *Cell*, Vol. 128, pp. 787-800, ISSN 0092-8674.

Misteli, T. (2010). Higher-order genome organization in human disease. *Cold Spring Harbour Perspectives in Biology*, Vol. 2, a000794, ISSN 1943-0264.

Molenaar, C.; Wiesmeijer, K.; Verwoerd, N.P.; Khazen, S.; Eils, R.; Tanke, H.J. & Dirks, R.W. (2003). Visualizing telomere dynamics in living mammalian cells using PNA probes. *EMBO Journal*, Vol. 22, pp. 6631-6641, ISSN 0261-4189.

Műnkel, C. & Langowski, J. (1998). Chromosome structure predicted by a polymer model. *Physical Review E*, Vol. 57, pp5888-5896, ISSN 1539-3755.

Nikiforova, M.N.; Stringer, J.R.; Blough, R.; Medvedovic, M.; Fagin, J.A. & Nikiforov, Y.E. (2000). Proximity of chromosomal loci that participate in radiation-induced rearrangements in human cells. *Science*, Vol. 290, pp. 138–141, ISSN 0036-8075.

Ondrej, V. ; Lukasova, E. ; Krejci, J. ; Matula, P. & Kozubek S. (2008). Lamin A/C and polymeric actin in genome organization. *Molecules and Cells*, Vol. 26, pp. 356-361, ISSN 1016-8478.

Parada, L.A.; McQueen, P.G. & Misteli, T. (2004). Tissue-specific spatial organization of genomes. *Genome Biology*, Vol. 12, pp. 1692-1697, ISSN 1474-7596.

Parada, L.A. & Misteli, T. (2002). Chromosome positioning in the interphase nucleus. *Trends in Cell Biology*, Vol. 12, No. 9, pp. 425-432, ISSN 0962-8924.

Pederson, T. (2000). Half a Century of "The Nuclear Matrix". *Molecular Biology of the Cell*, Vol. 11, No. 3, pp. 799-805, ISSN 1939-4586.

Sachs, R.; van den Engh, G.; Trask, B.J.; Yokota, H. & Hearst, J.E. (1995). A random-walk/giant-loop model for interphase chromosomes. *Proceedings of the National Academyof Sciences of the USA*, Vol. 92, pp. 2710-2714, ISSN 0027-8424.

Sadoni, N.; Langer, S.; Fauth, C.; Bernardi, G.; Cremer, T.; Turner, B.M. & Zink, D. (1999). Nuclear organization of mammalian genomes. Polar chromosome territories build up functionally distinct higher order compartments. *Journal of Cell Biology*, Vol. 146, pp. 1211–1226, ISSN 0021-9525.

Shi, Y.; Lan, F.; Matson, C.; Mulligan, P.; Whetstine, J.R.; Cole, P.A.; Casero, R.A. & Shi, Y. (2004). Histone demethylation mediated by the nuclear amine oxidase homolog LSD1. *Cell*, Vol.119, pp. 941-953, ISSN 0092-8674.

Trojer, P. &Reinberg, D. (2007). Facultative heterochromatin: Is there a molecular signature. *Molecular Cell*, Vol. 128, pp. 1-12, ISSN 1097-2765.

Van Driel, R.; Fransz, P.F. & Verschure, P.J. (2003). The eukaryotic genome : a system regulated at differenthierarchical levels. *Journal of Cell Science*, Vol. 116, pp. 4067-4075, ISSN 0021-9533.

Verschure, P.J.; van Der Kraan, I.; Manders, E.M. & van Driel, R. (1999). Spatial relationship between transcription sites and chromosome territories. *Journal of Cell Biology*, Vol. 147, pp. 13–24, ISSN 0021-9525.

Visser, A.E.; Jaunin, F.; Fakan, S. & Aten, J.A. (2000). High resolution analysis of interphase chromosome domains. *Journal of Cell Science*, Vol. 113, pp. 2585–2593, ISSN 0021-9533.

Volpi, E.V.; Chevret, E.; Jones, T.; Vatcheva, R.; Williamson, J.; Beck, S.; Campbell, R.D.; Goldsworthy, M.; Powis, S.H.; Ragoussis, J.; Trowsdale, J. & Sheer, D. (2000). Large-scale chromatin organization of the major histocompatibility complex and other regions of human chromosome 6 and its response to interferon in interphase nuclei. *Journal of Cell Science*, Vol. 113, pp. 1565–1576, ISSN 0021-9533.

Weierich, C.; Brero, A.; Stein, S.; von Hase, J.; Cremer, C.; Cremer, T. & Solovei, I. (2003). Three-dimensional arrangements of centromeres and telomeres in nuclei of human and murine lymphocytes. *Chromosome Research*, Vol. 11, pp. 485-502, ISSN 0967-3849.

Williams, R.R.; Broad, S.; Sheer, D. & Ragoussis, J. (2002). Subchromosomal positioning of the epidermal differentiation complex (EDC) in keratinocyte and lymphoblast interphase nuclei. *Experimental Cell Research*, Vol. 272, pp. 163–175, ISSN 0014-4827.

Woodcock, Ch.L. & Ghosh, R.J. (2010). Chromatin higher-order structure and dynaics. *Cold Spring Harbour Perspectives in Biology*, Vol. 2, a000596, ISSN 1943-0264.

Yokota, H.; van den Engh, G.; Hearst, J.E.; Sachs, R. & Trask, B.J. (1995). Evidence for the organization of chromatin in megabase pair-sized loops arranged along a random walk path in the human G0/G1 interphase nucleus. *Journal of Cell Biology*, Vol. 130, No. 6, 1239-1249, ISSN 0021-9525.

Zarebski, M.; Wiernasz, E., & Dobrucki, J.W. (2009). Recruitment of heterochromatin protein 1 to DNA repair sites. *Cytometry A*, Vol. 75, No. 7, pp. 619–625, ISSN 1552-4922.

Zhang, R.; Liu, S.T.; Chen, W.; Bonner, M.; Pehrson, J.; Yen, T.J. & Adams, P.D. (2007). HP1 proteins are essential for a dynamic nuclear response that rescues the function of perturbed heterochromatin in primary human cells. *Molecular and Cellular Biology*, Vol. 27, No. 3, pp. 949- 962, ISSN 0270-7306.

6

Mealybug as a Model for Studying Responses to High Doses of Ionizing Radiation

Kommu Naga Mohan[1], Jun Ge[2] and Jayarama S. Kadandale[1]
[1]*Centre for Human Genetics, Next to Aravind Mills, Biotechpark – Phase I, Electronic City, Bangalore,*
[2]*Baylor College of Medicine, Houston,*
[1]*India*
[2]*USA*

1. Introduction

Biological effects of exposure to different doses of ionizing radiation (IR) range from death within two days in case of high doses (>30 Gy) to various types of diseases such as cancers (due to gene mutations) in case of long-term exposure to sub-lethal doses (3-30 Gy). Further, IR-induced mutations when transmitted to the progeny can cause various types of cancer. At molecular level, IR results in double-strand breaks in DNA leading to loss of genetic material and cell death. To address the damaging consequences of exposure to radiation, recent research in radiation biology is focused on biological responses to IR in organisms showing resistance to high doses of ionizing radiation. For example, *Deinococcus radiodurans,* a bacterium that can survive radiation doses as high as 5,000 Gray can be potentially used for remediation of waste sites mixed with radioactive materials. However, investigation of radiation response in a higher eukaryotic organism that is resistant to high doses of ionizing radiation is desirable to offer treatment regimens to individuals exposed to IR in events of nuclear accidents. We propose here that mealybug (*Planococcus lilacinus,* Insecta; Coccoidea; Hemiptera) is an ideal model organism for studying responses to high doses of ionizing radiation. Among different organisms whose sensitivity to IR investigated so far, mealybug has been shown to tolerate radiation doses as high as 1100 Gy. Following exposure to such high doses, this insect has an extraordinary ability to survive and produce fertile offspring. Cytogenetic analysis of the embryos produced by irradiated parents showed that a number of tiny fragments of irradiated chromosomes are capable of metaphase alignment and anaphase segregation, suggesting that the centromere property is distributed throughout each chromosome. Taken together, mealybug has very efficient DNA repair machinery that ensures proper healing of double strand breaks induced by ionizing radiation. Because of the central role of telomeres in chromosome end-healing, the nature, distribution and maintenance of telomeric repeats in the mealybug were recently studied. Telomeric repeat has been shown to occur interstitially as well as at the ends of the mealybug chromosomes and telomerase which maintains these repeats is constitutively active. At radiation doses higher than 100 Gy, telomerase activity decreases, suggesting that telomerase-independent mechanisms may play a role in maintenance of telomeres. Based on these studies we propose that in the mealybug, features such as diffuse centromere, constitutive telomerase activity and interstitial telomeres are components of machinery that confers resistance to high doses of ionizing radiation.

2. Biological effects of ionizing radiation

Ionizing radiation (IR) is a stream of high energy particles (such as electrons, protons or α-particles) or any of the short wavelength electromagnetic radiation (such as X-rays, γ-rays or ultraviolet rays) that is capable of removing an electron from an atom or a molecule in the substance through which it passes. The most common biological effect of IR is ejection of an electron from water molecule, resulting in the formation of highly reactive species:

$$2H_2O \rightarrow e^{-}+H_2O+H_2O^{*}H_2O^{*} \rightarrow .OH+.HH_2O^{+}+H_2O \rightarrow .OH+H_3O^{+} \quad (1)$$

In the reaction above, the dot before a radical indicates unpaired electron and * indicates an excited species. These free radicals are highly reactive and can alter other molecules in the cell (Reactor Concepts Manual, NRC). DNA is one such important target which can also be directly ionized by IR. Cells experiencing IR-induced DNA damage respond in different ways, depending on the extent of damage. In cases where the extent of DNA damage is minimal, the cells successfully repair the damaged DNA and as a result, there is no recognizable biological effect. At the other extreme is when the damage is severe, leading to a failure in the cells' ability to repair the DNA and as a result, the cells undergo programmed cell death or apoptosis so that a potential damage from the entire tissue is prevented. In the middle of these two extremes are detectable DNA mutations that occur when repair results in a non-lethal DNA mutation that can be transmitted to the daughter cells produced during successive cell divisions and these mutations may induce cancer subsequently. A fourth possible consequence of exposure to IR is when cells experience irreparable DNA damage leading to replication and transcriptional errors that predispose their daughter cells to premature aging and cancer (reviewed in Maynard et al., 2009).

2.1 Acute and chronic exposure to ionizing radiation

Similar to the effects at cellular level, the biological effect of IR at the organismal level is also dependent on the severity of the dose received and the duration of exposure. Acute radiation exposure is usually for a short period of time but involves high dose of exposure as in case of nuclear explosions (instantaneous flashes), exposures because of handling of highly radioactive resources, accidents in laboratories or manufacturing units and high medical doses that could be either intentional or accidental. It is relatively more difficult to determine the effects of chronic exposure to IR and the results have been conflicting. For instance, based on studies on an Indian population that is exposed naturally to thorium-containing monazite mineral in sand (1.5-3.0 roentgen per year), Kochupillai et al. (1976) reported higher prevalence of abortions, mental retardation, Down syndrome, chromosomal and chromatid abnormalities. However in a separate study, no significant difference was observed between the high-level exposed population and control population (Cheriyan et al., 1999). In addition, a recent report on mitochondrial DNA sequence analysis revealed that there is an increased rate of mitochondrial DNA mutations in the 'exposed population' than in the control population (Forster et al., 2002). Clearly more detailed studies are required to be done to conclusively associate genomic effects of chronic exposure to IR.

Exposure to high doses of IR has been associated with leukemia and cancers of thyroid, breast, bladder, colon, liver, lung, esophagus, ovarian, multiple myeloma, and stomach. There have also been suggestions that ionizing radiation exposure may cause cancers of prostate, nasal cavity/sinuses, pharynx and larynx, and pancreas (ICRP, 1991).
Because of difficulties in interpretation of consequences of chronic exposure to low doses of IR (below 0.01 Gy), researchers depended mainly on models of the process by which lower doses of radiation causes cancer. Data on effects of exposure to IR at low but higher than the normal background levels gave mixed results on carcinogenesis or transgenerational effects, but were consistent with estimates of risk based on atomic bomb survivors (Cardis et al., 2005). These results suggest that there is a small increase in chances of developing cancers among individuals exposed to low levels of IR for extended periods of time. A collection of reports on consequences of chronic and acute exposures to IR led to the linear dose-response model that suggests an increase in the risk to develop cancer with any increase in dose, however small. Agencies such as Nuclear Regulatory commission, Environmental Protection Agency, National Academy of Sciences Committee lent support to the linear no-threshold model. According to this model, IR from background levels natural and man-made resources) can cause cancer in ~ 1% of the exposed population (BEIR VII report).

2.2 DNA damage due to exposure to IR

IR-induced DNA damages include single strand and double strand breaks that are repaired by different DNA repair machineries. Single-strand breaks are repaired by the same enzymes that mediate base-excision repair. These include DNA glycosylase, AP-endonuclease, DNA polymerase β and DNA ligase. Here, damaged nucleotides are excised and repaired using undamaged sequence from the complementary strand. Double-strand breaks are particularly hazardous because they create 'sticky' chromosome ends that cause

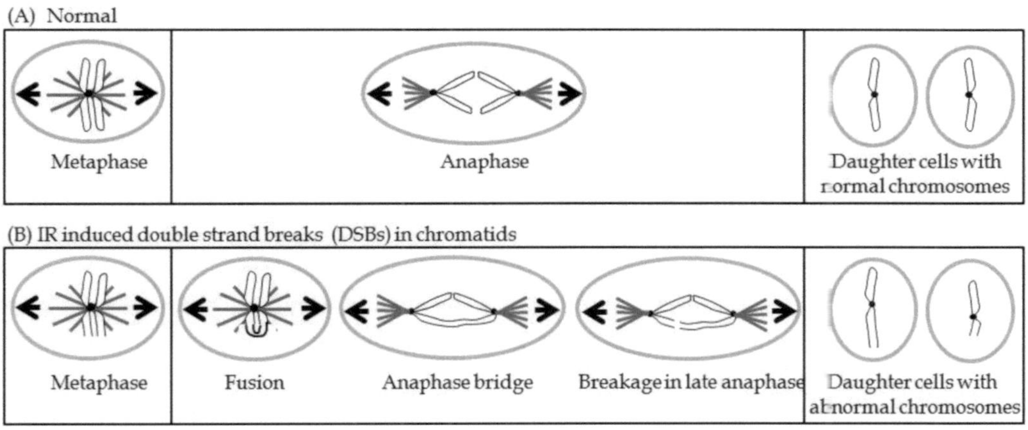

Fig. 1. IR-induced double strand breaks cause fusion-bridge-breakage-fusion cycles. IR induced double strand breaks in sister chromatids result in sticky ends that fuse and form anaphase bridges. Because of the force generated by spindle fibres, the sister chromatids in the bridge break unevenly leading to abnormal chromosomes at the end of cell division. One chromosome would have extra copies and the other would have lost a portion of the parent chromosome. Note that the daughter chromosomes also have sticky ends and participate in the fusion-bridge-breakage cycles again. These cycles result in complex genomic rearrangements.

fusion-bridge-breakage-fusion cycles (McClintock, 1938; reviewed in deLange, 2005; Fig.1) and lead to genomic rearrangements leading to cell death or transformation. DNA double-strand breaks can be repaired by non-homologous end joining (NHEJ; Moore & Haber, 1996), microhomology-mediated end joining (MMEJ; Liang et al., 1996) and homologous recombination (HR; Lundblad & Blackburn, 1993). In NHEJ, single strands of DNA ends showing microhomology are recognized and joined. If these overhangs are compatible, repair is usually accurate. A protein called Ku is essential for NHEJ. Other proteins such as DNA ligase IV, XRCC 4 participate in the process. MMEJ also depends on microhomology among the overhangs of the broken fragments of the chromosomes to be joined but these microhomologies are 'revealed' by chromosome end resection and therefore MMEJ usually results in deletion of the DNA sequences between microhomologies. In Homologous recombination, on the other hand uses the undamaged DNA sequence that is identical or highly related to the DNA sequence that has been damaged. Sister chromatids (after G2 phase) or a homologous chromosome (in G1 phase) serve as templates for recombination-mediated repair. Recombination-mediated repair uses the same machinery that is used in the process of meiotic recombination. Homologous recombination has been shown to be the process involved in the recovery from exposure to high doses of ionizing radiation in the radio-resistant bacterium *Deinococcus radiodurans* (see below).

3. Telomeres and sensitivity to IR

Unlike the genomes of bacteria which are circular with no 'exposed ends', eukaryote genome exists as linear chromosomal DNA molecules. This linearity also poses a problem in the completion of DNA replication, leaving a 20 nucleotide 'unfinished gap' in the lagging strand. Successive cell divisions lead to progressive shortening of the 3' ends of the chromosome ends. These ends of eukaryotic chromosomes are therefore 'exposed' and vulnerable to exonucleolytic attack yielding sticky ends which in turn mediate genomic rearrangements described above. In order to prevent loss of genetic material, and genomic rearrangements and, completion of replication at the chromosome ends, the ends of eukaryotic chromosomes are organized as telomeres (reviewed in Blackburn, 1991). Telomeres are specialized structures of DNA and proteins that have been characterized in a number of organisms. The nature of telomere-associated DNA sequences and the proteins associated with the telomeres are described in the next section. Telomere shortening was shown to result in increased sensitivity to ionizing radiation (Haber 1998; Ahmed & Hodgkin 2000). There is increasing evidence that enzymes involved in DSB repair also have a role at telomeres. Telomeres were shown to act as storehouses for proteins involved in DNA DSB repair (Ku70, Ku80, Sir2, Sir3, Mre11-RAD50-NBS1 complex) (Zhu et al. 2000). These proteins are relocated from telomeres to the damaged sites induced by IR (Martin et al. 1999; Mills et al. 1999). Genetic studies in Caenorhabditis elegans, yeast, and mice have shown that telomere shortening results in increased sensitivity to IR although DSB repair is unaffected (Haber 1998; Ahmed & Hodgkin 2000; Goytisolo et al. 2000). These results suggest a relationship between telomere length and sensitivity to IR.

3.1 Telomere structure

In several eukaryotes, telomeres are composed of simple tandem repeats at the end with complex subtelomeric repeats occurring next to these repeats (Blackburn 1991; Villasante et al. 2008; Kojima et al. 2002). However, there are exceptions to this general rule. In Bombyx mori where the insect-specific pentameric TTAGG repeat was first discovered (Okazaki et al. 1993), the telomeric repeats are interrupted by non-long terminal repeat retrotransposons such as SART and TRAS (Fujiwara et al. 2005), HeT-A, TAHRE, and TART elements are found at the chromosome termini in Drosophila (Mason & Biessmann 1995; Biessmann et al. 1992; Levis et al. 1993) and complex long tandem repeats in the chromosome termini of Chironomus (Lopez et al. 1996). Studies in various insect species have suggested that TTAGG repeat is an ancestral motif in Arthropoda and that the TTAGG repeats might have originated from the TTAGGG repeats, also found at vertebrate telomeres (Sahara et al. 1999; Frydrychova et al. 2004; Vitkova et al. 2005). In some organisms, telomeres form structures called telomere loops or T-loops containing double-stranded DNA and a 3′ single-stranded G-rich DNA overhang that is important for telomere maintenance (Fig. 2). The single-stranded DNA forms a circle and then disrupts a part of the double-stranded DNA in the telomere for base pairing with the C-rich strand, forming a displacement loop or D-loop (Griffith et al. 1999). The T-loop is held together by shelterin complex containing TRF1, TRF2, and POT1 that recognize TTAGG repeats and another three proteins TIN2, TPP1, and RAP1 that interconnect the first three proteins (reviewed in deLange 2005).

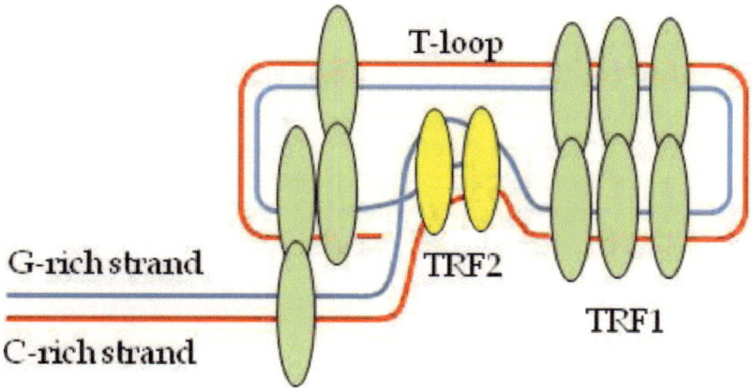

Fig. 2. Structure of telomeric DNA. The G-rich overhang (blue) forms a circle and then disrupts base pairing with the C-rich strand forming a displacement loop. The T-loop is held together by the sheltrin complex containing proteins such as TRF1 and TRF2.

3.2 Telomere maintenance

In almost all organisms containing simple tandem repeats, telomeres are maintained by telomerase, an enzyme with reverse transcriptase property by which it mediates chromosome end-healing (reviewed in Blackburn 1991). This enzyme is induced soon after cells are exposed to ionizing radiation (reviewed in Crompton 1997). In humans, the telomerase contains two units each of reverse transcriptase (TERT), telomerase RNA (TR or TERC), and dyskerin (DKC1) (Cohen et al. 2007). TERT folds with TERC and forms a mitten-like structure that allows the enzyme to wrap around the chromosomes to extend the telomeric repeats. Using the telomerase RNA that is complementary to the telomeric repeat unit, the reverse transcriptase subunit synthesizes multiple telomeric repeats by continuous elongation and translocation steps to produce a long single stranded DNA containing telomeric repeats (Fig. 3).

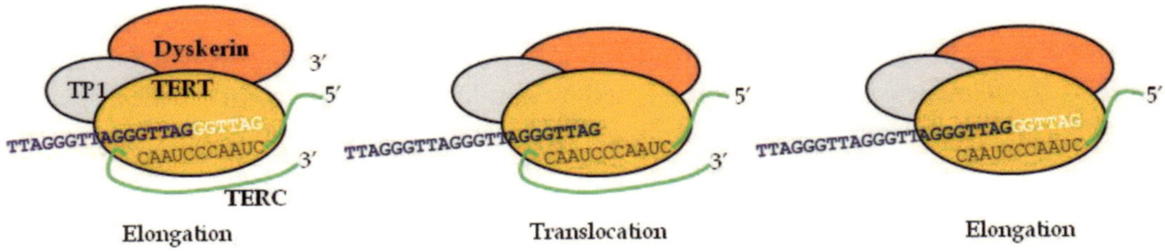

Fig. 3. Mechanism of elongation of telomeric repeats by telomerase. The RNA component of the telomerase pairs with the DNA to be extended and serves as a template for the elongation step. After elongation, there is a translocation step that brings the TERC to pair with the newly synthesized repeat unit to enable elongation again. Alternating elongation and translocation steps allows addition of multiple telomeric repeats.

In addition to telomerase-dependent mechanisms of chromosome end-healing, eukaryotic cells possess end-healing mechanisms that are telomerase-independent and are collectively called as alternative lengthening of telomeres (ALT). ALT, which was first demonstrated using telomerase-null yeast mutants, has been shown to be dependent on RAD52 that encodes a homologous recombination protein (Lundblad & Blackburn 1993). Such recombination-mediated chromosome end-healing mechanisms have also been demonstrated in human cells (Dunham et al. 2000). Additional studies have shown that in some cells lacking telomerase, there was a rapid increase as well as a decrease of telomeric repeats, suggesting that ALT may involve recombination at the telomeres (Murnane et al. 1994; Bryan et al. 1995). Unequal telomeric sister chromatid exchange and homologous recombination involving a damaged chromosome and an intact homologous chromosome have been proposed as the two mechanisms for recombination-mediated ALT (reviewed in Cesare & Reddel 2010). Unequal sister chromatid exchange results in two daughter chromosomes: one with longer telomeric repeats and another with shorter repeats. Preferential segregation of the chromosome with the longer telomeric repeats during cell division is expected, extending the cellular lifespan of one of the two daughter cells. Whereas there is no sufficient evidence for this model yet, down-regulation of homologous recombination promoting proteins such as the MRN complex (Jiang et al. 2005) and the SMC5-6 complex (Potts & Yu 2007) has been shown to result in telomere shortening, lending support to the homologous recombination model of ALT. NHEJ of two broken chromosomes, which does not involve recombination, is another telomerase independent mechanism that can restore chromosome integrity. In support of the evolutionary conservation of NHEJ as a means to maintain telomere integrity, a number of proteins involved in double-strand break DNA repair have also been shown to be involved in telomere maintenance (reviewed in Slijepcevic & Al-Wahiby 2005).

Several lines of evidence suggest that telomerase is absent in *Drosophila* species (reviewed in Pardue & DeBaryshe 1999). Telomerase activity is also not detectable in *B. mori* (Sasaki & Fujiwara 2000), a Lepidopteran insect, in which pentameric TTAGG repeats occur at the chromosome ends (Okazaki et al. 1993). Recent studies on cloning and expression analysis of the telomerase reverse transcriptase subunit from *B. mori* suggest that the reverse transcriptase gene is poorly transcribed and may be the reason for undetectable levels of its enzymatic activity (Osanai et al. 2006). It is however suggested that telomerase activity, even weak, is necessary for maintenance of telomeres in the domestic silk worm. In *Drosophila*, telomere length appears to be maintained by retrotransposition involving HeT-A, TAHRE, and TART elements (Biessmann et al. 1992). A role for proteins involved in doublestrand break (DSB) repair in telomere maintenance has been suggested from findings in which inactivation of proteins such as Ku, ATM results in telomere shortening (Boulton & Jackson, 1998; Hande et al. 2001). From the description above, it is evident that insects have evolved different mechanisms of maintenance of telomeres and telomerase-dependent mechanisms are not indispensible for maintenance of genomic integrity.

4. Model organisms for studying response to ionizing radiation

As noted above, exposure to a radiation dose of > 10 Gy (1000 rads) results in 95-100% death in humans within two days to two weeks of exposure depending on the radiation dose. Similar levels of exposure (> 12Gy) is also lethal in mice. From the limited information available in other mammals, we can consider that a radiation dose of 20-30 Gy can result in certain death in most mammals. In order to obtain insights into biological responses to IR and mechanisms that may confer resistance to exposure to radiation a variety of in vitro models using cell lines and different organisms showing resistance to high doses of IR have been sought. One example is *Deinococcus radiodurans,* an extremophilic bacterium that can survive exposures of IR up to 5000 Gy with no loss of viability (Murray, 1992). Analysis of *D. radiodurans* response to IR showed that there are 4-10 copies of the genome per cell in this bacterium.
Following exposure to IR, the bacterium is able to repair the DNA in 12-24 hours by reconnecting some chromosome fragments by a process called single-stranded annealing followed by homologous recombination (In: Clark et al. 2009). As a result, the entire genome is restored. Indeed homologous recombination and micro-homology mediated DSB repair is also observed in eukaryotes. It is however desirable that a eukaryotic multicellular organism that shows increased resistance to ionizing radiation is identified to study mechanisms conferring radiation resistance or a better response to ionizing radiation. Table 1 shows the lethal doses of IR in different eukaryotes. It is evident from the table that insects show a much higher level of resistance to IR than mammals. Among the three insect species shown, *Braconidae* show highest resistance. However, the nature of response has not been investigated either genetically or cytogenetically. Although the lethal dose has not been determined, *Planococcus citri* seems to be as resistant as Braconidae. It may also be noted that the radiation doses investigated in case of this insect are much higher than *Drosophila.*

Species	Lethal Dose (in Gy)	Taxonomic group
Homo sapiens	10	Mammals
Mus musculus	12	Mammals
Daenio rerio	40	Fish
Drosophila melanogaster	640	Insects
Planococcus citri	1200*	Insects
Braconidae	1800	Insects
Caenorhabditis elegans	800*	Worms
Escherichia coli	60	Bacteria
Deinococcus radiodurans	15000	Bacteria
Thermococcus gammatolerans	30000	Archaebacteria

Table 1. Minimum lethal dose of IR in different organisms. * denotes that at the indicated dosage, the organism survives and reproduces. Information on lethal dose (expected to be higher) is not known.

4.1 Mealybug as a model for studying responses to IR

The mealybugs (Planococcus species) are lecanoid coccids (Coccoidea; Homoptera) whose high degree of resistance to ionizing radiation has been investigated at both the cytological and genetical levels (Chandra 1963a, b). After exposure to radiation doses as high as 1200 Gy (120,000 rads), male insects not only survive but mate and reproduce to produce viable progeny. Cytological examination of the progeny obtained from such paternal irradiation revealed very small chromosome fragments that appear to retain their individual chromosome identity such as replication, metaphase alignment, and anaphase segregation. No apparently significant loss of genetic material was observed. These results suggest that the mealybug genetic system possesses very efficient DNA repair machinery that ensures protection of the many tiny chromosome fragments produced by irradiation and prevents from fusing to one another via rapid healing of broken chromosome ends. Because of these extraordinary features we propose that mealybug is an attractive model system to study mechanisms underlying resistance to high doses of IR.

4.2 Telomeric repeats, their distribution and telomerase activity in unirradiated and irradiated mealybugs

Because telomeric repeat lengths are critical for double-strand DNA repair and shortening of telomeric repeats is associated with increased sensitivity to IR, in a recent study, we examined the nature and distribution of telomeric repeats in the mealybug *Planococcus* lilacinus (Mohan et al. 2011).

In order to isolate telomeric sequences from the mealybug, a polymerase chain reaction was conducted on genomic DNA using insect telomeric repeat (TTAGG)-specific primers (Fig.4A.). The PCR products obtained were cloned and sequenced to analyze the telomeric

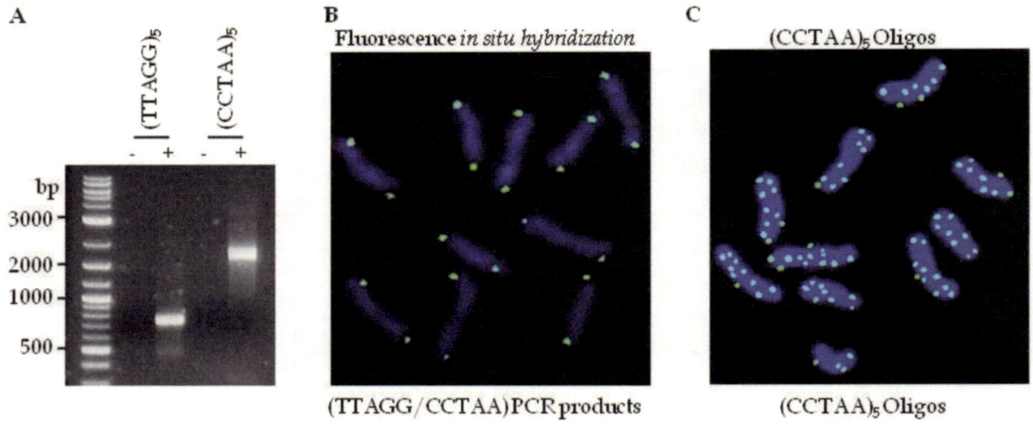

Fig. 4. Amplification and detection of TTAGG telomeric repeats in the mealybug Planococcus lilacinus. (A) Amplification of TTAGG sequences using $(TTAGG)_5$ and $(CCTAA)_5$ oligos. (B) Hybridization of metaphase chromosomes to a cloned TTAGG repeat sequence. The probe was labeled with digoxygenin (DIG) -UTP and hybridized to the chromosomes. Hybridization signals were detected using DIG-signal amplification kit. (C) Primed in situ hybridization of mealybug chromosomes with $(CCTAA)_5$ oligos. The oligos were hybridized to the chromosomes and an extension reaction was carried out with Taq DNA polymerase and DIG-labeling nucleotide mix. Signals were detected as in (B).

repeat-associated sequences. Two categories of sequences were identified: (1) Tandem repeats of TTAGG ranging from 100-200 repeats and (2) sequences containing 5 to 15 TTAGG repeats at the ends with a highly similar low-copy unrelated repeats of 400-450 bp. The unrelated low copy number repeats are interrupted by a variable number of TTAGG repeats. This sequence analysis suggested that whereas TTAGG repeats primarily occur at the ends of mealybug chromosomes, some of the repeats occur interstitially and are associated with certain low-copy repeats. Consistent with these two possibilities, FISH (fluorescent in situ hybridization) with a cloned TTAGG sequence (containing 96 TTAGG repeats) mainly hybridizes to the ends of mealybug chromosomes (Fig.4B) whereas primed in situ hybridization (PRINS) with TTAGG-specific oligos shows labeling at interstitial locations of the chromosomes (Fig.4C). Taken together, these results reveal a unique distribution of the telomeric repeats in the mealybug chromosome - they are distributed both terminally and interstitially. Because of this distribution, the interstitial telomeric repeats in this insect belong to the type I category in which short direct telomeric repeats interrupt precisely repetitive elements. Estimation by slot blot hybridization of different amounts of genomic DNA with $(TTAGG)_5$ probes suggest that the TTAGG repeats constitute ~0.03% of the mealybug genome (220 Mb; Mohan et al, 2002) and there are about 1275 TTAGG repeating units per each chromosome end. From these data, it may be noted that there is no significant difference in the length of telomeric repeats in the mealybug and other eukaryotic species and that resistance to high doses of IR is not due to unusual telomeric repeat lengths in this insect. To study the distribution of telomeric sequences in mealybug cells after exposure to IR, a series of experiments were performed in which gravid females (carrying embryos) were exposed to different amounts of IR, ranging from 30 Gy to 300 Gy. Whereas the number of chromosomes/fragments detectable under the microscope is relatively less at 30Gy (average number of fragments is about 14), at 300 Gy, the number of chromosome fragments is high and not estimable because of their tiny size. However at a radiation dose of 70 Gy, we were able to obtain an average chromosome/fragment number of 20 per metaphase, providing an adequate number of DSBs (about 10) per cell. This number of newly generated chromosome ends therefore provides a reasonable platform to study the distribution of TTAGG telomeric repeats in the context of IR and their terminal versus interstitial occurrence.

Analysis of chromosome fragments recovered after two days of exposure to a radiation dose of 70Gy by FISH showed that many of the chromosome ends induced by IR are labeled with the $(TTAGG)_{96}$ probe (Fig. 5A). Whereas in unirradiated chromosomes there are 20 labeled ends, in irradiated chromosomes with an average of 40 chromosome ends, 29 ends were labeled with the probe. It may be noted that at least some of the unlabeled chromosome ends may have extended TTAGG repeats but these repeats may not be long enough to be detectable by the FISH probe. Taken together, these results suggest that there is an efficient generation of TTAGG repeats at the recovered ends following IR.

To gain more insights into the origin of the telomeric signals in irradiated chromosomes, PRINS was performed on them. As shown in Fig. 5B, more than 95% of the ends of the irradiated chromosomes were labeled by PRINS. When the results in Fig. 5A and 5B are taken together, it is recognizable that there is an overlap in the signals between FISH and PRINS at half of the ends of irradiated chromosomes. It is also recognizable that there is a decrease in the number of interstitial signals in the irradiated chromosomes. Taken together, these results suggest that there is an extension of TTAGG repeats near the interstitial telomeric repeats that occur close to the DSBs induced by IR. It may be noted that there are no additional telomeric or centromeric probes available at present to further clarify on the involvement of interstitial telomeric repeats in the generation of FISH-detectable signals at the radiation-induced chromosome ends.

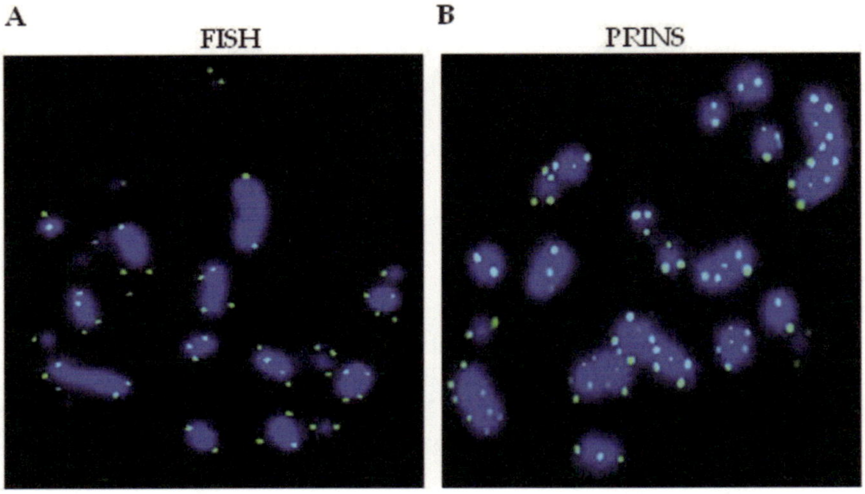

Fig. 5. Distribution of TTAGG telomeric repeat sequences in irradiated mealybug chromosomes. The chromosomes/chromosome fragments were subjected to FISH (A) and PRINS (B) as described in Fig.4.

4.3 Telomerase activity in unirradiated and irradiated mealybugs

For identification of components of machinery that maintains telomeric repeats in the mealybug, cell extracts were prepared from unirradiated and irradiated insects and telomerase assays were performed using TTAGG oligonucleotides as templates (Fig.6A and 6B). RNase-treated cell extracts were used as negative controls. As an internal size control, an end-labeled 30-bp oligonucleotide was used The results shown in Fig. 6A suggested that in both unirradiated and irradiated insects, cell extracts produced ladder-like extensions of the TTAGG oligonucleotides which is suggestive of telomerase activity both prior to and after irradiation. However, we observed approximately six times more telomerase activity in inirradiated insects than in the unirradiated insects. As one would expect for telomerase activity which is RNA-dependent, RNase-treated cell extracts did not produce detectable extensions. We also compared the levels of telomerase activity in insects exposed to different doses of ionizing radiation (30–300 Gy; Fig. 6B). Whereas there is an increase in the telomerase activity with an increase in the radiation dose up to 60 Gy, at higher levels there was a reduction in telomerase activity, and at 300 Gy, the telomerase activity approaches to the levels that are lower than that the unirradiated insects. Taken together, our results suggested that unlike in other organisms where telomerase activity ceases in adulthood, the mealybug telomerase is constitutively active. Such constitutive activity of telomerase was also reported in cockroaches, which are also radiation-resistant (Sasaki and Fujiwara 2000). It may be also noted that lower telomerase activity at doses >60 Gy highlights the possibility of occurrence of chromosome end-healing mechanisms that are telomerase-independent such as MMEJ-, NHEJ- and homologous recombination-mediated repairs (discussed in the section below).

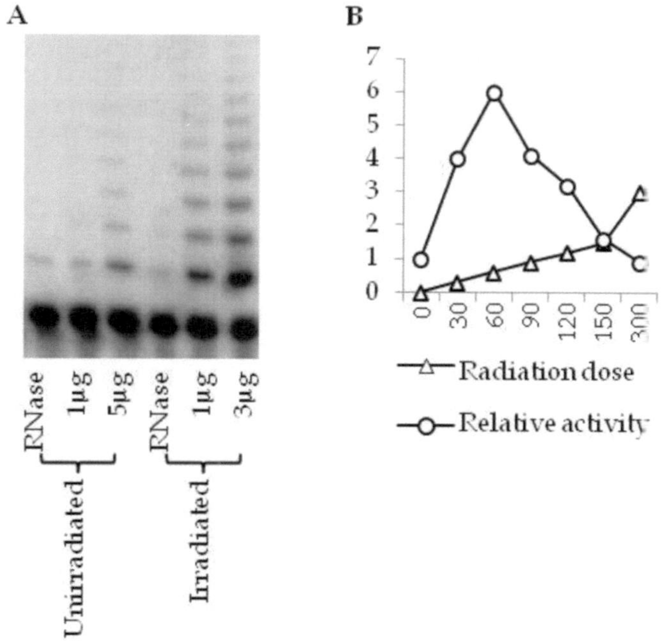

Fig. 6. Telomerase activity in mealybugs exposed to different doses of IR. (A). Telomerase activity in insects irradiated with a dose of 70 Gy is six-times the activity in unirradiated insects. (B) Progressive increase of telomerase activity was observed till 60 Gy. Exposure to higher doses was associated with a decrease in the level of activity.

5. Possible mechanisms of genomic stability in mealybugs exposed to high doses of ionizing radiation

From the results described here and the collection of prior data on the behavior of mealybug chromosomes after exposure to high doses of ionizing radiation, we suggest that there is highly efficient protection of radiation-induced chromosome ends that prevent the onset of fusion-bridge-breakage-fusion cycles and this machinery is excellently complemented by the centromere property in the mealybug chromosomes. As mentioned in the section 4.1, radiation-induced chromosome fragments which are very tiny and small, are capable of metaphase alignment and anaphase segregation. These results suggest that the centromere property is distributed throughout the chromosomes, or in other words the centromere is 'diffuse'. An obvious advantage of possessing diffuse centromere property is that there would not be a significant loss of genetic material among the collection of the chromosome fragments in the irradiated cells. These cells would be by and large viable and unless radiation-induced double strand breaks disrupt haplo-insufficient genes that are crucial for development or survival. Therefore, the combination of rapid chromosome end healing and diffuse centromere offer a great advantage to the mealybug in successful survival and reproduction upon exposure to high doses of IR.

What are the advantages in possessing constitutively active telomerase? As noted above, telomerase-mediated extension of telomeric repeats is one of the different mechanisms that operate in presumably all eukaryotes with the exception of Diptera. The existence of telomerase even before irradiation in mealybugs provides an opportunity to extend telomeric repeats whenever required for maintenance an optimal length of telomeric repeats (which is higher than the critical length) at newly broken ends of irradiated chromosomes. Such maintenance of optimal length creates a high probability to restore the original chromosome in the presence of a radiation-induced chromosome double-strand break. It may be noted that in cells containing critically short telomeres, the number of unprotected chromosome ends is higher and as a result restoration of the original chromosome is more unlikely and results in genomic instability. An additional advantage in efficient maintenance of optimal number of telomeric repeats is the availability of sufficient number of binding sites for the DSB repair proteins per cell. In this context, interstitial telomeric repeats when they occur at DSBs have the potential to serve as templates for formation of new repeats as part of protection of newly formed chromosome ends. An alternative mechanism is 'exposure' of the interstitial repeats akin to the MMEJ-mediated search for homology/micro homology by resection.

As discussed in section 4.3, telomerase activity decreases at doses higher than 60 Gy and at 300 Gy, the activity in irradiated insects is lesser than unirradiated insects. It is therefore clear the telomerase-dependent mechanism of end-healing may not be efficient at higher doses that mealybugs have been shown to survive. Therefore at these doses, chromosome end healing mechanisms that are telomerase-independent should be functional. These mechanisms are collectively called as alternative mechanisms of telomere maintenance (ALT). As mentioned previously, non-homologous end joining (NHEJ), microhomology-mediated end joining (MMEJ) and homologous recombination-mediated DNA repair are the three mechanisms which enable telomere maintenance. NHEJ and MMEJ, as mentioned before require microhomology between the damaged ends to be joined whereas homologous recombination requires a sequence homology from a homologous chromosome or a sister chromatid. In this context, it is conceivable that repeat sequences associated with the interstitial TTAGG repeats may play a role in one or more of the ALT pathways.

6. Conclusion

In summary, constitutively active telomerase, diffuse centromere and the unusual distribution of telomeric repeats together constitute components of efficient DNA repair machinery in the mealybug. In particular, interstitial telomeric repeats appear to be attractive candidate sequences that may help in the acquisition of additional telomeric repeats at DSBs either through telomerase-dependent or independent mechanisms. Direct cloning and analysis of healed chromosome ends induced by radiation and their comparison with 'intact' unirradiated homologs should shed more light on the role of interstitial telomeric sequences.

Future research on the isolation and analysis of telomerase encoding genes should provide important insights on regulation of telomerase and mechanisms underlying telomerase-dependent maintenance of telomeric repeats. A comparison of the DNA regulatory elements and analysis of the primary amino acid sequence should reveal any differences from the telomerase enzymes described in other insects. As in case of other eukaryotes, alternative lengthening of telomeres also appears to play an important role in the response to exposure to IR. In this context, isolation and analysis of isolation of genes involved in ALT pathways, their expression patterns and localization in cells following irradiation should provide more insights in the chromosome end-healing and DNA repair mechanisms in the mealybug. For a better understanding of the nature of biological response to IR, experiments such as mRNA expression profiling, protein profiling and protein modification profiling in insects before and after exposure to high doses of IR should provide more insights into the mechanisms conferring radiation resistance to the mealybug.

7. Acknowledgement

Research in KNM's lab is funded by the Department of Biotechnology, Government of India. We thank B. Sandya Rani and Pooja Swaroop Kulashreshta for FISH and PRINS studies, Prof. H. Sharat Chandra and Leonardo d'Aiuto for discussions. Part of the work discussed here (Figs. 4-6) was published in Chromosoma (vol. 120:165-175; please see the reference below).

8. References

Ahmed, S. & Hodgkin, J. (2000). MRT-2 checkpoint protein is required for germline immortality and telomere replication in C. elegans. Nature. (January 2000). 403(6766): 159-164.

BEIR VII report: Health Risks from Exposure to Low Levels of Ionizing Radiation in http://dels-old.nas.edu/dels/rpt_briefs/beir_vii_final.pdf

Biessmann, H.; Champion, L.E.; O'Hair, M.; Ikenaga, K.; Kasravi, B. & Mason, J.M. (1992). Frequent transpositions of Drosophila melanogaster HeT-A transposable elements to receding chromosome ends. EMBO J. (December 1992). 11(12):4459-4469.

Blackburn EH. (1991). Structure and function of telomeres. Nature. (April 1991). 350(6319): 569-573.

Boulton, S.J. & Jackson, S.P. (1998). Components of the Ku-dependent non-homologous end-joining pathway are involved in telomeric length maintenance and telomeric silencing. EMBO J. (March 1998). 17(6):1819-1828.

Bryan, T.M.; Englezou, A.; Gupta, J.; Bacchetti, S. & Reddel, R.R. (1995). Telomere elongation in immortal human cells without detectable telomerase activity. EMBO J. (September 1995). 14(17): 4240-4248.

Cardis, E.; Vrijheid, M.; Blettner, M.; et al. (2005). Risk of cancer after low doses of ionising radiation: retrospective cohort study in 15 countries. BMJ (July 2005) 331 (7508): 77

Cesare, A.J. & Reddel, R.R. (2010). Alternative lengthening of telomeres: models, mechanisms and implications. Nat Rev Genet. (May 2010). 11(5): 319-330.

Chandra, H.S. (1963a). Cytogenetic analysis following high dosage paternal irradiation in the mealybug Planococcus citri. I. Cytology of X1 embryos. Chromosoma (March 1963) 14: 310–329.

Chandra, H.S. (1963b). Cytogenetic analysis following high dosage paternal irradiation in the mealybug Planococcus citri. II. Cytology of X1 females and the problem of lecanoid sex determination. Chromosoma (March 1963). 14: 330–346.

Cheriyan, V.D.; Kurien, C.J.; Das, B.; Ramachandran, E.N.; Karuppasamy, C.V.; Thampi, M.V.; George, K.P.; Kesavan, P.C.; Koya, P.K. & Chauhan, P.S. (1999). Genetic monitoring of the human population from high-level natural radiation areas of Kerala on the southwest coast of India. II. Incidence of numerical and structural chromosomal aberrations in the lymphocytes of newborns. Radiat Res. (December 1999). 152(6 Suppl): S154-148.

Clark, D.P.; Dunlap, P.V.; Madigan, M.T.; Martinko, J.M. (2009). Brock Biology of Microorganisms. San Francisco: Pearson publishers. (2009). p281.

Cohen, S.B.; Graham, M.E.; Lovrecz, G.O.; Bache, N.; Robinson, P.J. & Reddel, R.R. (2007). Protein composition of catalytically active human telomerase from immortal cells. Science. (March 2007). 315(5820): 1850-1853.

Crompton, N.E. (1997). Telomeres, senescence and cellular radiation response. Cell Mol Life Sci. (July 1997). 53(7): 568-575.

de Lange, T. (2005) Shelterin: the protein complex that shapes and safeguards human telomeres. Genes Dev. (September 2005). 19(18):2100-2110. Review.

Dunham, M.A; Neumann, A.A; Fasching, C.L. & Reddel, R.R. (2000). Telomere maintenance by recombination in human cells. Nat Genet. (December 2000). 26(4): 447-450.

EMBO J. (October 1994). 13(20): 4953-4962.

Forster, L.; Forster, P.; Lutz-Bonengel, S.; Willkomm, H. & Brinkmann, B. (2002). Natural radioactivity and human mitochondrial DNA mutations. Proc Natl Acad Sci U S A. (October 2002) 99(21):13950-4.

Frydrychová, R.; Grossmann, P.; Trubac, P.; Vítková, M. & Marec F. (2004). Phylogenetic distribution of TTAGG telomeric repeats in insects. Genome. (February 2004). 47(1): 163-178.

Fujiwara, H.; Osanai, M.; Matsumoto, T. & Kojima KK. (2005) Telomere-specific non-LTR retrotransposons and telomere maintenance in the silkworm, Bombyx mori. Chromosome Res. (May 2005). 13(5): 455-467.

Goytisolo, F.A.; Samper, E.; Martín-Caballero, J.; Finnon, P.; Herrera, E; Flores, J.M.; Bouffler, S.D. & Blasco, M.A. (2000). Short telomeres result in organismal hypersensitivity to ionizing radiation in mammals. J Exp Med. (December 2000) 192(11):1625-1636.

Griffith, J.D.; Comeau, L.; Rosenfield, S.; Stansel, R.M.; Bianchi, A.; Moss, H. & de Lange T. (1999). Mammalian telomeres end in a large duplex loop. Cell. (May 1999). 14;97(4): 503-514.

Haber, J.E. (1998). The many interfaces of Mre11. Cell. (November 1998). 95(5): 583-586.

Hande, M.P.; Balajee, A.S.; Tchirkov, A.; Wynshaw-Boris, A. & Lansdorp, P.M. (2001). Extra-chromosomal telomeric DNA in cells from Atm(-/-) mice and patients with ataxia-telangiectasia. Hum Mol Genet. (March 2001). 10(5): 519-528.

International Commission on Radiological Protection, Recommendations: ICRP Publication No. 60, Pergamon Press, 1991, New York.

Jiang, W.Q.; Zhong, Z.H.; Henson, J.D.; Neumann, A.A.; Chang, A.C. & Reddel, R.R. (2005). Suppression of alternative lengthening of telomeres by Sp100-mediated sequestration of the MRE11/RAD50/NBS1 complex. Mol Cell Biol. (April 2005). 25(7): 2708-2721.

Kochupillai, N.; Verma, I.C; Grewal, M.S. & Ramalingaswami, V. (1976). Down's syndrome and related abnormalities in an area of high background radiation in coastal Kerala. Nature. (July 1976) 262(5563):60-61.

Kojima, K.K.; Kubo, Y. & Fujiwara, H. (2002) Complex and tandem repeat structure of subtelomeric regions in the Taiwan cricket, Teleogryllus taiwanemma. J Mol Evol. (April 2002) 54(4): 474-485.

Levis, R.W.; Ganesan, R.; Houtchens, K.; Tolar, L.A. & Sheen, F.M. (1993). Transposons in place of telomeric repeats at a Drosophila telomere. Cell. (December 1993) 17;75(6): 1083-1093.

Liang, F.; Romanienko, P.J.; Weaver, D.T.; Jeggo, P.A. & Jasin, M. (1996) Chromosomal double-strand break repair in Ku80-deficient cells. Proc Natl Acad Sci U S A. (August 1996). 93(17): 8929-8933.

Löpez, C.C.; Nielsen, L. & Edström, J.E. (1996) Terminal long tandem repeats in chromosomes form Chironomus pallidivittatus. Mol Cell Biol. (July 1996). 16(7): 3285-3290.

Lundblad, V. & Blackburn, E.H. (1993) An alternative pathway for yeast telomere maintenance rescues est1- senescence. Cell. (April 1993) 73(2): 347-360.

Martin, S.G.; Laroche, T.; Suka, N.; Grunstein, M. & Gasser, S.M. (1999). Relocalization of telomeric Ku and SIR proteins in response to DNA strand breaks in yeast. Cell. (May 1999) 97(5): 621-633.

Mason, J.M. & Biessmann H. (1995). The unusual telomeres of Drosophila. Trends Genet. (February 1995). 11(2): 58-62.

Maynard, S.; Schurman, S.H.; Harboe, C.; de Souza-Pinto, N.C. & Bohr, V.A. (2009). Base excision repair of oxidative DNA damage and association with cancer and aging. Carcinogenesis. (January 2009) 30(1):2-10.

McClintock, B. (1938). The Production of Homozygous Deficient Tissues with Mutant Characteristics by Means of the Aberrant Mitotic Behavior of Ring-Shaped Chromosomes. Genetics. (July 1938) 23(4):315-376.

Mills, K.D.; Sinclair, D.A. & Guarente, L. (1999) MEC1-dependent redistribution of the Sir3 silencing protein from telomeres to DNA double-strand breaks. Cell. (May 1999)97(5): 609-620.

Mohan, K.N.; Rani, B.S.; Kulashreshta, P.S. & Kadandale, J.S. (2011) Characterization of TTAGG telomeric repeats, their interstitial occurrence and constitutively active telomerase in the mealybug Planococcus lilacinus (Homoptera; Coccoidea). Chromosoma. (April 2011) 120(2): 165-175.

Moore, J.K. & Haber, J.E. (1996) Cell cycle and genetic requirements of two pathways of nonhomologous end-joining repair of double-strand breaks in Saccharomyces cerevisiae. Mol Cell Biol. (May 1996) 16(5):2164-2173.

Murnane, J.P.; Sabatier, L.; Marder, B.A. & Morgan, W.F. (1994). Telomere dynamics in an immortal human cell line.

Murray, R.G.E. (1992). The family Deino- coccaceae. In The Prokaryotes, Ed. A. Ballows, H.G. Truper, M. Dworkin, W. Harder amd K.H. Schleifer. Pp. 3732–44. New York: Springer-Verlag.

Okazaki, S.; Tsuchida, K.; Maekawa, H.; Ishikawa, H. & Fujiwara, H. (1993). Identification of a pentanucleotide telomeric sequence, (TTAGG)n, in the silkworm Bombyx mori and in other insects. Mol Cell Biol. (March 1993). 13(3):1424-1432.

Osanai, M.; Kojima, K.K.; Futahashi, R.; Yaguchi, S. & Fujiwara, H. (2006) Identification and characterization of the telomerase reverse transcriptase of Bombyx mori (silkworm) and Tribolium castaneum (flour beetle). Gene. (July 2006). 376(2) 281-289.

Pardue, M.L. & DeBaryshe, P.G. (1999). Telomeres and telomerase: more than the end of the line. Chromosoma. (May 1999). 108(2): 73-82.

Potts, P.R. & Yu, H. (2007). The SMC5/6 complex maintains telomere length in ALT cancer cells through SUMOylation of telomere-binding proteins. Nat Struct Mol Biol. (July 2007). 14(7): 581-590.

Reactor Concepts Manual, available from http://www.nrc.gov/reading-rm/basic-ref/teachers/09.pdf: Biological effects of radiation, UNRC technical training center.

Sahara, K.; Marec, F. & Traut, W. (1999). TTAGG telomeric repeats in chromosomes of some insects and other arthropods. Chromosome Res. (June 1999). 7(6): 449-460.

Sasaki, T. & Fujiwara, H. (2000). Detection and distribution patterns of telomerase activity in insects. Eur J Biochem. (May 2000). 267(10): 3025-3031.

Slijepcevic, P. & Al-Wahiby S. (2005). Telomere biology: integrating chromosomal end protection with DNA damage response. Chromosoma. 2005 Sep;114(4): 275-285.

Villasante, A.; de Pablos, B.; Méndez-Lago, M. & Abad, J.P. (2008) Telomere maintenance in Drosophila: rapid transposon evolution at chromosome ends. Cell Cycle. (July 2008) 7(14):2134-2138.

Vítková, M.; Král, J.; Traut, W.; Zrzavý, J. & Marec F. (2005). The evolutionary origin of insect telomeric repeats, (TTAGG)n. Chromosome Res. (February 2005). 13(2): 145-156.

Zhu, X.D.; Küster, B.; Mann, M.; Petrini, J.H. & de Lange T. (2000). Cell-cycle-regulated association of RAD50/MRE11/NBS1 with TRF2 and human telomeres. Nat Genet. (July 2000) 25(3): 347-352.

Part 2

Biological Process

7

Radiation-Sensitivity and Transcription Profiles in Various Mutant p53 Cells

Kumio Okaichi, Nanaka Izumi, Yuma Takamura,
Shoichi Fukui and Takashi Kudo
*Department of Radioisotope Medicine,
Atomic Bomb Disease and Hibakusha Medicine Unit,
Atomic Bomb Disease Institute,
Nagasaki University Graduate School of Biomedical Sciences, Nagasaki,
Japan*

1. Introduction

The tumor suppressor gene *p53* plays an important role in determining radiosensitivity. The normal *p53* gene product accumulates after exposure to ionizing radiation, and causes growth arrest or promotes cell death through the apoptosis pathway (Figure 1). Mutation of the *p53* gene is the most common genetic alteration observed in human cancers (Nigro et al. 1989). It has been widely reported that cells with mutant *p53* are more resistant to ionizing radiation or DNA-damaging agents (Fan et al. 1994; Wattel et al. 1994; Lee et al. 1993; Hamada et al. 1996). On the other hand, there have been reports of cells harboring mutant *p53* that are sensitive to ionizing radiation and anticancer drugs (Biard et al. 1994; Fan et al.

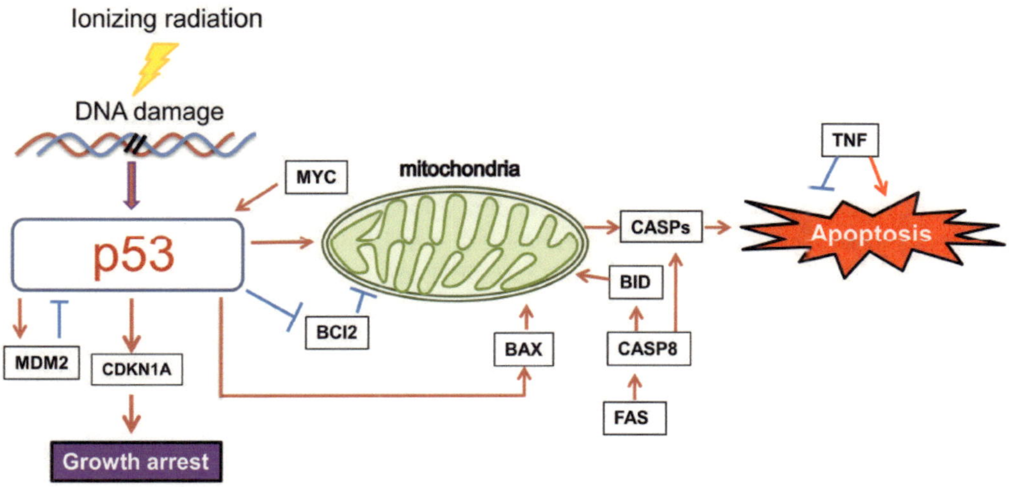

Fig. 1. Pathway of p53. From irradiation by ionizing radiation to growth arrest or apoptosis of the cell through p53.

1995), although specific details of the mutations were not discussed. Mutant forms of *p53* differ in their properties according to the points of the mutation. For example, Crook et al. using a large series of *p53* mutants, found that not all transcriptionally active mutants retained the ability to suppress transformation, and that some tumor-derived point mutations conferred both transforming and transactivating activity (Crook et al. 1994). Some mutant forms of the *p53* gene do not merely induce the functional equivalent of *p53* loss (Harvey et al. 1995). The radiosensitivity of cells may depend on the type of *p53* mutation they harbor. It is important to determine which mutations affect the radiosensitivity of tumor cells, because tumor cell radiosensitivity has substantial clinical relevance in the context of tumor radiotherapy.

2. Mutation in *p53*

We prepared 15 types of cells harboring mutant forms of *p53* (T123A, L130V, Q143A, V157F, H168R, R175H, I195T, C242F, G244C, G245S, R273H, C277F, R280T, R282W and E286K) to examine their radiosensitivity. First, we created various mutations in the *p53* gene using a QuikChange site-directed mutagenesis kit (Stratagene, La Jolla, CA) in accordance with the manufacturer's protocol and integrated them into the LacSwitch inducible mammalian expression system (Stratagene, La Jolla, CA) (Okaichi et al. 1999). These mutant *p53* genes were then transformed into the Saos-2 cell line, which is null for *p53*, and stable transformants were obtained.

3. Radiation sensitivity of cells harboring *p53* mutation

Radiation sensitivity may depend on the position at which mutation occurs in *p53*. We previously examined various *p53* mutants for radiation sensitivity (Okaichi et al. 2008). Cells were subjected to γ-ray irradiation and then plated onto dishes. Colonies were examined after about one month to calculate the surviving fraction. The cells with wild-type *p53* showed higher radiation sensitivity than Saos-2 (*p53*-null) cells. Some mutations also resulted in increased radiation sensitivity, but mutations including hot spot mutations (175H, 245S 273H and 282W) showed almost no alteration of radiation sensitivity compared with Saos-2. Other mutations conferred an intermediate level of radiation sensitivity (Okaichi et al. 2008).

We then compared the radiosensitivity of these mutants with the frequency of mutation at each point, which is correlated with the tendency for tumorigenesis. Figure 2 shows the relationship between the frequency of *p53* mutation in human cells and radiosensitivity of the *p53* mutants. We divided these mutants into three groups; R (resistant), M (medium) and S (sensitive). The 175H, 244C, 245S 273H and 282W transformants were placed in group R, which was radioresistant and included a high frequency of mutation at all hot spots. The 130V, 143A, 168R, 277F and 286K transformants were placed in group M, which showed medium radiosensitivity and a low frequency of mutation. The 123A, 157F, 195T, 242F and 280T transformants were placed in group S, which was radiosensitive and showed a relatively low frequency of mutation.

As the radiosensitivity of these cells may be related to the induced expression of various genes by each type of *p53* mutation, we investigated the genes whose expression appeared to be related to the radiosensitivity of cells bearing *p53* mutations.

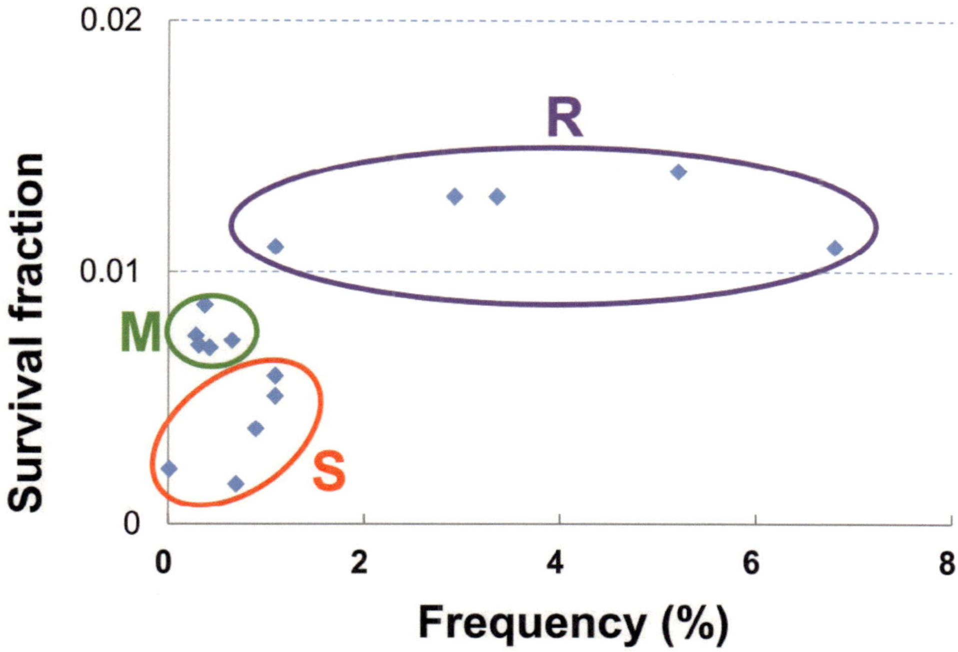

Fig. 2. Relationship between radiosensitivity of *p53* mutant cell lines and the frequency of *p53* mutation. The mutant cell lines were divided into three groups; R (radioresistant), M (medium radiosensitivity) and S (radiosensitive).

4. Transcriptional control in cells harboring *p53* mutation

We examined the expression of genes in cells harboring *p53* mutation that were subjected to γ-ray irradiation. For this we employed a DNA microarray (Gene Chip, Human Genome U133 Plus 2.0 Array; Affymetrix, Santa Clara, CA), containing over 54,000 probe sets, in accordance with the manufacturer's instructions. We extracted mRNA from each cell type 24 hours after irradiation at 6 Gy, and synthesized the cDNA. After we had synthesized, in turn, cRNA from the cDNA, we labeled the former with biotin and hybridized it with the DNA microarray. After staining and washing, we read the fluorescence using a scanner. The expression value (signal) of each gene was calculated and normalized using GeneSpring (Agilent Technologies, Santa Clara, CA) to adjust for minor differences between the experiments. In order to obtain the mean basal expression level of each gene, the signal values for unirradiated Saos-2 cells were used as the standard for the analysis. The change in value (signal log ratio) for each gene was calculated using Comparison Analysis in the software.

As radiosensitivity is intrinsically related to the apoptosis pathway, we summarized the genes associated with apoptosis, and these are shown in Table 1, where the cells harboring mutant *p53* cells are arranged from left to right according to their radiosensitivity, the cells harboring wild-type *p53* are located on the far left, and the parent cells, Saos-2, on the far right. We picked up genes showing an increase in gene expression of more than 2-fold, and indicated them by colored column. After irradiation, the cells with wild-type *p53* showed more than a 2-fold increase in the expression of 15 genes, whereas Saos-2 did not show any increase in the expression of apoptosis-related genes. Cells harboring mutant *p53* lacked expression of many genes that were induced in cells harboring wild-type *p53*, but showed induction of some genes that were not induced in the latter. The 245S mutant cell line showed a particularly marked increase in the expression of many TNF-associated genes upon irradiation. As the expression of TNF-associated genes inhibits apoptosis, this would explain the radioresistance of 245S cells. We also noticed that the expression of TNF-associated apoptosis-inducing genes, such as TNFSF9 (0.15), TNFSF10 (0.18) and TNFSF21 (0.22), was decreased by more than half, in the 245S mutant cell lines. We were unable to explain the radiosensitivity of other mutant cell lines upon induction of apoptosis-related genes.

		Sensitive					Medium					Resistant					
	Wild	195	123	242	157	280	277	168	286	130	143	244	273	245	282	175	Saos
CRADD	24.85	18.45	9.56	63.18	13.73	19.03	12.01	6.61	21.15	11.02	6.72	18.42	7.45	1.58	37.82	13.27	2.23
NFKBIA	4.35	2.95	2.86	1.73	2.54	4.72	4.92	3.33	6.15	2.52	1.87	2.54	3.44	0.82	0.88	4.21	2.43
PIK3R1	4.07	2.59	2.37	3.95	2.84	2.65	1.90	4.69	1.27	1.77	4.31	2.92	4.69	1.06	1.04	4.30	2.96
CASP1	27.65	87.42	9.50	0.52	0.37	8.97	1.51	1.29	4.27	3.97	4.75	5.19	2.04	0.49	9.59	4.81	1.47
CFLAR	2.89	2.00	1.51	3.52	3.07	1.47	1.82	2.39	1.51	1.69	2.89	2.58	2.51	3.71	1.81	1.48	2.29
IGF1R	3.51	4.16	1.52	5.86	4.24	2.77	1.81	3.83	0.92	1.70	3.94	3.39	3.19	1.04	0.89	1.79	2.26
MDM2	15.55	2.78	1.27	2.99	3.35	0.59	1.58	4.89	0.40	0.77	3.56	0.98	0.47	1.88	6.68	4.16	4.62
BCL2L1	6.52	2.62	0.90	3.44	0.70	0.69	1.76	1.03	3.58	2.11	1.44	5.89	0.62	2.73	6.90	0.84	2.04
CASP2	27.03	0.74	0.92	4.57	3.13	0.28	0.65	5.00	0.51	0.60	3.48	4.50	1.64	0.69	4.99	0.59	0.38
IKBKB	3.39	2.51	0.31	1.99	1.13	0.51	0.95	0.88	3.14	0.96	1.07	3.22	1.06	0.20	3.66	0.55	1.14
MCL1	3.24	2.09	0.56	2.28	0.95	0.55	1.10	1.13	1.22	1.72	1.86	2.50	1.96	0.51	2.41	1.20	1.37
FAS	2.16	0.44	3.79	0.36	1.02	0.19	0.51	0.90	0.33	0.70	2.07	0.44	1.08	1.49	0.45	2.29	1.10
TNFRSF10B	4.66	1.09	0.59	2.02	0.46	0.86	0.74	0.78	1.74	1.39	0.97	1.27	0.52	2.43	1.82	1.08	0.59
BIRC4	2.96	2.43	0.34	1.31	0.91	0.61	1.11	0.56	0.58	0.61	0.75	1.02	0.79	0.73	0.76	0.69	0.69
BAX	3.47	0.78	0.47	1.28	0.94	0.26	0.83	0.98	0.85	0.79	1.43	1.30	1.23	0.40	1.32	0.89	1.17
CASP4	0.21	1.35	0.56	3.79	1.33	1.97	0.87	7.81	0.97	13.87	9.53	0.72	2.73	1.24	2.63	0.44	0.98
BCL2L2	1.55	2.08	0.79	0.78	0.83	1.25	0.78	1.72	1.14	0.83	2.21	1.08	1.12	0.49	1.05	1.11	1.06
BCL2L11	0.74	0.39	0.95	1.15	1.28	0.43	0.34	1.12	2.01	0.63	1.30	1.19	1.75	0.95	2.42	1.22	1.26
MYC	0.34	0.53	0.39	0.21	0.25	0.34	0.44	0.71	0.12	1.95	0.77	0.22	0.45	2.04	0.10	0.41	1.19
NFKBIE	0.88	1.07	1.16	0.61	0.68	0.78	0.89	1.15	0.59	0.83	1.24	0.61	0.69	5.08	0.65	1.40	0.81
TNFRSF1A	1.63	0.84	0.85	0.67	0.64	0.66	0.94	0.55	0.76	0.61	0.61	0.85	0.84	2.39	0.54	0.61	0.85
TNFRSF25	0.32	0.06	1.13	1.03	1.03	0.16	0.91	1.27	0.20	2.49	1.03	1.53	0.98	5.92	2.29	0.37	0.25
TRAF2	0.75	0.88	0.96	1.36	0.62	0.81	0.57	1.03	0.98	0.58	0.99	0.85	1.28	2.31	0.98	0.99	1.08
TRAF3	1.14	1.60	1.06	1.03	0.72	0.79	0.92	0.55	0.94	0.77	0.98	0.87	0.67	2.03	1.31	1.28	1.01
BNIP3L	1.23	0.80	0.49	0.81	0.35	0.52	1.40	0.32	0.98	0.78	0.40	1.01	0.42	4.75	0.83	0.49	1.02
BID	1.09	0.79	1.18	0.66	0.80	0.62	1.43	1.77	1.19	1.83	2.01	0.69	0.77	0.36	0.85	1.32	0.92
TNFSF10	1.01	0.20	1.67	0.11	0.76	0.90	0.71	0.17	0.01	0.01	1.29	0.12	0.10	0.02	0.02	2.86	0.76

Table 1. Induction of gene expression in the apoptosis pathway by irradiation at 6 Gy. We listed the apoptosis genes whose expression was increased more than 2-fold in mutant cells. The numbers indicate the gene expression value in comparison with unirradiated Saos-2 cells. The colored columns indicate more than a 2-fold increase.

We speculated that certain genes might play an important role in making some cells radioresistant. In this connection, we listed those genes whose expression was increased more than 2-fold in radioresistant cells. Table 2 shows a list of genes whose expression was increased in more than 4 of the mutant strain of radioresistant cells. We paid attention to the level of gene expression in the radiosensitive mutant cells, because genes that play an important role in conferring radioresistance would be show lower levels of expression in radiosensitive cells. Expression of the genes CADPS2, DNPEP, NKTR, OVOS2, PSENEN, RASSF4, RBM14 and WTAP was not increased more than 2-fold in almost all of the radiosensitive cells. CADPS2 acts as a calcium sensor in constitutive vesicle trafficking and secretion (Cisternas et al. 2003), and DNPEP is an aspartyl aminopeptidase that catalyzes the sequential removal of amino acids from the unblocked N termini of peptides and proteins (Nakamura et al. 2011). NKTR plays an important role in NK-cell cytotoxicity (Anderson et al. 1993). OVOS2 is a member of the ovostatin family and possesses trypsin-inhibitory activity (Saxxena and Tayyab, 1997). PSENEN (Presenilin enhancer-2) is a component of the γ-secretase complex which catalyzes the final cleavage of amyloid precursor protein to generate the toxic amiloid β protein, the major component of plaques in the brain of Alzheimer disease patients, and protects embryos from apoptosis (Zetterberg et al. 2006). RASSF4 binds directly to activated K-Ras in a GTP-dependent manner via the effector domain, thus exhibits the basic nature of a Ras effector and plays an important role in Ras-dependent apoptosis (Eckfeld et al. 2004). RBM14 (CoAA) is a nuclear receptor coactivator protein at the interface of transcriptional coactivation and RNA splicing (Auboeuf et al. 2004). WTAP (Wilms' tumor 1-associating protein) is essential for embryonic development, and appears to exert an antiproliferative effect, inhibiting G_1-to-S phase cell cycle transition and also promoting apoptosis (Small et al. 2007).

		Sensitive					Medium					Resistant					
	Wild	195	123	242	157	280	277	168	286	130	143	244	273	245	282	175	Saos
ARHGDIA	22.6	1.78	2.33	5.98	0.91	1.93	4.59	3.27	6.46	0.76	4.02	31.3	2.02	1.79	11.3	3.56	11.2
ATP6V0E1	1.39	1.84	1.72	2.43	1.49	2.22	1.77	1.84	1.86	1.58	1.64	3.53	2.03	[illegible].24	1.23	2.02	2.54
C9orf86	3.04	1.49	2.00	2.57	1.79	2.97	2.34	1.84	4.49	0.61	1.73	3.79	2.07	[illegible].47	3.56	2.16	2.10
CADPS2	0.05	0.73	2.6	0.21	0.06	0.18	0.21	1.03	0.61	1.63	1.62	1.20	4.49	4.01	2.74	3.40	1.40
CUEDC1	4.57	0.70	1.82	2.22	2.26	1.67	2.63	2.34	3.00	1.12	2.54	4.73	5.14	2.60	1.99	3.47	3.00
DNPEP	1.78	2.91	1.52	1.98	1.55	1.19	3.02	3.62	1.79	4.78	3.19	3.33	2.61	1.38	2.73	3.50	2.62
EIF4G1	3.70	2.07	1.20	2.50	1.44	1.32	1.99	2.45	3.30	1.08	2.72	3.38	2.01	[illegible].77	3.26	2.20	3.87
FAM100B	2.79	2.95	1.37	2.51	1.77	1.79	2.17	1.72	3.16	2.48	1.56	3.06	2.43	[illegible].60	5.01	1.72	2.37
FAM129B	4.89	3.13	1.92	1.71	1.63	2.02	3.48	3.6	5.18	1.21	2.49	4.32	2.02	4.07	1.98	2.86	3.83
FUS	0.75	1.84	1.34	1.35	2.86	2.04	2.10	2.05	0.53	1.70	2.62	3.39	2.82	[illegible].38	0.93	2.73	1.37
HERC6	15.2	3.56	1.55	4.15	1.39	1.71	2.29	2.30	2.39	2.27	3.26	2.98	2.11	8.37	1.86	2.35	1.09
HOMER3	5.32	5.82	1.41	3.90	1.97	1.83	3.51	1.80	7.16	1.44	2.05	6.08	2.48	[illegible].82	9.04	2.11	1.83
HSF1	9.23	2.53	1.80	3.01	1.91	1.60	2.72	2.77	4.01	0.99	3.40	6.41	2.09	2.70	7.47	2.07	4.29
LDLR	4.34	2.4	0.32	2.15	0.21	0.95	3.78	0.24	1.60	0.07	0.60	4.36	2.66	[illegible].58	3.31	0.34	3.60
MYEF2	2.90	1.32	1.99	5.60	1.44	3.09	3.25	2.19	3.28	0.92	2.00	8.72	7.58	6.93	4.75	3.25	2.99
NKTR	0.66	1.64	1.54	2.65	1.68	1.85	2.07	1.85	1.04	2.00	1.58	2.12	3.96	2.70	1.12	2.59	1.34
OVOS2	0.56	0.80	0.65	1.96	1.24	0.33	0.19	1.87	3.03	4.3	2.32	2.01	2.83	7.44	2.93	0.27	1.04
PSENEN	4.48	1.44	2.74	1.99	1.41	1.86	1.68	2.61	2.94	1.24	2.65	2.29	2.76	[illegible].50	1.94	3.08	2.86
RASSF4	0.53	0.47	1.13	11.9	0.48	0.58	0.85	1.49	33.1	0.25	1.85	13.9	0.27	2.80	42.7	2.93	0.57
RBM14	10.0	1.35	1.26	3.90	1.81	0.76	1.52	2.42	6.92	0.99	3.61	7.01	2.60	[illegible].93	7.75	2.09	4.50
RHOU	0.34	0.36	4.82	4.11	1.00	0.92	0.42	2.66	1.13	2.15	5.27	3.48	3.39	5.72	1.92	7.63	1.32
SIRT6	2.24	1.71	3.02	1.73	1.33	2.38	1.37	1.82	2.48	1.29	1.87	2.14	2.14	2.37	1.53	3.07	1.96
SPAG9	2.13	1.73	1.03	2.41	3.38	1.72	2.04	3.19	1.12	1.46	3.33	2.14	2.64	7.77	1.14	2.09	2.69
TNRC6B	1.41	1.51	1.96	4.74	1.83	3.66	1.54	3.09	0.86	1.09	3.07	2.08	3.72	3.91	1.14	2.38	2.63
WTAP	0.79	0.80	3.32	1.97	1.89	1.68	2.46	2.02	0.99	3.93	3.15	2.99	2.43	2.69	2.27	4.07	1.71

Table 2. A list of the genes induced in radioresistant cells by irradiation at 6 Gy. We listed the genes that showed more than a 2-fold increase in almost all of the radioresistant cells. The numbers indicate the gene expression value in comparison with unirradiated Saos-2 cells. The colored columns indicate more than a 2-fold increase.

Among these genes, RASSF4 and WTAP are related to apoptosis, but exert a negative effect on radioresistant. As PSENEN blocks apoptosis, this gene may play an important role in radioresistance.
We approached this issue from the opposite perspective, and searched for genes that played an important role in conferring radiosensitivity. We listed genes showing more than a 2-fold increase in expression in radiosensitive cells. Table 3 shows a list of genes whose expression was increased in more than 4 of the radiosensitive mutant cell lines. Expression of CBR4, FOXP1, KPNA3, MFAP5, NEK3, TRIM2 and TRIM38 was not increased more than 2-fold in almost of all radioresistant cell lines. CBR4 (carbonyl reductase 4) is a mitochondrial NADPH-dependent quinine reductase that may be involved in the induction of apoptosis by cytotoxic 9, 10-phenanthrenequinone (Endo et al. 2008). FOXP1 is a forkhead transcription factor with functions in tissue and cell-type specific gene expression, and its gene is a direct target of p53-induced microRNA miR-34a (Rao et al. 2010). KPNA

		Sensitive					Medium					Resistant					
	Wild	195	123	242	157	280	277	168	286	130	143	244	273	245	282	175	Saos
ARNT	4.25	2.82	1.85	2.18	2.33	2.36	1.20	2.04	0.80	0.99	1.66	1.18	2.82	1.02	1.17	2.26	1.64
CBR4	0.87	2.09	2.69	1.56	2.43	2.64	1.99	2.72	1.04	2.80	2.33	1.07	1.87	1.94	1.16	2.12	1.07
CHD2	2.21	2.60	1.44	2.51	3.42	2.16	1.23	2.51	0.45	1.65	3.04	2.13	3.46	1.83	0.70	1.95	0.98
CRIM1	1.82	2.80	0.87	2.30	2.08	2.17	1.98	1.59	0.47	0.77	1.47	2.13	0.87	0.94	0.42	2.13	2.02
CSNK1A1	0.83	2.28	2.21	1.73	3.25	2.88	1.70	2.08	1.27	3.78	2.56	1.86	3.01	3.02	1.55	1.46	2.25
EPDR1	0.79	2.66	2.87	2.08	3.57	3.49	3.79	3.57	0.24	4.30	3.62	1.56	4.28	1.95	1.48	3.52	1.63
FOXP1	0.93	3.57	2.67	2.51	4.88	1.36	1.78	4.94	0.72	2.84	5.96	1.64	3.03	0.13	0.98	1.86	1.92
GNAS	0.90	2.94	1.40	2.45	3.81	2.22	1.95	2.60	1.89	5.39	2.46	2.76	2.47	1.16	1.94	1.89	2.10
IGF1R	3.51	4.16	1.52	5.86	4.24	2.77	1.81	3.83	0.92	1.70	3.94	3.39	3.19	1.04	0.89	1.79	2.26
KPNA3	1.24	2.92	0.79	2.07	2.69	2.07	1.84	1.45	0.66	0.96	1.80	1.29	1.59	2.03	0.92	1.20	2.06
KRAS	1.92	3.26	2.40	2.24	3.27	2.01	2.07	4.08	1.37	2.41	4.41	1.51	2.81	3.15	0.87	1.60	2.20
LYST	3.47	5.79	1.81	2.26	3.05	3.19	3.30	3.53	1.83	1.73	2.83	1.77	2.88	1.16	1.23	2.61	1.75
MARCKS	1.19	3.60	2.90	1.43	3.59	2.70	1.44	3.94	1.35	0.79	3.62	1.08	3.48	0.61	0.48	2.19	2.46
MFAP5	4.26	11.1	0.16	12.6	7.07	3.13	1.70	0.84	0.36	0.22	0.42	11.2	0.01	0.02	0.62	0.06	1.12
NBPF1	1.59	3.03	2.22	2.51	3.43	1.94	1.75	2.64	1.28	2.39	2.09	1.92	3.07	1.00	1.23	2.01	2.38
NBPF10	1.33	2.68	2.78	2.02	3.11	2.82	2.19	1.78	0.62	2.26	1.54	1.86	2.39	1.79	1.57	2.77	1.99
NEK3	1.37	2.93	1.23	2.10	2.45	2.84	1.72	2.64	0.89	0.90	1.99	2.36	1.83	0.52	1.06	0.89	1.39
PGF	1.11	1.94	2.53	2.02	3.22	3.12	1.37	2.46	0.88	2.05	2.65	1.51	1.99	2.67	1.46	3.45	1.68
PPP1R12A	2.40	2.32	2.11	2.32	2.83	3.16	2.24	3.53	1.23	1.24	3.84	1.65	2.68	1.77	1.13	2.63	1.92
QKI	2.33	3.30	2.96	1.69	5.23	2.53	1.83	3.39	1.94	9.80	3.07	1.85	3.27	0.91	1.94	2.54	2.50
RASAL2	4.52	5.9	1.16	2.73	2.79	2.44	2.89	2.19	1.61	0.89	2.36	3.06	3.16	0.89	1.32	1.72	2.14
SFRS11	2.01	3.99	3.54	1.96	7.18	2.86	1.47	4.95	0.53	1.96	4.51	1.98	5.37	1.31	0.95	6.45	2.95
SLC35E1	1.10	1.99	2.50	2.26	2.18	2.48	1.08	1.97	0.87	1.96	2.08	1.72	1.80	2.21	1.18	2.69	1.24
SPG21	1.35	2.13	3.53	2.21	3.25	3.73	1.93	2.72	1.16	2.61	1.97	1.60	1.88	2.43	0.84	3.25	1.62
SUMO2	1.72	2.41	2.55	1.87	2.28	2.77	1.54	2.10	1.63	1.30	2.53	2.30	1.96	1.88	1.35	2.83	1.64
TMEM165	0.99	2.51	3.18	3.07	3.07	2.5	1.76	4.25	0.7	5.08	3.87	1.39	4.51	1.51	1.68	3.62	2.75
TRIM13	2.78	3.09	1.61	3.27	3.82	3.22	2.02	3.22	1.12	0.99	2.12	3.26	1.88	3.99	1.20	1.42	0.92
TRIM2	1.67	3.04	0.47	2.53	2.05	2.13	4.37	0.82	0.73	1.14	0.57	1.23	1.36	0.06	0.49	0.42	2.32
TRIM38	1.87	2.15	1.71	2.45	3.57	2.54	1.51	2.75	1.03	3.26	2.86	1.95	1.59	2.92	1.75	1.78	1.93
TSPAN9	1.76	2.27	2.33	3.34	2.28	1.30	1.35	3.33	2.14	2.69	2.81	2.50	3.27	0.77	1.59	1.76	1.62
VGLL4	1.45	2.44	2.44	2.63	4.58	1.83	1.86	4.86	1.41	2.20	5.39	1.38	3.37	1.54	0.97	2.54	2.55
VPS13B	2.65	3.85	1.25	2.17	2.95	3.62	2.34	2.30	1.45	1.13	2.21	1.99	3.33	9.85	1.48	1.79	2.64
WNK1	1.69	2.33	2.48	2.03	3.01	1.00	1.29	2.98	1.22	1.99	3.21	2.22	1.72	4.99	1.52	1.84	1.34
WSB2	2.73	4.50	1.59	3.16	2.09	2.48	1.38	3.77	1.40	0.89	3.74	2.20	3.78	1.47	0.70	1.52	1.55
ZNF818	2.88	3.65	1.08	2.95	3.32	2.20	1.34	2.85	1.92	0.35	1.01	2.47	1.29	1.26	2.00	1.65	2.37

Table 3. A list of the genes induced in radiosensitive cells by irradiation at 6 Gy.
We listed the genes that showed more than a 2-fold increase of expression in almost all of the radiosensitive cells. The numbers indicate the gene expression value in comparison with unirradiated Saos-2 cells. The colored columns indicate more than a 2-fold increase.

(karyopherin-alpha) proteins are responsible for the transport of proteins into and out of the nucleus through the nuclear pore complex, and KPUNA3 contributes genetically to schizophrenia (Wei and Hemmings 2005). MFAP5 (microfibrillar associated protein 5), also known as a microfibril-associated protein (MAGP2), is a highly significant indicator of survival and chemosensitivity of the cells (Spivey and Banyard, 2010). NEK3 is a serine/threonine kinase that contributes to PRL-mediated breast cell cancer motility through mechanisms involving Rac1 activation and paxillin phosphorylation (Miller et al. 2007). TRIM (tripartite motif-containing) proteins are a family comprising more than 70 members in humans and contain conserved RING, G-box, coiled-coil, and SPRY domains, most of which are involved in protein ubiquitination, but only a few of them have been well studied. TRIM2 mediates the p42/p44 MAPK-dependent ubiquitination of Bim (Bcl-2-interacting mediator of cell death) in rapid ischemic tolerance, and suppression of TRIM2 expression stabilizes the level of Bim protein and blocks neuroprotection (Thompson et al. 2011). TRIM38 has E3 ubiquitin ligase activity and can be degraded during virus infection (Liu et al. 2011).

As CBR4 is involved in the induction of apoptosis, this gene may play an important role in radiosensitivity. However, the precise role of these genes in radiosensitivity remains unknown.

We have attempted to perform hierarchical clustering analysis of RNA expression in these mutant *p53* cell lines using Gene Tree software, but were unable to find any clear relationship between radiosensitivity and gene expression.

5. Conclusions

Ionizing radiation is used extensively in medical diagnostic and treatment protocols. With a better understanding of radiation induced molecular processes, it might become possible to identify the radiosensitivity of individuals before the start of radiation therapy, leading to individualization of radiation treatment. Radiation-induced transcriptional responses have been studied using DNA microarray (Kis et al. 2006; Jen and Cheung, 2006). Some previous studies have also examined cells harboring mutant *p53* using DNA microarray (Amandson et al. 2003; Scian et al. 2004), but they did not examine each type of mutation.

In the present study, we prepared 15 mutant *p53* cell lines, cells harboring wild-type *p53* and Saos-2 cells (null for *p53*). We examined the radiosensitivity of the mutant cell lines and classified them as R (resistant), M (medium) or S (sensitive). We then studied the radiation-induced transcriptional responses in these cell lines, and examined the relationship between their radiation-induced gene expression and radiosensitivity. We found some genes that appeared to have some correlation with radiosensitivity, for example PSENEN and CBR4. However, none of the genes directly determined the radiosensitivity of the cells. Further study will be needed to determine which of these genes is the main determinant of radiosensitivity.

Radiosensitivity may be determined by several genes working in collaboration. Mutation of *p53* leads not only to loss of function, but also gain of function. If such functions are related to growth arrest or DNA repair, then loss of function would confer radiosensitivity, and gain of function to radioresistance. On the other hand, if such functions are related to apoptosis, then loss of function would confer radioresistance and gain of function to radiosensitivity. Each mutation of *p53* may thus lead to loss of function and gain of some other function at the same time. This makes it very difficult to determine whether a certain mutation of *p53* leads to the radiosensitivity on the basis of transcriptional analysis alone. Recently it has been reported that many kinds of microRNAs related to tumorigenesis or apoptosis are regulated by *p53* (He et al. 2007; Suzuki et al. 2009). Thus p53 regulates not only mRNA but also microRNA. The regulation of microRNA in each mutant *p53* cell line would vary the degree of cell radiosensitivity. The available data suggest the importance of determining the type of mutation of *p53* and examining the regulation of overall transcriptional control in individual tumor cells in the context of radiotherapy.

6. Acknowledgments

This work was partly supported by the Global Center of Excellence (GCOE) of the Ministry of Education, Culture, Sports, Science and Technology of Japan.

7. References

[1] Amundson SA, Bittner M and Fornace Jr JF (2003) Functional genomics as a window on radiation stress signaling. Oncogene *22*: 5828-5833.

[2] Anderson SK, Gallinger S, Roder J, Frey J, Young HA and Ortaldo JR (1993) A cyclophilin-related protein involved in the function of natural killer cells. Proc Natl Acad Sci USA *90*: 542-546.

[3] Auboeuf D, Dowhan DH, Li X, Larkin K, Ko L, Berget SM and O'Malley BW (2004) CoAA, a nuclear receptor coactivator protein at the interface of transcriptional coactivation and RNA splicing. Mol Cell Biol *24*: 442-453.

[4] Biard DS, Martin M, Rhun YL, Duthu A, Lefaix JL, May E and May P (1994) Concomitant *p53* gene mutation and increased radiosensitivity in rat lung embryo epithelial cells during neoplastic development. Cancer Res *54*: 3361-3364.

[5] Cisternas FA, Vincent JB, Scherer SW and Ray PN (2003) Cloning and characterization of human CADPS and CADPS2, new members of the Ca^{2+}-dependent activator for secretion protein family. Genomics *81*: 279-291.

[6] Crook T, Marston N J, Sara EA and Vousden KH (1994) Transcriptional activation by p53 correlates with suppression of growth but not transformation. Cell *79*: 817-827.

[7] Eckfeld K, Hesson L, Vos MD, Bieche I, Latif F and Clark GJ (2004) RASSF4/AD037 is a potential Ras effector/tumor suppressor of the RASSF family. Cancer Res *64*: 8688-8693.

[8] Endo S, Matsunaga T, Kitade Y, Ohno S, Tajima K, El-Kabbani O and Hara A (2008) Human carbonyl reductase 4 is a mitochondrial NADPH-dependent quinine reductase. BBRC *377*: 1326-1330.

[9] Fan S, el-Deiry WS, Bae I, Freeman , Jondle D, Bhatia K, Fornace AJ, Jr Magrath I, Kohn KW and O'Connor PM (1994) *p53* gene mutations are associated with decreased sensitivity of human lymphoma cells to DNA damaging agents. Cancer Res *54*: 5824-5830.

[10] Fan S, Smith ML, Rivet DJn, Duba D, Zhan Q, Kohn KW, Fornace Jr AJ and O'Connor PM (1995) Disruption of *p53* function sensitizes breast MCF-7 cells to cisplatin and pentoxifylline. Cancer Res *55*: 1649-164954.

[11] Hamada M, Fujiwara T, Hizuta A, Gochi A, Naomoto Y, Takakura N, Takahashi K, Roth JA, Tanaka N and Orita K (1996) The *p53* gene is a potent determinant of

chemosensitivity and radiosensitivity in gastric and colorectal cancers. J Cancer Res Clin *122*: 360-365.

[12] Harvey M, Vogel H, Morris D, Bradley A, Bernstein A and Donehower LA (1995) A mutant *p53* transgene accelerates tumor development in heterozygous but not nullizygous *p53*-deficient mice. Nat Genet *9*: 305-311.

[13] He L, He X, Lim LP, de Stanchina E, Xuan Z, Ling Y, Xue W, Zender L, Magnus J, Ridzon D, Jakson AL, Linsley PS, Chen C, Lowe SW, Cleary MA and Hannan GJ (2007) A microRNA component of the p53 tumor suppressor network. Nature *447*: 1130-1134.

[14] Jen K-Y and Cheung VG (2006) Transcriptional response of lymphblastoid cells to ionizing radiation. Genome Res *13*: 2092-2100.

[15] Kis E, Szatmari T, Keszei M, Farkas R, Esik O, Lumniczky K, Falus A and Safrany G (2006) Microarray analysis of radiation response genes in primary human fibroblasts. Int J Radiation Oncology Biol Phys *66*: 1506-1514.

[16] Lee JM and Bernstein A (1993) *p53* mutations increase resistance to ionizing radiation. Proc Nat Acad Sci USA *90*: 5742-5746.

[17] Liu X, Lei X, Zhou Z, Sun Z, Xue Q, Wang J and Hung T (2011) Enterovirus 71 induces degradation of TRIM38, a potential E3 ubiquitin ligase. Virology *418*: 61-68.

[18] Miller SL, Antico G, Raghunath PN, Tomaszewski JE and Clevenger CV (2007) Nek3 kinase regulates prolactin-mediated cytoskeletal reorganization and motility of breast cancer cells. Oncogene *26*: 4668-4678.

[19] Nakamura Y, Inloes JB, Katagiri T and Kobayashi T (2011) Chondrocyte-specific microRNA-140 regulates endochondral bone development and targets Dnpep to modulate bone morphogenetic protein signaling. Mol Cell Biol *31*: 3019-3028.

[20] Nigro JM, Baker SJ, Preisiger AC, Jessup JM, Hostetter R, Cleary K, Bigner SH, Davidson N, Baylin S, Devilee P, Glover T, Collins FS, Weston A, Modali R, Harris CC and Vogelstein B (1989) Mutations in the *p53* gene occur in diverse human tumor types. Nature *342*: 705-708.

[21] Okaichi K, Wang L, Sasaki J, Saya H, Tada M and Okumura Y (1999) A point mutation of human *p53*, which was not detected as a mutation by a yeast functional assay, led to apoptosis but not p21Waf-1/Cip1/Sdi1 expression in response to ionizing radiation in a human osteosarcoma cell line, Saos-2. Int J Radiat Oncol Biol Phys *45*: 975-980.

[22] Okaichi K, Ide-Kanematsu M, Izumi N, Morita N, Okumura Y and Ihara M (2008) Variations in sensitivity to ionizing radiation in relation to *p53* mutant point. Anticancer Res *28*: 2687-2690.

[23] Rao DS, O'Connell RM, Chaudhuri AA, Garcia-Flores Y, Geiger TL and Baltimor D (2010) MicroRNA-34a perturbs B lymphocyte development by repressing the forkhead box transcription factor Foxp1. Immunity *33*: 48-59.

[24] Saxena I and Tayyab S (1997) Protein proteinase inhibitors from avian egg whites. Cell Mol Life Sci *53*: 13-23.

[25] Scian MJ, Stagliano KER, Ellis MA, Hassan S, Bowman M, Miles MF, Deb SP and Deb S (2004) Modulation of gene expression by tumor-derived p53 mutants. Cancer Res *64*: 7447-7454.

[26] Small TW, Penalva LO and Pickering JG (2007) Vascular biology and the sex of flies: regulation of vascular smooth muscle cell proliferation by Wilms' tumor 1-associating protein. Trends Cardiovasc Med *17*: 230-234.

[27] Spivey KA and Banyard J (2010) A prognostic gene signature in advanced ovarian cancer reveals a microfibril-associated protein (MAGP2) as a promoter of tumor cell survival and angiogenesis. Cell Adhesion & Migration *4*: 169-171.

[28] Suzuki H, Yamagata K, Sugimoto K, Iwamoto T, Kato S and Miyazono K (2009) Modulation of microRNA processing by p53. Nature *460*: 529-533.

[29] Thompson S, Pearson AN, Ashley MD, Jessick V, Murphy BM, Gafken P, Henshall DC, Morris KT, Simon RP and Meller R (2011) Identification of a novel Bcl-2-interacting mediator of cell death (Bim) E3 ligase, tripartite motif-containing protein 2 (TRIM2), and its role in rapid ischemic tolerance-induced neuroprotection. J Biol Chem *286*: 19331-19339.

[30] Wattel E, Preudhomme C, Hecquet B, Vanrumbeke M, Quesnel B, Dervite I, Morel P and Fenaux P (1994) *p53* mutations are associated with resistance to chemotherapy and short survival in hematologic malignancies. Blood *84*: 3148-3157.

[31] Wei J and Hemmings GP (2005) The KPNA3 gene may be a susceptibility candidate for schizophrenia. Neurosci Res *52*: 342-346.

[32] Zetterberg H, Cambell WA, Yang HW and Xia W (2006) The cytosolic loop of the γ-secretase component Presenilin enhancer 2 protects Zebrafish embryos from apoptosis. J Biol Chem *281*: 11933-11939.

8

Modulation of Gene Expression After Exposure to Ionizing Radiation

Mitsuru Nenoi and Vares Guillaume
National Institute of Radiological Sciences, Japan

1. Introduction

Regulation of gene expression is one of the fundamental mechanisms by which cells utilize the information in their DNA to obtain diverse characteristics, such as ability of differentiated cells to play specific roles and ability to respond to extracellular stresses including ionizing radiation (IR). The research on transcriptional regulation of genes after exposure to IR has a long history tracing back to early studies in late 1980's. In those studies, expression of individual genes was separately measured by a classical hybridization method. Recently, however, functional genomics approaches, such as microarray profiling, enable us to simultaneously monitor the expression of thousands of genes, and are now recognized as a firmly established methodology in radiation biosciences.
The porpose of investigation of transcriptional gene regulation by IR is primarily to gain an insight into how the human bodies respond to IR and eventually how radiation hazards develop. IR-induced hematopoietic death is an example where proapoptotic gene expression is enhanced in radiosensitive hematopoietic stem cells after exposure to high doses of IR, which causes untolerable loss of peripheral blood cells, ultimately resulting in the individual death. Secondary, the purpose of the reserach on transcriptional gene regulation is to search for biomarkers which indicate the quantitative IR-exposure records. Such biomarkers may help us estimate the exposed dose of public peolpe in the emergency cases. For these purposes, studies have been extensively carried out by lots of researchers, and plenty of insights were obtained.
In this review article, some of these studies, including those carried out in our laboratory, will be overviewed with an emphasis on the important roles of a tumor supressor p53 in transcriptional regulation after IR exposure and organ-dependence of transcriptional regulation *in vivo*. Methodology for the research on transcriptional regulation is also briefly touched. And the future perspectives of the research on transcriptional responses to IR will be discussed.

2. Studies with bacteria and yeasts

The SOS response is a well known bacterial response to DNA damage (d'Ari, 1985). There is a group of bacterial genes called SOS genes which are induced by severe DNA damages causing stoppage of DNA synthesis. The SOS genes play roles in DNA repair, cell cycle arrest and mutagenesis. Under normal growth conditions, the SOS genes are negatively regulated by

LexA repressor protein by binding to a 20 bp consensus sequence (the SOS box) in the operator region for the SOS genes. RecA, one of the SOS gene products, is expressed at certain levels even in the repressed state, because of the low affinity of LexA to its SOS box. When DNA damages are produced, DNA polymerase is blocked at the damages, generating single stranded (ssDNA) regions at replication forks. RecA forms a filament around the ssDNA regions, and becomes activated. The activated RecA facilitates dissociation of the LexA repressor from the operator. Then the SOS genes are transcriptionarily activated (Figure 1). First, the genes having SOS box with a lower affinity to LexA repressor protein (such as lexA, recA, uvrA, uvrB, and uvrD) are fully induced in response to mild DNA damages. Thus the first SOS repair mechanism to be induced is nucleotide excision repair (NER). However, if NER is not capable of fixing the damage, the LexA is further inactivated, so the transcription of genes containing SOS box with a stronger affinity to LexA (such as sulA, umuD, umuC - these are expressed late) is induced. SulA binds to FtsZ causing aborted cell division, which results in filamentation and the induction of UmuDC-dependent mutagenic repair.
Regulation of the SOS genes is utilized to detect the environmental mutagen and carcinogens (Oda et al., 1985). The umu-test is based on the ability of the DNA-damaging agents, most of which are potential carcinogens, to induce the umu operon. A bacterial strain containing a fusion gene of umuC and lacZ can be used to monitor the levels of umu operon expression by measuring the β-galactosidase activity. Using this strain, a wide range of environmental mutagens and carcinogens were shown to be inexpensively and sensitively screened.

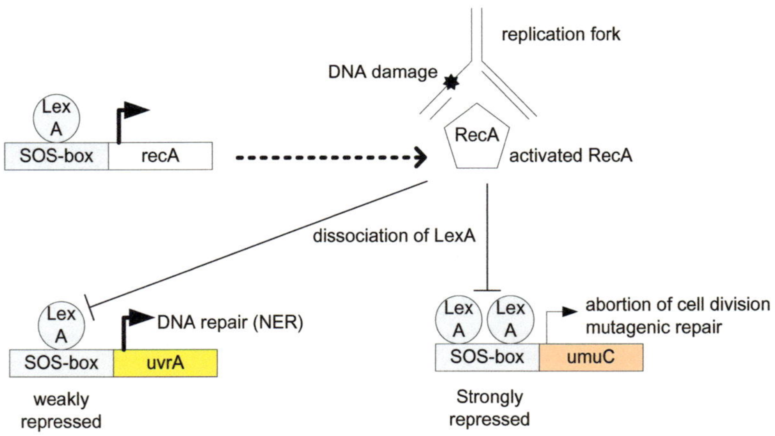

Fig. 1. Illustration of SOS response in bacterial cells.

3. Methodology in research on transcriptional regulation in mammalian cells

Transcriptional regulation of eukaryotic cells is usually mediated by binding of transcription factors (TF) to a DNA region (promoter) neighboring to the protein coding region of a gene. The activity of transcription is determined by a combination of basal TFs, such as RNA polymerase II, TATA binding protein (TBP), TBP-associated factor (TAF), and inducible TFs which act to either enhance or repress initiation of transcription by RNA polymerase II. Regulation of specific gene transcription is attributable to the inducible TFs. Therefore, identification of the inducible TF for IR-responsive genes provides important insight into how the cells respond to IR. First, the methodologies in research on transcriptional regulation is briefly introduced below.

3.1 Nuclear run-on analysis

Transcriptional activity of a gene is usually measured, as an amount transcription products (mRNA), by Northern blot analysis. However, mRNA level does not necessarily represent the transcription rate accurately. Instead, a nuclear run-on assay is performed to measure the transcription rate at a certain time point, which often differ from steady state mRNA levels revealed by Northern blot analysis. It is believed that the nuclear run-on assay is the most reliable method to measure transcription rates directly.

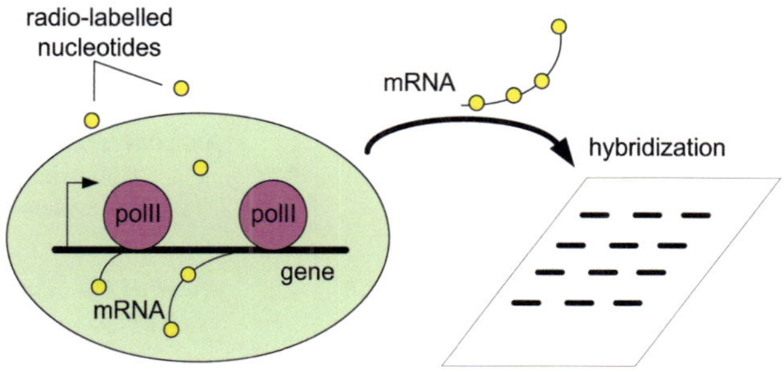

Fig. 2. Schematic illustration of the nuclear run-on assay.

As illustrated in Figure 2, the cell nuclei are isolated rapidly, and incubated in a short period of time in the presence of radio-labelled nucleotides, and the transcribed products are hybridized to a slot-blotted cDNA of interested genes.

3.2 Reporter gene analysis

Reporter genes are the genes that confer the cells characteristics easily identified and measured. In order to investigate the promoter activity, the researchers often attach reporter genes to a DNA fragment containing regulatory elements, and introduce them into cultured cells. The promoter activity can be measured by signal intensity produced by reporter genes as shown in Figure 3.

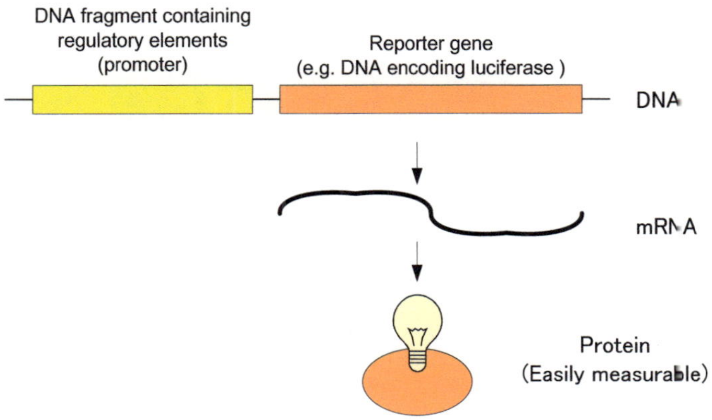

Fig. 3. Schematic illustration of reporter gene analysis.

3.3 Electrophoretic mobility shift assay

An electrophoretic mobility shift assay (EMSA) is an affinity electrophoresis technique used to determine if a protein is capable of binding to a given DNA sequence, as well as to identify the protein molecule. It is often performed to study binding of TFs to promoter region of interested genes. Electrophoresis of a DNA-protein mixture on a polyacrylamide gel can separate the protein-bound DNA from unbound DNA fragments, producing an additional band on the gel that represents the less mobile complex of DNA bound to protein (Figure 4). The DNA fragment is usually radio-labeled to be visible. An antibody that recognizes the protein is often added to the DNA-protein mixture to form an even larger complex with a greater shift. This method, called as a supershift assay, is used to unambiguously identify a protein present in the DNA-protein complex.

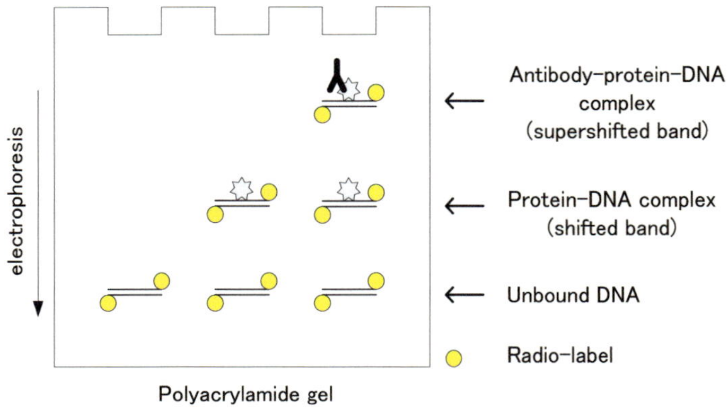

Fig. 4. Scematic illustration of electrophoretic mobility shift assay (EMSA)

3.4 DNase I footprinting assay

A DNase I footprinting assay is a molecular biological technique by which DNA-protein interaction can be detected and located using the fact that a DNA region bound to proteins is often protected from enzymatic cleavage (Figure 5). The method uses an enzyme, deoxyribonuclease I(DNnase I, for short) to cut the radioactively end-labeled DNA, followed by gel electrophoresis to detect the resulting cleavage pattern.

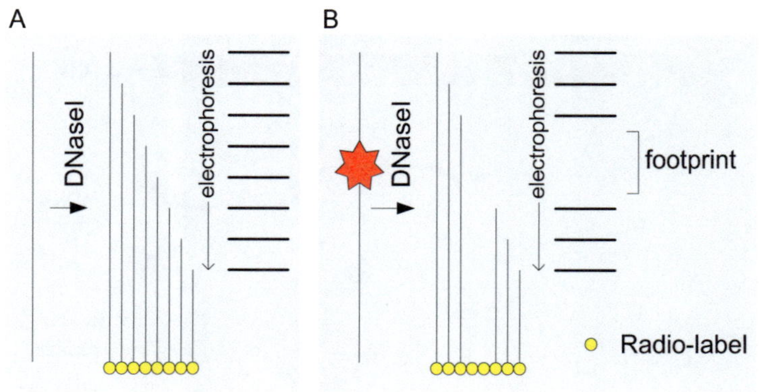

Fig. 5. Schematic illustration of DNase I footprinting assay

3.5 Chromatin immunoprecipitation assay

Chromatin Immunoprecipitation (ChIP) is a technique used to investigate the interaction between proteins and DNA in the cell, especially to determine whether specific proteins are interacted with specific genomic regions. It is applicable to investigate interaction of TFs with promoters *in vivo*. Protein and genomic DNA are temporarily crosslinked *in vivo*, and the DNA-protein complexes are fragmented by shear stresses as illustrated in Figure 6. Then the DNA fragments associated with the proteins of interest are selectively immunoprecipitated by using antibodies specific to those proteins. The associated DNA fragments are purified and their sequence is determined.

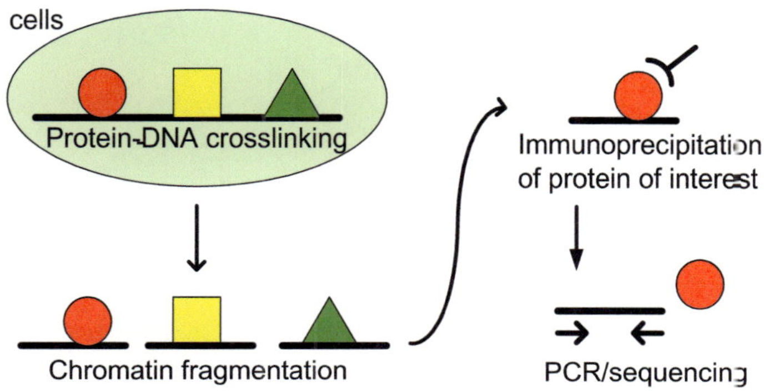

Fig. 6. Schematic illustration of Chromatin Immunoprecipitation assay

4. p53-mediated regulation of gene transcription after IR in cultured cells

Systematic studies on transcriptional regulation of mammalian genes have started in late 1980's. The representative studies were performed by Fornace and his colleagues (Fornace et al., 1988, Fornace et al., 1989). They screened a cDNA library for UV-induced genes by use of a hybridization subtraction method to obtain 20 differentially expressed genes including unidentified at that time. Some of these genes were co-ordinately regulated by DNA damaging agents and growth arrest signals, and were therefore called as gadd (**g**rowth **a**rrest and **D**NA **d**amage responsive) genes. The most widely characterized of these genes is *GADD45A*. The increase in *GADD45A* protein levels is correlated with a block in S phase progression as well as a stimulation of DNA repair (*Smith, et al., 1994, Chan, et al., 1995, Zhan, et al., 1994*). *GADD45A* gene was soon shown to respond to IR through a tumor supressor p53. The function of p53 in DNA damage responses was vigorously investigated at that time, and its pivotal roles as the guardian of the genome was established (Fei & El-Deiry, 2003). P53 functions primarily as a TF, and has been shown to be involved in multiple cellular pathways including cell cycle arrest, apoptosis, DNA repair, and so on, by controling many other genes. *CDKN1A* ($p21^{CIP1/WAF1}$) gene is another member of gadd genes which was identified among 20 clones originally isolated by cDNA subtraction. *CDKN1A* gene is induced after exposure to IR also through p53, and inhibits the cycln dependent kinase activity resulting in arrest of cell cycle progression at G1/S border (Hartwell & Weinert, 1989). Although both of *GADD45A* and *CDKN1A* were clearly demonstrated to be regulated through p53, involvement of other TFs has not been investigated. Elucidation of a set of TFs involved in gene regulation after IR should be pivotal for understanding of delicate mechanisms for biological responses to IR which include variety of cellular pathways. In the following sections, researches performed in our laboratory on regulation mechanisms for the *GADD45A* and *CDKN1A* genes are reviewed.

4.1 Regulation *GADD45A* gene

In human cells, *GADD45A* is up-regulated in response to IR. Figure 7 is a Northern blot analysis for induction of the *GADD45A* gene expression in human breast adenocarcinoma cell line MCF7 after exposure to IR. The induction was transient with the maximum induction at 2 h after irradiation.

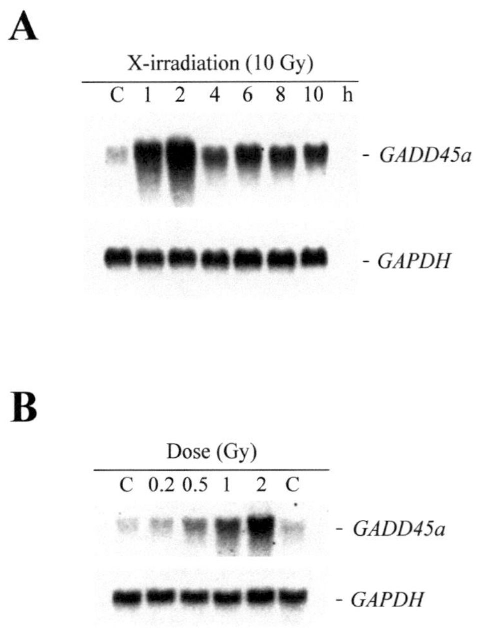

Fig. 7. Northern analysis for induction of the *GADD45A* gene expression after exposure to IR. (A) Total RNA isolated from the MCF-7 cells at various time intervals after irradiation with 10 Gy of X-rays was hybridized with a *GADD45A* gene-specific probe. The *GAPDH* gene was used as a control. (B) Induction of the *GADD45A* gene after irradiation with various doses of X-rays was analyzed. (Reproduced with permission of Elsevier B. V. from Ref (Daino et al., 2006))

To search for the DNA element functioning in transcriptional regulation of the *GADD45A* gene, a series of reporter gene constructs were designed. As the p53 recognition sequence was known to reside in the third intron, the third intron was fused to the 3′ end of the luciferase gene (*luc* +) (Figure 8A). MCF-7 cells were transiently transfected with these constructs, and luciferase activity was measured after treatment with or without X-rays. As shown in Figure 8B, constructs P/I through P6/I gave rise to a similar level of the basal luciferase activity, suggesting that there was no major DNA element functioning in transcriptional activity in the -2244 bp/-107 bp region. However, when the 5′ flanking region was truncated to -62 bp in P7/I, the luciferase activity was drastically decreased, suggesting that a regulatory element required for the basal expression of the *GADD45A*

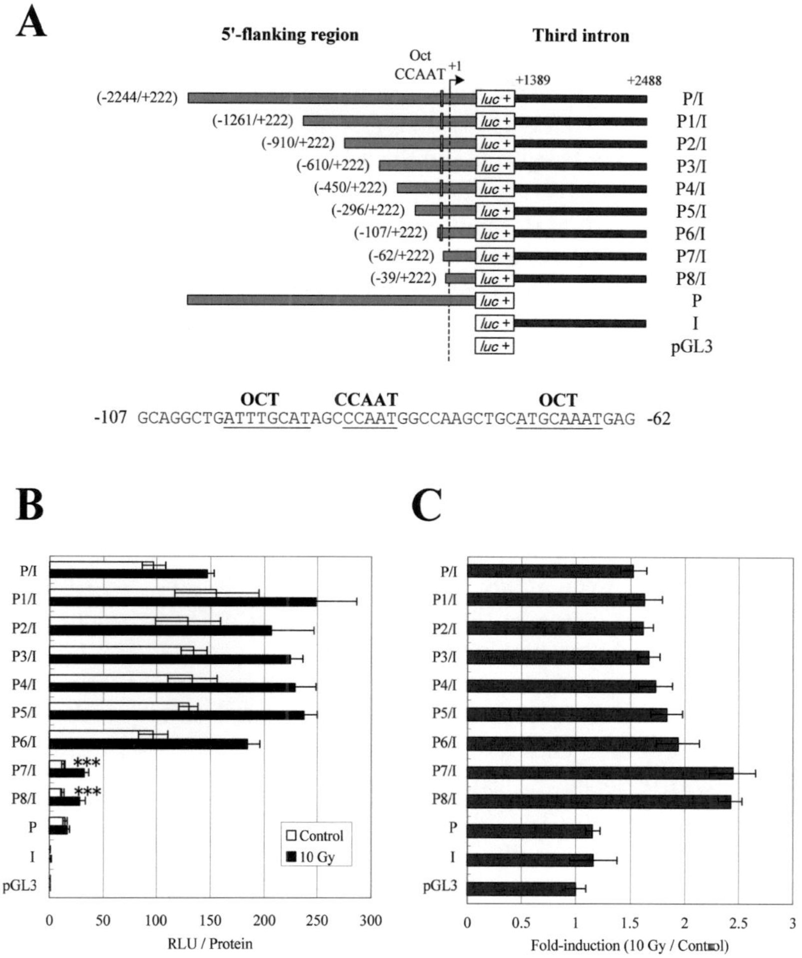

Fig. 8. Reporter gene analysis for the promoter activity of the *GADD45A* gene. (A) Various *GADD45A*-luciferase constructs are schematically illustrated. The light gray bars and the dark gray bars represent the 5′-flanking region and the third intron region of the *GADD45A* gene, respectively. The transcription start site is indicated by an arrow. The nucleotide sequence of the region, which reduced the basal promoter activity significantly as shown in (B), is presented at the bottom. Two OCT sites and a CCAAT box are underlined. (B) MCF-7 cells transiently transfected with reporter constructs were either irradiated with 10 Gy of X-rays (closed bars) or mock-treated as a control (open bars), and the luciferase activity in the cell lysates prepared 5 h after irradiation was measured. The relative light unit (RLU) per protein concentration is plotted. (C), Fold-induction of the luciferase activity after irradiation with 10 Gy of X-rays is plotted. Results are the mean of at least three independent experiments. Error bars represent the S.D. (*** $p < 0.001$ versus P/I, student's t-test). (Reproduced with permission of Elsevier B. V. from Ref (Daino et al., 2006))

gene was present in the region between -107 bp and -62 bp. There were two OCT sites and one CCAAT box in this region (Figure 8A). It was thought that these OCT and CCAAT elements functioned as the core promoter because a canonical TATA sequences could not be found. It can be seen, in Figure 8C, that fold-induction of luciferase expression after irradiation with 10 Gy of X-rays was not reduced even when the 5′-flanking region was deleted down to -39 bp. This result indicated the absence of major regulatory elements functioning in response to IR in the upstream region of the *GADD45A* gene.

When the region +1747 bp/+2488 bp in the third intron was deleted, the luciferase activity was not so much reduced. However, when the region +1625 bp/+2488 bp was deleted, the luciferase activity was decreased dramatically (data not shown), suggesting the presence of a DNA element, in +1625/+1746, that plays a role in the basal expression of the *GADD45A* gene. As AP-1 recognition element existed at the locus +1625 bp/+1631 bp, we constructed another reporter gene experiment with a plasmid (P6/I AP-1m) in which the AP-1 recognition sequence was mutated (Figure 9A). As shown in Figure 9B, the basal luciferase expression from P6/I AP-1m was drastically reduced compared with P6/I which contains wild type AP1 sequence. This result demonstrate the important role of AP-1. Deletion of the p53 site (P6/I p53Δ) also resulted in reduction of basal luciferase activity compared to that of the P6/I construct. Then it was suggested that both the AP-1 and p53 sites in the third intron of the *GADD45A* gene are critical for the basal expression. Concerning IR responsiveness, Figure 9C shows that deletion of the p53 site (P6/I p53Δ) diminished IR-response, while fold induction for the construct containing the mutated AP-1 site (P6/I AP-1m) was reduced, suggesting that the AP-1 site contributes to a part of p53-dependent transcriptional activation of the *GADD45A* gene after exposure to IR.

It was speculated that transiently transfected plasmid DNAs may not form tight chromatin complexes. Then, we used the recombinant adeno-associated virus (rAAV) vectors, which can stably integrate reporter genes into the genome of host cells. As a result, a remarkably higher fold-induction of luciferase expression (~3.4-fold, rAAV-P6/I) could be observed after IR irradiation as shown in Figure 9D. With this vector system, rAAV-P6/I AP-1m containing mutation at the AP-1 site gave rise to significantly decreased fold induction compared to the rAAV-P6/I construct (Figure 9E). Taken together, these results indicate that the AP-1 recognition site plays a significant role in the p53-dependent transcriptional activation of the *GADD45A* gene by IR.

When EMSA was performed with a DNA probe containing the AP-1 recognition sequence and nuclear extracts from MCF-7 cells, DNA-protein complexes were created. And these DNA-protein complexes disappeared when an anti-JunD antibody was added in the DNA-protein mixture. This result indicated the presence of JunD in the DNA-protein complexes.

When ChIP assay was carried out (Figure 10A), the binding of p53 and JunD to the third intron of the *GADD45A* gene was confirmed. However they were detected even in unirradiated cells. While the affinity of p53 to its site in the intron 3 was enhanced in X-ray-irradiated cells (Figure 10C), the bining of JunD to its recognition site was not increased in the X-ray irradiated cells (Figures. 10B and 10C). The binding of phospho-p53 (Ser20) to the p53 site was detected only in X-ray-irradiated cells (Figure 10C). These results indicated that JunD constitutively binds to the third intron of the *GADD45A* gene even in unstressed cells *in vivo*. JunD is considered to play an important role both in basal expression of the *GADD45A* gene and in transcriptional enhancement after IR irradiation via binding to the AP-1 site in the third intron.

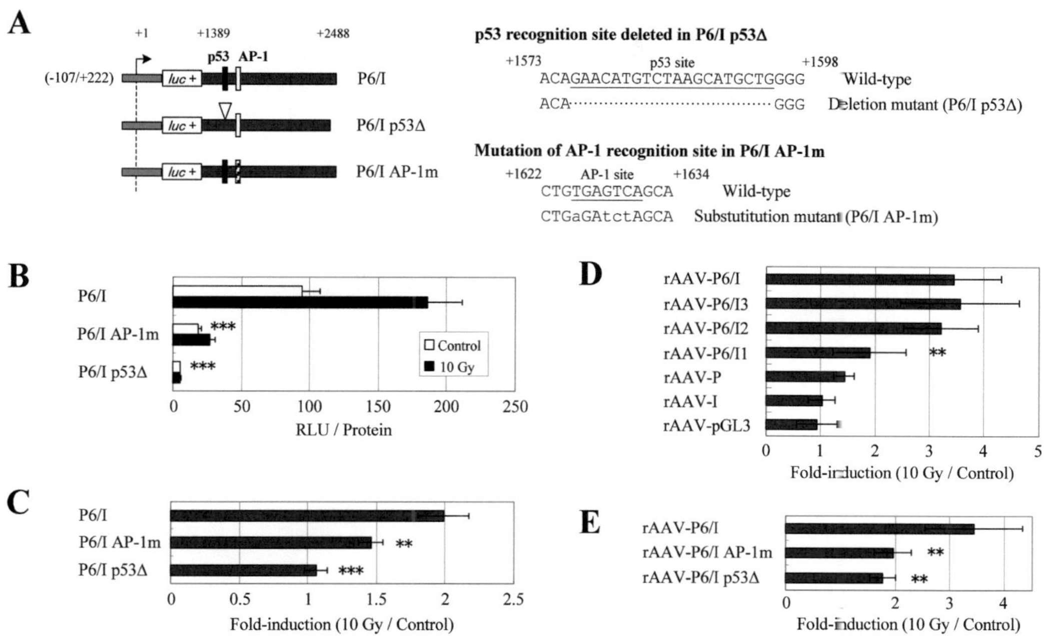

Fig. 9. Requirement of p53 and AP-1 recognition sites for both basal expression and transcriptional enhancement of the *GADD45A* gene after exposure to IR. (A) Mutant *GADD45A*-luciferase constructs are schematically illustrated. The light gray bars and the dark gray bars represent the 5'-flanking region and the third intron region of the *GADD45A* gene, respectively. The nucleotide sequences at the p53-deletion in P6/I p53Δ and the AP-1 mutation in P6/I AP-1m are shown on the right of the panel. The small letters represent the substituted nucleotides. The closed box and open box represent the p53 site and the AP-1 site, respectively. The arrowhead represents the site of p53-deletion, and the hatched box represents the mutated AP-1 site. (B) MCF-7 cells transiently transfected with reporter constructs were either irradiated with 10 Gy of X-rays (closed bars) or mock-treated as a control (open bars), and luciferase activity in the cell lysates prepared 5 h after irradiation was measured. The relative light unit (RLU) per protein concentration is plotted. (C) Fold-induction of the luciferase activity after irradiation with 10 Gy of X-rays is plotted. Results are the mean of at least three independent experiments. Error bars represent the S.D. (** $p < 0.01$, *** $p < 0.001$ versus P6/I, student's t-test). (D) and (E) MCF-7 cells transduced with reporter constructs by use of rAAV were either irradiated with 10 Gy of X-rays or mock-treated, and luciferase activity in the cell lysates prepared 5 h after irradiation was measured. Fold-induction of luciferase activity after irradiation with 10 Gy of X-rays is plotted. Results are the mean of at least three independent experiments. Error bars represent the S.D. (** $p < 0.01$ versus P6/I, student's t-test). (Reproduced with permission of Elsevier B. V. from Ref (Daino et al., 2006))

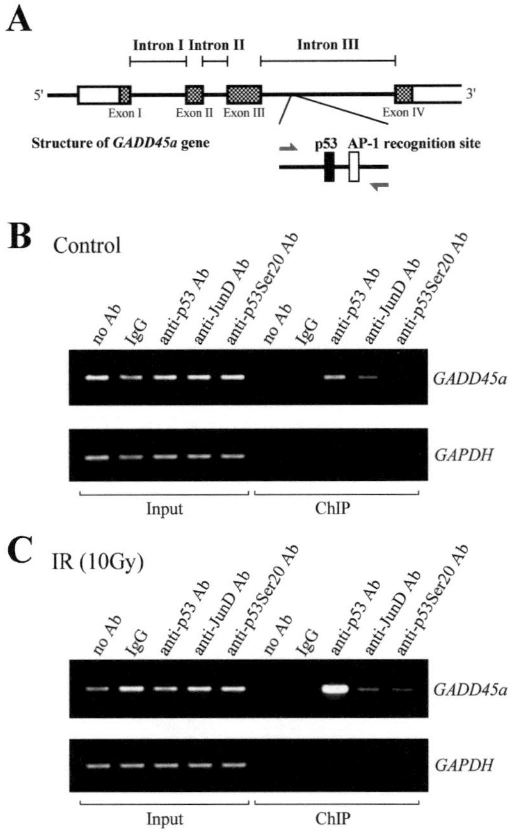

Fig. 10. Chromatin immunoprecipitation (ChIP) analysis for the in vivo binding of JunD at the third intron. (A) Schematic illustration of genomic structure of the *GADD45A* gene. PCR primers were designed to amplify the region containing both the p53 site and the AP-1 site, as represented by arrows. (B) ChIP analysis was performed with MCF-7 cells, unirradiated (B) or exposed to 10 Gy of X-rays with 1.5 h postincubated (C). Sheared chromatin was incubated without antibodies (no Ab) or with the IgG, anti-p53 antibody, anti-JunD antibody, and anti-phospho-p53 (Ser20) antibody. Immunoprecipitants were analyzed by PCR. The *GAPDH*_promoter containing neither an AP-1 nor p53 site was analyzed as a negative control. PCR was performed for 32 cycles, which was in the linear range of amplification, except for the case using the sample prepared from X-ray-treated cells with the anti-p53 antibody. (Reproduced with permission of Elsevier B. V. from Ref (Daino et al., 2006))

4.2 Regulation *CDKN1A* gene

Regulation of *CDKN1A* gene after exposure to IR occurs in transcription level as shown by a nuclear run-on analysis (Figure 11). While the induction kinetics of *CDKN1A* is similar to that of *GADD45A*, the fold-increase of *CDKN1A* is much higher, and is thought to be one of the most responsive genes to IR.

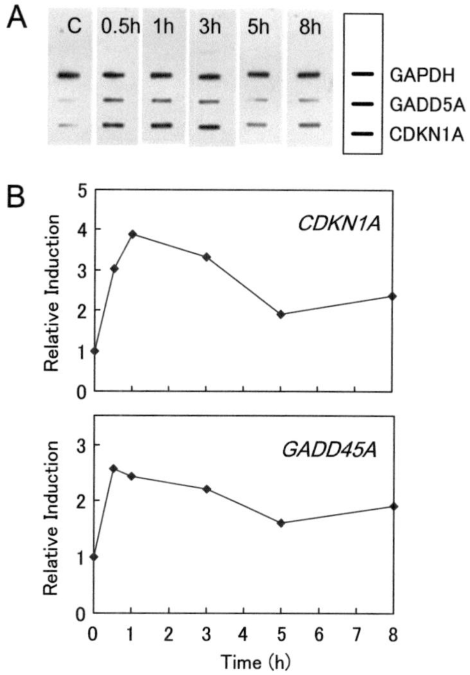

Fig. 11. Nuclear run-on analysis for transcription rates of the *CDKN1A* and *GADD45* genes in human myeloblastic leukemia cell line ML-1 after irradiation with 0.5 Gy of X-rays. (A) The resulting hybridization images. cDNA probes were blotted as shown in the right side panel. (B) The signal intensity was analyzed and plotted. (Reproduced with permission of Radiation Research Society of the USA from Ref (Daino et al., 2002))

The functional DNA region involved in the p53-mediated IR response was investigated by reporter gene analysis using AAV vectors, which can give rise to remarkably high responsiveness to IR as shown in Figure 12A. A series of reporter gene constructs with or without deletions in the 5′ flanking region of the *CDKN1A* gene was prepared as shown in Figure 12B, and were transduced into MCF-7 cells. The p53 recognition site at -2.2 kb was preserved in every construct. It can be seen, in Figure 12C, that the fold induction of luciferase activity increased linearly with radiation dose for every constructs. However, IR responsiveness of the constructs rAAV-del2, rAAV-del4 and rAAV-del5 was much reduced compared to that of rAAV-PLS which contains an intact 5′ flanking region of the *CDKN1A* gene, suggesting the presence of pivotal DNA elements for IR responsiveness in the regions -1962 bp/-1679 bp (rAAV-del2), -1398 bp/-1119 bp (rAAV-del4), and -1118 bp/-839 bp (rAAV-del5).

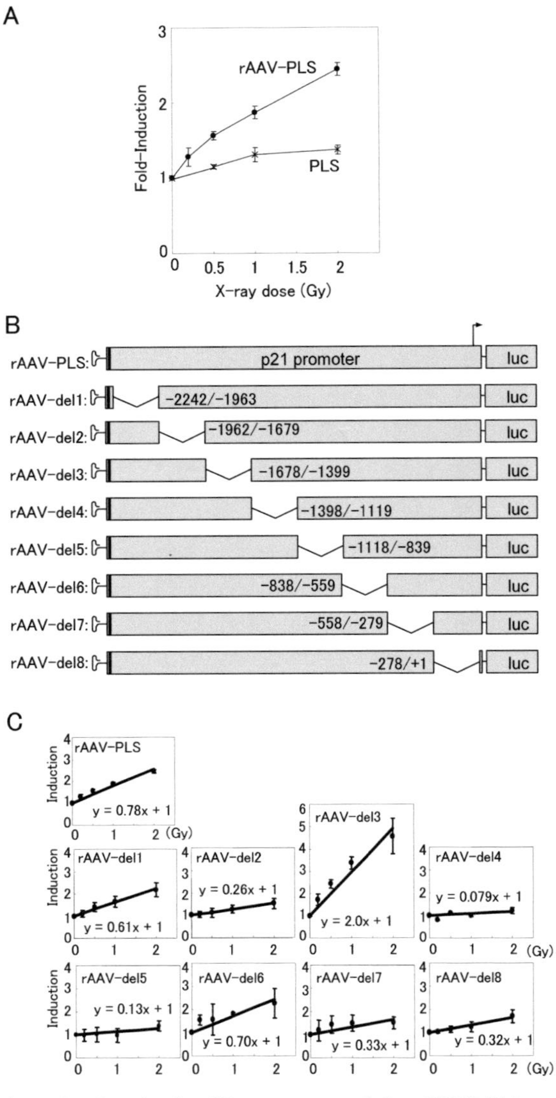

Fig. 12. DNA regions functioning in the IR response of the *CDKN1A* gene. A: Reporter gene constructs containing the luciferase gene under the control of the *CDKN1A* gene promoter were either transfected by electroporation (PLS) or transduced by rAAV vectors (rAAV-PLS) into MCF-7 cells. The cells were irradiated with X-rays and assayed for luciferase activity 5 h after irradiation or mock treatment. Results are the means of three independent experiments. Error bars represent the S.D. B: The reporter gene constructs with or without deletions in the 5′ flanking region are schematically represented. The TP53 site at –2.2 kb is indicated by a heavy line. These constructs were transduced into MCF-7 cells using rAAV vectors. C: X-ray response of each construct was assayed in the dose range of 0.2-2.0 Gy. Since the fold induction increased linearly with dose, the slope was used as an index of IR responsiveness. Results are the mean of three independent experiments. Error bars represent the S.D. (Reproduced with permission of Elsevier B. V. from Ref (Nenoi et al., 2009))

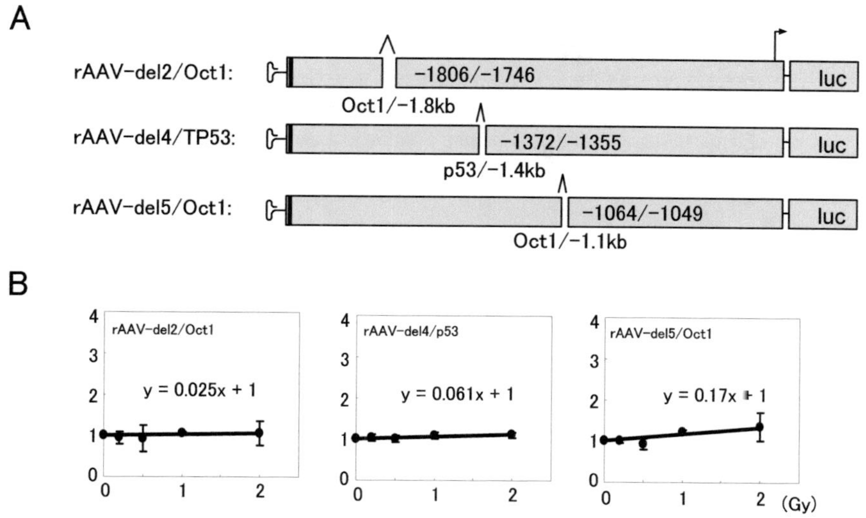

Fig. 13. Reporter vectors were constructed in which the Oct-1 site at –1.8 kb, TP53 site at –1.4 kb, and Oct-1 site at –1.1 kb were specifically deleted, resulting in rAAV-del2/Oct1, rAAV-del4/TP53, and rAAV-del5/Oct1, respectively (D). X-ray response of each construct was assayed at a dose range of 0.2-2.0 Gy (E). (Reproduced with permission of Elsevier B. V. from Ref (Nenoi et al., 2009))

The potential recognition sequences for TFs were searched in these regions, and Oct-1 recognition sequences were found at –1.8 kb and –1.1 kb. It was interesting to note that Oct-1 is the TF involved in regulation of IR-responsive genes *GADD45A*, *Prx1*, *Notch1*, and endothelial lipase, suggesting that Oct-1 may play a role in regulation of variety of IR-responsive genes by cooperating with p53. Then, mutant reporter vectors rAAV-del2/Oct1, rAAV-del4/TP53, and rAAV-del5/Oct1 were constructed in which the Oct-1 site at –1.8 kb, the p53 site at –1.4 kb, and the Oct-1 site at –1.1 kb, respectively, were specifically deleted (Figure 13A). Figure 13B shows that the IR responsiveness of these reporter constructs was mostly diminished, indicating that the recognition sequences for these TFs play pivotal roles in the IR response of the *CDKN1A* gene promoter.

EMSA analysis revealed that the DNA probes Oct-1/–1.8kb (Figure 14A) and Oct-1/–1.1kb (Figure 14B) could form complexes with nuclear proteins as indicated by arrows. These bands were supershifted by addition of anti-Oct-1 antibodies and competed away by addition of an excess amount of unlabeled oligonucleotides containing the Oct-1 consensus sequence. These results demonstrate Oct-1 could bind to its recognition sequences at -1.8 kb and -1.1 kb. However, binding of Oct-1 to these sites was not X ray-dependent. In accordance with the EMSA analysis, constitutive binding of Oct-1 to the sites at –1.8 kb and –1.1 kb was observed by ChIP analysis as shown in Figures 14C and 14E. In contrast, binding of p53 to the site at –1.4 kb was not constitutive but inducible after irradiation (Figure 14D).

Overall, it was considered that Oct-1 is constitutively bound to the sites at –1.8 kb and –1.1 kb and plays a cooperative role with p53 in induction of the *CDKN1A* gene promoter in response to IR.

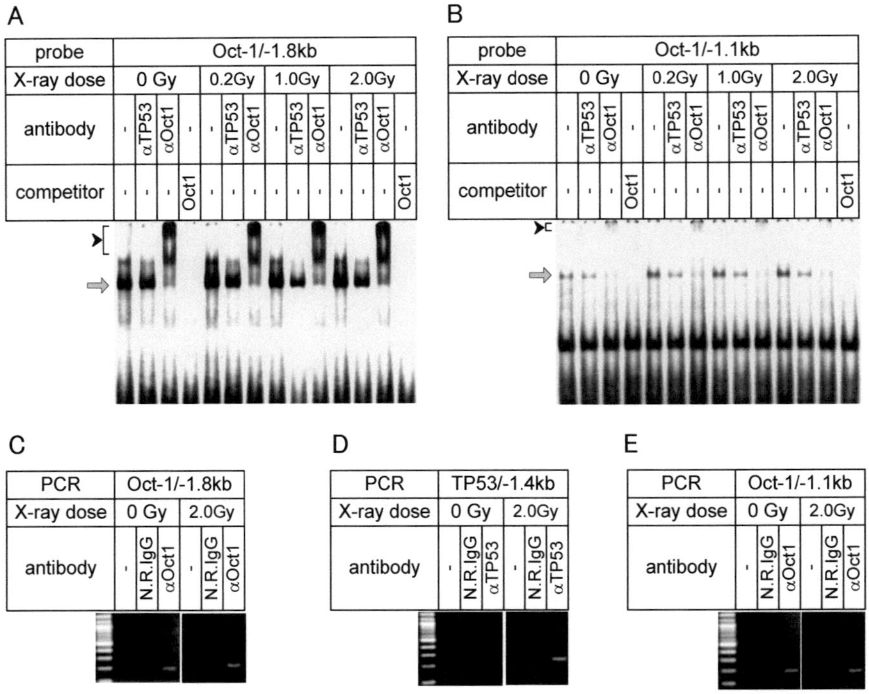

Fig. 14. DNA/chromatin binding of Oct-1. *In vitro* DNA binding of Oct-1 to the sites at -1.8 kb (A) and -1.1 kb (B) was analyzed by EMSA. Sequences of the probe and competitor oligonucleotides for Oct-1 are described in the text. Arrows indicate the DNA-protein complex containing Oct-1. Arrowheads indicate the supershifts caused by addition of anti-Oct-1 antibodies. *In vivo* chromatin binding of Oct-1 and TP53 was analyzed by ChIP. MCF-7 cells were either not irradiated or irradiated with 2.0 Gy of X-rays. After a period of 1.5 h, the chromatin was immunoprecipitated with anti-Oct-1 antibodies (C and E) or anti-p53 antibodies (D). Normal rabbit IgG was used as a control. Binding of the factors to chromatin was assessed using site-specific PCR. (Reproduced with permission of Elsevier B. V. from Ref (Nenoi et al., 2009))

5. *In vivo* studies

Regulated initiation of DNA replication (entry of S-phase) is one of the important cellular responses to IR. When DNA damages are produced by IR, multiple factors involved in cell cycle regulation such as CDKN1A which causes cell cycle arrest at G1/S are modulated. The so-called checkpoint regulation is thought to be necessary for sufficient repair of damaged DNA before DNA replication. However, the checkpoint regulation is absent in some of fully differentiated cells such as those *in vivo* organs. Thus, the cellular response to IR *in vivo* is much different from that in cultured cells, and it should have wide variety. In order to understand the difference in transcriptional responses to IR between tissues of mammals and also to reveal the underlying mechanisms for those differences, a microarray analysis of the global gene expression profile is a promising approach. There have been lots of *in vitro* studies investigating gene expression modulation after exposure to IR, where evident modulation of genes associated with DNA repair, stress response, cell cycle and apoptosis and the pivotal role played by p53 have been elucidated. However, few studies investigated the *in vivo* modulation of gene expression profiles after exposure to IR and its underlying mechanisms. Especially importance of studies with low dose and low dose-rate IR which people may encounter in social life should be emphasized.

5.1 Modulation of gene expression in the kidneys and testes of mice after exposure to IR

C57BL/6J mice were continuously irradiated with γ-rays for 485 days at the dose-rates in the range of 0.032 - 13 μGy/min. The lowest dose rates were only about 10 times higher than the normal background level. Gene expression profiles in the kidney and testis from irradiated and unirradiated mice were analyzed, and differentially expressed genes were identified after the standard procedure of quality control of the data. By hierarchical cluster analysis of differentially expressed genes (Figure 15), it was suggested that expression of the

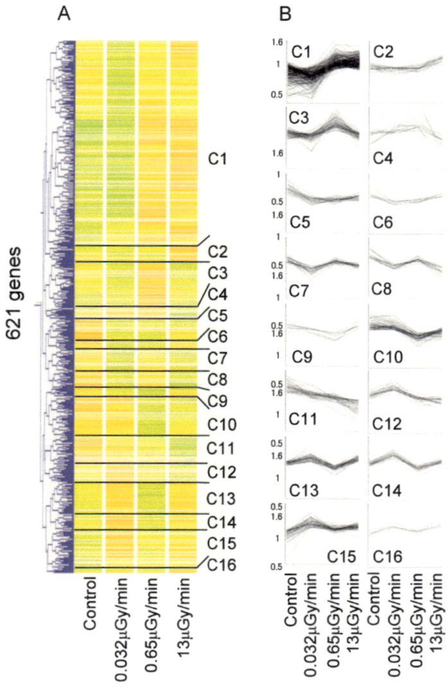

C

	Category	p-Value
C1	cytoplasm	2.8E-07
	mitochondrion	5.7E-04
	mitochondrion organization and biogenesis	0.0013
	energy pathways	0.0055
	organelle organization and biogenesis	0.0071
	cell organization and biogenesis	0.0071
	cytoplasm organization and biogenesis	0.0071
	transferase	0.031
C10	cell organization and biogenesis	1.4E-05
	cytoplasm organization and biogenesis	1.4E-05
	organelle organization and biogenesis	1.4E-05
	cytoplasm	0.0016
	mitochondrion	0.0061
	nucleosome	0.012
	mitochondrion organization and biogenesis	0.032
	GO SLIMS Cellular Component	0.040
C13	cytochrome-c oxidase activity	1.2E-05
	organelle inner membrane	6.5E-05
	mitochondrion organization and biogenesis	3.3E-04
	mitochondrion	3.3E-04
	energy pathways	7.1E-04
	cytoplasm	0.0086

Fig. 15. Hierarchical cluster analysis of genes whose expression was modulated after irradiation with a statistical significance in the kidney. A: Expression levels of 621 of significantly modulated genes are indicated by color. Green, yellow, and red colors represent low, medium, and high expression, respectively. On the basis of the expression pattern, genes were classified into 16 clusters. B: Expression profiles of genes in each of the 16 clusters are collectively shown. C: Gene Ontology categories significantly overlapped with the gene clusters are shown. No Gene Ontology categories significantly similar to the clusters other than C1, C10 and C13 were found. (Reproduced with permission of the Japanese Radiation Research Society from Ref (Taki et al., 2009))

genes involved in mitochondrial oxidative phosphorylation was elevated in the kidney after irradiation at the dose-rates of 0.65 μGy/min (the level of dose rate comparable to that in the spaceshuttle) and 13 μGy/min. A particular mitochondrial response may have occurred after low dose rate irradiation. It was reported that several cell cycle-related genes are up-regulated in the kidney after acute whole-body irradiation with 10 Gy of γ-rays (Zhao et al., 2006). However, we did not observe such gene modulation after irradiation at 13μGy/min We did not either extract clusters of significantly modulated genes that overlapped with the Gene Ontology category "cell cycle" (Figure 15C). These results suggested that modulation of the cell cycle-related genes *in vivo* following exposure to radiation may have a dose-rate threshold.

It was also demonstrated that alteration of the gene expression profile in the testis was largely different from that in the kidney. Gene Ontology categories "DNA metabolism", "response to DNA damage" and "DNA replication" were significantly overlapped to up-regulated genes in the testis in a dose rate-dependent manner (Figure 16). These results provide fundamental insight into the organ-specific responses to IR.

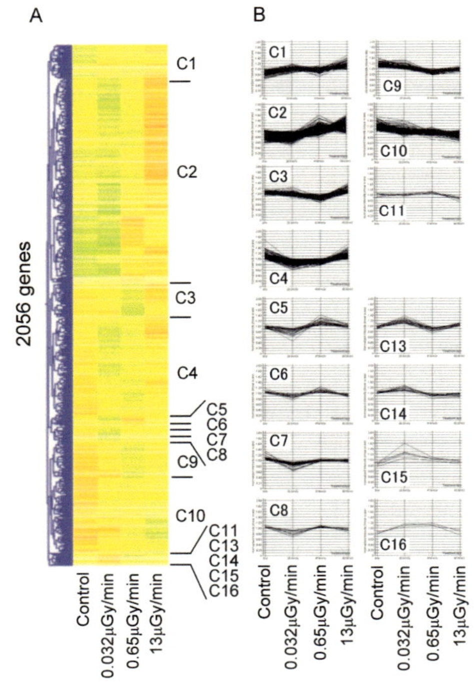

C

	Category	p-Value
C1	response to temperature	0.0013
C2	physiological process	0.0012
	protein biosynthesis	0.0088
	Tricarboxylic acid cycle intermediate metabolism	0.0077
	Chaperone cofactor dependent protein folding	0.0060
	aldehyde metabolism	0.0014
	glutathione metabolism	0.0077
	coenzyme metabolism	0.0058
	regulation of endocytosis	0.0095
	cellular physiological process	0.0003
	cytoskeleton-dependent intracellular transport	0.0049
	microtubule-based movement	0.0046
	organelle organization biogenesis	0.0078
	microtubule-based process	0.0015
	response to heat	0.0095
	response to hormone stimulus	0.0095
C9	DNA metabolism	0.0086
	DNA repair	0.0019
	response to DNA damage	0.0040
	response to endogenous	0.0049
C10	DNA replication	0.0024
	mitotic cell cycle	0.0056

Fig. 16. Hierarchical cluster analysis of genes whose expression was significantly modulated after irradiation in the testis. A: Expression levels of 2056 of significantly modulated genes are indicated by color. On the basis of the expression pattern, genes were classified into 16 clusters. The cluster 12 contained no gene. B: Expression profiles of genes in each of the 16 clusters are collectively shown. C: Gene Ontology categories significantly overlapped with the gene clusters C1, C2, C9 and C10 are shown. (Reproduced with permission of the Japanese Radiation Research Society from Ref (Taki et al., 2009))

5.2 Transcription factors

By use of microarray data, it is possible to pick up TFs whose recognition sequences are preferentially identified in the promoter of genes modulated by IR. These TFs are suggested

	E47	AP-4	Elk-1	NF-kappaB (p65)	Evi-1 (AGAY··)	Evi-1 (NGAT··)	CDP	S8	HFH-1	Brn-2	HSF1
unchanged	0.538	0.032	0.591	0.430	0.505	0.280	0.086	0.935	0.524	0.656	0.742
down-regulated	0.595	0.119	0.571	0.643	0.595	0.119	0.000	0.976	0.571	0.500	0.738
na	93.000	93.000	93.000	93.000	93.000	93.000	93.000	93.000	93.000	93.000	93.000
nb	42.000	42.000	42.000	42.000	42.000	42.000	42.000	42.000	42.000	42.000	42.000
P	0.556	0.059	0.585	0.496	0.533	0.230	0.059	0.948	0.507	0.607	0.741
Z	0.624	1.977	0.218	2.289	0.969	2.053	1.960	0.988	0.575	1.717	0.047
unchanged	0.538	0.032	0.591	0.430	0.505	0.280	0.086	0.935	0.524	0.656	0.742
up-regulated	0.346	0.077	0.346	0.692	0.308	0.192	0.000	0.808	0.308	0.423	0.538
na	93.000	93.000	93.000	93.000	93.000	93.000	93.000	93.000	93.000	93.000	93.000
nb	26.000	26.000	26.000	26.000	26.000	26.000	26.000	26.000	26.000	26.000	26.000
P	0.496	0.042	0.538	0.487	0.462	0.261	0.067	0.908	0.555	0.605	0.697
Z	1.726	1.004	2.217	2.365	1.787	0.896	1.548	1.989	2.366	2.147	1.997
down-regulated	0.595	0.119	0.571	0.643	0.595	0.119	0.000	0.976	0.571	0.500	0.738
up-regulated	0.346	0.077	0.346	0.692	0.308	0.192	0.000	0.808	0.308	0.423	0.538
na	42.000	42.000	42.000	42.000	42.000	42.000	42.000	42.000	42.000	42.000	42.000
nb	26.000	26.000	26.000	26.000	26.000	26.000	26.000	26.000	26.000	26.000	26.000
P	0.500	0.103	0.485	0.662	0.485	0.147	0.000	0.912	0.471	0.471	0.662
Z	1.996	0.556	1.806	0.419	2.306	0.829	#DIV/0!	2.381	2.117	0.618	1.691

	COUP-T	GR (NNNN··)	AP-1	GR (GGTAT··)	HNF-1	SREBP-1	AhR/Ar	Egr-3	Egr-2	Olf-1
unchanged	0.280	0.086	0.946	0.075	0.032	0.301	0.129	0.065	0.108	0.011
down-regulated	0.119	0.333	0.857	0.214	0.119	0.190	0.143	0.167	0.119	0.000
na	93.000	93.000	93.000	93.000	93.000	93.000	93.000	93.000	93.000	93.000
nb	42.000	42.000	42.000	42.000	42.000	42.000	42.000	42.000	42.000	42.000
P	0.230	0.163	0.919	0.119	0.059	0.267	0.133	0.096	0.111	0.007
Z	2.053	3.602	1.752	2.313	1.977	1.345	0.219	1.863	0.197	0.675
unchanged	0.280	0.086	0.946	0.075	0.032	0.301	0.129	0.065	0.108	0.011
up-regulated	0.192	0.269	0.769	0.077	0.038	0.500	0.346	0.192	0.115	0.115
na	93.000	93.000	93.000	93.000	93.000	93.000	93.000	93.000	93.000	93.000
nb	26.000	26.000	26.000	26.000	26.000	26.000	26.000	26.000	26.000	26.000
P	0.261	0.126	0.908	0.076	0.034	0.345	0.176	0.092	0.109	0.034
Z	0.896	2.488	2.755	0.028	0.155	1.887	2.567	1.989	0.114	2.617
down-regulated	0.119	0.333	0.857	0.214	0.119	0.190	0.143	0.167	0.119	0.000
up-regulated	0.192	0.269	0.769	0.077	0.038	0.500	0.346	0.192	0.115	0.115
na	42.000	42.000	42.000	42.000	42.000	42.000	42.000	42.000	42.000	42.000
nb	26.000	26.000	26.000	26.000	26.000	26.000	26.000	26.000	26.000	26.000
P	0.147	0.309	0.824	0.162	0.088	0.309	0.221	0.176	0.118	0.044
Z	0.829	0.556	0.924	1.495	1.139	2.685	1.965	0.270	0.046	2.252

p=0.01: Z=2.576

p=0.05: Z=1.960

[a] "Pro unaff.". "Pro down" and "Pro up" are the proportion of genes containing recognition sequence for the TF.

[b] "No unaff.". "No down" and "No up" are the number of genes contained in the unaffected. down-regulated and up-regulated gene groups. respectively.

[c] Po is the population proportion defined by Po=(Noxx•Proxx+Noyy•Proyy)/(Noxx+Noyy). where xx and yy is replaced either by "unaff.". "down" or "up".

[d] Z is the random variable defined by Z=|Proxx-Proyy|/sqrt(Po•(1-Po)•(1/Noxx+1/Noyy)). where xx and yy is replaced either by "unaff.". "down" or "up". If Z>1.960 then p<0.05. and if Z>2.576 then p<0.01.

[e] Evi-1 recognition sequence. AGAYAAGATAA

[e] Evi-1 recognition sequence. NGATANGANWAGATA

[g] GR recognition sequence. NNNNNNCNNTNTGTNCTNN

[h] GR recognition sequence. GGTACAANNTGTYCTK

Table 1. Statistical analysis of difference in the proportion of genes containing recognition sequences for TFs between unchanged, up-regulated and down-regulated gene groups. Boxes of yellow and red indicate that the difference in populationproportion is significant with p < 0.05 and p < 0.01, respectively. (Reproduced with permission of the Japanese Radiation Research Society from Ref (Vares et al., 2011)) to be functionally involved in gene modulations. Then, nucleotide sequences in the neighboring region of the up-regulated, down-regulated, and unaffected genes were retrieved from the Entrez Gene database, and recognition sequences for TFs were searched on the database using the TFSEARCH. As a result, 21 potential TF-binding sites with significantly different incidence between the three gene groups (up-regulated, down-regulated and unaffected gene groups) could be identified (Table 1).

The binding sites for sterol regulatory element-binding protein 1 (SREBP-1), aryl hydrocarbon receptor (AhR/Ar) and olfactory 1 (Olf-1) were suggested to be involved in up-regulation, while the binding sites for glucocorticoid receptor (GR) and hepatocyte nuclear factor 1 (HNF-1) were suggested to be involved in down-regulation of the genes. In addition, the binding sites for activating enhancer-binding protein 4 (AP-4), nuclear factor-kappaB (NFκB), and early growth response 3 (Egr-3) were correlated with modulation of gene expression regardless of the direction of modulation. Among them, GR and/or SREBP-1 are interesting TFs in association with the recent findings of a trend of elevated body weight in irradiated mice. GR is known to be associated to the metabolic syndrome. Similarly, SREBP-1 is also involved in metabolic process by mediating the final regulatory step in LDL metabolism. Alteration of TF activity of GR and/or SREBP-1 as well as the related signaling pathways after exposure to IR may be the underlying mechanisms for this metabolic process.

6. Conclusion

Studies of modulated gene expression after exposure to IR have historically played an important role in elucidating the molecular mechanisms underlying cellular IR response. Especially finding of transcriptional network derived from p53 made a deep impact on understanding of biological defence system as well as development of radiation hazards such as radiation carcinogenesis. Recent technologies enable us simultaneously monitor modulation of gene expression across essentially the entire genome. It was revealed that so many TFs are involved in gene regulation *in vivo* that gene expression is finely modulated depending on tissues. However no single consistent picture of IR response has not been obtained because variety of experimental models, experimental conditions and data analyses have been employed. As a future perspective in this research area, carefully worked out methods for data analysis, data mining and construction of signalling network are strongly need. In addition, protein researches, especially a comprehensive analysis of protein expression, modification, sub-cellular localization is required, because these data cannot be obtained from the researches on transcriptional regulation.

7. Acknowledgment

This article was prepared with a support by the Grant-in-Aid from the Ministry of Education, Culture, Sports, Science and Technology of Japan (21510060), the Budget for Nuclear Research from the Ministry of Education, Culture, Sports, Science and Technology of Japan based on screening and counseling by the Atomic Energy Commission, and under contract with the Aomori Prefectural Government, Japan.

8. References

Chen, I. T., Smith, M. L., O'Connor, P. M. & Fornace, A. J. Jr. (1995). Direct interaction of Gadd45 with PCNA and evidence for competitive interaction of Gadd45 and p21Waf1/Cip1 with PCNA. *Oncogene*. Vo.11, No.10, pp.1931-1937.

Daino,K., Ichimura,S. & Nenoi,M. (2002) Early induction of CDKN1A (p21) and GADD45 mRNA by a low dose of ionizing radiation is due to their dose-dependent post-transcriptional regulation. *Radiation Research*. Vol.157, No.4, pp.478-482.

Daino, K., Ichimura, S. & Nenoi, M. (2006). Both the basal transcriptional activity of the GADD45A gene and its enhancement after ionizing irradiation are mediated by AP-1 element. *Biochim Biophys Acta*. Vol.1759, No.10, pp.458-469.

d'Ari, R. (1985). The SOS system. *Biochimie*. Vol.67, No.3-4, pp.343-347.

Fei, P. & El-Deiry, W. S. (2003). P53 and radiation responses. *Oncogene*. Vol.22, No.37, pp.5774-5783.

Fornace, A. J. Jr, Alamo, I. Jr & Hollander, M. C. (1988). DNA damage-inducible transcripts in mammalian cells. *Proceedings of the National Academy of Sciences*. 8Vol.5, No.23, pp.8800-8804.

Fornace, A. J. Jr, Nebert, D. W., Hollander, M. C., Luethy, J. D., Papathanasiou, M., Fargnoli, J. & Holbrook, N.J. (1989). Mammalian genes coordinately regulated by growth arrest signals and DNA-damaging agents. *Molecular and Cellular Biology*. Vo.9, No.10, pp.4196-4203.

Hartwell, L. H. & Weinert, T. A. (1989). Checkpoints: controls that ensure the order of cell cycle events. *Science*. Vol.246, No.4930, pp.629-634.

Nenoi, M., Daino, K., Nakajima, T., Wang, B., Taki, K. & Kakimoto, A. (2009). Involvement of Oct-1 in the regulation of CDKN1A in response to clinically relevant doses of ionizing radiation. *Biochim Biophys Acta*. Vol.1789, No.3, pp.225-223.

Oda, Y., Nakamura, S., Oki, I., Kato, T. & Shinagawa, H. (1985). Evaluation of the new system (umu-test) for the detection of environmental mutagens and carcinogens. *Mutatation Research*. Vol.147, No.5, pp.219-229.

Smith, M. L., Chen, I. T., Zhan, Q., Bae, I., Chen, C. Y., Gilmer, T. M., Kastan, M. B., O'Connor, P. M. & Fornace, Jr A. J. (1994). Interaction of the p53-regulated protein Gadd45 with proliferating cell nuclear antigen. *Science*. Vol.266, No.5189, pp.1376-1380.

Taki, K., Wang, B., Nakajima, T., Wu, J., Ono, T., Uehara, Y., Matsumoto, T., Oghiso, Y., Tanaka, K., Ichinohe, K., Nakamura, S., Tanaka, S., Magae, J., Kakimoto, A. & Nenoi, M. (2009), Microarray analysis of differentially expressed genes in the kidneys and testes of mice after long-term irradiation with low-dose-rate γ-rays. *Journal of Radiation Research*. Vol.50, No.3, pp.241-252.

Vares, G., Uehara, Y., Ono, T., Nakajima, T., Wang, B., Taki, K., Matsumoto, T., Oghiso, Y., Tanaka, K., Ichinohe, K., Nakamura, S., Tanaka, S. & Nenoi M. (2011). Transcription Factor-recognition Sequences Potentially Involved in Modulation of Gene Expression after Exposure to Low-dose-rate γ-rays in the Mouse Liver. *Journal of Radiation Research*. Vol.52, No.2, pp.249-256.

Zhan, Q., Bae, I., Kastan, M. B. & Fornace, A. J. Jr. (1994). The p53-dependent gamma-ray response of GADD45. *Cancer Research*. Vol.54, No.10, pp.2755-2760.

9

The Role of MicroRNAs in the Cellular Response to Ionizing Radiations

Nicolas Joly-Tonetti and Jérôme Lamartine
Center of Genetics and Molecular and Cellular Physiology, CNRS, University Claude Bernard Lyon I, University of Lyon, Villeurbanne, France

1. Introduction

Ionising irradiation is a major issue in our society. Although public attention mainly focuses on it when industrial disasters or medical accidents occur, or in military contexts, it is also a permanent natural phenomenon and a powerful tool for diagnosis and therapy, notably in oncology. Understanding the action mechanisms and consequences of these radiations in living organisms is a major aim for public health and environmental conservation. Conceptions have begun to change over the past three decades. Previously, the central theory focused on the irradiated cells themselves, postulating that the biological effects are direct consequences of DNA damage when repair processes fail. More recently, however, there has been accumulating evidence that important biological consequences arise in cells with no direct radiation exposure (Kennedy et al., 1980; Nagasawa & Little, 1992; review in Wright, 2010). On the one hand, progeny cells over many generations present radiation-induced genomic instability, increasing mutation rates and chromosomal abnormality implicating reactive oxygen species. On the other hand, both neighbouring and remote cells are subject to a radiation-induced bystander effect, induced by signals from directly exposed cells *via* a variety of communication pathways. Both cell response phenomena point to an implication of epigenetic mechanisms, such DNA methylation, chromatin remodelling or small RNA regulation (Aypar et al., 2011).

MicroRNA was discovered nearly 20 years ago (Lee et al., 1993; Wightman et al., 1993). The growing family of small RNAs, including small interfering RNA (siRNA), piwi-interacting RNA (piRNA), repeat-associated-siRNA (rasiRNA) and heterochromatic small RNA (hcRNA), forms the most abundant class of endogenous RNA in metazoans, but has also been characterised in plants, unicellular algae (Zhao et al., 2007), DNA viruses (Pfeffer et al., 2004) and, controversially, in retroviruses (Klase et al., 2007). MicroRNAs are involved in numerous cellular processes such as development, viral infection, apoptosis and stress response, by regulating protein-coding gene expression at post-transcriptional and translational levels (review in Bushati & Cohen, 2007). They are also differentially expressed between tumour cells and normal counterparts in both benign proliferation and cancer (Lu et al., 2005). Due to this accumulating evidence, it is worthwhile studying microRNA response to ionising radiation, and this was started few years ago. This chapter will review microRNA studies relating to ionising radiation and discuss the perspectives in this field.

Before reviewing miRNA function in radiation response, we will begin with an account of the biogenesis of microRNAs, their mode of action in association with Argonaute protein, their function and the methods currently used to study them.

2. General information on microRNA: Biosynthesis, function and methods of investigation

2.1 MicroRNA biogenesis and microRNA ribonucleoprotein (miRNP) assembly

MicroRNA biogenesis is the cellular process producing active 22-nt long microRNA from large transcripts coded by the cell genome (review in Bartel, 2004; cf. fig. 1). Given differing microRNA profiles according to tissue and cell type and in response to environmental change, microRNA biogenesis can be seen to be a regulated process. Moreover, the bad correlation between the primary transcript synthesis rate, called pri-miR, and the mature microRNA reveals significant posttranscriptional regulation (Thomson et al., 2006).

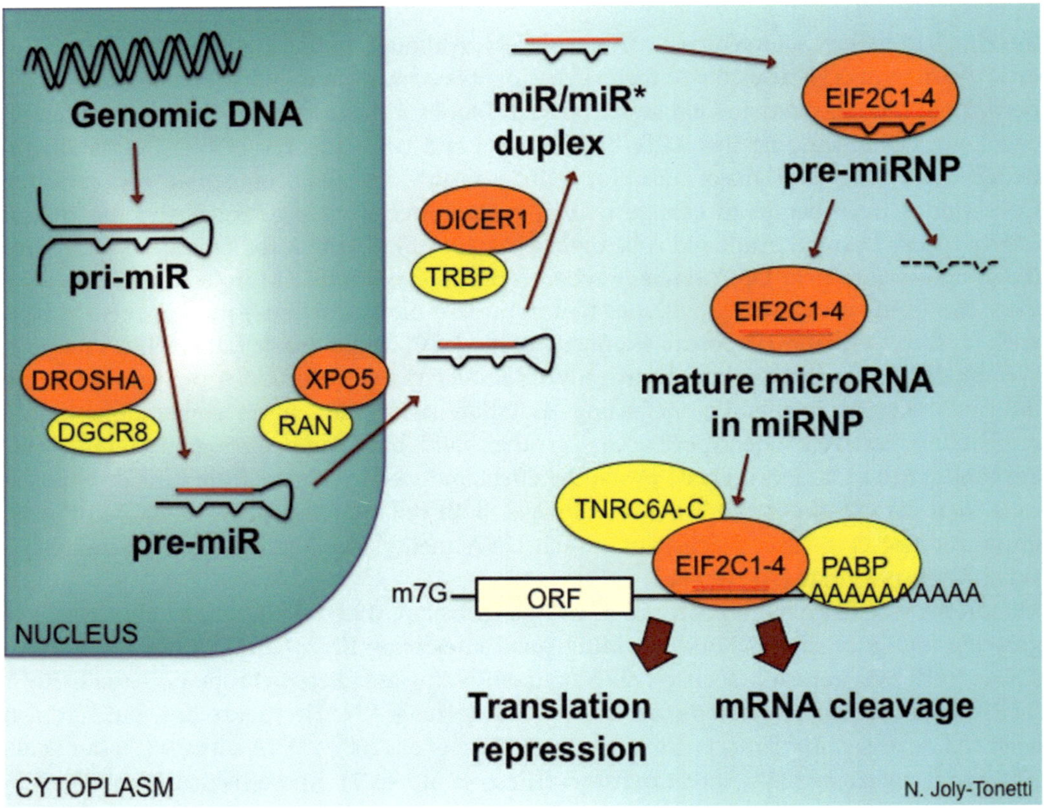

Fig. 1. MicroRNA biogenesis pathway

2.1.1 Pri-miR transcription and edition

Although a few microRNAs are transcribed by non-coding (nc) RNA associated with RNA polymerase III (Borchert et al., 2006), almost all are transcribed by RNA polymerase II (Lee et al., 2004). This primary transcript displays at least one stem-loop hairpin structure, presenting mismatches or wobbles, which is well conserved in all animal species. Like messenger RNA (mRNA) genes, pri-miR is capped and polyadenylated (Bracht et al., 2004; Cai et al., 2004). MicroRNAs are clustered on the genome and many are transcribed in polycistron RNA. One of the most frequently studied is the miR-17-92 cluster, which encompasses 6 stem-loops and plays an important role in non-solid and solid cancer (He et al., 2005; review in Mendell, 2008). Although few data are available about their transcription regulation and promoters, some microRNA are known to respond to conventional transcription factors, such as p53 (Raver-Shapira et al., 2007; Corney et al., 2007; Xiao et al., 2011) or MYC (O'Donnell et al., 2005), and are subject to chromatin remodelling (review in Sato et al., 2011).

The sequence transcribed in the pri-miR does not always correspond to the mature microRNA because of adenosine-to-inosine editing by the dsRNA adenosine desaminase ADAR1 or ADAR2 (review in Nishikura, 2010). Both enzymes catalyse adenine hydrolytic desamination in hypoxanthine (Bass & Weintraub, 1988) which pairs cytosine, creating a non-Watson-Crick pairing I:U. This wobble affects the secondary structure of pri-miR, with consequences for microRNA maturation and action. Depending on the site of edition on pri-miR, different consequences are possible, for instance impairment of microRNA maturation (Yang et al., 2006; Kawahara et al., 2007b) or modification of the complete target mRNA list (Kawahara et al., 2007a).

ADAR1 and ADAR2 tend to edit distinct sites and are subjected to regulation since ADAR2 inhibits ADAR1 activity (Kawahara et al., 2007a) and inflammation can increase ADAR1-isoform p150 expression (Yang et al., 2003). Moreover, ADAR1 can occur on pre-miR, in nucleus or cytoplasm. This regulated ADAR-drive edition may explain edition tissue-specificity: depending on the organ, only a few or all microRNA precursors will be desaminated at various sites (Kawahara et al., 2007a; Kawahara et al., 2007b). Although a pioneering general survey of microRNA precursors found editing in 6% of the pri-miRs examined (Blow et al., 2006), *in vitro* studies showed half of pri-miRs to be editable (Yang et al., 2006). This implies that inflammation may impact microRNA biogenesis and mRNA targeting.

Canonical maturation of animal pri-miR is then a sequential two-step process that take place in the nucleus and cytoplasm, performed by two bacterial homologue type-III RNase enzymes in partnership with other proteins, with a strong intermediate check-point, respectively producing a microRNA precursor (pre-miR) and the mature microRNA.

2.1.2 DROSHA - Dependent and independent pathways to pre-miR

Most microRNA genes are located in intronic regions of protein-coding genes or in non-coding regions of the genome (Rodriguez et al., 2004; Baskerville & Bartel, 2005). Based on pre-miR end characteristics, the endonuclease DROSHA has been identified as a pri-miR trimming enzyme: it identifies the pri-miR hairpin shape and produces what is called microRNA precursor (pre-miR), approximately 70 nucleotides in length (Lee et al., 2003). DROSHA acts in association with many proteins, leading to the formation of microprocessor complexes, including in particular DiGeorge syndrome critical region gene 8 (DGCR8), the homologue of *Drosophila melanogaster* pasha protein (Han et al., 2004; Gregory et al., 2004). The DROSHA-DGCR8 multimeric complex displays the most active *in vitro* pri-miR maturation activity. DGCR8 is therefore thought to be a molecular anchor (Han et al., 2006), positioning DROSHA at 10 bp from the hairpin base for trimming. In an alternative,

DROSHA-independent process, pre-miRs are generated during mRNA maturation by the formation of a short stem-loop intron, called a mirtron, then linearised by RNA lariat debranching enzyme (DBR1) (Okamura et al., 2007; Ruby et al., 2007). The resulting pre-miR structure is similar to that released by DROSHA. Recent studies have discussed coupling between microprocessor and spliceosome, trimming pre-miR into a larger intron before splicing and enhancing intron degradation (Morlando et al., 2008; Kataoka et al., 2009). In addition, other complex partners appear to be crucial for microRNA subset expression. DROSHA associates with DEAD-box helicase p68 (DDX5) and p70 (DDX17), which are involved in processing one subclass of pri-miR (Fukuda et al., 2007). Pri-miR maturation can also be influenced by various nuclear factors in response to stimuli, such as pri-miR-21 enhancement proceeding by R-SMAD in response to TGF-β and BMP in vascular smooth muscle cells (Davis et al., 2008) or pri-let-7 family excision inhibition by LIN28 in stem cells and tumours (Viswanathan et al., 2008). P53 also associates with DROSHA and helicase p68, enhancing maturation of growth-suppressive microRNA (Suzuki et al., 2009). For half of the microRNA family, DROSHA also cuts a microRNA-offset RNA (moRNA or moR) created from sequences immediately adjacent to pre-miR (Shi et al., 2009). These are systematically generated with RNA of early evolutionary origin (Langenberger et al., 2009), massively enriched in the nucleus (Taft et al., 2010), and sometimes more expressed than hairpin-sharing microRNA (Umbach et al., 2010); their function remains unknown.

2.1.3 Nuclear export checkpoint

Like smaller shuttling RNA (Izaurralde et al., 1997), pre-miR is actively exported by nucleocytoplasmic transport in a RAN-GTP dependent manner (Yi et al., 2003). The karyopherin family protein exportin-5 (XPO5) cargoes pre-miR sequence-independently (Bohnsack et al., 2004; Lund et al., 2004) and controls the quality of pre-miR: at one end, pre-miR needs at least a 18-nt double-stranded stem for effective nuclear export (Zeng & Cullen, 2004); and at the other end, the 5′ protruding end and 3′ long end inhibit shuttle (Lund et al., 2004; Zeng & Cullen, 2004). The terminal loop structure and the distance to the hairpin base are the only parts unchecked. Moreover, exportin-5 binding protects pre-miR against degradation (Zeng & Cullen, 2004; Lee et al., 2011). Pre-miR can be subject to edition by ADAR, before exportation by its two isoforms and then by p150 (Kawahara et al., 2007b).

2.1.4 Cytoplasmic generation and loading of microRNA duplex

In the cytoplasm, a class III RNAse III enzyme, DICER1, recognises pre-miR and cuts both strands of the imperfect duplex at about two helical turns (~22 nucleotides) from the base of the hairpin, creating a short RNA duplex with 3′ 2-nucleotide overhang ends (Hutvágner et al., 2001; MacRae et al., 2006). This cleavage, less reliable than with DROSHA, creates a 3′ polymorphism for 5′-strand-generated microRNA. DICER1 associates several proteins, like TRBP (Chendrimada et al., 2005), which contribute to his function. The imperfect duplex associates ATP-dependently (Kawamata et al., 2009; Yoda et al., 2010) with Argonaute protein to form an immature complex called microRNA ribonucleoprotein precursor (pre-miRNP) or RNA-induced silencing complex precursor (pre-RISC or pre-miRISC, with reference to mRNA cleavage siRNA). Central mismatch at position 10-11 (Tomari et al., 2007) and Hsc70/Hsp90 chaperone machinery (Iki et al., 2010; Iwasaki et al., 2010) contribute to duplex loading.

2.1.5 Strand selection and miRNP assembly

There are thermodynamic differences in base-pairing stability on either side of the duplex, and the strand with the greater 5′ stability will be preferentially sorted in the effecter complex, called miRNP or RISC. Mismatches in the seed region (nt 2-8) and 3′ mid region (nt 12-15) favour strand separation (Kawamata et al., 2009). Loading is highly sensitive to the presence of phosphate at the 5′ end, which completely prevents loading if lacking. The loaded strand is called the "guide" strand, and the other, "passenger", strand is discarded. However, the loaded strand is not always the same. Evolutionary pressure on both strand seed regions and their complementary sequences on target mRNA reveals that microRNA originally from opposite strands can regulate different target genes in a coordinated fashion (Okamura et al., 2008). In a former nomenclature, when there is a predominant loaded strand in terms of relative abundance, the mature sequence is named according to the miR form (e.g., miR-17 or let-7b) whereas the opposite arm, if loaded, is called star microRNA and named according to the miR* form (e.g., miR-17* or let-7b*) (Ambros et al., 2003a). However, miR/miR* nomenclature is phased out in favour of the -5p/-3p nomenclature, the suffixes 5p and 3p referring to the 5′ and 3′ sides of pre-miR, respectively (e.g., miR-28-5p or miR-28-3p). Argonaute proteins associate with several factors in the RISC, such as GW182-family proteins TNRC6 (Landthaler et al., 2008) or PABPC1 (Huntzinger et al., 2010). TNRC6 lead miRNP localisation to cytoplasmic P-bodies (Liu et al., 2005).

2.2 MicroRNA function

MicroRNAs are fundamental posttranscriptional regulators of gene expression (reviewed in Fabian et al., 2010). Although some studies reported that, in particular situations, microRNAs can activate protein translation (Vasudevan et al., 2007; Ørom et al., 2008; Henke et al., 2008) or can be reimported into the nucleus (Hwang et al., 2006), cytoplasm microRNA-mediated gene repression, by mRNA decay or translation repression, is the general mechanism of action. How one of these two mechanisms or both is brought into play, however, is not well understood. In addition, they have been described as reducing stochastic gene expression noise (Cui et al., 2007) and as possible modifiers of gene regulation paradigms, acting *in cis* and *trans* to regulate of MRE-containing RNA (Salmena et al., 2011).

2.2.1 Mechanistic aspects of microRNA repression

In animals, microRNA-mediated gene repression involves two mechanisms: translation inhibition of target mRNA and decay of mRNA after deadenylation. Although the initial dichotomy between animals and plants highlighted translation inhibition by imperfect matching in the former and mRNA decay by perfect matching in the latter, it is now clear that animal microRNAs mainly act on mRNA stability (Bagga et al., 2005; Guo et al., 2010) whereas plants exhibit translation inhibition mechanisms without notable effect on mRNA stability (Brodersen et al., 2008).

Recent debate has focused on three possible mechanisms of target mRNA translational inhibition: inhibition of translation initiation by disturbing mRNA circularisation or ribosome assembly, or ribosome drop-off during translation elongation. MicroRNA-mediated mRNA degradation is also still in debate.

Several studies revealed that mRNA that lack a functional m7GpppG or present a non-functional ApppG cap structure or whose translation is cap-independent are less sensitive to microRNA-mediated translational repression (Humphreys et al., 2005), possibly because miRNP can interfere with the cap recognition process at the initiation step (Pillai et al., 2005).

Another mechanism inhibiting initiation of translation may be inhibition of ribosome 80S assembly. Wang et al. (2008) mentioned 40S subunit enrichment in mRNP subjected to microRNA-mediated repression in rabbit reticulocyte lysates. Chendrimada et al. (2007) found eIF6, a 60S subunit inhibitor (Ceci et al., 2003), coimmunoprecipated with miRNP components and MOV10, the homologue of the *D. melanogaster* translational repressor Armitage. However, some contradictory data (Ding et al., 2008; Eulalio et al., 2008) highlight the importance to further investigate this mechanism of inhibition.

Mechanism of microRNA-mediated decay is a two-step process implicating poly(A) deadenylation and decapping and is dependent to GW182 family proteins. At the first step, GW182 interacts with PABP and interferes with mRNA "close conformation" loop by competition with eIF4G (Fabian et al., 2009). GW182 also recruits CCR4-NOT1 complex, including CAF1 exonuclease, to poly(A) removal (Behm-Ansmant et al., 2006). At the second step, GW182 recruits DCP1-DCP2 complex to induce decapping (Rehwinkel et al., 2005).

These various inhibition mechanisms are not mutually incompatible, and protein synthesis inhibition may be the first step or be reinforced by target mRNA decay (Fabian et al., 2009). However, nothing is known about how ionising radiation affects the process.

2.2.2 MicroRNA target recognition

MicroRNA has emerged as a protein synthesis regulator *via* translation repression and mRNA decay. In animals, microRNAs bind by base-pairing mainly to the 3′ untranslated region (3′UTR) of their target gene (Lee et al., 1993) on sites called microRNA-responsive element (MRE) (cf. fig. 2). The 5′UTR and coding region may also present MREs, but which are less efficient than in the 3′UTR (Gu et al., 2009). Except for particular MREs with good evolutionary conservation presenting almost perfect base-pairing on the entire microRNA, such as miR-196a and HOXB8 mRNA (Yekta et al., 2004), the canonical MREs constitute a sequence of perfect Watson-Crick matches between target mRNA 3′UTR and the microRNA 5′ proximal region (positions 2 to 7). Pairing for the rest of the microRNA is less stringent and more variable. This six-nucleotides sequence, called the seed region, is the strongest requirement for gene repression and rules microRNA annotation and sequence family composition. Accumulating evidence, however, has complicated this notion, as microRNA:mRNA interactions tolerate mismatch or GU wobble in the seed region (Didiano & Hobert, 2008), which can be helped by a 3′ compensatory site (Friedman et al., 2009). 3′ or centred regions of microRNA also play a role in the concert orchestrating mRNA regulation (Shin et al., 2010). Moreover, its position on 3′UTR can affect MRE activity: the functional target site is located away from the centre of long UTR and at least 15 nt from the stop codon (Grimson et al., 2007).

3′UTR sequence and structure also influence microRNA activity such as proximity to AU-rich elements (ARE) and presence of coexpressed microRNA-MRE. Secondary structure can hide MREs, depending on dsRNA binding protein (Kedde et al., 2010). Obviously, the presence of multiple MREs for the same microRNA strengthens its action. Many mRNA binding proteins (mRBP) have been shown to be involved in microRNA action: tristetrapolin (TTP) (Jing et al., 2005), APOBEC3G (Huang et al., 2007), DND1 (Kedde et al., 2007), PUF-family proteins (Nolde et al., 2007; Galgano et al., 2008) and HuR. The role of

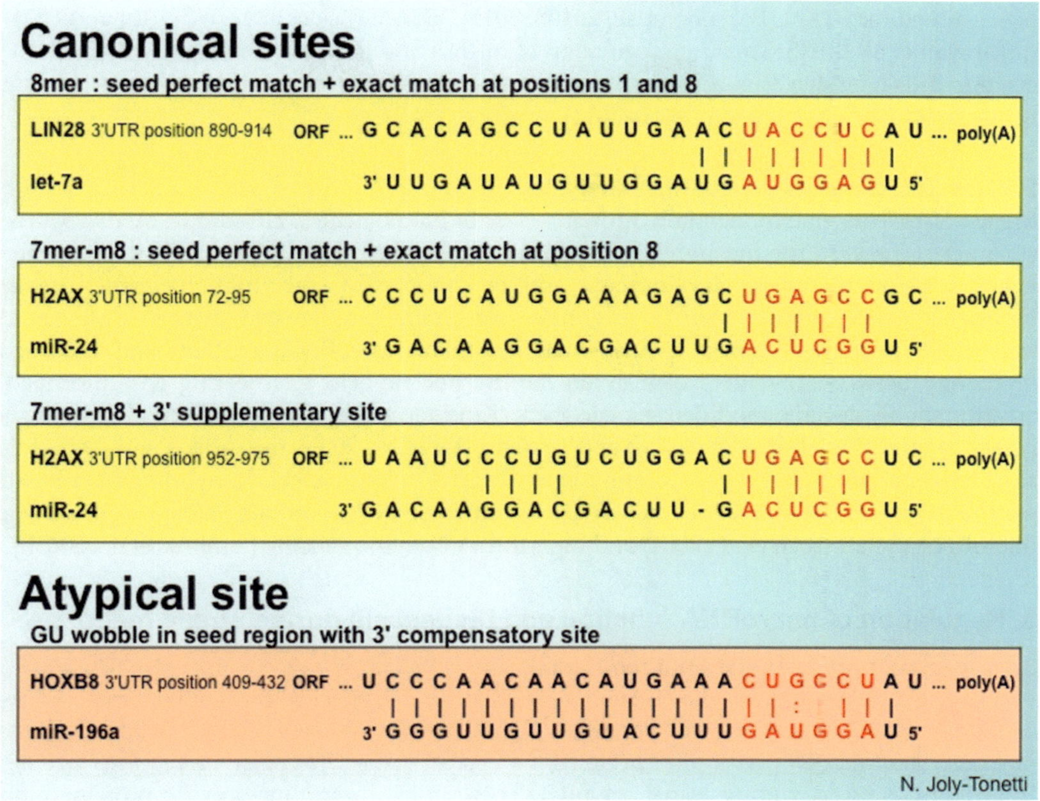

Fig. 2. Examples of microRNA-responsive elements (MRE). Seed region on microRNA and corresponding sequence on mRNA are highlighted in red.

mRNA binding protein could be ambivalent for microRNA activity, as in the case of HuR-related microRNA-mediated mRNA targeting: on the one hand, Bhattacharyya et al. (2006) showed that, in case of stress in Huh7 hepatoma cells, HuR competes with ARE-binding protein, originally directed at miR-122-mediated CAT1 mRNA repression, to address CAT1 messenger to the polysome; on the other hand, Kim et al. (2009) demonstrated that HuR binds MYC mRNA near a let-7-responsive element, enabling HuR-dependent let-7-mediated MYC mRNA decay and translation inhibition, probably by inducing secondary structural modifications of mRNA uncovering a let-7 MRE. This cell-type specific mRNA binding proteome highlights an important regulatory action mode of microRNA.

2.2.3 MicroRNA in the “competitive endogenous RNA” network

MicroRNA target regulation involves other parameters than microRNA:MRE interactions. It is now clear that the microRNA:target mRNA ratio is crucial for microRNA action: modulation of microRNA with weak expression has little impact on target mRNA; conversely, regulation of highly expressed microRNA has a more pronounced effect on the target (Arvey et al., 2010). Thus Seitz hypothesised that microRNA can be titrated by pseudo-target mRNAs, moving them away from their “true” target genes (Seitz, 2009). Another study identified pseudo-genes, the mRNA levels of which sponge microRNA targeting the corresponding gene, due to strong similarities between both 3’UTRs. This has been proved for PTEN (Poliseno et al., 2010b; 2011), VCAN (Lee et al., 2009; 2010) and CD44 (Jeyapalan et al., 2010). This transregulation of mRNA and long non-coding RNA (lncRNA) *via* microRNA and MREs acting as morphemes, constitutes a promising hypothesis for future investigations (Salmena et al., 2011).

2.3 Methods for global microRNA analysis

MicroRNAs form an exponentially growing class of biomolecules. Hundreds are discovered each year, from a growing number of species. Since alterations in microRNA expression levels were identified in development and in cancer and constitute powerful tumour biomarkers and potential targets, scientists were initially interested in high throughput microRNA profiling. Thanks to the small length of microRNA (~22 nt) and the high homology between members of a given family, nucleic acid engineering has improved, creating highly specific modified nucleic acids (Kloosterman et al., 2006). Several techniques have been developed: bead-based flow cytometry (Lu et al., 2005), microRNA serial analysis of gene expression (miRAGE) (Cummins et al., 2006) oligonucleotide hybridisation-based microarrays (Liu et al., 2004), RT-qPCR for mature (Chen et al., 2005) or precursor microRNA (Schmittgen et al., 2004) and high-throughput sequencing (Ambros et al., 2003b).

3. Regulation of microRNA function and biogenesis during stress response

3.1 Argonaute posttranslational regulation

Argonaute family proteins have been identified in all species presenting small RNA-mediated processes and are well conserved. They are phylogenetically classified into three families. The first comprises piwi proteins, named after the *Drosophila melanogaster* protein "P-element induced wimpy testis", or piwi. These animal germline-specific proteins bind piRNA and are essential to spermatogenesis and transposon control (Aravin et al., 2006; Brennecke et al., 2007). The second family consists of *Caenorhabditis elegans* secondary Argonaute (SAGO) proteins such as Sago-1 or Sago-2. That bind the 5′-triphosphate end of secondary siRNA to amplify the RNA interference signal in a retrocontrol loop fashion (Yigit et al., 2006). The third family, which is the largest and most universal, encompasses the Argonautes, named after and resembling the founding member AGO1 expressed in *Arabidopsis thaliana* (Bohmert et al., 1998; reviewed in Ender & Meister, 2010); this family form the universal core component of miRNP. The human genome encodes 4 Argonautes, called eukaryotic translation initiation factor 2C (eIF2C) subunits 1 to 4. EIF2C1, -3 and -4 are closely clustered on chromosome 1 whereas EIF2C2 is located on chromosome 8. Only EIF2C2-coded protein (known as eIF2C2, AGO2 or hAgo2 in the literature) has an experimentally determined cleavage activity (Liu et al., 2004). Argonaute proteins are overall bilobal ~90 kDa and share three characteristic domains. The first lobe incorporates the N-terminal Piwi-Argonaute-Zwille (PAZ) domain, binding the 3′-hydroxyl end of microRNA in a hydrophobic cleft (Yan et al., 2003; Lingel et al., 2004a; Ma et al., 2004). The second lobe contains the MID domain, binding the 5′-phosphate end of microRNA in a basic uridin- and adenine-specific pocket (Frank et al., 2010) and the C-terminal PIWI domain, an RNase H module which acts as an RNA double-strand-specific endonuclease that cleaves microRNA-targeted mRNA.

Like many other proteins, Argonaute proteins are substrates to posttranslational modifications. EIF2C2 is phosphorylable on several serine, tyrosine and threonine residues (Rüdel et al., 2011), in particular *via* a p38MAPK-MAPKAPK2 axis, that leads to its proper P-body localisation. Other phosphorylations impair microRNA association and suppress RISC activity (Rüdel et al., 2011). Argonaute proteins are also proline-hydroxylated, that is important for protein stability and effective RNA interference (Qi et al., 2008). It would be interesting to evaluate the impact of ionising radiation on such posttranslational modifications and on microRNA activity modulation.

3.2 DNA damage regulates microRNA transcription and maturation

DNA damage can regulate microRNA expression at the transcriptional level. MiR-34 family is p53-dependently induced by DNA damage. P53 binds to specific responding elements in the promoter of miR-34a and miR-34c gene and activates transcription of pri-miR-34a and pri-miR-34b/c (Chang et al., 2007; Corney et al., 2007; Raver-Shapira et al., 2007). MiR-34 family members directly repress the expression of several targets such as cyclin E2, BCL2, CDK4 and CDK6, involved in the regulation of the cell cycle, apoptosis and DNA repair. Ectopic expression of miR-34a induces apoptosis or cell cycle arrest, whereas inactivation of endogenous miR-34a inhibits p53-dependent apoptosis after DNA damage.

In addition to the miR-34 family, p53 regulates the transcription of several microRNA including miR-107 (Yamakuchi et al., 2010), miR-145 (Sachdeva et al., 2009) and the two homologous microRNA miR-192/215 (Braun et al., 2008). Ectopic expression of miR-192/215 induces cell cycle arrest. Their protein targets include transcripts that regulate G1/S and G2/M checkpoints. Altogether, these findings suggest that the DNA damage response (DDR) is able to activate a p53-dependent expression of specific microRNA that in turn play a key role in the cellular response to genotoxic stress.

Additional mechanisms by which DNA damage modulates microRNA processing and maturation have been recently described. It was reported that, in cells responding to DNA damage, p53 activated DROSHA-mediated processing of microRNA such as miR-16, miR-143 and miR-145, known to have a growth-suppressive function (Suzuki et al., 2009). More precisely, p53 interacts with the DROSHA/DGCR8 processing complex through an association with RNA helicase p68. DNA mutations in the p53 gene, frequently observed in tumours, can lead to decreased microRNA processing. Moreover, members of the p53 family (p53, p63, p73) could function as regulators of microRNA processing by direct transcriptional regulation of components of the microRNA processing complex (DROSHA-DGCR8, DICER-TRBP2, Argonaute proteins) (Boominathan. 2010).

4. MicroRNA and ionising radiation: The state of the art

MicroRNA is a new field of biological investigation. Studies of microRNA and ionising radiation in particular began less than five years ago and the number of articles published in the field is growing exponentially. Different strategies have been used in these studies: several teams tried to identify microRNA targeting proteins involved in signalling DNA double-strand breaks; others looked for miRNA biomarkers of radiation exposure or radiosensitivity. We will review the main results from these studies below.

4.1 MicroRNA and the DNA damage response: An overview

DNA double-strand breaks (DSB) caused by gamma-radiation activate a signal transduction process that leads to cell cycle arrest, followed by either apoptosis or DNA repair. Numerous proteins are involved in these pathways at different levels. Some are sensors or

mediators of DSB signalling: this is the case of H2AX, which is phosphorylated by ATM to further recruit partners such as BRCA1 and 53BP1 and form a scaffold around the DNA break site. The DNA damage signal is then transduced by ATM/ATR kinase to downstream pathways such as cell cycle, DNA repair or apoptosis (see figure 3). MicroRNA are involved in DDR regulation by targeting most of the proteins involved at each step of the DDR.
We will below briefly describe the core DDR pathway protein components and review the microRNA that can regulate DDR by targeting them.

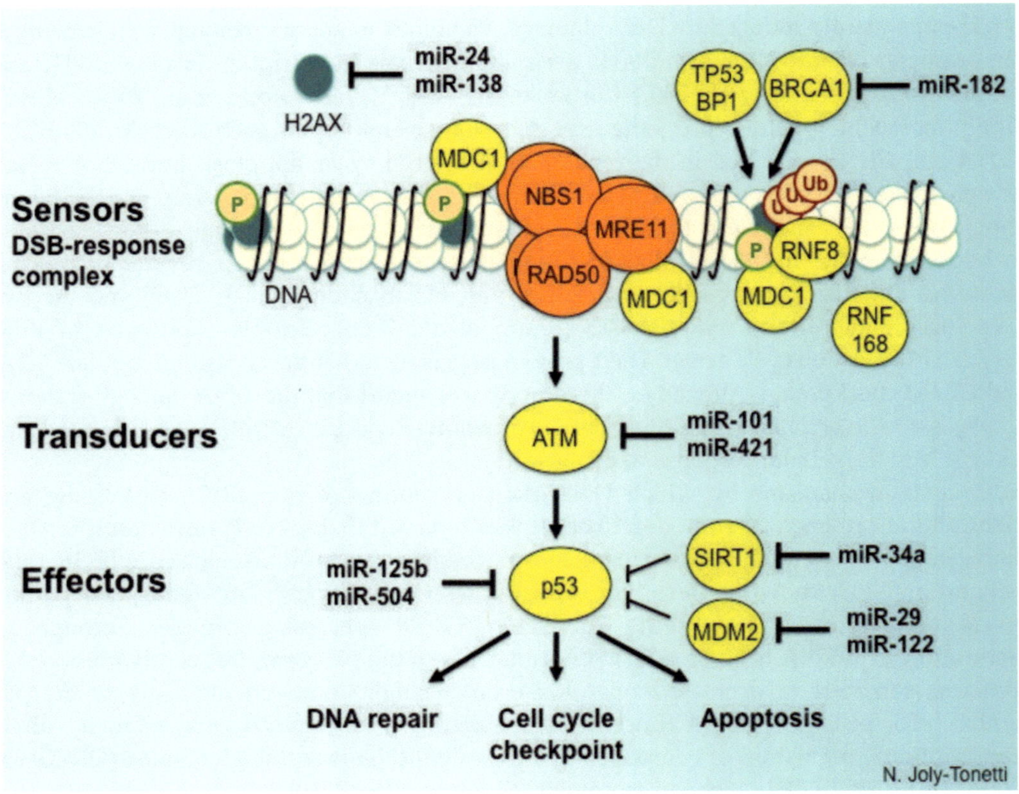

Fig. 3. MicroRNA involvement in the DSB response pathway.

4.1.1 Sensors/mediators of DSB signaling

The MRN complex composed of MRE11-RAD50-NBS1 proteins acts as a DSB sensor and recruits ATM. ATM-dependent phosphorylation of histone variant H2AX seems to be the initial signal for DDR protein accumulation. MDC1 binds to both the MRN complex and P-H2AX to further recruit ubiquitin ligase RNF8 and form a scaffold for accumulation of RNF168, BRCA1 and 53BP1 around the DSB site (cf. fig. 3).
MiR-24 has been identified as a regulator of H2AX: miR-24-mediated suppression of H2AX increases radiosensitivity in terminally differentiated hematopoietic cells (Lal et al., 2009). Another microRNA targeting H2AX has been recently identified (Wang et al., 2011): in a human osteosarcoma cell line, miR-138 reduced histone H2AX expression. Overexpression of miR-138 inhibited homologous recombination and enhanced cellular sensitivity to several DNA-damaging agents.

MiR-182 down-regulates BRCA1 expression. Antagonising miR-182 enhances BRCA1 protein levels and protects them from ionising radiation-induced cell death, while overexpressing miR-182 reduces BRCA1 protein level, impairs homologous recombination-mediated repair, and render cells hypersensitive to ionising radiation (Moskwa et al., 2011).

4.1.2 Transducers: ATM/ATR

ATM and ATR serine/threonine kinases transduce DNA damage signals to downstream proteins. ATM responds in particular to DSB, and ATR to single-strand break (SSB). ATM and ATR have global effects on many aspects of cell and organism function: a vast network of over 700 ATM/ATR targets that function in numerous signalling pathways has been revealed by a genomic approach (Matsuoka et al., 2007).

ATM expression is also under the control of microRNA: miR-421 and miR-101 are reported to suppress ATM expression by targeting the 3' UTR of ATM transcripts (Hu et al., 2010; Yan et al., 2010). Targeting ATM through miR-101 sensitises tumours to radiation (Yan et al., 2010). On the other hand, ATM has recently been described as a regulator of microRNA biogenesis in DDR: ATM directly binds and phosphorylates KRSP (KH-type splicing regulatory protein), enhancing interaction between KRSP and pri-miR and increasing microRNA processing (Liu & Liu, 2011; Zhang et al., 2011).

4.1.3 Effectors: p53 as a core regulator

DDR effectors include DNA repair, cell cycle checkpoints (G1/S, intra-S and G2/M) and apoptosis. Several proteins are specifically involved in these mechanisms; p53 transcription factor notably regulates all pathways. P53 is involved in the control of DNA repair through direct regulation of p48 (Hwang et al., 1999; Kastan & Bartek 2004). After DNA damage, the G1/S checkpoint is essentially controlled by the ATM/CHK2-p53/MDM2 pathway. ATM/CHK2 is activated by DSB to phosphorylate and stabilise p53, which in turn induces p21 expression. P21 silences the G1/S-promoting cyclin E-A/CDK2 complex through cyclin-dependent kinase inhibition, inducing G1 arrest. P53 is also involved in the G2/M checkpoint, but here it is the cyclinB/CDK1 complex which is silenced by p21, leading to late G2 arrest (Kastan & Bartek 2004). P53 also plays a central role in DSB-induced apoptosis through transcriptional activation of pro-apoptotic factors, such as Puma, BAX, Fax and NOXA (Haupt et al., 2003).

Recent studies have demonstrated that microRNAs interact with p53 and its networks at multiple levels. MiR-504 and miR-125b were both identified as direct regulators of p53 expression: ectopic expression of miR-504 reduces p53 expression and impairs p53 function, especially p53-mediated apoptosis and G1/S cell cycle arrest in response to stress (Hu et al., 2010). MiR-125b, a brain-enriched microRNA, also acts as a negative regulator of p53. Knocking down miR-125b leads to massive apoptosis in human cells during stress response (Le et al., 2009). In addition, p53 can be indirectly regulated by microRNA targeting its upstream regulators. MiR-34a targets SIRT1, a negative regulator of p53 and then exerts a positive regulatory effect on p53 to activate apoptosis (Yamakuchi et al., 2008). MiR-122 (Yamakuchi et al., 2008), miR-29 (Park al., 2009) and miR-605 (Xiao et al., 2011) play a similar role through direct or indirect regulation of MDM2, another p53 inhibitor.

4.2 Using microRNA as a biomarker of radiation exposure and radiosensitivity

MicroRNA profiles correlate more accurately than protein-coding gene transcriptome in cancer classification, especially for tumours that are poorly differentiated or of embryonic or unknown origin (Lu et al., 2005). Moreover, some microRNAs have also been identified as ubiquitous evolutionarily conserved stress players, such as miR-210 in hypoxia (Kulshreshtha et al., 2007; review in Chan & Loscalzo, 2010). As ionising radiation induces DNA-damage related and oxidative stress, it appears interesting to evaluate microRNA's potential as a biomarker. The goal is to use miRNA as a biomarker of radiation exposure or predictor of cell radiosensitivity. The pioneering work in this field investigated the response of TK6 human B-lymphoblastoid cell line to ionising radiation (Marsit et al., 2006). Noticing that TK6 cells exhibited greatly altered gene expression 4 hours and 6 days after 2.5 Gy gamma irradiation (Amundson et al., 2005), the authors reproduced the experimental conditions and analysed microRNA expression profile on oligonucleotide microarrays. They compared this response with 6-day folate deprivation, known to alter genomic methylation and impact cell viability (Stempak et al., 2005), and with 6-day sodium arsenite treatment, all three forms of stress being known to be carcinogenic in humans, although the link between the alterations and incidence of cancer remains to be clarified. The authors showed that gamma radiation did not usually alter microRNA expression, in contrast to a general increase after folate deprivation or sodium arsenite treatment. After 4 hours, only a dozen microRNAs, and after 6 days only a few, exhibited a non-significant >2-fold increase. In contrast, there was significant microRNA alteration after folate deprivation, confirmed *in vivo* in patients with low folate intake, making miR-222 a biomarker of low folate intake. In their model, the authors inclined towards both posttranslational and transcriptional response to radiation. Additional studies investigated potential biomarkers of ionising radiation exposure. Maes et al. (2008) crossed microRNA, mRNA and proteomic profiling, to analyse *in vitro* foreskin fibroblast response to low (0.1 Gy) and high dose (2 Gy) X-rays. SAM analysis was performed. Only a few microRNAs responded to radiation, showing weak modulation (≤2.6-fold change), confirmed by individual assays. By crossing microRNA profiles with transcriptome and proteome data, they identified potential target genes. Chaudhry et al. (2009) analysed 20 microRNAs by individual RT-qPCR. They explored two lymphoblast cells: TK6 B cell line and Jurkat acute lymphocytic leukaemia (ALL) T cell line, irradiated by 2 Gy gamma rays, and analysed their response 4, 8, 12 and 24h after irradiation. There was a 2- to 10-fold increase for a few miRNAs, with time- and cell line-dependent response. Only miR-142-5p, -3p and miR-145 were usually modulated between cell lines, but with different patterns. Small nuclear RNA RNU6-2 was used to normalise the qPCR data. Templin et al. (2011a) studied microRNA response in mouse blood 6h and 24h after 0.5, 1.5 and 6 Gy low-LET gamma irradiation and 0.1 and 0.5 Gy high-LET ^{56}Fe irradiation. Overall, they found 31 microRNAs differentially expressed; three-quarters, however, were specific to one condition and thus time-, dose- and radiation type-dependent: the results showed little overlap. Investigating blood microRNA response 4 hours after 1.25 Gy total-body X-ray irradiation in a cohort of 8 patients in leukaemia or lymphoma remission ahead of stem transplantation, Templin et al. (2011b) used RT-qPCR to identify 45 up-regulated microRNA, including 27 that were found in all patients. Their data selection, however, did not include a quantification cycle cut-off, which is essential.

By testing various cell lines, doses, radiation types and times after irradiation, numerous microRNA responding to ionising radiation were identified reviewed by Dickey et al. (2011). It seems that microRNA respond with modest modulation to ionising radiation, in a cell type-, dose- and time-dependent manner. Altogether, these data may indicate that microRNA are unsuitable as useful biomarkers of radiation exposure.

Another possible approach is to use microRNA as indicators of radiation sensitivity. It is well known that a given type of tumour can show varying radiosensitivity. Developing radiosensitivity prediction tools is important for effective tumour management. Lee et al. (2011) investigated the radiosensitivity of five cancer cell lines. Analysing the 3'UTR of EGFR, a membrane receptor frequently activated in cancer, they identified several miR-7-responsive elements. By modulating the miR-7 expression level by ectopic overexpression or by antisense inhibition, they were able to decrease or increase cell survival respectively.

4.3 MicroRNA and radiation-induced bystander effect

Another precursor study focused on the *in vivo* effect of radiation-induced genomic instability by the bystander effect (Koturbash et al., 2007). The authors irradiated rat skull hippocampus, protecting the rest of the body with a lead shield, and assessed the bystander effect in spleen at 7 months. They found global spleen cell genome demethylation, increasing transposable retro-element expression. Only one microRNA, miR-194, was significantly overexpressed, and even there the increase was less than 2-fold. There was an increase in miR-194 24 hours after irradiation, which was also less than 2-fold, in both spleen and plasma, suggesting a possible mechanism of plasma transportation.

5. Discussion

5.1 Towards a universal "radiomiR"?

There are a growing number of studies of microRNA response to ionising radiation. Results indicate that microRNA response depends mainly on cell system, LET, dose and time course after irradiation. Moreover, except in blood plasma, microRNA expression changes are too weak to be useful as exposure biomarkers. Despite some plasma blood microRNAs with a strong radiation response (Templin et al., 2011a), it has not been possible to identify a microRNA regulated independently of dose and time-course parameters. Unlike cancer or other diseases (Elmén et al., 2008) where microRNAs are powerful biomarkers (Brase et al., 2010; Heneghan et al., 2010) and potential therapeutic tools (Hausseker & Kay, 2010), there are no microRNAs that can serve as ubiquitous ionising radiation markers or radiosensitivity predictors: the universal "radiomiR" does not exist to date.

5.2 Critical review of microRNA expression studies

MicroRNA studies of ionising radiation have used large-scale analysis, mostly by RT-qPCR, compiling data on hundreds of microRNAs. Analysis of published results raises some questions. Firstly, a variety normalisation methods have been used, some of which are questionable. One or two small nuclear RNAs are used as reference gene, without justification by rigorous calculation (Vandesompele et al., 2002). Inconsistent choices of RT-qPCR normalisation in large-scale or individual assays can induce systematic bias in results and false interpretation. Secondly, the absence of a cut-off threshold in RT-qPCR analysis can lead to wide variation in estimations of microRNA abundance. The most common error in analysis of RT-qPCR results is the absence of a quantification cycle (Cq) threshold beyond which there is no repeatability. For example, for TaqMan Low Density Arrays, we and others (Gougelet et al., 2011) calculate a cycle threshold at Cq=32: above this limit, microRNA and control snRNA Cq repeatability is poor, whatever the cell type, radiation condition or time course, setting a technical limit to microRNA detection.

When studies identify microRNAs responding to radiation, little is done to identify the target gene and response mechanisms. In most studies, microRNA expression is inversely correlated with bioinformatically predicted target-mRNA or protein expression profile, without further validation by conventional methods, which are hard to implement because of weak post-radiation microRNA regulation. This may explain why gene ontology analysis of the identified regulated genes finds very divergent time- and dose-dependent microRNA-associated mechanisms (Maes et al., 2008).

Concerning the radiation-induced bystander effect, studies reported no overall change in microRNA profile. Only miR-194 was identified as up-regulated after ionising irradiation, with a less than two-fold increase whether a few hours or several months after exposure, in both blood and spleen, suggesting a possible mechanism of plasma transportation (Koturbash et al., 2007). By bioinformatic target analysis, the authors correlated DNA methyltranferase and methyl-binding protein down-regulation with the increase in miR-194 found in their results, but without further evaluation. To date, no miR-194 target has been experimentally confirmed in any organism. However, another study, analyzing sex-specific response, was not so clear about miR-194 up-regulation (Koturbash et al., 2008): the authors observed miR-34a up-regulation and presented this as a good response to ionising radiation, because of miR-34a's role downstream of p53. However, this role is not clear (Zenz et al., 2009) and if it has even been proved that p53 is up-regulated in this system (Koturbash et al., 2008) it can also have another role such as induction of senescence (Christoffersen et al., 2010) depending on temporally induced isoforms. Moreover, the spleen is a heterogeneous organ, including immune cells in the pulps, muscle and mesothelial cells of the capsule, and endothelial cells contouring spleen vessels. Whole-organ analysis averages individual cell-type responses, thus possibly masking microRNA regulation.

5.3 Future challenges

Technical tools for microRNA study are now powerful and cheap enough to allow elegant work on ionising radiation response. Scientists must now converge on analytical methods to produce reliable results. Radiation-induced genomic instability causes long-term effects, notably implicated in cancer development. One hypothesis suggests that ionising radiation induces potentially unstable chromosome regions (Suzuki, 1997) prone to delayed DNA breakage. As a majority of microRNA loci are on or near fragile chromosome sites (Calin et al., 2004), subject to deletion or rearrangement in cancer, and constitute powerful diagnostic, metastatic and prognostic markers (review in Koturbash et al., 2011), long-term surveys of irradiated cells will be crucial to understanding the steps of ionising radiation-mediated carcinogenesis. It now appears clear that crossing only microRNA expression profiling and mRNA and protein profiling restricts knowledge to a competitive-endogenous RNA hypothesis (Salmena et al., 2011). Absence of data on pseudo-genes and long non-coding RNA, sometimes some of the most regulated RNA between conditions (Panzitt et al., 2007), biases analysis of microRNA regulatory networks. Integration of all long and short RNA and protein profiling will need to bridge a major gap, using the most recent techniques of sequencing, to achieve a true characterisation of cell situation.

6. Conclusion

Little is known about the microRNA response to ionising radiation. It is now clear that "radiomiR" (i.e., a clear ubiquitous radiation biomarker and radiosensitivity predictor) does not exist. However, future studies, by increasing evidence about microRNA response to radiation, may specify cell-, dose-, time-, sex- and radiation type-related response profiles. Not only microRNA expression but also all aspects of the microRNA and mRNA-binding proteome pathways must be investigated to understand the epigenetic contributions to ionising radiation effects.

7. Acknowledgements

The authors thank EDF (Conseil de radioprotection), Ligue Nationale contre le Cancer (comité de Saône-et-Loire) and Agence Nationale de la Recherche (LODGRA 2008-2010) for funding this thematic.
The authors thank Odile Berthier-Vergnes, Ingrid Masse and Boris Guyot for helpful discussions and critical reading of the manuscript and Iain McGill for English editing.

8. References

Ambros, V., Bartel, B., Bartel, D.P., Burge, C.B., Carrington, J.C., Chen, X., Dreyfuss, G., Eddy, S.R., Griffiths-Jones, S., Marshall, M., Matzke, M., Ruvkun, G. & Tuschl, T. (2003a) A uniform system for microRNA annotation. *RNA (New York, N.Y.)*, 9, 277-279.

Ambros, V., Lee, R.C., Lavanway, A., Williams, P.T. & Jewell, D. (2003b) MicroRNAs and other tiny endogenous RNAs in C. elegans. *Current Biology: CB*, 13, 807-818.

Amundson, S.A., Do, K.T., Vinikoor, L., Koch-Paiz, C.A., Bittner, M.L., Trent, J.M., Meltzer, P. & Fornace, A.J., Jr. (2005) Stress-specific signatures: expression profiling of p53 wild-type and -null human cells. *Oncogene*, 24, 4572-4579.

Aravin, A., Gaidatzis, D., Pfeffer, S., Lagos-Quintana, M., Landgraf, P., Iovino, N., Morris, P., Brownstein, M.J., Kuramochi-Miyagawa, S., Nakano, T., Chien, M., Russo, J.J., Ju, J., Sheridan, R., Sander, C., Zavolan, M. & Tuschl, T. (2006) A novel class of small RNAs bind to MILI protein in mouse testes. *Nature*, 442, 203-207

Arvey, A., Larsson, E., Sander, C., Leslie, C.S. & Marks, D.S. (2010) Target mRNA abundance dilutes microRNA and siRNA activity. *Molecular Systems Biology*, 6, 363.

Aypar, U., Morgan, W.F. & Baulch, J.E. (2011) Radiation-induced epigenetic alterations after low and high LET irradiations. *Mutation Research*, 707, 24-33.

Bagga, S., Bracht, J., Hunter, S., Massirer, K., Holtz, J., Eachus, R. & Pasquinelli, A.E. (2005) Regulation by let-7 and lin-4 miRNAs results in target mRNA degradation. *Cell*, 122, 553-563.

Bartel, D.P. (2004) MicroRNAs: genomics, biogenesis, mechanism, and function. *Cell*, 116, 281-297.

Baskerville, S. & Bartel, D.P. (2005) Microarray profiling of microRNAs reveals frequent coexpression with neighboring miRNAs and host genes. *RNA (New York, N.Y.)*, 11, 241-247.

Bass, B.L. & Weintraub, H. (1988) An unwinding activity that covalently modifies its double-stranded RNA substrate. *Cell*, 55, 1089-1098.

Behm-Ansmant, I., Rehwinkel, J., Doerks, T., Stark, A., Bork, P. & Izaurralde, E. (2006) mRNA degradation by miRNAs and GW182 requires both CCR4:NOT deadenylase and DCP1:DCP2 decapping complexes. *Genes & Development*, 20, 1885-1898.

Bhattacharyya, S.N., Habermacher, R., Martine, U., Closs, E.I. & Filipowicz, W. (2006) Stress-induced reversal of microRNA repression and mRNA P-body localization in human cells. *Cold Spring Harbor Symposia on Quantitative Biology*, 71, 513-521.

Blow, M.J., Grocock, R.J., van Dongen, S., Enright, A.J., Dicks, E., Futreal, P.A., Wooster, R. & Stratton, M.R. (2006) RNA editing of human microRNAs. *Genome Biology*, 7, R27.

Bohmert, K., Camus, I., Bellini, C., Bouchez, D., Caboche, M. & Benning, C. (1998) AGO1 defines a novel locus of Arabidopsis controlling leaf development. *The EMBO Journal*, 17, 170-180.

Bohnsack, M.T., Czaplinski, K. & Gorlich, D. (2004) Exportin 5 is a RanGTP-dependent dsRNA-binding protein that mediates nuclear export of pre-miRNAs. *RNA (New York, N.Y.)*, 10, 185-191.

Boominathan L (2010). The tumor suppressors p53, p63 and p73 are regulators of microRNA processing complex. PLOS One, 5(5) :e10615.

Borchert, G.M., Lanier, W. & Davidson, B.L. (2006) RNA polymerase III transcribes human microRNAs. *Nature Structural & Molecular Biology*, 13, 1097-1101.

Bracht, J., Hunter, S., Eachus, R., Weeks, P. & Pasquinelli, A.E. (2004) Trans-splicing and polyadenylation of let-7 microRNA primary transcripts. *RNA (New York, N.Y.)*, 10, 1586-1594.

Brase, J.C., Wuttig, D., Kuner, R. & Sültmann, H. (2010) Serum microRNAs as non-invasive biomarkers for cancer. *Molecular Cancer*, 9, 306.

Braun, C.J., Zhang, X., Savelyeva, I., Wolff, S., Moll, U.M., Schepeler, T., Orntoft, T.F., Andersen, C.L. & Dobbelstein, M. (2008) p53-Responsive micrornas 192 and 215 are capable of inducing cell cycle arrest. *Cancer Res*, 68, 10094-104.

Brennecke, J., Aravin, A.A., Stark, A., Dus, M., Kellis, M., Sachidanandam, R. & Hannon, G.J. (2007) Discrete small RNA-generating loci as master regulators of transposon activity in Drosophila. *Cell*, 128, 1089-1103.

Brodersen, P., Sakvarelidze-Achard, L., Bruun-Rasmussen, M., Dunoyer, P., Yamamoto, Y.Y., Sieburth, L. & Voinnet, O. (2008) Widespread translational inhibition by plant miRNAs and siRNAs. *Science (New York, N.Y.)*, 320, 1185-1190.

Bushati, N. & Cohen, S.M. (2007) microRNA functions. *Annual Review of Cell and Developmental Biology*, 23, 175-205.

Cai, X., Hagedorn, C.H. & Cullen, B.R. (2004) Human microRNAs are processed from capped, polyadenylated transcripts that can also function as mRNAs. *RNA (New York, N.Y.)*, 10, 1957-1966.

Calin, G.A., Sevignani, C., Dumitru, C.D., Hyslop, T., Noch, E., Yendamuri, S., Shimizu, M., Rattan, S., Bullrich, F., Negrini, M. & Croce, C.M. (2004) Human microRNA genes are frequently located at fragile sites and genomic regions involved in cancers. *Proceedings of the National Academy of Sciences of the United States of America*, 101, 2999-3004.

Ceci, M., Gaviraghi, C., Gorrini, C., Sala, L.A., Offenhäuser, N., Marchisio, P.C. & Biffo, S. (2003) Release of eIF6 (p27BBP) from the 60S subunit allows 80S ribosome assembly. *Nature*, 426, 579-584.

Chan, S.Y. & Loscalzo, J. (2010) MicroRNA-210: a unique and pleiotropic hypoxamir. *Cell Cycle (Georgetown, Tex.)*, 9, 1072-1083.

Chang, T.C., Wentzel, E.A., Kent, O.A., Ramachandran, K., Mullendore, M., Lee, K.H., Feldmann, G., Yamakuchi, M., Ferlito, M., Lowenstein, C.J., Arking, D.E., Beer, M.A., Maitra, A. & Mendell, J.T. (2007) Transactivation of miR-34a by p53 broadly influences gene expression and promotes apoptosis. *Mol Cell*, 26, 745-52.

Chaudhry, M.A. (2009) Real-time PCR analysis of micro-RNA expression in ionizing radiation-treated cells. *Cancer Biotherapy & Radiopharmaceuticals*, 24, 49-56.

Chen, C., Ridzon, D.A., Broomer, A.J., Zhou, Z., Lee, D.H., Nguyen, J.T., Barbisin, M., Xu, N.L., Mahuvakar, V.R., Andersen, M.R., Lao, K.Q., Livak, K.J. & Guegler, K.J. (2005) Real-time quantification of microRNAs by stem-loop RT-PCR. *Nucleic Acids Research*, 33, e179.

Chendrimada, T.P., Finn, K.J., Ji, X., Baillat, D., Gregory, R.I., Liebhaber, S.A., Pasquinelli, A.E. & Shiekhattar, R. (2007) MicroRNA silencing through RISC recruitment of eIF6. *Nature*, 447, 823-828.

Chendrimada, T.P., Gregory, R.I., Kumaraswamy, E., Norman, J., Cooch, N., Nishikura, K. & Shiekhattar, R. (2005) TRBP recruits the Dicer complex to Ago2 for microRNA processing and gene silencing. *Nature*, 436, 740-744.

Christoffersen, N.R., Shalgi, R., Frankel, L.B., Leucci, E., Lees, M., Klausen, M., Pilpel, Y., Nielsen, F.C., Oren, M. & Lund, A.H. (2010) p53-independent upregulation of miR-34a during oncogene-induced senescence represses MYC. *Cell Death and Differentiation*, 17, 236-245.

Corney, D.C., Flesken-Nikitin, A., Godwin, A.K., Wang, W. & Nikitin, A.Y. (2007) MicroRNA-34b and MicroRNA-34c are targets of p53 and cooperate in control of cell proliferation and adhesion-independent growth. *Cancer Research*, 67, 8433-8438.

Cui, Q., Yu, Z., Purisima, E.O. & Wang, E. (2007) MicroRNA regulation and interspecific variation of gene expression. *Trends in Genetics: TIG*, 23, 372-375

Cummins, J.M. & Velculescu, V.E. (2006) Implications of micro-RNA profiling for cancer diagnosis. *Oncogene*, 25, 6220-6227.

Davis, B.N., Hilyard, A.C., Lagna, G. & Hata, A. (2008) SMAD proteins control DROSHA-mediated microRNA maturation. *Nature*, 454, 56-61.

Dickey, J.S., Zemp, F.J., Martin, O.A, Kovalchuk, O. (2011). The role of miRNA in the direct and indirect effects of ionizing radiations. *Radiation Environmental Biophysics* 50(4) : 491-499.

Didiano, D. & Hobert, O. (2008) Molecular architecture of a miRNA-regulated 3' UTR. *RNA (New York, N.Y.)*, 14, 1297-1317.

Ding, X.C., Slack, F.J. & Grosshans, H. (2008) The let-7 microRNA interfaces extensively with the translation machinery to regulate cell differentiation. *Cell Cycle (Georgetown, Tex.)*, 7, 3083-3090.

Elmén, J., Lindow, M., Silahtaroglu, A., Bak, M., Christensen, M., Lind-Thomsen, A., Hedtjärn, M., Hansen, J.B., Hansen, H.F., Straarup, E.M., McCullagh, K., Kearney, P. & Kauppinen, S. (2008) Antagonism of microRNA-122 in mice by systemically administered LNA-antimiR leads to up-regulation of a large set of predicted target mRNAs in the liver. *Nucleic Acids Research*, 36, 1153-1162.

Ender, C. & Meister, G. (2010) Argonaute proteins at a glance. *Journal of Cell Science*, 123, 1819-1823.

Eulalio, A., Huntzinger, E. & Izaurralde, E. (2008) GW182 interaction with Argonaute is essential for miRNA-mediated translational repression and mRNA decay. *Nature Structural & Molecular Biology*, 15, 346-353.

Fabian, M.R., Mathonnet, G., Sundermeier, T., Mathys, H., Zipprich, J.T., Svitkin, Y.V., Rivas, F., Jinek, M., Wohlschlegel, J., Doudna, J.A., Chen, C.-Y.A., Shyu, A.-B., Yates, J.R., 3rd, Hannon, G.J., Filipowicz, W., Duchaine, T.F. & Sonenberg, N. (2009) Mammalian miRNA RISC recruits CAF1 and PABP to affect PABP-dependent deadenylation. *Molecular Cell*, 35, 868-880.

Fabian, M.R., Sonenberg, N. & Filipowicz, W. (2010) Regulation of mRNA translation and stability by microRNAs. *Annual Review of Biochemistry*, 79, 351-379.

Frank, F., Sonenberg, N. & Nagar, B. (2010) Structural basis for 5′-nucleotide base-specific recognition of guide RNA by human AGO2. *Nature*, 465, 818-822.

Friedman, R.C., Farh, K.K.-H., Burge, C.B. & Bartel, D.P. (2009) Most mammalian mRNAs are conserved targets of microRNAs. *Genome Research*, 19, 92-105.

Fukuda, T., Yamagata, K., Fujiyama, S., Matsumoto, T., Koshida, I., Yoshimura, K., Mihara, M., Naitou, M., Endoh, H., Nakamura, T., Akimoto, C., Yamamoto, Y., Katagiri, T., Foulds, C., Takezawa, S., Kitagawa, H., Takeyama, K.-ichi, O'Malley, B.W. & Kato, S. (2007) DEAD-box RNA helicase subunits of the Drosha complex are required for processing of rRNA and a subset of microRNAs. *Nature Cell Biology*, 9, 604-611.

Galgano, A., Forrer, M., Jaskiewicz, L., Kanitz, A., Zavolan, M. & Gerber, A.P. (2008) Comparative analysis of mRNA targets for human PUF-family proteins suggests extensive interaction with the miRNA regulatory system. *PloS One*, 3, e3164.

Gougelet, A., Pissaloux, D., Besse, A., Perez, J., Duc, A., Dutour, A., Blay, J.-Y. & Alberti, L. (2010) Micro-RNA profiles in osteosarcoma as a predictive tool for ifosfamide response. *International Journal of Cancer. Journal International Du Cancer*.

Gregory, R.I., Yan, K.-P., Amuthan, G., Chendrimada, T., Doratotaj, B., Cooch, N. & Shiekhattar, R. (2004) The Microprocessor complex mediates the genesis of microRNAs. *Nature*, 432, 235-240.

Grimson, A., Farh, K.K.-H., Johnston, W.K., Garrett-Engele, P., Lim, L.P. & Bartel, D.P. (2007) MicroRNA targeting specificity in mammals: determinants beyond seed pairing. *Molecular Cell*, 27, 91-105.

Gu, S., Jin, L., Zhang, F., Sarnow, P. & Kay, M.A. (2009) Biological basis for restriction of microRNA targets to the 3′ untranslated region in mammalian mRNAs. *Nature Structural & Molecular Biology*, 16, 144-150.

Guo, H., Ingolia, N.T., Weissman, J.S. & Bartel, D.P. (2010) Mammalian microRNAs predominantly act to decrease target mRNA levels. *Nature*, 466, 835-840.

Han, J., Lee, Y., Yeom, K.-H., Kim, Y.-K., Jin, H. & Kim, V.N. (2004) The Drosha-DGCR8 complex in primary microRNA processing. *Genes & Development*, 18, 3016-3027.

Han, J., Lee, Y., Yeom, K.-H., Nam, J.-W., Heo, I., Rhee, J.-K., Sohn, S.Y., Cho, Y., Zhang, B.-T. & Kim, V.N. (2006) Molecular basis for the recognition of primary microRNAs by the Drosha-DGCR8 complex. *Cell*, 125, 887-901.

Haupt, S., Berger, M., Goldberg, Z. & Haupt, Y. (2003) Apoptosis - the p53 network. *J Cell Sci*, 116, 4077-85.

Haussecker, D. & Kay, M.A. (2010) miR-122 continues to blaze the trail for microRNA therapeutics. *Molecular Therapy: The Journal of the American Society of Gene Therapy*, 18, 240-242.

He, L., Thomson, J.M., Hemann, M.T., Hernando-Monge, E., Mu, D., Goodson, S., Powers, S., Cordon-Cardo, C., Lowe, S.W., Hannon, G.J. & Hammond, S.M. (2005) A microRNA polycistron as a potential human oncogene. *Nature*, 435, 828-833.

Heneghan, H.M., Miller, N. & Kerin, M.J. (2010) MiRNAs as biomarkers and therapeutic targets in cancer. *Current Opinion in Pharmacology*, 10, 543-550.

Henke, J.I., Goergen, D., Zheng, J., Song, Y., Schüttler, C.G., Fehr, C., Jünemann, C. & Niepmann, M. (2008) microRNA-122 stimulates translation of hepatitis C virus RNA. *The EMBO Journal*, 27, 3300-3310.

Hu, W., Chan, C.S., Wu, R., Zhang, C., Sun, Y., Song, J.S., Tang, L.H., Levine, A.J. & Feng, Z. (2010) Negative regulation of tumor suppressor p53 by microRNA miR-504. *Mol Cell*, 38, 689-99.

Hu, H., Du, L., Nagabayashi, G., Seeger, R.C. & Gatti, R.A. (2010) ATM is down-regulated by N-Myc-regulated microRNA-421. *Proc Natl Acad Sci U S A*, 107, 1506-11.

Huang, J., Liang, Z., Yang, B., Tian, H., Ma, J. & Zhang, H. (2007) Derepression of microRNA-mediated protein translation inhibition by apolipoprotein B mRNA-editing enzyme catalytic polypeptide-like 3G (APOBEC3G) and its family members. *The Journal of Biological Chemistry*, 282, 33632-33640.

Humphreys, D.T., Westman, B.J., Martin, D.I.K. & Preiss, T. (2005) MicroRNAs control translation initiation by inhibiting eukaryotic initiation factor 4E/cap and poly(A) tail function. *Proceedings of the National Academy of Sciences of the United States of America*, 102, 16961-16966.

Huntzinger, E., Braun, J.E., Heimstädt, S., Zekri, L. & Izaurralde, E. (2010) Two PABPC1-binding sites in GW182 proteins promote miRNA-mediated gene silencing. *The EMBO Journal*, 29, 4146-4160.

Hutvágner, G., McLachlan, J., Pasquinelli, A.E., Bálint, E., Tuschl, T. & Zamore, P.D. (2001) A cellular function for the RNA-interference enzyme Dicer in the maturation of the let-7 small temporal RNA. *Science (New York, N.Y.)*, 293, 834-838.

Hwang, B.J., Ford, J.M., Hanawalt, P.C. & Chu, G. (1999) Expression of the p48 xeroderma pigmentosum gene is p53-dependent and is involved in global genomic repair. *Proc Natl Acad Sci U S A*, 96, 424-8.

Hwang, H.-W., Wentzel, E.A. & Mendell, J.T. (2007) A hexanucleotide element directs microRNA nuclear import. *Science (New York, N.Y.)*, 315, 97-100.

Iki, T., Yoshikawa, M., Nishikiori, M., Jaudal, M.C., Matsumoto-Yokoyama, E., Mitsuhara, I., Meshi, T. & Ishikawa, M. (2010) In vitro assembly of plant RNA-induced silencing complexes facilitated by molecular chaperone HSP90. *Molecular Cell*, 39, 282-291.

Iwasaki, S., Kobayashi, M., Yoda, M., Sakaguchi, Y., Katsuma, S., Suzuki, T. & Tomari, Y. (2010) Hsc70/Hsp90 chaperone machinery mediates ATP-dependent RISC loading of small RNA duplexes. *Molecular Cell*, 39, 292-299.

Izaurralde, E., Kutay, U., von Kobbe, C., Mattaj, I.W. & Görlich, D. (1997) The asymmetric distribution of the constituents of the Ran system is essential for transport into and out of the nucleus. *The EMBO Journal*, 16, 6535-6547.

Jeyapalan, Z., Deng, Z., Shatseva, T., Fang, L., He, C. & Yang, B.B. (2011) Expression of CD44 3′-untranslated region regulates endogenous microRNA functions in tumorigenesis and angiogenesis. *Nucleic Acids Research*, 39, 3026-3041.

Jing, Q., Huang, S., Guth, S., Zarubin, T., Motoyama, A., Chen, J., Di Padova, F., Lin, S.-C., Gram, H. & Han, J. (2005) Involvement of microRNA in AU-rich element-mediated mRNA instability. *Cell*, 120, 623-634.

Kastan, M.B. & Bartek, J. (2004) Cell-cycle checkpoints and cancer. *Nature*, 432, 316-23.

Kataoka, N., Fujita, M. & Ohno, M. (2009) Functional association of the Microprocessor complex with the spliceosome. *Molecular and Cellular Biology*, 29, 3243-3254.

Kawahara, Y., Zinshteyn, B., Chendrimada, T.P., Shiekhattar, R. & Nishikura, K. (2007a) RNA editing of the microRNA-151 precursor blocks cleavage by the Dicer-TRBP complex. *EMBO Reports*, 8, 763-769.

Kawahara, Y., Zinshteyn, B., Sethupathy, P., Iizasa, H., Hatzigeorgiou, A.G. & Nishikura, K. (2007b) Redirection of silencing targets by adenosine-to-inosine editing of miRNAs. *Science (New York, N.Y.)*, 315, 1137-1140.

Kawamata, T., Seitz, H. & Tomari, Y. (2009) Structural determinants of miRNAs for RISC loading and slicer-independent unwinding. *Nature Structural & Molecular Biology*, 16, 953-960.

Kedde, M., van Kouwenhove, M., Zwart, W., Oude Vrielink, J.A.F., Elkon, R. & Agami, R. (2010) A Pumilio-induced RNA structure switch in p27-3′ UTR controls miR-221 and miR-222 accessibility. *Nature Cell Biology*, 12, 1014-1020.

Kedde, M., Strasser, M.J., Boldajipour, B., Oude Vrielink, J.A.F., Slanchev, K., le Sage, C., Nagel, R., Voorhoeve, P.M., van Duijse, J., Ørom, U.A., Lund, A.H., Perrakis, A., Raz, E. & Agami, R. (2007) RNA-binding protein Dnd1 inhibits microRNA access to target mRNA. *Cell*, 131, 1273-1286.

Kennedy, A.R., Fox, M., Murphy, G. & Little, J.B. (1980) Relationship between x-ray exposure and malignant transformation in C3H 10T1/2 cells. *Proceedings of the National Academy of Sciences of the United States of America*, 77, 7262-7266.

Kim, H.H., Kuwano, Y., Srikantan, S., Lee, E.K., Martindale, J.L. & Gorospe, M. (2009) HuR recruits let-7/RISC to repress c-Myc expression. *Genes & Development*, 23, 1743-1748.

Klase, Z., Kale, P., Winograd, R., Gupta, M.V., Heydarian, M., Berro, R., McCaffrey, T. & Kashanchi, F. (2007) HIV-1 TAR element is processed by Dicer to yield a viral micro-RNA involved in chromatin remodeling of the viral LTR. *BMC Molecular Biology*, 8, 63.

Kloosterman, W.P., Wienholds, E., de Bruijn, E., Kauppinen, S. & Plasterk, R.H.A. (2006) In situ detection of miRNAs in animal embryos using LNA-modified oligonucleotide probes. *Nature Methods*, 3, 27-29.

Koturbash, I., Boyko, A., Rodriguez-Juarez, R., McDonald, R.J., Tryndyak, V.P., Kovalchuk, I., Pogribny, I.P. & Kovalchuk, O. (2007) Role of epigenetic effectors in maintenance of the long-term persistent bystander effect in spleen in vivo. *Carcinogenesis*, 28, 1831-1838.

Koturbash, I., Zemp, F.J., Kutanzi, K., Luzhna, L., Loree, J., Kolb, B. & Kovalchuk, O. (2008) Sex-specific microRNAome deregulation in the shielded bystander spleen of cranially exposed mice. *Cell Cycle (Georgetown, Tex.)*, 7, 1658-1667.

Koturbash, I., Zemp, F.J., Pogribny, I. & Kovalchuk, O. (2011) Small molecules with big effects: the role of the microRNAome in cancer and carcinogenesis. *Mutation Research*, 722, 94-105.

Kulshreshtha, R., Ferracin, M., Wojcik, S.E., Garzon, R., Alder, H., Agosto-Perez, F.J., Davuluri, R., Liu, C.-G., Croce, C.M., Negrini, M., Calin, G.A. & Ivan, M. (2007) A microRNA signature of hypoxia. *Molecular and Cellular Biology*, 27, 1859-1867.

Lal, A., Pan, Y., Navarro, F., Dykxhoorn, D.M., Moreau, L., Meire, E., Bentwich, Z., Lieberman, J. & Chowdhury, D. (2009) miR-24-mediated downregulation of H2AX suppresses DNA repair in terminally differentiated blood cells. *Nature Structural & Molecular Biology*, 16, 492-498.

Landthaler, M. (2010) Chaperones get RISC loaded. *Molecular Cell*, 39, 161-162.

Langenberger, D., Bermudez-Santana, C., Hertel, J., Hoffmann, S., Khaitovich, P. & Stadler, P.F. (2009) Evidence for human microRNA-offset RNAs in small RNA sequencing data. *Bioinformatics (Oxford, England)*, 25, 2298-2301.

Le, M.T.N., Teh, C., Shyh-Chang, N., Xie, H., Zhou, B., Korzh, V., Lodish, H.F. & Lim, B. (2009) MicroRNA-125b is a novel negative regulator of p53. *Genes & Development*, 23, 862-876.

Lee, Y., Ahn, C., Han, J., Choi, H., Kim, J., Yim, J., Lee, J., Provost, P., Rådmark, O., Kim, S. & Kim, V.N. (2003) The nuclear RNase III Drosha initiates microRNA processing. *Nature*, 425, 415-419.

Lee, K.M., Choi, E.J. & Kim, I.A. (2011) microRNA-7 increases radiosensitivity of human cancer cells with activated EGFR-associated signaling. *Radiotherapy and Oncology: Journal of the European Society for Therapeutic Radiology and Oncology*.

Lee, R.C., Feinbaum, R.L. & Ambros, V. (1993) The C. elegans heterochronic gene lin-4 encodes small RNAs with antisense complementarity to lin-14. *Cell*, 75, 843-854.

Lee, D.Y., Jeyapalan, Z., Fang, L., Yang, J., Zhang, Y., Yee, A.Y., Li, M., Du, W.W., Shatseva, T. & Yang, B.B. (2010) Expression of versican 3'-untranslated region modulates endogenous microRNA functions. *PloS One*, 5, e13599.

Lee, Y., Kim, M., Han, J., Yeom, K.-H., Lee, S., Baek, S.H. & Kim, V.N. (2004) MicroRNA genes are transcribed by RNA polymerase II. *The EMBO Journal*, 23, 4051-4060.

Lee, D.Y., Shatseva, T., Jeyapalan, Z., Du, W.W., Deng, Z. & Yang, B.B. (2009) A 3'-untranslated region (3'UTR) induces organ adhesion by regulating miR-199a* functions. *PloS One*, 4, e4527.

Lee, S.J., Jiko, C., Yamashita, E. & Tsukihara, T. (2011) Selective nuclear export mechanism of small RNAs. *Current Opinion in Structural Biology*, 21, 101-108.

Lingel, A., Simon, B., Izaurralde, E. & Sattler, M. (2004) Nucleic acid 3'-end recognition by the Argonaute2 PAZ domain. *Nature Structural & Molecular Biology*, 11, 576-577.

Liu, Y. & Liu, Q. (2011) ATM signals miRNA biogenesis through KSRP. *Mol Cell*, 41, 367-8.

Liu, C.-G., Calin, G.A., Meloon, B., Gamliel, N., Sevignani, C., Ferracin, M., Dumitru, C.D., Shimizu, M., Zupo, S., Dono, M., Alder, H., Bullrich, F., Negrini, M. & Croce, C.M. (2004) An oligonucleotide microchip for genome-wide microRNA profiling in human and mouse tissues. *Proceedings of the National Academy of Sciences of the United States of America*, 101, 9740-9744.

Liu, J., Carmell, M.A., Rivas, F.V., Marsden, C.G., Thomson, J.M., Song, J.-J., Hammond, S.M., Joshua-Tor, L. & Hannon, G.J. (2004) Argonaute2 is the catalytic engine of mammalian RNAi. *Science (New York, N.Y.)*, 305, 1437-1441.

Liu, J., Rivas, F.V., Wohlschlegel, J., Yates, J.R., 3rd, Parker, R. & Hannon, G.J. (2005) A role for the P-body component GW182 in microRNA function. *Nature Cell Biology*, 7, 1261-1266.

Lu, J., Getz, G., Miska, E.A., Alvarez-Saavedra, E., Lamb, J., Peck, D., Sweet-Cordero, A., Ebert, B.L., Mak, R.H., Ferrando, A.A., Downing, J.R., Jacks, T., Horvitz, H.R. & Golub, T.R. (2005) MicroRNA expression profiles classify human cancers. *Nature*, 435, 834-838.

Lund, E., Güttinger, S., Calado, A., Dahlberg, J.E. & Kutay, U. (2004) Nuclear export of microRNA precursors. *Science (New York, N.Y.)*, 303, 95-98.

Ma, J.-B., Ye, K. & Patel, D.J. (2004) Structural basis for overhang-specific small interfering RNA recognition by the PAZ domain. *Nature*, 429, 318-322.

Macrae, I.J., Zhou, K., Li, F., Repic, A., Brooks, A.N., Cande, W.Z., Adams, P.D. & Doudna, J.A. (2006) Structural basis for double-stranded RNA processing by Dicer. *Science (New York, N.Y.)*, 311, 195-198.

Maes, O.C., An, J., Sarojini, H., Wu, H. & Wang, E. (2008) Changes in MicroRNA expression patterns in human fibroblasts after low-LET radiation. *Journal of Cellular Biochemistry*, 105, 824-834.

Marsit, C.J., Eddy, K. & Kelsey, K.T. (2006) MicroRNA responses to cellular stress. *Cancer Research*, 66, 10843-10848.

Matsuoka, S., Ballif, B.A., Smogorzewska, A., McDonald, E.R., Hurov, K.E., Luo, J., Bakalarski, C.E., Zhao, Z., Solimini, N., Lerenthal, Y., Shiloh, Y., Gygi, S.P. & Elledge, S.J. (2007) ATM and ATR substrate analysis reveals extensive protein networks responsive to DNA damage. *Science*, 316, 1160-6.

Mendell, J.T. (2008) miRiad roles for the miR-17-92 cluster in development and disease. *Cell*, 133, 217-222.

Morlando, M., Ballarino, M., Gromak, N., Pagano, F., Bozzoni, I. & Proudfoot, N.J. (2008) Primary microRNA transcripts are processed co-transcriptionally. *Nature Structural & Molecular Biology*, 15, 902-909.

Moskwa, P., Buffa, F.M., Pan, Y., Panchakshari, R., Gottipati, P., Muschel, R.J., Beech, J., Kulshrestha, R., Abdelmohsen, K., Weinstock, D.M., Gorospe, M., Harris, A.L., Helleday, T. & Chowdhury, D. (2011) miR-182-mediated downregulation of BRCA1 impacts DNA repair and sensitivity to PARP inhibitors. *Molecular Cell*, 41, 210-220.

Nagasawa, H. & Little, J.B. (1992) Induction of sister chromatid exchanges by extremely low doses of alpha-particles. *Cancer Research*, 52, 6394-6396.

Nishikura, K. (2010) Functions and regulation of RNA editing by ADAR deaminases. *Annual Review of Biochemistry*, 79, 321-349.

Nolde, M.J., Saka, N., Reinert, K.L. & Slack, F.J. (2007) The Caenorhabditis elegans pumilio homolog, puf-9, is required for the 3'UTR-mediated repression of the let-7 microRNA target gene, hbl-1. *Developmental Biology*, 305, 551-563.

O'Donnell, K.A., Wentzel, E.A., Zeller, K.I., Dang, C.V. & Mendell, J.T. (2005) c-Myc-regulated microRNAs modulate E2F1 expression. *Nature*, 435, 839-843.

Okamura, K., Chung, W.-J. & Lai, E.C. (2008) The long and short of inverted repeat genes in animals: microRNAs, mirtrons and hairpin RNAs. *Cell Cycle (Georgetown, Tex.)*, 7, 2840-2845.

Okamura, K., Hagen, J.W., Duan, H., Tyler, D.M. & Lai, E.C. (2007) The mirtron pathway generates microRNA-class regulatory RNAs in Drosophila. *Cell*, 130, 89-100.

Ørom, U.A., Nielsen, F.C. & Lund, A.H. (2008) MicroRNA-10a binds the 5'UTR of ribosomal protein mRNAs and enhances their translation. *Molecular Cell*, 30, 460-471.

Panzitt, K., Tschernatsch, M.M.O., Guelly, C., Moustafa, T., Stradner, M., Strohmaier, H.M., Buck, C.R., Denk, H., Schroeder, R., Trauner, M. & Zatloukal, K. (2007) Characterization of HULC, a novel gene with striking up-regulation in hepatocellular carcinoma, as noncoding RNA. *Gastroenterology*, 132, 330-342.

Park, S.Y., Lee, J.H., Ha, M., Nam, J.W. & Kim, V.N. (2009) miR-29 miRNAs activate p53 by targeting p85 alpha and CDC42. *Nat Struct Mol Biol*, 16, 23-9.

Pfeffer, S., Zavolan, M., Grässer, F.A., Chien, M., Russo, J.J., Ju, J., John, B., Enright, A.J., Marks, D., Sander, C. & Tuschl, T. (2004) Identification of virus-encoded microRNAs. *Science (New York, N.Y.)*, 304, 734-736.

Pillai, R.S., Bhattacharyya, S.N., Artus, C.G., Zoller, T., Cougot, N., Basyuk, E., Bertrand, E. & Filipowicz, W. (2005) Inhibition of translational initiation by Let-7 MicroRNA in human cells. *Science (New York, N.Y.)*, 309, 1573-1576.

Poliseno, L., Haimovic, A., Christos, P.J., Vega Y Saenz de Miera, E.C., Shapiro, R., Pavlick, A., Berman, R.S., Darvishian, F. & Osman, I. (2011) Deletion of PTENP1 Pseudogene in Human Melanoma. *The Journal of Investigative Dermatology*.

Poliseno, L., Salmena, L., Zhang, J., Carver, B., Haveman, W.J. & Pandolfi, P.P. (2010) A coding-independent function of gene and pseudogene mRNAs regulates tumour biology. *Nature*, 465, 1033-1038.

Qi, H.H., Ongusaha, P.P., Myllyharju, J., Cheng, D., Pakkanen, O., Shi, Y., Lee, S.W., Peng, J. & Shi, Y. (2008) Prolyl 4-hydroxylation regulates Argonaute 2 stability. *Nature*, 455, 421-424.

Raver-Shapira, N., Marciano, E., Meiri, E., Spector, Y., Rosenfeld, N., Moskovits, N., Bentwich, Z. & Oren, M. (2007) Transcriptional activation of miR-34a contributes to p53-mediated apoptosis. *Molecular Cell*, 26, 731-743.

Rehwinkel, J., Behm-Ansmant, I., Gatfield, D. & Izaurralde, E. (2005) A crucial role for GW182 and the DCP1:DCP2 decapping complex in miRNA-mediated gene silencing. *RNA (New York, N.Y.)*, 11, 1640-1647.

Rodriguez, A., Griffiths-Jones, S., Ashurst, J.L. & Bradley, A. (2004) Identification of mammalian microRNA host genes and transcription units. *Genome Research*, 14, 1902-1910.

Ruby, J.G., Jan, C.H. & Bartel, D.P. (2007) Intronic microRNA precursors that bypass Drosha processing. *Nature*, 448, 83-86.

Rüdel, S., Wang, Y., Lenobel, R., Körner, R., Hsiao, H.-H., Urlaub, H., Patel, D. & Meister, G. (2011) Phosphorylation of human Argonaute proteins affects small RNA binding. *Nucleic Acids Research*, 39, 2330-2343.

Sachdeva, M., Zhu, S., Wu, F., Wu, H., Walia, V., Kumar, S., Elble, R., Watabe, K. & Mo, Y.Y. (2009) p53 represses c-Myc through induction of the tumor suppressor miR-145. *Proc Natl Acad Sci U S A*, 106, 3207-12.

Salmena, L., Poliseno, L., Tay, Y., Kats, L. & Pandolfi, P.P. (2011) A ceRNA Hypothesis: The Rosetta Stone of a Hidden RNA Language? *Cell*.

Sato, F., Tsuchiya, S., Meltzer, S.J. & Shimizu, K. (2011) MicroRNAs and epigenetics. *The FEBS Journal*, 278, 1598-1609.

Schmittgen, T.D., Jiang, J., Liu, Q. & Yang, L. (2004) A high-throughput method to monitor the expression of microRNA precursors. *Nucleic Acids Research*, 32, e43.

Seitz, H. (2009) Redefining microRNA targets. *Current Biology: CB*, 19, 870-873.

Shi, W., Hendrix, D., Levine, M. & Haley, B. (2009) A distinct class of small RNAs arises from pre-miRNA-proximal regions in a simple chordate. *Nature Structural & Molecular Biology*, 16, 183-189.

Shin, C., Nam, J.-W., Farh, K.K.-H., Chiang, H.R., Shkumatava, A. & Bartel, D.P. (2010) Expanding the microRNA targeting code: functional sites with centered pairing. *Molecular Cell*, 38, 789-802.

Stempak, J.M., Sohn, K.-J., Chiang, E.-P., Shane, B. & Kim, Y.-I. (2005) Cell and stage of transformation-specific effects of folate deficiency on methionine cycle intermediates and DNA methylation in an in vitro model. *Carcinogenesis*, 26, 981-990.

Suzuki, K. (1997) Multistep nature of X-ray-induced neoplastic transformation in mammalian cells: genetic alterations and instability. *Journal of Radiation Research*, 38, 55-63.

Suzuki, H.I., Yamagata, K., Sugimoto, K., Iwamoto, T., Kato, S. & Miyazono, K. (2009) Modulation of microRNA processing by p53. *Nature*, 460, 529-533.

Taft, R.J., Simons, C., Nahkuri, S., Oey, H., Korbie, D.J., Mercer, T.R., Holst, J., Ritchie, W., Wong, J.J.-L., Rasko, J.E.J., Rokhsar, D.S., Degnan, B.M. & Mattick, J.S. (2010) Nuclear-localized tiny RNAs are associated with transcription initiation and splice sites in metazoans. *Nature Structural & Molecular Biology*, 17, 1030-1034.

Templin, T., Amundson, S.A., Brenner, D.J. & Smilenov, L.B. (2011a) Whole mouse blood microRNA as biomarkers for exposure to γ-rays and (56)Fe ion. *International Journal of Radiation Biology*.

Templin, T., Paul, S., Amundson, S.A., Young, E.F., Barker, C.A., Wolden, S.L. & Smilenov, L.B. (2011b) Radiation-induced micro-RNA expression changes in peripheral blood cells of radiotherapy patients. *International Journal of Radiation Oncology, Biology, Physics*, 80, 549-557.

Thomson, J.M., Newman, M., Parker, J.S., Morin-Kensicki, E.M., Wright, T. & Hammond, S.M. (2006) Extensive post-transcriptional regulation of microRNAs and its implications for cancer. *Genes & Development*, 20, 2202-2207.

Tomari, Y., Du, T. & Zamore, P.D. (2007) Sorting of Drosophila small silencing RNAs. *Cell*, 130, 299-308.

Umbach, J.L., Strelow, L.I., Wong, S.W. & Cullen, B.R. (2010) Analysis of rhesus rhadinovirus microRNAs expressed in virus-induced tumors from infected rhesus macaques. *Virology*, 405, 592-599.

Vandesompele, J., De Preter, K., Pattyn, F., Poppe, B., Van Roy, N., De Paepe, A. & Speleman, F. (2002) Accurate normalization of real-time quantitative RT-PCR data by geometric averaging of multiple internal control genes. *Genome Biology*, 3, RESEARCH0034.

Vasudevan, S., Tong, Y. & Steitz, J.A. (2007) Switching from repression to activation: microRNAs can up-regulate translation. *Science (New York, N.Y.)*, 318, 1931-1934.

Viswanathan, S.R., Daley, G.Q. & Gregory, R.I. (2008) Selective blockade of microRNA processing by Lin28. *Science (New York, N.Y.)*, 320, 97-100.

Wang, Y., Huang, J.W., Li, M., Cavenee, W.K., Mitchell, P.S., Zhou, X., Tewari, M., Furnari, F.B. & Taniguchi, T. (2011) MicroRNA-138 Modulates DNA Damage Response by Repressing Histone H2AX Expression. *Mol Cancer Res*, 9, 1100-11.

Wang, B., Yanez, A. & Novina, C.D. (2008) MicroRNA-repressed mRNAs contain 40S but not 60S components. *Proceedings of the National Academy of Sciences of the United States of America*, 105, 5343-5348.

Wightman, B., Ha, I. & Ruvkun, G. (1993) Posttranscriptional regulation of the heterochronic gene lin-14 by lin-4 mediates temporal pattern formation in C. elegans. *Cell*, 75, 855-862.

Wright, E.G. (2010) Manifestations and mechanisms of non-targeted effects of ionizing radiation. *Mutation Research*, 687, 28-33.

Xiao, J., Lin, H., Luo, X. & Wang, Z. (2011) miR-605 joins p53 network to form a p53:miR-605:Mdm2 positive feedback loop in response to stress. *EMBO J*, 30, 524-32.

Yamakuchi, M., Ferlito, M. & Lowenstein, C.J. (2008) miR-34a repression of SIRT1 regulates apoptosis. *Proc Natl Acad Sci U S A*, 105, 13421-6.

Yamakuchi, M., Lotterman, C.D., Bao, C., Hruban, R.H., Karim, B., Mendell, J.T., Huso, D. & Lowenstein, C.J. (2010) P53-induced microRNA-107 inhibits HIF-1 and tumor angiogenesis. *Proc Natl Acad Sci U S A*, 107, 6334-9.

Yan, D., Ng, W.L., Zhang, X., Wang, P., Zhang, Z., Mo, Y.-Y., Mao, H., Hao, C., Olson, J.J., Curran, W.J. & Wang, Y. (2010) Targeting DNA-PKcs and ATM with miR-101 sensitizes tumors to radiation. *PloS One*, 5, e11397.

Yan, K.S., Yan, S., Farooq, A., Han, A., Zeng, L. & Zhou, M.-M. (2003) Structure and conserved RNA binding of the PAZ domain. *Nature*, 426, 468-474.

Yang, W., Chendrimada, T.P., Wang, Q., Higuchi, M., Seeburg, P.H., Shiekhattar, R. & Nishikura, K. (2006) Modulation of microRNA processing and expression through RNA editing by ADAR deaminases. *Nature Structural & Molecular Biology*, 13, 13-21.

Yang, J.-H., Luo, X., Nie, Y., Su, Y., Zhao, Q., Kabir, K., Zhang, D. & Rabinovici, R. (2003) Widespread inosine-containing mRNA in lymphocytes regulated by ADAR1 in response to inflammation. *Immunology*, 109, 15-23.

Yekta, S., Shih, I.-H. & Bartel, D.P. (2004) MicroRNA-directed cleavage of HOXB8 mRNA. *Science (New York, N.Y.)*, 304, 594-596.

Yi, R., Qin, Y., Macara, I.G. & Cullen, B.R. (2003) Exportin-5 mediates the nuclear export of pre-microRNAs and short hairpin RNAs. *Genes & Development*, 17, 3011-3016.

Yigit, E., Batista, P.J., Bei, Y., Pang, K.M., Chen, C.-C.G., Tolia, N.H., Joshua-Tor, L., Mitani, S., Simard, M.J. & Mello, C.C. (2006) Analysis of the C. elegans Argonaute family reveals that distinct Argonautes act sequentially during RNAi. *Cell*, 127, 747-757.

Yoda, M., Kawamata, T., Paroo, Z., Ye, X., Iwasaki, S., Liu, Q. & Tomari, Y. (2010) ATP-dependent human RISC assembly pathways. *Nature Structural & Molecular Biology*, 17, 17-23.

Zeng, Y. & Cullen, B.R. (2004) Structural requirements for pre-microRNA binding and nuclear export by Exportin 5. *Nucleic Acids Research*, 32, 4776-4785.

Zeng, Y., Sankala, H., Zhang, X. & Graves, P.R. (2008) Phosphorylation of Argonaute 2 at serine-387 facilitates its localization to processing bodies. *The Biochemical Journal*, 413, 429-436.

Zenz, T., Mohr, J., Eldering, E., Kater, A.P., Bühler, A., Kienle, D., Winkler, D., Dürig, J., van Oers, M.H.J., Mertens, D., Döhner, H. & Stilgenbauer, S. (2009) miR-34a as part of the resistance network in chronic lymphocytic leukemia. *Blood*, 113, 3801-3808.

Zhang, C., Kang, C., Wang, P., Cao, Y., Lv, Z., Yu, S., Wang, G., Zhang, A., Jia, Z., Han, L., Yang, C., Ishiyama, H., Teh, B.S., Xu, B. & Pu, P. (2011) MicroRNA-221 and -222 regulate radiation sensitivity by targeting the PTEN pathway. *International Journal of Radiation Oncology, Biology, Physics*, 80, 240-248.

Zhao, T., Li, G., Mi, S., Li, S., Hannon, G.J., Wang, X.-J. & Qi, Y. (2007) A complex system of small RNAs in the unicellular green alga Chlamydomonas reinhardtii. *Genes & Development*, 21, 1190-1203.

10

Protein Ubiquitination in IR-Induced DNA Damage Response

Nur Yucer, Yi Shi and Yi Wang
Department of Biochemistry and Molecular Biology, Baylor College of Medicine, USA

1. Introduction

Ionizing radiation is a major source of double-strand DNA breaks, the most detrimental form of DNA damage that, if left unrepaired, leads to genomic instability and cancer. Eukaryotic cells have evolved a DNA damage response (DDR), an intricate network that can be rapidly activated to response to such challenge. Mutations in genes that encode DDR proteins are associated with many genetic diseases, suggesting the importance of DDR in maintaining genomic integrity (*1-5*).

DDR is a sophisticated signal transduction network comprised of proteins that are capable of sensing DNA damage, amplifying the signals and executing various cellular functions that have direct impact on cell fate. Of great importance for the understanding of DDR are series of studies that aim to identify proteome-wide substrates of checkpoint kinase ATM/ATR. These unbiased screens greatly expanded the DDR landscape and revealed the involvement of a broad spectrum of pathways that were not previously known (*6, 7*). One of the most intriguing observations made in these studies is the over-representation of proteins involved in ubiquitin proteasome system (UPS) pathways.

Ubiquitin (Ub) is a 76 amino acid protein that can be post-translationally covalently attached to the lysine residues of target proteins(*8*). Ubiquitin can influence the function of its target proteins by altering their enzymatic activity, abundance or localization (*9-14*). In this chapter, we will focus mainly on the role of ubiquitination in DDR signaling and cell cycle regulation. Following a brief introduction to principles of ubiquitination and ubiquitin-like modification, we will present an overview of enzymes and substrates that have been identified over the last two decades of intensive research. Finally, we will introduce recent advances in application of affinity purification and mass spectrometric-based identification of ubiquitinated proteins.

2. Principle of protein modification by ubiquitin and ubiquitin-like modification

2.1 Conjugation of Ub and UBLs through the three enzyme cascade

Ubiquitin is a highly conserved small protein that is expressed in all the eukaryotic cells and across different organisms. Ubiquitin was initially purified by Gideon Goldstein and co-workers from bovine thymus (*15*). The discovery of ubiquitin function in protein degradation was later made by Irvin Rose, Avram Hershko and his student Aaron Cieshanover when they were studying the ATP-dependent degradation of the tyrosine aminotransferase enzyme. Rose, Hershko and Ciechanover were awarded the 2004 Nobel Prize in Chemistry for their discovery (*16-19*).

Protein ubiquitination is an ATP-dependent reversible post-translational modification (PTM) that results in the formation of an isopeptide bond between carboxyl terminal of glycine (Gly) on ubiquitin and the epsilon amine of the lysine or, less commonly, the amino terminus or cysteine of the target protein (*8, 13, 17, 20-23*). Ubiquitination is carried out by a set of three enzymes: ubiquitin-activating enzymes (E1), ubiquitin-conjugating (E2) and ubiquitin ligase (E3) (*8, 18, 22*). Initially, ubiquitin is 'activated' (see below) in a two-step reaction by ubiquitin activating enzyme E1. The ubiquitin molecules are then transferred from E1 to the active site cysteine of an ubiquitin-conjugating enzyme E2 via a trans (thio) esterification reaction. Finally, an isopeptide bond between a lysine of the target protein and the C-terminal glycine of ubiquitin is established with the help of an ubiquitin ligase (E3), which recognizes and binds the target substrate and E2, thereby catalyzing the ubiquitin transfer from E2 to target protein **(Figure 1A)** (*8, 12-14, 21*). Human genome encodes two E1 activating enzymes, approximately 50 E2s and over 700 E3 ligases (*24, 25*).

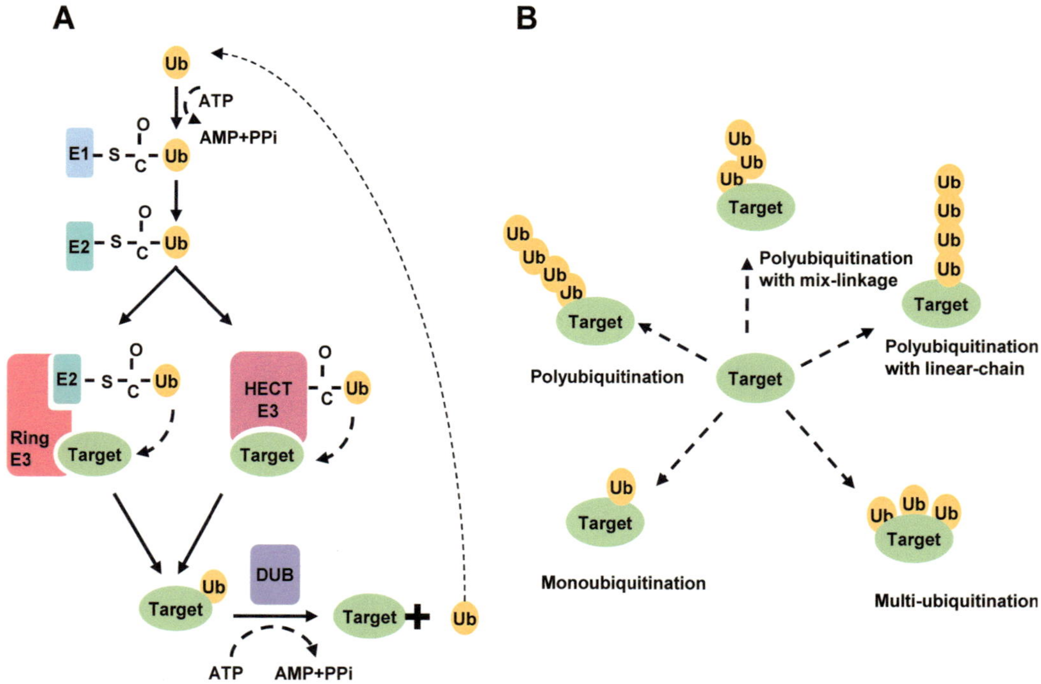

Fig. 1. An Overview of the Ubiquitination cascade and patterns. A) Schematic representation of the ubiquitination cascade as it is mediated by the E1, E2 and E3 enzymes and removed by DUB. B) Schematic representation of diferent ubiquination patterns.

The E1 activity represents an essential step during ubiquitin and ubiquitin-like protein (Ubl) conjugation, and the general mechanism of the E1-catalyzed reaction is well established (*17, 24-27*). In the first step of ubiquitin activation, the E1 enzyme catalyzes the adenylation of the Ub or Ubl C-terminus in an ATP-dependent process. In the second step, E1 forms a thioester between a conserved catalytic cysteine and the Ub or Ubl (*17, 18*). It has been long believed that a single E1 (Ube1) is responsible for activation and conjugation of all Ubls including ubiquitin, Nedd8 and SUMO. Recently, a divergent E1 in vertebrates and sea urchin, Uba6, was discovered. Human Uba6 charges a previously uncharacterized E2 (Use1) and specifically activates ubiquitin, but not other UBLs, *in vitro* and *in vivo* (*28, 29*).

Members of the ubiquitin-conjugating enzymes (E2s) family have highly conserved ubiquitin-conjugating catalytic (UBC) domain that provides a binding platform for E1, E3 and activated Ub/Ubl. Although E2s were initially thought to only be "ubiquitin carriers", recent studies revealed that E2s play an active role in determining ubiquitin chain length and linkage specificity(*30*). For instance, UBE2T specifically catalyzes the monoubiquitination of FANCD2, but not polyubiquitin chain formation due to lack of of extension activity(*30, 31*). UBE2N (Ubc13) mediates elongation of K63- specific polyubiquitin chains, yet it cannot initiate the ubiquitination (*30, 32-34*). UBE2C (UbcH10) is involved in the formation of K11-linked polyubiquitins, while UBE2K, UBE2R1 and UBE2G2 build the K48-linked ubiquitin chain (*30*).

In contrast to the E2s, whose catalytic sites are well conserved among its members, E3 ligases are structurally diverse and can be classified into two broad groups on the basis of their ubiquitin transfer mechanisms: HECT-type and adaptor-type. HECT-type E3 ligases are characterized by the HECT (Homologous to E6-associated protein C-Terminus) domain, which can form thioester bond with ubiquitin via an evolutionally conserved cysteine residue and play a direct catalytic role in final transfer of ubiquitin to a substrate (*35-37*).

On the other hand, adapter-type E3 ligases, including RING (Really Interesting New Gene) , U-box (modified RING motif without the full complement of Zn^{+2} – binding residues) domain containing E3 ligases, never form a covalent bond with ubiquitin (*38*). Rather, they function as scaffold proteins that facilitate the transfer of ubiquitin by recruiting ubiquitin conjugated E2 and substrate proteins into close proximity. The RING domain is a cysteine-rich sequence motif that is coordinated with two zinc atoms in "cross-brace" manner. Although U-box domain has very similar structure to RING domain, , the major difference between them is that U-box E3 ligase uses hydrogen bonds instead of zinc atoms (*38*).

Most the RING/U-box domain containing E3 ligases such as BRCA1, MDM2, RNF8 are multi-domain proteins with the exception of the Cullin–RING Ligases (CRL). Multi-domain E3 ligases are capable of catalyzing the reaction by themselves, albeit with relatively low efficiencies. The ligase activity is greatly enhanced by binding to a partner E3, e.g. BRCA1 with BARD1 and MDM2 with MDM4 (*39, 40*). On the other hand, the CRL are multi-components ubiquitin ligases and are arranged into elongated structure by a cullin scaffold protein and a catalytic RING subunit, RBX1 or RBX2 . Human genome encodes at least seven different cullin proteins: CUL1, CUL2, CUL3, CUL4A, CUL4B, CUL5 and CUL7 and two 'cullin homology domain' containing proteins: Parkin-like cytoplasmic protein (PARC) and subunit of the APC/C complex (APC2) (*41-44*). N-terminal domain of cullin binds substrate receptor, which holds an F box, SOCS box, VHL box or BTB box domain, directly or indirectly through adaptor proteins, such as SKP1 in SCF. C-terminal domain of cullin interacts with RBX RING domain, which brings to E2 ubiquitin conjugate enzyme into complex and promotes ubiquitin transfer (*41, 42*).

2.2 Different ubiquitination patterns have distinct functions

Attachment of single ubiquitin molecule on target protein (monoubiquitination) is sufficient to regulate various cellular processes such as endocytosis, DNA repair and replication, histone function, and transcription (*45-48*). Moreover, targeted proteins can be monoubiquitinated on several lysine residues, resulting in multi-ubiquitination (*49-51*). Alternatively, the targeted protein can be modified with a chain of ubiquitin molecules (polyubiquitination) **(Figure 1B)**. Since ubiquitin protein has seven lysine (K) residues (K6, K11, K27, K29, K33, K48 and K63) within its sequence, ubiquitin itself can be a target for ubiquitination. Since either lysine residue in ubiquitin can partake in chain formation, the reactions can produce structurally and functionally distinct polyubiquitin chains(*49-52*). K48-linked polyubiquitin chain with at least four ubiquitin molecules is well characterized as a signal for degradation by the proteasome (*53-55*). K63-linked polyubiquitin chains play crucial roles in DNA repair, protein trafficking and inflammation, and K6-linked chains are involved in DNA repair (*47, 52, 56-60*). The anaphase-promoting complex (APC/C) catalyzed K11 polyubiquitin chain also acts as efficient proteosomal targeting signal *in vitro* and *in vivo* (*61-63*). Another layer of complexity is introduced when several different lysines in the ubiquitin are conjugated and form mix-linkage or forked chains (*52*). There is also evidence for the linear polyubiquitin chain in which the C-terminal glycine of ubiquitin is bound to the α-amino group of the N-terminal methionine of another ubiquitin (*64*). Detailed characterization of these diverse ubiquitination patterns still awaits and will further extend our understanding of the functions of ubiquitin in the cell.

2.3 Principles of protein SUMOylation

Following the discovery of ubiquitin function, several structurally similar ubiquitin-like proteins (Ubls) have been identified as protein modifiers. Although Ubls do not share high sequence similarity, they all possess essentially the same three dimensional structure, and appear to be attached to substrates via similar sequential enzymatic cascade (*65*). Small Ubiquitin related MOdifier (SUMO)-protein is the best studied Ubl and functions as a critical regulator of many cellular processes, including nuclear transport, transcription, degradation, DNA repair and cell cycle control (*66-72*). SUMO conjugation to substrates occurs through a sequential enzymatic cascade involving a SUMO-specific E1 activating enzyme (SAE1/SAE2 heterodimer), an E2 conjugating enzyme (UBC9) and E3 ligases such as MMS21, PIAS1-4, RanBP2 or TOPORS (*70, 73, 74*). Similar to ubiquitin, the C-terminal glycine residue of SUMO is conjugated to lysine side chain of the target protein via an isopeptide bond, frequently, on the consensus sequence ψKxE (where ψ corresponds to a large hydrophobic amino acid, and x is any amino acid) (*70, 75, 76*). In mammals, three expressed forms of SUMO proteins have been well characterized: SUMO1, -2 and -3. SUMO2 and -3 have 95% homology with each other, so they are considered functionally equivalent and often referred to as SUMO2/3(*70, 75, 76*). Although SUMO2/3 is only 50% identical in sequence to SUMO1, all SUMO proteins share a common tertiary structure known as the ubiquitin superfold (*70, 71, 75*). Consistent with sequence divergences, SUMO1 and SUMO2/3 have distinct substrates specificity and also differ in their ability to form SUMO chains. SUMO2/3 possesses SUMO consensus sequence at their N-terminus and has the ability to form poly-SUMO chains, wheras SUMO1 lacks this consensus motif and is able to terminate the chain on SUMO2/3 polymers (*70, 71, 74-76*).

2.4 Reversing the ubiquitination and ubiquitin like modifications

Ubiquitination can be reversed by deubiquitination enzymes (DUBs). DUBs play four major roles in ubiquitin metabolism. First, DUBs engage in activation and processing of the ubiquitin precursor into mature ubiquitin monomers, probably co-translationally. Second, DUBs recycle the ubiquitin that was accidentally removed during the ubiquitination process. Third, DUBs restore monoubiquitin molecules from the unanchored polyubiquitin chains that is synthesized or released during the ubiquitination process. Finally, DUBs act as antagonist to E3 ligases by reversing the ubiquitination or ubiquitin-like modification of target proteins (*77, 78*).

Human genome encodes approximately 100 deubiquitinases and they can be divided into five major families: ubiquitin C-terminal hydrolases (UCHs), ubiquitin-specific proteses (USP), ovarian tumour proteases (OTUs), Ataxin-3/Josephins and JAB1/MPN/MOV34 metalloenzymes (JAMM/MPN+). The UCH, USP, OTU and Ataxin-3/Josephin families belong to Cys proteases, whereas the JAMM/MPN+ family members are zinc metalloproteases (*79*).

Protein SUMOylation is also a highly dynamic reversible process. The de-conjugation enzymes, also known as SUMO-specific proteases (SENP), cleave the isopeptide bond between SUMO and the target protein and are also responsible for processing of SUMO precursors into mature forms (*80*). In mammalian cell, SENP family contains six proteins (SENP1, SENP2, SENP3, SENP5, SENP6 and SENP7) (*81*). SENPs can be classified into three subgroups according to their functions. SENP1 and -2 can cleave both SUMO1 and SUMO2/3 from substrates and are involved in cellular maturation of the SUMO proteins. SENP3 and -5 remove the monomeric SUMO2/3 from substrates and SENP6 and -7 perform the editing and deconjugating of poly-SUMO-2/3 chains (*82, 83*).

3. An overview of DNA damage response signaling and modifications by ubiquitin and ubiquitin-like molecules

Exposure to endogenous and exogenous genotoxic stress triggers signaling pathways that allow the cell to slow down cell cycle progression and repair the damaged DNA. The framework of DDR can be defined in three categories: 1) sensors that continuously scan chromatin to detect the damage sites; 2) mediators or transducers that spread the signal, and 3) effectors that receive the signal and coordinate the response (*84*). Of crucial importance is the activation of the central checkpoint kinase Ataxia Telangiectasia Mutated (ATM) and ATM and Rad3-related kinase (ATR) that results in the phosphorylation of large number of substrates that execute diverse functions (*6, 7*). This is by far the most extensively studied pathway of the DDR.

Accumulating evidence suggest that phosphorylation is intimately coupled with other post-translational modifications including those by ubiquitination and ubiquitin-like molecules. Consistent with this hypothesis, proteins involved in ubiquitin proteasome systems are over-represented among the ATM/ATR substrates identified from proteomics screen (*6, 7*). Importantly, functional studies demonstrate that several E3 ligases acts at various stages of DNA damage response, including chromosome structure remodeling, signal amplification and repair, and their deregulation is associated with genetic diseases (*1, 25, 85-87*). In next section, we will describe the role of E3 ligases in DNA damage response pathway.

4. Modification by non-degradable Ubls in DNA damage signaling

The initial response to DSB is the sensing of DNA damage by multifunctional protein complexes that are able to bind directly to DNA. They include the MRE11/RAD50/NBS1

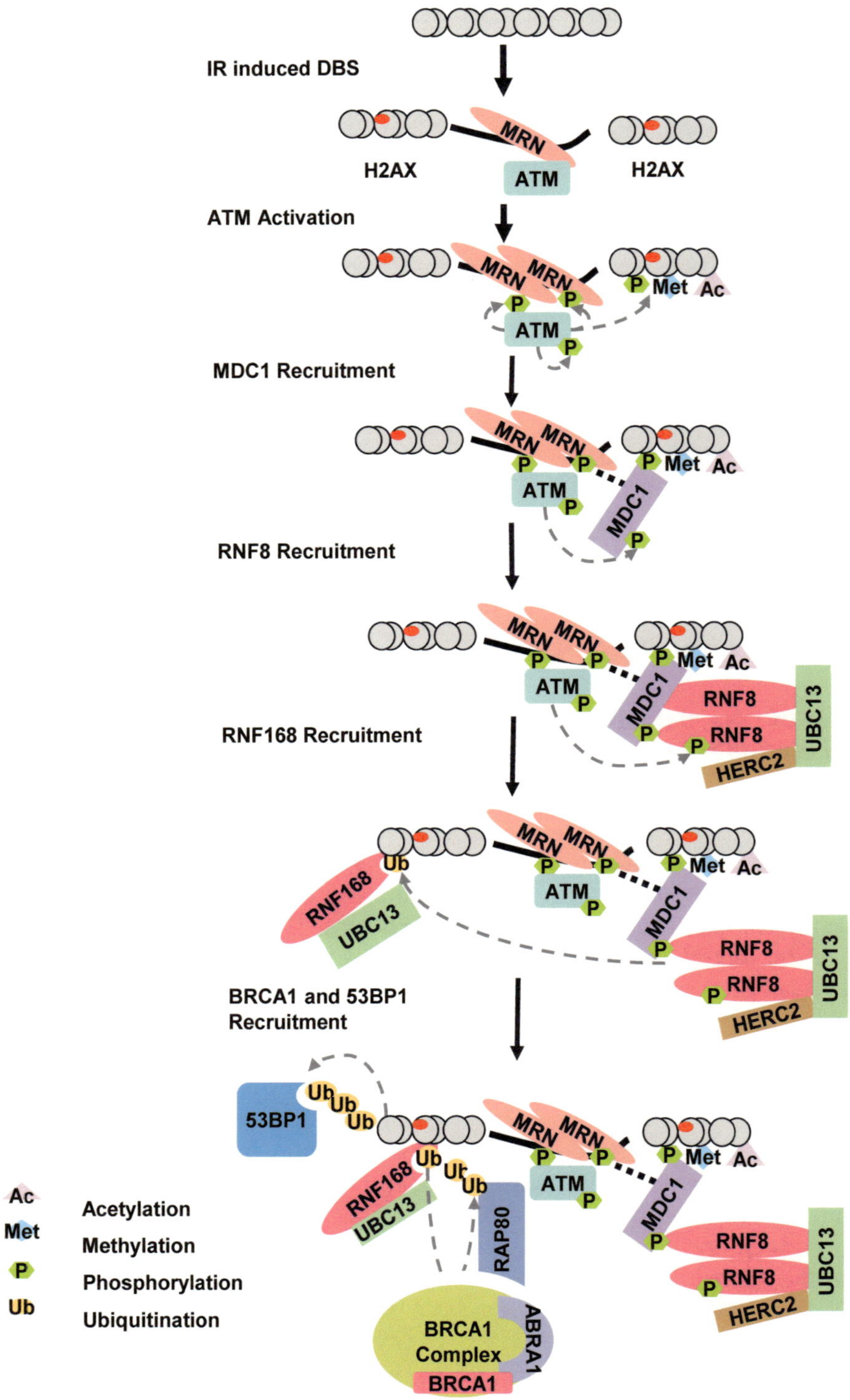

Fig. 2. Schematic representation of RNF8/RNF-168- mediatred ubiquitination at DSB and recruitment of DDR factors.

(M/R/N) complex, the Ku70/80/DNA-PKcs and perhaps ATM kinase itself (*88, 89*). Binding of M/R/N and Ku70/80 complexes onto DNA leads to the recruitment and activation of the Ser/Thr kinases ATM and DNA dependent kinase catalytic subunit (DNA-PKcs), respectively. Both kinases belong to Phosphatidylinositol -3-Kinase-like-Kinses (PIKK) family of serine–threonine kinases with sequence similarity to Phosphatidylinositol-3 Kinases (PI3Ks) and phoshorylate the histone variant H2AX on Ser139 residue (γ–H2AX) (*90, 91*). This phoshorylation marks the DNA damage sites and facilitates the recruitment of Mediator of DNA damage Checkpoint (MDC1). This initiates amplification of DDR signal by promoting the accumulation P53 Binding Protein 1(53BP1), RING Finger Protein 8 (RNF8), RING Finger Protein 168 (RNF168), RAP80 and Breast cancer susceptibility gene 1 (BRCA1) at damage sites (*92-97*). MDC1 and γ-H2AX dependent localization of RNF8 and RNF168 E3 ligases and ubiquitination of γ–H2AX are responsible for organizing the recruitment of BRCA1, 53BP1 and other DDR complexes and higher-order chromatin remodeling to facilitate local exposure of constitutive chromatin marks which leads to amplification of signal **(Figure 2)** (*58, 59, 96, 98*).

4.1 E3 ligases that are involved in the histone ubiquitination in the initial DNA damage signaling

RNF8 and RNF168 are substrates of checkpoint kinases that are recruited to sites of DNA damage

Following MDC1 phosphrorylation, RNF8 appears to be the first of the several E3s recruited to sites of DNA damage. RNF8 recognizes phosphorylated TQXF motifs on MDC1 through its FHA domain and cooperates with UBC13 to initiate the transfer of ubiquitins to histones (*58, 94, 96, 98*). Since UBC13 catalyzes the elongation of K63- specific polyubiquitin chain, RNF8/UBC13 mediated ubiquitination does not involve protein degradation (*58, 94, 96, 98, 99*). Importantly, accumulation of RNF8 on DNA damage sites is required for the recruitment of checkpoint mediators 53BP1 and BRCA1, suggesting that RNF8 lies upstream of these proteins. Recent studies also show that RNF8/UBC13-mediated ubiquitination contributes the recruitment of BRCA1 and 53BP1 to damage site, but it is not sufficient for persistent assembly of these factors (*58, 96, 100, 101*). Another RING type ubiquitin ligase, RNF168, recognizes RNF8/UBC13-mediated H2A and γ-H2AX ubiquitination by its Motif Interacting with Ubiquitin (MIU) domains to amplify and spread the regulatory K63-polyubiquitination on H2A and γ-H2AX (*1, 59*). In turn, RNF168-dependent ubiquitylation promotes the retention of BRCA1 and 53BP1 at DSB sites (*59*).

The importance of RNF8-mediated pathway in DNA damage response was demonstrated by the observation that biallelic mutations in RNF168 are associated with radiosensitivity, immunodeficiency, dysmorphic features, and learning difficulties (RIDDLE) syndrome (*1*), a disorder associated with defective DSB repair that shares some clinical features with other genome instability syndromes such as ataxia-telangiectasia (*102*). Cells derived from a RIDDLE patient are defective in DNA damage-induced ubiquitylation and exhibit impaired localization of 53BP1 and BRCA1 to DSBs, while MDC1 and NBS1, similarly to RNF8 and RNF168 depleted cells, remain unaffected (*1, 59, 102*).

HERC2 facilitates K63-polyubiquitination by stabilizing RNF8-UBC13 interaction

Recently, another ubiquitin ligase, HERC2 was shown to be involved in non-proteolytic histone polyubiquitination at DNA lesions (*103*). HERC2 interacts with the FHA domain of

RNF8 in phosphorylation-dependent manner. DNA damage-dependent phosphorylation of HERC2 on T4827 by PIKKs provides the docking site for FHA domain of RNF8 and allows the formation of a ternary MDC1-RNF8-HERC2 complex at DNA lesions (*103*). The exact function of HERC2 at DNA lesions is not entirely understood but it is suggested that HERC2 mediates and stabilizes the RNF8–UBC13 interaction and stimulates the K63-polyubiquitination. Another study showed that HERC2 promotes BRCA1 polyubiquitination thereby targeting BRCA1 for proteasome degradation (*104*). These findings are consistent with the overexpression of HERC2 in breast epithelial and carcinoma cells, implicating its role in breast cancer carcinogenesis.

BMI/RING1 E3 ligase contributes to γH2AX ubiquitination in a RNF8-independent manner

The polycomb repressive complex 1, which contains BMI1, RING1, and RING2, functions as an E3-ubuiquitin ligase. BMI1 and RING2 are recruited to sites of DSB where they contribute to the ubiquitination of γ-H2AX (*105*). In the absence of BMI1, ubiquitination-dependent recruitment of several proteins, including 53BP1, BRCA1, and RAP80, are impaired (*105, 106*). In contrast to RNF8, recruitment of BMI1 does not require H2AX, but depends on the FHA and BRCT domains of NBS1. Loss of BMI1 sensitizes cells to ionizing radiation to the same extent as loss of RNF8, and simultaneous depletion of both proteins revealed an additive increase in radiation sensitivity. Thus, it appears that RNF8 and BMI represent two distinct recruitment pathways for 53BP1 and BRCA1.

RNF20 and RNF40 regulate H2B mono-ubiquitination and homologous recombination

Recently, monoubiquitination of histone 2B (H2B), which is associated with transcription elongation, is shown to be a part of DDR in an ATM-dependent manner and required for timely damage repair (*107-109*). Heterodimeric RING-finger proteins RNF20 and RNF40, which are orthologs of the budding yeast protein Bre1 together with the ubiquitin-conjugating enzyme Rad6, catalyze H2B monoubiquitylation on lysine120 in the mammalian cells and on lysine 123 in yeast (*110*). Importantly, RNF20 and RNF40 physically interact with ATM and NBS1 (a member of the MRN complex) and are phosphorylated by ATM, which is essential for their recruitments to DNA damage site(*108, 111*). Either depletion of RNF20 /RNF40 or the mutation of H2B monoubiquitination sites lead to a significant decrease in RPA, RAD51 and BRCA1 localization on DSBs and interfere with Homologous Recombinational Repair (HRR) in human cells. Since the defect in HR causes the genomic instability by accelerating cancer-promoting mutations, dysfunctional RNF20/RNF40 increase the susceptibility to cancer (*112*).

4.2 E3s that are recruited to DSBs by ubiquitinated histones

Recruitment of BRCA1-A complex to DNA damage sites by RNF8/RNF168-mediated histone ubiquitination

The mechanism by which RNF8/RNF168 mediated-ubiquitination facilitates the recruitment of DDR factors is only partially resolved. RNF8/RNF168 can add K63-polyubiquitin chains on H2A and γ-H2AX that, in turn, act as molecular landing pads for downstream DDR factors with ubiquitin binding domains. The BRCA1 /BARD1 complex is one of the E3 complexes that are recruited to DNA damage sites in γ-H2AX-ubiquitination dependent manner.

The product of tumor suppressor gene BRCA1 plays a critical role in maintenance of genomic integrity by regulating DNA damage-induced cell cycle checkpoint activation, repair, chromatin remodeling, as well as transcriptional regulation and apoptosis (*5, 113*). BRCA1 protein contains two BRCT motifs at its C-terminus and RING domain at N-terminus (*114*). BRCT motif of BCRA1 functions as a phospho-protein binding domain and is required for its translocation and accumulation at DNA damage sites (*115-117*). RING domain of BRCA1 interacts RING domain of BARD1 (BRCA1-Associated Ring Domain protein 1) as a heterodimer and operates as an E3 ubiquitin ligase (*39, 118, 119*). The BRCA1/BARD1 complex can assemble with UBCH5C E2 ligase to promote the K6 ubiquitin chain *in vitro* (*120*). BRCA1/BARD1 heterodimer also forms at least four non-overlapping endogenous complexes with distinct functions. The "BRCA1-A" complex contains BRE/BRCC3/NBA1, FAM175A (Abraxas), FAM175B (Abraxas brother 1), and RAP80 (*100, 121, 122*). The RAP80 (Receptor Associated Protein 80 or UIMC1) is the component within BRCA1-A complex and is responsible for this recruitment of this complex onto the damage site by recognizing the K63 polyubiquitin chain through its ubiquitin interacting motifs (UIM) (*123, 124*).
Aside from RAP80 and BRCA1/BARD1, many components of the BRCA1-A complex possess structural domains implicating them in ubiquitination pathway. BRCC36 has a JAM domain characteristic of lysine 63-specific deubiquitinating enzymes, and Abraxas and BRE contain predicted Mpr-1/Pad1 N-terminal (MPN-) and ubiquitin E2 variant (UEV) domains respectively (*122, 125-128*). Possible targets of BRCC36 include RAP80 and di-ubiquitinated H2A/γH2AX. BRCC36 DUB activity and localization are regulated via its interactions within the context of each protein complex. In the BRCA1-A complex, ABRA1 and BRE are essential for BRCC36 DUB activity and localization at DNA damage site (*100, 122, 127, 128*). Although the exact function of BRCC36 deubiqutination at DNA damage site is not known, it has been suggested that BRCC36 catalyzes the deubiquitination of chromatin-associated proteins that lead to the chromatin remodeling and amplification of damage signal. In addition, deubiqutination by BRCC36 may contribute to the switch-off mechanism after the damage is repaired, which is required to resume normal DNA replication and cell proliferation. It is also speculated that BRCC36-driven deubiquitination not only removes the K63 polyubiquitin chain, but it may also promote BRCA1- dependent K6 ubiquitination.

Rad18 is another E3 ubiquitin ligase that accumulates at DSB sites in a manner dependent on RNF8

Rad18 is well-known for its function in DNA damage bypass and post-replication repair in yeast and vertebrates, where it promotes monoubiquitination of proliferating cell nuclear antigen (PCNA) at stalled replication forks (*129*). It was shown recently that RAD18 also participates in the signaling and homologous repair of DSB (*130*). The UBZ-like zinc finger domain on RAD18 preferentially binds to K63-linked Ub chains *in vitro*, and is required for Rad18 IR-induced foci (IRIF) formation, suggesting that RAD18 may be recruited to DSB through binding of its zinc finger domain to ubiquitinated proteins. RAD18 functions as an adaptor to facilitate homologous recombination through direct interaction with the recombinase RAD51C, although this function does not require its E3 ligase activity(*129-132*). Thus, the substrates of RAD18 in this process remain unknown.

4.3 SUMOylation system

Recent studies showed that the SUMO-conjugating system is operating in parallel and in cooperation with the ubiquitin-conjugating system for the assembly of 53BP1 and BRCA1.

SUMO1, SUMO2/3 and SUMO E3 ligases PIAS1 and PIAS4, as well as the E2 UBC9 also accumulate at DSB sites. Moreover, both 53BP1 and BRCA1 are possible substrates for SUMOylation, and the E3 Ub ligase activity of BRCA1 may also be subjected to regulation by SUMOylation (*73*).

5. Poly-ub-dependent protein degradation in cell cycle checkpoint control

Damaged DNA can lead to fixation of potentially harmful mutations during cell division. To this end, cells evolved checkpoints that can stop the cell cycle or slow down its pace to allow damage to be repaired. Two E3 ligases, the SCF and APC/C complexes that regulate normal cell cycle progression, also control cell cycle arrest. Another well-studied DNA damage checkpoint protein that is subjected to regulation by polyubiquitination-dependent degradation is the tumor suppressor protein p53, which regulates a wide variety of cellular processes, including transcription, apoptosis and senescence. Under normal growth conditions, p53 is kept at low levels by multiple E3 ligases through rapid turn-over; in response to DNA damage, p53 is rapidly stabilized and its level is maintained. Over the years, an increasing number of E3s and DUBs have been shown to be involved in this response. In this section, we will introduce the E3 ligases and DUBs that regulate p53 and substrates that are subjected to SCF and APC/C-mediated degradation for cell cycle controls.

5.1 The E3s and Dubs that keep the master tumor suppressor p53 under tight regulation

Tumor suppressor protein p53, also known as "the guardian of the genome", is a transcription factor that plays essential roles in multicellular organisms to regulate cell cycle progression, apoptosis, and senescence in response to DNA damage (*133-135*). Loss of function mutations in p53 are found in more than 50% of the tumors. In unstressed cell, p53 protein level is kept low predominantly through the ubiquitin-mediated proteosomal degradation. Several E3 ligases and their associated factors are involved in p53 ubiquitination and directly affect p53 levels, sub-cellular localization, and activity (*136, 137*). In addition, deubiquitination enzymes such as Herpes virus-Associated Ubiquitin Specific Protease (HAUSP, also known as USP7), and other ubiquitin-like (UBL) proteins also have impact on p53 abundance and activity (*136, 137*).

The identification of the human papilloma virus (HPV) E6-associated cellular protein E6AP was the first indication for the ubiquitination-dependent p53 regulation (*138*). Shortly after, RING type ubiquitin ligase Murine Double Minute 2 (MDM2) was characterized as the p53 E3 ligase that regulates its stability by promoting the formation of polyubiquitin chains on p53 and targeting it for ubiquitin-dependent proteasomal degradation (*139-141*). The importance of MDM2 in p53 regulation is highlighted by the *Mdm2* knock-out mice studies. The embryonic lethality of *Mdm2*-knockout mice, which is due to p53 induced apoptosis, is rescued by deletion of p53 (*142, 143*).

The MDM2 analogous MDM4 or MDMX, has emerged as another key regulator of p53. *Mdm4* was identified as a novel *p53* interacting protein in mouse and as a negative regulator of p53 by inhibiting p53 transcription activity (*144*). Importantly, MDM4 can form a heterodimer with MDM2 through their RING domains (*145, 146*) and while MDM4 lacks intrinsic ubiquitin-ligase activity, it can stimulate MDM2–dependent p53 ubiquitination and degradation *in vivo* (*40*). MDM4 also stabilizes MDM2, presumably by reducing MDM2 homodimers and thereby restricting MDM2 autoubiquitination activity (*146*). Like *Mdm2, Mdm4* deficiency in mice leads to early embryonic lethality due to cell arrest, and the phenotype is rescued by p53 loss and partially rescued by the loss of cyclin-dependent kinase inhibitor 1A (Cdk1a, the gene that encodes p21 which regulates proliferation) (*147, 148*).

The six key lysine residues (K370, K372, K373, K381, K382, and K386) located at the most C-terminus of p53 are well-characterized ubiqutination sites for MDM2 *in vitro* (*149*). However, knock-in mice studies, where all the lysine residues are replaced with arginines (p53-6KR mutant) *in vivo*, have shown that p53 expression levels are similar to wild type p53, implying that while these lysines are critical to regulate p53 function, they are not sufficient for degradation and alternative lysine residues may be involved in p53 stability (*150, 151*). Recently, it was shown that p53 could be ubiqutinated in the DNA binding domain *in vitro* and deletion mutant of DNA binding domain stabilizes p53 by decreasing its ubiqutination (*152*).

Many stress factors, such as DNA damage and oncogene activation, lead to p53 stabilization and increase in its transcriptional activity to regulate checkpoints especially the G1/S and apoptosis. In a widely accepted model, rapid p53 stabilization during early response to DNA damage occurs through phosphorylation of p53 at Ser15 and Ser20 and/or acetylation of p53 at C-terminal lysine residues instead of ubiqutination by DNA damage responsive enzymes, such as ATM/ATR, DNA-PK, CHK1, CHK2 and p300. These post-translational modifications (PTMs) cause p53 to dissociate from its negative regulators MDM2 and MDM4 and enhance the p53 transcriptional activity (*137, 153*). ATM also stimulates the phosphorylation of MDM2 and MDM4 at their C-terminus, which leads to their rapid degradation. ATM also activates another kinase c-Abl, to phosphorylate MDM2 at tyrosine 394 and MDM4 at tyrosine 99, which decrease their ability to inhibit p53 (*154-157*). Tumor suppressor ARF (known as p14ARF in humans and p19ARF in mouse) contributes to p53 stabilization by sequestering MDM2 and thus preventing MDM2 from physically interacting with p53 (*158*).

Active form of p53 is also known to be a transcription factor for its own major negative regulator MDM2 and lead to the accumulation of MDM2 by creating a negative feedback loop(*159, 160*). On the other hand, MDM2 contributes to the positive feedback loop to p53 stability by stimulating its and MDM4 ubiquitination for degradation (*161, 162*). Recent study also shows that another E3 ubiquitin ligase, RFWD3 (RING finger and WD repeat domain 3) positively regulates p53 stability as part of the late response to DNA damage. RFWD3, which is phosphorylated by ATM/ATR, is a potent E3 ligase for p53 that promotes its oligo-ubiquitination in vitro. RFWD3 forms a complex with MDM2 and p53, restricts MDM2 polyubiquitination activity in a way that p53 becomes resistant to 26S proteosomal degradation (*163*).

MDM2 also mediates ubiquitin-like modifications such as SUMOylation and Neddylation of p53 (*164, 165*). PIAS family of E3 ligases mediates p53 sumoylation at lysine 386 (K386) in vitro and MDM2 can fulfill this in an ARF-dependent manner. MDM2/ARF mediated p53 SUMOylation is associated with relocalization of p53 to subnuclear structures and modest increase in p53 transcriptional activity, the biological consequences of this phenomenon are not completely clear (*72, 165, 166*).

Although MDM2 is a major ubiquitin ligase of p53, other ubiquitin ligases also play a role in p53 ubiqutination, as p53 protein is still degraded at certain levels in Mdm2 deficient mice. Several p53 ubiquitin ligases have been identified including ARFBP1, COP1, E4F1, Ubc13,

Pirh2, TOPORS and WWP1 (*136, 137*) **(Figure 3).** These ubiquitin ligases can be categorized into two groups according to their ability to built K48 linked-polyubiquitin chain and non-proteasomal ubiquitin chain. The first group, which includes Pirh2, COP1, CARP1, CARP2 and ARFBP1, regulates the p53 protein level through the proteasomal degradation with the exception of TOROS, whose function is unknown (*166-171*). The main function of the proteins in second group (WWP1, MSL2 and E2 enzyme Ubc13) is to regulate the p53 localization (*172-174*).

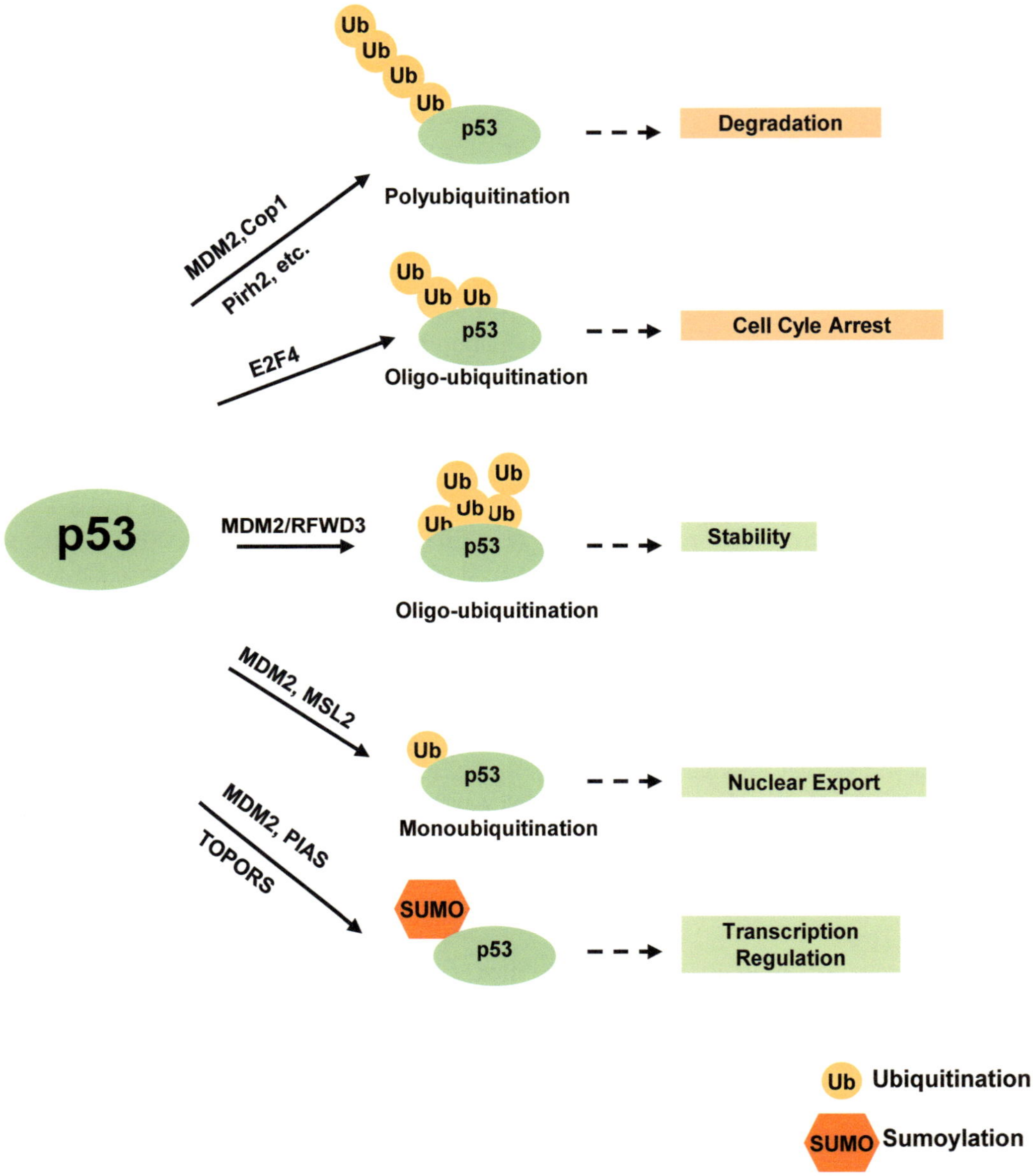

Fig. 3. MDM2-dependent and independent p53 ubiquitinationation patterns and their distinct functions in p53 biology.

In addition, transcription factor E4F1 was shown to ubiquitinate p53 mainly through the K48-linkage (*175*). E4F1 is considered as an atypical E3 ligase because it does not contain a HECT or RING domain. However, it has SUMO E3 ligase RanBP2 catalytic domain homologous region (aa 30-80), which provides its E3 ligase activity (*66, 175, 176*). E4F1 can catalyze the formation of mono-, di-, and tri-ubiqutination on the lysine cluster K319–K321 of p53 that are distinct from those targeted by MDM2 and these oligo-ubiquitin chains neither target p53 for degradation nor change the p53 nuclear/cytoplasmic ratio. E4F1-dependent ubiquitination instead increases p53 localization to the chromatin fraction selectively in p21 and cyclin G promoter region and regulates transcription of p53 target genes that are required for cell cycle arrest, but not of those involved in apoptosis. Furthermore, E4F1 mediated K320 ubiquitination competes for acetylation of K320 by PCAF that induce apoptosis in response to DNA damage (*175, 177*). This suggests that E4F1 dependent ubiquitination inhibits the PCAF-mediated acetylation of p53 and triggers an anti-apoptotic program in response DNA damage.

Ubiquitination is dynamic and reversible process through ubiquitin ligases and deubiquitinases (DUBs). Like ubiquitin ligases, deubiquitination enzymes (DUBs) regulate the fate and function of ubiquitin-conjugated proteins and have been implicated in several important pathways including cell growth and differentiation, development, oncogenesis and transcriptional regulation. DUBs activity is especially critical and tightly regulated when ubiquitination is used as a signal for non-proteasomal process and play an important role in p53 stability and activity. The first DUB shown to regulate p53 stability was HAUSP (*178*). HAUSP was originally characterized as Herpes virus-associated cellular factor and shown to interact and deubiquitinate p53 in vitro and in vivo (*178-180*). HAUSP belongs to the Ub-specific processing protease (UBP) subfamily of DUBs and contains conserved core enzymatic domain with characteristic Cys and His motifs (*181*). Beyond the catalytic domain, HAUSP has N-terminal and C-terminal extensions with no significant similarity to other members of UBP family and these extensions are important for substrate recognition and subcellular localization. HAUSP interacts and modulates not only p53 but also MDM2 and MDM4 protein level and function has greatly increased the complexity of its role within the p53-MDM2 pathway (*182, 183*). While overexpression of HAUSP results in the stabilization of p53 through de-ubiquitination and partial knockdown of HAUSP decreased p53 protein level, complete ablation of HAUSP led to an increase in p53 levels due to increase in MDM2 ubiquitination and proteasomal degradation (*184*). In the absence of stress, HAUSP is involved in stabilizing MDM2 and MDM4 by removing the ubiquitin molecules, however, MDM2 and MDM4 phosphorylated by ATM as a response to DNA damage and this phosphorylation inhibits MDM2 or MDM4 interaction with HAUSP and causes the MDM2 and MDM4 degradation in favor of p53 activation (*185*).

Besides the regulation of protein stability, DUBs have been shown to regulate chromatin structure and to facilitate both gene activation and gene silencing. HAUSP also stimulates p53 DNA binding activity and involves in its transcription function (*186*).

Another deubiquitinating enzyme USP2a was associated with MDM2 *in vivo* and shown to deubiquitinate MDM2, without reversing MDM2-mediated ubiquitination of p53. Ectopic expression of USP2a therefore leads to an increase in the levels of MDM2 and promotes p53 degradation. Downregulation of USP2a protein level with siRNA results in destabilization of MDM2 and thus elevated p53 protein expression and upregulation of p53 target genes. MDM2 specific deubiquitin ligase USP2a contributes to repression of p53 activity *in vivo* (*187*).

5.2 Substrates of CRLs that regulate cell cycle arrest

Many cellular factors that orchestrate cell cycle progression under normal conditions, including numerous UPS proteins, also play key roles in regulating damage-induced cell cycle arrest. Their functions in normal cell and in damage-induced cell mainly manage through synthesis-degradation and phosphorylation-dephosphorylation manner. Two well-studied E3 ligase families, the SCF and APC/C complexes polyubiquitinate these cellular factors and target them for proteasome-mediated degradation. Both SCF and APC/C are CRL-type E3 ligases composed of a scaffold protein (Cullins and APC2), a RING domain containing protein that mediates its interaction with E2 (RBX1 and APC11), adaptor proteins and co-activators that determine substrate specificity (F-box proteins, CDC20 or CDH1) (*188*). Importantly, activation of these E3 ligases is closely coupled with the activation of checkpoint kinases, which phosphorylate their substrates and facilitate the recognition by the F-box proteins in the SCF complex.

Chk1, a substrate of ATR/ATM, is an effector kinase that is part of both checkpoint initiation and termination

CHK1 is essential for cell viability and is phosphorylated and activated by ATR/ATM in response to replication block and DNA damage. Activation of CHK1 leads to the phosphorylation of a plethora of substrates including Cdc25A, Claspin, Kap1, which are critical for the activation of intra-S and G2/M checkpoint (*189*). CHK1 phosphorylation also promotes its own degradation mediated by the SCF E3 ligase cullin 4A/DDB1 or cullin 1 CUL1/Fbx6 (*190*). Therefore, CHK1 activity is fine-tuned by a negative feed-back loop to ensure timely termination of checkpoint signal and resuming cell cycle progression. In addition, the adaptor protein Claspin that facilitates ATR-mediated Chk1 phosphorylation, also contributes to down-regulation of Chk1 activity through its own proteolysis (*191*) (see below).

CDC25A is phosphorylated and degraded to prevent G1-S transition

Cdc25A is a phosphatase that is an essential regulator for CDK (Cyclin Dependent Kinase) activity during cell cycle progression (*189*). By removing the inhibitory phosphorylation of CDKs, Cdc25A drives cell cycle from G1 to S, and elevated Cdc25 levels resulting from their stabilization and persistence during cell cycle periods have significant oncogenic potential. In unperturbed cells, Cdc25A protein is maintained at modest levels during S and early G2 by the action of the ubiquitin ligase $SCF^{\beta TrCP}$, whose recognition of Cdc25A depends upon phosphorylation by a number of kinases (*192, 193*). At the end of mitosis and through G1, Cdc25A is eliminated via the APC/CCdh1 ubiquitin ligase. DNA damage induction of Chk1 activity increases the priming phosphorylation of Cdc25A, leading to greater $SCF^{\beta TrCP}$-induced degradation (*194*). Moreover, Cdc25A can also be stabilized by the ubiquitin hydrolase Dub3, which removes the polyubiquitin modifications that mark Cdc25A for proteasomal degradation (*195*). Importantly, Dub3 was able to transform NIH-3T3 cells and cooperated with H-Ras to promote growth and its overexpression is responsible for an abnormally high level of Cdc25A in a subset of human breast cancers (*195*).

CDT1 and Cdc6 are degraded upon DNA replication block to prevent replication

CDT1 is an essential protein required for licensing of replication origin in G1 and is degraded in S-phase to prevent re-replication and following DNA damage to prevent late origin firing (*196, 197*). In unperturbed cells, phosphorylation of CDT1 by cyclin-dependent kinase promotes its binding to and ubiquitination by SCF-Skp2 E3 ubiquitin ligase (*198-202*).

However, this pathway is not essential for the degradation of CDT1. It contains a second degradation signal at its N-terminus that is active after DNA damage, and is dependent on the loading of PCNA. Its ubiquitination and degradation requires CDT2 and DDB1, which are components of a CUL4 ubiquitin ligase (*200, 203*).

DNA damage-dependent ubiquitination of CDC6 has been attributed to two ubiquitin ligases, depending on the cell type and source of DNA damage. In p53-proficient cells treated with ionizing radiation, CDC6 is down-regulated through ubiquitination by the CDH1-associated form of the anaphase promoting complex, APC/C (*204*). In response to other forms of DNA damage, CDC6 is ubiquitinated by the HUWE1 enzyme irrespective of the p53 status of the cells (*205*).

Claspin is phosphorylated by CHK1 and PLK1, which mediate its degradation by $SCF^{\beta TrCP}$ and is involved in G2/M checkpoint-recovery from replication block

Claspin is an adaptor protein that facilitates ATR-mediated phosphorylation and activation of CHK1 in response to replication arrest and replication stress during normal cell cycle progression as a result of replication stress (*191, 206*). Claspin level is regulated throughout the cell cycle by ubiquitin mediated proteolysis. Two E3 ubiquitin ligases, the SKP1-CUL1-βTrCP and APC/CDH1 contribute to Claspin degradation (*207, 208*). Claspin is phosphorylated in a canonical DSGΦxSDSGFxS degron sequence, which is typical of $SCF^{\beta TrCP}$ substrates. Phosphorylation of Claspin is mediated by PLK1 and is essential for binding to βTrCP (*191, 209*). Thus, degradation of Claspin is essential for entering mitosis during normal growth and termination of checkpoint signaling to allow resumption of cell cycle. Claspin is also destabilized by the anaphase-promoting complex (APC) and thus remains unstable in G1. Whether APC/C-CDH1 also regulates Claspin stability after DNA damage is not clear.

Ubiquitination of Claspin is counteracted by 2 ubiquitin hydrolases. USP7 interacts with Claspin in vivo and is required to maintain steady-state levels of Claspin (*208*). Consistent with its role in stabilizing Claspin, USP7 also controls the recovery from replication arrest, presumably through regulating CHK1 phosphorylation. However, USP7 has no effect on APC/C-CDH1-mediated Claspin turn-over. USP28, a major DUB for maintaining 53BP1 protein level, has also been implicated in stabilization of several checkpoint proteins including that of Claspin (*210*).

6. Identification of new ubiquitinated proteins involved in DNA damage response

The key to a better understanding of how ubiquitination regulates biological processes is the identification of ubiquitinated substrates. An analogy can be drawn from recent systematic screens that identify multiple proteins that are phosphorylated by checkpoint kinase ATM/ATR. Taking advantage of a unique affinity of these kinases towards SQ motifs and the technical advancements in mass spectrometry, thousands of phosphorylated proteins were affinity purified with antibodies that recognize various pSQ motifs through "cross-reactivity" and identified by mass spectrometry (*6, 7*). Information derived from these studies tremendously expanded the landscape of DNA damage response pathways and provided valuable clues for follow-up functional characterization of these pathways. Unfortunately, such approach cannot be effectively applied to ubiquitination studies, as there is little information on any consensus motif that directs E3 ligases to their targets. Even less is known about how E3 activity is regulated after exposure to DNA damaging agents. Nevertheless, it is expected that improved sensitivity of mass spectrometry and new bioinformatics tools will allow for quantitative analysis of the ubiquitination events that correlate with changes that take place after DNA damage. In this section, we summarize recent proteome-wide approaches for affinity purification and mass spectrometry identification of ubiquitinated proteins.

6.1 Affinity-based purification coupled with Mass spectrometric analysis

Affinity purification of ubiquitinated proteins

The first and essential step in the identification of ubiquitinated substrates on a proteome-wide scale is enrichment by affinity purification; as the abundance of ubiquitinated species is generally low. Currently, there are at least three different strategies for large-scale ubiquitination profiling (Figure 4B-4D); Epitope Tagged Ubiquitin (Ub) expression systems (Figure 4B), Tandem Ubiquitin-Binding Domain (UBD) affinity purification (Figure 4C), Linkage-Specific Poly-Ubiqutin Antibody purification (Figure 4D), and Ubiquitination Signature Remnant Antibody purification (*211*).

Epitope tagged Ub strategy was initially applied to profile yeast ubiquitome by using an engineered yeast strain, which expresses epitope-tagged ubiquitin at endogenous level in ubiqutin null background. (*212*). Totally 110 ubiquitination sites were identified by using a LCQ mass spectrometer (*212*). A similar approach was adapted in the mammalian system, in which a tandem His-biotin-tagged Ub was stably expressed in HeLa cells for efficient substrate isolation. In this experiment, ~ 50 ubiquitination sites were identified with a LTQ-Orbitrap mass spectrometer (*213*). Recently, the identification of ~750 lysine ubiquitination sites from two human cell lines that stably express HA-Ub was reported, which stands, by far, as the largest collection of ubiquitination sites (*202*).

Major advantages of epitope tagged Ub strategy are to purify the substrates under protein denaturing conditions and to eliminate contaminant non-substrate proteins. However, it remains uncertain to which extent tagging Ub over-expression may interfere with normal cellular functions (*214, 215*). Substitution the endogenous Ub lysine mutants would allow to investigate linkage-specific substrates, but the presence of four Ub loci in the mammalian genome is discouraging to establish ubiquitin null cell lines and to analyze the substrates in that background. Recently, engineered cell line that stably expresses an exogenous K63R mutant Ub with a tetracycline-inducible RNAi that eliminates endogenous Ubiquitin expression was generated (*216*). This method may partially solve this problem and facilitates the profiling of linkage specific ubiquitination substrates in mammalian cells.

High-affinity Ub antibodies, once successfully generated, are powerful tools for ubiquitination profiling of endogenous substrates. Previous attempts to generate high affinity Ub antibody from mice and rabbits were unsuccessful due to sequence conservation among species. Recently, several linkage-specific monoclonal antibodies have been selected from phage display libraries and have been shown to be useful for immunoprecipitation and western blot (*217, 218*). Whether or not these antibodies can efficiently profile linkage-specific substrates remains unclear.

The Gly-Gly remnant antibody was generated against the Gly-Gly signature peptides that generated after trypsin digestion of ubiquitinated proteins (*198, 201*). Since it recognizes specifically the modified peptides, it is more efficient for ubiquitination site profiling than

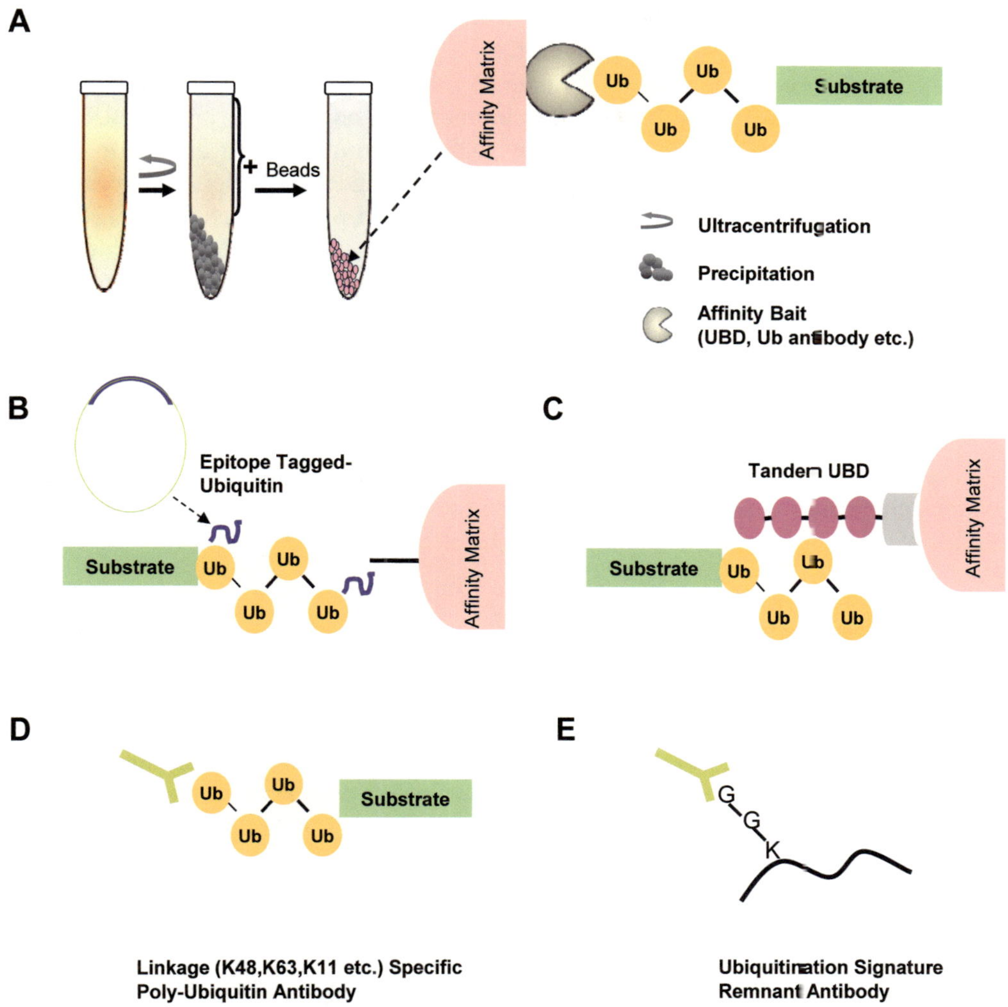

Fig. 4. A schematic presentation of ubiquitinated substrate purification strategies. A) Overall immunoprescipitation and purification procedure for ubiquitinated substrates. The affinity purification methods based on B) epitope-tagged ubiquitin, C) tandem ubiquitin-binding domain (UBD), D) linkage-specific poly-ubiquitin antibody or E) ubiquitination signature remnant antibody.

other methods that isolate the entire substrates. Using this reagent, hundreds of ubiquitination sites have been identified in human cells (*198, 201, 202*). Importantly, certain Ub-like modifications such as ISG15 or NEDD8 are also conjugated to their substrates through C-terminal Gly-Gly, as a result, the Gly-Gly remnant antibody does not distinguish ubiquitination from those modifications. It is necessary to couple with other purification method in order to obtain a homogenous population of ubiquitinated species for peptide IP (*211*) .

Ub binding domains (UBDs) are small structural entities that have affinity for Ub and can be used as natural affinity reagents to isolate endogenous poly-Ub proteins. Over 20 different UBDs have been discovered (*219-222*). The existence of a variety of UBDs implies their functional diversities in the UPS signaling network. Although all UBDs bind Ub, the binding affinity varies greatly with the Kd value in the range of several hundred mM to the a few μM (*219*). The inherent low affinity of a natural UBD is generally not sufficient for large-scale purification of substrates. The development of engineered tandem UBD with appreciable affinity in the range of antibody makes it an attractive reagent for substrate purification. We have applied such strategy by engineering a GST fusion with four tandem UBA domains of ubiquilin 1 (GST-qUBA), which allowed us to detect close to 300 lysine ubiquitination sites from human cells that were not treated with proteasome inhibitors (*223*). A complication for this approach is that purification has to be performed under non-denaturing conditions; thus, "contaminant proteins" (especially true for UBD proteins that bind poly-Ub chains) are inevitably co-purified and sequenced by MS. These co-purified proteins may not only be misidentified as Ub substrates, but also masks the low abundant true substrates. Moreover, UBDs are more favorable for poly-Ub chains than mono-Ub and this approach could be biased towards polyubiquitininated substrates. Whether UBDs have preference towards specific type of poly-Ub or even the length of the chain remains unclear. A better understanding of UBD and Ub interaction require systematic investigation of more varieties of UBDs and may facilitate the design of poly-Ub linkage-specific reagents with higher affinity and selectivity (*211, 223*).

Precautions to be taken in affinity purifications

One of the major challenges for large-scale substrate purification is to distinguish the true ubiquitinated substrate from contaminant proteins. There are mainly three types of protein contaminants for Ub affinity purification: 1) non-specific binders to the solid matrix (e.g., agarose beads) for affinity reagents, 2) precipitated proteins that are accumulated during incubation, and 3) specific binders to poly-ubiquitinated proteins and/or Ub itself (*223*). Although it is very difficult to completely eliminate them all, the amount of contaminants can be significantly reduced by 1) carefully removing the insoluble precipitations after high-speed centrifugation, 2) shortening the affinity incubation time, 3) increasing protein solubility with stronger detergent such as 0.1% SDS, and 4) washing the beads with higher concentration of salt (*223, 224*).

The second challenge during purification is to preserve the ubiquitination by restricting the DUB activity. The use of DUB inhibitors of both Cys alkylation (such as iodoacetamide, choloroacetamide, or N-ethylmaleimide) and zinc chelating chemical (1, 10-σ-phenanthroline) can alleviate the loss of poly-Ub and preserve the intact modified substrates (*126*). In addition, avoiding freezing-thawing cycles by using freshly prepared cell lysate can help to preserve ubiquitinated sites.

Identification by mass spectrometry

The ubiquitinated species, once enriched and purified, can be analyzed by mass spectrometry (MS). While the ubiquitinated peptides purified from Gly-Gly antibody can be directly sequenced, ubiquitinated proteins are first subjected to digestion by proteases, such as trypsin, which cuts at carboxyl-terminus of lysine and arginine. The affinity purified proteins can either be first separated on SDS-PAGE and in-gel digested for peptides to be extracted, or directly digested on affinity matrix and extracted. Although on-beads digestion is fast, convenient, and suffers less sample loss, the SDS-PAGE separation provides a number of advantages. SDS-PAGE not only adds another purification step that eliminates the interference from abundant proteins that potentially mask the signals of less abundant proteins, it also provides molecular size information for poly-ubiquitinted proteins that would migrate at higher molecular weight than their unmodified forms.

Ubiquitination can be identified by mass spectrometry with the detection of a mass shift of 114.043 Da- the ubiquitination signature that is derived from the di-glycine remnant of Ub after trypsin cleavage. Ubiquitination peptides (as well as sites) are identified by searching protein sequence database to match a tryptic peptide sequence with the addition of the ubiquitination signature mass on the particular amino acid, typically on Lys residues (*211, 223*).

The choice between peptide IP and protein IP has important implications. While peptide IP using Gly-Gly remnant antibody allows direct identification of Ub peptides and ubiquitination sites, these peptides may escape mass spectrometry detection if their sizes fall out of the mass range or they cannot be efficiently ionized. On the other hand, affinity-purified proteins that generate multiple peptides can be more readily detected, but cannot be distinguished from non-ubiquitinated proteins through association with ubiquitinated species. In this case, molecular weight information provided by SDS-PAGE can be used as a determinant.

6.2 Global Protein Stabilty (GPS) profiling of SCF Ub ligase substrates

Genetic screens are powerful tools for high throughput identification of biological targets through functional readouts. A strategy that combines global protein stability (GPS) profiling and genetic perturbation of E3 activity has been established to screen for substrates of the Skp1-cullin-F-box (SCF) ubiquitin ligase in mammalian cells (*225, 226*). In this approach, an HEK 293T reporter cell library expressing EGFP fused to ~8000 human open reading frames are established. In these cells, *Discosoma* sp. Red fluorescent protein (DsRed) and GFP fusion proteins are expressed under the control of the same promoter, serving as an internal control for protein expression. Thus, the GFP/DsRed ratio measured by FACS reflects the stability of the GFP-fusion protein, and the SCF targets can be detected in cells that show increased ratios in response to SCF inactivation. Since prolonged inhibition of SCF has the potential to alter cell cycle distribution, SCF is transiently reduced by lentiviral delivered dominant-negative Cul1. Among the >350 potential substrates identified, most known SCF targets are recovered and many previously unknown substrates involved in cell cycle, apoptosis, and signaling pathways are discovered (*225*). Although limited only to E3 ligases that regulate substrate stability, this approach has the potential for global discovery of any E3-substrate regulatory networks.

7. Concluding remarks

Protein modification by Ub and Ub-like molecules has emerged as a major regulatory mechanism for DNA damage signaling and repair. Accumulating evidence from last decade's research suggest that alteration of chromatin structure is a key step in the initiation and amplification of damage signals, as ubquitination of histones, such as H2A and H2AX, are substrates of several E3 ligases involved in DDR. Alternatively, it may also indicate that, while histones are previously known targets, other substrates remain unidentified due to technical difficulties. Likewise, two major E3 ligases (SCF and APC/C) appear to dominate the control of stability of most of proteins involved the cell cycle arrest. It is anticipated that technical advancement and availability of new affinity reagents will enable genome-wide, systematic characterization of UPS proteins in DDR.

8. References

[1] G. S. Stewart *et al.*, *Cell* 136, 420 (Feb 6, 2009).
[2] E. I. Palmero, M. I. Achatz, P. Ashton-Prolla, M. Olivier, P. Hainaut, *Curr Opin Oncol* 22, 64 (Jan, 2010).
[3] H. Joenje, K. J. Patel, *Nat Rev Genet* 2, 446 (Jun, 2001).
[4] S. P. Jackson, J. Bartek, *Nature* 461, 1071 (Oct 22, 2009).
[5] C. X. Deng, *Nucleic Acids Res* 34, 1416 (2006).
[6] J. J. Mu *et al.*, *J Biol Chem* 282, 17330 (Jun 15, 2007).
[7] S. Matsuoka *et al.*, *Science* 316, 1160 (May 25, 2007).
[8] C. M. Pickart, *Annu Rev Biochem* 70, 503 (2001).
[9] E. T. Yeh, L. Gong, T. Kamitani, *Gene* 248, 1 (May 2, 2000).
[10] C. M. Pickart, M. J. Eddins, *Biochim Biophys Acta* 1695, 55 (Nov 29, 2004).
[11] S. Jentsch, G. Pyrowolakis, *Trends Cell Biol* 10, 335 (Aug, 2000).
[12] M. Hochstrasser, *Nat Cell Biol* 2, E153 (Aug, 2000).
[13] M. Hochstrasser, *Science* 289, 563 (Jul 28, 2000).
[14] M. Hochstrasser, *Annu Rev Genet* 30, 405 (1996).
[15] G. Goldstein *et al.*, *Proc Natl Acad Sci U S A* 72, 11 (Jan, 1975).
[16] K. D. Wilkinson, M. K. Urban, A. L. Haas, *J Biol Chem* 255, 7529 (Aug 25, 1980).
[17] A. Hershko, H. Heller, S. Elias, A. Ciechanover, *J Biol Chem* 258, 8206 (Jul 10, 1983).
[18] A. Hershko, A. Ciechanover, *Annu Rev Biochem* 67, 425 (1998).
[19] A. Ciehanover, Y. Hod, A. Hershko, *Biochem Biophys Res Commun* 81, 1100 (Apr 28, 1978).
[20] K. Okumoto *et al.*, *Traffic* 12, 1067 (Aug, 2011).
[21] A. Hershko, *Cell* 34, 11 (Aug, 1983).
[22] M. H. Glickman, A. Ciechanover, *Physiol Rev* 82, 373 (Apr, 2002).
[23] K. Cadwell, L. Coscoy, *Science* 309, 127 (Jul 1, 2005).
[24] N. L. Lehman, *Acta Neuropathol* 118, 329 (Sep, 2009).
[25] C. Chen, A. K. Seth, A. E. Aplin, *Mol Cancer Res* 4, 695 (Oct, 2006).
[26] A. L. Haas, J. V. Warms, A. Hershko, I. A. Rose, *J Biol Chem* 257, 2543 (Mar 10, 1982).
[27] A. L. Haas, I. A. Rose, *J Biol Chem* 257, 10329 (Sep 10, 1982).
[28] B. A. Schulman, J. W. Harper, *Nat Rev Mol Cell Biol* 10, 319 (May, 2009).
[29] J. Jin, X. Li, S. P. Gygi, J. W. Harper, *Nature* 447, 1135 (Jun 28, 2007).
[30] Y. Ye, M. Rape, *Nat Rev Mol Cell Biol* 10, 755 (Nov, 2009).
[31] A. F. Alpi, P. E. Pace, M. M. Babu, K. J. Patel, *Mol Cell* 32, 767 (Dec 26, 2008).
[32] M. Windheim, M. Peggie, P. Cohen, *Biochem J* 409, 723 (Feb 1, 2008).
[33] M. D. Petroski *et al.*, *J Biol Chem* 282, 29936 (Oct 12, 2007).
[34] D. E. Christensen, P. S. Brzovic, R. E. Klevit, *Nat Struct Mol Biol* 14, 941 (Oct, 2007).
[35] S. E. Schwarz, J. L. Rosa, M. Scheffner, *J Biol Chem* 273, 12148 (May 15, 1998).
[36] M. Scheffner, O. Staub, *BMC Biochem* 8 Suppl 1, S6 (2007).
[37] J. M. Huibregtse, M. Scheffner, S. Beaudenon, P. M. Howley, *Proc Natl Acad Sci U S A* 92, 2563 (Mar 28, 1995).
[38] R. J. Deshaies, C. A. Joazeiro, *Annu Rev Biochem* 78, 399 (2009).
[39] Y. Xia, G. M. Pao, H. W. Chen, I. M. Verma, T. Hunter, *J Biol Chem* 278, 5255 (Feb 14, 2003).
[40] L. K. Linares, A. Hengstermann, A. Ciechanover, S. Muller, M. Scheffner, *Proc Natl Acad Sci U S A* 100, 12009 (Oct 14, 2003).
[41] W. Zachariae *et al.*, *Science* 279, 1216 (Feb 20, 1998).
[42] M. D. Petroski, R. J. Deshaies, *Nat Rev Mol Cell Biol* 6, 9 (Jan, 2005).
[43] A. Y. Nikolaev, M. Li, N. Puskas, J. Qin, W. Gu, *Cell* 112, 29 (Jan 10, 2003).
[44] J. Merlet, J. Burger, J. E. Gomes, L. Pintard, *Cell Mol Life Sci* 66, 1924 (Jun, 2009).

[45] A. Shilatifard, *Annu Rev Biochem* 75, 243 (2006).
[46] Y. Mosesson *et al.*, *J Biol Chem* 278, 21323 (Jun 13, 2003).
[47] T. T. Huang, A. D. D'Andrea, *Nat Rev Mol Cell Biol* 7, 323 (May, 2006).
[48] L. Hicke, *Nat Rev Mol Cell Biol* 2, 195 (Mar, 2001).
[49] F. Ikeda, I. Dikic, *EMBO Rep* 9, 536 (Jun, 2008).
[50] K. Haglund, I. Dikic, *EMBO J* 24, 3353 (Oct 5, 2005).
[51] C. Behrends, J. W. Harper, *Nat Struct Mol Biol* 18, 520 (May, 2011).
[52] H. T. Kim *et al.*, *J Biol Chem* 282, 17375 (Jun 15, 2007).
[53] J. S. Thrower, L. Hoffman, M. Rechsteiner, C. M. Pickart, *EMBO J* 19, 94 (Jan 4, 2000).
[54] C. M. Pickart, *FASEB J* 11, 1055 (Nov, 1997).
[55] V. Chau *et al.*, *Science* 243, 1576 (Mar 24, 1989).
[56] P. Xu *et al.*, *Cell* 137, 133 (Apr 3, 2009).
[57] W. Li, Y. Ye, *Cell Mol Life Sci* 65, 2397 (Aug, 2008).
[58] N. K. Kolas *et al.*, *Science* 318, 1637 (Dec 7, 2007).
[59] C. Doil *et al.*, *Cell* 136, 435 (Feb 6, 2009).
[60] Z. J. Chen, *Nat Cell Biol* 7, 758 (Aug, 2005).
[61] L. Jin, A. Williamson, S. Banerjee, I. Philipp, M. Rape, *Cell* 133, 653 (May 16, 2008).
[62] M. J. Garnett *et al.*, *Nat Cell Biol* 11, 1363 (Nov, 2009).
[63] T. Wu *et al.*, *Proc Natl Acad Sci U S A* 107, 1355 (Jan 26, 2010).
[64] K. Iwai, F. Tokunaga, *EMBO Rep* 10, 706 (Jul, 2009).
[65] O. Kerscher, R. Felberbaum, M. Hochstrasser, *Annu Rev Cell Dev Biol* 22, 159 (2006).
[66] A. Pichler, F. Melchior, *Traffic* 3, 381 (Jun, 2002).
[67] A. Pichler, A. Gast, J. S. Seeler, A. Dejean, F. Melchior, *Cell* 108, 109 (Jan 11, 2002).
[68] M. J. Matunis, E. Coutavas, G. Blobel, *J Cell Biol* 135, 1457 (Dec, 1996).
[69] R. Mahajan, C. Delphin, T. Guan, L. Gerace, F. Melchior, *Cell* 88, 97 (Jan 10, 1997).
[70] R. T. Hay, *Mol Cell* 18, 1 (Apr 1, 2005).
[71] J. R. Gareau, C. D. Lima, *Nat Rev Mol Cell Biol* 11, 861 (Dec, 2010).
[72] O. Bischof *et al.*, *Mol Cell* 22, 783 (Jun 23, 2006).
[73] J. R. Morris, *Biochem Soc Trans* 38, 92 (Feb, 2010).
[74] S. Bergink, S. Jentsch, *Nature* 458, 461 (Mar 26, 2009).
[75] K. A. Wilkinson, J. M. Henley, *Biochem J* 428, 133 (Jun 1, 2010).
[76] H. D. Ulrich, *Mol Cell* 32, 301 (Nov 7, 2008).
[77] K. D. Wilkinson, *FASEB J* 11, 1245 (Dec, 1997).
[78] F. E. Reyes-Turcu, K. H. Ventii, K. D. Wilkinson, *Annu Rev Biochem* 78, 363 (2009).
[79] D. Komander, M. J. Clague, S. Urbe, *Nat Rev Mol Cell Biol* 10, 550 (Aug, 2009).
[80] J. H. Kim, S. H. Baek, *Biochim Biophys Acta* 1792, 155 (Mar, 2009).
[81] D. Mukhopadhyay, M. Dasso, *Trends Biochem Sci* 32, 286 (Jun, 2007).
[82] L. N. Shen, C. Dong, H. Liu, J. H. Naismith, R. T. Hay, *Biochem J* 397, 279 (Jul 15, 2006).
[83] L. Gong, E. T. Yeh, *J Biol Chem* 281, 15869 (Jun 9, 2006).
[84] B. B. Zhou, S. J. Elledge, *Nature* 408, 433 (Nov 23, 2000).
[85] A. R. Meetei *et al.*, *Nat Genet* 35, 165 (Oct, 2003).
[86] I. Garcia-Higuera *et al.*, *Mol Cell* 7, 249 (Feb, 2001).
[87] S. de Sanjose *et al.*, *Int J Cancer* 106, 588 (Sep 10, 2003).
[88] R. S. Williams, J. S. Williams, J. A. Tainer, *Biochem Cell Biol* 85, 509 (Aug, 2007).
[89] B. L. Mahaney, K. Meek, S. P. Lees-Miller, *Biochem J* 417, 639 (Feb 1, 2009).
[90] E. P. Rogakou, D. R. Pilch, A. H. Orr, V. S. Ivanova, W. M. Bonner, *J Biol Chem* 273, 5858 (Mar 6, 1998).
[91] S. Burma, B. P. Chen, M. Murphy, A. Kurimasa, D. J. Chen, *J Biol Chem* 276, 42462 (Nov 9, 2001).

[92] M. Stucki *et al.*, *Cell* 123, 1213 (Dec 29, 2005).
[93] G. S. Stewart, B. Wang, C. R. Bignell, A. M. Taylor, S. J. Elledge, *Nature* 421, 961 (Feb 27, 2003).
[94] N. Mailand *et al.*, *Cell* 131, 887 (Nov 30, 2007).
[95] Z. Lou, C. C. Chini, K. Minter-Dykhouse, J. Chen, *J Biol Chem* 278, 13599 (Apr 18, 2003).
[96] M. S. Huen *et al.*, *Cell* 131, 901 (Nov 30, 2007).
[97] S. Bekker-Jensen, C. Lukas, F. Melander, J. Bartek, J. Lukas, *J Cell Biol* 170, 201 (Jul 18, 2005).
[98] V. Plans *et al.*, *J Cell Biochem* 97, 572 (Feb 15, 2006).
[99] R. M. Hofmann, C. M. Pickart, *Cell* 96, 645 (Mar 5, 1999).
[100] B. Wang, S. J. Elledge, *Proc Natl Acad Sci U S A* 104, 20759 (Dec 26, 2007).
[101] M. S. Huen *et al.*, *Mol Cell Biol* 28, 6104 (Oct, 2008).
[102] G. S. Stewart *et al.*, *Proc Natl Acad Sci U S A* 104, 16910 (Oct 23, 2007).
[103] S. Bekker-Jensen *et al.*, *Nat Cell Biol* 12, 80 (Jan, 2010).
[104] W. Wu *et al.*, *Cancer Res* 70, 6384 (Aug 1, 2010).
[105] I. H. Ismail, C. Andrin, D. McDonald, M. J. Hendzel, *J Cell Biol* 191, 45 (Oct 4, 2010).
[106] R. S. Gieni, I. H. Ismail, S. Campbell, M. J. Hendzel, *Cell Cycle* 10, 883 (Mar 15, 2011).
[107] K. Robzyk, J. Recht, M. A. Osley, *Science* 287, 501 (Jan 21, 2000).
[108] K. Nakamura *et al.*, *Mol Cell* 41, 515 (Mar 4, 2011).
[109] W. W. Hwang *et al.*, *Mol Cell* 11, 261 (Jan, 2003).
[110] J. Kim, S. B. Hake, R. G. Roeder, *Mol Cell* 20, 759 (Dec 9, 2005).
[111] L. Moyal *et al.*, *Mol Cell* 41, 529 (Mar 4, 2011).
[112] I. Cousineau, A. Belmaaza, *Cell Cycle* 6, 962 (Apr 15, 2007).
[113] P. M. Gilmore *et al.*, *Biochem Soc Trans* 31, 257 (Feb, 2003).
[114] W. W. Au, B. R. Henderson, *J Biol Chem* 280, 6993 (Feb 25, 2005).
[115] I. A. Manke, D. M. Lowery, A. Nguyen, M. B. Yaffe, *Science* 302, 636 (Oct 24, 2003).
[116] M. Rodriguez, X. Yu, J. Chen, Z. Songyang, *J Biol Chem* 278, 52914 (Dec 26, 2003).
[117] X. Yu, C. C. Chini, M. He, G. Mer, J. Chen, *Science* 302, 639 (Oct 24, 2003).
[118] L. C. Wu *et al.*, *Nat Genet* 14, 430 (Dec, 1996).
[119] P. S. Brzovic, P. Rajagopal, D. W. Hoyt, M. C. King, R. E. Klevit, *Nat Struct Biol* 8, 833 (Oct, 2001).
[120] F. Wu-Baer, K. Lagrazon, W. Yuan, R. Baer, *J Biol Chem* 278, 34743 (Sep 12, 2003).
[121] A. Malovannaya *et al.*, *Cell* 145, 787 (May 27, 2011).
[122] B. Wang *et al.*, *Science* 316, 1194 (May 25, 2007).
[123] B. Sobhian *et al.*, *Science* 316, 1198 (May 25, 2007).
[124] H. Kim, J. Chen, X. Yu, *Science* 316, 1202 (May 25, 2007).
[125] Y. Dong *et al.*, *Mol Cell* 12, 1087 (Nov, 2003).
[126] E. M. Cooper *et al.*, *EMBO J* 28, 621 (Mar 18, 2009).
[127] G. Shao *et al.*, *Proc Natl Acad Sci U S A* 106, 3166 (Mar 3, 2009).
[128] X. Hu *et al.*, *J Biol Chem* 286, 11734 (Apr 1, 2011).
[129] K. Watanabe *et al.*, *EMBO J* 23, 3886 (Oct 1, 2004).
[130] J. Huang *et al.*, *Nat Cell Biol* 11, 592 (May, 2009).
[131] L. Ting, H. Jun, C. Junjie, *DNA Repair (Amst)* 9, 1241 (Dec 10, 2010).
[132] C. Hoege, B. Pfander, G. L. Moldovan, G. Pyrowolakis, S. Jentsch, *Nature* 419, 135 (Sep 12, 2002).
[133] B. Vogelstein, D. Lane, A. J. Levine, *Nature* 408, 307 (Nov 16, 2000).
[134] A. J. Levine, *Cell* 88, 323 (Feb 7, 1997).
[135] D. P. Lane, *Nature* 358, 15 (Jul 2, 1992).
[136] A. Hock, K. H. Vousden, *Int J Biochem Cell Biol* 42, 1618 (Oct, 2010).
[137] C. L. Brooks, W. Gu, *FEBS Lett*, (May 27, 2011).

[138] M. Scheffner, J. M. Huibregtse, R. D. Vierstra, P. M. Howley, *Cell* 75, 495 (Nov 5, 1993).
[139] M. H. Kubbutat, S. N. Jones, K. H. Vousden, *Nature* 387, 299 (May 15, 1997).
[140] R. Honda, H. Tanaka, H. Yasuda, *FEBS Lett* 420, 25 (Dec 22, 1997).
[141] Y. Haupt, R. Maya, A. Kazaz, M. Oren, *Nature* 387, 296 (May 15, 1997).
[142] R. Montes de Oca Luna, D. S. Wagner, G. Lozano, *Nature* 378, 203 (Nov 9, 1995).
[143] S. N. Jones, A. E. Roe, L. A. Donehower, A. Bradley, *Nature* 378, 206 (Nov 9, 1995).
[144] D. Migliorini *et al.*, *Mol Cell Biol* 22, 5527 (Aug, 2002).
[145] S. Tanimura *et al.*, *FEBS Lett* 447, 5 (Mar 19, 1999).
[146] D. A. Sharp, S. A. Kratowicz, M. J. Sank, D. L. George, *J Biol Chem* 274, 38189 (Dec 31, 1999).
[147] H. A. Steinman, H. K. Sluss, A. T. Sands, G. Pihan, S. N. Jones, *Oncogene* 23, 303 (Jan 8, 2004).
[148] J. Parant *et al.*, *Nat Genet* 29, 92 (Sep, 2001).
[149] M. S. Rodriguez, J. M. Desterro, S. Lain, D. P. Lane, R. T. Hay, *Mol Cell Biol* 20, 8458 (Nov, 2000).
[150] K. A. Krummel, C. J. Lee, F. Toledo, G. M. Wahl, *Proc Natl Acad Sci U S A* 102, 10188 (Jul 19, 2005).
[151] L. Feng, T. Lin, H. Uranishi, W. Gu, Y. Xu, *Mol Cell Biol* 25, 5389 (Jul, 2005).
[152] W. M. Chan *et al.*, *Mol Cancer Res* 4, 15 (Jan, 2006).
[153] F. Toledo, G. M. Wahl, *Nat Rev Cancer* 6, 909 (Dec, 2006).
[154] V. Zuckerman *et al.*, *J Biol Chem* 284, 4031 (Feb 6, 2009).
[155] J. M. Stommel, G. M. Wahl, *Cell Cycle* 4, 411 (Mar, 2005).
[156] Z. Goldberg *et al.*, *EMBO J* 21, 3715 (Jul 15, 2002).
[157] L. Chen, D. M. Gilkes, Y. Pan, W. S. Lane, J. Chen, *EMBO J* 24, 3411 (Oct 5, 2005).
[158] C. J. Sherr *et al.*, *Cold Spring Harb Symp Quant Biol* 70, 129 (2005).
[159] X. Wu, J. H. Bayle, D. Olson, A. J. Levine, *Genes Dev* 7, 1126 (Jul, 1993).
[160] Y. Barak, T. Juven, R. Haffner, M. Oren, *EMBO J* 12, 461 (Feb, 1993).
[161] Y. Pan, J. Chen, *Mol Cell Biol* 23, 5113 (Aug, 2003).
[162] S. Fang, J. P. Jensen, R. L. Ludwig, K. H. Vousden, A. M. Weissman, *J Biol Chem* 275, 8945 (Mar 24, 2000).
[163] X. Fu *et al.*, *Proc Natl Acad Sci U S A* 107, 4579 (Mar 9, 2010).
[164] B. Di Ventura, C. Funaya, C. Antony, M. Knop, L. Serrano, *PLoS One* 3, e1507 (2008).
[165] L. Chen, J. Chen, *Oncogene* 22, 5348 (Aug 14, 2003).
[166] R. P. Leng *et al.*, *Cell* 112, 779 (Mar 21, 2003).
[167] W. Yang *et al.*, *J Biol Chem* 282, 3273 (Feb 2, 2007).
[168] P. Stehmeier, S. Muller, *DNA Repair (Amst)* 8, 491 (Apr 5, 2009).
[169] R. Rajendra *et al.*, *J Biol Chem* 279, 36440 (Aug 27, 2004).
[170] D. Dornan *et al.*, *Cancer Res* 64, 7226 (Oct 15, 2004).
[171] D. Chen *et al.*, *Cell* 121, 1071 (Jul 1, 2005).
[172] A. Laine *et al.*, *Mol Cell Biol* 26, 8901 (Dec, 2006).
[173] A. Laine, Z. Ronai, *Oncogene* 26, 1477 (Mar 1, 2007).
[174] J. P. Kruse, W. Gu, *J Biol Chem* 284, 3250 (Jan 30, 2009).
[175] L. Le Cam *et al.*, *Cell* 127, 775 (Nov 17, 2006).
[176] A. Pichler, P. Knipscheer, H. Saitoh, T. K. Sixma, F. Melchior, *Nat Struct Mol Biol* 11, 984 (Oct, 2004).
[177] L. Liu *et al.*, *Mol Cell Biol* 19, 1202 (Feb, 1999).
[178] M. Li *et al.*, *Nature* 416, 648 (Apr 11, 2002).
[179] R. D. Everett *et al.*, *EMBO J* 16, 1519 (Apr 1, 1997).
[180] N. Kon *et al.*, *Oncogene* 29, 1270 (Mar 4, 2010).
[181] S. M. Nijman *et al.*, *Cell* 123, 773 (Dec 2, 2005).

[182] C. L. Brooks, M. Li, M. Hu, Y. Shi, W. Gu, *Oncogene* 26, 7262 (Nov 8, 2007).
[183] E. Meulmeester *et al., Mol Cell* 18, 565 (May 27, 2005).
[184] M. Li, C. L. Brooks, N. Kon, W. Gu, *Mol Cell* 13, 879 (Mar 26, 2004).
[185] E. Meulmeester, Y. Pereg, Y. Shiloh, A. G. Jochemsen, *Cell Cycle* 4, 1166 (Sep, 2005).
[186] F. Sarkari, Y. Sheng, L. Frappier, *PLoS One* 5, e13040 (2010).
[187] L. F. Stevenson *et al., EMBO J* 26, 976 (Feb 21, 2007).
[188] S. K. Hotton, J. Callis, *Annu Rev Plant Biol* 59, 467 (2008).
[189] H. Zhao, J. L. Watkins, H. Piwnica-Worms, *Proc Natl Acad Sci U S A* 99, 14795 (Nov 12, 2002).
[190] Y. W. Zhang *et al., Mol Cell* 35, 442 (Aug 28, 2009).
[191] C. C. Chini, J. Chen, *J Biol Chem* 281, 33276 (Nov 3, 2006).
[192] N. Mailand *et al., EMBO J* 21, 5911 (Nov 1, 2002).
[193] L. Busino *et al., Nature* 426, 87 (Nov 6, 2003).
[194] L. M. Young, M. Pagano, *Cell Cycle* 9, 4613 (Dec 1, 2010).
[195] Y. Pereg *et al., Nat Cell Biol* 12, 400 (Apr, 2010).
[196] Y. Tatsumi *et al., J Cell Sci* 119, 3128 (Aug 1, 2006).
[197] S. Tada, *Front Biosci* 12, 1629 (2007).
[198] G. Xu, J. S. Paige, S. R. Jaffrey, *Nat Biotechnol* 28, 868 (Aug, 2010).
[199] J. R. Hall *et al., J Biol Chem* 283, 25356 (Sep 12, 2008).
[200] E. Guarino *et al., Nucleic Acids Res* 39, 5978 (Aug, 2011).
[201] W. Kim *et al., Mol Cell*, (Sep 7, 2011).
[202] J. M. Danielsen *et al., Mol Cell Proteomics* 10, M110 003590 (Mar, 2011).
[203] L. A. Higa *et al., Cell Cycle* 5, 1675 (Aug, 2006).
[204] A. Duursma, R. Agami, *Mol Cell Biol* 25, 6937 (Aug, 2005).
[205] J. R. Hall *et al., Mol Biol Cell* 18, 3340 (Sep, 2007).
[206] B. E. Gewurz, J. W. Harper, *Curr Biol* 16, R932 (Nov 7, 2006).
[207] A. Peschiaroli *et al., Mol Cell* 23, 319 (Aug 4, 2006).
[208] H. Faustrup, S. Bekker-Jensen, J. Bartek, J. Lukas, N. Mailand, *J Cell Biol* 184, 13 (Jan 12, 2009).
[209] I. Mamely *et al., Curr Biol* 16, 1950 (Oct 10, 2006).
[210] D. Zhang, K. Zaugg, T. W. Mak, S. J. Elledge, *Cell* 126, 529 (Aug 11, 2006).
[211] Y. Shi, P. Xu, J. Qin, *Mol Cell Proteomics* 10, R110 006882 (May, 2011).
[212] J. Peng *et al., Nat Biotechnol* 21, 921 (Aug, 2003).
[213] D. Meierhofer, X. Wang, L. Huang, P. Kaiser, *J Proteome Res* 7, 4566 (Oct, 2008).
[214] M. Tsirigotis *et al., Biotechniques* 31, 120 (Jul, 2001).
[215] M. Franco, N. T. Seyfried, A. H. Brand, J. Peng, U. Mayor, *Mol Cell Proteomics* 10, M110 002188 (May, 2011).
[216] M. Xu, B. Skaug, W. Zeng, Z. J. Chen, *Mol Cell* 36, 302 (Oct 23, 2009).
[217] K. Newton *et al., Cell* 134, 668 (Aug 22, 2008).
[218] M. L. Matsumoto *et al., Mol Cell* 39, 477 (Aug 13, 2010).
[219] L. Hicke, H. L. Schubert, C. P. Hill, *Nat Rev Mol Cell Biol* 6, 610 (Aug, 2005).
[220] I. Dikic, S. Wakatsuki, K. J. Walters, *Nat Rev Mol Cell Biol* 10, 659 (Oct, 2009).
[221] P. Young, Q. Deveraux, R. E. Beal, C. M. Pickart, M. Rechsteiner, *J Biol Chem* 273, 5461 (Mar 6, 1998).
[222] C. R. Wilkinson *et al., Nat Cell Biol* 3, 939 (Oct, 2001).
[223] Y. Shi *et al., Mol Cell Proteomics* 10, M110 002089 (May, 2011).
[224] A. Malovannaya *et al., Proc Natl Acad Sci U S A* 107, 2431 (Feb 9, 2010).
[225] H. C. Yen, S. J. Elledge, *Science* 322, 923 (Nov 7, 2008).
[226] H. C. Yen, Q. Xu, D. M. Chou, Z. Zhao, S. J. Elledge, *Science* 322, 918 (Nov 7, 2008).

11

X-Ray-Induced Radioresistance Against High-LET Radiations from Accelerated Neon-Ion Beams in Mice

Bing Wang et al.*
Research Center for Radiation Protection,
National Institute of Radiological Sciences, Inage-ku, Chiba,
Japan

1. Introduction

Radiation-induced adaptive response (AR) is the phenomenon of priming low-dose radiation-induced resistance to subsequent challenging radiation at higher doses. As investigations on the conditions essential for AR induction and the underlying mechanisms provide important scientific basis for radiation risk estimates and offer significant insight into the novel biological defense mechanisms regarding protection against radiation, study on AR is of great concern for both public health and academic research (United Nations Scientific Committee on the Effects of Atomic Radiation, 1994). Since the *editio princeps* of the AR concept introduced into radiation biology (Olivieri et al., 1984), this phenomenon has been demonstrated both *in vitro* and *in vivo* (Takahashi & Ohnishi, 2009; Mitchel, 2006; Vares et al., 2006). Though studies on the radiation-induced radioresistance in rodents *in vivo* could be retrospected to the later 1940s (Dacquisto, 1959), the first full-dress investigation on AR in mice started about 50 years later by Yonezawa and colleagues (Yonezawa et al., 1990). In a series of comprehensive studies, a variety of experimental condition combinations of the priming and challenging doses, the interval between the irradiations, and the mouse strain were testified and verified (Yonezawa, 2006), laying a cornerstone of *in vivo* model for AR in mice - successful establishment of the mouse models for AR induction by acute low liner energy transfer (LET) X-irradiations at the whole body level using survival from bone marrow death as the main endpoint, which is the so-called "Yonezawa Effect" in Japan (Takahashi & Ohnishi, 2009). The mouse models for AR induction (Yonezawa, 1996) were with great reliability and reproducibility in different laboratories (Nose et al., 2001; Otsuka et al., 2008;). The induced AR could be generally divided into two phenotypes according to the attitude of the priming dose and the interval between the priming and challenging irradiation, involving different mechanisms: AR induced 2 weeks after a priming irradiation at a dose from 0.3 to 0.5 Gy was due to radioresistance occurred in blood forming tissues which was a Trp53 dependent event (Otsuka et al., 2008; Horie et al., 2002), and interaction

* Kaoru Tanaka, Yasuharu Ninomiya, Kouichi Maruyama, Guillaume Varès, Yi Shang, Kiyomi Eguchi-Kasai and Mitsuru Nenoi
Research Center for Radiation Protection, National Institute of Radiological Sciences, Inage-ku, Chiba, Japan

between blood forming tissue and the central nervous system was thought to contribute to AR induced 2 months after a priming dose at 0.05 or 0.1 Gy (Yonezawa, 2000; Misonoh & Yonezawa, 1997).

AR, as one of the most important phenomena in radiation biology, has been studied for nearly three decades, however, almost all the investigations at the whole body level were performed using low LET X- or gamma-irradiations for both priming and challenging doses. With a significant increase in human activities dealing with space missions outside the earth's magnetic field, health-related concerns on exposures to space radiations become a hot topic to be addressed as the space radiation environment consists of energetic protons and heavy ions in galactic cosmic rays and solar particle events. Some important qualitative differences were found between the biological effects produced by the high LET particulate irradiations in space and the low LET photon exposures commonly encountered on earth (Horie et al., 2002). The accelerator-based experimental studies are important to quantify the health risks (Nelson, 2003) and of interest in the use of particle radiotherapy as well as for the exploration of mechanisms especially underlying high LET radiation effects. The present study is a part of the work in a series of investigations aiming to verify the possible existence of low LET irradiation-induced AR against the detrimental effects from the challenging irradiations with high LET heavy-ion irradiations in mammalians at a whole body level and explore its underlying mechanisms.

2. Materials and methods

2.1 Animals

C57BL/6J Jms strain female mice aged 5 weeks old purchased from SLC, Inc. (Japan), were maintained in a conventional animal facility under a 12 h light-12 h dark photoperiod (lights on from 8:00 a.m. to 8:00 p.m.). Animals housed in autoclaved cages with sterilized wood chips, were allowed free access to standard laboratory chow (MB-1, Funabashi Farm Co., Japan) and acidified water *ad libitum*. Animals were acclimatized to the laboratory conditions for 1 week before use. To avoid possible effects from the developmental condition of the animals, any mouse of 6 weeks old with a significantly higher or lower body weight (more or less than the mean $\pm$ 2 SD) was omitted from this study. Based on the preliminary experiment, in which at least 20 mice were used in each experimental group, in the present study at least 30 mice were used in each experimental group and the experiment was repeated twice. All experimental protocols involving mice were reviewed and approved (Experimental Animal Research Plan No. 07-2023) by The Institutional Animal Care and Use Committee of the National Institute of Radiological Sciences (NIRS), and were performed in strict accordance with the NIRS *Guidelines for the Care and Use of Laboratory Animals*.

2.2 Irradiation

X-rays were generated with an X-ray machine (Pantak-320S, Shimadzu, Japan) operated at 200 kVp and 20 mA, using a 0.50 mm Al + 0.50 mm Cu filter. An exposure rate meter (AE-1321M, Applied Engineering Inc, Japan) was used for the dosimetry. The dose rates for delivering the priming and challenging irradiations were about 0.3 Gy/min and 0.7 Gy/min, respectively. For high LET heavy-ion irradiations, the monoenergetic ion beams of neon particles were generated and accelerated by a synchrotron, the Heavy Ion Medical Accelerator in Chiba (HIMAC) at NIRS, Japan. The beam energy was 400 MeV/nucleon, corresponding to an average LET of 30 keV/micrometer. A challenging dose of 5.5 Gy was delivered at a rate about 1.5 Gy/min. The mice held in acryl containers were exposed to whole-body irradiation at room temperature.

2.3 The mouse model for induction of adaptive response

One set of the conditions essential for AR induction in the mouse model established by Yonezawa *et al* (Yonezawa et al., 1996) was adopted, verified and confirmed under the experimental conditions in our research facilities, and finally applied to the present study. In brief, the efficient priming dose of X-rays was 0.5 Gy. The timing for delivery of the priming dose and challenging dose were on postnatal ages of 6 and 8 weeks of the mice, respectively. The challenging dose of X-rays at 7.5 Gy was used based on the documented work and our previous report (Monobe et al., 2004; Mouthon et al., 1999; Monobe et al. 2003; Wang et al., 2010). The challenging dose of neon-ion irradiations at 5.5 Gy was determined by preliminary trials with the reference from the reported work and our previous study (Monobe et al., 2004; Monobe et al., 2003; Tomizawa et al., 2000; Wang et al., 2010).

2.4 Biological endpoints

The 30-day survival study: The number of deaths that occurred within the 30-day period after delivery of the challenging dose was recorded. Successful induction of AR was defined as the phenomenon that the priming dose induced significantly the suppression of the mortality caused by the challenging dose. The peripheral blood hemogram: The survivors of the 30-day survival study were anesthetized by CO_2 inhalation. The peripheral blood was collected from femoral artery and then the animals were killed by cervical dislocation. The differential blood cell count was done using a blood cell differential automatic analyzer (SYSMEX K4500, Sysmex Corporation, Japan). The data for each experimental group were from at least 6 mice. Micronucleus test: The bone marrow micronucleus test was carried out according to Schmid (Schmid, 2010) with minor modifications (Chaubey et al., 1993). Bone marrow smears prepared from both femurs were processed for the enumeration of micronucleated polychromatic erythrocytes (MNPCE) and micronucleated normochromatic erythrocytes (MNNCE). The slides were coded to avoid any observer bias. The micronuclei were scored using a light microscope at a magnification of 1000x. At least 5000 cells per mouse were counted and the data per experimental point were from at least 5 mice.

2.5 Statistical analysis

Statistical evaluation of the data was done using the c^2 test and Student's *t*-test when appropriate. Statistical significance was assigned to $P < 0.05$.

3. Results

3.1 Verification of the adaptive response model using low LET X-rays in delivery of both the priming and the challenging irradiations

As in our previous study (Wang et al., 2010), the reproducibility of the model for induction of AR in this system was verified and confirmed again by using X-rays as both the priming dose and the challenging dose. The priming dose at 0.5 Gy adopted from Yonezawa *et al*'s work (Yonezawa et al., 1996) and a challenging dose at 7.5 Gy which resulted in about 80% mortality in C57BL/6J Jms female mice when whole body exposed to X-rays at 8 weeks old, were delivered to the animals on postnatal ages of 6 and 8 weeks, respectively. Results (Fig. 1, drawn from a representative trial) showed that the priming dose significantly improved the survival from 16.7% to 80.0% in the 30-day survival study after the challenging dose, indicating clearly the successful induction of AR under the conditions in our experimental facilities. Serving also as a positive control, the verification work was performed in parallel to the following investigations with heavy neon-ion irradiations.

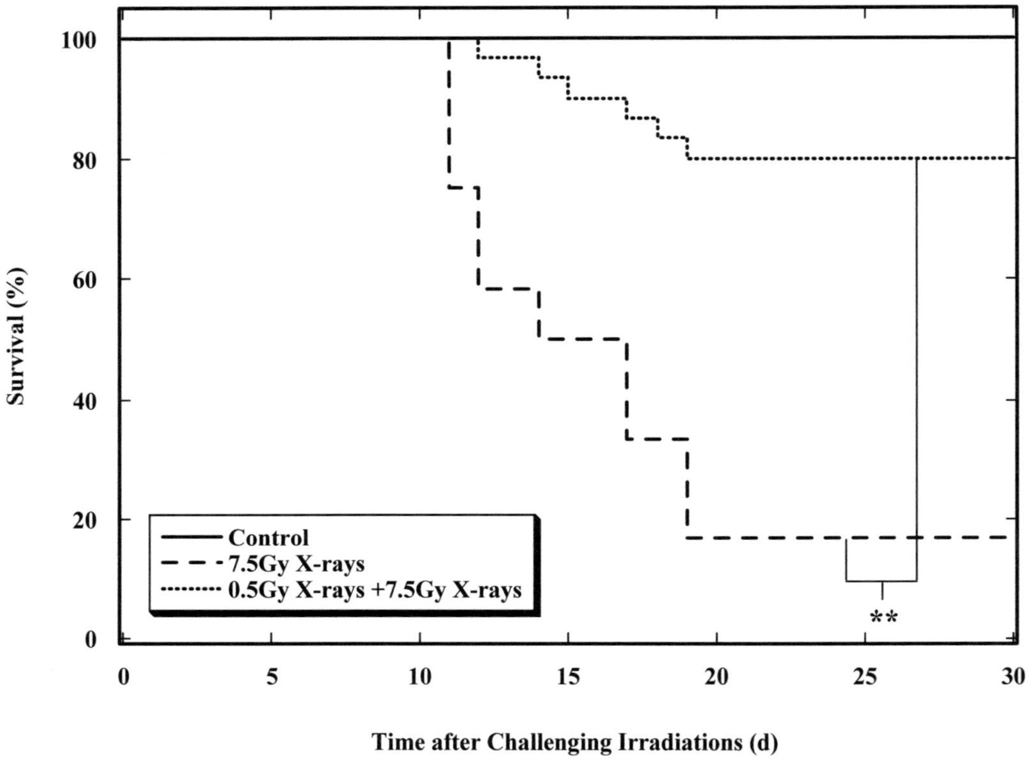

Fig. 1. Thirty-day survival study for the effect from priming low dose of X-rays (0.5 Gy) on the high challenging dose from X-rays (7.5 Gy) in young adult female mice.

The mouse death was recorded after the challenging dose. The solid, broken and dotted lines stand for the "Control" group receiving neither the priming nor the challenging irradiations, the "7.5Gy X-rays" group receiving only the challenging irradiations, and the "0.5Gy X-rays + 7.5Gy X-rays" group receiving both the priming and the challenging irradiations. Two asterisks indicate statistically significant differences at $P < 0.01$ between the two groups compared. This experiment was performed to serve as a positive control, demonstrating that Yonezawa et al's mouse model for induction of AR was successfully reproduced under the conditions in our experimental facilities.

3.2 Effects of the priming dose from low LET X-rays on the killing effect of the challenging dose from high LET heavy neon-ion irradiations

The 30-day survival study was applied to investigate the effects of the priming dose at 0.50 Gy from low LET X-rays on the challenging dose from high LET heavy-ion irradiations with neon beams in young adult female mice. A representative survival graph obtained using neon ion irradiations as the challenging irradiations at a dose of 5.5 Gy was shown in Figure 2. The priming dose markedly increased the survival rate from 26.1% to 56.7%. These data indicated that a priming dose at 0.50 Gy from low LET X-rays could induce radioresistance against the killing effects from subsequent challenging irradiations with heavy neon-ion beams.

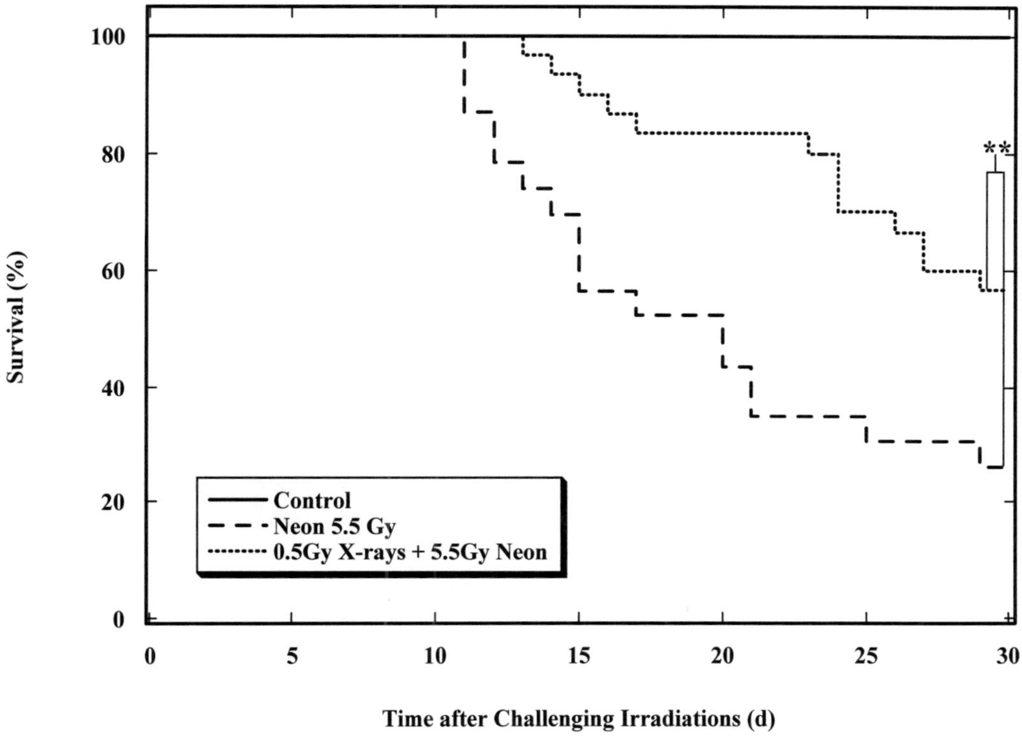

Fig. 2. Thirty-day survival study for the effects from priming low dose of X-rays (0.5 Gy) on the high challenging dose from 5.5 Gy neon-ion irradiations in young adult female mice.

The solid, broken and dotted lines stand for the "Control" group receiving neither the priming nor the challenging irradiations, the "5.5 Gy Neon" experimental group receiving only the challenging irradiations, and the "0.5Gy X-rays + 5.5Gy Neon" experimental group receiving both the priming and the challenging irradiations. Two asterisks indicate statistically significant differences at $P < 0.01$ between the two groups compared.

3.3 Effects of the priming dose from low LET X-rays on the detrimental effect induced by the challenging dose from low LET X-rays or high LET heavy neon-ion irradiations in the hematopoietic system in the survivors

As bone marrow failure was the main cause for the animal death, the residual damage in the hematopoietic system was also studied in the peripheral blood hemogram and the bone marrow cells in the survivors on the following day when the 30-day survival study was finished. Results showed that the priming dose significantly relieved the depression of peripheral blood platelet counts (Fig. 3). For other blood hemogram indexes, the mean value of leukocyte count in "0.5Gy X-rays + 7.5Gy X-rays" group was slightly higher than that of the "7.5Gy X-rays" group, and no significant effect on improvement of erythrocyte count and leukocyte count was observed (data not shown).

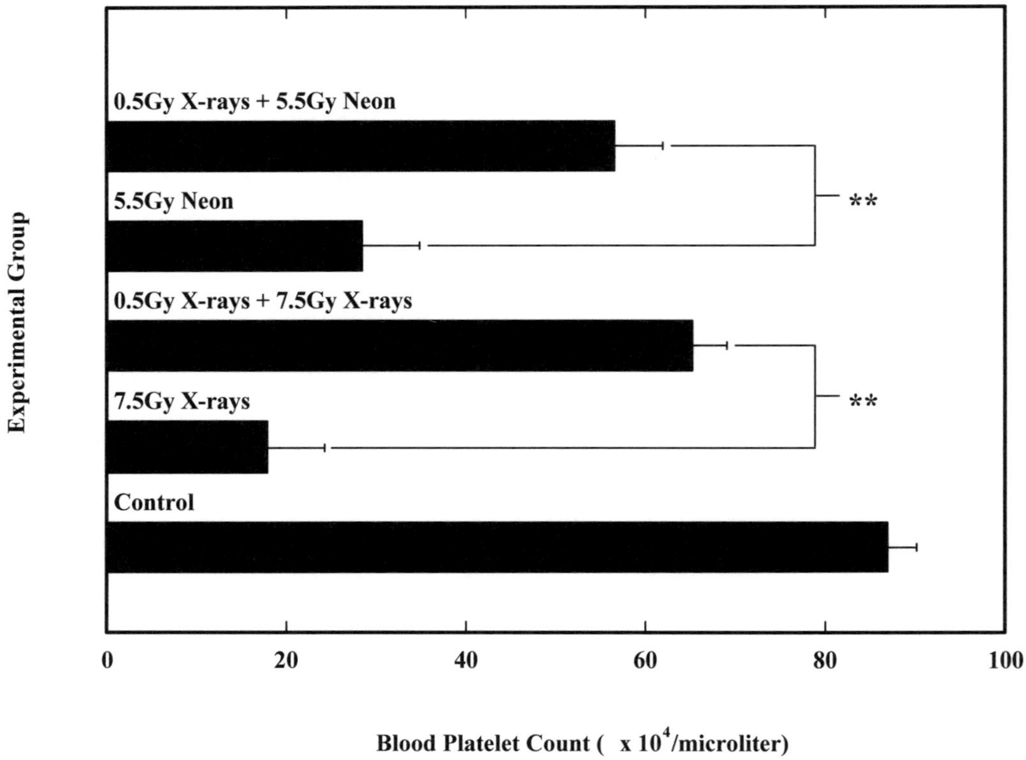

Fig. 3. Peripheral blood hemogram measurement in the survivors on the following day after the 30-day survival study.

Peripheral blood hemogram study in the survivors on the following day when the 30-day survival study was finished for the residual damage in young adult female mice receiving a high challenging dose from X-rays (7.5 Gy) or neon-ion irradiations (5.5 Gy) with or without a priming low dose of X-rays (0.5 Gy). The peripheral blood was collected from femoral artery of the survivors in the 30-day survival study. Two asterisks indicate statistically significant differences at $P < 0.01$ between the two groups compared.

Micronucleus test study showed that the priming dose markedly improved the ratio of polychromatic erythrocytes (PCEs) to the sum of PCEs and normochromatic erythrocytes (NCEs) (data not shown). The priming dose significantly reduced the occurrences of both micronucleated polychromatic erythrocytes (MNPCEs) (Fig. 4A) in PCEs and micronucleated normochromatic erythrocytes (MNNCEs) in NCEs (Fig. 4B). These data indicated that a priming dose at 0.50 Gy from low LET X-rays could induce radioresistance against the detrimental effects from subsequent challenging irradiations with X-rays or neon-ion beams in the hematopoietic system.

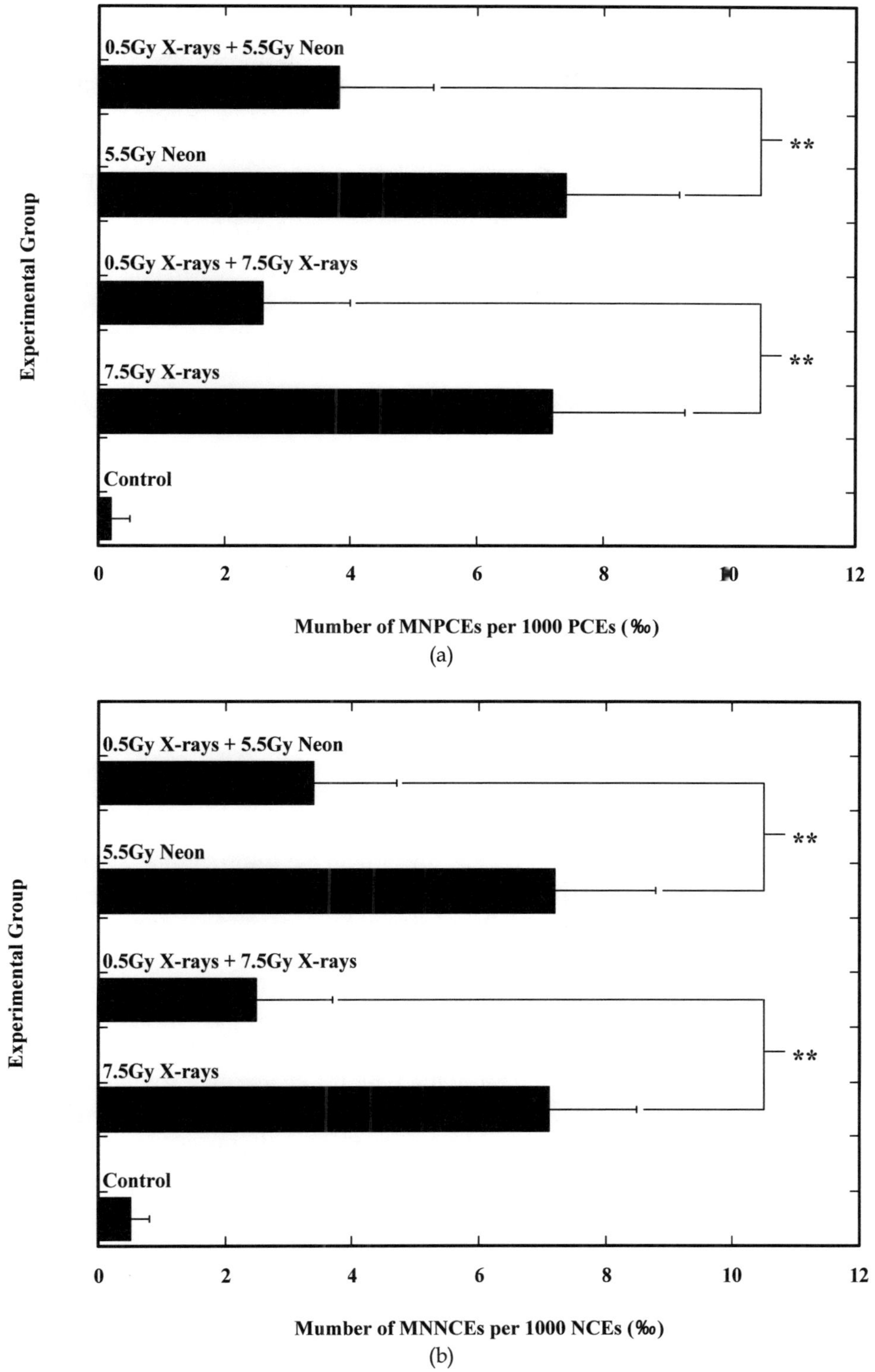

Fig. 4. Micronucleus test in the survivors on the following day after the 30-day survival study.

Micronucleus test study in the survivors on the following day when the 30-day survival study was finished for the residual damage in young adult female mice receiving a high challenging dose from X-rays (7.5 Gy) or neon-ion irradiations (5.5 Gy) with or without a priming low dose of X-rays (0.5 Gy). The "Control" group receiving neither the priming nor the challenging irradiations, the "7.5Gy X-rays" group receiving only the challenging irradiations, and the "0.5Gy X-rays + 7.5Gy X-rays" group receiving both the priming and the challenging irradiations from X-rays. The "5.5Gy Neon" group receiving only the challenging dose of 5.5 Gy from neon-ion irradiations, and the "0.5Gy X-rays + 5.5Gy Neon" group receiving both the priming of 0.5Gy from X-rays and the challenging dose of 5.5 Gy from neon-ion irradiations. Bone marrow smears were prepared from both femurs of the survivors in the 30-day survival study and processed for the enumeration of micronucleated polychromatic erythrocytes (MNPCE) and micronucleated normochromatic erythrocytes (MNNCE). Two asterisks indicate statistically significant differences at $P < 0.01$ between the two groups compared.

3.4 Effects of the priming dose from low LET X-rays on the reduced body weight gain induced by the challenging dose from low LET X-rays or high LET heavy carbon-ion irradiations in the survivors

The body weight was also scored in survivors in the 30-day survival study. For most of the time points measured, the body weight of either of the irradiated groups was markedly lower than that of the "Control" group after challenging irradiations. Results showed that the priming dose (0.50 Gy of X-rays) significantly suppressed the reduction in the body weight gain two weeks and three weeks, or at three weeks after the challenging dose of 7.5 Gy X-rays or 5.5 Gy of neon-ion irradiations (Fig. 5, Fig. 6). As the death cases were always with a low body weight and the significantly decreased reduction in the body weight was observed in the middle course of the 30-day survival study, the statistical significance was thought not to be of a markedly biological meaning. Of note, the body weight in animals that survived by AR was always significantly lower than the control group at the time from two weeks after the challenging irradiation to the end of the study, indicating the priming dose could rescue the animals from death but it could not completely prevent the damage from the challenging dose to the animals. In addition to the endpoints of the hematopoietic system, body weight is also a sensitive indicator.

The mouse body weight was recorded after the challenging dose. The solid, broken and dotted lines stand for the "Control" group receiving neither the priming nor the challenging irradiations, the "7.5Gy X-rays" group receiving only the challenging irradiations, and the "0.5Gy X-rays + 7.5Gy X-rays" group receiving both the priming and the challenging irradiations. One and two asterisks indicate statistically significant differences at $P < 0.05$ and $P < 0.01$ respectively between the "7.5Gy X-rays" group and the "0.5Gy X-rays + 7.5Gy X-rays" group.

The mouse body weight was recorded after the challenging dose. The solid, broken and dotted lines stand for the "Control" group receiving neither the priming nor the challenging irradiations, the "5.5Gy Neon" group receiving only the challenging irradiations, and the "0.5Gy X-rays + 5.5Gy Neon" group receiving both the priming and the challenging irradiations. One asterisk indicates statistically significant differences at $P < 0.05$ between the "5.5Gy Neon" group and the "0.5Gy X-rays + 5.5Gy Neon" group.

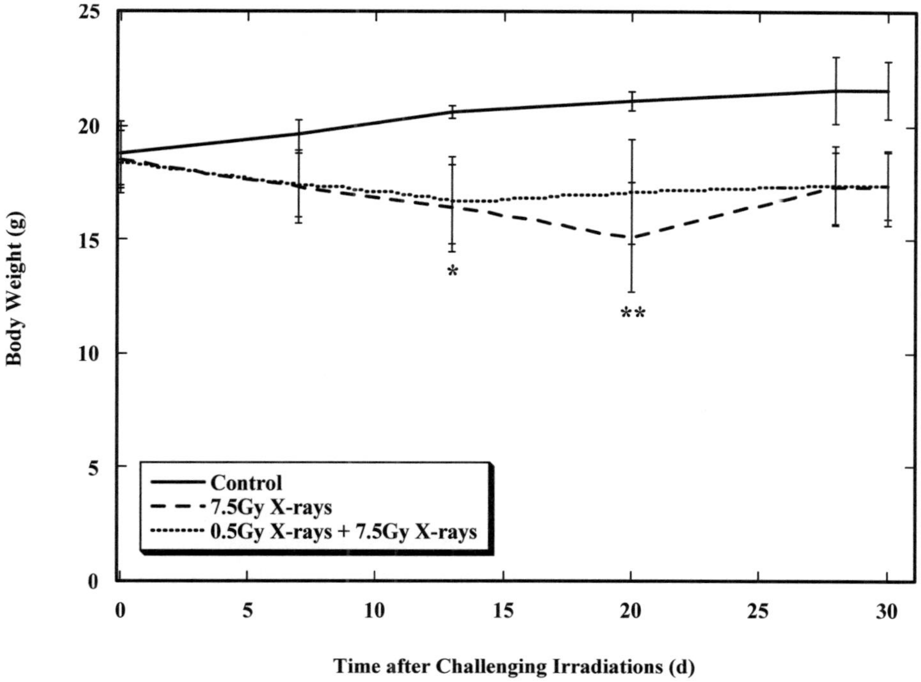

Fig. 5. Body weight study for the effects from priming low dose of X-rays (0.5Gy) on the high challenging dose from X-rays (7.5 Gy) in young adult female mice.

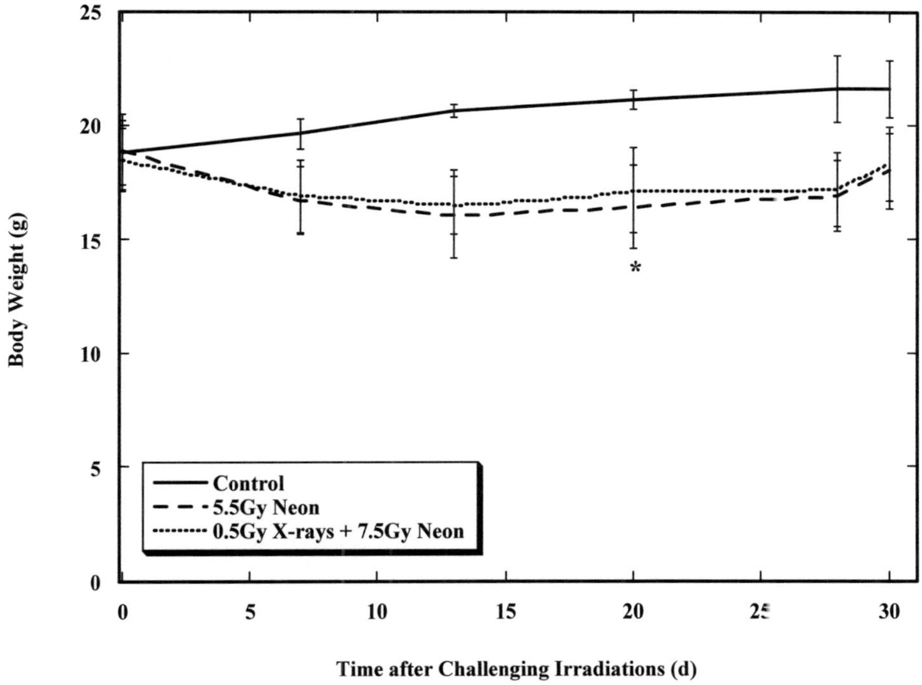

Fig. 6. Body weight study for the effects from priming low dose of X-rays (0.5Gy) on the high challenging dose from neon-ion irradiations (5.5 Gy) in young adult female mice.

4. Discussions

Low doses of low LET irradiations induce protective effects through possible mechanisms such as enhancement of antioxidative capacities, increase in cellular DNA double-strand break repair capacity leading to reduction of initial DNA damage in AR in mice *in vivo* (Otsuka et al., 2006), and reduction of cell death, chromosomal aberrations, mutations, and malignant transformation *in vitro*. These induced responses have been tightly conserved throughout evolution, suggesting that they are basic responses critical to life (Mitchel, 2006). Induction of AR may provide protection of the space mission crews and travelers (Mortazavi et al., 2003), the occupational radiation workers, and the normal tissues and organs of patients receiving heavy-ion radiotherapy. Considering the unavoidable exposure to high LET irradiations during varied human activities, it is crucial and of an on-the-spot inspection to study the possibility of triggering an AR against high LET irradiations in suitable biological models. In our series of investigations using AR model in mice, induction of AR was attempted using X-rays as priming irradiation in combination with challenging irradiations from four kinds of accelerated heavy ions of carbon, neon, silicon and iron with the LET values of about 15, 30, 55, and 200 keV/micrometer respectively. Results showed that a priming low dose of X-rays at 0.5 Gy significantly reduced the mortality from the high challenging dose of carbon, neon or silicon particles, but not iron particles. In addition, the priming low-dose of X-rays also markedly reduced some of the other detrimental effects in the survivors. In the present mouse model for AR induction, the mechanisms responsible for rescue of bone marrow death are mainly of priming irradiation-induced radioresistance to the challenging irradiation-induced Trp-53 derived-apoptosis in hematopoietic stem cells (Horie et al., 2002; Okazaki et al., 2007). Recovery of blood platelet counts after exposure is one of the most important factors for restoration of bone marrow death (Takeda et al., 1981). Results on blood platelet counts obtained in the present study are consistent with this conclusion.

The recovery ratio of the potentially lethal damage depended on the quality of radiations (Suzuki et al., 2000a) and cellular radiosensitivity correlated with the frequency of residual chromatin breaks (Suzuki et al., 2000b). Compared to X-rays, carbon ions induced higher rate of residual chromatin breaks per cell per Gy (Suzuki et al., 1998). Though high LET heavy-ion irradiations generally induces qualitatively different DNA damage such as clustered DNA damage than that of low LET irradiations, for certain cases, such as the results obtained on carbon, neon and silicon ion irradiations in our studies, the X-ray-induced biological defense mechanisms may be as effective countermeasures, being sufficient enough against the damages caused by challenging dose from these high LET irradiations. In fact, *in vitro* study showed that low doses of low LET X-irradiations were effective in reducing chromosomal aberrations induced by high LET irradiations from radon exposure (Wolff et al., 1991). Mechanism studies using cultured human fibroblasts reported that gene expression profiles following gamma-radiation and decays of high-LET-like ^{125}I shared the majority of genes in common, which indicated the involvement of similar pathways in signal transduction after radiation exposures of different modalities (Sokolov et al., 2006a) and DNA DSB may not be the major factor modulating changes in gene expression following irradiation (Sokolov et al., 2006b). These findings suggest that it may share at least to a large extent the same mechanisms against the damages caused by the high dose of challenging irradiations from low LET X-rays or from high LET carbon, neon and silicon ions. On the other hand, AR was not observable against, even at a small dose (LD5/30), the high LET challenging irradiations from iron ions (Wang et al., 2010). This indicates that the biological defense mechanisms induced by priming low LET X-rays are not, at least not sufficiently enough, capable of dealing with the damage caused by iron-ion irradiations, which seems probably qualitatively different from those induced by carbon, neon, or silicon ion irradiations. There is no a clear conclusion on if the killing effect from heavy ion irradiations at the whole body level is an event depending on either the LET values of the radiation or the ion species, or both. While *in vitro* studies suggested, for the same kind of ion species, heavy ion-induced cell killing effect is of LET dependency (Aoki et al., 2000). Study on the RBE values of carbon ion irradiations relative to X-rays for reproductive cell death showed LET dependent cell killing effects with the RBE values ranging from 1.06 to 1.33 for 13 keV/micrometer and from 2.00 to 3.01 for approximate 77 keV/micrometer irradiation (Suzuki et al., 2000c). For the different ion species, it was also reported that the cell killing effects varied with differences in the energy deposition track structures of the different ion sources, being of LET and ion species dependence: RBE for carbon ions increased steeply up to around 98 keV/micrometer (4.07), while the neon, silicon and iron ions showed maximum peaks around 180 keV/micrometer (3.03 to 3.39) (Tsuruoka et al., 2005). The mutation induction investigations also suggest that different ion species may cause qualitative and quantitative difference even if the LET values are similar (Suzuki et al., 2009; Tsuruoka et al., 2004; Suzuki et al., 2003). It should be noticed that in the cell culture system, though the enhanced clustering of ionization and DNA damage lowered the energy efficiency for producing damage resulting in low RBE values when the LET values of heavy ions exceeded 100 keV/micrometer (Mehnati et al., 2005), the fate for most of the cells hit by the heavy ions is death due to the unrepairable complexity of DNA damage. The biological sequences in vivo after iron-ion irradiations in our previous study should be the case thus AR could be not induced.

Depending on improvement in irradiation facilities, it will be possible to verify if this event depending on either LET or ion species or both when generation of heavy ion irradiations of the same ion species at different LET values and generation of different ion species at the same LET value are possible. To understand what extent low dose irradiations could be beneficially functional in humans, a better understanding of AR and other non-targeted effects other non-targeted effects is needed (Tapio & Jacob, 2007). Radiation quality dependent biological effects must be studied in the context of the mixed high- and low-LET radiations that are found in the space environment. In a recent study, a statistically significant increase in the yield of chromosomal aberrations in lymphocytes from cosmonauts at their first flight long-term space missions was observed. However, for cosmonauts involved in two or more space flights, the yield of aberrations at the end of the last mission was generally in the range of background frequencies measured before the first mission (Durante et al., 2003). This study suggests possible existence of AR induced by space irradiations *in vivo*. Further investigations on AR induction by low dose high LET heavy-ion irradiation and by mixed high- and low-LET radiations are needed. In addition to AR study, radiation effects from high LET irradiations in mammalians on such as development, behavior, mutation and carcinogenesis, are also of great interest to be studied on ground with heavy ions of different energy with suitable accelerators using well-designed animal experiments over extended periods of time (Curtis et al., 2004).

Taken together, in this series of studies, though the essential conditions vary on LET values and ion species, successful induction of AR is demonstrated for the first time using low LET X-rays as the priming dose and high LET heavy-ion irradiations as the challenging dose *in vivo* in the mouse model (Table 1.). The findings suggest that the efficacy of the biological defense mechanisms induced by priming dose of low LET X-rays depends on the quality (either the LET value or the ion particle species of the challenging irradiations, or both) of the challenging irradiations.

Priming Dose	Challenging Dose	Adaptive Response
X-Rays	Carbon	○
X-Rays	Neon	○
X-Rays	Silicon	○
X-Rays	Iron	×

Table 1. Induction of adaptive response by low LET X-rays again high LET heavy-ion irradiations.

5. Acknowledgments

This work was partially supported both by the Grant-in-Aid for Scientific Research (C), 21510060, 2009 from the Ministry of Education, Culture, Sports, and Science Culture, and a Research Project Grant (19B-258) with Heavy Ions at NIRS-HIMAC.

The authors thank all the staff of Experiment Support Group, Accelerator Division, Accelerator Engineering Corporation, for performing heavy-ion irradiations. The expert technical assistance and administrative support of Ms. Mikiko Nakajima, Ms. Nobuko Tsutsumi, Ms. Yasuko Morimoto, Ms. Satoko Idohara, Ms. Ichika Kishi, Ms. Kaori Tateno, Mr. Tatsuo Hayao, Mr. Toshiaki Kokubo, Ms. Maki Asano and Ms. Hiromi Arai, is gratefully acknowledged. We are also deeply grateful to Dr. Jun Misonoh, Dr. Hiroshi Tanooka, Dr. Takeshi Murakami and Dr. Isamu Hayata, for their continual encouragement and support throughout the study. Great appreciation is especially given to Dr. Takeo Ohnishi, Nara Medical University, for his constructive comments and continual encouragement throughout the study. Thanks are also due to the anonymous peer reviewers for providing the constructive comments that strengthened the presentation of this work.

6. References

Aoki, M.; Furusawa, Y. & Yamada, T. (2000). LET Dependency of Heavy-ion Induced Apoptosis in V79 Cells, *Journal of Radiation Research,* Vol.41, No.2, pp. 163-175.

Chaubey, R.C.; Bhilwade, H.N.; Joshi, B.N. & Chauhan, P.S. (1993). Studies on the migration of micronucleated erythrocytes from bone marrow to the peripheral blood in irradiated Swiss mice. *International Journal of Radiation Biology,* Vol.63, No.2, pp. 239-245.

Curtis, S.B.; Hazelton, W.D.; Luebeck E.G.; & Moolgavkar, S.H. (2004). From mechanisms to risk estimation--bridging the chasm. *Advances in Space Research*, Vol.34, No.6, pp. 1404-1409.

Dacquisto, M.P. (1959). Acquired radioresistance; a review of the literature and report of a confirmatory experiment. *Radiation Research*, Vol.10, pp. 118-129.

Durante, M.; Snigiryova, G.; Akaeva, E.; Bogomazova, A.; Druzhinin, S.; Fedorenko, B.; Greco, O.; Novitskaya, N.; Rubanovich, A.; Shevchenko, V.; Von Recklinghausen, U. & Obe, G. (2003). Chromosome aberration dosimetry in cosmonauts after single or multiple space flights. Cytogenet. *Genome Research*, Vol.103, No.1-2, pp. 40-46.

Held, K.D. (2009). Effects of low fluences of radiations found in space on cellular systems. *International Journal of Radiation Biology*, Vol.85, No.5, pp. 379-390.

Horie, K.; Kubo, K. & Yonezawa, M. (2002). p53 dependency of radio-adaptive responses in endogenous spleen colonies and peripheral blood-cell counts in C57BL mice. *Journal of Radiation Research*, Vol.43, No.4, 353-360.

Mehnati, P.; Morimoto, S.; Yatagai, F.; Furusawa, Y.; Kobayashi, Y.; Wada, S.; Kanai, T.; Hanaoka, F. & Sasaki, H. (2005). Exploration of "over kill effect" of high-LET Ar- and Fe-ions by evaluating the fraction of non-hit cell and interphase death. *Journal of Radiation Research*, Vol.46, No.3, pp. 343-50.

Misonoh, J. & Yonezawa, M. (1997). Dose ranges for radioadaptive response in mice on the viewpoint of aquired radio-resistance after low dose irradiation. In: *Health Effects of Low Dose Radiation: Challenges of the 21st Century*, British Nuclear Energy Society, (Ed.), 169-174, Thomas Telford Services Ltd, London

Mitchel, R.E. (2006). Low doses of radiation are protective in vitro and in vivo: Evolutionary origins. *Dose Response*, Vol.4, No.2, pp. 75-90.

Monobe, M.: Koike, S.; Uzawa, A. & Ando, K. (2003). Effects of beer administration in mice on acute toxicities induced by X rays and carbon ions. *Journal of Radiation Research*, Vol.44, No.1, pp. 75-80.

Monobe, M.: Koike, S.; Uzawa, A.; Aoki, M.; Takai, N.; Fukawa, T.; Furusawa, Y. & Ando, K. (2004). Radioprotective activities of beer administration for radiation-induced acute toxicity in mice. *Radiotherapy & Oncology*, Vol.73, No.Suppl 2, pp. S127-129.

Mortazavi, S.M.; Cameron, J.R. & Niroomand-rad, A. (2003). Adaptive response studies may help choose astronauts for long-term space travel. *Advances in Space Research*, Vol.31, No.6, pp. 1543-1551.

Mouthon, M.A.; Van der Meeren, A.; Gaugler, M.H.; Visser, T.P.; Squiban, C.; Gourmelon, P. & Wagemaker, G. (1999). Thrombopoietin promotes hematopoietic recovery and survival after high-dose whole body irradiation. *International Journal of Radiation Oncology Biology Physics*, Vol.43, No.4, pp. 867-875.

Nelson, G.A. (2003). Fundamental space radiobiology. *Gravitational and space biology bulletin*, Vol.16, No.2, pp. 29-36.

Nose, M.; Wang, B.; Itsukaichi, H.; Ohyama, H.; Hayata, I. & Yamada, T. (2001). Rescue of lethally irradiated mice from hematopoietic death by pre-exposure to 0.5 Gy X rays without recovery from peripheral blood cell depletion and its modification by OK432. *Radiation Research*, Vol.156, pp. 195-204.

Okazaki, R.; Ootsuyama, A. & Norimura, T. (2007). TP53 and TP53-related genes associated with protection from apoptosis in the radioadaptive response. *Radiation Research*, Vol.167, pp. 51-57.

Olivieri, G.; Bodycote, J. & Wolff, S. (1984). Adaptive response of human lymphocytes to low concentrations of radioactive thymidine. *Science*, Vol.223, No.4636, pp. 594-597.

Otsuka, K.; Koana, T.; Tauchi, H. & Sakai, K. (2006). Activation of antioxidative enzymes induced by low-dose-rate whole-body gamma irradiation: adaptive response in terms of initial DNA damage. *Radiation* Research, Vol.166, pp. 474-478.

Otsuka, K.; Koana, T.; Tomita, M.; Ogata, H. & Tauchi, H. (2008). Rapid myeloid recovery as a possible mechanism of whole-body radioadaptive response. *Radiation Research*, Vol.170, pp. 307-315.

Schmid, M. (2010). The micronucleus test. *Mutation Research*, Vol.174, pp. 532-536.

Sokolov, M.; Panyutin, I.G. and Neumann, R. (2006a). Genome-wide gene expression changes in normal human fibroblasts in response to low-LET gamma-radiation and high-LET-like 125IUdR exposures. *Radiation Protection Dosimetry*, Vol.122, No.1-4, pp. 195-201.

Sokolov, M.; Smirnova, N.A.; Camerini-Otero, R.D.; Neumann, R.D. & Panyutin, I.G. (2006b). Microarray analysis of differentially expressed genes after exposure of normal human fibroblasts to ionizing radiation from an external source and from DNA-incorporated iodine-125 radionuclide. *Gene*, Vol.382, pp. 47-56.

Suzuki, M.; Kase, Y.; Nakano, T.; Kanai, T. & Ando, K. (1998). Residual chromatin breaks as biodosimetry for cell killing by carbon ions. *Advances in Space Research*, Vol.22, No.12, pp. 1663-1671.

Suzuki, M.; Kase, Y.; Kanai, T. & Ando, K. (2000a). Change in radiosensitivity with fractionated-dose irradiation of carbon-ion beams in five different human cell lines. *International Journal of Radiation Oncology Biology Physics*, Vol.48, No.1, pp. 251-258.

Suzuki, M.; Kase, Y.; Kanai, T. & Ando, K. (2000b). Correlation between cell killing and residual chromatin breaks measured by PCC in six human cell lines irradiated with different radiation types. *International Journal of Radiation Biology*, Vol.76, No.9, pp. 1189-1196.

Suzuki, M.; Kase, Y.; Yamaguchi, H.; Kanai, T. & Ando, K. (2000c). Relative biological effectiveness for cell-killing effect on various human cell lines irradiated with heavy-ion medical accelerator in Chiba (HIMAC) carbon-ion beams. *International Journal of Radiation Oncology Biology Physics*, Vol.48, No.1, pp. 241-250.

Suzuki, M.; Tsuruoka, C.; Kanai, T.; Kato, T.; Yatagai, F. & Watanabe, M. (2003). Qualitative and quantitative difference in mutation induction between carbon- and neon-ion beams in normal human cells. *Biological effects of space radiation*, Vol.17, pp. 302-306.

Suzuki, M.; Tsuruoka, C.; Uchihori, Y.; Kitamura, H. & Liu, C.H. (2009). Radiation-quality dependent cellular response in mutation induction in normal human cells. *Journal of Radiation Research*, Vol.76, No.5, pp. 1189-1196.

Takahashi, A. & Ohnishi, T. (2009). Molecular mechanisms involved in adaptive responses to radiation, UV light, and heat. *Journal of Radiation Research*, Vol.50, No.5, pp. 385-393.

Takeda, A.; Yonezawa, M. & N. Katoh, (1981). Restoration of radiation injury by Ginseng. I. Responses of X-irradiated mice to Ginseng extract. *Journal of Radiation Research*, Vol.22, No.3, pp. 323-335.

Tapio, S. & Jacob, V. (2007). Radioadaptive response revisited. *Radiation and Environmental Biophysics*, Vol.46, No.1, pp. 1-12.

Tomizawa, M.; Miyamoto, T.; Kato, H. & Otsu, H. (2000). Relative biological effectiveness of carbon ions for causing fatal liver failure after partial hepatectomy in mice. *Journal of Radiation Research*, Vol.41, No.2, pp. 151-161.

Tsuruoka, C.; Suzuki, M. & Fujitaka, K. (2004). LET and ion-species dependence for mutation induction and mutation spectrum on hprt locus in normal human fibroblasts. *Biological effects of space radiation*, Vol.18, pp. 188-189.

Tsuruoka, C.; Suzuki, M.; Kanai T. & Fujitaka, K. (2005). LET and ion species dependence for cell killing in normal human skin fibroblasts. *Radiation Research,* Vol.163, pp. 494-500.

United Nations Scientific Committee on the Effects of Atomic Radiation (UNSCEAR). (1994). Adaptive responses to radiation in cells and organisms, In: *UNSCEAR 1994 Report: Sources and Effects of Ionizing Radiation*, UNSCEAR, (Ed.), 185-272, UNSCEAR, New York

Vares, G.; Wang, B.; Tanaka, K.; Nakajima, T.; Nenoi, M. & Hayata, I. (2006). Radiation-induced adaptive response with reference to evidence and significance: A review. *Indian Journal of Radiation Research*, Vol.3, pp. 16-34.

Wang, B.; Tanaka, K.; Varès, G.; Shang, Y.; Fujita, K.; Ninomiya, Y.; Nakajima, T.; Eguchi-Kasai, K. & Nenoi, M. (2010). X-Ray-induced radioresistance against high-LET radiations from accelerated heavy ions in mice. *Radiation Research*, Vol.41, pp. 151-161.

Wolff, S.; Jostes, R.; Cross, F.T.; Hui, T.E.; Afzal, V. & Wiencke, J.K. (1991). Adaptive response of human lymphocytes for the repair of radon-induced chromosomal damage. *Mutation Research,* Vol.250, No.1-2, pp. 299-306.

Yonezawa, M.; Misonoh, J. & Hosokawa, Y. (1990). Acquired radioresistance after small dose X-irradiation in mice. *Journal of Radiation Research*, Vol.31, No.3, pp. 256-262.

Yonezawa, M.; Misonoh, J. & Hosokawa, Y. (1996). Two types of X-ray-induced radioresistance in mice: Presence of 4 dose ranges with distinct biological effects. *Mutation Research*, Vol.358, pp. 237-243.

Yonezawa, M. (2000). Radioadaptive survival response in mice, In: *Biological Effects of Low Dose Radiation*, T. Yamada, C. Mothersill, B.D. Michael & C.S. Poten, (Eds.), 93-99, Elsevier Sciences, Amsterdam

Yonezawa, M. (2006). Induction of radio-resistance by low dose X-irradiation. *Yakugaku Zasshi* (in Japanese), Vol.126, No.10, pp. 833-840.

12

Radiation Toxins – Effects of Radiation Toxicity, Molecular Mechanisms of Action, Radiomimetic Properties and Possible Countermeasures for Radiation Injury

Dmitri Popov[1], Jeffrey Jones[2] and Vacheslav Maliev[3]
[1]*Advanced Medical Technologies & Systems, Richmond Hill, Ontario, L4E4W8,*
[2]*Baylor College of Medicine, One Baylor Plaza, Houston, Texas,*
[3]*Vladicaucasian Scientific Center of Russian Academy of Sciences, Biotechnology Dept., Kosta Hetagyrova, Vladicaucas, North Ossetia–Alania,*
[1]*Canada*
[2]*USA*
[3]*Russia*

1. Introduction

Radiation toxins induce high toxicity reactions after exposure

Acute Radiation Disease (ARD) or Acute Radiation Syndromes (ARS) are defined as the collective toxic clinical states observed from the acute pathological processes in various doses of irradiated mammals; to include: systemic inflammatory response syndrome (SIRS), toxic multiple organ injury (TMOI), toxic multiple organ dysfunction syndromes (TMODS), and finally, toxic multiple organ failure (TMOF). [2, 10, 18, 21] Moderate and high doses of radiation induces necrosis of radiosensitive cells with the subsequent formation of radiation toxins and their induced acute inflammatory processes. Radiation necrosis is the most substantial and most severe form of radiation induced injury, and when widespread, has grave therapeutic implications [1, 3, 53]. Low doses of radiation exposure induces apoptosis (controlled, programmed death of radiosensitive cells) without significant levels of specific radiation-induced toxin formation and with only low levels of inflammatory response [17, 50, 62].

Studying the trigger mechanism for radiation-induced lymphocyte death, N.I. Sorokina used the results of numerous experiments to show that ionizing radiation induces changes in the antigen phenotype of immature thymocytes in mice. This has the same type of effect as chemical differentiation inductors and thymotropin, which indirectly attests to the specific modifying effect of ionizing radiation [97, 98].

B.D. Zhivotovskiy described radiation-induced apoptosis and demonstrated a quantitative association between the pyknotic changes in the cell nuclei of thymocytes and production of postradiation chromatin decay products. The enzyme responsible for the decomposition of chromatin in irradiated cells is Ca/Mg-dependent endonuclease. Areas of endonuclease attack are distributed randomly and are not associated with the level of repetition of nucleotide sequences of DNA or the transcriptional activity of chromatin. This supports the conclusion that the radiation death of lymphoid cells and the general biological phenomenon of programmed cell death are identical in nature [95, 96]. Apoptosis, interphase cell death of irradiated lymphocytes, is considered one example of the biological phenomenon known as "programmed cell death". Interphase cell death occurs as the result of "switching on" a specific genetic program, the triggering of which evokes induction of specific genes that switch on the "cell-death program". [1] Apoptosis may occur as the result of many signals, among which include: various types of lympholytic agents, different chemical agents, and physical factors which include ionizing radiation [15,50].

But relatively high doses of radiation could induce other forms of programmed cell death, apoptosis and/or necrosis. Necrosis especially, initiates inflammation and formation of specific Radiation Toxins (RTs) [62, 64, 65, 66, 89, 90, 91]. Specific Radiation Toxins are playing an important role as the trigger mechanism for inflammation, cell lysis, and damage to vital cellular structures such as mitochondria, DNA, ion channels and cell membranes [89, 90, 91, 92, 93, 94]. The mechanism and action of Radiation Toxins could be compared to actions of different microbial toxins, snake and scorpion venoms, enteric bacterial toxins, natural and plant-borne toxins [6, 86, 87, 88]. At high doses of radiation, specific Radiation Toxins play an important role in the development of pathological processes, especially injury to the central nervous system [89, 90, 91]. Radiation induces changes in concentration of many chemical mediators which possess significant influence on the Central Neural System - acethylcholine, cholinesterase, adrenergic amines, glutamate, γ-Aminobutyric acid, aspartic acid and interact with receptors and ions channels.[4, 30]. Many natural toxins like scorpion venom induce a prolongation of action potentials caused by selective inhibition of sodium current inactivation and specifically affect voltage-gated sodium channels in excitable tissues.[6, 29, 36, 44, 49, 60, 82, 83, 84, 85]. "Excitotoxicity" is an important pathological mechanism which damages the central nervous system [27,105]. After high doses of radiation, some specific receptors such as the NMDA receptor and AMP receptor are over activated. "Excitotoxins" such as NMDA and kainic acid, which bind to these receptors, as well as pathologically high levels of glutamate, can cause excitotoxicity, by allowing high levels of calcium ions to enter the cell. Ca^{++} influx into cells activates a number of enzymes, including phospholipases, endonucleases, and proteases such as calpain. These enzymes go on to damage cell structures e.g. components of the cytoskeleton, membranes, and DNA. Excitotoxicity may be involved in stroke, traumatic brain injury and neurodegenerative diseases of the central nervous system (CNS) such as multiple sclerosis, Alzheimer's disease, amyotrophic lateral sclerosis (ALS), fibromyalgia, Parkinson's disease, and Huntington's disease.[27, 105]. It therefore seems logical that exicitotoxicity possibly plays a prominent role in CNS injury.

We also postulate that specific Radiation Neurotoxins damage cerebral blood vessels and can induce activation of specific neurotransmitters which may produce exitotoxicity and its consequences [100, 103, 104]. The general toxicity of different types of ionizing radiation is associated mainly with a formation of Specific Radiation Toxins - a group of Radiation Toxins (RT) – Specific Radiation Determinants (SRD). [100, 103, 104]. Radiation toxins concentrate initially in intercellular fluid then disperse into blood and lymph and thereby induce development of systemic injury of hematologic and lymphatic systems. The toxicity of RT SRD subsequently distributes to the cerebrovascular, cardiovascular, and gastrointestinal systems, depending on the magnitude o the exposure. The injury to cellular components of radiosensitive and pluripotent hematopoietic and other stem cells are induced by processes that are specific for different form of ARS. [100, 103, 104]. *Radiation Toxins*: The specific Radiation Toxins (group SRD) are composed from glycoproteins with high enzymatic activity including high proteolytic, lipidolytic, carbohydratelytic properties. RT are formed and accumulated in mammalian cells and migrate into blood and lymphatic circulation in the first hours after irradiation. Radiation Toxins, for the most part, are glycoproteins, pro-enzymes which are activated after relatively high doses of irradiation, i.e. after some radiation threshold for activation is exceeded. This group of glycoproteins has been isolated from cells of irradiated mammalian organisms, and possess the ability to induce inflammatory reactions, apoptosis, and necrosis similar to reactions mediated by radiation itself. The molecular weight of RT (SRD-group) range from 200-250 kDa. The SRD molecules have been isolated from blood, lymph and cells of irradiated animals but lymph has the highest concentrations measured. Irradiated animals develop the different forms of Acute Radiation Syndromes such as Cerebrovascular ARS, Cardiovascular ARS, Gastrointestinal ARS, depending on the level of radiation they were exposed to.

Acute Radiation Syndromes	**Type of Radiation Toxins**	**Grade ARS Mild**	**Grade ARS Moderate**	**Grade ARS Severe**	**Grade ARS Extremely Severe**
Cerebral form of ARS	cerebro-vascular R.T.	SRD-1.1	SRD-1.2	SRD-1.3	SRD-1.4
Cardio-vascular form of ARS	cardio-vascular R.T.	SRD-2.1	SRD-2.2	SRD-2.3	SRD-2.4
Gastro-intestinal form of ARS	gastro-intestinal R.T.	SRD-3.1	SRD-3.2	SRD-3.3	SRD-3.4
Hematological form of ARS	hemato-poietic R.T.	SRD-4.1	SRD-4.2	SRD-4.3	SRD-4.4

Table 1. Classification of radiation toxins (SRD group): Radiation Toxins are composed of toxic substances isolated from lymph of irradiated animals with different forms of Acute Radiation Syndromes.

RT Properties- In our experiments RT were isolated from central lymphatics of irradiated animals with different form of ARS and each of the RT possess differences in their radiomimetic properties. Cerebrovascular RT, and to a lesser extent, cardiovascular RT and gastrointestinal RT are neurotoxins. Hematopoetic RT - are strong hematotoxins. Hematopoetic RT can destroy red blood cells and initiate hemolysis, disrupt the blood clotting system and cause multi-organ degeneration and tissues damages. Hematopoietic RT possess important activity against pluripotent stem cells and blood marrow.

Component (%)	Toxic substance SRD-1	Toxic substance SRD -2	Toxic substance SRD-3	Toxic substance SRD-4.
Protein	44.9 ± 0.05	48.2 ± 0.03	50.1± 0.09	56.2 ± 0.12
Lipid	40.1± 0.04	39.6 ± 0.05	38.2± 0.04	30.1± 0.09
Carbohydrate	13.4 ± 0.09	11.3 ± 0.07	10.2 ± 0,03	10.1 ± 0,07
Mineral residue	1.6 ± 0.07	0.9 ± 0.02	1.3 ± 0.04	3.4± 0.17

Table 2. Biochemical composition of the toxic substances comprising Radiation Toxins. (SRD- Specific Radiation Determinants)

2. Material and methods

These studies were conducted at several different research institutions and laboratories listed as follows: Biotechnology Centre of Russian Academy of Science (North Osetia), Kazan All-Union Scientific Research Veterinary, Institute Belarussian Scientific and Research Institute for Radiobiology in Gomel, the St. Petersburg Veterinary Institute, the Advanced Medical Technology and Systems Inc., Ottawa-Richmond Hill, Ontario, Canada. The studies were approved by the Animal Care and Use Committee for ethical animal research equivalent, at each institution. A largest volume of purified Radiation Toxin was prepared from larger mammalian irradiated animals. Subsequently the RT were characterized chemically and biologically. The experimental design of later studies compared relative toxicity, potential for development of acute radiation syndromes, and potential lethality after intravenous or intramuscular injections of SRD containing Radiation Toxins. More recent studies focused on the immunogenicity of small doses of SRDs and the potential to use these agents as vaccine countermeasures against radiation injury.

2.1 Experimental animals

These experiments have employed a wide variety of experimental and agricultural animals which include : Black motley cattle, 2.5-3.0 years of age, live weight of 300-350 kg (n=134); Ukrainian pigs, 0.5-1.0 years of age, live weight 35-90 kg (n=142); "prekos" sheep, 3-12 months old, live weight 18-23 kg (n=156); mixed breed (mongrel) dogs, 2-4 years of age, live weight 6.0-6,5 kg (n=162), Chinchilla rabbits, 11-12 months old, live weight 3.5-3.7 (n=180), Latvian draft horses, 3-8 years of age, live weight 350-550 kg (n=32), Balb mice, 2-3 months old, live weight 20-22 g (n=2636), Wistar rats, 3-4 months old, live weight 180-220 g (n=4002).

In all animals, the pre-experiment blood profiles, body temperature, central venous and lymphatic pressure did not exceed the limits of normal variability, (which had been previously measured for five species of mammals. Throughout the entire period of each experiment, ranging from 1 day to 2 years, the control and experimental animals were maintained under identical conditions of feeding, housing, and care, corresponding to livestock maintenance requirements and standards.

2.2 Electromagnetic radiation exposure

The animals were irradiated in RUM-17, Puma, and Panorama devices. The dose rate varied from 30Gy to 100Gy

2.3 Experimental design

Four independent experiments were performed with administration (IV or IM) of Radiation Toxins (SRD RT) to healthy, radiation naive animals that induced development of clinical symptoms of the Acute Radiation Syndromes. The administration of injection of the toxin was administered into veins (IV), into a muscle (IM), into the abdomen (IP) or into the skin (SC).

Experiment N1. Administration of Cerebrovascular form of Radiation Toxins to radiation naive animals in doses 0.1 mg/kg; 0,5 mg/kg; 1 mg/kg; 2 mg/kg; 3 mg/kg up to 30 mg/kg initiates development of specific toxic reactions with symptoms mimicking the cerebral form of ARS. Injection of Cerebrovascular Radiation Toxins (SRD-1) in excess of 3 mg/kg doses to rats, rabbits, sheep resulted in lethality.

Experiment N2. Administration of radiation toxins of Cardiovascular Radiation Toxins (SRD-2) to radiation naive animals in doses 0.1 mg/kg; 0,5 mg/kg; 1 mg/kg; 2 mg/kg; 3 mg/kg up to 30 mg/kg produced specific toxic reactions with symptoms mimicking the Cardiovascular form of ARS.

Experiment N3. Administration of Gastrointestinal Radiation Toxins (SRD-3) group to radiation naive animals in doses 0.1 mg/kg; 0,5 mg/kg; 1 mg/kg; 2 mg/kg; 3 mg/kg up to 30 mg/kg produced development of specific toxic reactions with symptoms mimicking the GI form of ARS.

Experiment N4. Administration of Haemotopoietic Radiation Toxins (SRD-4) to radiation naive animals in doses 0.1 mg/kg; 0,5 mg/kg; 1 mg/kg; 2 mg/kg; 3 mg/kg up to 30 mg/kg produced specific toxic reactions with symptoms of the hematological form of ARS.

2.4 Radiation toxins description and isolation

The methods of immune depletion, affine immuno-lympho-plasmasorption, as well as direct extraction were used to refine and purify the specific Radiation Toxins from the central lymph of animals with Cerebrovascular, Cardiovascular , Gastrointestinal and Haemotopoietic forms of Radiation Toxins.[104, 110]

Specific Radiation Determinants (SRD)- a group of Radiation Toxins isolated from lymph of irradiated mammals, had been divided to four important groups: 1.Cerebrovascular neurotoxic RT (SRD-1); 2.Cardiovascular RT(SRD-2); 3.Gastrointestinal RT (SRD-3); 4.Hematopoetic RT (SRD-4). The radiomimetic properties of Radiation Toxins preparations were evaluated on the basis of their capacity to induce radiobiological effects in radiation-naive animals after they were administered parenterally. Specific Cerebrovascular, Cardiovascular, Gastrointestinal and Hematopoietic Toxins (isolated from the lymph of animals irradiated at doses inducing Cerebral, Cardiovascular, Gastrointestinal and Haemotopoietic clinical forms of ARS) were dissolved in an isotonic solution of NaCl. The doses of preparations of Radiation Toxins which were administered was based on computation of the amount of SRD per unit volume of central lymph and absorbed dose of radiation. Radiation toxins of SRD group possess highly toxic properties and induce neuro-vascular and hematotoxic reactions in short time after administration.

Neurotoxins of SRD-1 group isolated from L.S. (lymphatic system) of irradiated animals with Cerebrovascular ARS after intra-muscular or intra-venous single injection in doses 5 mg/kg, 10 mg/kg, 15 mg / kg, 30 mg/kg activate acute neuro-vascular reactions with signs of injury of central and peripheral nervous system and damage of vascular walls of central nervous system.
Neurotoxins of SRD-2 group isolated from L.S. of irradiated animals with cardio-vascular form of ARS after intra-muscular or intra-venous single injection in doses 5 mg/kg, 10 mg/kg, 15 mg/kg, 30 mg/kg activate acute cardio-vascular and neuro-vascular reactions with signs of injury of central and peripheral vascular system . Neurotoxins of SRD-3 group isolated from L.S. of irradiated animals with gastro-intestinal form of ARS after intra-muscular or intra-venous single injection in doses 5 mg/kg, 10 mg/kg, 15 mg/kg and 30 mg/kg induce acute gastrointestinal syndrome with signs of injury of vascular system of gastro-intestinal system and injury of epithelium of intestinal walls. Hematotoxin of SRD-4 isolated from L.S. of irradiated animals with Hematopoetic form of ARS after single dose injection with doses 5 mg/kg, 10 mg/kg, 15 mg/kg, 30 mg/kg activate hematotoxic reactions with activation red blood cells and white blood cell lysis and activation of apoptosis of hematopoietic cells progenitors.

3. Results

SRD-1 Radiation Toxins induced injury of Central Nervous System and development of Acute Cerebrovascular form of Radiation Syndrome

The Cerebrovascular form of Radiation Toxins SRD-1) were administered to radiation naive animals in doses of 0.1 mg/kg; 0,5 mg/kg; 1 mg/kg; 2 mg/kg; 3 mg/kg, 5 mg/kg, 10 mg/kg, 15 mg/kg and 30 mg/kg. The injection of Cerebrovascular Radiation Toxins (SRD-1) to rats, rabbits, sheep produced bioeffects similar to the cerebrovascular form of ARS. (see Table 3) After parenteral administration, the animals experienced a short period of extreme agitation, followed by a deep coma, with alterations in breathing patterns and subsequently cardiovascular collapse. The results of autopsy of their bodies demonstrated cerebral hemorrhagic strokes, cerebrospinal fluid with blood (reddish color), hemorrhagic lesions in brain tissue. Internal organs were filled with blood. Multiple petechiae were observed on serous membranes. Depending on doses of cerebrovascular Radiation Neurotoxins, death was registered in 15 min or up to 5 hours after injection.
Case #1: The species studied in the experiment- 4 month old lambs. Radiation neurotoxin (SRD-1) in single dose of 5 mg/kg of weight was administered intramuscularly to radiation-naïve sheep. The injected animals were dead within 20 minutes after injection, with a profound clinical picture of cerebrovascular ARS, Within 5 minutes there was central nervous excitation and nervousness, and within 15 minutes there were signs of coma, followed by death. Post-mortem histopathology revealed: Brain- red discoloration of cerobrospinal fluid (CSF) with inflammatory meninges; multiple petechiae and hemorrhage on meninges surfaces. Grey matter of the brain – hemorrhages. Liver- dark-red coloration, with multiple petechiae and hemorrhages. Kidneys- multiple petechiae and hemorrhages. Clinical Diagnosis: Acute Cerebrovascuar Shock / Acute Cerebrovascular form of ARS induced by single injection of radiation neurotoxin

Radiation Toxins	Symptoms	Toxic doses 5mg/kg	Toxic doses 10mg/kg	Toxic doses 15mg/kg	Toxic doses 30 mg/kg
Cerebrovascular Neurotoxin SRD-1.	Cerebral edema, hemorrhagic stroke	Death in one -three hours	Death in one hour	Death in 15-30 min	Death in 15 min
Cardiovascular SRD-2.	Cardiac insufficiency, hemorrhagic stroke.	Death in three days	Death in 14 hours	Death in 7 hours	Death in 2 hours
Gastro-Intestinal SRD-3.	Vomiting, diarrhea, hematochezia and melana.	Death in 10-15 days	Death in 5-7 days	Death in 2-3 days	Death in 24 hours
Hematopoietic SRD-4	Peripheral edema, ecchymosis, anemia, granulocytopenic fevers.	Death in 2-4 weeks.	Death in next 10 days.	Death in next 7-10 days.	Death in next 4-5 days.
Mixture of lower doses of toxins SRD 1+2+3+4	Cerebral edema, vomiting, hemorrhage, cardiac insufficiency.	Death in 4-5 hours	Death in 1-3 hours	Death in one hour	Death in one hour

Table 3. Radiation Toxins: Toxic properties of Radiation Toxins (SRD Group 1-4) after single injection and different forms of Acute Radiation Syndromes induced by each.

SRD-2 Radiation Toxins induced injury of the Cardiovascular System and development of symptoms mimicking the Acute Cardiovascular Radiation Syndrome

The Cardiovascular form of Radiation Toxins were administered to radiation naive animals in doses of 0.1 mg/kg; 0,5 mg/kg; 1 mg/kg; 2 mg/kg; 3 mg/kg, 5 mg/kg, 10 mg/kg, 15 mg/kg and 30 mg/kg. The injection of Cardiovascular (SRD-2) RT in toxic doses to rats, rabbits, sheep produced symptoms similar to the cardiovascular form of ARS. (see Table 3) In this experiment, the following symptoms were observed: a delayed, shorter duration and less extreme period of agitation than with SRD-1 RTs, followed by tachycardia, cardiac arrhythmias, tachypnea. Postmortem histopathological analysis revealed inflammatory changes in the cardiac muscle tissue.

Case #2: Species in experiment- 4 months old lamb. Radiation neurotoxin (SRD-2) in single dose 5 mg/kg of weight was administered intravenously. 14 hours after injection all the injected lambs were dead after with a clinical picture of cardiovascular shock: agitation, fever, agitation, tachycardia, arrhythmias, hypotension, hyperventilation, followed by lethargy, signs of coma and death. Post-mortal histopathology showed: Heart- damaged microvasculature, cardiac vasoconstriction , ischemic myocardium, contraction band necrosis, multiple cardiac hemorrhages. Kidneys: multiple petechies of haemorrhages, acute tubular necrosis and renal failure.

SRD-3 Radiation Toxins induced injury of the Gastrointestinal System and development of symptoms mimicking the Acute Gastrointestinal Radiation Syndrome

Administration of Gastrointestinal Radiation Toxins (SRD-3) were administered to radiation naive animals in doses of 0.1 mg/kg; 0,5 mg/kg; 1 mg/kg; 2 mg/kg; 3 mg/kg, 5 mg/kg, 10 mg/kg, 15 mg/kg and 30 mg/kg. Injection of SRD-3 in toxic doses to experimental animals produced different grades of the Gastrointestinal form of ARS. (see Table 3) Case #3: Experimental species: one year old sheep. After a single dose injection of 15 mg/kg of the SRD-3 Gastrointestinal radiation toxins (isolated from central lymph of animals with the gastrointestinal form of ARS) intravenously, the animals elicit the following: Clinical signs- decrease in appetite and increased peristalsis within hours after injection, followed by severe diarrhea, hematochezia, dehydration, and later followed by collapse over several days. Death occurred 5-10 days after toxin administration. Histopathology revealed- local vasoconstriction, damaged microvasculature, large vessel vasodilatation and hyperemia of the small bowel and colon. Hypertension was manifested approximately two hours after intravenous injection. Ultimately hypotension developed after days of vomiting, diarrhea and dehydration.

SRD-4 Radiation Toxins induced development of Acute Hematopoietic Acute Radiation Syndrome

The Hemotopoietic Radiation Toxins (SRD-4) were administered to radiation naive animals in doses of 0.1 mg/kg; 0,5 mg/kg; 1 mg/kg; 2 mg/kg; 3 mg/kg, 5 mg/kg, 10 mg/kg, 15 mg/kg and 30 mg/kg. Injection of SRD-4 to experimental animals resulted in erythrocytopenia, lymphocytopenia, leukocytopenia, and thrombocytopenia within days to weeks after injection. (see Table 3) The development of clinical features of the Acute Hemotopoietic Syndrome depended on the dose of SRD-4 Hematopoietic Radiation Toxins injected to radiation naïve animals. Autopsy of those animals that died showed acute or chronic hematotoxic reactions. The clinical signs were: short-term agitation within 2 hours after administration accompanied by a short-term leukocytosis which gave way to a progressive, profound leukopenia, mainly attributable to a decrease in the absolute number of lymphocytes, the minimal levels of which were measured between days 7 and 15 after injection. Total WBC was 1.2-1.6 1000/μl to 0.4-0.5 thousand/μl in sheep and 1.8-2.5 thousand/μl to 0.5-0.7 thousand/μl in cattle. The recovery of absolute and relative levels of leukocytes and lymphocytes was observed in some animals between days 30 and 60. Blood counts exhibited thrombocytopenia accompanied by progressive erythrocytopenia, which developed into profound anemia. An extensive blood analysis of the peripheral blood of the cattle showed that the processes induced by the SRD-4 injection and the processes occurring after irradiation were nearly identical. Analysis of the clinical reaction to SRD-4 administration, which was assessed on the basis of body temperature, and heart and respiration rate, established that all experimental animals showed reactions of the same type for all the tested doses and that sheep and horses were more sensitive to the administered preparation. Thus, in sheep, which received SRD-4 at the maximum dose tested, the body temperature increased by 1.5-2 ° C, reaching 41.2° 1 or 2 days before death. This core body temperature increase was accompanied by severe tachycardia and tachypnea, which were measured at 105-106/min. and 70-80/ minute, respectively. When SRD-4 was administered in the intermediate and maximum dose levels, both sheep and cattle showed changes of the same type but of lesser degree depending on the dose. The majority of experimental animals recovered between days 30-60 after injection of the preparation. Post-mortem examination of the animals that died showed changes characteristic of acute radiation sickness, accompanied by marked hemorrhage. Death often will occur secondary to overwhelming bacterial or fungal sepsis. Some sheep showed areas of skin epilation on the back and abdomen.

Case# 4. Species- sheep. Hematoxin, SDR-4, isolated from irradiated mammals with the Hematopoietic form of ARS, was injected via single dose to non-irradiated mammals. The SRD-4 RT induced significant changes in white blood cells (WBC) and red cells profile (RBC). Hematoxin injected to non-irradiated sheep in doses 5 mg/kg, 10 mg/kg, 30 mg/kg activated a complex reaction which included general inflammation, vascular endothelial cell injury, apoptosis and necrosis of blood progenitor/ stem cells. The levels of sheep erythrocytes, leukocytes, lymphocytes, thrombocytes significantly increased in first hours and day after single injection of radiation hematotoxin and could be a result of general inflammation reactions and stimulation of immune system. However the levels of erythrocytes, leukocytes, lymphocytes, thrombocytes significantly decreased after 72 hours after hematoxin administration and the minimal level of erythrocytes, leukocytes, lymphocytes were measured after 168- 360 hours following a single dose injection of the radiation-induced hematoxin. (see Tables 4- 7) Characteristics of the peripheral blood of large horn cattle after a single injection of Hematopoetic Radiation Toxin (SRD-4) were measured and levels of erythrocytes, leukocytes, lymphocytes significantly increased in first hours after a single dose injection of radiation hematotoxin and decreased after 72, 168, and 360 hours after administration. (see Tables 8- 11)

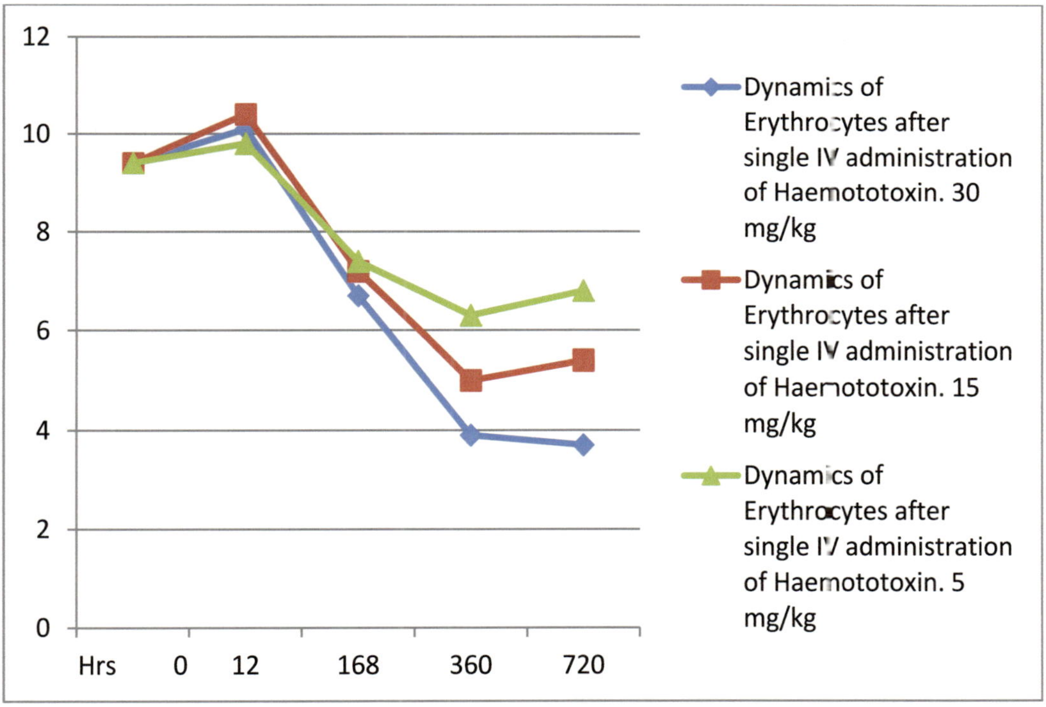

Fig. 1. Dose-dependent changes in sheep peripheral blood erythrocytes after a single injection of Hematopoietic Radiation Toxin (SRD-4).

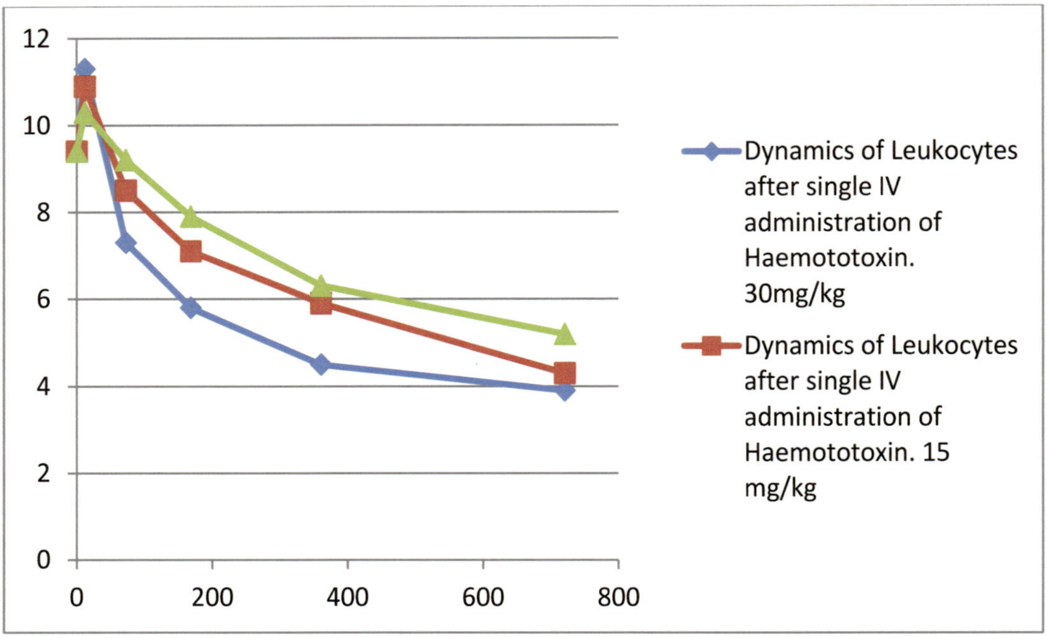

Fig. 2. Dose-dependent changes sheep peripheral blood leukocytes after a single injection of Hematopoietic Radiation Toxin (SRD-4).

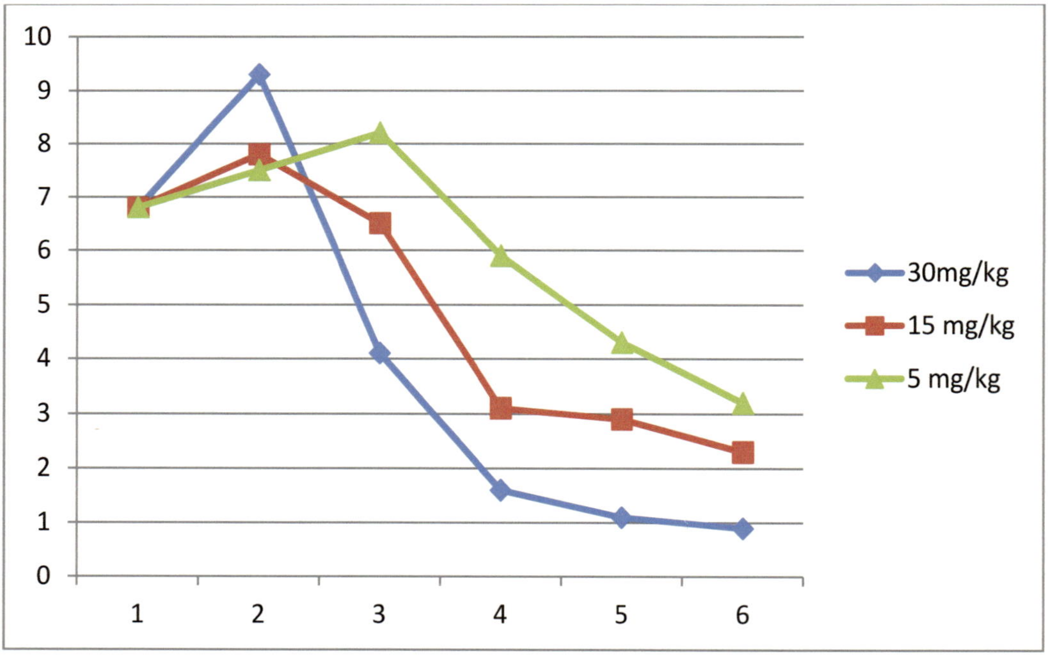

Fig. 3. Dose-dependent dynamics of sheep Lymphocytes after single IV administration of Hematotoxin (SRD-4).

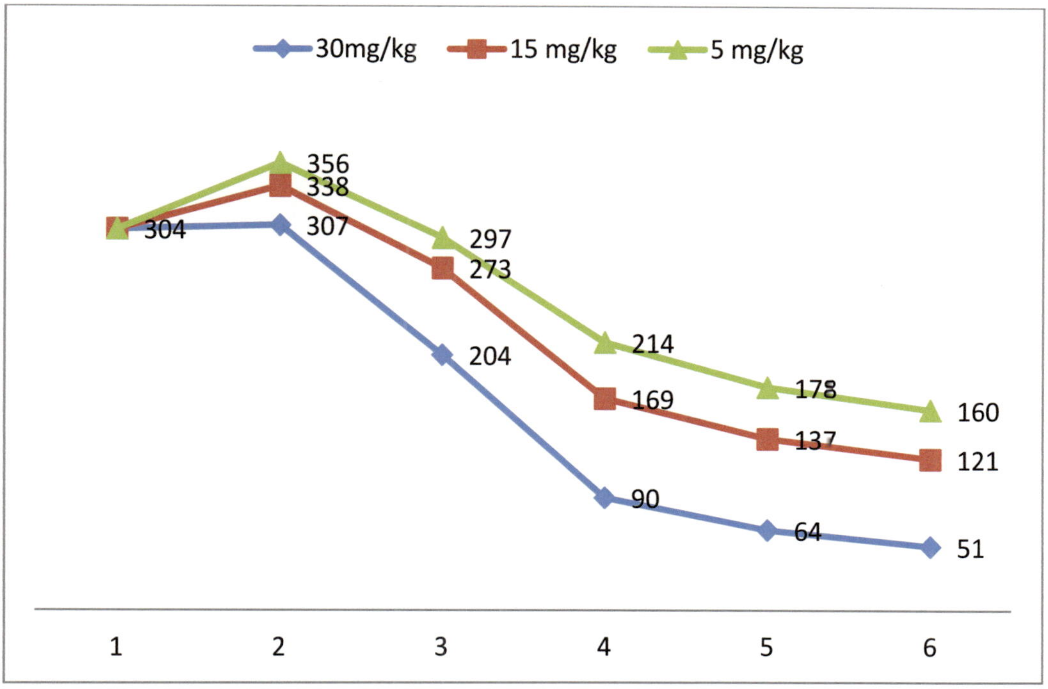

Fig. 4. Dose-dependent dynamics of sheep thrombocytes (platelets) after single IV administration of Hematotoxin (SRD-4).

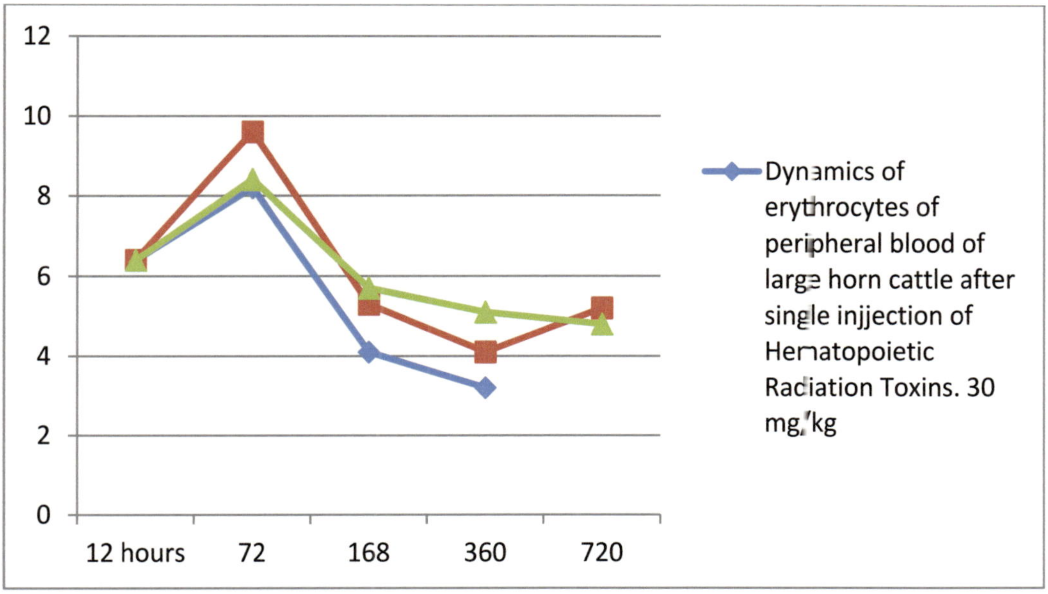

Fig. 5. Dose-dependent characteristics of large horn cattle peripheral blood erythrocytes after single injection of Hematopoetic Radiation Toxin (SRD-4).

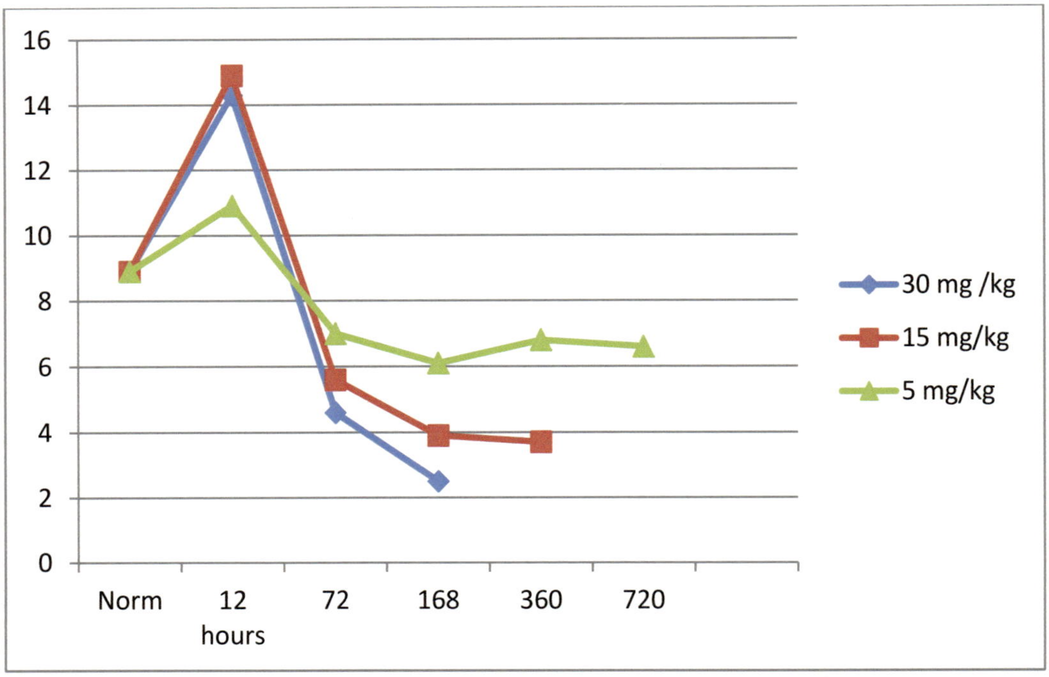

Fig. 6. Dose-dependent changes in cattle peripheral blood leukocytes after a single injection of Hematopoietic Radiation Toxin (SRD-4).

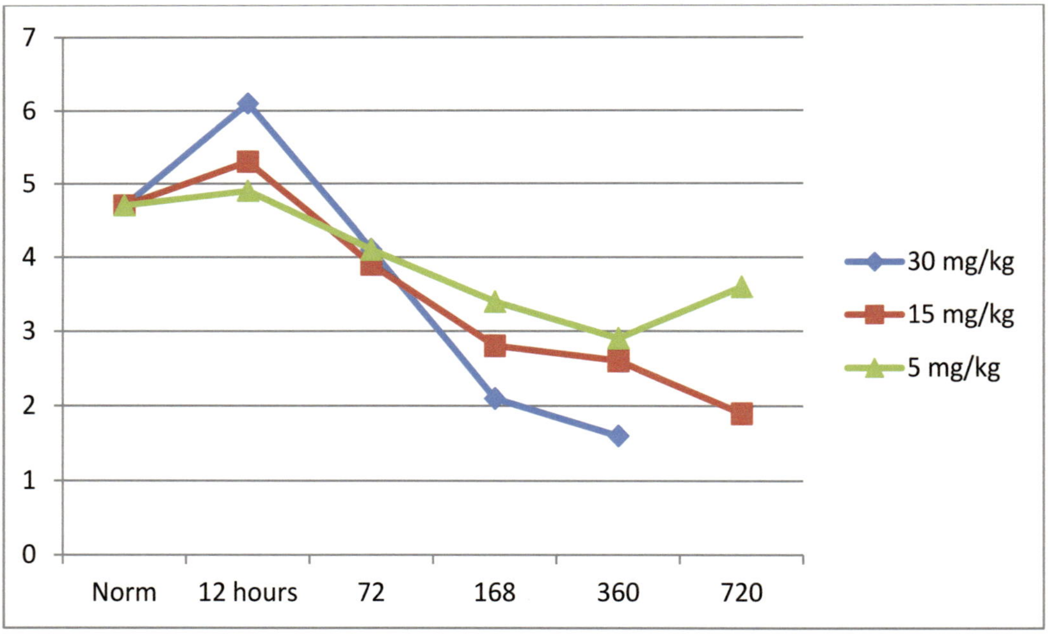

Fig. 7. Dose-dependent dynamics of cattle Lymphocytes after single IV administration of Hematotoxin (SRD-4).

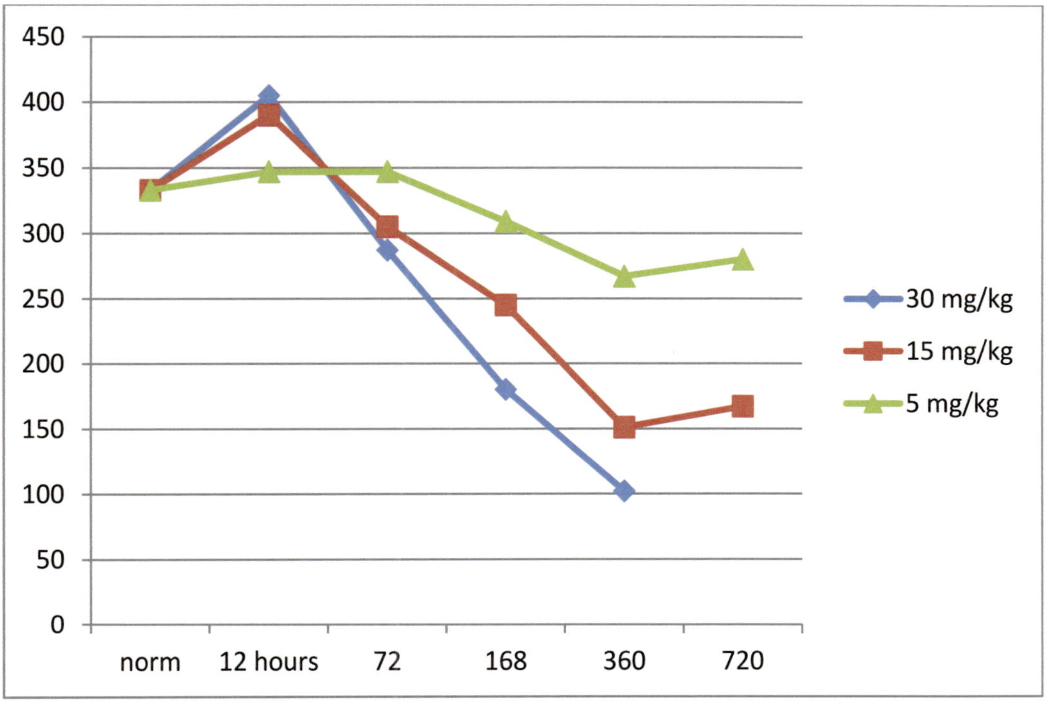

Fig. 8. Dose-dependent dynamics of cattle thrombocytes (platelets) after single IV administration of Hematotoxin (SRD-4).

4. Discussion

Radiation toxin SRD-1 (cerebrovascular radio-neurotoxin) activates pathophysiological processes and clinical symptoms / signs of central and peripheral nervous system injury with development of neurological and cognitive deficit, activation, followed by depression, of specific centers of the central nervous system, including autonomic nervous system pathways.

Radiation toxin SRD-2 (cardiovascular radiotoxin) activates pathophysiological processes and clinical symptoms / signs of cardiac and vascular inflammation and injury with endothelium cell, microvascular vessel and cardiovascular system tissue showing apoptosis and necrosis. Severe hypotension and tachycardia were important clinical indicators of cardiovascular acute radiation syndromes with the subsequent development of cardiac myocyte necrosis.

Radiation toxin SRD-3 (gastrointestinal radiotoxin) initiates symptoms signs of gastrointestinal system injury including anorexia and vomiting, induced by the toxic effects and damage to the vascular and lymphatic vessel network of the gastrointestinal system, as well as apoptosis of intestinal lining epithelial cells, including crypt / villi necrosis. Severe diarrhea and melena / hematochezia were important clinical indicators of the mimicked gastrointestinal acute radiation syndrome.

Radiation toxin SRD-4 (hematotoxin) induces development of red cell lysis and apoptosis of white blood cell and red cell progenitors. The action of the SRD-4 hematotoxin is described

in Tables 4-7 for sheep and 8-11 for cattle. Thrombocytopenia, lymphocytopenia, granulocytopenia , ecchymosis, hemorrhage and coagulopathy were important clinical signs of the mimicked hematopoietic acute radiation syndrome.
Other clinical indicators of the Radiation Toxins (RT) included cutaneous system involvement: including: cutaneous edemas, blustering, desquamation, hair loss, ulcer and necrosis.
Severe Acute Radiation Exposure sydromes will induce Toxic Multiple Organ Failure (TMOF) and Toxic Multiple Organ Involvement (TMOI) – e.g. pneumonitis, renal failure, renal hypo-perfusion, acute tubular necrosis, hepatic failure, etc. as an acute consequences of radiation toxemia. Similarly, bacterial toxemia induced by bacterial superantigens and radiation-induced toxemia are very similar in their clinical manifestation and similar in their pathophysiological actions. In a related matter, the molecular structure of radiation antigens (toxins) and bacterial superantigens, as well as different snake venom toxins have similarities.
Thus, if a researcher administers specific Radiation Toxins of one of the SRD groups to radiation naive animals, toxicity will be observed along a dose response curve, from doses as small as 0.1 mg/kg or 0.5 mg/kg up to the highest tested dose of 30 mg/kg. The observed toxicity will mimic the ARS syndrome specific to the exposure of the animals from which the SRDs were acquired. Thus the exogenously administered SRD initiates development of specific toxic reactions with symptoms of a specific ARS, hence the name-specific radiation determinant (SRD 1-4). The SRD-group molecules possess both toxic and antigenic properties. The molecular Koch's postulates, which are commonly applied to bacterial toxin pathogenecity, could also be applied to the biological effect produced by Radiation Toxins, and those postulates would mostly be satisfied to list RT as the causative agents [101, 106].

4.1 Comparative analysis of bacterial toxins vs radiation toxins (see table 12)

We compared lethal toxicity and clinicopathological features and characteristics for Radiation Toxins and Bacterial Toxins. Pathogenic bacteria which colonize the intestine, can invade intestinal epithelial cells and lymphatics, and/or produce one or more toxins that are important etiologies of diarrheal disease [32, 42, 48, 57, 86, 87, 88]. Bacteria produce toxins and the mechanisms of action of bacterial toxins can be classified into three important groups (see Table 12) : 1. Toxins with intrinsic enzymatic activity – cytotoxic bacterial toxins. 2. Toxins that binds to receptors that stimulate the actions of second messengers – cytotonic bacterial toxins (e.g. botulinum or cholera toxins) . 3. Toxins that damage eukaryotic cell membranes – membranotoxic bacterial toxins. [32, 42, 48, 57, 86,]. We postulate that radiation toxins possess high enzymatic activity, damage cell membranes, and also activate cell receptors. We postulate that radiation toxin mechanisms of action depend on the toxin type and dose of radiation toxins. We postulate that Radiation Toxins could initiate two different types of cell death – programmed (apoptosis) or necrosis. [90,91].
Escherichia Coli is an Enteric Gram-negative rods and their pathogenic strains contained the wide group of toxins with different properties. Escherichia Coli contain endotoxin - lipopolysaccharide, which is antigenic and virulence factor. [12, 48]. Escherichia Coli contains three groups of antigens – O, H and K. The O antigens – cell wall antigens - found in polysaccharide portion of LPS. The H antigens are associated with flagella and the K antigens are associated with the fimbriae or capsule. [12,48,78,79,87,88]. Enteroaggregative E.Coli – EAEC, produce three important toxins. EAEC heat stable enterotoxin is a 4.1-kDa protein. The

other EAEC heat -liable enterotoxin with molecular weight 120 kDa. The third EAEC enterotoxin is a heat liabile,108 kDa protein with the ability to induce a severe acute intestinal inflammation. [12,78,79,87]. Enterohemorrhagic Escherechia coli, EHEC contains a Shigella-like toxin. The Shiga toxin family contains Shiga-like toxin 1 (SLT-1) and Shiga-Like toxin 2 (SLT-2). EHEC strains produce SLT -2 or combination SLT 1 and SLT 2 [12]. Shiga toxin is a cytotoxic for endothelial and epithelial cells including human colonic and ileal epithelial cells, and B-lymphocytes [12]. Enteropathogenic Echerechia coli induce an important and distinctive histopathologic lesion in the intestine with destruction of microvilli – the classic histopathologic lesion - AE lesion [12]. There are two proteins responsible for development AE lesion – a 94-kDa protein and protein encoded by the eaeB gene. [12,78,79].

Toxins	Mechanism of Action	Symptoms
1. Enterotoxigenic E. Coli (ETEC).	1. ETEC - Three toxins: EA- ST, EA- HL, EA-HL	ETEC toxins induce hyper-secretion by intestinal mucosa cells, watery diarrhea
2. Enteropathogenic E. Coli (EPEC).	2. EPEC - Shiga-like toxins.	EPEC toxins causing destruction of microvilli and development focal lesions.
3. Enterohemorrhagic E. Coli (EHEC).	3. EHEC – verotoxin or Shiga-like toxin.	EHEC induce severe form of hemorrhagic colitis, acute renal failure or hemolytic uremic syndrome.
4. Enteroinvasive E. Coli (EIEC).		EIEC induce dysentery like syndrome
5. Enteroadherent E. Coli (EAEC).		EAEC cause traveler'diarrhea

Table 4. Bacterial Toxins and their mechanisms of action.

Enterotoxigenic Escherechia coli produce heat—stable (ST) and heat-labile enterotoxins (LT). STa is a cysteine-rich, 18-amino-acid peptide. ST is produced as a precursor that is cleaved by signal peptidase and translocated to the periplasm where is added three intra-molecular disulfide bonds crucial for toxin activity. A second proteolytic agent occurs extracellularly and produce biologically active and toxic ST [12,81,]. STa acts by binding to a protein intestinal epithelial receptor localized in brush border membrane. [12,78].
E.coli heat-stable enterotoxin b is a 71-amino-acid precursor protein secreted extracellularly and proteolytically processed to a toxic form as a 48-amino-acid protein [12, 79]. E. coli heat-labile enterotoxin I is closely related to CT, including its structure and enzymatic activity. [12]. Other Escherechia coli Toxins include Cytolethal distending toxin and Cytotoxic necrotizing factors [12, 78,79].
Staphylococcus aureus produces serologically distinct protein exotoxins named A, B, C1, C3, D, E. All of these exotoxins are small proteins (24-30 kDa) with important biological activities. All seven toxins stimulate local receptors in upper intestinal tracts [12, 86]. Staphylococcal enterotoxins are likely playing an important role in cytokine production and are most potent T-cell receptor activators and inducers of lymphocyte proliferation. This class of toxins could be classified as "superantigens", as well as the toxins produced by Streptococcus spp. and Clostridium perfringens. Superantigens presented directly to the T-

Clinical symptoms induced by Bacterial Superantigene	**Clinical symptoms induced by radiation toxins (SRD group) and by irradiation.**
Central Nervous System involvement Peripheral nervous system Confusion, Irritability, Lethargy	CNS involvement: Cerebro-Vascular Acute Radiation Syndrome. Radiation Induced Neurotoxin. Neurological and Cognitive Deficit, Vomiting. Severe form of brain microcirculation disorder. Clinical futures of hemorrhagic stroke.
Cardio-Vascular System involvement Systolic Blood Pressure :< 90 mmHg Tachycardia >100 beats/min, Hypotension, Hyperventilation, Loss of Sympathetic responsiveness .	Cardio-Vascular System involvement: Cardio-vascular Radiotoxin-induced Acute Radiation Syndrome.- Tachycardia, Bradicardia, Arrythmias, Hypertension followed by severe hypotension (Systolic Blood Pressure:< 80 mmHg), hyperventilation.
Gastro-Intestinal System Involvement Vomiting, diarrhea, hematochezia and stool mucous (depending on which enteric toxin).	Gastro-intestinal System Involvement: Gastro-Intestinal Acute Radiation Disease. Radiotoxin. Vomiting, diarrhea. Melena and hematochezia. Abdominal pain- moderate to severe.
Bacterial Superantigen hematotoxicity Thrombocytopenia, Platelet count < 100,000/mm3.- Thrombocytopenia., leukopenia, pancytopenia. Disseminated Intravascular Coagulation (DIC) Syndrome.	Hematopoietic Toxicity: Acute Radiation Hematopoetic Syndrome. Hematopoietic Radiation Toxin-Thrombocytopenia, Lymphocytopenia, Granulocytopenia. Pancytopenia.
Cutaneous involvement: diffuse maculopapular, petechial or ecchymotic rash, intense erythroderma, desquamation (e.g. meningococcemia or toxic shock syndrome).	Cutaneous System Involvement: Possibleaction of any type of Radiation Toxins. Swelling, edema, blustering, desquamation, hair loss, ulceration and skin necrosis
Multiple Organ Failure and Multiple Organ Involvement.	Multiple Organ Failure and Multiple Organ Involvement.
Renal failure, Renal hypo-perfusion; Oliguria, Acute tubular necrosis	Renal failure, renal hypo-perfusion, Acute tubular necrosis.
Hepatic failure and inflammation	Hepatic failure
Pulmonary System: Hyperventilation with Respiratory alkalosis, pulmonary hypertension and edema, hypoxemia, ARDS	Pulmonary System: Hyperventilation. Radiation induced pneumonitis and subsequent pneumonia, ARDS.

Table 5. Comparative analysis of clinical symptoms induced by radiation and/or Radiation Toxins – (SRD group 1-4)- induced ARS and those produced by Toxic Shock Syndrome induced by Bacterial Superantigene.

cell receptors by cells containing major histocompatability complex class II molecules, without processing by antigen-presenting cells [32,48,86]. Superantigens (SAgs) are a class of antigens which cause non-specific activation of T-cells resulting in polyclonal T cell activation and massive cytokine release. The large number of activated T-cells secrete large amounts of cytokines (the most important of which is TNF-alpha). [32,86]. SAg stimulation of antigen presenting cells and T-cells elicits a response that is mainly inflammatory, focused on the action of TH-1 T-helper cells. Some of the major products are IL-1, IL-2, IL-6, TNF-α, gamma interferon (IFN-γ), macrophage inflammatory protein 1α (MIP-1α), MIP-1β, and also monocyte chemoattractant protein 1(MCP-1). Clinical symptoms / signs induced by bacterial Superantigenes include central and peripheral nervous system involvement (confusion, lethargy, vomiting, anorexia), involvement of the cardiovascular system with development of hypotenison, tachycardia and loss of sympathetic responsiveness. Bacterial Superantigenes activate toxic processes within the gastrointestinal system including acute vomiting and diarrhea. The important signs and clinical picture of bacterial superantigen hematotoxicity with development thrombocytopenia, leukopenia, lyphocytopenia and development of coagulopathy. Bacterial Superantigenes could activate injury of cutaneus system and diffuse rash, erythrodermia, desquamation. [32,86] Radiation Toxins of SRD groups 1-4 can activate similar pathological processes, clinical picture and clinical signs as the bacterial superantigent toxicity and the three entities: 1) Acute Radiation Syndromes, 2) SRD RT- induced acute syndromes and Bacterial Toxin- induced sepstic reaction and toxicity can be compared. (see Table 5)

4.2 Comparative analysis of toxins produced by venom toxins and radiation toxins (see table 6)

Neurotoxic proteins can be isolated from various snake venoms. These proteins can potentially be a mix of neurotoxins ,which attack the nervous system; hemotoxins, which attack the circulatory system; plus cytotoxins, bungarotoxins and many other toxins that affect the body in different ways [6,29,49,60,61,82,83,84,85]. Almost all snake venom contains hyaluronidase, an enzyme that ensures rapid diffusion of the venom into the body of the bite-victim [82,84]. Viper venoms from the Elapidae, Hydrophiidae, Atractaspididae, Viperidae and Colubridae families contain at least 25 separate classes of biologically active compounds, including enzymes and non-enzymatic molecules. Venom- neurotoxins activate different reactions with nerve synapse - presynaptic and postsynaptic action; and induce anticholinesterase-induced flaccid paralysis. Myotoxins from Venom could induce systemic skeletal muscle damage. Hemotoxins induce the development of severe microcirculatory injury, damage to vascular membranes, thrombosis or bleeding. [6, 82, 84].

4.3 Radiation induced apoptosis and necrosis

Radiation induced Apoptosis and Necrosis as response of eukaryotic cells to influence of different types of radiation remain contraversion. Radiation induced cell death by triggering apoptosis pathways was described in many articles and supported by many scientists [11, 15, 17, 39, 51, 52, 53, 54, 55, 56; 57, 58, 59, 62, 63, 74, 75, 76, 78]. However some scientists and some institutions described processes developing in irradiated eukaryotic cells as necrosis [24;34;38] and some scientists describing a variety of complex mechanism and mentioned that radiation-induced cell death could developing under different mechanisms of pathogenesis which include and apoptosis and necrosis. [1, 34, 38, 43, 77] Activation of

Venom Toxins	Mechanism of Action	Symptoms
Polypeptide toxins.	Autonomic, Neurotoxic, Cardiotoxic, Myotoxic	Induce two forms of programmed cell death: cell apoptosis or cell necrosis. Presynaptic or postsynaptic actions. Induce excitation of some neurons. Neurotoxicity, Haemototoxity.
Phospholipase toxins.	Presynaptic neurotoxic, myolytic, procoagulant, cardiotoxic, necrotic.	Induce cerebrovascular toxicity, cardiovascular toxicity.
Enzyme toxins: Transferases, Oxydoreductases, Hydrolases, Carbon-nitrogen lyases, Metalloproteinases.	Disorder of haemostasis, necrotic, haemolytic.	Blood vessel injury, blood cell lysis, brain tissue damage, heart muscle damage, liver damage, kidney damage. Neurotoxicity. Haemototoxicity. Skin damage. Hair follicule damage.
The Group of Toxins: Lectins, Nerve growth factors, Phospholipase inhibitors, Complement inhibitors, Proteinase inhibitors.	Cardiotoxic, Neurotoxic.	Anticoagulant proteins. Platelet aggregation inducers. Platelet aggregation inhibitors.

Table 6. Venom toxins and mechanisms of action – biochemical and clinical classification.

cysteine proteases, cleavage of PARP - poly (ADP-ribosyl) ation of nuclear proteins causes the inner mitochondrial membrane to become permeable for ions and other small molecules during both types of cell death- apoptosis and necrosis. Many mechanisms of cell necrosis and apoptosis are identical – for example, degradation of the Nuclear matrix is a Common Element during Radiation-Induced Apoptosis and Necrosis [1,3,11,16,24,34,38,43,53,59,78].
Data from different sources described that the differences between apoptosis and necrosis are much less numerous than previously studies suggested. [1,78]. Joseph Dynlacht and his research group of scientists studied the outcome of human promyelocytic leukemia (HL60) cells irradiated with 10 or 50 Gy of X-rays and determined the mode of leukemic cell death. Different doses of X-rays induced different modes of cell death – cells irradiated with 10 Gy died by necrosis; cells irradiated with 50Gy died predominantly by apoptosis. [16].
The Cell Death Nomenclature Committee recommends the use of the appropriate diagnosis: apoptotic necrosis. [34]. Comparison and description of a cell's Necrosis and Apoptosis were provided by a group of scientists from different institutions and research groups [16,34]. Apoptosis is described as a normally active, programmed process of cell death devoid of inflammation and toxins. [16,34]. Necrosis possesses the characteristics of accidental or externally-induced cell death from the influence of different environmental factors and triggering development of inflammation and elaborating of toxins, release of pro-inflammatory active and toxic cellular contents into the intracellular and extracellular fluids and into the lymphatic systems and blood circulation [39,55,56,62]. Caspases are important molecular mechanisms and central components of the apoptosis process [39, 50, 51, 52, 54, 58, 62, 74, 75, 77, 78].

Apoptosis	Necrosis
Inflammation never present.	Inflammation always present.
Toxic substances never present.	Toxic substances always present.
Morphologically cells shrinks, become denser, condensation occur, original name - shrinkage necrosis. Mitochondria's structure and functions are not affected.	Morphologically cells swelling, lysis. Mitochondria's structure and functions are affected.
Karyorhexis – pyknotic nuclear fragments. DNA broken into segments.	Karyolysis, Pyknosis, Karyorhesis. Nuclear swelling, Chromatin granular, Chromatin flocculation, Types of radiation induced damage of DNA: Breaks of the strand, alteration to bases, destruction of sugar, cross-links and formation of dimmers
Caspase activation always present.	Caspase depended initiation of developing of necrosis possible. Caspase-8 initiation possible with apoptosis and necrosis.
Genetic control initiate apoptosis	No genetic control, environmentally-induced
The Budding phenomenon. Cell membrane not affected	The Blebbing phenomenon, increased membrane's permeability. Cell membrane affected by external and internal factors.

Table 7. Apoptosis and Necrosis: Comparison of Morphological and Biochemical features.

However, apoptosis is the principal mechanisms by which cells are physiologically eliminated. During apoptotic death, cellular dissociation by caspases of cell compartments occur and cell compartments and molecules are packaged into apoptotic bodies to avoid immune activation. [24,34,78]. Necrosis was considered in past times as passive, uncontrolled and unorganized mechanism of cell death. At the present time necrosis is considered as an alternative form of programmed cell death whose activation induces important biological consequences including immune reactions and induced inflammation. [43, 51, 52, 67, 74, 75, 77].

4.4 Inflammation induced by radiation

Radiation induces an important and aggressive inflammatory response in the irradiated tissues. [19, 20, 28, 30, 64, 65]. Acute Radiation Disease (ARD) or Acute Radiation Syndromes (ARS) were defined as a toxic, poisonous exposure with development of the acute pathological processes in irradiated animals: systemic inflammatory response syndrome (SIRS), toxic multiple organ injury (TMOI), toxic multiple organ dysfunction syndromes (TMOD), toxic multiple organ failure (TMOF) [2, 5, 18, 19, 21, 89, 90, 91]. Radiation-induced lung injury, radiation pneumonitis (RP), is a potentially fatal side-effect of thoracic radiation therapy [3, 4, 11, 15]. Radiation-induced thyroiditis: can occur after the use of radiation treatment and can result in lifelong hypothyroidism [18]. Irradiation of the liver with doses of 2,450-2,920 rads caused liver injury in 14 patients of 65 patients treated with radiation therapy [3, 28]. The clinical picture of radiation-induced hepatitis includes hepatomegaly, ascites, pleural effusion and alteration in liver function. Radiation injury of the central nervous system (CNS) has devastating and fatal clinical consequences and severely limits the dose that can be delivered in the radiotherapy of brain, head and neck, pulmonary and other tumors [19, 28, 30, 64, 65]. Hematopoetic system toxicity is a major limiting factor in the use of aggressive, combined modality therapy (chemotherapy + radiotherapy) in the treatment of malignant disease [14, 22, 40].

Permanent anemia and pancytopenia can be caused by the reduced capability for cellular proliferation due to the stem cells injury produced by radiation. [14, 69, 70, 72, 76]. Radiation induces an inflammatory response in the irradiated organs characterized by leukocyte infiltration and vascular permeability changes and vascular injury [90, 99]. Vascular injury is a key determinant of acute and chronic organ intoxication and dysfunction associated with irradiation of the different systems, including the gastrointestinal tract [9, 19, 20, 22,]. Radiotherapy is used in the treatment of pediatric brain tumors and is often associated with inflammation and behavior changing following irradiation-induced inflammation and injury in the developing brain [19, 20]. The consequences of irradiation microglial loss can be secondary to injury from pronounced inflammation [19, 20]. A prominent inflammatory response occurs following both ischemic and hemorrhagic forms of microcirculation disorders [19, 20].

4.5 Roles of radiation toxins in developing of cellular structure radiation-induced injury

Radiation Toxins contained a group of biologically-active compounds which induce complicated but specific mechanisms of action with tissues and cellular damage. Radiation Toxins include a group of glycoproteins with high enzymatic activity against proteins, lipids, carbohydrates. Radiation Toxins contain a group of peptides with selective affinity to number of receptors and specific interactions with receptors of mammalian organisms. Toxins isolated from the irradiated mammals inhibit or activate a group of an ion channels, acetylcholine receptors, nicotinic receptors, membranes, coagulant/anticoagulant pathways. Our experiments involved injections to radiation-naïve animals of toxic doses of Radiation Toxins (SDR- group) which damaged endothelial cells of blood and lymphatic vessels, and injury to blood microcirculation in the brain, lungs, gastrointestinal tract, cardiovascular system. Administration (IV or IM) of the Haemotopoietic Radiation Toxins to radiation-naïve animals induced development of anemia with reduction of hemoglobin concentration, reduction of red blood cell count and reduction of white cell blood count. Hemotopoietic Radiation Toxins induced the development of hemolytic anaemia with accelerated destruction of mature red cells outside the bone marrow and destruction of pluripotent stem cells inside of the bone marrow. Radiation-induced hemolytic anemia occurred with intravascular and extravascular hemolysis. As a result of stem cell injury erythropenia, leukopenia, thropmbocytopenia developed within days after irradiation. However, the acute hemolytic extravascular and intravascular anemia with lysis of erythrocytes, developed after high doses of radiation.

4.6 Development of radiation countermeasures

There are many sources of oxidative stress in the lives of workers, whether they work in nuclear power facilities, on the front lines of international conflicts, or in the reaches of outer space. The exposure dose can vary substantially, but at minimum will accelerate the aging of their organ systems, and at worse could result in acute exposure syndromes that may be fatal. A common thread of the oxidative stress exposures is ROS-binding to critical cellular organelles and molecules, which can result in cellular dysfunction, mutation of nucleic acids, or even apoptotic cell death [107]. Currently there are no proven countermeasures for these exposures, aside from a clinical agent, Amifostine, (Etyhol™), which is used to reduce mucositis and other side effects from radiation therapy dose in cancer patients.

The cytotoxic effects of different types of radiation may be the single most important clinicopathologic process by which oxidative damage is induced from reactive oxygen species and radiation toxicity induced by radiation toxins. Radiation toxins (SRDs) with high enzymatic activity and their ability to degraded a wide variety of extracellular proteins, lipids, carbohydrates and DNA molecules, induce damage of important intracellular compartments such as mitochondria, ion channels, DNA, as well as activating degradation of peptide bonds in important polypeptides in tissues and vascular endothelium. Yet the exact mechanism by which radiation toxins stimulate development of the ARS is poorly understood.

Countermeasures against radiation toxicity can be developed both in oral formulas [109] and parenteral agents, e.g. Manganese SuperOxide Dismutase (MnSOD)-liposomes [108, 109] which can reduce radiation exposure-induced biological effects. The oral formulas consist of a combination of antioxidants and chemopreventive, anti-progression agents, which can reduce the likelihood of mild toxicity in acute low dose radiation exposures and tumor induction in chronic low dose exposures. The combination of oral formulas with higher dose reactive oxygen species scavengers, like Mn-SOD can improve survival in animals exposed to acute high dose total body irradiation [109]. In addition to these measures, active immunization by low, non-toxic doses of radiation toxins, (SRDs) can also be employed to reduce radiation toxicity. [104, 110] To be effective, SRD immunization must occur no less than 30 days before irradiation, and can be effective up to three years or more. Active immunization by radiation toxins can significantly improve the survival rate (up to 60%) versus placebo-controlled irradiated animals. Our studies attempt to show the potential ability of specific antibodies to neutralize radiation toxins and thus substantially reduce the effects on radiation-induced neuro-, vascular, gastrointestinal, and hematopoietic toxicity. [104, 110] Antiradiation antibodies prevent the radiation-induced cytolysis of selected groups of cells that are sensitive to radiation. Anti-radiation antibodies derived from different phases of the Acute Radiation Syndrome can compete with and thus prevent cytolysis mediated by cytotoxic lymphocytes. The therapeutic benefit of neutralization of SRD radiation toxins could make hemopoetic stem cell transplantation more effective. Antiradiation vaccine and IgG antibodies have shown activity in animals against several different types of radiation include gamma, heavy ions, and neutron irradiation.[104, 110]

Preliminary and pilot studies in vitro, in animal models and recently in humans, are showing some promise for both efficacy and safety/tolerability. [108-110] With current unpredictability in the operation of nuclear power facilities, potential radio-biologic terrorist weapons, and human operations in extreme environments, it is important to develop countermeasures to radiation toxicity in order to protect the potentially exposed. The first step in developing these countermeasures is to fully understanding the mechanism of toxicity, as is attempted to be summarized in this chapter.

5. Acknowledgements

Carlos Montesinos, Kedar Prasad, Michael Epperly, Joel Greenberger.

6. References

[1] Akadi Y, et al. Radiation-induced Apoptosis and Necrosis in Molt-4 Cells: A Study of Dose-effect Ralathionships and Their Modification. International Journal of Radiation Biology. 1993, V.64, No 1, pp. 47-56.

[2] Azizova T.V. et al. Multi-organ Involvement and failure in selected accident cases with acute radiation syndrome observed at the Mayak Nuclear Facility. The British Institute of Radiology.

[3] Kumar A., Leeds N., Gregory N. et al. Malignant Gliomas: MR Imaging Spectrum of Radiation Therapy- and Chemotherapy-induced Necrosis of the Brain after Treatment.

[4] Ahmed, A. et al. Effect of Exposure to Electromagnetic Radiation from the Mobile Phone on Acetylcholinestrase Activity in the Hippocampus and Striatum of Young and Adult Male Rats , Med. J. Cairo Univ., 2006, 74(2),(Suppl.), 129–135.

[5] Berger M. and al. Medical management of radiation injuries: current approaches. Journal of Occupational Medicine. 2006. V.56. pp.162-172.

[6] Tilmisanuy A., Osman A. Snake, Scorpion and Envenomation: A Review. JKAU: Med. Sci., Vol.3, pp. 11- 18.

[7] Briggs S., Brinsfield K. Advanced Disaster Medical response Manual for Providers. Harvard Medical International, Boston, Ma.2003,

[8] Blakely W. Early biodosimetry response: recommendations for mass-casualty radiation accidents and terrorism. Armed Forces Radiobiology Research Institute, Uniformed Services University, Bethesda. pp. 1-26.

[9] Wong C.,Van der Kogel A. Mechanisms of Radiation injury to the Central Nervous System: Indications for Neuroprotection. Molecular Interventions. 10. Volume 4, Issue 5, 273-284, 2004.

[10] Casarez E. Basic principles of toxicology. BIOC, 2001.p.597.Centers For Disease Control and Prevention. Acute Radiation Syndrome: A fact Sheet for Physicians. June 29, 2005. www.bt.cdc.gov/radiation

[11] Chow, B., Li Y. and Wong, C. Radiation-induced apoptosis in the adult central nervous system is p53-dependent. Cell Death and Differentiation 7, 712–720 (2000).

[12] Sears C, Kaper J. Enteric Bacterial Toxins: Mechanisms of Action and Linkage to Intestinal Secretion. Microbiological Reviews, Mar. 1996, p.167-215.

[13] Netto D., Chiacchio S. et al. Humoral response and neutralization capacity of sheep serum inoculated with natural and cobalt 60 – irradiated Crotalus durissus terrificus venom.LAURETTI, 1968. Journal of Venomous Animals and Toxins. Vol. 8, No.2, 2002.

[14] Daniak N, Ricks R. The evolving role of haemotopoetic cell transplantation in radiation injury: potential and limitations. The British Institute of Radiology, 2005. Pp. 169-174.

[15] Dewey W et al. Radiation induced apoptosis: relevance to radiotherapy. International Journal Radiation Oncology Biology Physics. 1995, No 33, pp. 781-796.

[16] Dunlacht J., et al. Degradation of the Nuclear matrix Is a common Element during Radiation-Induced Apoptosis and Necrosis. Radiation Research, 1999, v.152. No 6, pp.590 - 603.

[17] Rio D., Pardo V. Monoamine neurotoxins-induced apoptosis in lymphocytes by a common oxidative stress mechanism: involvement of hydrogen peroxide

(H(2)O(2)) caspase-3, and nuclear factor kappa-B (NF- kappa B), P53, c-Jun transcription factors. Biochem Pharmacology Journal, 2002, 63, pp. 677-688.

[18] Fliedner T. et al. Multi-organ involvement as a pathogenic principle of the radiation syndromes: a study involving 110 cases histories documented in SEARCH and classified as the bases of hematopoietic indicators of effect. The British Institute of Radiology, 2005, p.1-6.

[19] Kalm M. Inflammation and behavior following irradiation-induced injury in the developing brain. Center for Brain Repair and Rehabilitation, Institute of Neuroscience and Physiology, Sahlgrenska Academy, University of Gothenburg, Sweden, 2009.

[20] Kleining T, Vink. R. Suppression of inflammation in ischemic and hemorrhagic stroke: therapeutic options. Current Opinion in neurology. 2009, No 22, pp. 294-301.

[21] Konchalovsky M. et al. Multiple Organ Involvement and failure selected Russian radiation accident cases re-visited. The British Institute of radiology. 2005,pp.26-29.

[22] Swarts S. et al. Radiation-induced cyto-toxicity, DNA damage and DNA repair: implications for cell survival theory. Radiation Environmental Biophysics. 1990, 29 (2), pp.93 - 102.

[23] Rakesh D. et al. Activation of a CrmA-insensitive, p35-sensitive pathway in Ionizing Radiation-induced Apoptosis. The American Society for Biochemistry and Molecular Biology. 1997, v.272, No. 3, pp. 1965 - 1969.

[24] Fink S., Cookson B. Apoptosis, Pyroptosis and Necrosis: Mechanic Description of Dead and Dying Eukaryotic Cells. Infection and Immunity. Apr.2005, p.1907 - 1916.

[25] Schultheiss, T., Kun, L., Ang, K. and Stephens, L. Radiation response of the central nervous system. *Int. J. Radiat. Oncol. Biol. Phys.* 31, 1093–1112 (1995).

[26] Tofilon, P.J. and Fike, J.R. The radioresponse of the central nervous system: A dynamic process. *Radiation Research,* 153, 357–370 (2000).

[27] Temple M., O'Leary D., Faden A. The role of glutamate receptors in the pathophysiology of traumatic CNS injury. Chapter 4 in Head Trauma: Basic, Preclinical, and Clinical Directions. Miller LP and Hayes RL, editors. Co-edited by Newcomb JK. John Wiley and Sons, Inc. New York. 2001. Pages 87-113.

[28] Trot K. Radiation Risks from Imaging of Intestinal and Abdominal Inflammation. Scandinavian Journal of Gastroenterology. 1994, vol. 29, No 203, pp. 43-47.

[29] Topchieva S., Mehrabova M., Influence of electromagnetic radiation on venom of Vipera Lebetina Obtuse. Radioprotection, 2008, vol. 43, n. 5.

[30] Van Der Kogel, A. Radiation-induced damage in the central nervous system: An interpretation of target cell responses. *Br. J. Cancer Suppl.* 7, 207–217 (1986).

[31] Hungerford J. Committee on Natural Toxins and Food Allergens. General Referee Reports: Journal of AOAC International. Vol., 89,No.1, 2006.

[32] Rudbach J., et al. Preparation and Properties of a National Reference Endotoxin. Journal of Clinical Microbiology. Jan. 1976. P.21-25.

[33] Luvisetto S. et al. Toxicity of botulinum neurotoxins in central nervous system of mice. J.Toxicon, 2003, March, 41(4), pp.475 - 481.

[34] Levin S. Toxicological highlights: Apoptosis, Necrosis, or Oncosis: What is your Diagnosis? A Report from the Cell Death Nomenclature Committee of the Society of Toxicologic Pathologists. Toxicological sciences. 1998, 41, 155 -156.

[35] MacCann D. Radiation poisoning: Current concepts in the Acute Radiation Syndrome. American Journal of Clinical Meidcine. V.3, No3. 2006, pp. 13-21.
[36] Ramirez-Dominguez M, Olamendi T. et al. Cn11, the first example of a scorpion toxin that a true blocker of Na+ currents in crayfish neurons. The Journal of Experimental Biology, 205, 869-876(2002).
[37] Molla M, Panes J. Radiation-induced intestinal inflammation. World Journal of Gastroenterology, 2007, No 13, pp. 3043-3046.
[38] Guido M. and Isabelle J. Apoptosis, Oncosis, and Necrosis . An Overview of Cell Death. American Journal of Pathology. 1995, v.146, No 1.
[39] Ozsahin M. et al., CD4 and CD8 T-lymphocyte Apoptosis Can Predict Radiation-Induced Late Toxicity: A Prospective Study in 399 Patients. Clinical Cancer Research, 2005, 11(20),October 15, 2005.
[40] Xue M. at al. Intracortical Hemorrhage Injury in Rats : Relationship Between Blood Fractions and Brain Cell Death. Journal of American Heart Association, 2000, No.31, pp.1721-1727.
[41] Nieder C. et al. Experimental concepts for toxicity prevention and tissue restoration after central nervous system irradiation. Radiation Oncology. 2007, No 2, p.23.
[42] Sagara-Ishijima N., Furuhama K. Toxic characteristics of the Synthetic Lipid A Derivative DT-5461 in Rats and Monkeys. Toxicological Sciences, 1999 ,49, 324-331.
[43] Nieder C., Andratschke N., Price R., Rivera, B., and Ang, K. Innovative prevention strategies for radiation necrosis of the central nervous system. Anticancer Research, 22,1017–1023 .
[44] Gilles N., Blanchet C. et al. A Scorpion α-Like Toxin That Is Active on Insects and Mammals Reveals an Unexpected Specificity and Distribution of Sodium Channel Subtypes in Rat Brain Neurons. The Journal of Neuroscience, October 15, 1999.
[45] Vladimiriov A. Biological membranes and non-programmed cell death. Soros educational Journal. V.6, No 9, 2000, p.2-9.
[46] Wasalenko J, McVittie T et al. Medical Management of the Acute Radiation Syndrome: Recommendations of the Strategic National Stockpile Radiation Working Group. 2004. Ann Int Med. No. 140, pp.1037-1051.
[47] Bonner W. Low-dose radiation: Thresholds, bystander effects, and adaptive responses. Proceeding of the national Academy of Sciences of United States of America. April 29, 2003, vol.10, no. 9, pp. 4973-4975.
[48] UenoY . Toxicology of microbial toxins. Pure & Appl. Chem., Vol.58. No.2, pp.339-350. 1986.
[49] Yu-Cheng X, Song-Ping L. Inhibition of Sodium Channels In Rat Dorsal Root Ganglion Neurons by Hainantoxin- IV, a novel Spider Toxin. Acta Biochimica et Biophysica Sinica, 2003, 35(1).
[50] Watters D. Molecular mechanisms of ionizing radiation-induced apoptosis. Immunology and Cell Biology (1999) 77, 263–271.
[51] Haimovitz-Friedman A. Radiation-induced signal transduction and stress response. Radiat. Res. 1998; 150: S102–8.
[52] Szumiel I. Monitoring and signaling of radiation-induced damage in mammalian cells. Radiat. Res. 1998; 150: S92–101.

[53] Kim Y, M.D. Radiation-induced Necrosis Deteriorating Neurological Symptoms and Mimicking Progression of Brain Metastasis after Stereotactic-guided Radiotherapy. Cancer Res Treat. 2007 March; 39(1): 16–21.

[54] Waterhouse, N.J. et al. (2003) Mitochondrial outer-membrane permeabilization in apoptosis.In: Cell Proliferation and Apoptosis. Hughes, D. and Mehmet, H., eds. 185–200. (Bios Scientific Publishers, Ltd. Oxford).

[55] Verheij M. et al. Radiation-induced apoptosis. Cell Tissue Res. 2000, Jul; 301 (1): 133-42.

[56] Rajesh P. et al. Apoptosis: molecular mechanisms and pathogenicity. EXCLI Journal 2009;8:155-181 ISSN 1611-2156

[57] Robertson J. and Orrenius S. Molecular Mechanisms of Apoptosis Induced by Cytotoxic Chemicals. Critical Reviews in Toxicology, 30(5):609–627 (2000).

[58] Riedl S. et al.Molecular mechanisms of caspase regulations during apoptosis. Molecular Cell Biology. 2004, v.5. p.897-907.

[59] Higuchi M. et al. Regulation of reactive oxygen species-induced apoptosis and necrosis by caspase 3-like proteases. Oncogene. 26 November 1998, Volume 17, Number 21, Pages 2753-2760.

[60] Tan P, Khan A., Brusic V. Bioinformatics for Venom and Toxins Science. Henry Stewart Publications. 1467-5463. Briefings in Bioinformatics. Vol. 4. No 1. 53–62. March 2003.

[61] Ji-Sheng C.. An Assessment of Toxins. Research Institute of Chemical Defense.P.O. Box 104. Beijing,102205, China.

[62] Kathryn D. Held. Radiation Induced Apoptosis. Massachusetts General Hospital. Harvard medical School.

[63] Wirth B. and Olander D. Fundamentals of Radiation Damage and Defects Production. US Berkeley, Nuclear Engineering. Ref: Sect 17.2, D.O. text.

[64] Molla M., Panes J. Radiation induced intestinal inflammation. World Journal of Gastroenterology. 2007. Pp.3043-3046.

[65] Molla M. et al. Influence of dose-rate on inflammatory damage and adhesion molecule expression after abdominal radiation in the rat. International Journal of Radiation Oncology and Biological Physics. 1999.45. pp.1011-1018.

[66] Linard C., et al. Acute induction of inflammatory cytokine expression after gamma-irradiation in the rat: effect of an NF-kappa B inhibitor. International Journal of Radiation Oncology and Biological Physics. 2004.58. pp.427-434.

[67] Ramanan S. et al. Role of PPARs in Radiation-Induced Brain Injury. PPAR Research. V.2010. article ID 234975. 12 pages.

[68] Schiller T., Chen Y. Modeling radiation-induced lung injury risk with an ensemble of support vector machines. Journal Neurocomputing , Volume 73. Issue 10-12, June, 2010.

[69] Baum S. et al. Radiation Induced Anemia in Rats Exposed repeatedly to mixed Gamma-Neutron radiation. Radiation Research.1970. 41. pp.492-499.

[70] Dritshilo A. Sherman D. Radiation and Chemical Injury in the Bone Marrow. Environment Health Prospectives. 1981, v.39, pp. 59-64.

[71] Wiebe E., and Rodrigues G. Radiation Induced Lung Injury. Strategies for reducing damages while optimizing therapeutic doses. 2006. Parchurst. Publisher for Oncology Exchange. V.5. No 2. pp.29-31.

[72] Wei-Chung L. et al. Protection against Radiation-Induced Bone Marrow and Intestinal Injuries by *Cordyceps sinensis,* a Chinese Herbal Medicine. Radiation Research, 2006. 166, 900–907.
[73] Wharton T. et al. Radiation Hepatitis induced by abdominal irradiation with the cobalt 60 moving strip technique. Presented as the Fifty-fourth Annual Meeting of the American Radium Society. 1973. pp.53-80.
[74] Elinger A., Thomson C. death by Design: Death by Design : Apoptosis, Necrosis and Autophagy. Current Opinion in Cell Biology. 2004, pp. 653-669.
[75] Dispersyn G. et al. Dissociation of cardiomyocite apoptosis and differentiation in infarct border zones. Europenean Heart Journal. 2002. 23. pp.849-857.
[76] Ohgushi M. et al. Molecular Pathway and Cell State Responsible for Dissociation-Induced Apoptosis in Human Pluripotent Stem Cells. Cell Stem Cell 7, 225–239, August 6, 2010 Elsevier Inc.
[77] Claudio A. et al. Caspase-dependent initiation of apoptosis and necrosis by the Fas receptor in lymphoid cells: onset of necrosis is associated with delayed ceramide increase. Journal of Cell Science 115, 4671-4683 © 2002 The Company of Biologists Ltd.
[78] Rastogi P. et al. Apoptosis: Molecular Mechanisms and Pathogenicity. EXCLI Journal 2009;8:155-181 ISSN 1611-2156. Received: June 29, 2009, accepted: July 27, 2009, published: August 03, 2009.
[79] Barnett N, editor. Biosecurity Reference: CFR-Listed Agent and Toxin Summaries. pp.1-150. Prepared by Sandia National Laboratory, Albuquerque, New Mexico 87185 and Livermore, California 94550.
[80] Simon J. Nuclear, biological, and chemical terrorism: Understanding the threat and designing responses. Int J Emergency Mental health 1999; 1: 81-89.
[81] Franz D, DVM, PhD, Colonel, Veterinary Corps. United States Army. Defense against Toxin Weapons. U.S. Army Medical Research and Materiel Command, pp. 1-36.
[82] Shivaji P. Snake venom neurotoxins: Pharmacological classification. Journal of toxicology. Toxin reviews. 2004, vol. 23, n°1, pp. 37-96 [60 page(s).
[83] Clemetson K, Morita T., Kini M. Scientific and standardization committee communications: classification and nomenclature of snake venom C-type lectins and related proteins. Posted on ISTH website,November 19, 2008.
[84] Viljorn C., Botes B. Snake Venom Toxins. The Journal of Biological Chemistry. Vol. 249, No. 2, Issue of January 25, pp. 366-372, 1974.
[85] Oren B., Ph.D. Venom Toxins as Ion Channel Research Tools. Modulator No.21 Fall 2006. 28-31. www.alomone.com
[86] Alouf, J. et al. 1991. The family of mitogenic, shock-inducing and superantigenic toxins from staphylococci and streptococci, p. 367–414. *In* J. E. Alouf and J. H. Freer (ed.), Sourcebook of bacterial protein toxins. Academic Press Ltd., London.
[87] Fiorentini, C.,et al. 1995. Escherichia coli cytotoxic necrotizing factor 1: evidence for induction of actin assembly by constitutive activation of the p21 Rho GTPase. Infect. Immun. 63:3936–3944.
[88] Gariepy, J., et al. 1987. Importance of disulfide bridges in the structure and activity of *Escherichia* coli enterotoxin ST1b. Proc. Natl. Acad. Sci. USA 84:8907–8911.

[89] Popov D., Maliev S. Antiradiation Vaccine: Immunological neutralization of Radiation Toxins at Acute Radiation Syndromes. 38th COSPAR Scientific Assembly. Held 18-15 July 2010, in Bremen, Germany, p.3.

[90] Popov D., Maliev S. Radiation Toxins: Molecular mechanisms and radiomimetic properties. 38th COSPAR Scientific Assembly. Held 18-15 July 2010, in Bremen, Germany, p.3.

[91] Popov D., Maliev S. Countermeasure development : Specific Immunoprophylaxis and Immunotherapy of Combined Acute Radiation Syndromes. 38th COSPAR Scientific Assembly. Held 18-15 July 2010, in Bremen, Germany, p.3.

[92] Jones, J. A., Riggs, P. K., Yang, T., Pedemonte, C. H., Clarke, M, Feedback, D., and Au, W. Ionizing radiation-induced bioeffects in space and strategies to reduce cellular injury and carcinogenesis ASEM 78(4) Section II suppl., pp. A67-78, 2007.

[93] Koniarek J, Thomas J., Vazquez V. Detection of microlesions induced by heavy ions using liposomes filled with fluorescent dye. Advances in Space Research , Volume 34, Issue 6, 2004, Pages 1373-1377. Space Life Sciences: Radiation Risk Assessment and Radiation Measurements in Low Earth Orbit.

[94] Koniarek J., Worgul B., Do heavy ions cause microlesions in cell membranes? Advances in Space Research, Volume 12, Issues 2-3, 1992, Pages 417-420.

[95] Zhivotovsky B., Samali, A., Gahm, A., and Orrenius, S Caspases: their intracellular localization and translocation during apoptosis. Cell Death Differ., 1999: 6, 644-651.

[96] Zhivotovsky, B., and Kroemer, G. Apoptosis and genomic instability. Nature Rev. Mol. Cell Biol. 2004: 5, 752-762.

[97] Sorokina N.I. et al. Molecular mechanisms of radiation death of lymphoid cells. Protein poly-ADP-ribosylation in thymocyte fractions. Radiobiologiia. ;23 (6):739-42.

[98] Filippovich I., Sorokina N. , Soldatenkov V., Alfyerova T., Trebenok Z. Effect of the inducers of cellular differentiation and of ionizing radiation on thymus lymphocytes: chromatin degradation and programmed cell death. Int J Radiat Biol Relat Stud Phys Chem Med. 1988 Apr. 53 (4):617-28.

[99] Bebeshko V et al. Biological markers of ionizing radiation. Atomic Medicine. N1 , (39) – I/II, 2004. pp.85 -104.

[100] Popov D., Kirchin V., Maliev V., Zlobin V. All Union conference of agriculture radiobiology (USSR). Obninsk, 1990. P. 136–137.

[101] Popov D., Kirchin V., Saphonova V, Usupova S. Republican Scientific Conference "Theoretical and practical questions of veterinary medicine" (USSR). Kazan, 1990. P. 34.

[102] Gel Filtration, Principles and Methods Handbook, USA. 181-0221-8, Amersham, Biosciences.

[103] Maliev V., Kirchin V., Popov D. Antiradiation vaccine. Herald of VSC of Russian Academy of Science. 2003. № 3. P. 12–28.

[104] Maliev V., Popov D., Casey R, Jones J. Mechanisms of Action for an Anti-Radiation Vaccine in reducing the biological impact of high dose and dose-rate, low-linear energy transfer Radiation Exposure. J. Radiation Biology, v.47.N3. pp. 286-291.

[105] Jaiswal M, Zech W, Goos M., Leutbecher C, Ferri A, Zippelius A, Carrì M, Nau R, Keller BU (2009). "Impairment of mitochondrial calcium handling in a mtSOD1 cell culture model of motoneuron disease". BMC Neuroscience 10: 64.

[106] Falkow S. Molecular Koch's postulates applied to bacterial pathogenicity--a personal recollection 15 years later. Nat Rev Microbiol. 2004 Jan; 2(1):67-72.

[107] Jones J, Karouia F, Casey R. "Ionizing Radiation as a Carcinogen". In: McQueen C, ed. Comprehensive Toxicology, Volume 14: Carcinogenesis. Oxford EN: Elsevier 2010

[108] Epperly, et al Modulation of Total Body Irradiation Induced Life Shortening by Systemic Intravenous MnSOD-Plasmid Liposome Gene Therapy. Radiation Research, 2008 170(4):437-444

[109] Epperly, M.W., Wang, H., Jones, J., Dixon, T., Montesinos, C.A., and Greenberger, J.S. Antioxidant-Chemoprevention Diet Ameliorates Late Effects of Total Body Irradiation and Supplements Radioprotection by MnSOD-Plasmid Liposome Administration. Radiat. Res. 2011 June 175(6):759-765

[110] Maliev, V, Popov,D, Jones, JA, Casey, RC. Mechanism of action for anti-radiation vaccine in reducing the biological impact of high-dose irradiation Advances in Space Research 2007 40:586-590

[111] Baum S. Erythrocyte stem cell kinetics in postradiation rat. Armed Forces Radiobiology Research Institute, Defense Atomic Support Agency, Bethesda, Maryland. May. 1966.

[112] Mihai R., Doina G. Membrane structural and functional modifications due to oxidative stress induction by irradiation and chemical agents. European Commission Center of Excellence, InterDisciplinary research and Applications based on Nuclear and Atomic Physics. Report WP5 ID RAHAP 25-02/2002. pp. 1-8.

[113] Benderitter M. et al. The cell membrane as a biosensor of oxidative stress induced by radiation exposure: a multiparameter investigation.Radiation Research: vol.159, No 4, pp.471-483.

[114] Dixon D., Darveau R. Lipopolysaccharide Heterogenity: Innate Host Responses to Bacterial Modificationof Lipid A Structure. J. DenRes 84(7): 584-595, 2005.

[115] Moroni M, Coolbaugh T., Lombardini E, Mitchell J, Moccia K, Shelton L, Nagy V, Whitnall M. Hematopoietic radiation syndrome in the Gottingen minipig. Radiat Res. 2011 Jul;176(1): 89-101. Epub 2011 Apr 26.

[116] Kudriashov I., Goncharenko E, Deev L, Gorskaia T, Samoĭlikova T. Depression of endogenous radiation-sensitizing agents--lipid toxic substances--as one of the mechanisms of radiation protection. Dokl Akad Nauk SSSR. 1970; 195(1):206-8.

[117] Smirnov D., Brady L., Halasa K, Morley M, Solomon S. and Cheun V.. Genetic variation in radiation induced cell death. Genome Research. August 15. 2011.

[118] Sanchez M., Benitez A., Ortloff L., Green L. Alterations in Glutamate Uptake in NT2-Derived Neurons and Astrocytes after Exposure to Gamma Radiation. Radiation Research. 2009 January; 171(1): 41–52.

13

Dynamical Aspects of Apoptosis

Antonio Bensussen and José Díaz
Theoretical and Computational Biology Group, Faculty of Science, Autonomous University of Morelos, Cuernavaca, Morelos, Mexico

1. Introduction

Cells are dynamical systems characterized by a high degree of complexity due to an intricate network of subcellular processes that sustain their correct performance. This network is formed by a series of subnetworks with modular functioning, and among them is the subnetwork of biochemical processes that produces cell death or apoptosis. In this chapter we deal with the dynamical behavior of the apoptotic pathway activated by ionizing radiation (IR). In the special case of the apoptotic process, mathematical modeling is required for understanding its nonlinear complex behavior. Although previous attempts to model apoptosis were often limited to small systems, or based on qualitative data only, recent models are based on experimental data that give support to the theoretical results.
In this form, mathematical models have shown that the interactions between the nuclear molecular components of the intrinsic apoptotic pathway drive the system into a persistent limit cycle in the respective phase space after the application of high doses of ionizing radiation. Experimental data confirm the existence of this limit cycle, showing that this dynamical feature of the system relies on the existence of a p53-Mdm2 negative feedback loop, which is induced by a switch-like activation of the ATM kinase.
At the cytoplasm, the molecular dynamics that leads to the activation of the caspases sets up a saddle point bifurcation in the *caspase 3-Bax* degradation rate diagram. In this region of bistability, the saddle point bifurcation can drive the system either to a stable point with a high concentration of caspase 3 (cell death) or to a stable point with a low concentration of caspase 3 (cell survival). Out of this region, the system dynamics is settled on by the existence of a single stable fix point, which determines cell surviving independently of the concentration of caspase 3, i.e., there exits a threshold mechanism that regulates apoptosis and it is based on bistability, ultrasensitivity and irreversibility.
Thus, the main goal of this chapter is to review and analyze the recent knowledge about these nonlinear dynamical aspects of IR-induced apoptosis, in order to understand how this complex subnetwork of molecular interactions integrates the information coded in the signals that control the cell survival, and executes the corresponding action to determine the correct death-life decision. The modeling of such decisions has a significant impact on the field of radiation biology and in the therapeutic aspects of inducing apoptosis in cancer cells by IR.

2. Cell cycle and apoptosis

In recent decades cancer research has gained great importance worldwide, not only for being one of the most important chronic diseases of mankind, but also by the great complexity of cellular process that it entails. A deeper insight into the causes of cancer progression, involves further research on the cell cycle because the malignant transformation starts as a deregulation of this cycle. In cancer, serious errors can be observed in DNA replication and repair simultaneously with an uncontrolled cell growth. Cancer cells lack the ability to respond to internal or external stimuli that induces death (Malumbres and Barbacid, 2009). It has seen that between 50 and 70% of all cancer types are related to the absence of the transcriptional factor p53 (Vousden, 2006), which serves to activate the cellular repairing response to DNA damage.

2.1 Regulation of the cell cycle progression

The cell cycle has four phases. During the first phase (G1) the cell acquires the necessary nutrients for replication of the genetic material. If external signals are received, or the cell is ready to continue, then Cdc25 is activated (Budirahardja and Gönczy, 2009). Cdc25 activates Cdk4-cyclin-D and Cdk6-cyclin-D heterodimers (Malumbres, 2009). These heterodimers phosphorylate the Retinoblastoma protein (pRb) inducing the release of the E2F group of transcription factors (Abukhdeir and Park, 2009). These factors are necessary to transcript genes that commit the cell to pass from G1 to S phase (Abukhdeir and Park, 2009) and to activate the first checkpoint, where is evaluated the genetic material integrity (Abukhdeir and Park, 2009). This test is performed by some specialized proteins such as ATM kinase (Ataxia-Telangiectasia Mutated), as well as some DNA binding proteins like the MRN complex (Pommier and Weinstein, 2006). If these proteins detect DNA damage, they induce cell cycle arrest through effector molecules such as Chk2, and p21Waf1/cip1 (Abukhdeir and Park, 2009). If the damage is severe, apoptosis is activated through the p53 pathway. If not damage in the DNA is detected; the cell induces degradation of cyclin D, activating the Cdk2-cyclin-E heterodimer, which is necessary to start the S phase (Malumbres and Barbacid, 2009).

The checkpoint commanded by ATM remains active during the S phase, in order to ensure the integrity of cell genome (Budirahardja and Gönczy, 2009). At the end of the S phase, the Cdk2-cycline-A homodimer is activated to conclude the S phase, and the cell enters into G2 phase (Malumbres and Barbacid, 2009). After the cell has entered G2 phase, the activation of Cdk1-cyclin-B and Cdk1-cyclin-A heterodimers, in response to a mitogenic stimuli, starts M phase (Mitosis) (Pommier and Weinstein, 2006). It is noteworthy that the DNA damage sensors commanded by ATM are both active during transition from G2 to M and during the M phase (Calonge and O'Connell, 2008). After cytokinesis, cell can either enter to G1 phase or go out of cell cycle and remain in G0 state indefinitely (Calonge and O'Connell, 2008).

2.2 p53 network in stress-free conditions

The p53 transcription factor essentially serves to regulate the expression of genes responsible for the cell cycle arrest, repair of DNA damage, senescence and apoptosis in the intrinsic pathway (Lacroix et al., 2006). It is known that expression of the gene encoding p53 is dependent of the p50 subunit of NF-kB transcription factor, and of other transcription factors like C/EPEβ-2, Ets-1, Pitx1, p73 and p53 itself (Baillat et al., 2006; Wang et al., 2006; Boggs and Reisman, 2007; Wu and Lozano, 1994). In stress-free human fibroblasts, p53 interacts with the anti apoptotic BCL-XL protein in the cytoplasm, and forms a heterodimer (Dogu & Díaz, 2009), and the remaining fraction of p53 is marked in the carboxyl-terminus with a nuclear import signal (NLS1), and is subsequently transported to the nucleus, where it will interact with the ubiquitin ligase Mdm2 (Hdm2 in humans). Mdm2 binds to p53, forming a complex that eventually will be degraded in the proteosome (Jänicke et al., 2008). It is noteworthy that at the basal level the Mdm2 gene promoter is activated by transcription factors other than p53, which have not yet been identified (Phelps et al., 2003). It is known that the *Mdm2* gene has two promoters (P1 and P2) that can lead to at least two isoforms of Mdm2 (Candeias et al., 2008). The p90Mdm2 isoform is responsible of the inhibition of *p53*, and the p76Mdm2 isoform is a p53 activator (Perry, 2004) and promotes the translation of p53 mRNA (Candeias et al., 2008). Under basal conditions p76Mdm2 expression is greater than p90Mdm2 (Phelps et al., 2004), and this fact could explain the origin of the basal levels of p53 throughout the cell cycle (Figure 1a). However, under cellular stress conditions, p53 induces the expression of p90Mdm2 in detriment of the p76Mdm2 isoform (Perry, 2004).

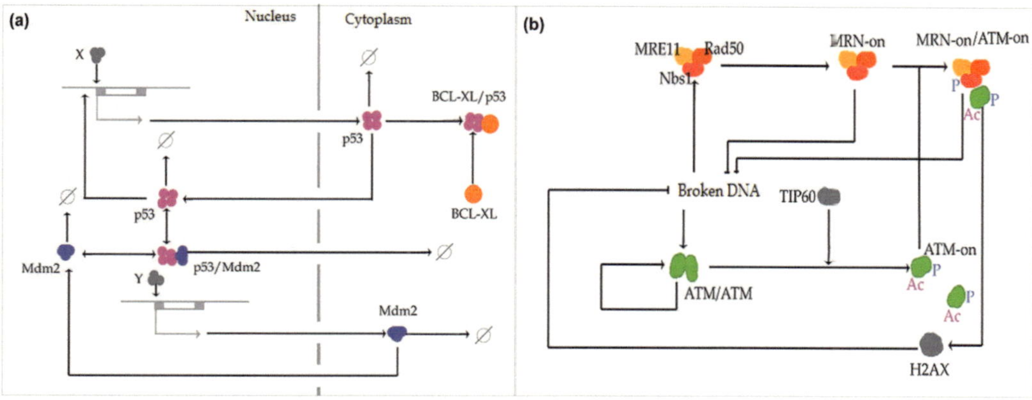

Fig. 1. (a) **p53 network at basal conditions**. p53 and other transcription factors, some of them no yet identified and marked as X, activates *p53* gene. Part of the novel p53 synthesized interacts with BCL-XL, forming heterodimers that will remain in the cytoplasm. The other part of p53 molecules is marked with an importation signal, enters the nucleus, and interacts with p90Mdm2 (represented by "Mdm2") forming a complex that degraded in the proteosome. "Y" represents a transcription factor set that activates p90Mdm2 in absence of genotoxic stress, and the empty set symbol represents protein degradation. (b) **ATM activation network in response to DNA damage.** DNA damage causes a nuclear topology change that induces the self-phosphorylation of ATM homodimers (represented by ATM/ATM), allowing its separation into active monomers (represented by ATM-on). The ATM activation is full after that TIP60 acetylates it. In the other side, if MRN complex (forming by Mre11, Rad50 and Nds1) detects broken DNA, it will be activated (represented by MRN-on) and then it will bind to broken DNA, in order to stop the damage propagation. After ATM activation, it will phosphorylate various substrates including Nbs1, and this permits ATM union to MRN-broken DNA complex, in order to start the DNA damage repair signaling pathway, through activation of effector proteins like H2AX.

2.3 Cellular response to genotoxic stress, and its different scenarios

The cell is able to respond to genotoxic stress produced by various endogenous and exogenous agents like ionizing radiation (IR), viral infection, heat, growth factor depravation and chemical DNA-damaging agents, among others (Plati et al., 2011). However, the mechanism and magnitude of the response will differ depending on both the intensity and nature of the aggression. In particular, if the cell undergoes genotoxic stress there are three possible scenarios, and multiple combinations of them. The first scenario is when the cell suffers minor levels of DNA damage and can repair it successfully. The second scenario is when the cell suffers a medium damage in its DNA, and the cell has to "decide" between life and death. In this scenario it is possible that the cell die for not being able to fully repair the damage after a finite period of time, or live for being able to repair it. The third scenario is when the cell suffers severe damage, and initiates the apoptotic death. For a cell can be able of executing one of these three different responses, DNA damage intensity must be first detected, and then, the cell must "make the choice" of the appropriate scenario.

2.4 Detection of DNA damage

In genotoxic stress conditions, in response to DNA damage, the change in the nuclear topology, together with histone acetyl transferase Tip60, and the MRN complex (consisting of Mre11, Rad50, and Nbs1), induces the activation of ATM kinase (Sun et al., 2005; Vissers et al., 2008; Tanaka et al., 2007) (Figure 1b). After its activation, ATM phosphorylates histone H2AX and other proteins with a BRCT domain (such as Nbs1, 53BP1, and MDC1) for their activation (Vissers et al., 2008; Pommier and Weinstein, 2006). These proteins will bind to the broken DNA to stop the spread of damage (Vissers et al., 2008). Similarly, ATM activates Chk2 kinase, partially responsible for promoting cell cycle arrest, because it induces phosphorylation of Cdc25A and Cdc25C (Yoda et al., 2008; Pommier and Weinstein, 2006). Chk2 also promotes the activity of the transcription factor E2F-1 (Yoda et al., 2008), which, in turn, regulates Chk2 and ASPP protein expression (Bergamaschi et al., 2004).

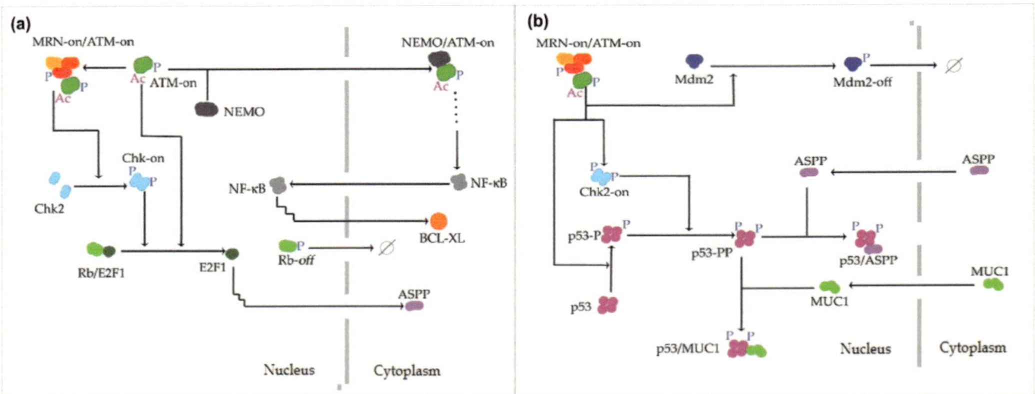

Fig. 2. (a) **Module of ATM effector functions.** When ATM is anchored to MRN-broken DNA complex, ATM phosphorylates Chk2 inactive monomers, to promote its binding into active homodimers. After that, free active ATM with Chk2 active homodimers will phosphorylate Rb protein to release E2F1. This transcription factor controls the pro-apoptotic protein synthesis, such as ASPP proteins. Free active ATM can bind to NEMO to induce NF-kB release in the cytoplasm. It is important because NF-kB is responsible to regulate the expression of proteins required to cell survival, like BCL-XL. In figure the stepped line represents the induction of protein transcription. (b) **The p53 activation by ATM and Chk2.** After the activation of ATM and Chk2, ATM inhibits Mdm2. Together with Chk2, ATM phosphorylates p53 in order to stabilize it. Phosphorylated p53 can bind either to MUC1 to promote the transcription of genes related with the cell cycle arrest and DNA damage repair, or can bind to ASPP proteins to promote the transcription of genes related with apoptosis.

In the nucleus, ATM interacts with NEMO forming a complex, which is ubiquitinated (Figure 2a), and transported to the cytoplasm where it induces the release of the transcription factor NF-kB (Schmid and Birbach, 2008). After that, NF-kB enters the nucleus and promotes the expression of various genes, including *BCL-XL* (Méndez et al., 2006; Chen et al., 2000). Simultaneously, ATM phosphorylates p53 nuclear stockpile, and, as a result, Mdm2 becomes less effective to recognize p53, delaying its degradation. Also, ATM phosphorylates Mdm2, inducing its autoubiquitination and degradation in the proteosome, and allowing an increase of p53 nuclear concentration over time (Dogu and Díaz, 2009). After ATM phosphorylation, Chk2 phosphorylates p53 in order to stabilize its structure (Pommier and Weinstein, 2006). At the end of this initial chain of phosphorylation, p53 can interact either with ASPP1 and ASPP2 proteins or with MUC1 (Pietsch et al., 2008). The MUC1 transmembrane glycoprotein is cleveaged in response to a genotoxic stimuli and the cytoplasmic segment is targeted to the nucleus, allowing its union with phosphorylated p53 and the induction of cell arrest (Pietsch et al., 2008). On the other hand, in response to severe DNA damage (Figure 2b), p53 will bind mainly with ASPP proteins. The concentration of the ASPP proteins increases gradually due to the activity of E2F-1 (Pietsch et al., 2008; Braithwaite, 2006), enhancing the pro-apoptotic function of p53. In this form, depending on the intensity of the stimulus, p53 can be oriented to transcribe pro-apoptotic genes or anti-apoptotic genes, depending on the accessory proteins to which it is attached.

2.4.1 Irreversible DNA damage, death by apoptosis

When DNA damage signal is either too strong (Figure 3a), or persistent, and if the cell has no energy to repair damage, p53 binds to ASPP proteins, and interacts with the histone acetyltransferase p300/CBP to form a complex (Patel et al., 2008). After that, p300/CBP acetylates p53 on lysine 373, allowing the selective recognization of the promoter sites of pro-apoptotic genes like *PUMA*, *BAX* and *BAK* by p53 (Pietsch et al., 2008). It is noteworthy that the formation of the p53-p300 complex is reversible, and p300 is separated from the complex by the removal of the acetyl group of p53 (Pediconi et al., 2009). The enzyme responsible for this step is the SIRT1 deacetylase (Tanno et al., 2006; Pediconi et al., 2009). SIRT1 is an enzyme that uses NAD+ as cofactor from cellular metabolism. However, in this scenario of severe DNA damage, the effect of SIRT1 is negligible, but will be important in the third scenario discussed below.

2.4.2 Soft DNA damage, cell cycle arrest and successful repair of damage

When the DNA damage signal is weak (Figure 3b), MUC1 nuclear concentration increases, and the union between p53 and nuclear MUC1 is enhaced. After that, the complex p53-MUC1 interacts with PCAF (p300/CBP associated factor) to form a new complex (Pietsch et al., 2008). PCAF acetylates p53 on lysine 320, allowing it to specifically recognize the promoters of genes related with the cell cycle arrest like p21WAF1/cip1, and DNA damage repair (Pietsch et al., 2008). Similarly, SIRT1 opposes PCAF-dependent phosphorylation of p53, blocking its transcriptional activity when the cell does not have energy to express proteins (Yamamori et al., 2009; Pediconi et al., 2009; Ikenoue et al., 2008). However, SIRT1 is not the only control point of p53. As discussed above, ATM, along with NEMO, activates NF-kB in its alternative pathway. This interaction is important, because NF-kB promotes the expression of anti-apoptotic genes like BCL-XL. NF-kB also competes with p53 for the cofactors: acetyltransferases p300, CBP and PCAF; which are required for the binding of p53 to DNA (Jänicke et al., 2008). The effect of this interaction leads to the indirect repression of the transcriptional activity of p53, given the fact that the concentration of the cofactors remains constant (Schmid and Birbach, 2008; Youle and Strasser, 2008).

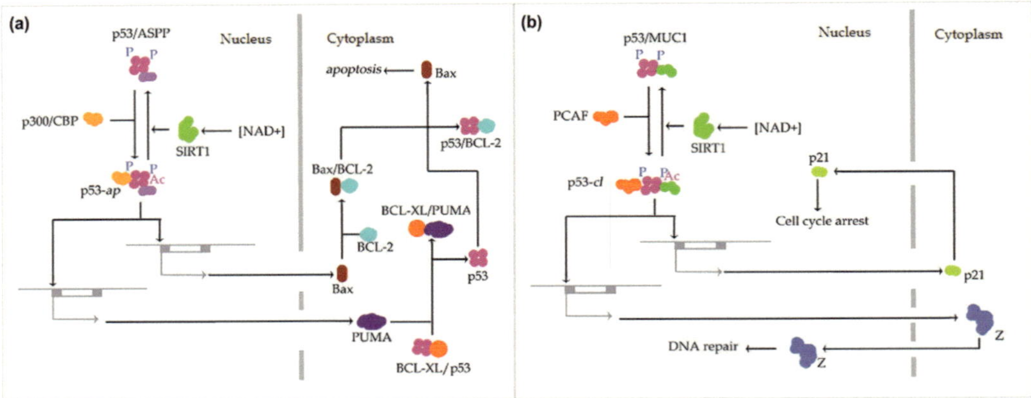

Fig. 3. (a) **The onset of apoptosis in its intrinsic pathway.** In response to severe DNA damage, p53 will bind to ASPP1 and ASPP2 (all of these are represented by ASPP). After that, p53-ASPP complex will interact with p300/CBP, and p300 acetylates p53-ASPP complex (represented by p53-ap). This allows p53-ap recognize the promoter side of apoptotic genes, like *PUMA* and *BAX*. The p53-ap transcriptional activity increases PUMA levels in the cytoplasm sufficiently to induce the release of p53 from p53/BCL-XL complexes. Free cytoplasmic p53 releases BAX in the cytoplasm starting C cytochrome-dependent apoptosis. (b) **Activation of cell cycle arrest and DNA damage repair pathways.** In response to medium intensity DNA damage, p53 will bind to MUC1. The p53-MUC1 complex interacts with PCAF (represented by p53-cl) to allow the expression of genes related with cell cycle arrest like *p21*, as well as the expression of genes related with DNA repair (in the figure are represented by "Z"). Another remarkable function of p53-cl is inducing the Wip1 and Mdm2 expression. Wip1 is a phosphatase that down-regulates the p53 network.

Among the p53-target genes that are activated in this pathway are WIP1 phosphatase and Mdm2 in its p90 isoform. The Wip1 function is to dephosphorylate Chk2, ATM, activated p53 and Mdm2 (Figure 4). This leads to inactivation of all of these proteins except Mdm2, which is activated (Braithwaite, 2006; Candeias et al., 2008; Kobet et al., 2000; Lahav et al., 2004; Jänicke et al., 2008; Bernards, 2004). The dephosphorylated form of Mdm2, activated WIP1, and new synthesized Mdm2, inhibit p53 reducing its nuclear concentration (Hu et al., 2007). Because the WIP1 function, the signal generated by ATM and Chk2 will blink while DNA damage is not repaired, since ATM activator proteins remain linked to broken DNA (Jänicke et al., 2008; Lu et al., 2008). When DNA damage is repaired nuclear topology is restored, and ATM activating proteins trigger stop signal. In this case, Wip1 definitely inactivates ATM, and effector molecules like Chk2. This encourages Mdm2 (p90 isoform) to efficiently suppress p53 activity (Figure 4). These interactions allow the cell to return to its basal state, and the levels of p90, Mdm2 and Wip1 also return to their basal value (Bernards, 2004).

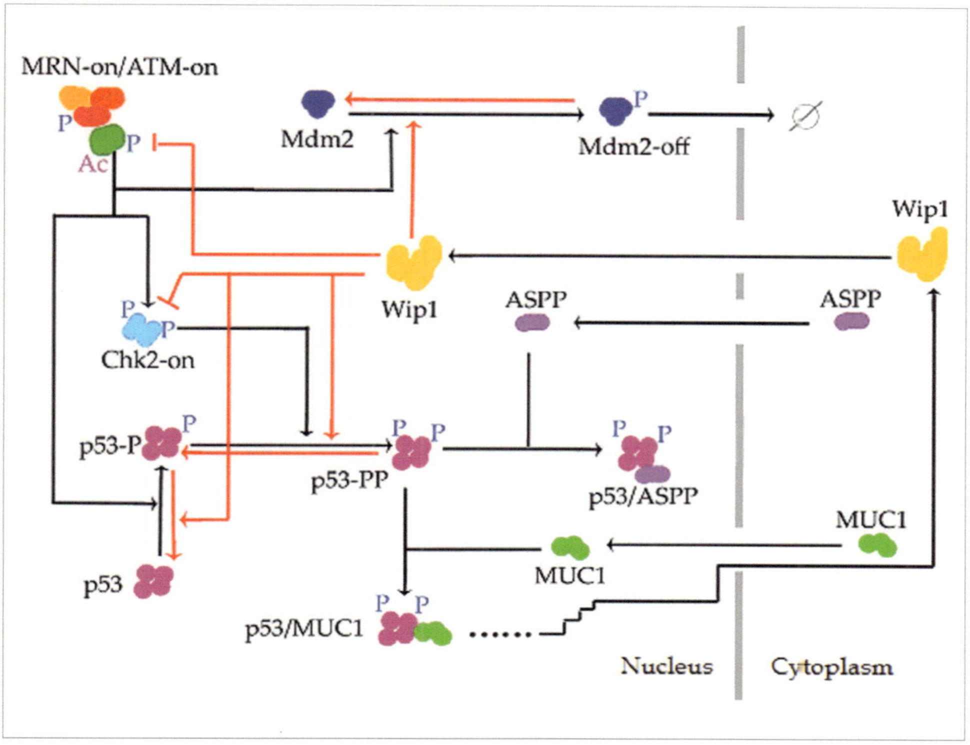

Fig. 4. **Mdm2 and Wip1 down-regulation of p53 network**. The main targets of Wip1 are ATM, Chk2 and Mdm2 phosphorylated. This results in a silencing of the ATM signal, and the Mdm2 activation. Mdm2 inhibits p53, in order to return its levels to normality, and restored the network equilibrium. In figure red lines represent the interactions that blocks Wip1.

2.4.3 Medium DNA damage, among life and death

The third stage is undoubtedly the most complicated of all, as it not only depends on the energy state of the cell, but also depends on the cell cycle phase in which the cell is. For example, if the cell is in S phase the cytoplasmic concentration of ASPP is higher than in other stages (Bermagaschi et al., 2004), and therefore this could lead more easily to apoptosis rather than cell cycle arrest. In this scenario, any type of delay suffered by the p53-dependent transcription process of the genes that control the DNA repair machinery and cell cycle arrest is of vital importance. These delays could be due to the effect of SIRT1 on p53, or to the competitive inhibition of p53 by other transcription factors, or to the action of the histone-acetyl transferases. If the transcription process of genes related to the DNA repair is slower than the normal, the cell will have fewer possibilities to survive. The intermittent signal of ATM produces a rise in the ASPP concentration, so it will be a question of time for the cell to enter in apoptosis. This could explain why after a long period of time if the cell is not repaired, it dies (Zhang et al., 2009; Yoda et al., 2001; Robinson et al., 2008; Shreeram et al., 2006; Wang et al., 2007; Bergamaschi et al., 2004; Braithwaite, 2006; Green and Kroemer, 2009; Toledo and Wahl, 2006; Liu et al., 2005; Haupt et al,, 2006; Strano et al., 2006; Zhang et al., 2008).

3. Nonlinear cell dynamics

Cells are complex networks of physicochemical processes that support their highly organized structure and function. Thus, each cellular process sustained by a cell involves different levels of cellular organization. Each level of organization can be represented as a cell subsystem (subnetwork) that function in a modular mode (Thieffry and Romero, 1999; Del Vecchio, Ninfa and Sotang, 2008). In this form, cells can be modeled as formed by a set of subsystems like the regulatory genetic network, the network of synthesis and distribution of proteins, the network of signaling pathways, and the metabolic network, among others. The information flow through the set of cellular subsystems that controls the cell response to environmental signals occurs according to the canonical schema of Figure 5. Tumor-inducing DNA or RNA viruses, and other agents like ionizing radiation (IR), can drastically modify this flow of information.

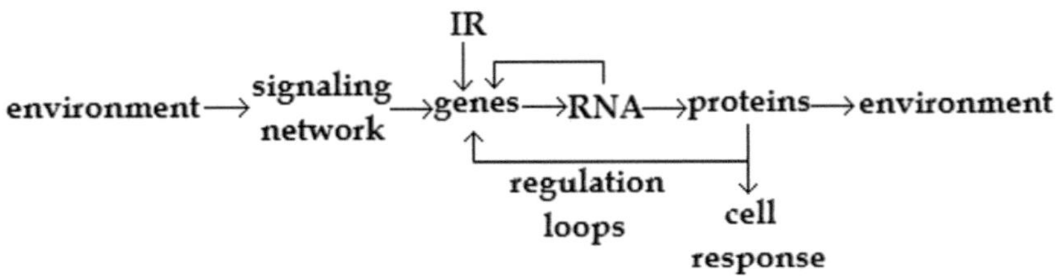

Fig. 5. **Flow of information between cell subsystems**. Changes in environmental conditions are sensed by cell signaling networks, which code and transmit this information to the nucleus. Coded information is decoded by the transcription machinery that, in response to this information, activates and inactivates a set of specific genes, giving raise to a specific distribution of effector proteins. These proteins are responsible for the cell specific response to the environmental conditions. Some of these proteins can be used to regulate gene expression, acting as specific transcription factors that conform complex regulatory loops inside the nucleus. Proteins can also be secreted to modify the cell environment. Tumor-inducing viral RNA and DNA can modify this flow of information by acting directly on the cell genome, drastically modifying the population of effector proteins and their function. Ionizing radiation (IR) can drastically interrupt this flow of information by producing severe damage to the cell DNA, and inducing apoptosis.

An important remark from Figure 5 is that the flow of information through cellular subsystems is due to a continuous flow of matter and energy, according to the respective laws of conservation. Taking into account the flow of matter at each point of the cell, the respective mass balance equation is:

$$\frac{\partial c_k(\mathbf{x},t)}{\partial t} = \sum_r \nu_{rk} \cdot \omega_r\left(c_1(\mathbf{x},t)c_2(\mathbf{x},t)...c_k(\mathbf{x},t)...c_s(\mathbf{x},t)\right) + D_k\nabla^2 c_k(\mathbf{x},t) \qquad (1)$$

which means that the local rate of variation of the concentration of the substance k (denoted by c_k) at point $\mathbf{x}$ at time t is equal to the net rate of diffusion of k inside the cell volume V (denoted by $D_k\nabla^2 c_k(\mathbf{x},t)$) plus the rate of formation/degradation of k due to the local chemical reactions inside V. In equation (3.1), ω_r represents the local rate of the chemical reaction r, which is a functional of the concentration of the reactive substances at point $\mathbf{x}$ at time t, and ν_{rk} is the stoichiometric coefficient of k in reaction r.

The reaction term of equation (1) can be rewritten as:

$$\sum_r \nu_{rk} \cdot \omega_r \left(c_1(\mathbf{x},t)c_2(\mathbf{x},t)...c_k(\mathbf{x},t)...c_s(\mathbf{x},t)\right) = f_k\left(c_1(\mathbf{x},t),c_2(\mathbf{x},t),...,c_k(\mathbf{x},t),...c_s(\mathbf{x},t)\right) \quad k=1,2,...,s \tag{2}$$

leading to:

$$\frac{\partial c_k(\mathbf{x},t)}{\partial t} = f_k\left(c_1(\mathbf{x},t),c_2(\mathbf{x},t),...,c_k(\mathbf{x},t),...c_s(\mathbf{x},t)\right) + D_k \nabla^2 c_k(\mathbf{x},t) \quad k=1,2,...,s \tag{3}$$

which is the well known form of the reaction-diffusion equation for the substance k. It indicates that the temporal variation of the concentration of k at point $\mathbf{x}$ at time t depends on the balance between the chemical processes in which this substance takes part, represented by the function f_k, and its diffusion rate in the cellular medium. Function f_k is generally a nonlinear function of the concentration of the reactive substances, and equation (3) usually has not an analytical solution. In a homogeneous medium the diffusive term in equation (3) is null, and f_k completely defines the entire system dynamics in the s-dimensional phase space, which is defined by the set of concentration values of the s reactive substances. The systems dynamics is represented by a trajectory in this space, defined by the column vector $\mathbf{c}(t) = \langle c_1(t)c_2(t)...c_s(t)\rangle$. In nonlinear systems this trajectory can have peculiar characteristics like high sensitivity to initial conditions, bifurcations and complex loops that represent a great variety of dynamical behaviors observed in biological systems like limit cycles, hysteresis, bistability, ultrasensitivity, among others. In not homogeneous medium the diffusive term of equation (3) produces a more complex dynamical behavior of the system, giving rise to phenomena like traveling waves, spirals and spatially located bursting of second messengers and proteins, among others.

In the subsequent sections it is discussed in depth the nonlinear dynamics of homogenous chemical systems.

3.1 Stability of nonlinear systems

The first problem concerning the dynamics of nonlinear systems is the determination the steady points of the system in its phase space.

A steady point is a column vector $\mathbf{c}^o = \langle c_1{}^o(t)c_2{}^o(t)...c_s{}^o(t)\rangle$ for which equation:

$$\dot{\mathbf{c}}(t) = \mathbf{f}(\mathbf{c}(t)), \text{ where } \mathbf{f}(\mathbf{c}(t)) = \begin{bmatrix} f_1(c_1(t),c_2(t),...c_k(t),...c_s(t)) \\ \vdots \\ f_s(c_1(t),c_2(t),...c_k(t),...c_s(t)) \end{bmatrix} \tag{4}$$

becomes cero. Once the set of steady points of the system is settled on, is necessary to determine their stability. Equation (4) subject to the initial condition $\mathbf{c}(0) = \mathbf{c}_0$ defines a **nonlinear dynamical system**.

The steady point $\mathbf{c}^o$ of a dynamical system is **Liapunov stable** if for each $\varepsilon > 0$ exits a $\delta > 0$ such that $\left\|\mathbf{c}(t) - \mathbf{c}^o\right\| < \varepsilon$ whenever $\left\|\mathbf{c}(0) - \mathbf{c}^o\right\| < \delta$, i.e., any trajectory that initiates at a distance δ of the steady point $\mathbf{c}^o$ remains at a distance ε of it for all positive time.

The steady point $\mathbf{c}^o$ of a dynamical system is attracting if exists a $\delta > 0$ such that $\lim_{t\to\infty} \mathbf{c}(t) = \mathbf{c}^o$ for any trajectory $\mathbf{c} = \mathbf{c}(t)$ whenever $\left\|\mathbf{c}(0) - \mathbf{c}^o\right\| < \delta$, i.e., any trajectory that initiates at a distance δ of the steady point $\mathbf{c}^o$ will converge to it eventually. In this case, the point $\mathbf{c}^o$ is an **attractor** of the dynamical system in the phase space. A steady point $\mathbf{c}^o$, which is Liapunov stable and attracting, is asymptotically stable.

A steady point $\mathbf{c}^o$, which is neither stable nor attracting, is unstable and is a **repulsor** in the phase space.

3.2 Phase plane analysis

In nonlinear systems the trajectories cannot be generally determined in an analytical form. However, it is possible to perform a qualitative analysis to find out the global behavior of the dynamical system in the corresponding phase space. As a vector can be assigned to each point of this space, according to equation (4), the vector field associated to the phase space can be drawn. By flowing this vector field, a phase point traces a solution **c**(*t*) of the dynamical system, corresponding to a trajectory winding trough the phase space.

It is of importance to point out the fact that if the function **f** of equation (4) is continuous and all its partial derivates $\frac{\partial f_i}{\partial c_j}$ $i,j = 1,2,\ldots,s$ are also continuous in **c** for a given subset $D \subset \Re^n$, then for every $\mathbf{c}_0 \in D$ the initial value problem of equation (4), has solution **c**(*t*) in some time interval (-*t*,*t*) around *t* = 0 and this solution is unique. A topological implication of this theorem is that two trajectories cannot intersect and, as consequence, chaos is ruled out of any 2-dimensional phase space but arises as a possible behavior of every s-dimensional dynamical system with s > 2 (Strogatz, 1994).

The phase space analysis of the dynamics of a nonlinear system takes into account the following aspects: 1) the number, position and stability of the steady points; 2) the arrangement of the trajectories near the steady points; and 3) the existence and stability of closed orbits.

In section 3.1 was presented the form in which the steady points are determined and how they can be classified according to their stability. The arrangement of the trajectories around steady points is determined by linearization of the original nonlinear system, and analysis of the behavior of the eigenvalues of the Jacobian matrix of the linearized system around each steady point. For example, considering a 2-dimensional phase space and a steady point $\mathbf{c}^o = \left\langle c_1^o, c_1^o \right\rangle$, a small perturbation from this steady state drives the nonlinear dynamical system $\begin{bmatrix} \dot{c}_1(t) \\ \dot{c}_2(t) \end{bmatrix} = \begin{bmatrix} f_1(c_1(t), c_2(t)) \\ f_2(c_1(t), c_2(t)) \end{bmatrix}$ into a new trajectory $\delta\mathbf{c}(t) = \left\langle \delta c_1(t), \delta c_2(t) \right\rangle$, where $\delta c_1(t) = c_1(t) - c_1^o$ and $\delta c_2(t) = c_2(t) - c_2^o$. In this form:

$$\begin{aligned} \delta\dot{c}_1 = \dot{c}_1 = f_1(c_1^o, c_2^o) + \left.\frac{\partial f_1}{\partial c_1}\right|_{(c_1^o, c_2^o)} \delta c_1 + \left.\frac{\partial f_1}{\partial c_2}\right|_{(c_1^o, c_2^o)} \delta c_2 + O\left(\delta^2 c_1, \delta^2 c_2, \delta c_1 \delta c_2\right) \\ = \left.\frac{\partial f_1}{\partial c_1}\right|_{(c_1^o, c_2^o)} \delta c_1 + \left.\frac{\partial f_1}{\partial c_2}\right|_{(c_1^o, c_2^o)} \delta c_2 + O\left(\delta^2 c_1, \delta^2 c_2, \delta c_1 \delta c_2\right) \text{ because } f_1(c_1^o, c_2^o) = 0 \end{aligned} \quad (5)$$

in a similar form:

$$\delta \dot{c}_2 = \dot{c}_2 = \left.\frac{\partial f_2}{\partial c_1}\right|_{(c_1^o, c_2^o)} \delta c_1 + \left.\frac{\partial f_2}{\partial c_2}\right|_{(c_1^o, c_2^o)} \delta c_2 + O\left(\delta^2 c_1, \delta^2 c_2, \delta c_1 \delta c_2\right) \quad (6)$$

which leads to the linearized dynamical system:

$$\begin{bmatrix} \delta \dot{c}_1(t) \\ \delta \dot{c}_2(t) \end{bmatrix} = \begin{bmatrix} \left.\frac{\partial f_1}{\partial c_1}\right|_{(c_1^o, c_2^o)} & \left.\frac{\partial f_1}{\partial c_2}\right|_{(c_1^o, c_2^o)} \\ \left.\frac{\partial f_2}{\partial c_1}\right|_{(c_1^o, c_2^o)} & \left.\frac{\partial f_2}{\partial c_2}\right|_{(c_1^o, c_2^o)} \end{bmatrix} \begin{bmatrix} c_1(t) \\ c_2(t) \end{bmatrix} = \mathbf{J}[c_1, c_2] \begin{bmatrix} c_1(t) \\ c_2(t) \end{bmatrix} \quad (7)$$

$\mathbf{J}[c_1, c_2]$ represents the Jacobian matrix of the linearized dynamical system in equation 6. The eigenvalues λ_1, λ_2 of $\mathbf{J}[c_1, c_2]$ can be calculated from the characteristic equation: $\left|\mathbf{J}[c_1, c_2] - \lambda \mathbf{I}\right| = 0$. Depending on the nature of the eigenvalues, it is possible to know the arrangement of the trajectories near each steady point of the nonlinear system (Figure 6). This linearization process can be extended to perform the phase space analysis of higher dimensional nonlinear dynamical systems (Edelstein-Kesher, 2005).

An important question that arises at this point is whether the behavior of the trajectories obtained from equation (7) accurately reflects the real behavior of the trajectories of the original nonlinear system. Andronov et al. (1973) show that this is the case. If the linearized system has a saddle, a node or a spiral at a given steady point, then the original nonlinear system really has also a saddle, a node or a spiral at that steady point. Furthermore, if a steady point is a stable saddle or node of the linearized dynamical system, then is also a stable saddle or node of the nonlinear system. In this case, the neglected nonlinear terms of equations (5) and (6) practically have not effect on the stability of these points when Re(λ) ≠ 0 for both eigenvalues. This kind of steady points is known as **hyperbolic points**, and they are not affected by the small nonlinear perturbations. The topological implication of this fact is that the vector field corresponding to a saddle or a node is not altered by small nonlinear perturbations and, as consequence, has **structural stability** (Strogatz, 1994).

When the eigenvalues of the Jacobian matrix are pure imaginary $\lambda = \pm i$, the steady point is a center. The trajectories around this point are closed orbits that are stable. However, the neglected nonlinear terms in equations (5) and (6) can produce an imperfect closure of the orbit, giving rise to a spiral. In this form, the vector field corresponding to a center is altered by small nonlinear perturbations that transform the center into a spiral and, as consequence, has not structural stability (Strogatz, 1994).

According to their stability, steady points of 2-dimensional dynamical systems can be classified into a: 1) **Robust case**, which includes repellers or sources, for which both eigenvalues have Re(λ) > 0; attractors or sinks, for which both eigenvalues have Re(λ) < 0 and saddles, for which one eigenvalues has Re(λ) > 0 and the other one has Re(λ) < 0; 2) **Marginal case**, which includes centers for which both eigenvalues are pure imaginary, and non-isolated steady points for which one eigenvalue has Re(λ) = 0 (Strogatz, 1994).

However, the phase space of a nonlinear system can exhibit another kind of closed orbits called **limit cycles**, which cannot be observed in linear systems. A limit cycle is an isolated

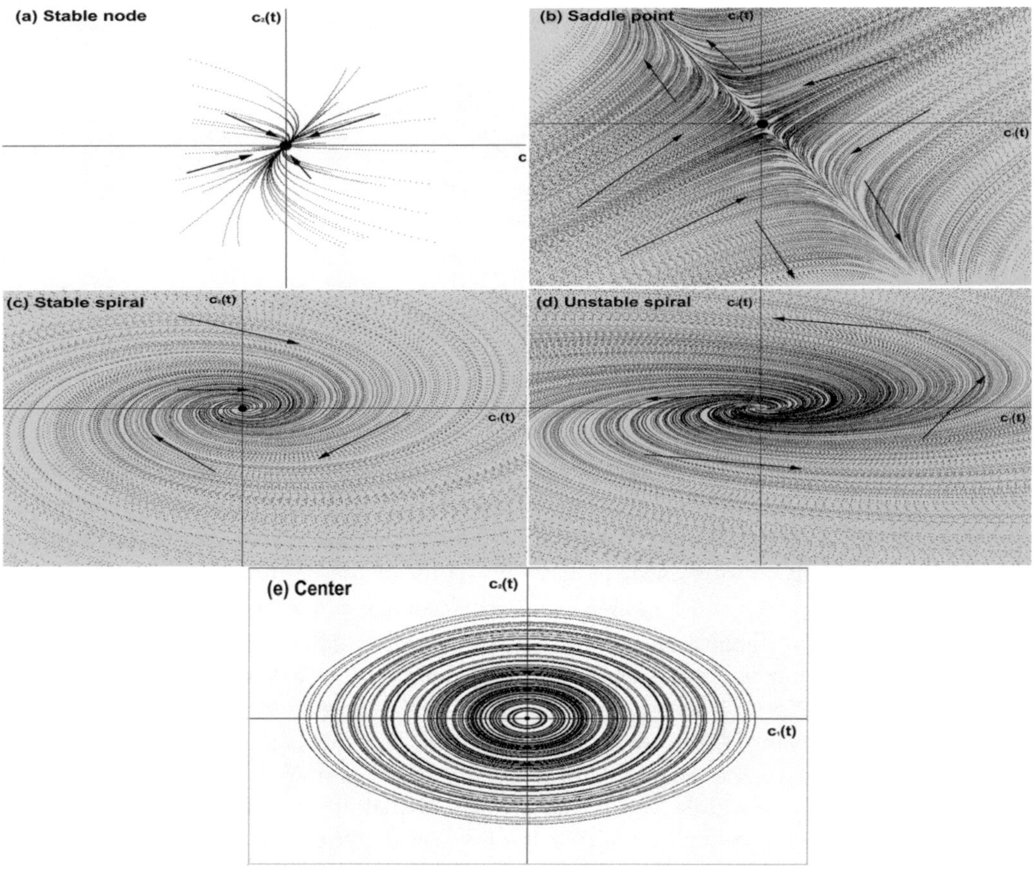

Fig. 6. **Classification of the steady points of a 2-dimensional linearized dynamical system.** (a) There is a stable node or attractor in the phase space when both eigenvalues λ_1 and λ_2 are real and negative (when both eigenvalues are positive, the steady point is a unstable node or repulsor); (b) There is a saddle point in the phase space when both eigenvalues λ_1 and λ_2 are real, but one of them is positive and the other is negative. The stable manifold is spanned by the eigenvector associated to the negative eigenvalue. The unstable manifold is spanned by the eigenvector associated to the positive eigenvalue. (c) There is a stable spiral in the phase space when both eigenvalues λ_1 and λ_2 are complex conjugated with negative real part. (d) On the contrary, if both eigenvalues λ_1 and λ_2 are complex conjugated with positive real part the spiral is unstable. (e) When λ_1 and λ_2 are pure imaginary the steady point is a center surrounded by a series of stable closed orbits. All figures show the flow of the dynamical system in the phase space spanned by the basis conformed by the variables $c_1(t)$ and $c_2(t)$. The black point represents the steady point of the dynamical system, and the arrows mark out the direction of the flow of the vector field.

trajectory for which neighbor trajectories can be only spirals that converge to it or diverge from it. If all the spirals converge into the limit cycle, this closed orbit is stable, otherwise is unstable. The existence of this kind of closed trajectories in the plane is settled down by the Poincaré-Bendixson theorem. According to this theorem, exits a trajectory C, which is either a closed orbit or a spiral that converges to a closed orbit as $t \to \infty$, confined inside a certain closed bounded region R of the plane. This theorem assumes 1) the existence of a vector field $\dot{\mathbf{c}} = \mathbf{f}(\mathbf{c})$ that is continuously differentiable on an open set of the plane containing R and 2) R does not contain any fixed point (Edelstein-Kesher, 2005). A consequence of this theorem is that in a 2-dimensional phase space any trajectory trapped into a closed bounded region R must converge to a limit cycle.

However, in higher dimensional systems the Poincaré-Bendixon theorem does not longer apply and trajectories can be trapped into a closed region of the phase space without converge into a limit cycle or settle down to a fixed point, and they could be attracted by a complex geometric object called strange attractor, which is a fractal set on which the motion is aperiodic and sensitive to very small changes in initial conditions. This sensitivity makes the motion unpredictable as t increases, giving rise to a chaotic dynamics (Strogatz, 1994).

3.3 Bifurcations

The qualitative features of the vector field of a biochemical dynamical system are strongly dependent on the set of parameters of its corresponding set of differential equations. As the value of one of these parameters changes, the qualitative features of the vector field undergo local variations around the steady points. This parameter-dependent change in the local topological structure of a vector field is known as bifurcation. They generally occur in a one-dimensional subspace, and the rest of the dimensions of the phase space are affected as consequence of the flow that can be attracted or repelled from this subspace (Strogatz, 1994; Edelstein-Kesher, 2005). Taking into consideration the imaginary plane, we can roughly classify bifurcations into two cases: 1) the eigenvalues of the Jacobian matrix are both real and bifurcations occur along the real axis as certain parameter α changes. This kind of bifurcation comprises the saddle-node bifurcation; the transcritical bifurcation, and the subcritical and supercritical pitchfork bifurcation. 2) The eigenvalues of the Jacobian matrix are complex conjugated. Bifurcations occur crossing the imaginary axis as certain parameter α changes. This kind of bifurcation comprises the supercritical and subcritical Hopf bifurcation.

In the first case, a) the saddle-node bifurcation causes local variations in the vector field around two points: a saddle and a node, as a bifurcation parameter α changes. These points become closer as parameter α varies until they collide and annihilate each other. This type of bifurcation has interesting applications in some models of biological processes that imply the presence of chemical switches; b) a transcritical bifurcation occurs when two steady points interchange their stability as the bifurcation parameter α varies. c) The normal form of an ordinary differential equation (ODE) that exhibits a subcritical pitchfork bifurcation is $\dot{c} = \alpha c + c^3$. When $\alpha < 0$ the steady point $c = 0$ is stable and there are two unstable points at $c = \pm\sqrt{-\alpha}$. When $\alpha > 0$ the only real steady point $c = 0$ becomes unstable. The normal form of an ODE that exhibits a supercritical pitchfork bifurcation is $\dot{c} = \alpha c - c^3$. When $\alpha < 0$ the only real steady point $c = 0$ is stable. For $\alpha > 0$ there is an unstable steady point at $c = 0$ and to stable steady points at $c = \pm\sqrt{\alpha}$.

In the second case, the presence of a Hopf bifurcation leads the system to a limit cycle. As the bifurcation parameter α varies, when a certain critical value α_c is reached the supercritical Hopf bifurcation drives the transformation of a stable spiral into an unstable spiral that converges to a stable limit cycle (Figure 7). The case of a subcritical Hopf bifurcation is more complicated. A typical example is when an unstable limit cycle shrinks to cero amplitude as the bifurcation parameter α reaches its critical value α_c, at which the cycle engulfs the node rendering it unstable and making the system to jump to a distant attractor when $\alpha > \alpha_c$ (Strogatz, 1994; Edelstein-Kesher, 2005). This new attractor could be a steady point, another limit cycle, infinity or a chaotic attractor (for higher dimensional systems).

4. Dynamical aspects of apoptosis

Ionizing radiation is a term applied to radiation that has energy enough to remove an electron from an atom or molecule. This ionization can produce free radicals chemically reactive. The effect of the ionizing radiation depends only on the energy of the particles (photons, electrons or α particles) that impacts a particular target. The effect of the ionizing radiation is measured in gray (Gy), which is the SI unit of **absorbed dose**, and is equivalent to the amount of radiation necessary to deposit one Joule of energy in one kilogram of matter. Ionizing radiation is used in the treatment of various types of cancer, under the assumption that the apoptotic network of cancer cells is disabled, making these cells more susceptible to the effects of single and double strand damage in their DNA (Qi et al., 2007).

4.1 Oscillatory nuclear dynamics of p53 in response to ionizing radiation

The response of a nonlinear dynamical system to an input is **digital**, when the output is conformed by a series of periodic discrete pulses or quanta, that have fixed form and size. The number of pulses increases with the strength of the input signal. The digital response corresponds to a limit cycle, which is generally driven by the onset of a **supercritical Hopf bifurcation** (Ma et al., 2005; see section 3.3 and Figure 7). In contrast, the response to a given input is **analog** when the strength of the output increases with the strength of the input signal (Lahav et al., 2004). Recent results show that the kind of radiation applied to a cell can produce different dynamical effects, either digital or analog, depending on the molecular subnetwork activated in response to the **double strand brakes** (DSB's) produced (Batchelor et al., 2011).

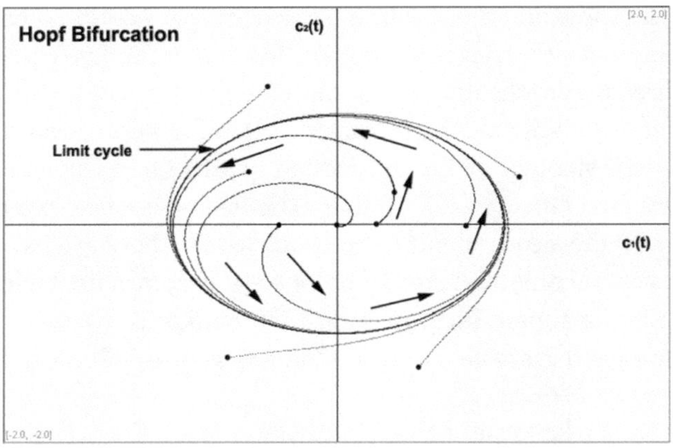

Fig. 7. **Supercritical Hopf bifurcation.** This kind of bifurcation transforms a stable spiral into an unstable spiral that converges to a stable limit cycle when the value of the bifurcation parameter α reaches some critical value α_c. The black points in the figure are different initial conditions of the dynamical system. The arrows mark out the direction of the flow of the vector field. The phase space is spanned by $c_1(t)$ and $c_2(t)$.

When γ-radiation or the radiomimetic drug neocarzinotastina (NCS) are applied to a cell, the output is digital. According to the results of Batchelor et al. (2008, 2011), the DSB's produced by these agents induce the onset of a limit cycle of fixed duration, amplitude and period (~ 5 h) in the p53-Mdm2 phase space. The number of pulses of both molecules increases as the damage increases. This series of pulses have the characteristic of an "all or none" response in single cells, which is trigged even by transient inputs (Batchelor et al., 2008).
A previous model by Ma et al. (2005) predicts this kind of digital dynamics under two basic facts: 1) the number of DSB's for a given doses is stochastic and follows a Poisson distribution with an average proportional to the radiation dose applied x and given by $35x$, which is consistent with the observed 30-40 DSB's per Gy. The repair dynamics is simulated with a Monte Carlo method; 2) the repairing of the DNA damage is a biphasic process, i.e., consists of a rapid and a slow phases of DSB fix, which are assumed to be produced by two different types of DSB's. The simple DSB's, which contain a break in each strand of DNA of a short segment of 10-20 bps in length, switches on the fast repairing process. The slow repairing process is trigged by complex DSB's, which contains breaks in each strand of DNA, together with other kind of alterations like base damage, base deletions, etc., within the same short segment. The fast and slow processes are represented by different chemical kinetics (Stwart, 2001).

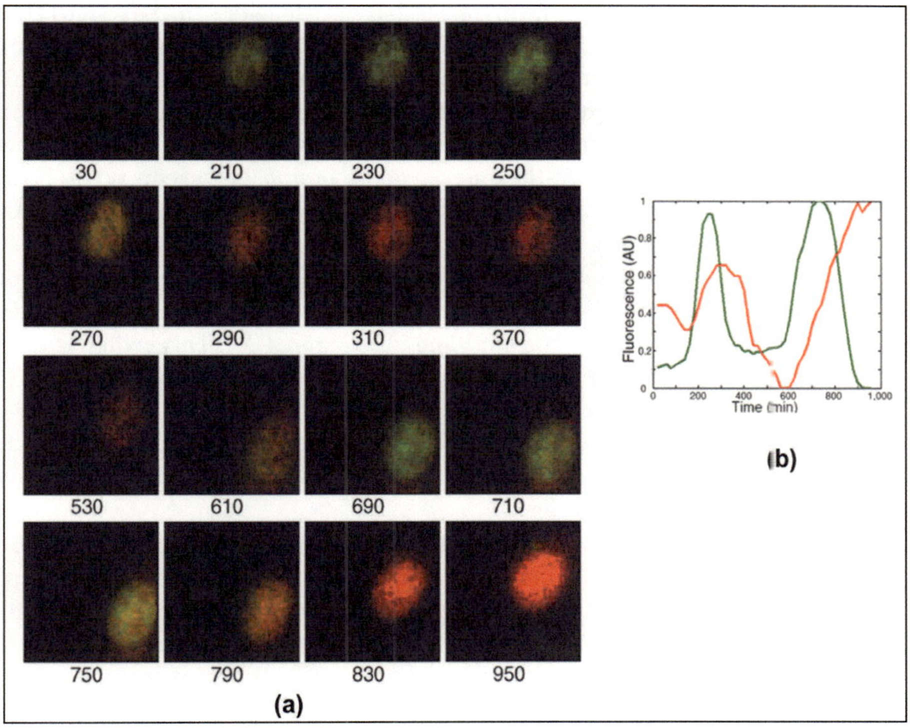

Fig. 8. (a) **Experiments with fluorescent p53 and Mdm2.** Two pulses of p53-CFP (green) and Mdm2-YFP (red) are shown in this panel. Time (in min) after irradiation is show below images. In the figure, a phase difference of ~ 120 min between the p53 and Mdm2 maximum fluorescence is observed. (b) **Levels of p53-CFP and Mdm2-YCP in the nucleus** of the cell of panel (a). AU, arbitrary units. Figure reproduced from Lahav et al. (2004), with authorization of Dr. U. Alon.

As mentioned above, the predicted dynamics from this model is a digital output, in which doses of the order of 5 Gy (~ 150 DSB's) applied in an on - off form produces pulses of p53 and Mdm2 with a period of ~ 7 hr and a phase difference of ~ 115 min. These results are consistent with the experimental results of Lahav et al. (2004) that are showed in Figure 8.

In order to study the behavior of the p53 and Mdm2 loop in individual MCF7 breast cancer cells, Lehav et al. (2004) fused p53 with the cyan fluorescent protein (CFP) under a zinc-inducible promoter, and Mdm2 with a yellow florescent protein (YFP) under the human Mdm2 native promoter. p53-CFP was completely functional in induced the transcription of Mdm2 after zinc induction. The p53 and Mdm2 levels were recorded using time-lapse fluorescent microscopy of the cell line clone expressing both fluorescent proteins after γ-irradiation at 20 minutes resolution during 16 hrs of growth (Figure 8). The results show that individual cells exhibit sustained oscillations in the p53-CFP and Mdm2-YFP nuclear levels in response to DSB's (Figure 8), but that the number of pulses varies from cell to cell. The width of each pulse was of 350 ± 160 min. The maximum of the first peak appeared 360 ± 240 min after DSB, and the time between successive peaks stabilized at a 440 ± 100 min. The time delay between the maximum levels of p53 and Mdm2 was ~ 100 min. However, some individual cells did not show pulses of p53 and Mdm2 after γ-irradiation.

The coincidence between the outputs from the models of Batchelor et al. (2008, 2011) and Ma et al., (2005) with their own and Lehav et al. (2004) experimental results, indicates that γ-irradiation produces DSB's that induces the onset of a limit cycle through a supercritical Hopf bifurcation in the p53-Mdm2 phase plane (Batchelor, 2008), and that the molecular mechanisms that underlay this dynamical behavior possibly lies on the p53-Mdm2 negative feedback loop (Figures 1a). However, the only existence of this loop seems not to be enough to explain the sustained oscillations observed in individual cells (Lehav et al., 2004; Ma et al., 2005; Batschelet et al., 2008 , 2011).

According to the results from the model of Ma et al., (2005), the presence of DSB's induces a switch-like behaviour of the ATM kinase, which dissociates from its inactive dimmer form into two activated monomers that phosphorylate p53 (Figures 2a). This switch is turned on by doses of γ-irradiation as low as 0.1 Gy (~ 3 DSB's) and saturates at 0.4 Gy (~ 14 DSB's). Furthermore, activated ATM monomers (represented by ATM*) close a positive feedback loop acting on the ATM inactive dimmer form to cleavage it and produce more activated monomers. The combination of these two feedback loops is able of producing the sustained oscillations observed in individual cells.

However, experimental and theoretical analysis of the p53-Mdm2 loop by Batchelor et al. (2011) show that the sustained oscillations observed in individual cells can also be explained by the action of a triple negative feedback loop. ATM* activates p53, which closes the first negative feedback loop with Mdm2. However, p53 also activates the transcription of *wip1* that closes a negative feedback loop with p53. Although, p53 also closes a negative feedback loop with ATM* mediated by Wip1 (Figures 4). Experimental and *in silico* results (Batchelor et al., 2008, 2011) show that the suppression of the negative action of Wip1 on ATM* switches p53 dynamics from repetitive pulses into a single analog pulse.

Probably, the positive ATM feedback loop together with a initial rapid degradation of Mdm2 (Batchelor et al., 2011) and the triple negative feedback loops mediated by p53, Mdm2 and Wip1 accounts for the onset and stability of the limit cycle that gives rise to the digital response to γ-irradiation observed in individual cells.

Digital response of the p53 machinery to DSB by γ-irradiation could be a timing mechanism that controls downstream events that determine if the DSB's can be repaired or not (Figure 3a), and, in consequence, if apoptosis should continue until cell death or not (Ma et al., 2005; Dogu and Díaz, 2009).

4.2 Non-oscillatory nuclear dynamics of p53 in response to ionizing radiation

The effect of UV is the cross link of consecutive pyrimidine bases producing the exposition of single strands of DNA (ss-DNA). In contrast to the effect of γ-irradiation, the treatment of cells with bursts of UV-radiation with different energy, ranging from 2 to 10 J/m2, produces, in each case, a single analog pulse whose amplitude and duration increases with the UV doses. (Batchelor et al., 2011).

Beside the difference in the type of damage produced by γ and UV irradiation to the DNA, digital and analog p53 dynamics are sustained by different feedback mechanisms. The presence of ss-DNA induces the activation of the ATR kinase, which phosphorylates Mdm2 on Ser-407, a rapid reversible process, and inhibits its activity on p53. In contrast, DSB's induces the activation of the ATM kinase, which phosphorylates Mdm2 on Ser-395, producing its rapid degradation and an initial sharp rise of p53 activity (Batchelor et al., 2011). In consequence, the initial inhibitory effect of ATR* on the p53-Mdm2 loop is less intense than the initial effect of ATM*. Furthermore, the restoration of this loop after the post-translational modification of Mdm2 by ATR* probably occurs in a time scale of minutes. In contrast, the restoration of the p53-Mdm2 loop after the rapid degradation of Mdm2 produced by ATM*, probably occurs in a time lapse of hours due to the process of translation of *mdm2*. Additionally, as Wip1 has not a detectable inhibitory action on ATR*, p53 has not a negative retroactivity on ATR* mediated by Wip1. In this form, there are only two persistent feedback loops acting on p53 after UV-irradiation: the p53-Mdm2 and the p53-Wip1. (Batchelor et al., 2008, 2011).

Additionally, in contrast to the switch-like activation of ATM in response to low doses of γ-irradiation, the rate of activation of ATR must be considered UV-doses dependent in order to computationally reproduce the analog dynamics of p53 (Batchelor et al., 2011). In this form, the analog response of p53 to UV-irradiation seems to be due to three facts: 1) rapid reversible inactivation of Mdm2 by ATR*; 2) absence of negative retroactivity of p53 on ATR* mediated by Wip1 and 3) a non switch-like dynamics of activation of ATR.

4.3 Heterogeneous oscillatory dynamics of p53 in cell populations

Previous experiments by Lahav et al. (2004) show that a population of cells from the same clone MCF7, exhibit damped oscillations of p53 and Mdm2, when their dynamics is recorded by assays based on the determination of the average or the relative values of the amount of proteins over populations of cells, like the immunoblot technique. When not such averaging techniques are used (see section 4.1), records of the dynamics of p53 and Mdm2 show sustained oscillations with wide heterogeneity in their amplitude, not only between isogenic cells, but also in individual cells (Figure 9).

Geva-Zatorsky et al. (2006) repeated the experiment mentioned in section 4.1, but with 30 h of continuous recording. They found that only 60% of the cells exposed to 10 Gy of γ-irradiation showed sustained oscillations of the p53-Mdm2 feedback loop. These oscillations have a relatively constant period of 5.5 ± 1.5 h, a high coefficient of variation (~ 0.7) in their amplitude, and a relatively constant peak width. The p53-Mdm2 delay is also relatively constant in these cells. In contrast, only 30% of the cells exposed to 5 Gy of γ-irradiation showed sustained pulses, indicating that the number of cells displaying oscillations increases with higher doses of γ radiation. Initially, the onset of the oscillations in different cells was synchronized with the pulse of γ-radiation applied. However, synchrony was lost due to the different frequency response exhibited by individual cells. Some cells stopped their oscillatory dynamics or changed their frequency of oscillation. Additionally, after cell division, sister cells exhibited oscillations of p53 and Mdm2 that were initially synchronized. However, after a lapse of time of ~ 11 h this correlation decreased in 50%.

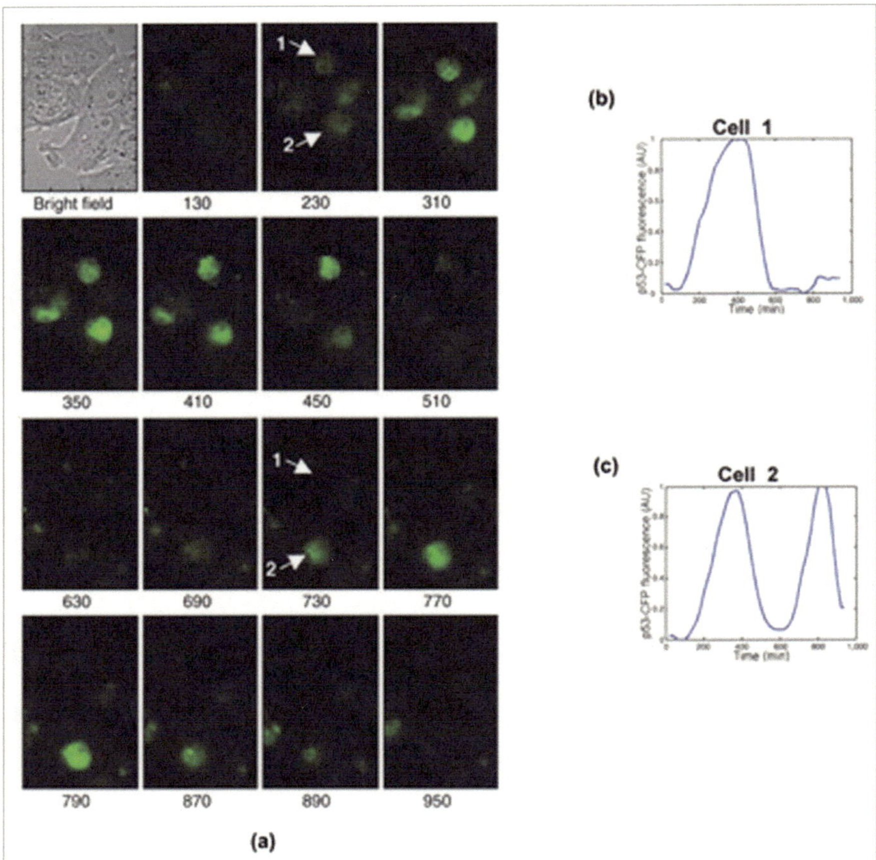

Fig. 9. (a) **p53-CFP (green)** in clonal MCF7+pU265+pU293 cells after 5-Gy γ-irradiation. Time (in min) after irradiation is shown below images. (b) and (c) **p53-CFP levels** (total CFP fluorescence in nuclei) from cells 1 and 2 (indicated by arrows in (a)), showing the heterogeneity in the pulses generated after γ-irradiation. AU, arbitrary units. Figure reproduced from Lahav et al. (2004), with authorization of Dr. U. Alon.

Geva-Zatorsky et al. (2006) work also reveals that after a treatment with 10Gy of γ-radiation, 40% of the cells showed a non-oscillatory response, or did not respond at all . The non-oscillatory response consisted of slow fluctuations of Mdm2-YFP and p53-CFP with only one, two or three peaks. Each fluctuation had duration of 8 - 12 h. Furthermore, a fraction of non-irradiated cells also show this kind of fluctuations.

4.4 Noisy oscillatory dynamics of p53

From section 4.3, it is clear that a deterministic approach for understanding the dynamics of apoptosis is not enough to account for the heterogeneity observed in the p53-Mdm2 oscillatory response of individual cells to IR. As we mentioned in section 3.3, the onset of a stable limit cycle in the p53-Mdm2 phase plane depends on the transition from a stable spiral to an unstable spiral (supercritical Hopf bifurcation), according to the Poincaré-Bendixson and Hopf theorems. Once the limit cycle is settled down, all nearby trajectories are trapped by it. In this form, any small perturbation of this cycle will be damped and the system will return slowly to its original rhythm of oscillation with constant amplitude and frequency. This is not the kind of dynamical behavior experimentally determined in MCF7 cells (Ma et al., 2205; Batchelor et al, 2011; Lahav et al., 2004; Geva-Zatorski et al., 2006). As consequence, deterministic models of the p53-Mdm2 oscillations can be considered only as approximate descriptions of reality if they take into account only the p53-Mdm2 negative feedback loop (Qi et al., 2007; Spencer and Sorger, 2011). As it was mentioned in section 4.1, better results are obtained from models that take into account the positive ATM feedback loop together with a initial rapid degradation of Mdm2 (Ma et al., 2005; Batchelor et al., 2011; Jolma et al., 2010) and the triple negative feedback loops mediated by p53, Mdm2 and Wip1. From these models, it is clear that in the nuclear compartment the activation of the dynamical characteristics of the DNA repairing molecular machinery conveys a trajectory that sharply leads to an unstable fixed point in the ATM*-p53-Mdm2 phase space, from which a limit cycle parallel to the p53-Mdm2 plane emerges through a supercritical Hopf bifurcation. However, there is also the possibility of the existence of a subcritical Hopf bifurcation, which can then lead to cycles of different IR-dose dependent amplitude (Qi et al., 2007). Nevertheless, these models cannot account for the variability of the pulses of p53 and Mdm2 observed in individual cells and in cell populations.

Cell systems are subject to intrinsic and extrinsic noise sources that can affect both the rate of production and the rate of degradation of proteins (Elowitz et al., 2002; Kaufmann and van Oudenaarden, 2006). Noise is a source of variability, even in cells from the same cell line. The intrinsic cell noise can insert a low frequency fluctuating component in the rate of production (degradation) of proteins producing a continuous stochastic variation in the corresponding number of molecules (Diaz, 2011). Geva-Zatorski et al. (2006) point out that the probable source of variability in the response of cells to γ-radiation is the stochastic variation in the rate of protein production, which generates fluctuations in the amplitude of the pulses without affecting their period. On the contrary, if the protein degradation rate is affected by noise, pulses exhibit variable amplitude and period. In this form, fluctuations in the rate of the transcription and translation processes of key apoptotic proteins after DSB damage can be reflected in the variable amplitude of the p53 and Mdm2 pulses, without affecting their rhythm of generation.

Furthermore, fluctuations can evoke the generation of spontaneous elevations in the basal concentrations of Mdm2, even in absence of γ-irradiation (Geva-Zatorski et al., 2006). Cancer cells have a unstable genetic system in which is possible that some control mechanism of apoptosis are modified or disabled in order to undergo cell death in response to cytotoxic stimuli. In consequence, any spontaneous fluctuations in the concentration of apoptotic nuclear proteins, like Mdm2, can be over amplified giving rise to the atypical dynamical features of cancer cells. For example, spontaneous increase in the concentration of Mdm2 can suppress the activity of p53 by ubiquitination and posterior degradation in the proteosome (Figure 1a), and disable the apoptotic mechanism of MCF7 cells during time lapses up to ~ 20 h (Geva-Zatorski et al., 2006).

After MCF7 cells division, p53 and Mdm2 oscillations continued with the same phase but sister cells lose ~ 50% of correlation in their oscillatory dynamics after 11 h (Geva-Zatorski et al., 2006). These results point to two important facts: 1) apoptotic information can be transferred to daughter cells and 2) sister cells can inherit similar levels of relevant apoptotic molecules, but noise generates fluctuations in protein production and degradation. As consequence, synchrony is lost after a lapse of time (Spencer and Sorger, 2011). In this form, the initial apoptotic timing synchronization caused by the γ.irradiation can be transferred to daughter cells but this correlation falls as time from division increases, until sister cells are no more correlated than a randomly chosen pair of cells (Spencer and Sorger, 2011).

4.5 Dynamics of cytoplasmic apoptotic proteins

Bistability is a broad dynamical feature of nonlinear biochemical systems, in which an unstable steady point separates two stable nodes. The transition between nodes requires energy to overcome the high energetic barrier among them. Generally, once the transition from one steady point to the other occurs, return to the original state requires a different pathway with different energy requirements, forming a **hysteresis** loop. Molecular switches are the common examples of this type of dynamics (Strogatz, 1994).

Bagci et al. (2006) work studied the role of Bax and Bcl2 synthesis and degradation rates in mitochondria-dependent apoptosis (Figure 3a), finding the existence of a saddle point bifurcation in the caspase 3-Bax degradation rate diagram. The Bax degradation rate of ~0.11 s^{-1} determines the upper limit to this region of bistability. Inside this region, the saddle point bifurcation can drive the system either to a stable point with a high concentration of caspase 3 (cell death) or to a stable point with a low concentration of caspase 3 (cell survival). Out of this region, the systems dynamics is settled on by the existence of a single stable fix point that determines cell surviving independently of the concentration of caspase 3. Cui et al. (2008) work points out that the activation of the "Bcl2 switch" upstream mitochondria induces apoptosis when the formation of the heterodimer Bak/Bax reaches a certain level. In this form, Bax level settles on the beginning of the apoptosis process (Bagci et al., 2006; Cui et al., 2008; Hua et al., 2005). An important conclusion from this work is that an abnormal state, like cancer, arises in cells when Bax degradation rate is above a threshold value, giving rise to a stable cell survival dynamics, i.e., cells do not die.

Fussenegger, Bailey and Varner (2000) work used mass-conservation law and kinetic rate laws to establish a model of caspases 8 and 9 activation by the extrinsic and intrinsic apoptotic pathways respectively. According to this model, in the case of the intrinsic pathway the release of cytochrome c starts 10 min after a stress signal like IR. The executioner caspases are active at 40% of its maximum value 1 h after the initiation of the stress response and 70% of them are active 2 h after. A negative mutation in p53 decreases the fraction of both activated caspase 9 and executioner caspases.

Dogu and Díaz (2009) model takes into account the role of proapoptotic proteins like p53 and Puma in the response to stress signals like IR (Figure 3a). This model shows that the overall kinetic behaviour of the p53 apoptotic network under genotoxic stress is mainly due to a possible p53-induced release of Bax from its complex with Bcl2 even in absence of Puma- However, the release of Puma in the cytoplasm increases the rate of liberation of Bax from the Bcl2/Bax cytoplasmic complex. In this form, this model proposes that Bax cytoplasmic level controls apoptosis by its continuous release from the Bcl2/Bax complex after DNA damage. If DNA is repaired or the damage is reduced enough before Bax reaches a threshold value, cells may divide or may have a Bax control mechanism that avoids the start of apoptosis. In this model PUMA plays the role of accelerator of the Bax level increasing process. The possible biological significance of this process may be to allow a lag between the starting of the DNA damage repair process and the onset of full apoptosis when the damage cannot be repaired in a pre-set moment of the cell cycle. After this time lag, Puma reaches a threshold concentration value that fires the complete and irreversible apoptotic process. From a dynamical point of view, this model states the stability and robustness of the network of interaction between p53 and the pro and anti-apoptotic proteins BclxL, Bcl2, Puma and Bax. The flow predicted by this model is formed by set of trajectories that converge into an attractor in the 4-dimensional phase space spanned by the concentration of the p53/Bcl-x_L, PUMA/Bcl-x_L, Bax/Bcl2 and p53/Bcl2 complexes. This flow is reasonably independent of the values of the rate constants of association and dissociation of these complexes, indicating its structural stability.

In this form, the cytoplasmic component of the apoptotic subnetwork exhibits a dynamics driven by a saddle-node bifurcation, which settles down a control point that determine the cell fate according to the information flow from the nucleus. This dynamics seems to be started by a Bcl2 switch that initiates the accumulation of Bax in the cytoplasm, until it reaches a threshold value. It seems that, although the cytoplasmic accumulation of Bax initiates the liberation of cytochrome c from mitochondria in a time lapse as short as ~ 10 min, the process is still reversible. However, once PUMA is present in the cytoplasm, the rate of liberation of Bax from the Bcl2/Bax complexes is sharply increased, and makes the apoptotic process completely irreversible due to the consequent increase of the levels of Bax over its threshold value. The high concentration of free Bax triggers the process of activation of caspase 3.

5. Conclusions

Apoptosis is a complex molecular process, which is capable of determine the cell fate according to the intensity of the DNA damage caused by IR (section 2.3). This network of molecular processes has a highly connected node, which is protein p53. Its key role in apoptosis is tightly regulated by a series of molecules like Wip1 and Mdm2, among others. DSB´s and ss-DNA can trigger the activation of either ATM or ATR (section 2.4). These kinases suppress the inhibitory activity of Mdm2 on p53, giving rise to a digital or to an analog response to IR, respectively (section 4.1 and 4.2).

Cancer cells show abnormalities in their genetic network, which includes the regulatory loops of apoptosis. As a result, the cancer cell response to IR is heterogeneous probably due to three facts: 1) IR produces random damage to DNA (Ma et al., 2005; Qi et al., 2007); 2) intrinsic noise adds some amount of randomness to the processes of transcription and translation of key apoptotic proteins (Geva-Zatorsky et al., 2006; Díaz, 2011); 3) regulatory genetic loops may be randomly disabled in cancer cells.

The effect of randomness is the generation of low frequency fluctuations in the levels of proteins like Mdm2. When these fluctuations coincide with the natural frequency of oscillation of the negative feedback loops that control p53 activation, the result is a series of pulses of different amplitude (Geva-Zatorsky et al., 2006). However, the rhythm of the oscillations remains relatively constant, suggesting that fluctuations principally affect the rate of synthesis of proteins. Models that include this randomness are successful in reproducing the heterogeneity of the pulses of p53 and Mdm2 in individual cells (Section 4.3).

However, there are not data about the oscillatory dynamics of p53 in normal cells after perturbation with IR, neither in other lines of cancer cells. In this form, it is not known if the oscillatory dynamics found in MCF7 cells represents a generalizaed dynamical behaviour of all cells under treatment with IR, or if it is only a particular feature of this cell line. However, variability in response to other proapototic tratments seems to be an universal feature of cells (Spencer and Sorger, 2011). A direct and very important consequence of this variability is that the killing fraction, i.e., the fraction of died cells from an initial living cell population after treatment with a proapototic agent, is generally much lower than 100%. This fact points to the necessity of combining at least two different treatments to completely erradicate cancer cells.

In order to gain a deeper insight into the mechanisms that generate this variability in cell response, is necessary to elaborate a more rigorousos quantitative approach to understand how key apoptotic processes are affected by stochasticity. Mathematical and computational models, combined with quantitative experimental research, are necessary tools to achieve this goal.

As we stated in this chapter, determinist models of apoptosis seems to have a limited power to explain the real dynamics of cell death. However, they are still useful tools because they incorporate in their core of equations a series of kinetic terms that represent the contribution of the different biochemical process at different times. Furthermore, from these models some general principles about the dynamics of apoptosis are beginning to emerge: 1) digital and analog responses are generated by different mechanisms of activation and regulation of p53; 2) digital response is generated by the nonlinear interaction of at least three negative feedback loops and an autocatalityc process, together with the rapid initial degradation of Mdm2 that generates a time delay in the restoration of its negative feedback loop with p53; 3) this nonlinear interactions conveys the evolution of the dynamical system to an unstable fixed point in the ATM*-p53-Mdm2 phase space, from which a limit cycle parallel to the p53-Mdm2 plane emerges through a supercritical Hopf bifurcation; 4) this limit cycle assures the stability of the oscillations, generating pulses of constant peak width, frequency and average amplitude; 5) as consequence, deterministic models predicts the onset of a clock-like mechanisms that determines the timing of cell death; 6) this nuclear clock may be someway synchronized with the cytoplasmic events that lead to the activation of executory caspases; 7) cytoplasmic biochemical processes generated by the interaction of pro-apototic and anti-apoptotic molecules like p53, PUMA, Bcl2, Bak, Bax , among others, settle down a saddle-node bifurcation, which determines either cell death or cell survival, depending on the levels of free Bax; 8) with respect to the nuclear analog response, deterministic models suggest that is due to the activation of only two negative feedback loops and the rapid reversible modification of Mdm2 by ATR*.

However, it is necessary to propose a broader deterministic model in order to understand: 1) how the nuclear Hopf bifurcation is linked to the cytoplasmic saddle-node bifurcation to determine the precise timing of cell death and 2) how the correct death-survival decision is taken according to the information processed in the nucleus.

Stochasticity can be added to this model in order to know if the fluctuations of p53 in the nucleus also influence death decisions in the cytoplasm, or if they are damped by the cytoplasmic feedback loops, avoiding lose of the correct timing of apoptosis. Mislay of this timing can produce cell death at random lapses of time.

Finally, the knowledge generated by stochastic models can be useful to improve the pro-apoptotic treatments that are used now day. These treatments can target nodes that are not practically affected by noise, allowing the increase of the killing fraction.

6. Acknowledgments

We thank Erika Juarez Luna for technical and logistical assistance. Financial support for this work was from CONACYT (Consejo Nacional de Ciencia y Tecnología) grant 105678.

7. References

Abukhdeir AM & Park BH (2008). p21 and p27: roles in carcinogenesis and drug resistance. Expert Rev Mol Med. 10:e19; doi: 10.1017/S1462399408000744.

Andronov AA, Leontovich EA, Gordon II, & Maier AG (1973). Qualitative theory of second order dynamic systems. J. Wiley. New York, NY. ISBN: 0470031956

Bagci EZ, Vodovotz Y, Billiar TR, Ermentrout GB & Bahar I (2006). Bistability in apoptosis: roles of Bax, Bcl-2, and mitochondrial permeability transition pores. Biophysical Journal 90:1546–1559.

Baillat D, Laitem C, Leprivier G, Margerin C & Aumercier C (2008). Ets-1 binds cooperatively to the palindromic Ets-binding sites in the p53 promoter. Biochemical and Biophysical Research Communications 378: 213–217.

Batchelor E, Loewer A, Mock C & Lahav G (2011). Stimulus-dependent dynamics of p53 in single cells. Molecular Systems Biology 7:488; doi:10.1038/msb.2011.20

Batchelor E, Mock CS, Bhan I, Loewer A & Lahav G (2008). Recurrent Initiation: A Mechanism for Triggering p53 Pulses in Response to DNA Damage. Molecular Cell 30: 277–289.

Bergamaschi D, Samuels Y, Jin B, Duraisingham S, Crook T & Lu X (2004). ASPP1 and ASPP2: Common Activators of p53 Family Members. Molecular and Cellular Biology 24: 1341-1350.

Bernards R (2004). Wip-ing out cancer.Nature Genetics 36: 319-320.

Boogs K & Reisman D (2007). C/EPBβ Participates in Regulating Transcription of the p53 Gene in Response to Mitogen Stimulation. The Journal of Biological Chemistry 282: 7982–7990.

Braithwaite AW (2006). Some p53-binding proteins that can function as arbiters of life and death. Cell Death and Differentiation 13: 984–993.

Budirahardja Y & Gönczy P (2009). Coupling the cell cycle to development. Development 136:2861-2872.

Calonge TM & O'Connell MJ (2008). Turning off the G2 DNA damage checkpoint. DNA Repair 7:136–140.

Candeias MM, Malbert-Colas L, Powell DJ, Daskalogianni C, Maslon MM, Naski N, Bourougaa K, Calvo F & Fähraeus R (2008). p53 mRNA controls p53 activity by managing Mdm2 functions. Nature Cell Biology 19: 1098-1105.

Chen C, Edelstein LC & Gélinas C (2000). The Rel/NF-kB Family Directly Activates Expression of the Apoptosis Inhibitor Bcl-xL. Molecular and Cellular Biology 20:2687–2695.

Cui J, Chen C, Lu H, Sun T, Shen P (2008). Two independent positive feedbacks and bistability in the Bcl-2 apoptotic switch. Plos One 3: e1469; doi:10.1371/journal.pone.0001469.

Daniel P. Teufel, Stefan M. Freund, Mark Bycroft, and Alan R. Fersht 2007 Four domains of p300 each bind tightly to a sequence spanning both transactivation subdomains of p53. PNAS, 104: 7009–7014.

Del Vecchio D, Ninfa AJ & Sontag ED (2008). Modular cell biology: retroactivity and insulation. Molecular Systems Biology 4:161; doi:10.1038/msb4100204

Díaz J (2011). Information flow in plant signaling pathways. Plant Signaling & Behavior 2011 6: 1-5.

Dogu Y & Díaz J (2009). Mathematical model of a network of interaction between p53 and Bcl-2 during genotoxic-induced apoptosis. Biophysical Chemistry 143:44–54.

Edelstein-Kesher L (2005). Mathematical Models in Biology. Classics in Applied Mathematics. Siam. Philadelphia, PA. ISBN: 978-0-898715-54-5.

Elowitz MB, Levine AJ, Siggia ED, Swain PS (2002). Stochastic Gene Expression in a Single Cell. Science 297: 1183-86.

Fussenegger M, Bailey JE, Varner J (2000) A mathematical model of caspase function in apoptosis. Nature Biotechnology 18:768–774.

Geva-Zatorsky N, Rosenfeld N, Itzkovitz S, Milo R, Sigal A, Dekel E, Yarnitzky T, Liron Y, Polak P, Lahav G & Alon U (2006). Oscillations and variability in the p53 system. Molecular Systems Biology 2:2006.0033; doi:10.1038/msb4100068.

Green DR & Kroemer G (2009). Cytoplasmic functions of the tumour suppressor p53. Nature 458: 1127-1130.

Haupt S, Alsheich-Bartok O and Haupt Y (2006) Clues from worms: a Slug at Puma promotes the survival of blood progenitors. Cell Death and Differentiation 13:913–915.

Hu W, Feng Z, Ma L, Wagner J, Rice JJ, Stolovitzky G & Levine AJ (2007). A Single Nucleotide Polymorphism in the MDM2 Gene Disrupts the Oscillation of p53 and MDM2 Levels in Cells. Cancer Research 67: 2757-2765.

Hua F, Cornejo MG, Cardone MH, Stokes CL & Lauffenburger DA (2005). Effects of Bcl2 Levels on Fas signaling-induced caspase-3 activation: molecular genetic tests of computational model predictions. The Journal of Immunology 175:985–995.

Ikenoue T, Inoki K, Zhao B & Guan K (2008). PTEN Acetylation Modulates Its Interaction with PDZ Domain. Cancer Research 68: 6908-6912.

Jänicke RU, Sohn D & Schulze-Osthoff K (2008). The dark side of a tumor suppressor: anti-apoptotic p53. Cell Death and Differentiation 15: 959–976.

Jolma IW, Ni XY, Rensing L, & Ruoff P (2010). Harmonic Oscillations in Homeostatic Controllers: Dynamics of the p53 Regulatory System. Biophysical Journal 98: 743–752.

Kaufmann BB & van Oudenaarden A (2007). Stochastic gene expression: from single molecules to the proteome. Current Opinion in Genetics & Development 17:107–12.

Kobet E, Zeng X, Zhu Y, Keller D & Lu H (2000). MDM2 inhibits p300-mediated p53 acetylation and activation by forming a ternary complex with the two proteins. PNAS 97:12547–12552.

Lacroix M, Toillon R & Leclercq G (2006). p53 and breast cancer, an update. Endocrine-Related Cancer 13: 293–325.

Lahav G, Rosenfeld N, Sigall A, Geva-Zatorsky N, Levine AJ, Elowitz MB & Alon U (2004). Dynamics of the p53-Mdm2 feedback loop in individual cells. Nature Genetics 36:147-150.

Liu Z, Lu H, Shi H, Du Y, Yu J, Gu S, Chen X, Liu KJ & Hu CA (2005). PUMA Overexpression Induces Reactive Oxygen Species Generation and Proteasome-Mediated Stathmin Degradation in Colorectal Cancer Cells. Cancer Res. 65:1647-1654.

Lu X, Nguyen T, Moon S, Darlington Y, Sommer M & Donehower LA (2008). The type 2C phosphatase Wip1: An oncogenic regulator of tumor suppressor and DNA damage response pathways. Cancer Metastasis Rev. 27:123–135

Ma L, Wagner J, Rice JJ, Hu W, Levine AJ & Stolovitzky GA (2005). A plausible model for the digital response of p53 to DNA damage. PNAS 102: 14266–14271.

Malumbres, M & Barbacid M (2009). Cell cycle, CDKs and cancer: a changing paradigm. Nature Reviews Cancer 3:153-166.

Marquez RT, Baggerly KA, Patterson AP, Liu J, Broaddus R, Frumovitz M, Atkinson EN, Smith DI, Hartmann L, Fishman D, Berchuck A, Whitaker R, Gershenson DM, Mills GB, Bast RC & Lu KH (2004). Selection of Potential Markers for Epithelial Ovarian Cancer with Gene Expression Arrays and Recursive Descent Partition Analysis. Clinical Cancer Research 11: 3291–3300.

Méndez O, Martín B, Sanz R, Aragüés R, Moreno V, Oliva B, Stresing V & Sierra A (2006) Underexpression of transcriptional regulators is common in metastatic breast cancer cells overexpressing Bcl-xL. Carcinogenesis, 27:1169–1179.

Michael Karin & Anning Lin (2002). NF-κB at the crossroads of life and death. Nature Immunology 3: 221-227.

Patel S, George R, Autore F, Fraternali F, Ladbury JE & Nikolova PV (2008). Molecular interactions of ASPP1 and ASPP2 with the p53 protein family and the apoptotic promoters PUMA and Bax. Nucleic Acids Research 36: 5139–5151.

Pediconi N, Guerrieri F, Vossio S, Bruno T, Belloni L, Valeria Schinzari V, Scisciani C, Fanciulli M & Levrero M (2009). hSirT1-Dependent Regulation of the PCAF-E2F1-p73 Apoptotic Pathway in Response to DNA Damage. Molecular an Cellular Biology 29:1989–1998.

Perry ME (2004). Mdm2 in the Response to Radiation. Molecular Cancer Research, 2: 9-19.

Phelps M, Darley M, Primrose JN & Blaydes JP (2003). p53-independent Activation of the hdm2-P2 Promoter through Multiple Transcription Factor Response Elements Results in Elevated hdm2 Expression in Estrogen Receptor a-positive Breast Cancer Cells. Cancer Research63: 2616–2623.

Pietsch EC, Sykes SM, McMahon SB & Murphy ME (2008).The p53 family and programmed cell death. Oncogene 27:6507–6521.

Plati J, Bucur O & Khosravi-Far R (2011). Apoptotic cell signaling in cancer progression and therapy. Integrative Biology 3: 279–296

Pommier Y & Weinstein JN (2006). Chk2Molecular InteractionMap and Rationale for Chk2 Inhibitors. Clin Cancer Res. 12: 2657-2661.

Qi JP, Shao SH, Xie J & Zhu Y (2007). A mathematical model of P53 gene regulatory network under radiotherapy. BioSystems 90: 698–706

Robinson RA, Lu X & Edith Yvonne Jones EY (2008). Biochemical and Structural Studies of ASPP Proteins Reveal Differential Binding to p53, p63, and p73. Structure 16 : 259–268.

Schmid JA & Birbach A (2008). IκB kinase β (IKKβ/IKK2/IKBKB)—A key molecule in signaling to the transcription factor NF-κB. Cytokine & Growth Factor Reviews 19:157–165.

Shreeram S, Demidov ON, Hee WK, Yamaguchi H, Onishi N, Kek C, Timofeev ON, Dudgeon C, Fornace AJ, Anderson CW, Minami Y, Appella E & Bulavin DV (2006). Wip1 Phosphatase Modulates ATM-Dependent Signaling Pathways. Molecular Cell 23: 757–764.

Spencer LS & Sorger PK (2011). Measuring and Modeling Apoptosis in Single Cells. Cell 144:926-939.

Stewart RD (2001). Two-Lesion Kinetic Model of Double-Strand Break Rejoining and Cell Killing. Radiation Research 156: 365–378.

Strano S, Dell'Orso S, Mongiovi AM, Monti O, Lapi E, Di Agostino S, Fontemaggi G & Blandino G (2006). Mutant p53 proteins: between loss and gain of function. Head and Neck 29: 488-496.

Strogatz S (1994). Nonlinear Dynanics and Chaos. With applications to Physics, Biology, Chemistry, and Engineering. Perseus Books Publishing, LLC. Cambridge, MA. ISBN: 978-0-7382-0453-6.

Sun Y, Jiang X, Chen S, Fernandes N & Price BD (2005). A role for the Tip60 histone acetyltransferase in the acetylation and activation of ATM. PNAS 102: 13182–13187.

Tanaka T, Huang X, Halicka HD, Zhao H, Traganos F, Albino AP, Dai W, Darzynkiewicz Z (2007). Cytometry of ATM Activation and Histone H2AX Phosphorylation to Estimate Extent of DNA Damage Induced by Exogenous Agents. Journal of the International Society for Analytical Cytology 71A: 648-661.

Tanno M, Sakamoto J, Miura T, Shimamoto K & Horio Y (2006). Nucleo-cytoplasmic shuttling of NAD+ -dependent histone deacetylase SIRT1. The Journal of Biological Chemistry, 282: 6823-6832.

Thieffry D & Romero D (1999). The modularity of biological regulatory networks. Biosystems 50:49-59.

Toledo F & Wahl GM (2006). Regulating the p53 pathway:in vitro hypotheses, in vivo veritas. Nature Reviews Cancer 6: 909-923.

Vissers JHA, Nicassio F, van Lohuizen M, Di Fiore P, Citterio E (2008). The many faces of ubiquitinated histone H2A: insights from the DUBs. Cell Division 3:8; doi:10.1186/1747-1028-3-8

Vousden, Karen H (2006). Outcomes of p53 activation - spoilt for choice. Journal of Cell Science 119: 5015-5020.

Wagner J, Ma L, Rice JJ, Hu W, Levine AJ & Stolovitzky G (2005). p53–Mdm2 loop controlled by a balance of its feedback strength and effective dampening using ATM and delayed feedback. IEE Proc.-Syst. Biol. 152:109-118.

Wang P, Yu J & Zhang L (2007). The nuclear function of p53 is required for PUMA-mediated apoptosis induced by DNA damage. PNAS 104:4054–4059.

Wang S & El-Deiry WS(2006) p73 or p53 Directly Regulates Human p53 Transcription to Maintain Cell Cycle Checkpoints. Cancer Research 66: 6982-6989 .

Wu H & Lozano G (1994). NF-κB Activation of p53 A POTENTIAL MECHANISM FOR SUPPRESSING CELL GROWTH IN RESPONSE TO STRES. The Journal of Biological Chemistry, 269:20067-20074.

Yamamori T, DeRicco J, Naqvi A, Hoffman TA, Mattagajasingh I, Kasuno K, Jung S, Kim C & Irani K (2009) SIRT1 deacetylates APE1 and regulates cellular base excision repair. Nucleic Acids Research 38: 832-845.

Yoda A, Toyoshima K, Watanabe YY, Onishi N, Hazaka Y, Tsukuda Y, Tsukada J, Kondo T, Tanaka Y & Minami Y (2008). Arsenic Trioxide Augments Chk2/p53-mediated Apoptosis by Inhibiting Oncogenic Wip1 Phosphatase. The Journal of Biological Chemistry 283: 18969–18979.

Youle RJ & Strasser A (2008). The BCL-2 protein family: opposing activities that mediate cell death. Nature Reviews in Molecular Cell Biology 9:47-59.

Zhang J, Tsaprailis G & Bowden GT (2008) Nucleolin Stabilizes Bcl-XL Messenger RNA in Response to UVA Irradiation. Cancer Research 68: 1046-1054.

Zhang X, Feng Liu F, Cheng Z & Wang W (2009). Cell fate decision mediated by p53 pulses. PNAS 106: 12245–12250.

14

Modelling Radiation Health Effects

Noriyuki B. Ouchi
Nuclear Science and Engineering Directorate,
Japan Atomic Energy Agency,
Japan

1. Introduction

Recent years, assessment of low-dose radiation risk have been of increasing importance in an area of public-health investigation because of an increase in many types of exposures, e.g. medical exposures, exposure of astronauts and the cosmic radiation exposure of aircraft crew, and for the radiation-protection viewpoints.

We still have many outstanding questions on health effects of radiation. Generally, the most demanded question is "how much radiation (dose) produces how much effect?" this is the so-called dose-response relationship (Fig.1), especially at low dose region (BEIR VII phase 2, 2006, UNSCEAR 2006 Report Vol. I, Annex A, 2006).

In general, dose-response relationship at low dose region (almost below 100mSv) have no statistical significance, consequently linear extrapolation from high dose region is used for explanation.

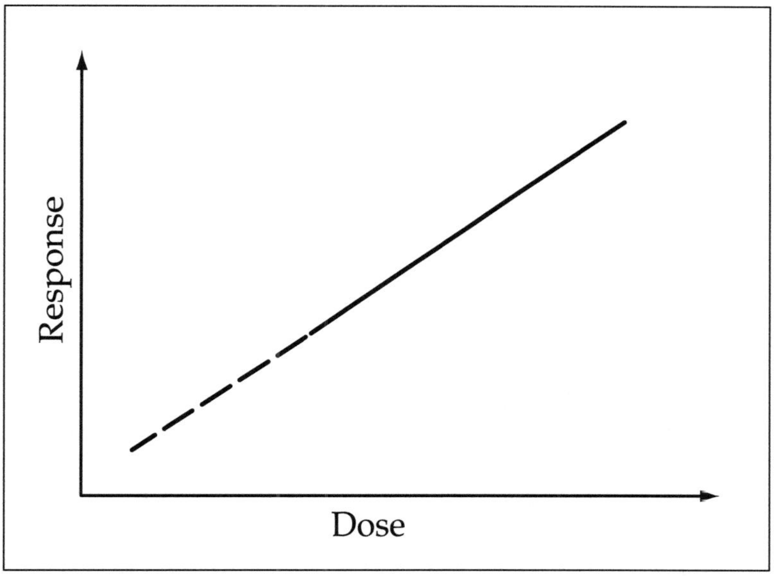

Fig. 1. Dose-response relationship of radiation effects. In general, cancer incidence is used for the response of radiation dose. Dashed line shows linear extrapolation from high-dose region.

At this point, we had better scientifically define the term "radiation effects". Well, then what is the radiation effect?

Generally speaking, biological radiation effects can be classified in two basic categories, deterministic effects and stochastic effects depending on radiation dose and exposure time, i.e., deterministic effects are consequences of acute high radiation dose. Examples of deterministic effects include skin erythema (reddening), radiation induced cataract, hair loss, radiation sickness (nausea, vomiting and diarrhoea), sterility and depression of red blood cell formation. These effects are due to cell killing, that is to say, they will occur when the cell killing is large enough to cause functional damage of tissue or organ. Therefore, there is a threshold for deterministic effect. In other words, deterministic effects are associated with intermediate to high dose region and the dose-response relationship is fitted well with linear non-threshold (LNT) model which is derived from epidemiological data, e.g., data from A-bomb survivors (Brenner et al., 2003).

On the contrary, stochastic effects are generally associated with long-term low dose exposures. As the name suggests, stochastic effects are described only by the probability. These phenomena (effects) are primarily many types of cancer. Here, we can now define the health effects of radiation at low dose region as carcinogenesis/tumorigenesis.

2. Radiation health effect problem at a glance

Historically, health effects of radiation exposure, especially by high-dose radiation, became apparent just after the diccovery of X-ray (BEIR VII phase 2, 2006). In 1895, Wilhelm Conrad Röentgen was investigating an electrical discharge generated from the various types of vacuum tube. During his experiments, he discovered a fluorescent effect on a small black cardboard screen covering vacuum tube and he thought that this fluorescent effect was a consequence of invisible emissions from his experimental equipment. He named these invisible emissions as X-rays, using the mathematical expression for unknown variable.

After the Röentgen's discovery of man-made radiation, Henri Becquerel discovered naturally occurring radiation emitted from uranium salts in 1896. He reported this discovery at 1896's proceedings of the French Academy of Sciences under the title "Sur les radiations émises par phosphorescence", [on the radiation emitted by phosphorescence], in english.

Furthermore, Marie and Pierre Curie succeeded in purification and concentration of uranium ore, pitchblende, and they found polonium. Marie also introduced new term "radio-active". Thus, both man-made and naturally occurring radiation were discovered within several years in the 1890's.

Because of the invisibility nature of radiation, unfortunately, many types of adverse health effects of radiation became apparent shortly after these early scientifical discoveries.

Most of the reported health effects in this period are the acute effects such as redness of the skin (erythema), hair loss, decrease in the number of white blood cells, tumorigenesis, and so on. In fact, many of radiation related researchers or radiologists had slow-healing skin lesions in their hands or even died by radiation induced diseases. The most famous example is the case of Thomas Edison's assistant, Clarence Dally. During Edison's development of fluoroscopy which is a machine using X-rays to take radiographs, Edison demonstrated his machine by the contribution of his assistant Dally. Thus, Dally died by tumor which was associated with too much exposure. As a consequence, scientists began to know about the potential of radiation to damage human health.

Finally, in 1915, the British Röentgen Society made probably the first adoption of radiation protection recommendations. Therefore, many international or national organizations for radiation health effects were established to meet the growing interest in the subject. At the early stage of radiation effects studies, the main purpose of study or recommendation was radiation protection for patients from diagnostic medical X-ray exposure and also for radiologists. However, stochastic effects of radiation exposure, particularly at low dose, also began to be recognized, during the first few decades of the twentieth century. In spite of the accumulation of knowledge concerning harmful indications of radiation to human body, biological mechanism for the effects of radiation on human body was still lacking the precise knowledge.

Generally speaking, only epidemiological study is the tool to reveal the long-term stochastic effects of radiation. In fact, we need well-controlled, large number and long-term epidemiological studies to investigate late effects (stochastic effects) of radiation health effects, because the numbers of such stochastic effects are too small to determine statistical significance. Although, there have been a small number of epidemiological studies to satisfy these requirements, we can give examples of such well known studies, A-Bomb survivors of Hiroshima and Nagasaki (Preston, 1998, Brenner et al., 2003), 100 years of British radiologists study (Berrington et al., 2001), study of Chernobyl accident (UNSCEAR 2008 Report Vol. II, Annex D, 2008) and international collaborative study of nuclear workers (Cardis et al., 2007, Thierry-Chef et al., 2007, Vrijheid et al., 2007). Adding to these epidemiological studies base on the "population" of considering subjects, more detailed study of biological radiation effects has started due to the birth of molecular biology, by discovery of DNA's molecular structure by Watson and Crick (Watson and Crick, 1953). In other words, it was the discovery of molecular structure as a "target" of radiation. It is clear that the study of radiation health effects was changed qualitatively and also quantitatively since the birth of molecular biology. It has been believed that the first target of radiation health effects (carcinogenesis) is DNA.

Currently, there are two major methods to investigate the radiation effects on humans, one is the epidemiology which aims to study population level radiation effects the other is the molecular biology which is intended to clarify the radiation effects at cellular or intracellular level. In general, it is known that carcinogenesis by low dose radiation will start from DNA damage by ionizing radiation. Then, these very small effects will appear on a cellular scale by accumulation of various intracellular biological responses with spending long time and finally grow to the tumor with clonal expansion of cancer cell. Fig. 2 is the known hierarchy of biological mechanisms of radiation effects.

Thus, the biological radiation effects are considered as chain reaction phenomena with a very wide scale; DNA damage (space: 10^{-9}m and time: 10^{-6}sec) to the tumorigenesis (space: 10^{-3} m and time: 10^{5} sec). This is one of the reasons why study of low dose radiation health effects is difficult (Adams & Jameson, 1980).

From the viewpoint of elucidation of dose-response relationship, we should connect small scale dynamics (e.g. DNA damage) with relatively quite large scale dynamics (e.g. tumorigenesis). As shown in Fig.2, we can see that many "subsystems" exist between DNA damage and tumorigenesis. One simple method to solve the problem is the integration of each subsystem's dynamics. However, it is easy to plan but it is very difficult to implement.

Because of the existence of intrinsic uncertainties originate from their own dynamics in each of the subsystems, integration or connection of subsystems will increase whole system's uncertainty (Jackson, 1991). This is the unpredictable nature of nonlinear systems.

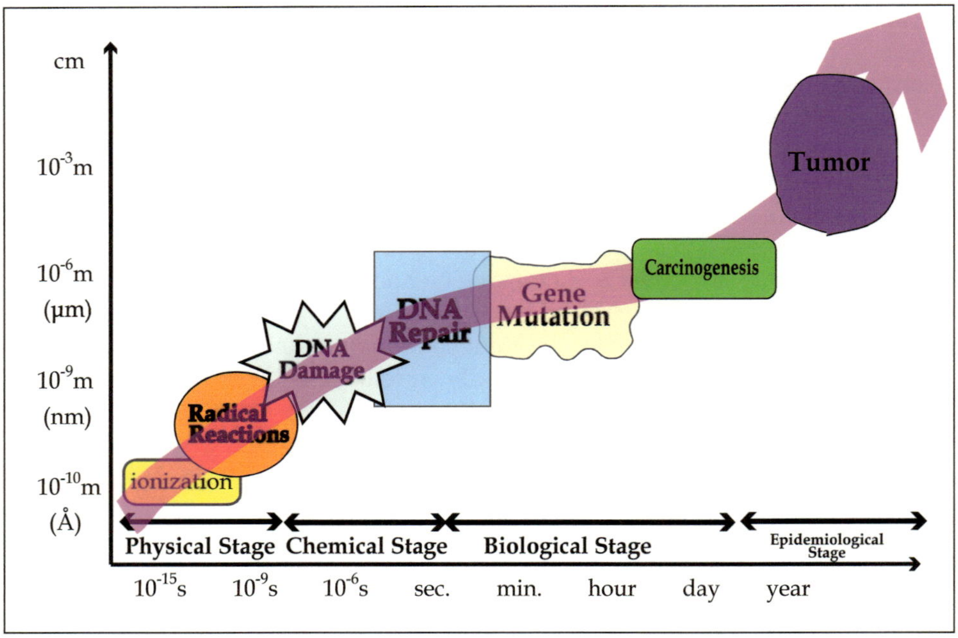

Fig. 2. Spatiotemporal order of radiation effects. Each event will progress to the tumor in a chain reaction as an arrow in the figure.

There are no other problems like radiation health effects study that has much broader spatiotemporal scale, it has at least twenty-four orders of magnitude in time and about ten orders of magnitude in space. Therefore, this will be a very challenging problem in the field of natural science.

Here, another practical difficulty of the problem is noted. In addition to the scaling complexity mentioned above, there is a statistical significance problem associated with sample sizes in low dose radiation health effects problem. For example, due to the inappreciable cancer incidence or infrequent cellular level effects at low dose radiation exposure, we need a large number of experimental data to detect mutations at low dose in a cell culture system, also huge number of data is needed to distinguish the net radiation effects from naturally occurring one in a epidemiological study, however it is practically difficult to get such a large number of data in both cases. To overcome these difficulties in the low-dose radiation effects' problem, it is considered a good approach to study the process of carcinogenesis using mathematical model which is based on biological mechanism. Next, we will show some examples of mathematical model approaches.

3. Modelling radiation effects – Dynamical system's view

As described in the previous section, radiation health effects study has much broader spatiotemporal scale. Here, let us begin our analysis by reducing this broad problem into more simple form. Practically, epidemiological problem concerning the population is going to be statistical problem if it turns out the precise incidence of carcinogenesis for each person. Then, what is the cancer incidence for each person? Tumor will be apparent only when it grows to a certain size however the incidence depends on the accuracy of diagnosis, it may define tumor growth at a certain size as one of the endpoints of this broad problem.

Generally speaking, tumorigenesis is thought to be a multistage process with gradual accumulation of mutations in a number of different genes (Fearon and Vogelstein, 1990). How many mutations are required to transform a single normal cell to a cancer cell? This question still has no answer (Sjöblom et al., 2006). Moreover, most of the combinations of mutations in tumors which taken from patients with same cancer type are almost different (Smith et al., 2002). However, biological pathway that transforms normal cell into cancer cell may not be a single pathway, there are well known three sequential state changes, i.e. initiation, promotion and progression (IPP) (Trosko, 1992). This is a so-called multistage carcinogenesis model. Moreover, it is a conceptual dynamics rather than an actual dynamics. The origin of these three processes is thought to be the result of the damage of the chromosomal DNA and failure of its repair. The initiation is defined basically as irreversible changes to target cell, followed by the gene mutations. The stage of promotion, neoplastic development, is believed to be the consequence of the damage of the specific gene expression, e.g., lipid metabolites, cytokines, so the neoplasm may get the enhancement of cellular growth potential, i.e., they lost the intracellular communication. The last stage progression is thoght to be a process that the cell acquires malignancy. Cleary, there are two different scale dynamics in this neoplastic process, one is intracellular change process via mutagenic changes and the other is extracellular change or cell group dynamics via cell-cell or cell-environment interactions. Here we may define, from the viewpoint of dynamical systems, "cancer cell" as the uncontrolled neoplastic cell which is developed by intracellular change and "tumor" as the aggregate of such neoplastic cells. In a clinical definition of cancer, there exists a concept "malignancy" which is based on invasion and metastasis; however, it would be a one of the aspects of the tumor growth process. Schematic explanation of tumorigenesis in association with IPP concept is shown in Fig. 3.

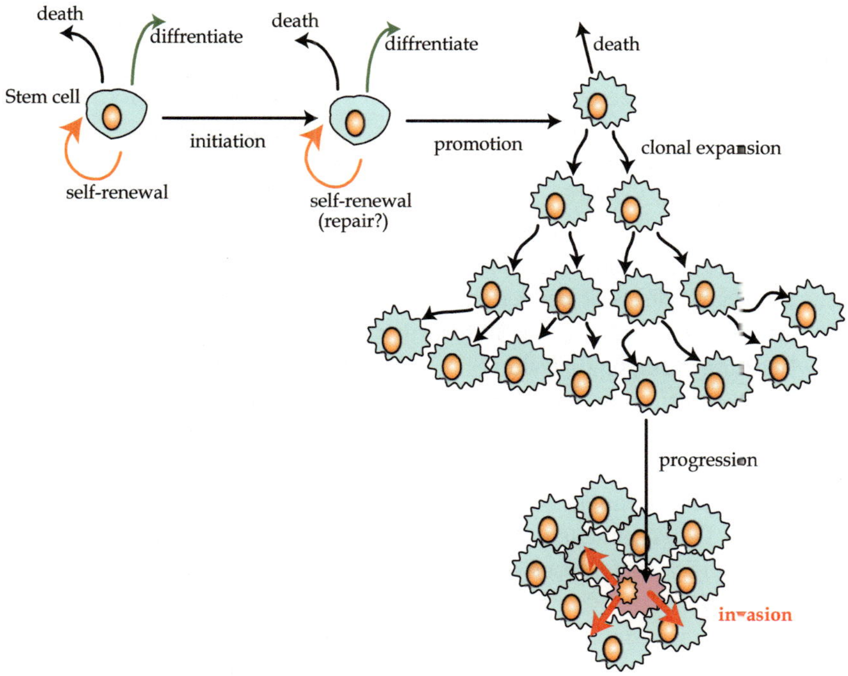

Fig. 3. Schematic explanation of tumorigenesis based on IPP concept.

This is a reduction from large spatiotemporal scale side, then, initial process is unwired for the purpose of modelling. How does radiation induce cancer? In other words, how does radiation induce mutagenic change in a cell? There are several possible pathways to the DNA damages from radiation. In general, radiation will ionize DNA by direct atomic collisions or by indirect manner, as described below.
When an ionizing radiation passes through a water molecule, ionization of water will occur:

$$H_2O \rightarrow \bullet H_2O^+ + e^-, \quad (1)$$

then ionized water molecule reacts with another water molecule and produce an hydroxyl radical ($\bullet OH$):

$$\bullet H_2O^+ + H_2O \rightarrow H_3O^+ + \bullet OH, \quad (2)$$

or $\bullet H_2O^+$ breaks up:

$$\bullet H_2O^+ \rightarrow H^+ + \bullet OH. \quad (3)$$

Ejected electron is trapped by polarizing water molecules and will produce hydrated electron e_{aq}^-:

$$e^- + nH_2O \rightarrow e_{aq}^-, \quad (4)$$

and e_{aq}^- will produce free radical $\bullet H$ according to:

$$e_{aq}^- + H_2O \rightarrow OH^- + \bullet H \quad (5)$$

or

$$e_{aq}^- + H^+ \rightarrow \bullet H. \quad (6)$$

Thus free radicals $\bullet OH$, $\bullet H$, and e_{aq}^- are produced by ionizing water molecules. These free radicals are thought to induce DNA damage indirectly; this is so called indirect action. Over the past decades, numerous studies have been made on the initial process of DNA damage using Monte Carlo track structure method (Nikjoo et al, 2006).
Currently, radiation induced DNA damages are roughly classified into four types, 1) base damage, 2) base release, 3) strand break, and 4) crosslink. Sometimes damage of the DNA is lethal to a cell so various repair mechanisms exist depending on the types of damage. For example, the necessary yields of DNA damage to kill 63% of irradiated cells are known, e.g., 1000 for the single strand break (ssb), 40 for the double strand break (dsb), and 150 for the crosslinks, where the number of lesions per cell per D_{37}, and D_{37} is dose of 37% survival (BEIR V, 1990). Scientific explanation about the rate of DNA repair needs further investigation. It depends on various factors, including cell type, age of the cell, extracellular environment and also the type of DNA damage. Generally, it is thought that these DNA damages are completely repaired, but sometimes there will be unrepaired DNA damages or incorrectly repaired DNA damages, and these *prognoses* of DNA damage will determine the cell fate. On a cell which has a large amount of DNA damage accumulation, senescence or apoptosis may occur. On the other hand, misrepaired DNA damage is so-called mutation. It is believed that the accumulation of mutations will induce neoplastic transformation. Schematic explanation of initial process of radiation effects which is concerned with DNA damage and mutation is shown in Fig. 4. As described above, detailed dynamics from DNA

damage to mutation still lacking scientific explanation, it is difficult to make mathematical models; it seems to be dominated by some probability.

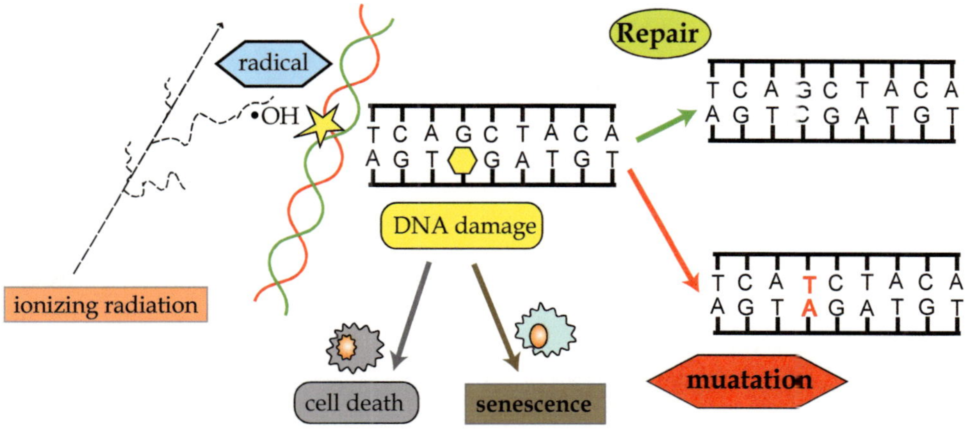

Fig. 4. Simplified schematic explanation of initial process of radiation effects, from DNA damage to mutation.

If tumor growth is set as an endpoint for the radiation health effects study, it is good to consider DNA damage, mutagenesis, carcinogenesis and tumorigenesis as the midpoints of the problem. Moreover, it is appropriate to divide roughly the dynamics into intracellular phenomena and extracellular or cell group phenomena because of the dynamics is thought to be dominated by stochastic in intracellular scale or deterministic over cellular scale. Hereafter, mathematical model of cellular scale are introduced.

4. Mathematical model

In fact, there are not so many mathematical models to investigate radiation induced cancer, but there are many types of mathematical models to study common cancer. Detailed mechanism of carcinogenesis is still not well defined concerning carcinogenesis which is induced not only by radiation but also by other carcinogens. Actually, there are not so many radiation specific effects on cancer; there are no diagnosis methods of radiation specific cancer. One of the radiation specific effects is its penetration to whole human body, or it will only appear as a specific type or character of DNA damages, not in cellular level.
Here, some examples of mathematical model of carcinogenesis are introduced even though most of them are not radiation-specific. However there are many mathematical models of carcinogenesis than radiation specific models, it may be helpful in modelling radiation carcinogenesis.

4.1 Modeling carcinogenesis/tumorigenesis in general

There are several types of mathematical models of tumor growth or carcinogenesis, e.g. simple temporal population based dynamics, models based on diffusion or reaction-diffusion type dynamics, individual cell based models, multi-scale models, and so on (Araujo and McElwain, 2004). These models may be classified into three categories, depending on the mathematical expression of the cell, 1) population based (no spatial structure), 2) spatial model which consider tumor as a one continuous density, 3) individual or single-cell based models. Adding to this classification, dynamical systems could be classified in two or more categories depending on its status: whether the variable (space, time) is continuous or discrete, whether the process is stochastic or deterministic. For example, reaction-diffusion model which is constructed by partial differential equations is deterministic and continuous, models of Cellular Automata (CA) with some rule are discrete and deterministic. Choice of the variable type and the process in constructing mathematical model is very important and it is dependent on the problem to solve. These are classification based on mathematics of the model. Likewise, the endpoint setting is very important for modelling process, because of the endpoint setting will affect the mathematical modelling strategies, i.e., top-down or bottom-up approaches. For the instance, if the cancer growth or age of cancer incidence is the endpoint to solve then reproducible model construction is the purpose of mathematical modelling. Therefore, constructed models often do not reflect specific biological mechanisms, or rather the model seems to be descriptive. On the contrary, the specific intracellular metabolic network is modeled to see some emergent behavior of considering system, or rather the model seems to be mechanistic. The former is also called top-down approach, the latter bottom-up approach.

Hereafter, some of these models will be introduced shortly. One of the classical models in this subject is Hill's diffusion based model (Hill, 1928). Hill's idea is that the diffusion of dissolved oxygen through tissues is an important factor by metabolic process. He did not apply his model to tumor growth explicitly, however his idea affect the later mathematical models of solid tumor growth.

Gatenby and Gawlinski (Gatenby and Gawlinski, 1996) made a simple Reaction-Diffusion based model with continuum cell population for the cancer invasion. They made a hypothesis which based on experimental evidences, that tumor-induced alteration of microenvironmental pH is the important key-role for mechanism of cancer invasion. They use simple reaction-diffusion equations with three field variables, $N_1(x,t)$, the density of normal tissue, $N_2(x,t)$, the density of neoplastic tissue, and $L(x,t)$, the excess concentration of H^+ ions, where x is for 1-D spatial coordinate and t is for time. It should be noted that this model include neither dynamics of early tumor formation, i.e. intracellular genetic changes, nor large-scale morphological features of tumor. That is to say, they modeled only microscopic scale population dynamics at the tumor-host interface. These dynamics can be written as following equations.

$$\frac{\partial N_1}{\partial t} = r_1 N_1 \left(1 - \frac{N_1}{K_1} - \alpha_{12}\frac{N_2}{K_2}\right) - d_1 L N_1 + \nabla \cdot \left(D_{N_1}[N_2]\nabla N_1\right) \tag{7}$$

$$\frac{\partial N_2}{\partial t} = r_2 N_2 \left(1 - \frac{N_2}{K_2} - \alpha_{21}\frac{N_1}{K_1}\right) + \nabla \cdot \left(D_{N_2}[N_1]\nabla N_2\right) \tag{8}$$

$$\frac{\partial L}{\partial t} = r_3 N_2 - d_3 L + D_3 \nabla^2 L \tag{9}$$

The first part in Eq. (1) is growth term of normal tissue based on competitive Lotka-Volterra type equation with competition strength parameter α_{12}, where r_1 is growth rate of normal tissue, K_1 the carrying capacity, and the second term $d_1 L N_1$ is death rate of normal tissue proportional to L, and the last part shows cellular diffusion with N_2 dependent diffusion coefficient $D_{N1}[N_2]$. Similarly, the first part in equation of neoplastic tissue growth (2) is growth term of neoplastic tissue with Lotka-Volterra competition parameter α_{21}, and second term is also cellular diffusion with N_1 dependent diffusion coefficient $D_{N2}[N_1]$ where r_2 is growth rate, K_2 the carrying capacity. Finally, equation (3) shows dynamics of excess H^+ ion. Production rate of excess H^+ ions is assumed to proportional to the neoplastic cell density N_2 and also diffuse chemically. The first and second part in equation (3) show excess H^+ ion production with rate r_3 and reabsorption term with rate d_3, and the last part is diffusion with coefficient D_3. With this model, they can predict a pH gradient extending from tumor-host interface, and benign to malignant transition which consistent with experimental data.

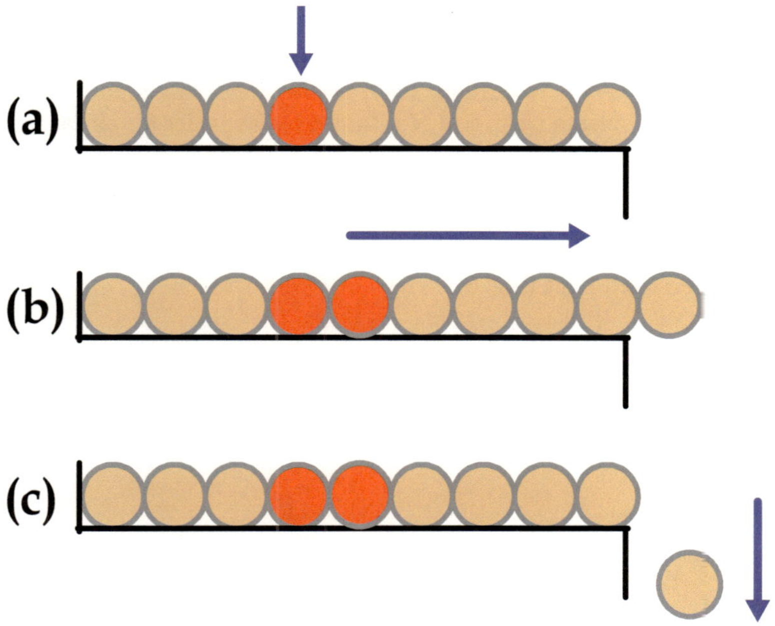

Fig. 5. The linear process. Redrawn from figure 1b of Novak et al. (2003).

Next, as an example of stochastic model, the linear process is introduced (Novak, Michor and Iwasa, 2003). The linear process represents stem cells and differentiated cells with rule based cell division process. The model also has spatial structure, so-called compartment. It is very simple model. As shown in Fig. 5, N cells (9 in Fig. 5) are placed in a linear array and at each time step, one cell is chosen for reproduction proportional to its "fitness". Then, the cell is replaced by two daughter cells and all cells beyond considering site are shifted one unit. The cell over the last edge falls off, it's a representation of apoptosis. Repeating this rule and time goes on. The term fitness is one of the key concepts of this model and it is derived from reproductive rate of cell. One of the purposes of this model is to obtain fixation probability, it is a concept of evolutional dynamics, i.e. the probability that one considering cell will take over the whole tissue. Despite the model has a spatial structure, the linear process could be categorized as a kind of population dynamics.

It is apparent that sufficient vacant space adjacent to neoplastic cells is needed for its growth. Thus the invasion is very important in case of insufficient vacant space for neoplastic cells as in specific organ cancer. Malignancy of neoplastic cell arises by acquirement of invasion and metastasis capability. On the modelling of tumorigenesis process, especially of solid cancer, existence of adjacent cells should be aware of. One of the important key dynamics of the invasion and metastasis is adhesion and microenvironmental adaptation. Moreover, in a stage of metastasis, we should consider a dynamics of respective cells rather than that of cell group's. Thus, it is thought that individual based cell model is applicable to model metastasis. The hybrid discrete continuum (HDC) model (Anderson and Quaranta, 2008) is one of such individual based cell models. The HDC model represents cells as points on a lattice which represents cell microenvironment, i.e. density of ECM and the concentrations of proteases and nutrients. The dynamics of cell microenvironment factors is defined by a set of continuous partial differential equations (PDEs).

Finally, multi-stage model proposed by Armitage and Doll (Armitage and Doll, 1954, Moolgavkar, 2004, Frank, 2007) is introduced. Armitage and Doll proposed multistage model in 1954 which was derived from excellent statistical findings about cancer mortality (Fisher and Holloman, 1953, Nordling, 1953). Fisher and Hollomon used statistics of stomach cancer in women from USA, and Nordling used statistics for all cancer in men from USA, United Kingdom, France, and Norway. Their data showed that the logarithm of cancer mortality increased in direct proportion to the logarithm the age between the ages of 25 and 74, the scale of the log-log plot showed sixth power of age; in other words, cancer mortality will increase with approximately the sixth power of age, $P=\alpha t^6$, where P is for cancer death rate, α the parameter, *t* is for age (Fig. 6).

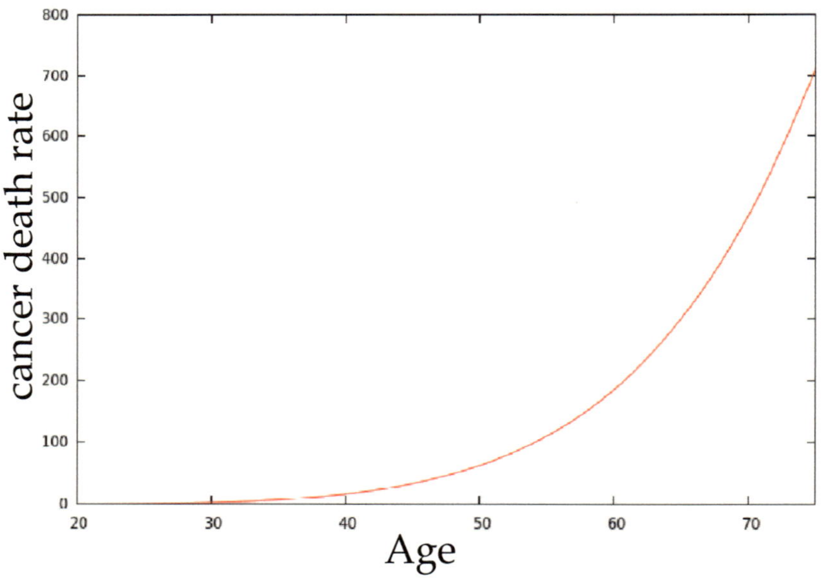

Fig. 6. Example plot of change in cancer mortality with age. Fisher and Hollomon, and Nordling used log-log scale for the plot of cancer mortality in their articles.

Thus, Armitage and Doll made a theory; human cancer is the end-result of several successive cellular changes and tested their theory by mathematical model (Fig. 7). Their model, so-called multistage model, has good agreement with many epidemiological data. Many followers of Armitage and Doll appear to refine multistage model to fit various specific epidemiological or biological data, thus there are many multistage variants (Frank, 2007).

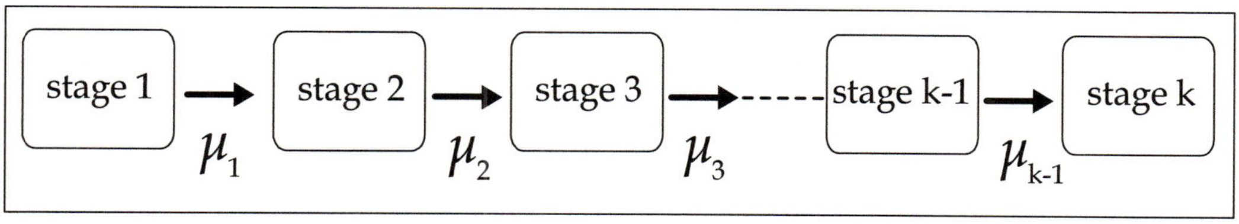

Fig. 7. Schematic explanation of Armitage-Doll multistage model. The cells acquire total (k-1) mutations with mutation rate μ_i where i=(1, k-1). The model supposes that an individual has a population of $N(t)$ normal cells at age t, then these cells acquire one mutation with a rate μ_1.Then cells with one mutation acquire a second mutation with rate μ_2, and so on. Cells with (k-1) are thought to be fully malignant cancer. Calculation of the population of malignant cells and its age are needed for statistical analysis.

The multistage theory could be summarized as follows,

$$P(t) = \frac{(Nu)^n t^{n-1}}{(n-1)!} \tag{10}$$

where P is cancer incidence, t is for age, n is the number of distinct carcinogenic events (number of stages -1), N is for number of cells at risk, u is for transformation rate per cell per unit time(Frank, 2007).

In spite of the multistage model was derived in 1954; it was about the same age as the discovery of DNA double helix, interestingly, it is thought to be a metaphor of biological fact: accumulation of carcinogenic mutations will induce cancer. Fig. 8 is a genetic multistage model of colorectal cancer (Fearon and Vogelstein, 1990). Today, carcinogenesis is known as the multistage genetic process with accumulation of mutagenic changes in the DNA. They derived mathematical structure from statistical data based on their mathematical knowledge. Their amazing modelling approach is instructive in constructing a model. However, the multistage model has some issues to be solved. For example, it seems to have no scientific evidence about the meaning of "stages" or number of mutations to fit the epidemiological data.

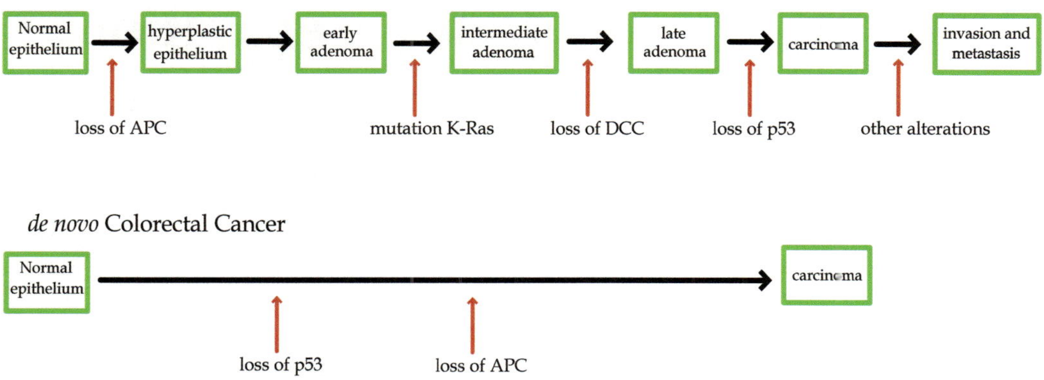

Fig. 8. Genetic multistage model of colorectal cancer. Most of colorectal cancer will follow upper pathway, however pathway like *de novo* type exists. Redrawn from figure 3 of Fearon and Vogelstein (1990).

4.2 Cellular large-*Q* Potts model

On the modelling of carcinogenesis process, it seems reasonable to introduce spatial structure to the model because of the fact that life and death of the cell is controled by its spatial strucuture, i.e., cell shape and cell size (Chen et al, 1997). These properties will not be ignored, because they will concern microenvironmental effects on the carcinogenesis process. Here we introduce one of the models of spatial cellular dynamics, called cellular large-*Q* Potts model (CPM) (Ouchi et al., 2003) which was first introduced by Graner and Glazier (Graner & Glazier, 1992, Glazier & Graner, 1993). The CPM is based on a biological hypothesis called differential adhesion hypothesis (DAH) first proposed by Steinberg (Steinberg, 1970). (Fig. 9). The problem *"How are tissues formed from populations of cells?"* is a one of the very famous questions in morphogenesis study (Gilbert, 2006, pp. 67). Steinberg showed certain cell types migrate centrally or peripherally depending on the accompanied cells, using cells derived from trypsinized embryonic tissues. From this observation, Steinberg made hypothesis that interacting cells will form an aggregate with the smallest interfacial energy because that is the thermodynamically stable. Therefore, cell adhesion strength of A-A is greater than A-B or B-B then sorting will occur. In the case of A-B strength is greater than or equal to strength of A-A or B-B then the aggregate will remain a randomly mixed structure. Finally, if the strength of A-B connection is so small compared to A-A or B-B connections, then A cells and B cells will make separate aggregates. It is considered that this mechanism is one of the principles of the dynamics of cell movement.

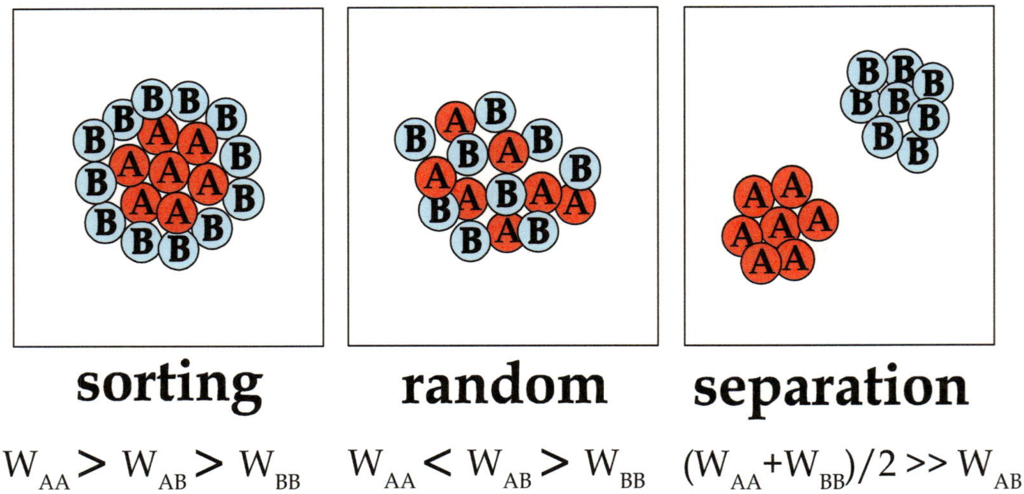

Fig. 9. Schematic representation of the DAH. Red circles with A, and Blue circles with B, represent two different types of cells. Mixture of these two types of cells finally makes spatial structure which is determined by their adhesive relationship. W_{xy} shows strength of cell adhesion between x and y.

4.2.1 Introduction of CPM

As a beginning we would like to introduce the CPM briefly. The CPM assigns a spin σ_{ij} to each lattice site, (*i, j*) and packed sites which have same spin number defines a cell, so the spin number is just a cell index which has no physical meanings. All movements of cells are determined by thermodynamic implementation of DAH, i.e. total energy of the system is minimized by Monte Carlo method. Each cell has an associated cell type τ. Fig. 10 shows schematic explanation of the formation rule of the cell of our model.

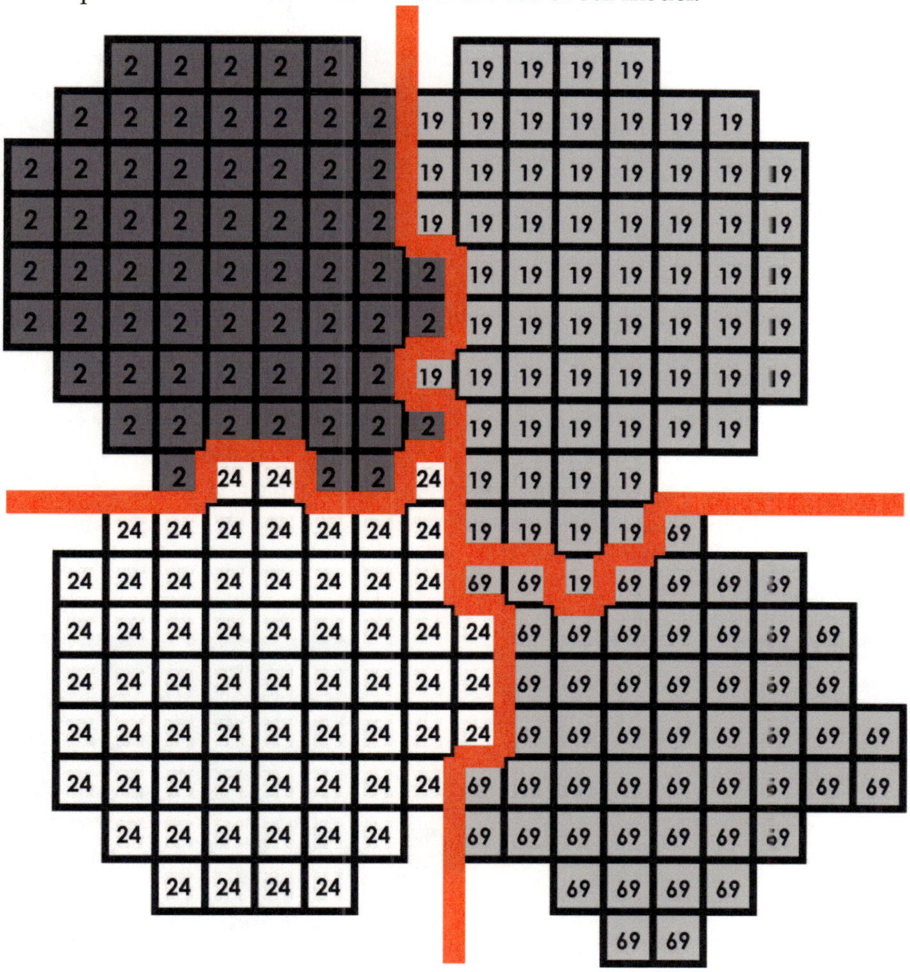

Fig. 10. Schematic explanation of the cell configuration of CPM. The numbers in a lattice shows σ_{ij} and each collection of σ_{ij} defines a cell. The colors of the each cell define the cell type, τ. A fat line in a figure shows the bonding sites of the cell. Redrawn from figure 2 of Ouchi (2007).

At the simulation, we calculate the total energy of the system H, from cell-cell or cell-substrate adhesion energy $J_{\tau,\tau}$ elastic bending energy of cell and surface tension energy. Thus, the total energy is

$$H = \sum_{(i,j)} J_{\tau\tau'}(1-\delta_{\sigma,\sigma'}) + \left\{ \lambda_1 \sum_{\sigma} (a(\sigma) - A_\tau)^2 + \lambda_2 \sum_{\sigma} (l(\sigma) - l_\tau)^2 \right\} (1 - \delta_{\tau,m}) \tag{11}$$

where $a(\sigma)$ and $l(\sigma)$ are respectively the area and perimeter of considering cell σ, λ_1 and λ_2 are the elastic parameter of cell surface tension and bending. In this equation, the symbol δ_{ij} represents so-called delta function,

$$\delta_{ij} = \begin{cases} 1 & i = j \\ 0 & i \neq j, \end{cases} \tag{12}$$

it takes 1 if the index $i=j$, or 0 if the index $i\neq j$. In our 2D simulation, the cell size is introduced as a target area A_τ and a target perimeter l_τ. At each step, we apply the Metropolis algorithm, choose a site at random and accept a proposed change in its spin value with a Boltzmann transition probability dependent on temperature $T > 0$:

$$P = \begin{cases} exp(-(\Delta H - H_0)/T & \Delta H > H_0 \\ 1 & \Delta H < H_0 \end{cases} \tag{13), (14}$$

where H_0 is a threshold for a spin flip adopted from Hogeweg (Hogeweg, 2000) and our temperature T is not a real world's temperature but just one of the controllable parameters. Fig. 11 is the detailed process of cell movement using this Metropolis algorithm. First we choose two adjacent sites located at cell surface randomly, and next changes one spin number to adjacent's spin number using above probability, and repeating again and again. Because of considering site with some spin number is a part of a cell with same spin number, spin number change represents cell deformation at cell surface. This is the cell movement.

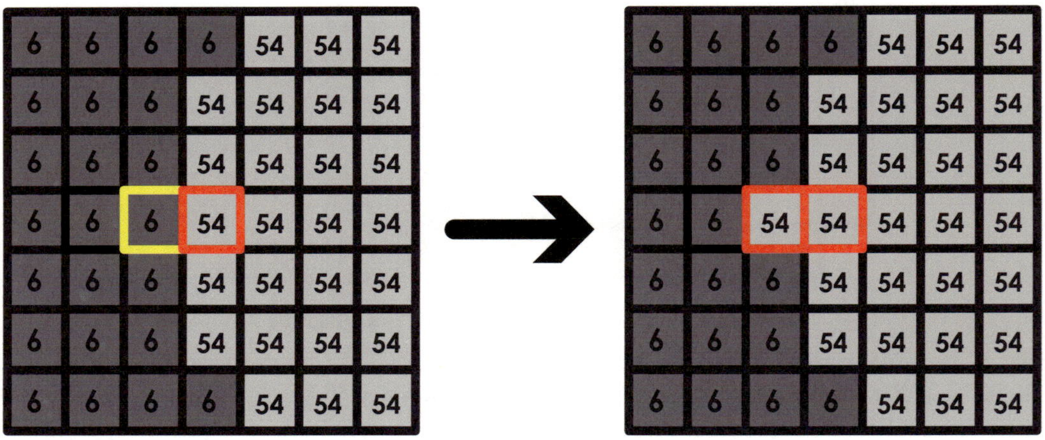

Fig. 11. Schematic explanation of the cell movement process of CPM. The site with spin number 6 is changed its number to 54. It shows that the cell 54 extend small part of its surface to adjacent cell 6.

Using this CPM formalism, we can simulate many types of cell group dynamics, e.g., the cell sorting, engulfment process and tissue rounding, moreover coupling with continuous field equations that depend on the solving problems, various cell group dynamics can be well simulated (Jiang, 2005, Turner & Sherratt, 2002) even for the entire life cycle of the slime mould Dictyostelium discoideum (Marée & Hogeweg, 2001). Fig. 12 and Fig. 13 are examples of the cell sorting and cell engulfment simulation. We used two cell types, endodermal cell and ectodermal cell, to simulate these cell rearrangement processes.

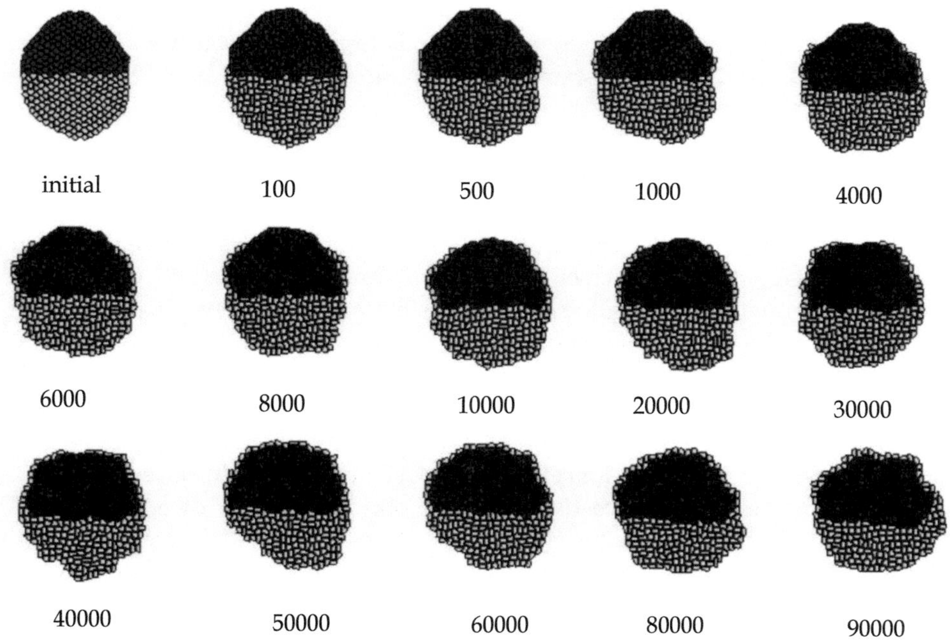

Fig. 12. Simulation of cell engulfment process using CPM. The numbers below each snapshot are simulation steps, Monte Carlo steps (MCS).

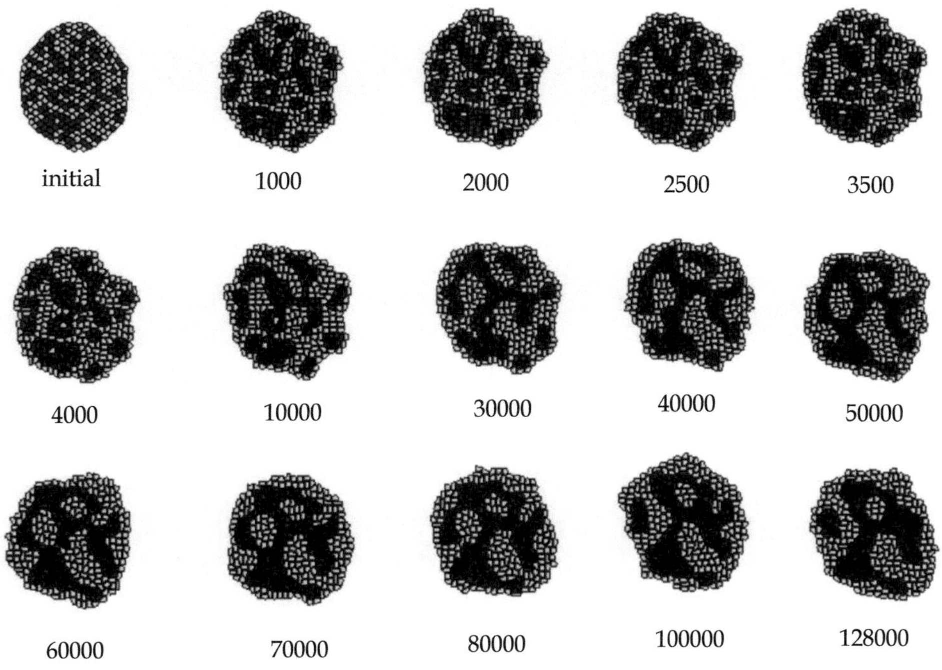

Fig. 13. Simulation of cell sorting process using CPM. The numbers below each snapshot are simulation steps, Monte Carlo steps (MCS).

The statistical calculations of mean square displacement (MSD) of cell, which is one of the characteristics of cell movement, agree with previous results obtained from corresponding experiments (Rieu et al., 2000). It seems that scientific validity of the CPM formalism is confirmed by these results.

4.2.2 Modelling of intracellular carcinogenesis process

Next, let us begin to model the cell transform process by introducing dynamics of cell type change in the CPM (Ouchi, 2007, Ouchi, 2011). Carcinogenesis process is commonly thought to be a multistage process, in which normal cells are transformed to cancer cells via accumulation of mutations. Modelling of cell transformation is achieved by employing cell type τ as one of the stages of multistage carcinogenesis. For the dynamics of cell transformation, three conceptual process, initiation, promotion and progression (IPP) of the cell are introduced in our model. Adding to these processes, cell killing and cell proliferation (cell division) are adopted to the model. All these dynamics are introduced as the probability of each process except for cell division. Cell division is introduced by conditional manner, i.e. if the cell size a exceeds the critical size a_c then cell division occur. Thus, adopted cell transformation dynamics (neoplastic process) can be drawn as Fig. 14, where P_1, P_2 and P_3 denote respectively the probabilities of initiation, promotion and progression, k_1, k_2, k_3 and k_4 are respectively the cell killing probabilities of corresponding cell types. According to the transformed cell types, this neoplastic process can be divided into four different cell stages, we call first cell stage as normal cell, second as initiated cell, third as promoted cell and final stage as cancer cell.

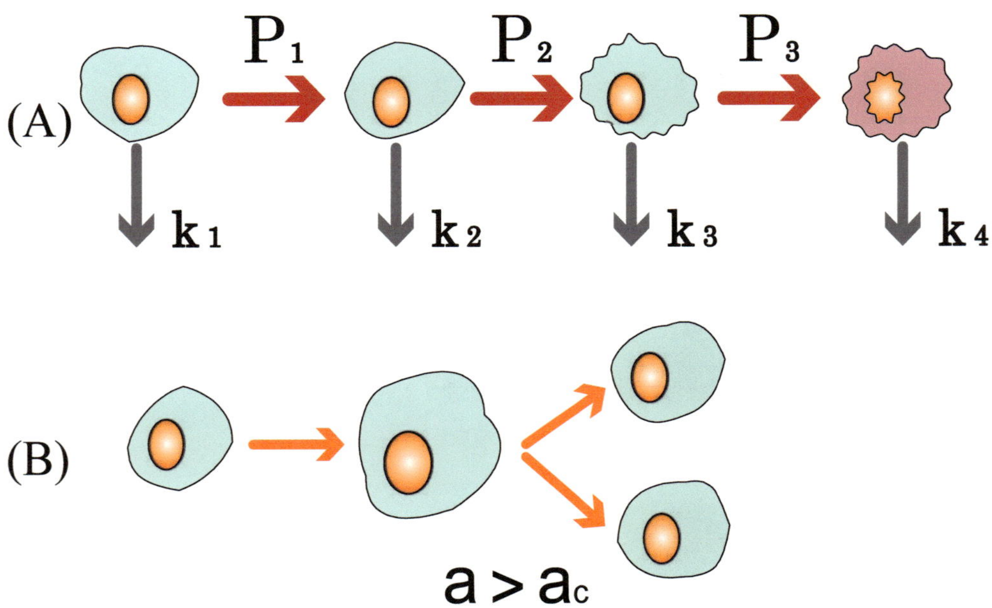

Fig. 14. Schematic explanation of cellular dynamics, intracellular (A) and cell division (B). In this figure, P_1, P_2 and P_3 are the mutation probabilities of each stage in the broadest sense, and k_1, k_2, k_3, and k_4 are the cell killing probabilities for each of the stages. If the cell area exceeds some threshold value, $a > a_c$, then cell division occurs (B).

By combination of intracellular dynamics and CPM which describes extracellular physical cell properties, mathematical model of tumorigenesis which has deformable sell shape and physical cell properties is introduced as shown in Fig. 15. The model has three specific physical properties, cell-cell or cell-medium adhision $J_{\tau\tau'}$, surface tension a, elastic bending l. The cells with diffrent cell type τ have different combination of physical properties ($J_{\tau\tau'}$, a, l), thus the cells will change its physical properties with the neoplastic transformation (Fig.15). As a concrete example, mutations in the genes involved in cell adhesion molecule change the physical adhesion of the cells. Therefore, adding to these physical parameters, the parameters of the model are the transformation probabilities P, the cell-kiling probabilities k, and the threshold cell size of cell division a_c. In this model, the effects of radiation will appear as the mutation increment or cell-killing increment. It is thoght that the radiation dose d and the mutation μ or cell kiling k have some relationship like $\mu=f(d)$ or $k=g(d)$, not a few experiments show linear relationship or linear quadratic rerationship depend on the situation. However, the clear relationship between dose and mutation frequency or cell-killing seem not to be obtained yet. Consequently, linear relationship between dose and mutation frequency or cell-killing is assumed: mutation increment means dose increment.
Generally, the physical parameters including the threshold cell size a_c could be determined with the aid of corresponding experiments, control parameters are thought to be the transformation probabilities P and the cell-kiling probabilities k. To show validity of the model described in this section, some simulation results are presented in the next section.

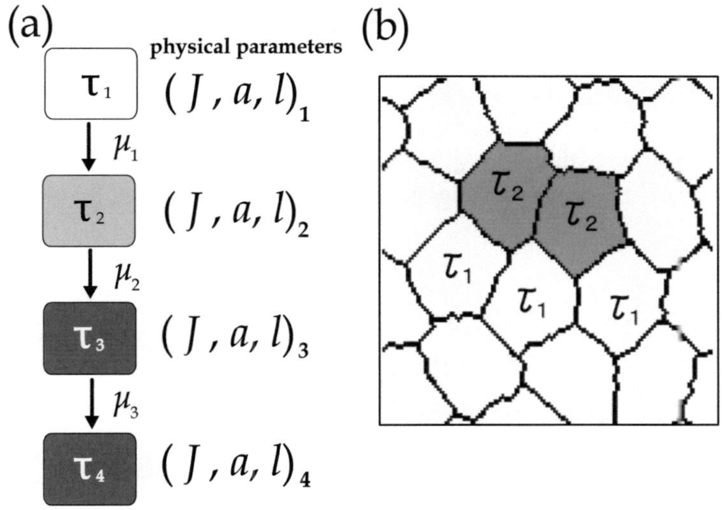

Fig. 15. Schematic explanation of the combination of intracellular dynamics and CPM. (a) The cells with diffrent cell type τ have different combination of physical parameters $(J, a, l)_i$, thus the cells will change its physical properties with the neoplastic transformation.
The actual spatial configuration of the model is shown in (b).

4.2.3 Calculation results

Before starting statistical calculations, we carefully checked that the chosen parameters k_i and P_i satisfied homeostasis of the cell number; the total number of the cell is in equilibrium. In most cases, we use 250 or 1000 or 4000 cells to calculate statistical quantities depending on the situation. These available cell numbers are limited by the computational resources. In all the calculation, we use randomly placed confluent normal cells mixed with small fraction (0.002%) of cells with initiation for initial condition of the system. Actually, changing this fraction has no influence on simulation results. In the simulation, as the transformation progresses, our cells are assumed to change softer and higher adhesion than the normal cells, for a first attempt of calculation. Figure 16 shows typical time evolution of the system starting from the initial condition, confluent normal cells mixed with a small number of cells with initiation. We can see the growth of the aggregate of cancer cells by using appropriate parameters (Ouchi, 2007, Ouchi, 2011).

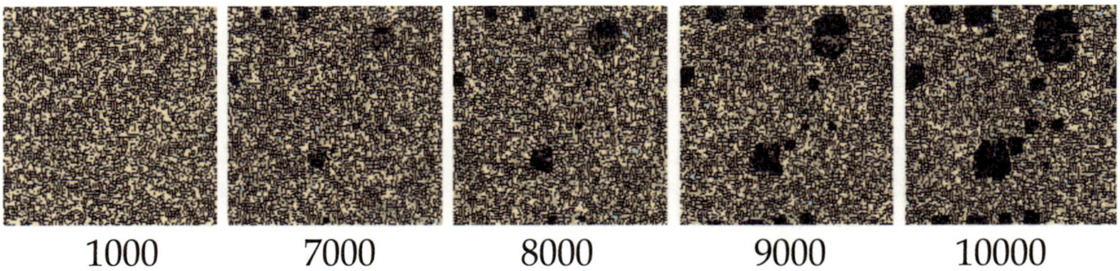

Fig. 16. Time evolution of the system. Time runs left to right. Numbers represent time steps. The initial condition is confluent normal cells mixed with a small number of cells with initiation. The system is consists of about 4000 cells. Darker cells represent higher stage of carcinogenesis, and black cells show cancer cells.

Next, population dynamics is shown in Fig. 17 (a) and Fig. 17 (b). The main diffrence between Fig. 17 (a) and Fig. 17 (b) is transformation parameters of the normal cell: P_1 and k_1 of (b) is five times gretaer than the (a)'s. In case of Fig.17 (a), no apparent tumorigenesis is seen, on the contrary, apparent tumorigenesis is seen in Fig.14 (b).

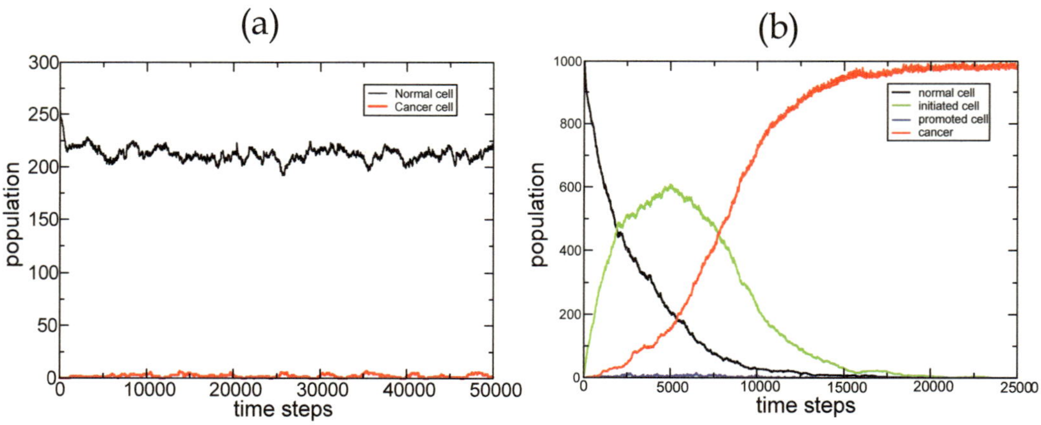

Fig. 17. Time series plots of the population for (a) normal cells (black line) and cancer cells (red line), and (b) normal cells (black line), initiated cells (green line), promoted cells (blue line) and cancer cells (red line). The saturation of the cancer cell number in (b) comes from the limitation of the system size.

Here, let us define a statistical quantity for criterion of tumourigenesis called "cancer growth rate" (*CGR*) : time averaged ratio of cancer cells to whole cells (Ouchi, 2011),

$$CGR = \langle \# \text{ of cancer cells/total } \# \text{ of cells} \rangle_t \quad (15)$$

where $\langle\rangle_t$ shows time averaging operation basically $t \to \infty$. Fig. 18 is plot of CGR calculation with changing initiation probability P_1 and fixed other parameters. Clearly two region can be seen, one is tumorigenesis region and the other is no tumor region, with some boundary value of $P_1 \approx 0.003$.

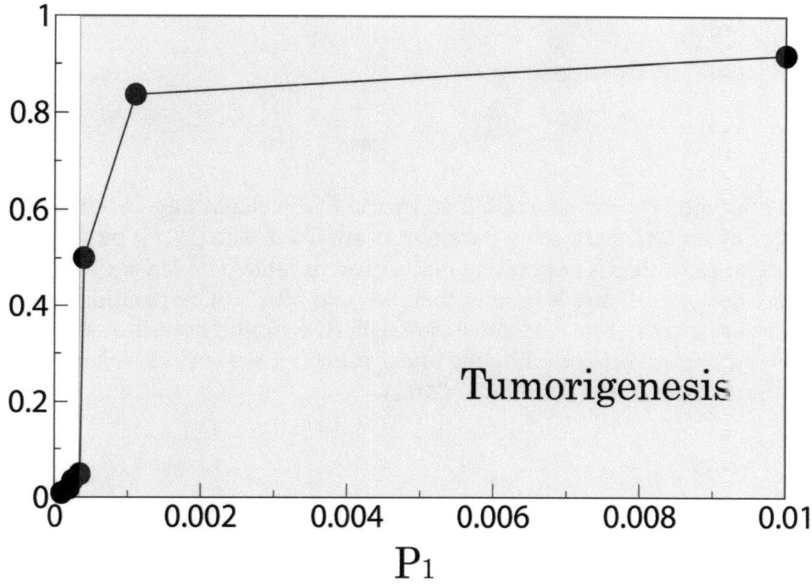

Fig. 18. Calculation of *CGR* with various P_1 values. It seems to have two regions, tumorigenesis and no tumor, depending on its P_1 value.

Finally, numerous calculations of *CGR* was performed to make phase diagram regarding the value of P_1 and k_1 of the model. Obtained phase diagram is shown in Fig. 19. If the calculated *CGR* is greater than 0.1, seems tumor to emerge, and thus *CGR*=0.1 is defined by the criteria of tumorigenesis. Tumor seems not to emerge if the *CGR* is less than 0.05. The tumorigenesis and non-tumorigenesis regions are separated by the linear function, k_1=--$\alpha P_1+\beta$, where α=3/2 and β=0.0006.

Generally, tumor growth curve analysis is very important for the cancer study and also important for the risk assesement of low dose radiation. Usually, empirical models are used for growth curve analysis to fit obtained experimental data. In case of tumorigenesis, well-known Gompertz function (Gompertz, 1825) is used to analyze the data. Here, the present model is based on a modelling of biological mechanisms, a kind of bottom-up approach, calculation of growth curve of our model seems interesting subject: whether the growth curve of our model fit a empirical growth curve or not. In conclusion, it seems that the

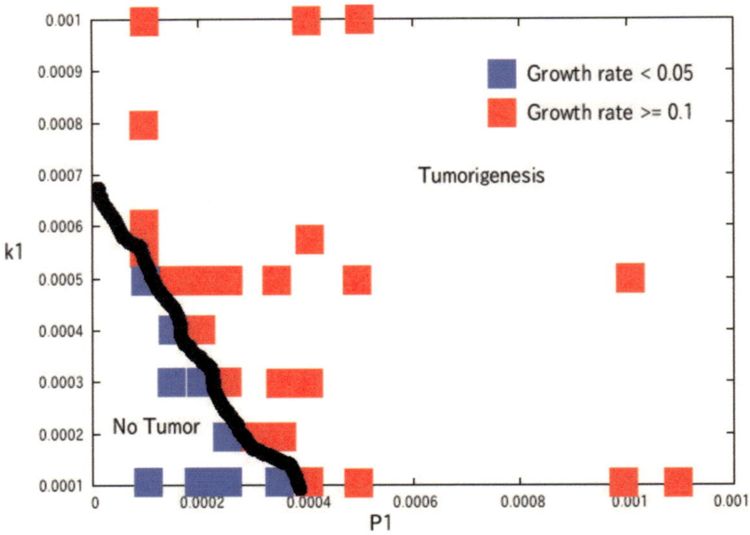

Fig. 19. Phase diagram of the model obtained by the CGR calculations in various combinations of P1 and k1, with other parameters are fixed. The (P_1, k_1) positions are painted by blue or red squares depending on its growth rates. The red squares show positions where the growth rate is greater than 0.1, and blue square positions are the positions where the growth rate is smaller than 0.05. The tumorigenesis and non-tumorigenesis regions are separated by the linear function, k_1=--$\alpha P_1+\beta$, where α=3/2 and β=0.0006. Redrawn from figure 5 of Ouchi (2011).

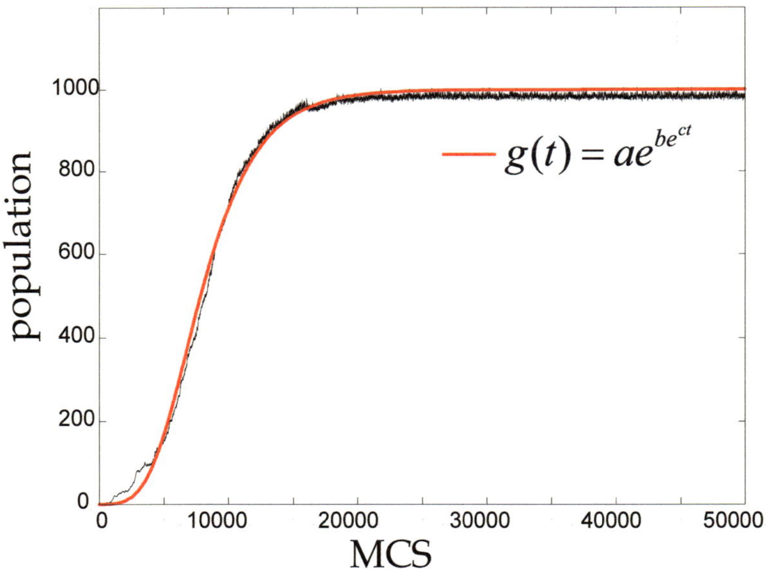

Fig. 20. Growth curve of cancer cells obtained from model calculation with parameters, P=(0.0005, 0.0001, 0.01), k=(0.0005, 0.001, 0.001, 0.001). Obtained growth curve seems to be fitted well with Gompertz function which is depicted in the figure by a red dotted line. The parameters of Gompertz function are a=1000, b=9.2, c=0.00033. Redrawn from figure 4 of Ouchi (2011).

model can reproduce the Gompertz curve very well as shown in Fig. 20. In a tumorigenesis region of our model, calculated tumor growth curves are well fitted with Gompertz curve with changing function's parameter (*a,b,c*). By changing model parameter *k*s and *P*s, the different growth curve of Gompertz form are obtained, ordered structure of phase space is observed when we treat Gompertz function's parameters (*a,b,c*) as a point of 3D phase space (unpublished data, NBO). However, further investigation are needed to clarify the relation between phase space structure and cancer in real world.

4.3 Modelling tumor invasion and metastasis

Here, importance of invasion and metastasis study is explained. Tumor invasion and metastasis is one of the cardinal features of malignant cancer. It occurs after carcinogenesis, and is in the late stage of cancer. Tumor invasion and metastasis is a tumor spreading process from local to other organs (Liotta, 1984). The tumor invasion is the expanding process of cancer cells into neighboring normal tissue, and the metastasis is transporting process of the malignant tumor from its primary site to other distant sites in the body (Molecular Biology of the Cell, 5th Ed. [CELL], 2008). This metastasis may occur by three main routes: 1) through the bloodstream, 2) through the lymphatic system or through both routes, and in addition, 3) direct seeding across body cavities, e.g. through the peritoneum. Thus, the process of the tumor invasion and metastasis concern many cell types, e.g., normal cell, cancer stem cell, immune cell, extracellular matrix (ECM). In fact, these cells have complex interaction with each other to promote the cancer progression and expansion. Taking these behaviors into consideration, it is clear that cell-cell and cell-microenvironment interactions play a very important role in the process of invasion and metastasis. Tumor cells must be able to detach the primary site by changing its adhesion, and in turn must be able to attach the secondary sites, in order to proliferate successfully. Therefore, the adhesion is one of the key players for the invasion and metastasis. Moreover, cancer cells change its genetic status even in a metastatic period (Santinelli et al., 2008), thus the malignant cancer cells are thought to be not in a steady state but more dynamic. Therefore, tumor cells can grow up by "niche" adaptation in cell microenvironment, in other words, tumor need microenvironmental adaptation for its growth. The tumor cells which expand in the primary site become invasive and spreading to surrounding sites, and detachment of invasive tumor cells leading to the colonization of other organs. Tumor cells traveling to the distant organ should have microenvironmental adaptation to make colonies, consequently the tumor cells keep always evolving with changing its gene expression. This is a reason why the cancer is called as a microevolutionary process (Merlo et al., 2006). For that reason, tissue structure should be needed to model the dynamics of invasion and metastasis. In the present, there are not so many models introducing spatial tissue structure. It is not easy, but the tissue structure can be modeled by CPM formalism. In tissue modeling, we need more cell types, which depending on the considering tissue, to construct the system. Fig. 21 shows simulation example of the tissue structures, epithelium (a) and so-called intestinal crypt (b). It is believed that the cancer stem cells exist at the crypt, study of crypt structure is important for colorectal cancer. We use epithelial cells, medium layer, lamina muscularis mucosae, lamina propria for the simulation. By using one of these structures as an initial configuration of the stage of carcinogenesis, tumorigenesis and invasion could be studied integratively with presented model.

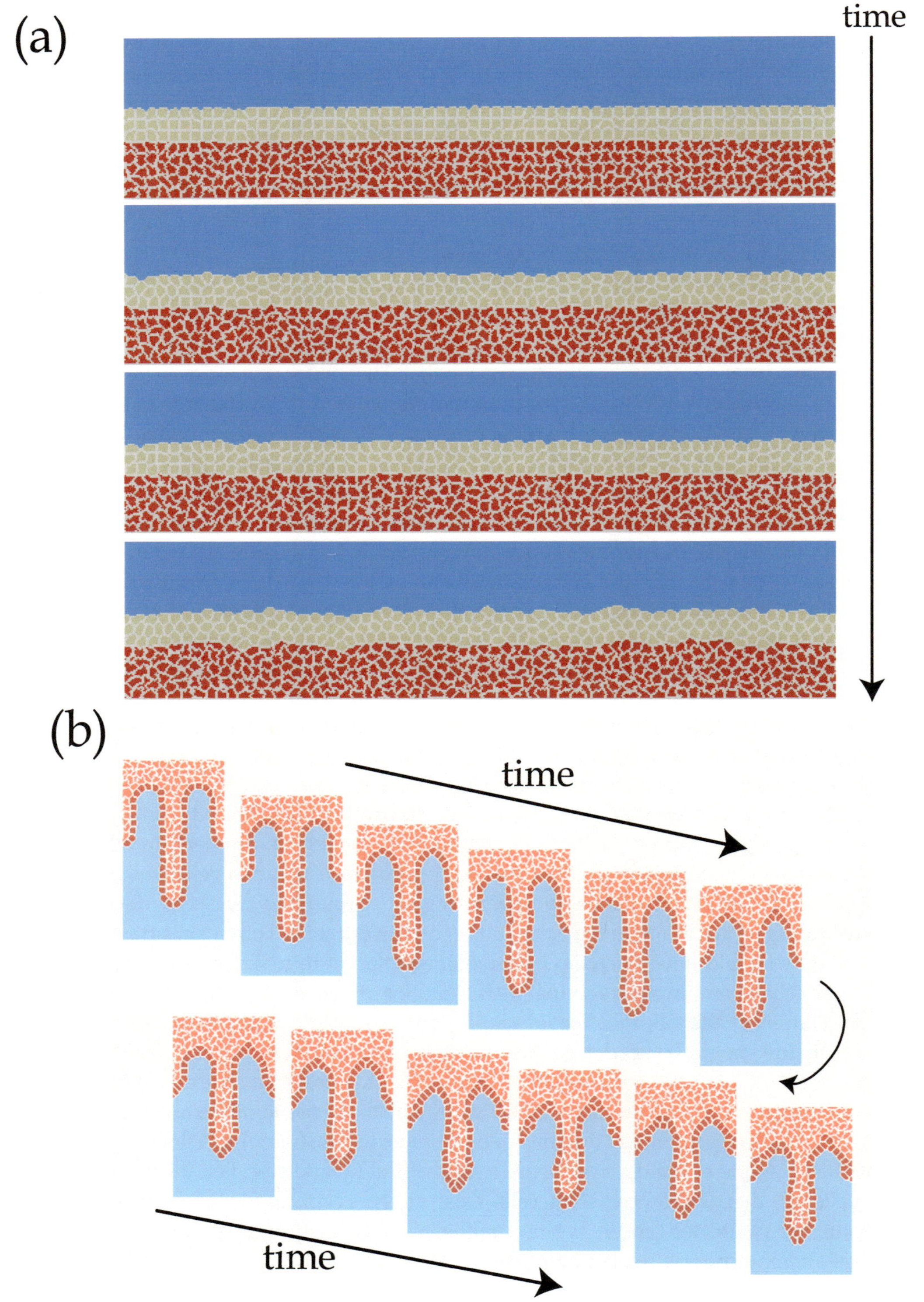

Fig. 21. Simulation examples of tissue structure using CPM.

5. Conclusion and perspectives of the problem

From the early days of the discovery of radiation, its health effects have been recognized. Although numerous investigations have been done to clarify the health effects of radiation, there still seems to be many questions to answer. Existence of low dose radiation specific phenomena, e.g. bystander effects or adaptive responses (see Matsumoto et al, 2007 for review) shows possibility of unknown biological mechanisms, however there should be a intrinsic uncertainty within a system. One of the difficulties not only for radiation biology but also for general biology is thought to be presence of exceptions. It is possible that most experiments show "positive" results but a few experiments will show the opposite results in a biological experiments, and perhaps that is the big difference between biology and physics. Following the example of Mendel's law, on the discovery of principle which is based on the biological mechanism will explain such kind of presence of exceptions with theoretically consistent. Here, one of the uncertainties of carcinogenesis is shown. Epidemiological data shows presence of the uncertainty about the age of cancer onset: ages of cancer onset are different among patients. By definition of "cancer onset" as an onset of neoplastic aggregate formation, detailed analysis of present model shows that cancer onset has initial condition dependency: very small differences in initial state will lead diffrent ages of cancer onset. (unpublished data, NBO). This uncertainty seems to be stochastic, moreover, some specific minimum number "cancer onset number", i.e. number of constituent cells of tumor, seems to be exist for the cancer onset (Fig. 22). However, a clarification of the cancer onset number problem needs further investigation.

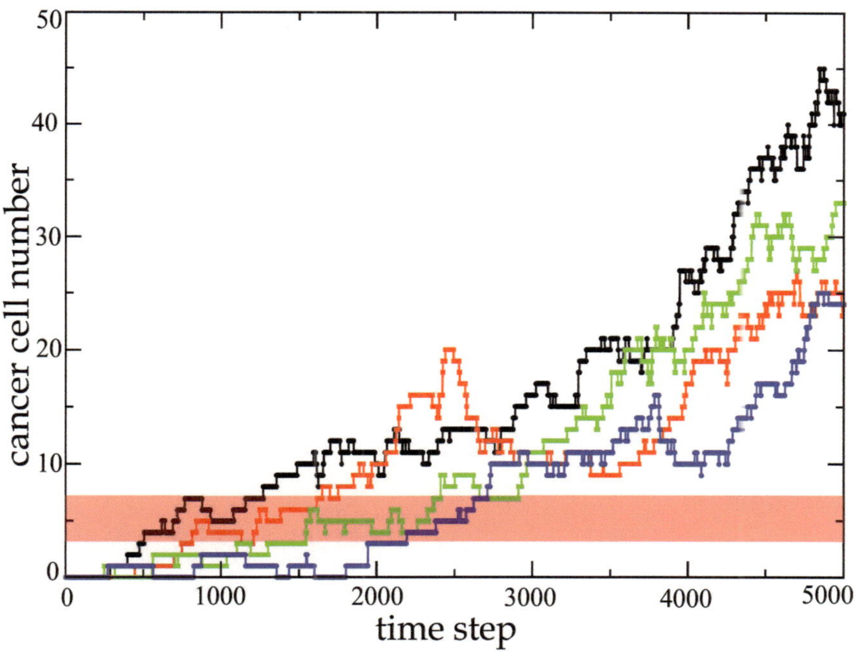

Fig. 22. Temporal plots of cancer cell population at early stage of tumorigenesis. Each of the lines shows the calculation starting from different initial conditions. It seems to overcome a specific number of cancer cell for further tumor growth: red shaded region highlights the estimated hurdle in tumor growth.

Next, as an another example of presence of exceptions, redundancy of the intracellular dynamics, e.g. metabolic network, is shown. It is thought that the redundancy is one of the factors of multi-pathway of carcinogenesis. Because of its redundancy, there seems to be many paths from some start point *A* to the goal *G* via intermediate point *B*, *C*, *D*..., thus even if some path *A*-*B*-*G* is down, then another path *A*-*C*-*G* will appear. This redundancy is the source of the stability of living systems. It is like a complicated subway route (Fig. 23). We can reach our destination even if some station is down by accident using another route to the destination. It may indicate that more detailed analysis needs to be carried out in the near future. Moreover, on the investigation of tumorigenesis process, we should aware of the existence of adjacent cells.

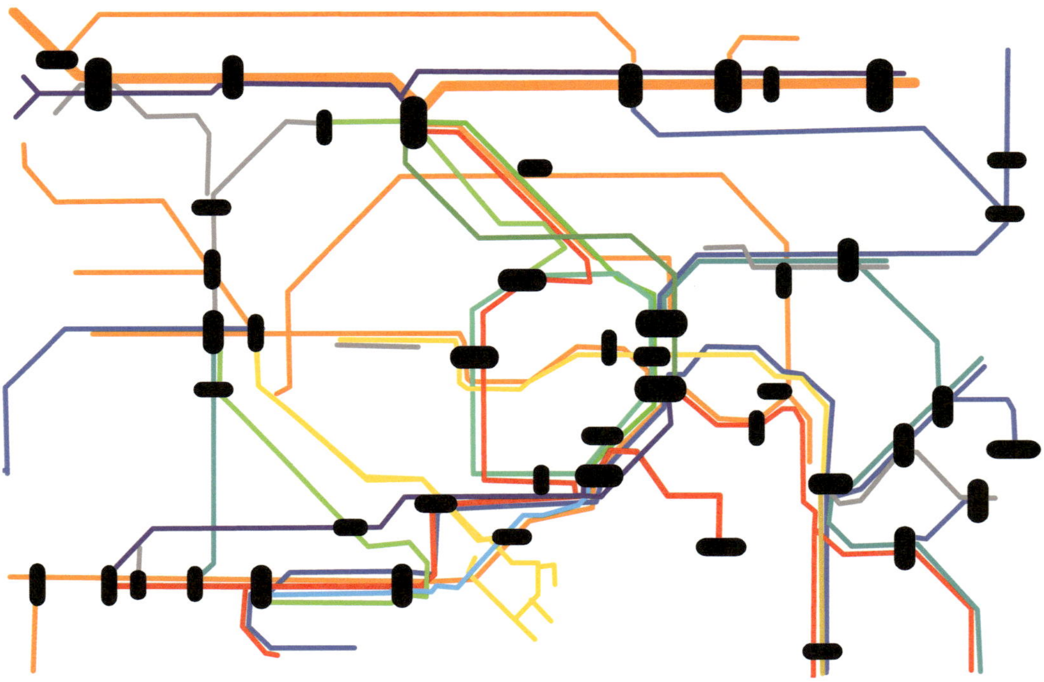

Fig. 23. Simplified subway map of some big city. Due to the redundancy of network, we can reach our destination even if some station is down by accident. Similar structure may be seen in intracellular metabolic network.

6. Acknowledgment

First, I would like to thank Dr. K. Saito for his continuous encouragement and useful discussions. Thanks are also due to Prof. M. Kai, Prof. M. Ootaki and Dr. O. Niwa for many suggestions and interesting discussions and holding very interesting and exciting small workshop about the mathematical model of carcinogenesis. Attending their workshop was author's best experience when he started this subject. The author also acknowledges all members of "SRB BENKYOKAI", it was very fascinating experience for me to attend the meeting and make interesting discussion. This work is partially supported by KAKENHI (18700294) and the Budget for Nuclear Research of the Ministry of Education, Culture,

Sports, Science and Technology, based on the screening and counseling by the Atomic Energy Commission.

7. References

Adams, G. E. and Jameson, D. G. (1980), Time Effects in Molecular Radiation Biology, *Radiat. Environ. Biophys.*, Vol. 17, pp. 95-113

Alberts, B., Johnson, A., Lewis, J., Raff, M., Roberts, K., and Walter, P. (2008), *Molecular Biology of the cell Fifth Ed.*, Garland Science

Anderson, A. R. A. and Quaranta, V. (2008), Integrative mathematical oncology, *Nat. Rev. Cancer*, Vol. 8, No. 3, pp. 227-234

Armitage P. and Doll R. (1954), The age distribution of cancer and a multistage theory of carcinogenosis, *Br. J. Cancer*, Vol.8, No.1, pp. 1–12 reprint version is available at Brit. J. Cancer's Web site: http://www.nature.com/bjc/journal/v91/n12/pdf/6602297a.pdf

Araujo, R. P. and McElwain, D. L. S. (2004), A History of the Study of Solid Tumour Growth: The Contribution of Mathematical Modelling, *Bulletin of Mathematical Biology*, Vol. 66, pp. 1039–1091

BEIR V (1990), *Health Effects of Exposure to Low Levels of Ionizing Radiation*, National Research Council of the National Academies, National Academies Press, USA

BEIR VII, Phase2 (2006), *Health risks from exposure to low levels of ionizing radiation*, National Research Council of the National Academies, National Academies Press, USA

Berrington A., Darby S. C., Weiss H. A., Doll R. (2001), 100 years of observation on British radiologists: mortality from cancer and other causes 1897–1997, *Br. J. Radiol.* Vol. 74, pp. 507–519

Brenner, D. J., Doll, R., Goodhead, D. T., Hall, E. J., Land, C. E., Little, J. B., Lubin, J. H., Preston, D. L., Preston, R. J., Puskin, J. S., Ron, E., Sachs, R. K., Samet, J. M., Setlow, R. B., and Zaider, M. (2003), Cancer risks attributable to low doses of ionizing radiation: Assessing what we really know, *Proc. Natl. Acad. Sci. USA*, Vol. 100, No. 24, pp. 13761-13766

Preston, L. E. and Abrahamson, S. (Ed.) (1998), *Effects of Ionizing Radiation*, Joseph Henry Press, USA

Cardis, E., Vrijheid, M., Blettner M., Gilbert E., Hakama M., Hill C., Howe G., Kaldor J., Muirhead C. R., Schubauer-Berigan M., Yoshimura T., Bermann F., Cowper G., Fix J., Hacker C., Heinmiller B., Marshall M., Thierry-Chef I., Utterback D., Ahn Y. O., Amoros E., Ashmore P., Auvinen A., Bae J. M., Bernar J., Biau A., Combalot E., Deboodt P., Sacristan A. D., Eklof M., Engels H., Engholm G., Gulis G., Habib R. R., Holan K., Hyvonen H., Kerekes A., Kurtinaitis J., Malker H., Martuzzi M., Mastauskas A., Monnet A., Moser M., Pearce M. S., Richardson D. B., Rodriguez-Artalejo F., Rogel A., Tardy H., Telle-Lamberton M., Turai I., Usel M., and Veress K. (2007), The 15-Country Collaborative study of cancer risk among radiation Workers in the Nuclear Industry: Estimates of Radiation-Related Cancer Risks, *Radiat. Res.* Vol. 167, No. 4, pp. 396-416

Chen, C. S., Mrksich, M., Huang, S., Whitesides, G. M., and Ingber, D. E. (1997), Geometric Control of Cell Life and Death, Science, Vol. 276, No. 5317, pp. 1425-1428

Fearon, E. R. and Vegelstein, B. (1990), A Genetic Model for Colorectal Tumorigenesis, *Cell*, Vol. 61, pp. 759-767

Fisher J. C. and Hollomon J. H. (1951), A hypothesis for the origin of cancer foci., *Cancer*, Vol. 4, pp. 916-918

Frank, S. A. (2007), *Dynamics of Cancer: incidence, inheritance, and Evolution*, Princeton University Press

Gatenby, R. A. and Gawlinski, E. T. (1996), A Reaction-Diffusion Model of Cancer Invasion, *Cancer Res.*, Vol. 56, No. 24, pp. 5745-5753

Gilbert, S. F. (2006), *Developmental Biology, 8th Ed.*, Sinauer Associates Inc.

Glazier, J. A. and Graner F. (1993), Simulation of the differential adhesion driven rearrangement of biological cells, *Phys. Rev. E*, Vol. 47, No. 3, pp. 2128-2154

Gompertz, B. (1825), On the nature of the function expressive of the law of human mortality, and on a new mode of determining the value of life contingencies. *Phil. Trans. B.*, Vol. 115, pp. 513–585

Graner, F. and , Glazier, J. A. (1992), Simulation of Biological Cell Sorting Using a Two-Dimensional Extended Potts Model, *Phys. Rev. Lett.*, Vol. 69, No. 13, pp. 2013-2016

Hill, A. V. (1928), The Diffusion of Oxygen and Lactic Acid through Tissues, *Proc. R. Soc. Lond. B 104*, 1 December, Vol. 104, No. 728, pp. 39-96

Hogeweg, P. (2000), Evolving Mechanisms of Morphogenesis: on the Interplay between Differential Adhesion and Cell Differentiation, *J. Theor. Biol.*, Vol. 203, No. 4, pp. 317-333

Jackson, E. A. (1991), *Perspectives of nonlinear dynamics, Vol. 2*, Cambridge University Press

Jiang, Y. , Pjesivac-Grbovic, J., Cantrell, C., and Freyer, J. P. (2005), A Multiscale Model for Avascular Tumor Growth, *Biophys. J.*, Vol. 89, 3884-3894

Liotta, L. A. (1984), Tumor invasion and metastases: role of the basement membrane, *Am. J. Pathol.*, Vol. 117, No. 3, pp. 339-348

Marée, A. F. M. and Hogeweg, P. (2001), How amoeboids self-organize into a fruiting body: Multicellular coordination in Dictyostelium discoideum, *Proc. Natl. Acad. Sci. USA*, Vol. 98, No. 27, pp. 3879-3883

Matsumoto, H., Hamada, N., Takahashi, A., Kobayashi, Y. and Ohnishi T. (2007), Vanguards of Paradigm Shift in Radiation Biology: Radiation-Induced Adaptive and Bystander Responses, *J. Radiat. Res.*, Vol. 48, No. 2., pp. 97-106

Merlo, L. M. F., Pepper, J. W., Reid, B. J., and Maley, C. C. (2006), Cancer as an evolutionary and ecological process, *Nat. Rev. Cancer*, Vol. 6, No. 12, pp. 924-935

Moolgavkar, S. H. (2004), Commentary: Fifty years of the multistage model: remarks on a landmark paper, *Int. J. Epidemiol.*, Vol. 33, No. 6, pp. 1182-1183

Nikjoo, H. ,Uehara, S., Emfietzoglou, D., and Cucinotta, F. A. (2006), Track-structure codes in radiation research, *Radiation Measurements*, Vol. 41, pp. 1052-1074

Nordling, C. O. (1953), A New theory on cancer-inducing mechanism, *Brit. J. Cancer*, Vol. 7, pp. 68-72

Nowak, M. A.m Michor, F., and Iwasa, Y. (2003), The linear process of somatic evolution, *Proc. Natl. Acad. Sci. USA*, Vol. 100, No. 25, pp. 14966-14969

Ouchi, N. B., Glazier, J. A., Rieu, J. P., Upadhyaya, A., and Sawada, Y. (2003), Improving the realism of Potts model in simulations of biological cells, *Physica A*, Vol. 329, pp. 451-458

Ouchi, N. B. (2007), Computer modeling of radiation effects, *Data Science Journal*, Vol. 6, Supplement, 10 May 2007, pp. S278-284

Ouchi, N. B. (2011), Growth Curve Analysis of Tumorigenesis using Cellular Level Cancer Model, *Radiat. Prot. Dosimetry*, Vol. 143, pp. 365-369

Rieu, J. P., Upadhyaya, A., Glazier, J. A., Ouchi, N. B., and Sawada, Y. (2000), Diffusion and Deformation of Single Hydra Cells in Cellular Aggregates, *Biophys. J.*, Vol. 79, pp. 1903-1914 Santinelli, A., Pisa, E., Stramazzotti, D., and Fabris, G. (2008), HER-2 status discrepancy between primary breast cancer and metastatic sites. Impact on target therapy, *Int. J. Cancer*, Vol. 122, No. 5, pp. 999-1004

Sjöblom, T., Jones, S., Wood, L. D., Parsons, D. W., Lin, J., Barber, T. D., Mandelker, D., Leary, R. J. , Ptak, J., Silliman, N., Szabo, S., Buckhaults, P., Farrell, C., Meeh, P., Markowitz, S. D., Willis, J., Dawson, D., Willson, J. K. V., Gazdar, A. F., Hartigan, J., Wu, L., Liu, C., Parmigiani, G., Park, B. H., Bachman, K. E., Papadopoulos, N., Vogelstein, B., Kinzler, K. W., and Velculescu, V. E. (2006), The Consensus Coding Sequences of Human Breast and Colorectal Cancers, *Science*, Vol. 314, No. 5797, pp. 268-274

Smith, G., Carey, F. A., Beattie, J., Wilkie, M. J. V. , Lightfoot, T. J., Coxhead, J., Garner, R. C. , Steele, R. J. C., and Wolf, C. R. (2002), Mutations in APC, Kirsten-ras, and p53 – alternative genetic pathways to colorectal cancer, *Proc. Natl. Acad Sci. USA*, Vol. 99, No. 14, pp. 9433-9438

Steinberg, M. S. (1970), Does differential adhesion govern self-assembly processes in histogenesis? Equilibrium configurations and the emergence of a hierarchy among populations of embryonic cells, *J. Exp. Zool.*, Vol.173, pp. 395-434

Thierry-Chef, I., Marshall, M., Fix, J. J., Bermann, F., Gilbert, E. S., Hacker, C., Heinmiller, B., Murray, W., Pearce, M. S., Utterback, D., Bernar, K., Deboodt, P., Eklof, M., Griciene, B., Holan, K., Hyvonen, H., Kerekes, A., Lee, M. C., Moser, M., Pernicka, F., and Cardis, E. (2007), The 15-Country Collaborative Study of Cancer Risk among Radiation Workers in the Nuclear Industry: Study of Errors in Dosimetry, *Radiat Res.* Vol. 167, No. 4, pp. 380-395

Trosko, J. E. (1992), Does Radiation Cause Cancer ?, *RERF Update*, Vol. 4, No. 1, pp. 3-5

Turner, S. and Sherratt, J. A. (2002), Intercellular adhesion and cancer invasion: a discrete simulation using the extended Potts model, *J. Theor. Biol.* Vol. 216, pp. 85-100

UNSCEAR 2006 Report Vol. I, Annex A (2006), *Epidemiological studies of radiation and cancer*, United Nations, NY
http://www.unscear.org/unscear/en/publications/2006_1.html

UNSCEAR 2008 Report Vol. II, Annex D (2008), *Health effects due to radiation from the Chernobyl accident*, United Nations, NY
http://www.unscear.org/unscear/en/publications/2008_2.html

Vrijheid, M., Cardis, E., Blettner, M., Gilbert, E., Hakama, M., Hill, C., Howe, G., Kaldor, J., Muirhead, C. R., Schubauer-Berigan, M., Yoshimura, T., Ahn, Y O., Ashmore, P., Auvinen, A., Bae, J. M., Engels, H., Gulis, G., Habib, R. R., Hosoda, Y., Kurtinaitis,

J., Malker, H., Moser, M., Rodriguez-Artalejo, F., Rogel, A., Tardy, H., Telle-Lamberton, M., Turai, I., Usel, M., and Veress, K. (2007), The 15-Country Collaborative Study of Cancer Risk among Radiation Workers in the Nuclear Industry: Design, Epidemiological Methods and Descriptive Results, *Radiat Res.*, Vol. 167, No. 4, pp. 361-379

Watson, J. D. and Crick, F. H. (1953), Molecular structure of nucleic acids; a structure for deoxyribose nucleic acid, *Nature*, Vol. 171, No. 4356, pp. 737–738 http://www.nature.com/nature/dna50/watsoncrick.pdf

Part 3

Health Effects

15

Ionizing Radiation Carcinogenesis

Otto G. Raabe
University of California Davis,
USA

1. Introduction

1.1 Ionizing radiation

Human exposure to ionizing radiation is a natural part of life on earth. These exposures occur every day from radiation associated with naturally occurring radionuclides in soil, air, and food, and also from cosmic rays. In addition, many people are exposed to dental and medical diagnostic procedures and therapy involving: x rays, gamma rays, charged particles, radionuclides, or other ionizing radiation sources. Others may be exposed in their workplace such as in laboratories, hospitals, underground mines, or nuclear power plants. Excessive exposures may lead to the development of cancer by promotion of ongoing carcinogenic biological processes or by independent cancer induction.

Ionization converts a neutral atom to a charged atom. The γ rays and x rays are ionizing electromagnetic radiation with energies above ultraviolet in the electromagnetic spectrum. The difference between x rays and γ rays is a matter of nomenclature. The γ rays are emitted from the nucleus of a radioactive atom while x rays originate in the orbital electrons of a energized atom. The α^{2+} particle has been identified as a positively charged helium nucleus with 2 protons and two neutrons emitted from the nucleus of certain radionuclides. The β^- radiation was identified as negatively charged electrons emitted from the nucleus of certain radionuclides. There are other types of ionizing radiation including most prominently uncharged neutrons emitted from the nucleus of atoms that ionize atoms via nuclear interactions, and positrons and that are positive beta particles, β^+. The measure of absorbed radiation dose in matter or tissue is the gray (Gy) equal to one joule of energy per kilogram of matter or tissue. To account for different types of ionizing radiation, such as gamma, alpha, and beta radiations, the dose from ionizing radiation is corrected for the theoretical relative biological effectiveness of the different radiations in causing cellular damage using a radiation weighting factor, w_R (Sv/Gy), and this is reported as the equivalent dose in sieverts (Sv). Hence, the measure of biological dose is the sievert (Sv), equal to the absorbed dose in Gy times the radiation weighting factor. The γ rays, x rays, and β radiation typically have a radiation weighting factor of 1, while α radiation has a radiation weighting factor of 20 and neutrons may have a radiation weighting factor from 5 to 20 depending on energy. The quantity of radioactivity is measured in becquerel (Bq), equal to one radioactive atomic event or nuclear disintegration per second.

Radiation induced cancer is a complex and not completely understood process involving multiple events including but not limited to cellular DNA damage, up and down regulation of genes, intercellular communication, tissue and organ responses, clonal expansion of altered cell lines, and possibly eventual malignancy. The current understanding of radiation carcinogenesis is informed by epidemiological studies of human populations exposed to elevated levels of ionizing radiation and controlled studies utilizing laboratory animals. The two major human epidemiological studies have led to sharply contrasting results. Studies of the atomic bomb survivors indicate a linear no-threshold dose-response relationship. Studies of the radium dial painters have led to a sharp threshold relationship. These contrasting findings occur because quite different mechanisms of ionizing radiation carcinogenesis are involved in brief acute exposures at relatively high dose rates and in protracted exposures. Brief acute high dose-rate exposures act by promoting ongoing biological processes that lead to cancer in later life and this promotion effect is proportional to the dose. Protracted exposures induce cancer with a latency that depended on the lifetime average dose-rate of exposure to the target organ. At low dose rates the latent period may exceed the natural lifespan leading to a lifetime virtual threshold for cancer induction. Another observation at lower dose rates is the amelioration of an ongoing pre-malignant process. These findings have important implications with respect to ionizing radiation safety standards.

1.2 Naturally occurring ionizing radiation and radioactivity

There are naturally occurring sources of ionizing radiation that have existed on the planet since its formation (Eisenbud & Gesell, 1997). Naturally occurring radiation sources include cosmic rays, cosmogenic radionuclides, and primordial radionuclides and their decay products. Cosmic rays consist primarily of extremely high energy (mean energy ~ 10 billion electron volts) particulate radiation (primarily protons) and high energy gamma rays. When the particulate radiations collide with the earth's atmosphere a shower of "secondary" radiations are produced, which include high energy electrons and photons. The average person's dose from cosmic radiation is 0.27 mSv per year or ~ 7% of natural background. Exposure to cosmic radiation increases with altitude as there is less atmosphere to absorb the radiation, so populations at high elevations receive higher cosmic doses. For example, people living in Leadville, Colorado, at 3,200 meters above sea level, receive ~ 1.25 mSv y^{-1} or five times the average exposure at sea level. A fraction of the secondary particulate cosmic radiation collides with stable atmospheric nuclei making them radioactive. These cosmogenic radionuclides contribute very little (~ 0.004 mSv y^{-1} or less than 1%) to natural background radiation. The majority of this component of natural background is from the formation of carbon-14 and tritium (hydrogen-3).

Terrestrial radioactive materials that have been present on earth since its formation are called primordial radionuclides. Population exposure from primordial radionuclides comes from external exposure, inhalation, and incorporation of radionuclides into the body. The decay chains of ^{238}U, whose half-life is 4.5 billion years (uranium series), and ^{232}Th, whose half-life is 14 billion years (thorium series), produce several dozen radionuclides that together with natural potassium-40, whose half life is 1.3 billion years, are responsible for the majority of the external terrestrial average equivalent dose rate of 0.28 mSv per year or ~9% of natural background. Traces of these primordial isotopes of uranium and thorium are found in all the soil and rock on the face of the earth. Some regions of the world with high concentrations of primordial radionuclides produce local equivalent dose rates of environmental gamma radiation as high as 25 mSv per year.

Radon, ^{222}Rn, and its decay products, which are constituents of the ^{238}U decay series (Fig. 1), are the most significant source of natural background radiation exposure. Once inhaled, the majority of the dose is deposited in the tracheobronchial region by its short-lived daughters rather than by ^{222}Rn itself. Radon concentrations in the environment vary widely due to differences in ^{238}U concentration in the soil and differences in ventilation and construction of buildings. All other factors being equal, buildings with less ventilation will tend to have higher radon concentrations and thus, higher level of background radiation exposure. Outdoor concentrations of radon and its decay products are normally low because of the rapid mixing and dilution of radon gas with ambient air.

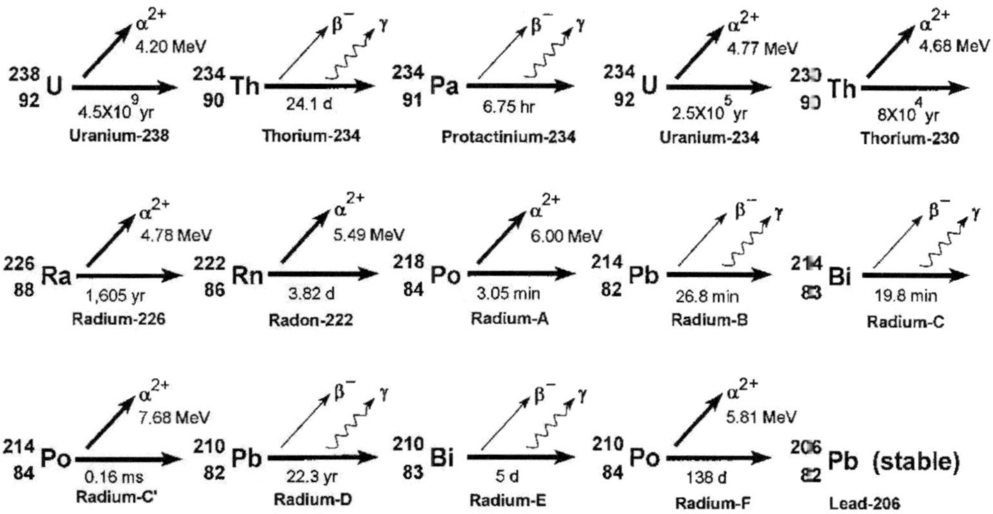

Fig. 1. Schematic illustration of the natural ^{238}U decay series as found in all soil and rock on the surface of planet Earth.

The second largest source of natural background radiation exposure comes from the ingestion of food and water that contain primordial radionuclides (and their decay products) of which ^{40}K is the most important. Altogether this pathway is responsible for an average effective dose rate of 0.4 mSv y^{-1}, or ~ 13% of natural background.

1.3 Typical human exposures to ionizing radiation

Every person who has ever lived has been exposed to ionizing radiation every minute of every day. Among the natural exposures are radiation emitted by nuclides naturally found on the earth (3% terrestrial background), internal exposure to natural radionuclides in food and water (5% internal background), cosmic radiation (5% space background), and radon and thoron gases released from the earth by the decay of naturally occurring radium isotopes (37% radon isotopes alpha irradiation of the lung). A major change has occurred here in the 21st Century from increased exposures that may occur through medical diagnostic use of x rays and radionuclides.

The average per capita "effective" radiation dose in the United States from naturally occurring (or "background") sources is about 3 mSv) per year. The "effective" dose is calculated using the International Commission on Radiological Protection (ICRP) methodology in which a tissue weighting factor is assigned to each organ or tissue and multiplied by the actual equivalent dose to the organ or tissue (ICRP 26, 1977). In the United States diagnostic medical radiology and nuclear medicine now add about 3 mSv of effective dose to ionizing radiation exposure per year per average person. The percent average contribution of various sources of ionizing radiation exposures of typical residents of the United States population in 2006 is illustrated in a pie-chart (Fig.2) prepared by the National Council for Radiation Protection and Measurement.

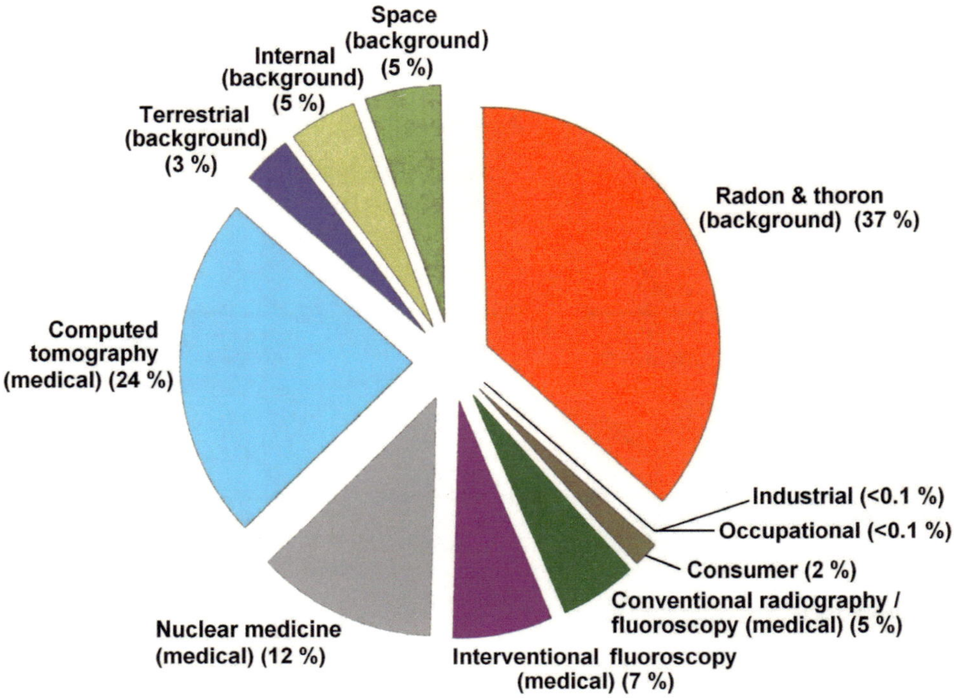

Fig. 2. Percent contribution of various sources of exposure to ionizing radiation to the average annual effective dose (6.2 mSv) per person in the United States population (about 300 million people) in 2006 (NCRP 160, 2009).

There are many places on the earth where natural background doses are much higher because of the natural presence of high concentrations of naturally occurring radionuclides in soil, rock, and water. Studies of these areas of naturally elevated ionizing radiation have not yielded any major deleterious effects associated with these natural exposures. For example, people in some areas of Ramsar, Iran, receive an annual ionizing radiation biological dose from natural background that is up to 260 mSv per year (over 40 times the USA national average). Chromosomal, hematological, and immunological studies and life span and cancer data show no untoward effects or responses (Ghiassi et al., 2002).

1.4 Naturally occurring radium isotopes and decay products

There are many naturally occurring radioactive nuclides that are continually being produced from decay chains in which the parent has a very long physical half life. For example, ^{226}Ra that is present in all the soil and rock on the earth at concentrations of about

25 Bq kg^{-1} is a product of the decay series headed by ^{238}U whose half life is comparable to the age of the earth (4.5 billion years). Consequently, it is also naturally found in ground water and the human body, usually in trace amounts (about 1 Bq per person). The ^{238}U decay series is shown in Figure 1. As shown in the figure, various types of ionizing radiation are emitted by the radionuclides in the chain including alpha radiation (α^{2+}), beta radiation (β^-) and gamma radiation (γ). In addition, similar or smaller amounts of natural ^{228}Ra, a decay product of natural ^{232}Th, are also found in soil and water. A similar ^{232}Th decay chain yields ^{228}Ra, ^{228}Ac, ^{228}Th, ^{224}Ra, ^{220}Rn, ^{216}Po, ^{212}Pb, ^{212}Bi, ^{212}Po, and stable ^{208}Pb (Eisenbud & Gesell, 1997).

As an analogue of calcium, radium that enters the body tends to incorporate into mineral bone where it and its progeny irradiate portions of the skeleton, primarily by alpha radiation emissions. Thorium isotopes tend to deposit on bone surfaces and irradiate surface cells, primarily by alpha radiation emissions. The radionuclide undergoing decay is referred to as the parent nuclide and the transformed nuclide is called the daughter or decay product. The daughter may be no more stable than its parent and may also be radioactive. Successive transformations will occur in a so-called decay chain, possibly yielding several radioactive progeny, until a stable nuclide is reached (Fig. 1).

In this context for ^{226}Ra, the word dose is used specifically to refer to the mean alpha radiation dose absorbed by the irradiated skeleton measured in units of energy deposited per unit mass of skeletal tissue (Gy). Dosage refers to the amount of material that enters the body or is administered (Bq). Burden refers to the amount retained in the body or skeleton with time post intake (Bq). Intake can refer to systemic intake into the blood or other organs of the body, as for the U.S. radium cases (Rowland, 1994), or it can refer only to that material that passes into the mouth or nose during ingestion or inhalation as in radiation protection practice (ICRP-30, 1979). Note that there is a really big difference between ingestion or inhalation intake and systemic intake, sometimes a factor of thousands. Cumulative dose, D, to an organ of the body is derived from the time integral of concentration (or dose rate) in that organ and requires a specified time limit to be meaningful.

2. Internally deposited radionuclides

2.1 Radium in people

Early in the last century people were exposed to high concentrations of ^{226}Ra and ^{228}Ra and their decay products in the luminous dial industry, in laboratory work, and in medical or private therapeutic use of radium. In particular, luminous-dial painters, who were mostly young women, were taught to tip their paint brushes on the tongue to make a sharp brush point; this procedure resulted in the ingestion of considerable radium leading to systemic uptakes of some of the ingested radium. These massive intakes of emulsions of pure radium salts resulted in life-time absorbed doses to the irradiated skeleton of some dial painters as high as a few thousand Sv, with consequent cases of severe injury to the skeleton and/or bone sarcoma as well as carcinomas of the head associated with retained ^{222}Rn and daughters in the nasal sinuses and mastoid spaces.

Detailed evaluations of these human cases have been described by Evans (1943), Evans et al. (1972), Evans (1974) and further documented in recent years (Rowland, 1994). Studies of individuals who had accidental intakes of massive amounts of radium have shown that the principal risk of exposure to large quantities of radium is direct damage to bone, bone cancer (but not leukemia), and cancers of the head caused by trapped decay of radon gas produced by decay of ^{226}Ra in the skeleton. It is interesting to note that no person among the

U.S. radium dial painters whose skeletal dose was less than 10 Gy (200 Sv for ionizing radiation weighting factor of 20 for alpha radiation) developed bone cancer because of exposure to ^{226}Ra or ^{228}Ra. An effective or virtual threshold dose, below which there is no radiation-induced cancer from exposure to radium, has been observed (Evans, 1974; Raabe et al., 1980; Raabe, et al., 1990; Raabe, 2010). In addition, serious non-neoplastic (non-cancer) radiation bone injury, such a bone osteodystrophy and fractures, occurred at even higher doses than did cancer.

Since deposition in and protracted irradiation of the skeleton follows systemic uptake, even a single brief intake initiates a period of chronic irradiation from radium and its retained decay products that depends in duration on skeletal retention or radioactive decay of the deposited radium. About 30% of radon-222 is retained from decay of radium-226, and almost all radon-220 is retained following the decay of radium-224 that forms from the decay products of radium-228. In the case of inhalation of radium (or thorium), irradiation of the lung is also an important aspect of the exposure. However, radium (or thorium) deposited in the lung is cleared by the lymphatic flow, coughed up and swallowed or expectorated, or enters the blood. Upon entry into the blood from the lung, inhaled radium is mostly excreted in the urine, while about 20% is deposited and tenaciously retained in the skeleton. The retention of radium in the human body, particularly the skeleton, has been the subject of considerable study and has been described in a mathematical form that allows the ionizing radiation dose to body tissues to be estimated from time of systemic intake until any time after intake (ICRP-20, 1973).

Fig. 3. Early 20th Century photo of young women painting clock dials with luminous paint containing radium. They tipped their brushes on their tongues and swallowed considerable quantities of ^{226}Ra or ^{228}Ra.

Lifetime studies were conducted in the 20th Century of the American radium dial painters (Fig. 3) and others with internally deposited radium isotopes (^{226}Ra and ^{228}Ra). Internally deposited radium chemically behaves somewhat like calcium and deposits somewhat uniformly in bone mineral in the skeleton. Radium is cleared from the body slowly so that a few percent of the intake can still be detected decades after intake. Alpha radiation from radium and its decay products lead to bone sarcoma in some exposed people at the highest doses. For ^{226}Ra (half life 1,600 years), the first decay product is ^{222}Rn gas much of which migrates in the blood to other parts of the body and is exhaled via the lung. Some ^{222}Rn gas sequesters in the sinus and mastoid regions of the head where alpha radiation from radon and its decay products lead to head carcinoma in some of the most highly exposed people. The most extensive studies of people exposed to radium were conducted at Massachusetts Institute of Technology (MIT) and Argonne National Laboratory (ANL). Results have been reported for over 2,000 people including about 1,700 who had significant intakes of radium, mostly to ^{226}Ra (Rowland, 1994). The nuclear physicist, Robley Evans, author of The Atomic Nucleus (1955) lead the early studies. Evans observed that only those people who received skeletal absorbed alpha radiation doses exceeding 1,000 cGy, developed bone sarcoma (Fig. 4), which he called a "practical threshold". He showed that the absence of sarcoma cases below 1,000 cGy was a statistically significant finding (Evans et al, 1972). This finding did not deter others from reorganizing the data into selective groups of cases and advancing a linear no-threshold model for radium induced bone sarcoma (Fig. 5).

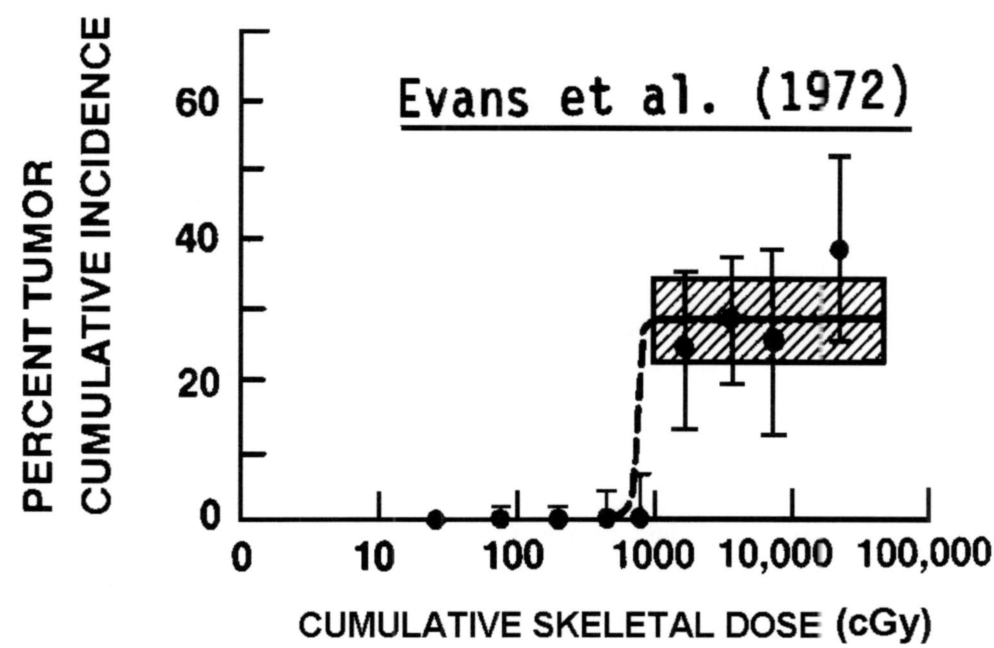

Fig. 4. Bone sarcoma incidence in people exposed to ^{226}Ra as a function of cumulative dose to the skeleton as reported by Evans et al. (1972).

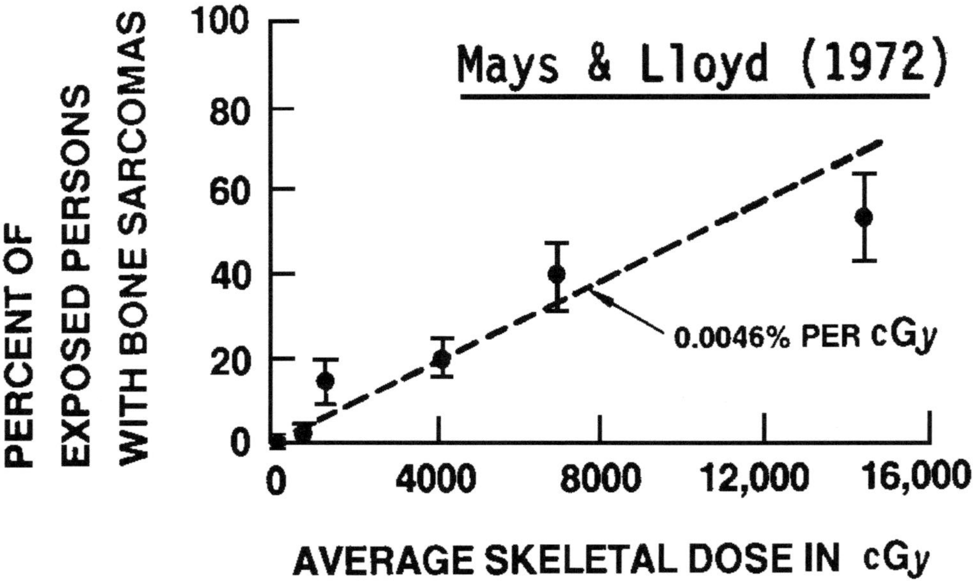

Fig. 5. Misleading linear dose-response model fit to bone sarcoma incidence in people who were exposed to ^{226}Ra as a function of cumulative dose by Mays and Lloyd (1972).

2.2 Radium-226 in beagle dogs

Since human data for exposure to internally deposited radionuclides are usually difficult to evaluate and readily confounded, in the late 1950's, the Division of Biology and Medicine of the United States Atomic Energy Commission conceived a series of carefully controlled scientific studies utilizing purpose-bred dog colonies of pedigreed beagles. These studies were conducted primarily at the University of Utah, the University of California, Davis (UC Davis), Hanford Laboratory, Richland, Washington, and Lovelace Foundation in Albuquerque, New Mexico.

The largest lifetime study of beagles exposed to ^{226}Ra was conducted at UC Davis from 1961 to 1989. Dogs were bred in a manner to maintain and randomize the gene pool and entered into the study over several years in a randomized-block design so that and all dosage levels and controls were represented contemporaneously in each treatment block. Exposure to ^{226}Ra, temporally simulating the dial painter exposures, were by eight intravenous injections two weeks apart in young adults from 435 to 540 days of age (Raabe 1989).

The distribution of deaths and causes are show in Fig. 6 (Raabe, 2010). The bone sarcoma deaths followed a tight lognormal power function of average radiation absorbed dose rate to skeleton with geometric standard deviation of only 1.22. Although the induced cancer response is a tight power function of dose rate, it is not a meaningful function of cumulative dose. For example, at an alpha radiation dose rate of about 0.1 Gy d^{-1} the median time to bone cancer death is about 1,200 days with a total cumulative dose of 120 Gy. In sharp contrast, at a alpha radiation dose rate of 0.003 Gy d^{-1} the median time to bone cancer death is about 4,000 days with a total cumulative dose of only 12 Gy. It would seem that the radiation is ten times more effective at the lower dose rate than at the higher dose rate, but that observation is misleading since it is the dose rate that controls the cancer induction and latent period. It is the lifetime average dose rate that controls the risk of radiation-induced cancer. Since the extrapolated bone cancer latent period exceeds the natural life span of the beagles at low dose rates, there is a virtual threshold for bone cancer from radium. None of the dogs in the lowest dose group developed bone cancer because they died first from causes associated with natural aging.

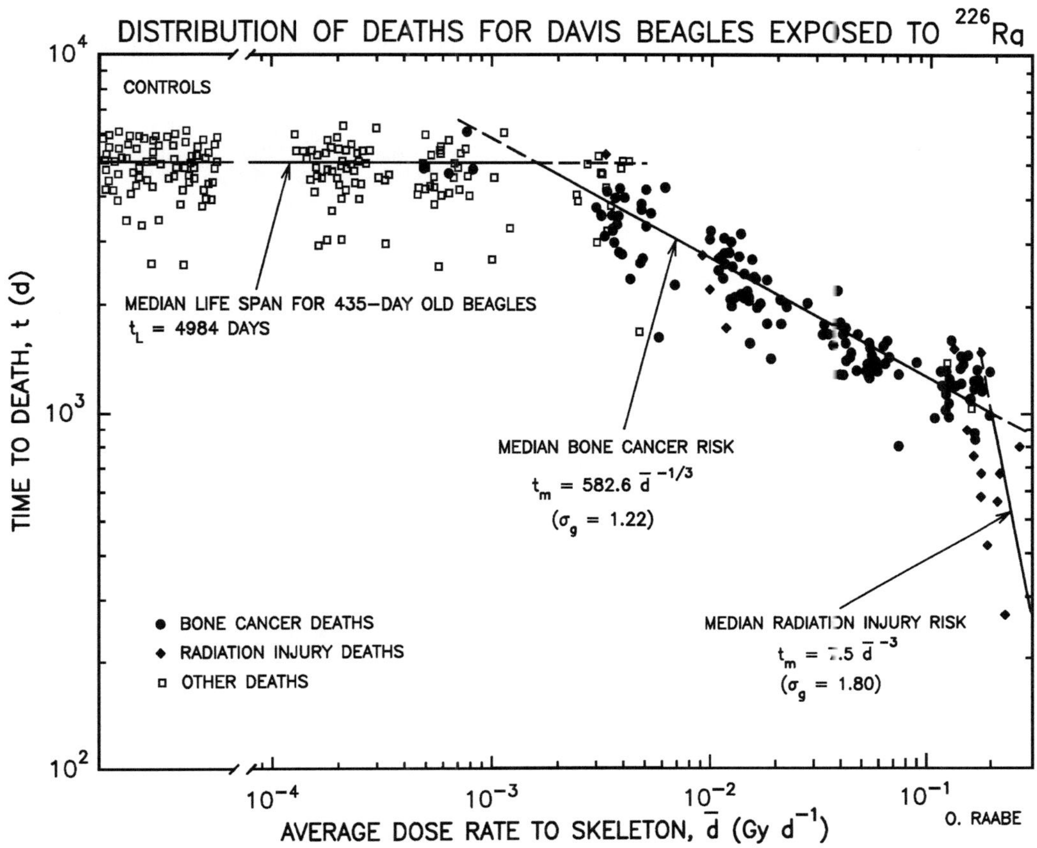

Fig. 6. Two-dimensional logarithmic representation of the data for radiation injury deaths, bone sarcoma deaths, and other deaths in beagles injected with ^{226}Ra at UC Davis, showing time from initial intake at age 435 days to death for each individual dog versus average dose rate to skeleton and fitted lognormal risk functions (Raabe, 2010).

The data shown in Fig. 6 is a two-dimensional representation of a three dimensional phenomenon. The three dimensions are lifetime average dose rate to the skeleton, time to death, and the frequency distribution of bone cancer deaths (Fig. 7). This three-dimensional phenomenon can be expressed in terms of the risk of death from various causes during the beagle lifetime (Fig. 8). The occurrence of beagle bone cancer deaths from ^{226}Ra-induced cancer is displayed as a mound rising out of a Euclidian plane. At low dose rates there are no cases of bone cancer observed because all of the beagles have died of causes associated with natural aging. At high dose rates there are radiation injury deaths.

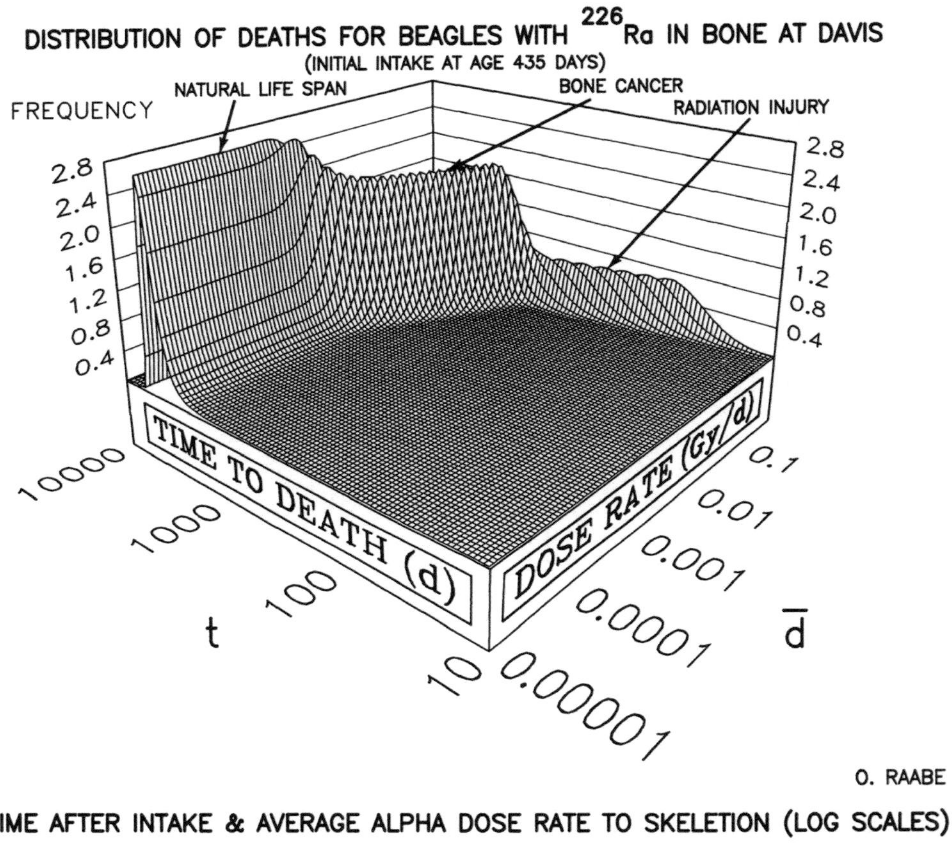

Fig. 7. Three-dimensional logarithmic representation of average-dose-rate/time/response relationships of Fig. 6, for beagles injected with ^{226}Ra at UC Davis, shown as the probability density distribution frequency of the combined risk of dying from causes associated with natural life span, radiation-induced bone sarcoma, and radiation injury, as a function of average dose rate to skeleton and elapsed time after intake at age 435 days (Raabe, 2010).

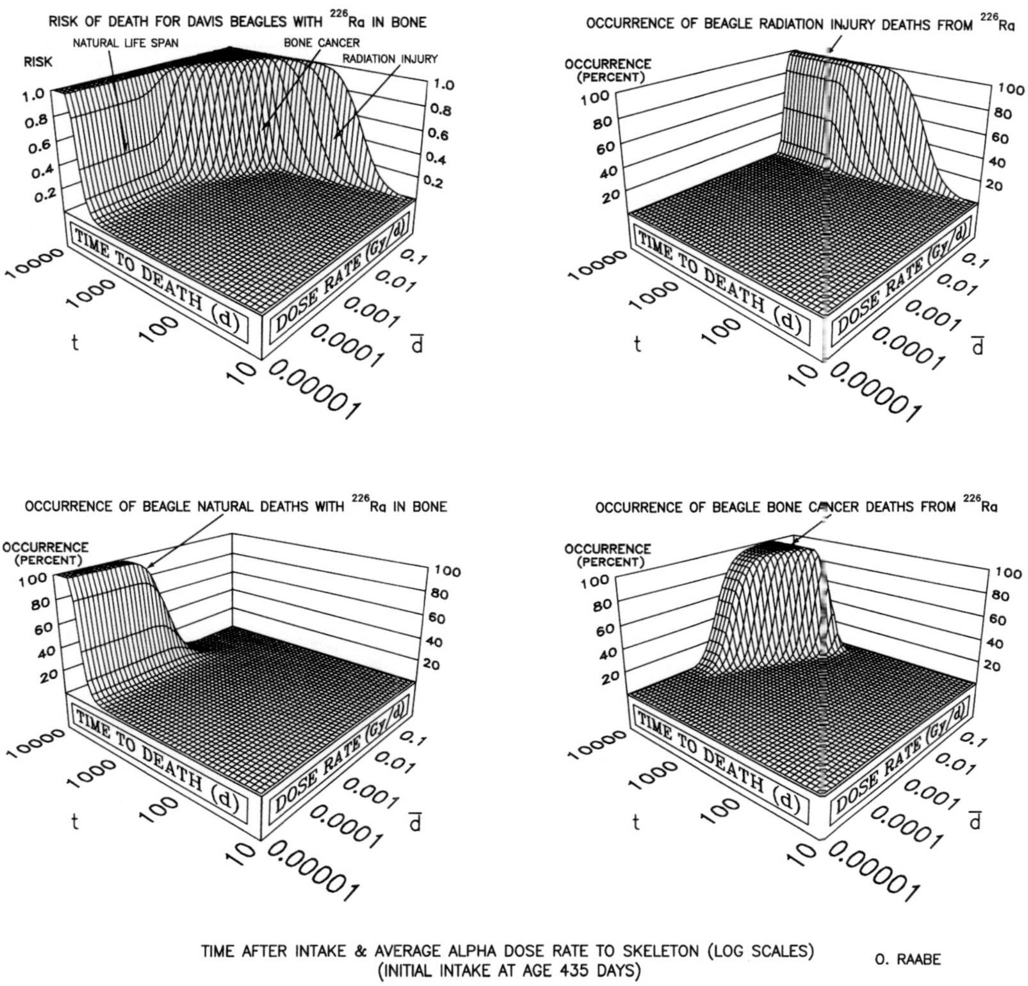

Fig. 8. Three dimensional representation from Fig. 6 of average-dose-rate/time/response relationships for beagles injected with ^{226}Ra at UC Davis shown in the top left panel as the combined risk of dying, and in the successive panels as the occurrence of deaths from radiation injury, natural life span, and radiation-induced bone sarcoma (Raabe, 2010).

The common but inappropriate practice of assuming that the cancer induction risk from exposure to ionizing radiation is proportional to cumulative radiation dose leads to misunderstandings about the true nature of the dose-response relationship. For example, the data from the ^{226}Ra beagle study can be plotted in the traditional fashion (Fig. 9). The common practice of fitting a linear no-threshold line starting at the zero-zero coordinates leads to an imprecise model that completely obscures the virtual threshold at low doses shown by the precise relationship to dose rate (Fig. 6).

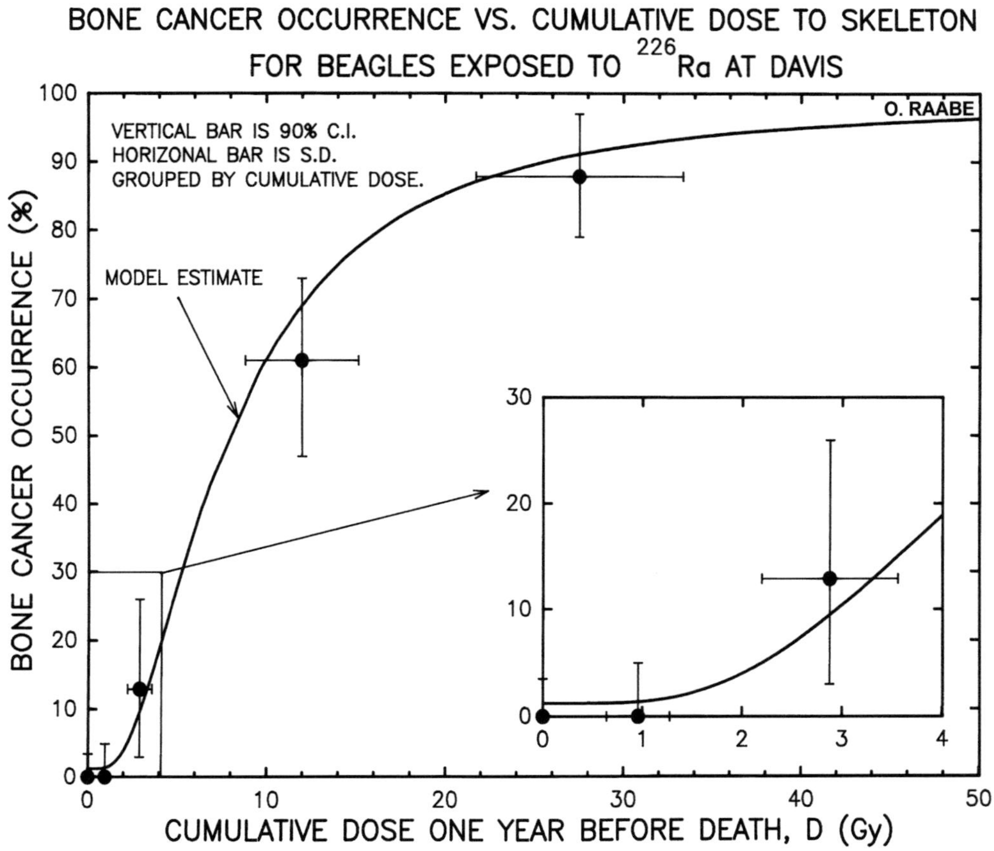

Fig. 9. Two-dimensional representation of the data for radiation-induced bone cancer deaths, in beagles injected with ^{226}Ra at age 435 days of age at UC Davis, showing lifetime bone sarcoma occurrence as percent of population versus calculated cumulative dose to skeleton one year before death. Note the life-span virtual threshold below 1 Gy as shown in the inset that would be obscured by a linear no-threshold (LNT) model of these data that starts at the origin with the only zero or near-zero risk at zero dose (Raabe, 2010).

2.3 Human cancer risk from radium-226

When a three-dimensional analysis is used for an interspecies comparison for radium-induced cancer, the different species display a similar relationship displaced in time based on natural life span (Fig. 10) (Raabe et al., 1980). The greater scatter of data for mice and people is the result of uncertainty and inaccuracy for the dose data compared to the precise beagle study. When the time scale is normalized with respect to natural lifespan for the species, all three sets of data overlay (Fig. 11). These results provided a basis for scaling cancer induction dose-response relationships among difference species. For example, the precise beagle data can be used to predict radium-induced bone cancer deaths in people by life-span normalization. The resulting prediction of median bone cancer risk for people falls almost perfectly at the median of the observed risk although the human data are quite scattered because of the dosimetric limitations (Fig. 12).

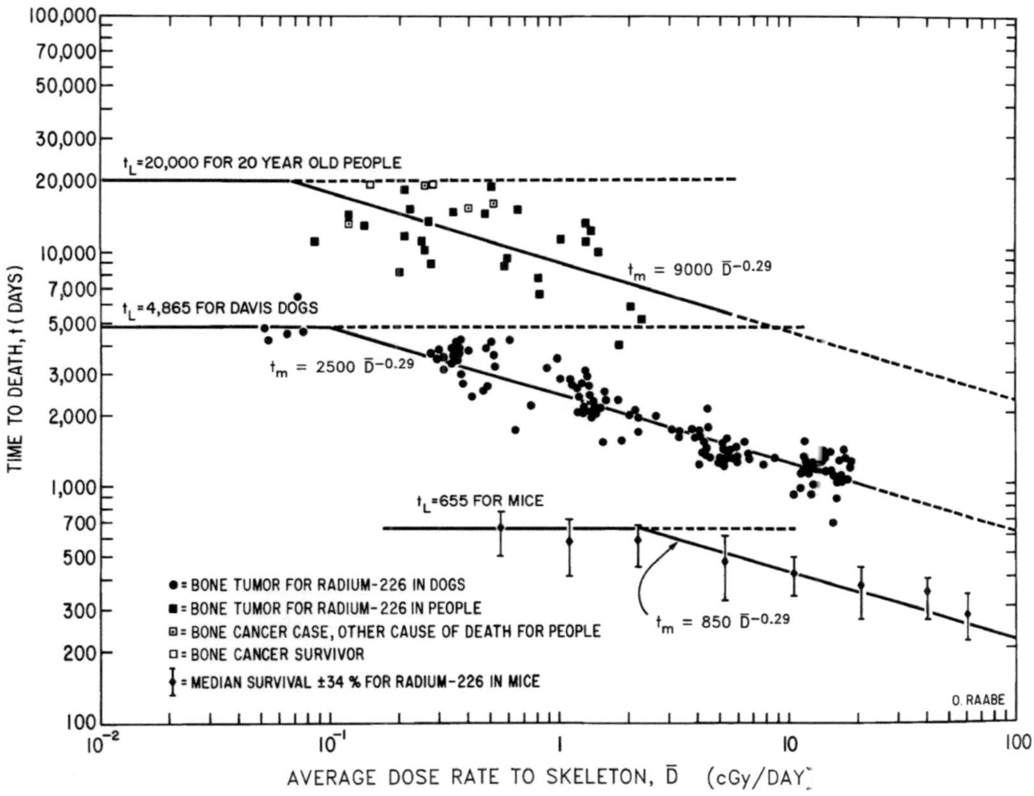

Fig. 10. Primary bone sarcoma deaths from ^{226}Ra in people, beagles, and female mice showing similar dose-response functions indicated by parallel median lines with negative slope of about one-third and t_L are the typical life spans for the three species (Raabe et al., 1980).

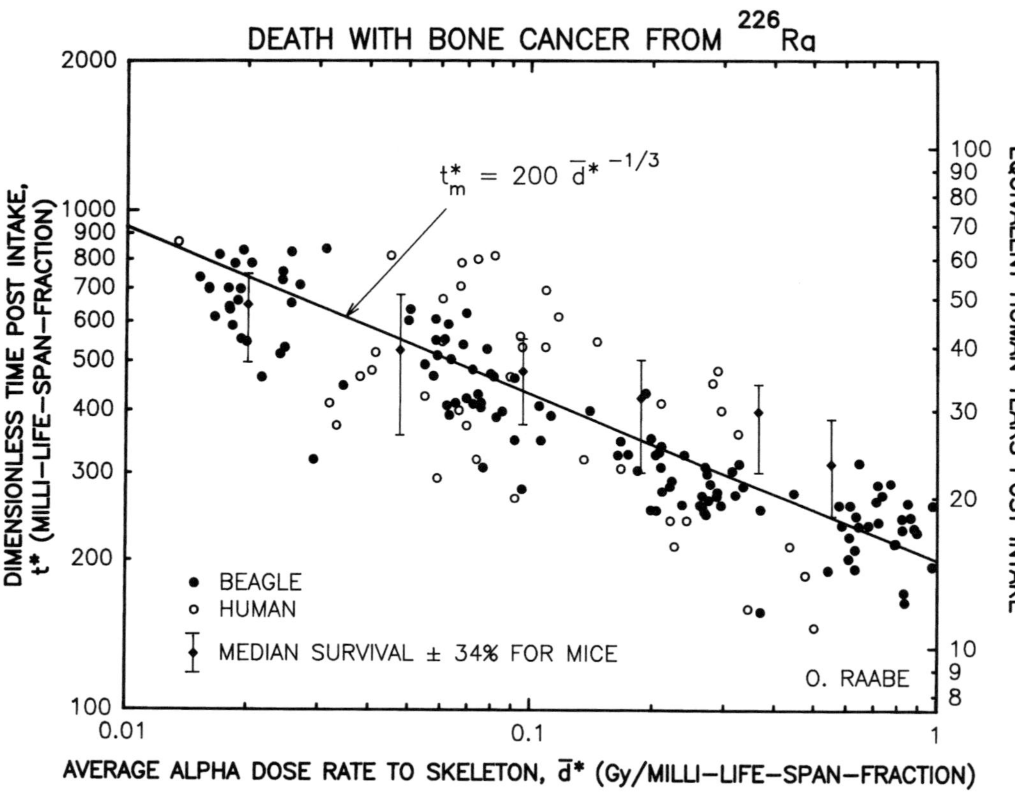

Fig. 11. Life span normalization with respect to dimensionless time and average alpha radiation absorbed dose rate to skeleton for fatal bone sarcoma in beagles, mice and people (Fig. 10) showing a single logarithmic regression line (determined for beagles) represents the median risk for all three species (Raabe, 2010).

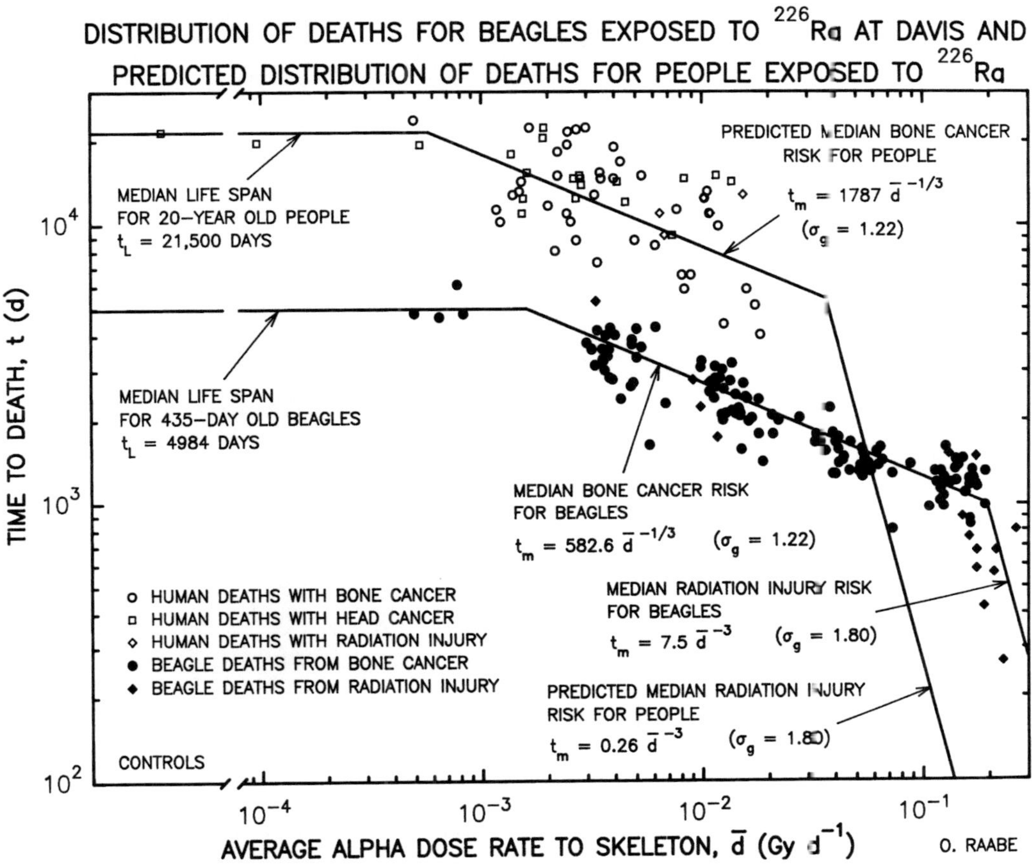

Fig. 12. Distribution of deaths for beagles and predicted distribution of deaths for people exposed to ^{226}Ra along with human data from the U.S. radium studies (Raabe, 2010).

Another form of alpha radiation-induced cancer occurs in people but not in laboratory animals. The radioactive decay of ^{226}Ra yields ^{222}Rn gas, much of which enters the blood stream and some of which sequesters in the nasal sinus and mastoid region of the head which leads to alpha irradiation of the internal tissues in these regions of the head from radon and its decay products. These deaths are somewhat delayed in time. The resulting three-dimensional frequency distribution of causes of death in people with internally deposited ^{226}Ra is shown in Fig. 13. Both head carcinoma and bone sarcoma display virtual threshold cancer risk relationships at low doses and dose rates.

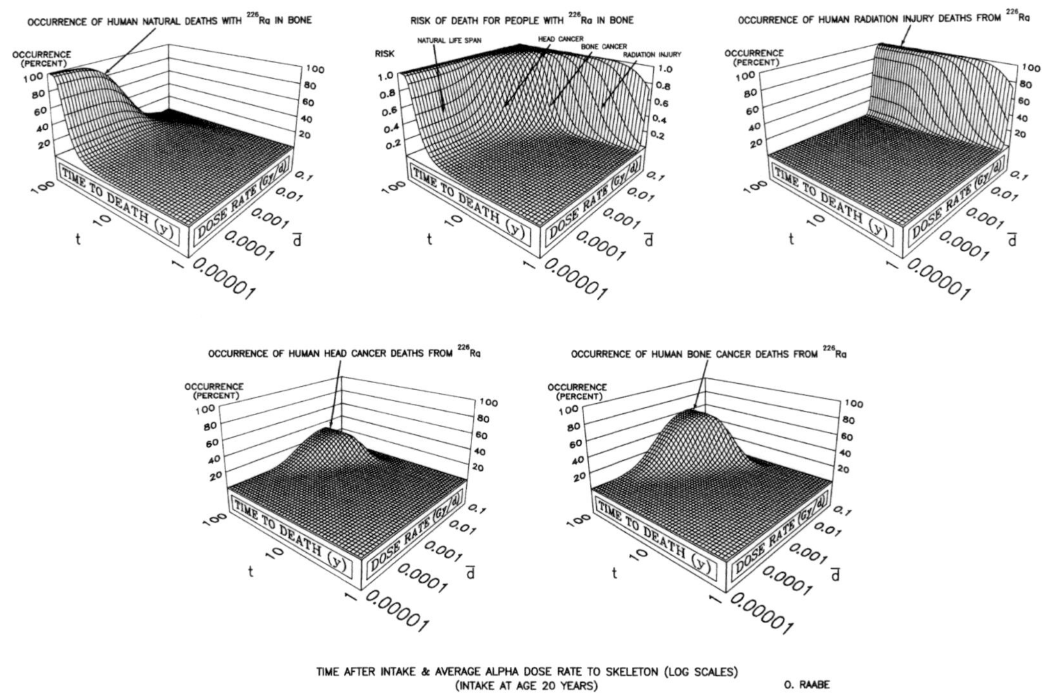

Fig. 13. Three-dimensional representation of the average-dose-rate/time/response relationships for people after intake of ^{226}Ra shown in the top middle panel as the cumulative risk of dying, in the top left panel as the occurrence of deaths associated with natural life span, in the top right panel as the occurrence of deaths from radiation injury to the skeleton, in the lower panels have the occurrence percent of deaths from radiation-induced head carcinoma and bone sarcoma, respectively (Raabe, 2010).

2.4 Strontium-90 in people

Radioactive ^{90}Sr (half life 29 years), a product of nuclear fission of uranium and plutonium, has been released into the atmosphere of the earth by nuclear weapons in Japan in 1945, by numerous atmospheric nuclear weapons tests up until 1966, and by the Chernobyl nuclear reactor accident in Russia in 1986. Strontium is chemically similar to calcium and deposits somewhat uniformly in bone mineral in the skeleton. Released to the environment, strontium is adsorbed and retained by clay in soil, and is found in small amounts in agricultural food products. The ^{90}Sr, a beta particle emitter, decays to ^{90}Y, a short-lived beta particle emitter that decays to stable ^{90}Zr. Everyone born in the Northern hemisphere after about 1950 probably has measurable trace amounts ^{90}Sr in bones and teeth.

2.5 Strontium-90 in beagle dogs

Because of the widespread exposure of people to ^{90}Sr from nuclear weapons testing in the 20th Century, the U. S. Atomic Energy Commission initiated controlled studies of ingested ^{90}Sr in purebred beagles. The largest lifetime study of beagles exposed to ^{90}Sr by ingestion was conducted at UC Davis from 1961 to 1989 in parallel with the study of injected ^{226}Ra.Dogs were bred in a manner to maintain and randomize the gene pool and entered into the study over several years in a randomized-block design so that and all dosage levels and controls were represented contemporaneously in each treatment block. Exposed beagles received measured amounts of ^{90}Sr in food via the mother during mid-gestation and via their food through adulthood at 540 days of age. The exposed young adult beagles in the study had bones and teeth that were uniformly labeled with ^{90}Sr. The controls were fed food with a high-dose mass equivalent of non-radioactive stable natural strontium, mostly ^{88}Sr (Raabe & Parks, 1993).

Unlike the short-range alpha radiation from ^{226}Ra and decay products in bone that irradiated primarily the bone mineral of the skeleton, the longer-range beta radiation from ^{90}Sr-Y irradiated the bone mineral, the bone marrow, tissue and tissues adjacent to bone and teeth. Induced cancer deaths were associated with four different types of cancer: (a) bone sarcoma, (b) periodontal carcinoma, (c) oral carcinoma, and (d) leukemia. Each of these displayed a virtual threshold relationship (Fig. 14). Although there were 2 cases of bone sarcoma deaths among the 80 controls, there were no cases among the 183 exposed beagles

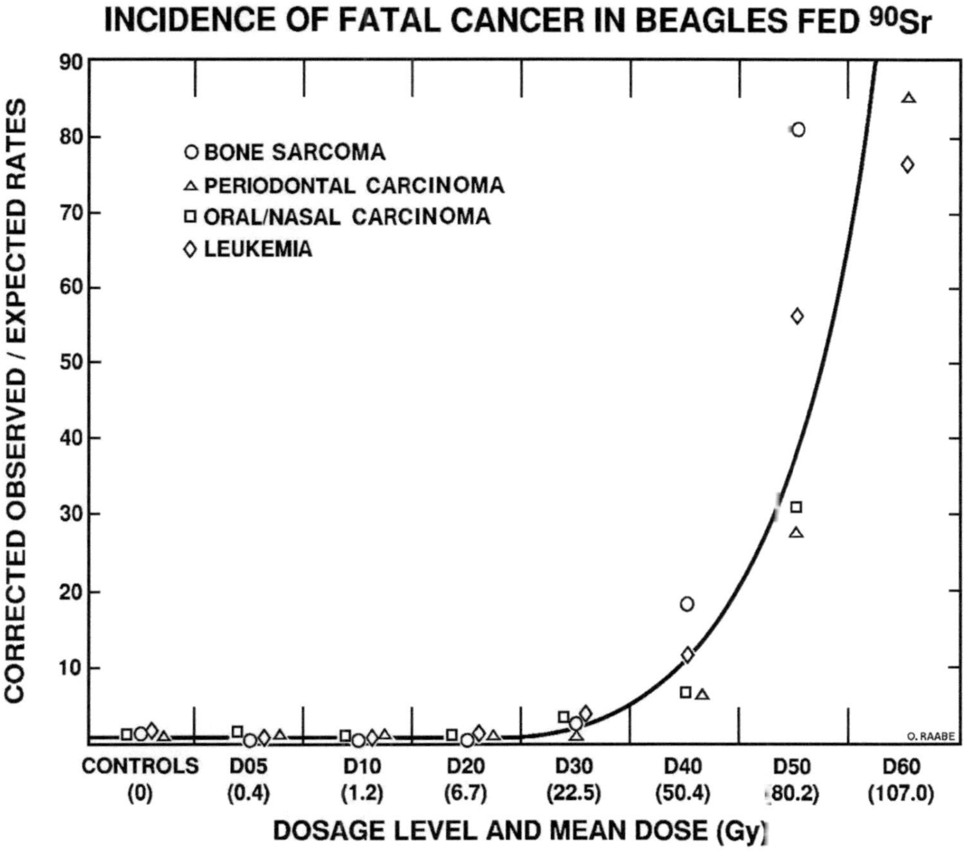

Fig. 14. Statistical evaluation by survival analysis (Peto et al. 1980) of the incidence of fatal leukemia, bone sarcoma, and carcinoma in beagles fed ^{90}Sr from before birth to adulthood at UC Davis demonstrating a life-span virtual threshold with all radiation-induced cancers occurring at calculated cumulative skeletal beta radiation doses above 20 Gy (20 Sv). The absence of bone sarcoma in the lowest three dosage groups is significantly less than those found in the controls ($p<0.047$)(Raabe, 2010).

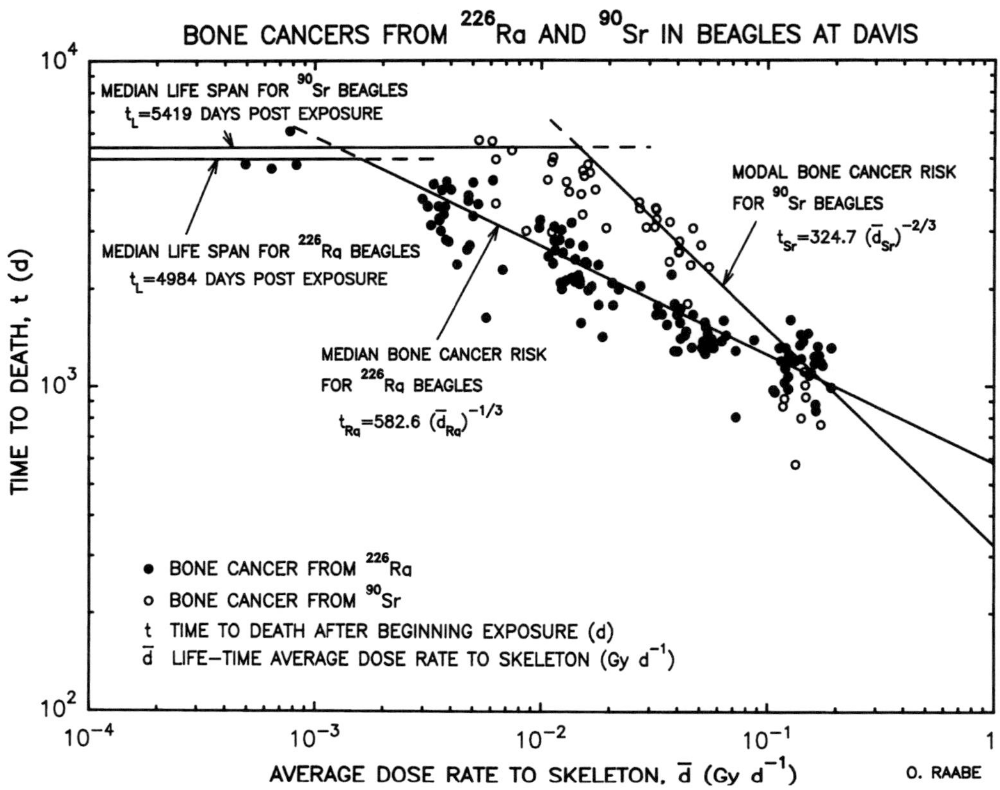

Fig. 15. Comparison of bone sarcoma in beagles at UC Davis from injected ^{226}Ra and ingested ^{90}Sr (Raabe, 2010).

in the three lowest dose groups. The chance of this is p=0.092 (Fisher's Exact Test). Among all 162 controls in the ^{226}Ra and ^{90}Sr studies there were 2 fatal bone sarcoma cases and two with bone sarcoma that died of other causes. With 4 bone sarcoma cases among 162 controls, the absence of bone sarcoma cases in the three lowest ^{90}Sr dose groups is a statistically significant reduction in tumor cancer incidence with p=0.047 (Fisher's Exact Test). The shape of the dose response relationship for ^{90}Sr-induced bone sarcoma is similar to that for ^{226}Ra except it follows a Weibull distribution and the power function slope is twice as steep as for ^{226}Ra indicating that an average of two beta particles are needed to equal the effect of one alpha particle for induction of bone cancer (Fig. 15).

2.6 Inhaled, ingested, and injected radionuclide studies in beagles

An analysis of 25 lifetime studies of inhaled, ingested, and injected internally-deposited radionuclides in beagles show remarkable consistency (Raabe, 2010). Three-dimensional models have been fit to selected data from life-time studies of internally deposited radionuclides in young adult beagles at four laboratories: University of California, Davis (Raabe and Abell, 1990), Lovelace Inhalation Toxicology Research Institute (ITRI) and University of Utah (Mauderly and Daynes, 1994), and Battelle Pacific Northwest Laboratory, PNL (Park et al., 1993). Dose response data were evaluated for beagles with skeletal burdens of ^{90}Sr after exposure by ingestion at Davis, by injection at Utah, and by inhalation at ITRI, for lung burdens of inhaled ^{144}Ce, ^{91}Y, and ^{90}Sr in fused aluminosilicate particles (FAP) at ITRI, inhaled ^{239}Pu dioxide at PNL, skeletal burdens of injected ^{226}Ra at Davis and Utah and inhaled ^{238}Pu at ITRI, and skeletal burdens of injected ^{228}Ra and ^{241}Am at Utah. Analyses were based on the mean organ absorbed doses to to the target tissues from parent and corresponding doses from decay products in their appropriate proportion, where all x ray and gamma emissions were ignored because of their minor contribution, and where beta emissions are also ignored in the cases where the primary exposures were from alpha radiation. Overall, there were separate injection studies of ^{226}Ra, ^{228}Ra, ^{224}Ra, ^{228}Th, ^{239}Pu, ^{249}Cf, ^{252}Cf, ^{241}Am and ^{90}Sr. There were separate inhalation studies of ^{239}Pu, ^{238}Pu, ^{90}Y, ^{91}Y, ^{90}Sr, and ^{144}Ce (Raabe, 2010).

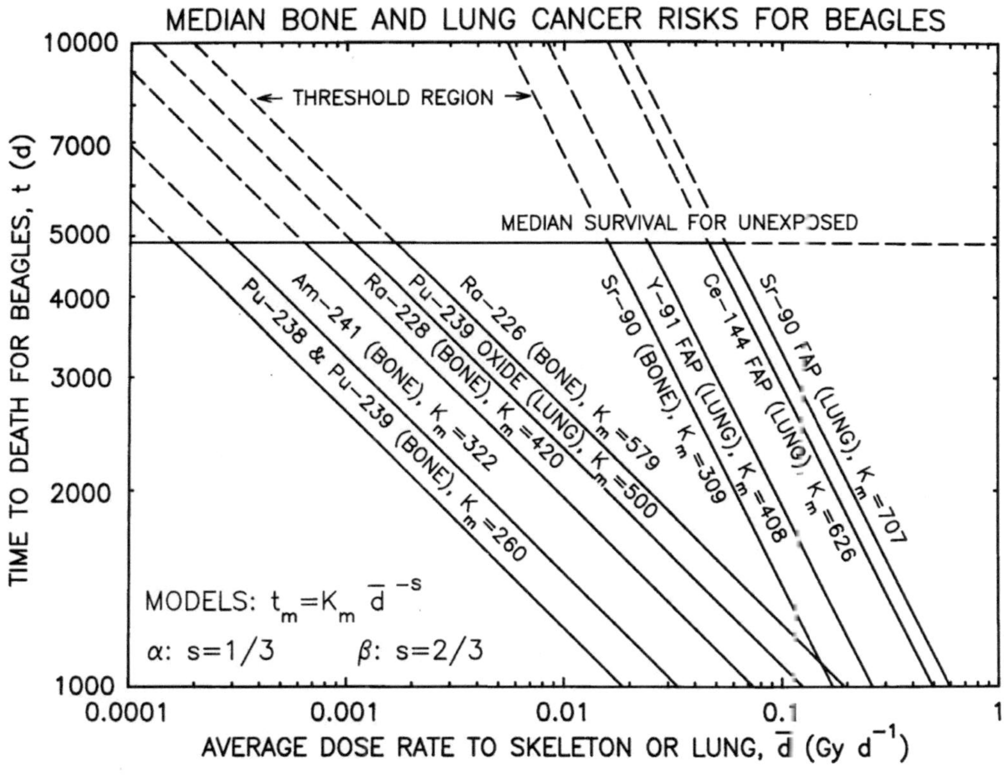

Fig. 16. Summary of bone sarcoma and lung carcinoma risk functions for beagles demonstrating similar target organ dose-rate/time/response patterns illustrating the life-span virtual thresholds at low dose rates. The positions of the lines vary because of inherent differences in irradiation of the target cells by the different radionuclides. Data were selected from life-time laboratory studies of internally deposited radionuclides in young adult beagles including skeletal deposits of ^{90}Sr after exposure by ingestion at Davis, by injection at Utah, lung deposits of inhaled ^{144}Ce, ^{91}Y, and ^{90}Sr in fused aluminosilicate particles (FAP) at ITRI, lung deposits of inhaled $^{239}PuO_2$ at PNL, skeletal deposits of injected ^{226}Ra at Davis, inhaled $^{238}PuO_2$ at ITRI, and skeletal deposits of injected ^{228}Ra and ^{241}Am at Utah (Raabe, 2010).

A selection of these results are shown in Fig. 16. All of the alpha radiation studies followed lognormal distributions with power function slope of negative one-third. All of the beta radiation studies followed Weibull distributions with power function slope of negative two-thirds. The different positions of these functions were shown to be associated with the efficiency of irradiation of the target sensitive tissue in each case. At low dose rates and associated low life-time cumulative doses there was a virtual threshold for ionizing radiation induced cancer. For young adult beagles, bone sarcoma induction from alpha-emitting or beta-emitting radionuclides was unlikely for cumulative doses below about 20 Sv delivered specifically to the sensitive tissues at bone surfaces. For inhaled alpha-emitting radionuclides lung carcinoma was unlikely for cumulative lung doses less than 10 Sv and for inhaled beta-emitting radionuclides lung carcinoma was unlikely for cumulative lung doses less than 5 Sv (Raabe 2010).

3. Radon exposures

3.1 Environmental radon

All of the earth's atmosphere contains low concentrations of radioactive airborne particles and gases associated with various naturally occurring radionuclides. Of particular interest are radioactive isotopes of the inert gas, radon. Three radioactive isotopes of the radon gas ^{219}Rn from the natural ^{235}U decay series, ^{220}Rn from the natural ^{232}Th decay series and ^{222}Rn from the natural ^{238}U decay series are produced in earth's crust and released by diffusion into the outdoor air. The ^{222}Rn with a decay half-life of 3.82 days is therefore the most important. In some cases ^{220}Rn, commonly called thoron, can be important also. The ^{238}U decay series that yields ^{222}Rn and its decay products is shown in Figure 1. The ^{222}Rn is the decay product of the naturally occurring radium isotope ^{226}Ra that occurs naturally in all the soils and rock on the surface of the earth at concentrations of about 25 Bq kg^{-1}. Consequently, ^{226}Ra is also naturally found in ground water and the human body, usually in trace amounts (about 1 Bq per person). When ^{226}Ra undergoes radioactive decay (half-life 1,600 y) it forms gaseous ^{222}Rn, which can percolate through and diffuse out of the soil or rock and into the air. This process has definite temporal limits since the half-life of ^{222}Rn is only about 3.8 d. However, enough radon reaches to the earth's atmosphere to provide an average outdoor concentration of about 10 Bq m^{-3} in outdoor air in the most populated part of the world. Much lower concentrations occur over the oceans and in cold polar locations. The ^{222}Rn decays in air to form radioactive, solid decay products that form radioactive aerosols. In addition to ^{226}Ra, similar or smaller amounts of ^{228}Ra, a decay product of ^{232}Th, are also found on the earth leading to the atmospheric release of the radon gas isotope, thoron, ^{220}Rn. However, the concentration of thoron and its decay products is usually negligible in comparison the ^{222}Rn and its decay products. Very high air concentrations of radon can be found in some mines and elevated levels can be found in most buildings.

When air containing radon gas is inhaled by a person, a small amount of this inert gas dissolves in body fluids. During its radioactive decay emitting an alpha particle and the following decay of its decay products some cells of the body are irradiated, but the level of whole body irradiation is quite small. Much more important is the irradiation of the lungs associated with the deposition in the bronchial airways of the radon decay products which are solid metal atoms rather than gases. While radon gas is inhaled and exhaled, the unattached decay products and those attached to small inhalable airborne particles can deposit in the bronchial airways of the lungs during normal breathing. For example, in the case of ^{222}Rn , the short-lived alpha-particle emitting isotopes of polonium, ^{218}Po and ^{214}Po, that deposit in the lungs can result in relatively high alpha-particle irradation of bronchial airways. The comparatively small dose associated with the beta-particle emitting decay products is insignificant compared to the alpha irradiation which is about 20 times more biologically effective than beta radiation per unit of energy. The equivalent dose to the bronchial airways from radon and thoron decay products in ambient air in homes for the average person in the USA has been estimated to be about 28 mSv per year (NCRP 160, 2009). In some areas of the USA, the annual dose to the bronchial airways from radon and thoron may be double this average value.

Given sufficient time and other favorable conditions, the decay products of ^{222}Rn will come into radioactivity equilibrium with the radon with each of the short-lived decay products having the same activity concentration in the air as the ^{222}Rn. However, this ideal equilibrium is rarely attained in air containing elevated levels of radon so that measurement of the radon gas concentration does not precisely indicate the concentration of decay products unless that state of disequilibrium is known. Since essentially all of the biologically important radiation dose to the respiratory epithelium is derived from the alpha-emitting radon decay products, their concentration is sometimes described in special units called the working level, WL. The WL unit is defined as any combination of the short-lived radon progeny in one liter of air that will result in the emission of 130,000 MeV of alpha particle energy. Air having a ^{222}Rn concentration of 3.7 kBq m^{-3} with the progeny in secular equilibrium would represent 1 WL. The exposure associated with a typical work month in a uranium mine for 170 h at 1 WL is called an exposure of 1 working level month, WLM. Dosimetric models indicate that the nominal dose to the bronchial epithelium associated with inhalation of radon decay product aerosols by a uranium miner is about 6 mGy/WLM. Assuming an alpha radiation weighting factor of 20, this yields about 120 mSv equivalent dose per WLM to the bronchial region of the lung.

Upon inhalation, the airborne particles containing radon decay products may deposit upon contact onto the surfaces of the respiratory airways. Because of their diffusivity, the very small molecular clusters may efficiently deposit in the head airways or in the trachea and bronchial airways of the lung. Other somewhat larger particles may reach the alveolar region of the lung, as well. Because of their short radioactive half-lives, they usually decay prior to being cleared from the respiratory tract and irradiate the respiratory epithelium. Of primary concern in this regard is the irradiation of the bronchial epithelium by the highly ionizing alpha radiation emitted by radium-A (^{218}Po) and radium-C' (^{214}Po). Since radon itself is an inert gas, it does not readily deposit in the respiratory airways during inhalation and is mostly exhaled. Occasionally an atom of radon gas may decay and emit alpha radiation in air present in the lung irradiating the epithelium, but the fraction of the dose contributed by the radon itself is small compared to that associated with the deposited decay product particles. Because naturally occurring radon decay products are in ambient air outdoors and within a building, the lung is the most highly irradiated organ of the body of a typical person from background radiation sources.

3.2 Radon and radon decay product dosimetry

The dosimetry methodology associated with estimating the alpha radiation dose to the lungs of people from exposure to atmospheres containing radon and thoron is quite complicated and the dosimetry estimates are subject to quite large uncertainties. Consider the case of radon gas entering an air space as it diffuses from the earth. Initially there are no decay products since they are produced as the radon atoms decay and convert from gas atoms to metallic atoms. This in-growth of ^{218}Po (commonly called Ra-A), ^{214}Pb (commonly called Ra-B), and ^{214}Bi (commonly called Ra-C) is shown graphically in Figure 17 for a concentration of radon equal to 3.7 kBq m^{-3} which at decay-product equilibrium is the level commonly called one working level (WL) associated with underground mining. While ^{218}Po comes to equilibrium with radon in about 20 minutes, it takes more than 2 hours for ^{214}Pb and ^{214}Bi to approach equilibrium. In that time more radon is entering the airspace, ventilation is changing the relative concentrations, and the metallic decay products are being surrounded by water molecules, are undergoing Brownian diffusion and are attaching to larger airborne particles and onto the surfaces of walls, floor, ceiling, furniture, and people. Measurements that are made of radon gas concentrations in a living space or underground mine do not describe the concentration of the airborne solid decay products that deliver more than 90% of the dose the bronchial airways of people in that room or underground mine.

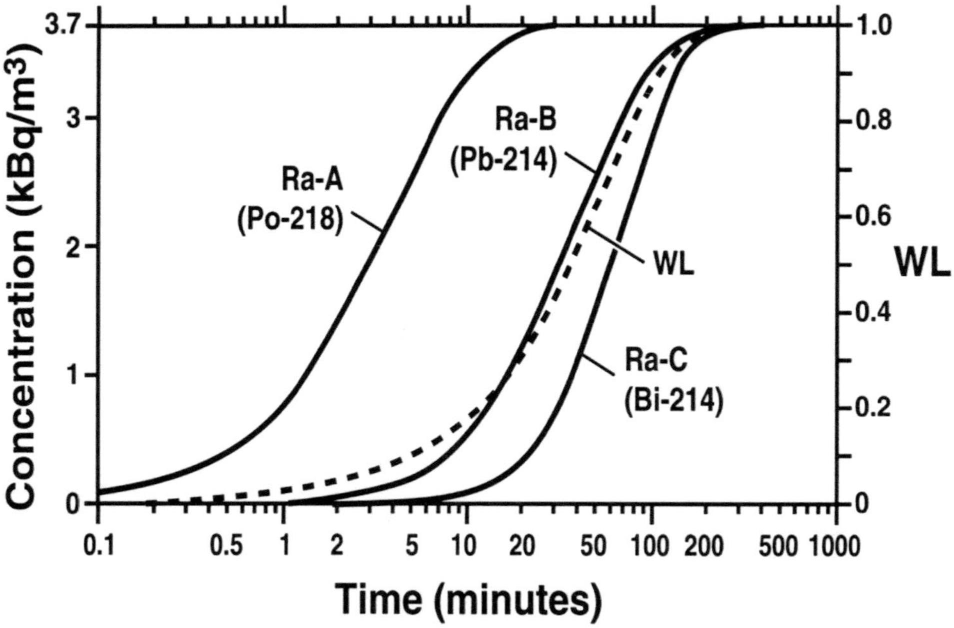

Fig. 17. Ingrowth of radon decay products as a function of time starting with 3.7 kBq m^{-3} of ^{222}Rn yielding one working level, WL.

As each radon gas atom in air decays by emitting alpha radiation it forms an atom of radioactive polonium, ^{218}Po (called radium-A, half-life about 3 minutes). This metallic atom quickly oxidizes and forms the center of a particulate molecular cluster of about 8 Å in diameter. When it decays emitting alpha radiation, a series of short-lived decay products are formed in air. These also form particulate clusters. This radon decay process is shown schematically in Fig 1. The long-lived lead isotope, ^{210}Pb (half-life 22.3 y) provides negligible radioactivity to the atmospheric aerosols and is, in effect, virtually non-radioactive compared to its short-lived progenitors. These radon decay product clusters represent the smallest airborne particles normally found in ambient air. In addition, these clusters attach upon contact to other, larger airborne particles to an extent that depends upon their concentration. Typically, more than 90% of radon decay product aerosols are associated with particles smaller than 0.5 μm in aerodynamic diameter.

Actually, the dosimetry is even more complicated. The deposition of airborne particles in the bronchial airways during inhalation is a function of the aerodynamic and diffusive properties of the airborne particles to which the radon decay products attach (ICRP 66, 1994). This depends on both the size distribution and concentration of the the airborne particles in the air. When the concentration of airborne particles is quite small, many of the decay products may not be attached to a larger particle (the so called unattached fraction). The unattached fraction efficiently attaches in all parts of the airways because of Brownian diffusion properties of very small particles causing somewhat efficient deposition in the upper airways of the nose and throat as well as the small bronchial airways. The irradiation of the living cells occurs primarily after the airborne particles deposit on the surface of the airways. If the radon decay products are associated with larger airborne parrticles, the deposition in the lung may be less because some of the inhaled particles may be exhaled. If the airborne particles are relatively large, they may deposit in the nose and throat and not readily reach the bronchial airways. Both the concentration and the aerodynamic particle size distribution of the airborne particles will affect the bronchial dose. Smoke in the air space can have a large effect on the bronchial radiation dose from radon decay products. Clearly, a simple measurement of the average concentration of radon in the air is insufficient to provide an accurate estimate of the lifetime average radiation dose rate to the bronchial epithelium of a person. Many reported studies only involve average radon measurements. Many reported studies only reported cumulative exposures to radon decay products in units such as working level months (WLM).

3.3 Lung cancer from exposure to radon and its decay products

Uranium mining in the 20th Century led to the clear realization that protracted and repeated exposures to high concentrations of radon can cause lung cancer. For example, miners working uranium mines in the Colorado Plateau region of the USA were found to have a high incidence of lung cancer depending on the level of exposure. People chronically exposed over extended period of employment to high levels of airborne radon decay products such as are found in uranium mines have developed bronchogenic carcinoma at incidence rates that significantly exceeded the expected rates in either smokers or non-smokers. However, almost all the lung cancer cases in the western U.S. mines occurred in smokers (Saccomanno et al., 1988). The reported excess relative risks of lung cancer as a function of cumulative dose expressed in working level months of exposure are shown in Figure 18. The miners did not have radon dosimeters and it could never be clear how long anyone worked in any of the areas within the mines, hence the working level month (WLM) values are estimates that are intended to represent cumulative radiation dose. As with all cumulative dose plots it is common to fit or draw a linear no-threshold function that begins at the origin obscuring the apparent virtual threshold. This is facilitated by the large dosimetric uncertainties (Lubin et al., 1995).

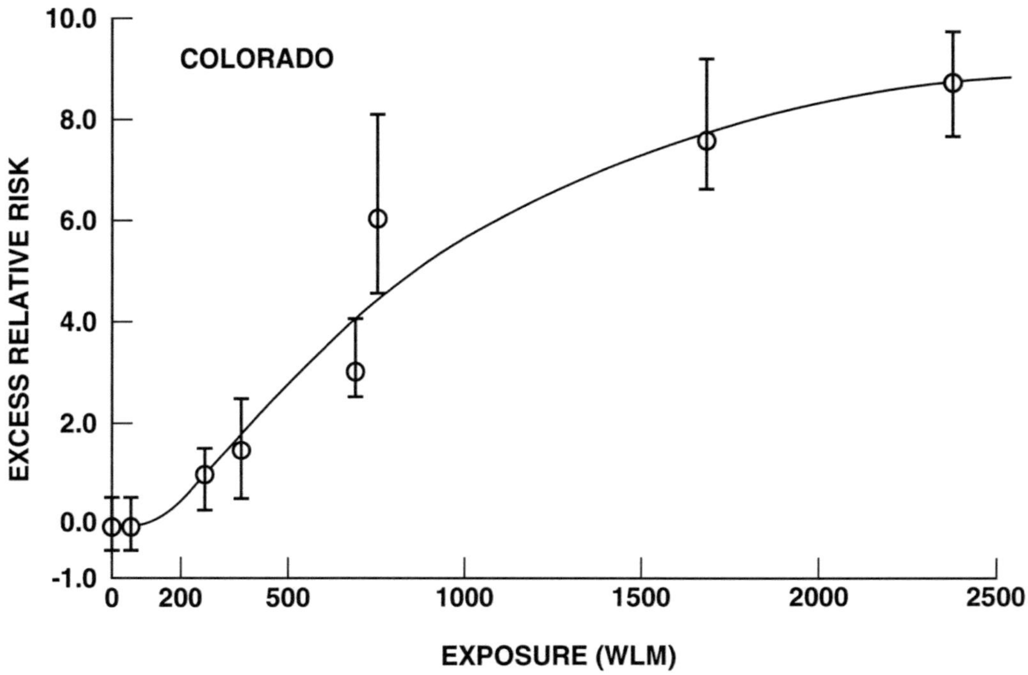

Fig. 18. Historical data of observed lung cancer risk as a function of exposure in working level months (WLM) among uranium miners from the Colorado Plateau in the United States (BEIR IV, 1988). Uranium miner data were routinely expressed in estimated cumulative doses in WLM so that the virtual threshold was badly obscured or ignored and linear no-threshold (LNT) models were assumed to apply in every case (Figure 19).

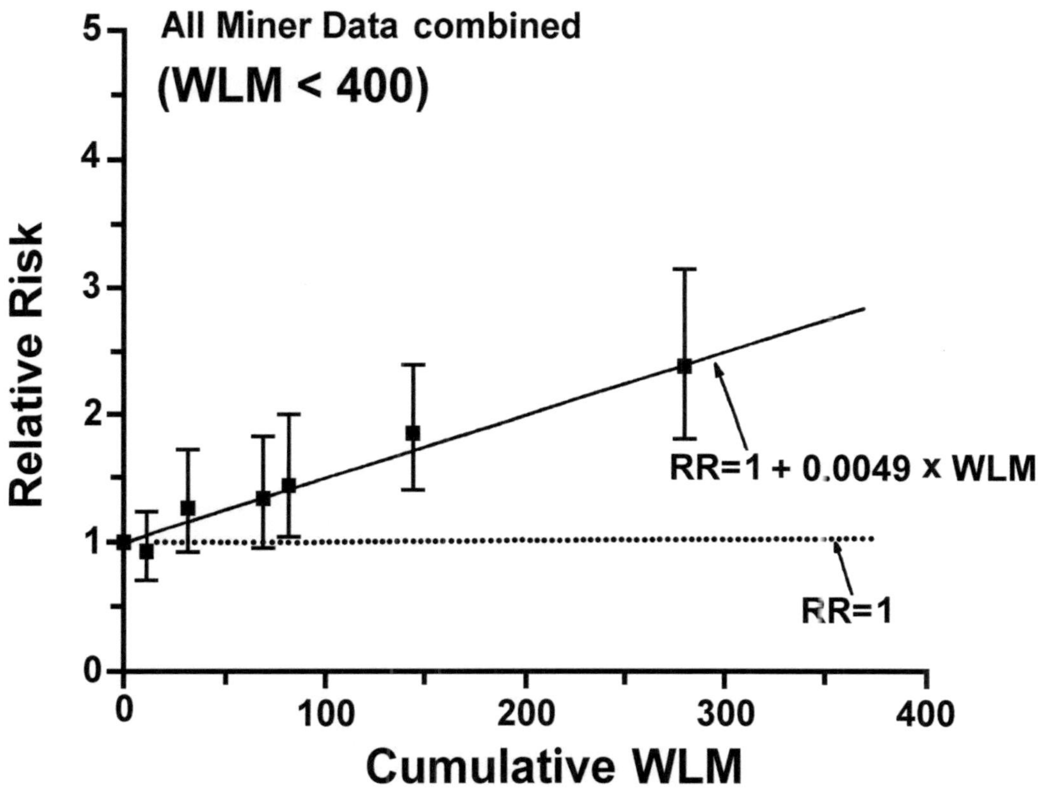

Fig. 19. Observed relative risk (RR) of lung cancer in underground miners versus cumulative airborne radon decay product exposure in working level months (WLM) for WLM less than 400 with a linear no-threshold risk model fit to the data (Lubin et al., 1994).

The reported epidemiological studies of radon in homes are not conclusive and are usually based on average radon concentration measurements rather than on decay product levels and the results were displayed using the standard linear no-threshold assumption. Also, the dosimetric uncertainties are usually not displayed. In Figure 20 the authors created a mathematical relative risk of 1 at a radon concentration exposure of zero but there is no such point in the real world. The major cause of lung cancer in the United States is known to be cigarette smoking (up to 95% incidence) so that cigarette dosimetry is actually more important than radon dosimetry in these types of studies.

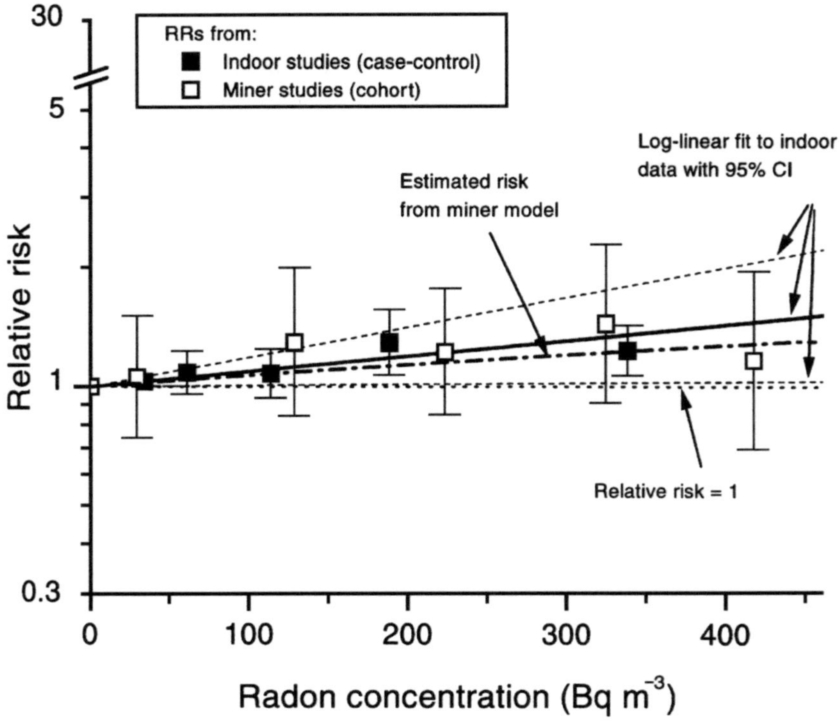

Fig. 20. Summary of calculated relative lung cancer risk mathematical meta-analysis of indoor radon studies along with miner studies (BEIR VI, 1999).

A careful ecological study was conducted of the relationship of lung cancer mortality as a function of measured air concentrations in homes in over 1,600 counties in the United States (Cohen, 1995). The remarkable results showed that the counties with homes having the higher radon concentrations tended to have the lowest lung cancer mortality rates (Figure 21). This relation was robust and was demonstrated for widely separated counties in various parts of the United States. His finding were in conflict with the prevailing linear no-threshold theory promulgated by the ICRP and United States Environmental Protection Agency (EPA). A scientific committee formed to review the radon risk data concluded that the results were faulty no matter how robust and reproducible because they were based on an ecological study that could not directly match cases and radon exposures (BEIR VII-2, 2006). Cohen argued to no avail that under the LNT hypothesis the distribution of the doses was irrelevant and his data proved the linear no-threshold model did not apply to exposure to radon in homes. Although it is mathematically possible that Cohen's ecological study is misleading, it seems highly unlikely that such an anomaly would uniformly appear everywhere in the United States. Cohen's result is substantiated by a simple comparison of the County by County radon zones and lung cancer mortality rates (Figure 22).

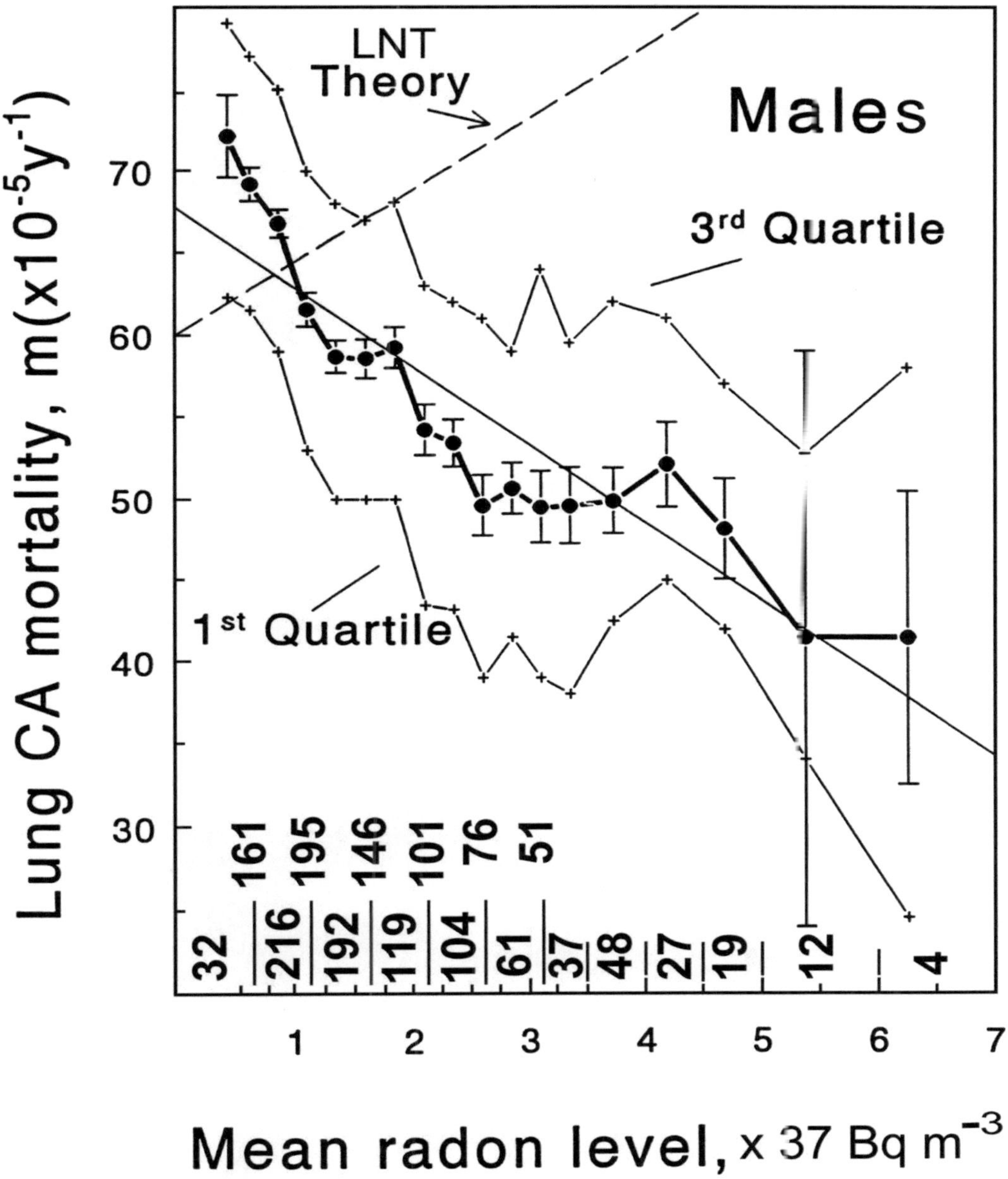

Fig. 21. Lung cancer mortality rates in men versus mean radon level for 1,601 counties spread throughout the continental United States as a function of mean radon level with the abscissa in units of 37 Bq m^{-3} (Cohen, 1995).

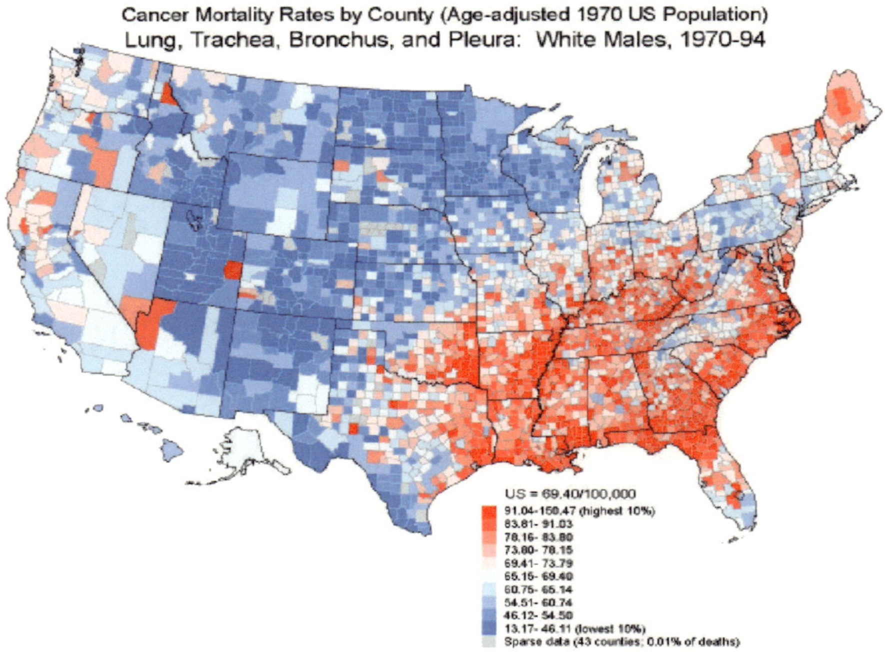

LUNG CANCER: Blue Low, Red High ↑

↓ RADON: Red HIGH, Yellow LOW

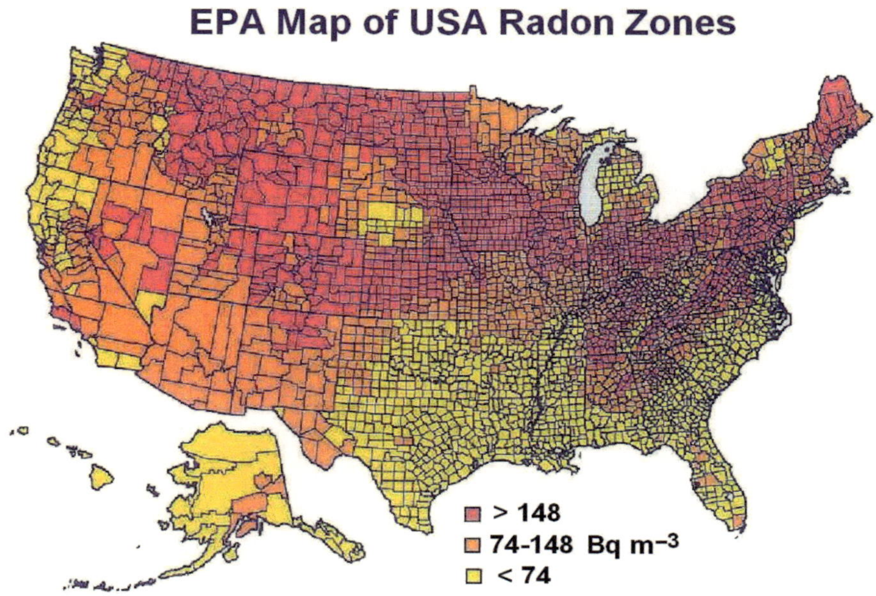

Fig. 22. Comparison of lung cancer mortality rates and radon concentration in homes among the counties of the United States of America.

A conditional logistic regression case-control study of lung cancer risk from residential radon was conducted in the State of Massachusetts in the United States by Thompson et al. (2008). Although their radon dosimetry was limited to average air concentration, they also evaluated cigarette smoking dose in a precise manner by stratifying the data with nine categories of smoking history. This ingenious approach led to results that agreed well with Cohen's ecological study (Cohen 1995). They found that lung cancer rates were the highest in those people with the smallest radon expose levels (<25 Bq m^{-3}). As radon levels were increased the lung cancer rate went down in a consistent manner which was statistically significant in one dose group ($p<0.05$ at 59-75 Bq m^{-3}) and statistically strong in three others ($p<0.1$). Radon induced lung cancer appeared to be present only for average radon air concentrations greater than 250 Bq m^{-3}.

Fornalski and Dobrzynski (2011) reanalyzed the data from 28 published studies of the relationship of lung cancer and radon exposure using Bayesian statistical methods. They concluded that for people exposed to radon concentrations less than 838 Bq m^{-3} in those 28 studies there is no evidence that there is an association between radon exposure and lung cancer incidence.

These studies suggest that radon exposures at low dose levels can interfere with ongoing neoplastic processes in the lung as may be associated with cigarette smoking. Mechanisms such as stimulation of DNA repair may be invoked by low level exposures to ionizing radiation helping to inhibit the damage associated with chemical carcinogens such as cigarette smoke exposures. Similarly, up-regulating and down-regulating of genes at lower ionizing radiation doses can interfere with ongoing carcinogenic biological processes and lead to reduced cancer development.

4. External ionizing radiation epidemiology

4.1 Typical study problems

Several external ionizing radiation epidemiology studies have been conducted. However, some of these studies involve technical problems that affect the reliability of the findings. A distortion of the possible carcinogenic response is associated with the use of cumulative dose as the radiation exposure measure rather than the more appropriate lifetime average dose rate. Several confounders such as exposure to toxic and carcinogenic chemicals are quite difficult to properly evaluate and are often ignored. In years past two carcinogenic agents, benzene and trichloroethylene, were commonly used as cleaning agents in laboratories including nuclear research laboratories. There were few if any records of these exposures or chemical doses. The most important confounder among the predominantly male radiation workers is cigarette smoking. Confounding occurs when the workers who are most likely to receive the highest exposures to ionizing radiation are also the most likely to have been smokers. It is possible that those workers at nuclear facilities (mostly men) who worked most directly in the radiation areas were more predominantly cigarette smokers than workers who had managerial, supervisory, or clerical assignments. This confounding from smoking is most likely important prior to 1960 before the health risks of smoking were well understood. This possibility is most prominent in studies of workers at Oak Ridge National Laboratory in the United States State of Tennessee where nuclear operations began in 1943. Some epidemiological studies of Oak Ridge workers have attributed lung cancer and other smoking-related illnesses to radiation exposure (Wing et al., 1991).

4.2 Atomic worker studies

Two major series and the most important are the ionizing radiation epidemiological studies of multinational radiation works in several countries of the world that have been conducted by Cardis and her numerous co-workers (Cardis et al. 1995, Cardis et al. 2005). The two studies were similar in approach, but the second was much larger and included subjects who were in the first study. In these studies emphasis was on workers who had available ionizing radiation dosimetry records of external photon exposure (x rays and gamma rays) but who did not have major internal exposure or neutron exposure. In the second study internally deposited radionuclide dose was specifically limited to ten percent of the total. Two cause of death risk types were separately evaluated: (a) leukemia excluding chronic lymphocytic leukemia and (b) all cancers excluding leukemia. Simple linear dose-response models of risk that was assumed to be proportional to cumulative external ionizing radiation dose were statistically evaluated. Radiation doses were lagged by 10 years for cancer cases and 2 years for leukemia. Hence, they assumed the cancer risk was proportional to the cumulative dose received prior to the lag period just before death.

If a person developed lung cancer, the radiation dose received in the prior 10 years was assumed to have not been a contributor to the risk. Typical lifetime doses in these studies were well below 1 Sv, so that the risk of radiation-induced cancer would be expected to be very near zero, based on the internal emitter studies discussed above.

In the first study (Cardis et al., 1994; Cardis et al., 1995) 95,673 nuclear industry workers (85.4% men) in the United States, United Kingdom, and Canada were monitored for external exposure to ionizing radiation prior to 1988. A linear cumulative dose response model with Poisson regression was used to estimate excess relative risk per Sv. Eleven dose categories were used. Also, it was assumed that radiation might increase cancer risk but it would not decrease cancer risk, so a one-tailed statistical test was used with significance at the 10% level instead of the usual 5% level. In this study about 99% of the deaths involved lagged doses less than 0.4 Sv. No evidence of association was found between radiation dose and mortality from all causes or from all cancers. However, mortality from leukemia, other than chronic lymphocytic leukemia, had a statistically significant association in a trend test with cumulative external radiation dose (estimated p=0.046, 119 deaths among 15,825 deaths). This estimated trend test p=0.046 is primarily based on 6 cases among 238 deaths above 400 mSv which by itself is not significant. The excess relative risk for leukemia excluding chronic lymphocytic leukemia was reported to be 2.18 per Sv (90% confidence interval 0.1 to 0.99). The excess relative risk for all cancer types excluding leukemia was minus 0.07 per Sv (90% confidence interval minus 0.4 to 0.3). These results may be somewhat affected by the use cumulative dose risk mathematical models rather than the more appropriate lifetime average dose rate models. Exposures to leukemia-causing organic chemicals may have occurred in the workplace.

In the second enlarged study (Cardis et al. 2005) 407,391 nuclear industry workers (90% men) in 15 countries who had available external exposure ionizing radiation dosimetry data were evaluated by a joint analysis group with 50 contributors. These 15 countries included Australia, Belgium, Canada, Finland, France, Hungary, Japan, South Korea, Lithuania, Slovak Republic, Spain, Sweden, Switzerland, United Kingdom, and the United States. Organ doses were estimated by dividing the recorded doses by organ dose bias factors. The colon and bone marrow doses were used for both all cancers excluding leukemia and for leukemia. Eleven dose groups were created. Analyses used a simple linear relative risk

Poisson regression model of the form where relative risk equals 1+βZ where Z is the cumulative equivalent dose in Sv; 95% likelihood based confidence intervals were calculated. Estimates of excess relative risk were stratified for sex, age, and calendar period, facility, duration of employment, and socio-economic status. A total of 24,158 (5.9%) died during the study period including 6,519 from cancer other than leukemia and 196 from leukemia other than chronic lymphocytic leukemia. The leukemia incidence was statistically limited and the risk per Sv was not significantly different from zero.

The key results of this second study for cancer other than leukemia (Cardis et al. 2005) are summarized in Fig. 23. None of the cohorts were found to have excess relative risk of cancer other than those in Canada. The overall combined excess relative risk per sievert for all cancer excluding leukemia was 0.97 (0.14 to 1.97 ninety-five percent confidence range). The excess relative risk per sievert for all cohorts for non-leukemia cancers excluding lung and pleural cancers, which may be related to smoking, was no longer significantly different from zero [0.59 (-0.29 to 1.70 ninety-five percent confidence range)]. Of the fifteen countries, the calculated excess relative risk of cancer that was observed in Canada appear to be anomalously high for some unknown reason. When the study analyzed only the other 14 countries the excess relative risk of cancer (other than leukemia) per sievert was no longer significantly different from zero [0.58 (-0.22 to 1.55 ninety-five percent confidence range)]. With this abridgement, there would be no significant cancer risks observed in this study.

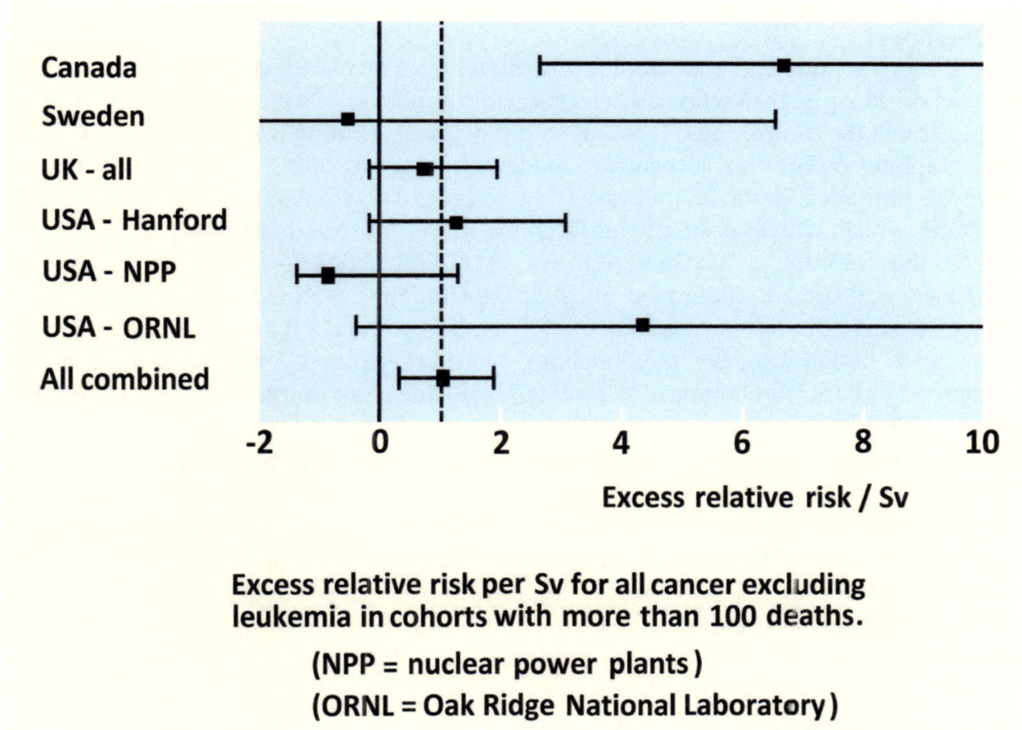

Fig. 23. Excess relative risk per Sv for all cancer types except leukemia in cohorts with more then 100 deaths in radiation worker study of 15 countries; NPP = nuclear power plants, ORNL = Oak Ridge National Laboratory in Tennessee, and Hanford = the Hanford Laboratory in Richland, Washington (Cardis et al., 2005).

5. Nuclear power plant accidents

5.1 Three Mile Island

The accident at the Three Mile Island in Dauphin County, Pennsylvania, USA, on March 28, 1979, drew attention to the possible exposures of people to radionuclides that might be released from a nuclear power during an accident involving the overheating of the nuclear fuel. Because of the design issue and instrument confusion, the plant operator of the Unit 2 reactor was led to think that the cooling water level was high when in fact it was low. Eventually, feed water pumps failed and the reactor overheated. Steam and water were vented through a relief valve and the reactor control rods scrammed in and the reactor shut down. Even though the fission chain-reaction process had been terminated, the heat generated by the highly radioactive fission decay products overheated and melted part of fuel elements and their containment tubes. Volatile radionuclides such as ^{131}I and ^{137}Cs were released from the reactor containment but because of high efficiency particle filters and activated charcoal filters, very little was released to the outdoors. The extent of the risk to people living near the reactor was not clear as the accident progressed. Fortunately, no member of the public received a serious exposure. The maximum total radiation exposure to the members of the public was estimated to be only a few μSv from small releases of radioiodine, ^{131}I. The Unit 2 nuclear reactor unit was damaged beyond repair. The twin Three Mile Island Unit 1 nuclear reactor is still operational.

5.2 Chernobyl

An extremely serious and catastrophic Chernobyl nuclear reactor accident in the Ukraine occurred on 26 April 1986 when a nuclear reactor number four went super-critical during a systems test of the control rods. This reactor did not have a strong containment building and used graphite carbon for a neutron moderator. The resulting super-critical reaction destroyed the fuel elements, ruptured the reactor vessel, and exposed the graphite moderator which, with exposure to air, began to burn. The seriously damaged containment allowed the resulting smoke and effluent of radionuclides to mix with the outdoor atmosphere and spread fallout over much of the Ukraine, Soviet Union, and other parts of Europe. Not only did the most volatile radionuclides such and ^{131}I and ^{137}Cs escape but also many other fission product radionuclides such as ^{90}Sr and ^{144}Ce were part of this widespread airborne contamination. Twenty-six power plant workers died of acute whole body radiation overdoses.

The large population exposure to long-term protracted ionizing radiation associated with the 1986 Chernobyl reactor accident in the Ukraine was predicted with the linear no-threshold model to result in a virtual epidemic of long term radiation-induced cancer (Anspaugh et al. 1988). Instead, there was no apparent major health effects of this widespread protracted exposure to ionizing radiation except for thyroid disease associated with very high acute radiation doses from short-lived (half life 8 days) radioiodine (^{131}I) contamination in cow's milk and dairy products (WHO 2006, Jaworowski 2010).

5.3 Fukushima Daiichi

Three of the six boiling water nuclear power plants at Okuma, Fukushima, Japan, underwent destructive meltdown of the nuclear fuel along with widespread releases of radioactive fission products to the environment as a result of massive flooding associated with the Tohoku earthquake and tsunami on March 11, 2011. Although plants safely shut during the earthquake, the tsunami that followed completely flooded the plants and shut down all sources of local and nearby electricity. Without electricity it was not possible to operate water pumps needed to cool the reactor fuel to prevent overheating associated with the natural radioactive decay of fission product radionuclides produced during normal operations.

When the quake occurred units 1, 2, and 3 were operational, but unit 4 had previously been defueled and units 5 and 6 had been shut down for maintenance. Ultimately the fuel elements in units 1,2, and 3, overheated and ruptured, The special zircaloy cladding reacted with steam to produce large quantities hydrogen gas which vented into the containment buildings and subsequently exploded injuring workers and damaging the building walls and roofs.

Ultimately, there were widespread releases to the outdoor air of smoke and volatile fission products from the damaged nuclear fuel. These releases lofted into the atmosphere and were diluted and carried for long distances. The most important human exposures were probably those associated with contamination of water and food. Two most prominent fission products released in this accident were short-lived (half life 8 days) radioiodine (^{131}I) and long-lived (half life 30 years) radiocesium (^{137}Cs). Both of these radionuclides are beta radiation emitters that can efficiently irradiate body organs if inhaled or ingested. Associated gamma rays contribute a more diffuse and less important portion of organ doses when these radionuclides are internally deposited in the body of a person.

Ingested ^{131}I primarily follows the grass to cow to milk route of human exposure. Ingested iodine tends to concentrate in the small thyroid gland where it can efficiently irradiate the thyroid tissue. The very efficient release of iodine from damaged nuclear fuel and the short radioactive half life of ^{131}I means that the risk of exposure is relatively short-lived after an accidental release but the acute doses to the thyroid gland can be quite large. From studies of children who ingested ^{131}I from the Chernobyl nuclear reactor accident, it has be observed that the risk of thyroid cancer is primarily associated with thyroid cumulative radiation doses larger than about 10 Sv.

Ingestion or inhalation of ^{137}Cs leads to very uniformly widespread irradiation of body because cesium is quite soluble in most of its chemical forms and behaves somewhat like potassium in the human body. Although ^{137}Cs has a long physical half-life, it has a relatively short biological half life on the order of about one month. For an acute single intake, very little remains in the human body after one year. In the environment, cesium tends to strongly bind by ion-exchange with clay in soils, somewhat limiting its uptake into plants.

In order to evaluate the biological risk, including carcinogenesis, associated with the intake of ^{137}Cs, fifty-four beagles were injected with graded dosages of ^{137}Cs as cesium chloride from 32.5 to 148 MBq per kg body weight and held for lifetime study at the Inhalation Toxicology Research Institute (Mauderly and Daynes, 1989)(Fig. 24). Because of the relatively short biological retention half-time of about 30 days (Boecker, 1969), about half of the whole body absorbed dose from ^{137}Cs beta rays and ^{137m}Ba daughter gamma rays was delivered in the first 30 days after the injection and nearly all of the dose was delivered within one year. Those deaths that occurred prior to one-half year after injection were associated with direct injury to the blood forming organs by beagles receiving the highest dosages of ^{137}Cs. There were only 12 unexposed controls assigned to this study, but the fate of 52 other controls assigned to contemporaneous studies was used for comparison to the exposed dogs. Those beagles that survived the acute injury phase in the very highest dose groups lived about as long as unexposed controls and had insufficient cases of radiation induced cancer to develop a meaningful three-dimensional model.

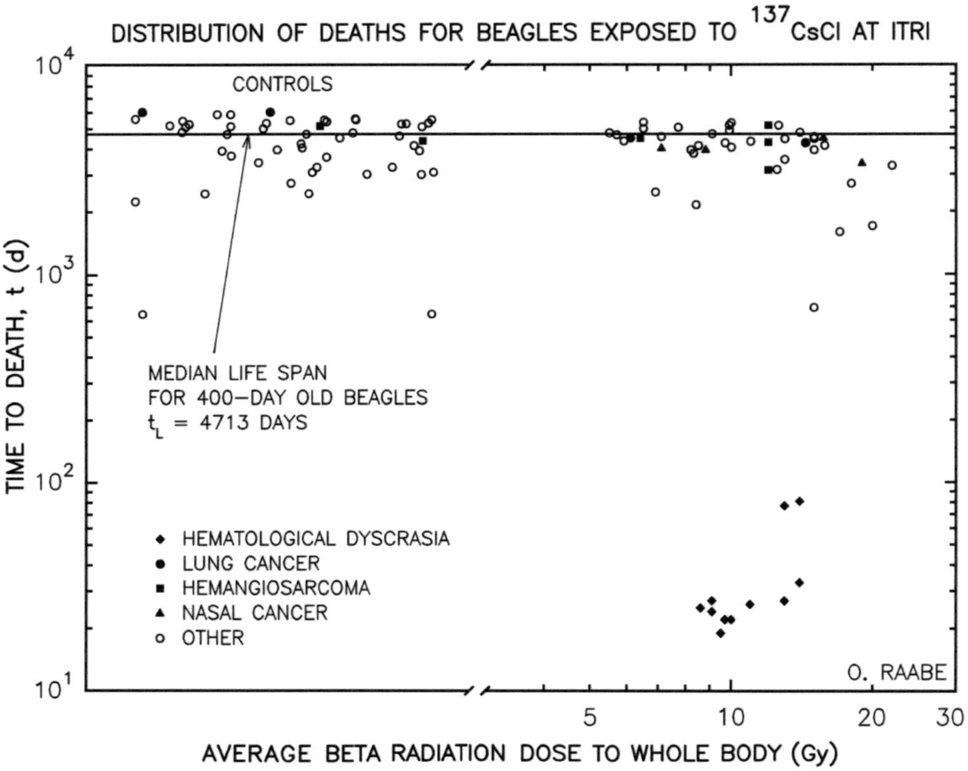

Fig. 24. Distribution of deaths in lifetime studies of beagles injected with $^{137}CsCl$ at ITRI (Raabe,1994).

A Peto trend test (Peto et al, 1980) of the cancer cases showed that two cancer types had a statistically significant occurrence: (a) nasal carcinomas reported in 4 exposed dogs and not observed in controls, and (b) hemangiosarcoma in different organs (4 cases) with one case in controls. The lowest lifetime cumulative whole body radiation dose associated with a death with nasal carcinoma was 7.7 Gy; this compares with the lowest skeletal beta radiation dose of 18.4 Gy associated with the occurrence of nasal carcinoma in beagles exposed to ^{90}Sr at Davis. There was no statistically significant incidence of lung, liver, or bone cancer, and no reported cases of leukemia. This result is consistent with the expected risk of radiation-induced cancer from protracted exposure as discussed above.

6. Summary of protracted and repeated exposure cancer induction risk

As presented in the previous sections, the studies of repeated and protracted exposures to ionizing radiation have shown that radiation induced cancer follows a sharply non-linear upward temporal pattern with a virtual threshold at lower life-time average dose rates (Raabe, 2010). The precision and time-delay of the cancer induction phenomenon indicate an underlying gradual biological process involving many altered cells associated with cellular deoxyribonucleic acid (DNA) mutations, clonal development depending on cell division cycles, cellular maturation, and average ionizing radiation dose rate over long latency periods. Studies described above of protracted exposures to ionizing radiation from internally deposited radionuclides in people and laboratory animals have demonstrated that cancer induction risk is actually a somewhat precise function of average dose-rate and that at lower average dose rates cancer latency can exceed natural life-span.

Because of the long latency at low dose rate that may exceed the natural life-span, the radiation-induced cancer risk associated with protracted exposures to ionizing radiation involves a life-span virtual threshold when the lifetime average dose rate is low and the cumulative dose to sensitive tissues is below about 10 Sv.

Life-span virtual thresholds for radiation-induced cancer risk should exist for other types of protracted and fractionated exposures including radon inhalation and external exposures associated with background levels of ionizing radiation from environmental radionuclides. The organs, tissues, and cells of the body cannot distinguish between ionizing radiation emitted by internally deposited radionuclides or entering the body from external sources. Low linear energy transfer (LET) gamma rays from external sources yield ionizing energetic electrons in the body and beta particles from internally deposited radionuclides are also ionizing energetic electrons. High LET neutrons from external sources yield ionizing energetic protons whose radiation quality weighting factors are similar to high LET ionizing alpha particles from internally deposited radionuclides (Raabe, 2011).

Another important finding is that two beta particles are about equal on the average to one alpha particle in the radiation induction process (Raabe 2010). This finding suggests that double strand damage to cellular DNA is involved in the cancer induction process. While one alpha particle hitting DNA may produce the necessary double-strand break, two beta particles almost simultaneously hitting the DNA are apparently needed to produce similar double-stand breaks. Also, the shape of the cancer-induction dose-response curve is not linear but rather sharply increasing as a function of cumulative dose delivered from virtually zero risk at low doses to high risk at high doses as a function of dose rate.

7. Single acute exposures to ionizing radiation

7.1 Atomic bomb survivor studies

A quite different phenomenon has been observed in studies of the Japanese survivors of atomic bombs detonated over Hiroshima, August 6, 1945, and over Nagasaki, August 9, 1945 (Manhattan Engineer District 1946). These Japanese survivors were exposed instantaneously to a large acute pulse of external gamma radiation (and some neutrons) delivered in about one minute. Many of these acute absorbed doses were several hundred times the normal annual radiation exposure from background radiation. All of the cells of the body were exposed to this large ionizing radiation pulse and those cells and body tissues were all affected in some way. Years later some of the highly exposed survivors developed cancer of the same types as occurred in the control population but somewhat earlier and at somewhat higher rates. In addition, the increases in these cancer cases were proportional to that one-minute whole-body absorbed radiation dose. This linear relationship is in sharp contrast with cancer induction observed in the protracted and repeated exposures described above. Except for some early cases of myeloid leukemia caused by acute radiation damage to the blood forming tissues of the body from the higher exposures, there were no apparent radiation induced cancer cases among the atomic bomb survivors (Raabe, 2011).

There has been a long-standing scientific disagreement concerning cancer induction associated with protracted and repeated exposures to ionizing radiation and the carcinogenic effects of the atomic bomb acute instantaneous exposures. The reason for this conflict is readily apparent from the data themselves. It is explained by the fact that the increased risk of cancer observed in the atomic bomb survivor studies was primarily the result of acute high dose-rate promotion of ongoing biological processes that lead to cancer in the Japanese people rather than actual de-novo cancer induction.

The Japanese survivor studies are very important since they have been extremely well done and represent a significant body of human data about the possible effects of acute exposure to ionizing radiation (RERF, 2008). Also, the currently accepted methodology for estimation of solid cancer risk associated with human exposure to ionizing radiation is based primarily on the detailed studies of the Japanese atomic-bomb survivors (ICRP-26, 1977; ICRP-60, 1991; BEIR VII Phase 2, 2006; NCRP 136, 2001; ICRP 103, 2007). The prevailing models of human ionizing radiation exposure risk and relative organ sensitivities (ICRP tissue weighting factors, W_T) are based primarily on those studies. Specifically, the concept that cancer risk as a linear function of cumulative dose and follows a linear no-threshold cancer risk model is based primarily on the acute high dose rate exposures received by the Japanese atomic bomb survivors.

7.2 Radiation promoted cancer from acute exposures

The Radiation Effects Research Foundation (RERF) and its predecessor Atomic Bomb Casualty Commission (ABCC) studied the development of solid malignant tumors in about 80,000 of the Japanese atomic bomb survivors (Pierce and Preston, 2000; Preston et al., 2003; Preston et al., 2004; Preston et al., 2007). The main study considered 79,972 survivors for which there were available calculated radiation doses. Of these, 44,636 survivors had calculated doses that exceeded 5 mSv. Those with less than 5 mSv doses were used as the control group (Preston et al., 2007). These about one-minute acute exposures involved very high-energy gamma radiation and some neutrons. Myeloid leukemia from high dose bone marrow exposures followed a different response course and is usually considered separately from the solid tumor incidence.

The traditional approach is to assume that the solid malignant tumors are the result of simple stochastic (isolated and random) initiating events in individual cells that occurred during that about one-minute exposure. This simplistic stochastic model of ionizing-radiation-induced cancer is based on the idea that a single cell is randomly altered by a unique ionizing radiation event causing a unique pre-malignant mutation in that cellular deoxyribonucleic acid [DNA] (Moolgavkar et al. 1988; Heidenreich et al. 1997). This single random event then ultimately leads to a clone of similar pre-malignant cells. Later, usually much later, a second random DNA alteration occurs in one of the clonal pre-malignant cells that produces a malignant cell that develops into a monoclonal malignant tumor. This process began with the single random (stochastic) cellular event. This process can be advanced by promoter agents including other ionizing radiation that presumably affect the clonal development, quantity, and maturation of the pre-malignant clonal cells (Heidenreich et al., 2007) . A cancer promoter is anything that advances the development of a malignancy other than a directly carcinogenic agent or an intrinsic component of the carcinogenesis process (Casarett, 1968). Mathematical models are readily constructed with unknown parameters and hypothetical modifying factors and process lag times designed to fit the data associated with acute ionizing radiation exposures (Heidenreich et al., 1997).

The internal emitter studies discussed above strongly suggest that double strand DNA damage or a related phenomenon is involved in the cancer induction process associated with ionizing radiation. In particular, two low LET beta particles were found to be required to match the radiation induction process associated with each alpha particle (Raabe 2010). Since the exposure of the Japanese atomic bomb survivors was primarily associated with low LET gamma radiation and associated energetic electrons, two hits at the same region of DNA in a target cell would be expected to be required for the induction of cancer. The resulting increase in cancer by induction in this two-hit process would follow a sharply increasing curvilinear power function of increasing cumulative absorbed dose. In fact, the increase in cancer among the atomic bomb survivors tended to follow a remarkably linear pattern as a function of absorbed ionizing radiation dose. Deterministic cancer promotion of those types found in the exposed population rather than simple isolated stochastic cancer induction better explains the increase of solid cancers in the atomic bomb survivor studies.

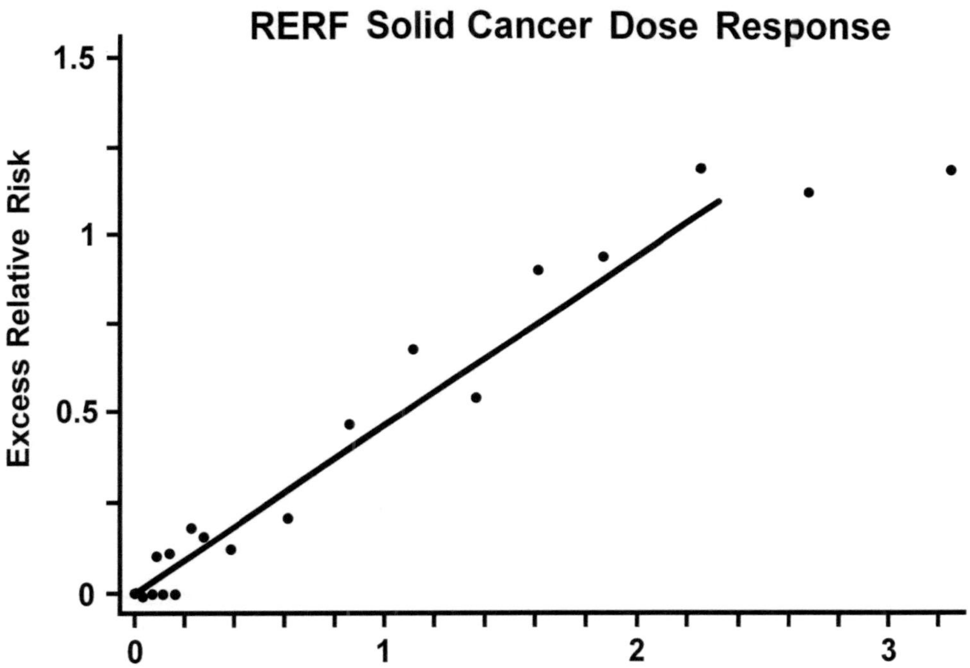

Fig. 25. Linear dose-response relationship of Excess Relative Risk for the promotion of solid cancer in Japanese survivors of the atomic bomb detonations at Hiroshima and Nagasaki in 1945 with respect to survivors who received low radiation exposures as reported by the Radiation Effects Research Foundation [RERF](Preston et al. 2007).

The studies of the atomic bomb survivors demonstrate a linear dose-response promotional effect related to the natural or existing biological processes that may eventually lead to cancer in the exposed population (Fig. 25). These processes involve years of cellular division, clonal expansion, and cellular maturation. The exposure to a sudden high dose of ionizing radiation delivered in about one minute at the time of the nuclear detonations may have advanced or stimulated the cellular changes that eventually lead to various typical types of cancer. Hence, some cancers may have appeared at an earlier time than otherwise would have occurred based on the existing underlying cellular and tissue processes (Fig. 26). This promotional effect was observed to advance cancer rates not only relatively soon after exposure but throughout the lives of the exposed individuals (Fig. 27). This behavior is proportional to the instantaneous dose just as would be expected for any phenomenon that involves augmentation of existing processes rather than a few random or "stochastic" changes in a few select cells.

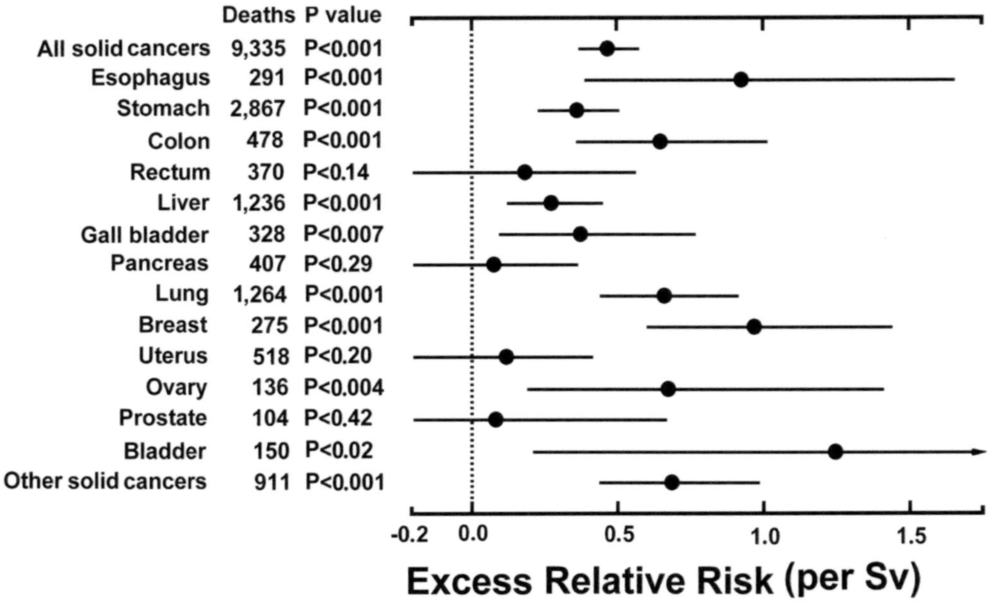

Fig. 26. Observed Excess Relative Risk by body organ for promoted solid cancer observed for one Sv of equivalent ionizing radiation dose in highly irradiated Japanese atomic bomb survivors compared to survivors who received low radiation exposure as reported by the Radiation Effects Research Foundation [RERF] (Preston et al. 2003).

Concerning the solid tumor incidence in the atomic bomb survivor studies, Pierce and Mendelsohn (1999) pose the question, "How could it be that the excess cancer rate might depend only on age and not on time since exposure or age at exposure?" Fig. 27 shows that the increase in malignant solid tumors in the atomic bomb survivors associated with their radiation exposure follows the same lifetime pattern irrespective of the age at exposure. The simple answer is that the normal progression of cancer incidence in the population was somewhat promoted by the radiation exposure without the actual independent induction of cancer. This promotion is not a single event stochastic process but rather the result of the almost instantaneous delivery of ionizing charged electrons produced in all the tissues by ionizing radiation from the atomic bombs. The tissues response is complex and unfocused. The cells of the tissue communicate among themselves as a result of the exposure in a process called bystander effect with various responses evoked even among the cells in these tissues that were not directly impacted (Hall 2003). Tissues respond to sudden ionizing radiation exposure with up-regulation of various genes and down-regulation of others (Snyder and Morgan, 2004; Coleman et al. 2005). This imprint is apparently not lost and carries forward throughout life promoting the existing biological processes that may lead to cancer in the exposed atomic bomb survivor population. This complex process suggests a systems biology phenomenon rather than a unique stochastic response (Barcellos-Hoff, 2008).

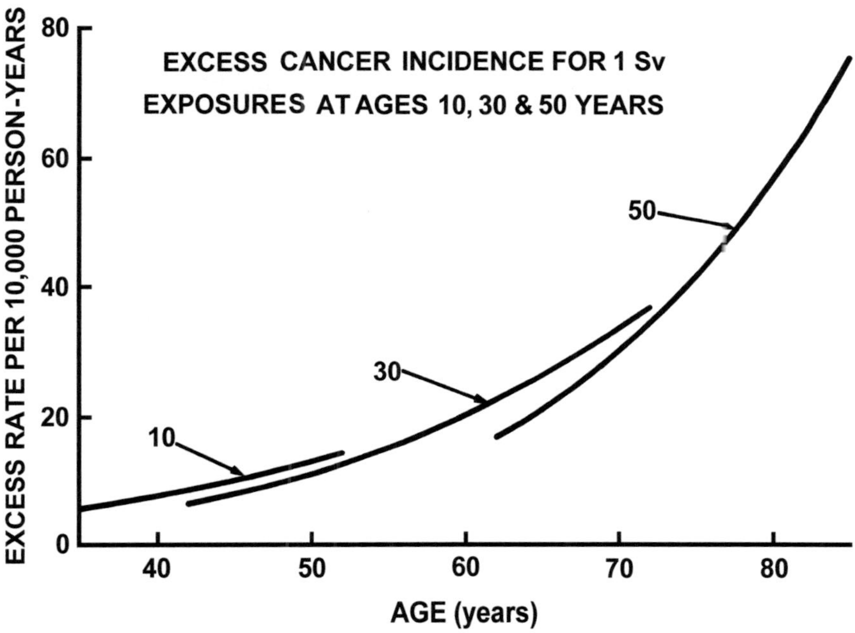

Fig. 27. Observed gender-averaged age-specific excess incidence rates at 1 Sv for most major solid cancers over the 1958-1987 follow-up period for ages at exposure 10 y, 30 y, and 50 for the Japanese atomic bomb survivors (Pierce and Mendelsohn, 1999). The excess rates appear to depend only on age and not on time since exposure or age at exposure as might be expected for radiation-induced promotion of the cancer types normally found in this population.

The Japanese atomic bomb survivor data are unique because they do not in any way predict the observed carcinogenesis associated with protracted exposures as occur in the case of internal emitters. In fact, the acute gamma ray exposures clearly represent a completely different mechanism of carcinogenesis from that which occurs with protracted exposure as with long-lived internal emitters (Raabe 2011). The resulting cancer promotion phenomena in the atomic bomb survivors should not be expected to describe the effects of similar exposures delivered uniformly or fractionated over a relatively long period of time. Since promotion is a relative process rather than an absolute process, it is not meaningful to try to create absolute risk estimates from relative response information.

The tissue-weighting factors (w_T) developed by the ICRP to calculate the so-called effective dose are actually a reflection of the convolution of the underlying incidences of different types of cancer in the control population and the relative promotional effect of the whole body exposure to gamma rays and some neutrons in the Japanese atomic bomb life-span studies (ICRP 26, 1977; ICRP 30, 1979; ICRP 60, 1991; ICRP 103, 2007). Cancers that were somewhat rare in this Japanese population, such as bone cancer, were assigned relatively low tissue weighting factors relative to the whole body cancer or assumed genetic risk (such as w_T = 0.01 for bone surfaces). Cancers that were somewhat common in this Japanese population, such as lung cancer, were assigned relatively high ratio values relative to the whole body cancer or genetic risk (such as w_T = 0.12 for whole lung) . These tissue-weighting factors (w_T) can be related to observed cancer promotion, but are unrelated to the cancer induction associated with protracted or fractionated exposure to ionizing radiation. The use of tissue-weighting factors (w_T) recommended by the ICRP based on the atomic bomb survivor studies is inappropriate for protracted or fractionated exposures.

The elaborate Radiation Effects Research Foundation (RERF) studies of the atomic bomb survivors have investigated in rigorous detail the effect of whole body irradiation by high-energy gamma rays (and some neutrons) delivered in about one minute. The A-bomb RERF life-span study clearly describes a meaningful linear dose model of promotion of ongoing biological processes that lead to increased cancer rates for brief high dose rate exposure to ionizing radiation. The relative risk values might be applicable to other brief high dose-rate ionizing radiation exposures as may occur in occupational exposures or in medical diagnosis and treatment (Hall and Brenner, 2008). However, there is still considerable uncertainty for acute doses less than about 0.05 Sv. Small acute doses may by beneficial as they may promote or stimulate DNA repair or other defensive cellular phenomena that reduce promotional cancer risks associated with ongoing cellular processes that might otherwise lead to cancer (Feinendegen 2005).

7.3 Acute medical exposures

7.3.1 Medical exposure concerns

Among the most common instantaneous exposures of people to ionizing radiation are those associated with diagnostic medical radiographs (X-rays). Modern computed tomography (CT) scans using x rays with higher radiation doses can provide precise three-dimensional images and are of great medical value. Fluoroscopic images that use nearly continuous irradiation are also valuable in certain medical procedures. The question as to whether these acute diagnostic medical exposures result in an increased cancer risk has been considered and even assumed in various studies. The promotion of cancer observed among the Japanese atomic bomb survivors is a reasonable basis for considering the possible cancer promotion associated with medical exposures to ionizing radiation, but protracted exposures do not seem to involve cancer risk at small doses and sometimes appear beneficial.

7.3.2 Neonatal medical exposures

Among the earliest studies of possible effects of medical exposures of patients associated with diagnostic X-rays are those associated with the Oxford Survey of Childhood Cancer (OSCC) of deaths in England, Wales, and Scotland from childhood cancer. These case-control studies were initiated and advanced by Alice Stewart, a British physician (Stewart et al., 1956; Stewart et al., 1958). The basic hypothesis of these studies is that childhood cancer was caused in part by fetal exposures from medical X-rays associated with diagnostic radiology of the pregnant mothers. Healthy control children were matched by sex, by birth location and, as closely as readily possible, by birth date. A comprehensive summary and statistical evaluation has been published (Mole 1990). The overall study considered a majority of the cancer deaths in of children in Britian under the age of 16 primarily spanning the years 1953 to 1978. Most of the major findings were based on children who were born between 1940 and 1969. The initial studies inquired from the mothers about X-ray films that might have been made during pregnancy. Later studies involved verification by review of medical records, but not all cases could be verified. There was no direct ionizing radiation dosimetry associated with these studies. Overall, there was a strong trend to a higher fraction of diagnostic radiology among the mothers of children who died of cancer compared to the healthy controls. Typical overall raw values of these ratios were 1.33 for children dying under age 6, 1.38 for children age 6 or older dying under age 10, and 1.27 for children age 10 or older dying under age 16. For births that occurred prior to 1957 the odds ratios were significantly above unity, while for later births they were not It may have been most difficult to verify the X-ray recollections of the mothers for those earlier years. The odds ratios for X-raying in Britian for the four birth years 1958 to 1961 were 1.27. 1.36 and`1.02 for cancer deaths at ages 0 to 5, 6 to 9, and 10 to15, respectively. The overall odds ratio for all ages 0 to 15 years was 1.23 with a 95% confidence interval from 1.04 to 1.48 and the associated excess lethal tumor rate from in utero x ray exposure was 0.00028 with a confidence interval from 0.48 to 0.00058 (Mole. 1990).

The findings from the Oxford Survey of Childhood Cancer tend to show an important trend, but they may not convincingly prove that in utero X-rays were the cause of childhood cancer. Much of the earlier data were based on patient-reported X-ray exposures. A mother whose child died of cancer is more likely to remember or readily believe that she received X-ray exposures than a mother of a healthy child. Early medical records may have been incomplete or difficult to locate.

In many cases the X-rays may have been associated with some medical complication or abnormality associated with the pregnancy especially in many cases where the mother was X-rayed more than once. Those medical abnormalities, if they existed may have directly contributed to later development of cancer in the child. Another possibility is that the in utero exposure promoted a pre-cancer process that led to earlier cancer development than would have otherwise occurred, but the low radiation doses involved in these studies was too low to correlate with the atomic bomb survivor studies promotional effects (Preston et al., 2007). Totter and McPherson (1981) point out that differences in population groups in the Oxford Survey and other factors can explain for the observed relationships.

Although no dosimetry was available to accurately estimate the maternal or in utero ionizing radiation doses associated with the X-ray exposures in these studies, Mole (1990) attempted to estimate the risk of childhood cancer per unit of absorbed ionizing radiation to the fetus using estimates made in 1966 by the Adrian Committee of the British Ministry of Health. He assumed that average dose from all obstetric radiography was 0.61 cGy. From his calculated excess lethal tumor rate he calculated and reported a risk coefficient for irradiation in the third trimester for childhood cancer deaths at ages up to 15 years equal to 0.00046 per cGy with a confidence interval from 0.8 to 0.00095 per cGy. This estimate is not valid because the Oxford Survey of Childhood Cancer does not provide any dose-response information for anyone in the study nor for the whole population at risk. More than 95% of the cases of childhood cancer are not explained by neonatal radiography. In 1958 there were 840,196 births in Britian and it was likely that at least 100,000 babies were irradiated in utero based on the estimates (Mole 1990). The number of deaths associated with cancer ages 0-14 born in 1958 was 977 and the data show that about 15% of those or about 150 had been irradiated in utero. Therefore the fraction of those irradiated that died of cancer in childhood was about 0.15%.

It might be noted, however, that the promotion of cancer observed in the atomic bomb survivor studies is a mechanism that could explain the observed results, but there are no statistically significant radiation promoted cancer risks for acute doses below about 10 cSv among the atomic bomb survivor data. (Preston et al. 2007). Totter and McPherson (1981) point out that differences in population groups can account for the observed misleading relationships since the sub-population of cancer cases does not have the same probability of being X-rayed as the sub-population of controls.

7.3.3 Medical diagnostic X-ray risks

Medical diagnostic radiology has become one of the highest sources of exposure to ionizing radiation in many countries of the world. This is largely associated with the increasing use of computed tomography (CT) images using x rays which involve much higher radiation doses than conventional X-ray images. Since these doses are delivered almost instantaneously, they can be expected to present similar cancer promotion risks as observed in the atomic bomb survivor studies discussed above. However, the radiation doses involved in the use of CT examination are typically well below the approximately 10 cSv dose above which the cancer promotion risk is statistically significant in the Japanese atomic bomb survivor studies (Preston et al 2007). Hence, attempt to extrapolate the cancer promotion risk associated with these small doses involves considerable imprecision and uncertainty. No one can be sure that the same linear no-threshold model applies precisely to individual organ exposures in radiology as applies to whole body exposures as occurred for the Japanese atomic bomb survivors. In addition, biological processes that affect the behavior of cells and tissues at lower doses may alter or cancel the promotional phenomena. For example, low acute doses may stimulate cellular DNA repair mechanisms.

Some investigators have created somewhat precise models of what may be occurring at low doses to predict lifetime cancer risks from the acute x ray exposures that may be associated with diagnostic radiology using linear no-threshold models based on the atomic bomb survivor studies of cancer promotion. These studies of diagnostic medical exposure to ionizing radiation usually refer to their studies as estimates of risk of radiation induced cancer, but in this chapter the acute exposure dose phenomenon has been be more correctly shown to be cancer promotion. In addition, some of these studies report their results as cancer risks, but they are not measured risks but calculated estimates subject to the theoretical use of the linear model and the approximate parameters of that model obtained from the atomic bomb survivor studies.

Brenner and associates (Brenner et al., 2000; Brenner, 2002; Brenner and Hall, 2007; Hall and Brenner, 2008) have published a series of papers providing estimates of lifetime fatal cancer as a function of dose from CT scans of various organs of the body as a function of exposure age utilizing data from the atomic bomb survivor studies and the risk models developed from those data by the United States National Research Council (BEIR V, 1990) and the International Commission on Radiological Protection (ICRP-60, 1991).

The calculated cancer risk depends on the organ irradiated and the sex and age of the irradiated person. The calculated overall lifetime attributable risk tends to be higher for children than for adults since children have many more years of life during which cancer might develop. Representative values for lifetime attributable risk in these studies are from

0.14% for babies to 0.02% for adults. These are typical calculated values using linear no-threshold models following from the separate organ exposure results at high doses from the atomic bomb survivor studies.

Based on BEIR V (1990), Brenner et al. (2000) summarizes observed relative increased risk of cancer for an exposure of Japanese atomic bomb survivors exposed to one Gy for males as a function of age at exposure (Fig. 28). This representation clearly shows cancer promotion. The total risk per year is about the same for everyone. Given an 85 year life span in this figure, the risk per year of life for a five year old boy is about 13% over 80 years = 0.16% per year. For a 25 year old man the risk per year of life if about 9.5% over 60 years = 0.16% per year. For a 45 year old man the risk per year of life if about 6.5% over 40 years = 0.16% per year. The total cancer risk per year is almost independent of age at exposure. This is not explained by any simple stochastic cancer induction model, but is explained by some sort of deterministic cellular reprogramming that promotes the cancer process in a somewhat linear way.

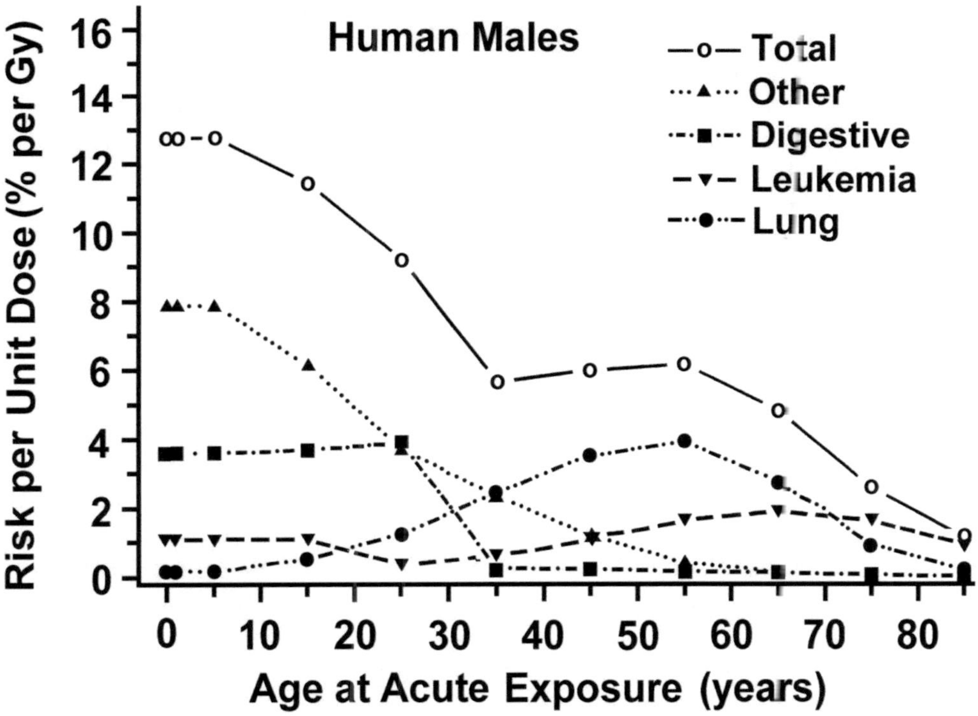

Fig. 28. Lifetime attributable cancer mortality risks for males by cancer type as a function of age at a single acute exposure of one Gy of γ radiation based on BEIR V (Brenner et al., 2000).

Addition evidence for promotion is seen in the lung cancer rates shown in Figure 28. Lung cancer in these males may be related to smoking history. For young boys up to age 15 there is no apparent promotion of lung cancer (total lifetime risk about 0.5%). For these non-smokers there was no ongoing lung cancer development to promote. But lung cancer risk per year climbs for older males many of whom are likely to have been smokers with a peak risk at age 55.

Berrington de Gonzalez and Darby (2004) estimated the hypothetical overall population cancer risks from diagnostic X-rays in the UK and 14 other countries based on frequency of diagnostic X-ray use, doses to different body organs, and linear no-threshold risk models based on the Japanese atomic bomb survival data. Their results led them to report that the cumulative risk of cancer for people up to age 75 from diagnostic X-ray exposures was indicated to be in the range from 0.6% to 1.8%.

In studies of 64,172 tuberculosis patients of whom 39% were exposed externally to highly fractionated x ray chest fluoroscopies, lung cancer deaths showed no evidence of cancer risk associated with the x ray exposures with the relative risk at a cumulative doses of 1 Sv being 1.00 [95% confidence interval 0.94-1.07] (Howe 1995). Also, studies of people exposed to unusually high levels of protracted external ionizing radiation associated with natural background (up to 260 mSv y^{-1}) have not detected increased cancer risks (Ghiass et al., 2002).

Overall, the attributable risk values estimated in these reports from individual X-rays and CT scans are hypothetical and unproved since the doses involved are much smaller than those for which statistically significant elevations in cancer rates have been observed in the atomic bomb survivors. The calculations depend on the unproved assumption that the promotion risk for acute exposures is perfectly linear down to a dose near zero, the linear no-threshold hypothesis. Various biological processes could compete with or nullify the ionizing radiation promotion process for these relatively small individual acute exposures. There seems to be considerable evidence there is no significant risk at very low acute doses, and some possible beneficial effects.

8. Discussion

An important observation is that the dose-response relationships for protracted exposures to internally deposited radionuclides are quite different from the observed effects of a single high dose rate exposure as experienced by the atomic bomb survivors in temporal manifestation and preciseness of the effect. In the internal emitter studies the time to cancer development depends on the lifetime average dose rate and the cases are not randomly distributed with respect to time but rather are tightly grouped over a predictable and narrow range of times. At low lifetime average dose rate the time required to develop cancer may exceed the natural life span yielding a virtual threshold for cancer induction.

Bone cancer and lung cancer are equally sensitive in the internal emitter studies with beagles, but there is no apparent induction of bone cancer in the atomic bomb survivors while there is a considerable increase in lung cancer. The main data are the relative increase that suggests radiation promotion was more effective than radiation induction.

Based on the linear relative risk relationship observed for acute exposure in the RERF studies of the atomic bomb survivors, the ICRP developed an absolute risk model for radiation-induced cancer as a function of absorbed cumulative ionizing radiation dose that was believed to apply to all forms of exposures to ionizing radiation. It is assumed that the linear model can predict cancer induction risk but needs a hypothetical dose and dose-rate effectiveness correction factor (DDREF) for protracted, fractionated, or low dose exposures.

In recognition that protracted or fractionated exposures have been observed to be less carcinogenic than the RERF data directly suggest, the ICRP reduced the calculated risk using a DDREFof 2, but still assumed lifetime LNT linearity of cancer induction was maintained (ICRP 103, 2007). This approach has led to simplistic linear cancer induction risk factors such as " 5% per Sv" (ICRP 60, 1991; ICRP 103, 2007).

The idea that such a risk factor is valid rests on the assumption that each ionizing-radiation-induced malignancy is the result of a simple isolated random or "stochastic" initial event that is equally probable in any exposed person. Although each malignant tumor is monoclonal, the biological processes that lead to that first tumor (and successive tumors) are apparently deterministic systemic processes.

Others have also adopted similar linear cancer induction risk models as a function of person-sievert based on the RERF studies. However, as discussed above, these cancer inductions models do not apply to non-acute exposures that depend on lifetime average dose rate rather than simply on cumulative dose.

These cancer induction risk models are not valid because of the confusion of cancer promotion associated with brief high dose-rate exposures and cancer induction associated with protracted exposures. Unfortunately, these misleading cancer induction risk values may be used to implement expensive cleanup standards for environmental radioactivity.

The LNT model does not predict the observed effects or lack thereof for protracted exposures to ionizing radiation exposure (Jaworowski 2010). The LNT model does not readily predict the results of a cumulative set of fractionated medical diagnostic x ray exposures of the lung where there was no indication of increased lung cancer with the observed relative risk at a cumulative doses of 1 Sv being 1.00 [95% confidence interval 0.94-1.07] contrasting with the high expected relative risk based on the atomic bomb survivors being 1.60 [95% confidence interval 1.27-1.99] (Howe 1995). Also, the large population exposures to long-term protracted ionizing radiation associated with the 1986 Chernobyl reactor accident in the Ukraine was predicted with LNT to result in a virtual epidemic of long term radiation-induced cancer (Anspaugh et al. 1988). Instead, there was no apparent major effect of this widespread protracted exposure to ionizing radiation except for thyroid disease associated with very high acute radiation doses from short-lived ^{131}I in milk (WHO, 2006; Jaworowski, 2010).

Failure to realize the fundamental differences between cancer promotion and cancer induction has been the source of scientific misunderstandings and risk estimate disagreements. A logical barrier has stood between the linear no-threshold (LNT) model of cancer promotion in the acute exposures associated with the Japanese survivors and the virtual threshold associated with induction of cancer associated with protracted exposures as received from long-lived internally deposited radionuclides in humans or animals. Further, it has led to a systematic overestimation of cancer induction risk from protracted exposures to ionizing radiation.

The underlying assumption in these current recommendations is that risk of radiation-induced cancer is proportional to cumulative dose without threshold. This assumption obviously conflicts with the induction of cancer observed in the cases of lifetime protracted and repeated exposures to ionizing radiation. An understanding of the source of this conflict of data interpretation is essential for sound estimates of cancer risk associated with exposures to ionizing radiation.

Radiation cancer induction is not a simple stochastic process in the sense that it is not the result of a random radiation event in a single cell that converts it into some kind of pre-malignant neoplastic cell. In reality, both radiation promoted and radiation-induced cancers are the result of complex biological processes involving multiple cellular events. These events include but are not limited to intercellular communication in irradiated tissues, up-regulation and down-regulation of genes, DNA mutations, cell division rates, clonal expansion of altered cells, and various responses to numerous specific radiation events such as DNA double-strand breaks. Induced cancer risk from protracted radiation exposures of body organs depends on the lifetime average dose rate to the irradiated organs rather than on some function of cumulative dose. The middle Twentieth Century radiation protection standards promulgated by the ICRP were properly focused on minimizing dose rate (ICRP 2, 1959).

9. Conclusion

Clearly the development of a radiation-induced malignant tumor from either protracted ionizing radiation exposures or acute exposures is not the result of a single random interaction of the ionizing radiation with an isolated cell. Hence, the term stochastic as used by the ICRP is not appropriate. The following conclusions indicate that major revisions of methodology and standards are needed and other currently accepted ionizing radiation risk models should be improved to provide more meaningful and realistic estimates of ionizing radiation cancer risk:
(1) Cancer induction risk associated with protracted or fractionated ionizing radiation exposure is a non-linear function of lifetime average dose rate to the affected tissues and exhibits a virtual threshold at low lifetime average dose rates. (2) Cumulative radiation dose is neither an accurate nor an appropriate measure of cancer induction risk for protracted or fractionated ionizing radiation exposure except for describing the virtual threshold for various exposures. (3) Cancer promotion risk for ongoing lifetime biological processes is a relative process as seen in the RERF studies of the Japanese atomic bomb survivors for brief high dose-rate exposures to ionizing radiation. It cannot be used to estimate cancer induction risk from protracted or fractionated ionizing radiation exposures over long times and at low dose rates.

10. References

Anspaugh, L.R.; Catlin R.J. & Goldman M. (1988). The Global Impact of the Chernobyl Reactor Accident. *Science* 242:1513–1519.

Barcellos-Hoff, M.H. (2008). Cancer as an Emergent Phenomenon in Systems Radiation Biology. *Radiat. Environ. Biophys.* 47: 33-38.

BEIR IV. (1988). *Health Risks of Radon and Other Internally Deposited Alpha-Emitters.* National Research Council, Washington DC: The National Academy Press.

BEIR V. (1990). *Health Effects of Exposure to Low Levels of Ionizing Radiation.* National Research Council, Washington DC: The National Academy Press.

BEIR VI (1999). *Health Effects of Exposure to Radon.* National Research Council, Washington DC: The National Academies Press.

BEIR VII Phase 2. (2006). *Health Risks From Exposure to Low Levels of Ionizing Radiation.* National Research Council, Washington DC: The National Academies Press.

Berrington de Gonzalez A. & Darby S. (2004). Risk of Cancer from Diagnostic X-rays: Estimates for the UK and 14 other countries. *Lancet* 363:345-51.

Boecker, B.B. (1969). Comparison of ^{137}Cs Metabolism in the Beagle Dog Following Inhalation and Intravenous Injection. *Health Phys.* 16: 785-788.

Brenner, D.J. (2002). Estimating Cancer Risks From Pediatric CT: Going From the Qualitative to the Quantitative. *Pediatr. Radiol.* 32:228-231.

Brenner, D.J.; Elliston. C.D.; Hall, E.J. & Berdon, W.E. (2000). Estimated Risks of Radiation-Induced Fatal Cancer From Pediatric CT. *American J. Radiology* 175:289-296.

Brenner, D.J. & Hall, E.J. (2007). Computed Tomography - An Increasing Source of Radiation Exposure. *N. Engl. J. Med.* 357:2277-2284.

Cardis, E. & International Agency for Research on Cancer Study Group on Cancer Risk Among Nuclear Industry Workers. (1994). Direct Estimates of Cancer Mortality Due to Low Doses of Ionising Radiation: An International Study. *Lancet* 344:1039-1043.

Cardis, E.; Gilbert, E.S.; Carpenter, L.; Howe, G.; Kato, I.; Armstrong, B.K.; Beral, V.; Cowper G.; Douglas, A.; Fix, J.; Kaldor, J.; Lave, C.; Salmon, L.; Smith, P.G.; Voelz, G.L. & Wiggs, L.D. (1995). Effects of Low Doses and Low Dose Rates on External Ionizing Radiation: Cancer Mortality Among Nuclear Industry Workers in Three Countries. *Radiat. Res.* 142:117-132.

Cardis, E.; Vrijheid, M.; Blettner, M.; Gilbert, E.; Hakama, M.; Hill, C,; Howe, G.; Kaldor, J.; Muirhead, C.R.; Schubauer-Berigan, M.;Yoshimura, T. & International Study Group.(2005). Risk of Cancer After Low Doses of Ionizing Radiation: Retrospective Cohort Study in 15 Countries. *British Med. J.* 331:77-80.

Casarett, A.P. (1968). *Radiation Biology*. Prentice-Hall, Englewood Cliffs, New Jersey.

Cohen, B.L. (1995). Test of the Linear-no Threshold Theory of Radiation Carcinogenesis for Inhaled Radon Decay Products. *Health Phys.* 68:157-174.

Coleman, M.A.; Yin, E.; Peterson, L.E.; Nelson,D.; Sorensen, K.;Tucker, J.D.; Wyrobek, A.J. Low-dose irradiation alters the transcript profiles of human lymphoblastoid cells including genes associated with cytogenetic radioadaptive response. (2005). *Radiat. Res.* 164:369-382.

Eisenbud, M. & Gesell, T.F. (1997) *Environmental Radioactivity From Natural, Industrial, and Military Sources*, Fourth Edition, Academic Press, San Diego, California.

Evans, R.D. (1955). *The Atomic Nucleus*. McGraw-Hill Book Company, New York.

Evans, R.D. (1943). Protection of Radium Dial Workers and Radiologists from Injury by Radium. *Journal of Industrial Hygiene and Toxicology* 25: 253-269.

Evans, R.D.; Keane, A.T. & Shanahan, M.M. (1972). Radiogenic Effects In Man of Long-term Skeletal Alpha-irradiation. In: Stover BJ, Jee WSS, eds. *Radiobiology of Plutonium.* Salt Lake City, UT: The J.W. Press; 431-468.

Evans, R.D. (1974). Radium In Man. *Health Phys.* 27: 497-510.

Finkel, M.P.; Biskis, B.P. & Jinkins. P.B. (1969) Toxicity of radium-226 in mice. In: *Radium Induced Cancer.* Vienna: International Atomic Energy Agency [IAEA-SM-118/11]; 369-391.

Feinendegen, L.E. (2005). Evidence For Beneficial Low Level Radiation Effects and Radiation Hormesis. *Br. J. Radiology* 78:3-7.

Fornalski, K.W. & Dobrzynski, L. (2011). Pooled Bayesian Analysis of Twenty-eight Studies on Radon Induced Lung Cancers. *Health Phys.* 101:265-273.

Ghiassi, M.; Mortazavi, S.M.J.; Cameron, J.R.; Niroomand-rad, A. & Karam, P.A. (2002). Very High Background Radiation Areas of Ramsar, Iran: Preliminary Biological Studies. *Health Phys.* 81:87-93.

Hall, E.J. (2003). The Bystander Effect. *Health Phys.* 85:31-35.

Hall, E.J. & Brenner, D.J. (2008). Cancer risks from diagnostic radiology. *Br. J. Radiology* 81:362-378.

Heidenreich, W.F.; Luebeck, E.G.; Moolgavkar, S.H. (1997). Some properties of the Hazard Function of the Two-Mutation Clonal Expansion Model. *Risk Anal.* 17:391-399.

Heidenreich, W.F.; Cullings, H.M.; Funamoto, S. & Paretske, H.G. (2007). Promoting action of radiation in the atomic bomb survivor Carcinogenesis Data? *Radiat. Res.* 168: 750-756.

Howe, G.R. (1995) Lung Cancer Mortality between 1950 and 1987 after Exposure to Fractionated Moderate-Dose-Rate Ionizing Radiation in the Canadian Fluoroscopy Cohort Study and a Comparison with Lung Cancer Mortality in the Atomic Bomb Survivors Study. *Radiat. Res.* 142: 295-304.

ICRP 2. (1959) *Report of Committee II on Permissible Dose for Internal Radiation.* The International Commission on Radiological Protection Publication 2. Pergamon Press, Headington Hill Hall, Oxford, UK.

ICRP 20. (1973) *Alkaline earth metabolism in adult man.* International Commission on Radiological Protection Publication 20. Pergamon Press, Elmsford, NY.

ICRP 26. (1977). *Recommendations of the International Commission on Radiological Protection.* International Commission on Radiological Protection Publication 26. *Annals of the ICRP* 1:3. Pergamon Press, Oxford.

ICRP 30. (1979). *Limits for Intakes of Radionuclides by Workers.* International Commission on Radiological Protection Publication 30 Part 1. *Annals of the ICRP* 2: 3-4. Pergamon Press, Oxford, UK.

ICRP 60. (1991). *1990 Recommendations of the International Commission on Radiological Protection.* International Commission on Radiological Protection Publication 60. *Annals of the ICRP* 21: 1-3. Pergamon Press, Elmsford, NY.

ICRP 66. (1994). *Human Respiratory Tract Model for Radiological Protection.* International Commission on Radiological Protection Pub. 66. Elsevier Science Ltd., Oxford.

ICRP 103. (2007). *The 2007 Recommendations of the International Commission on Radiological Protection.* Exeter, International Commission on Radiological Protection. Publication 103. Annals of the ICRP 37: 2-4. Elsevier, UK.

Jaworowski, Z. (2010). Observations on the Chernobyl Disaster and LNT. *Dose Response* 8: 148–171.

Lloyd, R.D.; Miller, S.C.; Taylor, G.N.; Bruenger, F.W.; Jee, W.S.S. & Angus, W. (1994). Relative Effectiveness of ^{239}Pu and Some Other Internal Emitters for Bone Cancer Induction in Beagles. *Health Phys. 67*: 346-353.

Lubin, J.H.; Boice, J.D.; Edling, C.; Horning, R.W.; Howe, G.; Kunz, E.; Kusiak, R.A.; Morrison, H.I.; Radford, E.P.; Samet, J.M.; Tirmarche, M.; Woodward, A.; Yao, S.X. & Pierce, D.A. (1994). *Radon and Lung Cancer Risk: A Joint Analysis of 11 Underground Miner Studies,* National Institutes of Health Publication No. 94-3644, United States Department of Health and Human Services, Washington, DC.

Lubin, J.H.; Boice, J.D.; Edling, C.; Horning, R.W.; Howe, G.R.; Kunz, E.; Kusiak, R.A.; Morrison, H.I.; Radford, E.P.; Samet, J.M.; Tirmarche, M.; Woodward, A.; Yao, S.X. & Pierce, D.A. (1995). Lung Cancer In Radon-Exposed Miners and Estimation of Risk From Indoor Exposure. *J. Natl. Cancer Inst.* 87: 817-827.

Manhattan Engineering District. (1946). *The Atomic Bombings of Hiroshima and Nagasaki.* Manhatten Engineer District, United States Army, Washington, DC.

Mauderly, J.L. & Daynes, R.A. (1994). *Biennial Report on Long-term Dose-response Studies of Inhaled or Injected Radionuclides 1991-93.* Inhalation Toxicology Research Institute, Albuquerque, New Mexico. Springfield, VA: National Technical Information Service; ITRI-139.

Mays, C.W. & Lloyd, R.D. (1972). Bone Sarcoma Incidence Versus. Alpha Particle Dose. In: Stover BJ, Jee WSS, eds. *Radiobiology of Plutonium.* Salt Lake City, UT: The J.W. Press, 409-430.

Ministry of Health. (1966). *Radiological Hazards to Patients.* Final Report. HMSO: London, UK.

Mole, R.H. (1990). Childhood Cancer After Prenatal Exposure To Diagnostic X-ray Examinations in Britian. *Br. J. Cancer* 62: 152-168.

Moolgavkar, S.H. Dewanji, A.; & Venzon, D.J. (1988). A Stochastic Two-Stage Model for Cancer Risk Asssessment. I. The Hazard Function and the Probability of Tumor. *Risk Anal.* 8:383-392.

NCRP 136. (2001). *Evaluation of the Linear-Nonthreshold Dose-Response Model for Ionizing Radiation.* National Council on Radiation Protection and Measurements Report No. 136. Bethesda MD.

NCRP 160. (2009). *Ionizing Radiation Exposure of the Population of the United States.* National Council on Radiation Protection and Measurements Report No. 160. Bethesda, MD.

Park, J.F. (1993). *Pacific Northwest Laboratory Annual Report for 1992 to the DOE Office of Energy Research: Part 1: Biomedical Sciences.* Pacific Northwest Laboratory, Richland, Washington. Springfield, VA: National Technical Information Service.

Peto, R.; Pike, M.C.; Day, N.E.; Gray, R.G.; Lee, P.N.; Parish, S.; Peto J.; Richards, S. & Wahredorf, J. (1980). *Guidelines For Simple, Sensitive Significance Tests for Carcinogenic Effects in Long-Term Animal Experiments, Supplement 2 In: Long-Term and Short-Term Screening Assays for Carcinogens: A Critical Appraisal.* IARC Monographs. Lyon, France: World Health Organization International Agency for Research on Cancer; Sup 2, Annex 311-426.

Pierce, D.A. & Mendelsohn, M.L.(1999). A Model For Radiation-Related Cancer Suggested by Atomic Bomb Survivor Data. *Radiat. Res.* 152: 642-654.

Pierce, D.A. & Preston, D.L. (2000). Radiation-Related Cancer Risks at Low Doses Among Atomic Bomb Survivors. *Radiat. Res.* 154: 178-186.

Preston, D.L.; Shimizu, Y.; Pierce, D.A.; Suyama, A. & Mabuchi, K (2003). Studies of Mortality of Atomic Bomb Survivors. Report 13: Solid Cancer and Noncancer Disease Mortality: 1950-1997. *Radiat. Res.* 160:381-407.

Preston, D.L.; Pierce, D.A.; Shimizu, Y.; Cullings, H.M.; Fujita, S.; Funamoto, S. & Kodama, K. (2004). Effect of Recent Changes in Atomic Bomb Sutvivor Dosimetry on Cancer Mortality Risk Estimates. *Radiat. Res.* 162:377-389.

Preston, D.L.; Ron, E.; Tokuoka, S.; Funamoto, S.; Nishi, N.; Soda. M.; Mabuchi, K. & Kodama, K. (2007). Solid Cancer Incidence in Atomic Bomb Survivors: 1958-1998. *Radiat. Res.* 168:1-64;

Raabe, O.G. (1987). Three-dimensional Dose-Response Models of Competing Risks and Natural Life Span. *Fundam. Appl. Toxicology* 8:465-473.

Raabe, O.G. (1989). Extrapolation and Scaling of Animal Data to Humans: Scaling of Fatal Cancer Risks From Laboratory Animals to Man. *Health Phys.* 57, Sup. 1:419-432.

Raabe, O.G. (1994). Cancer and Injury Risks From Internally Deposited Radionuclides. In: *Actualités sur le Césium,* Comité de Radioprotection, Electricité de France, Publication Numéro 8: 39-48.

Raabe, O.G.. (2010). Concerning the Health Effects of Internally Deposited Radionuclides. *Health Phys.* 98: 515-536.

Raabe, O.G. (2011). Toward Improved Ionizing Radiation Safety Standards. *Health Phys.* 101: 84-93.

Raabe, O.G.; Book, S.A. & Parks, N.J. (1980). Bone cancer from radium: Canine dose response explains data for mice and humans. *Science* 208:61-64.

Raabe, O.G. & Abell, D.L. (1990). *Laboratory for Energy-Related Health Research Final Annual Report, Fiscal Year 1989,* UCD 472-13. University of California, Davis, CA. Springfield VA: National Technical Information Service.

Raabe, O.G.; Rosenblatt, L.S. & Schlenker, R.A. (1990). Interspecies Scaling of Risk for Radiation-induced Bone Cancer. *Int. J. Radiat. Biol.* 57: 1047-1061.

Raabe, O.G. & Parks, N.J. (1993). Skeletal uptake and lifetime retention of ^{90}Sr and ^{226} Ra in beagles. *Radiat. Res.* 133:204-218.

RERF. (2008). 60-year Anniversary of ABCC/RERF, Annual Report 1 April 2007-31 March 2008 Radiation Effects Research Foundation Hiroshima, Japan.

Rowland, R.E. (1994). *Radium in Humans, A Review of U.S. Studies.* ANL/ER-3, Argonne, IL: Argonne National Laboratory.

Saccomanno, G; Huth, G.C.; Auerbach, O. & Kuschner, M. (1988). Relationship of Radioactive Radon Daughters and Cigarette Smoking in the Genesis of Lung Cancer in Uranium Miners. *Cancer* 62:1402-1408.

Snyder, A.R. & Morgan, W.F. (2004). Gene Expression Profiling After Irradiation: Clues To Understanding Acute and Persistent Responses? *Cancer Metastasis Reviews* 23:259-268.

Stewart, A., Webb, J., Giles, D. & Hewitt, D. (1956). Malignant Disease in Childhood and Diagnostic Irradiation In Utero, *Lancet* ii:447-448.

Stewart, A. , Webb, J., Hewitt, D. (1958). A Survey of Childhood Malignancies. *Br. Med. J.* 1:1495-1508.

Totter, J.R. & MacPherson, H.G. (1981). Do Childhood Cancers Result From Prenatal X-rays. *Health Phys.* 40:511-524.

Wing, S.; Shy, C.M.; Wood, J.L.; Wolf, S.; Cragle, D.L. & Frome, E.L. (1991). Mortality Among Workers at Oak Ridge National Laboratory. *J. American Med. Assoc.* 265:1397-2402.

WHO. (2006) *Health Effects of the Chernobyl Accident and Special Health Care Programmes.* World Health Organizion, Geneva, Switzerland.

16

Molecular Spectroscopy Study of Human Tooth Tissues Affected by High Dose of External Ionizing Radiation (Caused by Nuclear Catastrophe of Chernobyl Plant)

L. A. Darchuk[1], L. V. Zaverbna[2],
A. Worobiec[1] and R. Van Grieken[1]
[1]*University of Antwerp, Antwerp,*
[2]*National University of Medicine, Lviv,*
[1]*Belgium*
[2]*Ukraine*

1. Introduction

Ionizing radiation remains the most significant environmental factor, causing severe impacts on human health on the vast territory of Ukraine for 25 years.
Medical monitoring of the dental health of more than 1500 people who participated in cleaning work at the Chernobyl power station territory after the catastrophe has demonstrated an increasing of specific dental problems occurred after several years of participating at these works (Leus et al., 1998). The high level of hyperstenzy of hard dental tissues, the increase of enamel abrasion (first and second degrees) and the noncarious dental disease (cuneiform defects, pitting) are typical for around 63%, 36% and 47% of examined patients, respectively (Darchuk et al., 2008). Pathological dental abrasion, cuneiform defects, enamel erosion and enamel scissuras have been shown in Fig.1. Significant damages of parodontium (level II-III) by generalized severe periodontitis were detected for men with an acute radiation syndrome (Revenok, 1998). The main pathology was an intensified degeneration process while carious changes occurred, although these last were not directly related to irradiation.

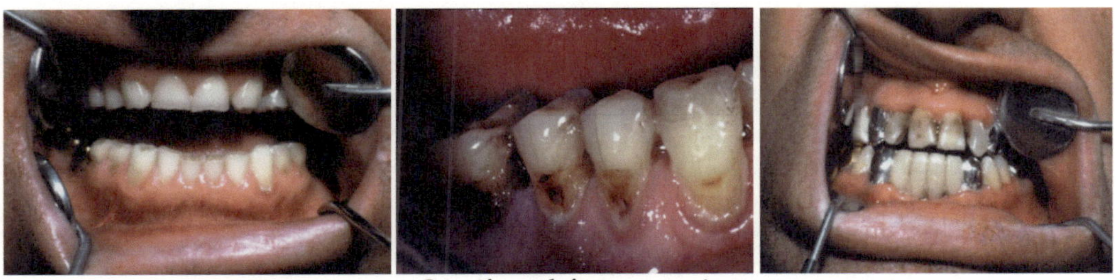

Pathological dental abrasion | Cuneiform defects, noncaries changes | Enamel erosion

Fig. 1. Dental defects which are typical of person undergone external ionizing radiation.

The influence of external ionizing radiation on dental tissues was analyzed by several scientists. Zhang et al. (2004) studied with SEM (Scanning Electron Microscopy) the effect of irradiation on the susceptibility of radiation caries, the structures of rats` tooth enamel and dentin. The collagen fibers and the resistance to acid after undergo external radiation were also investigated. Enamel structure changes were found after irradiation of 30 - 70 Gy. It was shown that dentin morphology changed, some collagen fibers vanished and resistance to acid was reduced after irradiation.

A study on the possibility of dose-dependent tooth-germ (for four-week-old dogs) damage produced by ionizing radiation has shown hyperemia and edemata on tooth-germ pulps from 1.3 Gy onward. Both of these dental diseases became more acute as the radiation dose increased from 1.3 Gy to 5.3 Gy. Possible damage to both the dentin and enamel was pointed out (Sobkoviak et al., 1977).

Dental analysis of rats which had undergone an exposure to 13.2 Gy during 100 days showed a temporary halting in tooth formation following exposure. The dentin which was formed immediately after the radiation was architecturally disarranged to the extent that no paralleling of dentinal tubules existed. In some instances, there was actual resorption of tooth structure with a fibrous replacement (English et al., 1954).

The goal of this work was to study structural changes of teeth taken from people which were exposed to high doses of external ionizing radiation. They worked for different periods: from several weeks to several months, on the territory of the Chernobyl reactor IV zone during the first year just after the catastrophe. The teeth were extracted according to medical recommendations because of oral surgery or mechanical damage due to accidents.

Infrared (IR) absorption and micro-Raman spectroscopy (MRS) as the most efficient techniques to analyze molecular structure of biological tissues (Ellis et al., 2006) have been applied to investigate tooth tissues. The main advantage of Infrared spectroscopy is a superior signal-to noise ratio, which is much less for Raman spectroscopy. Anyway Raman spectroscopy offers another benefit. The MRS is very useful for analysis of bone-like objects, which consist of mineral and organic substances, because Raman vibrational bands of crystalline materials are often sharp. So that the vibrational bands typical of organic matrix are more distinguishable and the identification of minerals is less doubtful.

Enamel, dentin, and cement belonging to 17 patients who absorbed high doses (0.5-1.7 Gy, high dose group, HDG) of ionizing radiation were studied. Only patients without the bone marrow affection were investigated. For comparison the same tooth tissues of 10 men who had not been exposed to radiation (control group, CG)) were investigated as well.

Tooth is composed by enamel, dentin, cement and root (Fig. 2) which are bioinorganic materials. The organic part of the dental tissues is represented by collagen (Ten Gate, 1998) which controls the calcification of bone and crystallization of apatite in bones (Castrom et al., 1956). Collagen is rich in glycine, proline, and hydroxyproline. Collagen molecules are composed by 3 polypeptide chains forming tubinal structure. Collagen is rich with glycine, praline and hydroxyproline (Neuberger, 1956). The composition of aminoacids depends on the dental tissues - the ratio of proline:hydroxyproline:glycine is about 11:11:55 for dental enamel and 29:21:62 for dentin (Geller, 1958). The mineral part of enamel, dentin and cement is composed by carbonate hydroxyapatite, with an approximate formula $Ca_{10}(PO_4)_{6-x}(OH)_{2-y}(CO_3)_{x+y}$, where $0 \le x \le 6$ and $0 \le y \le 2$ (Le Geros, 1991). The carbonate ion CO_3^{2-} can substitute for either PO_4^{3-} (B-substitution, a CO_3^{2-} ion substitutes for a single PO_4^{3-} ion) or OH^- (A-substitution, a CO_3^{2-} ion replaces two OH^- ions) (Le Geros, 1981; Elliott, 1994). Also the carbonate ion can be present in dental tissues as an adsorbed phase in the mineral lattice (Elliott, 1994).

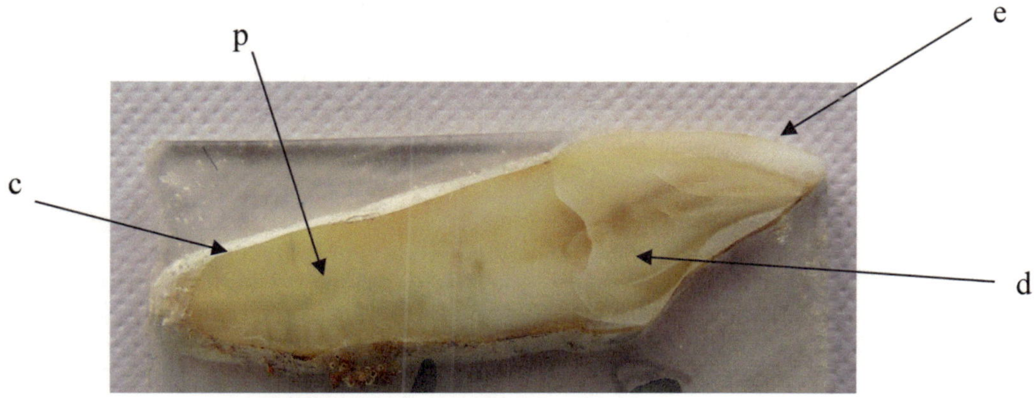

Fig. 2. Cross section of a tooth; e, enamel; d, dentin; c, cement; p, pulp cavity.

Tooth enamel, covering tooth external surface, is the hardest and most mineralized dental tissue. It contains about 96-98 wt.% of mineral matrix and around 2-4 wt. % of protein and water. Protein is rich in proline, glycine and glutamic acid.

Dentin makes up the bulk of the tooth and it is surrounded by cement at the root and by a thin protective layer of enamel. Dentin consists of about 70 wt.% of mineral matter, 20 wt.% of organic materials, and 10 wt.% of water (Ten Gate, 1998). The protein of dentin is composed mainly of Type I collagen and non-collagenous proteins (Le Geros, 1981; 1991).

Cement is a bonelike calcified connective tissue covering tooth roots (Ross et al., 2003) and its composition is similar to the composition of dentin. Cement consists of about 60-65 wt. % of mineral matter, around 23-26 wt.% of organic compounds (mainly collage) and about 12 wt.% of water.

Diversities of enamel, dentin and cement are provided by the nature of the tissues` embryogenesis. The tooth dentin and cement are grown up from mesenchyme while the tooth enamel is grown up from ectoderm, so that enamel almost does not consist of organic matrix.

2. Samples and experimental technique

2.1 Infrared absorption spectroscopy

The preparation of samples for infrared spectroscopy has some features because biological tissues consist of some amount of water. So that samples are dried to avoid overlapping of absorption bands arose from analyzed objects with a broad absorption bands typical of water.

For preparing samples teeth were washed with distilled water and then left for several hours in a drying chamber at 50° C. Afterwards enamel, dentin and cement were sawn down with diamond dental drill to get powder of tooth tissues for IR spectral analysis. Then 0.2 g of tooth tissue powder was thoroughly mixed with 50 µl vaseline oil and put in a cuvette of KBr.

All IR spectra were recorded with one-beam infrared spectrometer IKS-31. Ten accumulations of spectra has been applied for each measurement to improve signal-to-noise ratio. Infrared spectral analysis was done in a low vacuum camera (10^{-1} Pa) to minimize overlapping with broad absorption bands typical of atmospheric CO_2 and H_2O. The infrared intensity transmitted by the cuvette containing the sample (I_s) and that transmitted by an empty cuvette (I_o) were measured in the 600-4000 cm^{-1} spectral range with a resolution of 2-4 cm^{-1}. The spectral resolution depends on diffraction grate (number of grating groove per mm) applied for each spectral interval.

The absorption intensity $I_a(v)$, where v is wavenumber in cm^{-1}, of the investigated samples was obtained as $I_a = (1 - I_s/I_c)$, which is approximately proportional to the sample absorption. For a semiquantitative analysis, the $I_a(v)$ spectra were fitted using a standard procedure which describes the $I_a(v)$ spectrum as a sum of Lorentzian components (Burn, 1985).

2.2 Micro-Raman spectroscopy

For micro-Raman spectroscopy washed and dried teeth were put in polymer solution of Technoviz 4004 (Kulzer, Germany), which got solidified after 6-8 hours. Then the tooth samples were cut in slices with a thickness of 1÷1.5 mm with a diamond disk, Fig.2.

Raman spectra have been collected with a Renishaw InVia micro-Raman (Renishaw plc, GB) spectrometer. For vibrational excitation two laser were applied: argon laser (514.5 nm, Spectra Physica) with a maximum laser power of 50 mW as well as a diode laser (785 nm, Renishaw) with a maximum laser power of 300 mW. Raman spectra were obtained in the spectral range 100 – 3200 cm^{-1} at a resolution of 2 cm^{-1}. Measurement times between 10 and 30 s have been used to collect the Raman spectra with a signal-to-noise ratio of better than 100/1.

The slices of tooth were placed on the stage of the microscope, and the transverse cross section was oriented perpendicularly to the incident laser beam. A continuous laser beam was focused on a sample via 50x and 100x microscope objectives. The laser beam spot size depends on objectives. Spectra were obtained using a 100×(NA=0.95; theoretical spot size of 0.36 μm and 0.50 μm for 514.5 nm and for 785 nm laser, respectively) and 50×(NA=0.70; theoretical spot size of 0.45 μm and 0.68 μm for 514.5 nm and for 785 nm laser, respectively) magnification objectives.

Because of the heterogeneous nature of teeth, single point Raman microspectroscopy cannot adequately describe the microstructure of tissues and therefore spatial information is needed. Hence for the analysis of teeth, MRS mapping, with a spatial resolution of ~1 μm, allows to analyze tooth tissue zones such as individual cement lines, individual lamellae. Raman spectral mapping was employed to study the orientation of mineral and collagen components of bone tissues by Morris et al. (2004) and by Kazanci et al. (2006).

A computer controlled micro-Raman mapping was applied to analyze enamel, dentin, and cement. The microscope stage was XY-motorized and computer-controlled for point-by-point scanning with 0.1 μm resolution, 1 μm reproducibility and 90 mm × 60 mm maximum spatial range. Raman mapping data sets are collected by point-by-point scanning with 1 μm as a minimum step. A total of 20 images were obtained by using the motorized translation stage and the pixel reconstruction. Raman images were acquired in different rectangular areas of 60x60, 60x40, 50x50, 50x30 and 40x30 μm^2. The images were created with step size of 1 or 2 μm. The sample background fluorescence was subtracted with Wire 2.0 softwear (Renishaw, GB).

The Raman spectra, except MRS mapping, were normalized to the intensity of the strongest band at 962 cm^{-1} (v_1-PO_4 mode) of each spectrum.

3. Discussion

3.1 Enamel

Mineral component of dental tissues causes the presence in the IR spectra of bands assigned to the carbonate-ions (CO_3^{2-}) and the orthophosphate-ions (PO_4^{3-}). Collagen is presented by amide groups of proteins - ($CONH_2$). The amide absorption bands are well evident at 1660 cm^{-1} (amide I), 1540-1550 cm^{-1} (amide II), and band at 1240 cm^{-1} (amide III).

Infrared spectra of dental enamel taken from patients of both HDG and CG consisted of strong and wide absorption band at 900-1200 cm^{-1} which arose due to v_3 vibration mode of the phosphate-ion PO_4^{3-} (Fig. 3). This wide absorption band consisted of a set of bands at 950-970, 1000-1020, and 1050-1080 cm^{-1} in the IR spectra of enamel. A comparison of IR spectra of enamel belonging to patients from the control group and people exposed to high doses of ionizing radiation did not show significant changes of absorption band profile typical of phosphate-ion. Nevertheless it was detected that the intensity of the absorption band at 880 cm^{-1} which is typical for CO_3^{2-} ions was stronger in the spectra of irradiated enamel. So that the phosphate-carbon ratio in mineral matrix of enamel changed to increasing of carbonate content in enamel belonging to patients of HDG. Semiquantitative analysis was applied to estimate the area of the band assigned to the carbonate-ions. The area of the band was calculated with Origin softwear according to a procedure described by Burn (1985) and Bebeshko et al. (1998). This analysis showed increasing of the area of the band typical of the anion CO_3^{2-} (at 880 cm^{-1}) up to 0.98 ± 0.04 (a.u.) for people from HDG, in comparison with the area value of 0.84 ± 0.04 (a.u.) for patients of CG. Absorption bands at 1420 and 1482 cm^{-1} arose from the v_3-CO_3^{2-} vibration mode were not distinguished in the obtained IR spectra, because of their overlapping with the strong absorption band at 1440 cm^{-1} typical of vaseline.

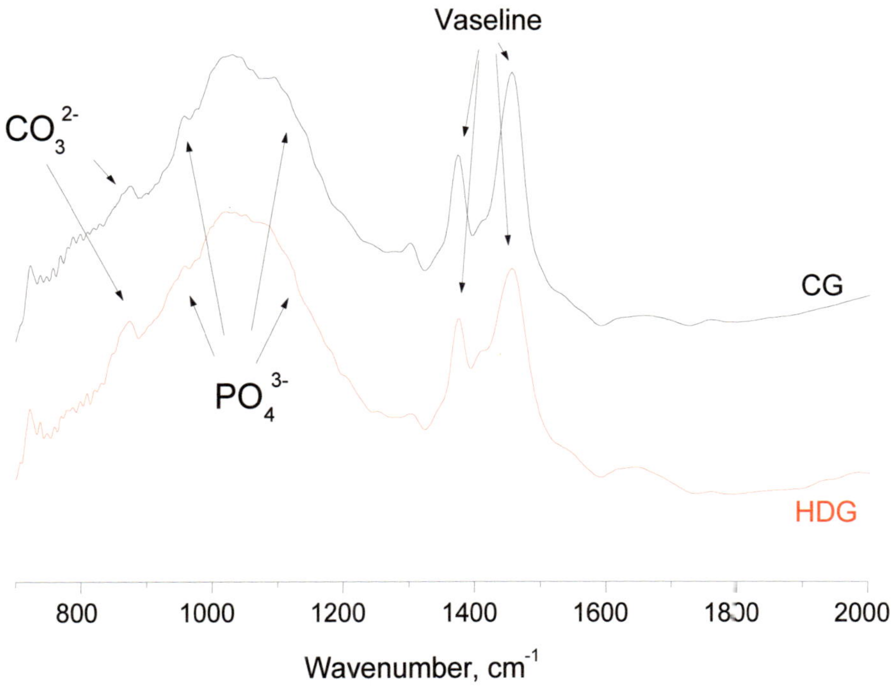

Fig. 3. Infrared spectra of enamel taken from patients exposed to high doses of ionizing radiation (HDG) and people from control group (CG).

Raman spectra of teeth enamel of HDG and CG were more informative than infrared spectra to detect transformations of the enamel mineral matrix. The Raman spectra of enamel obtained with 514 nm were as informative as those obtained with 785 nm laser beam excitation (Fig. 4), but the spectra with excitation at 514 nm often showed an intensive fluorescence background. The fluorescence should be easily subtracted, but if the fluorescence is huge and the intensity of Raman bands is very low compared to the fluorescence signal, the accuracy of analysis will be lower.

Hydroxyapatite single crystals belong to the crystal-class C_6 and they can be described with the Raman-active symmetry tensors (A, E_1, and E_2). The free PO_4^{3-} ion has four internal vibration modes: v_1 at 948 and 962 cm^{-1}, v_2 at 432 and 447 cm^{-1}, v_3 at 1028, 1053 and 1075 cm^{-1}, and v_4 at 580, 593 and 608 cm^{-1} (Fig. 4). Intensities of the Raman bands are invariant to the orientation of the crystal and depend only on the c-axis orientation (Tsuda et al., 1994; Nelson et al., 1982).

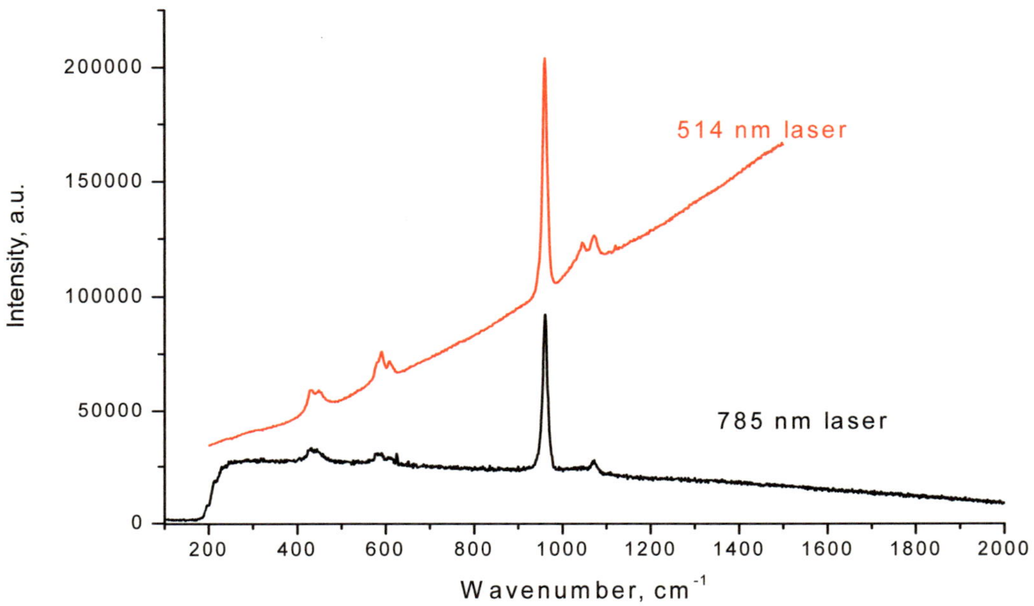

Fig. 4. Raman spectra of enamel recorded with laser beam excitation at 514 nm and at 785 nm.

In the case of enamel taken from patients exposed to ionizing radiation: for the v_4 vibration mode of PO_4^{3-}, the band at 580 cm^{-1} was not detected, the band at 450 and 608 cm^{-1} became very weak, but the band at 593 cm^{-1} did not change (Fig. 5) in comparison with the Raman spectra of enamel belonging to men from control group. These results can be explained as a breaking of the hydroxyapatite crystal symmetry.

The shoulder at 1028 cm^{-1} ($v_3(PO_4)$) was covered by a weak band at 1045 cm^{-1} ($v_3(PO_4)$, out of plane), which is typical of amorphous phosphate and a weak band at 1071 cm^{-1} (Fig. 5) which is assigned to carbonated apatite, type-B substitution (Elliott, 1994; Morris et al., 2004).

The increase of the carbonate content in the enamel is the result of a lack of equilibrium between phosphate and carbonate phases of $Ca_{10}(PO_4)_{6-x}(OH)_{2-y}(CO_3)_{x+y}$, induced by irradiation. The mentioned changes of the mineral matrix lead to a decrease of tooth enamel hardness.

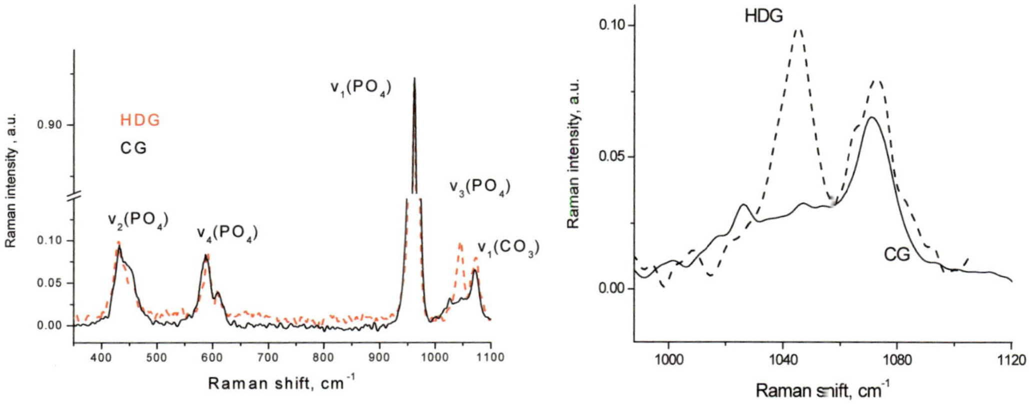

Fig. 5. Raman spectra in the range of internal vibrations of the mineral part of tooth enamel taken from patients of HDG and CG.

3.2 Dentin and cement

Dentin and cement differ from enamel and they look like coarse-fibered bone. IR spectra of dentin and cement were very similar; so only infrared absorption spectra of dentin will be discussed. Infrared spectroscopy of both dentin and cement taken from patients of HDG showed a decreasing content of the mineral matrix (intensities of phosphate bands are lower that those for CG) and changes in organic matrix (Fig. 6). The area of the absorption band at 880 cm^{-1}, typical of the anion CO_3^{2-} increased from 0.97 ± 0.05 (a.u.) for dentin taken from people of control group to 1.05 ± 0.05 (a.u.) for HDG. The absorption band of

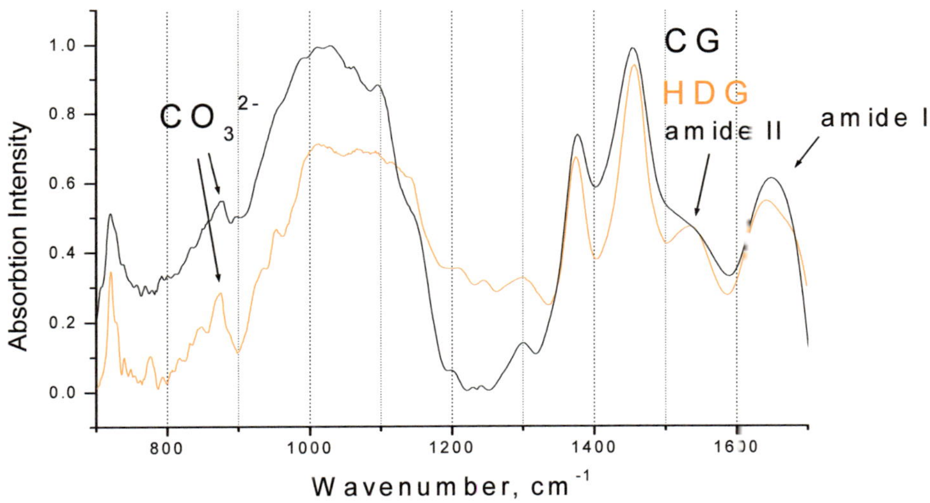

Fig. 6. Infrared spectra of dentin for patients exposed to a high doses of radiation (HDG) and people from control group (CG).

the amide III at 1240 cm-1 and amide II at 1550 cm-1 partly overlapped with strong absorption bands at 1380 and 1440 cm-1 assigned to vaseline oil, so that it was complicated to provide a semiquatitative analysis of areas of the bands assigned to amides. Anyway, reduction of the ratio between amide I and amide II amount was detected - ratio of amide I to amide II bands` area changed from 1.5 ± 0.05 (a.u.) for CG to 1.2 ± 0.05 (a.u.) for HDG. These changes of organic matter were the results of the collagen structure degradation caused by high doses of external radiation.

It can be seen in Fig. 7 that the Raman spectra of dentin and cement were also very similar: therefore mainly Raman spectra of dentin will be discussed.

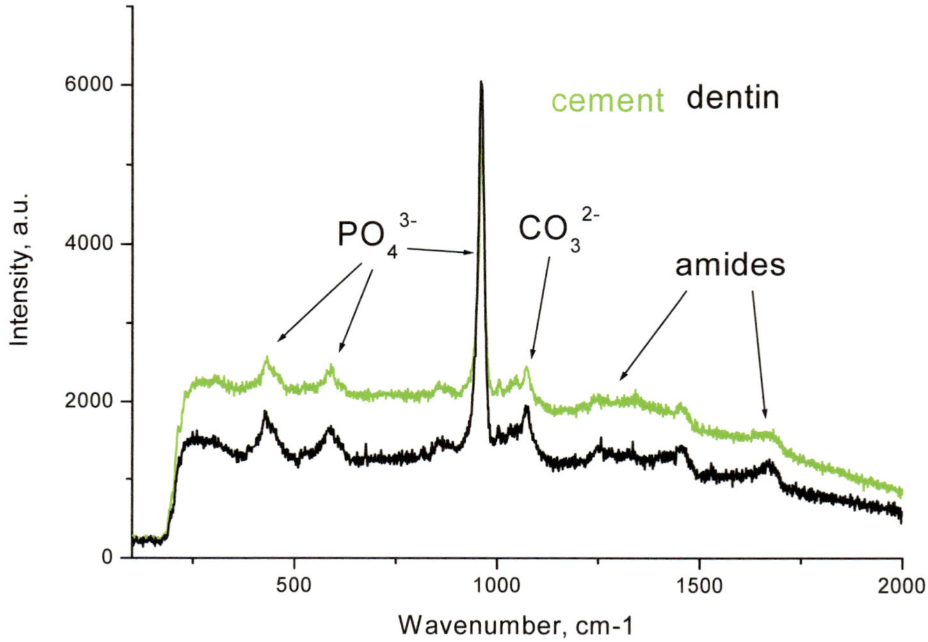

Fig. 7. Raman spectra of dentin and cement (control group) was recorded with 785 nm laser beam excitation.

In order to detect changes of the mineral matrix caused by high doses of external ionizing radiation, the full width at half maximum (FWHM) of the strongest band (at 961 cm-1) assigned to phosphate-ion has been estimated for both HDG and CG. FWHM bandwidth and area S of the PO_4^{3-} v_1 band were determined by curve of it with the procedure describing by Burn, (1985) and Bebeshko et al. (1998). It was found that the band could be well fitted with two Lorentian peaks at 955 cm-1 and 962 cm-1 (Table 1)

Center of peaks	955 cm-1		962 cm-1	
	FWHM (cm-1)	S (a.u.)	FWHM (cm-1)	S (a.u.)
Control group	18.1	29630	11.3	78210
High doses radiation group	18.3	37232	11.5	76138

Table 1. FWHM and area of the strongest band at 961 cm-1 calculated from Raman spectra of dentin belonging to HDG and CG patients

The extra peak at 955 cm^{-1} had to be involved into the calculation to curve the strong phosphate band and it is an additional v_1 peak assigned as amorphous or crystallographically disordered hydroxyapatite (Marcovich et al., 2004). For HDG this disordering is high and the area of the main peak at 962 cm^{-1} is less, this could be caused by ionizing radiation. For instance, the strongest band of the PO_4^{3-} in Raman spectra of both CG and HDG could be curved with only one peak at 961 cm^{-1} with FWHM of 10.1 cm^{-1}, so that crystallography of hydroxyapatite in enamel was less sensitive to high doses ionizing radiation.
The Raman spectra of dentin belonging to patients from HDG showed slight increase of the band at 1071 cm^{-1} ($v_1(CO_3^{2-})$ mode) which could be explained by substitution of the phosphate ion with carbonate (B-type substitution), Fig. 8.

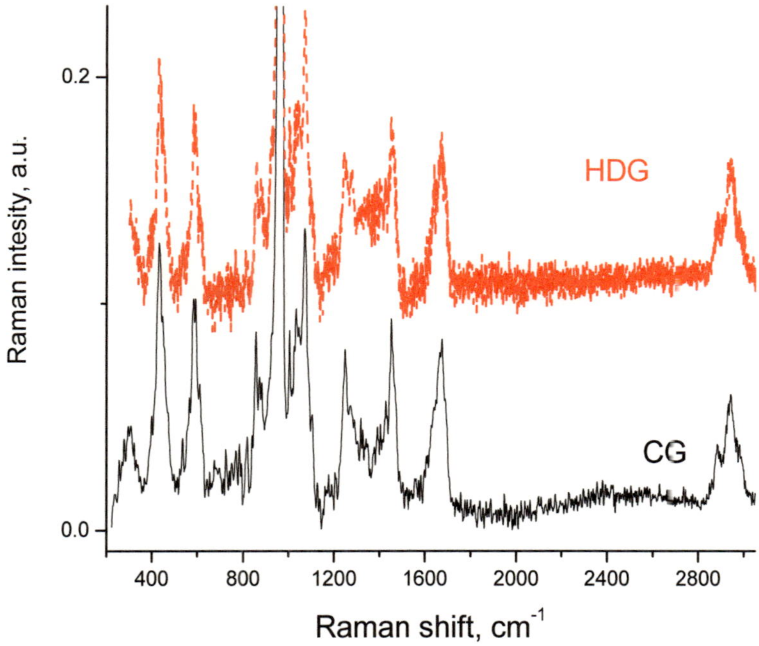

Fig. 8. Raman spectra of dentin taken from patients of HDG and CG.

The feature of dentin organic matrix is the presence of osteonectin which links the bone mineral (hydroxyapatite) and the collagen phases. Osteonectin is a specific noncollagenous protein and consists of two amino acids: glutamine (R=-CH_2-CH_2-CO-NH_2) and phenylalanine (R=-CH_2-C_6H_5). The band at 1004 cm^{-1} assigned to phenylalanine was observed in the Raman spectra of the dentine from patients of HDG and CG (Fig. 9). According to the Raman spectra of dentin taken from HDG, phenylalanine is affected by external ionizing radiation as shown by the band at 1004 cm^{-1} that splits into two narrow bands. The intensities of the bands typical of glutamine (at 815 cm^{-1} and 1330 cm^{-1}) were significantly lower in case of the HDG, which can be ascribed to the ionizing radiation influence. These results have shown a sensitivity of osteonectin to high doses of external radiation.

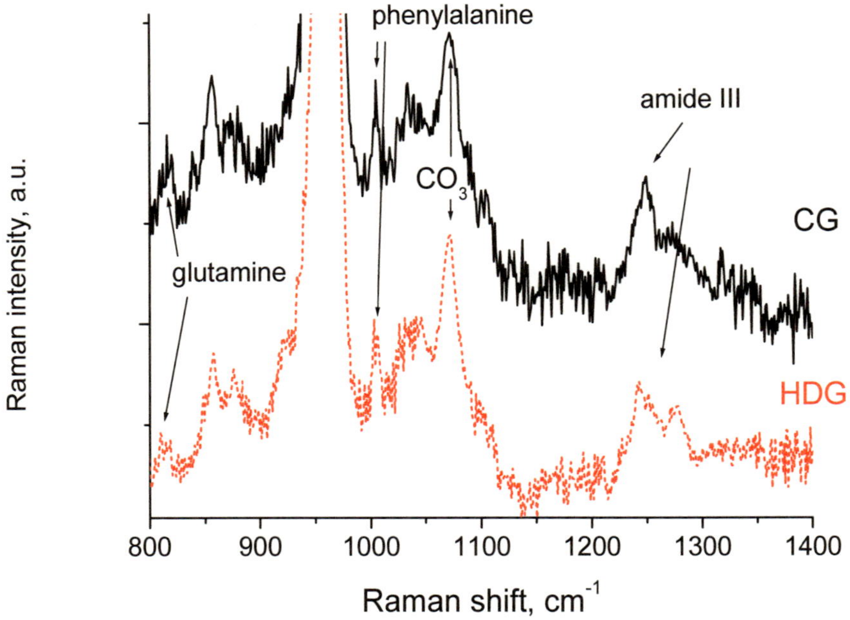

Fig. 9. Raman spectra of dentin taken from patients of HDG and CG in spectral region of osteonectin.

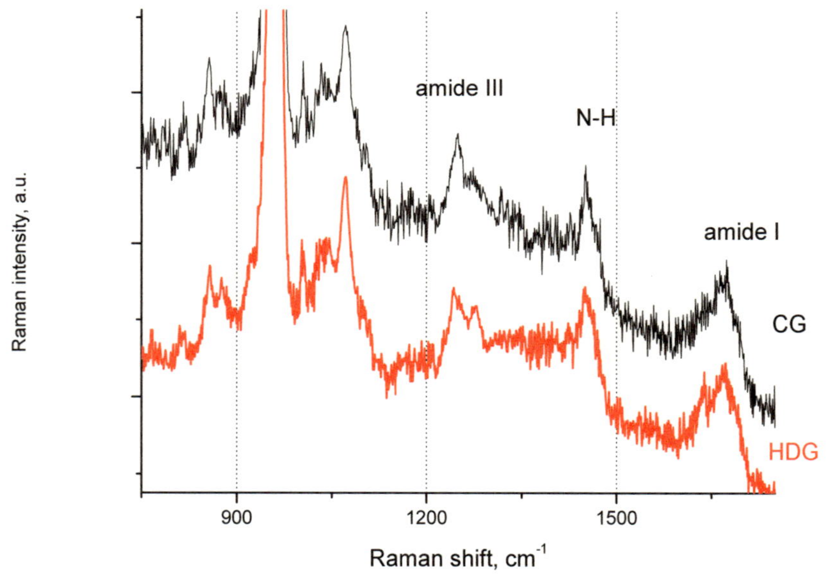

Fig. 10. Raman spectra of dentin belonging to patients of HDG and CG in spectral region of proteins.

The amide absorption bands are well evident in the Raman spectra at 1670 cm^{-1}(amide I) and at 1240 cm^{-1} (amide III) for both CG and HDG. A strong impact of high doses of ionizing radiation on the amide part of dentin and cement has been demonstrated by changes of amides` bands (Fig.10).

The wide band corresponding to amide I vibrations of collagen (peptide carbonyl coupled to the NH-out-of-plane band) is weakly structured with a main wide line at about 1670 cm^{-1} (poly-L-proline) and a shoulder at 1640 cm^{-1} for dentin and cement from patients of CG. For dentin and cement from patients of HFG, the wide band of amide I was split into two bands at 1245 and 1270 cm^{-1}. The split of the amide I band could be explained as breakdown of peptide bonds of collagen caused by ionizing radiation. The band at 876 cm^{-1} could be assigned to hydroxyproline which is an amino acid formed upon hydrolysis of connective-tissue proteins such as collagen (about 14) and elastin but rarely from other proteins.

High doses of radiation also provoked decreasing intensities of the hydroxyproline band at around 855 cm^{-1} and the proline band at around 875 cm^{-1} and the proline shoulder at 920 cm^{-1}. Changes of the proline band intensity are more significant compared to those in the hydroxyproline. Loss of nitrogen in collagen has been detected as a decreasing of the band at 1450 cm^{-1}. The changes of Raman bands typical of organic matrix of dentin and cement reflect a degradation of the collagen structure and breaking of collagen bonds affected by high doses of radiation.

So, high doses of external ionizing radiation lead to mineral composition changes and to alteration of the organic matrix of dentin and cement.

3.3 Mapping of dental tissues

Micro Raman analysis offers superior spatial resolution (0.6-1 µm) compared to the infrared microspectroscopy resolution which is not better than 5-10 µm. So that micro-Raman spectroscopy is very useful for the analysis of biological localities such as single cement lines, boundaries around microtracks, and human dentin tubules.

Micro-Raman mapping was performed for enamel, dentin and cement belonging to patients from both HDG and CG. Mapping of Raman spectra did not show any differences in spatial distribution of organic and mineral components, which could be caused by high doses of external ionizing radiation. Due to the Raman mapping it could be seen that osteonectin and amide III formed single lines (Fig. 11) which look like channels going through dental tissues, such as dentin and cement. The diameter of these lines was calculated according to the applied mapping step of 1 µm and was around 3-5 µm. We need to underline that the distribution of ostenectin was similar to the amide III distribution - the same location of maximum and minimum of their Raman bands intensities.

The applied mapping of dentin (from HDG) also found out several spots with size around 1 µm where the intensity of the strongest band typical of phosphate ion was incredibly low. The single Raman spectrum which was extracted from the mapping (Fig.12) showed that at this spot only organic matrix is present. The near absence of the bands typical of the phosphate-ion demonstrated disordering of the hydroxyapatite crystal. It is possible to suggest that this local resorption of mineral matrix was provoked by high doses of ionizing radiation, but only a few similar spots have been found.

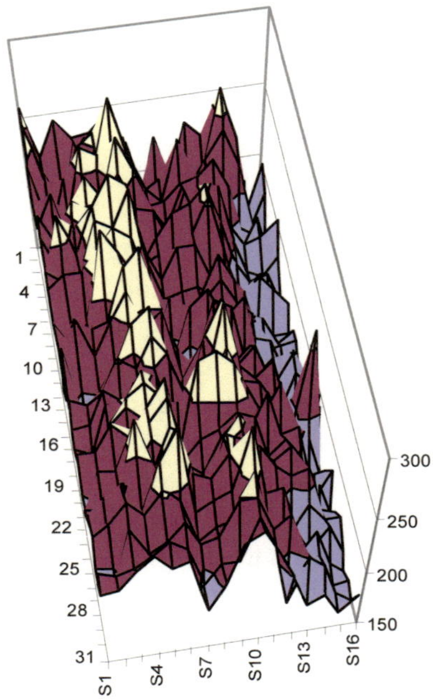

Organic matrix (according to the intensity of the band of amide III at 1260 cm^{-1}

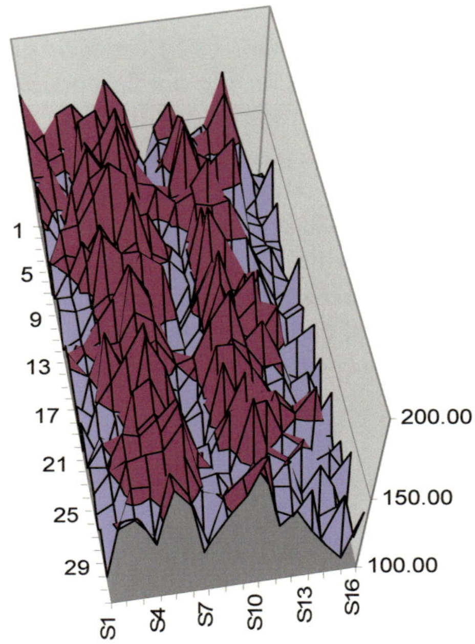

Distribution of osteonectin according to the intensity of the band at 1004 cm^{-1}

Fig. 11. Spatial distribution of amide III and osteonectin in dentin, the mapping has been done with step of 1 μm.

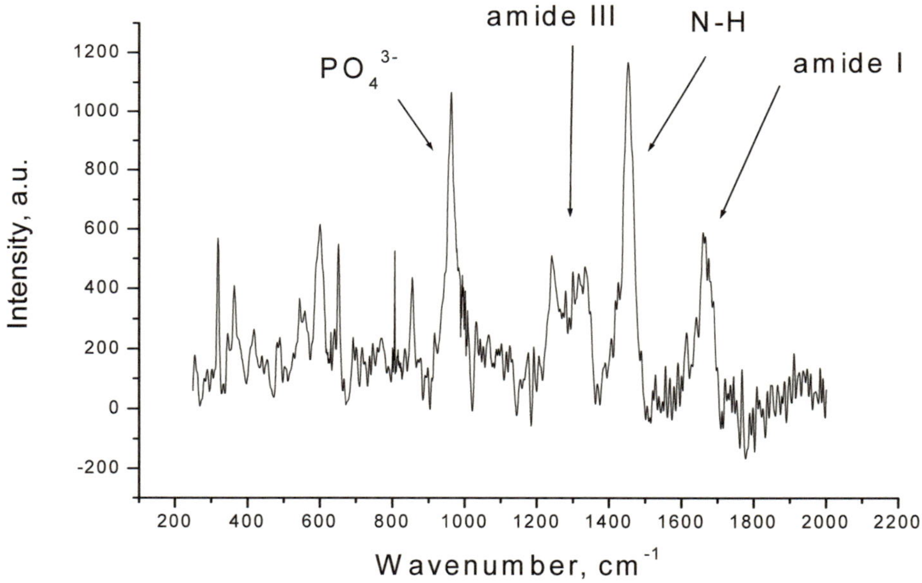

Fig. 12. Raman spectrum of resorbed area of dentin belonging to patient from HDG

Histological dentin consists of a calcified matrix with dental tubules. With Raman mapping it was demonstrated that some of these tubules go through the dentin-enamel junction and terminate in the enamel (Fig.12, taken from Darchuk et al., 2011). It was possible to estimate that the diameter of dentin channels was around 2 µm.

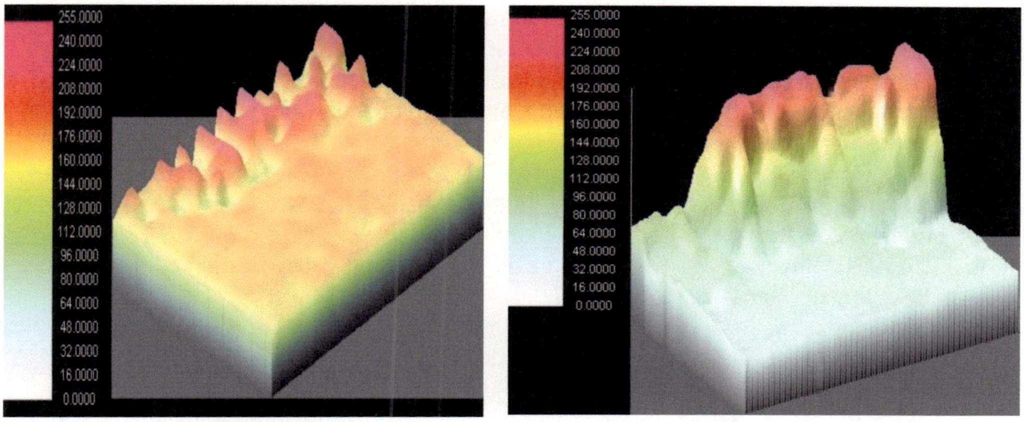

Inorganic matrix Organic matrix

Fig. 12. Spatial distribution of mineral (according to intensities of the band of hydroxyapatite at 962 cm^{-1}) and organic (according to intensities of the band of amide I at 1670 cm^{-1}) components on the boarder between enamel and dentin; area of mapping was 30x50 µm.

4. Conclusions

Thanks to the application of two different spectroscopic analytical methods, complementary information regarding the influence of high dose radiation on the dental tissues was achieved. The results of infrared and micro-Raman spectroscopy analysis of dental tissues taken from patients of CG and HDG were in good agreement but micro-Raman spectroscopy gave a lot of details about the pathological effects of high doses of external radiation on the human dental tissues.

The changes in mineral matrix of enamel and dentin (as well as cement) were caused by substitution of the phosphate ion with the carbonate ion which lead to lack of phosphate-carbonate equilibrium. An increase of amorphous phosphates in dentin and cement showed the breakage of the hydroxyapatite crystal symmetry brought about the decrease of dental tissues hardness.

Both, infrared and Raman spectroscopy have demonstrated changes in organic structure of tooth tissues. High doses of radiation effect resulted in destruction of the collagen chain of dentin and cement: chemical transformations of amino acids, breakdown of peptide bonds of collagen and loss of nitrogen in collagen. Micro-Raman spectroscopy demonstrated sensitivity of osteonectin to high doses of external radiation. Intercrystalline voids, appeared as a result of ionizing radiation effect, have been filling with low differentiation connective tissue which could not fix hydroxyapatite crystals well enough.

5. References

Bebeshko V.G., Darchuk-Korovina L.A., Liashenko L.A., Sizov F.F. & Darchuk S.D. (1998). Spectral diagnostics of the bone structure cnanges in the case of oncohematological disease and acute radiation syndrome, *J. Problems Osteol.* 1, 25-30 (in Russian).

Burn G. (1985). *Solid State Physics*. Burlington, MA: Academic Press, New York, 755 p., ISBN-13: 978-0121460709

Castrom D., Engsstrom A., & Finean J.B. (1956). The influence of collagen on the organization of apatite crystallization in bone. *Chem. Abstr*. 50, 12149 h.

Darchuk L.A., Zaverbna L.V., Bebeshko V.G., Worobiec A., Stefaniak E.A., & R. Van Grieken. (2008). Infrared investigation of hard human teeth tissues exposed to various doses of ionizing radiation from the 1986 Chernobyl accident. *Spectroscopy-Biomedical Application*, 22, 105-111.

Darchuk, L., Zaverbna, L., Worobiec, A., & Van Grieken, R. (2011) Structural features of human tooth tissues affected by high dose of external ionizing radiation after nuclear catastrophe of Chernobyl plant. *Microchemical Journal*, 97, 282-285.

Elliott J.C. (1994). Structure and Chemistry of the Apatites and Other Calcium Orthophosphates. Elsevier, Amsterdam, 387 p. ISBN 0-444-81582-1.

Ellis, D.I. & Goodacre, R. (2006) Metabolic fingerprinting in disease diagnosis: Biomedical applications of infrared and Raman spectroscopy. *Analyst* , 131, 875-885.

English, J.A., Scolack, C.A., & Ellinger F. (1954) Oral manifestations of ionizing radiation II. Effect of 200 kV. X-Ray on rat incisor teeth when administered locally to the head in the 1,500 R dose range. *J. Dent Res.*, 33, 377-388.

Geller J.H. (1958). Metabolism significance of collagen in tooth structure. *J. Dental Res.*, 4, 276-279.

Kazanci, M., Roschger P., Paschalis E.P., Klaushofer K., & Fratzl P. (2006) Bone osteonal tissues by Raman spectral mapping: Orientation–composition. *J. Struct. Biol.* 156, 489-496.

Le Geros R.Z. (1981). Apatites in biological systems. *Prog. Crystal Growth Charact.* 4, 1-5.

Le Geros R.Z. (1991). Calcium Phosphates in Oral Biology and Medicine. Ed. by H. Myers (S. Karger, Basel), New York, N.Y., 200 p., ISBN 978-3-8055-5236-3

Leus P.A., Dmitrieva N.I., & Beliasova L.V. (1998). Dental health of Chernobyl power station accident liquidators, in: *The Second Conf. Further Medical Consequents of Chernobyl Power Station Accident*, Ukraine, Kiev, May 1998, p.98 (in Russian).

Morris, M.D., & Finney, W.F. (2004) Recent developments in Raman and infrared spectroscopy and imaging of bone tissue. *Spectroscopy*, 18, 155–159.

Neuberger A. (1956). Metabolism of collagen under normal conditions. *Chem. Abstr.* 50, 12252f.

Nelson, D.G.A. & Williams B.E. (1982) Low-temperature laser Raman spectroscopy of synthetic carbonated apatites and dental enamel. *Australian Journal of Chemistry*, 35, 715-727.

Marcovic M., Fowler, B.O., & Tung, M.S. (2004). Preparation and comprehensive characterization of a calcium hydroxyapatite reference material. *NIST Journ. Research,* 109 (6), 553-568.

Revenok B.A. (1998) Oral cavity health of people which took part in liquidation of Chernobyl accident consequences, in: *The Second international Conf. Further Medical Consequents of Chernobyl Power Station Accident*, Kiev, Ukraine May 1998, p.352 (in Russian).

Ross, H. Michael, Gordon, I. Kaye, & Wojciech P. (2003) Histology: Text and Atlas. Lippincott Williams and Wilkins, 4th edition, 875 p., ISBN-13: 978-0683302424

Sobkowiak, E.M., Beetke, E., Bienengräber, V., Held, M., & Kittner K.H. (1977) The problem of damage to the tooth germ by ionizing radiation (in Germ). *Zahn Mund Kieferheilkd Zentralbl.*, 65 (1), 19-24.

Ten Cate, A.R. (1998). Oral Histology: Development, Structure, And Function. Saint Louis: Mosby-Year Book, 5th edition. ISBN 0-8151-2952-1

Tsuda, H. & Arends, J. (1994). Orientational Micro-Raman Spectroscopy on Hydroxyapatite Single Crystals and Human Enamel Crystallites. *J.Dent Res* , 73, 1703-1710.

Zhang, X., Li Y.J., Wang S.L., & Xie J.Y. (2004) Effect of irradiation on tooth hard tissue and its resistance to acid (in Chin.) *Zhonghua Kou Oiang Yi Xue Za Zhi*, 39, 463-466.

17

Ionizing Radiation in Medical Imaging and Efforts in Dose Optimization

Varut Vardhanabhuti and Carl A. Roobottom
Derriford Hospital and
Peninsula College of Medicine and Dentistry, Plymouth,
United Kingdom

1. Introduction

Medical-related radiation is the largest source of controllable radiation exposure to humans and it accounts for more than 95% of radiation exposure from man-made sources. Its direct benefits in modern day medical practices are beyond doubt but risks-benefits ratios need to be constantly monitored as the use of ionizing radiation is increasing rapidly. From 1980 to 2006, the per-capita effective dose from diagnostic and interventional medical procedures in the United States increased almost six fold, from 0.5 to 3.0mSv, while contributions from other sources remained static (NCRP report no 160, 2009).

This chapter will review radiation exposure from medical imaging initially starting from a historical viewpoint as well as discussing innovative technologies on the horizon. The challenges for the medical community in addressing the increasing trend of radiation usage will be discussed as well as the latest research in dose justification and optimization.

2. Sources and trends

Medical radiation is by far the largest artificial source of population exposure to ionising radiation, accounting for 90% of all doses from artificial sources (United Nations, 2010). This radiation burden is increasing. In the US, this increase has been primarily due to the increasing use of computed tomography (CT). Despite the fact that CT only accounts for 11% of the examinations it contributes 68% of collective dose. In comparison, conventional radiography constitutes 90% of the examinations but only 19% of collective dose (Mettler et al., 2009). From 1980s to 2006, there is an increase of approximately 6-fold in cumulative effective dose per individual in the US from 0.5 to 3mSV (NCRP report no 160., 2009).

As an overall exposure to all kinds of radiation to humans, medical-related radiation exposure now accounts for 48% (an increase from 15%) with background radiation remaining relative static at 50%. The remaining 2% of total radiation exposure came from consumer-related products and activities. These include cigarette smoking, building materials, commercial air travel, mining and agriculture.

	Collective Effective Dose (person-sievert)		Effective Dose per Individual in the U.S. Population (mSv)	
Exposure Category	Early 1980s	2006	Early 1980s	2006
Ubiquitous natural background	690,000	933,000	3	3.11
Medical Procedures	123,000	899,000	0.53	3
Consumer products	12,000 - 29,000	39,000	0.05 - 0.13	0.13
Industrial, security, medical, educational, and research exposures	200	1,000	0.001	0.003
Occupational exposures	2,000	1,400	0.009	0.005
Overall	835,000	1,870,000	3.6	6.2

Data from NCRP report no 160 (2006 data) and no 93 (1980s data).

Table 1. Changes in collective effective dose and effective dose per individual in the US population between early 1980s and 2006.

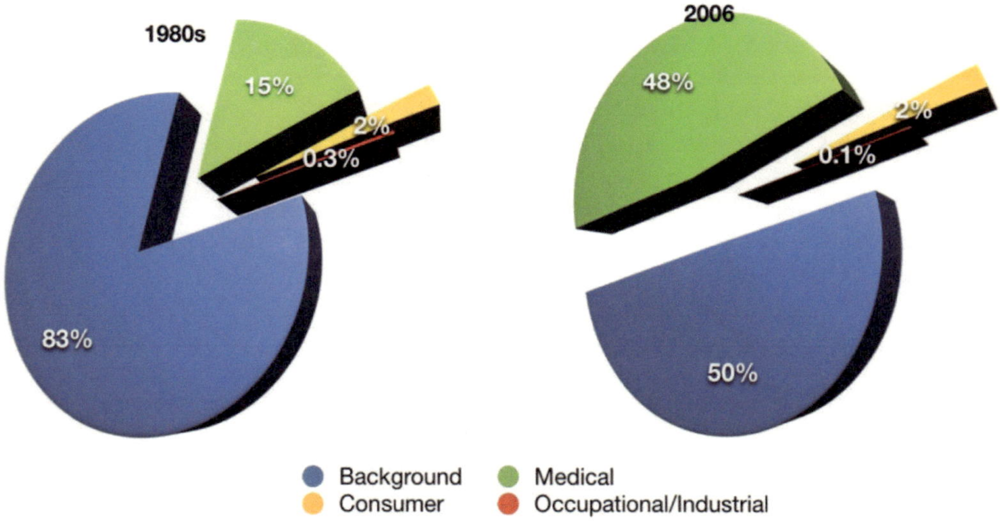

Fig. 1. Exposure of the US population in early 1980s and in 2006 according to NCRP report no 93 and 160.

From data in 2006, in radiology and nuclear medicine procedures, the biggest contribution as previously mentioned is from computed tomography (49%), followed by nuclear medicine studies (26%), then interventional procedures (14%). The remaining 11% came from radiographic and fluoroscopic studies despite accounting for more than 74% of total number of procedures. Not surprisingly, the per capita dose for CT is highest at 1.47 mSv. Data is summarised in the following table.

Type of Procedure	No. of Procedures (in millions)	Percentage of Total No. of Procedures	Collective Effective Dose (person-sievert)	Percentage of Collective Dose from Procedures	Per-Capita Dose (mSv)
Diagnostic radiographic and fluoroscopic studies	293	74%	100,000	11%	0.33
Interventional procedures	17	4%	128,000	14%	0.43
CT scanning	67	17%	440,000	49%	1.47
Nuclear medicine studies	18	5%	231,000	26%	0.77
Total	395	100%	899,000	100%	3.01

Data taken from NRCP report no 160

Table 2. Estimated number and collective effective doses from radiologic and nuclear medicine procedures in the US for 2006.

The overall trend is similar in the UK. A recent Health Protection Agency publication (Hart et al., 2010) showed a 28% increase in 2008 compared to 1997/8 in the United Kingdom. This increase is again mainly due to a doubling of the computed tomography (CT) examinations performed over this same period. The increase in radiation dose in the UK is thought to be modest and this is partly due to the number of examinations performed. A comparison with other European countries show that the 19 out of 20 most commonly performed examinations, UK shows a lower than European average in frequency (Aroua et al., 2010). Only barium enemas were shown to be more frequent than average compared to the European counterparts. In terms of dose, 17 out of 20 examinations in the UK are less on average dose compared to their European counterparts. This illustrates that even in a country like United Kingdom, the trend for increasing use of medically related ionising radiation continues apace despite the fact that there has been more traditional emphasis on

Country	Annual collective dose per caput (mSv)
Germany	1.52
Belgium	1.39
Switzerland	1.37
Iceland	1.28
France	1.11
Norway	0.94
Sweden	0.59
Netherlands	0.48
Lithuania	0.47
Denmark	0.46
Finland	0.35
United Kingdom	0.30

Table 3. Collective dose per caput in 2008 for the "top 20" examinations in Europe (from Aroua et al, 2010).

dose awareness. Table below shows annual collective dose per capita of European countries. The UK collective dose per caput from all medical and dental X-ray examinations stands at 0.3 mSv. This number can be compared with corresponding values of 2.2 mSv assessed for medical and dental X-rays in the USA in 2006 (NCRP, 2009) and the similar figure of 1.9 mSv (UNSCEAR., 2010) as the average for people living in Healthcare Level 1 (HCL1) countries. The UK collective dose per caput is clearly very low for a HCL1 country.

For CT, the UK annual per caput dose stands at 0.27 mSv. This number can be compared with corresponding levels of 1.5 mSv from CT in the USA in 2006 (NCRP., 2009) and 0.74 mSv from CT in Canada in 2006 (Chen and Moir., 2010). The UK per caput dose from CT is relatively low compared to North American countries.

At the time of writing, there is likely to be an even further increase in the use of CT compared to the level quoted in 2006. This is mainly due to rapid rise in the use of CT in emergency setting (White and Kuo, 2007; Street et al, 2009) and also the increasing use of PET/CT (Elliot A, 2009; Chawla et al, 2010; Devine et al, 2010). However, the dose increase may not be as high as recent concentrated efforts in the medical community have focused on strategies in dose reduction. Recent effective novel innovations have helped to reduce some of the dose burden (see later).

3. Risks

Ionizing radiation has long been known to increase the risk of cancer and is officially classified as a carcinogen by the International Agency for Research on Cancer (IARC). It is now on the official list of carcinogen by the World Health Organization (IARC list of carcinogen, 2011). The exact relationship between dose exposure and cancer induction, however, is complex and several issues merit further discussion.

First, perceived medical-related radiation has traditionally been quantified in comparison with radiation we all receive from background levels. It is worth noting that medical-related radiation exposure is not the same as background radiation. The exposure from background radiation is generally of mixed-energy particles (high and low LET-radiation) while the exposure from diagnostic medical procedures is generally of low-energy x-rays. Low-LET radiation deposits less energy in the cell along the radiation path and is considered less destructive per radiation track than high-LET radiation. Examples of low-LET radiation include X-rays and γ-rays (gamma rays), which are used in medical imaging. Low-LET radiation produces ionizations sparsely throughout a cell; in contrast, high-LET radiation transfers more energy per unit length as it traverses the cell and is more destructive per unit length.

Background radiation are naturally occurring from sources in soil, rocks, bricks and building material, and from radon gas which seeps out to homes. Radon is a colourless, odourless gas that emanates from the earth and as it decays also emits LET radiation. Average annual exposures worldwide to natural radiation sources (both high and low LET) would generally be expected to be in the range of 1–10 mSv, with 2.4 mSv being the present estimate of the central value (UNSCEAR, 2000). Of this amount, about one-half (1.2 mSv per year) comes from radon and its decay products. Some areas will have more radon gas background than others (e.g. Cornwall in the UK generally has twice the amount of background radiation than the rest of the UK).

After radon, the next highest percentage of natural ionizing radiation exposure comes from cosmic rays. This is followed by terrestrial sources, and "internal" emissions. Cosmic rays are particles that travel through the universe and are filtered by the earth atmosphere (therefore this varies from sea level to higher altitudes where less atmospheric filtration occurs). The Sun is a source of some of these particles. Other particles come from exploding stars called supernovas. The amount of terrestrial radiation from rocks and soils varies in different parts of the world. Much of this variation is due to differences in background radon levels. "Internal" emissions come from radioactive isotopes in food and water with uranium and thorium series of radioisotopes present in food and drinking water (UNSCEAR, 2000).

To quantify medical exposures (which are of predominantly low-LET radiation) to background radiation (which are of mixed-LET radiation) is uncertain due to this respect but it is now generally accepted that of the 2.4mSv total average background radiation of mixed-LET, the total average annual population exposure worldwide due to low-LET radiation would generally be expected to be in the range of 0.2–1.0 mSv, with 0.9 mSv being the present estimate of the central value. The pie chart below illustrates proportions of high and low-LET radiation of the background radiation worldwide (adapted from BIER VII, 2006).

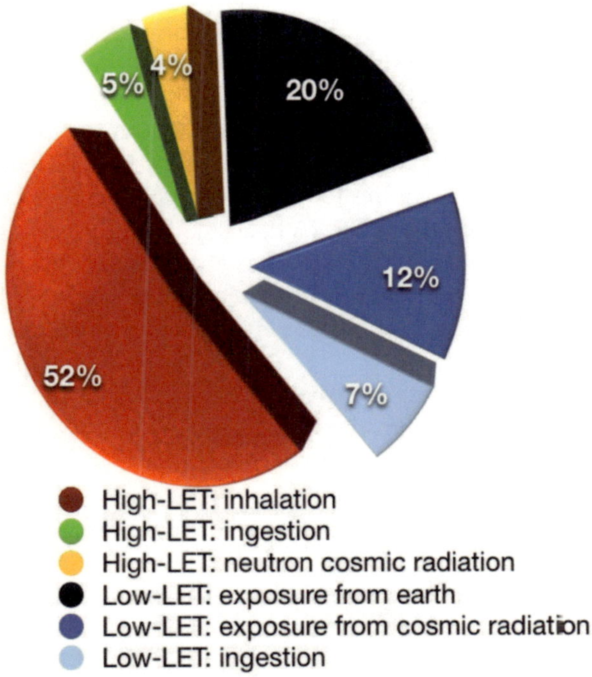

Fig. 2. Pie chart proportions of high and low-LET radiation of the background radiation worldwide (adapted from BIER VII, 2006).

Ubiquitous background radiation represents exposure to whole population (all ages, gender, health status) whereas medical radiation distributions are often skewed towards higher age groups and also sicker individuals. There may also be skewed in gender distribution (there is rightly a reluctance to image patients who are pregnant, for example). Therefore, one must acknowledge that the background radiation level is an approximation and will vary from individual to individual living at different locality having varying exposures.

Current radiation protection standards and practices are based on the premise that any radiation dose, no matter how small, can result in detrimental health effects, such as cancer and genetic damage. Further, it is assumed that these effects are produced in direct proportion to the dose received, i.e., doubling the radiation dose results in a doubling of the effect. These two assumptions lead to a dose-response relationship, often referred to as the linear no-threshold model, for estimating health effects at doses of interest. Although, this is of much benefit in practical terms in risk estimation in the context of radiation protection quantification, there is, however, substantial scientific evidence that this model is an oversimplification of the dose-response relationship. In particular, it results in an overestimation of health risks in the low dose range. Biological mechanisms including cellular repair of radiation injury, which are not accounted for by the linear, no-threshold model, reduce the likelihood of cancers and genetic effects. Therefore, it is now generally accepted that quantification of estimated health risk in the dose range similar to that of background radiation should be strictly qualitative and encompass a range of hypothetical health outcomes, including the possibility of no adverse health effects at such low levels (Burk Jr RJ 1996, updated in 2004).

Biological Effects of Ionizing Radiation (BEIR) VII committee defines low doses of ionizing radiation as less than 100mSv and agrees that at doses of 100 mSv or less, statistical limitations make it difficult to evaluate cancer risk. The current best estimation model for risks associated with low dose radiation exposure is that approximately one individual in 100 persons would be expected to develop cancer (solid cancer or leukemia) from a dose of 100 mSv while approximately 42 of the 100 individuals would be expected to develop solid cancer or leukemia from other causes (BEIR VII, 2006).

It is now generally accepted that at effective doses above 50 mSv the risk of cancer induction increases linearly with dose, this dose-response relation has not been demonstrated at doses below 50 mSv. Below 50 mSv no convincing epidemiological evidence exists currently for cancer risk induction.

There is, however, more convincing evidence to support risk of cancer induction using high dose radiation. A recent article from the USA estimated that CT scans performed in 2007 could result in 29,000 future cancers based on current risk estimations (Berrington de Gonzales et al., 2009). The risk to an individual patient is also high. For example a cardiac scan in a 20-year-old female can produce a lifetime cancer risk from that scan of approaching 1% (Einstein et al., 2007).

Specific effects of radiation are thought to be either deterministic and/or stochastic although the exact relationship is difficult to quantify with certainty. There is reasonable epidemiological evidence (though not definitive) from 30,000 A-bomb survivors that organ doses from 5 to 125 mSv result in a very small but statistically significant increase in cancer risk (Preston et al., 2007). Other low dose epidemiological studies from the occupational exposure of radiation workers are also generally in keeping with this trend with increasing cancer risk with increasing radiation dose (Muirhead et al., 2009 and Cardis et al., 2007). Of note, the third analysis of the National Registry for Radiation Workers in the UK (NRRW-3 from Muirhead et al., 2009) provides the most up-to-date and precise information on the risks of occupational radiation exposure based on cancer registrations as well as mortality from an enlarged cohort of 174,541 workers dating back to 1955. Data were available for 26,731 deaths during 3.9 million person-years of total follow-up period. It clearly shows a statistically significant increasing trend with dose in both mortality and incidence for all malignant neoplasms (leukaemia and several solid organs tumours). Table below summarises the findings from NRRW-3. Despite this, one must bear in mind that radiogenic excess cancer risk associated with medical-related radiation is in orders of magnitude smaller than the spontaneous cancer risk (Burk Jr RJ 1996, updated in 2004 and BEIR VII (2006).

	Leukaemia excluding CLL	All neoplasms excluding leukaemia	All malignant neoplasm excluding leukaemia, lung and pleura cancer
3rd NRRW Analysis			
Mortality	1.712 (0.06-4.29)	0.275 (0.02-0.56)	0.323 (0.02-0.67)
Incidence	1.782 (0.17-4.36)	0.266 (0.04-0.51)	0.305 (0.05-0.58)
2nd NRRW Analysis			
Mortality	2.55 (-0.03-7.16)	0.09 (-0.28-0.52)	0.17 (-0.26-0.70)
15-country Nuclear Worker Study			
Mortality	1.93 (0-7.14)	0.97 (0.27-1.80)	0.59 (-0.16-1.51)
Japanese A-bomb survivors			
BEIR VII: *Mortality*	1.4 (0.1-3.4)	0.26 (0.15-0.41)	-
BEIR VII: *Incidence*	-	0.43	-

Table 4. Comparison of estimates of ERR per Sv (and 90% CI) for cancer in NRRW, the 15-country nuclear worker and the Japanese A-bomb survivors.

In addition, there appears to be other determinants that also affect the risk of developing cancers. Among these, genetic considerations, age at exposure, sex and fractionation and protraction of exposure appear to play important roles.

Genetic considerations

There are 2 epidemiological studies that suggest that there is a subgroup of population who are more likely to develop cancer when exposed to radiation although in neither case has the genes responsible for the increased radiosensitivity been identified. Ronckers et al (2008) performed a retrospective study looking at 3,010 young women with scoliosis who regularly underwent radiographic follow-up to monitor disease progression between 1912 and 1965. They had found that there was a borderline but significant dose response in a subset of women with a family history of breast cancer in first- or second-degree relatives. Flint-Richter et al (2007) performed a case-control study looking at children who were epilated with x-rays for the treatment of tinea capitis and found that 1% of children developed meningioma with marked clustering in certain families suggestive of genetic susceptibility to the development of tumours after exposure to radiation. A meta-analysis by Jansen-van der Weide (2010) also supports increase risk of radiation in genetically susceptible group. They performed a meta-analysis from seven studies evaluating the effects of low-dose radiation exposure, such as mammography, on cancer risk in women with a familial or genetic predisposition. They had found that there is an increased risk with exposure to radiation that results in 1.3 times increased breast cancer risk. The risk is also higher in women who were exposed before the age of 20 or who were frequently exposed to radiation.

Age at exposure

Following on from the last point, there is convincing evidence to support a relationship between life-time attributable risk of cancer incidence and age at exposure. Graph below show analysis of lifetime attributable risk of radiation-induced cancer incidence derived from BEIR VI committee based on data of A-bomb survivors. This, in general, supports that children are more radiosensitive than adults. However, it is also true that social and environmental factors play a role as for some solid cancers, these risks do not decline with age. When looking at effective risks ratio (ERR), there is little difference in risk between 10 and 40 years of age, while for some cancers such as lung and bladder, there appears to be a significant increase in risk with increasing age of exposure.

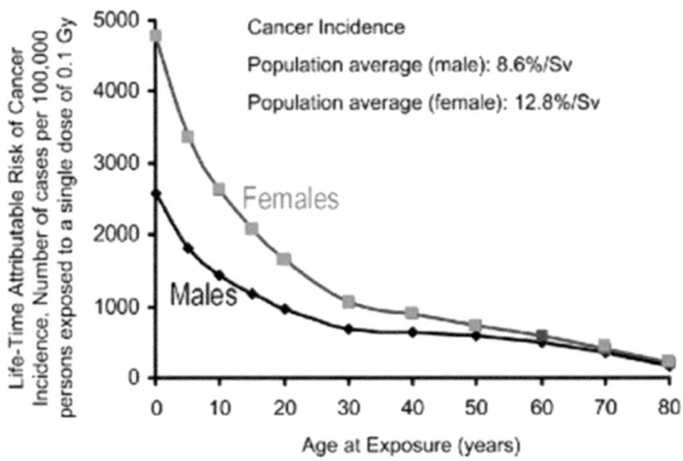

Fig. 3. Lifetime attributable risk of radiation-induced cancer incidence, as a function of age at exposure for males and females (data based on BEIR VII, 2006).
(with permission from Hricak et al, 2011)

Sex

BEIR VI report supports the notion that there is substantially higher lifetime attributable risk of cancer incidence in females compared with males. Figure 5 shows breast cancers risks are higher, but what is more notable is the risk for lung and bladder cancer in women is much higher than in men. This is, in spite of the fact that in 1945 Japan, men were heavy smokers while smoking was deemed as very uncommon in women.

Fractionation and protraction of exposures

It was previously thought that radiation risks per unit dose at low levels and at low dose rates were smaller than that of higher dose and dose rates. This is due to the perceived influence of DNA repair. A suggested dose and dose rate effectiveness factor (this is a multiplication factor use for low dose rates compared to high dose rates) by the ICRP was two, while BEIR VII suggested a value of 1.5. There are a couple of studies that have already been mentioned but are again worth mentioning in this regard. First the International Agency for Research on Cancer 15-country study looked at around 600,000 nuclear workers who were exposed to an average cumulative dose of 19mSv. The estimated ERR for this cohort for developing solid cancers was almost four times larger than that for the A-bomb

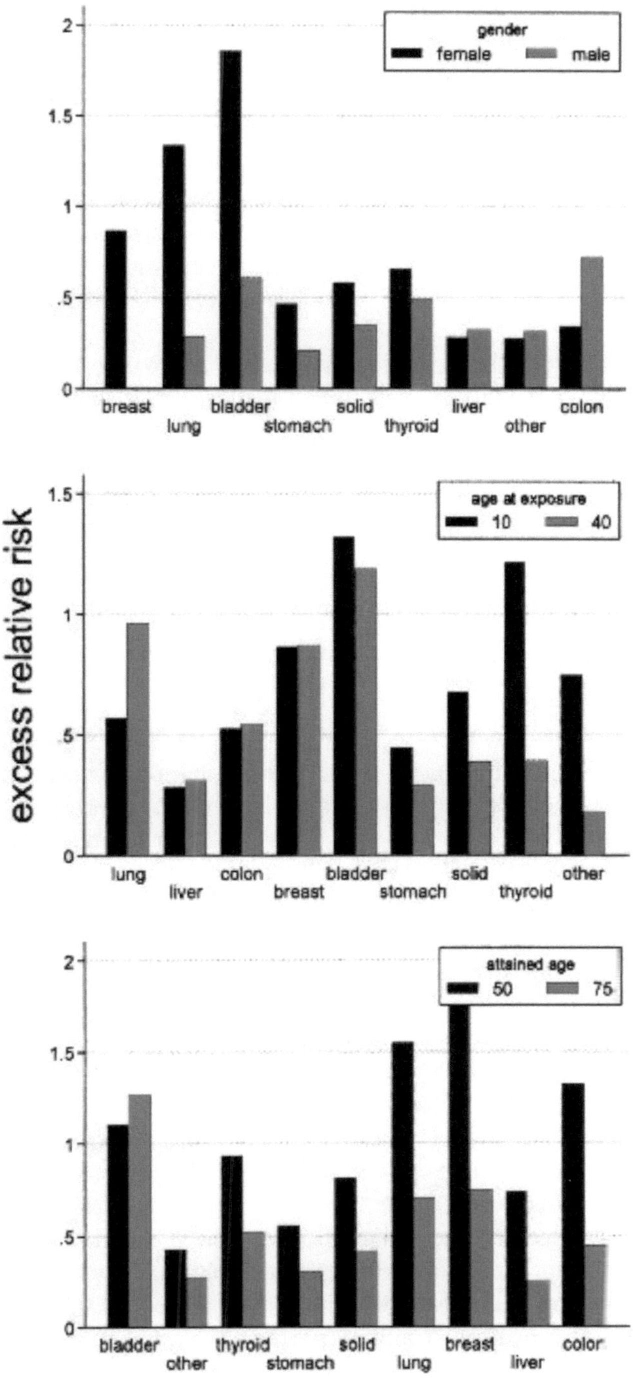

Fig. 4. Comparison of site-specific gender (top), age-at-exposure (middle) and attained-age (bottom) effects on standardised ERR_{1Gy} estimates for selected sites and all solid cancers. (with permission from Preston et al, 2007)

survivors. However, the study noted that there are likely to be important confounders. First, the results are likely to be skewed by the Canadian workers who were relatively few in number with high number of death rates. Second, the predominance of lung cancer suggests possible confounding effect of smoking. More recently, the update from the National Registry for Radiation Workers in the UK (NRRW-3 from Muirhead et al., 2009) followed cohorts with cumulative dose of 25mSv. Cancer risks do increase with dose and the estimated ERR per Sievert was very similar to that for the A-bomb survivor suggestive of a rather small reduction in cancer risk induction with dose protraction which is somewhat surprising. These 2 large studies in themselves suggest that it is not at all clear what the relationship is between dose fractionation and protraction with the risk of cancer induction.

Currently, there are several epidemiological studies following up patients who were exposed to CT at a young age in UK, Australia, Canada, France, Israel and Sweden. Results of such studies will shed more light into the precise influence of radiation and add to the existing body of evidence. This is likely to take time, however, due to the inherent nature of these studies.

4. Strategies in dose optimisations in Computed Tomography (CT)

The fact that CT accounts for most of the ionising radiation used in medical imaging is the reason for its focus on this chapter. Other imaging modalities will also be discussed but in lesser detail.

Dose elimination

Whilst, undoubtedly medical imaging using ionising radiation has several advantages, one must bear in mind that the best way to reduce radiation is to not perform the investigation at all. There is evidence that increasingly CT has been over utilised in various clinical settings meaning that unnecessary scans are being performed, or incorrect examinations are being performed without appropriate justification. The usual practice is to refer to clinical guidelines or appropriateness criteria to direct or justify an examination according to a clinical scenario. The American College of Radiology (ACR), the Royal College of Radiologists in the UK (RCR), and the European Commission all have published decision guidelines for the appropriate use of CT in different clinical scenarios. A retrospective study was performed in a level I trauma centre looking at appropriateness of scans (Hadley et al, 2006). It was found that 44% of the studies ordered would not have been indicated had the guidelines been rigorously followed. One recent innovative approach addressing this has been to incorporate these guidelines into computerised imaging order entry system. Pre-authorisation of CT examinations according to the ACR and RCR guidelines were utilised in an institution which showed significant deferral rate and substantial decrease in the use of CT and MRI. After reauthorisation was implemented, CT annual performance rates decreased from 25.9 examinations per 1,000 in 2000 to 17.3 per 1,000 in 2003 (Blachar et al, 2006). Despite being evidence-based and recommended for routine clinical usage, these guidelines still show poor uptake in general usage (Bautista et al, 2009). When a survey was performed looking at how physicians decide what the best imaging test to use for their patients, the use of ACR Appropriate criteria showed very low uptake in one institution (2.4%) compared with other available resources (e.g. Radiologist consult, specialty journal, UpToDate, Google, Pubmed, etc).

Dose reduction

CT dose reduction can broadly be divided into ways to reduce the total radiation emitted by the X-ray tube and ways to reduce scanning time. Ways to reduce scanning time include ECG gating in cardiac studies, or increasing the pitch of the scanner, for example.
Traditionally, the ways to reduce total dose have included X-ray beam filtration, X-ray beam collimation, X-ray tube current modulation and adaptation for patient body habitus (automatic exposure control), peak kilovoltage optimisation, improved detection system efficiency, low dose protocols for specific indications (e.g. CT KUB for renal stones).
Since the 1980s, a number of technical innovations have been responsible for dose reduction in CTs including the use of solid state scintillating detectors, electronic circuitry, multi-detector arrays, more powerful x-ray tubes and beam-shaping filters. More recently, a number of newer dose reduction techniques have gained widespread acceptance and these are as follows:

Automatic exposure control

Automatic exposure control (AEC) is one of the most important techniques in clinical practice to reduce radiation dose without compromising image quality. AEC is a broad term that encompasses not only tube current modulation (to adapt to changes in patient attenuation), but also determining and delivering the "right" dose for any patient (infant to obese) in order to achieve the diagnostic quality images. It is technologically possible for CT systems to adjust the x-ray tube current in real-time in response to variations in x-ray intensity at the detector (McCollough et al, 2005), much as fluoroscopic x-ray systems adjust exposure automatically. The modulation may be fully pre-programmed, occur in near-real time by using a feedback mechanism, or incorporate pre-programming and a feedback loop.

Tube current modulation

This is done by maintaining a constant image noise level through longitudinal and/or angular modulation of x-ray tube current according to patient size, shape and the resultant attenuation. This means that the tube current varies across different scan length. This is in contrast to fixed tube current methods where it is constant throughout the scan length meaning in effect that in certain areas, there are wasted radiation as one will not yield increased diagnostic capability.

Longitudinal (z) mA modulation

In the longitudinal (z-axis) modulation (AutomA) technique, the basic strategy is to provide predictable image quality to achieve a reliable diagnosis with the lowest necessary radiation dose depending on patient size and attenuation. This is done along the patient's longitudinal axis (i.e. shoulders to pelvis). For a specific patient anatomy and diagnostic task, a specific parameter (defined as noise index - NI) is prescribed by the user to specify the targeted image quality that represents the average noise in the centre of an image of a uniform water phantom. A 5% decrease in NI demands approximately a 10% radiation increase, whereas a 5% increase in NI decreases radiation dose by approximately 10% (Karla et al, 2004a). Therefore, an appropriate NI selection is imperative to control the balance between radiation dose and image quality. This is often recommended by a combination of manufacturer recommendation and clinical experience. This also varies depending on the type of scan performed. With a given NI, the AutomA automatically adjusts x-ray tube current in the scan to maintain the same noise level in all images regardless of patient size and attenuation. Previous studies have shown that in abdominal CT studies a NI of 10.5 to

15 leads to a reduction in radiation dose by 16.6-53.3% in comparison to that using a constant x-ray tube current (Karla et al, 2004b).

Angular (x,y) mA modulation

Angular (x,y) mA modulation addresses the variation in x-ray attenuation around the patient by varying the mA as the x-ray tube rotates about the patient (e.g. in the A.P. versus lateral direction). The operator chooses the initial mA value, and the mA is modulated upward or downward from the initial value within a period of one gantry rotation. As the x-ray tube rotates between the AP and lateral positions, the mA can be varied according to the attenuation information from the CT radiograph (i.e. Scout image), or in near real-time according to the measured attenuation from the 180° previous projection.

Using both angular and longitudinal mA modulation, significant dose reduction can be achieved. Although an approximately 50% reduction in dose has been found with automatic exposure control, the system is not foolproof. It seems that at the extremes (i.e. smaller and larger patient sizes), there needs to be adjustment of the noise level such that a higher noise level was recommended for a large sized patient to avoid a higher radiation exposure (Karla, 2004a), and a lower noise level for smaller patients was also suggested by a previous investigation (Karla, 2004b). Some of these studies have shown that there is an influence of patient weight on image quality and dose when a constant noise level is chosen for all patient sizes. Kalra et al (2004b) showed that smaller patients (in their study, defined as having weight less than 68 kg and corresponding to smaller transverse and anteroposterior diameters) had subjective image quality scores lower than larger patients (weights greater than 68 kg) despite using a fixed noise index parameter.

Adjusting noise level based on weight alone is also fraught with difficulties. Several studies have shown that weight is not the ideal factor for required dose calculations. This is because two patients with the same weight can have different regional dimensions and tissue attenuation properties on CT scanning, which can affect the image quality significantly (Karla et al, 2003). Schindera et al (2008) found that a phantom with increased anthropomorphic size received significantly increased skin and deep organ dose than a smaller sized phantom for fixed noise level. Some people have proposed a correction factor for patient size to find the optimal noise index using a combination of patient's weight, BMI or information body diameter from CT scout images (Li et al, 2011).

Tube Angle Start Position and Pitch

Various investigators have demonstrated that there are significant dose variations with a sinusoidal pattern on the peripheral of a CTDI 32 cm phantom or on the surface of an anthropomorphic phantom when helical CT scanning is performed, resulting in the creation of "hot" spots or "cold" spots (Svandi et al, 2009). Exploiting this in conjunction with adjustment of pitch can result in dose saving (Zhang et al, 2009). For example, at a pitch of 1.5 scans, the dose is usually lowest when the tube start angle is such that the x-ray tube is posterior to the patient when it passes the longitudinal location of the organ. For pitch 0.75 scans, the dose is lowest when the tube start angle is such that the x-ray tube is anterior to the patient when it passes the longitudinal location of the organ. For organs that have a relatively small longitudinal extent, dose can vary considerably with different start angles. While current MDCT systems do not provide the user with the ability to control the tube start angle, these results indicate that in these specific situations pitch 1.5 or pitch 0.75, small organs and especially small patients, there could be significant dose savings to organs if that functionality adjustment was available.

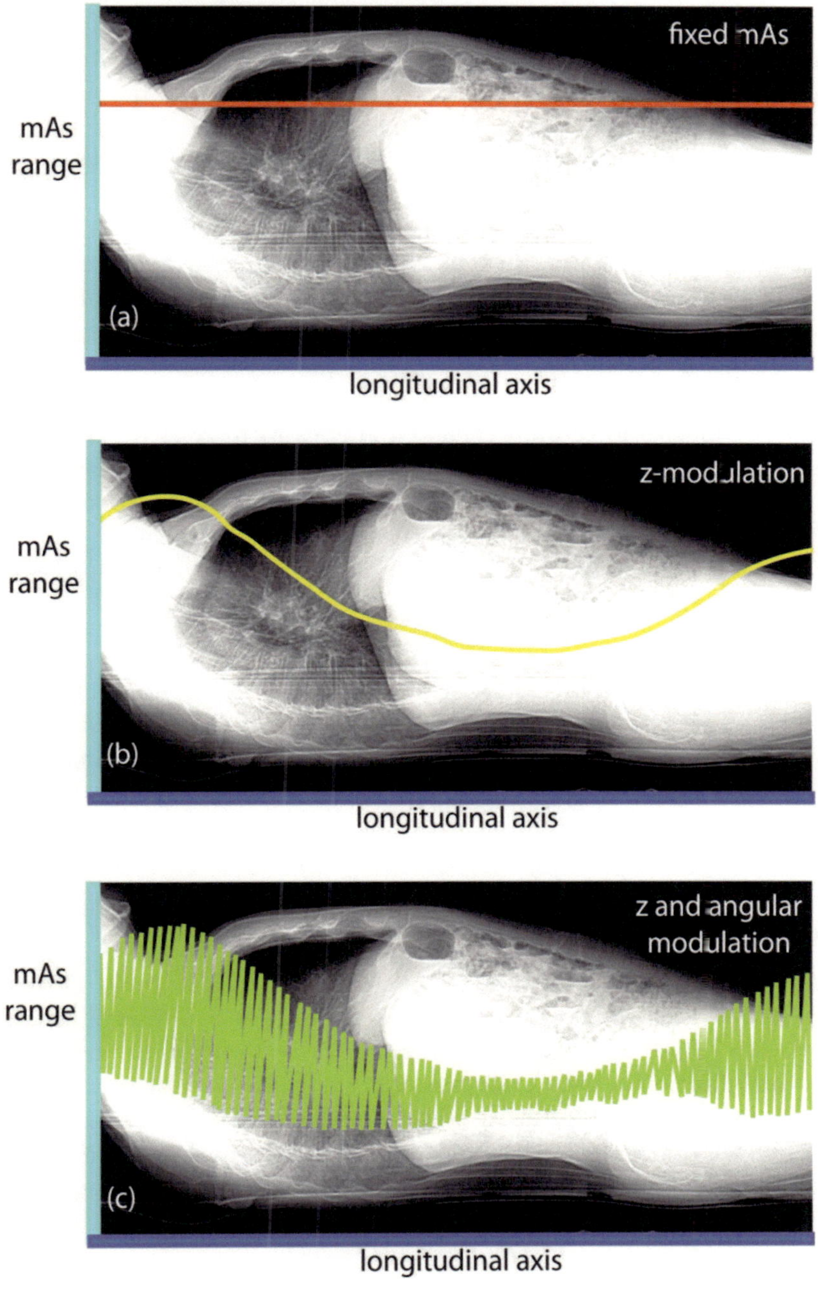

Fig. 5. Diagrammatic illustration of differences between fixed mA (a), z-modulation (b) and combined z and angular modulation (c). Note that the use of fixed mAs means that mA is too high for most parts of the scan. Alternatively, solely relying on z-axis modulation can also mean that tube currents are higher than fixed mA at the edges of the scan where the tube current is increased to account for shoulder girdles and bony pelvis (which inherently have higher attenuation than body scan).
(images are authors' own reproduction, modified from Hricak et al, 2011 with permission)

Tube voltage

There have been various studies looking at the adjustment of tube potential. The most reliable determinant when adjusting the tube voltage at present is the patient's BMI but factors such as fat distribution (e.g. around the thorax) should also be considered. The traditional setting for voltage has been at 120kV. Diagnostic images have been shown to be possible at 100 or even at 80kV (Hausleiter et al, 2010). This can be done according to the patient's size with published reports demonstrating up to 70% reduction in chest scans and 40% reduction in abdominal scans (Yu et al, 2010) by lowering the kV in selected groups to 100kV from 120kV. Lowering the voltage decreases penetration (therefore less useful in patients with high BMI) but also results in increased subject contrast as it approximates towards the k-edge of iodine (33 keV). This means that there is superior enhancement of iodine at lower potentials. As a result, there is improved conspicuity of hypervascular or hypovascular pathologies in contrast-enhanced studies as there is improved contrast-to-noise ratio. There is a trade-off in that there is increase image noise mainly due to higher absorption of low-energy photons by the patient. Therefore, this only works best in patients of smaller sizes and the improved contrast-to-noise is negated in larger patients. Dose is proportional to the current but to the square of tube voltage and so it remains more advantageous to lower the kV than the mAs.

Cardiac CT

Various innovations related to cardiac CT have markedly reduced the dose of this traditionally high dose examination and are worth discussing in more detail (Roobottom et al, 2010). The 2 main methods of data acquisitions in cardiac CT angiographic studies are inherently linked to the cardiac cycle (ECG-gating). These are retrospective and prospective gating. The former tends to be utilised when the heart rate is greater than 65 beats per minute (bpm) or irregular heart rate and the latter with regular heart rate below 65 bpm. Radiation reduction strategies for these are as follows.

ECG-linked tube current modulation

Traditionally, tube current is applied throughout the cardiac cycle in retrospective acquisition. But since the coronary arteries are best image at end diastole where they are most still (usually 75% of the RR interval), the tube current only needs to be maximum at this point. At all other points, the tube currents can be reduced (see Fig. 6 below). Phase data is still available for the rest of the cardiac cycle (although image will be affected by quantum mottle) and functional data can still be obtained. The traditional ECG-gated retrospective spiral acquisition, though results in good image quality, the dose still poses concerns. Thus sequential scanning as an alternative has been developed (Schoenhagen P, 2008).

Prospective axial gating

Axial acquisition (rather than spirally) on a pre-defined block (e.g. 4cm) can be used - so called 'step and shoot' technique (Stolzmann et al, 2008) - is a more radiation-efficient technique in that the tube current is only applied in that block and is zero outside that pre-determined window (see Fig 7, below). This results in a dramatic reduction in dose to a level where 5mSv and below can be regularly achieved. The current reconstruction technique limitation, however, determines that the heart rate needs to be below 65bpm. Therefore aggressive beta blockade should be utilised where possible to achieve this (unless there are contraindications). This technique is also prone to step artefacts. The heart rate also needs to be regular but the use of more 'padding' with centering of the acquisition to include end-systolic (35%) to end-diastolic (85%) phases can combat this to a certain extent. This also allows multi-phase analysis. Expanding phase data, however, increases dose penalty but still remains advantageous as this is still less than the use of retrospective spiral scanning, even with aggressive dose modulation.

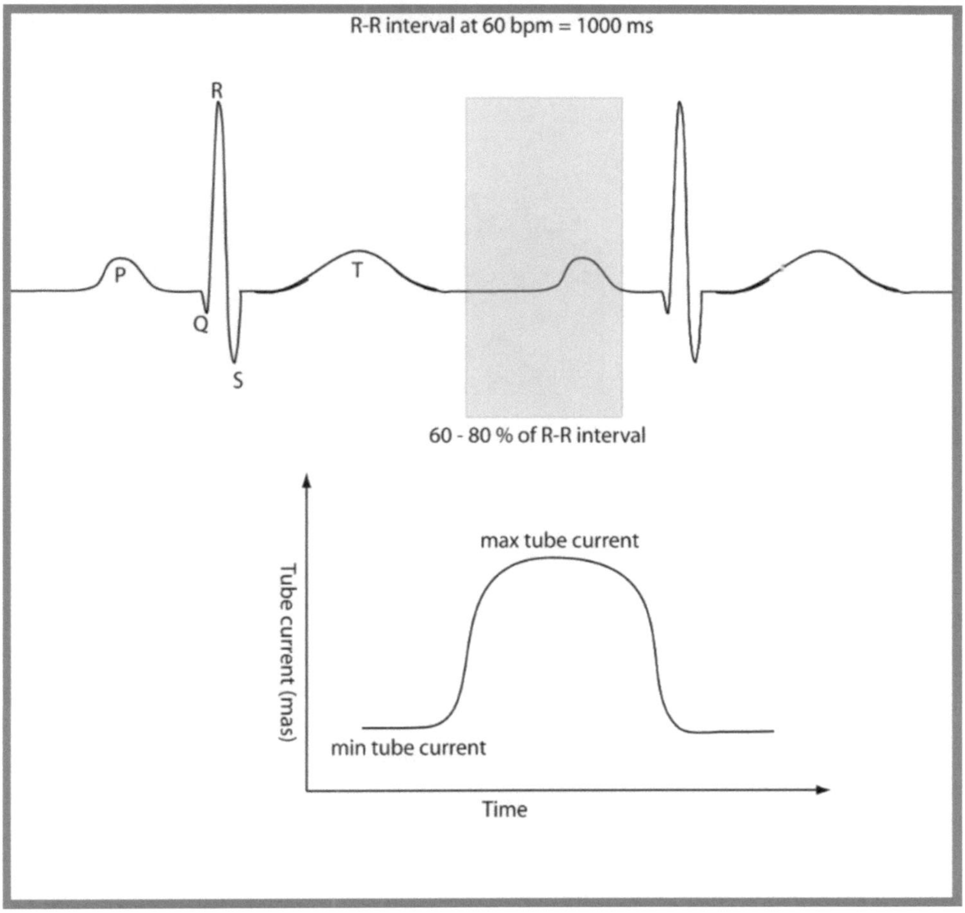

Fig. 6. ECG-linked dose modulation with maximum tube current during diastolic phase and minimum for the rest of the cardiac cycle.
(images are authors' own reproduction)

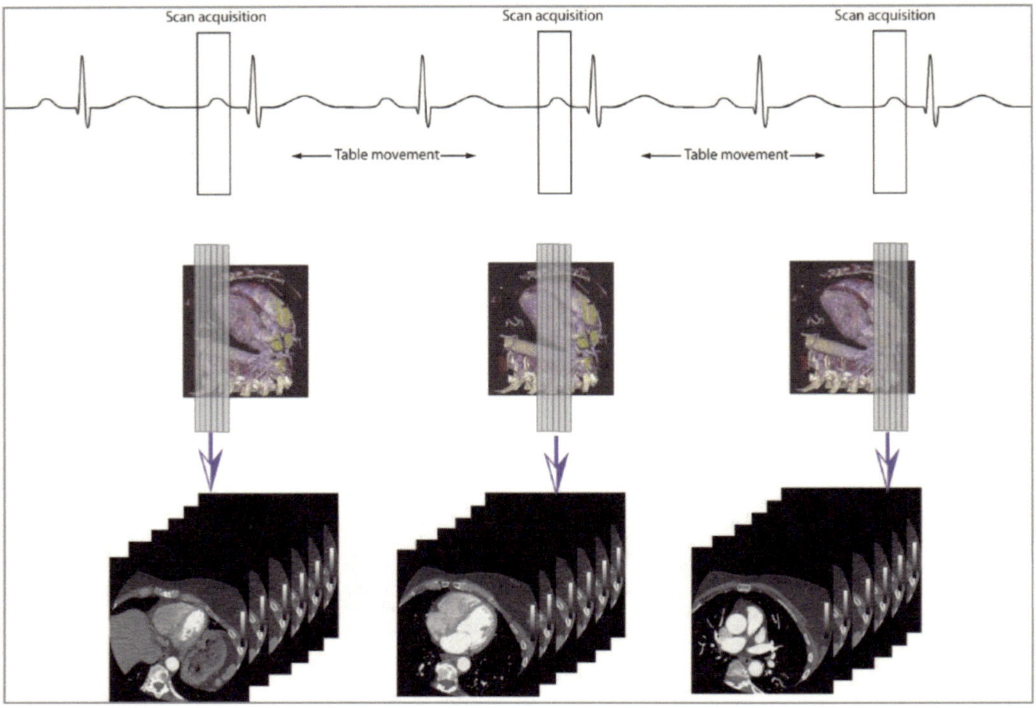

Fig. 7. The "step and shoot" technique with sequential axial scans covering the entire heart. (images are authors' own reproduction)

Scan length

Scan length should be kept at a minimum to lower the dose. In general CT, this is done by the operator from the scout view obtained on initial scanning. In certain studies, increase coverage is required but this is difficult to ascertain on the scanned projection radiograph (i.e. scanogram, scoutview or topogram). For example, increase coverage is needed in patients with vein grafts (e.g. internal mammary artery grafts require coverage from lung apex). At our institution, we performed the initial low dose unenhanced scan to look for coronary calcifications which also allows accurate delineation of the coronary arteries. The added information means that we do not have to perform overcoverage (and therefore eliminate unnecessary radiation) and we now routinely perform this at our centre (Roobottom et al, 2010).

High pitch spiral acquisition

High pitch ECG-gated acquisition using dual source CT scanner is sufficiently fast enough so that in a patient that has a slow heart rate, images of the whole heart can be acquired within a single heart beat. In particular, cardiac imaging have shown great efforts in dose reduction where a dose reduction of up to 90% has been achieved using novel acquisition techniques (Flohr et al, 2009 and McCollough et al, 2009).

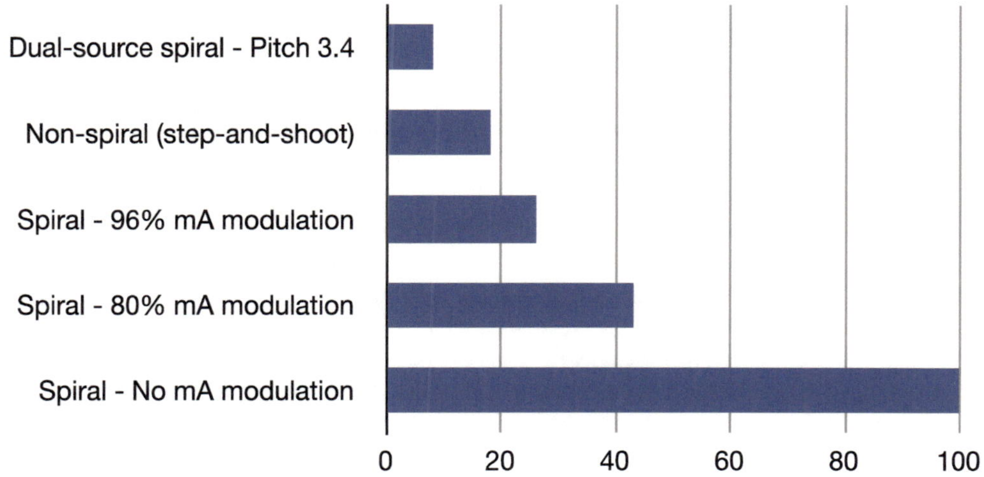

Fig. 8. Graphical illustration showing dose reduction in different techniques of cardiac CT angiography.
(modified from Hricak et al, 2011, with permission)

Displacement and shielding methods

There are several types of selective organ shielding available since 1997 (Hooper et al, 1997) with various studies focusing on the use of shields for the eyes, thyroid breasts and gonads. The evidence shows that they reduce dose (especially surface organ dose), however, whether they affect image quality is still a hugely contentious issue.

Breast shielding has been the most studied shielding method and is worth a more extensive mention for this reason. These are made of latex sheets consisting of bismuth and have the property of attenuating x-ray beam in the energy range used in CT. This can be up to about 50% depending on the tube voltage or filtration. The use of bismuth breast shield does not affect the mean energy of the x-ray beam, but reduces the overall intensity. There is fairly consistent agreement throughout the literature on its effectiveness at reducing radiation dose. Fricke et al (2003) found that bismuth breast shields for paediatric CT reduce dose by 29% in anthropomorphic phantoms. Coursey et al (2008) reported a dose reduction of 52% when the bismuth breast shield was employed after the acquisition of scout view using the automatic tube current modulation in a chest MDCT protocol. Foley et al (2011) showed that the use of breast displacement and lead shielding in a group of women undergoing cardiac CT for investigation of chest pain lead to significant reduction in mean breast surface dose without affecting coronary image quality. This varies with breast size but the mean reduction was around 24% of breast surface dose.

Other organ specific shields have also been studied. Mukundan et al (2007) reported a 42% lens dose reduction when using a 2-ply bismuth shield for axial brain and helical craniofacial paediatric CT protocols. Perisinakis et al (2005) also investigated the orbital bismuth shielding technique in children and found the average eye dose reduction of 38% and 33% for CT scans of the orbit and whole head in their Monte Carlo simulations as well as 34% and 20% for the entire and partial eye globe scans in paediatric patients, respectively.

However, despite some evidence supporting the use of shielding, routine use of these shields has been called into question. Geleijns et al (2006) and Vollmar et al (2008) reported that although dose reduction can be achieved with the use of shields, their use is also associated with significant artefacts and increase in image noise. Their findings were supported by Kalra et al (2009) indicating that as well as increasing noise, there is artifactually increased in attenuation values in the region immediately behind the shields. More recently, Lee et al (2011) showed that the use of thyroid shielding with cotton wool spacer reduces the dose to thyroid by up to 27% without affecting image noise although it was noted that there was noticeably increase in the attenuation values of the superficial neck structures such as the neck muscles. It was thought that this was likely related to a metal artefact caused by bismuth implanted within the shield itself.

Some of the streak artefacts in earlier studies can partly be explained by close apposition of the shields to the skin surface. This was the case for both Geleijns et al (2006) and Vollmar et al (2008). When a spacer or plastic shields are used the streak artefacts are noticeably eliminated (Hohl et al 2006, Karla et al, 2009 and Lee et al, 2011). However, the increase in attenuation appears to be a real issue (Karla et al, 2009, Lee et al, 2011). This could have implications when looking at coronary artery calcifications, renal cyst and adrenal mass characterisation, for example, where increase in attenuation is used specifically to characterise disease entities.

Another contentious issue concerns image quality. Although some of the studies have examined the effect on image quality, the robustness of these assessments has also been called into question. In particular, the image quality was not quantified objectively but rather only in qualitative terms with statements such as "we did not see any differences in quality between the shielded and unshielded lung" (Yilmaz et al, 2007). Some investigators have argued that when noise and image quality were analysed objectively and in a robust manner, studies appear to suggest that there is increase in image noise and deterioration of image quality (Geleijns et al, 2006, Vollmarr et al, 2008, Karla et al, 2009). Although, it must be noted that a more recent study by Lee et al (2011) looking at bismuth thyroid shield did not show significant differences in mean noise values. In general, there needs to be more evidence that robustly assess image quality prior to the routine use of shielding to be universally accepted.

Another issue concerns the effect on image reconstruction and wasted radiation. Some investigators argue that bismuth shielding has very different effects on patient dose for the frontal (AP) projection compared to the dorsal (PA) projection (Geleijns et al, 2010). The attenuation of the incoming and exiting beam ultimately has an effect of image quality and dose efficiency. It is argued that if the x-ray tube was in the dorsal position, the exiting x-ray beam is attenuated prior to reaching the detector and therefore this wastes unnecessary radiation, which may have been useful for image formation process. They argue that if the aim of bismuth shield is to reduce low energy photons that mainly deposit their energies on the surface of the patients, then this can be similarly achieved with lowering the tube current without the added artefacts.

The use of bismuth shielding in combination with automated tube current modulation has also been investigated. This is also fraught with danger as there is the potential for either a dose increase to the patient or unequal noise levels within/between images. The reason for this is that the AEC system may attempt to increase the dose to account for the extra patient attenuation. Some of this can be eliminated by placing the shields after the scanned projection radiograph (i.e. scanogram, scoutview or topogram), however, some manufacturer system performs continual monitoring of patient attenuation and adapts tube current in real time (e.g. when angular modulation is used - see earlier section). If this is the case, then the use of shielding can be detrimental and would actually increase the dose to the patient.

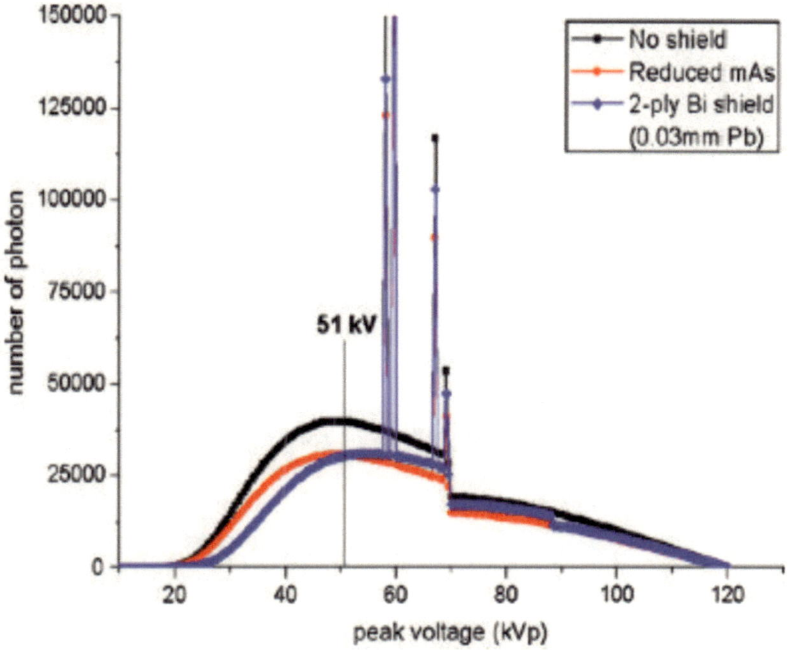

Fig. 9. Comparison of 120 kVp CT beam x-ray spectra with no shield, reduced mAs or 2-ply Bi shield. Note that the reduction of the low-energy photons is substantial in the 2-ply bismuth shield compared to the reduced tube current.
(with permission from Kim et al, 2010)

Patient factors

Increasingly, we are moving towards tailoring our examination according to individual patient to optimise image quality and minimising dose. Previous discussion regarding tube current modulation have already shed light on the benefits of this. In general, factors such as age, chest circumference, body mass index, and specific individual factors such as presence of stents and coronary calcification may influence the type of study being performed as well as dose limitation.

Patient scan length

Patient scan length is also worthy of consideration. Larger length scans produce larger dose. In the PROTECTION I trial, the average dose for a cardiac CT angiography was 12mSv with the average scan volume of 12cm (Hausleiter et al, 2006)). For cardiac scanning (at least for retrospective scanning), an increase in 1cm results in an increase of approximately 1mSv of added radiation. The volume of acquisition must therefore be tailored to suit each individual patient. Both radiologists and radiographers will need to have active role in selecting this.

Special patient groups

Paediatric patients

Usual dose reduction strategies such as tube current modulation, ECG-gating, etc as is used in adults can be similarly applied to children with further dose reduction. Adjustment of kV to 80kv is not an unreasonable approach as well as changing mAs values by employing weight-based specific protocols (Ben Saad et al, 2009, Lee et al, 2006, Young et al 2011).

Women's imaging and pregnancy

Efforts in dose reduction in pregnant patients owing to dose concerns to the foetus have previously been to not perform the study unless it is absolutely paramount. If this was deemed necessary, then various approaches have been adopted but mostly revolve around the use of shielding (see above). Some specific adjustment of protocols have been tried for certain clinical scenario such as for investigation of pulmonary embolism. Litmanovich et al (2009) compared reduced-dose pulmonary CT angiography (200 mA and 100 kV) with matched control group standard protocol (400 mA, 120 kv). The CT dose index, dose-length product, effective dose, image quality, and signal-to-noise ratio were assessed. There was a significant dose reduction of more than 65% using low dose protocols while maintaining diagnostic imaging quality. Though the dose to the chest has been substantially reduced, the dose to the foetus due to scatter radiation still poses concerns. Others have adopted a more simple approach to tackle this. Danova et al (2010) used lead aprons as shielding to the uterus when scanning thoracic CT achieving up to 34% reduction using the wrap around apron to cover for scatter radiation demonstrating that protective aprons are an effective dose reduction technique without additional costs and little effect on patient examination time.

Iterative reconstruction

Out of all the dose reduction strategies discussed, iterative reconstruction shows the most promise. CT workstations have used filtered back projection as the preferred method for producing images from the raw data acquired by the receptors. Filtered Back-Projection (FBP) algorithm has been used as the foundation of commercially available CT reconstruction techniques since the 1970's. Comparatively, it is robust and relatively undemanding on computer processing. It is still widely used today and considered acceptable for clinical diagnosis, but it does not provide optimal results for depiction of the patient as it makes many incorrect assumptions about the data. This is apparent in the inherent level of artefacts and noise in FBP images. To compensate for such noise, larger patient doses are required to overcome this.

Iterative reconstruction algorithms have been put forward as promising recent advances in CT technology but were in fact initially proposed by Shepp and Vardi back in 1982. Only recently, this has been shown to be superior to filtered back projection algorithms for noise reduction (Liu et al, 2007). Even though iterative reconstruction algorithms exhibit great advantages in situations of low signal to noise ratio, their use in real-time CT was previously limited by the time and computing power required to perform the iterations. Recently, however, due to improving computing technology, it is now possible to utilise various facets of iterative reconstruction to reduce the noise and thus achieving significant dose reduction.

- Adaptive Statistical Iterative Reconstruction (ASIR)

Adaptive statistical iterative reconstruction (ASIR) is a post-processing method marketed by GE (General Electric Medical systems, WI, USA) where images are obtained by applying adaptive statistical iterative reconstruction to filtered back projection images. The images are obtained in a low dose mode and the noise in the image is then reduced by applying ASIR. The amount of noise reduction applied can be varied from 1% to 100%. This technique allows modest (up to 40%) dose reduction with similar recorded levels of noise in the image. Several manufacturers have subsequently released similar technology and now all manufacturers are offering iterative reconstruction methods and a means for improved image quality and dose reduction.

There is increasing amount of research to suggest that utilising iterative reconstruction improves image quality, and thereby allows for lowering of kV and mAs and thereby reducing the dose (Prakash et al, 2010). Marin et al (2010) utilized ASIR at low kV and high mAs setting comparing with standard of care FBP technique scanning hepatic organs and found that there is improvement in both image quality and reduction in dose. Similarly, Singh et al (2010) found that abdominal scanning using ASIR compared with FBP yields significant benefit in terms of improved image quality. Due to improved contrast-to-noise ratio, lower radiation parameters can be utilised thereby achieving significant dose reduction. Similar results have been shown using chest CT in comparing ASIR with standard FBP. Leipsic et al (2011) showed that compared with FBP images, ASIR images had significantly higher subjective image quality, less image noise, and less radiation dose with around 30% reduction on average. Studies that have high dose burden such as CT colonography also shows much promise with the use of new ASIR technique. Flicek et al (2010) utilised ASIR in a pilot study using 18 patients undergoing CT colonography using an altered protocol of standard scan 50mAs (supine scan) and 25mAs (prone scan with 40% ASIR). The results show that the radiation dose can be reduced 50% below currently accepted low-dose techniques without significantly affecting image quality when ASIR is used. More recent investigations appear to confirm that using ASIR yields benefits in achieving dose reduction as well as improved image quality. Pontana et al (2011a) studied the utility of iterative reconstruction algorithm in comparison to FBP on 80 patients and found that iterative reconstructions provided similar image quality compared with the conventionally used FBP reconstruction at 35% less dose and also suggested that even higher dose reductions than 35% may be feasible by using higher levels of iterations. Significant noise reduction can also be achieved using the same dose/raw data (Pontana et al 2011b). More and more evidence are appearing in the literature for specific uses of ASIR in specific clinical setting (e.g. CT enterography in Crohn's patients - Kambadakone et al 2011; coronary CT angiography - Leipsic et al 2011) further emphasising its increasing usefulness in quest for improved image quality and dose reduction.

- Model-Based Iterative Reconstruction (MBIR)

Compared with ASIR, newer method so called 'Model-Based Iterative Reconstruction (MBIR)' has now been developed and instead of relying on a single model (as is used in ASIR), performs multiple iterations from multiple models. These models account for a complete three-dimensional assumptions that comprises of focal spot, beam shape, voxel size, and size of detector. In addition, MBIR also accounts for noise from photon flux as well as system noise from the scanner itself.

There is currently no literature research on the practical applications on MBIR as this has only recently been commercialised in December 2010. Preliminary work by the authors on phantoms (unpublished data) have revealed that MBIR shows the most reduction in terms of noise and is superior in terms of objective and subjective image quality and diagnostic confidence compared with ASIR and FBP technique. Dose can be further reduced by up to 80%. Studies are underway to see how image quality compares with traditional methods of FBP and ASIR. From the preliminary work, there is a distinct possibility of achieving body scanning at under 1mSv thus paving a way for significant dose reduction of up to 80% of current levels.

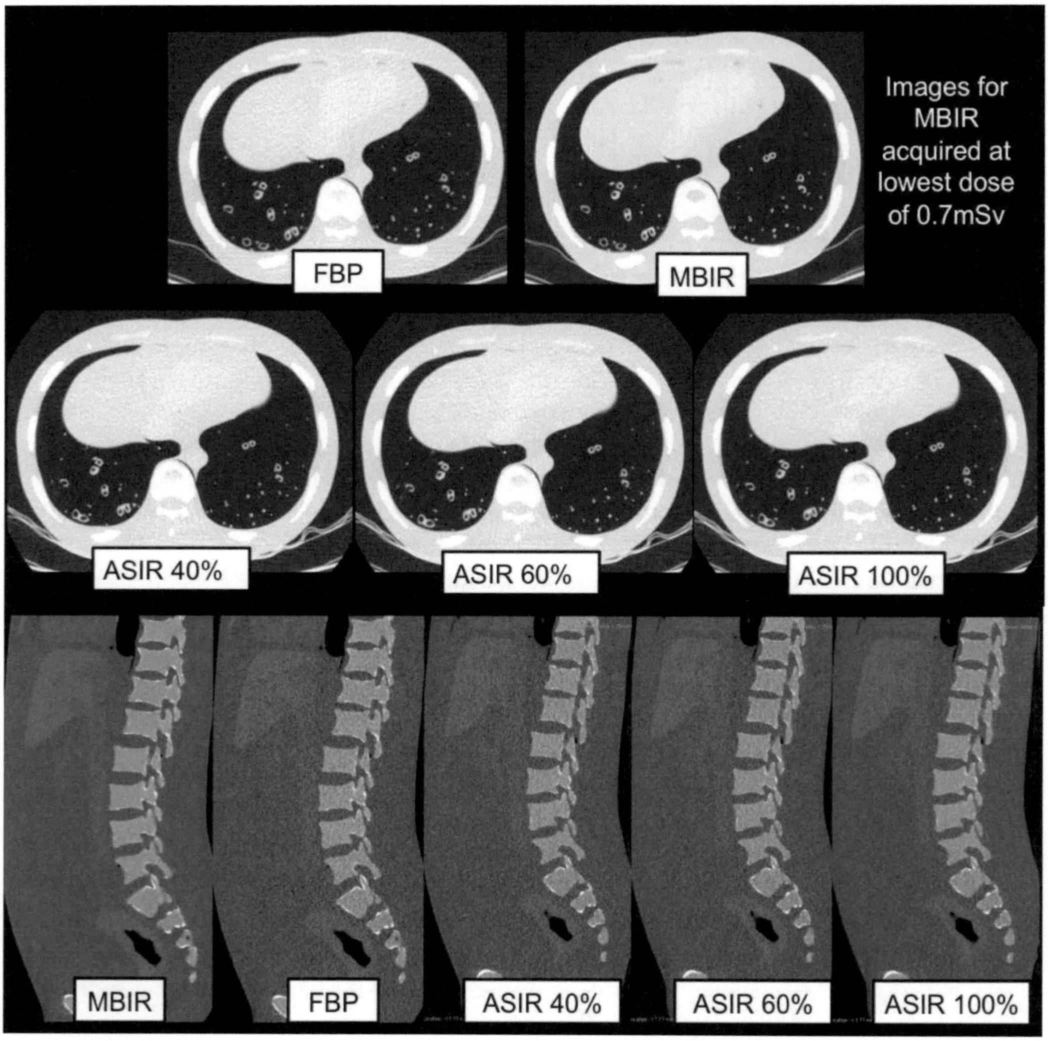

Fig. 10. Side-by-side comparison of scanned torso phantom showing different image quality between traditional filtered back projection, ASIR and MBIR. Note that despite MBIR being acquired at very low dose, this still shows remarkable superior image quality.
(images are authors' own work)

Garnet technology

Recently, Garnet-based detectors have been developed which are 100 times faster and have 25% less afterglow compared to the traditional GOS (Gd2O2S)-based detectors. Reduced inherent noise as well as greater contrast resolution should allow for lower voltage/current techniques even for patients with high BMI. Moreover, due to greater contrast resolution, it is hoped that this might aid more accurate detection of in-stent restenosis in cardiac CT angiography - something that is difficult to assess accurately previously due to streak artefact (Cademartiri et al, 2007; Haraldsdottir et al 2010).

320-Row detector

With new 320-row detectors now available, axial volumetric scanning of a 16cm segment range in a single 0.35s rotation with an acquisition configuration of 320x0.5mm is now possible. This has the advantage of preventing overranging, and offers high spatial and temporal resolution. This has the potential to reduce the dose considerably. It has been determined that axial volumetric scanning constitutes a dose saving of up to 55% (Kroft et al, 2010; Al-Mousily et al 2011).

5. Dose reduction in fluoroscopy

At the turn of the century, there is little awareness of the risk of associated with fluoroscopic procedures, and in particular fluoroscopically-guided intervention. This has in some parts been due to the lack of adequate measuring equipment, complex dose relationships, and false sense of perceived lack of risk. The mood has gradually shifted since 2 key publications in 2004 calling for better dose management for fluoroscopically-guided interventions (Hirshfeld Jr et al 2004; Miller et al, 2004). Since then the FDA in the US have included some of the recommended requirements into new safety-related regulations for manufacturers, and these include features such as last image hold, display of cumulative exposure time, cumulative air kerma, and real-time display of the air kerma dose rate.
Dose delivery in fluoroscopic procedure is complex interactions of numerous factors. There are some excellent review articles (Hirshfeld Jr et al 2004; Steckler et al, 2009) but in-depth discussion is beyond the scope of this chapter. The issue of adequate dose monitoring and effective dose estimations have previously been called to question. The latest dose management technology, which is a feature that is made available on all angiographic equipment in the US, is 'cumulative dose at reference point'. This actually refers to the cumulative 'free-in-air' air kerma and quantify total amount of radiation delivered to a specific point located a fixed distance from the x-ray source. Although, actual dose to patient is a complex, procedure-specific, and varies between operators and is dynamic process (patient's movement during procedure will affect dose received for example), the cumulative dose at reference point has been shown to correlate reasonably well with the absorbed dose at a specific skin site (Miller et al, 2003).

6. Dose reduction efforts in nuclear medicine

In the US, between 1972 and 2005, diagnostic nuclear medicine procedures increased 5- to 6-fold whereas the U.S. population increased by approximately 50%. Between 1996-2005, there was 5% annual growth in the number of nuclear medicine procedures while the growth of the U.S. population has been less than 1% annually. Between 1982 and 2005, the estimated per capita effective dose from in vivo diagnostic nuclear medicine increased by 550% and the collective effective dose increased by 720% (Mettler et al, 2008). In fact, the estimated 2005 per capita effective dose from diagnostic nuclear medicine (0.75 mSv) is greater than the total per capita dose from both diagnostic radiology and nuclear medicine examinations combined in 1982 (0.14 and 0.40 mSv, respectively). As might also be suspected, the estimated 2005 collective effective dose from diagnostic nuclear medicine (220,000 person Sv) is greater than the total per capita dose from both diagnostic radiology and nuclear medicine examinations in 1982 (32,600 and 92,000 person Sv, respectively).

Year	Procedures (Millions)	U.S. Population (Millions)	Exams per 1000 Population	Per-Capita Effective Dose (mSv)	Collective Effective Dose (Person-Sv)
1972	3.3	209.9	15.7		
1973	3.5	211.9	15.6		
1974		213.9			
1975	4.8	216.0	22.2		
1976		218.0			
1977		220.2			
1978	6.4	222.5	28.8		
1979		225.0			
1980	(5.8 to 6.4)	226.5			
1981	7.0	229.5	30.5		
1982	7.55 (7.4 to 7.7)	231.6	32.6	0.14	32,100
1983		233.0			
1984	6.3*	235.8	26.7		
1985	6.2*	237.9	26.2		
1986	6.7*	240.1	27.9		
1987	6.8*	242.3	28.1		
1988	7.1*	244.5	29.0		
1989	7.1*	246.8	28.9		
1990		249.7			
1991		252.1			
1992		255.0			
1993		257.7			
1994		260.3			
1995	10.2†	262.8	38.8		
1996	10.5†	260.3	40.3		
1997	10.9†	262.8	41.5		
1998	11.8†	265.2	44.5		
1999	12.6†	272.7	46.2		
2000	13.5†	282.1	47.9		
2001	14.5†	284.8	50.9		
2002	14.9†	287.9	51.8		
2003	15.7†	290.8	54.0		
2004	16.5†	294.7	56.0		
2005	(19) 17.2†	296.0	58.1	0.75	220,500

*FDA 1985 Radiation Experience Data 1980, Survey of U.S. Hospitals, DHEW Pub FDA 86-8253, National Technical Information Service, Springfield, VA.

†Data from IMV Inc Benchmark Report 2005 are patient visits, not procedures.

Table 5. Number of Nuclear Medicine Examinations Performed in the U.S. between 1972-2005.
(with permission from Mettler et al, 2008)

There has been a marked shift in the type of procedures with the studies of the brain and thyroid decreasing from a combined percentage of more than 56% of all procedures in 1973 to less than 4% in 2005. The most dramatic increase occurred in cardiac procedures increasing from 1% in 1973 to 57% in 2005. Cardiac studies are relatively high dose procedures and account for more than 85% of the effective dose to the patient population. Currently, more than 75% of all studies fall into 2 categories- cardiac and bone and these 2 types of examinations account for almost 95% of the collective effective dose. In 1982, the estimated number of nuclear medicine procedures was about 7.5 million. The per-capita effective dose from nuclear medicine was 0.14 mSv and the collective dose was 32,000 person Sv. By 2005, the estimated number of procedures had increased to about 19.6 million. The per-caput effective dose increased to about 0.75 mSv and the collective dose to about 220,000 person Sv.

	1973*		1982*		2005*	
Procedure	Number	(%)	Number	(%)	Number	(%)
Bone	125	(3.6)	1811	(24.5)	3450	(20)
Cardiac	33	(1.0)	950	(12.8)	9800	(57)
Lung	417	(11.9)	1191	(16.1)	740	(4)
Thyroid	460	(13.1)	677	(9.1)	—	(<2)
Renal	122	(3.5)	236	(3.2)	470	(3)
GI	535	(15.2)	1603	(21.7)	1210	(7)
Brain	1510	(43.0)	812	(11.0)	—	(<2)
Infection					380	(2)
Tumor	14	(0.4)	121	(1.6)	340	(2)
Other	294	(8.4)			—	(<2)
Total	3510	(100)	7400	(100)	17200	(100)

*National Council on Radiation Protection and Measurements.[2]
†IMV is for patient visits. Ratio of visits to procedures is about 1.14.

Table 6. Change (in Thousands) and Percentage of Total Examinations from 1972 in diagnostic nuclear medicine examinations.
(with permission from Mettler et al, 2008)

As with CT, dose reduction efforts have focused on improving technology with innovations such as increased sensitivity of single photon emission computed tomographic (SPECT) instrumentation. This is currently the principal modality for nuclear studies of the heart. Improvement in detectors such as the new cadmium-zinc-telluride detectors will facilitate higher speed/resolution gamma camera and reduce noise (Erlandsson et al, 2009). This equates to higher detector sensitivity, and the hope is that this will generally lead to less administered dose requirement over time (Peterson et al, 2011).

7. General notes on issues relating to dose monitoring

Modality-specific metrics are generally used for dose monitoring. These include administered radioactivity for radiopharmaceuticals, entrance skin kerma for x-rays, CT dose index for CT and dose-area product and cumulative air kerma for fluoroscopy. This can be confusing. In an ideal world, radiation safety should be assessed meaningfully across all imaging modalities using a single standard metric for assessment of radiation risk. This should be done not least because it would be more practical for clinicians, but also for ease of risk stratification for patients who are likely to undergo various imaging investigations of different modalities throughout their lifetime. Today, various methods exists, and more are under development to achieve that very goal but at the same time the validity of such endeavour are being questioned as standardised dose specification with a single quantity/unit may not necessarily encompass for both stochastic and deterministic effects. As it stands, no single unifying unit exists for dose monitoring for all medical-related procedure at present.

8. Conclusion

In summary, recent efforts in dose reduction in the medical community has arisen due to increasing awareness of radiation-related risks. As we perform more imaging, the risk to the population becomes more concerning. Trends to this effect show that this is increasing at an almost exponential rate. Fortunately, the medical community has made great strides both in utilising new technologies but also aided by closer examination of current practices. Some of the improvements can be implemented right away without much efforts and costs but rely on adherence to already available guidelines and decision making systems. Others rely on the use of new technologies that should make it easier for clinicians to both monitor accurately the dose delivered to the patient but also aid in dose reduction/optimization. In the future, medical imaging is likely to become intertwined with individualised medicine to such an extent that each patient will receive an appropriate test, at an appropriate dose level and according to his or her specific characteristics. Tailoring an investigation to the individual patient should become the norm rather than a one-size-fits-all solution. Several advances are on the horizon and further progress in computational power, new tube design, novel detectors and advances in medical physics and engineering will likely overcome current limitations which can only be beneficial.

9. References

Al-Mousily F, Shifrin RY, Fricker FJ, et al. (2011). Use of 320-detector computed tomographic angiography for infants and young children with congenital heart disease. *Pediatr Cardiol*, 32:426-32.

Aroua A, Olerud HM et al (2010). Collective doses from medical exposures: an inter-comparison of the "Top 20" radiological examinations based on the EC guidelines RP 154. Proceedings of the Third European IRPA Congress, June 2010, Helsinki, Finland.

Bautista, A.B., Burgos, A., Nickel, B.J., Yoon, J.J., Tilara, A.A. & Amorosa, J.K., (2009). Do clinicians use the American College of Radiology Appropriateness criteria in the management of their patients? *American Journal of Roentgenology*, 192(6), p. 1581.

BEIR VII report, (2006). Health Risks from Exposure to Low Levels of Ionizing Radiation: BEIR VII Phase 2, U.S. Nuclear Regulatory Commission. Washington, DC: National Academies Press.

Ben Saad M, Rohnean A, Sigal-Cinqualbre A et al (2009) Evaluation of image quality and radiation dose of thoracic and coronary dual-source CT in 110 infants with congenital heart disease. *Pediatr Radiol* 39:668-676

Berrington de Gonzalez A, Mahesh M, Kim K et al. (2009). Projected Cancer Risks From Computed Tomographic Scans Performed in the United States in 2007. *Arch Intern Med*, 169:2071-2077.

Blachar A, Tal S, Mandel A, Novikov I, Polliack G, Sosna J, Freedman Y, Copel L, Shemer J. (2006). Preauthorization of CT and MRI examinations: assessment of a managed care preauthorization program based on the ACR Appropriateness Criteria and the Royal College of Radiology guidelines. *J Am Coll Radiol*. Nov;3(11):851-9.

Burk Jr, R.J. (1996). Radiation risk in perspective: position statement of the Health Physics Society, *Health Physics Society Website. www. hps. org/documents/risk_ps010-1. pdf. Published January*. Updated in 2004.

Cardis, E., Vrijheid, M., Blettner, M., Gilbert, E., Hakama, M., Hill, C., Howe, G., Kaldor, J., Muirhead, C.R. & Schubauer-Berigan, M. (2007). The 15-Country Collaborative Study of Cancer Risk among Radiation Workers in the Nuclear Industry: estimates of radiation-related cancer risks, *Radiation research*, 167(4), pp. 396-416.

Cademartiri F, Schuijf JD, Pugliese , et al. (2007). Usefulness of 64-slice multislice com- puted tomography coronary angiography to assess in-stent restenosis. *J Am Coll Cardiol*, 49:2204–10.

Chawla SC, Federman N, Zhang D, et al. (2010). Estimated cumulative radiation dose from PET/CT in children with malignancies: a 5-year retrospective review. *Pediatr Radiol.* 40:681-686.

Chen J and Moir D (2010). An estimation of the annual effective dose to the Canadian population from medical CT examinations. *J Radiol Prot*, 30 (2), 131.

Coursey C, Frush DP, Yoshizumi T et al (2008) Pediatric chest MDCT using tube current modulation: effect on radiation dose with breast shielding. *AJR* 190:W54-W61

Danova D, Keil B, Kästner B, Wulff J, Fiebich M, Zink K, Klose KJ, Heverhagen JT. (2010). Reduction of uterus dose in clinical thoracic computed tomography. *Rofo.* Dec;182(12):1091-6.

Devine CE, Mawlawi O. (2010). Radiation safety with positron emission tomography and computed tomography. *Semin Ultrasound CT MR.* Feb;31(1):39-45. Review.

Einstein AJ, Henzlova MJ, Rajagoplana S. (2007). Estimating Risk of Cancer Associated With Radiation Exposure From 64-Slice Computed Tomography Coronary Angiography. *JAMA.* 298(3):317-323

Elliott A. (2009). Issues in medical exposures. *J Radiol Prot.* Jun;29(2A):A107-21.

Erlandsson, K., Kacperski, K., van Gramberg, D. & Hutton, B.F. (2009). Performance evaluation of D-SPECT: a novel SPECT system for nuclear cardiology, *Physics in medicine and biology*, 54, p. 2635.

Flicek, K.T., Hara, A.K., Silva, A.C., Wu, Q., Peter, M.B. & Johnson, C.D. (2010). Reducing the radiation dose for CT colonography using adaptive statistical iterative reconstruction: A pilot study, *AJR*. 195(1), pp. 126-31.

Flint-Richter, P. & Sadetzki, S. (2007). Genetic predisposition for the development of radiation-associated meningioma: an epidemiological study, *The Lancet Oncology*, 8(5), pp. 403-10.

Flohr TG , L eng S, Yu L, et al. (2009). Dual-source spiral CT with pitch up to 3.2 and 75 ms temporal resolution: image reconstruction and assessment of image quality. *Med Phys*, 3 6 (12): 5641 -5 653 .

Foley, S.J., McEntee, M.F., Achenbach, S., Brennan, P.C., Rainford, L.S. & Dodd, J.D. (2011). Breast Surface Radiation Dose During Coronary CT Angiography: Reduction by Breast Displacement and Lead Shielding, *American Journal of Roentgenology*, 197(2), pp. 367-73.

Fricke BL, Donnelly LF, Frush DP et al. (2003). In-plane bismuth breast shields for pediatric CT: effects on radiation dose and image quality using experimental and clinical data. *AJR* 180:407-411

Geleijns, J., Wang, J. & McCollough, C. (2010). The use of breast shielding for dose reduction in pediatric CT: arguments against the proposition, *Pediatric radiology*, pp. 1-4.

Geleijns J, Salvado AM, Veldkamp WJ et al. (2006). Quantitative assessment of selective in-plane shielding of tissues in computed tomography through evaluation of absorbed dose and image quality. *Eur Radiol* 16:2334-2340

Hadley, J.L., Agola, J. & Wong, P. (2006). Potential impact of the American College of Radiology appropriateness criteria on CT for trauma, *American Journal of Roentgenology*, 186(4), p. 937.

Hart, Wall, Hillier, Shrimpton. (2010). HPA-CRCE-012 - frequency and collective dose for medical and dental x-ray examinations in the UK, 2008. Health protection agency December 2010: Available from: http://www.hpa.org.uk/Publications/Radiation/CRCEScientificAndTechnicalReportSeries/HPACRCE012/. Accessed 5 May 2011.

Hausleiter et al. (2010) Image Quality and Radiation Exposure With a Low Tube Voltage Protocol for Coronary CT Angiography Results of the PROTECTION II Trial. *JACC Cardiovasc Imagingvol.* 3 (11) pp. 1113-23

Haraldsdottir et al. (2010) Diagnostic accuracy of 64-slice multidetector CT for detection of in-stent restenosis in an unselected, consecutive patient population. *European Journal of Radiology* vol. 76 (2) pp. 188-94

Hirshfeld JW Jr, Balter S, Brinker JA, Kern MJ, Klein LW, Lindsay BD, Tommaso CL, Tracy CM, Wagner LK, Creager MA, Elnicki M, Hirshfeld JW Jr, Lorell BH, Rodgers GP, Tracy CM, Weitz HH. (2004). American College of Cardiology Foundation; American Heart Association; American College of Physicians. ACCF/AHA/HRS/SCAI clinical competence statement on physician knowledge to optimize patient safety and image quality in fluoroscopically guided invasive cardiovascular procedures. A report of the American College of Cardiology Foundation/American Heart Association/American College of Physicians Task Force on Clinical Competence and Training. *J Am Coll Cardiol*. Dec 7;44(11):2259-82.

Hopper KD, King SH, Lobell ME et al. (1997). The breast: in-plane x-ray protection during diagnostic thoracic CT - shielding with bismuth radioprotective garments. *Radiology* 205:853-858

Hohl C, Wildberger JE, Suss C, et al. (2006). Radiation dose reduction to breast and thyroid during MDCT: effectiveness of an in-plane bismuth shield. *Acta Radiol*; 47:562-567

Hausleiter J, Meyer T, Hadamitzdy et al. (2009). Estimated radiation dose associated with cardiac CT angiography. *JAMA* ;301:500-7.

International Agency for Research on Cancer (IARC)
Web site: www.iarc.fr
IARC Carcinogen Monographs: http://monographs.iarc.fr. Accessed 6 Aug 2011.

Israel, G.M., Cicchiello, L., Brink, J. & Huda, W. (2010). Patient size and radiation exposure in thoracic, pelvic, and abdominal CT examinations performed with automatic exposure control, *American Journal of Roentgenology*, 195(6), p. 1342.

Jansen-van der Weide, M.C., Greuter, M.J.W., Jansen, L., Oosterwijk, J.C., Pijnappel, R.M. & de Bock, G.H. (2010). Exposure to low-dose radiation and the risk of breast cancer among women with a familial or genetic predisposition: a meta-analysis, *European radiology*, pp. 1-10.

Kambadakone, A.R., Chaudhary, N.A., Desai, G.S., Nguyen, D.D., Kulkarni, N.M. & Sahani, D.V. (2011). Low-Dose MDCT and CT Enterography of Patients With Crohn

Disease: Feasibility of Adaptive Statistical Iterative Reconstruction, *American Journal of Roentgenology*, 196(6), p. W743.

Kalra, M.K., Maher, M.M., Toth, T.L., Hamberg, L.M., Blake, M.A., Shepard, J.A. & Saini, S. (2004). Strategies for CT Radiation Dose Optimization1, *Radiology*, 230(3), p. 619. - 2004a

Kalra MK, Maher MM, Kamath RS, et al. (2004). Sixteen-detector row CT of abdomen and pelvis: study for optimization of Z-axis modulation tech-nique performed in 153 patients. *Radiology*; 233:241-249. - 2004b

Kalra MK, Maher MM, Prasad SR, et al. (2003). Correlation of patient weight and cross-sectional dimensions with subjective image quality at standard dose abdominal CT. *Korean J Radiol*; 4:234-238.

Kalra MK, Dang P, Singh S et al. (2009). In-plane shielding for CT: effect of off-centering, automatic exposure control and shield-to-surface distance. *Korean J Radiol* 10: 156-163

Kim, S., Frush, D.P. & Yoshizumi, T.T. (2010). Bismuth shielding in CT: support for use in children, *Pediatric radiology*, pp. 1-5.

Kroft LJ, Roelofs JJ, Geleijins J. (2010). Scan time and patient dose for thoracic imaging in neonates and small children using axial volumetric 320-detector row CT compared to helical 64-, 32-, and 16- detector row CT acquisitions. *Pediatr Radiol* ;40:294-300.

Lee, Y.H., Park, E., Cho, P.K., Seo, H.S., Je, B.K., Suh, S. & Yang, K.S. (2011). Comparative Analysis of Radiation Dose and Image Quality Between Thyroid Shielding and Unshielding During CT Examination of the Neck, *American Journal of Roentgenology*, 196(3), p. 611.

Lee T, Tsai IC, Fu YC. (2006). Using multi-detector row CT in neonates with complex congenital heart disease to replace diagnostic cardiac catheterization for anatomical investigation—initial experiences in technical and clinical feasibility. *Pediatr Radiol* 36:1273-1282

Leipsic, J., Nguyen, G., Brown, J., Sin, D. & Mayo, J.R. (2010). A prospective evaluation of dose reduction and image quality in chest CT using adaptive statistical iterative reconstruction, *AJR. American journal of roentgenology*, 195(5), pp. 1095-9.

Leipsic, J., Heilbron, B.G., and Hague, C. (2011). Iterative reconstruction for coronary CT angiography: finding its way. *Int J Cardiovasc Imaging*, Feb 27. (Epub ahead of print].

Leschka et al. (2008). Low kilovoltage cardiac dual-source CT: attenuation, noise and radiation dose. *Eur Radiol* ;18(9):1809-1817.

Li, J., Udayasankar, U.K., Tang, X., Toth, T.L., Small, W.C. & Kalra, M.K. (2011). Patient Size Compensated Automatic Tube Current Modulation in Multi-detector Row CT of the Abdomen and Pelvis, *Academic radiology*, 18(2), pp. 205-11.

Litmanovich D, Boiselle PM, Bankier AA, Kataoka ML, Pianykh O, Raptopoulos V. (2009). Dose reduction in computed tomographic angiography of pregnant patients with suspected acute pulmonary embolism. *J Comput Assist Tomogr*. Nov-Dec;33(6): 961-6.

Liu YJ, Zhu PP, Chen B, et al. (2007). A new iterative algorithm to reconstruct the refractive index. *Phys Med Biol*, 52:L5 - L13

Marin, D., Nelson, R.C., Schindera, S.T., Richard, S., Youngblood, R.S., Yoshizumi, T.T. & Samei, E. (2010). Low-tube-voltage, high-tube-current multidetector abdominal CT:

improved image quality and decreased radiation dose with adaptive statistical iterative reconstruction algorithm--initial clinical experience, *Radiology*, 254(1), pp. 145-53.

McCollough CH, Bruesewitz MR, Kofler JM. (2006). CT Dose Reduction and Dose Management Tools: Overview of Available Options; *Radiographics* March - April, 26:2: 503 - 512.

McCollough CH. (2005). Automatic exposure control in CT: are we done yet? *Radiology* Dec; 237(3): 755-756.

McCollough C H, Leng S, Schmidt B, Allmendinger T, Eusemann C, Flohr TG. (2009). Use of a pitch value of 3.2 in dual-source cardiac CT angiography: dose performance relative to existing scan modes [abstr]. In: Radiological Society of North America scientific assembly and annual meeting program. Oak Brook, Ill: Radiological Society of North America,; 485-486.

Mettler Jr, F.A., Bhargavan, M., Thomadsen, B.R., Gilley, D.B., Lipoti, J.A., Mahesh, M., McCrohan, J. & Yoshizumi, T.T. (2008). Nuclear medicine exposure in the United States, 2005-2007: preliminary results, *Seminars in nuclear medicine*, 38(5), pp. 384-91.

Mettler FA Jr, Bhargavan M, Faulkner K, et al. (2009). Radiologic and nuclear medicine studies in the United States and worldwide: frequency, radiation dose, and comparison with other radiation sources—1950-2007. *Radiology* 2009 ;2 53 (2): 520-5 31

Miller et al. (2004). Quality improvement guidelines for recording patient radiation dose in the medical record. *J Vasc Interv Radiol,* 15(5):423-429.

Miller DL, Balter S, Cole PE et al. (2003). Radiation doses in interventional radiology procedures: the RAD-IR study. II. Skin dose. *J Vasc Interv Radiol*, 14(8): 977-990.

Monson, R., Cleaver, J., Abrams, H.L., Bingham, E. & Buffler, P.A. (2005). *BEIR VII: health risks from exposure to low levels of ionizing radiation,* Washington DC: National Academies Press.

Muirhead, C.R., O'Hagan, J.A., Haylock, R.G.E., Phillipson, M.A., Willcock, T., Berridge, G.L.C. & Zhang, W. (2009). Mortality and cancer incidence following occupational radiation exposure: third analysis of the National Registry for Radiation Workers, *British journal of cancer*, 100(1), pp. 206-12.

Mukundan S, Wang PI, Frush DP et al. (2007). MOSFET dosimetry for radiation dose assessment of bismuth shielding of the eye in children. *AJR* 188:1648-1650

NCRP (2009). Ionizing radiation exposure of the population of the United States. NCRP Report 160. National Council on Radiation Protection and Measurements, Bethesda MD.

Perisinakis K, Raissaki M, Tzedakis A et al. (2005). Reduction of eye lens radiation dose by orbital bismuth shielding in pediatric patients undergoing CT of the head: a Monte Carlo study. *Med Phys* 32:1024-1030

Peterson, T.E. & Furenlid, L.R. (2011). SPECT detectors: the Anger Camera and beyond, *Physics in medicine and biology*, 56, p. R145.

Preston DL, Ron E, Tokuoka S, et al. (2007). Solid cancer incidence in atomic bomb survivors: 1958-1998. *Radiation research*, 168:1-64.

Pontana, F., Duhamel, A., Pagniez, J., Flohr, T., Faivre, J.B., Hachulla, A.L., Remy, J. & Remy-Jardin, M. (2011). Chest computed tomography using iterative reconstruction vs filtered back projection (Part 2): image quality of low-dose CT examinations in 80 patients, *European radiology*, 21(3), pp. 636-43. A

Pontana, F., Pagniez, J., Flohr, T., Faivre, J.B., Duhamel, A., Remy, J. & Remy-Jardin, M. (2011). Chest computed tomography using iterative reconstruction vs filtered back projection (Part 1): evaluation of image noise reduction in 32 patients, *European radiology*, 21(3), pp. 627-35. B

Prakash, P., Kalra, M.K., Ackman, J.B., Digumarthy, S.R., Hsieh, J., Do, S., Shepard, J.A. & Gilman, M.D. (2010). Diffuse lung disease: CT of the chest with adaptive statistical iterative reconstruction technique, *Radiology*, 256(1), pp. 261-9.

Pflederer T, Rudofsky L, Ropers D et al. (2009). Image quality in a low radiation exposure protocol for retrospectively ECG-gated coronary CT angiography. *AJR AM J Roentgenol*, 192:1045-50.

Preston DL, Ron E ,Tokuoka S, e t al. (2007). Solid cancer incidence in atomic bomb survivors: 1958-1998. *Radiat Res*, 168(1): 1-64 .

Ronckers, C.M., Doody, M.M., Lonstein, J.E., Stovall, M. & Land, C.E. (2008). Multiple diagnostic X-rays for spine deformities and risk of breast cancer, *Cancer Epidemiology Biomarkers & Prevention*, 17(3), p. 605.

Roobottom CA, Mitchell G and Morgan-Hughes G. (2010). Radiation-reduction strategies in cardiac computed tomographic angiography. *Clin Radiol*, vol. 65 (11) pp. 859-67

Schindera ST, Nelson RC, Toth TL, et al. (2008). Effect of patient size on radiation dose for abdominal MDCT with automatic tube current modulation: phantom study. *AJR Am J Roentgenol*, 190:W100-W105.

Schoenhagen P. (2008). Back to the future: coronary CT angiography using prospective ECG triggering. *Eur Heart J* ;29(2):153–4.

Singh, S., Kalra, M.K., Hsieh, J., Licato, P.E., Do, S., Pien, H.H. & Blake, M.A. (2010). Abdominal CT: Comparison of Adaptive Statistical Iterative and Filtered Back Projection Reconstruction Techniques, *Radiology*. Nov;257(2):373-83

Stecker MS, Balter S, Towbin RB, et al. (2009). Guidelines for patient radiation dose management. *J Vasc Interv Radiol*, 20(7 suppl): S263-S273.

Stolzmann P, Leschka S, Scheffel H, et al. (2008). Dual-source CT in step-and-shoot mode: noninvasive coronary angiography with low radiation dose. *Radiology*, 249(1): 71–80.

Savandi, A.S., Demarco, J.J., Cagnon, C.H., Angel, E., Turner, A.C., Cody, D.D., Stevens, D.M., Primak, A.N., McCollough, C.H. & McNitt-Gray, M.F. (2009). Variability of surface and center position radiation dose in MDCT: Monte Carlo simulations using CTDI and anthropomorphic phantoms, *Medical physics*, 36, p. 1025.

Street M, Brady Z, Van Every B, Thomson KR. (2009). Radiation exposure and the justification of computed tomography scanning in an Australian hospital emergency department. *Intern Med J*. Nov;39(11):713-9.

United Nations Scientific Committee on the Effects of Atomic Radiation. (2010). Sources and effects of ionizing radiation. Medical radiation exposures, annex A. 2008 Report to the General Assembly with Annexes. New York, NY: United Nations, 2010.

Vollmar SV, Kalender WA. (2008). Reduction of dose to the female breast in thoracic CT: a comparison of standard-protocol, bismuth-shielded, partial and tube-current-modulated CT examinations. *Eur Radiol* 18:1674-1682.

White CS, Kuo D. (2007). Chest pain in the emergency department: role of multidetector CT. *Radiology*. Dec;245(3):672-81. Review.

Yilmaz MH, Yasar D, Albayram S et al. (2007). Coronary calcium scoring with MDCT: the radiation dose to the breast and the effectiveness of bismuth breast shield. *Eur J Radiol* 61:139- 143.

Young, C., Taylor, A.M. & Owens, C.M. (2011). Paediatric cardiac computed tomography: a review of imaging techniques and radiation dose consideration, *European radiology*, 21(3), pp. 518-29.

Yu L,Li H, Fletcher JG, McCollough CH. (2010). Automatic selection of tube potential for radiation dose reduction in CT: a general strategy. *Med Phys*, 37(1): 234 -2 43.

Zhang, D., Zankl, M., DeMarco, J.J., Cagnon, C.H., Angel, E., Turner, A.C. & McNitt-Gray, M.F. (2009). Reducing radiation dose to selected organs by selecting the tube start angle in MDCT helical scans: a Monte Carlo based study, *Medical physics*, 36(12), pp. 5654-64.

18

Ultrasound Image Fusion: A New Strategy to Reduce X-Ray Exposure During Image Guided Pain Therapies

Michela Zacchino and Fabrizio Calliada
Fondazione IRCCS Policlinico "San Matteo", Radiology Department, Pavia - Piazzale Golgi, Pavia, Italy

1. Introduction

Many pain procedures cannot reliably be performed with a blind technique. Thus, imaging guidance is frequently mandatory, above all when the region of interest is deep and/or difficult to reach. In recent years new imaging techniques have been developed to improve diagnosis and to display greater anatomical details. Both Radiology and Pain Therapy have developed new and more accurate techniques in interventional pain linked to a better understanding of pathophysiology and mechanisms of pain.
There are many important anesthetic blocks performed under ultrasound guidance, but our experience is manly based on pudendal nerve and sacro-iliac joint infiltration.

2. Pudendal nerve

Chronic perineal pain syndrome, due to pudendal nerve impingement, has a specific etiology and taxonomy among other possible pain sources in the pelvic and perineal area. Typically patients present uni or bilateral pain in the perineum, which may be anterior (urogenital), posterior (anal) or mixed, with a history of local treatments failure (proctologic, urologic or gynecologic). Postural nature of the pain, exacerbated, if not entirely provoked, by the seated position, led to a therapeutic strategy based on peri-truncal anesthetic blocks (Robert et al., 1998). Chronic pelvic pain was estimated to affect approximately 15–20% of women aged 18–50 (Mathias et al., 1996). On the other hand, the prevalence in men is near 8% considering urological examinations, but only 1% during primary care consultations (Schaeffer, 2004). The pudendal nerve, a mixed (sensory and motor) nerve, supplies the anus, the urethral sphincters, the pelvic floor and the perineum, furthermore it provides for genital sensitivity. It arises from anterior rami of the second, third, and fourth sacral nerves on the ventral aspect of the piriformis muscle in the pelvic cavity and crosses the gluteal region, passing through the greater ischiatic foramen, into the infrapiriformis canal, accompanied by its artery. It is also surrounded by veins with plexiform appearance (pudendal neurovascular bundle). This bundle courses around the sacrospinous ligament just before the latter's attachment to the ischial spine, enters the perineum through the lesser ischiatic foramen and courses through the ischiorectal fossa and then through the pudendal

(Alcock's) canal. The Alcock's canal is the fascia tunnel formed by the duplication of the obturator internus muscle under the plane of the levator ani muscle on the lateral wall of the ischiorectal fossa. Subsequently, the pudendal nerve splits into three terminal branches: the dorsal nerve of the penis (or clitoris), the inferior rectal nerve, and the perineal nerve, providing the sensory branches to the skin of the penis (or clitoris), the perianal area, and the posterior surface of the scrotum or labia majora. It also innervates the external anal sphincter (inferior rectal nerve) and deep muscles of the urogenital triangle (perineal nerve) (Labat et al., 2008; Lefaucheur et al.,2007). Pudendal nerve impingements are possible in its proximal segment in the pinch between the sacrospinous and sacrotuberous ligaments at the ischial spine and when it crosses the inner border of the sacrotuberous ligament, which is thickened at the beginning of the falciform process. Another possible entrapment site may occur in the Alcock's canal as a result of a thickening of the obturator internus muscle fascia. Finally the pudendal artery may describe perineural curves or constrict the nerve trunk with its collateral branches. The vessels, often tortuous and dilated, narrow nerve's components within the vascular sheath (Robert et al., 1998; Lefaucheur et al., 2007; Labat et al., 1990).

Then, nerve decompression is made with different therapeutic conservative or surgical strategies (Robert et al., 2004; Amarenco et al., 1991). Peripheral nerve blocks approach was first described in 1908 (Benson and Griffis, 2005) and it is actually used by pain therapists. There are many different ways of placing needle: by a fluoroscopic, electroneuromyography (ENMG), computed tomography (CT) or ultrasound (US) guide. Pudendal nerve block is usually made with conventional fluoroscopic guidance with placement of the needle tip near the apex of the iliac crest (Calvillo et al., 2000) but fluoroscopy is unable to visualize the pudendal nerve in the anatomical plane formed by the sacrospinous and sacrotuberous ligament (interligamentous plane). Moreover, this technique exposed to ionizing radiations both patient and physician. CT guidance, first used 1999 (Thoumas et al., 1999), is also well established and documented (Fanucci et al., 2009; Robert et al., 2005; McDonald and Spigos, 2000). Using CT it is easy to recognize ischial spine, sacrospinous and sacrotuberous ligaments and pudendal bundle. Then, we can not only place needle tip in the interligamentous space (between the sacrospinous and sacrotuberous ligaments, as close as possible to the caudal portion of the ischial spine) but it is also possible to inject pudendal nerve at the entrance of Alcock's canal (a scan, at the level of the pubic symphysis, allows to identify pudendal bundle on the medial aspect of the obturator internus) (Fanucci et al., 2009). Despite all these benefits, this technique is performed without real-time visual control, and it leads to risks of unintended puncture of adjacent vessels. Ultrasonography allows the direct visualization of the ischial spine, sacrospinous and sacrotuberous ligament. Moreover color-Doppler improves pudendal artery's visualization. In a feasibility study (Rofaeel et al., 2008), it was shown that the pudendal nerve could be clearly visualized only in 12% of the patients. Pudendal nerve shows a diameter more or less between 4 mm to 6 mm (Mahakkanukrauh et al, 2005; Gruber et al., 2001; O'Bichere et al., 2000). All the structures of this size are hardly detected by US at a depth of 5.2-1.1 cm. Furthermore, the depth of the ischial spine from the cutaneous plane is usually more than 7 cm and in 30–40% of the cases, the pudendal nerve shows anatomical variants making the nerve visualization more difficult, especially when the pudendal nerve has dense or fatty tissue near itself, with a possible failure of the procedure. For all these reasons US is usually combined with intraoperative fluoroscopy with a concordance of the two methods in 82% of the

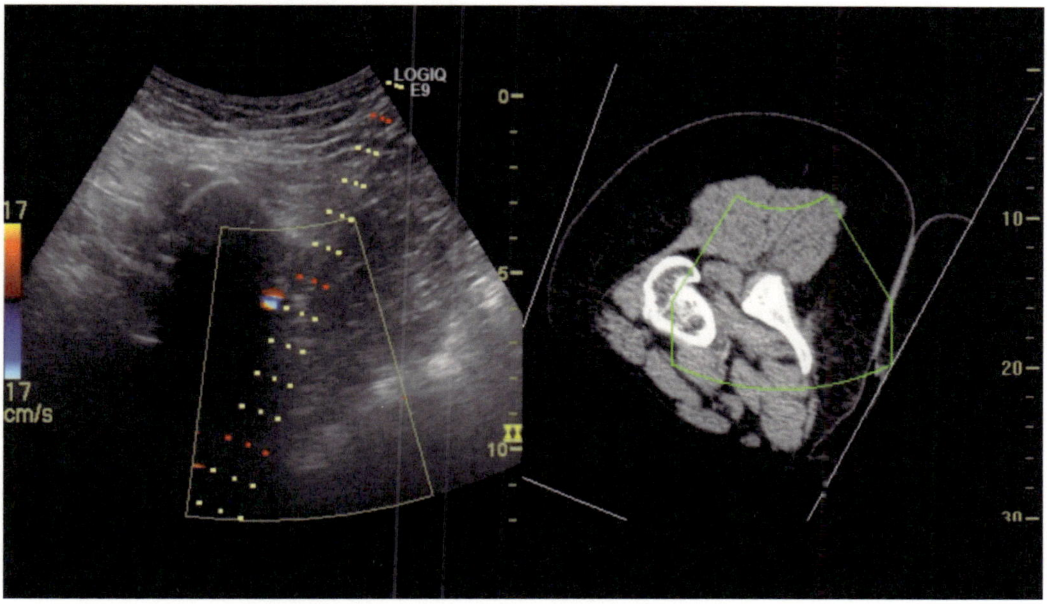

Fig. 1-a. the patient is in prone position. Fusion imaging of left side Alcok canal, pudendal artery (landmark for pudendal nerve) is visible in US side by color Doppler signal medially to ischium bone

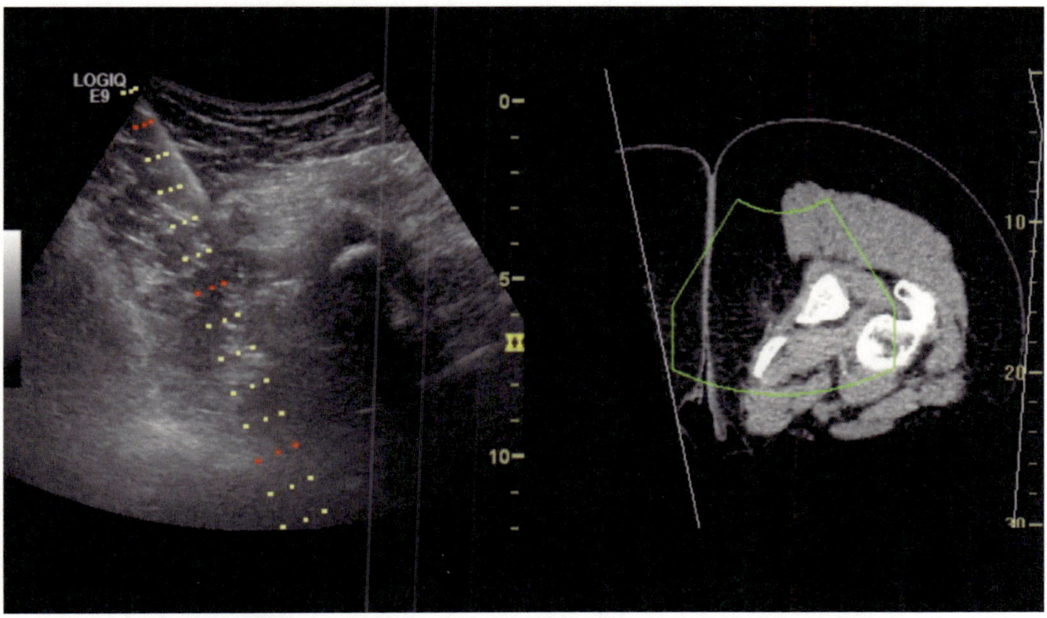

Fig. 1-b. Fusion imaging of right side Alcok canal. The spread of anesthetic solution is visible along the needle tract

procedures (Mahakkanukrauh et al, 2005; Gruber et al., 2001). Some authors describe an easier nerve visualization using a supplementary injection of quiescent solution, as reverse contrast, which should outline nerve's borders (Gray, 2006). Moreover US guidance ensures real-time needle advancement and confirmation of injection spread within the interligamentous plane (Peng and Tumber, 2008). Ultrasound guided block is a simple and reproducible approach avoiding ionizing radiations (Abdi et al; 2004).

3. The sacro-iliac joint

The sacro-iliac joint is a potential source of low back pain and/or buttock pain with or without lower extremity pain. Sacroiliac joint pain may be the result of direct trauma, unidirectional pelvic shear, repetitive and torsional forces, inflammation, or idiopathic onset. The prevalence of sacroiliac joint pain is estimated to range between 10% and 38% with 95% confidence intervals of 0% - 51% (Manchikanti 2009). It's a diarthrodial synovial joint but only the anterior portion and inferior one third of the SI joint is a true synovial joint, because of an absent or rudimentary posterior capsule, the SI ligamentous structure is more extensive dorsally, functioning as a connecting band between the sacrum and ilia. The anterior portion of the sacroiliac joint likely receives its major innervation from posterior rami of the L1-S2 roots but may also receive innervation from the obturator nerve, superior gluteal nerve, and lumbosacral trunk. The posterior aspect of the joint is supplied by the posterior rami of L4-S3, with major contribution from S1 and S2.

For this complex anatomy a clinically guided injection results actually intra-articular only in few cases. Particularly Rosenberg et al. demonstrated a success rate of 22%.

An image guide is required to reach the joint. Since 2000 Dussalt et al proved a fluoroscopy guide as a safe and rapid procedure. They rotated the fluoroscopic tube and reached the posterior aspect of the caudal end of the joint in 97% of the injections. Many authors used a CT guide with subsequent good results.

However radiation exposure under fluoroscopy guided SI joint injection ranges from 12-30 mGy/minute for the skin and 0.1-0.6 mGy/minute for the gonads, and using a CT guidance it may vary from 10 to 30 mGy/minute for the skin. Moreover ionizing radiation exposure leads to stochastic genetic and carcinogenic effects. Some authors have suggested using magnetic resonance (MR) imaging guidance to achieve intra-articular rates of up to 97% (Pereira, 2000). An MRI based procedure was found effective and safe, with the additional advantage to detect bone edema and others inflammatory signs. Even if both CT and MRI imaging are useful technique with an high contrast and spatial resolution, they takes too long with a cumbersome equipment. For all these reasons and to avoid radiation exposure Pekkafali et al. tried to perform SI joint injection using a sonographic guidance. They obtained a successful intra-articular injection rate of 60% for the first half of the procedures and about 93.5%in the last half. Pekkafali, Klauser, Migliore et al. showed that sonographically guided technique may be a valuable alternative in SI joint interventions, safe, rapid and reproducible, with the possibility to detect inflammatory signs using color-Doppler even if the result depends on radiologist's training. However a successful intra-articular injection using a sonographic guide depends not only on the radiologist experience in US performed anesthetic block, but also on the kind of patients and on the disease staging. In fact it's well known that patient 's body mass index influences sonographic scans, like the presence of bone spurs or articular space narrowing, due to the pathology itself.

For all these reasons both these anesthetic procedures requires a new strategy in the guidance of the block itself, to reach the region of interest avoiding radiation exposures, like fusion imaging technique.

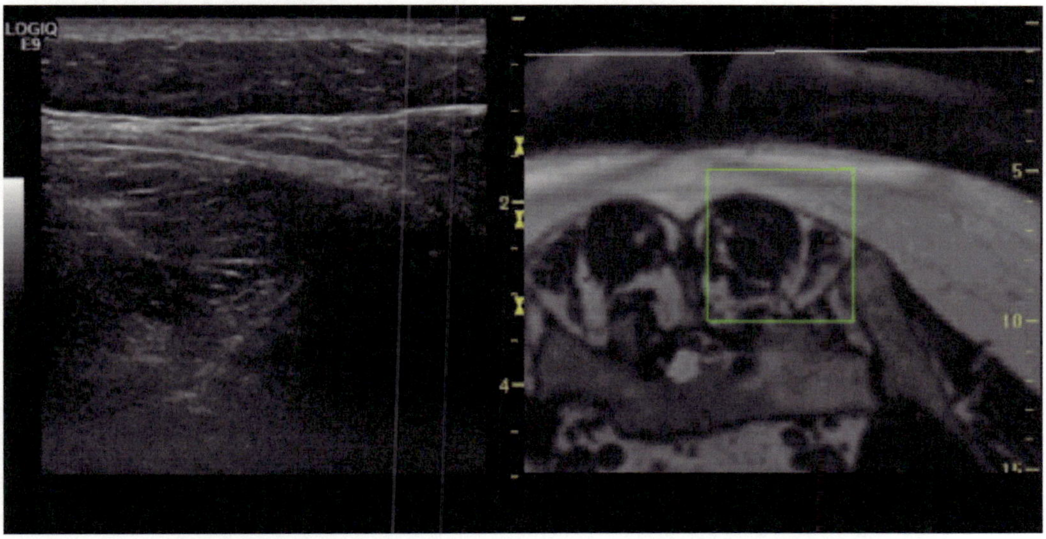

Fig. 2-a. Fusion imaging of right sacroiliac joint at second sacral foramen
Courtesy of Massimo Allegri, Pain Therapy Unit Policlinico S. Matteo Pavia

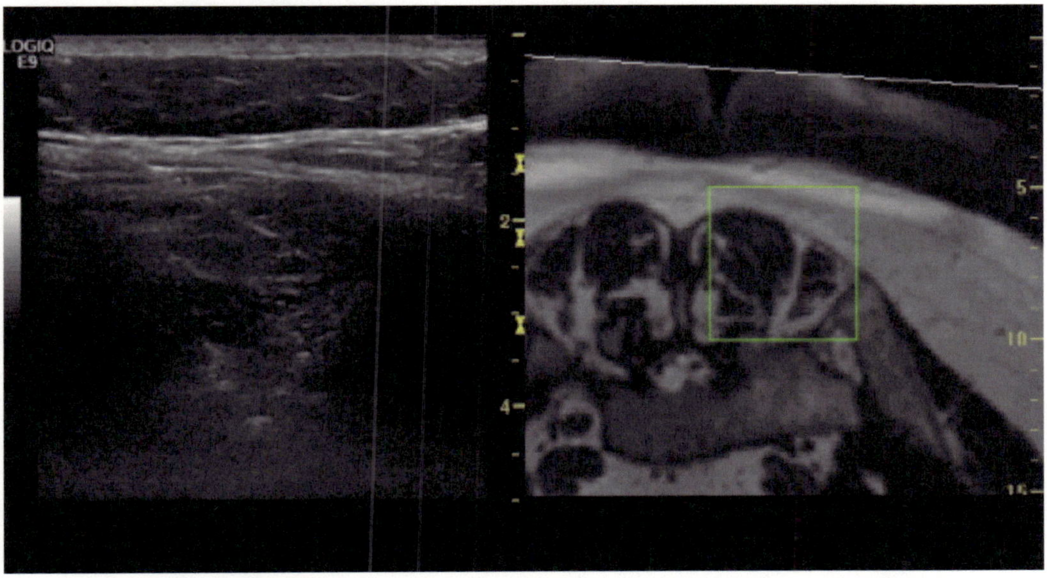

Fig. 2-b. the needle position is easily visible on the US side of Fusion Imaging. The needle tip is approaching the sacroiliac joint
Courtesy of Massimo Allegri, Pain Therapy Unit Policlinico S. Matteo Pavia

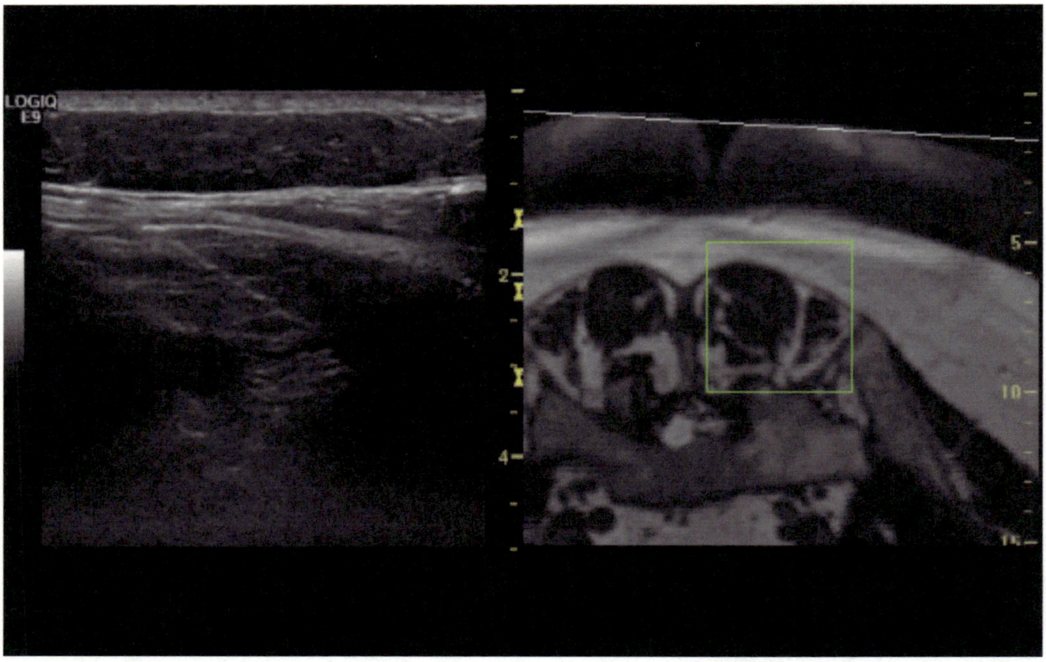

Fig. 2-c. the needle position is easily visible on the US side of Fusion Imaging. The needle tip is approaching the sacroiliac joint
Courtesy of Massimo Allegri, Pain Therapy Unit Policlinico S. Matteo Pavia

4. Fusion imaging

Fusion system can simultaneously show both ultrasound imaging with previous magnetic resonance (MR) and CT study. The basic concept of imaging fusion depends on the idea that an MR/CT study creates a 3D data set that represents the patient's anatomy. The 3D DICOM data set can be stored in any convenient imaging network location. On the other hand, US is the real-time modality, mostly used, that allows to use a feature called "Volume Navigation" to virtually project the patient's previous statics MR/CT data into the space where the clinician is performing a live, real-time, ultrasound. As the operator scans, the ultrasound system shows both the real-time ultrasound and the corresponding slice through the previous MR/CT 3D data. This is a real time reconstruction of an arbitrary slice through the MR/CT 3D data set morphed, with US data into a single unit so that the radiologist or pain therapist can intuitively use both imaging modalities simultaneously to provide diagnosis and treatment guidance. The images are presented in several formats. When presented side by side, the user can see the live ultrasound on one side of the display and the corresponding MR/CT on the other side (split screen image). Similarly when presented side by side the zoom factor of the MR/CT image and US should be varied simultaneously. In this view, the user can see not only the small live field of view (FOV) of ultrasound but also the rest of the MR/CT image that lies outside of the ultrasound scanning area but in the same plane of view. That adds the benefit of the MR/CT's wide FOV to the comparably narrow scope of ultrasound image. The images can also be presented in a mixed mode with MR/CT and US fused data from both modalities. Fusion imaging allows recording CT or

MRI data with US images in real time through an electromagnetic transmitter placed near the patient with two sensors closely bound to the probe. All these components are linked to a GPS system responsible to track the position of the ultrasound probe and to correlate with CT/MRI 3D data set. The critical phase of this procedure is the intermodality registration, involving the precise identification of multiple selected anatomical landmarks of the same patient on the different imaging modalities. Fusion imaging advantages are: 1. Ultrasound suffers from a relatively narrow FOV. MR/CT, by contrast, has an optimal contrast resolution and a wide FOV but lacks real time. Furthermore

- Ultrasound is well suited for many interventional procedures because of its real-time imaging.
- The fusion of CT/MRI and US allows using Color- Doppler or CEUS superimposed with CT/MRI.
- Fusing the images from different modalities could help to avoid multiple exposures to ionizing radiations, improving costs, and quality.
- To date, fusion imaging has been used for many diagnostic and interventional applications.

The first experiences in fusion imaging were carried out on liver imaging, because both biopsy or ablation therapy (radiofrequency) should require an imaging help.

Crocetti's group was the first to test the feasibility of fusion imaging, matching volumetric CT data of the calf livers containing internal targets, which simulated liver lesions, with real time US. The authors tried to reach each target using fusion imaging guidance and finally to perform a radiofrequency ablation. Their protocol consisted of various steps, but the most important is the registration time, during which ultrasound live examination was mixed with the previous CT scan, using radio-opaque markers applied to the calf liver capsule. Every marker was put down with a system of numbering. However during the procedure the authors used others target points (anatomic marker) to confirm the real concordance between both the modalities, such as portal vein. They found an high and consistent level of matching accuracy.

They used very small (1.5 mm) US undetectable target, reproducing a tiny lesion visible only at CT. The navigation system represented therefore the only guidance for the procedures.

However an important limitation of this study was the absence of breath and the absence of a real patient with a subsequent "lesion movement". They suggested solving this problem with the implementation of external electromagnetic position sensors, to patient body and using a breathing motion correction.

Crocetti however proved that fusion imaging could improve liver lesion interventional procedures, with a new possibility for all that lesions that are difficult to see on US B-mode scan.

At the present time, fusion imaging is realized with a new feature called volume navigation technology, which allows to mix US scan with CT/MRI DICOM data, and which uses a magnetic sensor to locate the probe position. Particularly the transmitter is fixed to, or near to, the patient's table as close as possible to the region of interest.

So the registration time becomes very important, like the choice of the target points, and it affects the success of the procedure. There is another important study carried out to analyze the accuracy of fusion imaging between CT and US to guide liver interventional procedures, according to several variables, that can impact the accuracy of this technique in clinical

conditions. Hakime et al. measured the different spatial locations of an established target lesion between virtual CT and US examination in real time.
They used some anatomical intra-hepatic and extra-hepatic landmarks with the help of some non-anatomical landmarks (cysts/calcification). On CT scan, they preferred to use portal phase during which tumor is much more visible such as the portal vein, an important anatomical target point. In this study the authors analyzed also breathing influence, that lead to a global distortion along the three axes, and they suggested to display respiration cycle, that could be helpful in non-anesthetized patients.
They found a greater mismatch for anterior-posterior axis versus the lateral one, maybe due to the pressure applied with the probe.
Another important result of this study is the variation in accuracy when patient was under general anesthesia, or when CT scan was performed several days before of the procedure. Both these conditions lead to more errors during registration time, probably because the patients took a different position with a subsequent distortion of the volume body. Moreover for patients under anesthesia was impossible to repeat the same apnea conditions.
In liver pathology diagnosis plays an important rule too. In fact in literature there are some reports using fusion imaging to optimize liver involvement. For example Jung's group first tested liver lesions characterization, vascularisation and perfusion using fusion imaging. Particularly they used contrast-enhanced US mixed with contrast-enhanced MD-CT in seventeen cases and with contrast-enhanced MRI in three cases. They preferred an arterial phase during the examination of Hepatocellular Carcinoma and of neuroendocrine tumors, a portal phase for the metastasis (above all for colo-rectal ones) and finally, for hemangiomas and focal nodular hyperplasia, the phase in which the tumors could be easily visualized. As target points they used above all vascular structure, both artery and veins. An additional registration was done when a lesion was visualized, particularly providing an adjustment of the lesion size.
They found a better characterization of liver lesions by matching different contrast-enhanced modalities, because this new feature could employ the advantages of the different imaging methods . Using an additional registration they marked even small lesions or they reached also a better characterization in cirrhotic livers, where important liver structure changes, due to the pathology itself, influenced a correct diagnosis. Even if one important limitation of this study was the number of cases, few, to prove fusion imaging, actual, diagnostic accuracy. However they found two others important advantages, such as the possibility to detect a lesion in patients with reduced renal function and to reduce radiation exposure during follow-up.
Stang's group too, evaluated colo-rectal liver metastasis using fusion-imaging technique. They found a better characterization of small hepatic lesions, with a higher rate of correctly classified nodules compared with CT imaging alone.
Fusion imaging was tested also in breast biopsy. MR imaging in fact has an high sensitivity in the detection of breast cancer even if it shows a variable specificity. Above all, there is a large overlap of MRI features of the lesions, during the enhancing phase, which influences patient's management. Moreover MRI guidance for biopsy shows many disadvantages above all regarding cost and time. Many groups of research try to reach MR suspicious with a second look US examination, for all these reasons Rizzatto and Fausto tested the feasibility of fusion guidance in breast imaging.
An important problem is the different position of the patient during US examination and during MRI scan, such as the mobility of breast tissue. In fact usually breast is examined

using clock position. To solve this problem different approaches were proposed, such as the use of algorithms to develope a model of deformation or the use of a redisegned bed that allows to lie in the prone position both during US and MRI examination. In the first case the registration phase in made using target points that are identified between the two images, obtained with different modalities, and then transformed. After US-MRI imaging coregistration, the software reconstructs a real time multiplanar MR image of the corresponding US examination. Particularly Rizzatto and Fausto performed their first MRI examination in the prone position, using precontrast and postcontrast phases, then they obtained another MRI data set with the patient in the supine position by using external fiducial markers at 9-, 12- and 3- o'clock. So they performed US live examination with the patient in the supine position. Using fusion imaging guidance for breast biopsy, the authors reached an important reduction of the time and costs.

About pain therapy Galiano's group first conducted a research in 2007 that analyzed a kind of fusion imaging. Galiano's group tried to improve US visualization of facet vertebral joint to pain therapists unfamiliar to US scan, that however would like to perform medial branch block with US guidance. They used CT reconstruction images to recognize the anatomical structures during real time US examination of the facet joints. Their research was conducted only on cadavers with subsequent good results. In fact, at the end of the study, they encouraged to start to perform anesthetic blocks in real patients under an US guide.

Klauser and Zacchino's groups published the first papers analyzing fusion imaging in pain therapy in 2010.

Klauser tested fusion imaging to guide sacro-iliac joint-infiltration, using ten cadavers and then ten patients. The basic concept of fusion imaging is always the same, so also Klauser's group chose anatomical target points to match US with CT. Particularly the landmarks were spinous process of the fifth lumbar vertebra, posterior superior and inferior iliac spine, first and second posterior sacral foramen and the sacro-iliac joint. They found a positive success rate, with an increasing learning curve in the second half of patients. Moreover CT scan provides further anatomical details about joint assessment, about eventual bone spurs and so on.

In this work too, registration time takes the longest time, but is the most important time of the entire procedure.

Zacchino's group tested the feasibility of pudendal nerve anesthetic block using fusion imaging. In this work US was matched with CT data, using as anatomical target points ischiatic spine, femoral head and coccyx. They performed anesthetic block in the Alcock's canal, which is difficult to reach using an US or CT guidance alone. Fusion imaging provides further anatomical details for this block, such as the possibility to visualize with the same technique both the pudendal vascular bundle, sacrotuberous and sacrospinous ligaments, independently from pudendal nerve depth. After the procedure the patient was immediately and completely pain free.

All these works carried out on different interventional procedures showed that fusion imaging is helpful to reach the region of interest, and that fusion imaging, combing CT spatial resolution and its wide field of view, with US real time, is essential to perform the procedure, avoiding possible complications (unintended puncture of vessels, hematoma, etc). Fusion imaging improves direct visualization of the region of interest matching CT/MRI advantages plus US advantages avoiding radiation exposure during the procedure, that is always important but above all when repeat the procedure is mandatory and in young patients. It avoids also US disadvantages, first of all, its dependence on patient body

mass index. However using fusion imaging too, requires a certain experience that lead to decrease procedure time and it improves the success rate.
Moreover it's important to place the patient in the same position, both during the CT/MRI previous acquisition and during the procedure, as Crocetti proved. Then it's important also to avoid patient's movement both during registration time and during the procedure.
Fusion imaging technique however is now applied to others different kind patology: a new report (Iagnocco 2011) evidences some possibilities in rheumatologic field, where fusion is used to investigate and monitor osteoarthritis and rheumatoid arthritis. This paper showed as fusion imaging should be useful for therapy monitoring matching MRI anatomical landmarks and US details.

5. References

Abdi S, Shenouda P, Patel N, Saini B, Bharat Y, Calvillo O. A novel technique for pudendal nerveblock. Pain Physician 2004;7:319–22.

Amarenco G, Le Cocquen-Amarenco A, Kerdraon J, Lacroix P, Adba MA, Lanoe Y. Perineal neuralgia. Presse Med 1991;20:71–4.

Benson JT, Griffis K. Pudendal neuralgia, a severe pain syndrome. Am J Obstet Gynecol 2005;192:1663–8.

Calvillo O, Skaribas IM, Rockett C. Computed tomography-guided pudendal nerve block. A new diagnostic approach to long-term anoperineal pain: a report of two cases. Reg Anesth Pain Med 2000;25:420–3.

Crocetti L., Lensioni R., De Beni S., Choon See T., Della Pina C., Bartolozzi C. Targeting Liver Lesions for Radiofrequency Ablation An Experimental Feasibility Study Using a CT–US Fusion Imaging System Invest Radiol 2008;43: 33–39

Dussault RG , Kaplan PA , Anderson MW . Fluoroscopy-guided sacroiliac joint injections. Radiology 2000 ; 214 (1): 273 - 277

Fanucci E., Manenti G., Ursone A., Fusco N., Mylonakou L., D'Urso S., Simonetti G. Role of interventional radiology in pudendal neuralgia: a description of techniques and review of the literature. Radiol Med 2009;114:425–36.

Gray AT. Ultrasound-guided regional anesthesia: current state of the art. Anesthesiology 2006;104:368–73.

Gruber H., Kovacs P., Piegger J., Brenner E. New, simple, ultrasound-guided infiltration of the pudendal nerve: topographic basics. Dis Colon Rectum 2001;44:1376–80.

Hakime A. Deschamps F., De Carvalho E. G. M., Teriitehau C., Auperin A., De Baere T. Clinical Evaluation of Spatial Accuracy of a Fusion Imaging Technique Combining Previously Acquired Computed Tomography and Real-Time Ultrasound for Imaging of Liver Metastases Cardiovasc Intervent Radiol 2011; 34:338–344.

Hirooka M, Iuchi H, Kumagi T et al (2006) Virtual sonographic radiofrequency ablation of hepatocellular carcinoma visualized on CT but not on conventional sonography. AJR Am J Roentgenol 186:S255–S260

Iagnocco A. et al. Magnetic resonance and ultrasonography real-time fusion imaging of the hand and wrist in osteoarthritis and rheumatoid arthritis. Rheumatology 2011; 50: 1409-1413

Jung EM, Schreyer AG, Schacherer D, Menzel C, Farkas S, Loss M, Feuerbach S, Zorger N, Fellner C. New real-time image fusion technique for characterization of tumor vascularisation and tumor perfusion of liver tumors with contrastenhanced ultrasound, spiral CT or MRI: first results. Clin Hemorheol Microcirc 2009;43:57–69.

Klauser A.S., De Zordo T., Feuchtner G. M., Djedovic G., Bellmann Weiler R., Faschingbauer R., Shirmer M., Moriggl B. Fusion of real-time US with CT images to guide sacro-iliac joint injection in vitro and in vivo. Radiology 2010; 256 (2): 547-553.

Labat J.J., Riant T., Robert R., Amarenco G., Lefaucheur J.P., Rigaud J. Diagnostic criteria for pudendal neuralgia by pudendal nerve entrapment (Nantes criteria). Neurourol Urodyn 2008;27:306–10.

Labat J.J., Robert R., Bensignor M., Buzelin J.M. Les névralgies du nerf pudendal (honteux interne). Considérations anatomo-cliniques et perspectives thérapeutiques. J Urol (Paris) 1990;96:239–44.

Lefaucheur J.P., Labat J.J., Amarenco G., et al. What is the place of electroneuromyographic studies in the diagnosis and management of pudendal neuralgia related to entrapment syndrome? Neurophysiol Clin 2007;37:223–8.

Mahakkanukrauh P., Surin P., Vaidhayakarn P. Anatomical study of the pudendal nerve adjacent to the sacrospinous ligament. Clin Anat 2005;18:200–5.

Manchikanti L., Boswell M. V., Singh V., Benyamin R. M., Fellows B., Abdi S., Buenaventura R. M., Conn A., Datta S., Derby R., Falco F. JE., Erhart S., Diwan S., Hayek S. M., Helm S. II, Parr A. T., Schultz D.M., Smith H.S., Wolfer L.R., Hirsch J.A. Comprehensive Evidence-Based Guidelines for Interventional Techniques in the Management of Chronic Spinal Pain. Pain Physician 2009; 12:699-802

Mathias S.D., Kuppermann M., Liberman R.F., Lipschutz R.C., Steege J.F. Chronic pelvic pain: prevalence, health-related quality of life, and economic correlates. Obstet Gynecol 1996;87:321–7.

McDonald J.S., Spigos D.G. Computed tomography-guide pudendal block for treatment of pelvic pain due to pudendal neuropathy. Obstet Gynecol 2000;95:306–9

Migliore A., Bizi E., Massafra U., Vacca F., Martin-Martin L.S., Granata M., Tormenta S. A new technical contribution for ultrasound-guided injections of sacro-iliac joints. Eur rev Med Pharmacol Sci 2010; 14(5): 465-9

Minami Y, Chung H, Kudo M et al (2008) Radiofrequency ablation of hepatocellular carcinoma: value of virtual ct sonography with magnetic navigation. AJR Am J Roentgenol 190(6): 335–341

O'Bichere A., Green C., Phillips R.K. New, simple approach for maximal pudendal nerve exposure: anomalies and prospects for functional reconstruction. Dis Colon Rectum 2000;43:956–60.

Pekkafahali MZ, Kiralp MZ, Basekim CC, Silit E, Mutlu H. OzturkE, et al. Sacroiliac joint injections performed with sonographic guidance. J Ultrasound Med 2003; 22:553-9.

Peng PW, Tumber PS. Ultrasound-guided interventional procedures for patients with chronic pelvic pain – a description of techniques and review of literature. Pain Physician 2008;11:215–24.

Pereira PL , Günaydin I , Trübenbach J , et al . Interventional MR imaging for injection of sacroiliac joints in patients with sacroilitis . Am J Roentgenol 2000 ; 175: 265 – 266

Rizzatto G, Fausto A. Breast imaging and volume navigation: MR imaging and ultrasound coregistration. Ultrasound Clin 2009;4:261–71.

Robert R, Bensignor M, Labat JJ, Riant T, Guerineau M, Raoul S, Hamel O, Bord E. Le neurochirurgien face aux algies périnéales: guide pratique. Neurochirurgie 2004;50:533–9.

Robert R, Labat JJ, Bensignor M, Glemain P, Deschamps C, Raoul S, Hamel O. Decompression and transposition of the pudendal nerve in pudendal neuralgia: a randomized controlled trial and longterm evaluation. Eur Urol 2005;47:403–8.

Robert R, Prat-Pradal D, Labat JJ, Bensignor M, Raoul S, Rebai R, Leborgne J. Anatomic basis of chronic perineal pain: role of the pudendal nerve. Surg Radiol Anat 1998;20:93–8.

Rofaeel A, Peng P, Louis I, Chan V. Feasibility of real-time ultrasound for pudendal nerve block in patients with chronic perineal pain. Reg Anesth Pain Med 2008;33:139–45.

Rosenberg JM JM, Quint TJ, de Rosayro AM. Computerized tomographic localization of clinically-guided sacroiliac joint injections. Clin J Pain 2000 ; 16 (1): 18 - 21 .

Schaeffer AJ. Etiology and management of chronic pelvic pain syndrome in men. Urology 2004;63:75–84.

Stang et al. Real-time Ulrasonography-Computed tomography fusion imaging for staging of hepatic metastatic involvement in patients with colo-rectal cancer. Initial results from comparison to US seeing separate CT images and to multidetector-row CT alone. Investigative radiology 2010; 45: 491-501

Thoumas D, Leroi AM, Mauillon J, Muller JM, Benozio M, Denis P, Freger P. Pudendal neuralgia: CT-guided pudendal nerve block technique. Abdom Imaging 1999;24:309–12.

Zacchino M., Allegri M., Canepari M., Minella C. E., Bettinelli S., Draghi F., Calliada F. Feasibility of pudendal nerve anesthetic block using fusion imaging technique in chronic pelvic pain. European journal of pain 2010; 4:329-333

19

Ionizing Radiation Profile of the Hydrocarbon Belt of Nigeria

Yehuwdah E. Chad-Umoren
Department of Physics, University of Port Harcourt, Choba, Port Harcourt, Nigeria

1. Introduction

Man and his environment are constantly bombarded with naturally occurring ionizing radiation. However, aside from the radiation occurring naturally in the human environment, there are those resulting from man's activities. These anthropogenic activities result in the elevation of the background ionizing radiation levels mainly through the depletion of the ozone layer so that an increase occurs in the amount of cosmic radiation reaching the earth (Foland et al, 1995). One crucial area of human activity is the hydrocarbon industry. Globally, the hydrocarbon industry is a strategic industry, contributing to the wealth of nations and the prosperity of individuals. Global economy is often appreciably impacted by the stability or otherwise of the industry. Furthermore, much of the world's population is dependent on the industry for its energy needs. Significantly also, the processes and techniques of oil and gas exploration, exploitation and usage contribute to the devastation, pollution and degradation of the environment.

In Nigeria, the hydrocarbon industry is reputed to be the highest user of radioactive substances in the country (Chad-Umoren and Obinoma, 2007). And the Niger Delta region of Nigeria plays host to the largest concentration of companies in this sector. The maze of crisis-crossing pipelines in the region (Fig.1) does not only transport oil and gas, but also dangerous radioactive substances so that the environment is not despoiled only during oil spillage, but is also polluted with harmful ionizing radiation. Also abundant in the region are numerous non-oil and gas industries and operations that have been attracted into the region by its oil and gas wealth.

2. History and status of the hydrocarbon industry in Nigeria

Oil was discovered in the Niger Delta of Nigeria in 1956 at Oloibiri in present day Bayelsa state and Shell D'Arcy Petroleum commenced production there in 1958. However, before this period, the German company, the Nigerian Bitumen company, had unsuccessfully drilled fourteen (14) wells in the area between 1908 and 1914 within the eastern Dahomey Basin (NNPC, 2004). By the year 2004, oil production had risen to 2.5 million barrels per day. It is estimated that Nigeria's recoverable crude oil reserves stand at about 34 billion barrels, while the nation's natural gas reserves stand at 163 trillion cubic feet (Tcf) making Nigeria

one of the top ten nations with natural gas endowments. The current development plan projects that by year 2020, production will increase to 4million barrels per day.
Though government policy had envisaged that gas flaring will end by 2010, about 50% of the natural gas is still wastefully flared (ECN, 2003), while 12% is re-injected to enhance oil recovery. An estimate by Shell shows that about half of the 2 Bcf/d of associated gas and gaseous by-products of oil exploitation is flared in Nigeria annually, while the World Bank's estimates put Nigeria's gas flaring at 12.5% of the world's total gas flared.

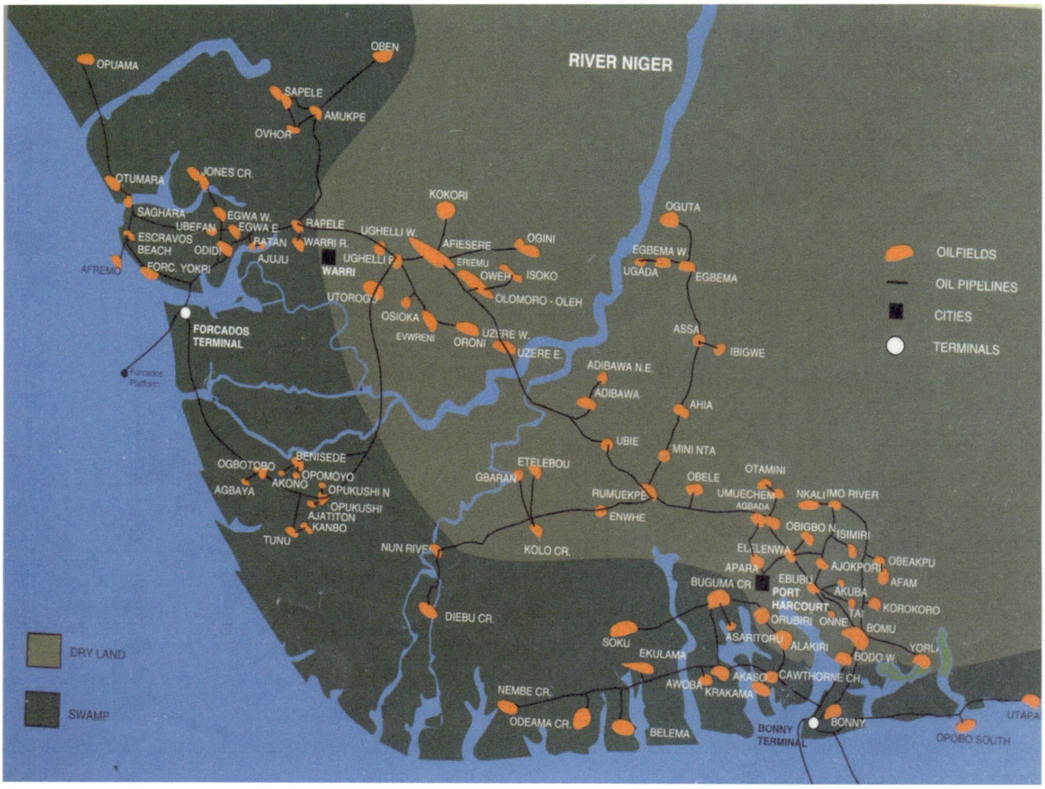

Fig. 1. A map showing network of pipes of a major oil company in the Niger Delta

3. Industrialization and ionizing radiation in the Niger Delta

A number of ionizing radiation surveys have previously been carried out in the Niger delta region to assess the impact of industrialization on the radiation profile of the region and a strong correlation has been established between high industrial activities and elevated background ionizing radiation (BIR) levels for parts of the region (Avwiri and Ebeniro, 1998; Chad-Umoren and Briggs-Kamara, 2010). Ebong and Alagoa (1992 a, b) studied the background ionizing radiation profile at the fertilizer plant at Onne, Rivers State. The study reported a significant increase in the background radiation indicating that the presence of the fertilizer plant had a negative impact on the radiation levels of the area. A survey of the external environmental radiation in the Trans-Amadi industrial area and at other sub-industrial areas of Port Harcourt showed an average value of 0.014 mRh^{-1} (Avwiri and

Ebeniro, 1998). Comparing this result to the standard background radiation level of 0.013mRh^{-1}, the study showed that there was a significant elevation in the radiation status of the surveyed areas which can be attributed to the industrial activities there
Chad-Umoren and Obinoma (2007) surveyed the background ionizing radiation patterns at the campus of the College of Education at Rumuolumeni in Rivers State. The work gave a low radiation dose equivalence of 0.745±0.085mSv/yr as the average for the campus. It was also suggested that further studies needed to be done to assess the impact of the presence of a nearby cement production plant on the radiation status of the college. Briggs-Kamara et al (2009) studied the ionizing radiation patterns of the Rivers State University of Science and Technology at Port Harcourt. The major focus of the study was to assess the impact of computer and photocopier operations on the environmental ionizing radiation patterns of the campus. Computers and photocopiers are widely used on Nigerian university campuses without any consideration for their health hazards. The study established that in deed these devices contributed to the radiation levels of the campus.
A study was also carried out to determine the radiation levels in solid mineral producing areas of Abia state (Avwiri et al, 2010). Solid minerals have their origin in the earth's crust where the primordial radionuclides such as ^{238}U, ^{232}Th and their progenies are found. Consequently, mining of solid minerals has the potential to impact the environmental ionizing radiation. The radiation exposure rate for the surveyed areas ranged between 14.7 μR/hr to 18.2 μR/hr indicating an elevation over the normal background radiation level of 11.4 μR/hr for the host communities. Considering this elevated radiation level, it was concluded that the mining activity had future radiological health hazards for the miners, the general populace and the environment.
The Niger delta region constitutes the hydrocarbon belt of Nigeria and our aim in this work is to study the impact of oil and gas and its ancillary services on the ionizing radiation profile of the region (Fig. 2). Furthermore, based on international practises, we have also suggested appropriate strategies for the control of the ionizing radiation from the hydrocarbon industry and therefore make the activities of the industry in Nigeria more environmentally friendly. And in conclusion we have suggested possible areas for further work.

4. Geology, physiography and evolution of the Niger Delta

The Niger-Delta forms one of the world's major hydrocarbon provinces. It is situated on the Gulf of Guinea on the west coast of Africa and in the southern part of Nigeria, lying between longitudes 4 - 9°E and latitudes 4 - 9° N (Fig. 1) with an estimated area exceeding over 200,000 square kilometres (Odigie, 2001). It is composed of an overall regressive clastic sequence, which reaches a maximum thickness of about 12km (Evamy et al, 1978) and is divided into three formations: Benin formation, Agbada formation and Akata formation. The Agbada formation consists of sand, sandstone and siltstones and is the principal host of Niger Delta petroleum (Beka and Oti, 1995). The Akata formation forms the base of the transgressive lithologic unit of the delta complex and is of marine origin. It is composed of thick shale sequence (potential source rock), turbidite sand (potential reservoir in deep water) and minor amounts of clay and silt (Avbovbov, 1998). The tectonic framework of the continental margin along the west coast of Equatorial Africa is controlled by Cretaceous fractured zones expressed as trenches and ridges in the deep Atlantic.

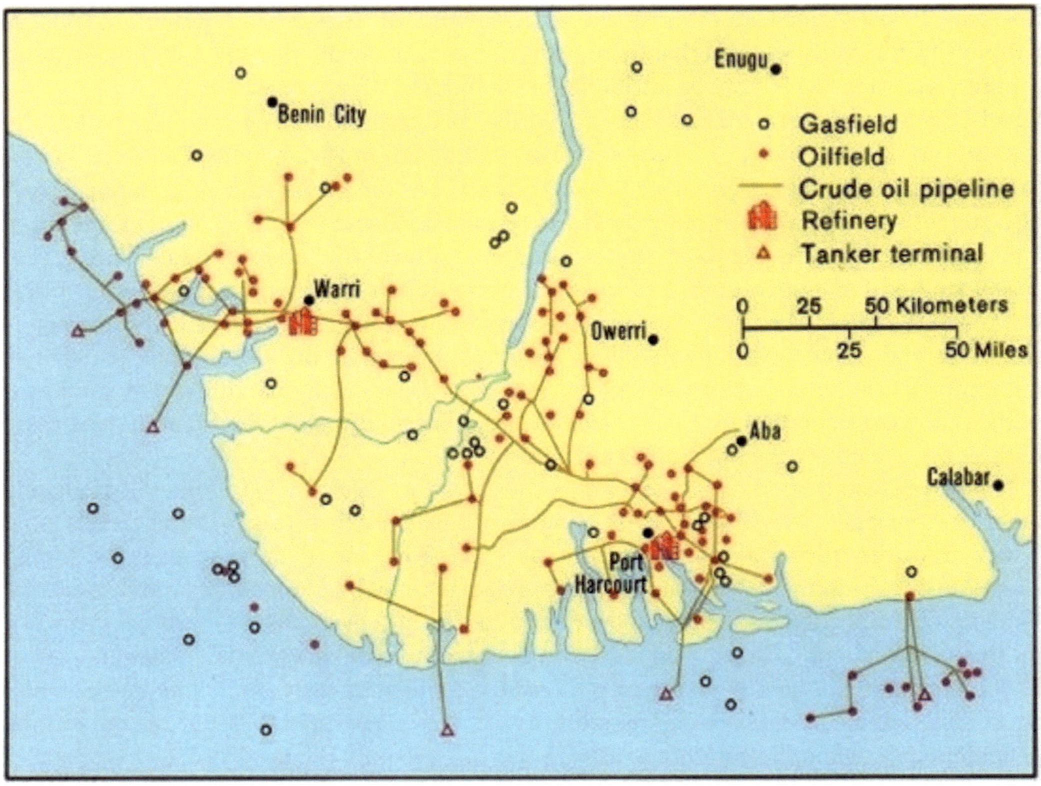

Fig. 2. Map of study area (Niger Delta)

5. Sources of ionizing radiation in the hydrocarbon industry

There are a number of sources for the ionizing radiation associated with the hydrocarbon industry. These include the earth's crust. The earth's crust is a natural source of oil and gas naturally occurring radioactive materials (NORM) as it contains the radionuclides ^{238}U and ^{232}Th. These radionuclides become mixed with the oil, gas and water in the process of oil exploration and exploitation. The radioactive decay chains of these naturally occurring parent radionuclides have very long half-lives and are ubiquitous in the earth's crust with activity concentrations that depend on the type of rock. The radioactive disintegration of ^{238}U and ^{232}Th produce several series of daughter radioisotopes of different elements and of different physical characteristics with respect to their half-lives, modes of decay, and types and energies of emitted radiation. A second source of NORM in the oil and gas industry is scale. Scales consist primarily of insoluble compounds of such elements like barium, calcium and strontium. These compounds precipitate from the produced water as a result of changes in temperature and pressure (IAEA, 2003). The radioactive substances found in scales are the radium isotopes (^{226}Ra and ^{228}Ra). Sulphate scales are the main types of scales in the oil and gas industry (OGP, 2008).

Sludge and scrapings are also sources of NORM in the hydrocarbon industry. Oil field sludge consists of dissolved solids which like scales precipitate from produced water due to variations in temperature and pressure. Generally, sludge is composed of oily, loose materials that may contain both silica compounds and a large amount of barium. The radionuclide found in sludge include radium, while ^{210}Pb and ^{210}Po are found in pipeline scrapings and in sludge that have accumulated in gas/oil separators and liquefied natural gas (LNG) storage tanks. Gas processing facilities are also another source of NORM. The radionuclide in this case is the radon gas.

In the Niger delta region of Nigeria the absence of infrastructural facilities for natural gas utilization has made gas flaring widespread in the region. As previously mentioned, about 50% of the natural gas is flared in the region thereby greatly contributing to the radon in the atmosphere of the region. Seawater injection systems also contain NORM, especially uranium. Seawater is used in the process of oil recovery from reservoirs and incidence of ionizing radiation due to this source is greatly increased in the case where a large volume of seawater is used (OGP, 2003). Incidence of ionizing radiation is further enhanced in the Niger delta of Nigeria as about 12% of the gas is re-injected into wells to improve oil recovery.

6. Effects of ionizing radiation

Radiation monitoring in the hydrocarbon region of Nigeria is important because of the health hazards and the environmental impact of ionizing radiation. Various health hazards of ionizing radiation have been documented. These include the following:

Erythema which is an increased redness of the skin as a result of capillary dilation.

Cancers: The skin is a radiosensitive part of the body and is also its most exposed part. Exposure to ionizing radiation leads to skin cancer. Also, various other kinds of cancers are linked to exposure to ionizing radiation. These include leukaemia, and cancers of the lung, stomach, oesophagus, bone, thyroid, and the brain and nervous system. Not all forms of cancer are traced to exposure to ionizing radiation.

Genetic Effects: Exposure of the reproductive cells to ionizing radiation can lead to miscarriage or genetic mutation. Genetic mutation affects the embryo causing deformity or death.

Sterility: Developing sperm cells known as gonads have very high radiosensitivity, exposure to ionizing radiation can therefore lead to sterility.

Cataracts: Opacities on the surface of the eye lens, called cataracts, result from the denaturing of the lens protein. One effect of ionizing radiation is the denaturing of the lens.

Atrophy of the Kidney: Exposure to ionizing radiation can lead to a condition known as atrophy of the kidney. In this case, the kidney and urinary tract waste or shrink with attendant loss of renal functions.

The level of damage experienced due to exposure to ionizing radiation is determined mainly by the radiation dose received and by such other factors as duration of exposure, nature of radiation and the sensitivity of the part of the body irradiated (ICRP, 1977, UNSCEAR, 1988). Radiation health hazards are dose-dependent. The overall effect of ionizing radiation in man is greatly enhanced when both radionuclide deposition and energy absorption by a specific organ or tissue occur together.

7. Ionizing radiation profile of the Niger Delta, Nigeria

Oil field development and gas exploitation activities have resulted in various forms of unsettling environmental activities in the Niger Delta region that have impacted the ecological, biophysical and socio-economic and political structure of the area (Abali, 2009).

Also, there is ample evidence globally that the hydrocarbon industry contributes to the elevation of the ionizing radiation of the areas where its operations are carried out.
The oil and gas industry, like the rest of the global community, places much emphasis on a safe work environment. Various indices may be adopted to assess what constitutes a safe and healthy work environment, among which is the incidence of certain levels of ionizing radiation in the given area. A study was conducted to evaluate the occupational ionizing radiation levels at 30 locations of oil and gas facilities in Ughelli in the Niger delta of Nigeria (Avwiri et al, 2009). The study was carried out during two time periods - during the production periods and during the off-production periods. Not unexpectedly, the study showed that radiation levels were higher during production periods than during the off-production periods (Fig. 3). The mean radiation levels workers in the oil fields were exposed to during both production and off-production periods are shown in Table 1. For example, the mean obtained for the radiation levels during the off-production periods had a range of 13.38±1.69μR/h (0.023±0.003mSv/wk) to 16.29±2.60μR/h (0.027±0.004mSv/wk) while the mean obtained for the radiation levels during the production periods had a range of 15.50± 1.65μR/h (0.026±0.003mSv/wk) to 19.14±3.16μR/h (0.32±0.005mS/wk). For the two periods under investigation, the Eriemu oil field had the highest radiation levels with a mean equivalent dose rate of $7.88 \pm 1.29 \mu R h^{-1}$, while the Kokori oil field had the highest percentage radiation deviation of 15.19%.

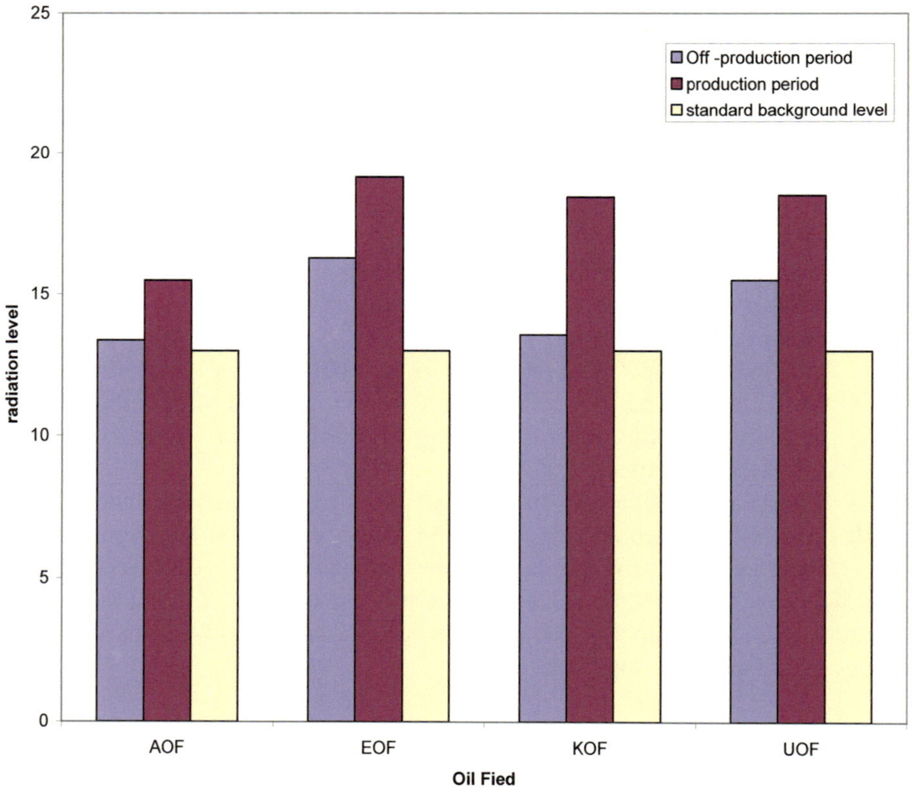

Fig. 3. Comparison of mean radiation levels (μR/h) in the fields with standard background level (Avwiri et al, 2009)

Field code	Surveyed oil and gas field	Mean radiation levels (μR/h)		Mean Deviation (%)	Mean dose equivalent rate (μSv/wk)
		Off - production	Production period		
AOF	Afiesere oil field	13.38 ±1.69	15.50±1.65	**7.34**	**6.43**±0.75
EOF	Eriemu oil field	16.29±2.60	19.14±3.16	**8.04**	**7.88**±1.29
KOF	Kokori oil field	13.57±1.80	18.43±2.68	**15.19**	**7.12±0.99**
UOF	Ughelli oil field	15.50±2.53	18.50±2.68	**8.82**	**7.57**±1.16
	MEAN	**14.69**±2.16	**17.89**±2.54	**9.85**	**7.25**±1.65

Table 1. Mean Exposure Rates in the Oil Fields.

We observe that though for both periods, the mean radiation levels are all within the safe radiation limit of 0.02mSv/wk as recommended by the UNSCEAR (1993) there is still a certain measure of occupational radiation health hazard in these locations for the following reasons:

1. The higher radiation levels during production periods compared to those during off-production periods show that the activities at the oil and gas facilities have led to the elevation of the radiation levels at the study locations. It further shows that the activities during the production periods contribute to higher occupational risk for the workers. Workers are therefore exposed to higher radiation risks during production periods and lower risks during off-production periods.
2. The observed disparity in the exposure rates between the production period and the off-production period was seen as implying that there was increased use of radionuclides, increased exposure to flowing crude oil (spillage) and gas flaring activities resulting in elevation of the ionizing radiation levels of the facilities and their immediate neighbourhoods (Avwiri et al, 2009).
3. The exposure rates at all the locations were generally higher than the standard background level of 13.0μR/hr (Fig. 3).
4. Very high exposure levels of 20μR/hr and above were common. The highest level of 26.00±0.5.1μR/h (0.044±0.009mSv/wk) was recorded at the Kokori oil field during production.
5. Even during off-production periods, high exposure levels were also recorded. For example, an exposure level of 22.00μR/h±2.1μR/h was obtained at the Eriemu natural gas compressor station vessel.
6. Generally, the radiation levels at the Natural Gas Compressor (NGC) facilities were found to be higher than the 0.02mSv/week recommended by the UNSCEAR (1993). The NGC facilities were therefore found to be the most unsafe work-environments as the highest radiation levels were obtained there. This may be attributed to the presence of radon at the facility as this radioactive gas is very much associated with the gas exploitation procedures of the hydrocarbon industry (OGP, 2008)
7. A previous study of the radiation around oil and gas facilities in Ughelli showed that for the oil fields the exposure dose rate had a mean range of 12.00± 0.1μRh (5.33±0.35μSv/wk) to 22.00±2.1μRh-1 (9.79±0.16) and 09.00±1.0 to 11.00±0.5μRh-1 in the host communities (Avwiri, et al., 2007)). Compared to the present investigations we observe that the radiation levels have increased appreciably.

A six-year impact assessment of ionizing radiation was published which reported the effect of oil spillage on the ionizing radiation profile of the oil spillage environment and the host communities in parts of Delta state (Agbalagba and Meindinyo, 2010). The study used a geographical positioning system (GPS) and a digilert nuclear radiation monitor. For the actual survey, 20 sites were chosen along with 6 host communities spread across the affected area and a control sample. The mean values obtained for the location ranged from 0.010 mRh^{-1} (0.532 mSv y^{-1}) to 0.019 mRh^{-1}(1.010mSv y^{-1}). In the area affected by oil spillage, the study gave a yearly exposure rate that ranged from 0.013±0.006 mRh^{-1}(0.692±0.080 $mSvy^{-1}$) to 0.016±0.005 mRh^{-1} (0.851±0.100 $mSvy^{-1}$); the values for the host communities ranged from 0.011 mRh^{-1}(0.585 $mSvy^{-1}$) to 0.015 mRh^{-1}(0.798 $mSvy^{-1}$). The exposure rate for the control was 0.010 mRh^{-1}(0.532 $mSvy^{-1}$).

The study established the impact of oil spillage on the radiation levels within the area and the host communities showing that the radiation levels of both the oil-spilled area and the host communities were elevated by the oil spillage. The radiation exposures were 55% and 33.3% respectively above the normal background level of 0.013 mRh^{-1}. Also, the mean equivalent dose rate for the study area although within the safe limit of 0.05 $mSvy^{-1}$ recommended by ICRP (1990) and NCRP (1993) was higher than the 0.0478 mSv/y normal background level.

Chad-Umoren and Briggs-Kamara (2010) assessed the ionizing radiation levels in parts of Rivers State, one of the states in the Niger delta region. The major thesis of that work was that since there is a near homogeneous distribution of oil and gas operations in the state, the radiation profile of the state will correlate with this, such that the state will exhibit a homogeneous radiation profile. The state was then delineated into three zones, namely, an upland college campus; a collection of rural communities and a group of industrial establishments forming the industrial zone. A Comparative analyses of the results for the three sub-environments showed that the highest dose equivalent of 1.332±0.076mSv/yr occurred in the industrial zone while the lowest value of 0.57±0.16mSv/yr was obtained in the rural riverine sub environment. The computed mean dose equivalent for the three zones also showed that the industrial sector had the highest mean value of 1.270±0.087mSv/yr, while the mean for the upland college campus environment was 0.745±0.085mSv/yr with the rural riverine communities having the lowest mean dose equivalent of 0.690± 0.170mSv/yr.

It is to be observed that the very high exposure rate of 0.0168mR/hr was recorded in this study and that it was obtained in an oil-activity-related environment.

This study revealed the following: Firstly, anthropogenic activities have great impact on the radiation levels of the environment. Secondly, the industrial environment of the state contributes the most to the radiation levels of the state (Tables 2, 3 and 4). Thirdly, the dose equivalent for the different components of the industrial environment are all higher than the European Council for Nuclear Research (CERN) recommended value of 1.0mSv/yr for the general population who are not engaged in nuclear radiation related occupations (CERN, 1995). Fourthly, the dose equivalents obtained for the other two environments are within the CERN regulations.

A national energy policy articulated by the national government of the federation of Nigeria encourages the efficient utilization of the nation's abundant oil and gas reserves (NEP, 2003). Recently the impact of this energy policy on the radiation profile of Ogba/Egbema/Ndoni, a Local Government Area in the central Niger delta state of Rivers state was studied (Ononugbo et al, 2011). The study area was divided into six zones (Table 5). To ensure that the original environmental characteristics of the samples were not tampered with, an *in situ* approach of background radiation measurement was used. The

Station	Name
1	Port Harcourt Refinery Company (PHRC) PHRC Junction, Alesa
2	Nigeria Ports Authority, Onne NAFCON(Fertilizer company) Onne
3	Bori/Onne Junction
4	Rumuokoro Junction
5	Choba (Wilbros)
6	Nkporlu Village
7	Arker Base
8	Mobil Area.

Table 2. Stations for experiment (Industrial environment) (Avwiri and Ebeniro, 2002)

Station	Counter	*mR/hr	S	μ	T	Tc	Remarks
1	3	0.0149	0.0008	0.0144	1.98	2.26	t<tc
	2	0.0154	0.0006	0.0144	5.27	2 26	t>tc
	1	0.0128	0.0008	0.0144	-6.32	-2.26	t<tc
2	3	0.0148	0.0008	0.0143	1.93	2.26	t<tc
	2	0.0159	0.0007	0.0143	7.23	2.26	t>tc
	1	0.0122	0.0010	0.0143	-6.04	2.26	t<tc
3	3	0.0145	0.0006	0.0146	-0.53	2.26	t<tc
	2	0.0156	0.0011	0.0146	6.04	-2.26	t>tc
	1	0.0138	0.0022	0.0146	-1.15	-2.26	t>tc
4	3	0.0151	0.0007	0.0145	2.71	2.26	t>tc
	2	0.0159	0.0011	0.0145	4.02	2.26	t>tc
	1	0.0124	0.0007	0.0145	-4.50	-2.26	t<tc
5	3	0.0150	0.0006	0.0143	3.69	2.26	t>tc
	2	0.0150	0.0006	0.0143	3.69	2.26	t>c
	1	0.0121	0.0007	0.0143	-9.94	-2.26	t<tc
6	3	0.0148	0.0009	0.0147	0.032	2.26	t<tc
	2	0.0160	0.0008	0.0147	4.30	2.26	t>tc
	1	0.0136	0.0019	0.0147	-2.00	-2.26	t>tc
7	3	0.0147	0.0003	0.0147	-0.109	-2.26	t>tc
	2	0.0155	0.0008	0.0147	3.35	2.26	t>tc
	1	0.0138	0.0006	0.0147	-4.688	-2.26	t<tc
8	3	0.0141	0.0009	0.0144	0.94	-2.26	t>tc
	2	0.0155	0.0012	0.0144	3.00	2.26	t>tc
	1	0.0135	0.0015	0.0144	-1.83	-2.26	t>tc
9	3	0.0130	0.0006	0.0140	-5.47	-2.26	t<tc
	2	0.0154	0.0006	0.0140	7.79	2.26	t>tc
	1	0.0135	0.0004	0.0140	-3.25	-2.26	t<tc
10	3	0.0155	0.0007	0.0152	1.26	2.26	t<tc
	2	0.0168	0.0013	0.0152	3.86	2.26	t>tc
	1	0.0132	0.0006	0.0152	-10.89	-2.26	t<tc

Table 3. T-test for stations at 5% confidence level and (n-1) degrees of freedom showing mean counter rate (*), mean background radiation (μ), standard deviation(s), computed t and critical t (tc) with sample size (n) of 10 (Industrial environment) (Avwiri and Ebeniro, 2002)

Station	Mean exposure rate, R (mR/hr)	Dose equivalent, D (mSv/yr)
1	0.0144±0.0007	1.261±0.064
2	0.0143±0.0008	1.253±0.073
3	0.0146±0.0013	1.279±0.114
4	0.0145±0.0008	1.270±0.073
5	0.0143±0.0019	1.253±0.166
6	0.0147±0.0012	1.288±0.105
7	0.0147±0.0006	1.288±0.053
8	0.0144±0.0012	1.261±0.105
9	0.0140±0.0005	1.226±0.044
10	0.0152±0.0009	1.332±0.076

Table 4. Dose equivalent, D computed for the data of Table 3 (Industrial environment) (Avwiri and Ebeniro, 2002; Chad-Umoren and Briggs-Kamara, 2010)

study used a digilert 50 and a digilert 100 nuclear radiation monitoring meters containing a Geiger-Muller tube capable of detecting α, β, γ and x-rays within the temperature range of -10°C to 50°C, alongside a geographical positioning system (GPS) which was used to measure the precise location of sampling. The survey was carried out between the hours of 1300 and 1600 hours, because the exposure rate meter has a maximum response to environmental radiation within these hours (Louis *et al*, 2005). At each of the selected sites and within the host communities four readings were taken at intervals of 5 minutes and the average value determined. The tube of the radiation meter was held at a standard height of 1.0m above the ground with its window facing first the oil installations and then vertically downwards (Chad-Umoren *et al*, 2006). A ^{137}Cs source of specific energy was used to calibrate the instrument to read accurately in Roentgens. Estimate of the whole body equivalent dose rate was done using the National Council on Radiation Protection and measurements (NCRP, 1993) recommendation:

$$1mRh^{-1}=(0.96\times24\times365/100)mSvy^{-1} \quad (1)$$

Table 5 compares the mean radiation exposure rates for the hydrocarbon industry facilities and the mean radiation exposure rates for the host communities. The lowest mean radiation exposure rate of $0.014mRh^{-1}$ was obtained at Obite/Ogbogu. This low level may be attributed to the concrete shielding of the gas plant in the area and the distance of the host community from the Idu flow station (about 2km). For the surveyed oil and gas facility locations, the highest mean exposure rate of $0.018\pm0.002mRh^{-1}$ was obtained at the gas plant; while for the host communities, the highest mean exposure rate of $0.017\pm0.001mRh^{-1}$ was recorded in Ebocha and Obrikom communities of zones A and B. This value is well above the mean field/site radiation exposure level. The percentage exposure rate difference is minimum at the treatment plant and maximum at the gas turbine in Obrikom/Omoku.

At location 5 of zone C, a high exposure rate of $0.034mRhr^{-1}$ was recorded. This could be because of both the presence of sharp sand (asphalt) along the river bank and the industrial waste discharged into the river.

A comparison of the measured radiation levels for the six zones with the normal background level (Fig. 4) shows that the computed mean effective equivalent dose rate for the locations in all the six zones are above the dose limit of $1mSvyr^{-1}$ for the general public and far below the dose limit of $20mSvyr^{-1}$ for radiological workers as recommended by the

Area Code	Industrial Site Surveyed	Host community	Mean Site Radiation levels (mRh^{-1})	Mean Host Community Radiation levels (mRh^{-1})
A	Ebocha Oil Gathering Center	Ebocha/Mgbede	0.016 ± 0.001	0.017 ± 0.001
B	OB/OB Gas Plant	Obrikom/ Ndoni	0.016 ± 0.001	0.017 ± 0.001
C	Gas Turbine	Obrikom /Omoku	0.018 ± 0.002	0.016 ± 0.003
D	Treatment plant (Arrival Manifold)	Obrikom /Ebegoro	0.014 ± 0.001	0.014 ± 0.001
E	Obite Gas Plant	Obite /Ogbogu	0.015 ± 0.001	0.014 ± 0.001
F	Idu Flow Station	Idu /Obagi	0.016± 0.001	0.015 ± 0.001

Table 5. Comparison of Exposure rate at Industrial sites and Host Communities (Ononugbo et al, 2011)

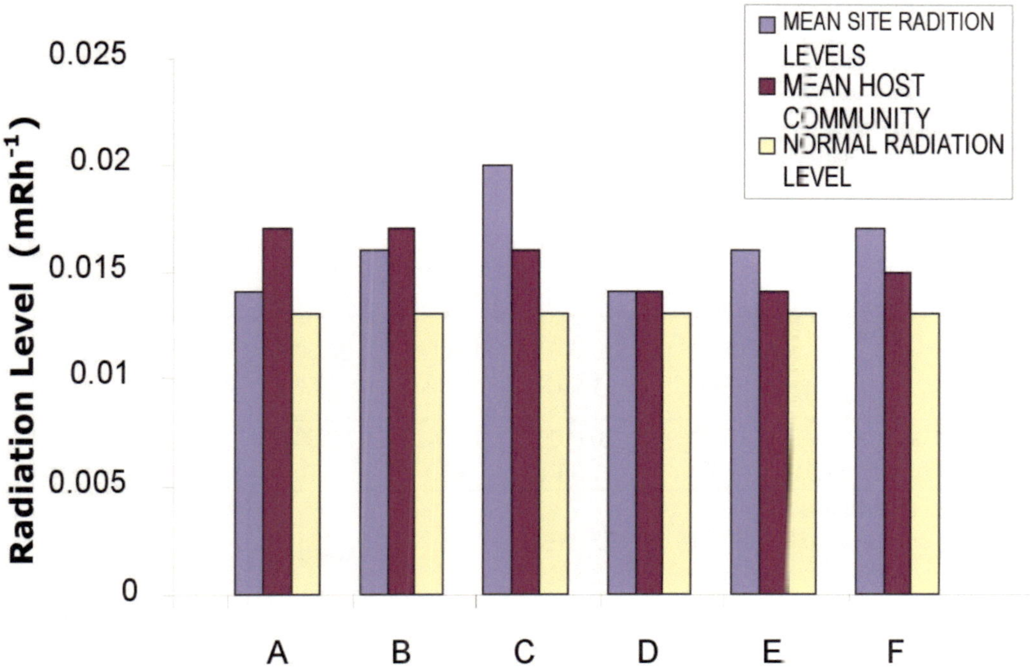

Fig. 4. Comparison of Measured Mean Radiation Levels with Normal Background Level (Ononugbo et al, 2011)

International Commission on Radiological Protection (ICRP, 1990). This indicates that while those exposed occupationally are safe, members of the host communities are not.

The results also show that although the values are within the range of values previously reported for the Niger Delta region (Arogunjo *et al*, 2004; Agbalagba and Avwiri, 2008 and Chad-Umoren and Briggs-Kamara, 2010), the radiation exposure rate for 71.7% of the sampled area (43 locations) exceeded the accepted ICRP background level of 0.013mRhr^{-1}.

S/N	Oil and Gas Facilities	Flow station installation radiation levels (μRh^{-1})					
		Afiesere Delta	**Ughelli West Delta**	**Etelebou Bayelsa**	**Kolo-Creek Bayelsa**	**Adibawa Rivers**	**Egbema Rivers**
1	Flow station entrance	08.00± 0.70	14.00± 2.30	12.00± 1.00	09.00± 1.10	19.00± 1.10	13.00± 1.10
2	Manifold	15.00± 2.00	19.00± 3.00	19.00± 2.80	12.00± 0.90	17.00± 2.20	16.00± 2.10
3	Flare monitoring meter units	12.00± 0.60	15.00± 2.10	16.00± 1.50	16.00± 1.00	16.00± 1.40	15.00± 2.20
4	Flare knock-out drum/vessel	14.00± 1.70	23.00± 4.20	18.00± 2.10	14.00± 1.60	25.00± 4.20	24.00± 3.20
5	Stack gas vent	15.00± 1.00	15.00± 1.00	21.00± 3.70	13.00± 0.80	19.00± 2.00	22.00± 3.20
6	Flare stack pathway	11.00± 0.80	17.00± 3.20	17.00± 1.80	24.00± 4.3.10	16.00± 2.20	17.00± 2.50
7	Flare stack point	19.00± 3.20	14.00± 1.40	14.00± 1.00	17.00± 2.10	14.00± 1.40	21.00± 3.60
8	Inflow delivery crude oil pipes	18.00± 1.60	19.00± 2.60	17.00± 3.00	24.00± 4.20	20.00± 2.00	21.00± 3.10
9	Inflow gas delivery pipes	15.00± 3.00	20.00± 4.00	13.00± 2.00	22.00± 3.00	20.00± 3.40	18.00± 2.40
10	Associate gas control meter	15.00± 2.00	20.00± 1.80	16.00± 3.10	20.00± 3.70	14.00± 0.60	19.00± 3.00
11	Natural gas compressor station (NGC)	21.00± 2.40	22.00± 2.10	Nil	Nil	Nil	Nil
	MEAN	**14.82± 1.74**	**18.00± 2.54**	**16.30± 2.20**	**17.10± 2.15**	**18.00± 2.02**	**18.60± 2.64**
Mean Rate	**Dose equivalent (μSv/wk)**	**6.70± 0.78**	**8.01± 1.13**	**7.25± 0.98**	**7.61± 0.96**	**80.01± 0.90**	**8.28± 1.17**

Table 6. Facilities exposure rate (Agbalagba et al, 2011)

A recent survey studied the gamma radiation profile of oil and gas facilities in six selected flow stations in the Niger delta region (Agbalagba et al, 2011). Of the different types of ionizing radiations, gamma rays are the most penetrating and therefore very hazardous. They emanate from radionuclides containing radon and may be ingested or inhaled by personnel in the course of routine repairs and maintenance of oil facilities. When inhalation occurs, the dust particles and aerosols containing radon become attached to the lungs so that the presence of the gamma rays emitted in the decay increases the risk of lung cancer and other hazards such as eye cataracts and mental problems to personnel and host communities (Laogun et. al., 2006).

The results are shown in Tables 6 and 7. The radiation levels for the facilities (Table 6) range from 08.00±0.70μRh^{-1} in Afiesere flow station entrance to 25.00±4.20μR^{-1} in Adibawa flare knockout vessel. The high value at the Adibawa knockout vessel can be attributed to the spill of associated crude and the exposure of the environment to effluent. The mean exposure rate for the flow stations range from 14.82±1.74μRh^{-1} (6.70±0.78μSv/wk) at Afiesere flow station to 18.60±2.64μRh^{-1} (8.28±1.17μSv/wk) at Egbema flow station, with field mean radiation value of 17.14±2.22μRh^{-1}. The high radiation levels recorded at the natural gas compressor stations (NGC) at Afiesere and Ughelli West flow stations may be attributed to the high concentration of radon in the natural gas and gas production facilities.

Table 7 compares the radiation exposure rate for both the flow stations and the host communities. The lowest radiation exposure rate of 10.00±0.70μRh^{-1} was obtained at Joinkrama 4 in Rivers State. This low radiation level may be attributed to the geology (underlying rock) of the area and to the distance of the host community from the flow station (~2.5km) which is the farthest among the six host communities. The highest average exposure rate (21.00±2.10μRh^{-1}) was obtained in Emeregha community in Afiesere field. This value is much higher than the mean field radiation exposure level.

S/N	Oil field/flow station	Host community	Mean fields Radiation levels μRh^{-1}	Mean Host community Radiation level μRh^{-1}	Exposure rate diff (%)
1	Afiesere	Emeragha	14.82.±1.74	21.00±2.10	17.3
2	Ughelli west	Ekakpamre	18.00±2.54	17.00±2.00	2.9
3	Etelebou	Nedugo	16.30±2.20	15.00±1.40	4.2
4	Kolo-creek	Imirigin	17.10±2.15	18.00±1.60	2.6
5	Adibawa	Joinkrama 4	18.00±2.02	10.00±0.70	28.6
6	Egbema	Egbema	18.60±2.64	14.00±1.10	14.1

Table 7. Comparison of Flow station and Host Communities (Agbalagba et al, 2011)

The presence of the radioactive gas, radon which is produced by the radioactive disintegration of radium-226, may be responsible for the variation of the radiation levels within the facilities. This is so because radon is known to be present in crude oil and gas products (IAEA, 2003). The dispersal of the gas is also favoured by wind direction at the point of liberation.

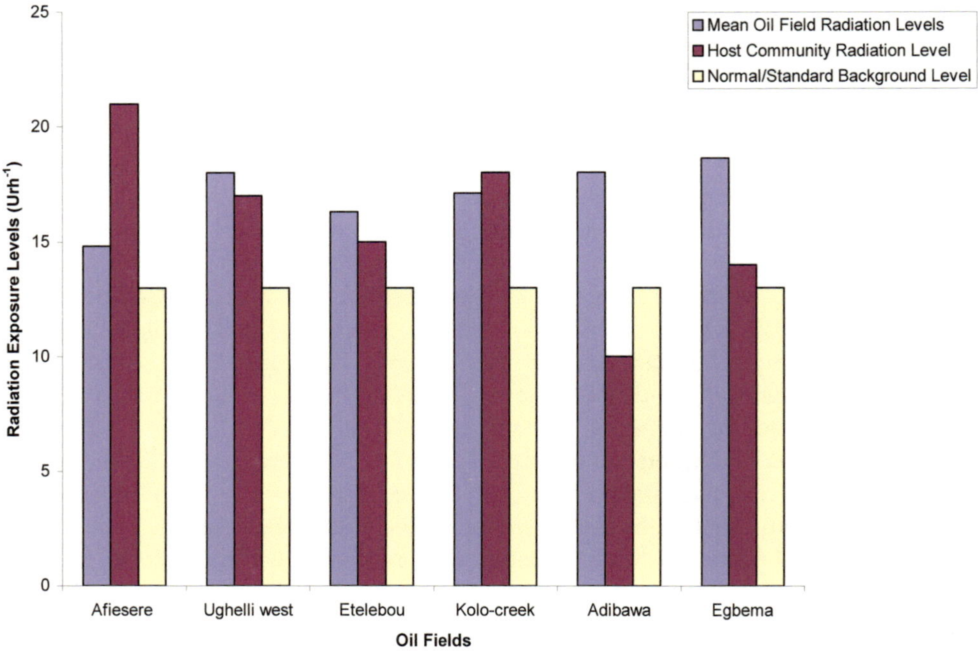

Fig. 5. Comparison of Mean Oil Fields and Host Communities Radiation Levels with Normal background Radiation Level (Agbalagba et al, 2011)

Fig.5 is a comparison of the mean radiation levels for the oil fields and host communities with the normal background ionizing radiation level. Overall, the results show that in all the six flow stations and their host communities except Joinkrama 4, the exposure rates exceed both the standard background levels and values reported previously by Ebong and Alagoa (1992a); Avwiri and Ebeniro (1998), but are in agreement with results reported in similar environments (Arogunjo et al., 2004; Laogun et al., 2006; Agbalagba et al., 2007). The differences in the exposure rates between host communities and flow stations can be attributed to the input materials and output substances (effluents) associated with the activities of the operating companies and dilution of radon as it transits to the host communities. The only exception, the Emeragha community, may be due to the existence of oil wells (wells 7, 10 and 14) within the community which may have enhanced radon concentration and hence the radiation level in the community.

Furthermore, the radiation levels recorded for gas flare facilities and natural gas compressor stations are fairly higher than those for other facilities. This is in agreement with the study of occupational radiation patterns during production and off-production periods (Avwiri et al, 2010). Also, the observation implies the presence of a high concentration of radon gas and heavy metals normally associated with natural and associated gas (Arogunjo et al. 2004; Laogun et al., 2006). However, this study indicates that the radon concentrations in the natural gas at these facilities in the region are low when compared to those obtained in countries like the USA, Great Britain and Canada (Laogun et al., 2006) where radon concentrations (radiation levels) pose enormous environmental challenges requiring government legislation for the control of naturally occurring radiation material (NORM) contamination in their petroleum industries.

The results of the survey did not indicate that there is in the short term the possibility of any health hazard either for the flow station attendants (staff) or the residents of the host communities since the highest average radiation exposure levels of 18.60±2.64μRh^{-1} (8.28±1.17μSv/wk) recorded at Egbema flow station and 21.00±2.10μRh^{-1}(9.35±0.93μSv/wk) obtained at Emeragha community are within the UNSCEAR recommended permissible limit of 200μSv/wk for public health and safety (UNSCEAR, 1993). However, the possibility of future health complications for the staff and host communities does exist as a result of prolonged cumulative dose intake from both direct and indirect radionuclides. Also, it is observed from Fig. 5 that the mean radiation exposure in all the oilfields is higher than the normal background which implies that the oil exploration and gas exploitation activities have impacted significantly on the background ionizing radiation levels of the area. According to Agbalagba et al (2011), this impact can be attributed to the underlying crude oil and gas being contaminated by radionuclide bearing rocks/elements (uranium, thorium and radium) which when drilled to the surface and exposed to the terrestrial environment at the oil installations, releases radon gas and other heavy metals which enhances the background levels of the area and other input materials."

S/N	SAMPLED AREA	GEOGRAPHICAL LOCATION	RADIATION LEVEL mRh^{-1}		AVE. RAD. VALUE mRh^{-1}	EQ. DOSE mSvy^{-1}
			RAD 50	RAD 100		
1	Crude Flow Pipe	NO5 32.297′ E005 53.780′	0.019	0.018	0.0185±0.004	0.9843±0.32
2	Natural Gas Compressor	NO5 26.021′ E005 52.940	0.025′	0.019	0.0220±0.008	1.170±0.43
3	Flow station entrance	NO5 26..057′ E005 52.926′	0.017	0.018	0.01759±0.007	0.931±0.37
4	Well 7	NO5 25.918′ E005 53.014′	0.021	0.024	0.2230±0.010	1.186±0.53
5	Pegging Manifold	N05 26.062′ E005 52.901″	0.019	0.021	0.020±0.008	1.064±0.43
6	Well 10	N05 25.671′ E005 52.930′	0.016	0.018	0.0170±0.006	0.9041±0.32
7	Flare Stack Site	N05 26.141 E005 52.653	0.024	0.025	0.0245±0.011	1.303±0.58
8	Well 5	NO5 25.701′ E005 52.608′	0.018	0.020	0.0190±0.009	1.011±0.48
9	Otorogu Gas Plant	NO5 25.701′ E005 52.608′	0.028	0.034	0.0310±0.010	1.649±0.53
10	***Otujeremi Town***	***NO5 25.865′ E005 52.567′***	***0.022***	***0.020***	***0.0210±0.007***	***1.117±0.37***
	MEAN FIELD LEVELS				**0.0213±0.008**	**1.134±0.44**

Table 8. Otorogu Oil and Gas Field

Tables 8 - 11 give the results of ionizing radiation monitoring for oil and gas fields located at Otorogu, Evwreni, Oweh, Uzere East and West, while Table 12 compares the ionizing radiation profile of the study fields with those of the host communities (Agbalagba et al, 2011). The radiation status of various areas and parts of the oil and gas facilities are given such as the crude flow pipe, flow station, gas plants, crude oil control valve, manifold, oil wells and host communities. In each specific area two Radalerts were used – Radalert 50 and Radalert 100 and the average radiation level computed.

S/N	SAMPLED AREA	GEOGRAPHICAL LOCATION	RADIATION LEVEL mRh^{-1}		AVE. RAD. VALUE mRh^{-1}	EQ. DOSE $mSvy^{-1}$
			RAD 50	RAD 100		
1	Camp site	NO5′ 22.720′ E006 02.962′	0.011	0.011	0.0110± 0.003	0.585± 0.16
2.	Well 13	NO5′ 22.615′ E006′ 02 640′	0.015	0.014	0.0145± 0.005	0.771± 0.27
3	Manifold	NO5′ 22.405′ E006′ 02.405′	0.019	0.013	0.0160± 0.006	0.851± 0.32
4	Well 1	N05′ 22.327′ E006′ 02.410′	0.017	0.014	0.055± 0.005	0.825± 0.27
5	Flow station Gate	N05′ 22.445′ E006. 02.470′	0.015	0.016	0.0155± 0.006	0.825± 0.32
6	L & S Tanga crude flow pipe	N05′ 22.428′ E006″ 02 500	0.015	0.014	0.0145± 0.005	0.771± 0.27
7	Gas vent (knockout drum)	N05′ 22.432′ E006 22.482′	0.020	0.022	0.0210± 0.009	1.117± 0.48
8	Flare stock site	N05″ 22.361′ E006″ 02.451′	0.021	0.018	0.0195± 0.008	1.0371± 0.83
9	Well 11	N05″ 22 .394′ E006′ 02.439′	0.014	0.014	0.0140± 0.005	0.771± 0.27
10	***Evwreni Community***	***N05′ 24.243′ E006″ 03.451′***	***0.017***	***0.014***	***0.0155± 0.007***	***0.8258± 0.32***
	MEAN FIELD LEVEL				**0.0160± 0.006**	**0.839± 0.34**

Table 9. Evwreni Oil and Gas Field

S/N	SAMPLED AREA	GEOGRAPHICAL LOCATION	RADIATION LEVEL mRh^{-1}		AVE. RAD. VALUE mRh^{-1}	EQ. DOSE $mSvy^{-1}$
			RAD 50	RAD 100		
1	Flow Station Gate	NO5′ 29.271′ E006 08.101′	0.016	0.012	0.0140±0.005	0.745±0.27
2.	Crude oil control valve	NO5′ 08.101′ E006′ 08′	0.019	0.019	0.0190± 0 007	1.011±0.37
3	Gas vent (knockout drum)	NO5′ 29.289′ E006′ 08.201′	0.017	0.016	0.0165±0.006	0.878±0.32
4	Flare stack site	N05′ 29.304′ E006′ 08.244′	0.016	0.018	0.017±0 005	0.904±0.27
5	NGC Station	N05′ 29.216′ E006. 08.132′	0.022	0.020	0.0210±0.008	1.117±0.43
6	L & S tango Crude flow pipe	N05′ 29.285′ E006″ 28 185′	0.016	0.014	0.0150± 0.006	0.798±0.32
7	Manifold	N05′ 28.185′ E006 07.720′	0.019	0.018	0.01850±0.008	0.984±0.43
8	Well 12	N05″ 29.666′ E006″ 06.567′	0.020	0.018	0.0190±0.007	1.011±0.37
9	Well 2	N05″ 29 .219′ E006′ 08.128′	0.018	0.023	0.0205±0.010	1.091±0.53
10	***Otor-Oweh community***	***N05′ 29.614′ E006″ 06.248′***	***0.012***	***0.014***	***0.0130±0.005***	***0.692±0.27***
	MEAN FIELD LEVEL				**0.0178±0.007**	**0.949±0.37**

Table 10. Oweh Oil and Gas Field

S/N	SAMPLED AREA	GEOGRAPHICAL LOCATION	RADIATION LEVEL mRh-1		AVE. RAD VALUE mRh-1	EQ. DOSE mSvy-1
			RAD 50	RAD 100		
1	Manifold	NO5′ 20.080′ E006 14.865 ′	0.016	0.015	0.0155±0.006	0.525±0.32
2.	Buster station	NO5′ 20.162′ E006′ 14 .781″	0.017	0.014	0.0155± 0.005	0.825±0.27
3	NGC Station	NO5′ 19.751′ E006′ 14.762′	0.016	0.019	0.0175±0.006	0.931±0.32
4	Flow station Gate	N05′ 19.627′ E006′ 14.655′	0.027	0.028	0.0275±0.013	1.463±0.69
5	L & S Tango crude flow pipe	N05′ 19.167′ E006. 14.642′	0.022	0.024	0.230±0.010	1.224±0.53
6	Flare knock out down	N05′ 19.601′ E006″ 14. 633′	0.020	0.018	0.01900± 0.008	1.011±0.43
7	Flare stack site	N05′ 19.584′ E006′ 14.566′	0.017	0.021	0.0190±0.007	1.011±0.37
8	Well 6	N05″ 19.251′ E006″ 15.960′	0.019	0.023	0.0205±0.009	1.277±0.64
9	Well 2	N05″ 19 .421′ E006′ 15.862′	0.022	0.026	0.0240±0.012	1.277±0.64
10	***Uzere community***	***N05′ 20.268′ E006″ 14.338′***	***0.016***	***0.019***	***0.0175±0.007***	***0.931±0.27***
	MEAN FIELD LEVEL				**0.0202±0.008**	**1.075±0.45**

Table 11. Uzere East and West Oil and Gas Field

Area Code	Oil and Gas Field	Host Community	Mean field dose rate (mSvy-1)	Host Community dose rate (mSvy-1)	Difference (%)
OUT	Otorugu	Otujeremi	1.134±0.31	1.117±0.37	1.51
EVN	Evwreni	Evwreni	0.839±0.34	0.612±0.16	22.70
OWT	Oweh	Otoweh	0.949±0.37	0.692±0.27	37.14
OLO	Olomoro-Oleh	Olomoro	0.943±0.37	0.931±0.27	1.29
UZE	Uzere West & East	Uzere	1.075±0.45	0.931±0.37	14.4

Table 12. Comparison of Studies fields and Host Communities Radiation Data

Table 12 shows that the highest mean dose rate for the oil fields, 1.131±0.31mSvy^{-1} was recorded at the Otorugu field and its host community, Otujeremi, recorded the highest dose rate, 1.117±0.37mSvy^{-1}, for the host communities. It would appear that the preponderance of gas operations in Otorugu as compared to the other fields has resulted in an increased release of radon gas into the atmosphere over the oilfield and its host community leading to significant impact on the radiation level of the oilfield and its host community.

8. Water and soil analyses

The Niger delta has a plethora of interconnected underground and surface waterways and this can have some impact on the dispersal of ionizing radiation in the environment as they have the potential to transport particulate radioactive matter. Sediments from different water bodies and industrial effluents discharged into the water bodies are transported from location to location. Furthermore, evaluating the water in the region for radioactivity becomes especially important as the water from creeks, streams and rivers in much of the region is also used for such activities as washing, bathing, cooking and drinking! Through any of this means radionuclide ingestion and consequent radiation contamination of the populace can attend epidemic proportions. Another route for radionuclide ingestion by the human population is through the consumption by man of fish and other sea foods that have been contaminated by radiation.

A study conducted in Delta state, another Nigeria Niger delta state, using Exploranium - The-Identifier GR-135 model, assessed the natural radioactivity concentration in river Forcados (Avwiri et al, 2008). 20 water samples were collected from 20 different locations separated a distance of 100m from each other. The study identified three radionuclides, namely ^{40}K, ^{226}Ra and ^{232}Th with respective average specific activity of 13.94±1.97Bq/l, 12.80±2.84Bq/l and 34.62±3.71Bq/l. The mean absorbed dose rates and dose equivalent were computed to be 9.90±1.61nGy/h and 0.084±0.003mSv/y respectively. The survey did not indicate any radionuclide concentration gradient as the values in the survey were found to be randomly distributed. This can be attributed to the water transportation mechanism and effluent discharge.

Also, the results show low concentration of the identified radionuclides in the water and may therefore not pose any radiological effects on the populace that use the water from this river. Furthermore, the values are comparable to previous surveys within the Niger delta region (Arogunjo et al, 2004), however, they are higher than the value of 1.2mBq/l (^{226}Ra) reported for Lake Ontario (Baweja et al, 1987) and 28 x 10^{-6} Bq/l (^{232}Th) for Lake Michigan. This disparity can be attributed to the activities of the hydrocarbon industry in the Niger delta.

In a survey of the gross alpha and beta radionuclide activity in the Ckpare Creek in Delta State, Nigeria, Avwiri and Agbalagba (2007) classified the creek into three zones and reported average alpha activities in the classified zones as 1.003±0.097, 4.261 ± 0.109 and 10.296±0.489Bql^{-1} respectively, and the beta activities as 0.129±0.100, 0 523+0.003 and 0.793+ 0.010Bql^{-1} respectively. These values are far above the 0.1Bq/l for alpha and 1.0Bq/l for beta WHO maximum recommended level for screening for drinking water (WHO, 2003). A previous study by Avwiri et al (2005) assessed the radionuclide contents of soil, sediments and water in Aba River in the Niger delta state of Abia state. In addition to these, the radionuclide content of the fish in the river was also evaluated. The work employed a rigorously calibrated sodium iodide (NaI) detector. The radionuclides identified were ^{226}Ra, ^{228}Ra and ^{40}K for all three study samples. For the soil samples the respective average specific activity levels obtained were 4.73±0.71Bq/kg, 4.26±0.9Bq/kg, and 118.75±10.54Bq/kg. The activity levels of the radionuclides for the water samples were 3.34±0.43Bq/l, 3.69±0.17Bq/l and 111.39±10.04Bq/l respectively. In the case of the sediments the study gave 11.56±2.35Bq/kg, 17.50±1.77Bq/kg and 253.42±21.10 respectively. And for the catfish from the river, the activity levels were 19.14±4.33Bq/kg, 22.58± 5.58Bq/kg and 63.42±14.09Bq/kg respectively. These results show significantly higher concentrations compared to a previous study (Jibiri et al, 1999). The operations of the Industrial Zone in Aba account for these significantly high values as they discharge their effluents directly into the river. An analogous study to assess the radiological impact of the petrochemical industry on the Aleto Eleme River showed a significant elevation of the ionizing radiation levels at the point of effluent discharge into the river (Avwiri and Tchokossa, 2006).

A study carried out to assess the level of natural radionuclides in borehole water in some selected wells in Port Harcourt, Rivers State showed that the mean specific activity of the resulting annual effective doses for ^{226}Ra, ^{228}Ra and ^{40}K were 3.51±2.22, 2.04 ± 0.29 at 23.03 ±4.37 and 0.36 ± 0.12, 0.51± 0.02 and 0.05 ±0.01mSy/y respectively (Avwiri et al, 2006). The results of this survey are within the range obtained elsewhere. Generally, public places showed the highest activity concentration due to poor sanitation.

Waste management in urban centres is important for environmental health. In many parts of the Niger delta region, wastes are sometimes collected and heaped up and no further action taken until they are scattered again by wind; disposed of indiscriminately or they may be disposed of in landfills. Landfills are openings in the soil used for waste disposal. They could be purpose-made or an abandoned pit or quarry. Many of the landfills in the Niger delta region are usually not differentiated and therefore will hold a mixture of different types of wastes, including those from the hydrocarbon industry. Landfills are an important source of groundwater pollution and may also result in the elevation of the ionizing radiation profile of the environment due to its radionuclide content. One survey conducted in the city of Port Harcourt, Rivers State, assessed the radionuclide content, the ionizing radiation level and associated dose rates of a landfill around the Eliozu area of the city (Avwiri et al, 2011). The Eliozu landfill is composed of different types of wastes including industrial wastes, chemical wastes, medical wastes, scraps, metals and other debris.

10 samples each of soil and water were collected from different parts of the landfill and analyzed for their radioactivity and radionuclide content using the gamma-ray spectrometer NaI (Tl) detector system. The results are shown in Table 13 for the soil samples and in Table 14 for the water samples.

The radionuclides found in both the soil and water samples were ^{232}Th, ^{238}U and ^{40}K. The study reported that for the soil samples, the mean activity concentration were 27.41 ± 9.97 Bq/kg for ^{238}U, 19.27 ± 8.14 Bq/kg for ^{232}Th and 326.08 ± 66.74 Bq/kg for ^{40}K (Table 7). And for the water samples, the mean activity concentration were 7.92 ± 2.69 Bq/l, 6.96 ± 2.37 Bq/l and 24.77 ± 8.33 Bq/l for ^{238}U, ^{232}Th, ^{40}K respectively (Table 8). The calculated absorbed dose rates for the soil had a range of 23.53 nGy.h^{-1} to 50.39 nGy.h^{-1} and a mean of 38.17 ±12.45 nGy.h^{-1} while for the water the range was 6.62 nGy.h^{-1} to 10.71 nGy.h^{-1} with a mean of 9.03±3.07 nGy.h^{-1}. The mean absorbed dose rate for the area is lower than the world's average of 55 nGy.h^{-1}for soil. It should however be observed that the upper limit of 50.39 nGy.h^{-1} is quite close to the world average.

		Soil (Bq/kg)				
S/No	**Sample**	**K-40**	**U-238 (Ra-226)**	**Th-232 (Ra-228)**	**Absorbed dose rates (nGy/hr)**	**Equivalent dose rate (mSv/yr)**
1	**A1**	570.08 ± 87.6	20.52 ± 5.2	18.95 ± 9.9	45.25 ± 12.5	0.3964 ± 0.1
2	**A2**	105.57 ± 24.9	18.94 ± 8.4	16.62 ± 6.7	23.53 ± 9.1	0.2061 ± 0.1
3	**A3**	404.95 ± 99.8	34.81 ± 13.6	19.33 ± 6.5	44.68 ± 14.3	0.3914 ± 0.1
4	**A4**	140.49 ± 35.8	23.87 ± 9.7	17.84 ± 6.5	27.92 ± 10.0	0.2446 ± 0.1
5	**A5**	254.55 ± 79.5	38.28 ± 11.4	29.54 ± 11.5	46.60 ± 15.3	0.4082 ± 0.1
6	**A6**	256.22 ± 68.6	26.54 ± 9.7	22.46 ± 8.7	36.97 ± 12.8	0.3239 ± 0.1
7	**A7**	527.91 ± 89.5	35.91 ± 11.4	19.41 ± 9.9	50.39 ± 15.2	0.4414 ± 0.1
8	**A8**	545.13 ± 87.5	26.41 ± 11.1	18.63 ± 7.8	46.52 ± 13.6	0.4075 ± 0.1
9	**A9**	323.64 ± 75.9	29.36 ± 9.3	13.43 ± 7.3	35.05 ± 12.0	0.3070 ± 0.1
10	**A10**	132.23 ± 18.7	19.44 ± 9.9	16.52 ± 6.5	24.80 ± 9.3	0.2173 ± 0.1
Mean Values		326 ± 66.7	27.41 ± 10.0	19.27 ± 8.1	38.17 ± 12.5	0.3344 ±0.1

Table 13. Radionuclide Concentration of Soil Samples ($BqKg^{-1}$) (Avwiri et al, 2011)

		Water (Bq/l)				
S/N	**Sample**	**K-40**	**U-238 (Ra-226)**	**Th-232 (Ra-228)**	Absorbed dose rates (nGy/hr)	Equivalent dose rates (mSv/yr)
1	**W1**	16.40 ± 7.3	8.24 ± 2.8	5.87 ± 2.0	7.80 ± 2.8	0.0683 ± 0.03
2	**W2**	26.74 ± 7.7	7.48 ± 3.3	7.44 ± 2.2	9.24 ± 3.2	0.0809 ± 0.03
3	**W3**	24.98 ± 9.8	9.32 ± 2.3	8.32 ± 2.5	10.54 ± 3.0	0.0923 ± 0.03
4	**W4**	19.84 ± 7.0	7.89 ± 3.0	7.89 ± 2.1	9.43 ± 3.0	0.0826 ± 0.03
5	**W5**	23.51 ± 9.7	8.04 ± 2.1	6.78 ± 2.3	8.91 ± 2.9	0.0781 ± 0.03
6	**W6**	32.08 ± 8.2	9.06 ± 3.2	8.29 ± 3.1	10.71 ± 3.8	0.0938 ± 0.03
7	**W7**	16.12 ± 7.2	9.41 ± 2.7	7.89 ± 2.5	9.92 ± 3.1	0.0869 ± 0.03
8	**W8**	27.16 ± 9.0	6.94 ± 3.5	5.20 ± 2.0	7.55 ± 3.2	0.0661 ± 0.03
9	**W9**	21.67 ± 7.4	5.99 ± 2.1	4.76 ± 2.1	6.62 ± 2.6	0.0580 ± 0.02
10	**W10**	39.15 ± 10.0	6.85 ± 1.8	7.34 ± 2.9	9.68 ± 3.[illegible]	0.0848 ±0.03
Mean Values		24.77± 8.3	7.92±2.7	6.96±2.4	9.03 ± 3.1	0.0791 ± 0.03

Table 14. Radionuclide Concentrations of Water Samples (Bql^{-1}) (Avwiri et al, 2011)

The determination of the equivalent radiation exposure (effective dose rate) for the immediate neighbourhood of the dumpsite indicate that the population in the vicinity of the site receives a dose that lies in the range of 0.2061 to 0.4414 $mSv.y^{-1}$ with a mean of 0.3344±0.1091 $mSv.y^{-1}$ from the soil and a dose that ranges from 0.0580 to 0.0938 $mSv.y^{-1}$ with a mean of 0.0791±0.0269 $mSv.y^{-1}$ from the water. The dumpsite therefore appears to pose minimal ionizing radiation risk to the environment and the population.

9. Ionizing radiation control strategies for Nigeria's hydrocarbon industry

The following strategies are proposed for the control and management of ionizing radiation in the region:

9.1 Identification of areas of radiation risks

Activities and locations with high ionizing radiation risks or potential for radiation hazards should be delineated. The reason for this is that radiation pollution and risks cannot be controlled if the location of the risk is unknown. Therefore, regular radiation monitoring in the region is to be sustained. Such radiation monitoring should include baseline surveys of oil and gas facilities and host communities, water and soil analyses, monitoring of well heads, production manifolds, storage tanks, flow stations, etc.

9.2 Establishment and enforcement of ionizing radiation limits for the hydrocarbon industry

Here such government agencies as the Nigeria Nuclear Regulatory Authority (NNRA) must enforce compliance with relevant laws by industry operators. Where existing laws are inadequate or obsolete they should be reviewed and appropriate legislations enacted.

9.3 Education and public enlightenment

Education of industry personnel and proper enlightenment of members of the public should form critical components of any ionizing radiation control strategy. Industry personnel, especially those most likely to be exposed to radiation, need to be trained to work safe and ensure that ionizing radiation is not spread.
Also, the general public, especially residents of communities that host oil and gas installations should be educated on the hazards of ionizing radiation.

9.4 Handling of ionizing radiation contaminated waste

Ionizing radiation contaminated waste generated by the hydrocarbon industry include sludge, scale, drilling pipes, storage tanks, etc. In the absence of appropriate control mechanisms, unintended spreading of ionizing radiation contamination with consequent contamination of areas of land and the environment and eventual exposure of the public can occur in the course of the handling, storage and transportation of ionizing radiation contaminated equipment or waste.
Used oilfield equipments should therefore be evaluated first to determine their radioactivity status before transportation; otherwise unintended spreading of ionizing radiation can occur when such materials as contaminated pipes and tanks are transported. Possible strategies for handling radioactive waste generated in the oil and gas industry include among others disposal in salt caverns, underground injection, and smelting (OGP, 2003).

Industry operators will have to study the Niger delta environment to determine the most appropriate strategies for the region.

9.5 Decommissioning of oil and gas production facilities

At some point in its productive cycle, an oil and gas facility such as a reservoir will become economically unviable for further exploitation and must therefore be properly disposed of rather than just abandoning it. This is the process of decommissioning. For example, available statistics (NNPC, 2004) show that there are 606 oil fields in the Niger Delta region, 251 of which are offshore, while the balance of 355 wells are on-shore. Also, of these, 193 wells are currently being exploited while 23 have been shut in or abandoned as no more viable and would therefore need to be properly decommissioned.

Among other reasons for decommissioning is the desire to limit the hazards of radioactive contamination. The steps in oilfield facility and equipment decommissioning include the following (IAEA, 2003):

i. Radiological assessment of equipments and facilities to ascertain radiation levels and therefore level of risk to personnel, the general populace and the environment.
ii. Submission of decommissioning plans, surveys and radiological reports on the equipments and facilities to relevant regulatory authorities such as the Nigeria Nuclear Regulatory Authority (NNRA) and such other agencies responsible for environmental health and safety especially as it relates to ionizing radiation.
iii. Decontamination of equipments and facilities to levels set by relevant regulatory bodies, especially if such equipments and facilities will become available for unrestricted public use.
iv. Radioactive waste management. This is the last step in the decommissioning procedure, where all hazardous radioactive wastes and the remaining contaminated items are disposed of in designated and approved radioactive waste disposal facilities.

10. Conclusion and suggestions for future research

It is obvious from this study that the Niger delta region of Nigeria is inundated with ionizing radiation as a result of the activities of the hydrocarbon industry and its subsidiary services. Though the profile in general does not indicate any immediate health complications both for the industry personnel and the general populace, the ionizing radiation patterns of the region is still crucial for the following reasons:

10.1. Low level radiations have long-term cumulative effects (Chang et al, 1997). Moreover for the hydrocarbon industry evidence exists on long term effect of low radiation dose. It has been observed that there is prevalence among hydrocarbon industry retirees and host community members of such diseases as eye cataracts, various kinds of cancer such as lung and bone cancer, leukaemia and mental disorder (UNDP, 2006; Otarigho, 2007). In the long term therefore, the current low radiation exposure could become a serious health hazard. It is therefore apparent that the present low radiation level in Nigeria's hydrocarbon belt should not be ignored.

10.2. Various research findings presented in this study show that in many specific locations, radiation levels exceed the expected normal background levels. The implication of this is that the ionizing radiation levels in the region are rising progressively so that the environment is gradually becoming unsafe for the general populace in the region.

10.3. Though the overall ionizing radiation profile for the region is that of low exposure rates, however evidence from the study indicates that in some of the locations the radiation levels exceed internationally recommended standards, especially for the general populace.
10.4. The activities of the industry are not abating. On the contrary, there appears to be more aggressive activities aimed at exploiting the abundant resources of the region with attendant increased introduction into the region of ionizing radiation. Increased oil and gas exploration, exploitation and production activities will result in increased radiation pollution due to increased radioactive waste generation and increased generation of such hydrocarbon industry sources of ionizing radiation as scales, sludge, formation water, etc.
10.5. The current practise of gas flaring poses the danger of radionuclide build-up in the atmosphere. As the work has shown, areas of high gas-related activities in the region have tended to experience marked ionizing radiation levels. Furthermore, when the flared gas comes in contact with rain water, it results in the radioactive pollution of the rain water. And like the water from the creeks and rivers, rain water in many parts of the Niger delta is an important source of water for various activities like bathing, washing, cooking and drinking.
10.6. It will be helpful to investigate the health challenges of the workers at the oil and gas facilities in the Niger delta to see if there is any correlation between ionizing radiation exposure and the health complaints of the workers.
10.7. Further research in the area should include:

i. Mathematical modelling of oil spill phenomenon in the area
ii. Mathematical modelling of impact of gas flaring in the area
iii. More extensive soil and water analyses of areas affected by oil spillage in the region so as to properly establish the impact of oil spillage especially considering that much of the population of the region are fishermen and rural subsistence farmers
iv. A study of the health challenges of oil and gas industry personnel and host community residents to determine any possible correlation between ionizing radiation and the prevailing health issues of these respective groups

11. References

Abali,B.K. (2009). Oil and Gas Exploration: What ONELGA suffers. Port Harcourt, B'Alive publications co.

Agbalagba, E. O , Avwiri, G. O and Chad-Umoren, Y. E. (2011). Radiological Impact of Oil and Gas Activities in Selected Oilfields in Production Land Area of Delta State, Nigeria (In press).

Agbalagba, O. E. and and Meindinyo, R. K. (2010). Radiological impact of oil spilled environment: A case study of the Eriemu well 13 and 19 oil spillage in Ughelli region of delta state, Nigeria. Indian Journal of Science and Technology, Vol. 3(9), 1001 - 1005

Agbalagba E.O., Meindinyo. R.K., Akpata, A.N. and Olali, S.A (2011). Assessment of Gamma-radiation Profile of Oil and Gas Facilities in Selected Flow Stations in the Niger- Delta Region of Nigeria. Indian Journal of Emerging Science and Technology (in press)

Arogunjo, M.A., I.P. Farai and I.A. Fuwape (2004). Impact of oil and gas industry to the natural radioactivity distribution in the delta region of Nigeria. Nigerian Journal of Physics, vol.16, 131-136.

Avbovbo, A. A. (1998). Tertiary Lithostratigraphy of Niger Delta: American Association of Petroleum Geologists Bulletin, vol. 62(2), 295-300

Avwiri, G. O. and Agbalagba, E.O. (2007). Survey of gross alpha and gross beta radionuclide activity in Okpare - Creek, Delta State, Nigeria. Asian Journal of Applied Science, vol. 7 (22), 3542 - 3542.

Avwiri, G.O., Chad - Umoren, Y.E., Enyinna, P.I. and Agbalagba, E.O. (2009). Occupational radiation Profile of oil and gas facilities during and off - production periods in Ughelli, Nigeria. Factia Universititatis, Series Working and Living Environmental Protection, Vol. 6(1).

Avwiri, G. O; Enyinna, P. I and Agbalagba, F. O.(2007). Terrestrial Radiation Around Oil and Gas Facilities in Ughelli, Nigeria. Journal of Applied Science, vol. 7, 1543 - 1546.

Avwiri, G. O., Owate, I. O. and Enyinna (2005). Radionuclide Concentration Survey of Soil, Sediments and Water in Aba River, Abia State, Nigeria. Scientia Africana, vol. 4(1&2), 67-72.

Avwiri, G.O. and Ebeniro, J.O. (1998). External Environmental Radiation in an industrial area of Rivers State, Nigeria. Nigerian Journal of Physics, vol. 10 103 - 107.

Avwiri, G.O., Enyinna, P. I. and Agbalagba, E.O. (2008). Assessment of natural radioactivity concentration and distribution in river Forcados, Delta State, Nigeria. Scientia Africana Vol. 4(1 & 2), 128-135

Avwiri, G.O., Enyinna, P.I and Agbalagba, E.O. (2010). Occupational radiation levels in solid minerals producing areas of Abia State, Nigeria. Scientia Africana. Vol. 9 (1).

Avwiri, G.O., and Ebeniro, J.O. (2002). A survey of the background radiation levels of the sub-Industrial areas of Port Harcourt, Global Journal of Pure and Applied Sciences. Vol. 8(2), 111 - 113.

Avwiri, G.O. and Tchokossa, P. (2006). Radiological Impacts of Natural radioactivity along Aleto River due to a Petrochemical Industry in Port Harcourt, Rivers state, Nigeria. Journal of Nigeria Environmental Society (JNES), Vol. 3 (3), 315 - 323.

Avwiri, G.O., Mokobia, C.E. and Tchokossa, P. (2006). Natural radionuclide in borehole water in Port Harcourt, Rivers state, Nigeria. Radiation Protection Dosimetry, Vol. 123(4), 509 - 514, Oxford Journals UK

Avwiri, G.O., Nte, F.U. and Olanrewaju, A.I. (2011). Determination of radionuclide concentration Of land fill at Eliozu, Port Harcourt, Rivers State. Scientia Africana (in press).

Baweja A. S, Joshi S.R. and Demayo A . (1987). Report on the National Radionuclide monitoring program, (1981-1984) Environment Canada, Ottawa, Scientific services No.156.

Beka, F. T., and Oti, M. N., (1995). The distal offshore Niger Delta: frontier prospects of a mature petroleum province, *in*, Oti, M.N., and Postma, G., eds., Geology of Deltas: Rotterdam, A.A. Balkema, p. 237-241.

Briggs-Kamara, M.A; Sigalo, F.B; Chad-Umoren, Y.E. and Kamgba, F.A. (2009). Terrestrial Radiation Profile of a Nigerian University Campus: Impact of Computer and Photocopier Operations. Journal Facta Universitatis: Working and Living Environmental Protection, vol 6 (1), 1 - 9

CERN (1995). Safety Guide for Experiments at European Council for Nuclear Research, CRN, Part III - Advice, 40, Ionizing Radiation. (http://cern.web.cern.../40).

Chad-Umoren, Y. E and Briggs-Kamara, M. A (2010). Environmental Ionizing Radiation Distribution in Rivers State, Nigeria. Journal of Environmental Engineering and Landscape Management, vol. 18(2), 154–161

Chad-Umoren, Y. E and Obinoma, O (2007). Determination of Ionizing Radiation Level of the Main Campus of the College of Education, Rumuolumeni, Rivers State, Nigeria. Int'l Journal of Environmental Issues, vol. 5 (1 and 2), pp 5 -10

Chad-Umoren, Y. E; Adekanmbi, M and Harry, S. O. (2006). Evaluation of Indoor Background Ionizing Radiation Profile of a Physics Laboratory. Journal Facta Universitatis: Working and Living Environmental Protection., Vol 3 (1), pp 1 - 8

Chang, W.P; Hwang, BF; Wang, D. and Wang, JD (1997). Cytogenetic Effect of Chronic Low-Dose, Low-Dose-Rate Gamma Radiation in Residents of Irradiated Buildings. Lancet. 350:330 - 333.

Ebong I.D.U. and Alagoa, K. D. (1992a). Estimates of gamma - ray background air exposure at a fertilizer plant. Discovery innovate, 4, 25-28.

Ebong, I.D.U. and Alagoa, K. D. (1992b) Fertilizer Impact in ionization radiation background at a production plant. Nig. J. Phys., vol. 4, 143 - 149.

Evamy, B.D., Haremboure, J., Kamerling, P., Knaap, W.A., Molloy, F.A., and Rowlands, P.H., (1978). Hydrocarbon habitat of Tertiary Niger Delta: American Association of Petroleum Geologists Bulletin, vol. 62, p. 277-298.

Foland, C.K., T.K. Kirland and K. VinniKoov (1995). Observed climate variations and changes (IPCC scientific Assessment) Cambridge University Press New York, Pp. 101 - 105.

IAEA (2003). Radiation Protection and the Management of Radioactive Waste in the Oil and Gas Industry. International Atomic Energy Agency Vienna, 2003 Safety Reports Series No.34

ICRP (1990). Recommendations of International Commission on Radiological Protection., Annals of ICRP-60, Pergamon press, Oxford.

ICRP (1977). Recommendation of the International Commission on Radiological Protection. ICRP Publication 261 (3).

Jibiri, N.A.O., Mbawonku, A.O., Oridale A.A. and Ujiagbadion C. (1999). Natural Radionuclide concentration levels in soil and water around cement factory, Ewekoro, Ogun State. Nigeria Journal of Physics, Vol. II, 12 - 16.

Laogun, A.A., N.O Ajayi and S.A. Agaja (2006). Variation in well head gamma radiation levels at the Nigeria Petroleum development company oil field, Ologbo Ede State, Nigeria. Nig. J. Phys, 18(1), 135-140.

Louis, E. A; Etuk, E. S. and Essien, U (2005). Environmental Radioactive Levels in Ikot Ekpene, Nigeria. Nig. J. Space Res. Vol 1, 80 – 87.

ECN(2003).National Energy Policy. The Presidency: Energy commission of Nigeria. Federal Capital Territory, Abuja, Nigeria, pp: 82

NNPC Yearly Bulletin (2004): History of the Nigerian Petroleum Industry, Nigeria pp 54 - 64

OGP (2008) Guidelines for the management of Naturally Occurring Radioactive Material (NORM) in the oil & gas industry, International Association of Oil & Gas Producers

Ononugbo, C. P; Avwiri, G. O. and Chad-Umoren, Y. E. (2011): Impact of Gas Exploitation on the Environmental Radioactivity of Ogba/Egbema/Ndoni Area, Nigeria. Energy and Environment (In press).

Otarigho, M.D., (2007). Impact of oil spillage on the people of Ughelli South Local Government Area, Delta State. J. Environ. Res. & Policies, vol. 2, 44 - 50.

United Nations Development Programme (UNDP) (2006). Niger Delta Human Development report: Environmental and Social Challenges in the Niger Delta. UN House, Abuja, Nigeria.

United Nations Scientific Committee on the Effects of Atomic Radiation (UNSCEAR, 1993). United Nations sources and effects of atomic radiation 1993, Report to the general Assembly with scientific Annexes, United Nations, New York 1993.992.

Part 4

Medical Uses

20

The Effects of Antioxidants on Radiation-Induced Chromosomal Damage in Cancer and Normal Cells Under Radiation Therapy Conditions

Maria Konopacka[1], Jacek Rogoliński[1] and Krzysztof Ślosarek[2]
[1]*Center for Translational Research and Molecular Biology of Cancer,*
[2]*Department of Radiotherapy and Brachytherapy Treatment Planning,*
Maria Sklodowska-Curie Memorial Cancer Center and
Institute of Oncology, Gliwice,
Poland

1. Introduction

Radiotherapy is the major form of treatment for many human cancer. During the course of treatment, the ionising radiation produces many biological effects not only in cancer but also in normal cells. Due to the risk of toxicity to normal cells, the radioprotectors are needed to reduce the normal tissue injury during the irradiation of tumours without influence on effectiveness on cancer treatment. Most of the compounds that showed the radioprotective capacity in laboratory studies failed because of their toxicity to the normal cells. The good radioprotector should be non-toxic and should selectively eliminate the cancer cells. A number of natural dietary ingredients show capacity to protect cells from damage induced by ionising radiation (Arora et al., 2008).

It is well known that antioxidant vitamins such as ascorbic acid and vitamin E protect cellular DNA and membranes from radiation-induced damage (Noroozi et al., 1998; Konopacka & Rzeszowska-Wolny, 2001; Kumar et al., 2002; Jagetia, 2007). Recently, several flavonoids, polyphenols and phenolic acids have become more popular as diet compounds due to their beneficial impact on human health. One of them is ferulic acid. It renders preferential radioprotection to normal tissue, but not to tumour cells under both ex vivo and in vivo conditions (Maurya et al., 2005; Maurya & Nair, 2006). Ferulic acid is present in many plant products such as giant fennel, green tea, coffee beans and grains. It is monophenolic phenylpropanoid that acts as an antioxidant against peroxyl radicals-induced oxidation in neuronal culture and in synaptosomal membranes (Kanski et al., 2002). It was found that ferulic acid reduced the number of radiation-induced DNA strand breaks and enhanced the DNA repair processes in peripheral blood lymphocytes but not influenced the level of radiation-induced damage in fibroblastoma tumour cells in mice (Maurya et al., 2005).

It has been also showed that the α-tocopherole can preferentially reduce the level of radiation-induced chromosomal damage in normal human cells, but in cancer cells it actually increases the level of damage (Jha et al., 1999; Kumar et al., 2002). Although there is much evidence about the modulating effects of antioxidants in cells directly irradiated the effects of low dose of scattered radiation are poorly studied.

Our previous study showed that the cells placed outside the irradiation field were exposed to very low doses of scattered radiation that induced the micronuclei (Konopacka et al., 2009) and decreased the cell viability (Rogolinski et al., 2009). The extent of micronuclei formation and apoptosis as well as the decrease of viability of cells exposed to scattered radiation was higher that could be predicted by dosimetric methodology based on the linear non-treshold model (LNT). In this situation, the radioprotection of normal tissues placed outside the irradiation field during radiotherapy of cancer seems to be very crucial. In present study, we compared the influence of vitamin C, vitamin E and ferulic acid on micronuclei formation in human cancer cells directly irradiated and in normal human cells placed outside the irradiation field during exposure. This study was performed to answer the question whether the antioxidants selected by us could preferentially protect normal cells but not cancer cells during exposure to radiation in conditions mimicking the radiotherapy. We tested the modulating effects of vitamin C, vitamin E and ferulic acid on micronuclei formation in directly irradiated cancer A549 cells as well as in normal BEAS-2B cells exposed outside of the radiation field.

2. Experimental procedures

2.1 Cell culture

Human lung carcinoma cells (A549 line) and normal human bronchial epithelial cells (BEAS-2B) were grown on DMEM/F12 medium supplemented with 10% fetal bovine serum (Immuniq) in a humidified atmosphere of 5% CO_2 at 37°C. Before irradiation the cells were trypsinized and 20 μl of cell suspension, containing approximately 10^3 cells, were transferred to an Eppendorf`s tubes and then the tubes were filled with medium up 0.5 ml so the cells were irradiated without presence of air.

2.2 Preparation of antioxidants

Vitamin C (Serva, Germany) was dissolved in culture medium and filter sterilized. Vitamin E (α- tocopherole, Sigma) and ferulic acid (Linegal) were dissolved in ethanol. Antioxidants were added to cultures 1 h before irradiation at final concentration from 1 to 100 μg/ml. Experiments included control cultures treated with ethanol alone at the same volume as volume of vitamin E or ferulic acid added to cultures.

2.3 Irradiation of cells

The cancer cells in Eppendorf`s tubes were placed in a special stand on 3 cm of depth in a water phantom and expose to 5 Gy. The normal cells in Eppendorf`s tubes were placed in a water phantom at distance of 4 cm outside the radiation field. These cells were exposed to scattered radiation at dose of 0.2 Gy. The tubes were placed horizontally in the water phantom in such a way that the A549 cancer cells were within the radiation beam field, whereas the normal BEAS-2B cells were placed 4 cm outside the beam field, as is presented in Fig.1.

Experiments were performed for electron (22 MeV) radiation generated in a linear accelerator Clinac series Varian Medical system, for 300 Mu/min accelerator mode and dose of 5 Gy in build-up (3 cm) depth in a water environment. After irradiation the cells were transferred into plastic dishes (50 mm diameter) and supplemented with up 5 ml of the culture medium.

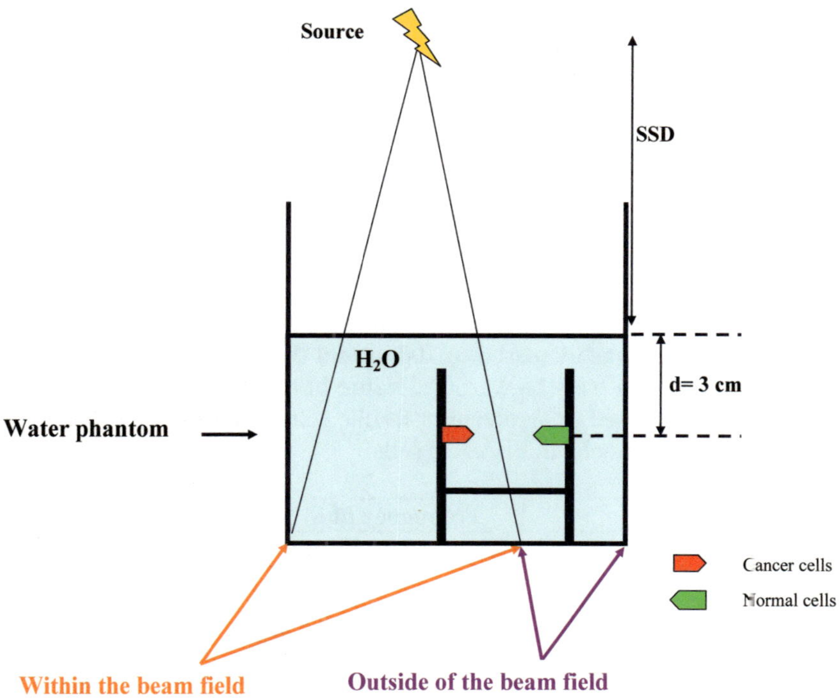

Fig. 1. The scheme of the irradiation set-up

2.4 Cytokinesis-block micronucleus test

The micronucleus test was performed according to the standard procedure (Fenech & Morley, 1985). After irradiation and transferring of cells to culture dishes, the cytochalasin B was added to medium to a final concentration 2 µg/ml and cells were incubated for 48h prior fixation. The cells were fixed *in situ* with a cold solution of 1 % glutaraldehyde (Sigma) in phosphate buffer (pH=7.5) and stained by Feulgen reaction. At least 500 binucleate cells were examined for the presence of micronuclei (MN-BN cells) under microscope.

2.5 Statistical analysis

Experiments were repeated three times. Means ± SD were calculated from experimental data and the Student`s t-test was used to determine the statistical significance of differences in the number of micronuclei between cells cultured and irradiated in the presence or absence of antioxidants.

3. Results

The effect of ferulic acid, vitamin C and vitamin E on the micronuclei formation in the normal human BEAS-2B cells is presented in Tab.1.

Concentration (μg/ml)	Frequency of MN-BN cells (%)		
	Ferulic acid	Vitamin C	Vitamin E
0		3.83 ± 0.44	
1	3.50 ± 0.32	3.44 ± 0.28	3.75 ± 0.43
0	3.33 ± 0.40	3.50 ± 0.42	3.80 ± 0.53
50	4.00 ± 0.62	3.85 ± 0.49	3.66 ± 0.44
100	4.98* ± 0.66	4.00 ± 0.55	4.10 ± 0.58
Ethanol		3.90 ± 0.42	

*- significantly different from control at $p < 0.01$.

Table 1. Influence of ferulic acid, vitamin C and vitamin E on the micronucleus formation in BEAS-2B cells. Means ±SD of three experiments are shown.

The background level of micronuclei was 3.83 ± 0.44 and none of the antioxidants at their concentration below 50 μg/ml caused any changes above this value. When the concentration of antioxidants was increased to 100 μg/ml the frequency of micronucleated cells was higher in comparison with background value but this difference was significant only for ferulic acid. Ethanol used as a solvent of ferulic acid and vitamin E did not change the spontaneous level of micronuclei in BEAS-2B cells.

Concentration (μg/ml)	Frequency of MN-BN cells (%)		
	Ferulic acid	Vitamin C	Vitamin E
0		2.10 ± 0.32	
1	2.18 ± 0.35	2.21 ± 0.33	2.25 ± 0.33
10	6.62* ± 0.55	2.26 ± 0.36	2.27 ± 0.20
50	9.16** ± 0.74	2.16 ± 0.29	2.20 ± 0.29
100	10.20** ± 0.98	2.23 ± 0.25	2.34 ± 0.42
Ethanol		2.16 ± 0.47	

Significantly different from control at: * $p < 0.01$, ** $p < 0.001$.

Table 2. Influence of ferulic acid, vitamin C and vitamin E on the micronucleus formation in A 549 cells. Means ±SD of three experiments are shown.

Tab.2 presents the effect of antioxidants on the micronuclei formation in tumour A549 cells. Addition of vitamin C or vitamin E at concentrations ranging from 1 up to 100 μg/ml did not cause any measurable changes in the spontaneous level of micronuclei (2.10 ± 0.32). Ferulic acid at the concentrations 10 μg/ml and higher increased significantly the number of micronucleated cells in comparison with the control cells incubated without antioxidants. Ethanol did not cause any changes above the background in A549 cells.

In the next experiments we tested the effect of antioxidants on the level of radiation-induced micronuclei in cells irradiated in a water phantom (see Material and Methods). Antioxidants were tested at concentrations 1, 10 and 25 μg/ml. The results of this experiment in normal BEAS-2B cells exposed to radiation outside the field are showed in Fig.2.

Vitamin C at the concentrations of 1 and 10 μg/ml protected cells from radiation-induced DNA damage but at concentration of 25 μg/ml it was not effective in reducing this damage. In contrast to ascorbic acid, vitamin E was effective as a radioprotector at concentration above 10 μg/ml. Ferulic acid showed the best protective effect against radiation-induced micronuclei in normal cells. It inhibited significantly the micronuclei formation in dose-dependent manner.

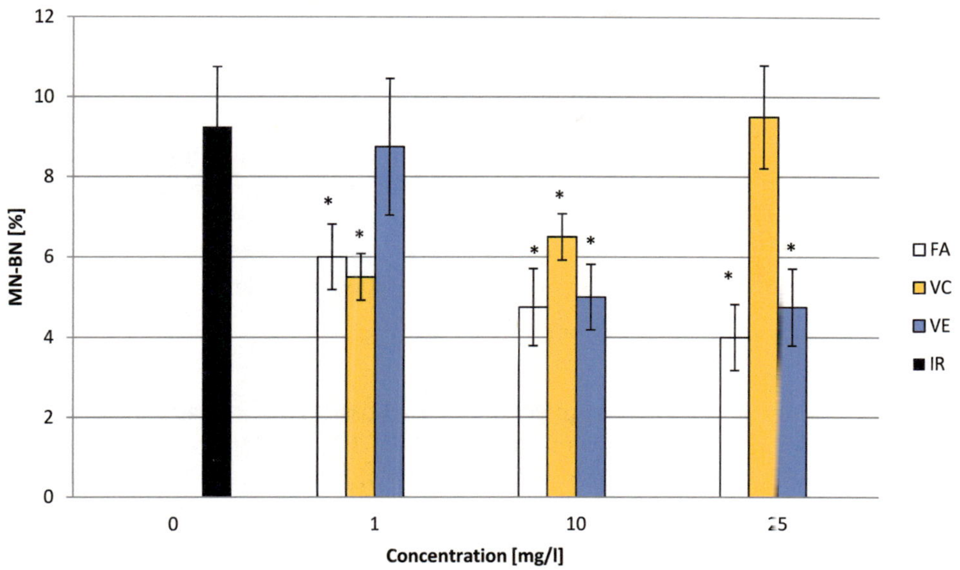

Values are means ±SD of three experiments.
* $p < 0.01$ refers to differences between irradiated only and preincubated with antioxidants cells (Student's t - test).

Fig. 2. Effect of ferulic acid, vitamin C and vitamin E on the micronuclei formation in BEAS-2B cells exposed to radiation outside the field.

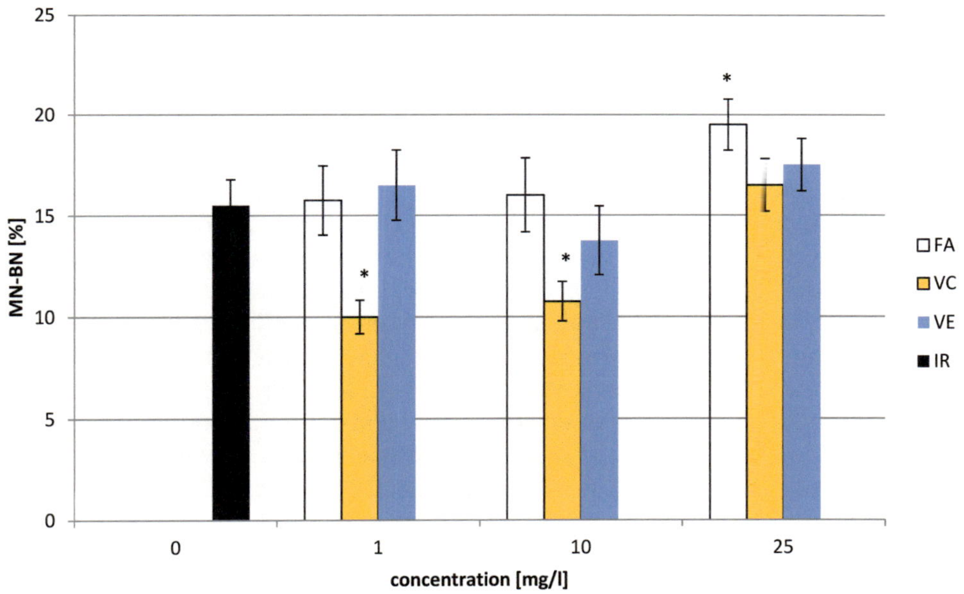

Values are means ±SD of three experiments.
* $p < 0.01$ refers to differences between irradiated only and preincubated with antioxidants cells (Student's t - test).

Fig. 3. Effect of ferulic acid, vitamin C and vitamin E on the micronuclei formation in A-549 cells irradiated in a beam axis.

Fig.3 shows the radioprotective capacity of antioxidants in cancer cells irradiated in beam of axle. Vitamin C at low concentration (below 10 µg/ml) diminished the level of radiation-induced micronuclei. At the concentration of 25 µg/ml we observed no effect of this vitamin on formation of micronuclei in irradiated cells in comparison with the cells irradiated without vitamin treatment. Vitamin E at all concentrations did not influence the level of radiation-induced chromosomal damage. In cells irradiated in the presence of ferulic acid at concentrations 1 or 10 µg/ml we did not observe any protective effect in reducing the level of radiation-induced micronuclei but at the highest concentration (25 µg/ml) ferulic acid significantly increased the number of radiation induced micronuclei.

4. Discussion

Radioprotectors should selectively protect the normal tissues during radiotherapy of cancer without inhibition of damaging of cancer cells. In present work we tested the radioprotective activity of known antioxidants, namely, ferulic acid, vitamin C and vitamin E, in normal cells versus cancer cells. To answer the question whether these antioxidants can preferentially protect the normal cells during radiotherapy, we performed experiments under conditions like those that would be utilized in radiation therapy treatment procedures: tumour cells were irradiated directly (in a beam axle), whereas the normal cells were placed outside the radiation field during exposure. Our results indicate that the radioprotective effect of vitamin C is concentration-dependent and similar in normal and cancer cells; vitamin C at low concentration diminished the radiation-induced micronuclei whereas at high concentration it enhanced the level of damage, that was stronger in cancer than in normal cells. It is known that high concentration of ascorbic acid can potentate the production of hydroxyl radicals from hydrogen peroxide via Fenton reaction, which enhances the level of radiation-induced DNA damage (Halliwell & Gutteridge, 1985). This action of vitamin C varied among cell types due to the differences in the intracellular concentration of the ascorbic acid ranging from 10 µM in serum blood cells to 700 µM in bone marrow cells (Umegaki et al., 1995). It has been showed that vitamin C supplemented with vitamin K3 was effective in killing of cancer cells via activation of DNase, which degrades tumour cell DNA and induces cell death (Jamison et al., 2004).

Our results indicate that ferulic acid and vitamin E are potentially very good radioprotectors of normal cells during radiotherapy because they reduced the number of radiation-induced micronuclei in normal cells and simultaneously they did not influence the damaging effect of radiation in cancer cells and at high concentration they enhanced the damaging effect of radiation in cancer cells. Moreover ferulic acid showed selectively clastogenicity expressed as a micronuclei formation in tumour cells. The number of micronucleated cancer cells increased over the background value in concentration-dependent manner.

Our observations are in agreement with published data that indicated that ferulic acid preferentially protected normal mouse bone marrow and blood cells but not cancer fibroblastoma cells in mice exposed to 4 Gy of γ-radiation (Maurya & Nair, 2006). It was also shown that it protected human lymphocytes against radiation-induced chromosomal damage (Prasad et al., 2006). Moreover, it was found that vitamin E selectively protected normal human fibroblasts but not human cervical cancer and ovarian carcinoma cells against radiation-induced chromosomal damage (Kumar et al., 2002) and cell cycle inhibition (Jha et al., 1999). It has been suggested that the preferential protection of normal cells against radiation can be connected with alteration of genes encoding the elements of the cell signalling pathways such as transcriptional factor E2F (Turley et al., 1997) and repair processes (Maurya & Nadir, 2006).

5. Conclusions

The results presented in this paper indicate that antioxidants such as ferulic acid and vitamin E protect to normal cells exposed to low dose of scattered radiation present outside the radiation field. Recently, there is increasing attention for the low dose radiation exposure including non-target phenomena such as bystander effect and low dose hypersensitivity and genetic instability (Morgan & Sowa, 2007) that can be responsible for the induction of secondary cancer. The protection of normal cells against these distant effects appears to be important element in radiotherapy and out to be taken into consideration in clinical practice. Ferulic acid and vitamin E seem to be the promising protectors of normal cells during radiotherapy.

6. Acknowledgment

This study was supported by a grant from Polish Ministry of Science and Higher Education, No. NN 402 4447 33.

7. References

Arora, R.; Kumar, R.; Sharma, A. & Tripathi, R.P. (2008). Radiomodulatory compounds of herbal origin for new frontiers in medicine, homeland security, management radiological incidents and space application. In: *Herbal Radiomodulators, applications in medicine, homeland defence and space*, R. Arora, (Ed.), 1 – 22, CABI North American Office, Cambridge

Fenech, M. & Morley, A.A. (1985). Measurement of micronuclei in lymphocytes. *Mutation Research* 147, 29-36

Halliwell, B. & Gutteridge, J.M.C. (1985). Free Radicals in Biology and Medicine, Clarendon Press, Oxford, pp 58-88

Jagetia, G.C. (2007). Radioprotective potential of plants and herbs against the effects of ionizing radiation. *Journal of Clinical Biochemistry and Nutrition* 40, 74-81

Jamison, J.M.; Gilloteaux, J.; Nassiri, M.R.; Venugopal, M.; Neal, D.R. & Summers, J.L. (2004). Cell cycle arrest and autoschizis in a human bladder carcinoma cell line following Vitamin C and Vitamin K3 treatment. *Biochemical Pharmacology* 15, 337-351

Jha, M.N.; Bedford, J.S.; Cole, W.C.; Edward-Prasad, J. & Prasad, K.N. (1999). Vitamin E (d-alpha-tocopheryl succinate) decreases mitotic accumulation in gamma-irradiated human tumor, but not in normal cells. *Nutrition and Cancer* 35, 189-194

Kanski, J.; Aksenova, M.; Stoyanova, A. & Butterfield, D.A. (2002). Ferulic acid antioxidant protection against hydroxyl and peroxyl radical oxidation in synaptosomal and neuronal cell culture systems in vitro: structure-activity studies. *Journal of Nutritional Biochemistry* 13, 273-281

Konopacka, M. & Rzeszowska-Wolny, J. (2001). Antioxidant vitamins C, E and beta-carotene reduce DNA damage before as well as after gamma-ray irradiation of human lymphocytes in vitro. *Mutation Research* 49, 1-7

Konopacka, M.; Rogoliński, J. & Ślosarek, K. (2009). Comparison of dose distribution of ionizing radiation in a water phantom with frequency of cytogenetic damage in a human bronchial cells. In: *World Congress on Medical Physics and Biomedical Engineering. IFMBE Proceedings,* Dossel Schlegel (Ed.), Munich, 7-12 September 2009, 25/3, pp 379-382

Kumar, B.; Jha, M.N.; Cole, W.C.; Bedford, J.S. & Prasad, K.N. (2002). D-alpha-tocopheryl succinate (vitamin E) enhances radiation-induced chromosomal damage levels in human cancer cells, but reduces it in normal cells. *Journal of American College of Nutrition* 21, 339-343

Maurya, D.K.; Salvi, V.P. & Nair, C.K. (2005). Radiation protection of DNA by ferulic acid under in vitro and in vivo conditions. *Molecular and Cellular Biochemistry* 280, 209-217

Maurya, D.K. & Nair, C.K. (2006). Preferential radioprotection to DNA of normal tissues by ferulic acid under ex vivo and in vivo conditions in tumor bearing mice. *Molecular and Cellular Biochemistry* 285, 181-190

Morgan, W.F. & Sowa, M.B. (2007). Non-targeted bystander effects induced by ionizing radiation. *Mutation Research* 616, 159-164

Noroozi, M.; Angerson, W.J. & Lean, M.E. (1998). Effects of flavonoids and vitamin C on oxidative DNA damage to human lymphocytes. *American Journal of Clinical Nutrition* 67, 1210-1218

Prasad, N.R.; Srinivasan, M.; Pugalendi, K.V. & Menon, V.P. (2006). Protective effect of ferulic acid on gamma-radiation-induced micronuclei, dicentric aberration and lipid peroxidation in human lymphocytes. *Mutation Research* 28, 129-134

Rogolinski, J.; Konopacka, M.; Sochanik, A. & Ślosarek, K. (2009). Scattering medium depth and cell monolayer positioning with respect to beam field affect cell viability. In: *World Congress on Medical Physics and Biomedical Engineering. IFMBE Proceedings,* Dossel Schlegel (Ed.), Munich, 7-12 September 2009, 25/3: pp 403-405

Turley, J.M.; Ruscetti, F.W.; Kim, S.J.; Fu, T.: Gou, F.V. & Birchenall-Roberts, M.C. (1997). Vitamin E succinate inhibits proliferation of BT-20 human breast cancer cells: increased binding of cyclin A negatively regulates E2F transactivation activity. *Cancer Research* 57, 2668-2675

Umegaki, K.; Aoki, S. & Esashi, T. (1995). Whole body X-ray irradiation to mice decreases ascorbic acid concentration in bone marrow: comparison between ascorbic acid and vitamin E. *Free Radicals in Biology and Medicine* 19, 493-497

21

Melatonin for Protection Against Ionizing Radiation

M. A. El-Missiry[1], A. I. Othman[1] and M. A. Alabdan[2]
[1]*Department of Zoology, Faculty of Sciences, Mansoura University, Mansoura,*
[2]*Department of Zoology, Faculty of Science,*
Princess Nora Bint AbdulRahman University, Riyadh,
[1]*Egypt*
[2]*Kingdom of Saudi Arabia*

1. Introduction

Radiations exist ubiquitously in the environment since the Earth's creation in soil, water and plants. Radiation exposure is a concern in the health industry and other occupations in the world. Apart from diagnostic, therapeutic and industrial purposes, humans also are exposed to ionizing radiations during air and space travel and exploration, background radiation, nuclear accidents, and nuclear terror attacks. Elevated radiation levels have been detected following Chernobyl on April 1986 at Ukraine, and recently Fukushima Daiichi Nuclear Power Plants on March 2011 at Japan. This raised the need for finding out efficient and reliable radioprotectors especially when a whole nation is exposed at high or even low levels for a prolonged period. The fallout and radioactivity cause concern during the weeks and months after the accidents. In addition, radiations are commonly used in a number of medical and industrial situations; however, their pro-oxidative effects limit their applications. Therefore, it is essential to protect humans from ionizing radiations by efficient pharmacological intervention. A valid approach to halt normal tissue radiotoxicity is the use of radioprotectors that when present prior to radiation exposure protect normal tissues from radiation effects. This view has also been used as a successful preventative measure for possible nuclear/radiological situation. From a practical point of view radioprotectors should perfectly have several criteria that relate to the ability of the agent to improve the therapeutic outcome. Ionizing radiation causes oxidative damage to tissues within an extremely short period, and possible protection against it would require the rapid transfer of smart antioxidants to the sensitive sites in cells. At this point, melatonin (N-acetyl-5-methoxytryptamine; MW= 232), an innate antioxidant produced mainly by the pineal gland, seems unique among antioxidants because of its multiple properties and reactions which reviewed and documented in several publications and summarised herein.

While ionizing radiation exposures, due to free radical generation, present an enormous challenge for biological and medical safety, melatonin is a potent radioprpotector. In several investigations, melatonin has been recognized for successful amelioration of oxidative injury and illness due to direct and indirect effects of ionizing radiation and against oxidative stress in several experimental and clinical settings. Furthermore, numerous studies have established that melatonin is a highly efficient free radical scavenger, broad antioxidant and stimulator of several antioxidants in biological systems. Because of its unique characteristics; melatonin has effects not only at the cell level but also within subcellular organelles and structures. The antioxidant and prophylatic properties of melatonin allow the use of radiation during radiotherapy to get better therapeutic outcomes. Several published articles documented that melatonin's anticancer and oncostatic effects make melatonin an excellent candidate and good choice to be used in routine radiotherapies, space travel and following nuclear accidents occupational settings where accidental exposure may occur. This article will review antioxidant features that put melatonin on top of potentially efficient pharmacological radioprotectors.

2. Ionizing radiation, free radicals & oxidative stress

Ionizing radiations are types of particle radiation (such as neutron, alpha particles, beta particles and cosmic ray) or electromagnetic (such as ultraviolet, X-rays and gamma rays) with sufficient energy to ionize atoms or molecules by detaching electrons from their valence orbitals. The degree and nature of such ionization depends on the energy of the individual particles or on frequency of electromagnetic wave. It is well known that exposure to ionizing radiation at sufficiently high doses results in various types of adverse biological effects. The biological effect of radiation involves direct and indirect actions. Both actions produce molecular changes that mostly need enzymatic repair. Indirect effect involves the production of reactive free radicals which produce oxidative mutilation on the key molecules. The environmental sources of oxidative attack include, in particularly, specific exposures of the organism to ionizing radiations like X-, γ- or cosmic rays and α-particles from radon decay as well as UVA and UVB solar light. Ionizing radiations prevalent in space, involve a broad range of radiation types and energies from cosmic and unpredictable solar sources, representing a very diverse range of ionization qualities and biological effectiveness. Linear energy transfer (LET) is a measure of the energy transferred to tissue or cells as an ionizing particle travels through it. The LET of the potential radiations can cover several orders of magnitude from <1.0 keV μm^{-1} to > several 100 keV μm^{-1} (Blakely and Chang 2007) Low LET radiation causes damage through reactive oxygen species (ROS) production mainly by the radiolysis of water present in living system.

From a chemical point of view, reactive oxygen species (ROS) and reactive nitrogen species (RNS) are oxygen and nitrogen containing molecules constitute the main category of free radicals which may be defined as any chemical moiety generated with an unpaired number of electrons in valancy orbital. ROS include oxygen-based free radicals, the superoxide anion (O_2•−), hydroxyl (•OH), alkoxyl (RO•), peroxyl (ROO•), and hydroperoxyl (HOO•). RNS include peroxynitrite (ONOO−), nitric oxide (NO•), and nitrogen dioxide (NO2•). ROS may be radical, such as, $O_2^{\bullet -}$ and •OH, or non-radical, such as, hydrogen peroxide (H_2O_2) and singlet oxygen ($^1O_2$). At high concentrations, free radicals can be harmful to living organisms. Some ROS damage biomolecules indirectly. For example, H_2O_2 and O_2•− initiate DNA and lipids damage by interaction with transition metal ion, in particular iron and copper, in the metal-catalysed Haber–Weiss reaction, producing •OH. It is the most electrophilic and reactive of the ROS, with a half-life of~10^{-9} s (Draganic and Draganic 1971). •OH can be produced by ultraviolet and ionizing radiations (Von Sonntag 1987). This radical is considered the most frequently damaging species. It has been estimated that the •OH is responsible for 60–70% of the tissue damage caused by ionizing radiations (Galano et al. 2011). Moreover, •OH has great ability to react with almost any molecule in the vicinity of where it is generated (Reiter et al. 2010). Chemical nature and reactivity of free radicals in biological systems has been recently reviewed (Galano et al. 2011). Once formed ROS and RNS can produce a chain reaction. The transfer of the free radical to a biological molecule can be sufficiently damaging to cause bond breakage or inactivation of key functions. The organic ROO• can transfer the radical from molecule to molecule causing damage at each encounter. Thus, a cumulative effect can occur, greater than a single ionization or broken bond.

A variety of external events, in particular, exposure to ionizing or ultraviolet radiation, can lead to an increase in the generation of ROS in comparison with available antioxidants leading to oxidative stress. Oxidative stress is caused by the presence of excessive amount of ROS which the cell is unable to counterbalance. This implies that the steady state balance of pro-oxidant/anti-oxidant systems in intact cells is shifted to the former. When excessive oxidative events occur, the pro-oxidants outbalance the anti-oxidant systems. Moreover, oxidative stress may result by overwhelming of antioxidant and DNA repair mechanisms in the cell by ROS. In radiation sickness oxidative stress is a factor as either cause or effect. The result is oxidation of critical cellular macromolecules including DNA, RNA, proteins and lipids eventually leading to cell death in severe oxidative stress. On the other hand, moderate oxidative stress may lead to activation of cytoplasmic/nuclear signal transduction pathways, modulation of gene and protein expression and alteration of DNA polymerase activity, affect the endogenous anti-oxidant systems by down-regulating proteins that participate in these systems, and by depleting cellular reserves of anti-oxidants (Acharya et al. 2010, Cadet et al. 2010, Little 2000).

3. Molecular biology effects of ionizing radiation due to free radical generation

As a matter of fact, ionizing radiation penetrating living tissue and can damage all important cellular components both through direct ionization and through generating ROS due to water radiolysis and induce oxidative damage. Radiation-induced oxidative stress was evaluated by three independent approaches; DNA damage, lipid peroxidation and protein oxidation.

3.1 DNA damage

Cells and their genomic constituent of the living organisms are continually exposed to oxidative attacks. Acute exposure to ionizing radiation can create oxidative stress in a cell and chronic exposure to this stress can result in permanent changes in the genome (Cooke et al. 2003). The main target of ionizing radiation has long since been indicated to be DNA which shows wide range of lesions. The oxidatively DNA damage commonly are apurinic/apyrimidinic (abasic) DNA sites, oxidized purines and pyrimidines, single strand (SSBs) and double strand (DSB) DNA breaks and non-DSB (Kryston et al. 2011). Other initial chemical events induced in DNA by ionizing radiation include cross-links, oxidative base modification (Hutchinson 1985) and clustered base damage (Goodhead 1994), sugar moiety modifications, and deaminated and adducted bases (Cooke et al. 2003, Sedelnikova et al. 2010, Sutherland et al. 2000, Ward 1994). The numbers of DNA lesions per cell that are detected immediately after a radiation dose of 1 Gy have been estimated to be approximately greater than 1000 base damage, 1000 SSBs, 40 DSBs, 20 DNA–DNA crosslinks, 150 DNA–protein cross-links and 160–320 non-DSB clustered DNA damage and defective DNA mismatch repair proteins (MMP) (Martin et al. 2010). Recently, it is suggested that radiation dose and the type of DNA damage induced may dictate the involvement of the MMP system in the cellular response to ionizing radiation. In particular, the literature supports a role for the MMP system in DNA damage recognition, cell cycle arrest, DNA repair and apoptosis (Martin et al. 2010). In addition, The DNA oxidation products are a direct risk to genome stability, and of particular importance are oxidative clustered DNA lesions, defined as two or more oxidative lesions present within 10 bp of each other (Sedelnikova et al. 2010).

The most common cellular DNA base modifications are 8-oxo-7,8-dihydroguanine (8-oxoGua) and 2,6- diamino-4-hydroxy-5-formamidopyrimidine. Both originate from the addition of the •OH to the C8 position of the guanine ring producing a 8-hydroxy-7,8-dihydroguanyl radical which can be either oxidized to 8-oxoGua or reduced to give the ring-opened FapyGua (Altieri et al. 2008, Kryston et al. 2011). The •OH interact with pyrimidines (thymine and cytosine) at positions 5 or 6 of the ring, and yield several base lesions. The most abundant and well known products, 5,6-dihydroxy-5,6-dihydrothymine (thymine glycol) and 5,6-dihydroxy-5,6-dihydrocytosine (cytosine glycol). It is generally accepted that 8-oxodG and thymine glycol are reliable biomarkers of high levels of oxidative stress and damage in the human body. These lesions ultimately are not lethal to the cell, but are considered to be highly mutagenic. Oxidized bases in DNA are potentially mutagenic and so are implicated in the process of carcinogenesis. It has been reported that X-radiation induced a significant increase in 8-OHdG concentration in mammary gland DNA (Haegele et al. 1998). High levels of 8-OHdG have been observed in normal human epidermis or purified DNA exposed to ultraviolet radiation (Wei et al. 1997). Thus, genome stability is crucial for maintaining cellular and individual homeostasis, but it is subject to many changes due to free radicals attack induced by the exposure to ionizing radiation. DNA breaks and fragments resulted from chromosomal damage appear as micronuclei in rapidly proliferating cells micronuclei frequency was markedly enhanced in bone marrow cells of mouse exposed to 5Gy radiation (Verma et al. 2010).

Comprehensive reviews have been appeared to give structural and mechanistic information on the radiation-induced damage to DNA (Cadet et al. 2010, Cadet et al. 2005). However, these authors showed that there is still a dearth of precise data on the formation of radiation-induced base injury to DNA in cells and tissues. This is because the determination of the radiation-induced base damage within DNA is achieved indirectly by methods utilizing hydrolysis of the biopolymer then followed by analysis of the free fragments. The situation is even more difficult for cellular DNA since highly sensitive assays are required to monitor the formation of very low amounts of injury, typically within the range of one modified base per 10^6 normal nucleotides.

Direct damage to DNA caused by ionizing radiation has been considered as a significant initiator of mutation and cancer. However, some reports suggest that extracellular and extranuclear targets may contribute to the genotoxic effects of radiation (Little 2000, Morgan 2003). In addition, it has been shown that irradiation of the cytoplasm produces gene mutations in the nucleus of the hit cells and that this process is mediated by free radicals (Wu et al. 1999). Recently, it is proposed that a possible extracellular signal-related kinase pathway involving ROS/RNS and COX-2 in the cytoplasmic irradiation-induced genotoxicity effect (Hong et al. 2010). Furthermore, it has been demonstrated that nitric oxide synthase (NOS) produces sustained high concentrations of nitric oxide (NO) in various mammalian cells after exposure to radiation (Matsumoto et al. 2001). In cytoplasmic-irradiated cells, 3-nitrotyrosine, a nitrosated protein product used as a marker of ONOO-, was significantly elevated and dramatically inhibited by L-NMMA, (NO inhibitor) implicating a critical role of RNS in the mutagenicity induced by cytoplasmic irradiation (Hong et al. 2010).

3.2 Lipid peroxidation

The direct and indirect destructive effects of ionizing radiation lead to peroxidation of macromolecules, especially those present in lipid-rich membrane structures, lipoproteins and chromatin lipids. Phospholipids in membranes and triglycerides in LDL are highly susceptible to free radical attacks. Once the process of lipid peroxidation is started, it proceeds as a free radical-mediated chain reaction involving initiation, propagation, and termination (Gago-Dominguez et al. 2005). The first step (initiation) in the lipid peroxidation process is the abstraction of a hydrogen atom, from a methylene group next to a double bond in polyunsaturated fatty acids. This produces a carbon centered radical which undergo rearrangement of the double bond to form a stable conjugated diene. In propagation step, carbon centered radicals react with oxygen to form new ROO• that react further with another neighboring lipid molecule forming a hydroperoxy group and a new carbon centered radicals. The lipid hydroperoxide will react further to form cyclic peroxide, cyclic endoperoxide, and finally aldehydes. The propagation phase can repeat many times until it is terminated by chain breaking antioxidants (Halliwell 2009, Reed 2011)

During lipid peroxidative pathway, several end products are formed such as malondialdehyde (MDA) and 4-hydroxy-2-nonenal (4-HNE), pentane and ethane, 2,3 transconjugated dienes, isoprostanes and cholesteroloxides (Catala 2009, Tuma 2002). These aldehydes are highly reactive and bind with DNA and proteins and form adducts which inhibit proteins functions and can disrupt nuclear events. The aldehydes are more diffusible than free radicals, thus injury can occur in distant locations. The aldehyde can be found in measurable concentrations in biological fluids and analytical methods used are sometimes complex and require sample preparation involving extraction and purification steps. Isoprostanes, prostaglandin like compounds, are generated from the free radical-initiated peroxidation of arachidonic acid. F2-isoprostanes are the most specific markers of lipid peroxidation and the most difficult to measure (Comporti et al. 2008). The mostly frequently lipid peroxidation markers used in free radical research are MDA and 4-HNE. Lipid peroxidation was assessed as thiobarbituric acid reactive substances (TBARS) in biological materials using thiobarbituric acid reaction method (Esterbauer and Zollner 1989, Moller and Loft 2010).

Several studies have examined the radiation-induced free radical damage evidenced by the elevation of lipid peroxidation levels. Lipid peroxidation-derived products have been implicated in the pathogenesis of oxidative stress-associated radiation sickness and diseases. Aldehydes showed significant increase with increasing doses of ionizing radiation in several organs (Bhatia and Manda 2004, Sener et al. 2003) and mitochondrial membranes (Kamat et al. 2000). A significant increase in DNA strand breaks and TBARS concentrations was found in rat brain exposed to 10Gy ionizing radiation (Undeger et al. 2004). Exposure to 5 Gy irradiation led to considerable elevation of MDA level in thymus, brain, jejunum liver and kidney of total body irradiated mice (Taysi et al. 2003, Verma et al. 2010) after 24 hrs of irradiation continued up to 48 hrs. Lipid peroxidation due to •OH attack were found to be in a radiation dose-dependent manner but no significant differences between radiation

resistant and radiation sensitive rats were detectable after whole-body-irradiation with x-rays at 2, 4, and 6 Gy. Among the subcellular organelles mitochondria are one of the key components of the cell injured by radiation-induced oxidative stress. In an interesting study, mitochondria from rat brain and liver was isolated then exposed to 450Gy gamma radiation. In this study there was considerable increase in lipid hydroperoxide (LOOH) and MDA in rat liver and brain mitochondria (Lakshmi et al. 2005). Recent research showed that increased free radicals due to radiation exposure damage membrane lipids, which results in cell lysis due to altered membrane fluidity (Gulbahar et al. 2009).
The ideas about the leading role of lipid peroxidation in radiation damage of cells and tissues arose from the damaging of cell membrane structures. Peroxidation of lipids can greatly alter the physicochemical properties of membrane lipid bilayers, resulting in severe cellular dysfunction. It causes the change in structure, fluidity and permeability of membranes and inactivates several membrane associated enzymes and protein receptors. In biological membranes, lipid peroxidation is also usually accompanied by oxidation of membrane proteins. In consequence, peroxidation of lipids may change the agreement of proteins in bilayers and by that interfere with their physiological role on the membrane function.

3.3 Protein oxidation

As defined earlier, ionizing radiation can interact and modify all cellular components both through direct ionization and through induction of ROS resulting in a variety of subtle and profound biological effects. Radiation-induced oxidative protein damage can be started by even quite low doses of radiation and can produce an alteration of the cellular redox balance, which lasts for substantial time after exposure and may contribute to changes in cell survival, proliferation, and differentiation(Shuryak and Brenner 2009). Several damages to the peptide chain or to the side-chains of amino acid residues have been identified, and some of their mechanisms of formation has been described (Griffiths et al. 2002).
Available data from various studies raveled that the most sensitive amino acids, cysteine, tryptophan, tyrosine and methionine, bear aromatic or sulphur-containing side-chains. Furthermore, protein oxidation can lead to hydroxylation of aromatic groups and aliphatic amino acid side chains, nitration of aromatic amino acid residues, nitrosylation of sulfhydryl groups, sulfoxidation of methionine residues, chlorination of aromatic groups and primary amino groups, and to conversion of some amino acid residues to carbonyl derivatives (Catala 2007). The fundamental mechanisms involved in the oxidation of proteins by ROS were described by studies in which amino acids, peptides, and proteins were exposed to ionizing radiations under conditions where •OH or a mixture of •OH and O_2•− are formed (Stadtman 2004). It has been demonstrated that the attack by •OH leads to an abstraction of a hydrogen atom from the protein polypeptide backbone and form a carbon-centered radical (Klaunig et al. 2011, Stadtman 2004). Oxidation due to radiation exposure can lead also to cleavage of the polypeptide chain and formation of cross-linked protein aggregates. Because the generation of carbonyl derivatives occurs by many different mechanisms, the level of carbonyl groups in proteins is widely used as a marker of oxidative protein damage (Guajardo et al. 2006). Studies performed with various tissues have revealed that radiation increases protein oxidation and carbonyl levels as well as produces structural and functional changes (Gulbahar et al. 2009).
In most reports describing the in vivo experiments on radiation sickness, the carbonyl levels were determined in tissue homogenates or soluble cytoplasmic proteins. Moreover, it is very

important to consider the carbonyl levels in different subcellular fractions, since they may show different susceptibility to oxidative damage, probably due to differences in their protein composition or activities of their antioxidant defenses, and, therefore, make different contribution to the impairment of cell functioning with radiation responses.
Numerous investigations showed that protein carbonyls, a marker of primary protein damage indicated a higher magnitude of damage in irradiated mice brains exposed to 1.5 Gy high-LET 56Fe beams (500 MeV/nucleon, 1.5 Gy). This effect was associated with impaired cognitive behavior of mice at day 30 post-exposure as well as apoptotic and necrotic cell death of granule cells and Purkinje cells (Manda et al. 2008). γ-irradiation of rats at a dose of 10 Gy caused increases in protein carbonyl groups in mitochondria and cytoplasm both in liver and spleen. Similar results have been obtained for homogenates of different tissues isolated from γ-irradiated gerbils and rats (Sohal et al. 1995). Post-irradiation accumulation of oxidized proteins in subcellular fractions, especially if occurring in nuclei, might probably affect not only the catalytic properties of enzymes but also the regulation of radiation-induced gene expression by interfering with the activation of transcription factors (Whisler et al. 1997). Among the nuclear proteins, histones are likely most susceptible to oxidative modification, due to high contents of lysine and arginine residues in their molecules. Information on the formation of radicals on peptides and proteins and how radical damage may be propagated and transferred within protein structures have been reviewed (Hawkins and Davies 2001).

4. Defenses against free radicals

Human and all of the aerobic organisms have a very efficient defense network of antioxidants against oxidative stress. An antioxidant can be defined as a molecule or an element that, when present at low concentrations compared to those of the oxidizable substrate, significantly combat, delays and inhibit oxidation of that substrate, thus, prevent free radicals from damaging healthy cells (Halliwell 1997, 2009). Under normal condition, cells have well coordinated and efficient endogenous antioxidant defense systems, which protect against the injurious effects of oxidants.
From the viewpoint of mechanistic functions, antioxidant defense mechanisms can be classified into the following five lines of defenses: preventing antioxidants, scavenging antioxidants, repair and de novo antioxidants, adaptive antioxidants, and finally cellular signaling messenger (Halliwell 1997, Niki 2010). The first line of defense is the preventing antioxidants which act by suppressing the formation of ROS and RNS by reducing H_2O_2 and lipid hydroperoxides that are generated during lipid peroxidation to water and lipid hydroxides, respectively, or sequestering pro-oxidant metal ions such as iron and copper by some binding of proteins (e.g., transferrin, metallothionein). The second line of defense can be described as the scavenging antioxidants which exist to intercept, or scavenge free radicals and remove active species rapidly before attacking biologically essential molecules. For example; superoxide dismutase (SOD) converts $O_2{\bullet}-$ to H_2O_2, while α-tocopherol and carotenoids are efficient scavenger of $^1O_2$ (Inoue et al. 2011). Many phenolic compounds and aromatic amines act as a free radical-scavenging antioxidant. There is a general agreement that electron transfer and hydrogen transfer are the main mechanisms involved in the reactions of melatonin with free radicals. The third line of defenses is various enzymes which function by repairing damages, clearing the wastes, and reconstituting the lost function. The adaptation mechanism is considered the fourth line of defense, in which

appropriate antioxidants are released at the right time and transported to the right site in right concentration. Some antioxidants constitute the fifth line of defense by functioning as a cellular signaling messenger to control the level of antioxidant compounds and enzymes (Niki and Noguchi 2000, Noguchi and Niki 2000).

5. Radioprotectors and mitigators of radiation induced injury

Generally, any chemical/biological agents given before to or at the time of irradiation to prevent or ameliorate damage to normal tissues are termed radioprotectors. While mitigators of normal tissue injury are agents delivered at the end of irradiation, or after irradiation is complete, but prior to the manifestation of normal tissue toxicity. The estimated time scale to use mitigators efficiently ranges from seconds to hours after radiation exposure. Agents delivered to improve established normal tissue toxicity are considered treatments which can be monitored over weeks to years after radiation exposure (Citrin et al. 2010). Since radiotherapy, occupational, accidental exposure to radiation or space travel and exploration can produce unwanted side effects, it is important to prevent such effects by the use of radioprotectors or mitigators. Ideally, radioprotective and mitigative agent should fulfill several characteristics that relate to the ability of the agent to improve the therapeutic results. First, the agent should have protective effects on the majority of organs and tissues. Second, the agent must reach all cells and organelles and can easily penetrate cellular membranes. Third, it must have an acceptable route of administration (preferably oral or alternatively intramuscular) and with minimal toxicity. Fourth, to be useful in the radiotherapy settings, radioprotectant should be selective in protecting normal tissues from radiotherapy without protecting tumor tissue. Finally, to a large extent radioprotectors should be compatible with the wide range of other drugs that will be prescribe to patients. Moreover, because free radicals are responsible for injury caused by ionizing radiation, therefore, for an agent to protect cells from primary free radical damage, the agent needs to be present at the time of radiation and in sufficient concentration to compete with radicals produced through radical-scavenging mechanisms (Citrin et al. 2010, Hosseinimehr 2007, Shirazi et al. 2007).

A large body of literature describes radioprotection or mitigation with a variety of agents after total body or localized exposures. A complete and comprehensive review of these agents is outside the scope of this chapter. Herein, we briefly highlight melatonin that have been described as radiation protectors and mitigators, and attempt to focus on it with demonstrated or anticipated usefulness for therapeutic radiation exposures. As defined above, an ideal radioprotectors need to have radical-scavenging properties and can also exert broad antioxidant activity. Whereas all antioxidants cannot afford full radioprotection, melatonin verify most of the criteria needed for efficient radioprotector , mitigators and treatment agent with antioxidant potential, radical scavenging characteristics and stimulator of intrinsic antioxidants.

6. Melatonin

6.1 Synthesis, distribution, and metabolism

Melatonin synthesis in the pineal gland has been reviewed in significant detail (Reiter 2003). In summary, pinealocytes take up L-tryprophane from blood. Via several enzymatic steps including tryptophan 5-hydroxylation, decarboxylation, N-acetylation and O-methylation, in that sequence, N-acetyl-5-methoxytryptamine (melatonin) is synthesized. It is secreted upon biosynthesis into the extracellular fluid to the general circulation from which it easily crosses various cellular membranes. It is secreted by the pineal gland and its levels have diurnal variation and also fluctuate with sleep stages. They are higher during night (Luboshitzky et al. 1999). The diurnal/nocturnal levels of blood melatonin can range between 8 ± 2 pg/mL (light phase) and 81 ± 11 pg/mL (dark phase). The synthesis and presence of melatonin have also been demonstrated in non-pineal tissues such as retina, Harderian gland, gastrointestinal track, testes, and human lymphocytes. Furthermore, the distribution of melatonin in the human being is very broad (Reiter 2003). Once synthesized, the majority of melatonin diffuses directly towards the cerebrospinal fluid of the brain's third ventricle, while another fraction is released into the blood stream where it is distributed to all tissues and body fluids (Cheung et al. 2006). It is found in serum, saliva (Cutando et al. 2011, Novakova et al. 2011), cerebrospinal fluids (Rousseau et al. 1999), and aqueous humor of the eye (Chiquet et al. 2006), ovarian follicular fluid, hepato-gastrointestinal tissues (Messner et al. 2001). Melatonin in the milk of lactating mothers exhibits a marked daily rhythm, with high levels during the night and undetectable levels during the day (Illnerova et al. 1993, Sanchez-Barcelo et al. 2011). Moreover, melatonin production is not confined exclusively to the pineal gland, but other tissues including retina, Harderian glands, gut, ovary, testes, bone marrow and lens also produce it (Esposito and Cuzzocrea 2010).

Melatonin has two important functional groups which determine its specificity and amphiphilicity; the 5-methoxy group and the N-acetyl side chain. In liver melatonin is metabolized by P- 450 hepatic enzymes, which hydroxylate this hormone at the 6- carbon position to yield 6- hydroxymelatonin which conjugated with sulfuric or glucuronic acid, to produce the principal urinary metabolite, 6-sulfatoxymelatonin. In the final stage, conjugated melatonin and minute quantities of unmetabolized melatonin are excreted through the kidney. In addition to hepatic metabolism, oxidative pyrrole-ring cleavage appears to be the major metabolic pathway in other tissues, including the central nervous system (Esposito and Cuzzocrea 2010).

A plethora of evidence suggests that melatonin mediates a variety of physiological responses through membrane and nuclear binding sites. In mammals, the mechanisms of action of melatonin include the involvement of high affinity G protein-coupled membrane receptors (MT1, MT2), cytosolic binding sites (MT3 and calmodulin), and nuclear receptors of the RZR/ROR family. Melatonin also has receptor-independent activity and can directly scavenge free radicals. A disulfide bond between Cys 113 and Cys 190 is essential for high-affinity melatonin binding to MT2 and possibly to MT1 receptors (Dubocovich and Markowska 2005). RZR/ROR family is expressed in a variety of organs. It presumably regulates the immune system and circadian cycles via the nuclear receptor and these also may be involved in its regulation of antioxidative enzymes (Cutando et al. 2011).

6.2 Melatonin and factors that determine antioxidant capacity

A number of criteria that characterize an ideal free radical scavenging antioxidant can be identified. First, because free radicals are highly reactive with very short half life time, therefore, an efficient antioxidant should be ubiquitous and present in adequate amounts in tissues and cells. Furthermore, the biological systems are heterogeneous in nature, which affects the action and efficacy of antioxidants. Both hydrophilic and lipophilic antioxidants act at respective site. Some antioxidants are present in free form, but others as metabolite or in bound form (Niki 2010). In contrast to other antioxidants that are either hydrophilic or lipophilic, melatonin is an amphiphilic small size molecule (Giacomo and Antonio 2007). These features of melatonin allow it to cross all morphophysiological barriers and to interact with toxic molecules throughout the cell and its organelles, thereby reducing oxidative damage to molecules in both the lipid and aqueous environments of cells. Numerous articles documented that melatonin is widely distributed and found in all body fluids, organs, cells and organelles. Recently, melatonin levels in the cell membrane, cytosol, nucleus, and mitochondrion was found to vary over a 24-hr cycle, although these variations do not exhibit circadian rhythms. The cell membrane has the highest concentration of melatonin followed by mitochondria, nucleus, and cytosol (Venegas et al. 2011).

Second, an efficient antioxidant should react with most of free radicals because as it is well known that free radicals are variable in their biological, chemical and physical properties that involved in the oxidative stress. In functional terms, have reported that melatonin exerts a host of antioxidant effects that can be described as a broad spectrum antioxidant (Karbownik and Reiter 2000). Initially, (Hardeland et al. 1993a, Hardeland et al. 1993b, Tan et al. 1993) are the first who documented that melatonin is a remarkably potent scavenger of the particularly reactive, destructive, mutagenic and carcinogenic •OH. It is documented that, melatonin is a more efficient •OH scavenger than either glutathione or mannitol, and that melatonin reacts at a diffusion-controlled rate with the •OH. Thus, melatonin is probably an important endogenously produced antioxidant. Extensive studies have established that melatonin is much more specific than its structural analogs in undergoing reactions which lead to the termination of the radical reaction chain and in avoiding prooxidant, carbon or oxygen centered intermediates (Hardeland et al. 2011, Poeggeler et al. 2002, Tan et al. 1993). Besides the •OH, melatonin in cell-free systems has been shown to directly scavenge H_2O_2, $^1O_2$ and HOCl with little ability to scavenge the O_2•–. Furthermore, melatonin scavenges nitric oxide (NO•) and suppresses the activity of its rate limiting enzyme, nitric oxide synthase (NOS), thereby inhibiting the formation of the ONOO-. In addition, melatonin scavenges a number of RNS including ONOO- and peroxynitrous acid (ONOOH). Moreover, melatonin has proven to scavenge alkoxyl, peroxyl radicals. The peroxyl radical (POO•), which is formed during the complex process of lipid peroxidation, is highly destructive to cells, because, once formed, it can propagate the process of lipid peroxidation. Therefore, agents that neutralize the POO• radical are generally known as chain breaking antioxidants, is important to maintaining the optimal function of not only cell membranes, but of cells themselves. It is estimated that each molecule of melatonin scavenged four POO• molecules (Pieri et al. 1994, Pieri et al. 1995), which would make it twice as effective as vitamin E, the principal well known chain-breaking lipid antioxidant and POO• scavenger. The most important products of melatonin's interaction with H_2O_2, N1-acetyl-N2-formyl-5-methoxykynuramine (AFMK), N-acetyl-5-methoxykynuramine(AMK), and 6-hydroxymelatonin, are also a highly efficient radical scavenger (Maharaj et al. 2003). The cascade of reactions where the secondary metabolites are also effective scavenges is believed to contribute to melatonin's high efficacy in reducing oxidative damage.

Very relevant to the development of this chapter is an alternate concept proposed to explain the protective effects of melatonin at the level of radical generation rather than detoxification of radicals already formed. It has been suggested that if melatonin is capable of decreasing the processes leading to enhanced radical formation, this might be achieved by low concentrations of the indole (Hardeland et al. 2011). Therefore, the main sources of free radicals should be investigated with regard to their modulation by melatonin and their sensitivity to the stressor. In addition to free radicals generated by leukocytes, mitochondria should be mentioned as main sources. In support of this suggestion, melatonin may also reduce free radical generation in mitochondria by improving oxidative phosphorylation, thereby lowering electron leakage and increasing ATP generation (Acuna-Castroviejo et al. 2001).

In addition to its direct free radical scavenging actions, melatonin influences both functional integrity of other antioxidative enzymes and cellular mRNA levels for these enzymes. Many studies documented the influence of melatonin on the activity and expression of the antioxidants both under physiological and under conditions of elevated oxidative stress. As an indirect antioxidant, melatonin stimulates gene expression and activity of SOD, thereby inducing the rapid conversion of $O_2{\bullet}^-$ to H_2O_2. The removal of $O_2{\bullet}^-$ by SOD also leads to a reduced formation of the highly reactive and damaging ONOO-. Catalase (CAT) and glutathione peroxidase (GSH-Px), enzymes that metabolizes H_2O_2 to H_2O, have also been shown to be stimulated by melatonin (Karbownik and Reiter 2000). In fact, it was demonstrated by several investigator that melatonin stimulates the rate-limiting enzyme, γ-glutamylcysteine synthase thereby increasing the level of an important endogenous antioxidant, glutathione (GSH) (El-Missiry et al. 2007, Othman et al. 2008, Urata et al. 1999) which, besides being a radical scavenger, is used by GSH-Px as a substrate to metabolize H_2O_2. In this process, GSH is converted to oxidized glutathione (GSSG). To maintain high levels of GSH, melatonin promotes the activity of glutathione reductase, which converts GSSG back to GSH. The possible intracellular mechanisms and pathways by which melatonin regulates antioxidant enzymes were reviewed (Rodriguez et al. 2004).

An added value of melatonin is that its metabolite N1-acetyl-N2-formyl-5-methoxykynuramine (AFMK) also has remarkable antioxidant properties and redox potential (Tan et al. 2001). It is formed when melatonin interacts with ROS, in particular, $^1O_2$ and H_2O_2. AFMK can be then enzymatically converted, by CAT, to N1-acetyl-5-methoxykynuramine (AMK). Cyclic 3-hydroxymelatonin (C3-OHM) is another product formed from melatonin by its interactions with free radicals, (Tan et al 1998), which can be further metabolized by free radicals to AFMK (Tan et al. 2003). All these findings indicate that AFMK is a central metabolite of melatonin oxidation especially in nonhepatic tissues.

Interestingly, melatonin was shown to prevent the loss of important dietary antioxidants including Vitamins C and E (Susa et al. 1997), bind iron and participate in maintaining iron pool at appropriate level resulting in control of iron haemostasis, thereby providing tissue protection (Othman et al. 2008). Furthermore, melatonin enhances antioxidant action of a-tocopherol and ascorbate against NADPH- and iron-dependent lipid peroxidation in human placental mitochondria (Milczarek et al. 2010).

The bioavailability is the main factor that determines the capacity of antioxidants in vivo. The antioxidants should be absorbed, transported, distributed, and retained properly in the biological fluids, cells and tissues (Cheeseman and Slater 1993, Niki 2010). Recently, it has been suggested that melatonin present in edible plants may improve human health, by virtue of its biological activities and its good bioavailability (Iriti et al. 2010). This could add a new factor to the of health benefits for patients associated to radiotherapy. Melatonin's interactions with other drugs that influence its bioavailability were summarized (Tan et al. 2007). Furthermore, melatonin can be administered by virtually any route, including orally, via submucosal or transdermal patches, sublingually, intranasally, intravenously (Reiter 2003).

6.3 Radioprotective effect of melatonin and its metabolites

The radiosensitivity of cells to ionizing radiation depends on several factors including the efficiency of the endogenous antioxidative defense systems to prevent oxidative stress. A number of natural and synthetic compounds have been investigated for their antioxidative as well as radioprotective potential. Most of the effective compounds were found to have some inconvenient side effects such as hypotention, hypocalcemia, nausea, vomiting and hot flashes. Furthermore, most of compounds must be given intravenously or subcutaneously which restrict their clinical application outside of controlled clinical situations. Thus, a need still exists for identifying a non-toxic, effective, and convenient compound to protect humans against radiation damage in accidental, occupational, clinical settings and space-travel. Melatonin received much attention for its unique antioxidative potential at a very low concentration compared with other antioxidants.

A number of in vitro and in vivo studies have reported that exogenously administered melatonin provides profound protection against radiation induced lipid peroxidation and oxidative stress (El-Missiry et al. 2007, Taysi et al. 2003). As indicated above, radiation-induced lipid peroxidation is a three steps free radical process. Melatonin inhibits lipid peroxidation by preventing the initiation phase of lipid peroxidation and interrupting the chain reaction. This is mainly due to melatonin ability to quench •OH and several other ROS and RNS. It has been reported to scavenge the several ROS, in particular •OH and the ONOO-•. When melatonin interacts with •OH it becomes indolyl (melatonyl) radical which has very low toxicity. After some molecular rearrangements, the melatonyl radical scavenges a second •OH to form cyclic 3-hydroxymelatonin (3-OHM). Thus, this reaction pathway suggests that 3-OHM is the footprint product of the interaction between melatonin with •OH. 3-OHM was also detected in the urine of both rats and humans. This provides direct evidence that melatonin, under physiological conditions, functions as an antioxidant to detoxify the most reactive and cytotoxic endogenous •OH (Tan et al., 1999). When rats were exposed to ionizing radiation which results in •OH generation, urinary 3-OHM increased significantly compared to that of controls (Tan et al. 1998). This provided direct evidence that radiation induced oxidative stress increases melatonin consumption in rats. The rapid decrease in circulating melatonin under conditions of excessive stress can be considered a protective mechanism for organisms against highly damaging free radicals; in this sense, melatonin can be categorized as a first line of defensive molecule (Tan et al. 2007). Along this line, the melatonin's metabolite AFMK protected against space radiation induced impairment of memory and hippocampal neurogenesis in adult C57BL mice (Manda et al. 2008). This study demonstrated that radiation exposure (2.0 Gy of 500 MeV/nucleon ^{56}Fe beams, a ground-based model of space radiation) significantly reduced the spatial memory of mice without affecting their motor activity. It is also reported that AFMK pretreatment significantly ameliorated radiation induced neurobehavioral ailments and reduced the loss of doublecortin and cell proliferation. Radiation exposure dramatically augmented the level of 8-OHdG in serum as well as DNA migration in the comet tail were impaired by AFMK pretreatment. In addition, radiation-induced augmentation of protein carbonyl content and 4-HNE + MDA and reduced the level of brain sulfhydryl contents was ameliorated by AFMK pretreatment (Manda et al. 2008). The ameliorating action of AFMK against radiation induced lipid peroxidation was attributed to free-radical scavenging property of AFMK. In vitro, AFMK showed a very high level of •OH scavenging potential which was measured by an electron spin resonance spin study of the 2-hydroxy-5,5-dimethyl-1-pyrrolineN-oxide (DMPO-OH) adduct. In this experiment, 10 Gy of X-ray for the radiolysis of water with different concentration of AFMK was used and intensity of spin adduct (DMPO-OH) were measured by ESR (Manda et al. 2007).

Extensive studies have established that pretreatment with melatonin at physiological dose 5 mg/kg or pharmacological dose 10 mg/kg bw significantly decreased MDA and NO• levels. The data documented that melatonin reduces tissue damage inflicted by irradiation when given prior to the exposure to ionizing radiation (Babicova et al 2011, Taysi et al. 2003, Verma et al. 2010). These authors explained that NO• is formed in higher amounts from L-arginine by inducible nitric oxide synthase (iNOS) during early response to ionizing radiation presumably as a part of signal transduction pathways (Babicova et al. 2011). Its cytotoxicity is primarily due to the production of ONOO-, a toxic oxidant, generated when the NO• couples with O_2•- (El-Sokkary et al. 2002). The processes triggered by ONOO- include initiation of lipid peroxidation, inhibition of mitochondrial respiration, inhibition of membrane pumps, depletion of GSH, and damage to DNA. Melatonin is reported to scavenge ONOO- both in vitro and in vivo (El-Sokkary et al. 1999, Gilad et al. 1997) and to inhibit iNOS activity thereby reducing excessive NO• generation.

It should emphasized that ionizing radiation causes oxidative damage to tissues within an extremely short period, and possible protection against it would require the rapid transfer of antioxidants to the sensitive sites in cells. At this point, melatonin seems unique among cellular antioxidants because of its physical and chemical properties; it can easily cross biological membranes and reach the cytosol, nucleus, and cellular compartments (Menendez-Pelaez and Reiter 1993). The effect of melatonin in maintaining normal hepatic and renal functions may be related to its ability to localize mainly in a superficial position in the lipid bilayer near the polar heads of membrane phospholipids (El-Sokkary et al. 2002).

The protective action of melatonin against lipid and protein oxidation as a factor modifying membrane organization may also be related to melatonin's ability to scavenge the oxidation-initiating agents, which are produced during the oxidation of proteins and lipids. Since membrane functions and structure are influenced by proteins in membranes and radiation is known to damage thiol proteins (Biaglow et al. 2003), it is possible that the protective action of melatonin against membrane damage may be related partially to the ability of melatonin to prevent lipid and protein oxidative damage (Le Maire et al. 1990). Changes in membrane structure and fluidity due to ROS reactivity after irradiation are also attributed to graded alterations in the lipid bilayer environment (Karbownik and Reiter 2000). It has been suggested that the protective role of melatonin in preserving optimal levels of fluidity of the biological membranes may be related to its ability to reduce lipid peroxidation (Garcia et al. 1997, Garcia et al. 2001).

Moreover, melatonin prevents inflammation and MDA caused by abdominopelvic and total body irradiation of rat (Taysi et al. 2003). Thus, the radioprotective effect of melatonin is likely achieved by its ability to function as a scavenger for free radicals generated by ionizing radiation. Furthermore, these findings may suggest that melatonin may enable the use of higher doses of radiation during therapy and may therefore allow higher dose rates in some patients with cancer. In another study, pretreatment with melatonin (10mg/kg bw) for 4 days before acute γ-irradiation significantly abolished radiation induced elevations in MDA and protein carbonyl levels in the liver and significantly prevented the decrease in hepatic GSH content, GST, and CAT activities. Moreover, preirradiation treatment with melatonin showed significantly higher hepatic DNA and RNA contents than irradiated rats. The levels of total lipids, cholesterol, triglycerides (TG), high density lipoproteins (HDL), low density lipoproteins (LDL), total proteins, albumin, total globulins, creatinine, and urea, as well as the activities of AST, ALT, and GGT in serum were significantly ameliorated when melatonin was injected before irradiation. The protection evidenced by normalization of the clinical parameters was associated with and attributed to decreased lipid peroxidation in the presence of melatonin. This data indicate that melatonin has a radioprotective impact against ionizing-radiation induced oxidative stress and organ injury (El-Missiry et al. 2007).

At sufficiently high radiation doses, GSH becomes depleted, leaving highly reactive ROS, beyond the immediate and normal needs of the cell, to react with critical cellular biomolecules and cause tissue damage. The concentration of intracellular GSH, therefore, is the key determinant of the extent of radiation-induced tissue injury. Thus, interest has been focused on melatonin, which acts as an antioxidant and is capable of stimulating GSH synthesis. It has been shown that melatonin enhances intracellular GSH levels by stimulating the rate-limiting enzyme in its synthesis, γ-glutamylcysteine synthase, which inhibits the peroxidative enzymes NOS and lipoxygenase. Experimentally, melatonin is demonstrated to increase hepatic GSH content, (El-Missiry et al. 2007, Sener et al. 2003), and hence to inhibit formation of extracellular and intracellular ROS (Reiter et al. 2004). It is also proposed that prevention of GSH depletion is the most efficient method of direct protection against radiation-induced oxidative toxicity. A significant decrease in hepatic GST activity and increase in serum GGT activity was recorded after exposure to 2&4 Gy (El-Missiry et al. 2007, Sridharan and Shyamaladevi 2002), and 10 Gy gamma irradiation (Samanta et al. 2004). Melatonin treatment at a dose level of 10mg/Kg bw before irradiation significantly countered radiation-induced decrease in the activities of these enzymes in the liver and serum. Furthermore, melatonin increases the activity of GSH-Px (Barlow-Walden et al. 1995) and superoxide dismutase (Antolin et al. 1996). These findings support the conclusion that melatonin affords radioprotection by modulating antioxidative enzyme activities in the body (El-Missiry et al. 2007).

Al the level of clinical markers, like cholesterol, TG, LDL, and free radicals are also risk factors that tend to damage arteries, leading to cardiovascular diseases. Pre-irradiation treatment with melatonin is found to reduce serum cholesterol, TG, and LDL levels in serum, indicating modulation of lipid metabolism in cells. It is well reported that antioxidants reduce oxidation susceptibility of HDL (Schnell et al. 2001) and control hyperlipidaemia (Mary et al. 2002). This might potentiate antiatherogenic effects of antioxidants including melatonin. Therefore, it is suggested that preirradiation treatment with melatonin appears to be hypolipidemic which might potentate its beneficial use in occupational, clinical, and space settings (El-Missiry et al. 2007).

Ionizing irradiation is among reproductive harmful agent and is widely identified to affect testicular function, morphology and spermatogenesis. Irradiation of the testes can produce reversible or permanent sterility in males. In an experiment, rats were subjected to sublethal irradiation dose of 8 Gy, either to the total body or abdominopelvic region using a ^{60}Co source. In this experiment, melatonin pretreatment resulted in less apoptosis as indicated by a considerable decrease in caspace-3 immunoreactivity. Electron microscopic examination showed that all spermatogenic cells, especially primary spermatocytes, displayed considerably inhibited of degeneration in the groups treated with melatonin before total body and abdominopelvic irradiation (Take et al. 2009).

An extensive literature implicates cellular DNA as the primary target for the biological and lethal effects of ionizing radiation. Melatonin has the ability to protect the DNA of hematopoietic cells of mice from the damaging effects of acute whole-body irradiation (Vijayalaxmi et al. 1999). The radioprotective ability of melatonin was investigated in the

Indian tropical rodent, Funambulus pennanti during its reproductively inactive phase when peripheral melatonin is high and the animal is under the influence of environmental stresses. Exogenous melatonin with its anti-apoptotic and antioxidant properties additively increased the immunity of the squirrels, by protecting their hematopoietic system and lymphoid organs against X-ray radiation induced cellular toxicity (Sharma et al. 2008).

Human keratinocytes is the main target cells in epidermal photodamage. Protection against UVR-induced skin damage was manifested by suppression of UV-induced erythema by topical pretreatment with melatonin with / without combination of vitamins C and E (Bangha et al. 1997, Dreher et al. 1998). Melatonin increased cell survival of HaCaT keratinocytes and ensured keratinocyte colony growth under UV-induced stress and showed decrease of UV-induced DNA fragmentation. Also, transcription of several classical target genes which are up-regulated after UV-exposure and play an important role in the execution of skin photodamage were down-regulated in HaCaT keratinocytes by melatonin pretreatment. It has been previously reported that melatonin reduces UVB-induced cell damage and polyamine levels in human skin fibroblasts (Lee et al. 2003). Furthermore, it was reported that melatonin increases survival of HaCaT keratinocytes by suppressing UV-induced apoptosis (Fischer et al. 2006). The molecular mechanisms underlying protective effects of melatonin on human keratinocytes and human fibroblasts upon UVB induced-apoptotic cell death was investigated (Cho et al. 2007). In this study, cDNA microarray analysis was perform from HaCaT keratinocytes, exposed to 100 mJ /cm2 and pretreated with melatonin for 30 min. Data showed that melatonin inhibits the expression of apoptosis related protein-3, apoptotic chromatin condensation inducer in the nucleus, and glutathione peroxidase 1 in UVB-irradiated HaCaT cells. The inhibitory effect of melatonin upon UVB irradiation is likely to be associated with antioxidant capacity of melatonin, thereby suggesting that melatonin may be used as a sunscreen agent to reduce cell death of keratinocytes after excessive UVB irradiation.

Radiotherapy plays an important role as part of the multimodality treatment for a number of malignancies in children. In young children, significant growth arrest was demonstrated with fractionated doses of 15 Gy and above as well as, in children less than one year of age, with doses as low as 10 Gy (Robertson et al. 1991). Proliferating chondrocytes, distal metaphyseal vessels, and epiphyseal vasculature are main targets for radiation-induced injury of bone growth plate (Kember 1967). Proliferating chondrocytes show marked cytological changes and cell death with a single fraction of 5 Gy. Melatonin with its antioxidant capacity protected against the hypoxic conditions of chondrocytes and promoted continued proliferation despite exposure to radiation. Moreover, in vitro studies showed that melatonin is capable of promoting osteoblast proliferation directly and stimulating these cells to produce increased amounts of several bone matrix proteins such as bone sialoprotein, alkaline phosphatase, osteocalcin (Roth et al. 1999) and procollagen type I c-peptide (Nakade et al. 1999), responsible for bone formation. Osteoprotegerin, an osteoblastic protein that inhibits the differentiation of osteoclasts is also increased by melatonin in vitro (Koyama et al. 2002). This data may support the radioprotective effect of melatonin on bone growth. The effects of fractionated radiotherapy combined with radioprotection by melatonin compared with fractionated radiotherapy alone in preserving the integrity and function of the epiphyseal growth plate from radiation damage in a weanling rat model was investigated. Data revealed that melatonin is more protective for bone growth protection than amifostine and a potential exists to implement the use of melatonin in an effort to maximize the radiotherapeutic management of children with less long-term morbidity than previous clinical experience (Yavuz et al. 2003).

Recent studies have documented that radiation in general and radiotherapy in particular has effects on brain function, such as thinking, memory and learning ability (Hsiao et al. 2010). Because cognitive health of an organism is maintained by the ability of hippocampal precursors to proliferate and differentiate, radiation exposures have been shown to inhibit neurogenesis and are associated with the onset of cognitive impairments. In recent investigation, on the protection by melatonin against the delayed effects of cranial irradiation on hippocampal neurogenesis melatonin maintained adult hippocampal neurogenesis and cognitive functions after irradiation (Manda and Reiter 2010). In this study the pretreatment with melatonin showed a significantly higher count of microtubule binding protein doublecortin and the proliferation marker Ki-67 positive cells compared with irradiated only animals. The protection was achieved by a single intraperitoneal injection of 10 mg melatonin/kg bw prior to irradiation. These protective effects were accompanied by significant control of oxidative stress indicated by reduction in the count of immunohistochemical localization of DNA damage and lipid peroxidation using the anti-8-hydroxy-2-deoxyguanosin the anti-hydroxynonenal. This indicated that melatonin minimize cell death.

6.4 Melatonin modulates apoptosis in radiotherapy and space radiation

Recent studies have showed the exposure to heavy ions such as ^{56}Fe or ^{12}C particle can induce detrimental physiological and histological changes in the brain, which lead to behavioral changes, spatial learning, and memory deficits. During space travel, astronauts are exposed to high-LET galactic cosmic rays at higher radiation doses and dose rates than humans received on Earth is one of the acknowledged showstoppers for long duration manned interplanetary missions. Hadrontherapy is an innovative modality of high precision tool for radiotherapy which consists in using hadrons (mainly protons or carbon ions) to irradiate tumors. This technique is used in certain cases to treat patients whose tumors are resistant to conventional X-ray radiotherapy. Given cancer therapy and space radiation protection, there is a demand for reliable agent for the protection of the brain against oxidative stress induced by heavy-ion radiation. It is well known that ionizing radiations can induce apoptosis. It is well known that oxidative stress is a mediator of apoptosis by compromising the fine balance between intracellular oxidant and their defense systems to produce abnormally high levels of ROS. Melatonin supplementation at 1,3&10mg/Kg bw reduced irradiation-induced oxidative damage, and stimulated the activities of SOD & CAT together with total antioxidant capacity in brain of rats exposed to heavy-ion radiation. Furthermore, pretreatment with melatonin significantly elevated the expression of Nrf2 which regulates redox balance and stress. In addition, pre-irradiation treatment with melatonin mitigated apoptotic rate, maintained mitochondrial membrane potential, decreased cytochrome C release from mitochondria, down-regulated Bax/Bcl-2 ratio and caspase-3 levels, and consequently inhibited the important steps of irradiation-induced activation of mitochondrial pathway of apoptosis (Liu et al. 2011). Studies performed by other investigators documented that melatonin pretreatment inhibited the cerebellum cell apoptosis after mice received 2 Gy ^{56}Fe particle irradiation (Manda and Reiter 2010) and that decreased apoptosis by melatonin was associated with a reduction in Bax/Bcl-2 ratio in mice splenocytes exposed to 2 Gy X-ray irradiation (Jang et al. 2009). Along this line, melatonin supplementation suppresses NO-induced apoptosis via induction of Bcl-2 expression in immortalized pineal PGT-β cells (Yim et al. 2002). A similar pathway of inhibitory effects of melatonin on apoptosis induced by ischemic neuronal injury has been determined (Sun et al. 2002).

Relevant to this context, it has been recently proposed that ERK MAPK plays a central role to determine whether cells will live or die in response to apoptotic stimuli. It is well documented that the apoptotic signaling activated during UVB stress mainly converges at the mitochondrial level into intrinsic pathway and supporting evidence consider this pathway might be the principle target of melatonin to prevent apoptosis in human leukocytes (Radogna et al. 2008) as well as in other tumor cell lines and in vivo models (Acuna-Castroviejo et al. 2007). An in vitro stress model for the cell protection and antiapoptotic functions of melatonin was studied using U937 cells exposed to UVB radiation (Luchetti et al. 2006). Melatonin sustained the activation of the survival-promoting pathway ERK MAPK (extracellular signal-regulated kinase) which controls the balance between survival and death-promoting genes throughout the MAPK pathway, and is required to antagonize UVB induced apoptosis of U937 cells. This kinase was found to modulate the antioxidant and mitochondrial protection effects of melatonin that may find therapeutic applications in inflammatory and immune diseases associated with leukocyte oxidative stress and accelerated apoptosis (Luchetti et al. 2009, Luchetti et al. 2006). Recently, it is reported that redox-sensitive components are included in the cell protection signaling of melatonin and in the resulting transcriptional response that involves the control of NF-κB, AP-1, and Nrf2. Through these pathways, melatonin stimulates the expression of antioxidant and detoxification genes, acting in turn as a glutathione system promoter (Luchetti et al. 2010).

It is suggested that ionizing radiation produces oxidative stress due formation of mitochondrial ROS resulting in calcium influx into the mitochondria with opening of the mitochondrial permeability transition pore (MPTP) (Andrabi et al. 2004, Halestrap 2006) and depolarization of the mitochondrial membrane potential as the end result of radiation-induced mitochondrial damage and cell apoptosis. The different regulatory mechanisms of apoptosis and their modification by treatment with melatonin were tested in different cells after irradiation. It is found that the mitochondrial pathway was strongly influenced by melatonin by reducing mitochondrial ROS generation and calcium release as well as inhibiting the opening of the MPTP as shown in rat brain astrocytes (Jou et al. 2004), mouse striatal neurons (Andrabi et al. 2004) and rat cerebellar granule neurons (Han et al. 2006). Moreover, the prevented decreases in the mitochondrial membrane potential resulted from irradiation suggests that melatonin, due to its physiochmeical characters crosses the blood–brain barrier and biological membranes to easily reach mitochondria, stabilizes oxidative stress-mediated dysfunction and integrity of mitochondria by preserving its membrane potential and increasing the efficiency of mitochondrial electron transfer chain and ATP synthesis (Acuna-Castroviejo et al. 2001). When melatonin treated cultured keratinocytes were irradiated with UVB radiation (50 mJ/cm2), there were less cell leaky, more uniform shape and less nuclear condensation as compared to irradiated, nonmelatonin-treated controls (Fischer et al. 2006). Exogenous melatonin with its anti-apoptotic and antioxidant properties additively increased the immunity of the animals, by protecting their hematopoietic system and lymphoid organs against X-ray radiation induced cellular toxicity (Sharma et al. 2008). These findings strongly highlight melatonin as a potential antiaopoptotic neuroprotective and mitigative agent against the space radiation hazards and the side effect risk in hadrontherapy. Consistent with all the overwhelming experimental evidence described above it may be concluded that melatonin can efficiently protect against and mitigate radiation induced oxidative stress. The majority of the published works on its ROS scavenging action coincide on the conclusion that melatonin is excellent for this task and make melatonin efficient pharmacological radioprotectors.

7. Toxicity and biosafety of melatonin

The melatonin doses chosen in several studies were between 5 and 15 mg/kg bw, which are rather minimal effective doses as reported in animal studies. Whereas, pharmacological studies in rats of up to 250 mg/kg doses did not indicate any adverse effects. In addition, human volunteers who ingested a single oral dose of 1-300 mg and 1 g of melatonin daily for 30 days did not report any adverse side effects. In a study, none of the 15 weanling rats administered with 5-15 mg/kg of melatonin died during the 6-wk observation period (Yavuz et al. 2003). In addition, ip treatment with melatonin for 45 days did not show abnormal singes (El-Missiry et al. 2007). All of these observations provide support for the non-toxic nature of melatonin (Cheung et al. 2006).

8. Conclusion

Apart from nuclear accidents, radiation has been used increasingly in medicine and industry to help with diagnosis, treatment, and technology. However, radiation hazards present an enormous challenge for the biological and medical safety. The deleterious effects of ionizing radiation in biological systems are mainly mediated through the generation ROS in cells as a result of water radiolysis leading to oxidative stress. •OH considered the most damaging of all free radicals generated in organisms, are often responsible for biomolecular damage caused by ionizing radiation. Oxidative stress greatly contributes to radiation-induced cytotoxicity and to metabolic and morphologic changes in animals and humans during occupational settings, radiotherapy, and space flight. Melatonin is an indoleamine hormone synthesized from tryptophan in pinealocytes. It is distributed ubiquitously in organisms and in all cellular compartments, and it easily passes through all biological membranes. Several studies have indicated that melatonin may act as a scavenger of ROS such as hydroxyl radical, alkoxyl radical, hypochlorous acid and singlet oxygen. A number of in vitro and in vivo studies have reported that exogenously administered melatonin provides protection against radiation induced oxidative stress in different species. Its ability to reduce DNA damage, lipid peroxidation, and protein damage may originate from its function as a preventive antioxidant (scavenging initiating radicals directly or indirectly). Furthermore, this indoleamine manifests its antioxidative properties by upregulation of endogenous antioxidant defense mechanisms, increases the efficiency of the electron transport chain thereby limiting electron leakage and free radical generation, protects the integrity of the mitochondria and promotes ATP synthesis. Furthermore, several metabolites that are formed when melatonin neutralizes damaging reactants are themselves scavengers suggesting scavenging cascade reaction that greatly increase the efficacy of melatonin in preventing oxidative damage. Several observations provide support for the non-toxic nature of melatonin. The radioprotective and mitigative effects of melatonin against cellular damage caused by oxidative stress and its low toxicity make this innate antioxidant a potential drug in situations where the effects of ionizing radiation are to be controlled.

9. References

Acharya MM, Lan ML, Kan VH, Patel NH, Giedzinski E, Tseng BP, Limoli CL. 2010. Consequences of ionizing radiation-induced damage in human neural stem cells. Free Radic Biol Med 49: 1846-1855.

Acuna-Castroviejo D, Escames G, Rodriguez MI, Lopez LC. 2007. Melatonin role in the mitochondrial function. Front Biosci 12: 947-963.

Acuna-Castroviejo D, Martin M, Macias M, Escames G, Leon J, Khaldy H, Reiter RJ. 2001. Melatonin, mitochondria, and cellular bioenergetics. J Pineal Res 30: 65-74.

Altieri F, Grillo C, Maceroni M, Chichiarelli S. 2008. DNA damage and repair: from molecular mechanisms to health implications. Antioxid Redox Signal 10: 891-937.

Andrabi SA, Sayeed I, Siemen D, Wolf G, Horn TF. 2004. Direct inhibition of the mitochondrial permeability transition pore: a possible mechanism responsible for anti-apoptotic effects of melatonin. FASEB J 18: 869-871.

Antolin I, Rodriguez C, Sainz RM, Mayo JC, Uria H, Kotler ML, Rodriguez-Colunga MJ, Tolivia D, Menendez-Pelaez A. 1996. Neurohormone melatonin prevents cell damage: effect on gene expression for antioxidant enzymes. FASEB J 10: 882-890.

Babicova A, Havlinova Z, Pejchal J, Tichy A, Rezacova M, Vavrova J, Chladek J. 2011. Early changes in L-arginine-nitric oxide metabolic pathways in response to the whole-body gamma irradiation of rats. Int J Radiat Biol.

Bangha E, Elsner P, Kistler GS. 1997. Suppression of UV-induced erythema by topical treatment with melatonin (N-acetyl-5-methoxytryptamine). Influence of the application time point. Dermatology 195: 248-252.

Barlow-Walden LR, Reiter RJ, Abe M, Pablos M, Menendez-Pelaez A, Chen LD, Poeggeler B. 1995. Melatonin stimulates brain glutathione peroxidase activity. Neurochemistry International 26: 497-502.

Bhatia AL, Manda K. 2004. Study on pre-treatment of melatonin against radiation-induced oxidative stress in mice. Environ Toxicol Pharmacol 18: 13-20.

Biaglow JE, Ayene IS, Koch CJ, Donahue J, Stamato TD, Mieyal JJ, Tuttle SW. 2003. Radiation response of cells during altered protein thiol redox. Radiat Res 159: 484-494.

Blakely EA, Chang PY. 2007. A review of ground-based heavy ion radiobiology relevant to space radiation risk assessment: Cataracts and CNS effects. Advances in Space Research 40: 1307-1319.

Cadet J, Douki T, Ravanat JL. 2010. Oxidatively generated base damage to cellular DNA. Free Radic Biol Med 49: 9-21.

Cadet J, Douki T, Gasparutto D, Ravanat JL. 2005. Radiation-induced damage to cellular DNA: measurement and biological role. Radiation Physics and Chemistry 72: 293-299.

Catala A. 2007. The ability of melatonin to counteract lipid peroxidation in biological membranes. Curr Mol Med 7: 638-649.

Catala A. 2009. Lipid peroxidation of membrane phospholipids generates hydroxy-alkenals and oxidized phospholipids active in physiological and/or pathological conditions. Chem Phys Lipids 157: 1-11.

Cheeseman KH, Slater TF. 1993. An introduction to free radical biochemistry. Br Med Bull 49: 481-493.

Cheung RT, Tipoe GL, Tam S, Ma ES, Zou LY, Chan PS. 2006. Preclinical evaluation of pharmacokinetics and safety of melatonin in propylene glycol for intravenous administration. J Pineal Res 41: 337-343.

Chiquet C, Claustrat B, Thuret G, Brun J, Cooper HM, Denis P. 2006. Melatonin concentrations in aqueous humor of glaucoma patients. Am J Ophthalmol 142: 325-327 e321.

Cho JW, Kim CW, Lee KS. 2007. Modification of gene expression by melatonin in UVB-irradiated HaCaT keratinocyte cell lines using a cDNA microarray. Oncol Rep 17: 573-577.

Citrin D, Cotrim AP, Hyodo F, Baum BJ, Krishna MC, Mitchell JB. 2010. Radioprotectors and mitigators of radiation-induced normal tissue injury. Oncologist 15: 360-371.

Comporti M, Signorini C, Arezzini B, Vecchio D, Monaco B, Gardi C. 2008. F2-isoprostanes are not just markers of oxidative stress. Free Radic Biol Med 44: 247-256.

Cooke MS, Evans MD, Dizdaroglu M, Lunec J. 2003. Oxidative DNA damage: mechanisms, mutation, and disease. FASEB J 17: 1195-1214.

Cutando A, Aneiros-Fernandez J, Lopez-Valverde A, Arias-Santiago S, Aneiros-Cachaza J, Reiter RJ. 2011. A new perspective in Oral health: Potential importance and actions of melatonin receptors MT1, MT2, MT3, and RZR/ROR in the oral cavity. Arch Oral Biol.

Draganic IG, Draganic ZD. 1971. The Radiation Chemistry of Water. Academic Press, New York.

Dreher F, Gabard B, Schwindt DA, Maibach HI. 1998. Topical melatonin in combination with vitamins E and C protects skin from ultraviolet-induced erythema: a human study in vivo. Br J Dermatol 139: 332-339.

Dubocovich ML, Markowska M. 2005. Functional MT1 and MT2 melatonin receptors in mammals. Endocrine 27: 101-110.

El-Missiry MA, Fayed TA, El-Sawy MR, El-Sayed AA. 2007. Ameliorative effect of melatonin against gamma-irradiation-induced oxidative stress and tissue injury. Ecotoxicol Environ Saf 66: 278-286.

El-Sokkary GH, Omar HM, Hassanein AF, Cuzzocrea S, Reiter RJ. 2002. Melatonin reduces oxidative damage and increases survival of mice infected with Schistosoma mansoni. Free Radic Biol Med 32: 319-332.

El-Sokkary GH, Reiter RJ, Cuzzocrea S, Caputi AP, Hassanein AF, Tan DX. 1999. Role of melatonin in reduction of lipid peroxidation and peroxynitrite formation in non-septic shock induced by zymosan. Shock 12: 402-408.

Esposito E, Cuzzocrea S. 2010. Antiinflammatory activity of melatonin in central nervous system. Curr Neuropharmacol 8: 228-242.

Esterbauer H, Zollner H. 1989. Methods for determination of aldehydic lipid peroxidation products. Free Radic Biol Med 7: 197-203.

Fischer TW, Zbytek B, Sayre RM, Apostolov EO, Basnakian AG, Sweatman TW, Wortsman J, Elsner P, Slominski A. 2006. Melatonin increases survival of HaCaT keratinocytes by suppressing UV-induced apoptosis. J Pineal Res 40: 18-26.

Gago-Dominguez M, Castelao JE, Pike MC, Sevanian A, Haile RW. 2005. Role of lipid peroxidation in the epidemiology and prevention of breast cancer. Cancer Epidemiol Biomarkers Prev 14: 2829-2839.

Galano A, Tan DX, Reiter RJ. 2011. Melatonin as a natural ally against oxidative stress: a physicochemical examination. Journal of Pineal Research 51: 1-16.

Garcia JJ, Reiter RJ, Guerrero JM, Escames G, Yu BP, Oh CS, Munoz-Hoyos A. 1997. Melatonin prevents changes in microsomal membrane fluidity during induced lipid peroxidation. FEBS Lett 408: 297-300.

Garcia JJ, Reiter RJ, Karbownik M, Calvo JR, Ortiz GG, Tan DX, Martinez-Ballarin E, Acuna-Castroviejo D. 2001. N-acetylserotonin suppresses hepatic microsomal membrane rigidity associated with lipid peroxidation. Eur J Pharmacol 428: 169-175.

Giacomo CG, Antonio M. 2007. Melatonin in cardiac ischemia/reperfusion-induced mitochondrial adaptive changes. Cardiovasc Hematol Disord Drug Targets 7: 163-169.

Gilad E, Cuzzocrea S, Zingarelli B, Salzman AL, Szabo C. 1997. Melatonin is a scavenger of peroxynitrite. Life Sciences 60: PL169-174.

Goodhead DT. 1994. Initial events in the cellular effects of ionizing radiations: clustered damage in DNA. Int J Radiat Biol 65: 7-17.

Griffiths HR, et al. 2002. Biomarkers. Mol Aspects Med 23: 101-208.

Guajardo MH, Terrasa AM, Catala A. 2006. Lipid-protein modifications during ascorbate-Fe2+ peroxidation of photoreceptor membranes: protective effect of melatonin. J Pineal Res 41: 201-210.

Gulbahar O, Aricioglu A, Akmansu M, Turkozer Z. 2009. Effects of radiation on protein oxidation and lipid peroxidation in the brain tissue. Transplant Proc 41: 4394-4396.

Haegele AD, Wolfe P, Thompson HJ. 1998. X-radiation induces 8-hydroxy-2'-deoxyguanosine formation in vivo in rat mammary gland DNA. Carcinogenesis 19: 1319-1321.

Halestrap AP. 2006. Calcium, mitochondria and reperfusion injury: a pore way to die. Biochem Soc Trans 34: 232-237.

Halliwell B. 1997. Antioxidants: the basics--what they are and how to evaluate them. Adv Pharmacol 38: 3-20.

Halliwell B. 2009. The wanderings of a free radical. Free Radic Biol Med 46: 531-542.

Han YX, Zhang SH, Wang XM, Wu JB. 2006. Inhibition of mitochondria responsible for the anti-apoptotic effects of melatonin during ischemia-reperfusion. J Zhejiang Univ Sci B 7: 142-147.

Hardeland R, Poeggeler B, Balzer I, Behrmann G. 1993a. A hypothesis on the evolutionary origins of photoperiodism based on circadian rhythmicity of melatonin in phylogenetically distant organisms. Gutenbrunner, C., Hildebrandt, G., Moog, R. (Eds.), Chronobiology & Chronomedicine. 1: 113-120.

Hardeland R, Reiter RJ, Poeggeler B, Tan DX. 1993b. The significance of the metabolism of the neurohormone melatonin: antioxidative protection and formation of bioactive substances. Neurosci Biobehav Rev 17: 347-357.

Hardeland R, Cardinali DP, Srinivasan V, Spence DW, Brown GM, Pandi-Perumal SR. 2011. Melatonin--a pleiotropic, orchestrating regulator molecule. Prog Neurobiol 93: 350-384.

Hawkins CL, Davies MJ. 2001. Generation and propagation of radical reactions on proteins. Biochim Biophys Acta 1504: 196-219.

Hong M, Xu A, Zhou H, Wu L, Randers-Pehrson G, Santella RM, Yu Z, Hei TK. 2010. Mechanism of genotoxicity induced by targeted cytoplasmic irradiation. Br J Cancer 103: 1263-1268.

Hosseinimehr SJ. 2007. Trends in the development of radioprotective agents. Drug Discov Today 12: 794-805.

Hsiao KY, Yeh SA, Chang CC, Tsai PC, Wu JM, Gau JS. 2010. Cognitive Function before and after Intensity-Modulated Radiation Therapy in Patients with Nasopharyngeal Carcinoma: A Prospective Study. International Journal of Radiation Oncology Biology Physics 77: 722-726.

Hutchinson F. 1985. Chemical changes induced in DNA by ionizing radiation. Prog Nucleic Acid Res Mol Biol 32: 115-154.

Illnerova H, Buresova M, Presl J. 1993. Melatonin rhythm in human milk. J Clin Endocrinol Metab 77: 838-841.

Inoue S, Ejima K, Iwai E, Hayashi H, Appel J, Tyystjarvi E, Murata N, Nishiyama Y. 2011. Protection by alpha-tocopherol of the repair of photosystem II during photoinhibition in Synechocystis sp. PCC 6803. Biochim Biophys Acta 1807: 236-241.

Iriti M, Varoni EM, Vitalini S. 2010. Melatonin in traditional Mediterranean diets. J Pineal Res 49: 101-105.

Jang SS, Kim WD, Park WY. 2009. Melatonin exerts differential actions on X-ray radiation-induced apoptosis in normal mice splenocytes and Jurkat leukemia cells. Journal of Pineal Research 47: 147-155.

Jou MJ, Peng TI, Reiter RJ, Jou SB, Wu HY, Wen ST. 2004. Visualization of the antioxidative effects of melatonin at the mitochondrial level during oxidative stress-induced apoptosis of rat brain astrocytes. J Pineal Res 37: 55-70.

Kamat JP, Devasagayam TP, Priyadarsini KI, Mohan H. 2000. Reactive oxygen species mediated membrane damage induced by fullerene derivatives and its possible biological implications. Toxicology 155: 55-61.

Karbownik M, Reiter RJ. 2000. Antioxidative effects of melatonin in protection against cellular damage caused by ionizing radiation. Proc Soc Exp Biol Med 225: 9-22.

Kember NF. 1967. Cell survival and radiation damage in growth cartilage. Br J Radiol 40: 496-505.

Klaunig JE, Wang Z, Pu X, Zhou S. 2011. Oxidative stress and oxidative damage in chemical carcinogenesis. Toxicol Appl Pharmacol 254: 86-99.

Koyama H, Nakade O, Takada Y, Kaku T, Lau KH. 2002. Melatonin at pharmacologic doses increases bone mass by suppressing resorption through down-regulation of the RANKL-mediated osteoclast formation and activation. Journal of Bone and Mineral Research 17: 1219-1229.

Kryston TB, Georgiev AB, Pissis P, Georgakilas AG. 2011. Role of oxidative stress and DNA damage in human carcinogenesis. Mutat Res 711: 193-201.

Lakshmi B, Tilak JC, Adhikari S, A. DTP, andJanardhanan KK. 2005. Inhibition of lipid peroxidationinduced by g-radiation and AAPH inrat liver and brain mitochondria by mushrooms. CURRENT SCIENCE 88: 484-488.

Le Maire M, Thauvette L, Deforesta B, Viel A, Beauregard G, Potier M. 1990. Effects of Ionizing-Radiations on Proteins - Evidence of Nonrandom Fragmentations and a Caution in the Use of the Method for Determination of Molecular Mass. Biochemical Journal 267: 431-439.

Lee KS, Lee WS, Suh SI, Kim SP, Lee SR, Ryoo YW, Kim BC. 2003. Melatonin reduces ultraviolet-B induced cell damages and polyamine levels in human skin fibroblasts in culture. Exp Mol Med 35: 263-268.

Little JB. 2000. Radiation carcinogenesis. Carcinogenesis 21: 397-404.

Liu Y, Zhang L, Zhang H, Liu B, Wu Z, Zhao W, Wang Z. 2011. Exogenous melatonin modulates apoptosis in the mouse brain induced by high-LET carbon ion irradiation. J Pineal Res.

Luboshitzky R, Lavi S, Lavie P. 1999. The association between melatonin and sleep stages in normal adults and hypogonadal men. Sleep 22: 867-874.

Luchetti F, Betti M, Canonico B, Arcangeletti M, Ferri P, Galli F, Papa S. 2009. ERK MAPK activation mediates the antiapoptotic signaling of melatonin in UVB-stressed U937 cells. Free Radic Biol Med 46: 339-351.

Luchetti F, Canonico B, Curci R, Battistelli M, Mannello F, Papa S, Tarzia G, Falcieri E. 2006. Melatonin prevents apoptosis induced by UV-B treatment in U937 cell line. J Pineal Res 40: 158-167.

Luchetti F, Canonico B, Betti M, Arcangeletti M, Pilolli F, Piroddi M, Canesi L, Papa S, Galli F. 2010. Melatonin signaling and cell protection function. FASEB J 24: 3603-3624.

Maharaj DS, Limson JL, Daya S. 2003. 6-Hydroxymelatonin converts Fe (III) to Fe (II) and reduces iron-induced lipid peroxidation. Life Sciences 72: 1367-1375.

Manda K, Reiter RJ. 2010. Melatonin maintains adult hippocampal neurogenesis and cognitive functions after irradiation. Prog Neurobiol 90: 60-68.

Manda K, Ueno M, Anzai K. 2007. AFMK, a melatonin metabolite, attenuates X-ray-induced oxidative damage to DNA, proteins and lipids in mice. J Pineal Res 42: 386-393.

Manda, K.Ueno, M. Anzai, K. 2008. Space radiation-induced inhibition of neurogenesis in the hippocampal dentate gyrus and memory impairment in mice: ameliorative potential of the melatonin metabolite, AFMK. J Pineal Res 45: 430-438.

Martin LM, Marples B, Coffey M, Lawler M, Lynch TH, Hollywood D, Marignol L. 2010. DNA mismatch repair and the DNA damage response to ionizing radiation: making sense of apparently conflicting data. Cancer Treat Rev 36: 518-527.

Mary NK, Shylesh BS, Babu BH, Padikkala J. 2002. Antioxidant and hypolipidaemic activity of a herbal formulation--liposem. Indian J Exp Biol 40: 901-904.

Matsumoto H, Hayashi S, Hatashita M, Ohnishi K, Shioura H, Ohtsubo T, Kitai R, Ohnishi T, Kano E. 2001. Induction of radioresistance by a nitric oxide-mediated bystander effect. Radiat Res 155: 387-396.

Menendez-Pelaez A, Reiter RJ. 1993. Distribution of melatonin in mammalian tissues: the relative importance of nuclear versus cytosolic localization. J Pineal Res 15: 59-69.

Messner M, Huether G, Lorf T, Ramadori G, Schworer H. 2001. Presence of melatonin in the human hepatobiliary-gastrointestinal tract. Life Sciences 69: 543-551.

Milczarek R, Hallmann A, Sokolowska E, Kaletha K, Klimek J. 2010. Melatonin enhances antioxidant action of alpha-tocopherol and ascorbate against NADPH- and iron-dependent lipid peroxidation in human placental mitochondria. J Pineal Res 49: 149-155.

Moller P, Loft S. 2010. Oxidative damage to DNA and lipids as biomarkers of exposure to air pollution. Environ Health Perspect 118: 1126-1136.

Morgan WF. 2003. Is there a common mechanism underlying genomic instability, bystander effects and other nontargeted effects of exposure to ionizing radiation? Oncogene 22: 7094-7099.

Nakade O, Koyama H, Ariji H, Yajima A, Kaku T. 1999. Melatonin stimulates proliferation and type I collagen synthesis in human bone cells in vitro. J Pineal Res 27: 106-110.

Niki E. 2010. Assessment of antioxidant capacity in vitro and in vivo. Free Radic Biol Med 49: 503-515.

Niki E, Noguchi N. 2000. Evaluation of antioxidant capacity. What capacity is being measured by which method? IUBMB Life 50: 323-329.

Noguchi N, Niki E. 2000. Phenolic antioxidants: a rationale for design and evaluation of novel antioxidant drug for atherosclerosis. Free Radic Biol Med 28: 1538-1546.

Novakova M, Paclt I, Ptacek R, Kuzelova H, Hajek I, Sumova A. 2011. Salivary Melatonin Rhythm as a Marker of the Circadian System in Healthy Children and Those With Attention-Deficit/Hyperactivity Disorder. Chronobiol Int 28: 630-637.

Othman AI, El-Missiry MA, Amer MA, Arafa M. 2008. Melatonin controls oxidative stress and modulates iron, ferritin, and transferrin levels in adriamycin treated rats. Life Sciences 83: 563-568.

Pieri C, Marra M, Moroni F, Recchioni R, Marcheselli F. 1994. Melatonin: a peroxyl radical scavenger more effective than vitamin E. Life Sciences 55: PL271-276.

Pieri C, Moroni F, Marra M, Marcheselli F, Recchioni R. 1995. Melatonin is an efficient antioxidant. Arch Gerontol Geriatr 20: 159-165.

Poeggeler B, Thuermann S, Dose A, Schoenke M, Burkhardt S, Hardeland R. 2002. Melatonin's unique radical scavenging properties - roles of its functional substituents as revealed by a comparison with its structural analogs. J Pineal Res 33: 20-30.

Radogna F, Cristofanon S, Paternoster L, D'Alessio M, De Nicola M, Cerella C, Dicato M, Diederich M, Ghibelli L. 2008. Melatonin antagonizes the intrinsic pathway of apoptosis via mitochondrial targeting of Bcl-2. J Pineal Res 44: 316-325.

Reed TT. 2011. Lipid peroxidation and neurodegenerative disease. Free Radic Biol Med 51: 1302-1319.

Reiter RJ. 2003. Melatonin: clinical relevance. Best Pract Res Clin Endocrinol Metab 17: 273-285.

Reiter RJ, Manchester LC, Tan DX. 2010. Neurotoxins: free radical mechanisms and melatonin protection. Curr Neuropharmacol 8: 194-210.

Reiter RJ, Tan DX, Gitto E, Sainz RM, Mayo JC, Leon J, Manchester LC, Vijayalaxmi, Kilic E, Kilic U. 2004. Pharmacological utility of melatonin in reducing oxidative cellular and molecular damage. Polish Journal of Pharmacology 56: 159-170.

Robertson WW, Butler MS, Dangio GJ, Rate WR. 1991. Leg Length Discrepancy Following Irradiation for Childhood Tumors. Journal of Pediatric Orthopaedics 11: 284-287.

Rodriguez C, Mayo JC, Sainz RM, Antolin I, Herrera F, Martin V, Reiter RJ. 2004. Regulation of antioxidant enzymes: a significant role for melatonin. J Pineal Res 36: 1-9.

Roth JA, Kim BG, Song F, Lin WL, Cho MI. 1999. Melatonin promotes osteoblast differentiation and bone formation (vol 274, pg 22041, 1999). Journal of Biological Chemistry 274: 32528-32528.

Rousseau A, Petren S, Plannthin J, Eklundh T, Nordin C. 1999. Serum and cerebrospinal fluid concentrations of melatonin: a pilot study in healthy male volunteers. J Neural Transm 106: 883-888.

Samanta N, Kannan K, Bala M, Goel HC. 2004. Radioprotective mechanism of Podophyllum hexandrum during spermatogenesis. Molecular and Cellular Biochemistry 267: 167-176.

Sanchez-Barcelo EJ, Mediavilla MD, Reiter RJ. 2011. Clinical uses of melatonin in pediatrics. Int J Pediatr 2011: 892624.

Schnell JW, Anderson RA, Stegner JE, Schindler SP, Weinberg RB. 2001 Effects of a high polyunsaturated fat diet and vitamin E supplementation on high-density lipoprotein oxidation in humans. Atherosclerosis 159: 459-466.

Sedelnikova OA, Redon CE, Dickey JS, Nakamura AJ, Georgakilas AG, Bonner WM. 2010. Role of oxidatively induced DNA lesions in human pathogenesis. Mutat Res 704: 152-159.

Sener G, Jahovic N, Tosun O, Atasoy BM, Yegen BC. 2003. Melatonin ameliorates ionizing radiation-induced oxidative organ damage in rats. Life Sciences 74: 563-572.

Sharma S, Haldar C, Chaube SK. 2008. Effect of exogenous melatonin on X-ray induced cellular toxicity in lymphatic tissue of Indian tropical male squirrel, Funambulus pennanti. Int J Radiat Biol 84: 363-374.

Shirazi A, Ghobadi G, Ghazi-Khansari M. 2007. A radiobiological review on melatonin: a novel radioprotector. J Radiat Res (Tokyo) 48: 263-272.

Shuryak I, Brenner DJ. 2009. A model of interactions between radiation-induced oxidative stress, protein and DNA damage in Deinococcus radiodurans. J Theor Biol 261: 305-317.

Sohal RS, Agarwal S, Sohal BH. 1995. Oxidative stress and aging in the Mongolian gerbil (Meriones unguiculatus). Mech Ageing Dev 81: 15-25.

Sridharan S, Shyamaladevi CS. 2002. Protective effect of N-acetylcysteine against gamma ray induced damages in rats--biochemical evaluations. Indian J Exp Biol 40: 181-186.

Stadtman ER. 2004. Role of oxidant species in aging. Curr Med Chem 11: 1105-1112.

Sun FY, Lin X, Mao LZ, Ge WH, Zhang LM, Huang YL, Gu J. 2002. Neuroprotection by melatonin against ischemic neuronal injury associated with modulation of DNA damage and repair in the rat following a transient cerebral ischemia. Journal of Pineal Research 33: 48-56.

Susa N, Ueno S, Furukawa Y, Ueda J, Sugiyama M. 1997. Potent protective effect of melatonin on chromium(VI)-induced DNA single-strand breaks, cytotoxicity, and lipid peroxidation in primary cultures of rat hepatocytes. Toxicol Appl Pharmacol 144: 377-384.

Sutherland BM, Bennett PV, Sidorkina O, Laval J. 2000. Clustered DNA damages induced in isolated DNA and in human cells by low doses of ionizing radiation. Proc Natl Acad Sci U S A 97: 103-108.

Take G, Erdogan D, Helvacioglu F, Goktas G, Ozbey G, Uluoglu C, Yucel B, Guney Y, Hicsonmez A, Ozkan S. 2009. Effect of melatonin and time of administration on irradiation-induced damage to rat testes. Braz J Med Biol Res 42: 621-628.

Tan DX, Manchester LC, Terron MP, Flores LJ, Reiter RJ. 2007. One molecule, many derivatives: a never-ending interaction of melatonin with reactive oxygen and nitrogen species? J Pineal Res 42: 28-42.

Tan DX, Poeggeler B, Reiter RJ, Chen LD, Chen S, Manchester LC, Barlow-Walden LR. 1993. The pineal hormone melatonin inhibits DNA-adduct formation induced by the chemical carcinogen safrole in vivo. Cancer Lett 70: 65-71.

Tan DX, Manchester LC, Reiter RJ, Plummer BF, Hardies LJ, Weintraub ST, Vijayalaxmi, Shepherd AM. 1998. A novel melatonin metabolite, cyclic 3-hydroxymelatonin: a biomarker of in vivo hydroxyl radical generation. Biochem Biophys Res Commun 253: 614-620.

Tan DX, Hardeland R, Manchester LC, Poeggeler B, Lopez-Burillo S, Mayo JC, Sainz RM, Reiter RJ. 2003. Mechanistic and comparative studies of melatonin and classic antioxidants in terms of their interactions with the ABTS cation radical. J Pineal Res 34: 249-259.

Tan DX, Manchester LC, Burkhardt S, Sainz RM, Mayo JC, Kohen R, Shohami E, Huo YS, Hardeland R, Reiter RJ. 2001. N1-acetyl-N2-formyl-5-methoxykynuramine, a biogenic amine and melatonin metabolite, functions as a potent antioxidant. FASEB J 15: 2294-2296.

Taysi S, Koc M, Buyukokuroglu ME, Altinkaynak K, Sahin YN. 2003. Melatonin reduces lipid peroxidation and nitric oxide during irradiation-induced oxidative injury in the rat liver. J Pineal Res 34: 173-177.

Tuma DJ. 2002. Role of malondialdehyde-acetaldehyde adducts in liver injury. Free Radic Biol Med 32: 303-308.

Undeger U, Giray B, Zorlu AF, Oge K, Bacaran N. 2004. Protective effects of melatonin on the ionizing radiation induced DNA damage in the rat brain. Experimental and Toxicologic Pathology 55: 379-384.

Urata Y, Honma S, Goto S, Todoroki S, Iida T, Cho S, Honma K, Kondo T. 1999. Melatonin induces gamma-glutamylcysteine synthetase mediated by activator protein-1 in human vascular endothelial cells. Free Radic Biol Med 27: 838-847.

Venegas C, Garcia JA, Escames G, Ortiz F, Lopez A, Doerrier C, Garcia-Corzo L, Lopez LC, Reiter RJ, Acuna-Castroviejo D. 2011. Extrapineal melatonin: analysis of its subcellular distribution and daily fluctuations. J Pineal Res.

Verma S, Gupta ML, Dutta A, Sankhwar S, Shukla SK, Flora SJ. 2010. Modulation of ionizing radiation induced oxidative imbalance by semi-fractionated extract of Piper betle: an in vitro and in vivo assessment. Oxid Med Cell Longev 3: 44-52.

Vijayalaxmi, Meltz ML, Reiter RJ, Herman TS, Kumar KS. 1999. Melatonin and protection from whole-body irradiation: survival studies in mice. Mutat Res 425: 21-27.

Von Sonntag C. 1987. The Chemical Basis of Radiation Biology. Taylor & Francis, London.

Ward JF. 1994. The complexity of DNA damage: relevance to biological consequences. Int J Radiat Biol 66: 427-432.

Wei H, Cai Q, Rahn R, Zhang X. 1997. Singlet oxygen involvement in ultraviolet (254 nm) radiation-induced formation of 8-hydroxy-deoxyguanosine in DNA. Free Radic Biol Med 23: 148-154.

Whisler RL, Chen M, Beiqing L, Carle KW. 1997. Impaired induction of c-fos/c-jun genes and of transcriptional regulatory proteins binding distinct c-fos/c-jun promoter elements in activated human T cells during aging. Cell Immunol 175: 41-50.

Wu LJ, Randers-Pehrson G, Xu A, Waldren CA, Geard CR, Yu Z, Hei TK. 1999. Targeted cytoplasmic irradiation with alpha particles induces mutations in mammalian cells. Proc Natl Acad Sci U S A 96: 4959-4964.

Yavuz MN, Yavuz AA, Ulku C, Sener M, Yaris E, Kosucu P, Karslioglu I. 2003. Protective effect of melatonin against fractionated irradiation-induced epiphyseal injury in a weanling rat model. J Pineal Res 35: 288-294.

Yim SV, et al. 2002. Melatonin suppresses NO-induced apoptosis via induction of Bcl-2 expression in PGT-beta immortalized pineal cells. Journal of Pineal Research 33: 146-150.

22

Radiosensitization with Hyperthermia and Chemotherapeutic Agents: Effects on Linear-Quadratic Parameters of Radiation Cell Survival Curves

Nicolaas A. P. Franken et al.*
Academic Medical Centre, University of Amsterdam,
Laboratory for Experimental Oncology and Radiobiology (LEXOR),
Centre for Experimental Molecular Medicine, Department of Radiation Oncology,
The Netherlands

1. Introduction

Radiosensitization effects of hyperthermia and chemotherapeutic agents as currently exploited in the clinic are discussed with respect to the linear quadratic parameters of dose-survival curve presentations. Studies of different human tumour cell lines show that a synergistic interaction can be obtained between hyperthermia, chemotherapy and radiation and that this interaction is more likely to occur in cell lines which are relatively sensitive to chemotherapy. The influence of modifying agents on radiation dose survival curves can adequately be analysed with the use of the linear-quadratic model: $S(D)/S(O)= \exp-(\alpha D+\beta D^2)$. The linear parameter, α, represents lethal damage from single particle events and describes the low dose area while the quadratic parameter, β, indicating sub lethal damage (SLD) dominates the effectiveness in the high dose region (Barendsen, 1990, 1994, 1997).

The linear-quadratic model is based on well accepted biophysical concepts, involving the assumption that lethal damage can be induced by single-particle tracks and by interaction of damage from multiple particles. It has been found to describe the low-dose region of the survival curves up to 6 Gy rather accurately. Furthermore the LQ-model has been shown to describe adequately dose fractionation effects for normal tissue tolerance and for experimental tumours. The LQ-model has also the advantage that it requires only two parameters to describe radiation dose-survival curves. It allows the separate analysis of changes in effectiveness in the low dose range, mainly determined by the linear term and in the high dose range determined mainly by the quadratic term (Barendsen, 1982; Joiner &

* Suzanne Hovingh[1], Arlene Oei[1], Paul Cobussen[1], Judith W. J. Bergs[3], Chris van Bree[1], Hans Rodermond[1], Lukas Stalpers[1], Petra Kok[1], Gerrit W. Barendsen[1,2] and Johannes Crezee[1]
Academic Medical Centre, University of Amsterdam,
[1]*Laboratory for Experimental Oncology and Radiobiology (LEXOR),*
Centre for Experimental Molecular Medicine, Department of Radiation Oncology, The Netherlands
[2]*Department of Cell Biology and Histology, Amsterdam, The Netherlands*
[3]*Institute of Molecular Biology and Tumor Research,Philipps-University Marburg, Emil-Mannkopff-Str. 2, Marburg, Germany*

Kogel, 2009). An additional advantage of the LQ model is that its parameters can be discussed in terms of specific mechanisms of cell inactivation by radiation (Barendsen, 1990, 1994). Linear-quadratic analyses of hyperthermia- or chemotherapy-induced radiation sensitisation have been reported for exponentially growing and plateau phase human tumour cells in culture and for different experimental rodent cell lines (Franken et al., 1997a, 1997b; van Bree et al., 1997, 2000; Castro Kreder et al., 2004; Bergs et al., 2006, 2007abc). When the additional treatment results in increases of the value of the α-parameter, this indicates that this treatment radiosensitizes at clinically relevant doses. When the additional treatment influences the value of the β-parameter it indicates that the additional treatment has an effect on the repair of sublethal damage.

In order to determine the linear and quadratic parameters and the effects of the different agents on these parameters clonogenic assays (Franken et al., 2006) were conducted of cells after ionizing radiation only and after combined hyperthermia or chemotherapy with radiation treatment. Radiation dose survival curves have been obtained and analysed according to the LQ formula: $S(D)/S(0)= \exp(-(\alpha D+\beta D^2))$. (Barendsen, 1982; Barendsen et al., 2001; Franken et al., 2004, 2006). The effects of the different agents on the linear parameter, α and quadratic parameter, β will be described. If significant changes are derived, the values of α/β ratio's are also of interest, because these ratio's show whether the influence of the dose per fraction, dose fractionation and dose rate in radiation treatments is larger or smaller due to the combined treatments. These ratio's may increase with increasing α or decreasing β. A change in the α/β ratio might be clinically of interest for the selection of fractionation schedules.

2. Hyperthermia

Hyperthermia refers to heat treatments if cells or malignancies in which the temperature is elevated in the range of 39°C to 45°C. It is used in combination with chemo- and/or radiotherapy since it is has been shown to enhance the anti-cancer effects of both therapies (Gonzalez Gonzalez et al., 1995; Van Der Zee et al., 2002; Crezee et al. 2009). Many *in vitro* studies on the combination of hyperthermia and radiation have shown a synergistic interaction between the two modalities, especially at higher temperatures (above 42°C) (Dewey et al., 1978; Roti Roti, 2004; Raaphorst et al., 1991). This interaction is believed to result from inhibition of repair of radiation-induced DNA damage by hyperthermia (Kampinga et al., 2001; Hildebrandt et al., 2002). The sequence of combined radiation and hyperthermia treatment is important. Optimal sensitization is obtained when radiation and hyperthermia are applied simultaneously or with a short interval (Hall & Giaccio, 2006). In the clinic this is not always possible. In our experiments hyperthermia was applied immediately after radiation treatment.

Despite the clinical goal to reach (cytotoxic) temperatures as high as 43 °C, tumour temperature distributions are in practice heterogeneous. In large areas of the tumour temperatures are often lower than 43°C. Nonetheless, good results have been obtained in locally advanced cervical cancers with tumour temperatures below 43 °C (van der Zee et al., 2000). Mild temperatures have more subtle effects than high temperatures, such as tumour-reoxygenation (Dewhirst et al., 2005; Bergs et al., 2007abc). Recently it has been shown that hyperthermia (42 °C for 1h) transiently breaks down the BRCA2 protein (Krawczyk et al., 2011). In this paragraph the effects of hyperthermia treatment for 1h at 41 or 43 °C on the linear quadratic parameters are summarized. Several different cell types have been studied.

2.1 Effect of hyperthermia treatment on radiosensitivity of RKO cells

The RKO cells, derived from human colon cancer, are relatively sensitive to hyperthermia treatment. Hyperthermia treatment for 1h at 43°C decreases the relative survival to less than 0.01 and combination with radiation doses in excess of 5 Gy always resulted in a situation in which no colony formation was observed. Treatment of cells with 41°C hyperthermia (1h) alone had little effect and resulted in a surviving fraction of 0.8 ± 0.4 in immediately plated (ip) cells and of 0.9 ± 0.1 in delayed plated (dp) cells. When cells were treated at 41°C for 1h immediately prior to irradiation, a significant ($p < 0.001$) enhancement of cellular radiosensitivity was observed both in ip (figure 1A) and dp (figure 1B) cells.

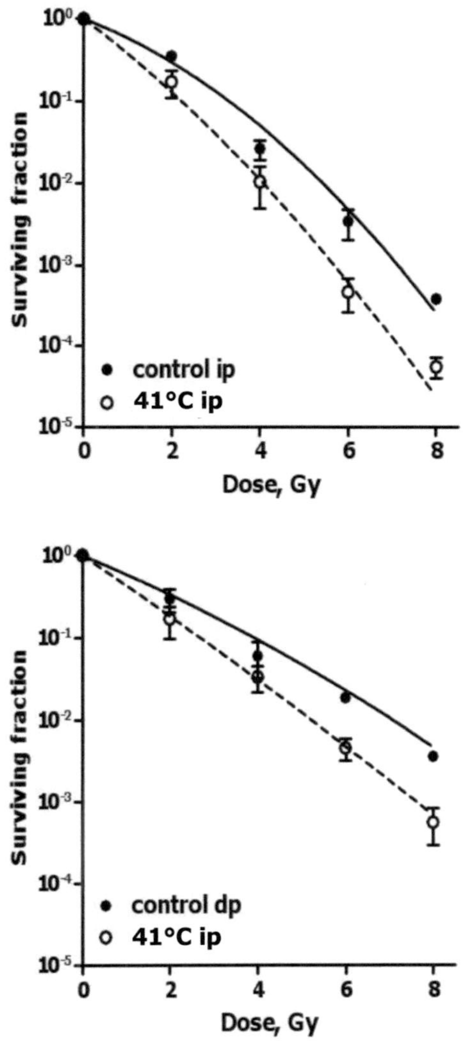

Fig. 1. Radiation survival curves of confluent cultures of RKO cells (human colon cancer cells) plated immediately after irradiation, ip (top) or 24h after irradiation, dp (below) with or without hyperthermia pre-treatment at 41°C for 1h. Means with standard errors of at least three experiments are shown

The effects of hyperthermia on the LQ parameters are summarized in table 1. The value of the linear parameter α increased by a factor 1.7-1.8 while the value of the β parameter even increased with a factor as high as 2.5-7.0. One must bear in mind that the quadratic component in this cell line is very small and small changes can have a large effect on the numerical values of β.

2.2 Effect of hyperthermia treatment on radiosensitivity of SW-1573 cells

SW-1573 cells are derived from a human lung tumour and are much less sensitive to hyperthermia treatment than RKO cells. Studies were performed to evaluate whether pretreatment with hyperthermia at 41°C or at 43°C in SW-1573 cells was able to enhance the radiosensitivity of these cells. Hyperthermia treatment at 41°C for 1h without radiation did not result in a decrease of the surviving fraction for ip and dp cells as compared to radiation alone. One hour hyperthermia treatment at 43 °C decreased survival to 0.5 ± 0.1 for ip and to 0.4 ± 0.2 for dp cells. Pre-treatment of cells at 41°C for 1h did not alter cellular radiosensitivity of both ip and dp cells (figure 2A). However, 1h treatment at 43°C resulted in a significant ($p < 0.001$) radiation enhancement both in ip and dp cells (figure 2B). In table I the values of the linear-quadratic parameters for radiation alone and for combined treatments are given. Hyperthermia treatment for 1 h at 41°C did result in an increase of the value of β by a factor 1.3-1.8 while the value of α even decreased. Hyperthermia treatment for 1 h at 43°C result in an increase of the value of α by a factor 2.3-4.4 while the value β increased with a factor 1.8-2.0.

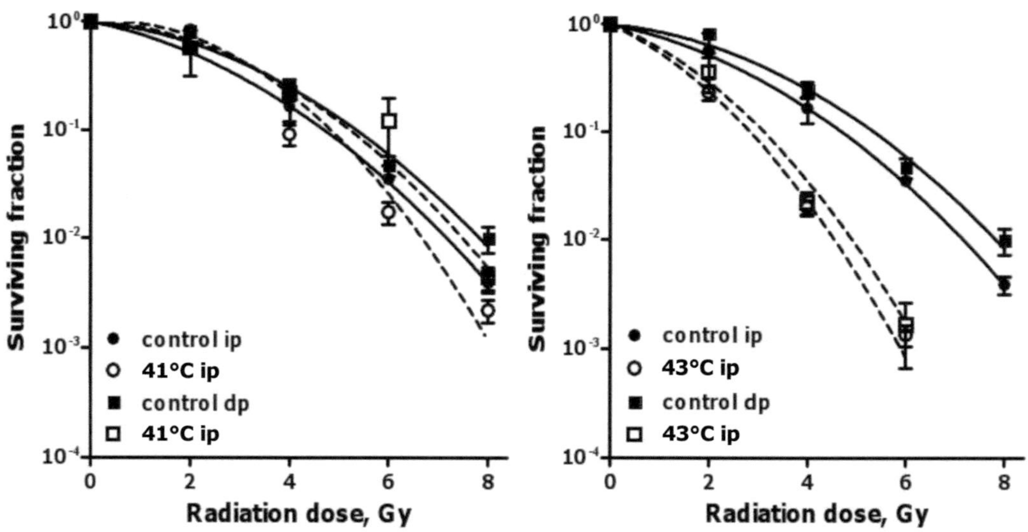

Fig. 2. Radiation survival curves of confluent cultures of SW-1573 cells (human lung tumour cells) plated immediately after irradiation (ip) or 24h after irradiation (dp) with or without hyperthermia pre-treatment at 41°C (left) or at 43°C (right) for 1h. Means with standard errors of at least three experiments are shown.

Cells	Treatment	α (Gy⁻¹)	β (Gy⁻²)	α/β	α-enhanc factor	β-enhanc factor
RKO ip	sham	0.55 ± 0.09	0.02 ± 0.01	27.5 ± 14.1		
	HT 41 1h	0.93 ± 0.09*	0.05 ± 0.02	18.6 ± 7.7	1.7 ± 0.3	2.5 ± 1.6
RKO dp	sham	0.47 ± 0.09	0.01 ± 0.01	47.0 ± 47.6[1]		
	HT 41 1h	0.83 ± 0.08*	0.07 ± 0.02	11.9 ± 3.6	1.8 ± 0.4	7.0 ± 7.3
SW1573 ip	sham	0.21 ± 0.02	0.06 ± 0.02	3.5 ± 1.2		
	HT 41 1h	0.06 ± 0.02	0.11 ± 0.03	0.6 ± 0.2	0.3 ± 0.1	1.8 ± 0.8
	HT 43 1h	0.49 ± 0.04*	0.12 ± 0.03	4.1 ± 1.1	2.3 ± 0.3	2.0 ± 0.8
SW1573 dp	sham	0.09 ± 0.02	0.06 ± 0.02	1.5 ± 1.6		
	HT 41 1h	0.05 ± 0.02	0.08 ± 0.02	0.6 ± 0.6	0.5 ± 0.3	1.3 ± 0.6
	HT 43 1h	0.40 ± 0.04*	0.11 ± 0.03	3.6 ± 1.1	4.4 ± 1.1	1.8 ± 0.8

Sham= is radiation only; ip=immediately plated; dp=delayed plated. * Significant from sham p<0.05.
[1]The α/β has a large variation because of the high uncertainty of the value of β.

Table 1. Values of the linear-quadratic parameters α and β, α/β and enhancement factors from cells treated with ionizing radiation only and after combined radiation and hyperthermia treatment

3. Cisplatin

Cisplatin is a widely used anti-cancer drug, often combined with radiotherapy (Gorodetsky et al., 1998). Chemo-radiation application based on cisplatin has now become the standard treatment for, among others, locally advanced cervical carcinoma (Duenas-Gonzalez et al., 2003) and locally advanced non-small cell lung cancer (NSCLC) (Loprevite et al., 2001). There have been many studies on the radiation sensitizing effect of cisplatin, but results vary from a clear cisplatin-induced radiosensitization (Begg et al., 1986; Nakamoto et al., 1996; Bergs et al., 2006, 2007ab) to only an additive effect on cell survival (Fehlauer et al., 2000). Cisplatin and radiation have in common that their cellular target is DNA (Rabik & Dolan, 2007).

Cisplatin causes DNA damage by the formation of inter- and intrastrand adducts (Crul et al., 2002). The cisplatin-DNA adducts can cause cell cycle arrest, inhibition of DNA replication and transcription, and eventually apoptosis (Myint et al., 2002). Repair inhibition of DNA has also been implicated (Lawrence et al., 2003) The most important repair pathways reported to be involved in cisplatin-induced DNA damage repair are nucleotide excision repair (NER) and/or homologous recombination (HR) (Haveman et al., 2004; De Silva et al., 2002). An additional route for the repair of cisplatin-DNA interstrand adducts is the post-replication/translation repair pathway which helps the cell to tolerate or bypass the lesion (Dronkert & Kanaar, 2001). Irradiation causes repairable (potentially lethal) and non-repairable (lethal) lesions to the DNA which are induced independently. The ultimate effect of the repairable lesions depends on competing processes of repair and misrepair. The repair of the potentially lethal damage (PLDR) is reflected by the difference in survival between immediately and delayed plated cells. Inhibition of PLDR is implicated to play a role in cisplatin-induced radiation sensitization (Bergs et al., 2006). More specifically, cisplatin-induced radiation sensitization has been shown to occur through inhibition of the

non-homologous end joining (NHEJ) pathway and recombinational repair (Myint et al., 2002; Haveman et al., 2004; Dolling et al., 1999).
In this paragraph the radiation sensitization of cisplatin on the lung tumour cell line SW1573 and the cervical tumour cell line Siha is described as changes in linear and quadratic parameters. In figure 3 survival curves are shown for SW1573 lung tumour cells after radiation alone and after radiation combined with cisplatin treatment (1 µM for 1 h). Cisplatin was added to the cultures just before radiation. The survival curves are obtained directly (ip) and 24 h after (dp) treatment to determine potentially lethal damage repair. A slight, but statistically significant effect of cisplatin on the radiosensitivity was only

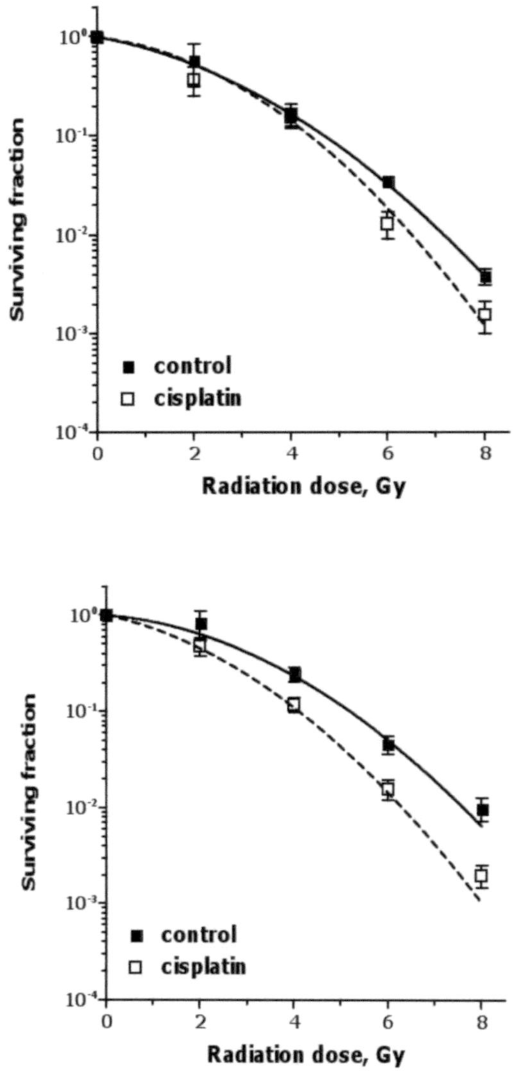

Fig. 3. Radiation survival curves of confluent cultures of SW-1573 cells plated immediately after irradiation, ip (top) or 24 h after irradiation, dp (below) with or without 1µM cisplatin treatment for one hour. Means with standard errors of at least three experiments are shown.

observed in delayed plated cells (p = 0.02). This was also described by an increase in the α- and β-value (table 2). Only for the delayed plated cells an increase with a factor of 2.5 for the value of α was obtained by cisplatin treatment. For both plating conditions an increase with a factor of 1.2 was obtained for the value of β. In the table 2 also the effects on the linear and quadratic parameters of different plating conditions are presented as well as a 1 h incubation with 1 or 5 μM cisplatin and a continuous incubation with cisplatin during the complete duration of the clonogenic assay. It is obvious that the cervical tumour cells Siha are more radiosensitzed with 1 μM continous cisplatin incubation than the SW1573 lung tumour cells.

Cells	Treatment	α (Gy^{-1})	β (Gy^{-2})	α/β	α-enhanc factor	β-enhanc factor
SW1573 ip	**sham**	**0.21 ± 0.09**	**0.061 ± 0.016**	**3.4 ± 1.7**		
	1 μM cisplatin (1h)	**0.21 ± 0.08**	**0.072 ± 0.018**	**2.9 ± 1.3**	**1.0 ± 0.6**	**1.2 ± 0.4**
SW1573 dp	**sham**	**0.10 ± 0.09**	**0.063 ± 0.016**	**1.6 ± 1.5**		
	1 μM cisplatin (1h)	**0.25 ± 0.09***	**0.077 ± 0.017**	**3.3 ± 1.4**	**2.5 ± 2.4**	**1.2 ± 0.4**
SW1573 ppi	**sham**	**0.37 ± 0.12**	**0.014± 0.034**	**26.4 ± 64.8[1]**		
	1 μM cisplatin (cont)	**0.41 ± 0.08**	**0.019 ± 0.025**	**21.6 ± 28.7[1]**	**1.1 ± 0.4**	**1.4 ± 3.8**
	5 μM cisplatin (cont)	**0.58 ± 0.20***	**0.030 ± 0.008***	**19.3 ± 8.4**	**1.6 ± 0.7**	**2.1 ± 5.2**
Siha ppi	**sham**	**0.41 ± 0.04**	**0.01 ± 0.01**	**41.0 ± 41.2[1]**		
	1 μM cisplatin (cont)	**0.81 ± 0.12***	**0.02 ± 0.02**	**40.5 ± 41.0[1]**	**2.0 ± 0.4**	**2.0 ± 2.8**

Sham is radiation only; ip=immediately plated; dp=delayed plated; ppi=plated prior to irradiation. * Significant from sham $p<0.05$. α/β values show that with SW1573 cells the quadratic term is affected more than the linear term, while with Siha cells only the linear term is significanty increased. [1] The α/β has a large variation because of the high uncertainty of the value of β.

Table 2. Values of the linear-quadratic parameters α and β and enhancement factors from SW1573 and Siha cells treated with ionizing radiation only and after combined radiation cisplatin (1 μM for 1h; 1 μM continuously; 5 μM continuously) treatment

4. Gemcitabine

Gemcitabine (dFdC, Difluorodeoxycytidine) is a deoxycytidine analogue with clinical activity in non-small cell lung cancer (NSCLC) and pancreatic cancer (Fosella et al., 1997; Manegold et al., 2000; Castro Kreder et al., 2004). It requires phosphorylation to its active metabolites, gemcitabine-diphosphate (dF-dCTP) and gemcitabine-triphosphate (dF-dCTP), with the initial phophorylation by deoxycytidine kinase (dCK) being the rate limiting step (Heinemann et al., 1992; Shewach et al., 1994). The dF-dCTP inhibits ribonucleotide reductase which regulates the production of deoxynucleotides necessary for DNA synthesis and repair (Plukett et al., 1995). The depletion of the deoxynucleotides leads to an increased incorporation of the dF-dCTP into DNA, blocking DNA synthesis (masked chain termination). After incorporation of the dF-dCTP into the DNA an increase in the number of DNA single-strand breaks, chromosome breaks and micronuclei has been observed (Auer et al., 1997).

Both in vitro and in vivo studies have shown that gemcitabine is a potent radiosensitizer (Shewach et al., 1994; Rockwell & Grindey, 1992; Lawrence et al., 1996; Latz et al., 1998; Gregoire et al. 1998; Milas et al., 1999; van Putten et al., 2001, Wachters et al., 2003, Castro Kreder et al., 2003). However, in an early study in non-small cell lung cancer patients, concurrent gemcitabine and radiotherapy resulted in unacceptable pulmonary toxicity related to the large volume of radiation delivered to the lung (Scalliet et al., 1998). More recent ongoing phase I trials show that concurrent gemcitabine at lower doses and radiotherapy is feasible without severe pulmonary toxicity (Manegold et al., 2000; Blackstock et al., 2001). Its unique mechanism of action, its lack of overlapping toxicity and its favourable toxicity profile define gemcitabine as an ideal candidate for combination therapy (Manegold et al., 2000). Currently many randomized studies are ongoing in which gemcitabine is combined with radiotherapy.

Gemcitabine radiosensitization is studied in a gemcitabine sensitive and resistant human lung tumour cells, SWp and SWg, resp., and in gemcitabine sensitive and resistant human ovarian tumour cells, A2780 and AG6000, resp. (Van Bree et al., 2002; Bergman et al., 2000). Gemcitabine was given 24 h before radiation treatment. The SWp is in fact similar to the SW1573 cell line which has been described above. It is called here SWp to distinguish it from SWg, the gemcitabine resistant counterpart which has been developed by van Bree et al. (2002). The lung tumour cells have different sensitivities to radiation alone as compared to the ovarian cancer cells (Van Bree et al., 2002; Bergman et al., 2000).

In table 3 the linear and quadratic parameters of the different cell lines obtained after analyses of the radiation dose survival curves for radiation alone and after combined radiation and gemcitabine treatment are summarized. SWp and SWg were almost equally sensitive to ionizing radiation alone with respect to the low dose region described by the α-value of the linear quadratic formula (Table 3). A slight increase in survival was observed in SWg cells in the high dose region which was reflected by a slightly lower β-value of the linear-quadratic formula (0.040 ± 0.006 vs 0.055 ± 0.008). The human ovarian carcinoma cell line A2780 and its gemcitabine-resistant variant AG6000 were equally sensitive to ionizing radiation. The surviving fractions of the different cell lines after incubation with gemcitabine alone are: SWp 10 nM: 0.52 ± 0.06; SWg 10 μM: 0.95 ± 0.03, 100 μM: 0.24 ± 0.11; A2780 2 nM: 0.82 ± 0.08, 10 nM: 0.21 ± 0.08; AG6000 20 μM: 0.62 ± 0.07, 50 μM: 0.22 ± 0.04.

As can be observed in figure 4 and table 3 radiosensitization is observed with gemcitabine-sensitive as well as in gemcitabine resistant cells. For the resistant cells much higher gemcitabine doses are needed for the radiation sensitization to result in similar cytotoxicity. Both gemcitabine-sensitive cell lines SWp and A2780 are sensitized by incubation with 10 nM of gemcitabine for 24 h before irradiation while the SWg and AG6000 are not sensitized with this dose of gemcitabine. The sensitization is described by an increase in the α-values with factors of 3 and 1.4 respectively, whereas the β-values are not significantly altered. Higher concentrations of Gemcitabine (50 and 100 nM resp.) are required to sensitize gemcitabine-resistant AG6000 and SWg cells to irradiation. For the SWg cells, the radiosensitization was reflected by an increase by a factor of 2.25 in the value of β, whereas in the AG6000 only the α-value was increased by factor of 1.3.

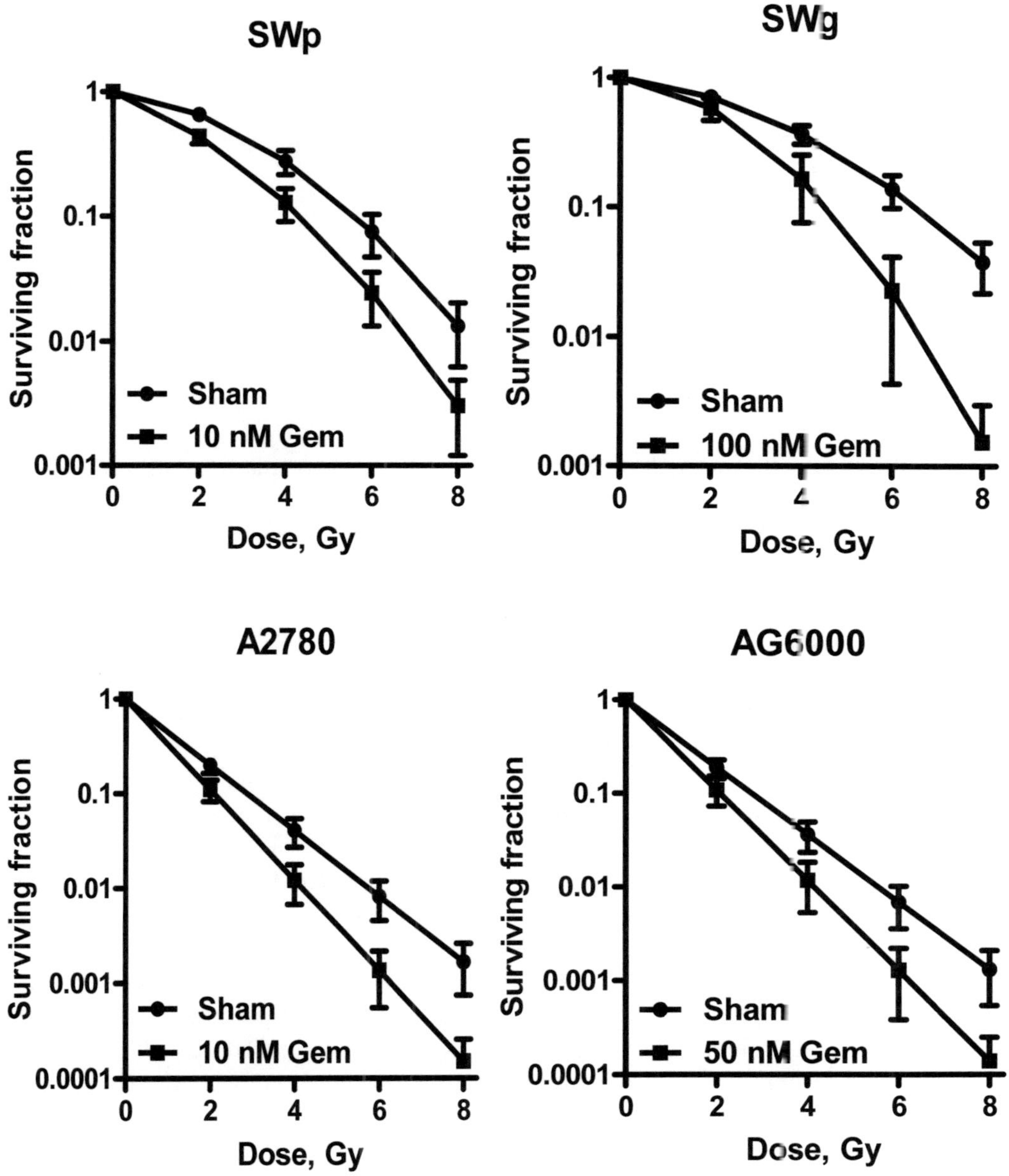

Fig. 4. Radiation sensitization after 24 h incubation with different concentrations of gemcitabine in gemcitabine sensitive SWp and resistant SWg lung tumour cells and in gemcitabine sensitive A2780 and resistant AG6000 ovary cancer cells. Surviving fractions are corrected for gemcitabine toxicity alone (for values see text). Cells are plated immediately after irradiation. Means with SEM of at least three separate experiments are shown.

Cells	Treatment	α (Gy⁻¹)	β (Gy⁻²)	α/β	α -enhanc factor	β -enhanc factor
SWp	sham	0.10 ± 0.03	0.055 ± 0.008	1.8 ± 0.6		
	10nM gemcitabine	0.30 ± 0.06*	0.053 ± 0.007	5.7 ± 1.4	3.0 ± 2.8	0.96 ± 0.2
SWg	sham	0.09 ± 0.02	0.040 ± 0.006	2.3 ± 0.6		
	100 μM gemcitabine	0.09 ± 0.03	0.090 ± 0.041†	1.0 ± 0.6	1.0 ± 0.5	2.3 ± 1.1
A2780	sham	0.80 ± 0.10	na			
	10nM gemcitabine	1.10 ± 0.15*	na		1.4 ± 0.3	
AG6000	sham	0.83 ± 0.13	na			
	50μM gemcitabine	1.11 ± 0.20†	na		1.3 ± 0.3	

significant difference with *P<0.01, †P<0.05, na is not applicable

Table 3. Values of the linear-quadratic parameters α and β and enhancement factors from cells treated with ionizing radiation only and gemcitabine-sensitized radiation dose survival curves of gemcitabine-sensitive (SWp and A2780) and gemcitabine-resistant (SWg and AG6000) cells.

5. Halogenated pyrimidines

Incorporation of halogenated pyrimidines (HPs), chloro-, bromo- and iodo-deoxyuridine (CldUrd, BrdUrd, IdUrd) into DNA is known to sensitise cells to ionizing radiation (Franken et al. 1997ab, 1999ab; van Bree et al. 1997; Iliakis et al. 1999; Miller et al. 1992ab). The induced radiosensitisation increases with the degree of thymidine-replacement. The mechanism of radiation sensitisation by the HPs has been suggested to be either an increase in the amount of DNA damage induced by radiation, an influence on repair of sublethal damage (SLD), and/or an enhanced expression of potentially lethal damage (PLD) (Jones et al. 1995; Franken et al. 1997). Since different processes are involved in these phenomena several mechanisms might contribute to the radiosensitisation.

HPs have been suggested to provide an advantage in radiotherapy as radiosensitisers of cells in rapidly growing tumours, in particular in clinical conditions in which critical normal tissues show limited proliferation and as a consequence take up less HP. Labelling depends on the growth fraction, cell loss, cell cycle time and potential doubling time. Of special importance for sensitisation is the rate at which non-cycling cells are recruited into the proliferative compartment during exposure to HPs and a course of radiotherapy. However, even in rapidly growing tumours, cells may, after proliferative cycles, move into a non-proliferative stage. This might compromise the degree of radiation sensitisation if resting cells are less affected by HPs, or are better able to cope with additional damage by repair of PLD.

Here the results of radiosensitization after incubation with 4 μM IdUrd for 72 h are presented. IdUrd-induced radio sensitisation was obtained in all studied cell lines, SW1573, RUCII (Rat urether carcinoma), R1 (Rat rhabdomyosarcoma) and V79, in exponentially growing and in plateau-phase cells. Values of α and β derived by linear-quadratic analyses of survival curves of exponentially growing cells and plateau-phase cells are presented in table 4. Survival curves of SW cells and V79 cells are given in figure 5. The plating conditions of the V79 cells, i.e. exponentially growing cells plating before or after irradiation (ppi or dp resp.), and plateau phase cells plated immediately or 6-24 h delayed after irradiation (ip or dp resp.) had no influence on the factor of increase of the α-value. It is shown that the value of the linear parameter, α can be enhanced by a factor of 1.9 to 7.5 and that in general low values of α are enhanced more than higher values of α. The value of β is less enhanced and the enhancement factor ranges from 0.7 to 2.4.

The direct comparison between immediate and delayed plating of plateau-phase cells and between plateau phase and exponentially growing cells shows significant quantitative differences. The data on the linear and quadratic parameters described here provide various new insights in the interpretation of radiosensitisation of delay plated plateau-phase cells. It is demonstrated that in delay plated HP-sensitized plateau phase cells PLD is not abolished.

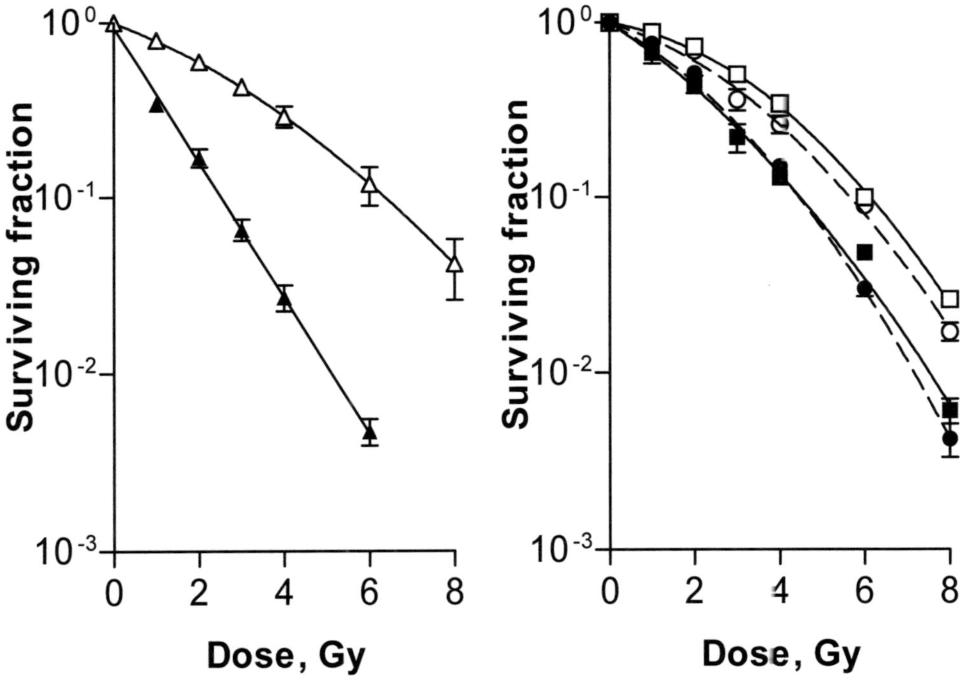

Fig. 5. Radiation dose-survival curves of exponentially growing SW-1573 cells (left) without IdUrd (open triangles) and after incubation with 4 μM IdUrd (closed triangles) and plateau-phase cells (right) plated immediately after irradiation (dashed lines) and plated 24 h after irradiation (solid lines) without IdUrd (open symbols) and after incubation with 4 μM IdUrd (closed symbols). Each point represents the mean value of 3 different experiments ± sem.

Cell line	α (Gy^{-1}) control	β (Gy^{-2}) control	α (Gy^{-1}) IdUrd-sens	β (Gy^{-2}) IdUrd-sens	α/β control	α/β IdUrd-sens	α - enhanc factor	β - enhanc factor
SW 1573 cells Exp growing ip	0.22 ± 0.01	0.022 ± 0.001	0.83 ± 0.06	na	10.0 ± 0.6	na	3.8 ± 0.3	na
SW 1573 cells Plateau phase ip	0.17 ± 0.03	0.042 ± 0.004	0.31 ± 0.03	0.047 ± 0.005	4.1 ± 0.8	6.6 ± 1.0	1.8 ± 0.4	1.1 ± 0.2
SW 1573 cells Plateau phase dp	0.09 ± 0.02	0.046 ± 0.002	0.37 ± 0.04	0.033 ± 0.006	2.0 ± 0.4	11.2 ± 2.4	4.1 ± 1.0	0.7 ± 0.1
RUCII cells Exp growing ppi	0.008 ± 0.007	0.025 ± 0.001	0.06 ± 0.02	0.026 ± 0.001	0.3 ± 0.3	2.3 ± 0.8	7.5 ± 7.0	1.0 ± 0.1
R1 cells Exp growing ppi	0.23 ± 0.01	0.068 ± 0.003	0.44 ± 0.05	0.075 ± 0.016	3.4 ± 0.2	5.9 ± 1.4	1.9 ± 0.3	1.1 ± 0.2
V79 cells Exp growing ip	0.18 ± 0.02	0.017 ± 0.003	0.38 ± 0.04	0.023 ± 0.007	10.6 ±2.2	16.5 ± 5.3	2.1 ± 0.3	1.4 ± 0.5
V79 cells Exp growing ppi	0.15 ± 0.02	0.013 ± 0.003	0.29 ± 0.03	0.016 ± 0.004	11.5 ± 3.1	18.1 ± 4.9	1.9 ± 0.3	1.2 ± 0.4
V 79 cells Plateau phase ip	0.09 ± 0.03	0.026 ± 0.004	0.17 ± 0.02	0.062 ± 0.005	3.5 ± 1.3	2.7 ± 0.4	1.9 ± 0.7	2.4 ± 0.4
V 79 cells Plateau phase dp	0.07 ± 0.02	0.020 ± 0.002	0.30 ± 0.03	0.024 ± 0.004	3.5 ± 1.1	12.5 ± 2.4	4.3 ± 1.3	1.2 ± 0.2

Means with SEM of at least three separate experiments are shown. ip=immediately plated after irradiation; dp=delayed plated after irradiation; ppi=plated prior to irradiation; na=not applicable.

Table 4. Values of the linear-quadratic parameters α and β and enhancement factors of several cell lines treated with ionizing radiation only and after sensitization with iododeoxyuridine (incubation with 4 μM IdUrd for 72 h)

6. PARP-1 inhibitors

The effect of the Parp-1 inhibitor NU-1025 on the linear and quadratic parameters was tested in Mouse embryonic fibroblasts. Parp1 also known as Poly (ADP-ribose) polymerase is an enzyme which is involved in the single strand break (SSB) repair of the DNA. The DNA SSB induced by ionizing radiation are mostly repaired by the base excision repair system, BER, wheras the DNA DSB are repaired by non homologous endjoining NHEJ or homologous recombination, HR. Inhibiting Parp-1 activity reduces the single strand break repair (Bouchard, 2003). Besides its role in BER Parp-1 is involved in many nuclear processes like DNA replication, transcription, DSB repair, apoptosis and genome stability (Rouleau et al. 2010; Bouchard et al. 2003, Löser et al. 2010). Recently it was hypothesized that cells deficient in BRCA2 or BRCA1 are particularly sensitive to inhibition of Parp-1 (Rouleau et al. 2010, Krawczyck et al. 2011). During DNA replication SSBs are induced. In the absence of Parp-1 these SSB are transformed in DSB. These DSB are repaired with homologous recombination (HR). Therefore cells deficient in HR (e.g. BRCA1 or BRCA2 tumours) might be sensitive to Parp-1 inhibitors. As Parp-1 is involved in many DNA repair processes, Parp-1 inhibitors might work well as radiosensitizers (Löser et al. 2010). As can be observed in figure 6 a modest sensitization effect by the Parp-1 inhibitor NU-1025 was obtained and increase of the value of α in the repair deficient cell line was larger than in the repair proficient cell line, 1.4 vs 1.2 respectively (table 5). The radiation dose survival curves of these MEF cells did not show a shoulder and therefore the quadratic parameter, β, could not be determined.

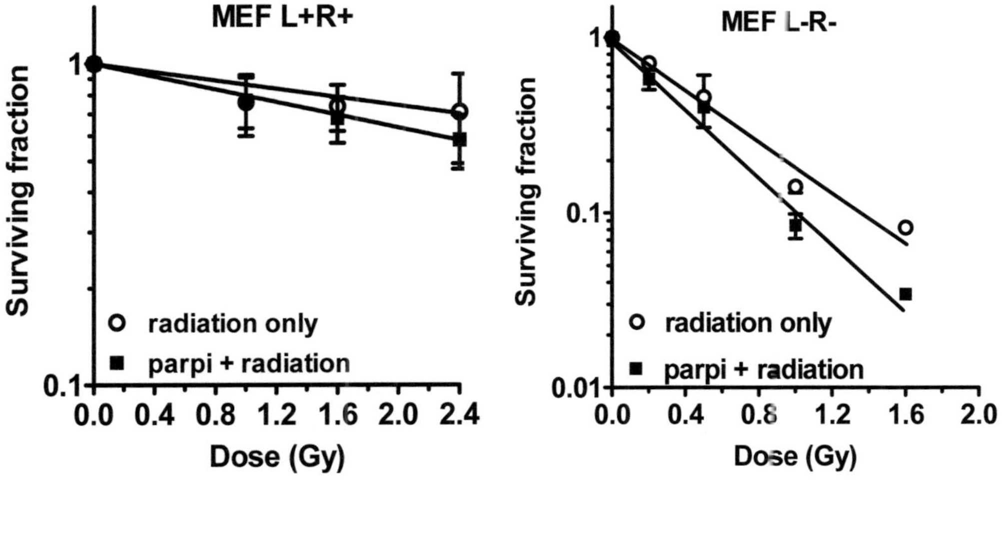

Fig. 6. Radiation dose-survival curves of Mouse embryonic fibroblasts (A: LigIV+/+, Rad54+/+) A and LigaseIV deficient and RAD54 deficient (B: LigIV-/-, Rad54-/-) mouse embryonic fibroblasts. Open circles radiation only curves, closed squares: radiation + Parp inhibitor. Cells were treated with 100 μM Nu-1025 for 24 h before irradiation.

MEF Cells	Treatment with Parp-i	α (Gy^{-1})	β (Gy^{-2})	α -enhanc. factor
LigIV+/+,Rad54+/+	No	0.28 ± 0.01	na	
LigIV+/+,Rad54+/+	yes	0.33 ± 0.03	na	1.2
LigIV-/-,Rad54-/-	no	1.59 ± 0.18	na	
LigIV-/-,Rad54-/-	yes	2.28 ± 0.42	na	1.4

Na= not applicable.

Table 5. Values of the linear parameter, α, and the enhancement factors. The quadratic parameter β could not be determined in these MEF cells.

7. Discussion and conclusion

In most cases an increase of the α-component was observed which corresponds to an enhanced (potentially) direct lethal damage (PLD) at low doses. The β-component, which is assumed to depend on the interaction of sublethal lesions (SLD), was rarely affected by the studied radiosensitization agents. Moreover, it appeared that more radioresistant cell lines were more sensitised than the radiosensitive lines. Furthermore it can be concluded that radiosensitization is also dependent on cell cycle stage like plateau or exponentially growing phase or post treatment plating conditions.

Hyperthermia is an excellent radiosensitizer which can already be effective at mild temperatures. One hour hyperthermia treatment at 41°C without radiation had only a small cytotoxic effect in both the heat sensitive and the heat resistant cell line. This is in agreement with the general idea of cell kill induction at temperatures ≥42°C for 1h or more (Dewhirst, 2005). Hyperthermia treatment at 43°C for 1h did not have a large cytotoxic effect in heat resistant SW-1573 cells. Radiosensitization by 41°C temperature hyperthermia was observed in RKO, but not in SW-1573 cells. The ability of mild temperatures (in the range of 40-42°C) hyperthermia to increase radiosensitivity of human tumor cells has been shown to be cell line dependent (Ruy et al., 1996; Franken et al., 2001; Bergs et al., 2007ab; Larsson & Ng, 2003; Murthy et al., 1977). In a study by Xu et al. (1999) 41.1°C pre-treatment of cells for 1h did not induce radiosensitization whereas treatment for 2h or more resulted in radiosensitization, in the hyperthermia resistant, but not in the hyperthermia sensitive cell line (Xu et al., 1999]. However, simultaneous treatment of the sensitive cell line with 1h 41.1°C hyperthermia and radiation did increase cellular radiosensitivity (Xu et al., 2002). An important mechanism of mild hyperthermia induced radiosensitization *in vivo* is the reoxygenation of tumors by an increase in blood flow (Vujaskovic et al., 2004; Oleson & Robertson, 1995; Song et al., 1995). Recently it was demonstrated that the BRCA-2 protein is transiently inhibited by mild hyperthermia (Krawzcyk et al., 2011). Also translocation of the Mre11 DSB repair protein from the nucleus to the cytoplasm has been implicated (Xu et al., 2002, 2007). However, disappearance of Mre11 protein foci at the sites of irradiation induced DNA double strand breaks by 41°C pre-incubation of cells was not observed (Krawzcyk et al., 2011; Bergs, 2007a). A role for mitotic catastrophe occurring as a result of G2/M checkpoint abrogation has also been suggested (Mackey & Ianzini, 2000). It has been shown that radiosensitization by 41-43°C hyperthermia correlates with an increased number of chromosomal fragments, but not of color junctions, at 24h after treatment compared to radiation alone (Bergs et al., 2008).

It is shown that cisplatin causes radiosensitization as measured by clonogenic survival, but only after allowing a potentially lethal damage repair (PLDR) time of 24 hours. These results are in agreement with those of Wilkins et al. (1993) who investigated the effect of cisplatin and radiation on PLDR in confluent cultures of two different brain tumor cell lines. Wilkins et al. (1993) also observed no radiosensitization by cisplatin in immediately plated cells whereas a cisplatin-induced radiosensitization was seen in cells plated eight hours after irradiation. Their results indicate that the radiosensitizing effect of cisplatin occurs through the inhibition of post-irradiation recovery. The strongest inhibition of PLDR was achieved when cisplatin was administered shortly before or after irradiation (Wilkins et al., 1993). In our experiments, cells were irradiated while cisplatin was present in the medium.

Results from studies using exponentially growing cell cultures vary from a cisplatin-induced radiosensitization (Loprevite et al., 2001; Begg et al. 1986; Nakamoto et al. 1996; Huang et al. 2004) to only an additive effect (Gorodetsky et al., 1998; Loprevite et al., 2001; Britten et al., 1996; Monk et al., 2002). The effect of cisplatin treatment on radiosensitivity may depend on the cell type used. Loprevite et al (2001) observed synergism in a squamous lung carcinoma cell line when exposed to cisplatin, whereas an adenocarcinoma of the lung was not sensitized by cisplatin. Even cell lines derived from a single biopsy can differ in the response to cisplatin and radiation combination therapy (Britten et al., 1996).

Although dependence on cell cycle phase (Meyn et al., 1980; Krishnaswamy & Dewey, 1993), cisplatin incubation time and the sequence of treatment modalities have been implicated (Gorodetsky et al., 1998; Meyn et al., 1980; Krishnaswamy & Dewey, 1993), there is currently no consensus to account for the varying response of cells to cisplatin and radiation.

The mechanism of cisplatin induced radiosensitization might be due to the inhibition of the DNA repair, NHEJ and HR, pathways (Myint et al., 2002; Dolling et al., 1998). The Ku protein complex, which plays an important role in NHEJ, was demonstrated to show a reduced ability to translocate on DNA containing cisplatin-DNA adducts compared with undamaged DNA. This resulted in a decreased interaction between Ku and DNA-PK_{cs} (Turchi et al., 2000) However, the biochemical processes that cisplatin undergoes in the cell are complex and the intracellular fate of cisplatin may be linked to copper transport (Muffia & Fojo, 2004). Therefore, other processes such as the formation of peroxy complexes inside the cell might be involved in cisplatin-induced radiosensitization (Dewit, 1987). Bergs et al. (2006) demonstrated an increase in the induction of apoptosis after combined treatment as compared to radiation or cisplatin alone at 24 h after treatment. This was confirmed in several other studies (Kumala et al., 2003; Guchelaar et al. (1998). These apoptotic effects observed by Bergs et al. (2006) correlated with clonogenic survival. Fujita et al. (2000) also observed an inhibitory effect of the combination of cisplatin and radiation on the survival of lung tumor cells and ascribed this effect on the induction of tumor cell apoptosis.

In conclusion, a radiosensitizing effect of cisplatin on cell survival is observed in confluent cultures when cells were replated after a 24 hour incubation period during which PLD repair could take place. In contrast, cisplatin did not induce a significant radiosensitization after immediate plating.

Several studies have shown that gemcitabine is a potent sensitizer of ionizing radiation (Shewach et al., 1994; Gregoire et al., 1999; Ostruszka & Shewach, 2000). Among other proposed mechanisms of action, the effect of gemcitabine on cell cycle distributions may be the most important (Milas et al., 1999; Van Putten et al., 2001). In our studies, both gemcitabine-sensitive cell lines SWp and A2780 could be sensitized to irradiation when cytotoxic gemcitabine-treatments were given. The radiosensitization was accompanied by a clear arrest of cells in early S phase which has been argued to be vital for gemcitabine-induced radiosensitization (Latz et al., 1998). Both cell lines showed an increase in α-value indicating the efficacy of gemcitabine-induced radiosensitization in the clinically relevant dose range. Although the gemcitabine resistant cells still could be sensitized only much higher gemcitabine doses were necessary to reach an effect. In the resistant ovarian carcinoma cell line AG6000 this was demonstrated by an increase in the value of α. In contrast with this change, in the gemcitabine resistant lung tumour cell line an increase in the β-value was obtained, the α-value was not affected. In both gemcitabine-resistant cell lines the sensitivity to ionizing radiation alone was not altered. It is reported that gemcitabine resistant tumours are cross-resistant to related drugs like Ara-c (Ruiz van haperen et al., 1994; Peters et al., 1996). In both gemcitabine-resistant cell line, AG6000 and SWg, this was indeed the case (Van Bree et al., 2002). Moreover, the AG6000 cells were also more resistant to cisplatin and taxoids (Bergman et al., 2000). However, no altered sensitivity was found in SWg cells for cDDP, paclitaxel, MTX and 5 FU, while AG6000 cells were 2.5-fold more sensitive to MTX (Bergman et al., 2000). These findings indicate that patients previously treated with gemcitabine may receive additional radiotherapy with or without cDDP or paclitaxel.

The HP-induced-radiosensitisation is mainly due to an increase in the linear parameter α. The quadratic parameter, β, is only rarely influenced. Different mechanisms involved in the radiosensitisation induced by halogenated pyrimidines have been described. Wang et al. (1994) suggested that in exponentially growing cells increased DNA damage production was the major component of radiosensitisation while in plateau-phase cells radiosensitisation occurred through inhibited repair and/or enhanced fixation of potentially lethal damage. The increase of the α values for exponentially growing cells as found in our study, indicates an increase in the number of directly lethal events due to the HPs. This is in agreement with observations of Webb et al. (1993) and Jones et al. (1995) which suggest that an important mechanism of radiosensitisation involves an increase of effective DNA double strand breaks. Miller et al. (1992ab) have suggested that radiation-induced damage in cells which have HPs incorporated into the DNA after low-LET irradiation resembles the damage produced by high-LET radiation. In plateau-phase cells plated immediately after irradiation the increase of α might be due to the same mechanism as involved in exponentially growing cells. In these cells also an increase of β was observed indicating that accumulation of sublethal lesions contributed significantly (Barendsen 1990). Due to the immediate plating after irradiation this sublethal damage might be fixated.

Greatest increases in α were found in delayed plated plateau-phase cells. This radiosensitisation can be interpreted as an enhanced fixation of potentially lethal damage due to immediate DNA damage and/or to damaged DNA repair function in these cells expressed during the interval before delayed plating. The value of β in these cells returned to values as found in cells not containing HPs. This demonstrates that sublethal damage has been repaired in HP-containing plateau-phase cells.

Because Parp-1 is implicated in several DNA repair processes, Parp-1 inhibitors might be good radiosensitizers. Several studies have already demonstrated the radiosensitizing effect of Parp-1 inhibitors (Albert et al. 2007; Löser et al. 2010; Krawczyk et al. 2011). Loser et al concluded that the effects of Parp-1 inhibitors are more pronounced on rapidly dividing and/or DNA repair deficient cells. In our study at time of treatment most of the cells in culture had accumulated in G1 phase. Therefore radiosensitization effects are modest. However, the increase of value of the linear parameter, α, of the repair deficient cells was more increased after the parp-1 inhibition than of the repair proficient cells.

8. Acknowledgements

We thank Dr. G. Iliakis for sending us (with permission from Dr. F. Alt) the mouse embryonic fibroblast cells. The authors are thankful for financial support from several foundations. The Maurits and Anna de Kock and the Nijbakker Morra foundations are acknowledged for sponsoring laboratory equipment. The Dutch Cancer Foundation (grant # UVA 2006-3484, # UVA 2008-4019) and the Stichting Vanderes are acknowledged for financing personnel support

9. References

Auer, H.; Oehler, R., Lindner, R., Kowalski, H.; Sliutz, G., Orel, L., Kucera, E., Simon, M. M., & Glössl, J. (1997). Characterisation of genotoxic properties of 2',2'-difluorodeoxycytidine. *Mutation Research*, Vol.393, pp. 165-173.

Barendsen, G.W. (1982). Dose fractionation, dose rate and iso-effect relationships for normal tissue responses. *International Journal of Radiation Oncology, Biology and Physics*, Vol.8, pp. 1981-1997.

Barendsen, G.W. (1990). Mechanisms of cell reproductive death and shapes of radiation dose-survival curves of mammalian cells. *International Journal of Radiation Biology*, Vol.57, pp. 885-896.

Barendsen, G.W. (1994). The relationship between RBE and LET for different types of lethal damage in mammalian cells: biophysical and molecular mechanisms. *Radiation Research*, Vol.139, pp. 257-270.

Barendsen, G.W. (1997). Parameters of linear-quadratic radiation dose-effect relationships: dependence on LET and mechanisms of reproductive cell death. *International Journal of Radiation Biology*, Vol.71, pp. 649-655.

Barendsen, G.W.; van Bree, C. & Franken, N.A.P. (2001). Importance of cell proliferative state and potentially lethal damage repair on radiation effectiveness: implications for combined tumor treatments (review). *International Journal of Oncology*, Vol.19, pp257-256.

Begg, A.C.; Van der Kolk, P.J., Dewit, L. & Bartelink, H. (1986). Radiosensitization by cisplatin of RIF1 tumour cells in vitro. *International Journal of Radiation Biology*, Vol.50, pp. 871-884.

Bergman, A.M.; Giaccone, G.; Van Moorsel, C.J.A.; Mauritz, R.; Noordhuis, P.; Pinedo, H.M.; & Peters, G.J. (2000). Cross-resistance in the 2',2'-difluorodeoxycytidine (Gemcitabine)-resistant human ovarian cancer cell line AG6000 to standard and investigational drugs. *European Journal of Cancer*, Vol.36, p. 1974-1983.

Bergman, A.M.; Pinedo, H.M.; Jongsma, A.P.; Brouwer, M.; Ruiz van Haperen, V.W.T.; Veerman, G.; Leyva, A.; Eriksson, S. & Peters, G.J. (1999). Decreased resistance to gemcitabine (2',2'-difluorodeoxycitidine) of cytosine arabinoside-resistant myeloblastic murine and rat leukemia cell lines: role of altered activity and substrate specificity of deoxycytidine kinase. *Biochemical Pharmacology*; Vol.57, pp. 397-406.

Bergs, J.W.J.; Franken, N.A.P., ten Cate, R., Van Bree, C. & Haveman, J. (2006). Effects of cisplatin and gamma-irradiation on cell survival, the induction of chromosomal aberrations and apoptosis in SW-1573 cells. *Mutation Research*, Vol.594, pp. 148-154.

Bergs, J.W.J. (2007a). Hyperthermia, cisplatin and radiation trimodality treatment: In vitro studies on interaction mechanisms. PhD. Thesis, University of Amsterdam.

Bergs, J.W.J.; Haveman, J., ten Cate, R., Medema, J.P., Franken, N.A.P. & Van Bree, C. (2007b). Effect of 41 degrees C and 43 degrees C on cisplatin radiosensitization in two human carcinoma cell lines with different sensitivities for cisplatin. *Oncology Reports*, Vol.18, pp. 219-226.

Bergs, J.W.J.; Franken, N.A.P., Haveman, J., Geijsen, E.D., Crezee, J. & Van Bree, C. (2007c). Hyperthermia, cisplatin and radiation trimodality treatment: a promising cancer treatment? A review from preclinical studies to clinical application. *International Journal of Hyperthermia*, Vol.23, pp. 329-341. Review.

Bergs, J.W.J.; ten Cate, R., Haveman, J., Medema, J.P., Franken, N.A.P. & Van Bree, C. (2008). Chromosome fragments have the potential to predict hyperthermia-induced radiosensitization in two different human tumor cell lines. *Journal of Radiation Research*, Vol.49, pp. 465-472.

Blackstock, A.W.; Lesser, G.J., Fletcher-Steede, J., Case, L.D.. Tucker, R.W., Russo, S.M., White. D.R. & Miller, A. (2001). Phase I Study of twice-weekly gemcitabine and concurrent thoracic radiation for patients with locally advanced son-small cell lung cancer. *International Journal of Radiation Oncology Biology Physics*, Vol.51, pp. 1281-1289.

Bouchard, V.J.; Rouleau, M. & Poirier, G.G. (2003). Parp1, a determinat of cell survival in response to DNA damage. *Experimental Hematology*, Vol.31, pp. 446-454.

Britten, R.A.; Peacock, J. & Warenius, H.M. (1992). Collateral resistance to photon and neutron irradiation is associated with acquired cis-platinum resistance in human ovarian tumour cells. *Radiotherapy and Oncology*, Vol.23, pp. 170-175.

Britten, R.A.;.Evans, A.J, Allalunis-Turner, M.J. & Pearcey, R.G. (1996). Effect of cisplatin on the clinically relevant radiosensitivity of human cervical carcinoma cell lines. *International Journal of Radiation Oncology, Biology, Physics*, Vol.34, pp. 367-374.

Castro Kreder, N.; Van Bree, C., Franken, N.A.P. & Haveman, J. (2003). Colour junctions as predictors of radiosensitivity: X-irradiation combined with gemcitabine in a lung carcinoma cell line. *Journal of Cancer Research and Clinical Oncology*, Vol. 129, pp. 597-603.

Castro Kreder, N.; Van Bree, C., Franken, N.A.P. & Havenman, J. (2004). Effects of gemcitabine on cell survival and chromosome aberrations after pulsed low dose-rate irradiation. *Journal of Radiation Research*, Vol.45, pp. 111-118.

Crezee J.; Barendsen, G.W., Westermann, A.M., Hulshof, M.C.C.M., Haveman, J., Stalpers, L.J.A., Geijsen, E.D. & Franken, N.A.P. (2009). Quantification of the contribution of hyperthermia to results of cervical cancer trials: In regard to Plataniotis an Dale (Int J Radiat Oncol Biol Phys 2009;73:1538-1544). *International Journal of Radiation Oncology, Biology, Physics*, Vol.75, p. 634.

Crul, M.; Van Waardenburg, R.C., Beijnen, J.H. & Schellens, J.H. (2002). DNA-based drug interactions of cisplatin. *Cancer treatment reviews*, 2002;28:291-303

De Silva, I.U.; McHugh, P.J., Clingen, P.H. & Hartley, J.A. (2002). Defects in interstrand cross-link uncoupling do not account for the extreme sensitivity of ERCC1 and XPF cells to cisplatin. *Nucleic Acids Research,* Vol.30, pp. 3848-3856.

Dewey, W.C.; Sapareto, S.A. & Betten, D.A. (1978). Hyperthermic radiosensitization of synchronous Chinese hamster cells: relationship between lethality and chromosomal aberrations. *Radiation research,* Vol.76, pp.48-59.

Dewhirst, M.W.; Vujaskovic, Z., Jones, E. & Thrall, D. (2005). Re-setting the biologic rationale for thermal therapy. *International Journal of Hyperthermia,* Vol.21, pp. 779-790.

Dewit, L. (1987). Combined treatment of radiation and cisdiamminedichloroplatinum (II): a review of experimental and clinical data. *International Journal of Radiation Oncology, Biology, Physics,* Vol.13, pp. 403-426.

Dolling, J.A., Boreham, D.R., Brown, D.L., Raaphorst, G.P. & Mitchel, R.E. (1999). Cisplatin-modification of DNA repair and ionizing radiation lethality in yeast, Saccharomyces cerevisiae. *Mutation Research* Vol.433, pp. 127-36.

Dronkert, M.L. & Kanaar, R. (2001). Repair of DNA interstrand cross-links, *Mutation Research* Vol.486, pp. 217-247.

Duenas-Gonzalez, A.; Cetina, L. & de la Mariscal, I.G.J. (2003). Modern management of locally advanced cervical carcinoma. *Cancer Treatment Reviews* Vol.29, pp. 389-399.

Fehlauer, F.; Barten-Van Rijbroek, A.D., Stalpers, L.J., Leenstra, S., Lindeman, J., Tjahja, I., Troost, D., Wolbers, J.G., van der Valk, P. & Sminia. P. (2000). Additive cytotoxic effect of cisplatin and X-irradiation on human glioma cell cultures derived from biopsy-tissue. *Journal of Cancer Research and Clinical Oncology,* Vol.126, pp. 711-716.

Fossella, F.V.; Lipmann, S.C., Shin DM, *et al.* (1997). Maximum-tolerated dose defined for single-agent gemcitabine: a Phase I dose-escalation study in chemotherapy-naïve patients with advanced non-small-cell lung cancer. *Journal of Clinical Oncology,* Vol.15, pp. 310-316.

Franken, N.A.P., Van Bree, C., Kipp, J.B.A. & Barendsen, G.W., (1997a). Modification of potentially lethal damage in irradiated chinese hamster V79 cells after incorporation of halogenated pyrimidines. *International Journal of Radiation Biology,* Vol.72, 101-109.

Franken, N.A.P.; Van Bree, C., Streefkerk, J.O., Kuper, I.M.J.A., Rodermond, H.M., Kipp, J.B.A. and Barendsen G.W. (1997b). Radiosensitization by iodo-deoxyuridine in cultured SW-1573 human lung tumor cells: Effects on α and β of the linear-quadratic model. *Oncology Reports,* Vol.4, pp. 1073-1076.

Franken N.A.P.; Ruurs, P., Ludwików, G., Van Bree, C., Kipp, J.B.A., Darroudi, F. & Barendsen, G.W. (1999a). Correlation between cell reproductive death and chromosome aberrations assessed by FISH for low and high doses of radiation and sensitization by iododeoxyuridine in human SW-1573 cells. *International Journal of Radiation Biology,* Vol.75, pp. 293-299.

Franken, N.A.P.; van Bree, C., Veltmaat, M.A.T., Ludwików, G., Kipp, J.B.A. & Barendsen, G.W. (1999b) Increased chromosome frequencies in iodo-doxyuridine-sensitized human SW-1573 cells after γ-irradiation. *Oncology Reports* Vol 6, pp. 59-63.

Franken, N.A.P.; Van Bree, C., Veltmaat, M.A., Rodermond, H.M., Haveman, J. & Barendsen, G.W. (2001). Radiosensitization by bromodeoxyuridine and hyperthermia: analysis of linear and quadratic parameters of radiation survival curves of two human tumor cell lines. *Journal of Radiation Research,* Vol.42, pp. 179-190.

Franken N.A.P.; ten Cate, R., Van Bree, C. & Haveman, J. (2004). Induction of the early response protein EGR-1 in human tumour cells after ionizing radiation is correlated with a reduction of repair of lethal lesions and an increase of repair of sublethal lesions. *International Journal of Oncology,* Vol.24, pp. 1027-1031.

Franken, N.A.P.; Rodermond, H.M., Stap, J., Haveman, J. & Van Bree, C. (2006) Clonogenic assay of cells in vitro. *Nature Protocols,* Vol.1, pp. 2315-2319.

Froelich, J.J.; Schneller, F.R. & Zahn, R.K. (1999). The influence of radiation and chemotherapy-related DNA strand breaks on carcinogenesis: an evaluation. *Clinical Chemistry and Laboratory Medicine,* Vol.37, pp. 403-408.

Fujita, M.; Fujita, T., Kodama, T., Tsuchida, T. & Higashino, K. (2000). The inhibitory effect of cisplatin in combination with irradiation on lung tumor cell growth is due to induction of tumor cell apoptosis. *International Journal of Oncology,* Vol.17, pp. 393-397.

Gonzalez Gonzalez, D.; Van Dijk, J.D., & Blank, L.E. (1995). Radiotherapy and hyperthermia. *European Journal of Cancer,* Vol. 31A, pp.1351-1355.

Gorodetsky, R.; Levy-Agababa, F., Mou, X. & Vexler, AM. (1998). Combination of cisplatin and radiation in cell culture: effect of duration of exposure to drug and timing of irradiation. *International journal of Cancer,* Vol.75, pp. 635-642,.

Gregoire, V.; Hittelman, W.N., Rosier, J.F. & Milas, L. (1999). Chemo-radiotherapy: radiosensitizing nucleoside analogues. *Oncology Reports,* Vol.6, pp. 949-957.

Groen, H.J.M.; Sleijfer, S., Meijer, C, *et al.* (1995). Carboplatin- and cisplatin-induced potentiation of moderate dose radiation cytotoxicity in human lung cancer cell lines. *British Journal of Cancer,* Vol.72, pp. 1406-1411.

Guchelaar, H.J.; Vermes, I., Koopmans, R.P., Reutelingsperger, C.P. & Haanen, C. (1998). Apoptosis- and necrosis-inducing potential of cladribine, cytarabine, cisplatin, and 5-fluorouracil in vitro: a quantitative pharmacodynamic model, *Cancer Chemotherapy & Pharmacology,* Vol.42, pp. 77-83.

Hall, E.J. & Giaccia, A.J. (2006). Radiobiology for the Radiobiologist sixth Edition, Lippincott Williams & Wilkins. Chapter 28, Hyperthermia, pp. 469-490.

Harima, Y.; Nagata, K., Harima, K., Ostapenko, V.V., Tanaka, Y. & Sawada, S. (2001). A randomized clinical trial of radiation therapy versus thermoradiotherapy in stage IIIB cervical carcinoma. *International Journal of Hyperthermia,* Vol.17, pp. 97-105.

Hatzis, P.; Al-Madhoon, A.S., Jullig, M., Petrakis, T.G., Eriksson, S. & Talianidis, I. (1998). The intracellular localization of deoxycytidine kinase. *Journal of Biological Chemistry,* Vol.273, pp. 30239-30243.

Haveman, J.; Castro Kreder, N., Rodermond, H.M., Van Bree, C., Franken, N.A.P., Stalpers L.J., Zdzienicka, M.Z., & Peters. G.J. (2004). Cellular response of X-ray sensitive hamster mutant cell lines to gemcitabine, cisplatin and 5-fluorouracil, *Oncology Reports* Vol.12 pp. 187-192.

Haveman, J.; Rietbroek, R.C., Geerdink, A., Rijn van, J. & Bakker, P.J.M. (1995). Effect of hyperthermia on cytotoxicity of 2',2'-difluorodeoxycytidine (gemcitabine) in cultured SW1573 cells. *International Journal of Cancer,* Vol.62, pp. 627-630

Heinemann, V.; Xu, Y.Z., Chubb S, Sen, A., Hertel, L.W. & Grindey, G.B. & Plunkett, W. (1992). Cellular elimination of 2',2'-difluorodeoxycytidine 5'-triphosphate: a mechanism of self-potentiation. *Cancer Research,* Vol.52, pp. 533-539.

Hildebrandt, B.; Wust, P., Ahlers, O., Dieing, A., Sreenivasa, G., Kerner, T , Felix, R. & Riess, H. (2002). The cellular and molecular basis of hyperthermia. *Critical reviews in oncology/hematology,* Vol.43, pp. 33-56.

Huang, H.; Huang, S.Y., Chen, T.T., Chen, J.C., Chiou, C.L. & Huang, T.M. (2004). Cisplatin restores p53 function and enhances the radiosensitivity in HPV16 E6 containing SiHa cells. *Journal of Cellular Biochemestry,* Vol.91, pp. 756-765.

Iliakis, G., Kurtzman, S., Pantelias, G. & Okayasu, R. (1989). Mechanism of radiosensitisation by halogenated pyrimidines: Effect of BrdU on radiation induction of DNA and chromosome damage and its correlation with cell killing. *Radiation Research,* Vol.119, pp. 286-304.

Iliakis, G., Wang, Y., Pantelias, G.E. & Metzger, L. (1992). Mechanism of radiosensitisation of halogenated pyrimidines. Effect of BrdU on repair of DNA breaks, interphase chromatin breaks and potentially lethal damage in plateau-phase CHO cells. *Radiation Research,* 1Vol.29, pp. 202-211.

Iliakis, G., Wright, E and Ngo, F.Q.H. (1987). Possible importance of PLD repair in the modulation of BrdUrd and IdUrd-mediated radiosensitisation in plateau-phase C3H10T1/2 mouse embryo cells. *International Journal of Radiation Biology,* Vol.51, pp. 541-548.

Iliakis, G.; Pantelias, G & Kurtzman, S., (1991). Mechanism of radiosensitisation by halogenated pyrimidines: Effect of BrdU on cell killing and interphase chromosome breakage in radiation sensitive cells. *Radiation Research,* Vol.25, pp. 56-64.

Joiner, M. & Van der Kogel, A. (2006). Basic Clinical Biology. Fourth Edition.

Jones, G.D.D., Ward, J.F., Limoli,C.L., Moyer, D.J. & Aguilera, J.A. (1995). Mechanisms of radiosensitization in iododeoxyuridine-substituted cells. *International Journal of Radiation Biology,* Vol.76, pp. 647-653.

Kampinga, H.H. & Dikomey, E. (2001). Hyperthermic radiosensitization: mode of action and clinical relevance. *International journal of radiation biology,* Vol.77, pp. 399-408.

Kinsella, T.J., Mitchell, J.R., Russo, A., Morstyn, G. & Glatstein,E. (1984). The use of halogenated thymidine analoge as clinical radiosensitisers rationale, current status*and* future prospects: nonhypoxic cell sensitisers. *International Journal of Radiation Oncology, Biology and Physics,* Vol.10, pp. 1399-1406.

Krawczyk, P.M., Eppink, B., Essers, J., Stap, J., Rodermond, H.M., Odijk, H., Zelensky, A., Van Bree, C., Stalpers, L.J., Buist, M.R., Soullié, T., Rens, J., Verhagen, H.J. M., O'Connor, M., Franken, N.A.P., ten Hagen, T.L.M., Kanaar, R. & Aten, J.A. (2011) Temperature-controlled induction of BRCA2 degradation and homologous recombination deficiency sensitizes cancer cells to PARP-1 inhibition. *Proceedings of the National Academy of Science,* Vol. 108, pp. 9851-9856.

Krishnaswamy, G. & Dewey, W.C. (1993). Cisplatin induced cell killing and chromosomal aberrations in CHO cells: treated during G1 or S phase. *Mutation Research,* Vol.293, pp. 161-172.

Kumala, S.; Niemiec, P., Widel, M., Hancock, R. & Rzeszowska-Wolny, J. (2003). Apoptosis and clonogenic survival in three tumour cell lines exposed to gamma rays or chemical genotoxic agents. *Cellular and Molecular Biology Letters,* Vol.8, pp. 655-665.

Larsson, C. & Ng, C.E. (2003). p21+/+ (CDKN1A+/+) and p21-/- (CDKN1A-/-) human colorectal carcinoma cells display equivalent amounts of thermal radiosensitization. *Radiation research,* Vol.160, pp. 205-259.

Latz, D.; Fleckenstein, K., Eble, M., Blatter, J., Wannenmacher, M. & Weber, K.J. (1998). Radiosensitizing potential of gemcitabine (2',2'-difluoro-2'-deoxycytidine) within the cell cycle in vitro. *International Journal of Radiation Oncology, Biolology, Physics,* Vol.41, pp. 875-882.

Lawrence, T.S.; Blackstock, A.W. & McGinn, C. (2003). The mechanism of action of radiosensitization of conventional chemotherapeutic agents. *Seminars in radiation oncology,* Vol.13, pp. 13-21.

Lawrence, T.S.; Davis, M.A. & Normolle D.P. (1995). Effect of bromodeoxyuridine on radiation-induced DNA damage and repair based on DNA fragment size using pulsed-field gel electrophoresis. *Radiation Research,* Vol.144, pp. 282-287.

Loprevite, M.; Favoni, R.E., de Cupis, A., Pirani, P., Pietra, G., Bruno, S., Grossi, F., Scolaro, T. & Ardizzoni, A. (2001). Interaction between novel anticancer agents and radiation in non-small cell lung cancer cell lines. *Lung Cancer* Vol.33, pp. 27-39.

Löser, D.A.; Shibata, A., Shibata, A.K., Woodbine, L.J. Jeggo, P.A. & Chalmers, A.J. (2010). Sensitization to radiation and alkylating agents by inhibitors of poly(ADP-ribose) polymerase is enhanced in cells deficient in DNA double strand break repair. *Molecular Cancer Therapeutics,* Vol.9, pp. 1775-1787.

Mackey, M.A. & Ianzini, F. (2000). Enhancement of radiation-induced mitotic catastrophe by moderate hyperthermia. *International Journal of Radiation Biology,* Vol.76, pp. 273-280.

Manegold, C.; Zatloukal, P., Krejcy, K. & Blatter, J. (2000). Gemcitabine in non-small lung cancer (NSCLC). *Invesigationalt New Drugs,* Vol.18, pp. 29-42.

Meyn, R.E.; Meistrich, M.L. & White, R.A. (1980). Cycle-dependent anticancer drug cytotoxicity in mammalian cells synchronized by centrifugal elutriation. *Journal of the National Cancer Institute,* Vol.64 , pp. 1215-1219.

Milas, L.; Fujii, T., Hunter, N., Elshaikh, M., Mason, K., Plunkett, W., Ang, K.K. & Hittelman, W. (1999). Enhancement of tumor radioresponse in vivo by gemcitabine. *Cancer Research,* Vol.59, pp. 107-114.

Miller, E.M.; Fowler, J.F. & Kinsella, T.J. (1992a) Linear-quadratic analysis of radiosensitisation by halogenated pyrimidines, I. Radiosensitisation of human colon cancer cells by iododeoxyuridine. *Radiation Research,* Vol.131, pp. 81-89.

Miller, E.M.; Fowler, J.F. & Kinsella, T.J. (1992b). Linear-quadratic analysis of radiosensitisation by halogenated pyrimidines. II. Radiosensitisation of human colon cancer cells by bromodeoxyuridine. *Radiation Research,* Vol.131, pp. 90-97.

Monk, B.J.; Burger. R.A., Parker, R., Radany, E.H., Redpath, L. & Fruehauf, J.P. (2002). Development of an in vitro chemo-radiation response assay for cervical carcinoma. *Gynecologic Oncology,* Vol.87, pp. 193-199.

Muggia, F.M. & Fojo, T. (2004). Platinums: extending their therapeutic spectrum. *Journal of Chemotherapy,* Vol.16 Suppl 4, pp. 77-82.

Munch-Petersen, B.; Cloos, L., Tyrsted, G. & Eriksson, S. (1991). Diverging substrate specificity of pure human thymidine kinases 1 and 2 against antiviral dideoxynucleosides. *Journal of Biological Chemistry,* Vol.266, pp. 9032-9038.

Murthy, A.K.; Harris, J.R. & Belli, J.A. (1977). Hyperthermia and radiation response of plateau phase cells. Potentiation and radiation damage repair. *Radiation Research,* Vol.70, pp. 241-247.

Myerson, R.J.; Roti Roti, J.L., Moros, E.G., Straube, W.L. & Xu, M. (2004). Modelling heat-induced radiosensitization: clinical implications. *International Journal of Hyperthermia,* Vol.20, pp. 201-12.

Myint, W.K., Ng, C. & Raaphorst, P. (2002). Examining the non-homologous repair process following cisplatin and radiation treatments, *International Journal of Radiation Biology* Vol.78, pp. 417-424.

Nakamoto, S.; Mitsuhashi, N., Takahashi, T., Sakurai, H. & Niibe, H. (1996). An interaction of cisplatin and radiation in two rat yolk sac tumour cell lines with different radiosensitivities in vitro. *International journal of radiation biology,* Vol.70, pp. 747-53.

Oleson, J.R. & Robertson, E. (1995). Special Lecture. Hyperthermia from the clinic to the laboratory: a hypothesis. *International Journal of Hyperthermia,* Vol.11, pp. 315-322.

Ostruszka, L.J. & Shewach, D.S. (2000). The role of cell cycle progression in radiosensitization by 2′,2′-difluoro-2′-deoxycytidine. *Cancer Research,* Vol.60, pp. 6080-6088.

Peters, G.J.; Ruiz van Haperen, V.W., Bergman, A.M., Veerman, G., Smitskamp-Wilms, E., Van Moorsel, C.J.A., Kuiper, C.M. & Braakhuis, B.J.M. (1996). Preclinical combination therapy with gemcitabine and mechanisms of resistance. *Seminars in Oncology,* Vol.23 (Suppl. 10), pp. 16-24.

Plunkett, W., Huang, P &, Gandhi, V. (1995). Preclinical characteristics of gemcitabine. *Anti-Cancer Drugs,* Vol.6 (Suppl. 6), pp. 7-13.

Raaphorst, G.P.; Azzam, E.I. & Feeley, M. (1988). Potentially lethal radiation damage repair and its inhibition by hyperthermia in normal hamster cells, mouse cells, and transformed mouse cells. *Radiation research,* Vol.13,pp. 171-182.

Raaphorst, G.P.; Feeley, M.M., Danjoux, C.E., DaSilva, V. & Gerig, L.H. (1991). Hyperthermia enhancement of radiation response and inhibition of recovery from radiation damage in human glioma cells. *International Journal of Hyperthermia,* Vol.7, pp. 629-641.

Raaphorst, G.P.; Heller, D.P., Bussey, A. & Ng, C.E. (1994). Thermal radiosensitization by 41 degrees C hyperthermia during low dose-rate irradiation in human normal and tumour cell lines. *International Journal of Hyperthermia,* Vol.10, pp. 263-270.

Raaphorst, G.P.; Wang, G., Stewart, D. & Ng, C.E. (1996). Concomitant low dose-rate irradiation and cisplatin treatment in ovarian carcinoma cell lines sensitive and resistant to cisplatin treatment. *International Journal of Radiation Biology,* Vol.69, pp. 623-631.

Raaphorst, G.P.; Yang, D.P. & Niedbala, G. (2004). Is DNA polymerase beta important in thermal radiosensitization? *International Journal of Hyperthermia,* Vol.20, pp. 140-103.

Rabik, C.A. & Dolan, M.E. (2007). Molecular mechanisms of resistance and toxicity associated with platinating agents. *Cancer treatment reviews,* Vol.33, pp. 9-23.

Rockwell, S. & Grindey, G.B. (1992). Effect of 2',2'-difluorodeoxycytidine on the viability and radiosensitivity of EMT6 cells in vitro. *Oncology Research,* Vol.4, pp. 151-155.

Roti Roti, J.L.(2004). Introduction: radiosensitization by hyperthermia. *International Journal of Hyperthermia,* Vol.20, pp. 109-114.

Roti Roti, J.L.; Kampinga, H.H., Malyapa, R.S., Wright, W.D., Vander Waal, R.P. & Xu M. (1998). Nuclear matrix as a target for hyperthermic killing of cancer cells. *Cell stress & chaperones,* Vol.3, pp. 245-255.

Rots, M.G.; Willey, J.C., Jansen, G., Van Zantwijk, C. H., Noordhuis, P., DeMuth, J. P., Kuiper, E., Veerman, A. J., Pieters, R. & Peters, G. J. (2000). mRNA expression levels of methotrexate resistance-related proteins in childhood leukemia as determined by a standardized competitive template based RT-PCR method. *Leukemia,* Vol.4, pp. 2166-2175.

Rouleau, M.; Patel, A., Hendzel, M.J., Kaufmann, S.H. & Poirier, G.C. (2010). PARP inhibition: PARP1 and beyond. *Nature Reviews Cancer,* Vol.10, pp. 293-301.

Ruiz van Haperen, V.W.; Veerman, G., Eriksson, S., Boven, E., Stegmann, A.P., Hermsen, M., Vermorken, J.B., Pinedo, H.M. &, Peters, G.J. (1994). Development and molecular characterization of a 2',2'-difluorodeoxycytidine-resistant variant of the human ovarian carcinoma cell line A2780. *Cancer Research,* Vol.54, pp. 4138-4143.

Ryu, S.; Brown, S.L., Kim, S.H., Khil, M.S. & Kim J.H. (1996). Preferential radiosensitization of human prostatic carcinoma cells by mild hyperthermia. *International Journal of Radiation Oncology, Biology, Physics,* Vol.34, pp. 133-138.

Scalliet, P., Goor, C., Galdermans D, *et al.* (1998). Gemzar (Gemcitabine) with thoracic radiotherapy - a phase II pilot study in chemo-naïve patients with advanced non-small-cell lung cancer (NSCLC) (Abstract). *Proc ASCO,* Vol.17, pp.499a.

Shewach, D.S.; Hahn, T.M., Chang, E., Hertel, L.W., & Lawrence, T.S. (1994). Metabolism of 2′,2′-difluoro-2′-deoxycytidine and radiation sensitization of human coloncarcinoma cells. *Cancer Research,* Vol.54, pp. 3218-3223.

Song, C.W.; Shakil, A., Osborn, J.L. & Iwata, K. (1996). Tumour oxygenation is increased by hyperthermia at mild temperatures. *International Journal of Hyperthermia,* Vol.12, pp. 367-373.

Spasokoukotskaja, T.; Arner, E.S.J., Brösjo, O., Gunvén, P., Julisson, G., Liliemark, J. & Eriksson, S. (1995). Expression of deoxycytidine kinase and phosphorylation of 2-chlorodeoxyadenosine in human normal and tumor cells and tissues. *European Journal of Cancer,* Vol.31, pp. 202-208.

Storniolo, A.M.; Enas, A.H., Brown, C.A., Voi, M., Rothenberg, M.L. & Schilsky, R. (1999). An investigational new drug treatment program for patients with gemcitabine-results for over 3000 patients with pancreatic carcinoma. *Cancer,* Vol.85, pp.1261-1268.

Turchi, J.J.; Henkels, K.M. & Zhou Y. (2000). Cisplatin-DNA adducts inhibit translocation of the Ku subunits of DNA-PK. *Nucleic Acids Research,* Vol.28, pp. 4634-4641.

Twentyman, P.R.; Wright, K.A. & Rhodes, T. (1991). Radiation response of human lung cancer cells with inherent and acquired resistance to cisplatin. *International Journal of Radiation Oncology, Biology, Physics,* Vol.20, pp. 217-220.

Van Bree, C.: Van Der Maat, B., Ceha, H.M., Franken, N.A.P., Haveman, J. & Bakker, P.J. (1992). Inactivation of p53 and of pRb protects human colorectal carcinoma cells against hyperthermia-induced cytotoxicity and apoptosis. *Journal of Cancer Research and Clinical Oncology,* Vol,125, pp. 549-55.

Van Bree, C.; Savonije, J.H., Franken, N.A., Haveman, J. & Bakker, P.J. (2000). The effect of p53-function on the sensitivity to paclitaxel with or without hyperthermia in human colorectal carcinoma cells. *International Journal of Oncology,* Vol.16, pp. 739-744.

Van Bree, C.; Castro Kreder, N., Loves, W.J., Franken, N.A.P., Peters, G.J. & Haveman, J. (2002). Sensitivity to ionizing radiation and chemotherapeutic agents in gemcitabine-resistant human tumor cell lines. *International Journal of Radiation Oncology, Biology, Physics,* Vol.54, pp. 237-44.

Van Bree, C.; Franken, N.A.P., Bakker, P.J., Klomp-Tukker, L.J., Barendsen, G.W. & Kipp, J.B. (1997). Hyperthermia and incorporation of halogenated pyrimidines: radiosensitization in cultured rodent and human tumor cells. *International Journal of Radiation Oncology, Biology, Physics,* Vol.39, pp. 489-496.

Van Der Zee, J.; Gonzalez Gonzalez, D., Van Rhoon, G.C., Van Dijk, J.D., Van Putten, W.L. & Hart, A.A. (2000). Comparison of radiotherapy alone with radiotherapy plus hyperthermia in locally advanced pelvic tumours: a prospective, randomised, multicentre trial. Dutch Deep Hyperthermia Group. *Lancet,* Vol.355, pp. 1119-1125.

Van Der Zee, J.; Treurniet-Donker, A.D., The, S.K., Helle, P.A., Seldenrath. J.J., Meerwaldt, J.H., Wijnmalen, A.L., Van den Berg, A.P., Van Rhoon, G.C., Broekmeyer-Reurink, M.P. & Reinhold, H.S. (1988). Low dose reirradiation in combination with hyperthermia: a palliative treatment for patients with breast cancer recurring in previously irradiated areas. *International Journal of Radiation Oncology, Biology, Physics,* Vol.15, pp. 1407-1413.

Van Der Zee, J. & González, G.D. (2002). The Dutch Deep Hyperthermia Trial: results in cervical cancer. *International Journal of Hyperthermia.* Vol.18, pp. 1-12. Erratum in: *International Journal of Hyperthermia* (2003) Vol.19, p. 213.

Van Putten, J.W.G.; Groen, H.J.M., Smid, K., Peters, G. J. & Kampinga, H. H. **(2001).** Endjoining deficiency and radiosensitization induced by gemcitabine. *Cancer Research,* Vol.61, pp. 1585-1591.

Vujaskovic, Z. & Song, C.W. (2004). Physiological mechanisms underlying heat-induced radiosensitization. *International Journal of Hyperthermia,* Vol.20, pp. 163-174.

Wang, Y.; Pantelias, G.E. & Iliakis, G., 1994, Mechanism of radiosensitization by halogen ated pyrimidines: the contribution of excess DNA and chromosome damage in BrdU radiosensitization may be minimal in plateau cells. *Internat ional Journal of Radiation Biology,* Vol.**66,** 133± 142.

Warters, R.L. & Axtell, J. (1992). Repair of DNA strand breaks at hyperthermic temperatures in Chinese hamster ovary cells. *International Journal of Radiation Biology,* Vol.61, pp. 43-48.

Webb, C.F.; Jones, G.D.D., Ward, J.F., Moyer, D.J., Aguilera, J.A. & Ling, L.L. (1993). Mechanisms of radiosensitisation in bromodeoxyuridine-substituted cells. *International Journal of Radiation Biology,* Vol.64, pp. 695-705.

Wilkins, D,E.; Ng, C.E. & Raaphorst, G.P. (1996). Cisplatin and low dose rate irradiation in cisplatin resistant and sensitive human glioma cells. *International Journal of Radiation Oncology, Biology, Physics,* Vol.36, pp. 105-111.

Wilkins, D.E.; Heller, D.P. & Raaphorst, G.P. (1993). Inhibition of potentially lethal damage recovery by cisplatin in a brain tumor cell line. *Anticancer Research,* Vol.13, pp. 2137-2142.

Xu, M.; Myerson, R.J., Xia, Y., Whitehead, T., Moros, E.G., Straube WL, Roti Roti, J.L. (2007). The effects of 41 degrees C hyperthermia on the DNA repair protein, MRE11, correlate with radiosensitization in four human tumor cell lines. *International Journal of Hyperthermia,* Vol.23, pp. 343-351.

Xu, M.; Wright, W.D., Higashikubo, R., Wang, L.L. & Roti Roti, J.L. (1999). Thermal radiosensitization of human tumour cell lines with different sensitivities to 41.1 degrees C. *International Journal of Hyperthermia,* Vol.15, pp. 279-290.

Xu, M.; Myerson, R.J., Straube, W.L., Moros, E.G., Lagroye, I., Wang, L.L., Lee, J.T. & Roti Roti, J.L.(2002). Radiosensitization of heat resistant human tumour cells by 1 hour at 41.1 degrees C and its effect on DNA repair. *International Journal of Hyperthermia,* Vol.18, pp. 385-403.

23

Phage-Displayed Recombinant Peptides for Non-Invasive Imaging Assessment of Tumor Responsiveness to Ionizing Radiation and Tyrosine Kinase Inhibitors

Hailun Wang, Miaojun Han and Zhaozhong Han
Department of Radiation Oncology and Cancer Biology,
Vanderbilt-Ingram Cancer Center,
School of Medicine, Vanderbilt University, Nashville, TN,
USA

1. Introduction

Recent studies have resulted in a variety of therapeutic options for cancer. However, tumor patients, even a same patient at different disease stages, respond(s) to a treatment protocol with different efficacy. A concept of personalized medicine has been developed to deliver individually tailored treatment upon the unique responsiveness of each patient. Currently, it is still challenging to predict treatment outcomes due to the genetic complexity and heterogeneity of cancers, which underlies the varied responses to treatment. To efficiently design treatment strategies and monitor the outcomes of therapies for individual patient, tumor responsiveness to a specific treatment regimen needs to be assesses in a time- and cost-efficient manner.

Non-invasive imaging technologies have demonstrated great potentials in diagnosis and treatment management by monitoring individual patient's disease condition and progression. Currently, anatomic and functional imaging modalities have been generally applied to detect, stage, and monitor tumors. Compared to the anatomic imaging that measures tumor size, functional or molecular imaging provides more information on tumor metabolism, biomarker expression, cell death or proliferation and thus is more relevant to the imaging assessment of tumor responsiveness to a treatment regimen, especially when the treatment affects tumors through blocking the tumor progression instead of shrinking the tumor size. Discovery of novel probes that specifically binds to tumor-limited targets with sound biological relevance is a limiting factor to develop such functional imaging modality.

Compared to antibody (~150 kD), peptide is in a much smaller size (1-2 kD) that enables an improved tissue penetration, faster clearance from circulation system, and less immunogenic property that are expected for a imaging probe, especially in the repeated assessment of treatment responsiveness in solid tumors. Advances in phage display-related technologies have facilitated the discovery and development of peptide derivatives as imaging probes for a variety of tumors. By using one example of HVGGSSV peptide that has been discovered and tested for non-invasive imaging assessment of tumor response to ionizing radiation (IR) and receptor tyrosine kinase (RTK) inhibitors in multiple tumor types, this review demonstrates that phage-displayed peptides hold potentials in personalized medicine by facilitating molecular imaging, discovery of diagnostic biomarker or therapeutic target, and tumor-targeted drug delivery.

2. Advancement in radiation therapy of cancer

Cancer is the leading cause of death worldwide, deaths from cancer worldwide are projected to continue rising, with an estimated 12 million deaths in 2030 (World Health Organization). Currently, radiotherapy is one of the most important modalities for the treatment of cancers. Over 60% of cancer patients received radiotherapy as part of their treatments.

2.1 Radiotherapy

Radiotherapy is the medical use of ionizing radiation as part of cancer treatment to control malignant cells. Radiation therapy may be used to treat localized solid tumors, such as cancers of the skin, tongue, larynx, brain, breast, lung, prostate or uterine cervix. It can also be used to treat leukemia and lymphoma. It works by damaging the DNA of cancerous cells through the use of one of two types of high energy radiation, photon or charged particle. This damage is either direct or indirect ionizing the atoms which make up the DNA chain. Indirect damage happens as a result of the ionization of water by high energy radiations, such as X-ray or gamma ray, forming free radicals, notably hydroxyl radicals, which then damage the DNA and form single- or double-stranded DNA breaks. Direct damage to DNA occurs through charged particles such as proton, boron, carbon or neon ions. Due to their relatively large mass, charged particles directly strike DNA and transfer high energy to DNA molecules and usually cause double-stranded DNA breaks. The accumulating damages to cancer cells' DNA cause them to die or proliferate more slowly. To minimize the damage to normal cells, the total dose of radiation therapy is usually fractionated into several smaller doses to allow normal cells time to recover. In clinics, to spare normal tissues from the treatment, shaped radiation beams are aimed from several angles of exposure to intersect at the tumor, providing a much larger absorbed dose in the tumors than in the surrounding tissues. Brachytherapy, in which a radiation source is placed inside or next to the cancer area, is another technique to minimize exposure to healthy tissues during treatment of cancers in the breast, prostate and other organs. It is also common to combine radiotherapy with surgery, chemotherapy, hormone therapy or immunotherapy to maximize treatment efficiency.

2.2 Radiosensitizer

Besides the rapid advances in radiotherapy technologies, the increased understanding of cancer biology and signaling networks behind radiotherapy has led to the development of newer chemotherapy agents that help to increase radiation treatment efficiency. Pathways targeted for radiosensitization include DNA damage repair, cell cycle progression, cell survival and death, angiogenesis, or modulation of tumor microenvironment. For example, hypoxia is one general characteristic associated with fast-growing solid tumors. It stimulates tumor malignant progression and induces HIF-1a. A few studies have found that low oxygen levels in tumors are associated with a poor response to radiotherapy (Overgaard, 2007). Well-oxygenated cells show an approximately 2-3 fold increases in radiosensitivity compared to hypoxic cells (Dasu and Denekamp, 1998). This discovery results in the development of a family of drugs - oxygen radiosensitizers. From initial attempts to increase oxygen delivery to the tumor by using hyperbaric oxygen in radiotherapy (Mayer et al., 2005), to later use oxygen mimetics/Electron-affinity agents, such as nitroimidazoles (Brown, 1975), or transition metal complexes, such as cisplatin (Liu et al., 1997), oxygen radiosensitizers significantly increase the radiotherapy efficiency. Currently, attention has been given to hypoxic cytotoxins, a group of drugs that selectively or preferably destroys cells in a hypoxic environment. These classes of compounds, such as mitomycin (De Ridder et al., 2008; Moore, 1977), are different from classic radiosensitizer in that they can be converted to cytotoxic agents under low oxygenation states, and they provide valuable adjuncts to radiotherapy. Recently, a wide variety of drugs that influence the DNA damage and repair pathways are being evaluated in conjunction with radiation. It includes topoisomerase inhibitors (e.g. camptothecin, topotecan), the hypoxia-activated anthraquinone AQ4N, and alkylating agents such as temozolomide. Proteins involved in tumor malignant progresses are also drawn attention as attractive targets of radiosensitizers, such as HIF-1a (Palayoor et al., 2008), survivin (Miyazaki et al.), Ras (Cengel and McKenna, 2005), epidermal growth factor receptors and related kinases (Sartor, 2004; Williams et al., 2004). Inhibitors for receptor tyrosine kinases such as vascular endothelial growth factor have been extensively studied and applied to improve the therapeutic efficacy of radiotherapy (Vallerga et al., 2004).

3. Assessment of tumor responses to radiotherapy

Different types of cancers possessed different mutations. Even the same type of cancers, they show different growth characteristics at different locations and in different patients. The heterogeneity of cancers underlies the different responses of cancers to the same treatment. Currently, cancer response is measured by imaging assessment of tumor volumes or by repeated biopsy. The whole processes are time consuming and inefficient. The recent advancement in imaging technologies has revolutionized medical diagnosis and prognosis. From the macroscopic anatomical sites down to a functional assessment of processes within tumors, imaging provide us a method to evaluate tumor response to irradiation treatment in a non-invasive, reliable and repeatable way (Lowery et al., 2011). So far, a few biomarkers have been explored for imaging to predict patients' outcomes after radiation treatment.

3.1 Cell metabolism

Cell metabolism is the earliest biomarker being studied after radiation treatment. Positron emission tomography (PET) has been used to evaluate tumor metabolism. ^{18}F-fuorodeoxyglucose (FDG) is the most common PET tracer for metabolism study. FDG, a glucose analog, is taken up by high-glucose-consuming cells, such as cancer cells. But FDG cannot be further metabolized during glycolysis and it becomes trapped and rapidly accumulates within the cell. As a result, the distribution of ^{18}F-FDG is a good reflection of the location of cancer cells. It is routinely used for the staging of cancer and for the monitoring of therapy (Allal et al., 2004).

3.2 Cell proliferation

The development of proliferation probes for PET imaging has enabled the *in vivo* evaluation of cell proliferation (Shields et al., 1998). Among those probes, nucleoside-based imaging probes (3'deoxy-3'-^{18}F-fluorothymidine, FLT) or amino acids based imaging probes are gaining popularity. ^{18}F-FLT is a pyrimidine analog that, after uptake into the cell, is phosphorylated by thymidine kinase 1. The phosphorylated ^{18}F-FLT can not leave the cell and result in the intracellular accumulation of radioactivity. Thymidine kinase 1 is a principal enzyme in the salvage pathway of DNA synthesis and exhibits increased activity during the S phase of the cell cycle. ^{18}F-FLT uptake, therefore, reflects cellular proliferation. Amino acid metabolism is increased in fast proliferating tumor cells. Among the 20 essential amino acids, l-[^{11}C]MET, [^{18}F]fluorotyrosine, l-[^{11}C]leucine, and [18F]fluoro-α-methyl tyrosine have been widely used in the detection of tumors (Laverman et al., 2002). Changes in l-[^{11}C]MET uptake have already been shown to reflect response to radiotherapy treatment in patients suffering from a wide variety of tumors (Team, 2005b).

3.3 Tumor vasculature and hypoxia

Although being characterized as vasculature-rich structures, tumors often develop regions of hypoxia due to the leaky and disorganized tumor blood vessels. Low oxygen environment will promote tumor angiogenesis, metastasis and render tumors resistant to radiation treatment (Tatum et al., 2006). Therefore, the tumor vasculature structure and oxygen level are valuable biomarkers for prognosis after treatment. ^{18}F-fluoromisonidazole is the most widely used PET tracer for detecting tumor hypoxia. After uptake in cell, it is reduced and binds selectively to macromolecules under hypoxic conditions (Team, 2005a). One recent study indicates that ^{18}F-fluoromisonidazole uptake is correlated with radiation treatment outcome in Head and neck cancer (Thorwarth et al., 2005). As to the tumor vasculature, several studies have been proposed using two different techniques - quantified power Doppler sonography or Dynamic contrast-enhanced MRI (DCE-MRI). And both showed promising results (Hormigo et al., 2007; Kim et al., 2006; Mangla et al., 2010).

3.4 Apoptosis

Since its recognition as one of the major forms of cell death after radiation, apoptosis is being increasingly studied as a biomarker of cellular radiosensitivity and a prognosis marker for radiotherapy outcome. During the apoptosis process, phosphatidylserine (PS) flips from the inner leaflet of the cell membrane to the exterior of the cells. Annexin V, a cellular protein of the Annexin family, binds to the exposed PS. To date, Annexin V has been fluorinated for PET and radioiodinated for SPECT. Annexin V labeled with 99mTc has demonstrated significant uptake in patients suffering form myocardial infarction (Narula et al., 2001). Studies assessing quantitative ^{99m}Tc-Annexin V uptake in human tumors and their relationship to radiotherapy outcome are underway.

4. Phage-displayed peptides as novel imaging probes for assessing tumor response to treatment

Recently, advances in phage display-related technologies facilitate the use of small peptide derivatives as probe molecules for recognition and targeting tumors. Phage display enables discovery and optimization of affinity probes for the known tumor-specific biomarkers.

Furthermore, this technology makes it possible to *de novo* discover novel imaging probes, and eventually identify novel diagnostic markers or therapeutic targets of cancer. In vivo screening against heterogeneous tumor targets have generated a diverse group of peptides for cancer-targeted delivery of imaging or therapeutic agents.

4.1 Principle of phage display

A phage is a type of viruses that infect bacteria. Typically, phages consist of a protein capsid enclosing genetic materials. Due to its simple structure, phages have been developed into a powerful tool in biological studies. Phage display was originally invented by George P. Smith in 1985 when he demonstrated the display of exogenous peptides on the surface of filamentous phage by fusing the DNA of the peptide on to the capsid gene of filamentous phages (Smith, 1985) (Fig. 1). This technology was further developed and improved to display large proteins such as enzymes and antibodies (Fernandez-Gacio et al., 2003; Han et al., 2004). The connection between genotype and phenotype enables large libraries of peptides or proteins to be screened in a relative fast and economic way. The most common phages used in phage display are M13 filamentous phage and T7 phage (Krumpe et al., 2006; Smith and Petrenko, 1997). The functional moiety on the phage surface can be short peptides, recombinant proteins, engineered antibody fragments or scaffold proteins. Screening can be conducted on the purified organic or inorganic materials, cells, or tissues.

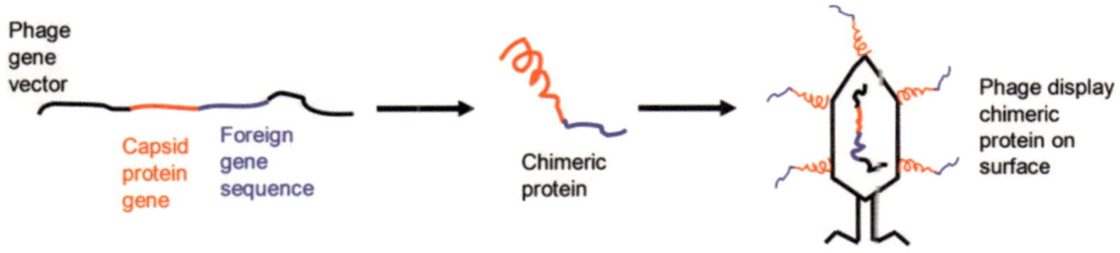

Fig. 1. Schematic illustration of phage display. Foreign gene sequences encoding short peptides, recombinant proteins or large antibody fragments can be fused with capsid protein genes with recombinant DNA technologies. As a result, the recombinant phages express the foreign peptides or proteins on the phage surface for affinity-based selection. The affinity-selected phages can be replicated in bacterial host for further rounds of selection or DNA-sequencing to identify the affinity peptides or proteins expressed on the phage surface.

4.2 Applications of phage display

4.2.1 General applications

The application of phage display technology include determination of binding partners of organic (proteins, polysaccharides, or DNAs) (Gommans et al., 2005) or inorganic materials (Hattori et al., 2010; Whaley et al., 2000). The technique is also used to study enzyme evolution *in vitro* for engineering biocatalysts (Pedersen et al., 1998). Phage display has been widely applied in drug discovery. It can be used for finding new ligands, such as enzyme inhibitors, receptor agonists and antagonists, to target proteins (Hariri et al., 2008; Pasqualini et al., 1995; Perea et al., 2004; Ruoslahti, 1996; Uchino et al., 2005). Invention of antibody phage display revolutionized the drug discovery (Han et al., 2004). Millions of different single chain antibodies on phages are used for isolating highly specific therapeutic antibody leads. One of the most successful examples was adalimumab (Abbott Laboratories), the first fully human antibody targeted to TNF alpha (Spector and Lorenzo, 1975).

4.2.2 *In vivo* phage display and its application in clinical oncology

Because isolating or producing recombinant membrane proteins for use as target molecules in phage library screening is often facing insurmountable obstacles, innovative selection strategies such as panning against whole cells or tissues were devised (Jaboin et al., 2009; Molek et al., 2011; Pasqualini and Ruoslahti, 1996). Due to cells inside the body may express different surface markers and possess different characteristics from cell lines in culture, *in vivo* phage bio-panning was developed to identify more physiologically relevant biomarkers (Fig. 2) (Pasqualini and Ruoslahti, 1996). Since its invention, *in vivo* phage display has been used extensively to screen for novel targets for tumor therapy. Majority of those studies focused on analyzing the structure and molecular diversity of tumor vasculature and selecting tumor stage- and type-specific markers on tumor blood vessels (Arap et al., 2002; Rajotte and Ruoslahti, 1999; Sugahara et al., 2010; Valadon et al., 2006). Recently, the use of this technique was expanded to the field of discovering new biomarkers for evaluation of cancer treatment efficacy. (Han et al., 2008; Passarella et al., 2009).

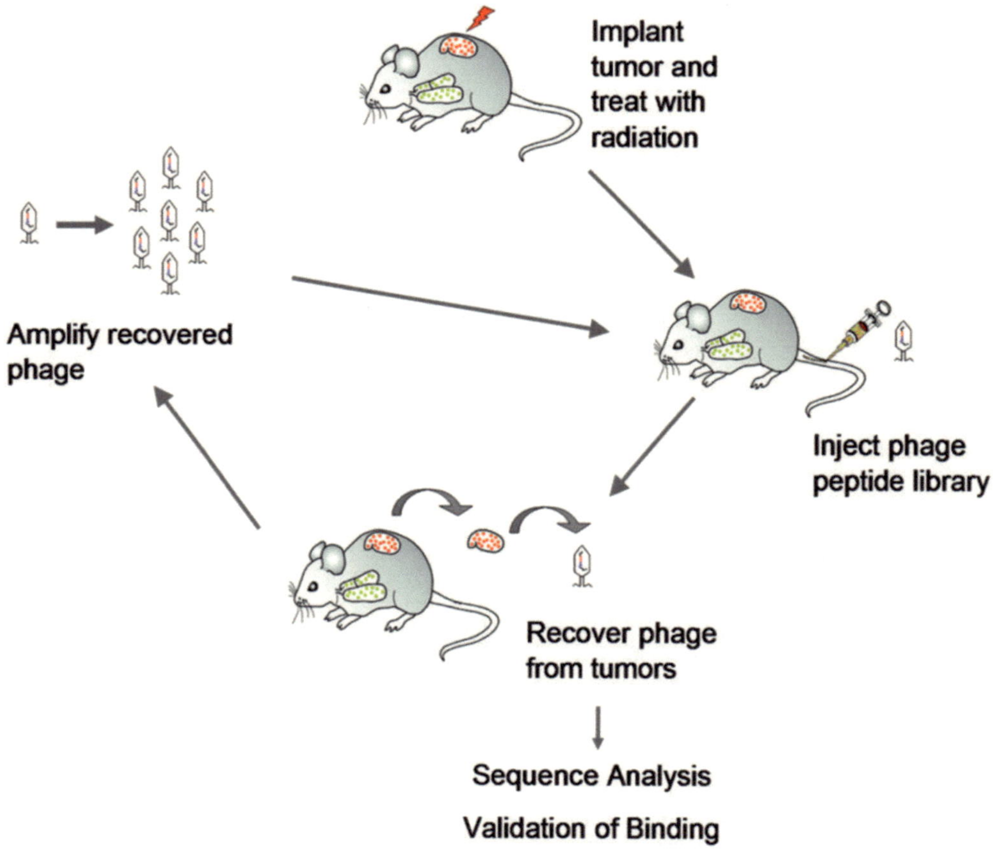

Fig. 2. *In vivo* phage display for screening peptides specifically target to radiation- or drug-treated tumors.

4.3 Peptides as probes for tumor targeted imaging

4.3.1 Advantages of peptide vs. antibody for tumor targeting

Antibodies, especially monoclonal antibodies, have been successfully utilized as cancer-targeting therapeutics and diagnostics due to their high target specificity and affinity. However, due to antibody large size (150 kDa) and limited tissue permeability, non-specific uptake into the reticuloendothelial system, and immunogenicity, most antibody-based therapeutics are of limited efficacy (Lin et al., 2005; Stern and Herrmann, 2005). In contrast to antibodies, peptides are much smaller molecules (1-2 kDa). Peptides have favorable biodistribution profiles compared to antibody, characterized by high uptake in the tumor tissue and rapid clearance from the blood. In addition, peptides have increased capillary permeability, allowing more efficient penetration into tumor tissues. Also peptides are easy to make and safe to use, they will not elicit an immune response (Ladner et al., 2004). With all these advantages, peptides have been increasingly considered as a good tumor targeted imaging probe (Aloj and Morelli, 2004; Okarvi, 2004; Reubi and Maecke, 2008).

4.3.2 Peptide as imaging probe

To date, a large number of peptides derived from natural proteins have already been successfully identified and characterized for tumor targeting and tumor imaging, such as integrin (RGD), somatostatin, gastrin-releasing peptide, cholecystokinin, glucagon-like peptides-1 and neuropeptide-Y (Cai et al., 2008; Hallahan et al., 2003; Korner et al., 2007; Miao and Quinn, 2007; Reubi, 2003; Reubi, 2007). A list of a few tumor homing peptides isolated using phage display technique is shown in Table 1.

Tumor Types	Tumor- targeting peptides
Prostate carcinoma	IAGLATPGWSHWLAL (Newton et al., 2006) ANTPCGPYTHDCPVKR (Deutscher et al., 2009) R/KXXR/K (Sugahara et al., 2009)
Colon carcinoma	CPIEDRPMC (Kelly et al., 2004)
Breast carcinoma	EGEVGLG (Passarella et al., 2009)
Hepatocellular carcinoma	AWYPLPP (Jia et al., 2007) AGKGTAALETTP (Du et al., 2010)
Pancreatic carcinoma	KTLLPTP (Kelly et al., 2008)
Head and Neck Cancer	SPRGDLAVLGHKY (Nothelfer et al., 2009)
Osteosarcoma	ASGALSPSRLDT (Sun et al., 2010)
Fibrosarcoma	SATTHYRLQAAN (Hadjipanayis et al., 2010)
Esophageal Cancer	YSXNXW and PXNXXN (Zhivotosky and Orrenius, 2001)
Bladder Cancer	CSNRDARRC (Ginestier et al., 2007)

Table 1. Phage display-derived tumor-targeting peptides

For use as *in vivo* imaging probes, peptides can be directly or indirectly labeled with a wide range of imaging moieties according to the imaging modality. For instance, near-infrared (NIR) fluorescent dyes or quantum dots have been labeled for optical imaging (Fig. 3), several radionuclides have been employed for positron emission tomography (PET) or single-photon emission computed tomography (SPECT), and paramagnetic agents have been used for magnetic resonance imaging (MRI) (Frangioni, 2003; Reubi and Maecke, 2008). Peptides can also be conjugated to other tumor targeted polymers or nanoparticles and dramatically increase their tumor targeted selectivity and efficiency (Hariri et al., 2010; Lowery et al., 2010; Passarella et al., 2009).

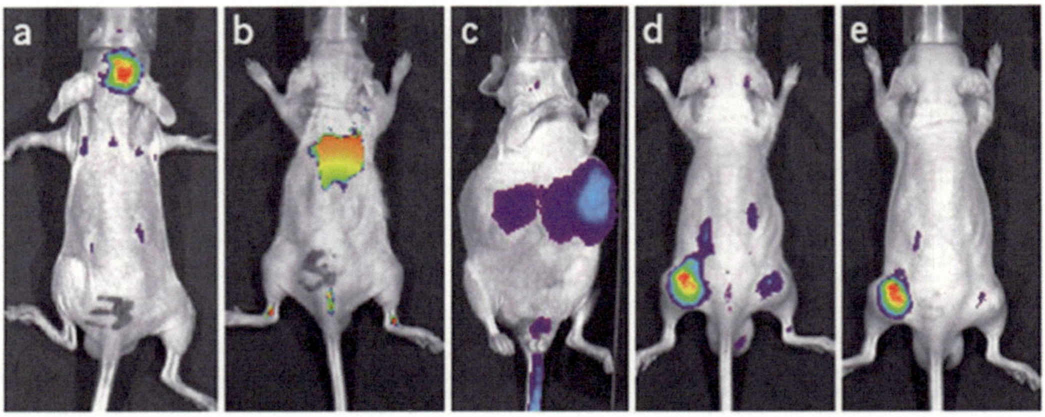

Fig. 3. HVGGSSV peptide labeled with near-infrared (NIR) fluorescent dyes specifically located to radiation-treated tumors. a) brain tumor (D54 human glioblastoma cell), b) lung tumor (H460 cell), c) colon cancer liver metastasis (HT22 cell), d) prostate cancer subcutaneous model (PC-3 cell), and e) breast cancer subcutaneous model (MDA-MB-231 cell). (Adapted from Han et al., 2008).

5. HVGGSSV peptide as one imaging probes to detect tumor response to radiation and tyrosine kinase inhibitor (TKI) *in vivo*

5.1 Discovery of HVGGSSV peptide

In our recent studies, we employed *in vivo* phage display technique and intended to identify peptides that will specifically home to radiation or drug treatment responsive tumors (Han et al., 2008; Passarella et al., 2009). During the studies, we first treated tumors in mice with radiation and tyrosine kinase inhibitors. Then a peptide phage library was injected from the tail vein of tumor bearing mice for tumor binding screening. After several rounds of *in vivo* screening and enrichment of phages isolated from the treated tumors (Fig. 2), one phage clone, encoding HVGGSSV peptides, was identified preferentially target to treatment responsive tumors. The binding preference of those phages were confirmed by fluorescence labeled phage or peptide imaging (Han et al., 2008; Passarella et al., 2009).

5.2 HVGGSSV peptide as imaging/targeting probe for radiation responsive tumors

To explore HVGGSSV peptide's clinical application in noninvasive imaging of tumor response to treatment, fluorescent labeled HVGGSSV peptide were used to target human tumors in several mouse models. Optical imaging studies indicated that the signal intensities of peptide binding within tumors correlate to the overall efficacy of treatment regimens on tumor growth control in multiple tumor models that had been treated with a variety of RTK inhibitors with or without combination of radiation (Han et al., 2008). SPECT/CT provides high spatial resolution and sensitivity in functional imaging. We

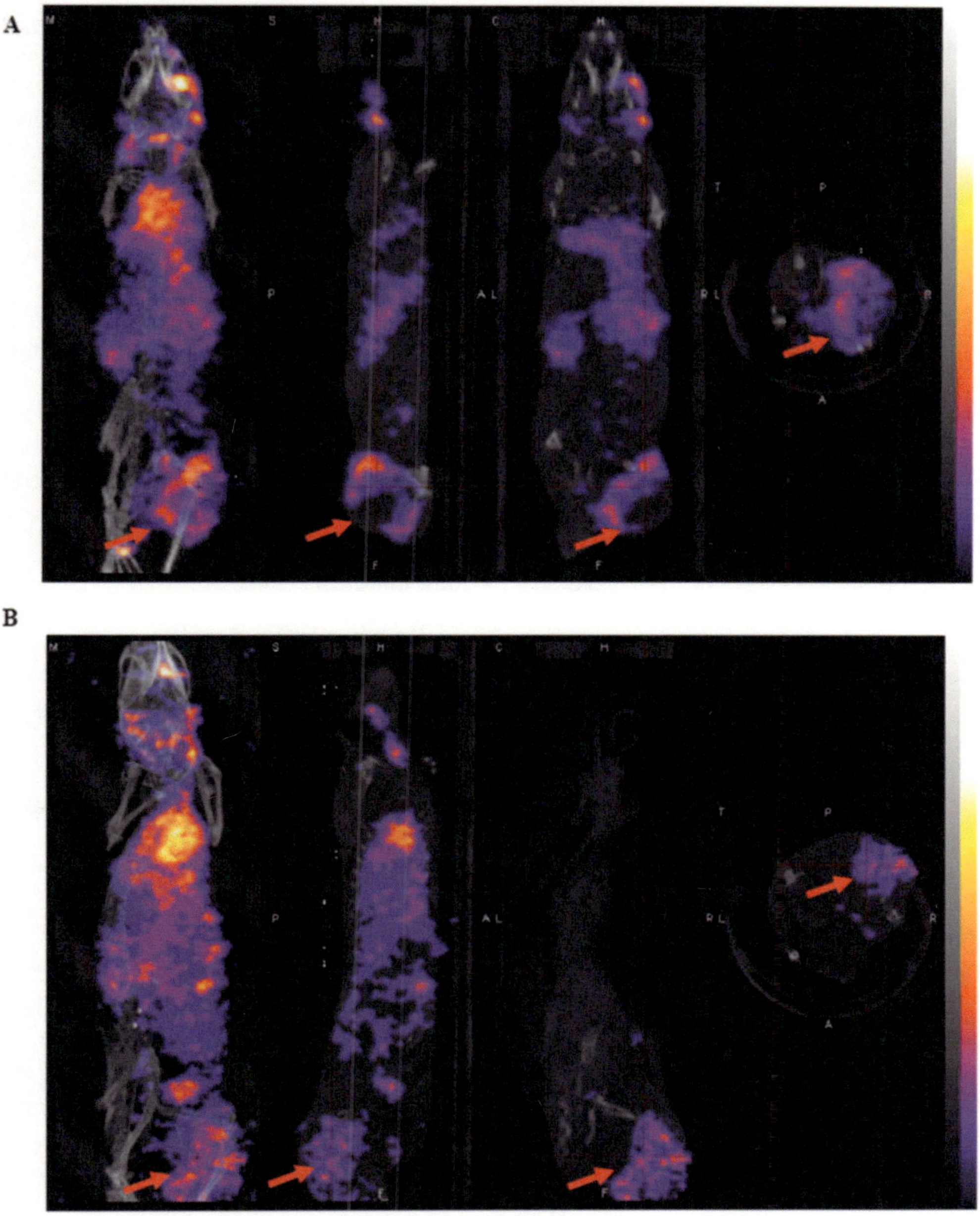

Fig. 4. SPECT/CT imaging of the HVGGSSV peptide within LLC tumors after treatment. The biotinylated HVGGSSV peptide was complexed with iodine-125-labeled streptavidin. The implanted tumor was treated once with radiation (5 Gy) alone (**A**), or combination of Sunitinib (40 mg/Kg) and radiation (5 Gy) (**B**) before intravenous administration of the imaging probe. Shown are 3D virtual rendering (3D-VR) images (far left) and hybrid SPECT/CT fusion images in the coronal, sagittal, and transaxial planes (the second to the fouth from the left, respectively) acquired 4 hours after the imaging probe administration. The LLC tumors was pointed with arrows, high resolution images enable spatially localizing the radiation-responding cells within the peripheral and central regions of the tumors.

employed this imaging modality to detect tumor response to radiation by using the HVGGSSV peptide. The mice were treated with radiation alone or combination of radiation and one TKI - Sunitinib (40 mg/Kg,). After the treatment, the HVGGSSV peptide complexed with ^{125}I-labeled streptavidin was selectively targeted to the tumors treated with radiation or radiation combined with Sunitinib. High resolution SPECT/CT images (Fig. 4) also showed that majority of the imaging probes were located in the peripheral area of the tumors that were treated with radiation alone. However, treatment with radiation and Sunitinib extended the imaging probe binding to both the peripheral and central parts of the subcutaneous tumors. This data might reflect the radiosensitization effect of Sunitinib.

The tumor targeting potential of HVGGSSV peptide has been further explored in several drug delivery studies. In these studies, HVGGSSV peptide has been conjugated to different nanoparticles, such as liposome, FePt, and nanoparticle albumin bound (nab) (Hariri et al., 2011; Hariri et al., 2010; Lowery et al., 2011), and selectively targeted those nanoparticles to irradiated tumors. One study also showed >5-fold increase in paclitaxel levels within irradiated tumors in HVGGSSV-nab-paclitaxel-treated groups and significantly increase tumor growth delay as compared with controls (Hariri et al., 2010).

5.3 The biological basis of the HVGGSSV peptide imaging

5.3.1 Peptide receptor identification

To understand the physiology underlines peptides binding, we need to identify the molecular targets of peptides. However, peptides are usually unstable. Their surface charges and structures will change dramatically in different environment. And peptides usually interact with their targets with low binding affinity due to their small sizes. Therefore, traditional affinity purification methods are of little use because of high background of non-specific binding. To date, there are very few identified receptors for peptides in contrast to the great number of discovered cancer targeting peptides (Sugahara et al., 2009). New strategies are needed for identifying peptide's receptors.

5.3.2 TIP-1 as a molecular target of HVGGSSV peptide

In our recent studies of one peptide (HVGGSSV), we utilized a phage cDNA library screening to search for peptide's receptors. Because several rounds of phage display screening can significantly enrich the low-affinity or low-abundance proteins, we successfully identified a PDZ protein - TIP-1 as the target of HVGGSSV peptide (Wang et al., 2010). Through the PDZ domain, TIP-1 binds to the classic C-terminal PDZ motif within the HVGGSSV peptide. One TIP-1-specific antibody that inhibited the in vitro interaction between TIP-1 and the HVGGSSV peptide attenuated the peptide's accumulation within irradiated tumors. Imaging with TIP-1-specific antibody recapitulated the pattern of peptide imaging in tumor-bearing mice. Mutation in the classic PDZ binding motif of the HVGGSSV peptide destroyed the specific binding within irradiated tumors. These results also demonstrated the potentials of screening phage-displayed cDNA library in discovery of molecular targets of the peptides with a simple structure and low affinity.

5.3.3 The biological relevance of TIP-1 relocation onto tumor cell surface to the radiation response of tumor cells

With a TIP-1 specific antibody, it was further identified that radiation induced translocation of the basically intracellular TIP-1 protein onto the cell surface in a dose-dependent manner.

The treatment-induced TIP-1 expression on the cell surface is detectable in the first few hours after the treatment and before the onset of treatment associated apoptosis or cell death. Majority of the cells expressing TIP-1 on the cell surface are the live albeit such cells are less potent in proliferation and more susceptible to subsequent radiation treatment (Wang et al., 2010). The increased susceptibility to the subsequent irradiation might explain why the peptide binding is predictive in assessing tumor overall responsiveness in the early stage of treatment course. The treatment-inducible TIP-1 translocation before the onset of cell apoptosis or death further suggests potentials of the HVGGSSV peptide in non-invasive imaging assessment of tumor response to radiation and tyrosine kinase inhibitors.

6. Perspectives

The development of imaging technologies revolutionizes medial diagnosis and clinical management. Functional molecular imaging becomes one critical part of personalized medicine. Peptide, with its small sizes and versatile structures, is increasingly recognized as a promising imaging probe to predict the outcomes of radiotherapy and other medical treatments. Although some disadvantages associated with peptides, such as its degradation inside human body and its low affinity with its targets, with chemical modification to improve its stability and association with nanoparticles to increase its binding affinity, peptides will play a major role in the future molecular imaging.

7. References

Allal, A. S., Slosman, D. O., Kebdani, T., Allaoua, M., Lehmann, W., and Dulguerov, P. (2004). Prediction of outcome in head-and-neck cancer patients using the standardized uptake value of 2-[18F]fluoro-2-deoxy-D-glucose. International journal of radiation oncology, biology, physics *59*, 1295-1300.

Aloj, L., and Morelli, G. (2004). Design, synthesis and preclinical evaluation of radiolabeled peptides for diagnosis and therapy. Current pharmaceutical design *10*, 3009-3031.

Arap, W., Kolonin, M. G., Trepel, M., Lahdenranta, J., Cardo-Vila, M., Giordano, R. J., Mintz, P. J., Ardelt, P. U., Yao, V. J., Vidal, C. I., *et al.* (2002). Steps toward mapping the human vasculature by phage display. Nat Med *8*, 121-127.

Brown, J. M. (1975). Selective radiosensitization of the hypoxic cells of mouse tumors with the nitroimidazoles metronidazole and Ro 7-0582. Radiation research *64*, 633-647.

Cai, W., Niu, G., and Chen, X. (2008). Imaging of integrins as biomarkers for tumor angiogenesis. Current pharmaceutical design *14*, 2943-2973.

Cengel, K. A., and McKenna, W. G. (2005). Molecular targets for altering radiosensitivity: lessons from Ras as a pre-clinical and clinical model. Critical reviews in oncology/hematology *55*, 103-116.

Dasu, A., and Denekamp, J. (1998). New insights into factors influencing the clinically relevant oxygen enhancement ratio. Radiother Oncol *46*, 269-277.

De Ridder, M., Van Esch, G., Engels, B., Verovski, V., and Storme, G. (2008). Hypoxic tumor cell radiosensitization: role of the iNOS/NO pathway. Bulletin du cancer *95*, 282-291.

Deutscher, S. L., Figueroa, S. D., and Kumar, S. R. (2009). Tumor targeting and SPECT imaging properties of an (111)In-labeled galectin-3 binding peptide in prostate carcinoma. Nuclear medicine and biology *36*, 137-146.

Du, B., Han, H., Wang, Z., Kuang, L., Wang, L., Yu, L., Wu, M., Zhou, Z., and Qian, M. (2010). targeted drug delivery to hepatocarcinoma in vivo by phage-displayed specific binding peptide. Mol Cancer Res *8*, 135-144.

Fernandez-Gacio, A., Uguen, M., and Fastrez, J. (2003). Phage display as a tool for the directed evolution of enzymes. Trends in biotechnology *21*, 408-414.

Frangioni, J. V. (2003). In vivo near-infrared fluorescence imaging. Current opinion in chemical biology *7*, 626-634.

Ginestier, C., Hur, M. H., Charafe-Jauffret, E., Monville, F., Dutcher, J., Brown, M., Jacquemier, J., Viens, P., Kleer, C. G., Liu, S., *et al.* (2007). ALDH1 is a marker of normal and malignant human mammary stem cells and a predictor of poor clinical outcome. Cell Stem Cell *1*, 555-567.

Gommans, W. M., Haisma, H. J., and Rots, M. G. (2005). Engineering zinc finger protein transcription factors: the therapeutic relevance of switching endogenous gene expression on or off at command. Journal of molecular biology *354*, 507-519.

Hadjipanayis, C. G., Machaidze, R., Kaluzova, M., Wang, L., Schuette, A. J., Chen, H., Wu, X., and Mao, H. (2010). EGFRvIII antibody-conjugated iron oxide nanoparticles for magnetic resonance imaging-guided convection-enhanced delivery and targeted therapy of glioblastoma. Cancer Res *70*, 6303-6312.

Hallahan, D., Geng, L., Qu, S., Scarfone, C., Giorgio, T., Donnelly, E., Gao, X., and Clanton, J. (2003). Integrin-mediated targeting of drug delivery to irradiated tumor blood vessels. Cancer cell *3*, 63-74.

Han, Z., Fu, A., Wang, H., Diaz, R., Geng, L., Onishko, H., and Hallahan, D. E. (2008). Noninvasive assessment of cancer response to therapy. Nature medicine *14*, 343-349.

Han, Z., Karatan, E., Scholle, M. D., McCafferty, J., and Kay, B. K. (2004). Accelerated screening of phage-display output with alkaline phosphatase fusions. Comb Chem High Throughput Screen *7*, 55-62.

Hariri, G., Wellons, M. S., Morris, W. H., 3rd, Lukehart, C. M., and Hallahan, D. E. (2011). Multifunctional FePt nanoparticles for radiation-guided targeting and imaging of cancer. Annals of biomedical engineering *39*, 946-952.

Hariri, G., Yan, H., Wang, H., Han, Z., and Hallahan, D. E. (2010). Nanoparticle Albumin Bound Paclitaxel Retargeted to Radiation Inducible TIP-1 in Cancer. Clin Cancer Res.

Hariri, G., Zhang, Y., Fu, A., Han, Z., Brechbiel, M., Tantawy, M. N., Peterson, T. E., Mernaugh, R., and Hallahan, D. (2008). Radiation-guided P-selectin antibody targeted to lung cancer. Ann Biomed Eng *36*, 821-830.

Hattori, T., Umetsu, M., Nakanishi, T., Togashi, T., Yokoo, N., Abe, H., Ohara, S., Adschiri, T., and Kumagai, I. (2010). High affinity anti-inorganic material antibody generation by integrating graft and evolution technologies: potential of antibodies as biointerface molecules. J Biol Chem *285*, 7784-7793.

Hormigo, A., Gutin, P. H., and Rafii, S. (2007). Tracking normalization of brain tumor vasculature by magnetic imaging and proangiogenic biomarkers. Cancer cell *11*, 6-8.

Jaboin, J. J., Han, Z., and Hallahan, D. E. (2009). Using in vivo biopanning for the development of radiation-guided drug delivery systems. Methods in molecular biology (Clifton, NJ *542*, 285-300.

Jia, W. D., Sun, H. C., Zhang, J. B., Xu, Y., Qian, Y. B., Pang, J. Z., Wang, L., Qin, L. X., Liu, Y. K., and Tang, Z. Y. (2007). A novel peptide that selectively binds highly metastatic hepatocellular carcinoma cell surface is related to invasion and metastasis. Cancer letters *247*, 234-242.

Kelly, K., Alencar, H., Funovics, M., Mahmood, U., and Weissleder, R. (2004). Detection of invasive colon cancer using a novel, targeted, library-derived fluorescent peptide. Cancer research *64*, 6247-6251.

Kelly, K. A., Bardeesy, N., Anbazhagan, R., Gurumurthy, S., Berger, J., Alencar, H., Depinho, R. A., Mahmood, U., and Weissleder, R. (2008). Targeted nanoparticles for imaging incipient pancreatic ductal adenocarcinoma. PLoS medicine *5*, e85.

Kim, D. W., Huamani, J., Niermann, K. J., Lee, H., Geng, L., Leavitt, L. L., Baheza, R. A., Jones, C. C., Tumkur, S., Yankeelov, T. E., *et al.* (2006). Noninvasive assessment of tumor vasculature response to radiation-mediated, vasculature-targeted therapy using quantified power Doppler sonography: implications for improvement of therapy schedules. J Ultrasound Med *25*, 1507-1517.

Korner, M., Stockli, M., Waser, B., and Reubi, J. C. (2007). GLP-1 receptor expression in human tumors and human normal tissues: potential for in vivo targeting. J Nucl Med *48*, 736-743.

Krumpe, L. R., Atkinson, A. J., Smythers, G. W., Kandel, A., Schumacher, K. M., McMahon, J. B., Makowski, L., and Mori, T. (2006). T7 lytic phage-displayed peptide libraries exhibit less sequence bias than M13 filamentous phage-displayed peptide libraries. Proteomics *6*, 4210-4222.

Ladner, R. C., Sato, A. K., Gorzelany, J., and de Souza, M. (2004). Phage display-derived peptides as therapeutic alternatives to antibodies. Drug discovery today *9*, 525-529.

Laverman, P., Boerman, O. C., Corstens, F. H., and Oyen, W. J. (2002). Fluorinated amino acids for tumour imaging with positron emission tomography. European journal of nuclear medicine and molecular imaging *29*, 681-690.

Lin, M. Z., Teitell, M. A., and Schiller, G. J. (2005). The evolution of antibodies into versatile tumor-targeting agents. Clin Cancer Res *11*, 129-138.

Liu, T. Z., Lin, T. F., Chiu, D. T., Tsai, K. J., and Stern, A. (1997). Palladium or platinum exacerbates hydroxyl radical mediated DNA damage. Free radical biology & medicine *23*, 155-161.

Lowery, A., Onishko, H., Hallahan, D. E., and Han, Z. (2010). Tumor-targeted delivery of liposome-encapsulated doxorubicin by use of a peptide that selectively binds to irradiated tumors. J Control Release.

Lowery, A., Onishko, H., Hallahan, D. E., and Han, Z. (2011). Tumor-targeted delivery of liposome-encapsulated doxorubicin by use of a peptide that selectively binds to irradiated tumors. J Control Release *150*, 117-124.

Mangla, R., Singh, G., Ziegelitz, D., Milano, M. T., Korones, D. N., Zhong, J., and Ekholm, S. E. (2010). Changes in relative cerebral blood volume 1 month after radiation-temozolomide therapy can help predict overall survival in patients with glioblastoma. Radiology *256*, 575-584.

Mayer, R., Hamilton-Farrell, M. R., van der Kleij, A. J., Schmutz, J., Granstrom, G., Sicko, Z., Melamed, Y., Carl, U. M., Hartmann, K. A., Jansen, E. C., *et al.* (2005). Hyperbaric oxygen and radiotherapy. Strahlenther Onkol *181*, 113-123.

Miao, Y., and Quinn, T. P. (2007). Alpha-melanocyte stimulating hormone peptide-targeted melanoma imaging. Front Biosci *12*, 4514-4524.

Miyazaki, A., Kobayashi, J., Torigoe, T., Hirohashi, Y., Yamamoto, T., Yamaguchi, A., Asanuma, H., Takahashi, A., Michifuri, Y., Nakamori, K., *et al.* Phase I clinical trial of survivin-derived peptide vaccine therapy for patients with advanced or recurrent oral cancer. Cancer science *102*, 324-329.

Molek, P., Strukelj, B., and Bratkovic, T. (2011). Peptide phage display as a tool for drug discovery: targeting membrane receptors. Molecules (Basel, Switzerland) *16*, 857-887.

Moore, H. W. (1977). Bioactivation as a model for drug design bioreductive alkylation. Science (New York, NY *197*, 527-532.

Narula, J., Acio, E. R., Narula, N., Samuels, L. E., Fyfe, B., Wood, D., Fitzpatrick, J. M., Raghunath, P. N., Tomaszewski, J. E., Kelly, C., *et al.* (2001). Annexin-V imaging for noninvasive detection of cardiac allograft rejection. Nat Med *7*, 1347-1352.

Newton, J. R., Kelly, K. A., Mahmood, U., Weissleder, R., and Deutscher, S. L. (2006). In vivo selection of phage for the optical imaging of PC-3 human prostate carcinoma in mice. Neoplasia (New York, NY *8*, 772-780.

Nothelfer, E. M., Zitzmann-Kolbe, S., Garcia-Boy, R., Kramer, S., Herold-Mende, C., Altmann, A., Eisenhut, M., Mier, W., and Haberkorn, U. (2009). Identification and characterization of a peptide with affinity to head and neck cancer. J Nucl Med *50*, 426-434.

Okarvi, S. M. (2004). Peptide-based radiopharmaceuticals: future tools for diagnostic imaging of cancers and other diseases. Medicinal research reviews *24*, 357-397.

Overgaard, J. (2007). Hypoxic radiosensitization: adored and ignored. J Clin Oncol *25*, 4066-4074.

Palayoor, S. T., Mitchell, J. B., Cerna, D., Degraff, W., John-Aryankalayil, M., and Coleman, C. N. (2008). PX-478, an inhibitor of hypoxia-inducible factor-1alpha, enhances radiosensitivity of prostate carcinoma cells. International journal of cancer *123*, 2430-2437.

Pasqualini, R., Koivunen, E., and Ruoslahti, E. (1995). A peptide isolated from phage display libraries is a structural and functional mimic of an RGD-binding site on integrins. The Journal of cell biology *130*, 1189-1196.

Pasqualini, R., and Ruoslahti, E. (1996). Organ targeting in vivo using phage display peptide libraries. Nature *380*, 364-366.

Passarella, R. J., Zhou, L., Phillips, J. G., Wu, H., Hallahan, D. E., and Diaz, R. (2009). Recombinant peptides as biomarkers for tumor response to molecular targeted therapy. Clin Cancer Res *15*, 6421-6429.

Pedersen, H., Holder, S., Sutherlin, D. P., Schwitter, U., King, D. S., and Schultz, P. G. (1998). A method for directed evolution and functional cloning of enzymes. Proceedings of the National Academy of Sciences of the United States of America *95*, 10523-10528.

Perea, S. E., Reyes, O., Puchades, Y., Mendoza, O., Vispo, N. S., Torrens, I., Santos, A., Silva, R., Acevedo, B., Lopez, E., *et al.* (2004). Antitumor effect of a novel proapoptotic peptide that impairs the phosphorylation by the protein kinase 2 (casein kinase 2). Cancer research *64*, 7127-7129.

Rajotte, D., and Ruoslahti, E. (1999). Membrane dipeptidase is the receptor for a lung-targeting peptide identified by in vivo phage display. J Biol Chem *274*, 11593-11598.

Reubi, J. C. (2003). Peptide receptors as molecular targets for cancer diagnosis and therapy. Endocrine reviews *24*, 389-427.

Reubi, J. C. (2007). Targeting CCK receptors in human cancers. Current topics in medicinal chemistry *7*, 1239-1242.

Reubi, J. C., and Maecke, H. R. (2008). Peptide-based probes for cancer imaging. J Nucl Med *49*, 1735-1738.

Ruoslahti, E. (1996). RGD and other recognition sequences for integrins. Annual review of cell and developmental biology *12*, 697-715.

Sartor, C. I. (2004). Mechanisms of disease: Radiosensitization by epidermal growth factor receptor inhibitors. Nature clinical practice *1*, 80-87.

Shields, A. F., Grierson, J. R., Dohmen, B. M., Machulla, H. J., Stayanoff, J. C., Lawhorn-Crews, J. M., Obradovich, J. E., Muzik, O., and Mangner, T. J. (1998). Imaging proliferation in vivo with [F-18]FLT and positron emission tomography. Nature medicine *4*, 1334-1336.

Smith, G. P. (1985). Filamentous fusion phage: novel expression vectors that display cloned antigens on the virion surface. Science *228*, 1315-1317.

Smith, G. P., and Petrenko, V. A. (1997). Phage Display. Chemical reviews *97*, 391-410.

Spector, R., and Lorenzo, A. V. (1975). Myo-inositol transport in the central nervous system. Am J Physiol *228*, 1510-1518.

Stern, M., and Herrmann, R. (2005). Overview of monoclonal antibodies in cancer therapy: present and promise. Critical reviews in oncology/hematology *54*, 11-29.

Sugahara, K. N., Teesalu, T., Karmali, P. P., Kotamraju, V. R., Agemy, L., Girard, O. M., Hanahan, D., Mattrey, R. F., and Ruoslahti, E. (2009). Tissue-penetrating delivery of compounds and nanoparticles into tumors. Cancer Cell *16*, 510-520.

Sugahara, K. N., Teesalu, T., Karmali, P. P., Kotamraju, V. R., Agemy, L., Greenwald, D. R., and Ruoslahti, E. (2010). Coadministration of a tumor-penetrating peptide enhances the efficacy of cancer drugs. Science *328*, 1031-1035.

Sun, X., Niu, G., Yan, Y., Yang, M., Chen, K., Ma, Y., Chan, N., Shen, B., and Chen, X. (2010). Phage display-derived peptides for osteosarcoma imaging. Clin Cancer Res *16*, 4268-4277.

Tatum, J. L., Kelloff, G. J., Gillies, R. J., Arbeit, J. M., Brown, J. M., Chao K. S., Chapman, J. D., Eckelman, W. C., Fyles, A. W., Giaccia, A. J., *et al.* (2006). Hypoxia: importance in tumor biology, noninvasive measurement by imaging, and value of its measurement in the management of cancer therapy. International journal of radiation biology *82*, 699-757.

Team, T. M. R. (2005a). [18F]Fluoromisonidazole. Molecular Imaging and Contrast Agent Database (MICAD) [Internet], http://www.ncbi.nlm.nih.gov/books/NBK23099/.

Team, T. M. R. (2005b). l-[methyl-11C]Methionine. Molecular Imaging and Contrast Agent Database (MICAD) [Internet], http://www.ncbi.nlm.nih.gov/books/NBK23696/.

Thorwarth, D., Eschmann, S. M., Scheiderbauer, J., Paulsen, F., and Alber, M. (2005). Kinetic analysis of dynamic 18F-fluoromisonidazole PET correlates with radiation treatment outcome in head-and-neck cancer. BMC cancer *5*, 152

Uchino, H., Matsumura, Y., Negishi, T., Koizumi, F., Hayashi, T., Honda, T., Nishiyama, N., Kataoka, K., Naito, S., and Kakizoe, T. (2005). Cisplatin-incorporating polymeric micelles (NC-6004) can reduce nephrotoxicity and neurotoxicity of cisplatin in rats. Br J Cancer *93*, 678-687.

Valadon, P., Garnett, J. D., Testa, J. E., Bauerle, M., Oh, P., and Schnitzer, J. E. (2006). Screening phage display libraries for organ-specific vascular immunotargeting in vivo. Proc Natl Acad Sci U S A *103*, 407-412.

Vallerga, A. K., Zarling, D. A., and Kinsella, T. J. (2004). New radiosensitizing regimens, drugs, prodrugs, and candidates. Clin Adv Hematol Oncol *2*, 793-805.

Wang, H., Yan, H., Fu, A., Han, M., Hallahan, D., and Han, Z. (2010). TIP-1 translocation onto the cell plasma membrane is a molecular biomarker of tumor response to ionizing radiation. PloS one *5*, e12051.

Whaley, S. R., English, D. S., Hu, E. L., Barbara, P. F., and Belcher, A. M. (2000). Selection of peptides with semiconductor binding specificity for directed nanocrystal assembly. Nature *405*, 665-668.

Williams, K. J., Telfer, B. A., Brave, S., Kendrew, J., Whittaker, L., Stratford, I. J., and Wedge, S. R. (2004). ZD6474, a potent inhibitor of vascular endothelial growth factor signaling, combined with radiotherapy: schedule-dependent enhancement of antitumor activity. Clin Cancer Res *10*, 8587-8593.

Zhivotosky, B., and Orrenius, S. (2001). Assessment of apoptosis and necrosis by DNA fragmentation and morphological criteria. Curr Protoc Cell Biol *Chapter 18*, Unit 18 13.

24

Biosafety in the Use of Radiation: Biological Effects Comparison Between Laser Radiation, Intense Pulsed Light and Infrared and Ultraviolet Lamps in an Experimental Model in Chicken Embryos

Rodolfo Esteban Avila, Maria Elena Samar, Gustavo Juri, Juan Carlos Ferrero and Hugo Juri
Department of Cell Biology, Histology and Embryology, Faculty of Medical Sciences, National University of Cordoba, Argentina

1. Introduction

Physics is the science that studies the interaction between matter and energy. This discipline studies the general properties of bodies, the forces that modify the transfer of energy and the interaction between particles. Physics now has many branches, and one of them is the Biomedical Physics.

Biomedical Physics in Medicine applies the principles and methods of physics. This will generate new knowledge and progress towards new horizons in the management of certain diseases. Thus the study of radiation led to progress in the field of diagnosis and therapy of some diseases.

Radiation it's called all energy that propagates as a wave through space. Radiation can be classified as ionizing radiation (cosmic rays, gamma rays, x-rays) and non-ionizing radiation. The concept of non-ionizing radiation includes ultraviolet (UV), infrared (IR) and others lasers.

Non-ionizing radiations are those that radiation interaction with matter does not generate ions due to its energy content is relatively low.

Non-ionizing radiation (laser, ultraviolet and infrared) are commonly used in medicine for diagnosis and therapy. However, they have a deleterious effect on organogenesis (organ formation).

1.1 Ultraviolet radiation (UV)

UV radiation is part of the natural light. According to its wavelength is recognized groups A (400 - 320 nm), B (320 - 290 nm) and C (290 - 200 nm).

UV radiation sources are natural (the sun) and artificial (hospitals, industries, cosmetics, etc.).

UVC radiation does not reach the surface because it is retained by the layer of ozone in the stratosphere.

The natural radiation that reaches us is UVA and UVB. The UVC is the most dangerous to health because of its higher energy.
As for the benefits of using ultraviolet (UV), is recognized for use in phototherapy of patients with psoriasis, vitiligo and other skin diseases. In addition, ultraviolet rays have a bactericidal action, allowing its use in pressure ulcers.
However, involvement of UV radiation in different pathologies has led to the approach of concepts such as photo-aging, accelerated aging process due to modification by UV radiation, DNA (deoxyribonucleic acid) and lipids membrane and induction of programmed cell death (apoptosis) of epithelial cells and the activation of enzymes degrading the collagen in the skin.
While studying the effects of ultraviolet radiation on the apoptotic induction and angiogenesis the results are mixed for use in the therapy of skin diseases including cancer and photocarcinogenesis, understood as the mechanism of DNA mutation due to the alteration of repair when damaged by UV radiation.
Ultrviolet radiation is involved in the alterations of protein synthesis, the immunosuppressive properties and Its relationship to skin cancer and change in the synthesis of melanin.
Ultraviolet radiation B are related to the induction and progression of cutaneous melanoma in mice, so its use helps to know the specific immune processes committed to these cancers, and thus to develop new treatments for melanoma. This type of radiation has also been used experimentally to elucidate the mechanisms of cellular radioresistance in ovarian cell lines.

1.2 Infrared radiation (IR)

Other natural rays are infrared (IR) or infrared radiation (the prefix infra means below, correspond to an emission of energy in the form of electromagnetic waves in the spectrum located immediately after the red zone). These rays have wavelengths between 800 and 0.25 nm and cause heating of the exposed regions, being the less penetrating wavelength formerly called them "calorific rays."
As for the infrared rays have a wavelength between 800 and 0.25 nm, cause heating of the exposed areas and those are more penetrating radiation of shorter wavelength. Special lamps are used commercially to produce 14,000 to infrared.
Its heating effect allows blood vessels to dilate and increase blood flow.
They are also used in conjunction with hyperthermia and photodynamic therapy to treat tumor in the ablation of organs, and therapy of processes related to inflammation of nerves, muscles and joints.
It has been postulated that IR produce a higher activity of phagocytosis and metabolic reactivity in infectious and has been used in phototherapy of patients with certain skin tumors.
However, the intracellular level, has shown that infrared radiation acting microtubule disrupting the centrosome (organoids) in mammalian cells in vitro.

1.3 Laser radiation

The laser light (artificial) could be defined as an amplification process that culminates in the production of a light. In turn, electricity is the band of the electromagnetic spectrum including UV radiation, visible light and IR radiation. The laser light is not normally found in nature and is of extraordinary intensity.

The best known property is the emission of beams of highly monochromatic, coherent and directional light.
There are different laser for medical use, classified as high-power-like carbon dioxide (CO2), for surgical use, and low power-such as Helium-Neon (He-Ne) for clinical use.
The term laser means in English Light Amplification by Stimulated Emission of Radiation.
The laser is a device or devices that produce a very special light (visible or invisible depending on their wavelength), created by man and acts like solid matter.
Einstein laid the theoretical basis for his work on the Quantum Theory of Radiation in 1917, by which the energy (light) can be converted into mass and mass into energy.
As light is reflected, the laser is absorbed, burning, and changes its direction through different lenses. As solid mass you can cut, melt, burn, and transmit.
The use of laser energy is not without several problems as the high price of the apparatus and the need for prior learning to use. In addition there is one type of laser that fits all applications, but each type has it sitting indications, many of which are unique to it.
From the medical point of view the use of laser can be used to remove tumors, seal blood vessels to reduce blood loss, sealed lymph vessels to reduce swelling and decrease the spread of tumor cells and nerve endings sealed to reduce postoperative pain.
Aesthetic laser therapy treatments can be used to remove warts and moles and to remove tattoos. It is also used to treat stretch marks, cellulite, sagging skin, acne sequels.It is based on the fundamental physiological properties of the laser, anti-inflammatory, spasmolytic and antiphlogistic effects and bio-stimulants. (Andreu & Valiente Zaldivar 1996).
Lasers and biomedical use:

- Solid lasers: the most used is Neodymium-Yag laser. In ophthalmology it is used to coagulate tissue. It is also used to treat hyperpigmentation of the skin
- Gas lasers: they are the most widely used therapeutic and amongst these the helium-neon laser, used in beauty treatments, therapies reductive, etc. (red), the argon laser used in dermatology (bluish green) and the laser CO2 used in surgery as a scalpel.

The Helium-Neon laser is used to treat various conditions "satisfactory" mainly osteomioarticular (rehabilitation), skin disorders and wound healing.
Among the lasers used in medical practice are the 308nm excimer. This type of laser emits 85% of ultraviolet (UV) B and 15% of UV radiation type A.
Contraindications: The laser should not be used in patients affected by neoplasia (cancer). Nor can it be used in the presence of acute infections and in patients treated with photosensitizing drugs or remedies. Not recommended laser surgery in people with pacemakers or pregnant women.
In a previous study, we irradiated chick embryos and new born chickens with He-Ne laser, infrared and ultraviolet radiation, finding post irradiation histopathologic changes. (Samar et al., 1993, Avila et al., 1994 , Samar et al., 1995)

1.4 Intense Pulsed Light (IPL)

In addition to the laser, since 1995 it also has been available a device that emits at 308 nm, called intense pulsed light (IPL) which is basically a XeCl lamp that has proved to be a useful tool for the treatment of the changes of the skin. The IPL provides high energy pulsed excitation consisting of 85% of ultraviolet radiations UV B and 15% of radiations UV A. This technology is also known as "Photoderm". Although the active medium of this lamp is also XeCl, the emission is polychromatic but also non-coherent; therefore is not a laser. However, the use in medicine and the risks associated with it are comparable to the medical lasers of the high energy (class 3b and 4) and therefore their use should be subject to the same safety guidelines. (Spencer & Hadi, 2004). (Town, et al. 2007).

The IPL is used in a similar way to the excimer laser in the treatment of various pathologies of the skin: psoriasis, vitiligo, etc. Clinically it is also used to stimulate the regeneration of the cartilage in degenerative processes. It has been postulated that its action is based on the activation of the cell division, collagenous and elastic fibers formation, regeneration of blood vessels, cicatrisation of bone tissue and reepithelization of damaged tissue. (Schoenewolf et al. 2011)

Nevertheless, there are some discrepancies in results obtained with animal experimentation on the medical use of not-ionizing radiations. (Chan et al. 2007)

However, the biological effect of the 308 nm IPL has not been investigated using the chick embryos as sensor.

1.5 Objective

Different studies led us to establish preliminary results concerning the action of He-Ne laser, ultraviolet, infrared, LIP on the different tissues of chick embryos and in salivary glands and tongue of newly hatched chicks.

We also carry out the observation of effects that these types of radiation have on the chorioallantoic membrane of chick embryos, particularly on the formation of new blood vessels.

Moreover, the exposure time and the amount of non-ionizing radiation doses used in animal experiments and in medical practice are not yet fully known. Thus, our experimental design established in embryonated eggs and newly hatched chicks will contribute to the quantitative analysis (exposure and dose) of radiation mentioned above.

The purpose of this chapter describes the biological effects comparison between laser radiation, intense pulsed light and infrared and ultraviolet lamps in an experimental model in chicken embryos.

The chick embryo is a good model to evaluate direct effects of ionizing radiation because of easy handling and availability ITS.

Describe the most important results with different sources that emit non-ionizing radiation.

2. Materials and methods

2.1 Experimental model of chicken embryo

We used the chick embryo model as a mechanism for measuring biological effects of radiation on tissues, and this is an interesting method to be easily replicable.

Also to compare the results obtained by applying the radiation we have established in previous studies, the sequence of morphological changes, biochemical and histochemical occurring during differentiation and growth of the tongue, stomach, and ovaries both mesonephros "in ovo" as "in vitro" from 7 to 21 days post-birth immediately.

The problem groups were irradiated for 5 minutes through a window opened in the egg shell, and the eggs were maintained aseptically for 24hs in an incubator.We used the following radiations: intense pulsed light (excimer laser wavelength 308 nm) He-Ne laser (power 5mV, wavelength 632.8 nm), UV germicidal lamp and IR lamp OSRAM infraphil.

The control group were irradiated and only opened a window in the eggshell. All controls and aseptically problems remained in an incubator at 37 ° C for 24 hours post-irradiation.

The samples taken was fixed in Bouin and were processed by routine histological technique and stained with: hematoxylin and eosin; conventional Histochemistry: PAS for the demonstration of glycoproteins, Alcian Blue at pH 2.5 and 1.0 for the demonstration of glycosaminoglycans sulfated and sulfated; Toluidine blue at pH 3.3 to demonstrate basophilic and metachromatic substances, alcohol-resistant.

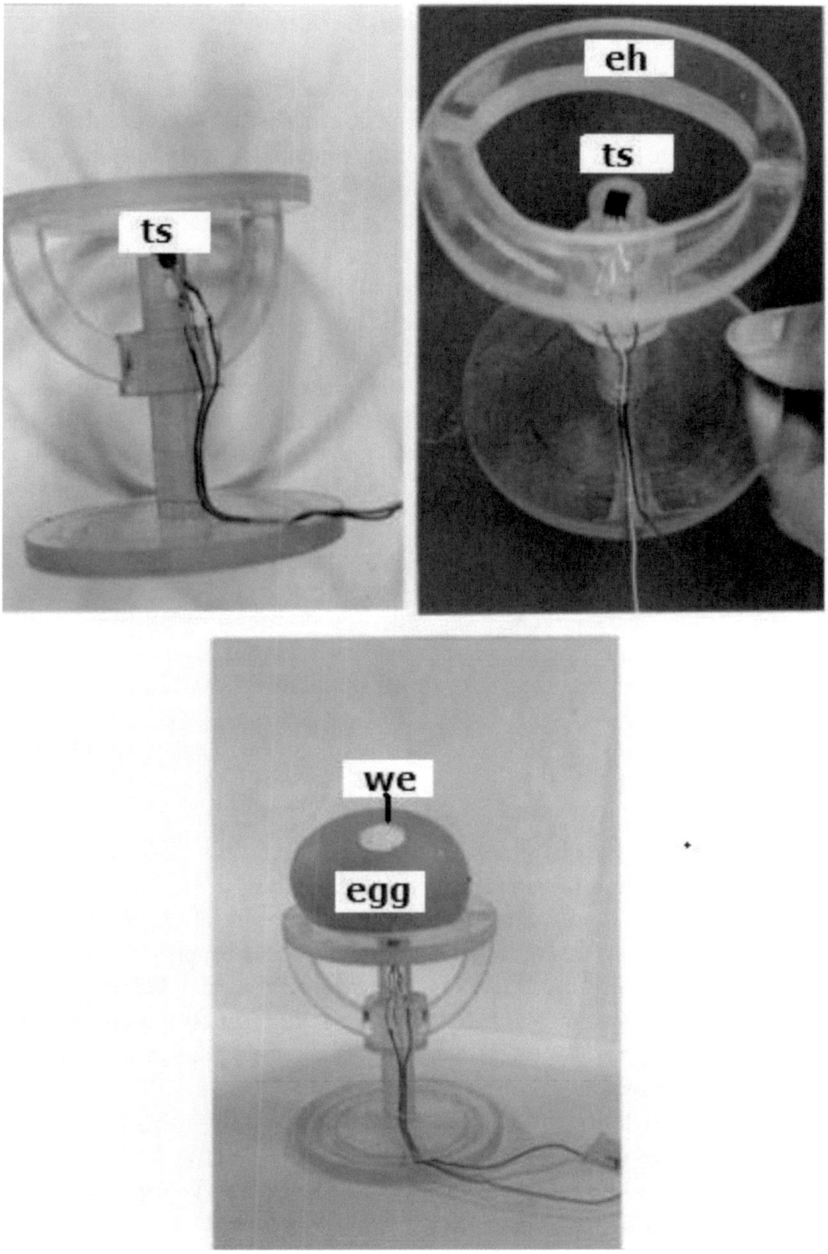

Sample (egg) with thermic sensor (ts). Egg holder (eh). Window in the egg shell (we).

Fig. 1. The system adaptable to the experimental requirements.

2.2 Infrared radiations

In 1996 we published design of a thermic detection system applied to chick embryonated eggs irradiated with infrared rays (Avila et al. 1996).

Infrared radiations are widely used in medical therapeutics. It has been argued that doses and periods of time employed in experimental animals are higher than those used in clinic. Thus, we considered of interest to analyse aspects of dosimetry and thermic effects of infrared rays with current methods in medical practice, using in ovo chick embryo as a model of easy control.

To this end we designed a system to measure temperatures and their acquisitions and software for its handling. The system consists of: a) thermic points: thermocouples or termistores adaptable to the experimental requirements and calibrated with a greater precision within a range of ten degrees around the incubation temperature; b) acquirer circuit of thermic data (hardware): it generates a time base that varies with the thermic sensor. Software: the PC XT or AT detects changes in the time base by means of a programs' in a Turbo Pascal; c) storage and analysis of data allows, through a menu the scale selection, time of program data to be acquired, storage and recovery of the diskette information and graphic impression; d) chick's embryonated eggs.

This system allows to measure temperature distribution in small physical spaces with little disturbance of the system to be measured in irradiated bodies, to analyse variations of the temperatures in time and to secure a greater confidence and automatism to obtain the required data. Figure 1.

2.3 Experimental Intense Pulsed Light (IPL) system with and without different filters (Avila et al. 2009)

To carry out the studies of the effects of the radiation on the different tissues in embryos of chicken we designed the experimental plan that is shown in the Figure 2.

The system consists of an Intense Pulsed Light (IPL) radiation source of 308 nm from DEKA model: Excilite (XeCl Excimer Light) wave length of 308 nm, capable to emit a radiation density of 4,5 J/cm2. The emission spectrum of this lamp was characterized in this work and is shown in the Figure 2.

Besides we used a filter set:

Acetate: Colorless, Orange, Green, Blue and Yellow.

Cellophane: Colorless, Orange, Green, Blue, Yellow and Red

A spectrometer (E1) Ocean Optics model HR4000 was used to characterize the emission of the IPL and the filters used in this work. The measurements of energy were carried out with an energy meter (E2) Scientech model 756 and a sensor (E3) model 362 with complete range (200nm to 1m). With this system we measured the total emitted energy per surface unit, mJ/cm2, on each sample with the different filters used and also the variation of the emission spectrum of the IPL after passing through the filter.

Optical characteristics of the light emitted by the XeCl lamp were determined, Intense Pulsed Light (IPL), using the experimental plan shown in Figure 2. The typical emission spectrum of this lamp was obtained by means of this system We determined that 85% of UVB radiation emitted by the IPL corresponds to 307.94 nm and only 15% to UVA centered in 367,5 nm. This implies that the effects seen on the tissue samples were due mainly by only one emission with very similar characteristics to the effects that could be caused by a laser radiation of this wave length.

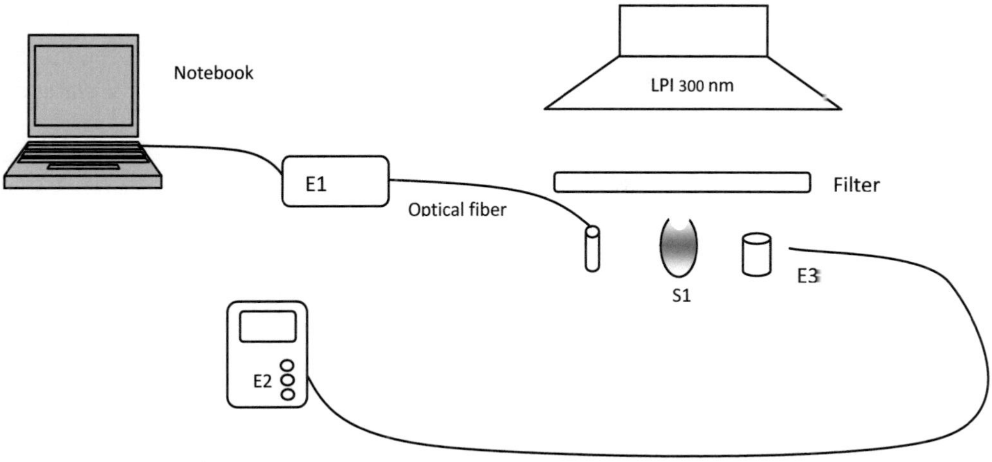

E1: Spectrometer Ocean Optics HR4000, E2: Energy meter, E3: Energy sensor, S1: sample (egg), LPI: IPL source (308 nm). (Avila et al, 2009)

Fig. 2. Scheme of the experimental setup

In Table 1 we shown the light intensity transmitted in the different units by the IPL with the different filters used of acrylic (Ac) and cellophane (Cel) with the exposure time that are normally given for skin treatment. The selection criteria used for the filters to work with were those which showed a transmittance grater than 30%. This transmittance percentage guarantees the possibility to observe the effects of radiation on tissues.

Filter	Exposure time / s	Transmitted Intensity (mW)	Transmitted Fluence (mJ/cm²)	Filter absorbance	Filter transmittance (%)
None	20	18.1	147.50	0.00	100
F1: Green acetate	20	0.46	3.74	1.59	2.5
F2: Blue acetate	20	0.00	0.00	2.00	0.0
F3: Clear acetate	20	0.00	0.00	2.00	0.0
F4: Yellow acetate	20	0.00	0.00	2.00	0.0
F5: Orange acetate	20	6.70	54.60	0.43	37.0
F6: Red Cellophane	10	3.29	13.40	1.04	9.1
F7: Clear Cellophane	10	18.0	73.34	0.30	49.7
F8:Yellow Cellophane	10	2.2	8.96	1.21	6.07
F9: Blue Cellophane	10	0.48	1.95	1.87	1.32
F10: Orange Cellophane	10	9.2	37.50	0.60	25.0
F11:Green Cellophane	20	5.25	42.8	0.53	30.0

Table 1. Intensity of the light from the XeCl lamp (IPL) transmitted through the different filters, measured at 30 cm from the lamp. (Avila et al. 2009).

Once we characterized the IPL emission and the filter absorption we proceeded to radiate the chicken embryos by previously removing their white membrane (below the shell of the egg). We first carried out a set of measurements without using any filters and later using the filters that were previously selected.
Before starting the radiation process in the different samples we determined the absorption spectrum of the albumen of the egg and of the white membrane. From these results we determined that the white membrane should be removed before the radiation of each sample since it has a very large absorption coefficient to the study wave length and would not permit to determine the effects of the radiation in the embryo tissues.
The acetate filters (transparent and orange) and cellophane filters (transparent orange and green) allow to the passage of the radiations producing biological effects in the organs of the chicken embryo, for this reason it is not recommended to be used as a filter.
The acetate filters (green, blue and yellow) and cellophane filters (red, yellow and blue) do not allow to the passage of radiation, without showing any biological effects in the chicken embryo.

3. Results

3.1 He-Ne laser irradiation

3.1.1 Chick embryo mesonephros

A study on the effects of He-Ne laser irradiation on glomeruli and renal tubules of the chick embryo mesonephros at 7 days of in ovo development was made. (Avila et al, 1992)
To this purpose, He-Ne laser irradiation (potency: 5 mW, wavelength: 632.8 nm) was beamed for 5 minutes through a window opened in the egg shell, and the eggs were maintained aseptically for 24 h in an incubator.
Mesonephros were dissected out and processed for hematoxylin/eosin, periodic acid-Schiff, and Alcian blue at pH 2.5 and 1.0, toluidine Blue at pH 3.8.
In controls, the glomeruli were formed by coiled capillaries with a homogeneous basement membrane, the proximal tubules presented a high cubic epithelium with acidophilic cytoplasm and a developed brush border. Distal and collecting segments were lined by cubic epithelium. An alcianophilic and PAS-positive reaction stood out at the membrane coats of the proximal tubules and tubular and glomerular basement membranes. No glomerular alterations were observed during the experiment. However, there was a marked enlargement of the tubular interstitium, with edema and lymphohistiocytic inflammatory infiltration. Some tubular cells desquamated to the lumen. Other cells presented a raveled apical surface. Some nuclei were dispersed out, and mixing with chromatin, formed diminutive granules. Pyknotic nuclei were seen occasionally. Epithelial necrosis and cytoplasmic debris in the lumen were also noted. For mucins, some zones showed brush border coats of the proximal tubules as discontinuous.

3.1.2 Chick embryo ovary

We published in 1992 the structural changes induced by He-Ne laser on the chick embryo ovary. (Avila et al. 1992)
The morpho-histochemical alterations that occur in the chicken ovary at 7 days of incubation after irradiation with He-Ne laser of a potency of 5 mw and at a wavelength of 632.8 nm were studied.

The embryos were irradiated for 5 minutes through a window opened in the eggshell and aseptically maintained in incubator for 24 hours. The gonads were dissected out and processed for the following techniques: H/E, PAS, Alcian blue, and Toluidine blue.
Controls: The ovaries were formed by a germinative or superficial epithelium, with germ and epithelial cells, and by primary sex cords compressed between them, although separated by a reduced stroma. The cords contained germ cells The surface coat of the germinative epithelium presented a thin layer of PAS positive, alcianophilic at pH 2.5 and orthochromatic material. Basement membranes and intercord extracellular substance were also PAS positive. Problems: Disorganization of the tissue structure was well manifest in irradiated gonads, accompanied by negativization of the histochemical reactions. A lymphocytic infiltration was also found. No structural alterations were observed in germ or epithelial cells.
It is concluded that laser radiations would act producing decrease of the mucosubstances associated to the plasma membrane and basement membrane.
They would also provoke the appearance of an inflammatory mononuclear infiltration.

3.1.3 Newborn chicken

3.1.3.1 Structural and cytochemical modifications in the lingual glands of the newborn chicken irradiated with He-Ne laser (Avila et al. 1997)

Despite the increasing and successful use of laser in Medicine and Odontology, the possible iatrogenic and otherwise deleterious side effects of this radiation remain mostly unknown. In previous studies, it was shown that both the embryonic and the post-hatched chicken constitute reliable experimental models for this type of studies.
Hence, the purpose of the present work was to analyze the structural and cytochemical alterations of the lingual glands of the newborn chicken irradiated with low energy He-Ne laser.
This laser produced regressive structural changes of the glands towards the embryonic stage as well as hyperplasia of the reserve glandular basal cells. Furthermore, a decrease in the glycoprotein content and a rise in the sulfated glycosaminoglycans were also found.
These results corroborate the pathogenic effects of the He-Ne- laser on the experimental model employed and, at the same time, emphasize the importance of considering, regarding clinical applications, possible previous neoplastic alterations as well as adverse reactions which might appear once laser therapy has been installed.

3.2 Intense Pulsed Light (IPL)

3.2.1 Tongue of the chicken embryo

3.2.1.1 Biological effects on the lingual cartilage of the chicken embryo

In 2009 we described the biological effects produced by intense pulsed light (Xe-Cl) on the cartilage of the tongue chick embryo using various filters. (Avila et al. 2009)
The Laser used correctly in the medical practice offers clear advantages compared with traditional therapies. The improvement and even the elimination of many significant skin lesions can be achieved with reduced risks to patients. However, it is important to keep security measures and understand the possible effects on an experimental model.
The chick embryo is a good model to evaluate the direct effects of non-ionizing radiation for its easy handling and availability. The purpose of this communication is to show our

histological findings in organs of the chick embryo with and without protective barrier to be subjected to radiation excimer.
We used the following emitter: intense pulsed light (excimer Xe-Cl laser of 308 nm wavelength). It was irradiated embryos through an open window on eggshells. Aseptically the eggs were kept for 24 hours in an incubator. The protective barriers were used with and without colored glass, latex, cellophane, paper, polycarbonate of different colors and thicknesses.
The tissue changes observed are consistent with possible side effects of these fotothermical radiations we warned about possible side effects when they are applied indiscriminately. We believe it is important to explore different means to safeguard the safety of operators and patients. Figure 3.

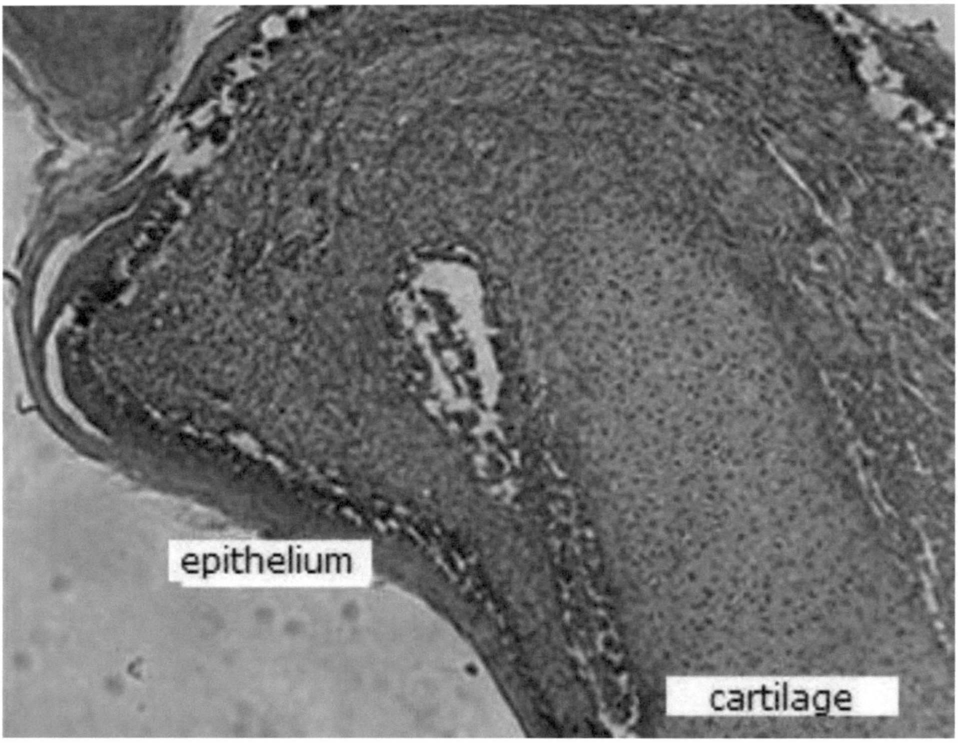

Fig. 3. Tongue of chicken 15 days of incubation in "ovo" irradiated with Intense Pulsed Light (Xe-Cl) 308 nm wavelength. Stain: hematoxylin and eosin. 5 x . The most striking findings consisted degenerative changes at the level of cartilage necrosis following groups glandular mucosa nature, accompanied by leukocyte infiltration, vascular wall thickening at the expense of the tunica media, perivascular edema.

3.2.2 Chick embryo wing

3.2.2.1 Histopathologic findings in epithelium and stroma of the chick embryo wing irradiated with intense pulsed light of 308 nm

Intense Pulsed Light (IPL) is used in medical practice enabling improved significantly even elimination of many skin lesions of patients. However, it is important to keep safety measures and know the possible effects on an experimental model. The chick embryo is a good model to evaluate the direct effects of ionizing radiation is its easy handling and availability.

The purpose of this communication is to show our histopathological findings in chick embryo wings with and without barrier protection when subjected to intense pulsed light radiation.

We used intense pulsed light of 308 nm wavelength Excilite model DEKA brand, Luz Excimer (XeCl) Wavelength 308 nm. Embryos were irradiated through an open window in the eggshell without barriers (controls) or barriers (problems). The eggs were maintained aseptically for 24 hours in an incubator. The barrier used was transparent cellophane without color and a green.

The most outstanding results obtained without a barrier, with clear cellophane and green were epithelial haematic fibrinous exudate and epithelial hyperplasia. In the stroma are varying degrees of vasocongestion, erythrocyte extravasation, and focal hemorrhage and edema, mononuclear infiltrate. It is concluded that tissue changes observed are consistent with possible changes produced by these collateral photothermal radiation, which warns of possible adverse effects when these are applied indiscriminately. Figure 4.

We believe it is important to study the means to ensure the safety of operators and patients.

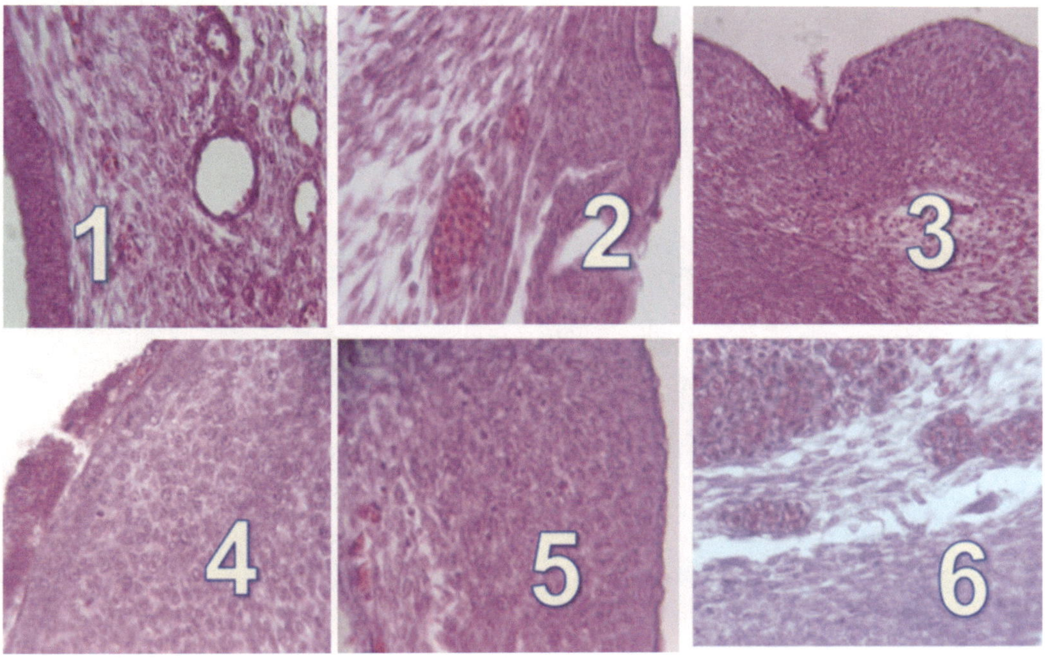

1. Control. 10 X H / E. 2. Problem:epithelial hyperplasia. 40 X. H / E. 3. Problem: Extravasation erythrocytes. 40 X H / E. 4. Problem: Fibrinous exudate 40 X. H / E. 5. Problem epithelial hyperplasia, edema and mononuclear infiltrate (3), which migrate into the epithelium sectors 40 XH / E. 6. Issue: Stromal with vessels congestion and epithelial hyperplasia. 10 X. H / E

Fig. 4. Epithelium and stroma of the chick embryo wing 15 days irradiated with (problem) or without (control) intense pulsed light of 308 nm.

3.2.3 Heart of the chick embryo

3.2.3.1 Biological effects on the heart of the chick embryo

The 308 nm excimer laser is a new application in cardiology for the treatment of congenital heart malformations, vascular and ischemic cardiomyopathy. (Spencer & Hadi, S. M. 2004). This type of laser emits 85% of ultraviolet (UV) type B and 15% of UV radiation type A. However, the literature does not describe the changes that occur in myocardial cells and surrounding embryonic tissue when exposed to this type of laser.
Moreover, the intense pulsed light therapy with high-energy UVB spectrum is used in a manner similar to the excimer laser in different conditions. Intense Pulsed Light (IPL), is based on generating a polychromatic light source, high intensity incoherent.

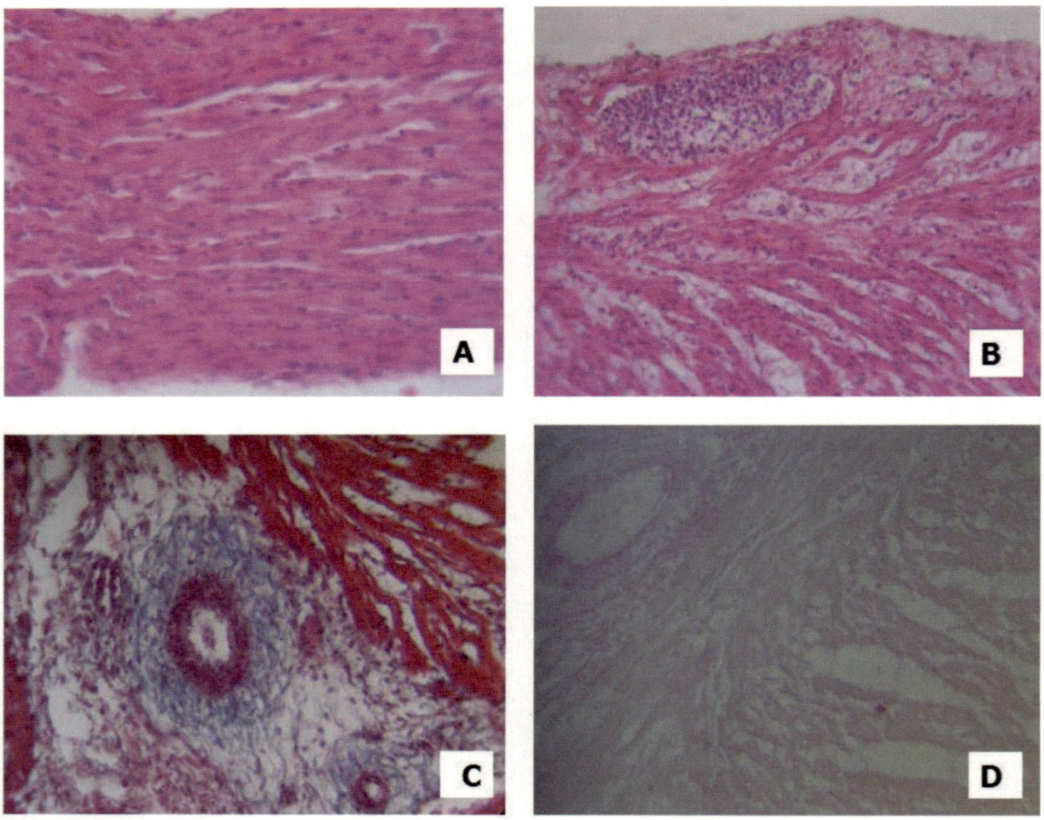

A. Control: cardiomyocytes. Coloration hematoxylin / eosin. B. Problem: disorganized tissue with interstitial edema, vasocongestion, exudate. Pericardium with leukocyte exudate. Coloration hematoxylin / eosin. C. Problem: Interstitium with blood vessels. Cardiomyocytes. Masson's Trichrome Staining. D. Problem: Interstitium and cardiomyocytes. Reduction of glycoproteins. PAS Staining

Fig. 5. Heart of the chick embryo 15 days days irradiated with (problem) or without (control) intense pulsed light of 308 nm.

4. Discussion and conclusion

The chick embryo is a good model to evaluate direct effects of ionizing radiation is its easy handling and availability.
We irradiated chick embryos and newly hatched chicks with He-Neon laser and infrared radiation (IR) and ultraviolet (UV) radiation found histopathological changes post. On the other hand, in the databases searched found no experimental studies on the biological effects of radiation from pulsed light of 308 nm (XeCl Excimer Light) in chick embryo organs. So also did experiments with intense pulsed light corroborating deleterious effects on organs of chicken embryos.
On the other hand, in the databases searched found no experimental studies on the biological effects of radiation from pulsed light of 308 nm (XeCl Excimer Light) in chick embryo organs. So also did experiments with intense pulsed light corroborating deleterious effects on organs of chicken embryos.
Authors studied the Influence of UV-B radiation on embryonic development of chickens Hampshire breed and considered that it is possible to state that short-lasting UV radiation appealing can have positive influence on organisms, which can be used in medicine for preventive end treating purposes. (Veterany et al. 2004)
Some authors (Schroeder et al. 2008) imply that IRA-radiation is capable of altering gene expression which brings forward a pro-aging phenotype of the skin. Apart from minimising exposure to natural IRA and responsible use of artificial IRA sources, the questions arises how a protection against the detrimental effects of IRA can be achieved.
The XeCl (IPL) lamp used in these experiments present a very monochromatic spectrum of emission of 307,94 nm and its effects could be very similar to those corresponding to a XeCl laser. The IPL is normally used for skin treatment for which the spectroscopic characterization carried out lets us determine the effects of this wave length on the tissues. (Nahavandi et al., 2008) The filters used basically acted only as intensity attenuators and they let us determine which optic material is more convenient for the protection of the personnel that uses this type of IPL.
It is because of this that we investigated the histopathological changes produced on an organ of the oral cavity of the chicken, the tongue and especially those centered in the cartilage, being similar to those described in prior publications produced in different organs of the embryo of chicken radiated with ultraviolet radiation, infrared and HeNe Lasers. (Samar et al., 1993, Avila et al., 1994 , Samar et al., 1995)
Clinically it is used as cartilage stimulus for its regeneration in the degenerative processes. It's been postulated that its action is based on the cell multiplication, the formation of collagen and elastic fibers, the vessel regeneration, the scar formation of bony tissues and the re-epithelization of the damaged tissue. (Baumanna et al., 2006 and Andreu & Zaldivar 1996).
Others authors. (Baumanna et al., 2006) studied the influence of laser radiation in human osteoartritic condrocites using different wave lengths (laser diode of 690nm and laser Nd: YAG of 1064 nm), power densities and exposition times and they observed that uaing a specific set of parameters (2 W; 16W/cm2; 60 s; 120 J) they observed an increase in the synthesis of the matrix (material of cartilage originating from 36 patients). They also describe that, using a very high power density, but a constant energy density, a reduction of the rate was produced (28%) of synthesis of the matrix.

By using only the filters with a transmittance grater than 30%, we could guarantee the possibility to observe the biological effects in tissue radiation.
Many countries have published safety guidelines for the use of lasers and most of them are harmonized with the international standards of the International Electrotecnic Commission (IEC). The maximum limit permitted of the exposure with the (MPE) laser used in international safety standards of the IEC are based on guidelines of the International Commission in the protection against the not-ionizing radiation (ICNIRP). (Moseley 1994; Sliney 2006; Parker,2007).
The CIS 825-1 norm (1993) is set for the manufacturers; However, it also offers some limitations oriented on the safety for users. All commercial lasers must exhibit the risk classification indicated.
We repeat that the use of IPL in spite of having similar biological effects than UV Lasers, do not have an equal standard safety classification.
The study of different types of radiation in our experimental model allowed to obtain information regarding the behavior of cells and tissues. We believe it is important to study the various means to ensure the safety of patients and operators. Also allow those responsible for advising on the lamps to regulate medical and aesthetic use.
We conclude that these tissue changes are compatible with photothermal side effects of this radiation that warn us about possible adverse effects of dose and time used experimentally.

5. References

Andreu, M. I. & Valiente Zaldivar, C. (1996). Efectos biológicos de la radiación láser de baja potencia en la reparación hística. *Rev Cubana Estomatol*, Vol. 33, No. 2, pp. 60-63. 9.

Avila, R. E., Ferraris, R. & Samar, M. E. (2001). Efectos biológicos inmediatos de las radiaciones ultravioletas en el mesonefros del embrión de pollo. *Rev Med Cordoba*, No. 89, pp. 9-11.

Avila, R. E., Plivelic, T., Samar, M. E. & Benavidez, E. (1996). Diseño de un sistema de detectores térmicos aplicado en huevos embrionados de pollo, irradiados con rayos infrarrojos. *Rev Fac Cienc Med Univ Nac Cordoba*, No. 54, pp. 13-18.

Avila, R. E., Samar, M. E., De Fabro, S. P., Leguina, M. & Juri, H. O. (1994). Histologic changes produced by non ionizing radiation in the chick embryo. *Rev. Fac. Cien. Med. Univ. Nac. Cordoba*, Vol 52 No.1, pp. 27-30.

Avila, R. E., Samar, M. E., de Fabro, S. P. & Plivelic, T. S. (1997). Structural and cytochemical modifications in the lingual glands of the newborn chicken irradiated with He-Ne laser. *Acta Odontol Latinoam*, Vol. 10, No. 2, pp. 81-88.

Avila, R. E., Samar, M. E., Fabro, S. P. de Juri, H. & Leguina, M. (1994) Alteraciones histológicas por radiaciones no ionizantes en el embrión de pollo. *Rev FacCienc Med Univ Nac Cordoba*, Vol. 52, No. 1, pp. 27-30.

Avila, R. E., Samar, M. E., Juri, H., Fabro, S. P. de, Centurión, C. & Sánchez Mazzaferro, R. (1992). Structural changes induced by He-Ne laser on the chick embryo ovary. *Rev Fac Cienc Med Univ NacCordoba*, No. 50, pp. 7-10.

Avila, R. E., Samar, M.E., Juri, H. & Fabro, S. P. de. (1992). Effects of He Ne laser irradiation on chick embryo mesonephros. *J Clin Laser Med Surg*, No. 10, pp. 287-290.

Avila, R. E.; Samar, M. E.; Juri, H.; Femopase, G. A.; Hidalgo, M.; Ferrero, J. C.; Rinaldi, C.; Fonseca, I. & Juri, G. (2009). Biological Effects Produced by Intense Pulsed Light (Xe-Cl) on the Cartilage of the Tongue Chick Embryo Using Various Filters. *Int. J. Morphol., 27(4)*:1003-1008.

Baumanna, M., Bjorn, J., Rohdec E., Bindigd, U., Mullerc, G. & Schellera, E. (2006). Influence of wavelength, power density and exposure time of laser radiation on chondrocyte cultures – An in-vitro investigation. *Med Laser Appl*, No. 21, pp. 191–198.

Chan, H. H., Yang, C. H., Leung, J. C., Wei, W. I. & Lai, K. N. (2007). An animal study of the effects on p16 and PCNA expression of repeated treatment with high-energy laser and intense pulsed light exposure. *Lasers Surg. Med.*, Vol. 39, No. 1, pp. 8-13.

Moseley, H. (1994). Ultraviolet and laser radiation safety. *Phys. Med. Biol.*, No. 39, pp. 1765-1799.

Nahavandi, H., Neumann, R., Holzer, G. & Knobler, R. (2008). Evaluation of safety and efficacy of variable pulsed light in the treatment of unwanted hair in 77 volunteers. *J. Eur. Acad. Dermatol. Venereol.*, Vol. 22, No. 3, pp. 311-315.

Parker, S. (2007). Laser regulation and safety in general dental practice. *Br Dent J.* Vol. 12, No. 202(9), pp. 523-532.

Samar, M. E., Avila, R. E., de Ferrais, M. E., Ferraris, R. V. & de Fabro, S. P. (1993). Embryogeny of human labial glands: a structural, ultrastructural and cytochemical study. *Acta Odontol Latinoam*, Vol. 7, No. 2, pp. 23-32.

Samar, M. E., Avila, R. E., Juri, H., Centurión, C. & Fabro, S. P. de. (1993). Histologycal changes produced by He-Ne laser on different tissues from chick embryo. *J Clin Laser Med Surg*, No. 11, pp. 87-89.

Samar, M. E., Avila, R. E., Juri, H., Plivelic, T. & Fabro, S. P. de. (1995). Histopathological alterations induced by He-Ne laser in the salivary glands of the posthatched chicken. *J Clin Laser Med Surg*, Vol. 13, No. 4, pp. 267-272.

Samar, M. E.; Avila, R. E. & Juri, H. (1993). Histological changes produced by He-Ne laser on different tissues from chick embryo. *J. Clin. Laser Med. Surg., No. 11, pp.* 87-89.

Schroeder, P., Haendeler, J. & Krutmann. (2008). The role of near infrared radiation in photoaging of the skinJ. *Exp Gerontol*, Vol. 43, No. 7, pp. 629-632.

Schoenewolf, N. L., Barysch, M. J. & Dummer, R. (2011). Intense pulsed light. *Curr Probl Dermatol.* No. 42, pp. 166-172.

Sliney D. H. (2006) Risks of occupational exposure to optical radiation. *Med Lav, Vol.* 97, No. 2, pp. 215-20.

Spencer, J. M. & Hadi, S. M. (2004). The excimer lasers. *J. Drugs Dermatol.*, No. 5, pp. 522-525.

Town, G., Ash, C., Eadie, E. & Moseley, H. (2007). Measuring key parameters of intense pulsed light (IPL) devices. *J. Cosmet. Laser Ther.*, Vol. 9, No. 3, pp. 148-160.

Veterány, L., Hluchý, S. & Veterányová, A. (2004). The Influence of Ultra-Violet Radiation on Chicken Hatching. *J Environ Sci Health A Tox Hazard Subst Environ Eng*. Vol. 39, No. 9, pp. 2333-2339.

25

Alternatives to Radiation Investigations in Orthodontics

Shazia Naser-Ud-Din
School of Dentistry, University of Queensland, Brisbane, Australia

1. Introduction

There has been a massive paradigm shift in investigatory tools at present. A greater surge of interest in soft tissue analysis, real time imaging and functional recordings to enhance our understanding of the complexity of the biological system is underway; with high emphasis on non-invasiveness and cost effectiveness. Current studies are looking at the possibilities of using non-ionizing alternatives to derive the required parameters to diagnose, treatment plan and evaluate the orthodontic case. This is in stark contrast to some decades ago where static hard tissue investigations were considered gold standard.

This chapter will explore the different investigations traditionally used for Orthodontic treatment planning and evaluation of treatment objectives. Since the introduction of lateral cephalometrics in 1931 by Broadbent and Hoffrath this 2 dimensional investigation has been considered the ultimate. However, it has several key limitations - radiation exposure particularly for growing children, 2 dimensional representation of a 3 dimensional living body with static image output. The need to provide a more holistic view to the case assessment with new investigatory tools like 3D imaging, ultrasonography, video recordings and EMG are perhaps the way for the future to assess growing living functioning body.

At the time of writing there is a wide array of imaging technology with innovative developments occurring at a phenomenal pace. Nearly every time one searches the world wide web something new is presented. The new hype- 3D video softwares are considered to be the next big thing [1]. The imaging industry finds wide application in defence, architecture, resource industries and general entertainment. In the medical field, it has immense application for diagnosis and potential to generate understanding not previously possible with 2D investigations. The classic examples are MRI and CT scans [2-6].

In orthodontics, the focus is more based on the external features and soft tissues and 3D reproductions are invaluable not only for diagnosis in general medical scenarios, but more so for treatment evaluation and long-term growth changes. However, the majority of published work tends to come from Japan followed by European and USA researchers [7-9].

In this chapter a brief historical background of the different imaging techniques namely Lateral cephalometrics, sEMG, Ultrasonography & 3D imaging as used in contemporary investigations along with important findings from recent studies from author's research involving such technology are discussed.

2. Lateral cephalometrics

Lateral cephalometric techniques have dominated in mainstream orthodontic, orthognathic and dentofacial growth studies for over seven decades. It is considered a versatile tool and has served the profession well since it was first introduced in 1931 by Broadbent [10]. Cephalograms have improved our understanding of growth through seminal research [11,12]. It is an important diagnostic tool utilized extensively in orthodontics. Recent literature search brought out 28,200 items related to "lateral cephalometrics in Orthodontics" while a similar search on PubMed found 808 documents on related topics. This highlights the volume of work done in the field.

There are two major limitations with cephalometrics; primarily, the additive radiation dosage of progressive films and, secondly, it is a two dimensional representation of the three dimensional craniofacial region. Moreover, the validity of cephalometric analyses has been questioned[13] and there have been concerns about the application of such simple analyses in diagnosis and treatment planning. Errors generally are attributed to orientation, geometry and association. A meta-analysis was conducted on six studies for random and systematic error along with repeatability and reproducibility [14] with good repeatability found for limited landmarks namely menton (Me) posterior nasal spine (PNS) anterior nasal spine (ANS) sella (S) pterygomaxillary fissure(Ptm) point A deepest point on the anterior maxillary margin (A) and point B deepest point on the anterior symphysis region of the mandible(B).

A number of computer software sites such as the Dolphin® (Los Angeles CA, USA), Mona Lisa® (Tidbinbilla, Canberra, Australia) (Fig.2) etc. are available to analyse the lateral cephalogram with multiple analyses at a click or a mouse scroll. However, the computer analyses have also been criticized for error through loss of contrast, related to pixel size and poor calibration. Overall, the trained eye can effectively trace and resolve finer details up to 0.1mm or smaller [15] as compared to 2.48mm accuracy recorded from Active Appearance Models (AAM) which is a computer program designed to locate landmarks [16]. Such technology needs refinement with mathematical model in order to provide a higher level of accuracy and reproducibility with work in progress by most of the softwares.

Furthermore, the lateral cephalogram has been supplemented with other radiographic conventional images such as panoramic radiography, periapicals, and hand wrist films. The later is now being replaced by the cervical maturation index (CMI) which can be derived from standard lateral cephalograms reinforcing the trend towards minimizing radiation exposure in growing individuals.

There is a galaxy of analyses to choose from. One needs to be mindful of the analyses being a composite of several unique dimensions often with a signature eponym. So it is imperative to choose specific dimensions of relevance to the study at hand. Prudent selection of analysis proves to be time efficient. Figure 1 depicts the basic linear, angular and proportional analysis for majority of diagnostic features commonly required for treatment planning in Orthodontics.

The common range and standard deviations for the different variables from a recent study are given in Table 1. These reflect the general spread for the variables elicited in Fig.1(part of the research project of Dr S Naser-ud-Din for the fulfillment of PhD University of Adelaide 2009)[18].

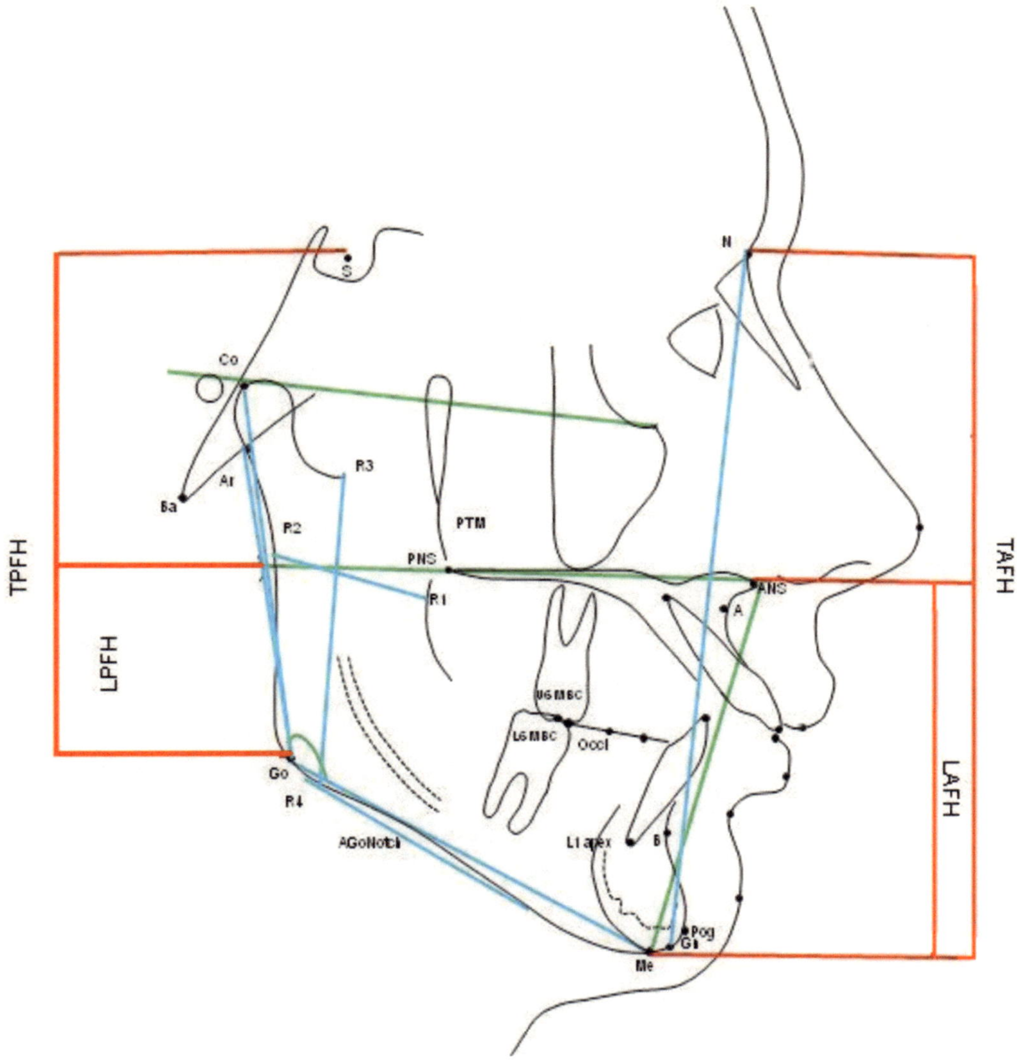

Fig. 1. **Lateral Cephaolmetric with Linear (Blue)-** 9 linear (Co-Go, Ar-Go, Go-Me, Masseter Length, R1-R2, R3-R4, N-Me, N-Gn, AGoNotch).**Angular (Green)** -6 angular (GoAngle, UGoAngle, LGoAngle, Max-Man, Frankfort Horizontal to Mandibular plane, ODI) **& Proportional (Red) analysis-**5 proportional (NGn-ArGo, LAFH/TAFH,LPFH/TPFH, ArGo-GoMe, Jarabak Ratio)[17]

		Variables	Abbreviations	Mean	SD	Range
Linear Variables	1	Condylion-Gonion	Co-Go	64.3mm	(6.0)	55.4-79.7
	2	Ante-Gonial Notch	Ago-Notch	1.9mm	(0.6)	0.9-3.2
	3	Articulare-Gonion	Ar-Go	45.8mm	(5.2)	35.3-55.4
	4	Gonion-Menton	Go-Me	73.7mm	(4.6)	65.4-80.8
	5	Nasion-Gnathion	N-Gn	107.7mm	(6.8)	96.9-121.3
	6	Ricketts R1-R2	R1-R2	29.2mm	(3.9)	21.7-36.2
	7	Ricketts R3-R4	R3-R4	48.0mm	(6.5)	39.6-58.0
	8	Masseter Length	ML	56.3mm	(5.6)	44.6-65.8
	9	Nasion-Menton	N-Me	110.4mm	(7.4)	99.2-126.6
Angular Variables	1	Maxillary - Mandibular Plane	Max-Man	21.3°	(6.2)	10.2-36.5
	2	Frankfort Horizontal -Mandibular Plane	FH-Man	21.8°	(6.3)	13.2-36.1
	3	Upper Gonial Angle	Go Angle U	49.9°	(5.5)	41.6-63.6
	4	Lower Gonial Angle	Go Angle L	68.3°	(5.2)	60.1-82.1
	5	Gonial Angle	Go Angle	118.2°	(7.8)	102.7-134.6
	6	Overbite Depth Indicator	ODI	73°	(7.8)	56.5-92.3
Proportional Variables	1	Nasion-Gnathion/ Articulare Gonion	NGn/ArGo	2.2	(0.2)	1.8-2.7
	2	Articulare Gonion / Gonion Menton	ArGo/GoMe	0.7	(0.1)	0.6-0.9
	3	Lower Anterior Face Height	LAFH/TAFH	57.1%	(3.9)	47.1-63.3
	4	Lower Posterior Face Height	LPFH/TPFH	73.4%	(6.4)	59.1-84.2
	5	Jarabak Ratio	J Ratio	83.6%	(6.5)	71.0-95.7

Table 1. Common distribution for the variables from cephalometrics. (Part of the doctorate project Dr S Naser-ud-Din University of Adelaide Australia, 2009)

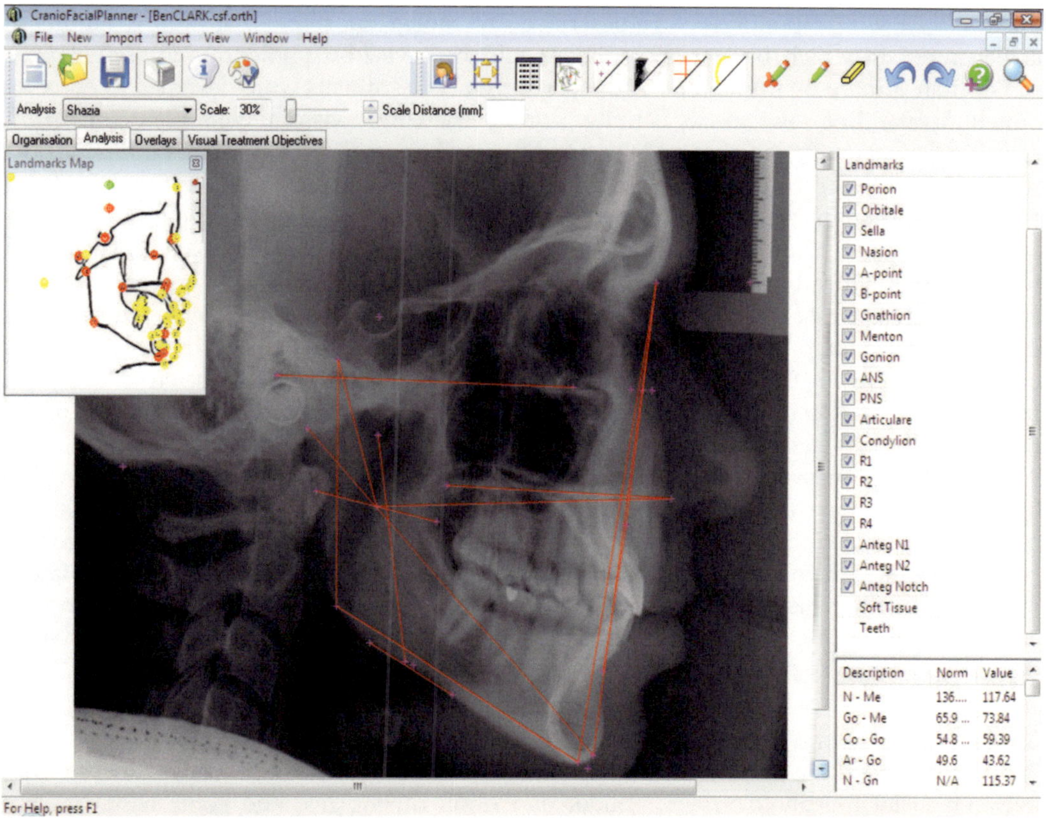

Fig. 2. MonaLisa ® Program for custom made cephalometric analysis. For efficient and standardized analysis computer based software packages are widely utilized.

3. sEMG (surface Electro Myo Graphy)

Our understanding of the neurophysiology of the muscles of mastication has increased substantially in the past 60 years. With the introduction of electrophysiological tools in the 1950's [19], an array of experimental protocols have become available around the globe. Although, the majority of studies have used animal models due to the invasiveness of the procedures, results from such studies still provide valuable data that is extrapolated into humans. The earliest EMG studies reported on muscles of mastication during basic mandibular movements were from Carlsöö(1952), Göpfert & Göpfert (1955) in Moller(Moller 1966[20]. These preliminary studies improved our fundamental understanding of the complex masticatory system. Other work of historical interest is from Perry & Ahlgren[21,22].
EMG registers signals of muscle contractility through action potentials delivered by the motoneurones. Highly refined bipolar surface electrodes are sensitive to these electrical signals and, once amplified, are visible as the EMG recordings.
In the human model, indirect methods are employed and are preferably, non-invasive. Surface electromyography (sEMG) is a tool that is used extensively to explore the neural circuitry. Although, it has its limitations (discussed later) is overall considered user-friendly, non-invasive and has extensive applications in neurophysiology.

3.1 Development of sEMG devices

The first chewing apparatus was described by Moller in 1966 [20]. It was a landmark achievement in which bipolar recordings could be saved on photographic oscilloscopes rather than printed by inkwriters. Therefore, several experiments were designed to enhance our understanding of muscle activity during chewing, postural activities, full effort clenching, swallowing patterns and even facial morphology. This innovative apparatus only required an EMG recorder with electrodes. The electrical activity was recorded on three channels. Although, an advanced methodology for its time, Moller's apparatus had the serious limitation of un-standardized positioning of the subject. This produced a lack of reproducibility, albeit the reproducibility factor poses a challenge even with contemporary advancements. Other methodological issues arose from interference from neighbouring potentials, rejection of common voltage, distortion of amplitude and uneven distribution of electrical activity [20]. Over time, several global studies have been conducted which have improved designs and chewing devices [23,24] leading to an enhanced understanding of the reflexes in the oro-facial regions.

The periodontal-masseteric reflex was first described by Goldberg 1971 [25] highlighting the importance of not only PMR but also the role of the masseter muscle and central connections, due to the short latency. This study concluded that there was a central connection located in the mesencephalic nucleus that was responsible for the excitatory masseteric reflex evoked by PMR and gingival receptors. One can appreciate that, even with a very simplistic methodology and manual tapping of the teeth, Goldberg was able to deduce these important findings.

3.2 EMG studies and significance in orthodontics

Work in this field has resulted in few conclusive findings. A canonical correlation analysis between facial morphology, age, gender and EMG during rest and contraction has not found a statistically significant correlation [26]. EMG and bite force have been studied extensively[20,27,28] with consistent findings of reduced force levels in dolichofacial patterns.

Different vertical facial types, both in adults and children, produce differences in EMG responses recorded over the course of a day [29]. Masseter muscle EMG activity was found to be consistently longer in short vertical dimension facial types as compared to high mandibular plane angle individuals. The variables of bite force, muscle efficiency and mechanical advantage in children with vertical growth patterns [30] were negatively correlated with muscle efficiency and vertical proportions.

Morimitsu *et al.* documented recordings of EMG from masseter muscles which showed positive correlations with linear cephalometric measurements; particularly, the muscle activity was significantly related to mandibular dimensions such as (Co-Go) and body length (Go- Me) [31]. In fact, the muscle activity increased with decrease in vertical proportions such as mandibular plane angle. Furthermore, they compared the muscle activities with anterior-posterior skeletal base relationships indicated by Sella Nasion Point A (SNA) and Sella Nasion Point B (SNB) variables. Morimitsu *et al.* strongly advocated the use of EMG for examining masticatory activity, but one could argue that the reproducibility with electrode placement at subsequent visits may prove to be a challenge [32].

A comprehensive study by Tuxen *et al.* [33] compared masseter muscle fibre types along with function from EMG and facial morphology. They concluded that even with intensive analysis of different dimensions, linear regression analyses failed to show any significant association highlighting, once again, the complexity of the craniofacial region.

Muscle spindle reflexes are stronger in short face height individuals[17,34] which may explain the phenomenon with bruxism and brachyfacial tendency. Moreover, this information may translate into better designs for splints and Orthodontic appliances.

A recent publication [35] has utilized EMG to analyse masseter muscle differences in individuals who brux in their sleep. The authors believe their system can help diagnose different types of bruxism and is aptly termed the innovative bruxism analysing system. Once again, the literature supports the use of sEMG as a useful mode of investigation.

3.3 Limitations & methodological issues

EMG has provided insight into the neuromuscular system but there are inherent limitations which must be minimized in order to derive meaningful data. This has been reviewed thoroughly [36]. EMG accuracy may be enhanced by the use of intramuscular electrodes compared with sEMG electrodes. However, sEMG electrodes with built-in amplifiers can assist in reducing cross-talk and movement artifact. Passive surface electrodes require skin preparation and monitoring of resistance below 10Ω for clear data recording. Contemporary sEMG involves the use of bipolar electrodes, which suppress noise. The amplification should counter distortion, preferably 10 x higher than the electrode-to-skin electrical resistance. The filter level is dependent upon skin thickness and the frequencies recorded. It should not eliminate any of the frequencies within the range of recording and the choice lies between upper (high-cut or low pass) and lower (low-cut or high pass).

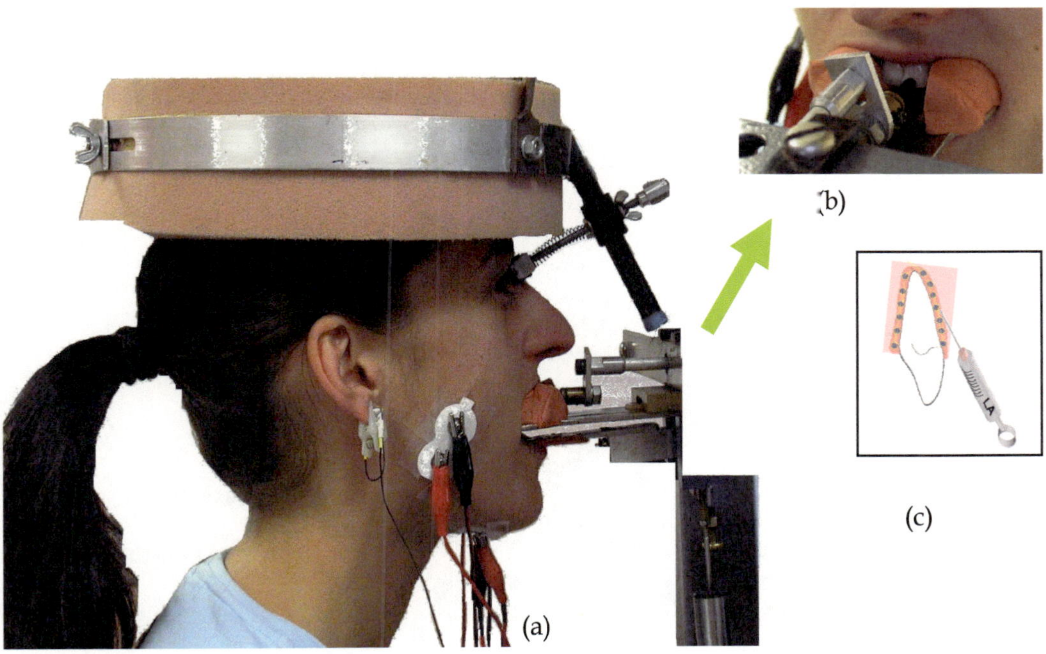

Fig. 3. Subject seated comfortably throughout the experiment session. sEMG recorded bilaterally from the masseter and the digastric muscles: (a) biting on a custom made impression; (b) left central incisor mechanically stimulated by an orthogonal probe;(c) local anaesthetic was infiltrated to eliminate periodontal mechanoreceptor input during the spindle trials. An ear clip provided the electrical ground.

In addition, the quality of data production is dependent on the recording and display devices, the sophistication of the computer programs to reliably identify the initiation of muscle activity, the type of ground used, and obviating the use of long leads which contribute to noise and, therefore, contamination of the data. Several different types of ground have been utilized in human studies such as forehead, wrist, elbow, ear lobe and lip clips. Currently the ear lobe is selected due to ease of placement and distance from the experiment site (Fig 3).

Minimizing artifacts is essential for clear data recording and minimizing movements from adjacent muscles is very helpful. In the study by author[34], the customized nose rest and the head halo provided stabilization during the chewing and static phases of the experiments (Fig. 3).

Cross-talk is a phenomenon commonly experienced in sEMG recordings where adjacent muscle activity leads to electrical volume added to the data. Cross-talk can be minimized by standardizing the experimental conditions. Moreover, double deferential techniques can also assist in eliminating cross-talk [37-43]. This is the prime cause of criticism in telemetry sEMG recording where cross-talk and noise can compromise the results [44].

Finally, correct processing of the data is essential and custom-made software programs are routinely used. IZZY® [45], in Fig. 4 provides an array of systematic offline analysis which allows the data from sEMG to be full wave rectified and further processed as CUSUM (Cumulative Sum).

Normalization (Fig.5) is a standardization process for data acquisition which allows comparison between different subjects and with the same subject data on different occasions. It means normalizing sEMG levels to the percentage of an individual subject's MVC (maximum voluntary contraction) for each muscle group, hence reducing the variability between records.

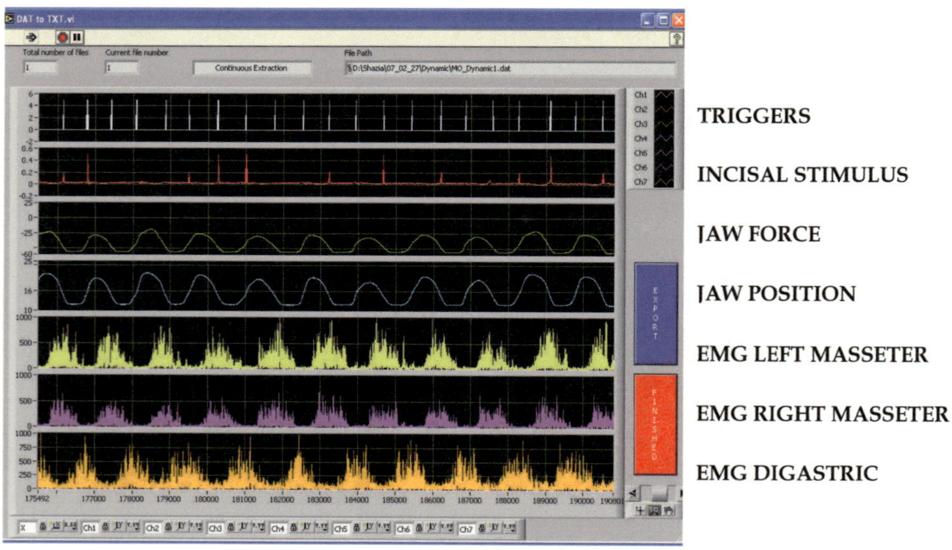

Fig. 4. Recording sEMG screen shot of the computer generated triggers (white), profile of stimulus (red), Jaw force (green), Jaw position in opening and closing (blue) and raw sEMG recorded from left masseter (yellow), right masseter (pink) and anterior digastric (orange). Total of seven channels were operating at the time of the experiments.

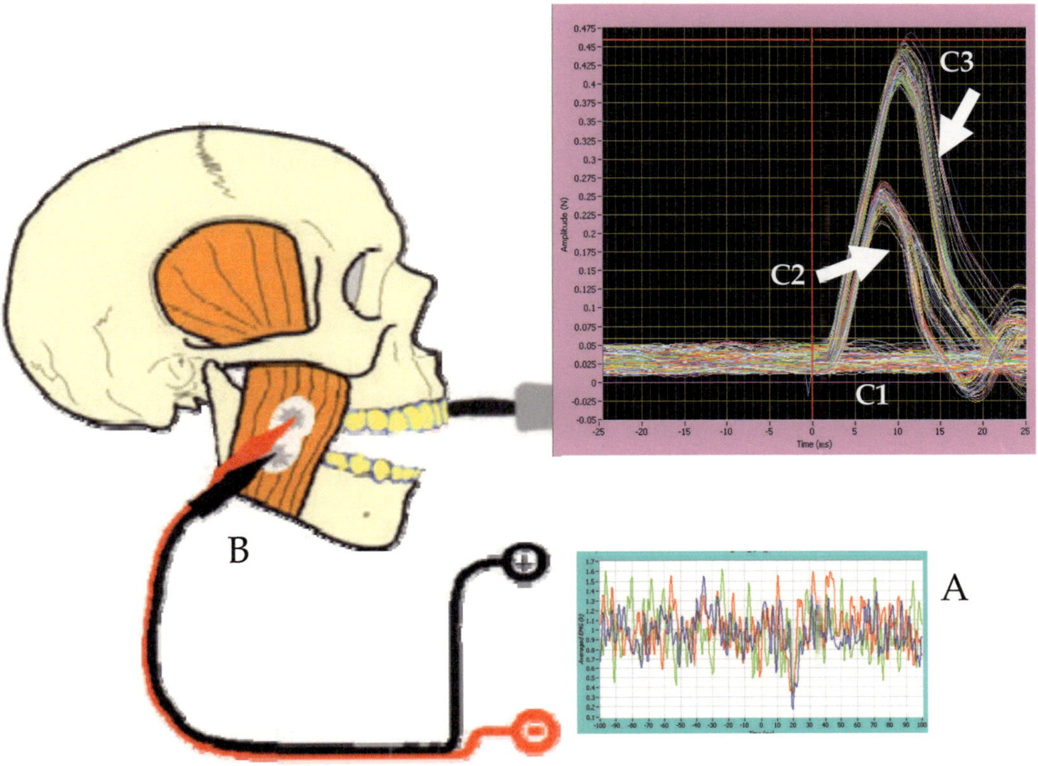

Fig. 5. sEMG (A) was recorded from the masseter (B) during experimental conditions with normalization of the raw scan in A to C1, C2 and C3 (control, with low tap and high force tap to the central incisors).

4. US (Ultra Sonography)

Ultrasonography is developing rapidly as the mainstream investigatory tool.
Ultrasonography (US), as the name implies, utilizes sound waves within the frequency of 2-18MHz that bounce off the tissue to provide the depth and density of the image conventionally captured on the screen as a real time image. It was developed for medical use in the 1940's with parallel developments across the Atlantic in the USA and Sweden. It has since been widely accepted as the safest investigating tool and is applied widely in obstetrics. However, it has applications across several medical disciplines including physiotherapy therapeutic outcomes for muscular-skeletal conditions. It is now easily available infact handheld devices which are easy to use and with higher resolution are making way into the clinical set ups and for field work. The 3D graphics along with real time color coding will indeed provide greater understanding and acceptance in the future to evaluate functioning muscles. There has been a major paradigm shift with realization that form and function are inextricably intertwined in Orthodontics. Our understanding of the hard tissue relationship has been comprehensive, with extensive clinical trials both cross sectional and longitudinal over the past several decades. The majority of studies have used

standardized cephalograms that reflect our current concepts. However, the same cannot be stated for soft tissues and, more specifically, the muscles that envelope the skeletal bases. Hence, the interest with non-invasive soft tissue real time evaluation with US.

The hypothesis was that if such a simple non-invasive investigation can provide diagnostic data then it would obviate the need for radiation investigations, in particular for randomized controlled clinical trials (RCCT) that are currently considered the highest caliber of evidence based literature. Such study designs require several radiographs which are difficult to justify ethically. Hence, non-invasive US could prove vital for long term studies.

US is advancing from 2D to 3D imaging whereby a series of 2D images are collated and rendered into 3D. The main limitation is cost as it needs a specialized probe and the image acquisition speed is relatively slow, especially in moving tissues such as contracting muscles. Further advances with colour Doppler technology are of value for imaging tortuous vessels. Generally speaking, conventional US serves well and is good for imaging muscles and soft tissues, but has poor acoustic impedance for bone.

Muscles of mastication have been studied extensively with US which has been considered to be a valuable, precise technique for analysing muscle shapes [46-49].

Furthermore, US is superior to radiographs for soft tissue evaluation and definitely overcomes the radiation hazards. The surface topography for the masseter muscle is excellent due to its superficial anatomical position Fig. 6 &7); however, the volume and cross-section assessments have been difficult [50]. Hence, the present study sought to evaluate these dimensions with a simplified approach and using arithmetical formulas [51].

A comprehensive meta analysis was recently presented by Serra in 2008[50] which reported masseter muscle thickness in contraction to range from 5-14.1mm [52,53]. One could argue that this wide variation would be due to racial or ethnic diversity not to mention the time of recording be the muscle relaxed or tensed in contraction. US has recorded thickness variations in masseter muscle contraction and relaxation [49] during function which would otherwise have been impossible with snap exposure type investigations.

US has been extended to determine association between several variables such as the masticatory muscle thickness, TMD and bite forces [54]. The study found a positive correlation between masseter muscle thickness and posterior facial dimensions. The study concluded that muscle thickness is related to vertical facial dimensions and bite force. Previous studies have had similar findings in large samples where not only negative relations between muscle thickness and anterior facial heights and mandibular lengths were noted but positive relationships existed with intergonial width and bizygomatic facial width according to anthropometric measurements[55,56]. Likewise, a Swedish study consistently found a relationship between thin masseter and longer faces in females [57]. This study found US to be a reliable and accurate method for muscle assessment.

Hatch and associates have clearly indicated the importance of multivariate analyses of the masticatory system for comprehensive diagnosis in dentistry [58]. They went so far as to include blood glucose levels along with bite force and cross-sectional area of the masseter muscle.

Overall, US has been documented as a reproducible and reliable investigating tool for masseter muscle assessment [59,60] and could produce useful, non-radiological information to enhance orthodontic practice.

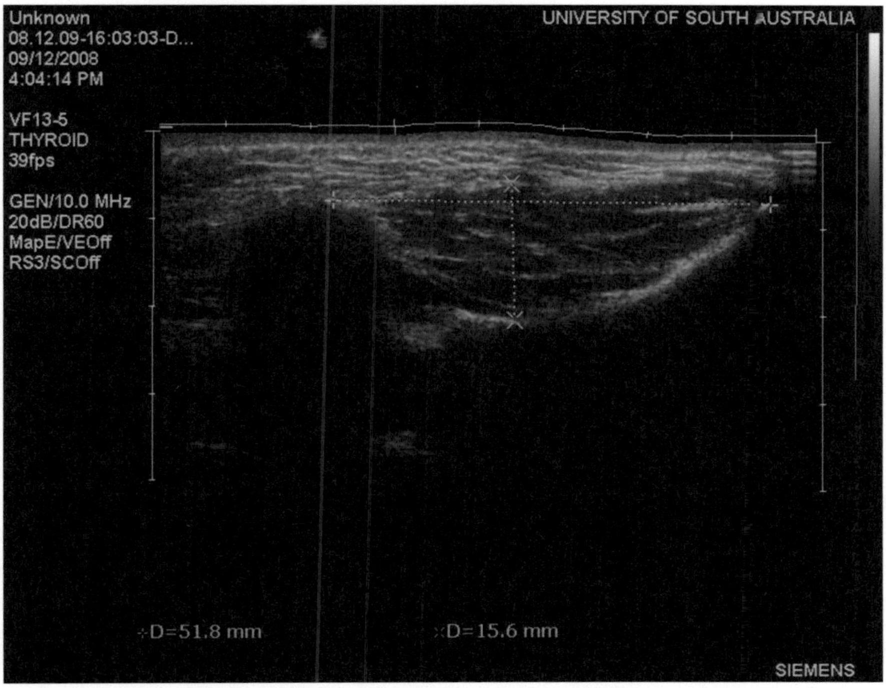

Fig. 6. US Scan of the left masseter showing (dotted lines) the length (horizontal line) and the thickness (vertical line)

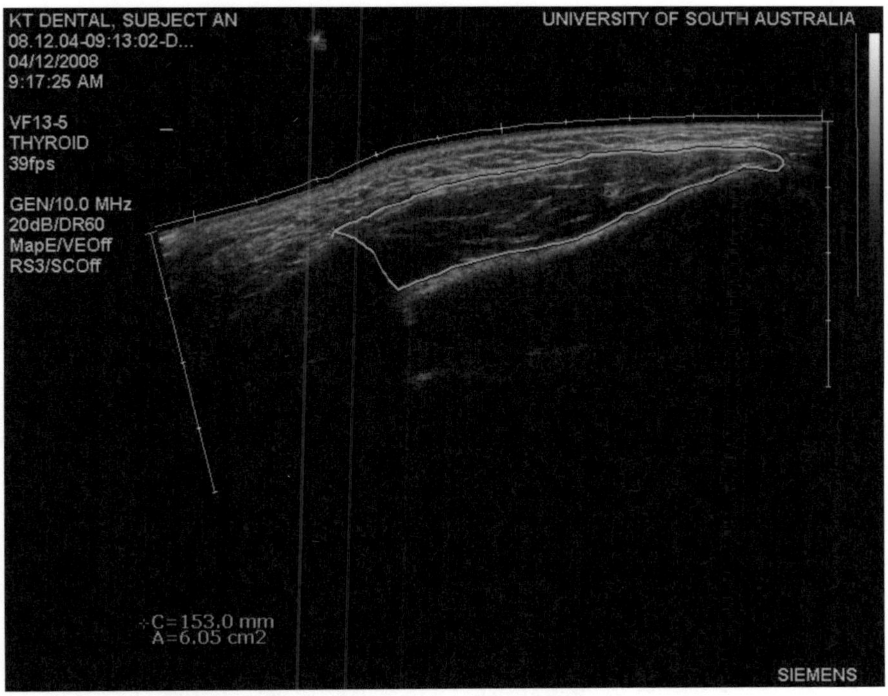

Fig. 7. Area measured manually by tracing the circumference of the masseter muscle (B).

5. 3D imaging

3D imaging has come a long way in the last 20 years, particularly since its introduction six decades ago [Thalmann-Degen 1944][61]. Primarily developed for application in industry [62] as an effective and non-invasive method, it quickly found its way into clinical applications, particularly in orthodontics [63] and oral surgery. 3D applications in restorative dentistry with the Cerec technique and CAD-CAM are already in main stream clinical application and boast high accuracy at the level of 70 microns [64]. Scanning of comparably smaller dimensions of hard surfaces with accurate reproduction has evolved over the past few years[7]. A similar level of precision is desired for facial soft tissue imaging. It's only a matter of time and refinement of technology until 3D imaging becomes a routine part of private orthodontic practice [65]. With innovative robotic wire bending, tailormade appliances both labial and lingual (Insignia, SureSmile[66] and Incognito-3M Unitek USA) 3 D imaging is gaining an integral component of how Orthodontics shall be delivered in 21st century.

In orthodontics, 3D imaging was tested for the first time in 1981 with optical surface scans [67-69]. A major change came with the use of a hand-held scanner making the scan portable (McCallum *et al.;* 1996 in Moss[70]) followed by a probe to record 3D coordinates [55]. Imaging is sensitive to the surface acquisition and sufficient data are required to appreciate the subtle changes that occur in soft tissues over a period of time.

There are various types of imaging techniques ranging from stereo photogrammetry, to 3D laser scanning, vision-based scans like Moiré tomography along with the latest, safest and most cost- effective structured light 3D imaging [9]. Innovative approaches are under way to integrate the conventional lateral cephalogram with stereo photogrammetry and digital fusion. It is claimed that such techniques will obliterate errors through image sensors and 3D orientation [71].

Structured light creates a superficial shell-like reproduction of the face enabling the digitized topology of the face to be displayed in 3D without any ionizing radiation [71]. It is a simple and cost effective method of generating three-dimensional images with minimal time required for exposure, usually within one second.

The structured light (SL) technique provides reasonable accuracy when following certain protocols such as the exposure should be frontal with deviations of up to +/-15° only because with increasing profile view there is proportionally reduced accuracy. Linear measurements can be erroneous up to 1mm and if that is acceptable to the operator and duly accounted for it should not be of much concern. Likewise, the smaller inter-distance between the two cameras creates a limited field of view leading to diminished accuracy in z-coordinate measurements [72].

Overall, SL is considered a simple, cost effective and readily applied 3D imaging system [73]. It is based on stereoimaging and triangulation to produce a 3D image. From a light grid or pattern which is usually horizontal Fig.8. The mean absolute error for linear measurements with SL 3D imaging in the current study was 0.53mm [18,74,75].

3D imaging is valuable in assessing the growth changes in soft tissues over time, because previous investigating methods were not able to adequately assess soft tissue changes related to growth. A large scale growth study with 3D optical surface scanning [76] found significant changes in the vertical facial dimension with increase in the cross sectional cohort age range 5-10 years old. Moreover, they found a significant increase in dimension in the masseter muscle mass across the age range. Hence, age-related changes can be appreciated and quantified with 3D imaging. Furthermore, a longitudinal study analysing facial

morphology changes also found similar results with an increase in the vertical dimension being gender dependent with greater significance in males [8].
Similarly, growth changes have been evaluated with 3D facial morphometry and Fourier analyses depicting changes in profile [77]. Another area where 3D imaging is used extensively concerns pre- and post-orthodontic treatment effects on soft tissues. The very controversial and anecdotal references often made regarding functional orthopaedic appliances has been backed with evidence [78-80]. Yet another highly contentious issue in orthodontics has been the extraction and non-extraction debate, where the treatment decision pendulum keeps swinging by the decade. A study in London [81] used 3D optical surface scans for patients with fixed appliances; one group with extractions and another without. The results were conclusive and laid the age-long controversy to rest with no significant facial soft tissue changes discernable in either study group.

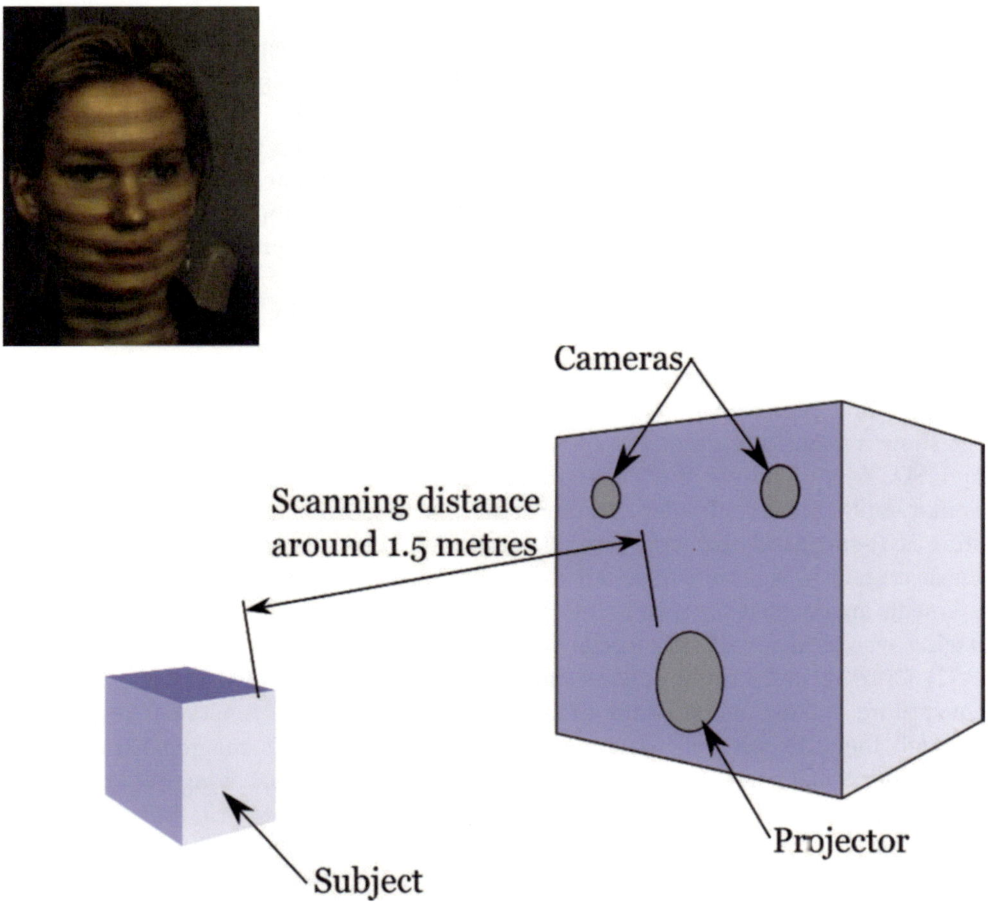

Fig. 8. Capturing an image with the 3D structured light projector placed 1.5m from the subject. The grid of light with horizontal pattern is flashed on the subject's face for 1sec. Triangulation helps to generate the 3D effect.

Work is underway to fabricate user-friendly analysis methods especially as 3D imaging is relatively contemporary and lacking well-established quantification and comparison techniques. 3D studies conducted with Procrustes analysis have determined the translation of soft tissues in monozygotic and dizygotic twins [70]. Another study has used the Procrustes registration method in conjunction with 3D CT scan and Virtual Reality Modelling Language (VRML) to assess the registration error associated with such techniques and found it to be within +/- 1.5mm in most parts [82]. A South Korean study [83] found negligible error of 0.37mm and less than 0.66% magnification with laser scanners. They claim that the soft tissue rendering was highly reliable and reproducible. However, one would argue about the safety of such imaging as the study was not conducted on humans but mannequins.

Unique to anthropometry and 3D imaging is the ability to measure and evaluate the transverse dimensions of the face. Although, a large list of measurements are presented from the frontal aspect in the Farkas textbook [55] for diagnostic purposes one should consider selected ones.

Two horizontal planes commonly used for facial widths are the bizygomatic diameter, inter-zygonion (zy-zy), and mandibular or lower face width, inter-gonion (go-go). Generally, such evaluations assist in picking up asymmetry greater than 2mm [84]. These dimensions form indices for facial proportions. Thus, the Facial Index was calculated with horizontal (zy-zy) as a percentage of vertical from nasion and gnathion (n-gn). Similarly, the (go-go) width is used in the Jaw Index with stomion to gnathion as the lower face proportion (Fig.9).

The midline landmarks were used for vertical proportions which were Index of lower jaw to facial height and Index of Jaw to facial height. These proportions are related to the lateral cephalogram and the correlations can be beneficial substitutes in progressive evaluation over the course of the treatment.

The curvilinear measurements were included because they are more biologically meaningful than straight line representations of complex 3D structures (Fig.10). For example, our findings show a significant difference ($p < 0.0001$) at 95% CI and with a standard error of 0.52 and SD = +/- 1.65 for the linear mandibular depth compared to the curvilinear mandibular depth. The mandibular curvilinear arc extended from left tragus to subnasale to right tragus (t-sn-t) and did not have an equivalent linear counterpart[75]. The use of curvilinear measures is gaining application in tooth morphology studies where curvilinear measurements are deemed superior to the conventional linear ones [85,86].

3D imaging was used for the superficial soft tissue measurements and transverse indices (Figs.9-11). Once again the findings were similar to past work. Currently, the imaging is still a single capture in time and presents limited insight into the true functional assessment. Even though there are several existing packages available for 3D imaging it has yet to become a routine diagnostic tool. There are several reasons for not having 3D imaging in mainstream clinical set ups. Primarily the cost, followed by *not completely realistic* imaging which still needs refinement for accurate and life-like reproduction of the face. Also, the data acquired from such analyses needs to be unique and provide information that cannot be generated by other more conventional means. Finally, the product needs to be user-friendly and easy to apply by the clinician or auxiliary staff. Above all, the suggestion that 3D imaging could be a surrogate for lateral cephalograms with predictive correlations obviating the need for serial radiography and radiation exposure, could make it an attractive choice.

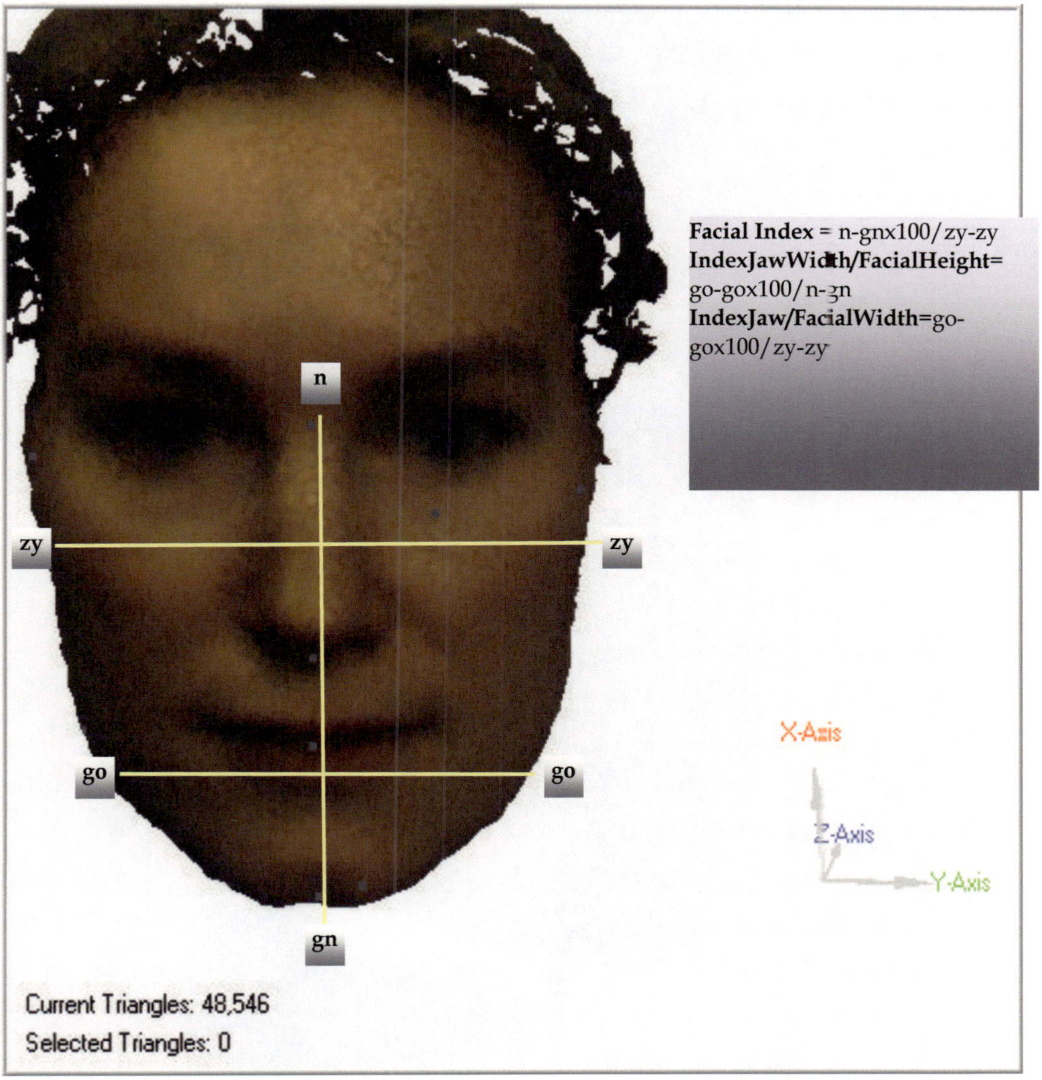

Fig. 9. Anthropometric Indices of Farkas[55] applied to 3D images for transverse and vertical assessments.

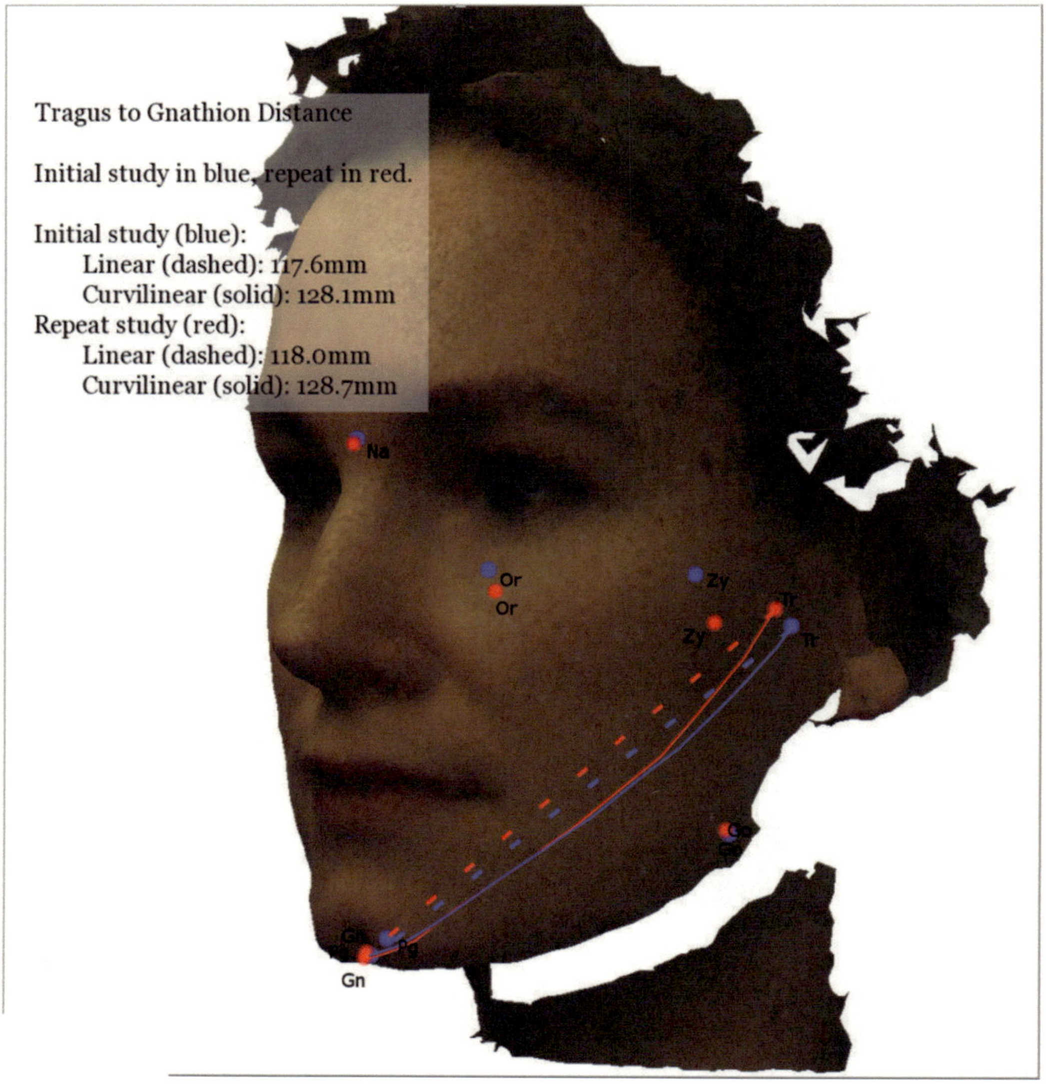

Fig. 10. Curvilinear and linear mandibular measurements for the left side of the subject. There is a clear advantage with 3D images over photographs.

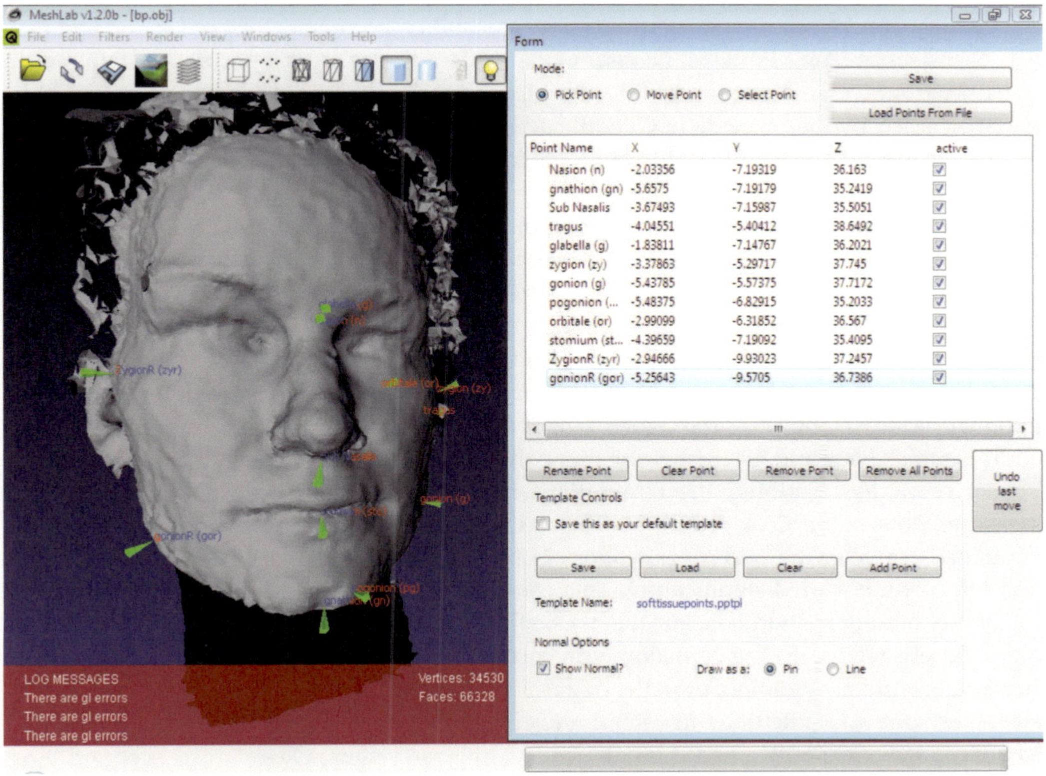

Fig. 11. Analysis of 3D scanned images using MeshLab® shows point selection for computing distances between the landmarks.

Author's study has found high correlation of certain lateral cephalometric variables, particularly the vertical ones, with 3D imaging indices. However, before it enters into mainstream clinical practice, the following criteria need to be fulfilled.

a. *Reliability* is considered a composite of repeatability and reproducibility [87] and different 3D imaging techniques have been scrutinized for accuracy and reliability. Kusnoto and Evans [88] used a surface laser scanner (Minolta Vivid 700) to assess various objects ranging from cylinders to cast analyses and facial models. They found accuracy in the range of 1.9 +/- 0.8mm for facial imaging which was not as high as for cast analyses 0.2 +/- 0.2mm. Hence, one can conclude that the larger the surface to scan and the greater the data acquisition, the higher the error rate with current 3D scans. Likewise, the suggestions for improvement and exploring more user-friendly options in 3D technology have been highlighted in a review article addressing these issues [89].
b. *Standardization* is equally essential for the translation of the predictive equations across the board with different imaging packages. This has been norm with many software cephalometric packages currently in use and could become an overall package.

c. *Cost Effectiveness* and ease of accessibility with perhaps even free downloads such as used in this study can enhance its wider applicability. Such off-line computer analyses can provide flexibility and effective access [90]. However, one has to be wary of the quality of such softwares; reproducibility, reliability, accuracy of reproduction and ease of manipulation. Perhaps custom-made software packages specifically tailored to orthodontic case assessment [85,86] with higher resolution and reproducibility would be welcomed by the profession. Overall such endeavour will aid in patient education and serve as a vital teaching tool to students to appreciate subtle changes during treatment and growth.
d. *Safety* Efforts are being made to assess mandibular growth with 3D MRI [91]. However, there is real concern about the justification of such expensive investigations and is often under fire for overuse in certain countries. Clearly, for the conventional orthodontic patient such scans would not be routinely recommended.

Multiple investigations used in the past such as CT, MRI and radiographs can be replaced with carefully selected diagnostic and evaluating tools which provide maximum information with minimal hazard along with the added bonus of being cost-effective. The current work has addressed this issue and future investigations should be geared to validate it.

Recent work has explored the correlations between anthropometric and cephalometric measurements of vertical profile [92]. The study found nasion and menton to be highly correlated between the two investigation, but landmarks such as subnasale, supradentale and infradentale varied significantly. This could be due to the overlying soft tissue and its variability among individuals. For this particular reason we chose landmarks that would be more closely related to the cephalometric counterparts such as glabella, nasion, gnathion, subnasale, gonion, zygonion, orbitale, tragus and stomion.

However, due to difficulties in lateral projections inherent to the Structured Light (SL) technique we consistently had difficulty in locating the tragus. Overall, our best and most highly repeatable landmarks were soft tissue subnasale (y-axis) with Dahlberg's statistic (DS) = 0.009, followed by zygonion (y-axis) DS=0.004, glabella (z-axis) with DS=0.005 and stomion (z-axis) DS=0.01. The above mentioned landmarks provide a reliable array of reference points to include in future 3D assessments. However, the majority of the 3D landmarks were within an acceptable range. Perhaps the repeatability could be enhanced with refined software with greater matrix size and pixel ratio [87]. This would further refine the accuracy of the prediction equations and improve overall reliability.

Currently, work is in progress in Liverpool where Brook *et al.* are evaluating 3D imaging for better assessment of surface contours of teeth [85,86,93]. Precision and accuracy with e-tools will improve the way we measure. Thus, more studies will be needed to explore landmark assessment and measurement.

6. Prediction equations

Predictive equations were developed from multiple linear regressions which were significant[18,75]. Unfortunately very few variables qualified as the projection from one medium of investigation to the other was diverse, highlighting that each has its own individual data set that is novel and there is very little overlap. However, those that showed somewhat strong correlations were developed with ß- weights which act as constants for a given equation.

Predictive Equation	R^2	MSE	SIG
Linear Variables			
R3-R4 = 25.44932 + 0.48688*t-go	0.33	5.46	*
Masseter = -37.9501 + 0.81574*FaceHt	0.88	2.19	*
N-Gn = 75.76050 + 0.20701*go-gn + 0.10290*ManDepthLinear	0.42	3.83	*
Masseter = -45.1784 + 0.67128*FaceHt + 0.20006*Mandiblewidth	0.92	1.98	*
Go-Me = 58.13533 + 0.06039*Mandibulararc	0.17	4.06	
Co-Go = 28.11536 + 0.76730*t-go	0.58	5.14	*
Ar-Go = -8.69762 + 2.50677*ManDepthLinear + -1.93239*ManDepthCurvilinear	0.67	4.16	*
Angular Variables			
GoangleU = 71.06816 + -0.36734*ManIncli	0.49	4.13	
Goangle = 144.2894 + -0.46133*ManIncli	0.47	5.33	
Proportional Variables			
AFH = 52.67724 + 0.11825*Indexjawwidthfacialheight	0.01	3.06	
AFH = 52.71398 + 0.18845*JawIndex	0.07	2.96	
AFH = 54.06144 + -0.03102*Indexjawwidthfacialheight + 0.19646*JawIndex	0.07	3.17	
JarabakR = 109.0283 + -0.20828*Indexjawfacialwidth	0.11	6.62	
JarabakR = 69.87241 + 0.24662*IndexJaw-FH	0.13	6.52	
JarabakR = 101.1132 + -0.27080*Indexjawfacialwidth + 0.30918*IndexJaw-FH	0.32	6.2	

Table 2. Predictive Tables for the various 3D imaging variables that can provide information pertaining to lateral cephalometric variables. Starred equations show significance at the level of $p < 0.05$. Linear measurements have the statistically significant predictive equations. Gray bars are the ß-weights.

We used simple methods to produce prediction equations rather than the more resource intensive Besaynian or Monte Carlo factor models with the aim of, decreasing the complexity and making it more clinically applicable. The concept of generating predictive equations from 3D images for lateral cephalometric variable estimation has been presented for the first time. Although, it can be debated that with a small sample size the value of such prediction is questionable, it does set the scene for future work[75].

Compared to the reported Eigenvalues from Benington *et al* [94] our values were lower. Also their study was the first to document the masseter volume measurements with high technology 3D US scanning. Likewise, the current study by author did measure masseter volume, albeit indirectly, utilizing a simple, portable US unit that could find application in clinical settings[75]. Similarly, when compared with Radsheer [95], our values were slightly lower but their study included, 121 adults and previous work with 329 subjects aged 7-22 from a Greek population [56]. Our result differences could be attributed to the sample size and also the population from which they are selected.

The method error indicated that there were certain landmarks that had reduced repeatability. This may be explained by background noise and lack of clear definition in some scans. Palpation is generally advantageous in precise positioning of soft tissue gonion, but it is considered a difficult point overall to determine even in anthropology [55]. Nasion (n) and subnasale (sn) were relatively easily identified on the 3D images. For a better view, the image was rotated upwards by 30°. This is an advantage of the scans because repositioning the live patient can be difficult and sometimes embarrassing. The accuracy and repeatability of 3D Imaging has been documented as exceptional [96] and, as with most recordings, in order to eliminate inter-operator variability all measurements should be made by the same investigator. The angular measurements are dependent on head position and need to be standardized across the sample [55].

Baumrind eloquently stated that the "craniofacial complex system needs to be evaluated from different perspectives as no single view would do justice in its evaluation and future treatment planning" [97]. 3D imaging has extended not only to facial surface analyses but is extending to cephalometry as well [98]. Subtle soft tissue changes have been studied extensively in recent work particularly with pre- and post-treatment outcomes [8,64,80,89,91,99,100] . Similar interest with mandibular growth changes[91] and four-dimensional analyses for the TMD [100] are gaining momentum in diagnostic imaging. A common critique with new imaging techniques has been the reliability which is of concern and can be addressed with the development of self –calibrating measuring systems [64].

New directions in technology will soon be evident with the need to enhance the quality of 3D images with more real-life effect. This would assist in patient education and teaching sessions for dental students. Moreover, programs with substantial mouse manipulation create undue hand and wrist fatigue whereas touch screen options would be ergonomically valuable.

Clinical applications of 3D imaging are vast and numerous. From treatment planning to pre- and post-treatment evaluation, the images can be manipulated from any direction providing in-depth analysis of the case without patient recalls and inconvenience of anthropometry. It is becoming a vital tool in orthognathic surgery planning, patient education, "*tele-orthodontics*" [7] to manage global movement of patients, and where interdisciplinary treatments are sought. It is an effective recording method for various facial asymmetries [96] and severe craniofacial dysostoses, hemifacial microsomias [101], and where simple photographs have serious limitations. The non-invasive nature of 3-D imaging will make its application widely accepted for sequential evaluation in growing children and on controls for research [96].

However, the idea of obviating lateral cephalograms entirely may not yet be the case because radiographs still provide valuable skeletal and dental information vital for initial treatment planning. Moreover, we do acknowledge the limitations of prediction because as with any prognostic approach, particularly in a biological system, there will be variations. However, a method that can assist the clinician in reducing the number of radiographic investigations would surely be welcomed by the public and profession alike.

7. Other investigations

a. Genetic profiling

Research is ever expanding in this novel frontier and will offer the possibility of diagnosis in Orthodontics. It already is promising in screening individuals with high risk of root resorption [102].

Predictive and presymptomatic testing is already creating a lot of media interest and long term implications with financial issues related to health insurance. Most importantly the possibility of genetic discrimination may occur.

Recent work from Hong Kong is uncovering the genetic etiology of Class III malocclusion[103,104].
The genetic profiling is evolving and has to be refined to be reliable.
Incomplete penetrance of the genetic code may prove challenging when considering probable malocclusion in the offspring. However, some conditions that have had high genetic predisposition e.g Class III skeletal patterns and anterior open bites (AOB) could provide useful genetic information with the probability of expression.
However, caution has to be exercised as malocclusion is multifactorial and with global migrations miscegenation is increasing with leading to genetic dilution so to speak.

b. Heat sensitive Scans

Moss has shown heat sensitive scans can be utilized to determine the "hot spots" of growth particularly following Orthodontic intervention[79]. This is a simple non-invasive investigation albeit costly.

c. Image Fusion

The latest of combining various investigations for a holistic evaluation process has been aptly described as "Image fusion" [105] assisting in facial triad analysis (namely soft tissues, skeletal basis and dentition) . The aim of this amalgamation is to recreate a virtual head for treatment planning, long term follow up and documentation. Moreover, it may provide a realistic prediction model in the future.

d. MRI

MRI (Magnetic Resonance Imaging) although radiation free is a costly investigation, with implication for time and quality of life (unbearably loud at the moment) . MRI usage is limited in Orthodontics needs to be justified and avoided when possible.
Previous work [3] has affirmed that lower face height and masseter muscle thickness have inverse correlation. Generally, the increased lower face height individuals will present with classic cephalometric characteristics such as large gonial angles, steep mandibular planes and reduced posterior face height. The masseter muscle cross-section has been measured with MRI's and found to be significantly smaller by nearly 30% in the dolichofacial types [3].
A recent study has compared volume of masseter muscle derived from MRI with vertical facial dimensions and found volume to be more relevant than cross-sectional area [106]. Moreover, the posterior face height was significantly correlated to masseter muscle volume. Although, MRI has shown high resolution for TMJ scans[107] it needs to refine to a small field of view [108] for acceptance as an investigation tool.

e. Videofluoroscopy

Even though not entirely radiation free Videofluoroscopy is showing promising results in the investigation of Obstructive Sleep Apnoea (OSA)[109]. There is the issue for 2D instead of 3D as found in CBVT scans, perhaps image fusion would add to its limitations.

f. Videos

Increasingly videos are being recorded to provide a real time assessment on Orthodontic patient smiles, speech and expressions which were not possible with static shots in repose and unnatural smiles. Many of the softwares for diagnosis and data storage are adding video as an important tab to achieve patient motions.

8. Conclusion

Even though the findings from this study are not clinically applicable it has highlighted the importance of finding cost effective and non-invasive methods of investigations for

Orthodontic case assessment. There is now a fast moving trend towards 3D imaging and indeed the lateral cephalometric may phase out in years ahead.
In orthodontics there is a strong association between form and function [110]. This has been studied extensively and proved with strong correlation that exists between the soft tissue and underlying osseous foundation. However, there are not many studies from the soft tissue perspective, particularly, the muscles of mastication including the masseter.
There is definitely scope for US for assessment of masseter muscle in real-time. This has been documented widely in the literature with masseter muscle thickness demonstrating differences in posterior crossbite cases and their relationship to TMDs [5,54,111,112]. But there is concern that some findings can be controversial and others produce non-conclusive results. For example, some studies [113] found the crossbite side had thinner masseter muscle. Obviously, 3D US is superior in function and would be better for oro-facial applications [94] provided it is cost-effective. One can also anticipate the rapid advancement in technology and 4D US may well be the future for diagnosis and treatment evaluation.
Our understanding of malocclusion has come a long way but it is multidimensional, multifaceted and highly variable and needs to be analyzed from several different perspectives. The emphasis is shifting from static, radiation-based investigations which are a snap shot in time to dynamic real-time recordings of soft tissue function. The fourth dimension in combination with conventional diagnostic tools could provide a comprehensive single diagnostic package that would not only assist in complete records diagnosis and evaluation but provide predictions for future expected changes.

9. References

[1] Gibson C: 3D video software the next big thing?, The Adeladian Adelaide, 2009

[2] Raadsheer MC, Van Eijden TM, Van Spronsen PH, et al: A comparison of human masseter muscle thickness measured by ultrasonography and magnetic resonance imaging. Arch Oral Biol 39:1079-84, 1994

[3] van Spronsen PH, Weijs WA, Valk J, et al: A comparison of jaw muscle cross-sections of long-face and normal adults. J Dent Res 71:1279-85, 1992

[4] Hussain AM, Packota G, Major PW, et al: Role of different imaging modalities in assessment of temporomandibular joint erosions and osteophytes: a systematic review. Dentomaxillofac Radiol 37:63-71, 2008

[5] Jank S, Rudisch A, Bodner G, et al: High-resolution ultrasonography of the TMJ: helpful diagnostic approach for patients with TMJ disorders ? J Craniomaxillofac Surg 29:366-71, 2001

[6] Weijs WA, Hillen B: Correlations between the cross-sectional area of the jaw muscles and craniofacial size and shape. Am J Phys Anthropol 70:423-31, 1986

[7] Hajeer MY, Millett DT, Ayoub AF, et al: Applications of 3D imaging in orthodontics: part II. J Orthod 31:154-62, 2004

[8] Kau CH, Richmond S: Three-dimensional analysis of facial morphology surface changes in untreated children from 12 to 14 years of age. Am J Orthod Dentofacial Orthop 134:751-60, 2008

[9] Kau CH, Richmond S, Incrapera A, et al: Three-dimensional surface acquisition systems for the study of facial morphology and their application to maxillofacial surgery. Int J Med Robot 3:97-110, 2007

[10] Broadbent BH: A new x ray technique and its application to orthodontics. Angle Orthod:45-66, 1931
[11] Bjork A, Skieller V: Contrasting mandibular growth and facial development in long face syndrome, juvenile rheumatoid polyarthritis, and mandibulofacial dysostosis. J Craniofac Genet Dev Biol Suppl 1:127-38, 1985
[12] Skieller V, Bjork A, Linde-Hansen T: Prediction of mandibular growth rotation evaluated from a longitudinal implant sample. Am J Orthod 86:359-70, 1984
[13] Vig PS, Spalding PM, Lints RR: Sensitivity and specificity of diagnostic tests for impaired nasal respiration. Am J Orthod Dentofacial Orthop 99:354-60, 1991
[14] Trpkova B, Major P, Prasad N, et al: Cephalometric landmarks identification and reproducibility: a meta analysis. Am J Orthod Dentofacial Orthop 112:165-70, 1997
[15] Mankovich NJ, Samson D, Pratt W, et al: Surgical planning using three-dimensional imaging and computer modeling. Otolaryngol Clin North Am 27:875-89, 1994
[16] Rueda S, Alcaniz M: An approach for the automatic cephalometric landmark detection using mathematical morphology and active appearance models. Med Image Comput Comput Assist Interv Int Conf Med Image Comput Comput Assist Interv 9:159-66, 2006
[17] Naser-Ud-Din S, Sowman PF, Sampson WJ, et al: Masseter length determines muscle spindle reflex excitability during jaw-closing movements. Am J Orthod Dentofacial Orthop 139:e305-13, 2011
[18] Naser-Ud-Din S: Analysis & Correlation Study of Human Masseter Muscle with EMG, US & 3D Imaging Orthodontics. Adelaide, University of Adelaide, 2009, pp 214
[19] Hultborn H: State-dependent modulation of sensory feedback. J Physiol 533:5-13, 2001
[20] Moller E: The chewing apparatus. An electromyographic study of the action of the muscles of mastication and its correlation to facial morphology. Acta Physiol Scand Suppl 280:1-229, 1966
[21] Perry HT, Harris SC: Role of the neuromuscular system in functional activity of the mandible. J Am Dent Assoc 48:665-73, 1954
[22] Ahlgren J: An intercutaneous needle electrode for kinesiologic EMG studies. Acta Odontol Scand 25:15-9, 1967
[23] Huck NL, Abbink JH, Hoogenkamp E, et al: Exteroceptive reflexes in jaw-closing muscle EMG during rhythmic jaw closing and clenching in man. Exp Brain Res 162:230-8, 2005
[24] van der Glas HW, van der Bilt A, Abbink JH, et al: Functional roles of oral reflexes in chewing and biting: phase-, task- and site-dependent reflex sensitivity. Arch Oral Biol 52:365-9, 2007
[25] Goldberg LJ: Masseter muscle excitation induced by stimulation of periodontal and gingival receptors in man. Brain Res 32:369-81, 1971
[26] Fogle LL, Glaros AG: Contributions of facial morphology, age, and gender to EMG activity under biting and resting conditions: a canonical correlation analysis. J Dent Res 74:1496-500, 1995
[27] Ringqvist M: Isometric bite force and its relation to dimensions of the facial skeleton. Acta Odontol Scand 31:35-42, 1973
[28] Ingervall B, Thilander B: Relation between facial morphology and activity of the masticatory muscles. J Oral Rehabil 1:131-47, 1974

[29] Ueda HM, Miyamoto K, Saifuddin M, et al: Masticatory muscle activity in children and adults with different facial types. Am J Orthod Dentofacial Orthop 118:63-8, 2000

[30] Garcia-Morales P, Buschang PH, Throckmorton GS, et al: Maximum bite force, muscle efficiency and mechanical advantage in children with vertical growth patterns. Eur J Orthod 25:265-72, 2003

[31] Morimitsu T, Nokubi T, Nagashima T, et al: [Relationship between orofaciocranial morphologic factors and electromyographic activities of the masticatory muscles]. Nihon Ago Kansetsu Gakkai Zasshi 1:162-71, 1989

[32] Turker KS: Electromyography: some methodological problems and issues. Phys Ther 73:698-710, 1993

[33] Tuxen A, Bakke M, Pinholt EM: Comparative data from young men and women on masseter muscle fibres, function and facial morphology. Arch Oral Biol 44:509-18, 1999

[34] Naser-Ud-Din S, Sowman PF, Dang H, et al: Modulation of masseteric reflexes by simulated mastication. J Dent Res 89:61-5, 2010

[35] Yoshimi H, Sasaguri K, Tamaki K, et al: Identification of the occurrence and pattern of masseter muscle activities during sleep using EMG and accelerometer systems. Head Face Med 5:7, 2009

[36] Turker KS: Reflex control of human jaw muscles. Crit Rev Oral Biol Med 13:85-104, 2002

[37] Kubota K, Masegi T: Muscle spindle supply to the human jaw muscle. J Dent Res 56:901-9, 1977

[38] Yemm R: The orderly recruitment of motor units of the masseter and temporal muscles during voluntary isometric contraction in man. J Physiol 265:163-74, 1977

[39] Eriksson PO, Thornell LE: Relation to extrafusal fibre-type composition in muscle-spindle structure and location in the human masseter muscle. Arch Oral Biol 32:483-91, 1987

[40] Rowlerson A, Mascarello F, Barker D, et al: Muscle-spindle distribution in relation to the fibre-type composition of masseter in mammals. J Anat 161:37-60, 1988

[41] Morimoto T, Inoue T, Masuda Y, et al: Sensory components facilitating jaw-closing muscle activities in the rabbit. Exp Brain Res 76:424-40, 1989

[42] Thexton AJ: Mastication and swallowing: an overview. Br Dent J 173:197-206, 1992

[43] Soboleva U, Laurina L, Slaidina A: The masticatory system--an overview. Stomatologija 7:77-80, 2005

[44] Ueda HM, Ishizuka Y, Miyamoto K, et al: Relationship between masticatory muscle activity and vertical craniofacial morphology. Angle Orthod 68:233-8, 1998

[45] Brinkworth RS, Turker KS: A method for quantifying reflex responses from intra-muscular and surface electromyogram. J Neurosci Methods 122:179-93, 2003

[46] Bakke M, Tuxen A, Vilmann P, et al: Ultrasound image of human masseter muscle related to bite force, electromyography, facial morphology, and occlusal factors. Scand J Dent Res 100:164-71, 1992

[47] Bertram S, Rudisch A, Bodner G, et al: The short-term effect of stabilization-type splints on the local asymmetry of masseter muscle sites. J Oral Rehabil 28:1139-43, 2001

[48] Close PJ, Stokes MJ, L'Estrange PR, et al: Ultrasonography of masseter muscle size in normal young adults. J Oral Rehabil 22:129-34, 1995

[49] Kubo K, Kawata T, Ogawa T, et al: Outer shape changes of human masseter with contraction by ultrasound morphometry. Arch Oral Biol 51:146-53, 2006

[50] Serra MD, Duarte Gaviao MB, dos Santos Uchoa MN: The use of ultrasound in the investigation of the muscles of mastication. Ultrasound Med Biol 34:1875-84, 2008
[51] Naser-Ud-Din S, Sampson WJ, Dreyer CW, et al: Ultrasound measurements of the masseter muscle as predictors of cephalometric indices in orthodontics: a pilot study. Ultrasound Med Biol 36:1412-21, 2010
[52] Prabhu NT, Munshi AK: Measurement of masseter and temporalis muscle thickness using ultrasonographic technique. J Clin Pediatr Dent 19:41-4, 1994
[53] Kiliaridis S, Katsaros C, Karlsson S: Effect of masticatory muscle fatigue on cranio-vertical head posture and rest position of the mandible. Eur J Oral Sci 103:127-32, 1995
[54] Pereira LJ, Gaviao MB, Bonjardim LR, et al: Ultrasound and tomographic evaluation of temporomandibular joints in adolescents with and without signs and symptoms of temporomandibular disorders: a pilot study. Dentomaxillofac Radiol 36:402-8, 2007
[55] Farkas L: Anthropometry of the Head and Face (ed second edition). New York Raven Press Ltd, 1994 pp. 405
[56] Raadsheer MC, Kiliaridis S, Van Eijden TM, et al: Masseter muscle thickness in growing individuals and its relation to facial morphology. Arch Oral Biol 41:323-32, 1996
[57] Kiliaridis S, Kalebo P: Masseter muscle thickness measured by ultrasonography and its relation to facial morphology. J Dent Res 70:1262-5, 1991
[58] Hatch JP, Shinkai RS, Sakai S, et al: Determinants of masticatory performance in dentate adults. Arch Oral Biol 46:641-8, 2001
[59] Emshoff R, Bertram S, Brandlmaier I, et al: Ultrasonographic assessment of local cross-sectional dimensions of masseter muscle sites: a reproducible technique? J Oral Rehabil 29:1059-62, 2002
[60] Satiroglu F, Arun T, Isik F: Comparative data on facial morphology and muscle thickness using ultrasonography. Eur J Orthod 27:562-7, 2005
[61] Burke PH, Beard FH: Stereophotogrammetry of the face. A preliminary investigation into the accuracy of a simplified system evolved for contour mapping by photography. Am J Orthod 53:769-82, 1967
[62] Da Silveira AC, Daw JL, Jr., Kusnoto B, et al: Craniofacial applications of three-dimensional laser surface scanning. J Craniofac Surg 14:449-56, 2003
[63] Ireland AJ, McNamara C, Clover MJ, et al: 3D surface imaging in dentistry - what we are looking at. Br Dent J 205:387-92, 2008
[64] Kopp S, Kuhmstedt P, Notni G, et al: G-scan--mobile multiview 3-D measuring system for the analysis of the face. Int J Comput Dent 6:321-31, 2003
[65] Mah J: 3D imaging in private practice. Am J Orthod Dentofacial Orthop 121:14A, 2002
[66] Sachdeva RC: SureSmile technology in a patient--centered orthodontic practice. J Clin Orthod 35:245-53, 2001
[67] Arridge S, Moss JP, Linney AD, et al: Three dimensional digitization of the face and skull. J Maxillofac Surg 13:136-43, 1985
[68] Moss JP, Linney AD, Grindrod SR, et al: Three-dimensional visualization of the face and skull using computerized tomography and laser scanning techniques. Eur J Orthod 9:247-53, 1987
[69] Aung M, Sobel DF, Gallen CC, et al: Potential contribution of bilateral magnetic source imaging to the evaluation of epilepsy surgery candidates. Neurosurgery 37:1113-20; discussion 1120-1, 1995

[70] Moss JP: The use of three-dimensional imaging in orthodontics. Eur J Orthod 28:416-25, 2006

[71] Quintero JC, Trosien A, Hatcher D, et al: Craniofacial imaging in orthodontics: historical perspective, current status, and future developments. Angle Orthod 69:491-506, 1999

[72] Lee JY, Han Q, Trotman CA: Three-dimensional facial imaging: accuracy and considerations for clinical applications in orthodontics. Angle Orthod 74:587-93, 2004

[73] Nguyen CX, Nissanov J, Ozturk C, et al: Three-dimensional imaging of the craniofacial complex. Clin Orthod Res 3:46-50, 2000

[74] Vallance S: *(personal communications)* Structured Light technique with Mona Lisa Imaging System. Adelaide, 2009

[75] Naser-ud-Din S, Thoirs K, Sampson WJ: Ultrasonography, lateral cephalometry and 3D imaging of the human masseter muscle. Orthod Craniofac Res 14:33-43, 2011

[76] Nute SJ, Moss JP: Three-dimensional facial growth studied by optical surface scanning. J Orthod 27:31-8, 2000

[77] Darwis WE, Messer LB, Thomas CD: Assessing growth and development of the facial profile. Pediatr Dent 25:103-8, 2003

[78] Bourne CO, Kerr WJ, Ayoub AF: Development of a three-dimensional imaging system for analysis of facial change. Clin Orthod Res 4:105-111, 2001

[79] Clark W: Twin Block Functional Therapy Applications in Dentofacial Orthopaedics (ed Second). London, Mosby, 2002

[80] Sharma AA, Lee RT: Prospective clinical trial comparing the effects of conventional Twin-block and mini-block appliances: Part 2. Soft tissue changes. Am J Orthod Dentofacial Orthop 127:473-82, 2005

[81] Ismail SF, Moss JP, Hennessy R: Three-dimensional assessment of the effects of extraction and nonextraction orthodontic treatment on the face. Am J Orthod Dentofacial Orthop 121:244-56, 2002

[82] Ayoub AF, Xiao Y, Khambay B, et al: Towards building a photo-realistic virtual human face for craniomaxillofacial diagnosis and treatment planning. Int J Oral Maxillofac Surg 36:423-8, 2007

[83] Baik HS, Lee HJ, Lee KJ: A proposal for soft tissue landmarks for craniofacial analysis using 3-dimensional laser scan imaging. World J Orthod 7:7-14, 2006

[84] Nechala P, Mahoney J, Farkas LG: Maxillozygional anthropometric landmark: a new morphometric orientation point in the upper face. Ann Plast Surg 41:402-9, 1998

[85] Smith R, Zaitoun H, Coxon T, et al: Defining new dental phenotypes using 3-D image analysis to enhance discrimination and insights into biological processes. Arch Oral Biol, 2008

[86] Brook AH, Pitts NB, Yau F, et al: An image analysis system for the determination of tooth dimensions from study casts: comparison with manual measurements of mesio-distal diameter. J Dent Res 65:428-31, 1986

[87] Kneafsey LC, Cunningham SJ, Petrie A, et al: Prediction of soft-tissue changes after mandibular advancement surgery with an equation developed with multivariable regression. Am J Orthod Dentofacial Orthop 134:657-64, 2008

[88] Kusnoto B, Evans CA: Reliability of a 3D surface laser scanner for orthodontic applications. Am J Orthod Dentofacial Orthop 122:342-8, 2002

[89] Papadopoulos MA, Christou PK, Athanasiou AE, et al: Three-dimensional craniofacial reconstruction imaging. Oral Surg Oral Med Oral Pathol Oral Radiol Endod 93:382-93, 2002

[90] Fuhrmann RA, Schnappauf A, Diedrich PR: Three-dimensional imaging of craniomaxillofacial structures with a standard personal computer. Dentomaxillofac Radiol 24:260-3, 1995

[91] Cevidanes LH, Franco AA, Gerig G, et al: Assessment of mandibular growth and response to orthopedic treatment with 3-dimensional magnetic resonance images. Am J Orthod Dentofacial Orthop 128:16-26, 2005

[92] Budai M, Farkas LG, Tompson B, et al: Relation between anthropometric and cephalometric measurements and proportions of the face of healthy young white adult men and women. J Craniofac Surg 14:154-61; discussion 162-3, 2003

[93] Khalaf K, Robinson DL, Elcock C, et al: Tooth size in patients with supernumerary teeth and a control group measured by image analysis system. Arch Oral Biol 50:243-8, 2005

[94] Benington PC, Gardener JE, Hunt NP: Masseter muscle volume measured using ultrasonography and its relationship with facial morphology. Eur J Orthod 21:659-70, 1999

[95] Raadsheer MC, van Eijden TM, van Ginkel FC, et al: Contribution of jaw muscle size and craniofacial morphology to human bite force magnitude. J Dent Res 78:31-42, 1999

[96] Hartmann J, Meyer-Marcotty P, Benz M, et al: Reliability of a Method for Computing Facial Symmetry Plane and Degree of Asymmetry Based on 3D-data. J Orofac Orthop 68:477-90, 2007

[97] Baumrind S: Taking stock: a critical perspective on contemporary orthodontics. Orthod Craniofac Res 7:150-6, 2004

[98] Togashi K, Kitaura H, Yonetsu K, et al: Three-dimensional cephalometry using helical computer tomography: measurement error caused by head inclination. Angle Orthod 72:513-20, 2002

[99] Martensson B, Ryden H: The holodent system, a new technique for measurement and storage of dental casts. Am J Orthod Dentofacial Orthop 102:113-9, 1992

[100] Terajima M, Endo M, Aoki Y, et al: Four-dimensional analysis of stomatognathic function. Am J Orthod Dentofacial Orthop 134:276-87, 2008

[101] Takashima M, Kitai N, Murakami S, et al: Volume and shape of masticatory muscles in patients with hemifacial microsomia. Cleft Palate Craniofac J 40:6-12, 2003

[102] Low E, Zoellner H, Kharbanda OP, et al: Expression of mRNA for osteoprotegerin and receptor activator of nuclear factor kappa beta ligand (RANKL) during root resorption induced by the application of heavy orthodontic forces on rat molars. Am J Orthod Dentofacial Orthop 128:497-503, 2005

[103] Xue F, Wong R, Rabie AB: Identification of SNP markers on 1p36 and association analysis of EPB41 with mandibular prognathism in a Chinese population. Arch Oral Biol 55:867-72, 2010

[104] Xue F, Wong RW, Rabie AB: Genes, genetics, and Class III malocclusion. Orthod Craniofac Res 13:69-74, 2010

[105] Plooij JM, Maal TJ, Haers P, et al: Digital three-dimensional image fusion processes for planning and evaluating orthodontics and orthognathic surgery. A systematic review. Int J Oral Maxillofac Surg 40:341-52, 2011

[106] Boom HP, van Spronsen PH, van Ginkel FC, et al: A comparison of human jaw muscle cross-sectional area and volume in long- and short-face subjects, using MRI. Arch Oral Biol 53:273-81, 2008

[107] Toll DE, Popovic N, Drinkuth N: The use of MRI diagnostics in orthognathic surgery: prevalence of TMJ pathologies in Angle Class I, II, III patients. J Orofac Orthop 71:68-80, 2010

[108] Antonio GE, Griffith JF, Yeung DK: Small-field-of-view MRI of the knee and ankle. AJR Am J Roentgenol 183:24-8, 2004

[109] Johal A, Sheriteh Z, Battagel J, et al: The use of videofluoroscopy in the assessment of the pharyngeal airway in obstructive sleep apnoea. Eur J Orthod 33:212-9, 2011

[110] Naini FB, Moss JP: Three-dimensional assessment of the relative contribution of genetics and environment to various facial parameters with the twin method. Am J Orthod Dentofacial Orthop 126:655-65, 2004

[111] Emshoff R, Brandlmaier I, Bodner G, et al: Condylar erosion and disc displacement: detection with high-resolution ultrasonography. J Oral Maxillofac Surg 61:877-81, 2003

[112] Landes CA, Sterz M: Evaluation of condylar translation by sonography versus axiography in orthognathic surgery patients. J Oral Maxillofac Surg 61:1410-7, 2003

[113] Kiliaridis S, Mahboubi PH, Raadsheer MC, et al: Ultrasonographic thickness of the masseter muscle in growing individuals with unilateral crossbite. Angle Orthod 77:607-11, 2007

Part 5

Detection and Measurement

26

Nanoscale Methods to Enhance the Detection of Ionizing Radiation

Mark D. Hammig
University of Michigan, USA

1. Introduction

The dominant modern methods by which ionizing radiation is sensed are largely based on materials that were developed decades ago, in the form of single crystal scintillators (such as sodium iodide (NaI(Tl))) or semiconductors (silicon (Si) or high purity germanium (HPGe) principally), and gas-filled counters (e.g. ^{3}He for neutron detection). These legacy materials have survived and flourished because they have delivered adequate performance for many medical imaging, military, and plant-monitoring applications, and there was no low-cost replacement materials that delivered equivalent or superior performance. Thus, most of the research effort throughout the latter half of the 20th century was focused on the implementation of single-crystal solid and gas-filled detectors into various radiation-detection niches.

The elevation of the concern over the threat presented by secreted radiological and nuclear-weapons has, however, driven a need for superior detection media. If one wishes to have the ubiquitous deployment of radiation sensors, then existing materials are unsuitable because they have either poor efficiency (Si), demanding logistical burdens (liquid nitrogen associated with HPGe), or unduly high cost (cadmium zinc telluride (CZT)). For simple *counting* applications, emerging solutions exist in the form of boron-coated straw based neutron detectors (Lacy, et al., 2010) and liquid scintillators; however, if high resolution imaging of the radiation field is to be accomplished, then one prefers a material that makes a highly accurate conversion of the particle's energy into the information carriers in the detection media.

The dominant sensing technologies of *ionizing* radiation depend on its namesake; that is, they sense the non-equilibrium charge states induced in the interaction media. Thus, either charges, in the form of electron-ion or electron-hole pairs, can be sensed by their effects on the electric field created by the surrounding device architecture, or the light that accompanies the radiative-recombination of that charge is monitored using scintillation photon detectors. Unfortunately, in the process of creating that charge, information is lost to phonon-creation which doesn't participate in signal formation.

One of the main motivations that drives the development of nanostructured materials, whether nanoscintillators or nanosemiconductors, is that the phonon-assisted loss-processes can be suppressed to a larger degree than is possible in single-crystal materials, such that more of the incident information is converted into the information carriers (charge, light) that participate in signal formation.

As we will discuss, realizing those properties requires careful control over: (1) the structure, size, and uniformity of the nanoparticles themselves, (2) effective coupling of the nanoparticles into a colloidal solid through which the information can flow unimpeded, and (3) suitable device design such that the information can be coupled into a readout circuit. Given the challenging tasks associated with deploying nanostructured media, one might ask whether one can eschew the charge-conversion modality altogether.

If the loss of information to heat is a challenge that requires careful control over the energy-band structure of the solid, why doesn't one just measure the heat directly with a sensitive thermometer? This, in-fact, is the approach that delivers the best energy-resolution currently achievable, in the form of microcalorimeters. In fact, the resolution is several orders-of-magnitude better than that produced by HPGe or the best nanocrystalline detectors. However, the temperature variations that accompany radiation impact are sufficiently slight that small detection-volumes bounded to superconducting readout circuits are required, whether one uses transition edge sensors (Ullom et al., 2007) or magnetic microcalorimetry (Boyd et al., 2009). Useful interaction rates can be achieved by multiplexed arrays of the underlying sensor, but the microcalorimetry technologies magnify the cost and logistical concerns associated with using HPGe detectors.

The second motivation for employing nanoparticle approaches is thus that high energy-resolution performance can be achieved with a lower-cost, larger-area alternative to detectors based on single-crystal materials. Although one can deploy high-vacuum, precisely controlled chamber-based equipment (e.g. molecular beam epitaxy) to create and grow nanoparticles, the most common approach is one of wet-chemistry, in which atmospheric pressure or low-vac processes are used to create the colloidal dispersions. Furthermore, the resulting solutions can then be deposited into solids of various form by the utilization of self-assembly during the solvent drying phase. Thus, the initial capital cost is far less than that required in a fabrication methodology based on microelectronic processing equipment, such as single-crystal growth furnaces. For instance, the current cost of deploying a cubic centimeter of the leading room-temperature single crystal semiconductor, CZT, is roughly $1,000, while we estimate that an equivalent volume of a nanostructured lead-selenide detector can be deployed for less than $10, based on labor and materials costs. In Section 2, we discuss some of those processing steps required to make the nanocrystalline detectors, and we show that they can produce excellent energy response utilizing large detection volumes.

As will be shown, experimental results reveal that structuring *semiconductors* in a solid nanocrystalline composite can suppress the heat-loss mechanism relative to charge-creation processes, resulting in better estimates of the initial energy of the radiation than those produced by single-crystalline materials. In principle, the uncertainty in the energy-measurement process can be reduced the size of the band-gap, which serves as the fixed increment that governs the number of charge-carriers created. However, we currently measure ~1 keV energy uncertainty rather than typical band-gap values of 1 eV, and substantial improvement must therefore be achieved to turn the principle into practice. In fact, the precision of the nanosemiconductor is such that much of the uncertainty is governed by the electronic noise of the readout circuit, but if that is quenched, then one expects the measurement uncertainty to be governed by the non-uniformities in the underlying material.

One might ask: can one do better even than the uncertainty produced by the *electronic* band-gap? Microcalorimetry provides one pathway because information-carrier creation is governed by the phonon energy structure. One can also utilize sensors which gauge the momentum of the incident particle, and therefore avoid the energy-loss pathways inherent to energy-conversion devices, as summarized in (Hammig et al., 2005). In fact, *mechanical* radiation detectors are close cousins to microcalorimeters because small detection volumes must be cooled to quench competing thermal noise sources, and arrays of sensors must be deployed to realize large detection efficiencies. However, holigraphic methods can be brought to bear to read-out an array of vibrating elements in parallel.

When capturing the incident radiation and transforming its physical parameters into a measurable quantity, modern sensing technologies of *neutral* particles depend on an abrupt modality, in which the impinging photon or neutron is converted into a secondary charge-carrier in a point interaction, which subsequently generates heat, charge, or momentum during its slowing in the detection medium. One might prefer to avoid the point conversion altogether and take advantage of the wave-mechanics of the incident particle. In Section 3, we will discuss some of the proposed methods by which radiation can be guided and bent such that its inherent information is not processed through inefficient information-carrier creation processes.

2. Nanosemiconductors

2.1 Why not nanoscintillators?

For nanostructured media, the dominant modality currently being investigated by the radiation-detection community consists of composite arrays of nanoparticles that scintillate upon their excitation. The physics motivation is that in comparison to single crystalline materials, nanostructured scintillators exhibit: (1) enhanced light emission due to the suppression of non-radiative loss processes, and (2) more rapid emission kinematics due to a higher degree of intradot charge coupling, both of which yield higher imaging performance. Nanoscintillators consisting of lead-iodide (Withers et al., 2009), rare-earth-doped fluorides, such as CaF_2(Eu) and LaF_2(Eu) (Jacobsohn et al., 2011), and rare-earth oxides, such as $LuBO_3$ (Klassen et al., 2008) among many other materials are currently being studied, many exhibiting accelerated decay responses and high light conversion efficiency on a nanoparticle basis.

The main challenge presented by the composite media is the optical self-absorption in the nanoparticles, the matrix, and the innumerable interfaces as the photons meander their way to the readout surface. To date, no nanocrystalline scintillator detector has exhibited comparable characteristics to the best single-crystal media, whether composed of NaI(Tl) or brighter scintillators such as $LaBr_3$(Ce). More to the point, one doesn't anticipate that an optimized light-emission media will ever produce a greater number of information carriers than a semiconductor equivalent because of the losses associated with converting the charge into scintillation light- and back again, at the photocathode or photodiode readout.

Thus, our initial investigations have been into nanosemiconductor composites, in which we have determined if one can overcome their expected limitations; namely, (1) charge loss during the slowing-down of the charged-particle in the matrix, (2) charge trapping during subsequent electron and hole transport, and (3) small detector volumes.

Regarding the latter point, small thin detectors typically accompany nanosemiconducting devices because of the inevitable colloidal defects that accumulate as a greater number of layers are cast onto the sample. Furthermore, biasing the device such that a high electric field is realized throughout the volume can, *a priori*, be challenging because of either poor

Schottky junctions or low-resistivity substrates. One might reasonably anticipate that depleting several centimetres of charge might be impossible without high bias voltages. Fortunately, these reasonable expectations are not met and sizable detector volumes can be realized over which rapid, clean charge transport can be achieved, as we will discuss next.

2.2 Cadmium telluride NC detector responses

Lead chalcogenide (PbS, PbSe, and PbTe) quantum dots have favorable properties for their use in NC applications. In contrast to a colloidal cadmium salt, such as CdTe, the large bulk Bohr radii of its excitons (e.g. 46 nm for PbSe) enables strong quantum confinement in relatively large NC structures. Since the conductivity improves sharply as the degree to which the particles are both mono-disperse and close-packed (*cf.* Remacle et al., 2002), larger particles result in better charge transport because the growth techniques are approaching the precision corresponding to the addition of an atomic layer. Another relative advantage of lead chalcogenides compared to CdTe is the nearly identical and small effective masses of the electrons and holes that can be attributed to the symmetric structure of the conduction and valence bands, which gives rise to relatively simple and broadly separated electronic energy levels that in turn, leads to much slower intraband relaxation (Schaller et al., 2005). Moreover, previous studies on lead chalcogenide NCs have shown that these materials have unique vibration states, weak electron-phonon coupling, and negligible exchange and Coulomb energies; thus, they have temperature-independent energy band-gap (e.g. Du et al., 2002; Olkhovets et al., 1998).

Our main focus therefore lies with the lead salts. However, the cadmium chalcogenides have, at present, more mature fabrication recipes and more importantly, their bulk band-gap is higher- 1.44 eV for CdTe compared to 0.26 eV for PbSe. This latter point is important because for low (< ~1 eV) band-gap semiconductors the fluctuations in the thermal noise compete vigorously with the small energy depositions that accompany radiation interaction, and they can swamp the measurement so that x-rays and low-energy particles are unresolvable.

2.2.1 Detector fabrication

CdTe is a II-VI semiconductor material that has been extensively investigated as a single-crystal semiconductor radiation detector. NCs of cadimum chalcogenides are also widely studied for the detection of optical photons. In fact, the fabrication methods are geared toward thin films consistent with the stopping of non-ionizing poorly-penetrating radiation. Although less favorable than the lead chalcogenides for the intended application, the synthesis processes are more fully characterized, and we have therefore investigated both systems.

One method to produce monodisperse NCs involves the mixing of all the reaction precursors in a vessel at a low temperature and then moderately heating the solution to grow the particles. An accelerated chemical reaction induced by increasing the temperature of the solution gives rise to supersaturation, which is relieved by the nucleation burst. The subsequent control of the solution temperature will induce further growing of NCs (Kim et al., 2009).

For instance, NCs in the ~3 nm range can be grown in 20 minutes, whereas larger particles can take 24 hours or more, the size judged during growth via the UV-photoluminescence.

For instance, a synthesized NC dispersion may exhibit green-colored photoluminescence of 540 nm average wavelength, which corresponds with a NC particle size of 3 - 5 nm. According to the chemistry of the synthetic process, the resulting dispersion contains NCs with a CdTe crystalline core surrounded by deprotonated –OH and –COOH groups of the thioglycolic acid; thus, the NCs have negative charge (Rogach, 2002).

For electronic testing, assemblies composed of ~4 nm NCs were prepared by two methods: the layer-by-layer (LBL) method and via drop-casting the NC dispersion on aluminum and gold-metalized glass substrates. The LBL method is a well-known method for efficiently depositing NC colloidal dispersions into high quality and stable thin-film layers on the substrate, while preserving the distinctive optoelectrical and magnetic properties of the size-quantized states of the NCs. For the LBL deposition, poly(diallyl dimethyl ammonium chloride) (PDDA), was induced as a polycation used for adsorbing nanoparticles. Thioglycolic-acid (TGA)-stabilized CdTe nanocrystal solution was spread on the PDDA layer and adsorbed. By an alternatively adsorbing procedure, a bilayer consisting of a polymer/nanocrystal composite was developed and the cycle was repeated from 1 to 30 times, to obtain a multilayer film of the desired thickness.

Unfortunately, in order to stop the high-energy charged particles associated with ionizing radiation one desires a thickness of at least 10's of micrometers if not several centimeters, which requires repetitive and lengthy casting procedures; for instance, roughly 80 hours of aqueous layer-by-layer deposition is required to achieve 10's of micrometers thickness. This can be enabled by robotic dipping tools, but for characterization studies, one can utilize sub-micrometer layers to evaluate the material.

The result of spin-casting the NC assembly onto a glass 4" diameter wafer is shown in Fig. 1A. The varying color reflects the non-uniform thickness deposited towards the outside of the wafer. If the uniform, inner-section of the assembly is diced and bounded by metallic electrodes, then 1 x 1 cm^2 detectors can be realized, as shown in Fig. 1B. In order to determine the thickness of the resulting NC assembly, we examine the cross-section of the layer and utilize energy dispersive x-ray spectroscopy to identify the constituent layers, as shown in Fig. 2. Note that the gold, indium, and cadmium signals overlap but a careful examination of the various structures reveals that the layer is only 130 nm thick, achieved after 48 drops of the LBL process was repeated (cf. Fig. 2C).

2.2.2 Spectroscopic ion measurements

A priori, one expects only slight energy depositions within a 130 nm layer, particularly for weakly ionizing electrons. Even if one exposes the assembly to 5.485 MeV alpha particles (from ^{241}Am), the energy deposited from the transmission of the ion through the layer is less than 30 keV. This is enough energy to produce a measurable signal, as shown in Fig. 3, which is very promising because gamma-rays, for instance, can produce energy depositions that are 10 times as big if there is enough volume to fully stop the products of the interaction.

Thus, even though the cadmium salts may not be as favorable as their lead equivalents- at least in their physics properties- they represent a promising material upon which nanostructured ionizing radiation detectors can be developed. Furthermore, detectors fabricated using the LBL method are more reproducible in their behavior than those that are solidified using gross drop- or spin-casting methods.

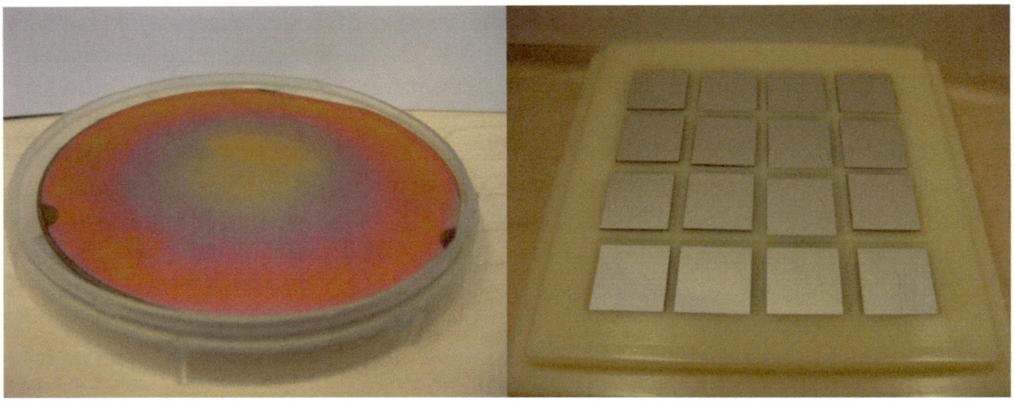

(A) (B)

Fig. 1. (A) Gold-metalized glass wafer upon which a colloidal dispersion of CdTe NCs is deposited via the LBL method utilizing spin-casting. (B) The 1 x 1 cm^2 detectors after indium evaporation and dicing.

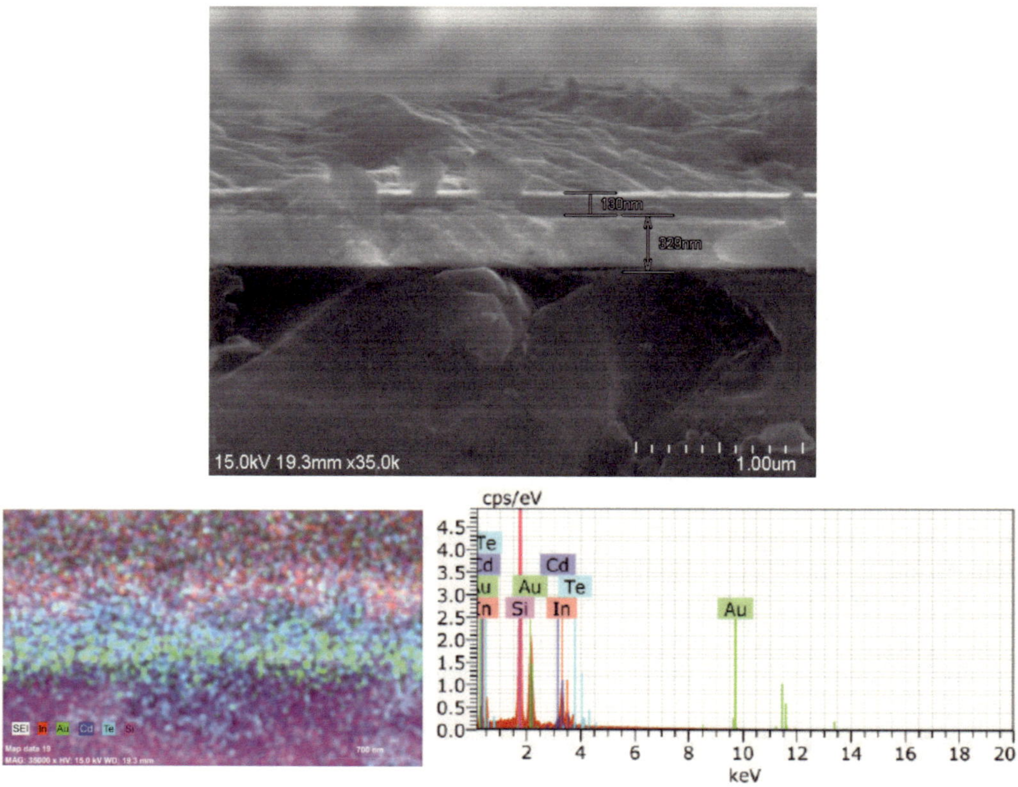

Fig. 2. (A) SEM micrograph of a cross-section of the NC CdTe detectors, showing the 329 nm gold layer and 130 nm CdTe assembly, assignments verified using energy dispersive x-ray spectroscopic mappings of (B) and (C).

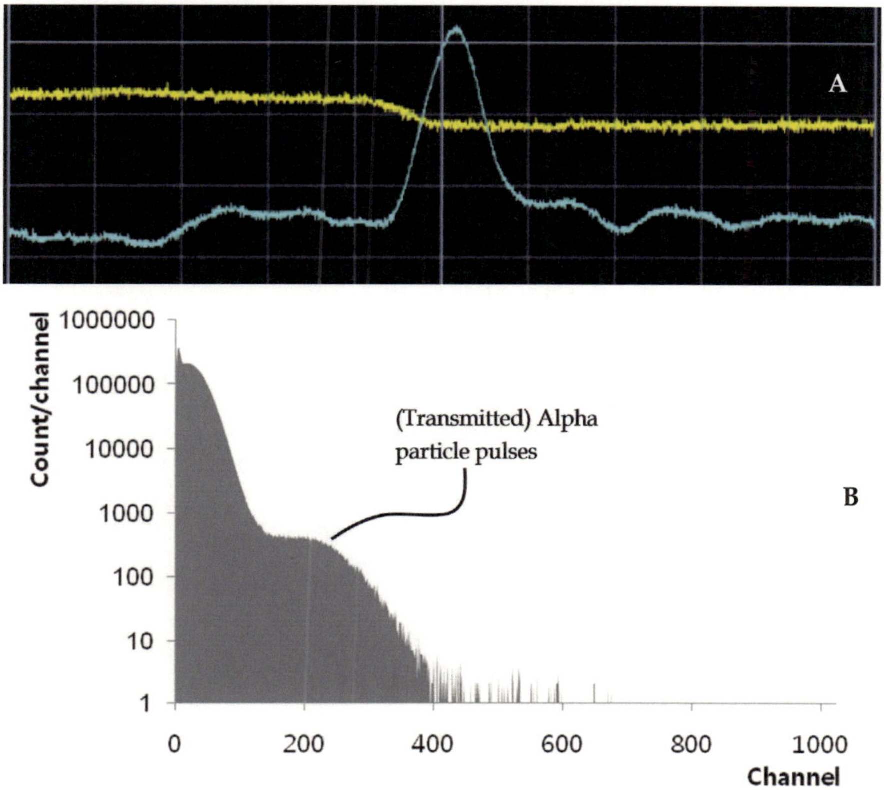

Fig. 3. (A) Preamp (yellow) and shaped (blue) signals derived from the passage of a 5.485 MeV alpha particle through a 130 nm CdTe nanocomposite detector, in which the induced charge signal is measured with a Ortec 142A preamp. The time per division is 10 μs and the volts per division are 200 mV for the preamp and 5 V for the shaped pulse. (B) An example alpha particle spectrum in which the alpha pulses are shown as a hump above the noise.

Nevertheless, instead of using the dozens of hours required to grow detectors with good intrinsic detection efficiency, one can produce exceptional detector behavior using chemistry based on fast-drying organic solvents and through deposition procedures that result in detectors that can be realized in minutes instead of hours. Furthermore, the lead salts, with their higher atomic numbers and densities, have inherntly higher stopping power to either charged or neutral ionizing radiation, as shown next.

2.3 Evaluation of the relative efficiencies of competing materials

Fig. 4A shows that when one compares various common single-crystal detection media and the lead chalcogenides under study, PbSe and PbTe in both single-crystal and nanocrystalline form are superior in terms of their combined mass densities and atomic numbers, Z, both of which combine to yield a higher specific detection efficiency.

The nanocrystalline forms of PbSe and PbTe are consistent with some of our current colloidal solids, in which spherical 10 nm PbSe nanoparticles or spherical 5 nm PbTe dots are spun cast into a face-centered-cubic (FCC) nanocrystalline solid with an interstitial para-MEH-PPV polymer of order 1. In realized detectors, we vary the relative blending of the polymer and NCs such that higher polymer orders are typically employed, and we grow rods and cubes, as well as spheres. Furthermore, the geometry of the solid deviates from a perfect FCC structure to a degree depending on the size dispersion of the colloidal solution; nevertheless, the reduction in densities between the single-crystal lead chalcogenides and the nanocrystalline forms reflects the values that are achieved when the ordering and spacing are optimized for spherical NCs. As shown in the figure, even in nanocrystalline form, the lead chalcogenides maximize the probability of interacting with impinging high-energy photons, and more generally, they exhibit superior characteristics for the efficient stopping of primary or secondary charged particles.

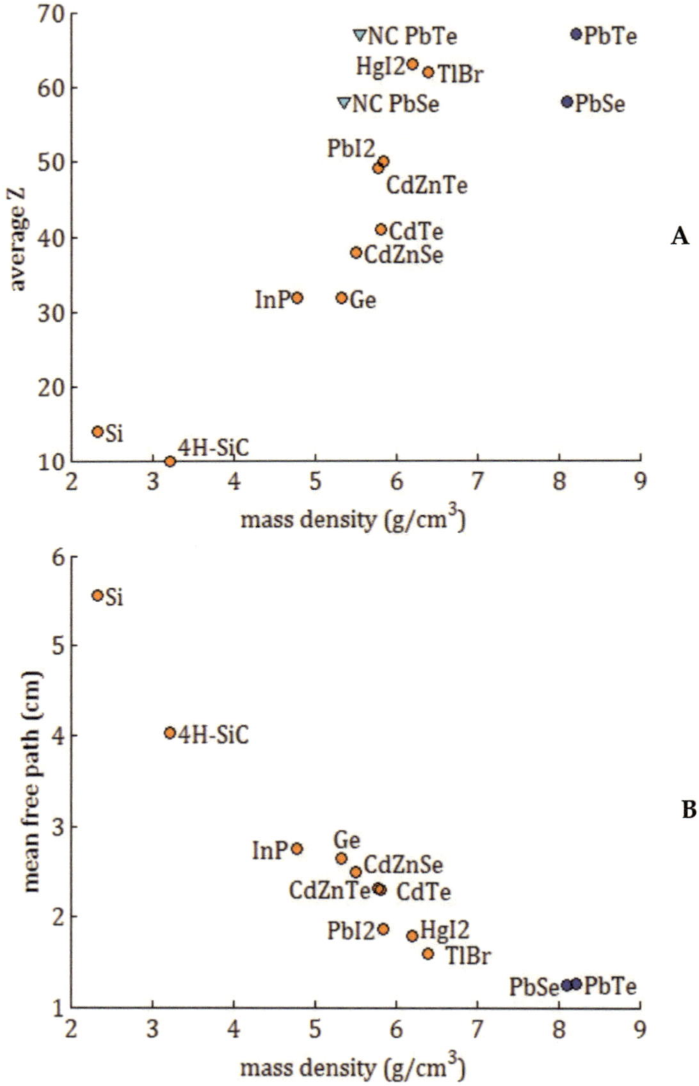

Fig. 4. (A) Relationship between the average atomic number Z and the mass density for various semiconducting materials. (B) The mean free path for a 662 keV photon interacting with single-crystal forms of the semiconducting media (Hammig & Kim, 2011).

As an example, Fig. 4B shows the photonic mean free path, as derived from the density and the macroscopic *photoelectric-absorption* linear attenuation coefficient at a photon energy of 662 keV, which is consistent with the gamma-ray emission from ^{137}Cs. A comparison of PbSe or PbTe with CZT- the most common room-temperature gamma-ray detection medium- shows that the mean free path in CZT is 1.85 times longer than that of either lead salt. On a path-length normalized basis therefore, the lead chalcogenides can form a highly efficient gamma-ray detection medium. Nevertheless, the question is: in a field in which device thicknesses are measured in nanometers to hundreds of nanometers, can the solid be formed into electronic structures with enough active thickness and lateral extent to take advantage of the inherently high interaction probabilities and stop charged particles that may range over hundred of micrometers? As described in Section 2.41, the drop-, dip- and spun-cast solids can be formed into thicknesses which yield highly efficient detector configurations.

One not only desires high probabilities of interaction with the incident quanta, but the multiplicity of the subsequent charge-creation should be maximized and the trapping of the charge-ensemble should be minimized so that the information conversion is optimized. As summarized in Section 2.42, the charge transport can vary across the lateral area as well as in depth because of the varying field or mobilities that accompany different packing regions. Nevertheless, one can bias the detector such that the charge velocities are saturated, resulting in uniform collection across the ensemble.

As shown in Section 2.43, even with a colloidal solid that is not optimized from a charge-separation and transport perspective, the resulting energy-resolving behavior is comparable to or better than state-of-the-art single-crystal semiconductors- Si, CZT, and high-purity germanium operated at liquid nitrogen temperatures- which reflects the promise of the nanostructured approach to semiconductor device design, extending beyond the sensing application reported.

2.4 Lead selenide NC detector responses

2.4.1 Detector fabrication

The chemical recipe used to form the colloidal dispersion, the subsequent characterization of the nanoparticles, and their formation into solid assemblies is described in (Kim & Hammig, 2009). Colloidal PbSe NCs were synthesized through the high-temperature solution-phase routes (Murray et al., 2001), as illustrated in the Fig. 5. Lead oleate and selenide-dissolved trioctylphosphine (TOP-Se) were used as the organo-metallic precursor and diphenyl ether (DPE) as the high-temperature organic solvent. The lead oleate precursor was prepared by dissolving lead acetate, $Pb(CH_3COO)_2$, into a mixture of DPE, TOP and oleic acid, heated up to 85 °C for 1 hour and cooled down. Mixed with 1 M TOP-Se solution, precursors were rapidly injected into rigorously stirred DPE at various reaction conditions (T_{inj}) in an inert environment. PbSe NCs were grown in T_{gr} for 3 - 6 minutes and cooled.

In the fabrication of the NC assembly detector, one forms a mold in an inert substrate which acts as a basin into which the colloidal dispersion is dripped. Furthermore, both the top and bottom metal contacts must be accessible in order to connect each face of the detector to the readout electronics. Although we have used silicon, printed circuit board, glass, and plastic substrates, we typically use glass substrates for thin detectors (10's of micrometers to 100's of micrometers) and employ plastic substrates for millimeter or centimeter thick devices. Fig. 6 shows the typical process if a glass substrate is used, which makes use of standard photolithography processes.

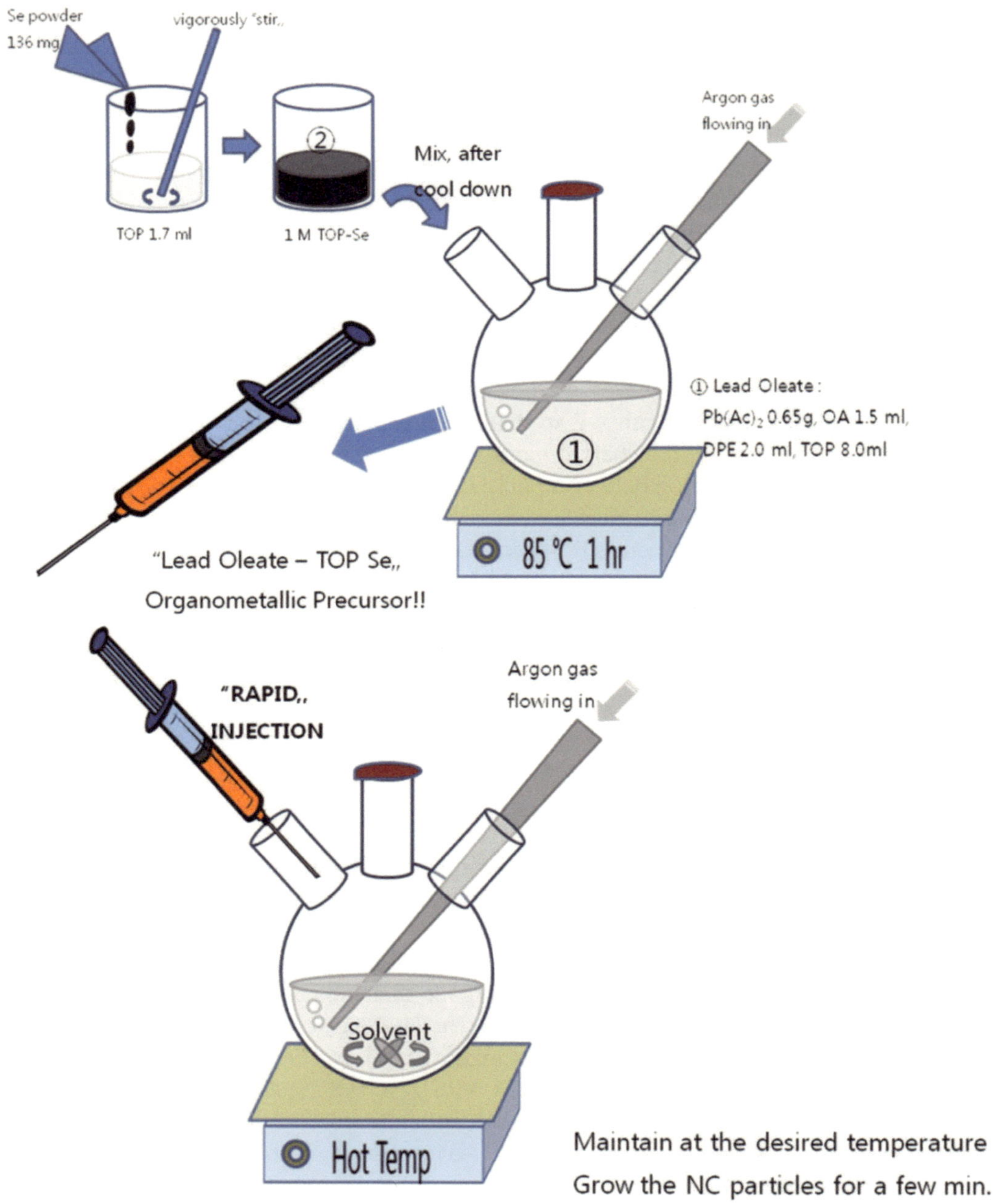

Fig. 5. Schematic of the PbSe NC dispersion synthesis.

Several fabricated substrates are shown in the Fig. 7. The lines stretching horizontally and vertically are due to the dicing of the substrate. Blue colored tape is attached to the backside, before the substrate is diced. The glass wafer-based substrates were designed in a circular shape to make the spin-casting procedure more efficient. Once the metalized molds are realized, the NC assembly can be deposited to form the active region of the sensor.

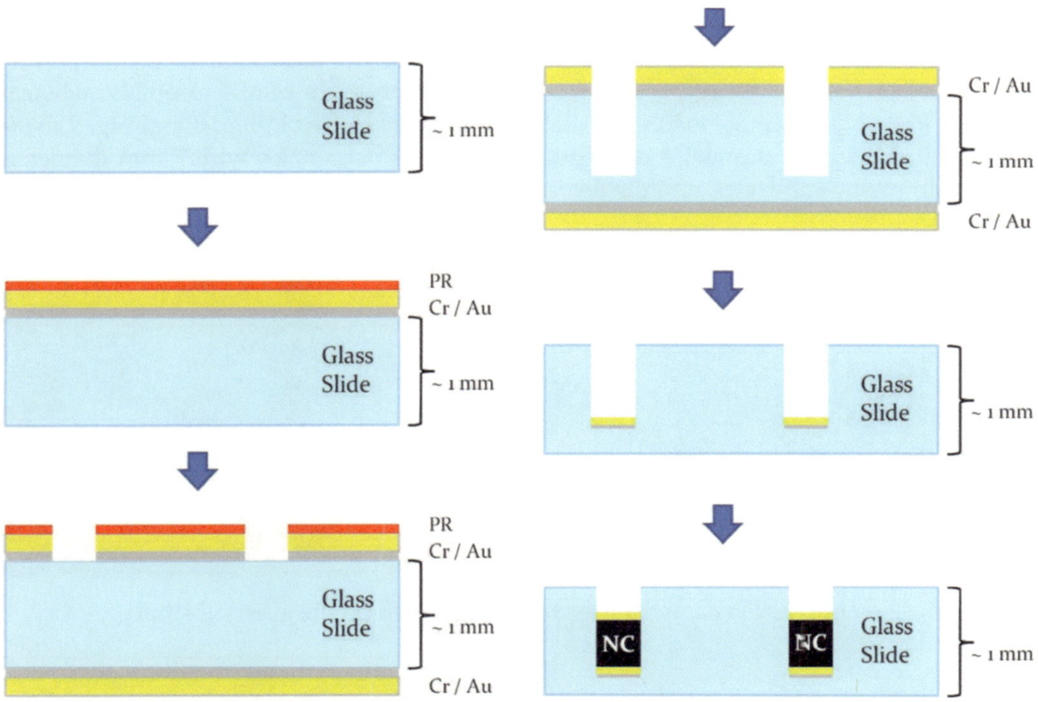

Fig. 6. Schematic of glass wafer-based NC assembly substrate fabrication procedure.

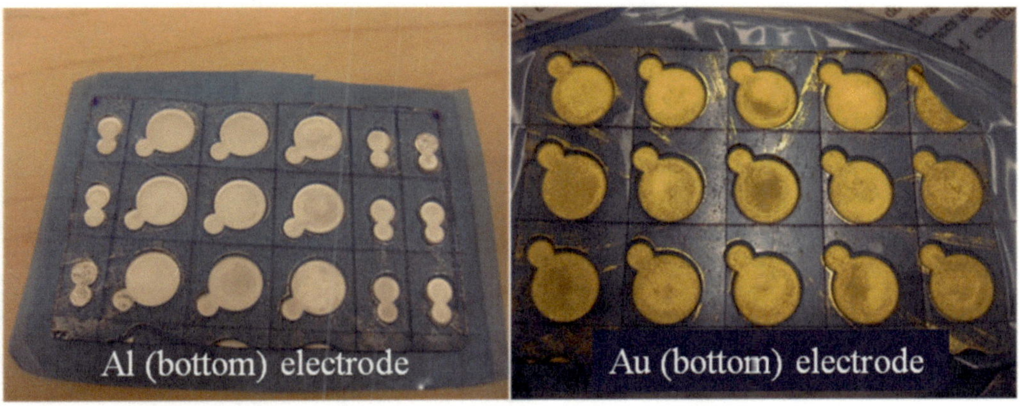

Fig. 7. Glass wafer-based NC assembly substrate with Al and Au bottom contacts.

Synthesized NC particles inherently self-assemble, if coordinated with organic ligands properly. NC particles dispersed in the chloroform solvent can be deposited on the substrate using many methods, but we have focused on drop- and spin-casting. Drop-casting is the process of depositing a droplet of the NC solution on the substrate and repeating the procedure, while allowing time for the solvent to evaporate between applications. Spin-casting is the process of dripping a droplet of the solution on a spinning substrate, which allows more uniform spreading of the NC solution as well as more rapid drying.

In order to realize suitable charge transport, the PbSe NC solution is mixed with a conductive polymer, for instance, para-MEH-PPV, which is used as a hole transporting agent. Greater polymer concentrations formed a brighter, more yellow assembly, whereas the pure PbSe NC assembly exhibits a dark assembly, as shown in Fig. 8 and Fig. 9. Note that Fig. 9 shows the assemblies deposited into ~5 mm deep holes with 7 mm diameters milled into plastic substrates, and the detectors therefore form active depths that are several millimeters thick.

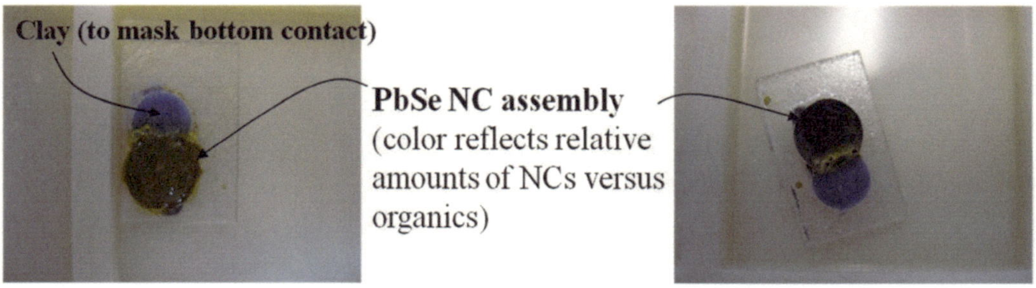

Fig. 8. PbSe NC/MEH-PPH composite assembly deposited on the glass substrate.

Fig. 9. PbSe NC/MEH-PPH composite assembly deposited on plastic substrates, before the top metal contact is formed (left) and after gold evaporation (right).

Various metal contacts can be used to abut the NC assembly, in which the band-structure can vary significantly with the particle size. In general, the MEH-PPV is p-type and we therefore use metal contacts with higher work functions, such as Pt (5.6 eV) and Au (5.2 eV), for the ohmic contact, and we form the rectifying contact with lower work function metals, such as In (3.8 eV) and Al (4.2 eV). Various combinations of samples were made in order to study the effects of the metal contacts, including Pt/NC/Pt, Pt/NC/In, Au/NC/Au, Au/NC/In, etc.. Nevertheless, the behavior of the resulting detector does not depend strongly on the bounding Schottky barrier; rather, the charge creation and transport is more dependent on the uniformity of the colloidal solid underlying the electrodes, some features of which will be described next.

2.4.2 Charge transport characteristics

As mentioned in Section 2.2, a close-packed solid with a high degree of size uniformity results in high conductivity. The size-dispersion in the NC solution depends on procedures used to prepare the liquid dispersion of coordinated-NCs. One can achieve monodisperse (particle size with less than 5 % uncertainty) solutions either directly from the nucleation and growth procedure- with finely controlled reaction conditions- or by employing size-selective precipitation procedures. For instance, Fig. 10 shows the size separation for CdTe NCs that can be elicited using incompatible and compatible solvents along with centrifugation, into a size-stratified solution (Fig. 10A). The separated layers (Fig. 10B) then exhibit different particle size via their size-dependent band-gap in their color or in their excitation spectra. The size can be further refined by repeated application of the procedures if needed.

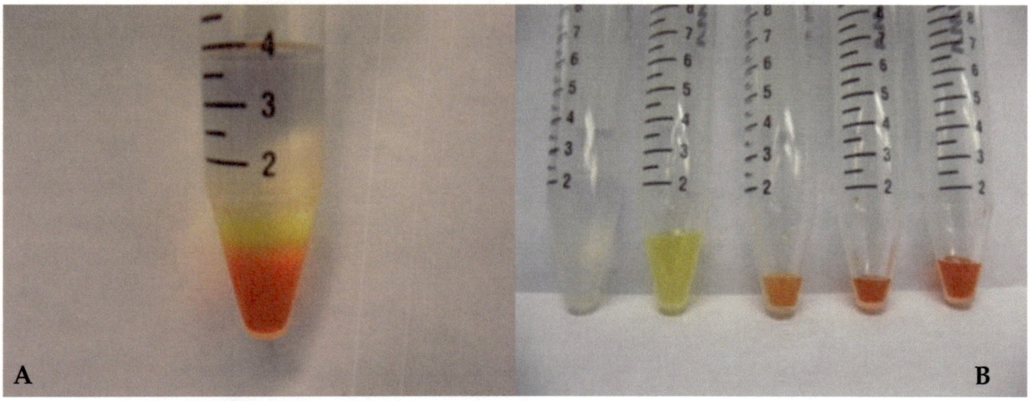

Fig. 10. (A) The separation of a CdTe NC dispersion (with size range in the 3-5 nm range) into different precipitation bands following centrifugation. (B) The separation of the bands into different vials through pipetting.

The importance of monodispersity is not only in the electrical characteristics but in the mechanical ordering properties (Coe-Sullivan et al., 2005). A large (> 10 %) distribution in size results in random packing, but narrow distributions that approach the uncertainty due to the addition of an atomic layer (< 5 %) result in high degrees of order.

Over microscopic areas therefore, monodispersity and the coordination chemistry can be employed to realize close-packed assemblies. However, when one examines the solid over larger volumes, then defects are apparent, ranging from quantum dot (QD) crystal defects, such as vacancies and dislocations, to large-scale cracks, such as those shown in Fig. 11. The large-scale cracking, which we see in all of our samples is a natural consequence of stress-relief during drying of the solvent. The formation of domains of differing degrees of NC order can be reflected in the charge transport characteristics.

For instance, if the bias if held low enough (a value depending on the diode design), then Fig. 12 shows that following the interaction of a single quanta, the holes move in fits and starts, as reflected in the induced charge pulses (in yellow) along with their amplified and shaped counterparts (in blue). This reflects an underlying variation in the drift velocity as the charge cloud traverses different depths of the active volume, with can be caused by either variations in the electric field or the mobility. Although these variations could potentially be problematic if one were to use pulses such as those shown in Fig. 12 to extract the energy deposited by the radiation, one can increase the voltage such that all of the slow-drift regions are eliminated, and the pulse rises sharply and cleanly, such as that shown in Fig. 13. One explanation for this behavior is that the charge-velocities saturate across the various domains in the NC detector and the transport variations are therefore of no consequence.

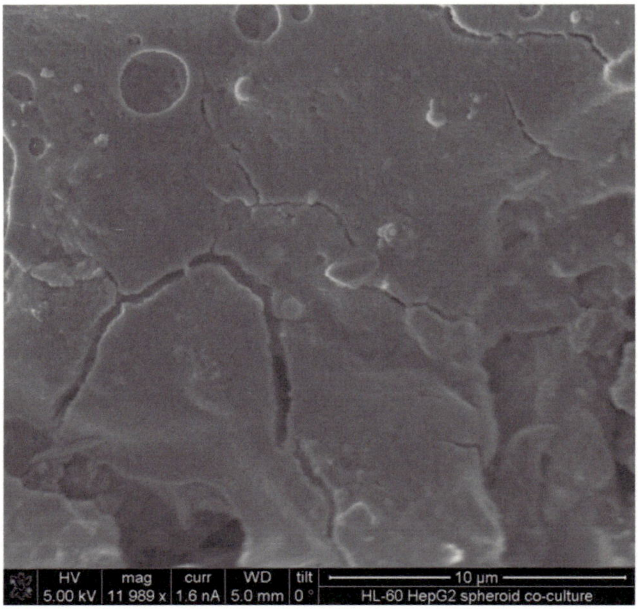

Fig. 11. The cracked and stacked dried mud-puddle look of the NC assembly at a scale of 10's of micrometers.

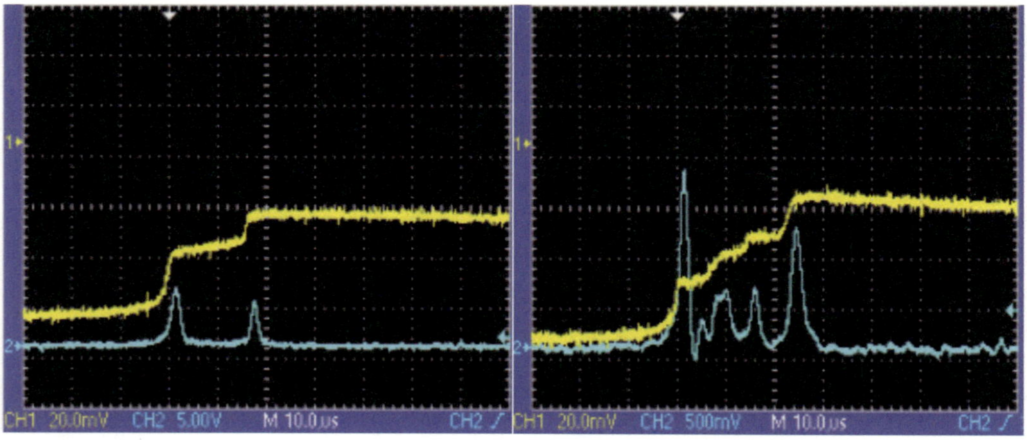

Fig. 12. Two examples of preamp (yellow) and shaped (blue) signals derived following the stopping of an ^{241}Am alpha particle within an under-based colloidal PbSe detector.

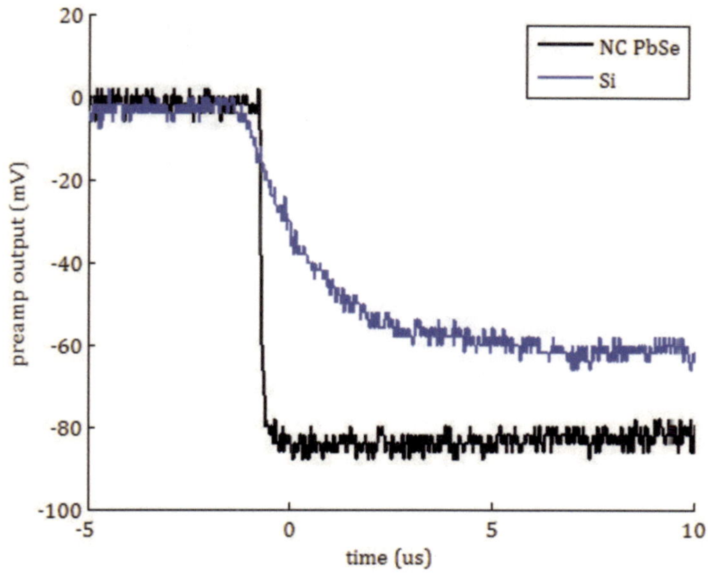

Fig. 13. The pulse shape derived from the charge-sensitive preamplifier (Ortec 142a) when an alpha particle from ^{241}Am is impinged upon a silicon PIN or NC PbSe detector, both operated at 0 V applied bias. The PbSe nanoparticles are spherical with a mean diameter of 7.4 nm (± 0.5 nm).

Thus, although the dispersion in not monodisperse and the solid is not uniform, the promise of the nanocrystalline approach is revealed in the fact that the results are comparable to or better than state-of-the art single crystal detectors, as will be shown next.

2.4.3 Spectroscopic measurements

Fig. 14A corresponds to a mixed spectrum derived from alpha particles- emerging from a *thin-film* ^{241}Am alpha particle source and ^{133}Ba gamma-rays impinging upon the 1 x 1 cm detector shown in Fig. 14A. The alpha source, placed 3.7 cm away from the detector surface, attenuates the 5.49 MeV particles to the energy range shown, and the ^{133}Ba source was placed external to the box housing the detector, 5 cm distant. If one focuses on those spectral features whose appearance correlates to the presence of the ^{133}Ba source, then we can identify the features using both the pulse amplitude and the frequency of occurrence as follows.

Although the material need not be linear, the pulse amplitudes of the full-energy gamma-ray peaks were linearly related to the energies identified in Fig. 14B. Second, the areas under the spectral features correspond to the expected values. For instance, if one takes the gamma-ray emission yield as well as the photoelectric absorption probabilities into account, the area of the 383.85 keV peak should be 12 % of that under the 356.02 keV peak, compared with a measured value of 11 %. The sizes of the x-ray escape features diminish for thicker detectors; however, the thin detector was utilized to characterize the charge-creation characteristics, minimally adulterated by charge-loss considerations, so that the inherent physical properties could be better estimated.

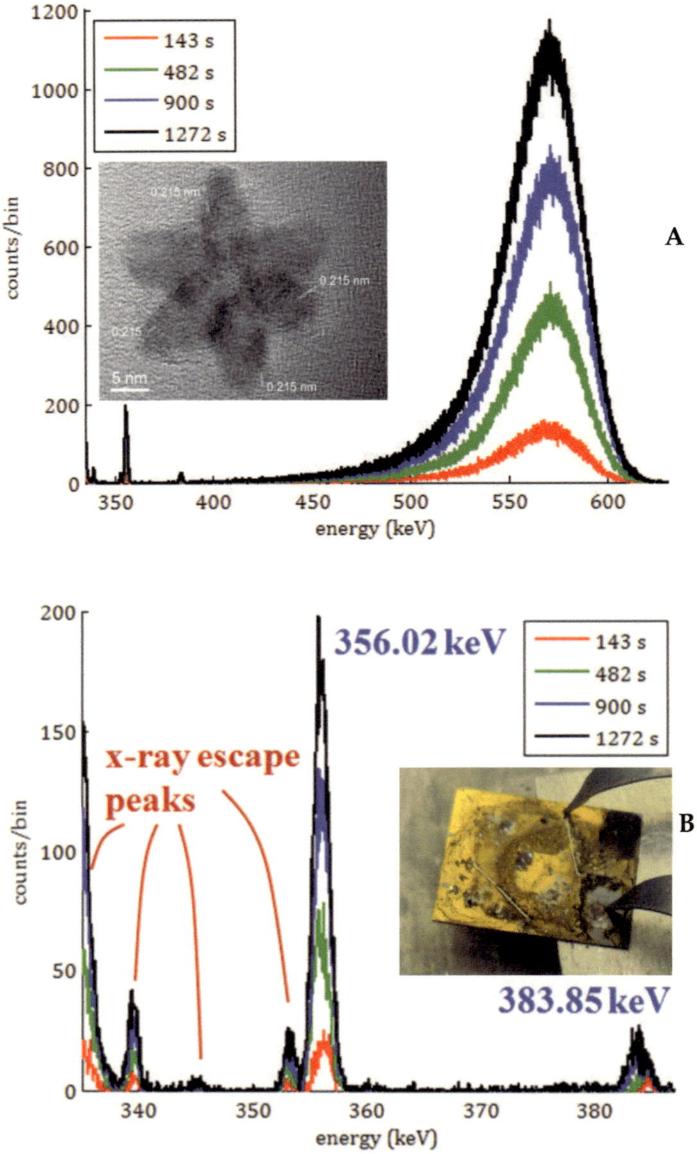

Fig. 14. (A) Energy spectrum derived from ^{133}Ba gamma-rays and ^{241}Am alpha particles, attenuated through 3.7 cm of air, impinging upon a 1 x 1 cm thin (10's of micrometers) composite assembly of para-MEH-PPV and star-shaped PbSe nanopartiles, accumulated for various durations shown in the legend. The inset shows a TEM micrograph of a typical PbSe NC in the assembly. (B) Typical ^{133}Ba spectrum derived from a thin detector, in which the Pb and Se x-ray escape peaks are prominent.

As reported in (Hammig & Kim, 2011), Fig. 15 shows the spectral widths compared with three state-of-art detectors, all operated at room temperature except for the high purity germanium (HPGe) detector, which was cooled with liquid nitrogen.

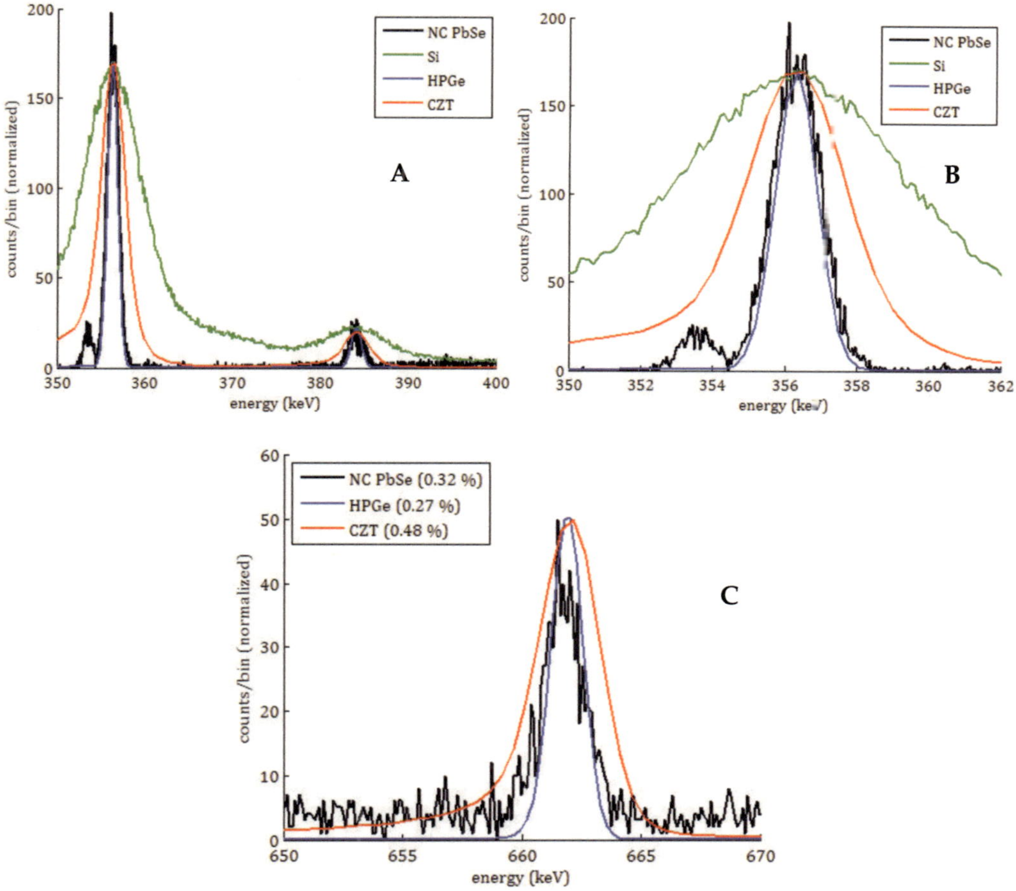

Fig. 15. (A) Spectral comparison between high-resistivity silicon (green), CZT (red), HPGe (blue), and the PbSe nanocomposite (black), when exposed to gamma-rays emitted from ^{133}Ba. The different measurement periods (e.g. 21.2 min for NC PbSe, 24 hr for HPGe) account for the differences in curve smoothness, and the curves were normalized to the number of counts in the PbSe peak. Note that the HPGe and NC PbSe traces largely overlap except for the x-ray escape peak from the NC PbSe detector. (B) Close-up of the 356.02 keV peak. (C) Spectral comparison between CZT (red), HPGe (blue), and the PbSe nanocomposite (black), when impinged upon by gamma-rays emitted from ^{137}Cs. The energy resolutions are shown in the legend. The measurement period for the NC PbSe spectrum was 46.4 min. (Hammig & Kim, 2011).

As shown in Figs. 15A and 15B, the energy resolution of the PbSe nanocomposite is superior to silicon and CZT and comparable to HPGe, yielding an energy resolution of 0.42 % (1.5 keV) at 356 keV, compared with a resolution of 0.96 % (3.4 keV) for CZT and 0.39 % (1.4 keV) for HPGe. Fig. 15C shows an equivalent comparison between the detectors when impinged upon by 661.7 keV gamma-rays from ^{137}Cs. The ^{137}Cs source was more than one hundred times less intense than the ^{133}Ba source and the peak was therefore less well sampled. More importantly, a slight gain drift in the detector dominated the peak width shown in the figure. Nevertheless, the ^{137}Cs results buttress those derived from ^{133}Ba

gamma-rays; namely, the nanocomposite detector provides an energy estimate with similar fidelity to that produced by HPGe. The results imply that the conductive polymers are not only accomplishing the goal of realizing good charge transport through the device, but they are not substantially participating in the charge loss of the recoil electrons.

2.5 Summary

Nanostructured cadmium and lead salts have been extensively studied as advanced media for optoelectronic devices. The sensitivity of their charge-transport properties to the surface conditioning of the nanoparticles and their degradation under ambient conditions has discouraged their use in a semiconductor configuration; rather, most of the efforts geared toward nuclear radiation detection have concentrated on nanostructured *scintillation* materials. Furthermore, the slight thicknesses that accompany the traditional spin-, dip-, and drop-casting methods have argued against their use as sensors of more highly-penetrating radiation.

However, we have shown that detectors with deep depletion regions can be constructed such that the deposited charge can be effectively drifted and the energy of the interacting quanta can therefore be measured. Furthermore, the accuracy of that measurement is comparable to those estimates provided by the best single-crystalline semiconducting materials, providing evidence for enhanced charge multiplication as the degree of strong confinement is increased.

Nevertheless, if one does employ thin detectors, then they are susceptible to the escape of the primary or secondary charged particles induced by the radiation's interaction, particularly for neutral particles that interact with roughly equal probability throughout the interaction depth of thin stopping media. The low energy tailing in the spectral features that are inherent to this energy loss can negatively impact the ability of the sensor to extract the physical characteristics of the impinging quantum, but more importantly, small volume detectors are inefficient for highly penetrating particles.

For neutral species, one might prefer to avoid the abrupt modality altogether, in which the particle is first converted into secondary charge particles which then confer the incident particle's information via electromagnetic interactions. If one can take advantage of the wave mechanics of the impinging quanta, then they can be guided into the detection volume such that the momentum and energy information is fully preserved, the potential of which is discussed in the next section.

3. Extending the bandwidth of wave-guiding media via self-assembled metamaterials

If one can inject the incident particle's energy into mechanical vibrational modes, such as the lever modes discussed in (Hammig, 20005) or the whispering-gallery modes discussed in (Zehnpfennig et al., 2011), then one can eskew the charge-conversion process altogether, as well as the inefficiencies inherent to the charge creation and collection processes. Utilizing the wave-mechanics of the radiation proceeds most simply from long wavelength electromagnetic waves, where microwave and optical light guiding and cloaking have already been demonstrated, to the shorter wavelengths that accompany ionizing radiation. The science of guiding electromagnetic waves is aided by the development of metamaterials, which permit one to exercise control over the permittivity and permeability of the transmission medium. If one can reduce the sizes of the nanoscale features in a metamaterial, then the effective medium approximation can be extended to increasingly small wavelengths such that UV, x-ray, and potentially, gamma-ray wave-guiding can be elicited. Such a development has consequnce for not only sensing applications but it also provides an alternative radiation-shielding modality to the existing materials, and provides the promise of light-weight, high efficiency shielding systems.

There has been no *fundamental* advance in the science of nuclear-radiation shielding since the dawn of the nuclear era. Existing shielding materials depend on *particle*-interaction models, in which one maximizes the number of high cross-section scattering sites that can be placed between the radiation source and the protected asset. However, the advent of nanostructured materials provides means through which novel physics can be realized. In particular, they provide a physical pathway through which the wave mechanics of electromagnetic radiation can be controlled via the spatial mapping of the permittivity and permeability. Such control can be realized over electromagnetic waves ranging from microwaves to optical photons. Extending that capability to shorter wavelengths requires either correspondingly smaller nanostructures, or new modalities for controlling the indices of refraction.

Invisibility cloaks generally operate on two operational principles. In the first, transformational optics- in which optical conformal mapping is used to design and grade a refractive index profile- can be used to render an object invisible by precisely guiding the flow of electromagnetic radiation around the object, in a push-forward mapping cloak or a conformal mapping cloak (H. Chen et al., 2010). The underlying mechanism stems from the invariance of Maxwell's equations to coordinate transformations, the effect of which is to only change constitutive parameters and field values. Thus, the space about a cloaked object can be removed from the space that is transformed by the coordinate transformation.

If one utilizes a metamaterial with a negative index of refraction, then one can also cloak an object by cancelling the wave scattering from the target. Optical negative index metamaterials (NIMs) typically require a plasmonic material, which is comprised of a metal-dielectric composite comprised of nanoscale-sized elements. For instance, Valentine et al. utilized a structure consisting of alternating layers of Ag and MgF_2 in a cascaded fishnet structure, in order to realize a 3D optical metamaterial with very low losses (Valentine et al., 2011). The unit cell dimensions were 860 x 565 x 265 nm and the material successfully exhibited the desired refractive index at optical wavelengths of 1200 – 1700 nm; thus, the nanostructural dimensions were approximately one-half to one third of the wavelength. Note that the regular structure of the fishnet structure, shown in Fig. 16b, was patterned using focused ion-beam (FIB) milling, which is capable of precise nanoscale cutting with high aspect ratios.

Another example is shown in Fig. 16a, in which an optical cloak for the transverse magnetic waveguide mode is comprised of isotropic dielectric materials and delivered broadband and low-loss invisibility in the 1400-1800 wavelength range. The hole diameter shown in Fig. 16a is 110 nm, approximately one-tenth the wavelength, and was again realized with FIB milling.

In fact, most of the cloaking demonstrations to date have used a material-*removal* fabrication modality in which the composite structure is built-up via deposition procedures, and then electron beam lithography or FIB milling is used to realize the structure by etching the underlying composite. Unfortunately, both procedures are slow and costly when implemented over large areas. More importantly, they do not generally allow the precise patterning of materials to the levels required to realize wave optics with x- or gamma-rays.

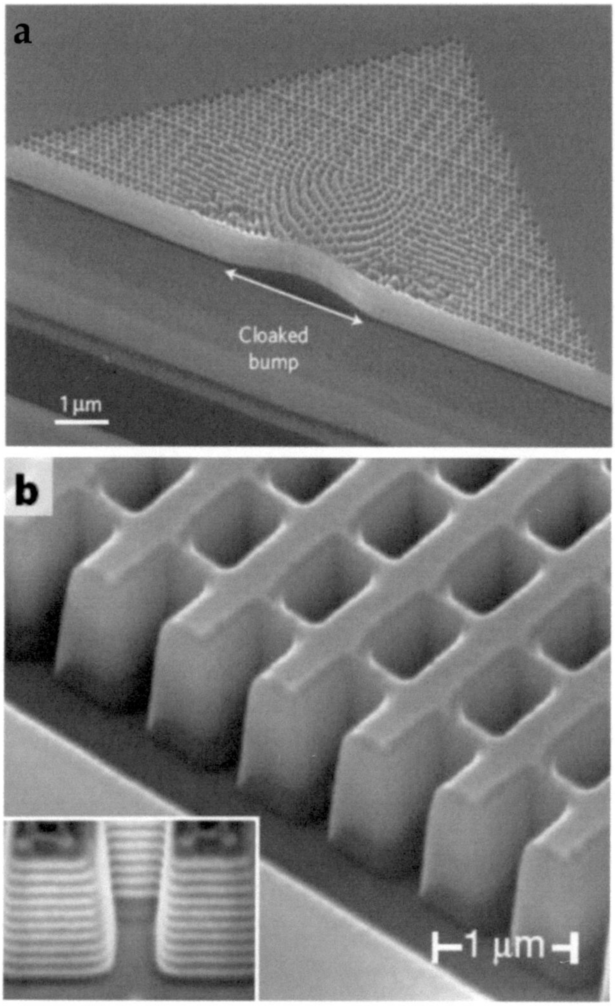

Fig. 16. (a) The carpet cloak design that transforms a mirror with a bump into a virtually flat mirror. Note the gradient index cloak above the bump, the pattern fabricated in a SOI wafer where the Si slab serves as a 2D waveguide. (b) SEM image of fabricated 21-layer fishnet structure, consisting of alternating layers of 30 nm silver and 50 nm magnesium fluoride (Valentine et al., 2011).

That is, the greatest challenge to achieving wave-guiding effects at x- or gamma-ray wavelengths, is that for existing metamaterials, the material loss and fabrication difficulties are substantial. As suggested above, the size of the nanostructure is generally sub-wavelength, so that the index of refraction of the material is governed by the effective medium approximation. The wavelength of a 1 keV x-ray is 1.23 nm; thus, a nanostructured metamaterial must have an ordered nano-crystalline structure in the nanometer range, which is beyond the capabilities of standard e-beam tools, and one must therefore deviate from the material-removal modality.

Fortunately, the desired size-scale can be realized with self-assembly, in which the current lower limit is in this nanometer scale and thus potentially applicable to x-ray cloaking. A combination of sol preparation and electrophoretic deposition has been used to synthesize a variety of oxide nanorod arrays, such as TiO_2, SiO_2, Nb_2O_5, V_2O_5, $BaTiO_3$ Pb(Zr,Ti)O_3, Sn-doped In_2O_3 (ITO), and $Sr_2Nb_2O_7$ (Kao, 2004). We have grown nanowires from CdTe and PbSe semiconductors, upon which metallic shells can be deposited. There has been a myriad of different self-assembled nanostructures, but they do not approach the process uniformity and control that accompanies batch-processing methods (cf. Ko et al., 2011) , and the deviations in the uniformity would diminish any wave-guiding achieved from a designed array of plasmonic structures of similar size. Nevertheless, some wave guiding can occur from an array of proper materials. Thus, using fabrication techniques similar to that that have been successfully employed to make high-performance charge-conversion devices, one can conceivably change the materials in the nanostructured array in order to excercise control over the wave mechanics of high energy photons.

4. Conclusion

Nanostructured media allow us to measure ionizing radiation with greater precision than can be achieved with single-crystal media because one has greater experimental control over the detector's governing parameters- the size and shape of the nanoparticles- than can be equivalently achieved in a nominally single-crystal media, in which defects and impurities can readily spoil the performance of the sensor. More importantly, one can yield fundamentally superior detection characteristics because one has a mechanism through which the informational processes- such as electron creation- can be controlled relative to the loss processes- such as phonon excitation. Fortuitously, this superior physics is also accompanied by a less expensive fabrication modality.

Nanostructured materials have therefore been projected as the next-generation material for the detection of ionizing radiation, the adoption of which depends on overcoming the expected difficulties associated with using a composite material comprised of innumerable interfaces at which information can be lost. As we have shown in this chapter, methods exist that can quench interface trapping so that the promise of the material can be realized today.

5. References

Bahl, G.; et al., Stimulated optomechanical excitation of surface acoustic waves in a microdevice. *Nat. Commun.* 2:403 doi: 10.1038/ncomms1412 (2011).

Chen, H.; C. Chan & P. Sheng. Transformation optics and metamaterials, *Nature Materials* Vol. 9, (2010).

Coe-Sullivan, S.; et al., "Large-Area Ordered Quantum-Dot Monolayers via Phase Separation during Spin Casting", *Adv. Funct. Mater.* 15 (2005).

Du, H.; et al., Optical properties of colloidal PbSe nanocrystals, *Nano Lett.*, vol. 2, no. 11, pp. 1321-1324, (2002).

Hammig, M.D; D. K. Wehe and J. A. Nees, , The measurement of sub-brownian lever deflections. *IEEE Trans. Nuc. Sci.* 52, 3005-3011 (2005).

Hammig, M.D.; G. Kim, Fine Spectroscopy of Ionizing Radiation via a Colloidal Array of Blended PbSe Nanosemiconductors (submitted, 2011).

Jacobsohn, L.G.; et al., Fluoride Nanoscintillators, Journal of Nanomaterials, Vol. 2011 (2011).

Kim, G.; J. Huang and M.D. Hammig, An Investigation of Nanocrystalline Semiconductor Assemblies as a Material Basis for Ionizing-Radiation Detectors. *IEEE Trans. Nuc. Sci.* 56 841-848 (2009).

Kim, G; M.D. Hammig, Development of Lead Chalcogenide Nanocrystalline (NC) Semiconductor Ionizing Radiation Detectors. *2009 IEEE Nuc. Sci. Sym. Conf. Record,* 1317-1320 (2009).

Klassen, N.V.; et al., Advantages and problems of nanocrystalline scintillators, IEEE Trans. on Nuc. Sci., Vol. 55, 1536-1541 (2008).

Lacy, J; et al., Boron-Coated Straw Detectors: a Novel Approach for Helium-3 Neutron Detector Replacement, *2010 IEEE Nuclear Science Symposium and Medical Imaging Conference (2010 NSS/MIC)*, p 3971-5, (2010).

Murray, C.B.; et al., Colloidal synthesis of nanocrystals and nanocrystal superlattices, *IBM J. Res. & Dev.*, vol. 45, no. 1, pp. 47-56, Jan. 2001.

Olkhovets, A.; et al., Size-dependent temperature variation of the energy gap in lead-salt quantum dots, *Phys. Rev. Lett.*, vol. 81, pp. 3539-3542, (1998).

Remacle, F. et al., Conductivity of 2-D Ag Quantum Dot Arrays: Computational Study of the Role of Size and Packing Disorder at Low Temperatures, *J. Phys. Chem.*, vol. 106, pp. 4116-4126, (2002).

Rogach, A.L; et al., II-VI semiconductor nanocrystals in thin films and colloidal crystals, *Colloids Surf. A,* vol. 202, iss. 2-3, pp. 135-144, (2002).

Schaller, R. D., et al., Breaking the phonon bottleneck in semiconductor nanocrystals via multiphonon emission induced by intrinsic nonadiabatic interactions, *Phys. Rev. Lett.*, vol. 95, 196401, (2005).

Ullom, J.N.; et al., Multiplexed microcalorimeter arrays for precision measurements from microwave to gamma-ray wavelengths. *Nucl. Inst. and Meth. A* 579 161-164 (2007).

Withers, N.; et al., Lead-iodide-based nanoscintillators for detectino of ionizing radiation, Proc. Of SPIE Vol. 7304, 73041N (2009).

Zehnpfennig, J.; et al., Surface optomechanics: calculating optically excited acoustical whispering gallery modes in microspheres, *Optics Express*, vol. 19, 14241 (2011).

Glass as Radiation Sensor

Amany A. El-Kheshen
[1]*National Research Centre, Glass Research Department,*
[2]*Chemistry Department, Faculty of Science, Taif University,*
[1]*Egypt*
[2]*KSA*

1. Introduction

1.1 What is glass? [Noel, 1998]

Let's begin with what glass is not—it is not a crystal. The atoms in a crystal are organized in a regular, repetitive lattice so you need only locate a few atoms in order to predict where all of their neighbors are (Fig.1*a*). The atoms are so neatly arranged that, except for occasional crystal defects, you can predict positions for thousands or even millions of atoms in every direction. This spatial regularity is called **long-range order**.

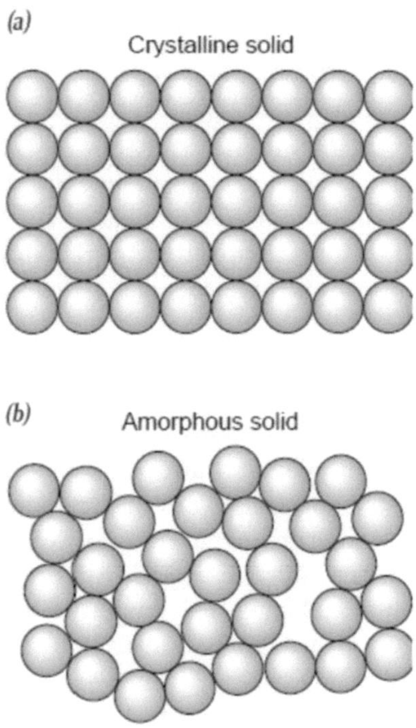

Fig. 1. (a), a crystalline solid with long-range order, (b), an amorphous solid, no long-range order.

Glass is an **amorphous solid**, a material without long-range order (Fig. 1*b*). Locating a few glass atoms tells you nothing about where to find any other atoms. The atoms in glass are arranged in the random manner of a liquid because glass is essentially a super-stiff liquid. Its atoms are jumbled together in a sloppy fashion but they can't move about to form a more orderly arrangement. Glass arrives at this peculiar amorphous state when hot liquid glass is cooled too rapidly. If molten glass were an ordinary liquid, it would begin to solidify abruptly during cooling once it reached its freezing temperature. At that point, its atoms would begin to arrange themselves in crystals that would grow in size until there was no liquid left. That's what's normally meant by freezing. However some liquids are slow to crystallize when you cool them slightly below their freezing temperatures. While they may be cold enough to grow crystals, they must get those crystals started somehow. If crystallization doesn't start, material's atoms and molecules will continue to move about and it will behave as a liquid. When that happens, the liquid is said to be **supercooled**. Supercooling is common in liquids that have difficulties forming the initial **seed crystals** on which the rest of the liquid can crystallize. Because almost all of the atoms in a seed crystal are on its surface, it has a relatively large surface tension and surface energy. Below a certain critical size, a crystal is unstable and tends to fall apart rather than grow. However once the first seed crystals manage to form, a process called nucleation, the rest of the supercooled liquid may crystallize with startling rapidity. Just below its freezing temperature, the atoms in glass don't bind to one another long enough to form complete seed crystals, and nucleation takes almost forever. At somewhat lower temperatures, seed crystals begin to nucleate, but glass's large viscosity (thickness) prevents these crystals from growing quickly. The glass remains a supercooled liquid for an unusually long time. At even lower temperatures, glass becomes so viscous that crystal growth stops altogether. The glass is then a stable supercooled liquid. At this temperature range, glass still pours fairly easily and can be stretched or molded into almost any shape.

However, when you cool the glass still further, it becomes a ***glass***. Here the word *glass* refers to a physical state of the material—a type of amorphous solid. To distinguish this use of the word glass from the common building material, it is italicized. Glass, the material, becomes a *glass*, the state, at the **glass transition temperature** (Tg). Below Tg, the atoms in the glass rarely move past one another; they continue to jiggle about with thermal energy but they don't travel about the material.

1.2 A quick journey with glass [Gerry, M., 2002]

The production of glass has been occurring naturally for millions of years, but the discovery of manufactured glass leads us to live a luxury life.

Natural glass has existed since the beginnings of time, formed when certain types of rocks melt as a result of high-temperature phenomena such as volcanic eruptions, lightning strikes or the impact of meteorites and then cool and solidify rapidly. The earliest man-made glass objects, mainly non-transparent glass beads, are thought to date back to around 3500 B.C., which found in Egypt. Phoenician merchants and sailors spread this new art along the coasts of the Mediterranean Sea.

After 1500 B.C., Egyptian craftsmen are known to have begun developing a method for producing glass pots by dipping a core mould of compacted sand into molten glass and then turning the mould so that molten glass adhered to it. While still soft, the glass-covered mould could then be rolled on a slab of stone in order to smooth or decorate it. The earliest

examples of Egyptian glassware are three vases bearing the name of the Pharaoh Thoutmosis 2I (1504-1450 B.C.).
The first instance of glass being used for windows in buildings, especially in Britain, was during the period of the Romans. In the 1st century A. D., it is known that the Romans used glass for a variety of reasons including mosaic tiles, decorating pottery and as windows. It seems that the glass window became more popular with the advent of churches and other places to worship. Most of the earliest examples of Roman window glass are colored, suggesting that they were glass windows for churches, for example. Around the world, the need for glass windows (and the manufacture of the glass) did not really take hold until the 11th century.

1.3 It only looks like magic (A world of glass)

1.3.1 First types of glass

The first types of glass invented centuries ago, scientists made that glass by mixing materials, melting and then cooling to form a solid. There are many different types of glass with different chemical and physical properties. Each can be made by a suitable adjustment to chemical compositions, but the main types of glass are:

- Commercial Glass also known as soda-lime glass
- Lead Glass
- Borosilicate Glass
- Glass Fiber

Glasses may be devised to meet almost any imaginable requirement. For many specialized applications in chemistry, pharmacy, the electrical andelectronics industries, optics, or the comparatively family of materials known as glass ceramics? Glass is a practical material for the engineer to use.

1.3.1.1 Commercial glass

Most of the glass we see around us in our everyday lives in the form of bottles and jars, flat glass for windows or for drinking glasses is known as commercial glass or soda-lime glass, as soda ash is used in its manufacture.
The main constituent of practically all commercial glass is sand. Sand by itself can be fused to produce glass but the temperature at which this can be achieved is ≥ 1700oC. Adding other minerals and chemicals to sand can considerably reduce the melting temperature.
The addition of sodium carbonate (Na_2CO_3), known as soda ash, to produce a mixture of 75% silica (SiO_2) and 25% of sodium oxide (Na_2O), will reduce the temperature of fusion to about 800oC. However, a glass of this composition is water-soluble and is known as water glass. In order to give the glass its stability, other chemicals like calcium oxide (CaO) and magnesium oxide (MgO) are needed.
Commercial glass is normally colorless, allowing it to freely transmit light, which is what makes glass ideal for windows and many other uses. Additional chemicals have to be added to produce different colors of glass.
Most of commercial glasses have roughly similar chemical compositions of:
70wt% - 74wt% SiO_2 (silica)
12wt% - 16wt% Na_2O (sodium oxide)
5wt% - 11wt% CaO (calcium oxide)
1wt% - 3wt%MgO (magnesium oxide)
1wt% - 3wt% Al_2O_3 (aluminium oxide)

Flat glass is similar in composition to container glass except that it contains a higher proportion of magnesium oxide.
Within these limits the composition is varied to suit a particular product and production method. The raw materials are carefully weighed and thoroughly mixed, as consistency of composition is of utmost importance in making glass.
Nowadays, recycled glass from bottle banks collections, known as cullet, is used to make new glass. Using cullet has many environmental benefits, it prevents pollution by reducing quarrying, and because cullet melts more easily, it saves energy.
Almost any proportion of cullet can be added to the mix (known as batch), but in the right condition. Although the recycled glass may come from manufacturers around the world, it can be used by any glassmaker, as the container glass compositions are very similar. It is however important, that glass colors are not mixed and that the cullet is free from impurities, especially metals and ceramics.

1.3.1.2 Lead glass

Commonly known as lead crystal, lead glass is used to make a wide variety of decorative glass objects.
It is made by using lead oxide instead of calcium oxide, and potassium oxide instead of all or most of the sodium oxide. The traditional English full lead crystal contains at least 30% lead oxide (PbO) but any glass containing at least 24% PbO can be described as lead crystal. Glass containing less than 24% PbO, is known simply as crystal glass. The lead is locked into the chemical structure of the glass so there is no risk to human health.
Lead glass has a high refractive index making it sparkle brightly and a relatively soft surface so that it is easy to decorate by cutting and engraving which highlights the crystal's brilliance making it popular for glasses, decanters and other decorative objects.
Glass with even higher lead oxide contents (typically 65%) may be used as radiation shielding because of the well-known ability of lead to absorb gamma rays and other forms of harmful radiation.

1.3.1.3 Borosilicate glass

Most of us are more familiar with this type of glass in the form of ovenware and other heat-resisting ware, better known under the trade name Pyrex.
Borosilicate glass, the third major group, is made mainly of silica (70-80wt %) and boric oxide (7-13wt %) with smaller amounts of the alkalis (sodium and potassium oxides) and aluminum oxide. This type of glass has relatively low alkali content and consequently has good chemical durability and thermal shock resistance. As a result it is widely used in the chemical industry, for laboratory apparatus, for ampoules and other pharmaceutical containers and as glass fibers for textile and plastic reinforcement.

1.3.1.4 Glass fiber

Glass fiber has many uses from roof insulation to medical equipment and its composition varies depending on its application.
For building insulation, the type of glass used is normally soda lime. For textiles, an alumino-borosilicate glass with very low sodium oxide content is preferred because of its good chemical durability and high softening point. This is also the type of glass fiber used in the reinforced plastics to make protective helmets, boats, piping, car chassis, ropes, car exhausts and many other items.

In recent years, great progress has been made in making optical fibers which can guide light and thus transmit images round corners. These fibers are used in endoscopes for examination of internal human organs, changeable traffic message signs now on motorways for speed restriction warnings and communications technology, without which telephones and the internet would not be possible.

1.3.2 Special glass

Types of special glass change according to human needs and development. Earlier, the special types of glass were:

- Vitreous silica
- Aluminosilicate glass
- Alkali-barium silicate glass
- Technical Glass
- Glass Ceramics
- Optical glass
- Sealing glass

1.3.2.1 Vitreous silica glass

Silica glass or vitreous silica is of considerable technical importance as it has a very low thermal expansion. This glass contains tiny holes and is used for filtration. Porous glasses of this kind are commonly known as Vycor.

1.3.2.2 Alumino-silicate glass

An important type of glass, aluminosilicate, contains 20wt% aluminium oxide (alumina-Al2O3), often including calcium oxide, magnesium oxide and boric oxide in relatively small amounts, with very small amounts of soda or potash. It is able to withstand high temperatures and thermal shock and is typically used in combustion tubes, gauge glasses for high-pressure steam boilers, and in halogen-tungsten lamps capable of operating at temperature as high as 750°C.

1.3.2.3 Alkali-barium silicate glass

Without this type of glass, watching TV would be very dangerous. A television produces X-rays that must be absorbed; otherwise they could in the long run cause health problems. The X-rays are absorbed by glass with minimum amounts of heavy oxides (lead, barium or strontium). Lead glass is commonly used for the funnel and neck of the TV tube, while glass containing barium is used for the screen.

1.3.2.4 Technical glass

Technical is the term given to a range of glasses used in the electronics industry.

Without borate glass, the computer revolution would not have been possible as it's vitally important in producing electrical components. This type of glass contains little or no silica and is used for soldering glass, metals or ceramics as it melts at the relatively low temperature (450-550oC), below that of normal glass, ceramics and many metals.

Glass of a slightly different composition is used for protecting siliconsemi-conductor components against chemical attack and mechanical damage, known as passivation glass.

Chalcogenide glass, similar semi conductor effect, is a type of glass that can be made without the presence of oxygen. Some of them have potential use as infrared transmitting materials and as switching devices in computer memories because their conductivity changes abruptly when particular voltage values are exceeded.

1.3.2.5 Glass ceramic

Some of these "Glass ceramics", formed typically from lithium aluminosilicate glass, are extremely resistant to thermal shock and have found several applications where this property is important, including cooking ware, mirror substrates for astronomical telescopes and missile nose cones.
An essential feature of glass is that it does not contain crystals. However, by deliberately stimulating crystal growth in glass, it is possible to produce a type of glass with a controlled amount of crystallization that can combine many of the best features of ceramics and glass.

1.3.2.6 Optical glass

Optical glasses are found in scientific instruments, microscopes, and fighter aircraft and most commonly in spectacles.
The most important properties are the refractive index and the dispersion. The index is a measure of how much the glass bends light. The dispersion is a measure of the way the glass splits white light into the colors of the rainbow. Glass makers use the variations in these characteristics to develop optical glasses.

1.3.2.7 Sealing glass

A wide variety of glass compositions are used to seal metals for electrical and electronic components. Here the available glasses may be grouped according to their thermal expansion which must be matched with the thermal expansions of the respective metals so that sealing is possible without excessive strain being induced by differing levels of expansion.
For sealing to tungsten, in making incandescent and discharge lamps, borosilicate alkaline earths or aluminous silicate glasses are suitable. Sodium borosilicate glasses may be used for sealing to molybdenum and the iron-nickel-cobalt (Fernico) alloys, are frequently employed as a substitute. The amount of sodium oxide permissible depending on the degree of electrical resistance required. With glasses designed to seal to Kovar alloy, relatively high contents of boric oxide (approximately 20%) are needed to keep the transformation temperature low and usually the preferred alkali is potassium oxide so as to ensure high electrical insulation.
Where the requirement for electrical insulation is paramount, as in many types of vacuum tube and for the encapsulation of diodes, a variety of lead glasses (typical containing between 30% and 60% lead oxide) can be used.

1.3.2.8 Colored glase (Weyl, 1959)

Color is the most obvious property of a glass object. It can also be one of the most interesting and beautiful properties. Although color rarely defines the usefulness of a glass object it almost always defines its desirability.

1.3.2.8.1 The colored glass recipe

The earliest people who worked with glass had no control over its color. Then, through accident and experimentation glass makers learned that adding certain substances to the glass melt would produce spectacular colors in the finished product. Other substances were discovered that, when added to the melt, would remove color from the finished project.
The Egyptians and the Romans both became expert at the production of colored glass. In the eighth century, an Arab chemist known as "Gerber" recorded dozens of formulas for the production of glass in specific colors. Gerber is often known as the "father of chemistry" and he realized that the oxides of metals were the key ingredients for coloring glass.

1.3.2.8.2 The glass color palette

Once the methods of colored glass production was discovered, an explosion of experimentation began. The goal was to find substances that would produce specific colors in the glass. Some of the earliest objects made from glass were small cups, bottles and ornaments.

Religious organizations were among those who provided incentive to the early glass artisans. Stained glass windows became very popular additions to churches and mosques over 1000 years ago. These artists needed a full palette of colors to make a realistic stained glass scene. This search for a full palette fueled research and experimentation to produce a vast array of colors.

1.3.2.8.3 Colors of duration

Then, another problem was discovered. Many of the glass colors did not stand up to year-in, year-out exposure to the direct rays of the sun. The result was a stained glass scene of deteriorating beauty. Some colors darkened or changed over time, while others faded away. Research and experimentation continued in an effort to meet the need for colors of duration. Eventually a full palette of fairly stable colors was achieved.

I.3.2.8.4 Metals used to color glass

The recipe for producing colored glass usually involves the adding of a metal to the glass. This is often accomplished by adding some powdered oxide, sulfide or other compound of that metal to the glass while it is molten. The table below lists some of the coloring agents of glass and the colors that they produce. Manganese dioxide and sodium nitrate are also listed. They are discoloring agents - materials that neutralize the coloring impact of impurities in the glass.

Metals Used to Impart Color to Glass	
Yellow	Cadmium Sulfide
Red	Gold Chloride
Blue-Violet	Cobalt Oxide
Purple	Manganese Dioxide
Violet	Nickel Oxide
Yellow-Amber	Sulfur
Emerald Green	Chromic Oxide
Fluorescent Yellow, Green	Uranium Oxide
Greens and Browns	Iron Oxide
Reds	Selenium Oxide
Amber Brown	Carbon Oxides
White	Antimony Oxides
Blue, Green, Red	Copper Compounds
White	Tin Compounds
Yellow	Lead Compounds
A "decoloring" agent	Manganese Dioxide
A "decoloring" agent	Sodium Nitrate

Table 1.

1.4 What types of radiation are there?

The radiation one typically encounters is one of four types: alpha radiation, beta radiation, gamma radiation, and x radiation. Neutron radiation is also encountered in nuclear power plants and high-altitude flight and emitted from some industrial radioactive sources.

1.4.1 Alpha radiation

Alpha radiation is a heavy, very short-range particle and is actually an ejected helium nucleus. Some characteristics of alpha radiation are:

Most alpha radiation is not able to penetrate human skin.

Alpha-emitting materials can be harmful to humans if the materials are inhaled, swallowed, or absorbed through open wounds.

A variety of instruments has been designed to measure alpha radiation. Special training in the use of these instruments is essential for making accurate measurements.

A thin-window Geiger-Mueller (GM) probe can detect the presence of alpha radiation.

Instruments cannot detect alpha radiation through even a thin layer of water, dust, paper, or other material, because alpha radiation is not penetrating.

Alpha radiation travels only a short distance (a few inches) in air, but is not an external hazard. Alpha radiation is not able to penetrate clothing. Examples of some alpha emitters: radium, radon, uranium, thorium.

1.4.2 Beta radiation

Beta radiation is a light, short-range particle and is actually an ejected electron. Some characteristics of beta radiation are:

Beta radiation may travel several feet in air and is moderately penetrating.

Beta radiation can penetrate human skin to the "germinal layer," where new skin cells are produced. If high levels of beta-emitting contaminants are allowed to remain on the skin for a prolonged period of time, they may cause skin injury.

Beta-emitting contaminants may be harmful if deposited internally.

Most beta emitters can be detected with a survey instrument and a thin-window GM probe (e.g., "pancake" type). Some beta emitters, however, produce very low-energy, poorly penetrating radiation that may be difficult or impossible to detect. Examples of these difficult-to-detect beta emitters are hydrogen-3 (tritium), carbon-14, and sulfur-35.

Clothing provides some protection against beta radiation. Examples of some pure beta emitters: strontium-90, carbon-14, tritium, and sulfur-35.

1.4.3 Gamma and X- radiation

Gamma radiation and x rays are highly penetrating electromagnetic radiation. Some characteristics of these radiations are:

Gamma radiation or x rays are able to travel many feet in air and many inches in human tissue. They readily penetrate most materials and are sometimes called "penetrating" radiation.

X rays are like gamma rays. X rays, too, are penetrating radiation. Sealed radioactive sources and machines that emit gamma radiation and x rays respectively constitute mainly an external hazard to humans.

Gamma radiation and x rays are electromagnetic radiation like visible light, radio waves, and ultraviolet light. These electromagnetic radiations differ only in the amount of energy they have. Gamma rays and x rays are the most energetic of these.

Dense materials are needed for shielding from gamma radiation. Clothing provides little shielding from penetrating radiation, but will prevent contamination of the skin by gamma-emitting radioactive materials.
Gamma radiation is easily detected by survey meters with a sodium iodide detector probe.
Gamma radiation and/or characteristic x rays frequently accompany the emission of alpha and beta radiation during radioactive decay.
Examples of some gamma emitters: iodine-131, cesium-137, cobalt-60, radium-226, and technetium-99m.

2. Effect of radiation on different types of glass

2.1 Silicate glass

Jiawei Shenga (2009) studied the effect of UV-laser irradiation on the soda-lime silicate glass by preparing Commercial soda-lime silicate glass substrates composed of (wt%): 73.2 SiO_2, 15.3 Na_2O, 1.3 Al_2O_3, and 10.2 CaO, they found that As shown in Fig. 1, the as-quenched base glass was colorless and had no measurable absorptions in the visible region. In addition, no ESR signal was observed in the blank samples. After laser irradiation, two characteristic absorption bands in the visible region with maxima at about 620, and 430 nm, respectively, were observed. More precise peak positions were determined to be 627 and 431 nm for these two bands, respectively, with the assistance of Gaussian resolution [**Friebele 1991, Marshall 1997, Zhang 2007**] (Fig. 2,). As a result, the glass showed slight brown. The induced color was unstable, as indicated by the decease of the peak intensity after 50 h room temperature storage following the irradiation (Fig. 2). These results were similar to the case using X-ray radiation [**Sheng, 1970**]. Radiation may cause the displacement of lattice atoms or electron defects that involve changes in the valence state of lattice or impurity atoms. The ionizing radiation produces electron-hole pairs in the glass structure. Accordingly, new optical absorption bands were developed [**Sheng, 1970**]. In general, these absorptions are associated with either oxygen deficiency or oxygen excess in the glass network. The most fundamental radiation-induced defects in glass are the nonbridging oxygen hole center (NBOHC: ^Si–O*), the E' center (^Si*), the peroxy radical (POR: ^Si–O–O*), and the trapped electrons (TE), where the notation "^" represents three bonds with other oxygen in the glass network and "*" denotes an unpaired electron. According to current knowledge, absorption bands of 627 and 431 nm in the soda-lime silicate glass were identified mainly as absorption of NBOHCs [**Bishay, 1970, Wong, 1967, Sheng, 2007**]. The main absorptions of the E0 center and the POR are in the far UV region, which have less effect on the glass visible color. Fig. 3 shows the ESR spectrum of the glass after the irradiation. Two distinctive signals at g ¼ 1.99w2.00 and g ¼ 1.992, respectively, were observed. As expected, the g ¼ 1.99w2.00 was identified as the defects of NBOHCs, correlating bands of 627 and 431 nm. The signal of g ¼ 1.992 might attribute to the defect of E' center. The spectra of the induced absorption with varying exposure time are presented in Fig. 4. The peak heights of the induced absorption increased with the laser irradiation time, and the peak positions absorption band were relatively constant, suggesting that only variations of their populations of defects be effected by the irradiation time. As shown in Fig. 4 (insert), the absorptions at 627 or 431 nm increased rapidly during the first 30 min radiation. The increase was then slow down and no more increase was observed after 60 min radiation. Irradiation energy is another important factor that affects the induced absorption of glass. As can be seen in Fig. 5, the induced absorption increased when the energy density was increased, while the induced absorption peak positions were relatively constant.

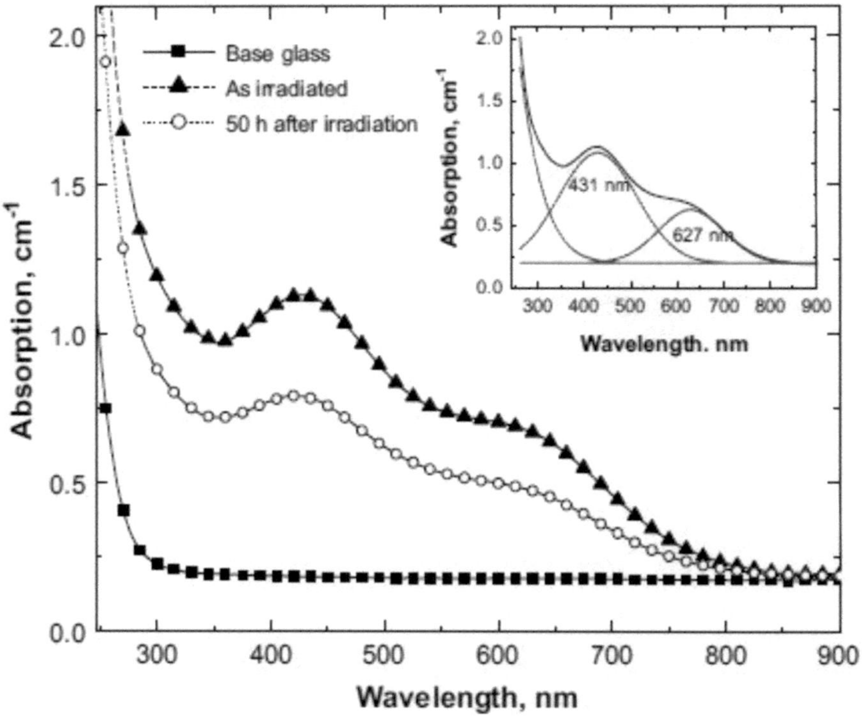

Fig. 2. Optical absorption in soda-lime silicate glass after UV-laser irradiation at 75 mJ/cm^2 for 10 min (insert Gaussian resolution of induced absorption)

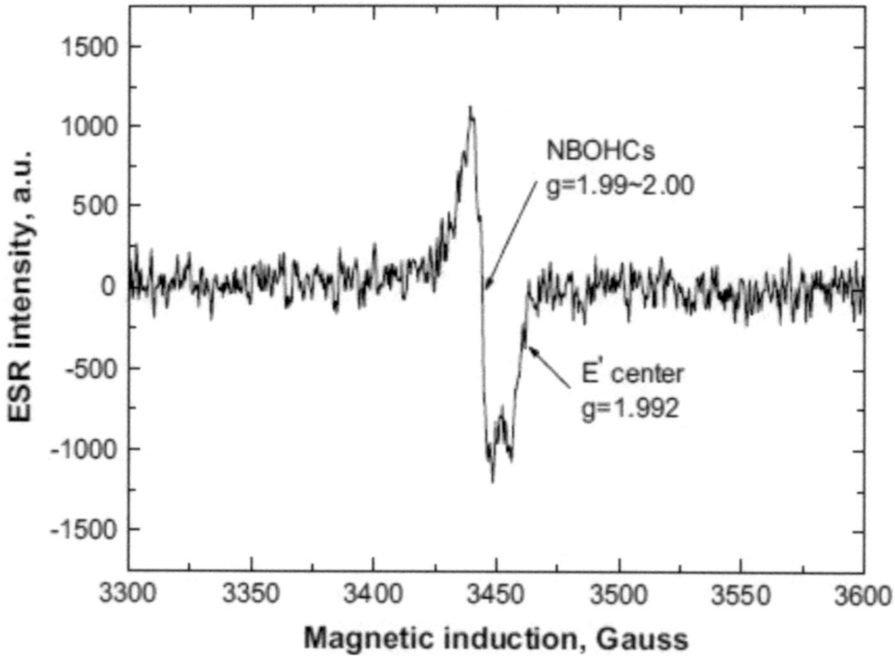

Fig. 3. ESR spectra of glass after irradiation at 75 mJ / cm^2 for 10 min

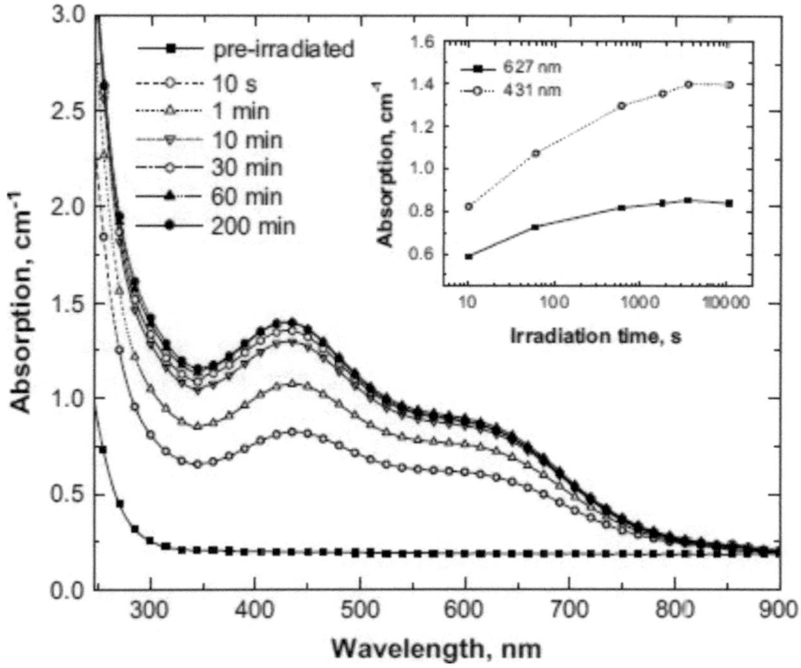

Fig. 4. Radiation induced optical Absorption with varied irradiation time (75 mJ / cm^2) (insert: Optical absorptions affected by irradiation time)

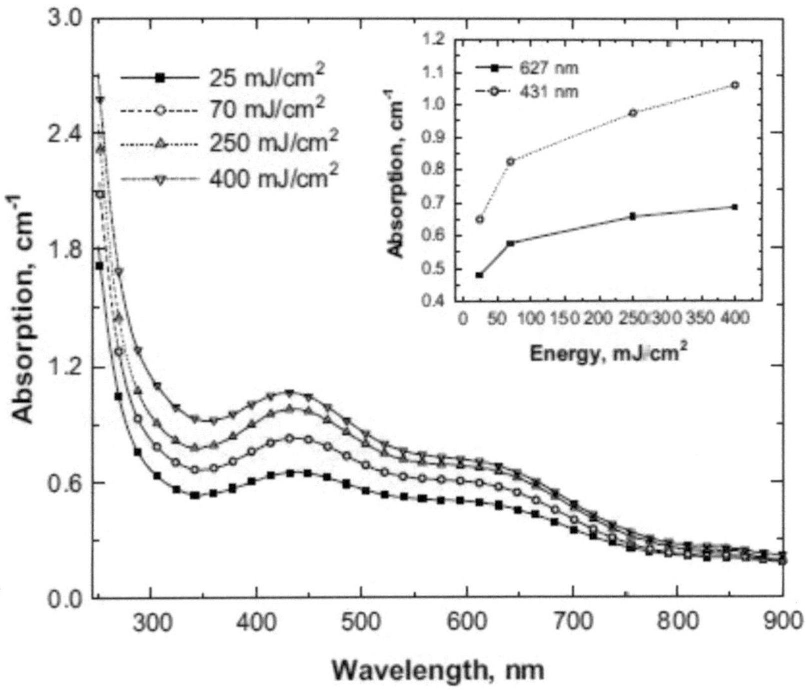

Fig. 5. Optical absorption affected by irradiation energy (100 shots) (insert: Optical absorption affected by irradiation energy).

Gusarov et al (2010) studied the effect of gamma irradiation on Silica-soda-lime glass with a composition $75SiO_2$–$22Na_2O$–3CaO, wt.% (labelled R1) and the same glass doped with 0.05 wt.% CeO_2 (labelled R7) were melted from high purity raw components using a technology which allowed one to keep the concentration of impurities of Fe ions at a few ppm and other transition metals below 1 ppm. Experimental samples were prepared as plates 25×25mm2, ~1mm thick, and polished for optical measurements. The glass samples contained a small number of air bubbles with diameters up to ~0.1mm, however transmission measurements at different parts on several samples showed that their influence was below the detection level. All transmission spectra were measured from 195 to 3300 nm with a commercial double-beam UV–Vis–NIR spectrophotometer. Irradiations have been performed in the Brigitte gamma irradiation facility at SCK CEN at 70, 200, and 350 ◦C, 25 kGy/h. They found that the addition ofCeO_2 results in significant transmission degradation in the UV (Fig. 6)

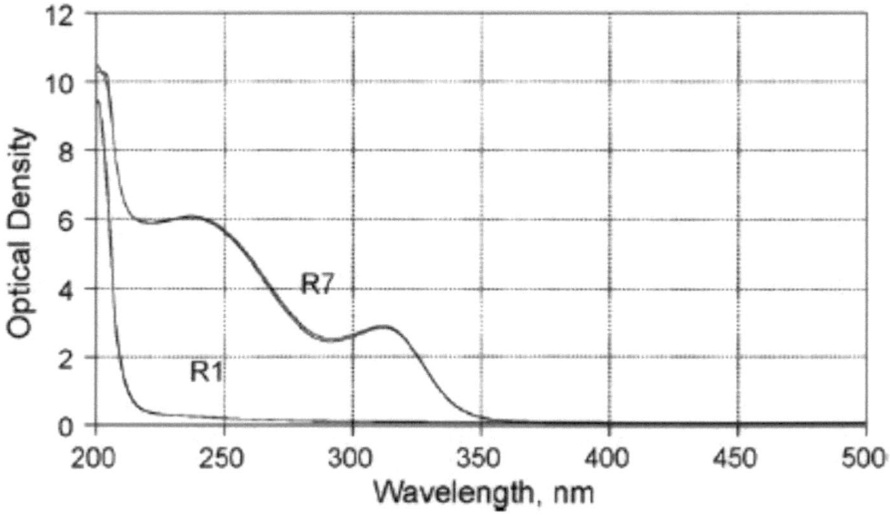

Fig. 6. Optical absorption spectra of R1-5 and R7-5 glass samples in the UV range. For each glass two spectra are plotted: before and after annealing for 24 h at 350 ◦C

This degradation is due to the overlapping optical absorption bands of Ce^{3+} and Ce^{4+} ions, with maxima at 313 and 240 nm (3.96 and 5.15 eV), respectively [**Arbuzov, 1990**]. Above 350 nm both materials show the same high transmission with no observable optical absorption bands. No detectable changes in transmission were observed after annealing R1 and R7 control glass samples for 24 h at 200 ◦C. Annealing for 24 at 350 ◦C does not change the transmission of R1. For R7 a weak band with amplitude of ~0.3cm−1 at 275 unappeased. This effect can be tentatively attributed to thermally-stimulated charge transfer between Ce and Fe ions. However, the amplitude of this thermally induced band is significantly less than the amplitude of bands produced by irradiation.

2.2 Phosphate glass doped with cobalt oxide

Sodium phosphate glass with different CoO percentage was prepared [El-Batal, 2010], A 60Co gamma cell (2000 Ci) was used as a c- ray source with a dose rate of 1.5 Gy/s (150 rads/s) at a room temperature of ~30 °C. The investigated glass samples were subjected to

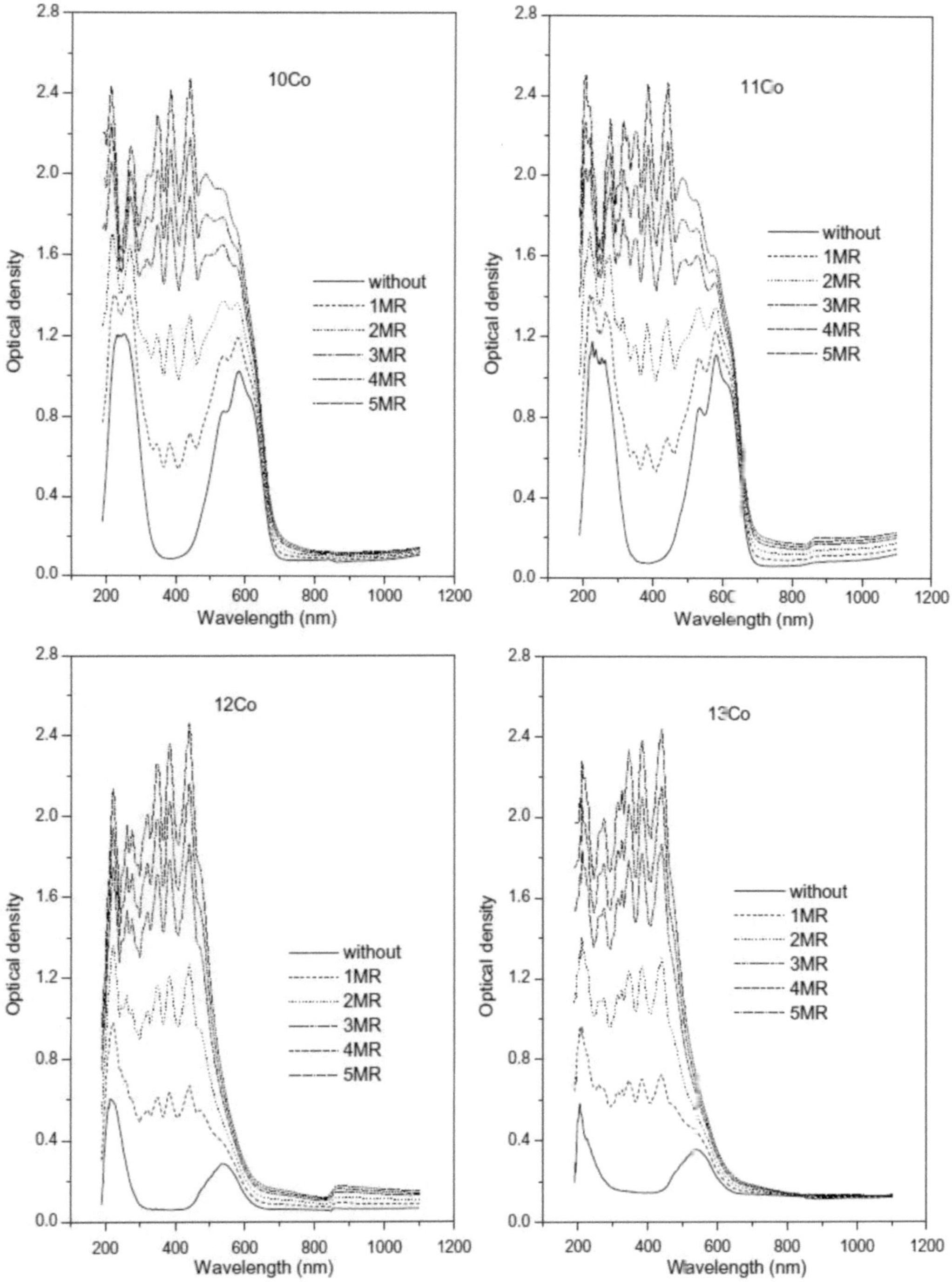

Fig. 7. Absorption spectra of CoO-doped sodium phosphate glasses with constant 0.25% CoO with additional 5% or 10% of Na_2O or 5% or 10% P_2O_5 before and after gamma irradiation.

the same gamma dose every time. Using a Fricke dosimeter, the absorbed dose in water was utilized in terms of dose in glass. No cavity theory corrections were made. Each glass sample was subjected to a total dose of 5 × 104 Gy (5MR). The UV–visible absorption of glasses containing CoO-doped sodium metaphosphate glasses. The glass containing 0.05% CoO (glass/Co) reveals before gamma irradiation a strong ultraviolet charge transfer absorption with two small peaks at about 205 and 235 nm and the visible spectrum shows a small broad band centered at about 530 nm. On subjecting this glass to successive gamma irradiation, the UV absorption progressively increases with marked splitting to three component peaks at about 205, 275, and 310 nm and followed by subsequent closely-connected four peaks at about 340, 400, 450, and 500 nm. With glass containing higher concentrations P0.5% CoO), the optical absorption spectra reveal the same strong ultraviolet absorption with two peaks at about 205 and 235 nm and the visible region shows a very broad band increasing in intensity with the increase of CoO content and exhibiting finally three small peaks at about 500, 540, and 595 nm. With continuous gamma irradiation, the absorption spectra reveal numerous connected peaks extending from 205 up to 400 nm exhibiting continuous growth with gamma irradiation and the distinct broad visible absorption is observed to be finally unaffected by progressive irradiation at high CoO content . The rate of increase with the first dose of irradiation is observed to be very high especially in the band at about 500 nm and the optical density is seen to be slowly increased with progressive irradiation.

2.2.1 Condition of cobalt in glass

Although cobalt can exhibit different oxidation states in many inorganic complexes but in glasses melted under normal atmospheric conditions, cobalt ion exists in the divalent state with two possible coordination forms, namely the octahedral and tetrahedral **[Bates 1962, Bamford 1977, Paul 1990, Aglan 1955**]; The only known ion having the 3d7 configuration is Co^{2+}. Early, **Aglan 1955**, and later other scientists **[El-Batal,2003]** have studied and interpreted the absorption spectra of glasses containing Co^{2+} in terms of equilibrium between octahedral and tetrahedral coordination forms of Co^{2+} ions depending on the type and composition of glass and condition of melting. The energy level diagram for d7 system in octahedral coordination predicts that the spectrum of Co^{2+} in octahedral symmetry will consist essentially of three bands corresponding to spin-allowed transitions $^4\Gamma_4 \rightarrow 4C5$, $^4\Gamma_4 \rightarrow {}^4\Gamma_2$ and $^4\Gamma_4 \rightarrow {}^4\Gamma_4$ (P) together with several weak lines corresponding to spin-forbidden transitions. However, the band corresponding to the transition $^4\Gamma_4 \rightarrow {}^4\Gamma_2$ is expected to occur only with low intensity as it corresponds to a forbidden two electron jump. **Bates, 1962** has assumed that in low alkali glasses, Co^{2+} ions are present in octahedral symmetry probably with a rhombic distortion arising from the Jahn- Teller effect.

The energy diagram for d7 system in tetrahedral symmetry predicts that the spectrum of Co^{2+} in four coordination consists essentially of three bands **[Sreekanth, 1998]** corresponding to the spin-allowed transition $^4\Gamma_4 \rightarrow {}^4\Gamma_5$, $^4\Gamma_2 \rightarrow {}^4\Gamma_4$ and $^4\Gamma_2 \rightarrow {}^4\Gamma_5$ (P) together with several weak lines and bands corresponding to spin-forbidden transitions. The splitting of the 4Γ4 (P) bands has been attributed to L, S interactions (i.e. a departure from Russel–Sanders coupling) which is also quite large in the free ion. The high intensity of the tetrahedrally coordinated band is a consequence of the mixing of the 3d-orbitals with 4P-orbitals and ligand orbital **[ElBatal, 2008]**.

2.3 Lead silicate glass

Lead silicate and borate glasses glasses **[El-Kheshen, A., 2008]** have been extensively studied these last decades mainly because of their presence and importance in a broad range of technological applications. These glasses have been used in the electronic and optical technologies, such as for electron multiplier **[Anderson,1979]**, micro-channel plated **[Wiza, 1979]**, non-linear optical and magneto-optical devises **[George, 1999]**. Also, high lead silicate glasses can find application as radiation shielding materials **[Friebele, 1991]**.

A base lead silicate glass of the chemical composition PbO 60 wt%, SiO_2 40 wt% (71.24 PbO mol%, 28.76 SiO_2 mol%) was prepared by **[Azooz , 2009]** .Batches containing the base glass composition to which was added 0.1% of one of the 3d-transition metal oxides: TiO_2, V_2O_5, Cr_2O_3, MnO_2, Fe_2O_3, CoO, NiO or CuO. An Indian 60Co gamma cell (2000 ci) was used as a gamma ray source with a dose rate of 1.5Gy s−1.

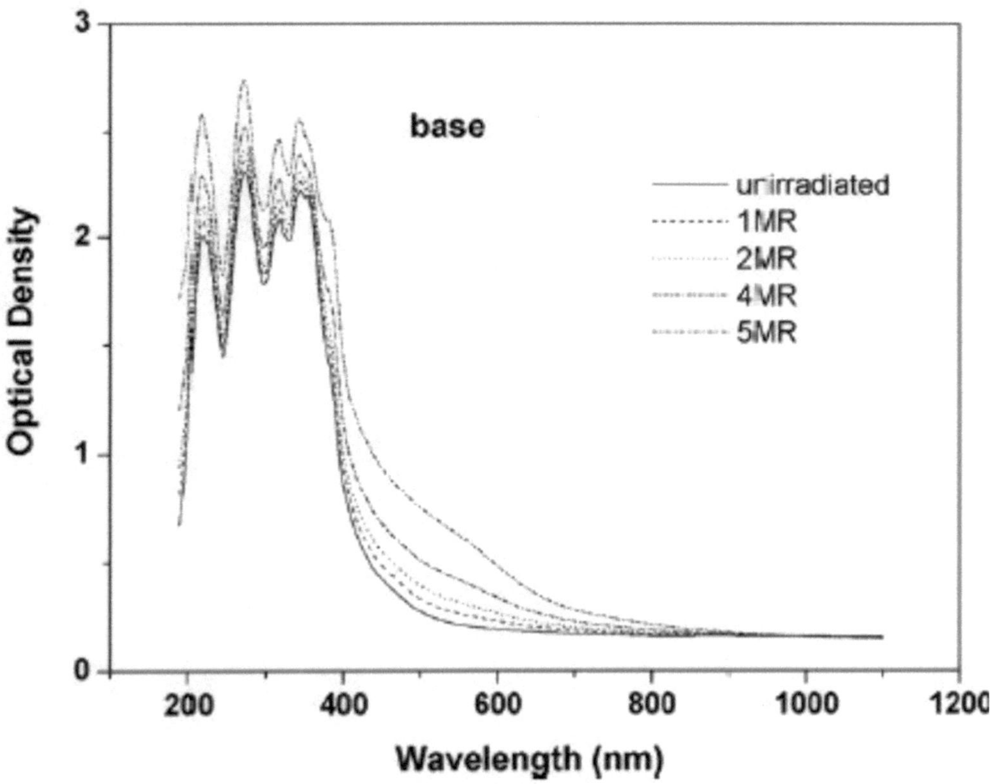

Fig. 8. The optical absorption spectra of the parent glass before and after different doses of gamma rays

Experimental results indicate that the effect of successive gamma irradiation is mainly concentrated in the changes of the intensities of the already observed UV bands which are assumed to be due to trace iron impurities (mainly Fe^{3+} ions) and to the Pb^{2+} ions. The main pronounced difference is the sharp growth of the kink at about 390nmto a pronounced peak and the quite resolution of a broad visible band centered at about 500 nm. It seems that the presence of high content of the heavy metal oxide (PbO), shows some shielding towards successive gamma irradiation and only limited changes are observed. This behavior is related to the presence of heavy Pb^{2+} ions which retard the movement or transfer of released electrons during the irradiation process **[Singh, 2004]**.

The realization of the effect of gamma irradiation can be interpreted by assuming that when the lead silicate host glass is subjected to ionizing radiation, electron-hole pairs are produced, which then become individually trapped at various intrinsic defect sites in the glass structure. Also, the presence of trace iron impurities in the host glass can easily trap electrons or holes and further extrinsic defects may be formed by photochemical process. It has been shown by several authors **[Bishay, 1977]** that the principal feature of the spectra of Pb-containing glasses is a band at 1.59 eV with a width of 0.6 eV which has been correlated with Pb^{3+} in silicate glasses. **[Friebele, 1977]** assumed that in oxide glasses this band is observed only when the Pb concentration is less than ≈25 mol%. A possible explanation for this is the assumption that Pb2+ is incorporated in glass as an ionic modifier for concentration of PbO ≤25-30 mol%, while at higher concentrations, PbO^{2+} is covalently bonded in PbO_4 unit **[Barker, 1965]** and/or PbO_3 unit **[Fayon, 1999]** as a glass former. **[Friebele, 1991]** further added that two bands with peak energies of 2.36 and 3.31 eV are present in high lead glasses and can be associated with Pb3+ on a network forming unit. Regarding the SiO2 as a partner in the composition, **[Bishay, 1977]** and **[Friebele 1991]** have classified the origin of the UV induced bands and attributed them to electron centers while they related to the visible induced bands to positive holes. Later, **[Shkrob, 2000]** have assumed that irradiation of alkali silicate glasses results in the formation of metastable spin centers such as oxygen hole centers (OHC1 and OHC2), silicon peroxy radicals and a silicon dangling electron center (E/center).

2.3.1 Cr doped lead silicate glass

This chromium-doped glass reveals before irradiation (Fig. 9) five strong ultraviolet absorption bands at about 205, 270, 310, 340 and 380nm followed by a strong visible band at about 440 nm and a medium band at about 480 nm and finally a very broad band centered at about 635 nm. This observed spectrum represents collective presence of absorption bands due to trace iron impurities (205, 270, 310 nm), Pb^{2+} ions (340 nm) and hexavalent chromium ions Cr^{6+} (380 nm) while the visible bands due to trivalent chromium bands (440 and 650 nm) and a new mixed band (470 nm). These collective specific absorption bands are identified to the respective mentioned ions in various glass systems by several authors **[Ghoneim, 1983]**.With progressive gamma irradiation, the six successive ultraviolet and near visible absorption bands slightly increase in intensity followed by slight increase and the presence of chromium ions seems to retard the effect of gamma irradiation this mentioned region. The visible band at about 470 nm shifts to longer wavelength at 480 nm while broad band highly increases in intensity and shifts from 650 to 620 nm.

It can be assumed that some of the present Cr^{6+} ions capture liberated electrons during gamma irradiation and are converted to induced Cr^{3+} ions producing absorption in the same position of the trivalent chromium in the following photochemical reaction:

$$Cr^{6+} + 3e^- \rightarrow Cr^{3+}$$

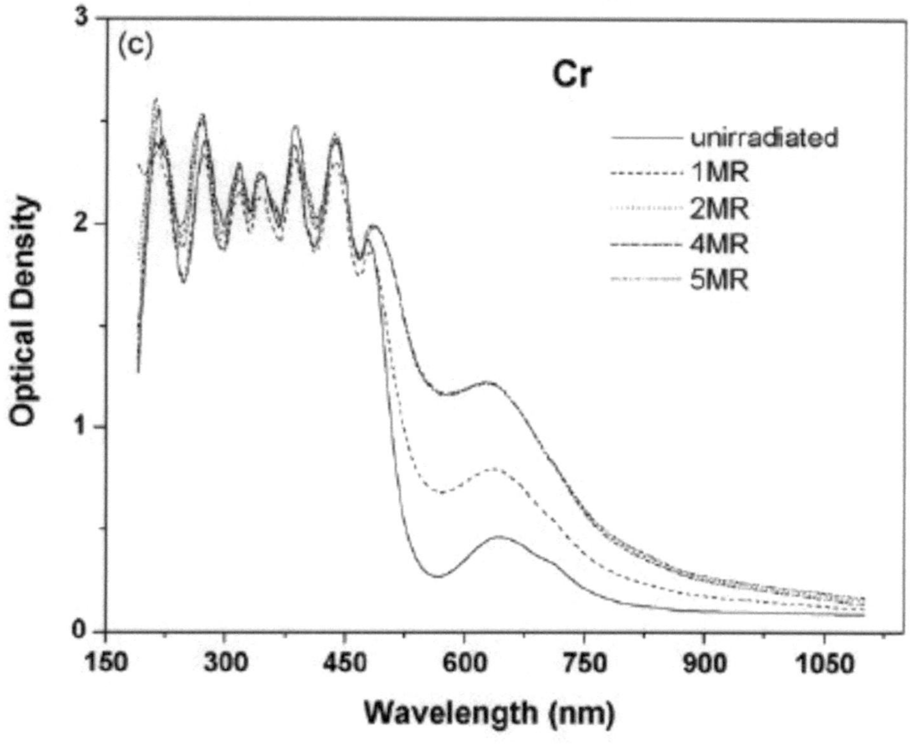

Fig. 9. The optical absorption spectra of the Cr-doped lead silicate glasses before and after different doses of gamma rays.

2.4 Radiation processing for glass coloration / discoloration

Glass is particularly susceptible to radiation-induced coloration/ discoloration due to its amorphous, non-crystalline structure. The nature of the optical changes varies, but usually consists of coloration in the visible light region and the formation of absorption bands in the infrared and/or ultraviolet regions. The optical density is almost always increased. It is well documented, for example, that high doses of gamma radiation turn glass various shades ranging from deep brown to pale amber **[Prasil, 1991].** Clear glass discolors when exposed to gamma irradiation, such as from a Cobalt-60 source. Ordinary flint, borosilicate, and lead glass undergo a color change from clear to light amber, brownish to black, depending on the amount of energy absorbed. Milk glass, when exposed to gamma irradiation, yields a grayish color, depending on energy absorbed and any swirling effect is highlighted, probably due to concentrations of color in the glass that have not mixed uniformly. The actual mechanism of the formation of "color centers" has been described **[Prasil, 1991].** The final color created is dependent on the chemical composition of the glass and can be altered by selection of additives (e.g., cerium ions can reduce browning; manganese ions induce an amethyst color). Final color is a combination of original glass color and the effects of color center formations.

Color from irradiation is considered metastable and heat, for example, is known to reverse the effect. Depending on the type of color centers formed and the hardness of the glass, varying amounts of heat or energy penetration are required to reverse color. Glass with a low diffusion rate (high softening point) is more stable. Combining all of these factors gives a very unique decorative coloration to a glass item. Very striking color contrasts can evolve, depending upon the color of the paints utilized on an item, the chemical composition, and impurities in the glass. At low doses, the color intensity increases linearly but eventually saturates at high doses. The final product is not radioactive or contaminated in any sense.

The general physical properties of colored glass include:

1. **Color stability:** There is a slight loss of color during the first several days after processing, and thereafter the rate of fade is slight. The process can actually be reversed, causing the glass to become clear, by placing the unit in an oven at approximately 300°F for a couple of hours. Hence, for applications where the product is exposed to high heat or low heat on a routine basis (e.g., dishwashers), this may not be ideal.
2. **Ultraviolet absorption:** There is a small amount of ultraviolet absorption characteristic which changes as the glass is further discolored.
3. **Physical properties:** Other than the color change, the normal physical properties of glass are not affected **[Dietz 1976, Prasil, 1990]**.

The advantages of irradiation for decorative glass coloration include:

1. Processing is done in shipping carton. There is no need to unpack or handle individual pieces.
2. The process is simple, clean, and completely safe to the end product and to the consumer.
3. Although economics would tend to favor large volumes, smaller volumes can also be economically attractive. The irradiation process is not dependent on processing large runs.
4. There is no waste product or breakage associated with the process

3. Colored glass as radiation sensor [Dunn, T.M. 1960]

This wonderful material (colored glass), with its beautiful appearance, has another use rather than decoration which is: radiation sensor. Some other instruments were used as radiation detectors but now colored glass has the same action, due to transition elements doping in glass. There are a number of properties shared by the transition elements that are not found in other elements, which results from the partially filled *d* shell. These include:

- The formation of compounds whose color is due to *d* - *d* electronic transitions
- The formation of compounds in many oxidation states, due to the relatively low reactivity of unpaired *d* electrons.
- The formation of many paramagnetic compounds due to the presence of unpaired *d* electrons. A few compounds of main group elements are also paramagnetic (e.g. nitric oxide, oxygen).

Color in transition-series metal compounds is generally due to electronic transitions of two principal types.

Charge transfer transitions. An electron may jump from a predominantly ligand orbital to a predominantly metal orbital, giving rise to a ligand-to-metal charge-transfer (LMCT) transition. These can most easily occur when the metal is in a high oxidation state. For example, the color of chromate, dichromate and permanganate ions is due to LMCT transitions. Another example is: Mercuric iodide, HgI_2, is red because of a LMCT transition. As this example shows, charge transfer transitions are not restricted to transition metals. A metal-to ligand charge transfer (MLCT) transition will be most likely when the metal is in a low oxidation state and the ligand is easily reduced. *d-d* transitions. An electron jumps from one d-orbital to another. In complexes of the transition metals the *d* orbitals do not all have the same energy. The pattern of splitting of the *d* orbitals can be calculated using crystal field theory. The extent of the splitting depends on the particular metal, its oxidation state and the nature of the ligands. In octahedral complexes, *d-d* transitions are forbidden and only occur because of coupling in which a molecular vibration occurs together with a *d-d* transition. Tetrahedral complexes have somewhat more intense color because mixing *d* and *p* orbitals is possible when there is no centre of symmetry, so transitions are not pure *d-d* transitions. The molar absorptivity (ε) of bands caused by *d-d* transitions are relatively low, roughly in the range 5-500 $M^{-1}cm^{-1}$ (where M = mol dm^{-3}). Some *d-d* transitions are spin forbidden. An example occurs in octahedral, high-spin complexes of manganese (2), which has a d^5 configuration in which all five electron has parallel spins; the color of such complexes is much weaker than in complexes with spin-allowed transitions. In fact many compounds of manganese (2) appear almost colorless. The spectrum of $[Mn(H_2O)_6]^{2+}$ shows a maximum molar absorptive of about 0.04 $M^{-1}cm^{-1}$ in the visible spectrum.

A characteristic of transition metals is that they exhibit two or more oxidation states, usually differing by one. Main group elements in groups 13 to 17 also exhibit multiple oxidation states. The "common" oxidation states of these elements typically differ by two. For example, compounds of gallium in oxidation states +1 and +3 exist in which there is a single gallium atom. No compound of Ga(2) is known: any such compound would have an unpaired electron and would behave as a free radical and be destroyed rapidly. The only compounds in which gallium has a formal oxidation state of +2 are dimeric compounds, such as $[Ga_2Cl_6]^{2-}$, which contain a Ga-Ga bond formed from the unpaired electron on each Ga atom. Thus the main difference in oxidation states, between transition elements and other elements is that oxidation states are known in which there is a single atom of the element and one or more unpaired electrons. The maximum oxidation state in the first row transition metals is equal to the number of valence electrons from titanium (+4) up to manganese (+7), but decreases in the later elements. In the second and third rows the maximum occurs with ruthenium and osmium (+8). In compounds such as $[MnO_4]^-$ and OsO_4 the elements achieve a stable octet by forming four covalent bonds. The lowest oxidation states are exhibited in such compounds as $Cr(CO)_6$ (oxidation state zero) and $[Fe(CO)_4]^{2-}$ (oxidation state −2). These complexes are also covalent. Ionic compounds are mostly formed with oxidation states +2 and +3. In aqueous solution the ions are hydrated by (usually) six water molecules arranged octahedrally.

3.1 Why do many glasses turn green, brown or black when exposed to ionizing radiation? [El-Kheshen A. A., 1999]

The phenomenon of radiation-induced color change is called 'activation of color centers'. The details are quite complex and involve an alteration of the orbital distribution of an atom's valence (outermost) electrons, causing the atom to absorb photons of a different frequency (color) after irradiation than before. What we describe as the reflected, transmitted, and emitted colors of a material is a consequence of the outermost shell of electrons, the valence electrons of atoms. In the following discussion, we will consider primarily reflected light.' White' light is actually a mixture of photons of many different frequencies (colors), and color is typically described by wavelength. The valence shell of an atom, described by classical physics as comprised of up to eight discreet electrons, is described by quantum physics to behave as a singular 'thing'. The valence shell will resonate with (it will capture, or absorb) photons of certain discrete energies. This resonance causes the valence shell to become excited, or 'pumped', to a higher energy meta-stable 'level' or state. The atom has a 'ground' (un-excited) state, and may have multiple excited states. Upon absorption of radiation of energy, electrons may be stripped from the atom, resulting in multiple degrees of ionization, each capable of multiple states of excitation. For example, a hypothetical atom with eight electrons in its valence shell has its ground state and shall have three meta-stable excited states. Upon capture of a photon of energy, one of the valence electrons is stripped resulting in the atom becoming a singly-ionized ion. This particular ion shall have two meta-stable excited states, and upon capture of a photon of energy a second of the valence electrons is stripped resulting in a doubly-ionized ion capable of four meta-stable states. And so on. For any mono-atomic atom, and for the shared valence electron orbitals of any compound, there are these discreet states capable of absorbing (and emitting) discreet quanta of energy which is shuttled around as photons. Now, let's work with a real example: Let's assume that we have a 'white' light beam composed of red, blue and yellow photons (the primary transmissive colors). We direct this beam onto a puddle of red anthracine dye. Anthracite dyes have tri-cyclic ('aromatic') structures which are capable of resonating with discreet spectra of frequencies [El-Batal, 2008]. By slight modifications of their molecular structures the valence orbitals of their constituent carbon atoms can be altered, which results in slightly different resonant characteristics. The different anthracine dyes thus appear to our eyes as having different colors. We have selected a red one. The blue and yellow photons in our white light resonate with (are captured and absorbed by) the red dye molecule, and the red photons are scattered (reflected). Some of this scattered red light is directed toward our eyes, and we thus see the puddle of dye as 'red'. If we put the puddle of red dye on a clear film, the red photons will be seen also to pass through the dye.

Now, let's examine a piece of colored glass. The materials which are added to glass as pigments are selected because their valence shell electrons are capable of selectively absorbing photons of certain frequencies, while passing and reflecting others. Cobaltic oxide for example, will absorb red and yellow but not blue, thus cobalt glass is blue. Transparent glasses in general are a mixture of alkaline and transition metal oxides, most of which are selected precisely because they *won't* interact with visible light. Put a piece of such a transparent glass in a high neutron-flux or high radiation (such as γ-ray) environment, and they will 'become colored' as the glass becomes physically altered at the atomic level **[Moustafa , 2010].**

3.2 Can the activation of color centers be reversed? Can the colored glass be made clear again? [Radek, 2009]

Yes. Glass, even at room temperature, is a liquid: the interatomic bonds are weak and constantly breaking and reforming. These bonds are in fact an interaction of (or, sharing of) the valence electrons of adjacent atoms. (In metals this sharing of electrons results in electrical conduction. For example, in a length of copper wire, the individual electrons at one end of the wire will, in theory, eventually migrate to the other end of the wire, due to random motion and without the application of an outside force). The inter-atomic bonding structure within the bulk of a material places physical constraints upon the valence electrons. Therefore, if you activate a color center (by any means) the alteration in the valence shell will be either stable (unchanging with time) or meta-stable (will change gradually with time). The stronger the interatomic structure (or 'lattice' in the case of a true crystal which, unlike a glass, is a solid) the more stable the change. The change can however be reversed by weakening the inter-atomic bonds which will allow the formation of new, lower energy bonds (atoms will break their initial bonds and reform bonds with other neighbors). This can be accomplished simply by the application of heat. In practice, the temperature required for complete color deactivation in an amorphous material (such as a glass) is its annealing temperature. Therefore, simply annealing a piece of glass will deactivate the color centers. Having worked extensively with researchers who conduct experiments which activate color centers by both of these means, I have frequently had need to deactivate color centers. An unexpected consequence of color center activation is the formation of strained inter-atomic bonds, and unless the piece is annealed, it may spontaneously fracture! As a rule, the more a piece is colored, the more strain the piece is under, and the more it needs to be annealed. Note that the color doesn't 'cause' the strain. Rather, color and strain are two entirely separate issues sharing the same cause. Finally, it's worth pointing out that color center activation is now so well understood that unusually colored gemstones are now being created by the process. In some cases, semiprecious irradiated stones are fraudulently sold as precious gem stones, and the difference can only be distinguished by a professional gemologist. Many materials when exposed to radiation will darken or change color. For example you can occasionally find a discarded bottle that had been left outside for many years that has become darkened due to exposure to sunlight. Co-60 emits gamma rays, which are light waves but with a much higher energy than visible light. Many hospitals use a Co-60 source to sterilize medical supplies after they have been packaged. So why does gamma radiation change the color of glass? We need to look at the structure of glass and at why certain materials absorb different colors of light to understand the effect. Glass is made of silicon and oxygen. Oxygen atoms are hungry for more electrons than they are naturally given, so each oxygen atom will share an electron from the silicon atom. This shared electron creates a very strong attraction between the oxygen and silicon, and is the reason why glass is a very hard material.

As you can tell by looking out a window, visible light goes right through glass. For light to be absorbed in a material, the light wave has to interact with an electron in the material and basically "shake it loose" from its atom. In glass the electrons are so tightly held by the silicon and oxygen atoms that visible light cannot shake one of those electrons loose, so light travels right through without any absorption. But gamma rays, which have a million times as much energy as visible light, can shake a lot of things loose in glass. When a gamma ray enters glass, it can actually knock a silicon atom out of its place, shoving it in-between other atoms, and leaving a hole, or vacancy, in the material. The electrons surrounding the hole will no longer be tightly bound to the atoms, and these "loose" electrons will be able to interact with lower energy light (visible light). When visible light is transmitted through the glass, specific colors of the light can interact and be absorbed by the electrons near the vacancy, and the transmitted light will have a color that is the complement of the absorption. These holes, or defects, are often called "color centers" because they impart a color to the glass. If you heat the irradiated glass (called annealing) the atoms will begin to shake so much that the displaced silicon atom will eventually go back to its original place and the glass will become clear again.

4. References

Aglan, M.A., Moore, H., J. Soc. Glass Technol. 39 (1955) 35T.

Anderson, L.P., Grusell, E., Berg, S., J. Phys. E 12 (1979) 1015.

Arbuzov, V. Belyankina, N.,"Spectroscopic and photochemical properties of cerium in silicate glass", Fiz. Khim. Stekla 16 (1990) 593–604.

A world of glass, http://www.britglass.org.uk/Index.html.

Azooz, M.A., El-Batal, F.H., " Gamma ray interaction with transition metals-doped lead silicate glasses", Materials Chemistry and Physics 117 (2009) 59–65.

Barker, R.S., McConckey, E.A., Richardson, D.A. Phys. Chem. Glasses 6 (1965) 24–29.

Bates, T., in: D. Mackenzie (Ed.), "Modern Aspects of the Vitreous State", vol. 2, Butterworth, London, 1962, p. 195.

Bamford, C.R., "Colour Generation and Control in Glass", Elsevier, New York, 1977.

Bishay A. Radiation induced color centers in multicomponent glasses. J Non-Cryst Solids 1970;3:54–114.

Bishay, A., Maklad, M. ,Phys. Chem. Glasses 7 (1966) 149–156.

Dietz, George R., "Radiation Coloring of Glass" Presentation to Society of Glass Decorators, Annual Meeting October 11- 13, 1976, Pittsburgh, PA.

Dunn, T.M., Wilkins, R.G. "Modern Coordination Chemistry. New York: Interscience., Chapter 4, Section 4, "Charge Transfer Spectra", pp. 268-273, 1960.

El-Batal, F., Ouis, M., Morsi, R. Marzouk, S.," Interaction of gamma rays with some sodium phosphate glasses containing cobalt", Journal of Non-Crystalline Solids 356 (2010) 46–55.

ElBatal, F.H., ElKheshen, A.A., Azooz, M.A., Abo-Naf, S.M., Opt. Mater. 30 (2008).

El-Batal,. Khalil, M.M.I, Nada, N., Desouky, S.A. Mater. Chem. Phys. 82 (2003) 375.

El Kheshen, A. A. El-Batal, F., Marzouk,S."UV-visible, Infrared and Raman spectroscopic and thermal Studies of tungsten doped lead borate glasses and the effect of ionizing gamma irradiation", Indian Journal of Pure & Applied Physics, 46, 4, 225-238, 2008.

El-Kheshen, A. A., "Effect of γ- rays on some glasses containing transition metals, A PhD thesis, Ain-Shams University, Egypt, 1999.

Fayon, F., Landron, C., Sakwari, K., Bessada, C., Massiot, D., J. Non-Cryst. Solids 243 (1999) 39.

Friebele EJ. Radiation effects in optical and properties of glass. New York, Westerville; 1991.

Friebele, E.J. in: D.R. Uhlmann, N.J. Kreidl (Eds.), Optical Properties of Glasses, The American Ceramic Society,Westerville, OH, 1991, p. 205.

Friebele, E.J. Proc. Int. Congr. Glass, 11th, Progue, vol. 3, 1977, pp. 87–95.

George, H.B., Vira, C., Stehle, C. , J. Meyer, S. Evers, D. Hogon, S. Feller,M. Affatigato, Phys. Chem. Glasses 40 (1999) 326.

Ghoneim, N.A., El-Batal, H.A., Zahran, A.H., Ezz ElDin, F.M. Phys. Chem. Glasses 24 (1983) 83.

Gerry, M., Alan M., "Glass: a world history, published by; University of Chicago Press", 2002.

Gusarov, A., Huysmans, S., Berghmans, F., "Induced optical absorption of silicate glasses due to gamma irradiation at high temperatures", Fusion Engineering and Design 85 (2010) 1–6.

Jiawei S, Yanfen W, Xinji Y, Jian Z., "UV-laser irradiation on the soda-lime silicate glass", International Journal of Hydrogen Energy, 34 (2009) 1123 – 1125.

Marshall CD, Speth JA, Payne SA. Induced optical absorption in gamma, neutron and ultraviolet irradiated fused quartz and silica. J Non-Cryst Solids 1997;212:59–73.

Moustaffa, F.A., El-Batal, F.H., Fayad, A.M., El-Kashef, I.M.Absorption Studies on Some Silicate and Cabal Glasses Containing NiO or Fe_2O_3 or Mixed NiO + Fe_2O_3", Acta physica polonica A, Vol. 117, 2010.

Noel, C.,S., "The Glass and Glazing Handbook; Standards Australia, "SAA HB125–1998.

Paul, . A. Chemistry of Glasses, second ed., Elsevier, New York, 1990.

Prasil, Z , Marlind, T. "Two colors out of one" Glass Decoloration #125, beta-gamma, 1991, 2-3.

Prasil, Z. and Marlind, T. "Radiation Coloration of Glass-State of the Art. Glass Decoloration #249, beta-gamma 1991, 2-3.

Prasil, Z., Schweiner, E., Pesek, M. "Radiation Modification of Physical Properties of Inorganic Solids", Radiation Physics and Chemistry, Vol. 35, Nos. 4-6, pp.509-513, 1990.

Radek , P. Vojtěch E., Viktor G., Mariana K., Martin M., Ondřej, S., Ladislav S., " A comparison of natural and experimental long-term corrosion of uranium-colored glass", J. of Non-Crystalline Solids, Volume 355, Issues 43-44, , Pages 2134-2142, 2009.

Sheng J., "Photo-induced and controlled synthesis of Ag nanocluster in soda-lime silicate glass". Int J Hydrogen Energy 2007;32(13). 2062–65.

Sheng J, Li J, Yu J. The development of silver nanoclusters in ion-exchanged soda-lime silicate glasses. Int J Hydrogen Energy 2007;32(13):2598–601).

Shkrob, I.A., Tadjikov, B.M., Trifunac, A.D. J. Non-Cryst. Solids 262 (6) (2000) 35.

Singh, N., Singh, K.J., Singh, K., Singh, H., Nucl. Instrum.Methods Phys. Res., B 255 (2004) 305.

Sreekanth , R.P., Murali, A., Lakshmara Rao, J. J. Alloy. Compd. 281(1998) 99.

Weyl, W.A "Coloured Glasses, reprinted by Dawsons of Pall Mall, London, 1959.

Wiza, J.L. Nucl. Instrum. Methods 62 (1979) 587.

Wong J, Angell CA. Glass: structure by spectroscopy. NewYork: Marcel Dekker; 1976.

Zhang J, Dong W, Qiao L, Li J, Zheng J, Sheng J. Silver nanocluster formation in the soda-lime glass by X-ray irradiation and annealing. J Cryst Growth 2007;305:278–84

Formation and Decay of Colour Centres in a Silicate Glasses Exposed to Gamma Radiation: Application to High-Dose Dosimetry

K. Farah[1,2], A. Mejri[1], F. Hosni[1],
A. H. Hamzaoui[3] and B. Boizot[4]
[1]National Center for Science and Nuclear Technology, Sidi-Thabet,
[2]ISTLS, 12, University of Sousse,
[3]National Centre for Research in Materials Science, Borj Cedria, Hammam-Lif,
[4]Laboratory of Irradiated Solids, UMR 7642 CEA-CNRS - Polytechnic School, Route de Saclay, Palaiseau,
[1,2,3]Tunisia
[4]France

1. Introduction

The interactions of ionizing radiation with glass matrix produce ionization, excitation, and atomic displacement. The main modification induced during γ, X rays, or electron irradiation on the glass structure is the creation of stable defects and the changes of the valence state of lattice atoms or of the incorporated impurities in glass. Some of the modified electronic configurations or defects cause preferential light absorption. Thus, glass becomes coloured and consequently these defects are called "colour centres". These centres are of many types and depending on the glass composition (Yokota, 1954, 1956) and are associated with optical absorption bands and EPR signals.

The change of optical properties of glasses when subjected to ionizing radiation has been investigated by many authors due to wide applications of this kind of material. The earlier studies were focused on ways to prevent the darkening in glasses used in reactor or hot-cell windows and optical devices (Friebele, 1991). Recently, many studies have been concentrated on the application of the irradiation induced colour to develop recyclable colour glasses which is of great interest in the glass industry from the economical and environmental point of view (Sheng et al., 2002a).

In addition to the applications mentioned above, the ionizing radiation induced colour centres in some glasses have been found wide application in radiation dosimetry (Farah et al., 2010; Fuochi et al., 2008, 2009; Mejri et al., 2008). This material is very interesting for dosimetry and very useful for many applications such as food irradiation, sterilization of medical devices, radiation treatment of industrial and municipal waste-water and radiation processing of materials.

In the present work, we investigated the effect of the irradiation dose and thermal annealing on the formation and the decay of the induced colour centres in gamma irradiated silicate glass. The activation energies of the two colour centres were calculated from Arrhenius equation. In order to evaluate its potential as radiation-sensitive material in high dose dosimetry, the main dosimetric properties were also studied in details.

2. Materials and methods

2.1 Glass composition

The glass samples were obtained from the same glass sheets purchased from the local market and were cut into pieces of 11 x 30 x 3 mm^3 dimensions for optical measurements. The chemical composition of the glass samples were determined by the Prompt Gamma Activation Analysis technique (Anderson et al., 2004) in the Budapest Neutron Centre (wt%: 68.52 SiO_2, 13.77 Na_2O, 8.19 CaO, 4.34 MgO, 1.003 Al_2O_3, 0.588 K_2O, 0.105 Fe_2O_3 and about 3.5 % of other components). To avoid grease contamination on glass surface, which may affect the absorbance measurements, the samples were carefully cleaned with ethyl alcohol. A thermal treatment at 300°C for 1 h was used to eliminate any spurious optical signal. Then the samples were wrapped in aluminium papers and stored in a dust free dark place to avoid any possible light effect.

2.2 Irradiation sources and procedures

Irradiations of the glass samples were done at the Tunisian semi-industrial ^{60}Co gamma irradiation facility at the dose rate of about 2 kGy/h (Farah et al. 2006). Dosimetry was done using Fricke and Ethanol-ChloroBenzene chemical dosimeters and the traceability was established with alanine/EPR dosimetry system in terms of absorbed dose to water traceable to Aérial Secondary Standard Dosimetry Laboratory (SSDL), Strasbourg-France (Aérial, 2011). All the irradiated samples were stored in the dark in a room where the temperature was maintained between 20 and 25°C and humidity 45-60% R.H.

2.3 Optical absorption measurements

When exposed to gamma radiation, glass turns to a brown colour in a quantifiable and reproducible manner. Measurement of the change in absorbance with calibrated spectrophotometers at specified wavelengths provides a method for accurately determining absorbed dose.

Optical absorption spectra were taken with a Perkin Elmer spectrophotometer Lambda 20 in the range 350-800 nm. The optical spectra of non irradiated samples were measured with reference to air. All optical spectra of the samples after irradiation were measured against to non irradiated sample in order to obtain the net induced changes of absorption. Genesis 5 spectrophotometer and Käfer MFT 30 thickness gauge were used to measure the specific absorbance changes produced in glass (i.e. absorbance divided by dosimeter thickness). An electrical furnace and a freezer were used to reach the desired temperatures.

2.4 Temperature control during irradiation

To control and maintain the glass samples at the desired temperature during gamma irradiation, Julabo refrigerating circulator type F25-EC, with ultra purified water for the temperature range 5-90°C or a mixture of water and ethylene glycol for the range -25°C to +50°C as coolant liquid, was used. Samples were placed inside an aluminium cylinder in which the coolant liquid was circulating thus maintaining the desired temperature. Before irradiation, the set-up was kept for 20 minutes to reach the equilibrium temperature. Temperature fluctuation inside the aluminium cylinder during irradiation was within ±1°C.

2.5 Method and conditions of calibration

In order to minimize the contribution of influence quantities to the overall uncertainty and to ensure similar irradiation conditions both for calibration and routine dosimetry during the production run, full in-plant calibration of the glass samples was performed by irradiating them at the Tunisian semi-industrial cobalt-60 gamma irradiation facility together with transfer standard dosimeters in the Risø HDRL calibration phantoms (Sharpe&Miller, 1999). Four glass samples were placed inside the phantom, together with four alanine transfer standard dosimeters in their small 3 mm thick polyacetal holders as shown. The phantoms containing the alanine pellets and the glass samples were then taped on cartons of simulated product (dummy product box) placed on the test site facing the cobalt source. Care was taken to minimize self shielding effects between the dosimeters. The distance of the dummy product box from the source pencils was such as to reduce the spatial variations in the radiation field over the surface area of the dosimeter package to negligible small values.
The alanine pellets were sent back after irradiation for evaluation to the Aérial SSDL. The absorbed doses, as measured by the alanine dosimeters, were then used to establish the calibration curves for the glass.
Temperature strips were placed in the dosimeter packages during the calibration irradiations to record maximum temperature. Temperatures were found to vary around a mean value of 26°C with a maximum variation of +3°C.

3. Results and discussion

3.1 Effect of gamma radiation dose

Before gamma irradiation, the glass used in this study was transparent. When irradiated, two induced bands have been observed at 410 and 600 nm leading to the development of the brown colour. The intensities of the overall absorption spectra are observed to increase progressively with increasing doses between 1 and 1200 kGy (Fig.1). It is obvious that the broad band at 600 nm is less sensitive to radiation than that at 410 nm.
The induced optical absorption by gamma irradiation in the visible range of this silicate glass is due to the generation of two Non Bridging Oxygen Hole Centres (NBOHCs) (≡Si-O°): HC1 at 410 nm and HC2 at 600 nm (Griscom, 1984).
Fig.2 shows the two bands separated in the region between 375 and 800 nm of the absorption spectrum of glass irradiated to 10 kGy. The induced absorption spectra were well modeled through a Gaussian shape with correlation coefficients (R^2) better than 0.99.
Table 1 shows the results of the best fit with two Gaussian bands (correlation coefficients (R^2) better than 0.99) of the induced absorption spectra in the region between 380-800 nm. It was found that the gamma radiation dose had no influence on the absorption band position. The band peak positions the FWHM were relatively constant and only the heights and the area under the bands changed, suggesting that only an increase of the number of induced colour centres were affected by the increasing of the irradiation dose.
These results show that this glass is radiation-sensitive material and the induced colour centres may be used for dose determination in large dose range.

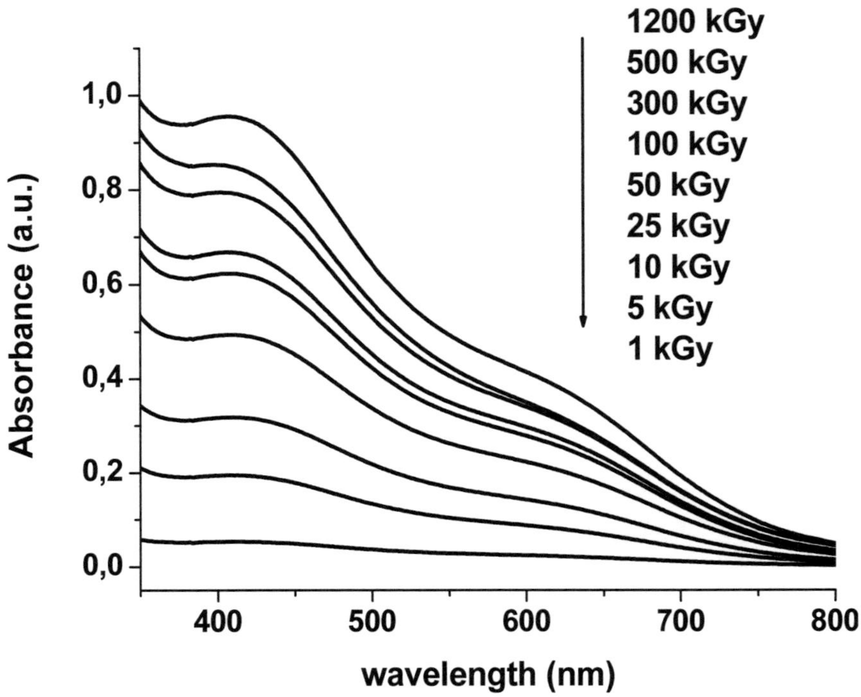

Fig. 1. Optical absorption spectra for the gamma irradiated glass samples in the range 1-1200 kGy

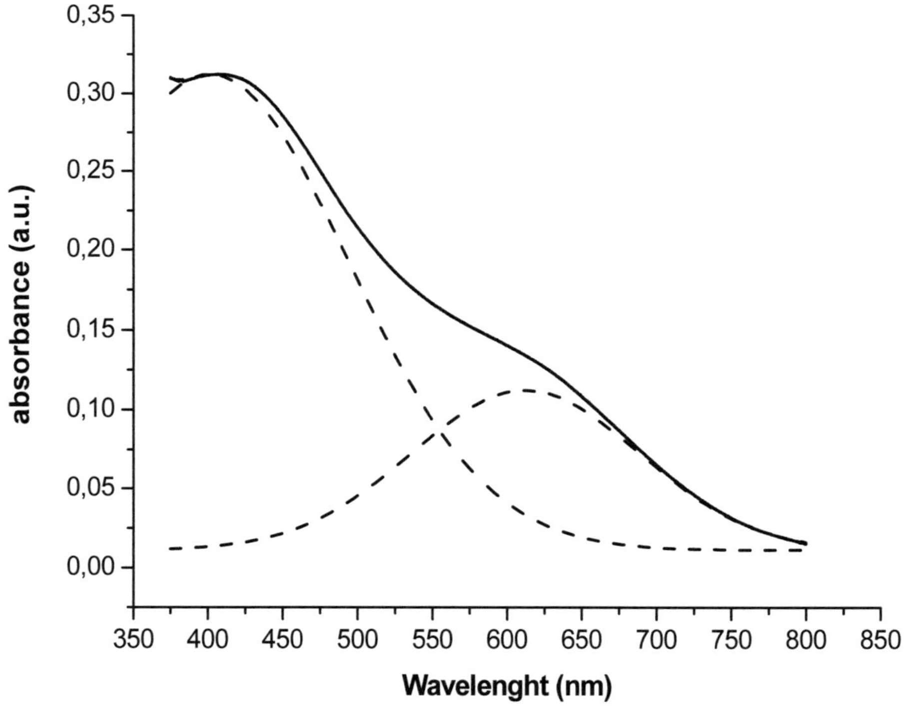

Fig. 2. Induced optical absorption with band separation of glass irradiated at 10 kGy.

Dose (kGy)	λ (nm)	FWHM (nm)	A
1	402	163.93	10.57
	598	174.93	4.50
10	405	162.42	62.31
	599	182.78	29.04
100	403	156.12	140.48
	602	174.82	43.93
1200	403	165.50	200.88
	603	182.07	60.31

Table 1. Results of band separation for the induced absorbance spectra

Where λ is the peak wavelength of band (nm), FWHM is the full width at half-maximum of the band (nm) and A is the area of the band

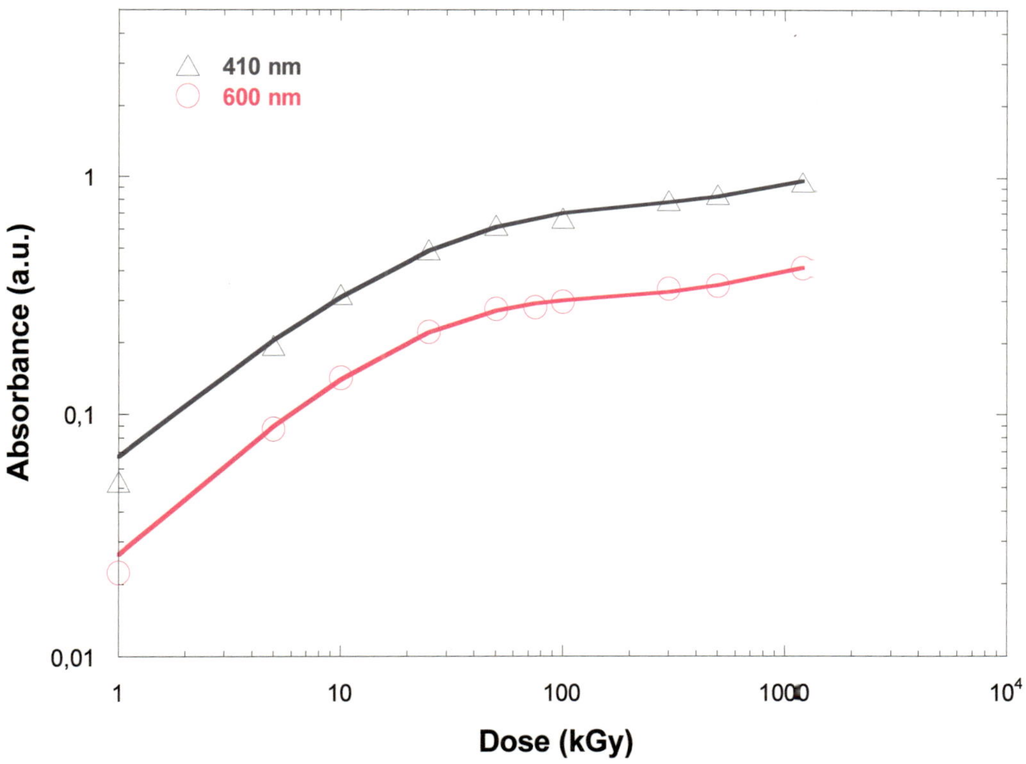

Fig. 3. Dose response curves of gamma irradiated silicate glass

The dose response curves shown in Fig.3 are successfully fitted following Mashkov equation (Eq.1) (Mashkov et al.,). The fitting parameters are given in Table 2.

$$Q(D) = Qc\,(kD)^b + Qa\,[1 - \exp[-(kD)^b]] \quad (1)$$

Where Q (D), Qa (D) et Qc (D) represent the measured quantities (Optical Absorption) proportional respectively to the total concentration of the colour centres for an accumulated dose D, the concentration of the colour centres created by an extrinsic process of activation and the concentration of the colour centres created during an intrinsic process of the rupture, k is a rate constant and b is a number between 0 and 1. The nonlinear dose dependence of the specific absorbance as function of the dose can be interpreted in terms of two different processes involved in the creation of colour centres: the creation of colour centres induced by activation of precursory defects which saturate with the dose because their concentration in glass is limited, and an "unlimited" creation of new colour centres during an intrinsic process of Si-O-Si bond rupture (Boizot, 1997).

Fitting parameters	Wavelengths	
	410 nm	600 nm
R2	0.995	0.996
k (kGy^{-1})	0.049 ± 0.007	0.052 ± 0.005
b	0.72 ± 0.06	0.86 ± 0.07

Table 2. Fitting parameters for Figure 3

The results of fitting our experimental data by Mashkov Equation showed that the values obtained for the k parameter (defects formation rate or rate constant) are equal to 0.049 and 0.052 kGy^{-1} for the two bands at 410 and 600 nm respectively. The similarity of this parameter for the two bands suggests that the related optical transitions correspond to the same type of NBOHCs in different configurations.

3.2 Activation energy

Figures 4 and 5 show the kinetics, for the absorption band at 410 nm, of temperature annealing performed between (-20 °C) and 150 °C of the glass samples irradiated at 30 kGy. The annealing process can be described by a sum of two first order decay kinetic functions ($\exp(-t/\tau)$) where t is the annealing time and τ is an appropriate time constant.

The activation energy characteristic of the annealing process was calculated from the Arrhenius equation. The obtained values for the fast and slow components of the 410 nm and 600 nm bands are presented in Table 3. The similarity of the activation energy values for both bands may suggest that the related optical transitions correspond to the same type of NBOHC's in different configurations (Griscom, 1984). The HC1 centre is a hole trapped in the 2p orbital of one non-bridging oxygen (NBO), analogue to NBOHC in silica glass, to which is correlated the absorption band at 410 nm. The HC2 centre is a hole trapped on two or three NBO's bonded to the same silicon to which is correlated the absorption band at 600 nm (Suszynska& Macalik, 2001).

By the mean of real-space multigrid electronic structure calculations, Jin and Chang proposed a diffusion mechanism of interstitial oxygen ions generated from O_2 under the UV irradiation with activation energies of 0.27 eV for O^- and 0.11 eV for O^{2-} (Jin & Chang, 2001). The value of 0.27 eV corresponding to the O^- diffusion activation energy in the glass network under UV irradiation is identical, within experimental error, to our values of activation energies calculated from Arrhenius plots of the slow component τ corresponding to the long-time isothermal annealing. Approximately the same value of activation energy was found by Tsai and al. for the long-time slow thermal annealing of radiolytic atomic hydrogen in OH containing amorphous silica (Tsai et al., 1989).

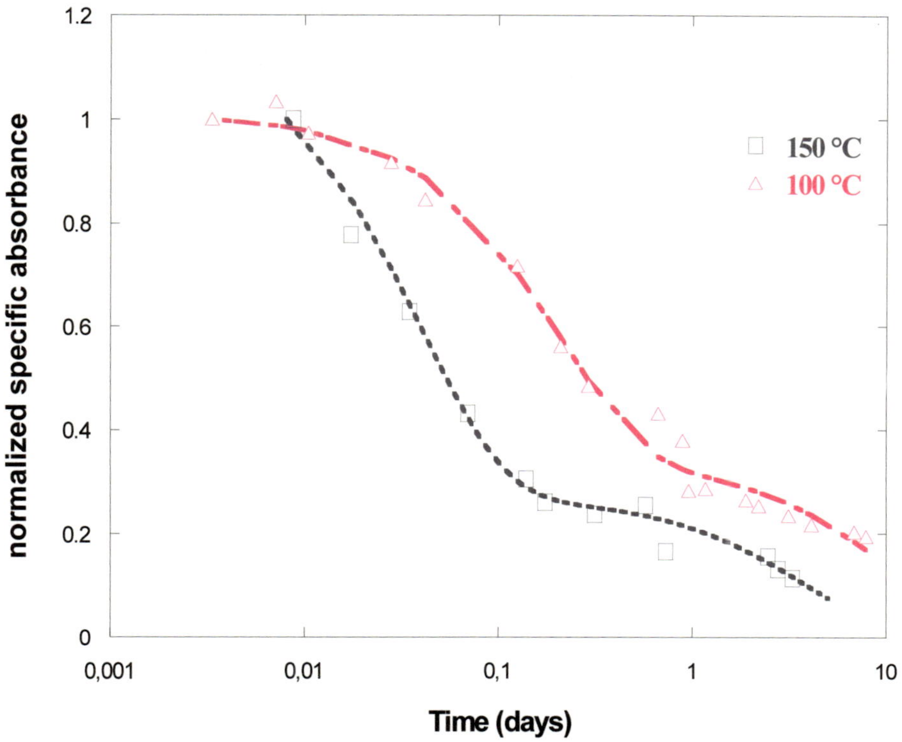

Fig. 4. Isothermal annealing of glass samples at high temperatures.

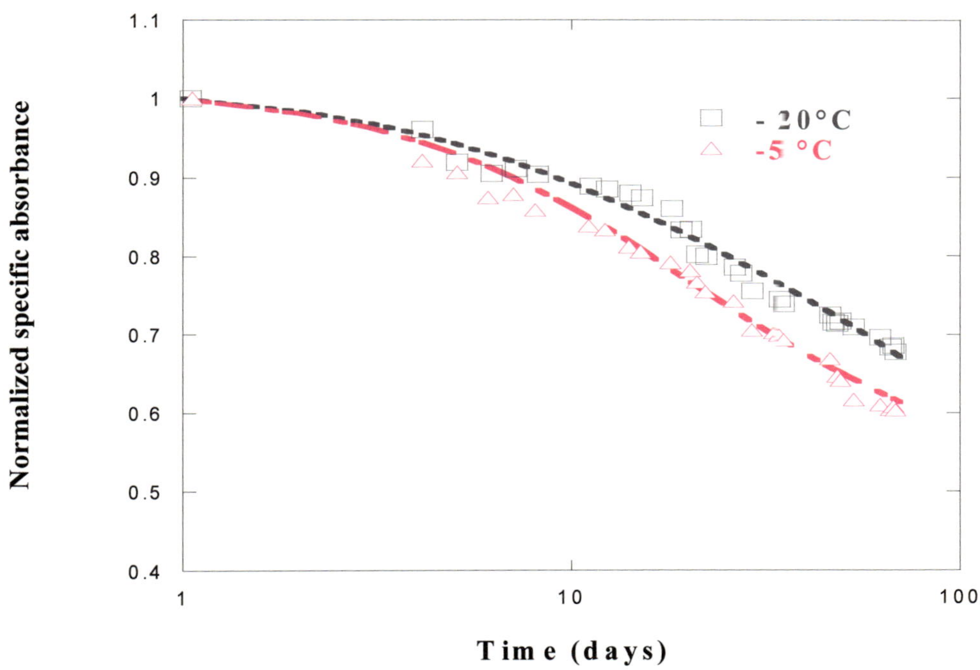

Fig. 5. Isothermal annealing of glass samples at low temperatures.

	Activation energy (eV)	
Wavelength(nm)	Fast component (τ_1)	Slow component (τ_2)
410	0.251± 0.022	0.310 ± 0.015
600	0.261± 0.022	0.353 ± 0.030

Table 3. Activation energy for the fast and slow components of both bands.

3.3 Application to high dose dosimetry

3.3.1 Room temperature fading behaviour

Five replicate glass dosimeters were irradiated with ^{60}Co gamma rays to 30 kGy. The changes of absorbance were followed up to 535 days. After each measurement glass samples were stored in the dark at room temperature. Fig. 6 presents the fading behaviour of the 410 nm absorption band at room temperature respectively for the long-term and the short-term period. A strong fading can be observed in the first 9 days followed by a slow fading up to 535 days. The 600 nm absorption band showed similar behaviour.
The data in Fig. 6 were fitted using first-order kinetics based on the data after the first 10 days. The coefficient of correlation (R^2) was 0.99.

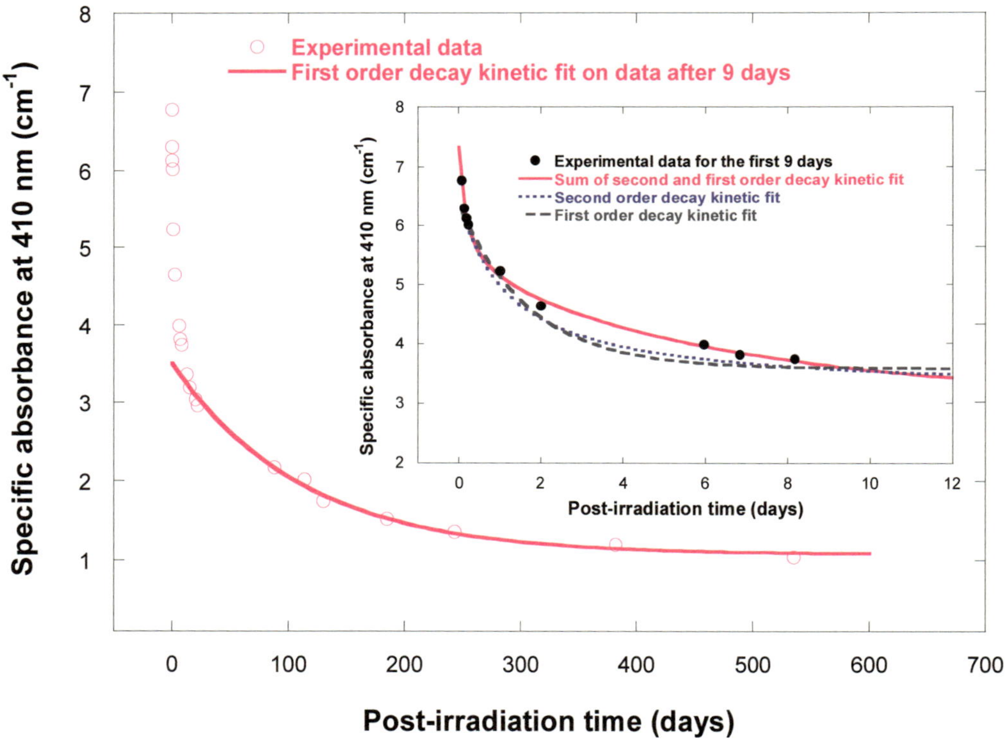

Fig. 6. Post-irradiation stability at room temperature of response of glass samples irradiated at 30 kGy. Inset: kinetics of the fading behaviour at room temperature for the short-term period.

The stability of radiation induced colour centres at room temperature was mainly controlled by the initial strong fading process. This initial fading process seems to follow neither a simple first-order nor second-order kinetics, but it can be well described by a sum of a first order decay kinetics function (exp $(-t/\tau)$) and second order decay kinetics function $[1+ k (t/\kappa)]^{-1}$, where t is the annealing time, τ and κ are appropriate time constants and k is a fitting parameter.

The thermal decay of colour centres after irradiation is mainly controlled by the diffusion of species in glass and it is generally attributed to diffusion - limited processes (Agnello & Boizot, 2003; Sheng et al., 2002c). Agnello and Boizot speculate that reaction of mobile species induced by irradiation with the induced colour centres is in the origin of their thermal decay (Agnello& Boizot, 2003). These mobile species are for example oxygen ions or anions and cations impurities released by broken bonds. In fact, a large number of studies have demonstrated that alkalis ions are movable and diffusible under irradiation. The association with NBO confines the alkali ion to local motion, whereas the absence of a co-ordinating NBO allows the alkali ion to explore more easily its environment (Ojovan & Lee, 2004).

By the combination of (MAS NMR) and (XPS) spectroscopy investigations, (Boizot and al., 2000) provided additional evidence in favour of a long range migration of sodium in glass.

Sheng et al., 2002c), gave evidence that the long-term fading process of colour centres induced by X-rays in soda-lime silicate glass was dominated by a first order kinetics, while, both the first and the second order kinetics played role in the short-term fading process (Sheng et al., 2002). Indeed, each NBOHC induced by gamma irradiation is surrounded by electrons and other NBOHCs. They assumed that the recombination of NBOHCs is controlled by the diffusion of electrons in the glass network (reaction 4) and/or by reaction of NBOHC with neighboring NBOHCs (reactions 2 and 3):

$$\equiv \mathrm{Si}-\mathrm{O}^{\circ}+\mathrm{e}^{-} \rightarrow \equiv \mathrm{Si}-\mathrm{O}^{-} \qquad (2)$$

$$2(\equiv \mathrm{Si}-\mathrm{O}^{\circ}) \rightarrow \equiv \mathrm{Si}-\mathrm{O}-\mathrm{O}-\mathrm{Si} \equiv \qquad (3)$$

It is obvious that the reaction (2) dominate the recombination process at short-term range because of the small initial distance between NBOHCs (Waite, 1957). Sheng and al. demonstrated that in addition to reaction (2), reaction (3) played also role in the short-term recombination process, while the long-term recombination process was dominated only by reaction (2).

3.3.2 Effect of post irradiation heat treatments on the room temperature fading

The effect of post irradiation heat treatments on the glass response fading was studied in the temperature range of 60-150 °C using sets of three glass samples. After gamma irradiation with 30 kGy absorbed dose, dosimeter sets were immediately submitted to the different heat treatments for 20 minutes, which was found to be the best treatment time, and stored after irradiation in the dark at room temperature. Optical absorbance measurements were carried out up to two months. The specific absorbance values were normalized to the first measurements taken 5 minutes after the heat treatments. The results are presented in Fig. 7 and Table 4. The best results have been obtained with heat treatments at 150 °C (20 min). This procedure is very effective for the removal of unstable entities responsible for the initial strong fading. The standard deviation of glass dosimeters response measurements is about 0.5 % (1σ) within the first two hours after irradiation. The response decay of irradiated glass dosimeters is about 8 % between the first 24 h and 20 days. This means that glass dosimeters can be evaluated either within the first two hours or just after one day after irradiation and heating.

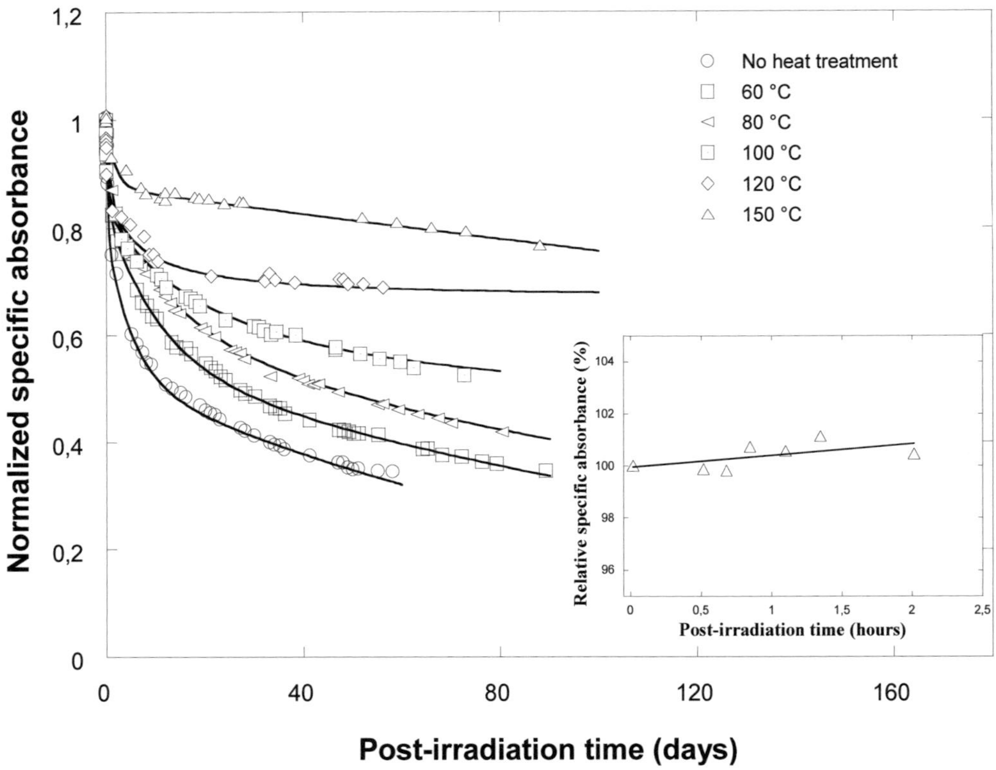

Fig. 7. Post-irradiation stability at room temperature of glass samples after 30 kGy irradiation and different heat treatments for 20 minutes. Inset: behaviour of glass samples, heated at 150 °C for 20 min, within the first two hours after the heat treatment.

Storage time	Reduction of relative specific absorbance (%)							
	Room temperature	-5 °C	-20 °C	60 °C	80 °C	100 °C	120 °C	150 °C
24 h	25	5	5	17	16	15	12	7
20 days	54	40	39	49	45	35	29	15
60 days	65	66	65	60	54	45	31	19

Table 4. Effect of thermal treatments on fading of glass samples after an irradiation of 30 kGy with gamma rays

3.3.3 Dose response curves

In order to find out the useful dose range for this silicate glass, the response curve (specific absorbance versus dose) was measured in the dose range 0.5 kGy– 87 kGy. Irradiations were carried out at the 60Co Gamma cell 220 excel of the Egyptian Radiation Technology Centre with the dose rate of 4.78 kGy/h at controlled temperature of 34°C. All data for the dose response curve were taken 24 h after irradiation. The reported specific absorbance was measured at 410 nm (Figure 8). The specific absorbance shows a rapid growth up to 40 kGy.

At higher doses the specific absorbance continued to grow slowly up to 87 kGy which was the upper dose level of the present experiments. The glass response had not yet reached saturation at this dose level. The dose response curve taken after irradiation and heat treatment at 150°C for 20 min for the dose range of 0.5–87 kGy is shown as inset of the Fig.8. The dose response curves shown in Fig.8 are successfully fitted following Mashkov equation.

T °C	First-order kinetics τ (h)	Second order kinetics κ (h)
Room temperature	189.6	30.9
-5	1128	50
-20	2184	117
60	169.5	7.5
80	168	0.43
100	110.4	0.024
130	6.25	0.0006
150	3.15	0

Table 5. Fading time constants for temperature range from -20 to 150 °C

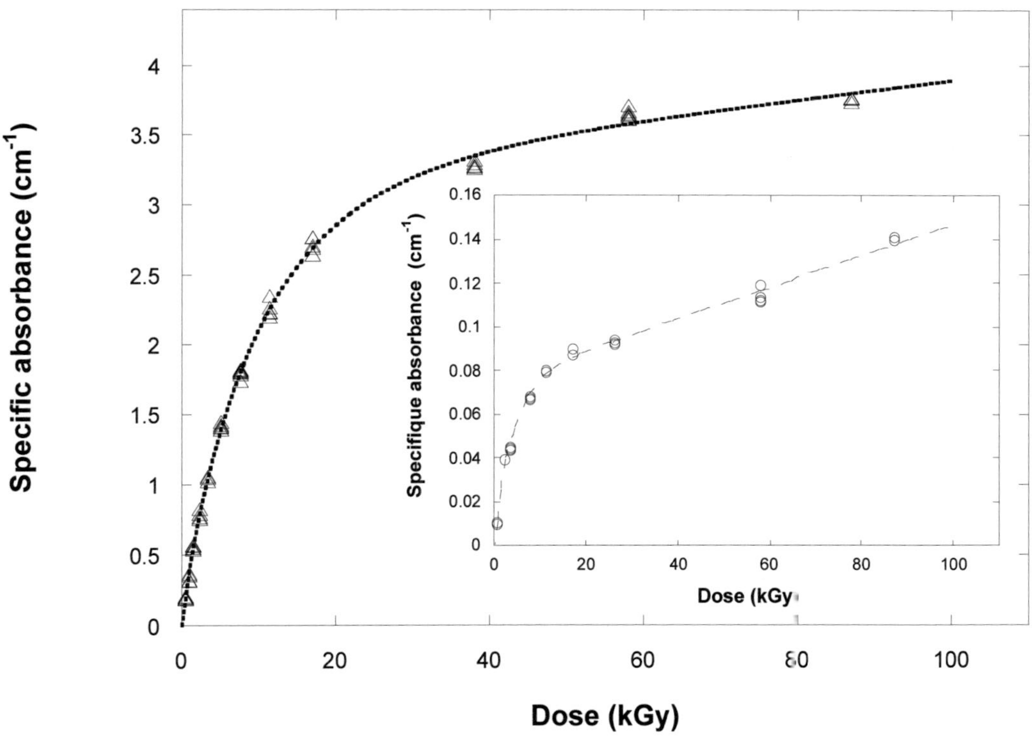

Fig. 8. Response curve for gamma irradiated glass in the dose range of 0.5-87 kGy (dose to water), absorbance at λ = 410 nm, dose rate: 4.78 kGy/h. Inset: response curve after heat treatment at 150°C for 20 min.

3.3.4 Effect of temperature during irradiation

The response of the majority of dosimeters is affected by the temperature during irradiation (Abdel-Fattah and Miller, 1996, Farah et al., 2004). In large gamma ray irradiators, the dosimeter temperature can reach 60°C for high doses. This effect should be carefully investigated especially if the dosimeters are going to be used for dose measurement at temperatures different than that for which they were calibrated.

In order to study the effect of temperature during irradiation on response of the glass dosimeters, samples were irradiated to an absorbed dose of 5 kGy at temperatures in the range from -3°C to +80°C, temperature range used during the usual irradiation processing. After irradiation, dosimeters were stored for 24 h in dark at room conditions (25 ± 3°C, 40-60% R.H.) then specific absorbance was measured at 410 nm. Figure 9 shows the specific absorbance, normalized at -3°C, plotted as a function of temperature during irradiation. The response of glass dosimeter decreases from -3°C to +80° C and the temperature coefficient of optical absorption was negative corresponding to a mean value of (-0.53 ± 0.02) %°C-1 and (-0.45 ± 0.04) % °C-1 respectively for 410 nm and 600 nm. These results extend the published data of (Zheng et al., 1988) and (Zheng, 1996). These authors observed, for glass dosimeters irradiated with 3 kGy, a negligible variation of absorbance at the wavelength of 500 nm between 0 and 50 °C followed by a fast decrease between 50 and 80 °C.

These differences can be explained by the fact that the colour centre studied by (Zheng et al., 1988) is probably due to free electrons trapped at some imperfection in the glass structure. While the induced absorption bands observed in our glass can be attributed to a trapped holes (NBOHCs colour centres).

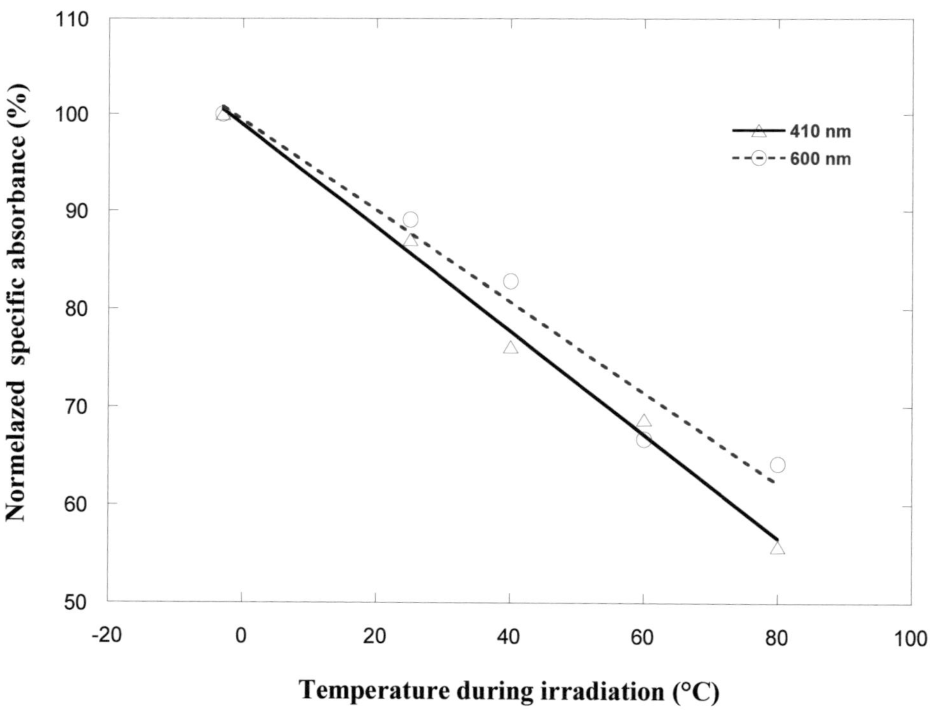

Fig. 9. Effect of temperature during irradiation on the response of glass samples irradiated at 5kGy and measured at 410 and 600 nm absorption bands

3.3.5 Relative humidity effect during post-irradiation storage

In order to investigate the effect of Relative Humidity (R.H.) during post-irradiation storage on the response of glass dosimeter, sets of three replicate glass samples were irradiated using 60Co gamma rays at a dose of 7 kGy. Following irradiation, glass samples were stored in the dark under different extreme R.H. conditions, dried condition (about 0%), moist condition (in water) and room condition (between 40-60%). The specific absorbance was measured at 410 and 600 nm up to 22 days. As seen in Figs. 10 and 11, small differences of specific absorbance have been observed between the glass dosimeters stored in the two extreme R.H. conditions (0 and 100%). Relative differences of specific absorbance compared to the ambient conditions of glass dosimeters stored in extreme R.H. conditions (0 and 100%) are significant for the two wavelengths 410 and 600 nm.

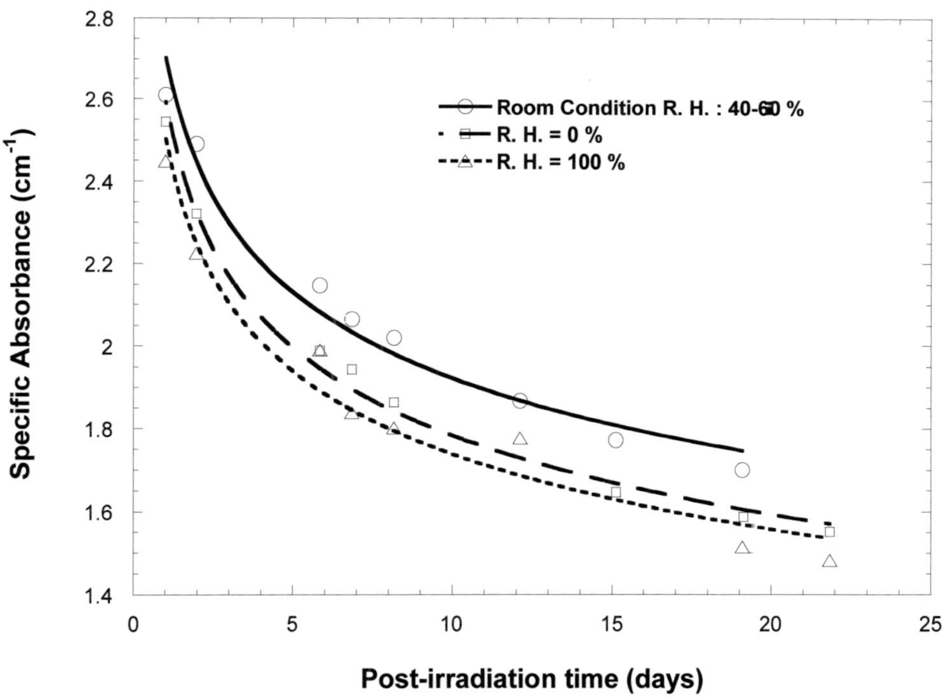

Fig. 10. Post-irradiation stability of irradiated glass to 7 kGy and stored at a different humidity conditions (410 nm band).

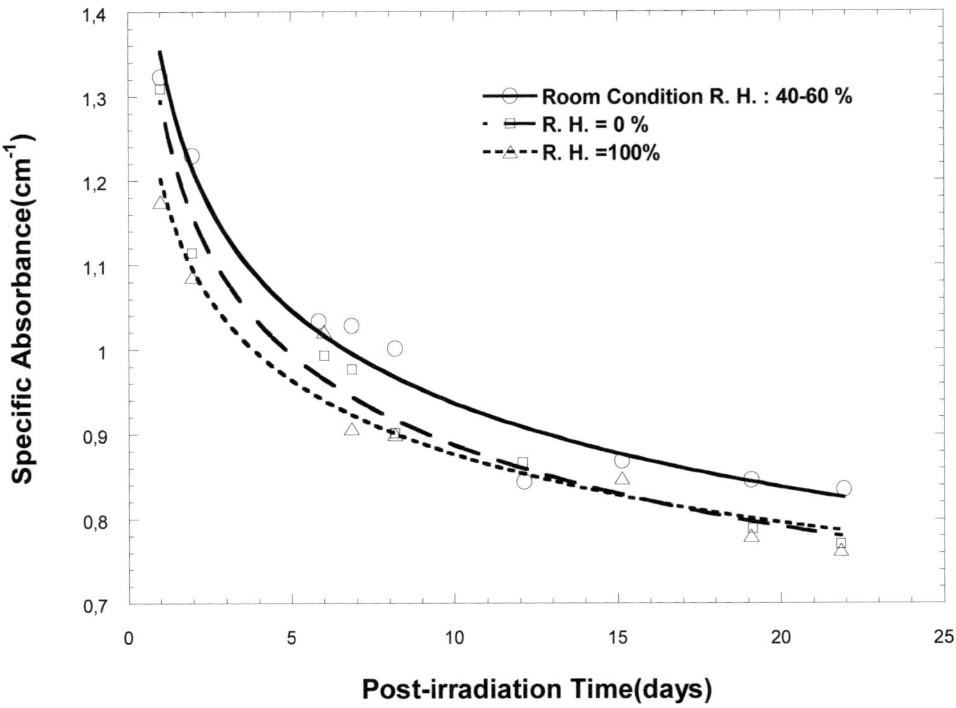

Fig. 11. Post-irradiation stability of irradiated glass to 7 kGy and stored at a different humidity conditions (600 nm band).

3.3.6 Dose rate effect

In the semi-industrial applications that depending on plant design, products may be treated in different positions with different dose rates, whereas, dosimeters were calibrated at a fixed dose rate. In order to investigate the effect of the dose rate on the response of glass dosimeter, several groups of glass samples were irradiated to different doses in the range of 0.5-50 kGy at the following dose rates: 1 and 6 kGy/h. Glass samples were submitted to a thermal treatment at 150°C for 20 minutes immediately after irradiation. Specific absorbance measurements were carried out after 24 hours. As seen in Fig. 12, the dose rate effects did not appear to be significant in the dose range of 0.5-20 kGy. Only a weak change of response within 1-2% was observed. This result is in good agreement with published data (Engin et al., 2006; Zheng et al., 1988). Above 20 kGy, glass samples exhibit significant dose rate dependence within 8-11%. In fact, the absorbance increases within 8–11% when the dose rate decreases. This observation is in agreement with the data published by (Ezz-Eldin et al., 2008). For low dose rate the rate of electrons production is low, thus the ejected electrons has enough time to annihilate a glass defect or forming non-bridging oxygen centers which causes an increase in the glass absorbance.

For the high dose rate, the rate of electrons production become high, which gives a better chance for their fast recombination rather than for annihilation resulting in less absorbance.

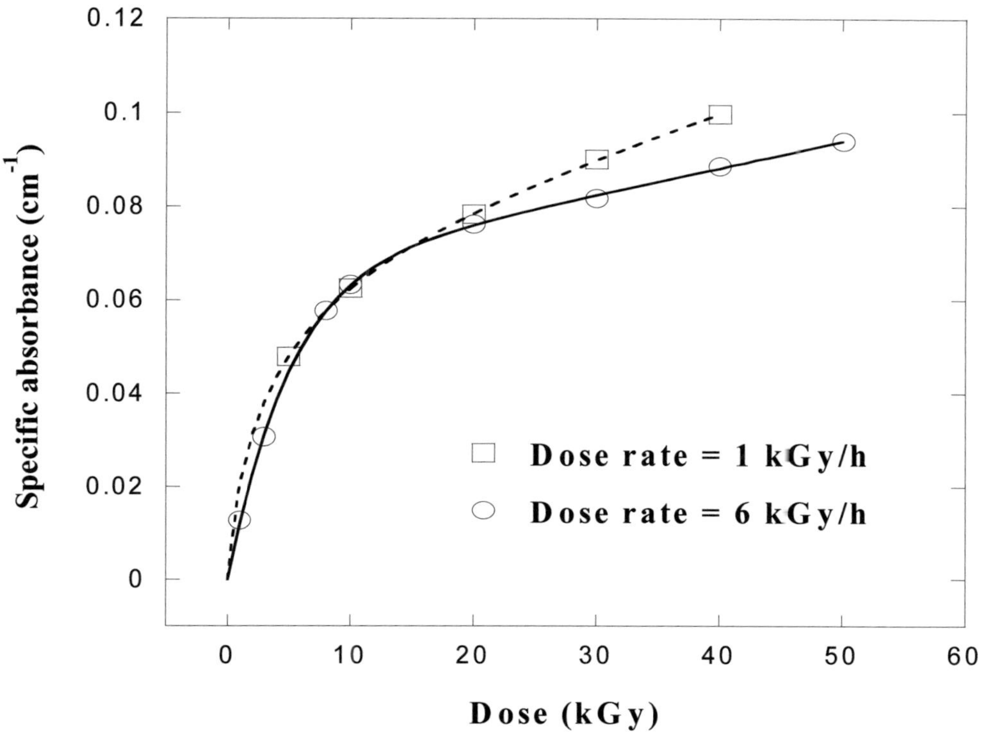

Fig. 12. Dose rate effect on the response of glass samples irradiated at 1 and 6 kGy/h and measured at 410 nm absorption band.

3.3.7 Reuse and reproducibility of glass dosimeter

Few investigations were found in the literature (Sheng, 2002b) about the reuse of glass dosimeters, by thermal bleaching of the radiation induced coloration.

In order to study the possibility of re-using glass dosimeter, three sets of glass samples, each containing three, were irradiated, at room temperature, with 60Co gamma rays at three different doses 1, 5 and 20 kGy.

Measurement of the specific absorbance was taken 30 min after irradiation. Then the irradiated samples were submitted to a heat treatment of 300°C for 30 min, sufficient to remove the radiation induced colour centres, before performing a new irradiation. This procedure was repeated six times giving good reproducibility as it can be seen from the results shown in Fig. 13. The standard deviation of the measurements was found to be lower than 4% (1σ).

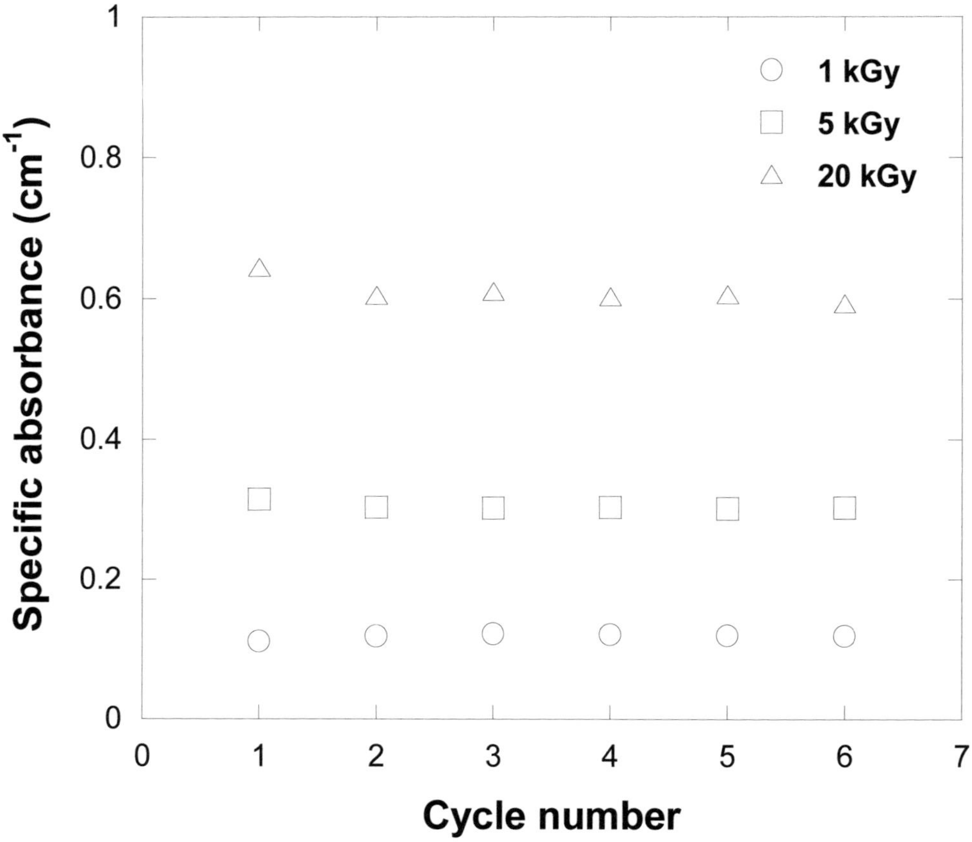

Fig. 13. Reproducibility of the results by reuse of glass dosimeter.

3.4 In-plant calibration of glass dosimeter for gamma irradiation

3.4.1 Calibration conditions

The in-plant calibration of glass dosimeters was done in the Tunisian semi-industrial cobalt-60 gamma irradiation facility rigorously following the procedure described in section 2.5 Method and condition of calibration. The phantoms, containing the alanine transfer standard dosimeters and the glass samples, were fixed on dummy product boxes and irradiated to nominal doses from 0.5 to 17 kGy at the dose rate of 1 kGy/h.

All dosimeters were read 24 hours after irradiation. Figs. 14 and 15 represent the response functions where the specific absorbance of the irradiated glasses vs dose to water, as measured by the alanine dosimeters, is reported. Here the mean values of the specific absorbance for each group of glass samples are plotted.

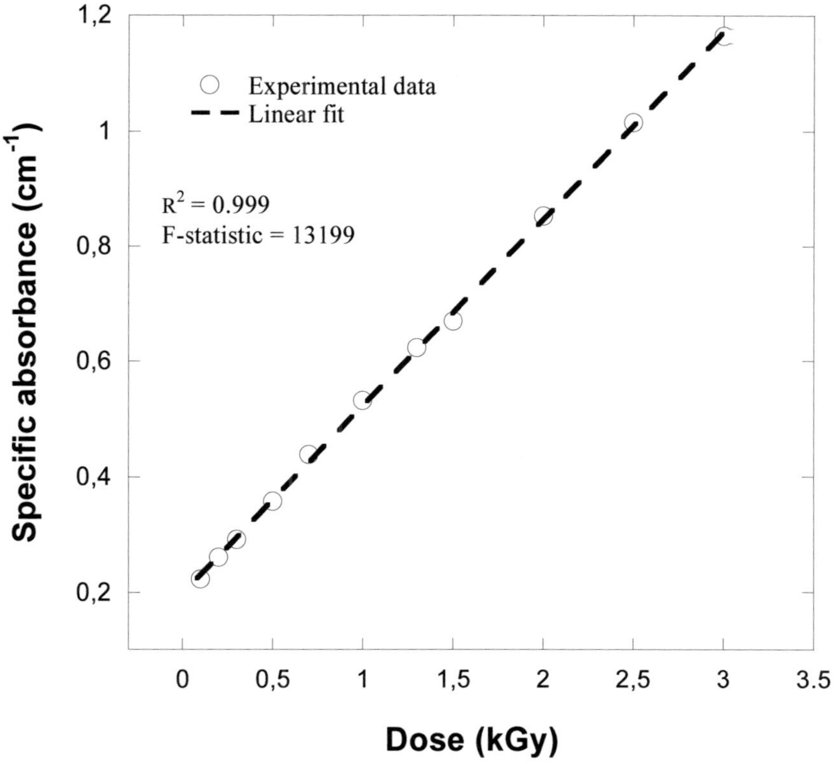

Fig. 14. Calibration curve of glass dosimeter in the dose range 0.1-3 kGy (dose to water). Y = 0.1982 +0.3245.X equation is used for linear fit.

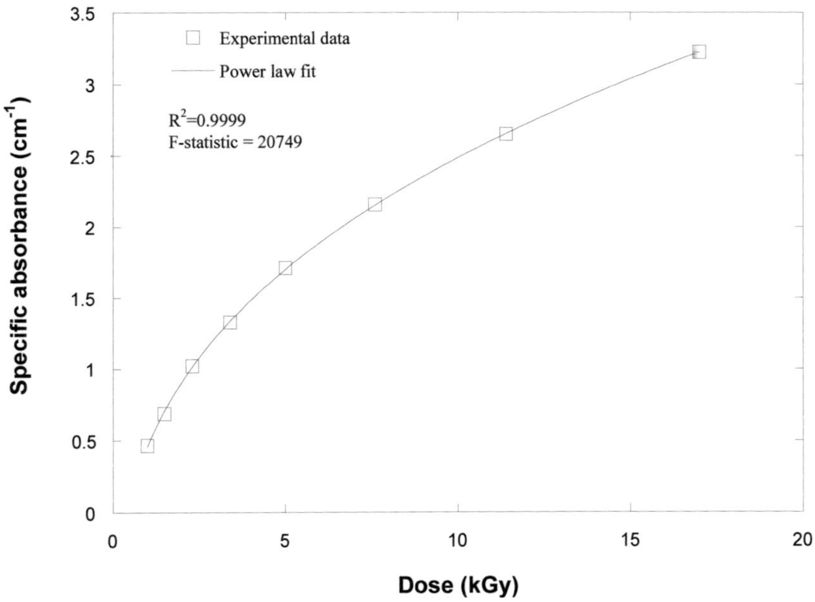

Fig. 15. Calibration curve of glass dosimeter in the dose range of 1-17 kGy (dose to water). Y = - 1.262 + 1.717.X 0.34 equation is used for power law fit.

3.5 Estimation of uncertainty

The procedure outlined in ISO/ASTM Standard 51707 (ISO/ASTM, 2002) was followed to estimate the overall uncertainty in the measurement of absorbed dose from the glass dosimeter. Three components of uncertainty were taken into account. These included, (1) uncertainty in absorbed doses reported by the Aérial, (2) uncertainty in the spectrophotométric analysis of dosimeter response and measurement of dosimeter thickness (Eq. 4), and (Eq. 5) goodness of fit of the power law function to the calibration data.

$$CV_{overall}(\%) = \sqrt{\frac{\sum_{i}(n_i - 1)S_{i-1}^2/\bar{k}^2}{\sum_{i}(n_i - 1)}} \times 100\% \tag{4}$$

Where:
C.V. (%) = overall coefficient of variation
Si-1= sample standard deviation for ith set of measurement,
(ni-1) degrees of freedom for ith set of data,
$\bar{k}$ = average value of specific absorbance for ith set of measurement,
ni = number of replicate measurements for ith set of data.

$$u_f = \sqrt{\frac{\sum_{i=1}^{i=n}(y_i - f(x_i))^2}{n - m}} \tag{5}$$

Where:
uf = fit standard uncertainty
yi = y data value
f (xi) = predicted y value using the power law function
n = number of data points
m = number of coefficient fitted
Results of this analysis are given in Table 6. The components of uncertainty combined in quadrature gave an estimate of overall uncertainty at a 95% confidence level of ± 7.58% (2σ) which is in basic agreement with the expected uncertainty for the routine use of these glass dosimeters and acceptable for the intended applications.

Uncertainty	Type A (%)		Type B (%)
Reference Dose (Aérial)		2.50	
Uncertainty in calibration curve	2.43		.48
Overall uncertainty (%)		3.79	
Expanded overall uncertainty (%) (k=2)		7.58	

Table 6. Components of uncertainty.

3.6 Field trial irradiations

In order to check the applicability and the performance of glass for routine application, field trial irradiations were carried out with the same irradiation arrangement for used the in-plant calibration. Five sets of three glass dosimeters and five sets of three Gammachrome dosimeters were irradiated, together in the polystyrene phantom in close contact, without shielding each other, to nominal doses of 0.2, 1, 1.5, 2.5, and 3 kGy. Four sets of three glass dosimeters and four sets of three Amber Perspex dosimeters were irradiated, together in the polystyrene phantom in close contact, without shielding each other, for the nominal doses: 1, 3, 7, and 15 kGy.

Dose determination for the irradiated glass was done using the established in-plant calibration curves of Figs.14 and 15. The evaluation of the absorbed dose with Harwell Perpsex dosimeters was done using our own calibration function traceable to the Aérial SSDL.

Linear regressions were applied to compare the absorbed dose results obtained in the gamma plant, measured by glass dosimeters and by the Gammachrome and Amber Perspex dosimeters (Figs. 16 and 17). The values of slope lines and correlation coefficients calculated from the data of Figs. 16 and 17 indicate the good agreement.

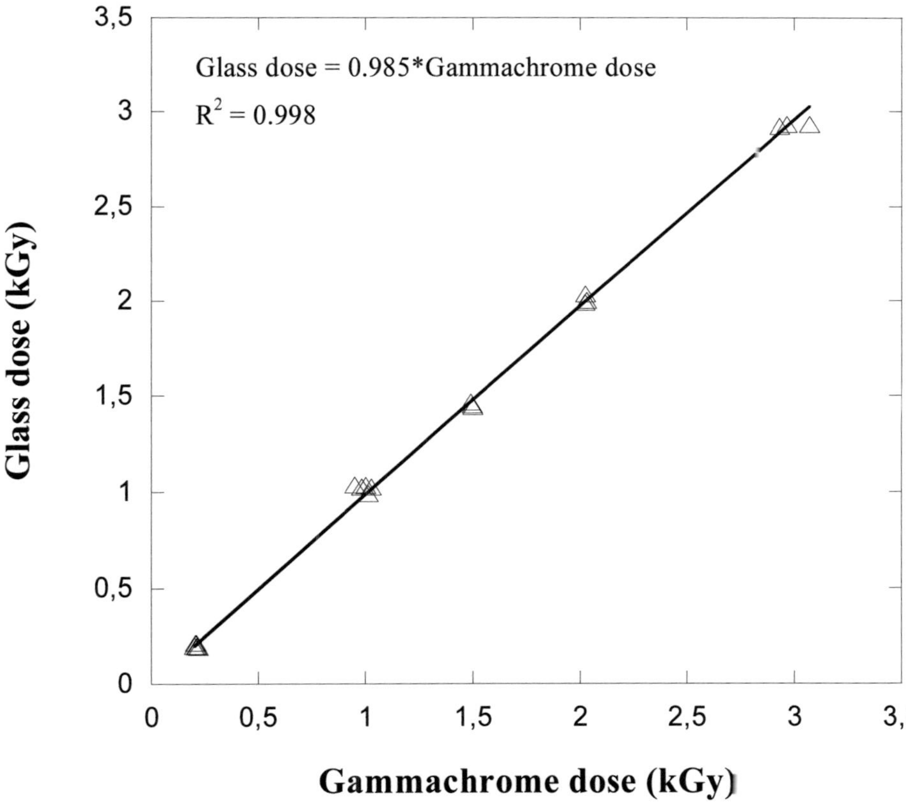

Fig. 16. Comparison of doses measured by Gammachrome dosimeters and glasses irradiated together during production runs at the gamma plant.

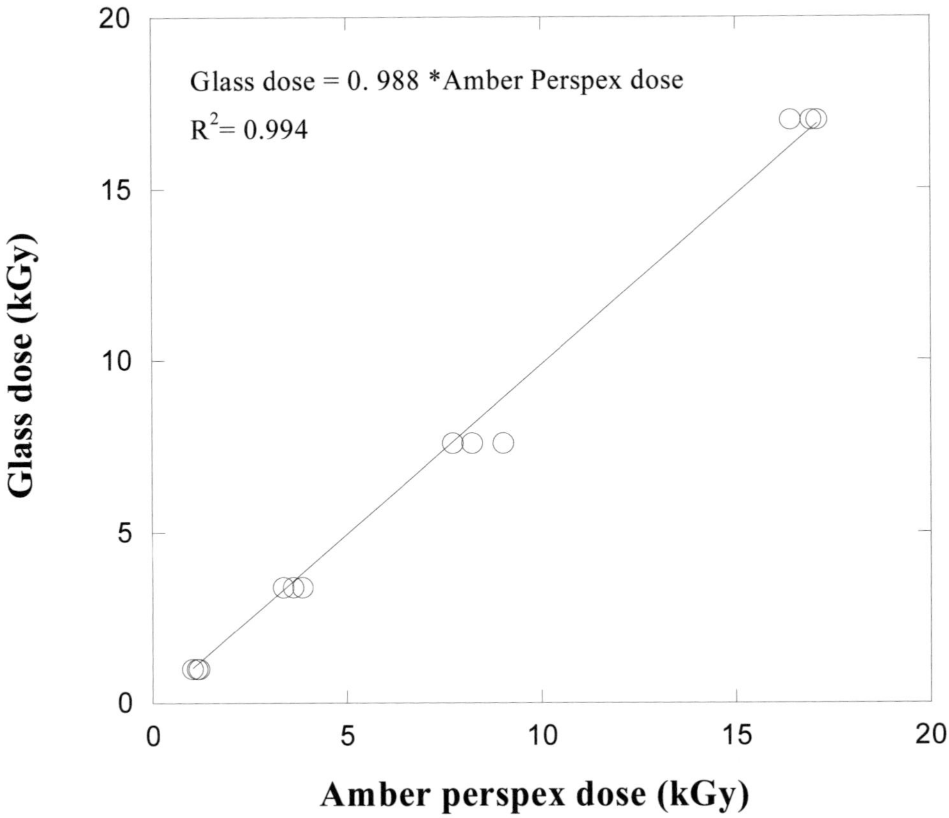

Fig. 17. Comparison of doses measured by Amber Perspex dosimeters and glasses irradiated together during production runs at the gamma plant.

4. Conclusion

The induced optical absorption by gamma irradiation in the visible range of this silicate glass is due to the generation of two Non Bridging Oxygen Hole Centres: HC_1 at 410 nm and HC_2 at 600 nm. The similarity of the activation energy values of the annealing process for both bands suggests that the related optical transitions correspond to the same type of NBOHCs in different configurations. The increase of gamma irradiation dose had no influence on the band peak positions and the FWHM of the induced colour centres and only the heights and the areas under the bands changed.

The nonlinear growth of the absorbance with increasing of dose can be explained by the competition of two different colour centres creation processes. The first is an intrinsic process corresponding to the rupture of the bond ≡ Si-O-Si ≡ in which the concentration of colour centres increases linearly with the dose. The second one is an extrinsic process corresponding to the activation of a fixed number of pre-existing precursors which must therefore saturate.

These colour centres were unstable at room temperature or at heating conditions. Most induced colour disappeared after being treated at 300 °C for 20 minutes and completely disappeared at 500 °C. The initial strong post-irradiation decay can be explained by reaction of NBOHC with neighboring NBOHCs because of the small initial distance between them. At long-term stage the slow decay is only controlled by the diffusion of impurities and electrons in the glass network.

The present study reveals that the optical absorbance response of this type of glass to γ can be quantified reasonably well in the 0.1-50 kGy range. The tests conducted on this type of glass show its good performances in the production irradiators and if the glass samples are calibrated in the plant they are reliable dosimeter material that can be used as a routine dosimeter for measuring doses between 0.1-17 kGy ensuring that the environment conditions are carefully controlled during the use of this dosimeter. The effect of influence quantities on the glass response could be minimized by performing the in-plant calibration. Even if the fading effect can give rise to significant errors during the routine use of these dosimeters, it is not of great importance if calibration curves and routine dose evaluation are done at the same time interval.

These results indicate that this material is very interesting for dose evaluation in many radiation processing applications.

5. References

Abdel-Fattah, A., Miller, A. (1996). Temperature, Humidity and Time. Combined effects on Radiochromic film dosimeters. *Radiat. Phys. Chem.* 47, 611–621.

Aérial. (2011). Available from http://www.aerial-crt.com

Agnello, S. and Boizot, B. (2003). Transient visible-UV absorption in beta irradiated silica. *J. Non-Cryst. Solids* 332, 84.

Anderson, D.L., Belgya, T., Firestone, R.B., Kasztovsky, Zs., Lindstom, R.M., Molnár, G.L., Révay, Zs., and Yonezawa, C. (2004). Handbook of Prompt Gamma Activation Analysis (Kluwer Academic Publishers, Dordrecht), The Netherlands.

Boizot, B. (1997). Défauts d'irradiation dans la silice amorphe a-SiO_2, 1997. Rapport Bibliographique CEA-R-5749.

Boizot, B., Petite,G., Ghaleb, D., Pellerin, N., Fayon, F., Reynard, B., Calas, G. (2000). Migration and segregation of sodium under β-irradiation in nuclear glasses. *Nucl. Instrum. and Meth. B* 166-167, pp 500-504.

Engin, B., Aydas, C., Demirtas, H. (2006). ESR dosimetric properties of window glass. *Nucl. Instrum. and Meth. B* 243, 149-155.

Ezz-Eldin, F.M., Mahmoud, H.H., Abd-Elaziz, T.D., El-Alaily, N.A. (2008) Response of commercial window glass to gamma doses. *Physica B* 403, 576.

Farah,K., Kuntz, F., Kadri,O., Ghedira, L. (2004). Investigation of the effect of some irradiation parameters on the response of various types of dosimeters to electron irradiation. *Radiat. Phys. Chem.* 71, 337–341.

Farah, K., Jerbi, T., Kuntz, F., and Kovacs, A. (2006). Dose measurements for characterization of a semi-industrial cobalt-60 gamma-irradiation facility. Radiat. Meas. 41, 201.

Farah K, Mejri A, Hosni, F., Ben Ouada , H., Fuochi P G, Lavalle, M, Kovács, A. (2010). Characterization of a silicate glass as a high dose dosimeter. Nuclear Instruments and Methods in Physics Research A, 614, pp. 137–144.

Friebele, E.J. (1991). Radiation effects. In: Optical Properties of Glass, Uhlmann. , D.R., Kreidl, N.J. (Eds.), 205, American Ceramic Society.

Fuochi P G, Corda, U, Lavalle, M, Kovács, A, Baranyai, M, Mejri, A and Farah, K (2008). Commercial window glass tested as possible high dose dosimeter, Electron and gamma irradiation. Proc. 10th Int. Conf. on Astroparticle, Particle, Space Physics, Detectors and Medical Physics Applications Ed M Barone et al (Villa Olmo, Italy, 8-12 October 2007) pp 70-74.

Fuochi P G, Corda, U, Lavalle, M, Kovács, A, Baranyai, M, Mejri, A and Farah, K. (2009).

Dosimetric properties of gamma-and electron-irradiated commercial window glasses Nukleonika, 54, pp. 39-43.

Griscom, D.L. (1984). Electron spin resonance studies of trapped hole centers in irradiated alkali silicate glasses: A critical comment to current models for HC1 and HC2. J. Non-Cryst. Solids 64, 229.

ISO/ASTM Standard. (2002). Guide for Estimating of Uncertainties in Dosimetry for Radiation Processing, ISO/ASTM Standard 51707. American Society for Testing and Materials, Philadelphia, PA.

Jin, Y.G. and Chang, K.J. (2001). Mechanism for the Enhanced Diffusion of Charged Oxygen ions in SiO2. *Phys. Rev. Lett.* 86 (9), 1793.

Mashkov, V.A., Austin, W. R., Zhang, L., and Leisure, R.G. (1996). Fundamental role of creation and activation in radiation-induced defect production in high-purity amorphous SiO2. Phys. Rev. Lett., 76 (6), 2926.

Mejri A, Farah K, Eleuch H, and Ben Ouada, H. (2008). Application of commercial glass in gamma radiation processing Radiat. Meas. 43 1372 – 76.

Ojovan, M.I. and Lee, W.E. (2004). Alkali ion exchange in γ-irradiated glasses. *J. Nucl. Mater.* 335, 425.

Sharpe, P.H.G. and Miller, A. (1999). Guidelines for the calibration of routine dosimetry systems for use in radiation processing. NPL report CIRM 29, National Physical Laboratory, (Teddington, UK).

Sheng, J. (2002a). Easily recyclable coloured glass by X-Ray irradiation induced coloration. *Glass Technology*, 43, 238.

Sheng, J, Kadono, K., Utagawa, Y., Yazawa,T. (2002b). X-ray irradiation on the soda-lime container glass. Appl. Radiat. Isotop. 56, 61.

Sheng, J., Kadona, K., Yazawa, T. (2002c). Fading behaviour of X-ray induced color centers in soda-lime silicate glass. *Appl. Radiat.Isotop.* 57, 813.

Suszynska,M. and Macalik, B. (2001). Optical studies in gamma irradiated commercial soda-lime silicate glasses. *Nucl. Instrum. and Meth. B* 179, 383.

Tsai, T.E., Griscom, D.L. and Friebele, E.J. (1989). Medium-range order and fractal annealing kinetics of radiolytic atomic hydrogen in high-purity silica. *Phys. Rev. B* 40 (9) 6374.

Waite, T.R. (1957). Diffusion-limited annealing of radiation damage in germanium. *Phys. Rev.* 107 (21), 471.

Yokota, R. (1954). Colour centres in alkali silicate and borate glasses. Phys. Rev. 95, 1145.

Yokota, R. (1956). Colour centres in alkali silicate glasses containing alkaline earth ions. Phys. Rev. 101, 522.

Zheng, Z., Honggui, D., Jie, F., Daochuan, Y. (1988). Window glass as a routine dosimeter for radiation processing. *Radiat. Phys. Chem.* 31, 419-423.

Zheng, Z. (1996). Study on the possibility of reading two kinds of data from one glass detector. *Radiat. Phys. Chem.* 50, 303-305.

29

Optical Storage Phosphors and Materials for Ionizing Radiation

Hans Riesen and Zhiqiang Liu
The University of New South Wales, Australia

1. Introduction

Over recent years there has been significant progress in the development of optical storage phosphors and materials for ionizing radiation, resulting in important applications in the fields of dosimetry and computed radiography. The aim of this Chapter is to provide some background information as well as an overview of a range of modern storage phosphors and materials and their applications in dosimetry and computed radiography. Dosimetry is the measurement of absorbed dose of ionizing radiation by matter or tissue and is measured in the SI unit of gray (Gy) or sievert (Sv), respectively, where 1 Gy and 1 Sv are equal to 1 joule per kilogram. It is noted here that the non-SI units of rad (absorbed radiation dose; for matter) and rem (Roentgen equivalent in man; for tissue) are still heavily used; their conversions are given by 1 Gy=100 rad and 1 Sv=100 rem. A dosimeter allows us directly or indirectly to measure exposure, kerma, absorbed or equivalent dose and associated rates of ionizing radiation. A dosimetry system consists of the actual dosimeter and an associated reader. A dosimeter is characterized by its accuracy and precision, linearity, energy response, dose and dose rate dependence, spatial resolution and directional dependence.

It is well documented that even low doses of ionizing radiation can induce cancer. Thus personal radiation monitoring, i.e. dosimetry applied to people, has become highly important in recent years as the number of potential radiation sources is ever increasing. Radiation monitoring will also become more important in medical diagnostics and in radiation therapy as regulations become stricter. Nuclear accidents, such as the recent events in Fukushima, Japan, have highlighted the need for inexpensive and effective radiation monitoring systems that can be deployed to a large population. We note here that, of course, the exposure to natural background radiation is, to a great extent, unavoidable. For example, the average annual radiation dose per capita from natural and medical sources is 2.0 mSv/year and 2.8 mSv/year in Australia and the United Kingdom, respectively. The amount of radiation exposure is a function of geography and building standards. For example, in European homes there is a relatively high level of radon as the internal walls are based on clay bricks. This is reflected in the fact that the average equivalent dose by radiation from radon progeny is 0.2 mSv/year and 1.4 mSv/year in Australia and the United Kindom, respectively.

Ionizing radiation has a wide range of applications such as in the sterilization of medical hardware and food, in medical diagnostics and radiotherapy and in technical and scientific imaging (Andersen et al., 2009; Dyk, 1999; Eichholz, 2003; Mijnheer, 2008; Podgorsak, 2005;

Riesen & Piper, 2008). For safety and quality assurance, reliable radiation monitoring systems (dosimetry) are mandatory for these applications (Mijnheer, 2008). Dosimetry systems that are currently used for evaluating dose and its distribution include (Dyk, 1999; Mijnheer, 2008; Podgorsak, 2005; Schweizer et al., 2004):

1. Ionization chambers and electrometers for the determination of radiation dose. The standard chambers are either cylindrical or parallel plates. Active volumes for cylindrical chambers are between 0.1 and 1 cm^3 i.e. they are relatively bulky. Ionization chambers allow an instant readout, are accurate and precise and are recommended for beam calibration. Disadvantages include the fact that they have to be connected by cables and are relatively bulky i.e. they, in general, do not allow point measurements.
2. Standard radiographic films are based on radiation sensitive silver halide emulsions in thin plastic films. Exposure to ionizing radiation leads to a latent image in the film that can be read out by measuring the optical density variations within the film. Since the films are very thin they do not perturb the beam. However, the photosensitive silver halide grains in the film are rapidly saturated with a low number of X-ray photons i.e. the dynamic range of the blackening process is very limited. Nevertheless, due to the thin film thickness which limits scattering effects, the spatial resolution of this method is still one of the highest up to date. A dark room processing facility is required to work with radiographic film and they also need proper calibration.
3. Radiochromic film is based on polymer films that contain dye molecules that polymerize upon exposure to ionizing radiation. For example, GafChromic® film yields a blue coloration due to the radiation induced polymerization to polyacetylene. The film can be read out by a standard scanner.
4. Thermoluminescence is based on thermally excited phosphorescence of a range of systems. Sensitive materials include $CaSO_4$:Dy, CaF_2:Dy and LiF:Mg,Ti. The most commonly used thermoluminescence (TL) dosimeters in radiotherapy include the tissue-equivalent LiF:Mg,Cu,P and LiF:Mg,Ti. TL dosimeters can come in all forms and sizes and allow high spatial resolution e.g. they are suitable for point dose measurements. However, TLD requires relatively complex reader units and hence instant readout is not possible. Moreover, the information is destroyed upon readout and can, in general, easily be lost.
5. Silicon diode and MOSFET (metal-oxide-semiconductor-field-effect-transistor) based dosimetry offer the advantages of small size, instant readout and high sensitivity but require that cables are connected. Also, their sensitivity changes with accumulated dose i.e. results are not very reproducible.
6. Optically stimulated (or photostimulated) luminescence (OSL or PSL) systems are related to TLD but require higher stimulation energy. In a typical OSL (PSL) material, such as BaFBr/Eu^{2+}, red light stimulates the recombination of electrons and holes that were created by exposure to ionizing radiation, yielding broad blue luminescence light. OSL materials offer high spatial resolution and sensitivity but are susceptible to ambient light and therefore data can be inadvertently lost. An OSL material that has been used in vivo fibre based dosimetry is the Al_2O_3:C system by Landauer. In this case, the blue emission is stimulated by green light (e.g. Nd:YAG laser at 532 nm). However, the size of the crystal required for this system is still relatively large (2x0.5x0.5 mm^3) and the material loses its information under ambient lighting conditions.
7. Electronic Portal Imaging Devices (EPIDs), based on amorphous silicon, are used for real time acquisition of megavolt images during patient treatment and can be, to some

extent, used to undertake dosimetry. Disadvantages of these devices include the fact that they are not linear with dose and dose rate and they are also relatively bulky and need to be connected with cables etc.

Storage phosphors can also be used as two-dimensional dosimeters, i.e. some spatial information about radiation exposure can be derived. This, of course, is based on computed radiography which is another important application of storage phosphors. The basic principle of computed radiography is illustrated in Figure 1.

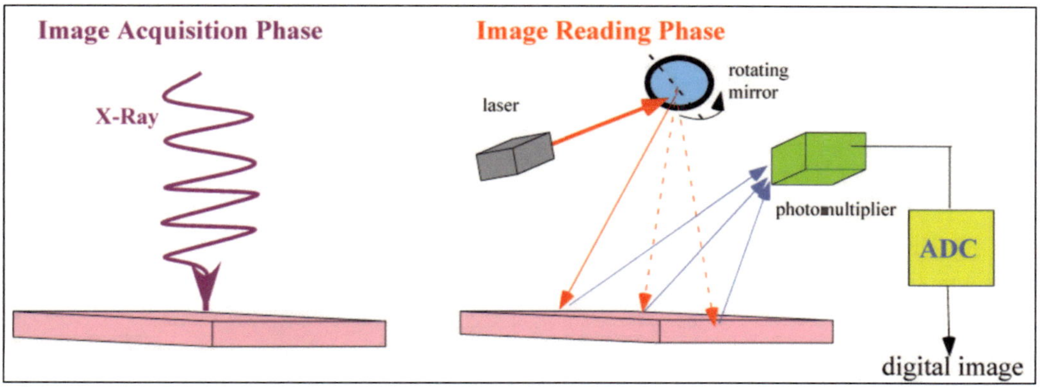

Fig. 1. Principle of computed radiography. An object on top of an imaging plate is exposed to X-rays. The storage phosphor renders a latent image of the object via the distribution of absorbed dose. The absorbed dose is measured in the reading phase and the signal is digitized. The readout for the conventional storage phosphor is undertaken by the so-called "flying-spot" method.

Since its introduction in the early 1990s, the method of computed radiography has gained significant momentum since it allows a reduction of the dose to as low as 18% in comparison with screen-film technology. In conventional computed radiography (CR), the latent image on an imaging plate (comprising of an X-ray storage phosphor) formed by exposure to ionizing radiation, is read out by photostimulated (optically stimulated) emission by using the so-called "flying-spot" method. In this method a focused red helium-neon laser beam is scanned across the imaging plate and the resulting photostimulated emission in the blue-green region of the visible spectrum as measured by a photomultiplier is converted into a digital signal pixel-by-pixel. In contrast to the screen-film method, computed radiography phosphors enable a dynamic range of up to eight orders of magnitude. However, the spatial resolution of imaging plates in CR is not yet on a par with screen-film technology due to the relatively large crystallites/grain size required in the commercially used photostimulable X-ray storage phosphor, BaFBr(I):Eu^{2+} (Nakano et al., 2002; Paul et al., 2002).

In this chapter, an overview of some optical storage phosphors and materials is given and storage mechanisms and applications are briefly discussed with emphasis on a novel class of photoluminescent storage phosphors.

2. Optical storage phosphors and materials

We define an optical storage phosphor or material to be a system that undergoes some electronic or structural change that allows an optical readout of radiation dose. This definition encompasses thermoluminescence where glow curves are measured upon heating the thermoluminescent material that was exposed to ionizing radiation, photostimulated (optically stimulated) luminescence where the stored energy is released upon stimulation by light, radiochromic film where chemical species are created that can be read out by absorption spectroscopy, or photoluminescent or radio-photoluminescent phosphors where the radiation induced centres are metastable or stable and can be repetitively read out by photoexciation. Silver-based emulsions are also optical storage media in the broader sense, as the darkening of the film can be read out quantitatively by the change of optical density. However, such films are not discussed here in detail. Our definition of what constitutes an optical storage phosphor or material is schematically depicted in Figure 2.

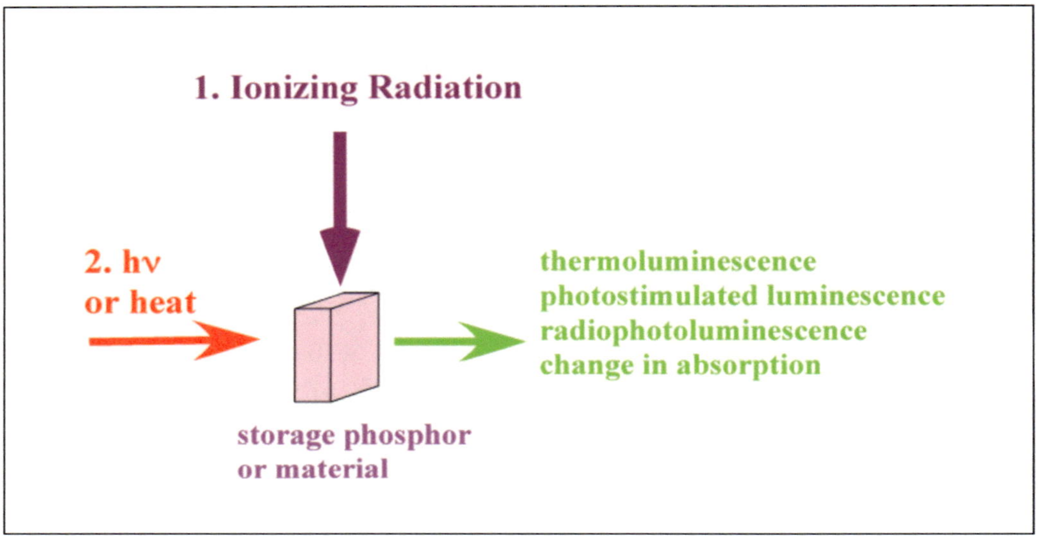

Fig. 2. General principle of an optical storage phosphor. Ionizing radiation leads to electronic and/or structural modifications in a material and the associated changes can be measured by luminescence or absorption spectroscopy.

The general mechanism of materials that exhibit thermoluminescence or photostimulated luminescence is schematically illustrated in Figure 3. Upon exposure to ionizing radiation, electron-hole pairs are created and the electrons and holes are subsequently trapped at defect sites within the crystal lattice or self-trapped such as by the creation of the V_k-centre. For example, electrons can be trapped at anion-vacancies such as F, Cl or I vacancies in halide crystals, forming the so-called F-centres (F stands for Farbe which is the German word for colour) which display oscillator strengths of close to unity. Holes can be self-trapped by the well-known V_k centres, e.g. a hole can be shared between two adjacent halide ions, leading to significant lattice distortion, which in turn may immobilise the hole to a certain extent. In thermoluminescent materials the electrons are liberated by heating the material since the electron traps are relatively shallow and the conduction band is thermally accessible. In photostimulable (optically stimulable) materials the electron trap is deeper and higher energies are required to liberate the electron. Upon the liberation of the electron it can combine with the hole leading to emission of light, either directly through the recombination process or by transferring excitation energy to an activator centre such as Eu^{2+} or the like.

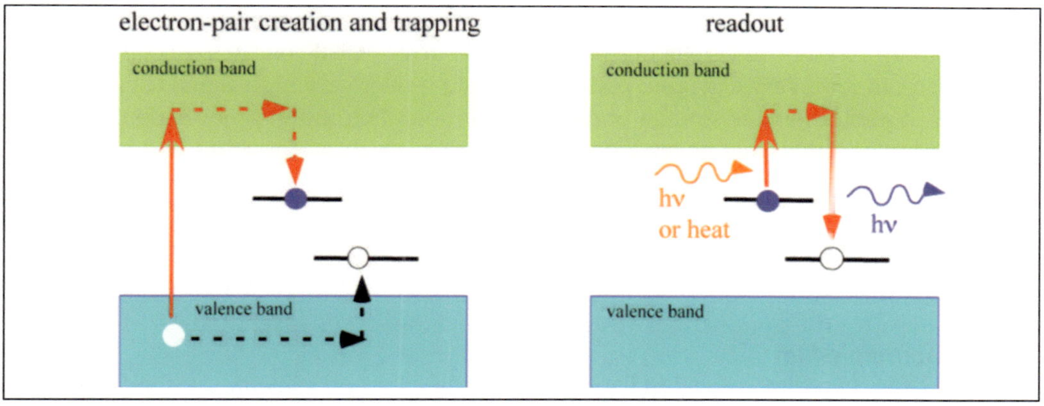

Fig. 3. Schematic diagram of the main mechanism of materials that display thermoluminescence or optically stimulated luminescence after exposure to ionizing radiation. Electrons and holes created by ionizing radiation are trapped out. Subsequently, heat or light can liberate electrons back into the conduction band and after recombination with the hole light is either emitted directly or by the transfer of the exciton energy to an activator such as a transition metal ion or a rare earth ion.

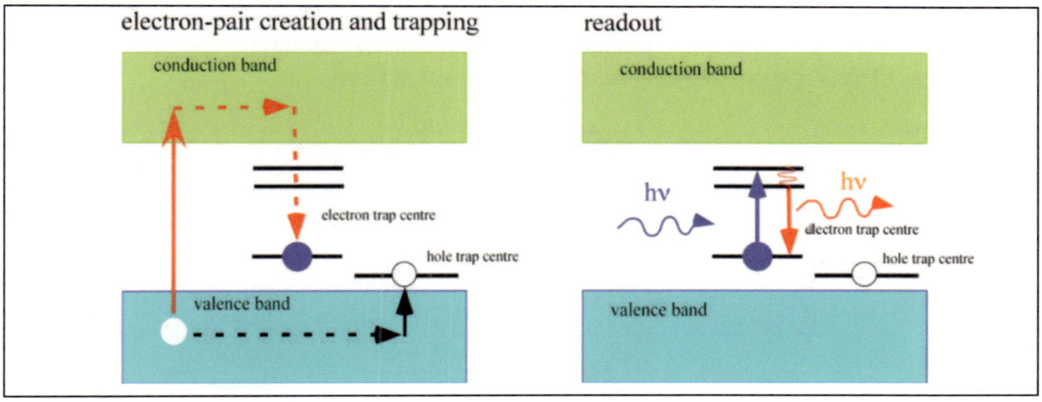

Fig. 4. Schematic and simplified diagram of the main mechanism of materials that display radiophotoluminescence after exposure to ionizing radiation. Electrons are localized in deep traps, e.g. transition metal ions, F-centres, rare earth ions. The electron trap centres can then be repetitively photoexcited from the electronic ground state to an electronically excited state. The electronically excited state is then deactivated by the emission of light and the centre induced by ionizing radiation is conserved.

In materials that display photoluminescence or radiophotoluminescence, electrons promoted to the conduction band by ionizing radiation migrate to deep traps. This is illustrated in Figure 4. These deep traps may be rare earth or transition metal ions doped as impurities into a wide bandgap host. The reduced trap can then be photoexcited without that the electron is promoted back into the conduction band. Naturally, if high light intensities are used, the electron may be promoted back into the conduction band by a two-step photoionization process. It is also possible that the excited state is close to the conduction band and hence some electrons will be liberated thermally from these excited states. However, in contrast to photostimulable materials, the signal in radiophotoluminescent materials does not decrease under ideal conditions i.e. the material can be read out for a very extended period of time to render very high signal to noise ratios. We define a photoexcitable storage phosphor as a radiophotoluminescent material that can be restored to its initial state by processes such as two-step photoionization.

A major advantage of optical storage materials for ionizing radiation in comparison with electronic devices is the fact that they are always in the accumulation mode and no battery or electric power supply is needed. This is particularly important in dosimetry applications where it is required that a dosimeter is on at all times.

2.1 Radiochromic film

Since the 1980s, two-dimensional radiation detectors employing radiochromic films have been widely applied in ionizing radiation dosimetry and medical radiotherapy (Soares, 2007; 2006). Radiochromic films consist of a thin polyester base impregnated with radiation-sensitive organic microcrystal monomers. Upon ionizing radiation, the film emulsions undergo a colour change through a chemical reaction (usually polymerization) and the radiation response signal can be read out by measuring the absorbance or optical density using spectrometers at certain wavelengths. Currently, the most popular radiochromic film in medical applications are the polydiacetylene based GafChromic® films designed by International Specialty Products (ISP). GafChromic® film was first introduced by Lewis (Lewis, 1986) and further developed by McLaughlin et al. at the National Institute of Standards and Technology (NIST) in the United States. Table 1 lists some current commercial GafChromic® films.

Application	Film type	Active Layer Thickness (μm)	Dose Range (Gy)	Spatial Resolution (dpi)	Energy Range
Radiotherapy	HD-810	6.7	10 ~ 2500	10,000	MV
	MD-55-2	67	10 ~ 100	5,000	MV
	RTQA2	17	0.02 ~ 8	5,000	MV
	EBT-2	30	0.01 ~ 40	5,000	kV-MV
	EBT-3	30	0.01 ~ 40	5,000	kV-MV
Radiology	XR-QA2	25	0.001 ~ 0.2	5,000	kV
	XR-CT2	25	0.001 ~ 0.2	5,000	kV
	XR-M2	25	0.001 ~ 0.2	5,000	kV
	XR-RV3	17	0.01 ~ 30	5,000	kV

Table 1. List of commercial GafChromic® films.

With different film implementations, the GafChromic® film exhibits a variety of dose responses in the energy range from keV to MeV, targeting different areas of medical radiotherapy. For example, with only a 6.7 μm thick active layer, the transparent GafChromic® HD-810 film is predominantly employed in high dose radiation measurements with doses up to 2500 Gy. In contrast, by adding a yellow dye into the sensitive layer, the latest GafChromic® EBT film is designed mainly for radiotherapy applications. Many studies have been devoted to the characterization and performance assessment of the GafChromic® films including the dose response functions (sensitivity), film stability and image resolution (Arjomandy et al., 2010;

Rink et al., 2008; Secerov et al., 2011). In addition to GafChromic® films, other commercial radiochromic films include FWT-60 from Far West Technology Inc., the B3 dosimetry lines manufactured by the Gex Corporation and the SIRAD (Self-indicating Instant Radiation Alert Dosimeter) developed by Gordhan Patel at JP Laboratories (Mclaughlin et al., 1991). Compared with conventional silver halide based photographic films, the radiochromic film offer many advantages in medical dosimetry and radiography, such as outstanding spatial resolution (>1200 lines/mm for GafChromic® film (Secerov et al., 2011), accurate and precise dose measurements, weak energy dependence and relatively easy handling.

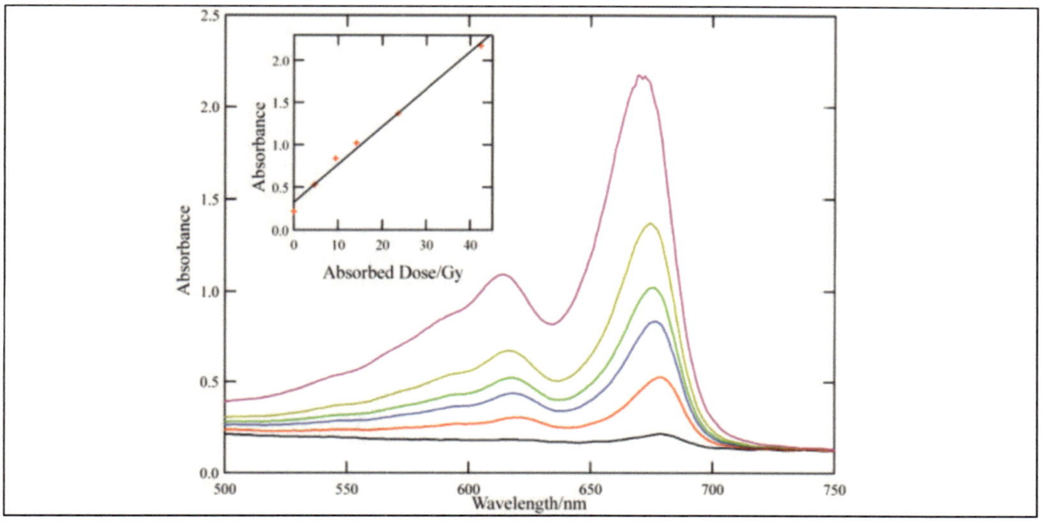

Fig. 5. Absorption spectra of GafChromic® film HD-810 as a function of high X-ray dose (40 kVp). The inset shows a plot of the absorbance at 676 nm of the film as a function of the absorbed dose. The absorption spectra were measured on a Cary-50 UV-Vis absorption spectrometer.

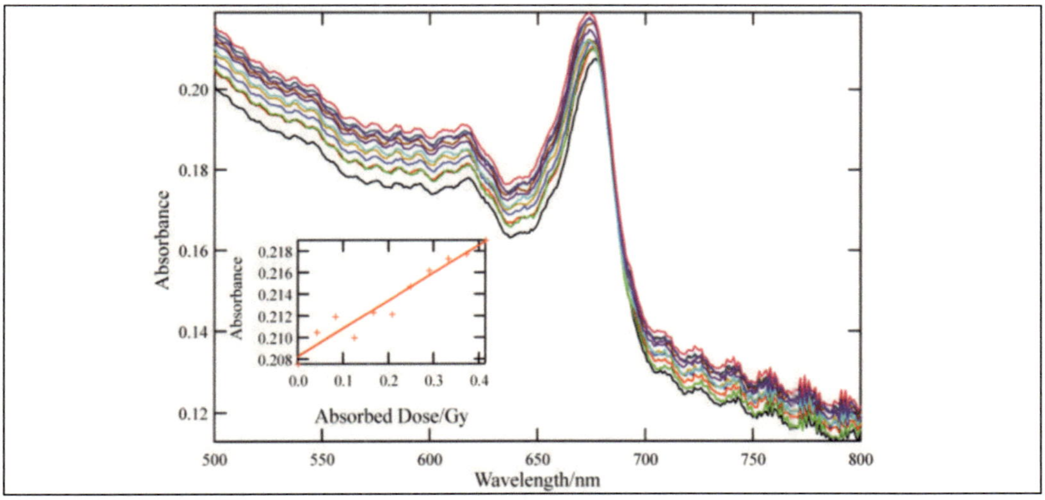

Fig. 6. As in Figure 5 but for low X-ray dose and 60 kVp radiation.

Figures 5 and 6 illustrate the X-ray dose dependence of the absorption spectra of GafChromic® film HD-810 at high and low doses, respectively. The optical absorbance at 676 nm of the film increases linearly in the dose range of 0.04 Gy to 2 Gy of X- irradiation. Weak energy dependence is another desirable property for a two-dimensional radiation detector in order to obtain accurate values of absorbed doses.
Due to the high atomic number components in silver-based film emulsions, the conventional photographic films suffer from a strong photon energy dependence of the response to dose. This has been significantly improved by radiochromic film dosimetry. Buston et al. (Butson et al., 2010) examined the energy dependence of the GafChromic® EBT2 film dose response in the X-ray range of 50 kVp to 10 MVp and a 6.5% ± 1% variation in optical density to absorbed dose response was obtained. The near energy-independence of radiochromic film has made it an ideal technology in some medical applications where a broad energy range of radiation sources is delivered. Furthermore, unlike the photographic films, the radiochromic films do not require subsequent chemical processing and can produce repetitive responses over a relatively long period of time. Despite the numerous advantages, special attention has to be paid to the environmental factors such as temperature, humidity and UV exposure in the irradiation and readout processes of the radiochromic films. Abdel-Fattah and Miller (Abdel Fattah & Miller, 1996) have investigated the effects of temperature during irradiation on the dose response of the FWT-60-00 and B3 radiochromic film dosimeters. At temperatures between 20 and 50 °C and a relative humidity between 20 and 53%, the temperature dependences of FWT-60-00 and Risø B3 dosimeters were found to be 0.25 ± 0.1% per °C and 0.5 ± 0.1% per ° C, respectively. For the GafChromic® films, it was suggested to be best stored in the dark at the temperature of ~22°C (Mclaughlin et al., 1991).

2.2 Thermoluminescent materials

Thermoluminescence (TL) is the emission of light produced by heating of a material. The phenomenon of thermoluminescence has been known for a long time and has been reported by Robert Boyle as early as 1663 when he observed thermoluminescence from diamond (Horowitz, 1984). At the end of the 19th century Borgman (Borgman, 1897) reported that X-rays and radioactive substances can induce thermoluminescence in particular materials. However, it was only in 1953 when Daniels et al. suggested the application of thermoluminescence in the field of dosimetry (Daniels et al., 1953). The early work was mostly involving the tissue equivalent LiF but also Al_2O_3. From the 1960s onwards thermoluminescence was regarded to be a viable alternative to film dosimetry and LiF became the TLD-100 standard after Harshaw purified the material and refined the crystal growth to make the properties of the material more reproducible. LiF is still in use today but the main key is to select a material for a specific monitoring application. Up to this date dosimetry based on thermoluminescence is still the main mode of personal radiation monitoring, with a plethora of suitable materials for a wide range of applications.
The mechanism of thermoluminescence induced by radiation is generally described in the following way. Upon ionizing radiation, free electron and hole pairs are created due to the photoelectric effect. The electrons and holes are then captured at different defect centres (electron and hole traps) within the material. During the subsequent heating process, the trapped electrons are released and recombine with the trapped holes, resulting in emission of light with the intensity proportional to the absorbed ionizing radiation dose (see Figure 4

above). Compared with other radiation dose measurement techniques, thermoluminescence dosimetry offers many advantages, such as a relatively small dosimeter size, high sensitivity over a wide range of surface dose levels from μGy to several Gy, long-term information storage and versatile applications in different types of radiation including X-rays, γ and β rays and neutrons.

There are a large number of materials exhibiting thermoluminescence, mostly insulators or semiconductors containing lattice defects resulted from substitutional impurities or ionizing irradiation. However, not all the materials with thermoluminescence properties are suitable for radiation dosimetry. There are several requirements for the practical use of thermoluminescence materials in radiation measurements (Kortov, 2007). For example, it is desirable that the material has a simple glow curve with the main peak occurring at around 200 °C. If thermoluminescence is also induced at lower temperatures, the thermal energy at room temperature may be high enough to liberate trapped electrons and fading can occur. On the other hand, when the glow curve is complex on the high-temperature side, the stored energy remaining after the radiation measurement has been performed, has to be dissipated (Yen et al., 2007). Figure 7 shows the glow curve of LiF:Mg, Cu, P (Bos, 2001; McKeever et al., 1993) after exposed to 0.4 Gy ^{60}Co •-radiation. From this figure it follows that dosimeters based on this material should not be subject to temperatures above 80 °C, otherwise fading starts to occur. In addition to a simple glow curve, a linear dose response, high luminescence efficiency, low photon energy dependence and long term stability are also desirable for the application of thermoluminescent materials in radiation dosimetry (Azorin et al., 1993).

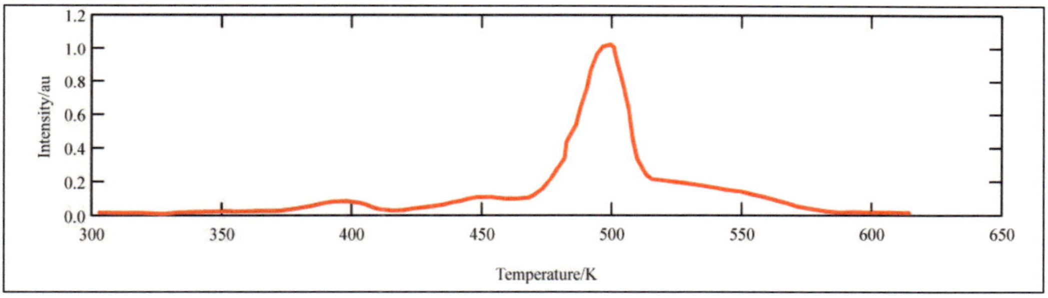

Fig. 7. Glow curve of LiF:Mg, Cu, P (TLD-100H). Thermal annealing procedure: 10 min at 240 °C with a heating rate of 2.1 °C /s followed by cooling in air. The absorbed dose was 0.4 Gy from a ^{60}Co source. The main dosimetric peak occurs at ~200 °C.

Table 2 lists some typical commercial thermoluminescence dosimeters with their applications and useful dose ranges. These commercial dosimeters generally exhibit emission at 380 - 480 nm, which falls off in the spectral response range of common photomultipliers. The thermoluminescence peak is located from 180 °C to 260 °C, offering an easy readout process. With an effective atomic number (Z_{eff}) close to that of soft biological tissue, LiF-based dosimeters are the most widely used thermoluminescence phosphors in personal radiation monitoring due to its very low photon energy dependence (Azorin et al., 1993). However, the glow curves of LiF-based dosimeters consist of at least seven peaks (Satinger et al., 1999). In contrast, CaF_2:Mn and Al_2O_3:C dosimeters offer simpler glow curves and higher sensitivity (Kortov, 2007). Nakajima et al. (Nakajima et al., 1978) first introduced the LiF:Mg, Cu, P (TLD-100H) which shows a 15 times greater sensitivity than that of the LiF:Mg, Cu (TLD-100). Furthermore, to improve the sensitivity of the dosimeters to thermal neutrons, ^{6}Li and ^{7}Li isotope based LiF dosimeters were designed to measure thermal neutrons in a mixed radiation field (Azorin et al., 1993).

Material	Dosimeter	Z_{eff}	Application	Useful Range
LiF:Mg, Ti	TLD-100	8.2	medical physics	10 μGy - 10 Gy
CaF_2:Dy	TLD-200	16.3	Environmental	0.1 μGy - 10 Gy
CaF_2:Tm	TLD-300	16.3	Low dose gamma rays	0.1 μGy - 10 Gy
CaF_2:Mn	TLD-400	16.3	Environmental	0.1 μGy - 100 Gy
Al_2O_3:C	TLD-500	10.2	Personnel	0.05μGy - 10 Gy
^{6}LiF:Mg, Ti	TLD-600	8.2	Neutron and gamma rays	10 μGy - 10 Gy
^{7}LiF:Mg, Ti	TLD-700	8.2	Gamma rays and thermal neutron	10 μGy - 10 Gy
LiB_4O_7:Mn	TLD-800	7.4	High dose and neutron	450 · Gy-10^4 Gy
$CaSO_4$:Dy	TLD-900	15.5	Environmental	1 μGy – 10^3 Gy
LiF:Mg, Cu, P	TLD-100H	8.2	Personnel	1 μGy - 20 Gy
^{6}LiF:Mg, Cu, P	TLD-600H	8.2	Personnel	1 μGy - 20 Gy
^{7}LiF:Mg, Cu, P	TLD-700 H	7.4	Personnel	1 μGy - 20 Gy

Table 2. Examples of commercial thermoluminescence dosimeters.

One of the better performers is $CaSO_4$:Dy which is approximately 10 times more sensitive than LiF:Mg,Ti (TLD-100). This material offers a simple processing cycle and is typically suspended in a Teflon disc. Annealing of only 2 hours at a temperature of ~280°C is required to erase the residual radiation induced thermoluminescence before the material can be reused. In contrast, LiF:Mg,Ti requires annealing of 5 hours at 300°C, followed by 20 hours at 80°C. However, CaF_2 and $CaSO_4$ based thermoluminescence materials are exceptionally light (UV) sensitive, and fading is enhanced significantly. These materials have to be handled, used and stored in opaque containers to avoid belaching of the thermoluminescence signal. Moreover, the reading accuracy depends on the heating rate and multiple readings are not possible as the stored energy is lost upon heating. Importantly, TL dosimeters cannot work in test-accumulation mode (multiple irradiation and reading). Also, the irradiation memory is dependent on the operation temperature. The precision on reuse is also lowered (±2% above 1 mSv and higher below this value) and the long term response retention is restricted (the best materials show 5% loss at room temperature for one year). A major drawback of thermoluminescence based dosimetry systems is the requirement of relatively complex reading instrumentation including a well-controlled flowing gas thermoelectric heating system (±1°C). Current thermoluminescence dosimeter manufacturers include the Solid Dosimetric and Detector Laboratory in China, Nemoto in Japan, the Institute of Nuclear Physics in Poland and Saint-Gobain Crystals and Detectors in the USA.

2.3 Photostimulable (optically stimulable) storage phosphors

As is illustrated in Figure 3, the mechanism of materials that show thermoluminescence and photostimulated (or optically stimulated) luminescence (PSL or OSL) is, to a great extent, the same but OSL (PSL) materials provide significantly deeper traps for the electrons and/or holes created upon exposure to ionizing radiation and hence thermal energies are not high enough to liberate the trapped electrons. Thus, in contrast to thermoluminescence, optically stimulated luminescence is an all-optical technique that does not require heating of the storage material. This allows much simpler and compact designs of reader units. The application of optically stimulable phosphors in dosimetry was first suggested in the 1950s. It was then initially mostly used for archaeological and geological dating, e.g the dating of quartz until the development of sensitive anion-deficient Al_2O_3:C. Photostimulable materials also became very important with the implementation of BaFBr:Eu^{2+} in computed radiography.

2.3.1 Al_2O_3:C

The Al_2O_3:C system has been extensively covered in a book by Bøtter-Jensen et al. (Botter-Jensen et al. 2003) and an elegant overview of the main features of this material is also given in a recent article by Akselrod (Akselrod, 2010). Anion-deficient α-Al_2O_3:C was firstly developed as a thermoluminescence material (Akselrod et al., 1990) but it was soon realized that it can serve as an OSL phosphor with excellent properties. In terms of its thermoluminescence properties, it is about 40-60 times more sensitive than LiF based TLD-100. However, the sensitivity is strongly dependent on the heating rate, complicating routine dosimetry measurements. In contrast, Al_2O_3:C appears to be an ideal material for optically stimulated luminescence. Neutral and charged oxygen vacancies (F, F^+) play an important part as luminescence centres. In particular, the presence of F^+ centres is important as they are the recombination centres for electrons, yielding excited F-centres. Al_2O_3:C displays a linearity of light output as a function of radiation dose over seven orders of magnitude. The long luminescence lifetime of 35 ms of the emitting F-centres allows simple time-dependent readout schemes and this has been implemented in so-called pulsed optically stimulated luminescence measurements (Akselrod et al., 1998) where the emitted light is only measured between the stimulating light pulses in order to reduce the requirement for heavy optical filtration. The system is now fully commercialized by Landauer with a suite of dosimetry products that allow static and dynamic monitoring of ionizing radiation. Landauer also offers a portable reader unit. A drawback of the Al_2O_3:C material is the fact that it has to be kept in light-tight containers, otherwise the stored energy is rapidly erased by visible light.

2.3.2 BaFBr(I):Eu^{2+}

Excellent and detailed reviews of this material have been published by von Seggern (von Seggern, 1999) and Schweizer (Schweizer, 2001). BaFBr(I):Eu^{2+} is the most widely used storage phosphor in computed radiography, with a number of major companies that offer imaging plates and associated readers for medical imaging.

Current storage phosphors contain centres for the capture of X-ray generated electrons and holes. For the BaBrF:Eu^{2+} storage phosphor both the $F^+(Br^-)$ and the $F^-(F^-)$ defects can act as electron storage centres whereas the Eu^{2+} acts as the hole trap. Upon X-ray irradiation of the X-ray storage phosphor BaBrF:Eu^{2+} electron-hole pairs are created. The electrons and holes are trapped at anion vacancies and the Eu^{2+}, respectively, creating F-centres and Eu^{2+}/V_K-centres (a V_K-centre is a hole shared between two neighbouring halide anions). However, some electron-hole pairs recombine spontaneously after their creation without being trapped and lead to spontaneous emission (scintillation). Upon excitation of the F-centres at 2.1 eV and 2.5 eV for the F(Br⁻) and F(F⁻) centres, respectively, the electrons recombine with the holes to form excitons. When the excitons are deactivated its excitation energy is resonantly transferred to the activator, Eu^{2+}, which in turn, leads to the broad $4f^65d \rightarrow 4f^7$ emission at about 390 nm. It was believed initially (Takahashi et al., 1984) that the hole combines with Eu^{2+} to form Eu^{3+}. However, there is evidence that this is not the case as follows, for example, from the fact that the Eu^{2+} photoluminescence does not decrease upon X-irradiation (Hangleiter et al., 1990; Koschnick et al., 1991). The exact mechanism of the electron–hole recombination is still controversially discussed; some experiments are at variance with a recombination of the electron-hole pairs via the conduction band. An alternative mechanism proposes a spatial correlation of the electron and hole traps and it is believed that recombination occurs via tunnelling of the electrons.

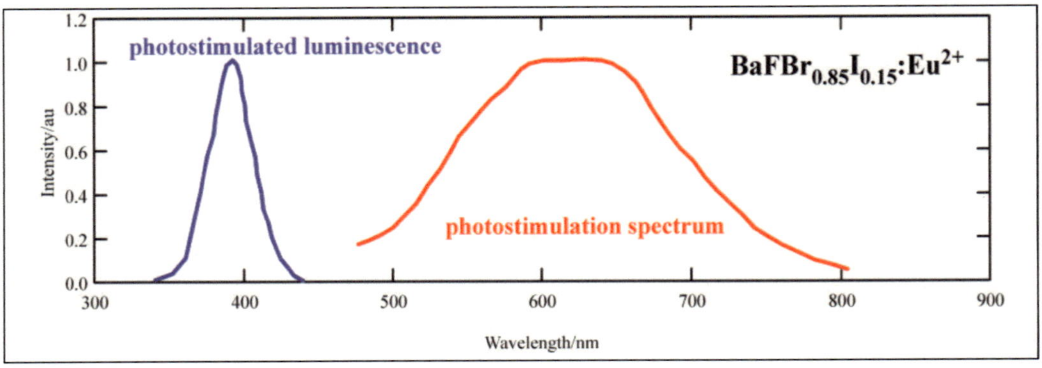

Fig. 8. Photostimulated luminescence and photostimulation spectra of $BaFBr_{0.85}I_{0.15}:Eu^{2+}$.

2.3.3 $CsBr:Eu^{2+}$

$CsBr:Eu^{2+}$ displays significant photostimulated luminescence after X-ray exposure (Leblans et al., 2001; Schweizer et al., 2000) and its figure-of-merit as a storage phosphor is as high as for $BaFBr(I):Eu^{2+}$ (Hackenschmied et al., 2002; Schweizer, 2001). In particular, this material can be grown in the form of needle crystals by vacuum deposition (Weidner et al., 2007), allowing an improved lateral resolution in comparison with $BaFBr(I):Eu^{2+}$. The sensitivity of this material appears to be strongly dependent on thermal treatment (Hackenschmied et al., 2003) and hydration (Appleby et al., 2009). In $CsBr:Eu^{2+}$ the holes are trapped as V_K-centres in the vicinity of Eu^{2+}, and the electrons are trapped in F-centres and hence the spectroscopy of this system is very much related to the $BaFBr:Eu^{2+}$ spectra with minor dielectric shifts of the spectra. However, poor radiation hardness appears to be a major problem with this material. In particular, the photostimulated luminescence deteriorates rapidly upon exposure to high X-ray dose and this deterioration is correlated with a reduction of the photoexcited Eu^{2+} fluorescence that has been attributed to an agglomeration of Eu^{2+} leading to luminescence quenching. (Zimmermann et al., 2005). This conclusion has been reached since no increase in Eu^{3+} luminescence is observed.

2.3.4 BeO

With near tissue equivalence (Z_{eff} = 7.1), relatively high sensitivity and low cost, beryllium oxide (BeO) has been well-documented as a thermoluminescence dosimetry (TLD) material with a range of applications particularly in medical fields (McKeever, 1988; McKeever et al., 1995; Vij & Singh, 1997). One of the most popular forms of BeO-based TLD is Thermalox® 995. Modifications of the material can be found as BeO:Li (Yamashit et al., 1974), BeO:Na (Yamashit et al., 1974) and $BeO:TiO_2$ (Milman et al., 1996). Depending on the preparation methods and the type of radiation source, the glow curves of BeO-based TLD materials exhibit the main dosimetric peaks between 160 °C and 280 °C. Compared with other commercial thermoluminescence dosimetric systems, the main disadvantage of the BeO-based materials is the rapid light-stimulated thermoluminescence fading. Under laboratory fluorescent light, rapid fading of 50 % in 30 min for the thermoluminescence peak at 167 °C has been reported in several types of commercial BeO ceramic dosimeters (Crase & Gammage, 1975). Based on the characteristic light-sensitive thermoluminescence of BeO, Albrecht and Mandeville (Albrecht & Mandeville, 1956) first investigated optically stimulated luminescence of the X-ray irradiated BeO by using visible photons. Bulur and Göksu (Bulur & Göksu, 1998) examined the detailed dosimetric properties of Thermalox® 995 as optically stimulated luminescence dosimeter, such as the OSL signal, the stimulation spectrum, the dose dependence, multireadability and short-term fading. Upon optical stimulation in a broad band region from 420 nm to 550 nm with maximum at 435 nm (Bulur & Göksu, 1998), the stimulated luminescence of BeO is generally assumed to be in the same region as that of the thermoluminescence, i.e. UV region with main peak at around 335 nm (McKeever et al., 1995). The exact origin of the optically stimulated luminescence of BeO is not clear yet, although a possible link to the thermoluminescence peaks at 220 °C and 340 °C has been proposed (Bulur & Göksu, 1998). The linear dose response of the optically stimulated luminescence of BeO was reported from 1 µGy up to a few Gy, covering more than six orders of magnitude (Sommer et al., 2008). In practical applications of BeO in radiation dosimetry, the stimulation is usually realized by employing blue light-emitting diodes and the detection of the light can be easily achieved with a photomultiplier. Optical filters are usually used in front of the photomultiplier to discriminate the stimulation light from photostimulated luminescence. The main problem associated with BeO materials is the relatively low reproducibility but in particular the very high toxicity.

2.4 Photoluminescent storage phosphors

Radiophotoluminescent materials and photoluminescent storage phosphors are based on an alteration of the luminescence spectrum of a material upon exposure to ionizing radiation. The phenomenon was reported as early as 1912 by Goldstein (Goldstein, 1912) but it was only in the 1950s that it was realized that the effect could be used for dosimetry (Schulman et al., 1953; Schulman, 1950; Schulman & Etzel, 1953; Schulman et al., 1951). Schulman's system comprised of a silver-activated aluminophosphate glass. Upon exposure to ionizing radiation, electrons are promoted to the conduction band and migrate to the silver activation sites in this system, yielding a reduction of the silver atoms. This reduction results in orange luminescence that can be excited at 365 nm. The main difference to photostimulable materials is the fact that in photoluminescent storage phosphors the radiation-induced change persists upon photoexcitation. For example, if stable colour centres are induced, they can be repetitively read out by photoexcitation whereas in optically stimulable materials *one* radiation induced centre, e.g. electron-hole pair, is annihilated upon the absorption of *one* photon. A study from 1984 suggests the application of rare earth ion co-doped $CaSO_4$ to produce a radiophotoluminescent material at room temperature that also exhibits thermoluminescence i.e. a combined TL/RPL dosimetry system (Calvert & Danby, 1984). In particular, Eu^{2+} ions if illuminated by ultraviolet radiation display radiophotoluminescence after exposure to ionizing radiation. However, the low sensitivity turned out to be a very limiting factor for this material.

2.4.1 Nanocrystalline BaFCl:Sm^{3+}

Nanocrystalline BaFCl:Sm^{3+}, as prepared by a facile co-precipitation method, is a very efficient photoluminescent storage phosphor for ionizing radiation (Liu et al., 2010; Riesen & Kaczmarek, 2007). The mechanism is based on the reduction of Sm^{3+} to Sm^{2+} by trapping electrons that are created upon exposure to ionizing radiation in the BaFCl host; subsequently, the Sm^{2+} can be efficiently read out by measuring the intraconfigurational *f-f* transitions, such as the $^5D_0$-$^7F_0$ line at around 688 nm, by excitation into the parity allowed $4f^6 \rightarrow 4f^55d$ transition at around 415 nm. The latter wavelength is ideal for efficient excitation by blue-violet laser diodes or LEDs as the transition is electric dipole allowed and thus very intense.
The photoluminescence and excitation spectra of nanocrystalline BaFCl:Sm^{3+} after exposure to 50 mGy of 60 kVp X-ray are illustrated in Figure 9. The spectra are dominated by the Sm^{2+} species that is created by the ionizing radiation. The efficiency of this class of storage phosphors can be optimized by the preparation of core-shell nanoparticles comprising of a BaFCl core and a BaFCl:Sm^{3+} shell. The $Sm^{3+} \rightarrow Sm^{2+}$ conversion efficiency of these nanoparticles is drastically higher than that of microcrystals obtained by conventional high temperature (HT) sintering. This is illustrated in Figure 10 where the 415 nm excited luminescence spectra of HT and nanocrystalline BaFCl:Sm^{3+} are compared.

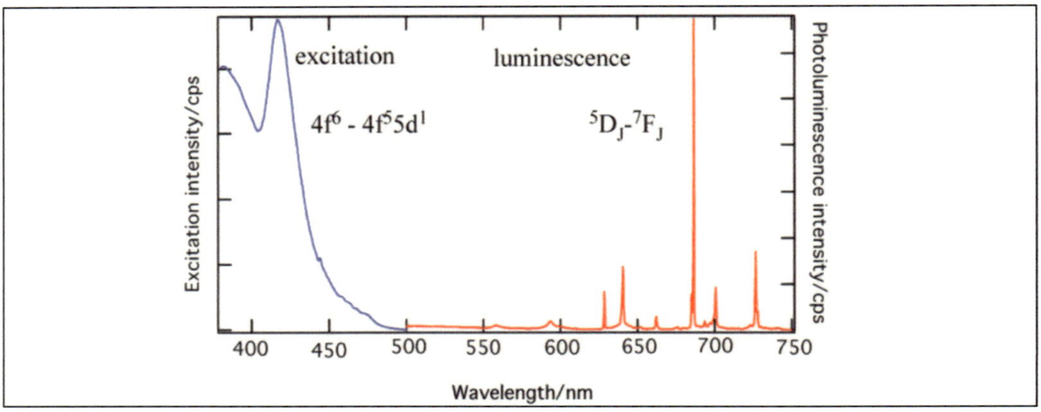

Fig. 9. Photoluminescence and excitation spectra of nanocrystalline BaFCl:Sm^{3+} after exposure to 50 mGy of 60 kVp X-ray radiation. All the pronounced transitions are due to Sm^{2+} which are absent before irradiation. The excitation spectrum is not corrected for the effect of a BG-18 glass filter used for the excitation light.

The HT sample was exposed to a 240,000 times higher radiation dose (12000 Gy) than the nanocrystalline material (50 mGy). It is noted here that there are significant differences in the $^4G_J \rightarrow {}^6H_J$ emission lines for the two materials, indicating a significantly different local environment for the Sm^{3+} ions in the nanocrystalline material. It is also clear that both materials provide a range of sites for the Sm^{3+} ions and in the case of the HT material it is straightforward to achieve high site selectivity when changing the excitation wavelength in the vicinity of 400 nm. Nevertheless, after the reduction to Sm^{2+} the resulting $^5D_J \rightarrow {}^7F_J$ emission lines perfectly coincide for both materials.

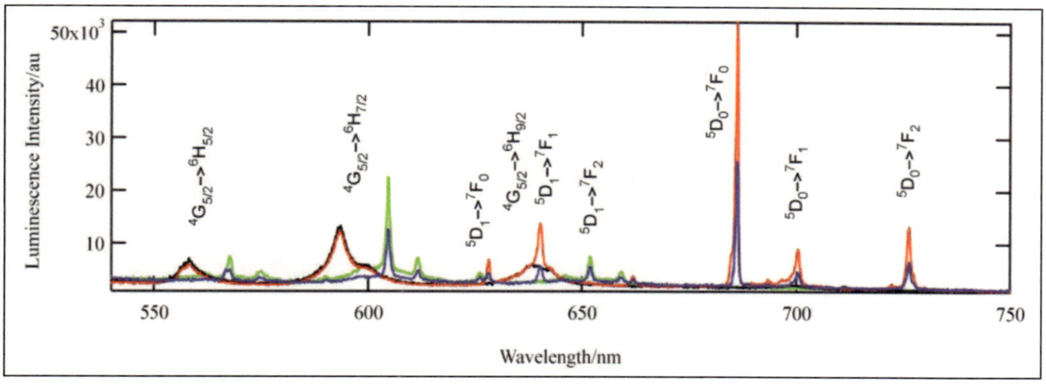

Fig. 10. Photoluminescence of high temperature (green and blue lines) and nanocrystalline (black and red lines) BaFCl:Sm^{3+} before and after exposure to 12000 Gy and 50 mGy, respectively. Prominent $^4G_J \rightarrow ^6H_J$ (Sm^{3+}) and $^5D_J \rightarrow ^7F_J$ (Sm^{2+}) transitions are denoted. Note that the unexposed samples (green and black lines for HT and nano sample, respectively) do not exhibit any emission lines due to Sm^{2+}.

The much higher conversion efficiency (~500,000 times) of the nanocrystalline material manifests itself also in cathodoluminescence where the Sm^{3+} and Sm^{2+} peaks dominate in the spectra of the HT sintered and nanocrystalline samples, respectively (Stevens-Kalceff et al., 2010). The nanocrystalline material has been characterized with a range of techniques such as electron microscopy and high resolution synchrotron powder X-ray diffraction. The diffraction lines in the powder X-ray diffraction pattern can be indexed to the tetragonal matlockite-type structure with space group P4/nmm. The resulting lattice parameters ($a = b$ = 4.395(1) Å, and c = 7.227(8) Å) indicate a small expansion of the crystal lattice in comparison with the literature values (Beck, 1976). The average volume-weighted column length of the crystallites is estimated to be 160 nm by applying the Williamson-Hall procedure (Williamson & Hall, 1953). This average size is consistent with electron microscopy micrographs and renders the nanocrystals for applications where extremely small probes for ionizing radiation are required. For example, if bonded or embedded to/in the end of an optical fibre the material can be used for remote sensing of CW or pulsed ionizing radiation; examples of such applications include real time in vivo monitoring of dose in radiotherapy and in food irradiation and the monitoring of hot labs and/or nuclear reactor environments. In such applications a selected Sm^{2+} emission line, e.g. $^5D_0 \rightarrow ^7F_0$, is monitored with a narrow bandpass filter (e.g. 1 nm FWHM) with gated detection 180⁰ out of phase from the excitation light pulse. The latter is facilitated by the relatively long lifetime of 2 ms of the $^5D_0$ excited state and dramatically reduces the lower detection limit by elimination of excitation light, fluorescence due to other impurities and surface defects, and Raman scattered light.

The core-shell nanocrystalline storage phosphor displays a relatively linear response to ionizing radiation up to a surface dose of ca. 10 Gy in the 50 keV X-ray region as is illustrated in Figure 11. Moreover, the lower detection limits for 50 keV and 1 MeV radiation are about 100 nGy and 10 μGy, respectively. This results in an impressive dynamic range of about seven orders of magnitude for radiation detection. A range of photon-gated spectral hole-burning experiments at low temperatures has also been conducted (Liu et al., 2010). The stored energy in the phosphor material can be released by a two-step photoionization process when higher light powers are used. However, the storage phosphor can only be fully restored after exposures of up to 10-20 mGy. An example of the reproducibility of the nanocrystalline BaFCl:Sm^{3+} for dosimetry applications is illustrated in Figure 12. In this example the Sm^{2+} photoluminescence signal was measured as a function of accumulated dose (60 kVp X-ray irradiation) up to 1 mGy. The dosimeter material was then bleached by a two-step photoionization and the experiment was repeated. As follows from this figure, the phosphor can be fully restored in this range of absorbed doses.

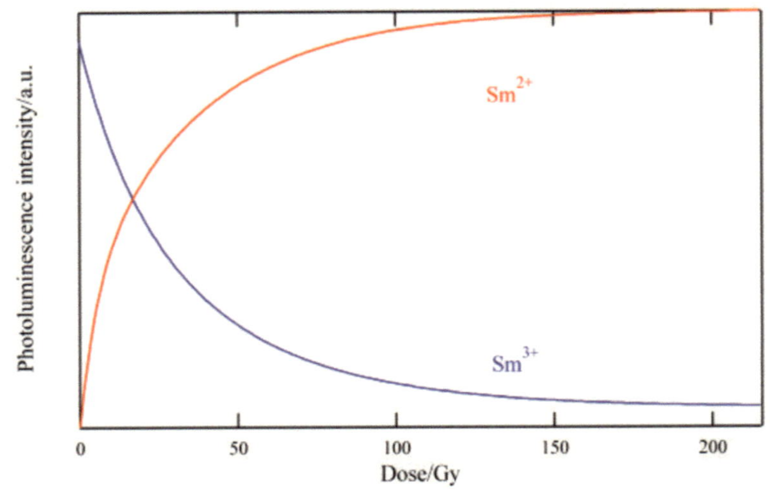

Fig. 11. Dependence of the Sm^{2+} and Sm^{3+} photoluminescence signals of nanocrystalline BaFCl:Sm^{3+} on absorbed 60 kVp X-ray dose.

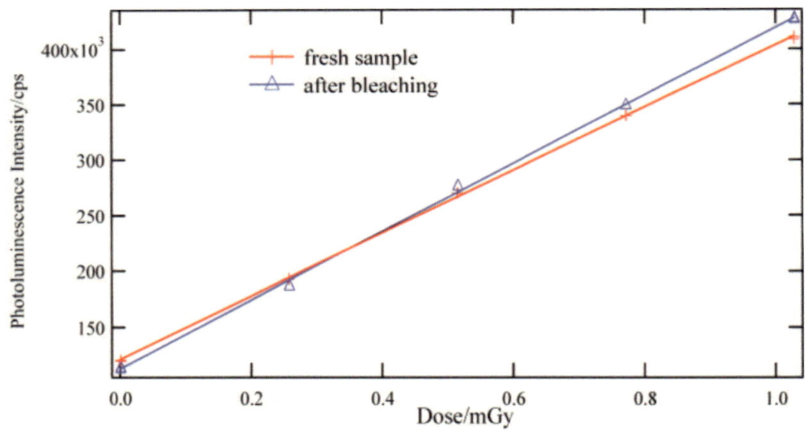

Fig. 12. Reproducibility of nanocrystalline BaFCl:Sm^{3+} in dosimetry.

Figure 13 illustrates the dose dependence of the Sm^{2+} photoluminescence signal upon exposure to Cs-137 γ-rays (662 keV). The phosphor's response is independent on the dose rate and fairly linear up to the maximum exposure of ca ~10 mGy.

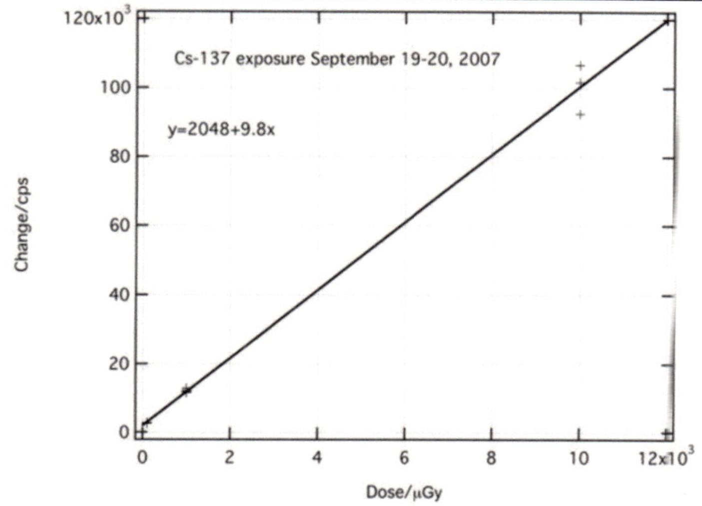

Fig. 13. Dependence of the Sm^{2+} photoluminescence intensity on absorbed dose of Cs-137 γ-rays (662 keV)

As can be expected, due to the presence of heavy atoms in the $BaFCl:Sm^{3+}$ material, the phosphor's quantum efficiency is subject to a relatively strong energy dependence. This is illustrated in Figure 14. As a consequence, if used in dosimetry, filters need to be applied. However, the strong energy dependence together with a filter arrangement allows the determination of the energy of the ionizing radiation, a desirable feature

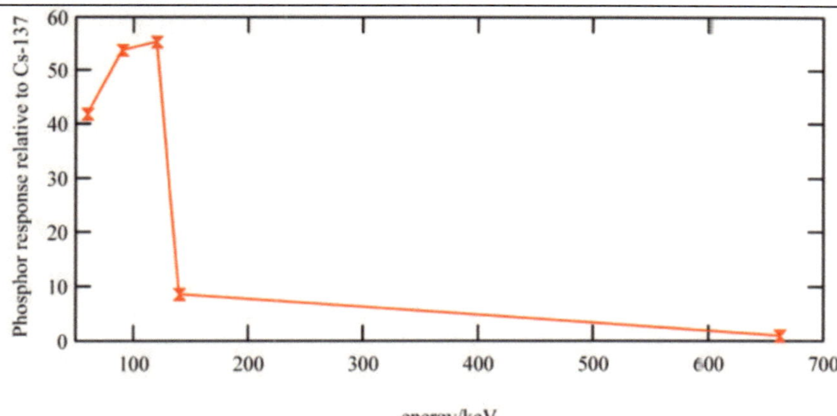

Fig. 14. Energy dependence of the conversion efficiency of nanocrystalline $BaFCl:Sm^{3+}$ relative to Cs-137 γ-rays (662 keV).

The lifetime of the excited ^5D_J multiplet is about 2 ms which is a severe limitation in computed radiography when the conventional "flying-spot" readout scheme is applied. However, the long lifetime allows for gated excitation and luminescence light with a phase shift of 180°. The gated readout facilitates a very sensitive detection scheme since fluorescence of impurities is blocked from reaching the light detector. A big advantage of the nanocrystalline $BaFCl:Sm^{3+}$ phosphor is the submicron particle size that increases the spatial resolution which, in the commercial BaFBr(I):Eu^{2+} is limited by the large grain size that induces significant light scattering in the readout process. The nanocrystalline BaFCl:Sm^{3+} also allows a higher packing density and the light scattering in the readout process can be expected to be a lesser problem. This can be tested by determining the modulation transfer function (MTF). The MTF can be determined following the procedure outlined in Figure 15 (Boone, 2001; Cunningham & Fenster, 1987; Cunningham & Reid, 1992; Samei et al., 1998; Samei et al., 2001). In particular, in the first step the edge spread function (ESF) is measured, by imaging a lead edge. The line spread function (LSF) is then calculated by differentiating the ESF and subsequently the MTF is obtained by Fast Fourier Transformation of the LSF. Thin films of nanocrystalline BaFCl:Sm^{3+} were applied to a plastic substrate by mixing the phosphor with a small amount of polyvinyl acetate. A 2D reader unit that images the excited photoluminescence of the BaFCl:Sm^{3+} storage plate by an electron-multiplying CCD (EMCCD) camera was employed to image the plate in parallel mode. The system relies on gated excitation-detection by pulsing the LEDs and mechanically gating the collimated luminescence light by a mechanical chopper wheel. This prevents fluorescence of impurities from reaching the detector. The 2D reader unit is illustrated in Figure 15 where the LED light sources, the film platform, the lens system, the chopper wheel and the EMCCD camera can be recognized.

Fig. 15. 2D reader for imaging plates based on nanocrystalline BaFCl:Sm^{3+}. The image plate is illuminated by two collimated and pulsed 405 nm mounted blue LEDs. The emitted light is collimated then passed through an aperture, that is opened with a 180° phase shift with respect to the blue LED pulses, by a mechanical light chopper wheel. The light is then refocused, filtered and detected by an EMCCD camera.

Radiographs were analyzed according to the mathematical steps shown in Figure 16. In particular, the line spread function (LSF) is obtained from the edge spread function (ESF) by differentiation and the MTF by Fast Fourier Transformation (FFT) of the LSF.

$$LSF(x) = \frac{d}{dx} ESF(x) \tag{1}$$

$$MTF(k) = \int_{-\infty}^{\infty} LSF(x)\exp(-i2\pi kx)dx \quad (2)$$

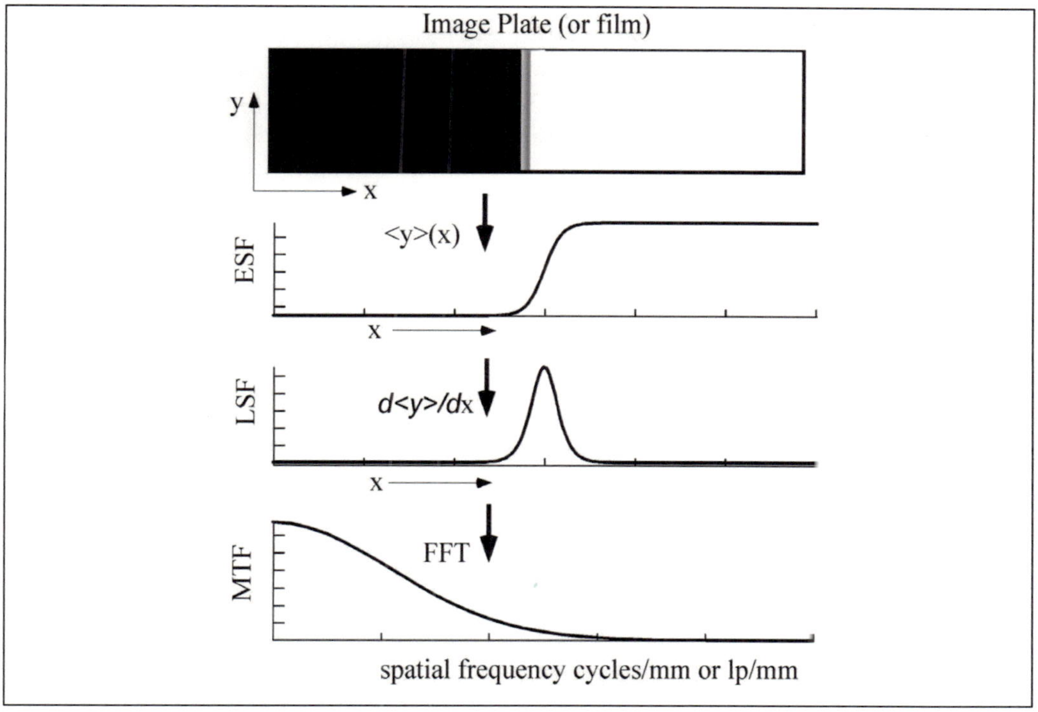

Fig. 16. Procedure to determine the MTF of an optical storage phosphor.

The resulting MTF is illustrated in Figure 17 and, for the purpose of comparison, the MTFs of the Kodak Insight dental film and Kodak GP and HR imaging plates in conjunction with Kodak Ektascan Model-400 and CR9000 computed radiography reader units are shown. The MTF points for the $BaFCl:Sm^{3+}$ imaging plates are calculated from raw data whereas for the smooth green line the ESF is based on a fit of a sigmoid to the ESF and the solid blue line is based on smoothing the ESF with the Savitzky-Golay procedure in 4th order using 37 points.

	$BaFCl:Sm^{3+}$ imaging plate	Kodak CR9000 reader with CR-GP plate	Kodak Ektascan Model-400 reader with GP plate	Kodak Ektascan Model-400 reader with HR plate	Kodak Insight With 2D reader	Kodak Insight with plustek OpticFilm 7400
MTF=0.5	2.4	1.3	1.22	1.92	2.8	4.7
MTF=0.1	5.4	3.1	3.27	4.35	18	18

Table 3. Comparison of MTF values of $BaFCl:Sm^{3+}$ imaging plates with Kodak Computed radiography systems and conventional Kodak Insight dental film.

In Table 3 a comparison of MTF data is compiled, in particular the spatial frequency is summarized at MTF values of 0.5 and 0.1. It is obvious that conventional dental film exhibits by far the best values. However, the BaFCl:Sm^{3+} based imaging plates are significantly better than all the commercial computed radiography systems shown in Table 3. It follows that the nanocrystalline BaFCl:Sm^{3+} storage phosphor yields imaging plates with a better MTF in comparison with commercially available computed radiography systems. It is noted here that a better MTF is highly desirable as it is the main drawback of computed radiography based on currently used storage phosphors. In particular in medical imaging a good MTF is required in order to detect, for example, micro-calcifications in soft tissue. High resolution imaging is also important for a range of scientific and technical imaging applications.

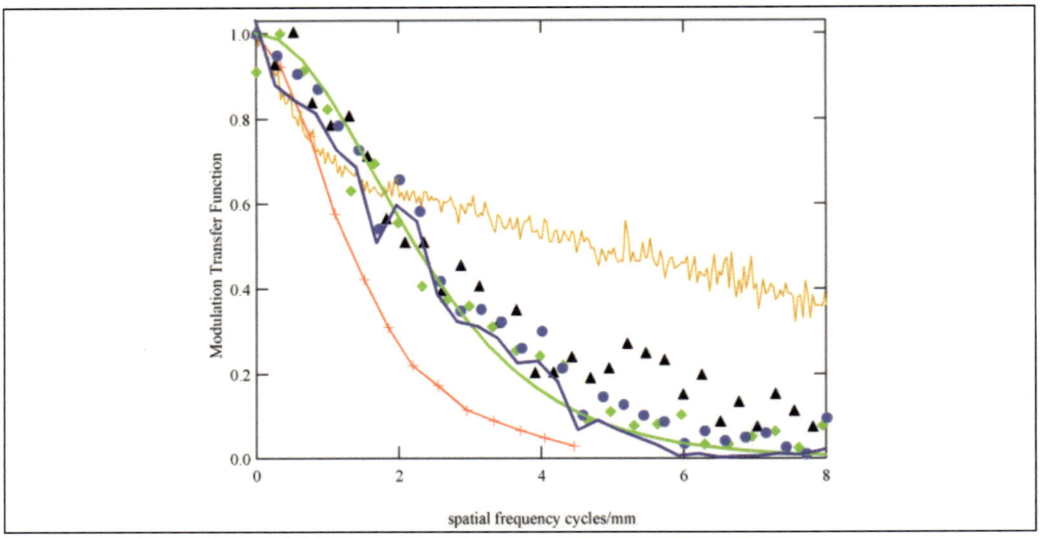

Fig. 17. MTF data for BaFCl:Sm^{3+} based imaging plate (blue solid circles, black solid triangles, green diamonds and green trace). The smooth green line is based on a fit of the ESF to a sigmoid. The blue line is the result of a Savitzky-Golay smoothing procedure applied to the ESF. The red trace with markers is the MTF of the Kodak CR9000 computed radiography system. The beige trace is the MTF of Kodak Insight dental film as read out by the 2D reader unit illustrated in Figure 15.

3. Conclusion

Over the last 60 years an impressive range of optical storage phosphors and materials has been developed and investigated for a plethora of applications, including personal radiation monitoring, quality control in radiation therapy and food irradiation, and for applications in computed radiography both for medical and technical applications. Although full electronic devices, both for point and 2D detection, are becoming increasingly more sensitive and inexpensive, many applications for optical storage phosphors prevail, possible for many years to come, as electronic solid state devices are not flexible and hence for many applications not practical. Importantly, storage phosphor based dosimeters are always accumulating and independent of any power supply. Optical storage phosphors have a significant potential to facilitate much lower radiation doses in medical imaging, a highly desirable outcome.

4. Acknowledgment

The Australian Research Council is acknowledged for financial support of our research into photoluminescent X-ray storage phosphors (ARC Discovery Project DP0770415, ARC Linkage Project LP110100451). We also like to thank Dosimetry & Imaging Pty Ltd, and especially A. Ujhazy, for supporting this project. The Australian Synchrotron is acknowledged for time on the powder X-ray diffraction and X-ray absorption beamlines.

5. References

AbdelFattah, A.A. & Miller, A. (1996). Temperature, Humidity and Time Combined Effects on Radiochromic Film Dosimeters. *Radiation Physics and Chemistry,* Vol. 47, No. 4, 611-621.

Akselrod, M.S., Kortov, V.S., Kravetsky, D.J. & Gotlib, V.I. (1990) Highly Sensitive Thermoluminescent Anion-Defective α-Al_2O_3:C Single Crystal Detectors. *Radiation Protection Dosimetry,* Vol. 32, No. 1, 15-20.

Akselrod, M.S., Lucas, A.C., Polf, J.C. & McKeever, S.W.S. (1998). Optically Stimulated Luminescence of Al_2O_3. *Radiation Measurements,* Vol. 29, No. 3-4, 391-399, 1350- 4487.

Akselrod, M.S. (2011). Fundamentals of Materials, Techniques, and Instrumentation for OSL and FNTD Dosimetry, *AIP Conference Proceedings*, Sydney, Australia, September 2010.

Albrecht, H. & Mandeville, C. (1956). Storage of Energy in Beryllium Oxide. *Physical Review,* Vol. 101, No. 4, 1250-1252.

Andersen, C.E., Nielsen, S.K., Greilich, S., Helt-Hansen, J., Lindegaard, J.C. & Tanderup, K. (2009). Characterization of a Fiber-Coupled Al_2O_3:C Luminescence Dosimetry System for Online in Vivo Dose Verification During ^{192}Ir Brachytherapy. *Medical Physics,* Vol. 36, No. 3, 708-718, 0094-2405.

Appleby, G., Zimmermann, J., Hesse, S., Karg, O. & von Seggern, H. (2009). Sensitization of the Photostimulable X-ray Storage Phosphor CsBr: Eu^{2+} Following Room-Temperature Hydration. *Journal of Applied Physics,* Vol. 105, No. 7, 0735111-0735115, 0021-8979.

Arjomandy, B., Tailor, R., Anand, A., Sahoo, N., Gillin, M., Prado, K & Vicic, M. (2010). Energy Dependence and Dose Response of Gafchromic EBT2 Film Over a Wide Range of Photon, Electron, and Proton Beam Energies. *Medical Physics,* Vol. 37, No. 1942-1947.

Azorin, J., Furetta, C. & Scacco, A. (1993). Preparation and Properties of Thermoluminescent Materials. *physica status solidi (a),* Vol. 138, No. 1, 9-46, 1521-396X.

Beck, H. (1976). A Study on Mixed Halide Compounds MFX (M= Ca, Sr, Eu, Ba; X= Cl, Br, I). *Journal of Solid State Chemistry,* Vol. 17, No. 3, 275-282, 0022-4596.

Boone, J.M. (2001). Determination of the Presampled MTF in Computed Tomography. *Medical Physics,* Vol. 28, No. 356-360.

Borgman, I.L. (1897). Thermoluminescence Provoquee par les Rayons de M. Roentgen et les Rayons de M. Becquerel. *Comptes Rendus,* Vol. 124, No. 895-896,

Bos, A. (2001). High Sensitivity Thermoluminescence Dosimetry. *Nuclear Instruments and Methods in Physics Research Section B: Beam Interactions with Materials and Atoms,* Vol. 184, No. 1-2, 3-28, 0168-583X.

Bøtter-Jensen, L., McKeever, S.W.S. & Wintle, A.G. (2003). *Optically Stimulated Luminescence Dosimetry, Elsevier Science, Amsterdam; Boston; London.*

Bulur, E. & Göksu, H. (1998). OSL From BeO Ceramics: New Observations from an Old Material. *Radiation Measurements,* Vol. 29, No. 6, 639-650, 1350-4487.

Butson, M.J., Yu, P.K.N., Cheung, T. & Alnawaf, H. (2010). Energy Response of the New EBT2 Radiochromic Film to X-ray Radiation. *Radiation Measurements,* Vol. 45, No. 7, 836-839, 1350-4487.

Calvert, R. & Danby, R. (1984). Thermoluminescence and Radiophotoluminescence From Eu- and Sm-Doped $CaSO_4$. *physica status solidi (a),* Vol. 83, No. 2, 597-604, 1521- 396X.

Crase, K. & Gammage, R. (1975). Improvements in the Use of Ceramic BeO for TLD. *Health Physics,* Vol. 29, No. 5, 739-746, 0017-9078.

Cunningham, I. & Fenster, A. (1987). A Method for Modulation Transfer Function Determination from Edge Profiles with Correction for Finite-Element Differentiation. *Medical Physics,* Vol. 14, No. 4, 533–537.

Cunningham, I. & Reid, B. (1992). Signal and Noise in Modulation Transfer Function Determinations Using the Slit, Wire, and Edge Techniques. *Medical Physics,* Vol. 19, No. 1037-1044.

Daniels, F., Boyd, C.A. & Saunders, D.F. (1953). Thermoluminescence as a Research Tool. *Science,* Vol. 117, No. 3040, 343-349, 0036-8075.

Dyk, J.V. (1999). *The Modern Technology of Radiation Oncology: A Compendium for Medical Physicists and Radiation Oncologists,* Medical Physics Publishing, 9780944838389, Madison, Wisconsin.

Eichholz, G.G. (2003). Dosimetry for Food Irradiation. *Health Physics,* Vol. 84, No. 5, 665, 0017-9078.

Goldstein, E. (1912). Concerning the Emission Spectra of Aromatic Bonds in Ultra Violet Light, in Cathode Rays, Radium Radiation and Positive Rays. *Pysikalische Zeitschrift,* Vol. 13, No. 188-193.

Hackenschmied, P., Zeitler, G., Batentschuk, M., Winnacker, A., Schmitt, B., Fuchs, M., Hell, E. & Knüpfer, W. (2002). Storage Performance of X-ray Irradiated Doped CsBr. *Nuclear Instruments and Methods in Physics Research Section B: Beam Interactions with Materials and Atoms,* Vol. 191, No. 1-4, 163-167, 0168-583X.

Hackenschmied, P., Schierning, G., Batentschuk, M. & Winnacker, A. (2003). Precipitation-induced Photostimulated Luminescence in CsBr: Eu^{2+}. *Journal of Applied Physics,* Vol. 93, No. 9, 5109-5112, 0021-8979.

Hangleiter, T., Koschnick, F. & Spaeth, J. (1990). Temperature Dependence of the Photostimulated Luminescence of X-irradiated BaFBr: Eu^{2+}. *Journal of Physics: Condensed Matter,* Vol. 2, No. 32, 6837-6846.

Horowitz, Y.S. (1984). *Thermoluminescence and Thermoluminescent Dosimetry,* CRC Press, 9780849356650 Boca Raton, FL.

Kortov, V. (2007). Materials for Thermoluminescent Dosimetry: Current Status and Future Trends. *Radiation Measurements,* Vol. 42, No. 4-5, 576-581, 1350-4487.

Koschnick, F.K., Spaeth, J.M., Eachus, R.S., McDugle, W.G. & Nuttall, R.H.D. (1991). Experimental Evidence for the Aggregation of Photostimulable Centers in BaFBr:Eu^{2+} Single Crystals by Cross Relaxation Spectroscopy. *Physical Review Letters,* Vol. 67, No. 25, 3571-3574, 0031-9007.

Leblans, P.J.R., Struye, L. & Willems, P. (2001). New Needle-Crystalline CR Detector, *Procccedings of SPIE* San Diego, CA, USA, February 2001.

Lewis, D.F. (1986). A Processless Electron Recording Medium, *Proceedings of SPSE'86 Symposium*, Arlington, VA.

Liu, Z., Massil, T. & Riesen, H. (2010). Spectral Hole-Burning Properties of Sm^{2+} ions. Generated by X-rays in BaFCl: Sm^{3+} Nanocrystals. *Physics Procedia,* Vol. 3, No. 4, 1539-1545, 1875-3892.

McKeever, J., Walker, F. & McKeever, S. (1993). Properties of the Thermoluminescence Emission from LiF (Mg, Cu, P). *Nuclear Tracks and Radiation Measurements,* Vol. 21, No. 1, 179-183, 0969-8078.

McKeever, S.W.S. (1988). *Thermoluminescence of Solids,* Cambridge University Press, 0521368111, Cambridge, New York.

McKeever, S.W.S., Moscovitch, M. & Townsend, P.D. (1995). *Thermoluminescence Dosimetry Materials: Properties and Uses,* Nuclear Technology Publishing, 1870965191, England.

Mclaughlin, W.L., Yundong, C., Soares, C.G., Miller, A., Vandyk, G. & Lewis, D.F. (1991). Sensitometry of the Response of a New Radiochromic Film Dosimeter to Gamma-Radiation and Electron-Beams. *Nuclear Instruments & Methods in Physics Research Section a-Accelerators Spectrometers Detectors and Associated Equipment,* Vol. 302, No. 1, 165-176, 0168-9002.

Mijnheer, B. (2008). State of the Art of in vivo Dosimetry. *Radiation Protection Dosimetry,* Vol. 131, No. 1, 117-122, 0144-8420.

Milman, I., Sjurdo, A., Kortov, V. & Lesz, J. (1996). TSEE and TL of Non-Stoichiometric BeO-TiO_2 Ceramics. *Radiation Protection Dosimetry,* Vol. 65, No. 1-4, 401, 0144-8420.

Nakajima, T., Murayama, Y., Matsuzawa, T. & Koyano, A. (1978). Development of a New Highly Sensitive LiF Thermoluminescence Dosimeter and its Applications. *Nuclear Instruments and Methods,* Vol. 157, No. 1, 155-162, 0029-554X.

Nakano, Y., Gido, T., Honda, S., Maezawa, A., Wakamatsu, H. & Yanagita, T. (2002). Improved Computed Radiography Image Quality from a BaFI: Eu Photostimulable Phosphor Plate. *Medical Physics,* Vol. 29, No. 592-597.

Paul, J.R.L., Willems, P. & Alaerts, L.B. (2002). New Needle-Crystalline Detector For X-Ray Computer Radiography (CR). *The e-Journal of Nondestructive Testing,* Vol. 7, No. 12, 1435-4934.

Podgorsak, E. (2005). *Radiation Oncology Physics: A Handbook for Teachers and Students,* International Atomic Energy Agency, 9201073046, Vienna.

Riesen, H. & Kaczmarek, W.A. (2005). *Radiation Storage Phosphor & Applications,* International PCT Application PCT/AU2005/001905, International Publication Number WO 2006/063409.

Riesen, H. & Kaczmarek, W.A. (2007). Efficient X-ray Generation of Sm^{2+} in Nanocrystalline BaFCl/Sm^{3+}: A Photoluminescent X-ray Storage Phosphor. *Inorganic Chemistry,* Vol. 46, No. 18, 7235-7237, 0020-1669.

Riesen, H. & Piper, K. (2008). *Apparatus and Method for Detecting and Monitoring Radiation,* International PCT application.

Rink, A., Lewis, D.F., Varma, S., Vitkin, I.A. & Jaffray, D.A. (2008). Temperature and Hydration Effects on Absorbance Spectra and Radiation Sensitivity of a Radiochromic Medium. *Medical Physics,* Vol. 35, No. 4545-4555.

Samei, E., Flynn, M.J. & Reimann, D.A. (1998). A Method for Measuring the Presampled MTF of Digital Radiographic Systems Using an Edge Test Device. *Medical Physics,* Vol. 25, No. 102-113.

Samei, E., Seibert, J.A., Willis, C.E., Flynn, M.J., Mah, E. & Junck, K.L. (2001). Performance Evaluation of Computed Radiography Systems. *Medical Physics,* Vol. 28, No. 361- 371.

Satinger, D., Horowitz, Y.S. & Oster, L. (1999). Isothermal Decay of Isolated Peak 5 in 165degC/15 Minute Post-Irradiation Annealed LiF:Mg,Ti (TLD-100) Following Alpha Particle and Beta Ray Irradiation. *Radiation Protection Dosimetry,* Vol. 84, No. 1-4, 67-72.

Schulman, J., Shurcliff, W., Ginther, R. & Attix, F. (1953). Radiophotoluminescence Dosimetry System of the US Navy. *Nucleonics (U.S.) Ceased publication,* Vol. 11, No. 10, 52-56.

Schulman, J.H. (1950). *Dosimetry of X-rays and Gamma Rays.*, Naval Research Laboratory. Report No. 3736, Washington, DC.

Schulman, J.H., Ginther, R.J., Klick, C.C., Alger, R.S. & Levy, R.A. (1951). Dosimetry of X-Rays and Gamma-Rays by Radiophotoluminescence. *Journal of Applied Physics,* Vol. 22, No. 12, 1479-1487, 0021-8979.

Schulman, J.H. & Etzel, H.W. (1953). Small-Volume Dosimeter for X-Rays and Gamma-Rays. *Science,* Vol. 118, No. 3059, 184-186, 0036-8075.

Schweizer, S., Rogulis, U., Assmann, S. & Spaeth, J.M. RbBr and CsBr Doped with Eu^{2+} As New Competitive X-ray Storage Phosphors, *Proceedings of the Fifth International Conference on Inorganic Scintillators and their Applications,* 1350-4487, Riga-Jurmala, Latvia, August 2000.

Schweizer, S. (2001). Physics and Current Understanding of X-Ray Storage Phosphors. *physica status solidi (a),* Vol. 187, No. 2, 335-393, 1521-396X.

Schweizer, S., Secu, M., Spaeth, J.M., Hobbs, L., Edgar, A. & Williams, G. (2004). New Developments in X-ray Storage Phosphors. *Radiation Measurements,* Vol. 38, No. 4-6, 633-638, 1350-4487.

Secerov, B., Dakovic, M., Borojevic, N. & Bacic, G. (2011). Dosimetry Using HS GafChromic Films the Influence of Readout Light on Sensitivity of Dosimetry. *Nuclear Instruments & Methods in Physics Research Section a-Accelerators Spectrometers Detectors and Associated Equipment,* Vol. 633, No. 1, 66-71, 0168-9002.

Soares, C.G. (2006). New Developments in Radiochromic Film Dosimetry. *Radiation Protection Dosimetry,* Vol. 120, No. 1-4, 100-106, 0144-8420.

Soares, C.G. (2007). Radiochromic Film Dosimetry. *Radiation Measurements,* Vol. 41, No. Supplement 1, S100-S116, 1350-4487.

Sommer, M., Jahn, A. & Henniger, J. (2008). Beryllium Oxide as Optically Stimulated Luminescence Dosimeter. *Radiation Measurements,* Vol. 43, No. 2-6, 353-356, 1350- 4487.

Stevens-Kalceff, M.A., Riesen, H., Liu, Z., Badek, K. & Massil, T. (2010). Microcharacterization of Core Shell Nanocrystallites. *Microscopy and Microanalysis,* Vol. 16, No. S2, 1818-1819, 1435-8115.

Takahashi, K., Kohda, K., Miyahara, J., Kanemitsu, Y., Amitani, K. & Shionoya, S. (1984). Mechanism of Photostimulated Luminescence in BaFX:Eu^{2+} (X=Cl,Br) Phosphors. *Journal of Luminescence,* Vol. 31-32, No. PART 1, 266-268.

Vij, D.R. & Singh, N. (1997). Thermoluminescence Dosimetric Properties of Beryllium Oxide. *Journal of Materials Science,* Vol. 32, No. 11, 2791-2796, 0022-2461.

von Seggern, H. (1999). Photostimulable X-ray Storage Phosphors: A Review of Present Understanding. *Brazilian journal of physics,* Vol. 29, No. 2, 254-268, 0103-9733.

Weidner, M., Batentschuk, M., Meister, F., Osvet, A., Winnacker, A., Tahon, J.P. & Leblans, P. (2007). Luminescence Spectroscopy of Eu^{2+} in CsBr: Eu Needle Image Plates (NIPs). *Radiation Measurements,* Vol. 42, No. 4-5, 661-664, 1350-4487.

Williamson, G. & Hall, W. (1953). X-Ray Line Broadening From Filed Aluminium and Wolfram. *Acta Metallurgica,* Vol. 1, No. 1, 22-31, 0001-6160.

Yamashit, T. Yasuno, Y. & Ikedo, M. (1974). Beryllium-Oxide Doped with Lithium or Sodium for Thermoluminescence Dosimetry. *Health Physics,* Vol. 27, No. 2, 201-206, 0017- 9078.

Yen, W.M., Shionoya, S. & Yamamoto, H. (2007). *Phosphor Handbook, (second edition),* CRC, 0849335647, Boca Raton, FL.

Zimmermann, J., Hesse, S., von Seggern, H., Fuchs, M. & Knüpfer, W. (2005). Radiation Hardness of CsBr: Eu^{2+}. *Journal of Luminescence,* Vol. 114, No. 1, 24-30, 0022-2313.

Ionizing Radiation Induced Radicals

Ahmed M. Maghraby
National Inst. of Standards (NIS) – Radiation Dosimetry Department, Giza, Egypt

1. Introduction

When ionizing radiation passes through a material it imparts some of its energy to that material. The imparted energy may be high enough to cause a break in bonds inside the molecules or between molecules or both, in such cases, free radicals are created. The type, lifetime, extent, fate and origin of those free radicals may differ according to several factors; most of them are beyond the subject of this chapter. However, the abundance of ionizing radiation induced radicals in a material is directly proportional (unless saturated) to the amount of ionizing radiation received by that material, hence it could be a method for determination of radiation doses passively by accurate evaluation of the extent of free radicals created.
Study of radiation induced radicals is not always related to radiation dose assessment; sometimes it is of great importance to investigate what type of radicals are produced when a specific material is exposed to a specific radiation dose, and for how long those induced radicals can persist. This could be of specific interest specially when dealing with environments of high radiation levels, for example: in space environments. Behavior of radiation induced radicals may lead to further understanding of molecular interactions or molecular dynamics, or may lead to a decision on the preference of a material for a specific design from the material science point of view.

2. Use of EPR for studying radiation induced radicals

Electron Paramagnetic Resonance (EPR) is the major universal technique for investigation of radiation induced radicals. The first observation of an electron paramagnetic resonance peak was made in 1945 when Zavoisky detected a radio frequency absorption line from a $CuCl_2.2H_2O$ sample (Zavoisky, E., 1945), The first EPR study of radiation damage in materials of biological (organic) interest powders was made by Gordy in 1955 (Gordy, W., et al., 1955), Papers on irradiated dimethylglyoxine and α-alanine of Miyagawa and Gordy (Miyagawa, I., and Gordy, W., 1959), and malonic acid by McConell (McConell, H., et al., 1960) soon published and confirmed that it was possible to investigate radicals produced by irradiating single crystals, in a similar way to that which had been used to study paramagnetic ions. Brady et al (Brady et al, 1968) suggested using EPR dosimetry and the additive re-irradiation method to obtain dose estimates from accidental overexposures, where human teeth were used for the first time as radiation dosimeter. In 1962, irradiated L-

α-alanine was suggested as a possible dosimeter material in the high-dose range by (Bradshaw, W., et al., 1962) after which great efforts were made to establish dosimetry systems based mainly on alanine/EPR systems (Deffner, U., and Regulla, F., 1980, Nette, H., et al., 1993).

3. Basic principles of EPR

Electron, as a rotating charge can be considered as a very tiny magnet, and hence the single electron which moves freely in absence of external magnetic field has only two orientations if placed in the field of external magnet: aligned parallel or anti-parallel. Those two cases reflect the two energy states arise after applying external magnetic field: state of lower energy when the moment of the electron, μ, is aligned with the magnetic field and a higher energy state when μ is aligned against the magnetic field.
The two states are designated by the projection of the electron spin, m_s, on the direction of the magnetic field. Because the electron is a spin 1/2 particle, the parallel state has $m_s = -1/2$ and the antiparallel state has $m_s = +1/2$. The energy difference between these two states, caused by the interaction between the electron spin and the applied magnetic field (B_0), is shown in the following relation:

$$\Delta E = g\ \mu_B\ B_0\ \Delta m_s = g\ \mu_B\ B_0 \quad (1)$$

Where μ_B is Bohr magneton (the natural unit of the electron's magnetic moment), g is the g-factor, and $\Delta m_s = \pm 1$.
So, if this electron gains energy of ΔE transition occurs between the two spin states, if this energy is in the form of photons, then:

$$\Delta E = h.\ v = g\ \mu_B\ B_0 \quad (2)$$

Where h is Planck's constant and v is the frequency of the electromagnetic radiation.
Now, resonance may occur either by scanning frequency at a constant magnetic field or by scanning magnetic field while the frequency of the magnetic field was held constant and the later is easier from the practical point of view.

4. Radiation induced radicals in biological molecules

Study of radiation induced radicals in biological molecules or molecules of biological origin is of high concerns in order to understand their impacts on functional and/or structural changes of such molecules after exposure to ionizing radiation. In the following sections, there are two examples for studying radiation induced radicals using EPR in biological molecules: bovine hemoglobin as a biological molecule of animal origin, and chitosan as a biological molecule of plant origin.

4.1 Bovine hemoglobin

4.1.1 Structure

Structure of bovine Hb is shown in Figure (1). It is composed of two pairs of non-identical subunits, alpha and beta. Each alpha-beta pair is more closely associated than they are with each other, but the overall arrangement is roughly tetrahedral (http://www.bmb.uga.edu/wampler/tutorial/prot4.html) (Marta et al., 1996).

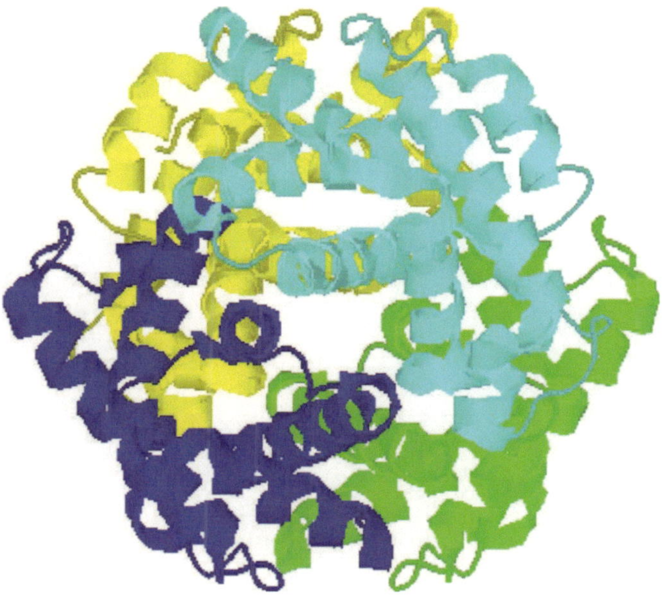

Fig. 1. Bovine Hb is composed of two pairs of non-identical subunits, alpha and beta. Each alpha-beta pair is more closely associated than they are with each other, but the overall arrangement is roughly tetrahedral.

4.1.2 EPR spectra of bovine hemoglobin

Fig. 2 represents the EPR spectra of unirradiated (solid line) and 743Gy gamma irradiated sample (dotted line). Major features of Bovine Hb EPR spectrum are comparable to the human one (Ikeya, 1993). About four features comprise the spectrum of g-factor equal to 5.91017, 4.27507, 2.14737 and 2.00557, respectively. The first two signals are due to Fe (III) of high spin form (S ¼ 5/2). The first signal (S1) is associated with oxidized heme iron, which clarifies its indication to methemoglobin (MetHb), in which a water molecule replaces O2 ion as a ligand of iron (Wajnberg and Bemski 1993). The second signal (S2) is corresponding to non-heme Fe(III) ions at sites endowed with rhombic symmetry, which is not associated with species involved in blood, such as transferrin which causes an EPR signal near g = 4.3 (Ślawska-Waniewska et al., 2004). From Fig. 2, it is clear that S2 is greater than S1, while this is not true in case of human Hb spectrum (Ikeya, 1993). This means higher nonheme iron content in bovine Hb than that of human. The nature of bovine Hb itself or the deterioration of its molecular structure and hence the decomposition of heme during sample preparation may give reason for the increase of non-heme iron content.

Third signal group (S3) is associated with low spin derivatives of ferrihemoglobin called "hemichrome", copper proteins and some transition-metal complexes (Rachmilewitz et al., 1971). Hemichromes are low spin derivatives of ferric Hb brought about through discrete reversible or irreversible changes of protein conformation (Venkatesh et al., 1997). The purity of our sample is about 95%, hemichromes are not easily to be separated from Hb molecules but the rest which is about 5% may contain some of these hemichromes that produce S3. Changes in normal Hb under the effect of time, pH and protein denaturants such as urea or salicylate can form different kinds of hemichromes with different endogenous ligands; hence hemichromes form the primary step to the destructive pathway for denaturation (Venkatesh et al.,1997).

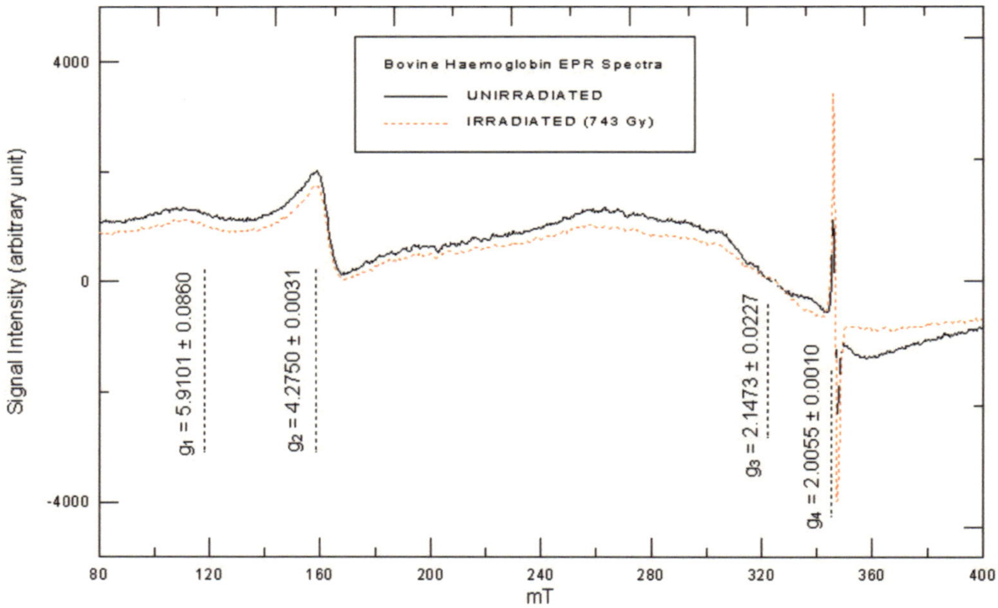

Fig. 2. EPR spectra of unirradiated and 743Gy gamma irradiated bovine Hb samples recorded at room temperature.

With regard to the fourth signal (S4), it appears as a singlet with no hyperfine structures as shown in Fig. 2. This signal is attributed to free radicals in hemoglobin formed by the degradation of blood constituents (Miki et al., 1987). Several investigators agree that at least two different kinds of radicals are formed on the protein (Kelman et al.,1994; Gunther et al., 1995). Although the formation of peroxyl radicals is well proven (Ikeya, 1993; Miki et al.,1987), this kind of radical constitutes only a fraction of the total concentration of radicals (Svistunenko et al.,1997a, b). The globin-based free radical (HB(Fe(IV)=O)) was suggested to be major contributor for S4 (Svistunenko et al., 1997a,b). Many investigations revealed that it is the tyrosine (Tyr) radical (shown in Fig. 3) (Svistunenko et al.,1997a,b, 2002, 2004; Svistunenko, 2005).

4.1.3 Radiation-induced changes in the bovine Hb EPR spectra

The EPR spectrum of 743 Gy gamma irradiated sample is shown in Fig. 2; from the figure it is clear that no new radicals have emerged, and no remarkable changes in the intensity of the first two signals were recorded. The unchanged intensity of the first signal (with all the irradiation doses) suggests no net change in MetHb, which may be explained as follows: The manipulated sample is in powder form, so any enzymatic reaction that can lead to change in MetHb content is excluded. The presence of oxygen is mandatory for MetHb production by other pathways such as the oxidation of heme iron by the electron transfer from Fe(II) to O2 creating Fe(III) and superoxide radicals (O_2^-) (Misra and Fridovich, 1972) During irradiation, wrapped samples were prevented from molecular oxygen in air. So, MetHb production through the second pathway is prevented; while the first pathway may be blocked by the removal of oxygen molecules from the bovine Hb sample during preparation in its powder form. The non-significant change in intensity of S2 by irradiation (Fig. 1), reflects the stability of non-heme iron content as radiation doses increase up to 743 Gy. It is clear from

Fig. 1 that S3 suffered apparent significant decrease upon irradiation; which ensures the decrease of the net amount of hemichromes and reflects the crosslinking processes following Hb irradiation at high doses. Results showed that the most obvious radiation-induced change in bovine Hb EPR spectrum is the significant increase in S4 even for very low dose (4.95 Gy). This may be due to the increase in the production of free radicals in Hb protein (peroxyl and tyrosyl radicals) and reflects the high sensitivity of Hb protein to irradiation.

4.2 Chitosan

Chitosan is a natural polysaccharide, that can be prepared on an industrial scale by deacetylation of much more abundant chitin and its chemical structure is a copolymer of β-(1-4)-D-glucosamine and N-acetyl- β-(1-4)-D-glucosamine (Jaroslaw, M., et al., 2005). Due to unique biophysical and chemical properties of polysaccharides, such as biocompatibility, biodegradability, nontoxicity and nonantigenicity (Bin Kang, et al., 2007), a broad spectrum of applications has been emerged in different modern fields: water treatment (Kurmaev E.Z. et al., 2002), chromatography, additives for cosmetics, textile treatment for antimicrobial activity (Le Hai et al., 2003), novel fibers for textiles, photographic papers, biodegradable films, biomedical devices, and microcapsule implants for controlled release in drug delivery. Also, its nano-ordered hydrogel is a potential responsive material for biochips and sensors for the development of PC- controlled biochips, and is used also in some attempts as a gene delivery system for curing of some hereditary diseases (IAEA, 2004). Tissue engineering and adsorption of metal ions as well as dyes removal are some of the many applications of chitosan (Jayakumar, R., et al., 2005).

4.2.1 Chitosan EPR spectral features

Figure (3) represents the EPR spectrum of chitosan-A. There are two overlapped singlet signals, the first signal (S_1) is of g-factor = 2.00725 ± 0.00018 with peak-to-peak line width

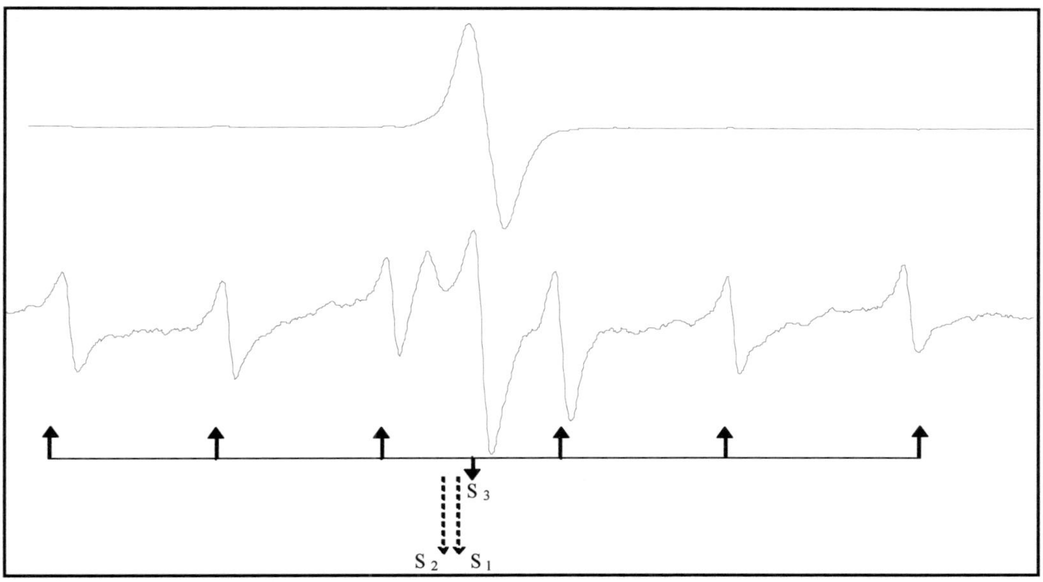

Fig. 3. EPR spectra of chitosan: lower represents the unirradiated sample spectrum, and the upper represents 30 kGy irradiated sample spectrum.

(W_{PP}) = 1.08 ± 0.072 mT, while the second signal (S_2) is of g-factor = 2.00966 ± 0.00073, while its W_{PP} = 5.70 ± 0.139 mT. EPR lines of Mn^{2+} (I = 5/2), ions are well recognized signal (S_3) with sextet hyperfine structure were observed (g_{iso}=2.01147±0.0002, a_{av}=9.06±0.35 mT). This signal is attributed to the presence of Mn^{2+} in crab shells from which chitosan-A was extracted (Maghraby, A., 2007). No new radicals have been emerged after irradiation up to 30 kGy, while some slight broadening in line width values for S_1 was noticed.

5. EPR radiation dosimetry

Radiation dosimetry using EPR utilizes the produced radicals as a result of the passage of ionizing radiation through matter, as extent of those radicals proportionate directly to the amount of ionizing radiation. Radiation dosimetry using EPR possesses some advantages which enabled the use of EPR in different fields of ionizing radiation dosimetry.

5.1 Industrial applications

5.1.1 Food irradiation

Food irradiation is a means of microbiological contamination reductions through which, food in packages or in bulk are exposed to a high radiation dose which is enough to destroy microbiological contents (Maghraby, A., 2007, Miyagusku, L., et al., 2007). The aim of food irradiation is sterilization and preservation of food during food transfer from region to another or to increase its quality locally before use (IAEA, 1978). The use of irradiation alone as a preservation technique can play an important role in cutting losses and reducing the dependence on chemical pesticides. Each year a few hundred thousand tones of food products and ingredients are irradiated worldwide. International food stuff trade especially seafood is growing up which increases the risk of the transfer of some diseases; this led the World Health Organization to legalize radiation treatment of food (WHO, 1988, 1994). International health and safety authorities have endorsed the safety of irradiation for all foods up to a dose level of 10 kGy, however, recent evaluation of an international expert study group appointed by FAO, IAEA, and WHO showed that food treated according to good manufacturing practices (GMPs) at any dose above 10kGy is also safe for consumption(ICGFI, 1999).

Irradiated food could be identified using different methods such as Gas chromatography, mass spectrometry, thermoluminescence (Schreiber, G., et al., 1993a, Schreiber , G., et al., 1993b, Schreiber , G., et al., 1993 c, Ziegelmann, K., et al., 1999, Parlato, A. et al., 2007, D'Oca, M. C. et al., 2008), and Electron paramagnetic resonance (EPR) spectroscopy (Desrosiers, M., 1989, Sünnetcioğlu, M., et al., 1999) which is characterized by its non-destructive detection of radiation-induced radicals, hence, it is a major technique for the investigation of irradiated food and the determination of the dose delivered accurately, which results in accepting or refusing of food transfer.

Mineral nutrients and Irradiated vegetables were identified using EPR (Lakshmi Prasuna, C.P, 2008, Maghraby, A., and Maha Anwar Ali. 2007), EPR spectral investigations have been carried out on some vegetable samples and the presence of various paramagnetic metal ions in various oxidation states is indicated. In almost all the fibrous vegetable samples, the free radical signal corresponding to cellulose radical is observed. However, in some vegetables like carrot, though the free radical signal cannot be distinguished, the free radical assignments are made depending on the organic radicals. The reason is that, the organic free radicals are generated due to CO group in the organic radicals (Lakshmi Prasuna, C.P,2008).

5.1.2 Drugs and medical products irradiation

Use of ionizing radiation in sterilization of medicinal products, such as catheters, syringes, drug and drug raw materials, is a new technology alternative to heat and gas exposure sterilization (Jacobs, 1995; Reid, 1995; Tilquin and Rollmann, 1996; Boess and Bögl, 1996). The advantages of sterilization by irradiation include high penetrating power, low measurable residues, small temperature rise and the fact that there are fewer variables to control (Fauconnet et al., 1996; Basly et al., 1997). Thus, sterilization can be carried out on the finally packaged product and is applicable to heat-sensitive drugs.

The regulations of radiosterilization differ among countries. In other radiation sources such as X-rays, fast electrons and UV illumination. Irradiation produces new radiolytic products. To prove the safety of radiosterilization, it is important to determine physical and chemical features of the radiolytic products and elucidate the mechanism of radiolysis. Thus, it is desirable to establish a method to discriminate between irradiated and unirradiated drugs. Electron spin resonance (ESR) spectroscopy appears to be well suited for determination of free radical concentrations in complex media and so, it can be used to detect and distinguish between irradiated drugs from unirradiated ones (Gibella et al., 1993; Signoretti et al., 1994; Miyazaki et al., 1994; Onori et al., 1996).

As a cephalosporin antibiotic, duricef is used in the treatment of nose, throat, urinary tract and skin infection that are caused by specific bacteria, including staph strep and E. coli, effects of gamma radiation on cephalosporins with various substitutive groups have been reported in different papers (Miyazaki et al., 1994; Onori et al., 1996; Jacobs, 1983; Yũrũs and Korkmaz, 2005; Gibella et al., 2000), but low-and high-temperature kinetic features, structure and activation energies of the radical species involved in the formation of their ESR spectra were investigated in none of these papers, except of few published work (Yũrũs, S and Korkmaz, M., 2005). Eleven of the studied cephalosporins with various substitutive groups have been shown to exhibit, interestingly ESR spectra of unresolved doublet appearence (Onori et al., 1996; Yũrũs and Korkmaz, 2005). This conlusion was considered as an indication relevant to the origin of radicalic species produced in irradiated cephalosporins. Species originating from substitutive groups and cephem ring are expected to be responsible from ESR spectra of irradiated cephalosporins, but the ultimate patterns of the latter are believed to be determined by the relative weights of those species. Thiamphinicol sterilization using irradiation to doses up to 25 kGy was investigated using EPR spectroscopy, and it was found that standard dose used for sterilization (25 Gy) did not produce any detectable changes in the physico-chemichal properties of Thiamfinicol.

5.1.3 Other applications

An electron paramagnetic resonance (EPR) investigation of a series of glasses irradiated at room temperature with β or X radiation sources has been made in order to predict the long term behavior of glasses used in the nuclear waste disposal, in particular, no paramagnetic defects associated with aluminum ions are detected in these irradiated glasses (Bruno Boizot, et al., 1998).

Applications in polymer research and industries are numerous, for example, Poly (vinyl chloride) is one of the most used thermoplastic polymers, thanks to its low production cost, easy processing, excellent mechanical properties, high compatibility with additives and possibility to be recycled. Particularly interesting is its employment in vital single use medical devices such as catheters, cannulae, urological products and flexible tubes for extracorporeal connections. The sterilization of these products is performed by electron beam or γ irradiations, a termination step, in which the active centers are deactivated. At 25 and 50 kGy oxygen appears to saturate all radicals and the EPR spectrum shows only one component, associated with peroxyl radicals. At the 100 and 150 kGy irradiation doses the EPR spectrum shows more structure and it comprises overlapping signals from the peroxyl and polyenyl radicals not yet oxidized (Baccaro, S. et al., 2003, Costa, L. et al., 2004).

5.2 Medical applications

5.2.1 Alanine dosimetry

5.2.1.1 Introduction

The natural amino acid L-α-alanine (see Figure (4)) has attracted considerable interest for use in radiation dosimetry (Bradshaw, W., etal., 1962, Regulla, D., and Deffiner, U., 1982, Nam, J., and Regulla, D., 1989, Ciesielski, B., and Wielopolski, L., 1994) and has been formally accepted as a secondary standard for high-dose and transfer dosimetry (Nett, H., ate al., 1993). The EPR powder spectrum from amorphous alanine pellets or other types of disordered alanine samples has been used for this purpose. The peak-to-peak amplitude of the central line of this powder spectrum is commonly used for monitoring the radiation dose (Eirik Malinen, et al., 2003a).

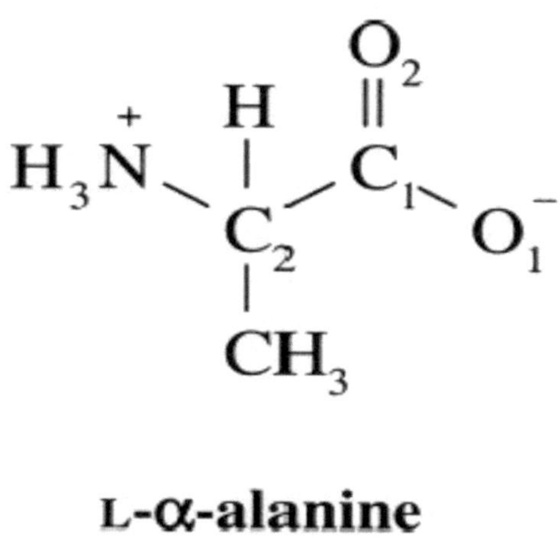

Fig. 4. Molecular structure of L-α- alanine.

It was found that at least three different radicals are formed and stabilized in alanine after X irradiation at room temperature as shown in Figure (5), (Sagstuen, E., et al., 1997a, Sagstuen, E., et al., 1997b, Mojgan Z. Heydari, et al., 2002). Two of these, the R1 radical (Structure 1) and a species formed by net hydrogen abstraction from the central carbon atom (radical R2, Structure 2), appear to occur in comparable relative amounts (55–60 and 30–35%, respectively) (Mojgan Z. Heydari, et al., 2002). The third species (denoted R3) is a minority species (5–10%), and an unambiguous structure assignment has not been made (Sagstuen, E., et al., 1997b). Structure 3 below, was suggested as a possible candidate for R3 (Eirik Malinen, et al., 2003b).

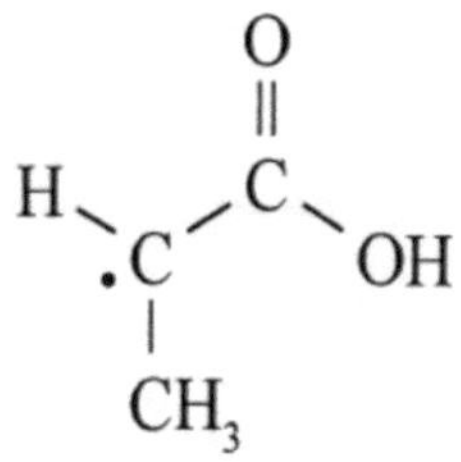

Structure 1

R1 (SAR)

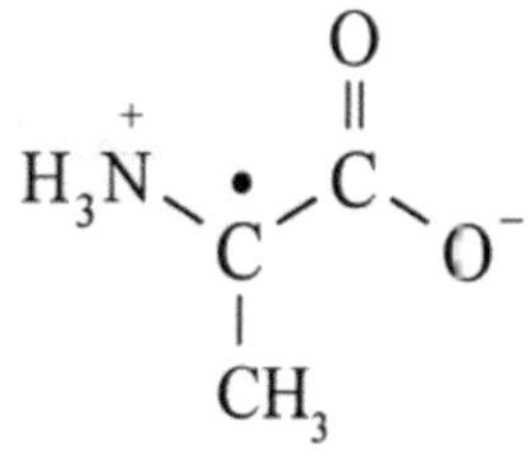

Structure 2

R2

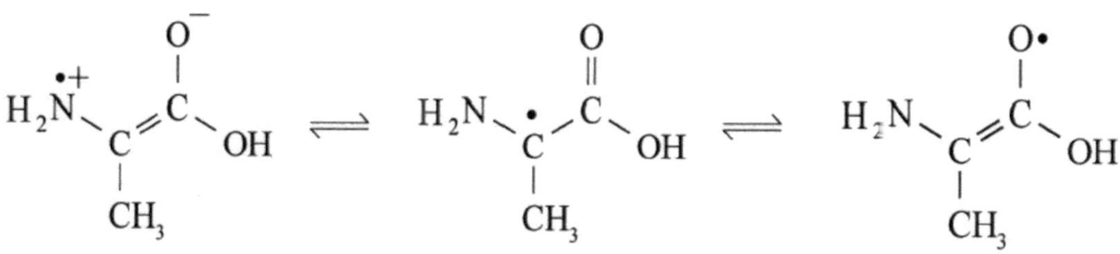

Structure 3

R3

Fig. 5. Three structures of alanine radicals, R1 or (SAR) is the Stable Alanine Radical, R2, and R3 in its three forms (Eirik Malinen, et al., 2003a and Eirik Malinen, et al., 2003b).

5.2.1.2 Dosimeters structure and shape

Alanine dosimeters often are composed of a binder and the alanine itself, the ratio and the binder type differs from type to type. The pellets are now in routine use as dosimeters for medical applications because of their robustness they are used as reference dosimeters sent by mail as a service offered by many laboratories (Maghraby, A., 2003).

Transfer alanine dosimeters which are distributed by the IAEA for intercomparison purposes were made of 70% per weight DL-α-alanine + 30% per weight polystyrene (Mehta, K., and Girzikowsky, R., 2000).

There are a lot of variations in alanine dosimeter shapes, varies according to different reasons, like the purpose of use, resolution required, and achievable sensitivity.

5.2.1.3 Applications

Experimental procedures of the use of alanine as dosimeter for brachytherapy purposes was described by De Angeles, C., et al., 1999. Also, a device for in vivo measurements of the rectal dose in radiotherapy for prostate cancer was developed (Daniela Wagner, et al., 2008) as shown in Figure (6). It was possible to insert this device without clinical complications and without additional rectal discomfort for the patients. For irradiations of alanine dosimeter probes under reference conditions, deviations of less than 1% in reference to the national German standard were achieved. In the absence of metallic implants, the relative deviations between measured and applied dose values at the anterior rectal wall are less than or equal to 1.5% for the in vivo measurements.

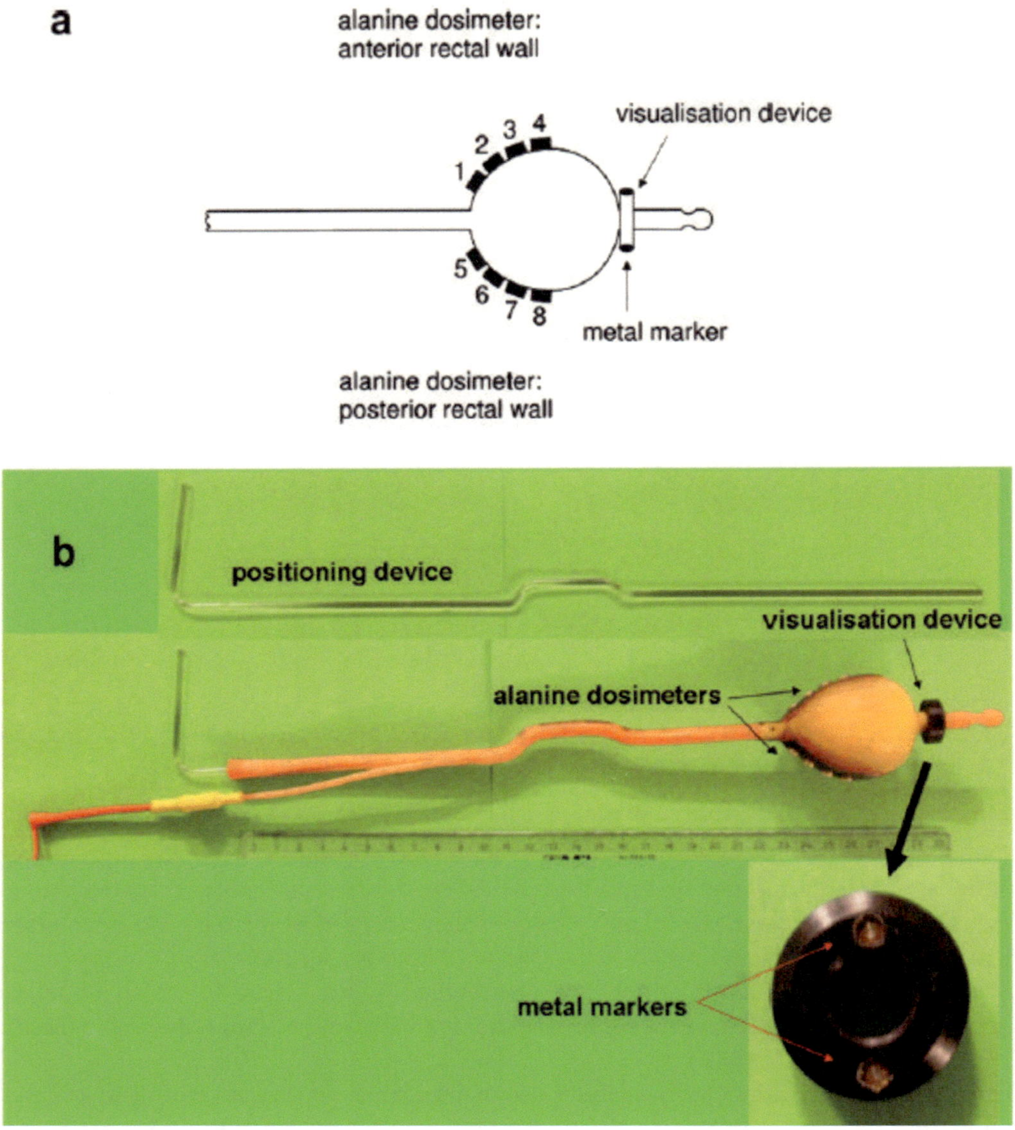

Fig. 6. (a) Sketch of the used rectal balloon equipped with alanine dosimeters. Four alanine dosimeter probes were placed at the anterior rectal wall (1–4) and: four alanine dosimeters at the posterior rectal wall (5–8). Note the visualization device on top of the balloon consisting of two metal markers. (b) Picture of one rectal balloon equipped with alanine dosimeters. For illustration, the positioning device was inserted. At the top of the rectal balloon, the visualization device with metal markers was added (Daniela Wagner, et al., 2008).

At the posterior rectal wall, relative deviations of up to 3.5% may occur. The dominant contribution to the overall uncertainty for the in vivo measurements was the positioning of the dosimeter probes in the patient's body and their corresponding localization in the Computed Tomography (CT) data. Therefore it is expected that improving the probe positioning in the patient's body (e.g. by an increased visibility in the radiographic images) will lead to more accurate results.
The method presented in this study turned out to be useful for in vivo quality control of the irradiations. The relative deviation between the dose determined by the ESR measurements and the planned dose determined by the treatment planning systems (TPS) was shown to be within the 5% limit recommended by the ICRU (ICRU, 1976) for doses above 0.7 Gy (Daniela Wagner, et al., 2008).
Alanine was used also for the in-phantom dose mapping in neutron capture therapy (NCT) (Baccaro, S., et al., 2004), Recoil proton dose can be measured by means of alanine detectors, after subtraction of gamma dose contribution as evaluated from the dose measured with TLD-300. To this purpose, it is necessary to study the sensitivity of alanine to proton recoils and the coefficient for converting the recoil proton dose in alanine to that in tissue.

5.2.1.4 Influencing factors

Time passes before starting measurements after irradiation of alanine dosimeters is a key factor for acquiring reliable measurements, it was found that there is complex dependence on the time conjugated with the received dose range (Nagy, V., and Desrseirs, M. 1996). However in one of recent studies (Anton, M., 2008), the postirradiation behavior is highly dependent on the dosimeter structure and constituents.
In addition to the time dependence, extensive studies on ambient environmental conditions effects on the alanine dosimeters sensitivity and response were performed (Nagy, V. et al., 2000, Sleptchonok, O.F., et al., 2000, Anton, M., 2005). It was found that the effect of storage conditions is dependent on the type of the dosimeter. Effect of higher temperatures also was studied (Maltar-Strmečkia, N. and Rakvin. B. 2005). Same studied parameters (relative humidity and postirradiation behavior) were carried out for a new alanine film dosimeters (Ruth M.D. Garciaa, et al., 2004).
Some other influencing factors were studied: the angular response of alanine dosimeters either in the form of pellets or powder (Jean-Michel Dolo, and Tristan Garcia. 2007), and the grain size (Tristan Garcia and Jean-Michel Dolo. 2007).

5.2.1.5 Energy dependence

Energy dependence of alanine dosimeters were investigated (Eva Stabell Bergstrand, 2003). It was found that the alanine response is 0.8% lower for high energy X-rays than for Co-60 gamma rays, this result indicates a small energy dependence in the alanine response for the high-energy photons relative to Co-60 which may be significant. Another recent study has been performed to reveal the response of alanine dosimeters to high energy photons 8 and 16 MV relative to the response of alanine dosimeter irradiated to the reference beam quality (Co-60), a confirmation of these results was made using EGSnrc package (Anton, M., et al., 2008). Additional study of energy dependence of Dl-alanine to 10 MV high energy photons has revealed the decrease in alanine response with respect to the Co-60 reference beam quality (Borgonove, A.F., et al., 2007).

5.2.1.6 Alanine derivatives

Some researchers tried to use different alanine substitute in order to take the advantages of alanine as solid-state dosimeter with minimum drawbacks, however, the positive results are not mandatory, for example: Polycrystalline phenyl-alanine and perdeuterated L-a-alanine (L-a-alanine-d4) were studied as potential high-energy radiation-sensitive materials (RSM) for solid state/EPR dosimetry. It was found that phenyl-alanine exhibits a linear dose response in the dose region 0.1–17 kGy. However, phenyl-alanine is about 10 times less sensitive to γ-irradiation than standard L-a-alanine irradiated at the same doses. Moreover, the EPR response from phenyl-alanine is unstable and, independent of the absorbed dose, decreases by about 50% within 20 days after irradiation upon storage at room temperature. γ-irradiated polycrystalline perdeuterated L-a-alanine (CD3CD(NH2)COOH) has been studied at room temperature by EPR spectroscopy. By spectrum simulations, the presence of at least two radiation induced free radicals, R1 (CH3C*(H)COOH) and R2 ($H3N^{+}$-C*(CH3)COO^{-}), was confirmed very clearly. Both these radicals were suggested previously from EPR and ENDOR studies of standard alanine crystals (Veselka Ganchevaa et al., 2006). Also, Minidosimeters of 2-methylalanine (2MA) were prepared and tested as potential candidates for small radiation field dosimetry (Bruno T. Rossi, et al., 2005). To quantify the free radicals created by radiation, a K-Band (24 GHz) EPR spectrometer was used. X-rays provided by a 6 MV clinical linear accelerator were used to irradiate the minidosimeters in the dose range of 0.5–30 Gy. The dose–response curves for both radiation sensitive materials displayed a good linear behavior in the dose range indicated with 2MA being more radiation sensitive than L-alanine. Moreover, 2MA showed a smaller LLD (Lower Limit of Detection) value. The proposed system minidosimeter/K-Band spectrometer was able to detect 10 Gy EPR spectra with good signal-to-noise ratio (S/N). The overall uncertainty indicates that this system shows a good performance for the detection of dose values of 20 Gy and above, which are dose values typically used in radiosurgery treatments (Chen,F., et al., 2007).

Alanine-in-glass dosimeters were prepared by packing pure polycrystalline L-*a*-alanine directly as supplied by the manufacturer in glass tubes. These dosimeters exhibited a linear dose response in the dose range from 0.1 to 20 Gy. These positive properties favor the polycrystalline alanine-in-glass tube as a radiation dosimeter (Anan M. Al-Karmi, and M.A. Morsy, 2008).

A new generation of self-calibrated alanine dosimeters were developed, and a regular international intercomparison is held for evaluation of radiation dos using these dosimeters (Gancheva, V. et al., 2008), these dosimeters are consists of RSM (α-alanine, sugar, other ones), Mn^{2+}/MgO as internal EPR intensity standard (IES) and a binder. Necessity to assurance of very good homogeneity of dosimeter material; and the cost of IES present in the amount of some percent in each self-calibrated dosimeter are of the main shortcomings of this technique. Also, it was found that addition of gadolinium to the alanine dosimeters in definite amounts helps to improve the Linear Energy Transfer (LET) sensitivity for γ photons (because of its high atomic number, Z = 64) and thermal neutrons as well (because of its high thermal neutron cross section) (Marrale, M., et al., 2007).

5.2.2 Other organics

The search for new EPR dosimeter is non stopping, the search for organic dosimeter ensures the minimum energy dependence at least at 0.5 MeV and higher photon energies, and hence tissue equivalency of the dosimeter (Maghraby, A., and Tarek, E., 2006), for example: sulfanilic acid possesses several good features of the good dosimeter and is characterized by its simple spectrum. Although its sensitivity is less than that of alanine, it could be pressed into pellets purely without need to a binder, and hence more homogeneity could be achieved. Sulfanilic acid is nearly tissue equivalent which enables its use in radiation therapy dosimetry, also it is isotropic and its detection limit is about 100±30 mGy. Sulfanilic acid EPR signal intensity shows noticeable stability for a sufficient time, which enables its use as a transfer dosimeter. Sulfanilic acid deserves further studies in order to be established as a common radiation dosimeter using EPR (Maghraby, A., and Tarek, E., 2006).

One of the most promising organic dosimeters ever since the discovery of alanine at 1962 (Bradshaw, W., et al., 1962), is the Lithium format (Gustafsson, H., et al., 2008). It is characterized by its simple spectrum (one narrow peak spectrum) and its high sensitivity (six times more than alanine), beside its obvious stability. Lithium formate also possesses tissue equivalency more than that of alanine, with suitable microwave dependence and modulation amplitude dependence (Tor Arne Vestad, et al., 2004). Lithium formate may replace alanine in the near future specially for its clinical uses.

5.3 Inorganic dosimeters

A wide variety of inorganic dosimeters were investigated, some of these dosimeters processed higher sensitivity than alanine with reasonable stability, and some examples are following:

5.3.1 Combined TL and EPR dosimeters

Beryllium oxide ceramics was investigated for TL and EPR dosimetry. Ceramics were doped with lithium and neodymium ions. TL and EPR signals associated with Li centers whose amplitude is proportional to the absorbed dose are observed. A complete anneal of the EPR signal takes place in the temperature range of the TL recordings (Kortov, V., et al., 1993b).

Onori, S., et al., have used the well known thermoluminescent material; calcium sulphate: Dy, for EPR high dose assessment (Onori, S., et al., 1998). Three EPR signals are detectable for $CaSO_4$: Dy phosphor, two out of the three signals were studied for high dose applications since one of them showed saturation at about 1 kGy. The concentration of both centers, $(SO_3)^-$ and Ca-vacancy, increases with dose at least up to 10^7 Gy or more. In both cases, the dose-effect relationship is not linear, being supralinear for $(SO_3)^-$ center and sub-linear for $(Ca)^{2-}$ center. $(Ca)^{2-}$ center is very stable over time independently of dose, while $(SO_3)^-$ center is not stable over time and can be used for a first dose estimation soon after irradiation.

The effect of grinding on powder form of clear fused quartz was studied by EPR and TL techniques. The minimum detectable dose (MDD) for using EPR was about 2 Gy for the clear fused quartz in powder form, which is 2 times greater than the bulk form. The EPR signal for of background varied inversely with particle size and was quite high for sizes lower than 38 μm, while for a Co-60 irradiated samples (about 22 Gy), the EPR intensity of the coarse powder varied directly with particle size. Thus, the intensity of a particle size of 20-38 μm was very low (Ranjbar, A., et al., 1999).

Feasibility of reading LiF thermoluminescence dosimeters by EPR was studied by Breen, S., (Breen, S., and Battista, J., 1999). EPR signals can be observed in irradiated polycrystalline LiF rods, but only near liquid nitrogen temperatures. The magnitude of this signal was measured by two methods: direct measurement of the radiation-induced signal, and curve fitting. The direct measurement showed that the radiation-induced signal increased linearly with dose; however, the technique suffered from low sensitivity; at low doses (below 20 Gy) the signal was barely discernible above the noise in the spectrum.

The curve-fitting method isolated three peaks in the EPR spectra of TLD-100. One component increased linearly with dose, although the spectrum of unirradiated TLD-100 possessed a large contribution from this component. This behavior is similar to that observed in the peak isolated by direct measurement. A second component was uncorrelated with dose. The third component decreased with dose, although this part of the spectrum was modeled poorly, perhaps due to the presence of another (fourth) signal in the spectrum. Due to the experimental difficulty in interpreting these spectra, and the low sensitivity of the application, EPR is not recommended as a substitute for thermoluminescent dosimetry of polycrystalline LiF.

5.3.2 Other inorganic compounds

Sulfamic acid possesses high-sensitivity to gamma radiation and is able to detect radiation doses below 5.0 Gy which would be advantageous for a possible use in medical applications, beside other good dosimetric properties like almost energy independence, narrow line and simple spectrum of well-defined radicals, also predictable decay kinetics which enable its use as a transfer dosimeter (Maghraby, A., 2007).

Ferrous ammonium sulfate was used as a high dose dosimeter in the range from 33.5 to 546 kGy (Juárez-Calderón, J. et al., 2007). Also, Sulphur trioxide anion in $K_2CH_2(SO_3)_2$ and carbon dioxide anion in irradiated sodium formate ($NaHCO_2$) were suggested as inorganic alternatives to the EPR/alanine dosimeter by Keizer, P.N., (Keizer, P.N., et al. 1991). These two systems have a four-fold sensitivity advantage over alanine. The radicals sulphur trioxide anion and carbon dioxide anion are, moreover, found in a wide variety of matrices, and it may be possible to find one in which they are even stronger. Sulphur trioxide anion was found to be completely stable over time, while carbon dioxide anion was not so stable against decay as sulphur trioxide, for example the signal decayed by 30 % over the first six weeks when irradiated to 10 kGy. The main drawback in inorganic dosimeters is their non-water equivalence.

Dosimetric properties of magnesium sulphate were investigated (Morton, J., et al., 1993). On irradiation with Co-60 γ rays, the stable SO_3^- is produced whose EPR signal amplitude increases linearly with dose up to about 10^5 Gy Advantages and disadvantages of the SO_3^- radical system were compared with α-alanine. The studied dose range was between 0.25 Gy and 50 Gy, focusing on the region below 10 Gy. Storage as well as measurement of the encapsulated samples was at room temperature, sample spectra are reproducible several weeks later. It was seen that the signal enhancement for magnesium sulphate is about 80%, quite significant for the lower dose range.

Magnesium oxide was suggested as a combined thermostimulated luminescence TSL-EPR detector for ionizing and ultraviolet (UV) radiations (Kortov, V., et al., 1993a).

Dose dependence of the EPR signal intensity caused by Fe^{+3} ions impurities is linear from 1 to 10^{4} Gy for X irradiation. Temperature range of measurements for the EPR signal is expanded, allowing measurements of radiation dose at higher temperatures.

In Physikalisch-Technische Bundesanstalt (PTB), Schneider, C., (Schneider, C., 1994), tested the SO_3^- radical in an anhydrous $MgSO_4$ matrix, it shows a three times stronger EPR signal than the alanine radical at the same dose. Because the SO_3^- spectrum has negligible high frequency noise components, the pattern recognition of the derivative spectral shape out of the noise background is better. Peak-to-peak evaluation at low doses (<10 Gy) is therefore more accurate for this simple single line spectrum. Discussion of EPR dosimeter materials cannot circumvent the problem of water equivalence. At low energies (<50 keV) the photon response of a martial is proportional to Z^4 up to Z^5, so, for instance, with sulphur nuclei instead of oxygen in a dosimeter, the response will differ by a factor of more than 16. In practice, the mass collision stopping power (the quotient energy loss over beam pathlength, ($\Delta E/\Delta l$) of alanine is about 3% lower than that of water in the electron energy range 0.1-1.0 MeV, whereas for $MgSO_4$ it is about 17% lower. Therefore, in the search for alternative EPR dosimeter materials, it is desirable that a matrix be found which combines spectral advantages (as shown by the SO_3^- radical for example) with a low Z value for its constituent nuclei.

Electron paramagnetic resonance (EPR) studies have been made on lithium metaitanate (Li_2TiO_3) ceramics irradiated by gamma rays of Co-60 source at ambient and Liquid Nitrogen Temperatures (LNT). The EPR spectra have been found strongly dependent on irradiation dose and temperature. The radiation defects induced by gamma irradiation at LNT were found to disappear almost completely by heating up to 255 k. in contrast to this, the radiation defects produced by irradiation at ambient temperature showed tolerance of elevated temperatures up to about 600 k (Grišmanovs, V., et al., 2000).

The sulphur trioxide anion looks interesting, so, new inorganic dosimeters based on barium and strontium dithionates ($BaS_2O_6.2H_2O$ and $SrS_2O_6.4H_2O$) were developed by Bogushevich, S., (Bogushevich, S., and Ugolev, I., 2000). The minimum detection limit was found to be 0.05 Gy at the measurements uncertainty of ± 30%. The dependence for γ irradiated dosimeters was linear up to 50 kGy for barium dithionates, and 80 kGy for strontium dithionates. Radical ions SO^-_3 in barium and strontium dithioniates are stable at 20°C. Studies have shown that, at temperatures below 35°C, the signal is stable within ± 5% for at least two years. Similarly, Barium sulphate (BaSO4) was irradiated by γ-rays and analyzed with electron spin resonance (ESR) to study radiation induced radicals for materials as radiation dosimeter (Sharaf, M.A. and Gamal M. Hassan. 2004).

Ammonium dithionate has been investigated as a potential dosimeter material (Danilczuk, M., et al., 2008). The radical signal in irradiated polycrystalline samples is a structureless narrow line. Ammonium dithionate was found to be more sensitive an l-α-alanine by a factor of seven at he same spectrometer settings. The results indicate that the ammonium dithionate can be applied as a dosimeter for situations when a material more sensitive than l-α-alanine is needed.

5.4 Retrospective and emergency dosimetry

5.4.1 Teeth enamel dosimetry

The EPR spectrum of irradiated tooth enamel (Shin Toyoda, et al., 2008) contains a multitude of signals that can be divided into two categories, radiation-induced and radiation insensitive signals (Figure (7)). This approach is an approximation because the intensity of the so-called non-radiation sensitive EPR spectral components from tooth enamel are also slightly affected by irradiation, which is evident after irradiation with doses above one hundred Gray. However, these EPR spectral components can be considered as radiation insensitive in the application range of retrospective dosimetry.

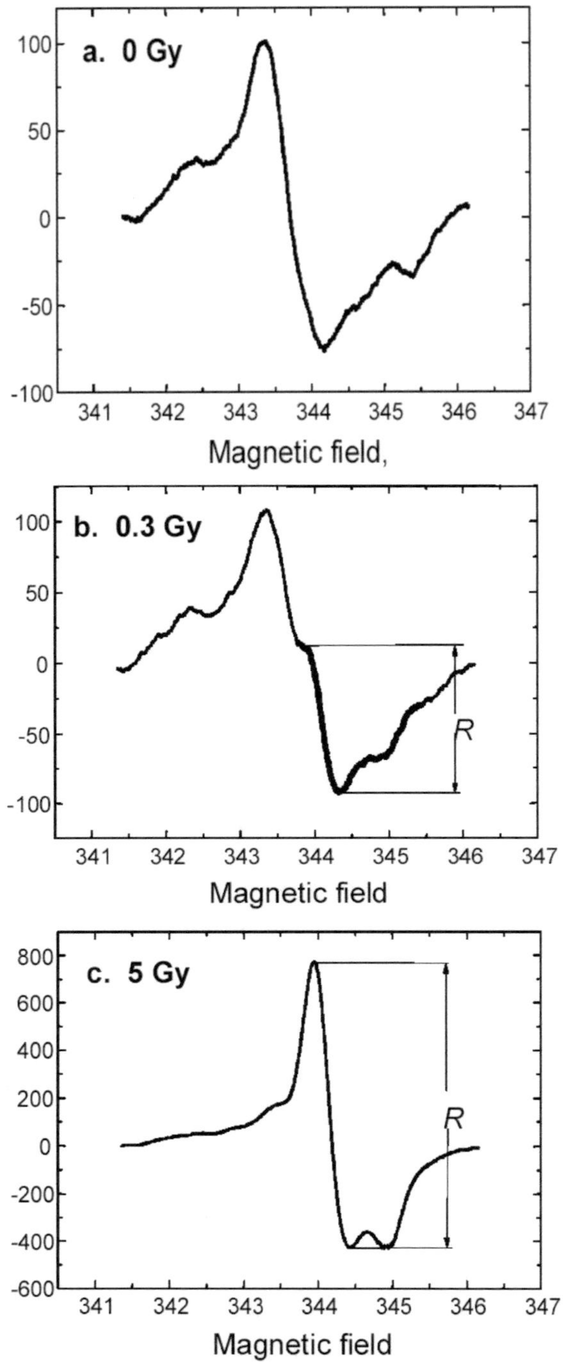

Fig. 7. EPR spectra of tooth enamel after irradiation with different doses: a- 0 Gy; b- 0.3 Gy; c- 5 Gy. The dosimetric component of the spectrum after irradiation with 0.3 Gy (in the middle – b) is in dark. R is the peak-to-peak amplitude used for EPR reconstruction (IAEA, 2002).

The majority of radiation-induced radicals in tooth enamel are carbonate derived, i.e.,CO_2^-, $CO3^-$, CO^-, $CO3^{3-}$, but also radicals derived from phosphate, i.e., $PO4^{2-}$, and oxygen, i.e.,O^- and O^{3-} were identified. The identification of radicals was based on EPR and ENDOR (Electron Nuclear Double Resonance) measurements of irradiated synthetic hydroxyapatite doped with ^{13}C (Callens, F., et al., 1998). Not all radiation-induced radicals are thermally stable, e.g., the CO3-radical, with g-value of the EPR signal ranging from 2.0060 to 2.0122, decays completely at room temperature during the first two weeks after irradiation (Callens, F., et al., 1998, Cevc, P., et al., 1972, Romanyukha, A., et al., 1996). For dose reconstruction the asymmetric EPR signal with $g_\perp$=2.0018 and $g_{||}$=1.9971 (signal maximum at g=2.0032 and minimum at g=1.9971) is used. The signal is predominantly derived from stable CO_2^- radical (IAEA, 2002).
Two methods have been used to assess the absorbed dose of irradiated enamel by EPR: additive re-irradiation and the use of calibration curve, in the additive re-irradiation method: the sample is incrementally irradiated to construct a response curve specific to the sample in question (Pavlenko, A et al., 2007) as shown in Figure (8). The other method uses a universal calibration curve (EPR signal Intensity versus absorbed dose) generated using a large blended sample pool of enamel material designed to average the sample-to-sample variances. For doses greater than a few hundred mGy, the variation in EPR signal intensity from sample to sample for tooth enamel is about 10%, however, dose reconstruction using the universal calibration curve method is much less time-consuming and is non destructive (Desrosiers, M., and Schauer, D.A., 2001). Some other techniques are not widely used (Lanjanian, H., et al., 2008). Some consideration to be taken into account, such as the internal irradiation of human body by radioactive cesium isotopes (Borysheva, N., et al., 2007), many other influencing parameters have been investigated (El-Faramawy, N., 2008).

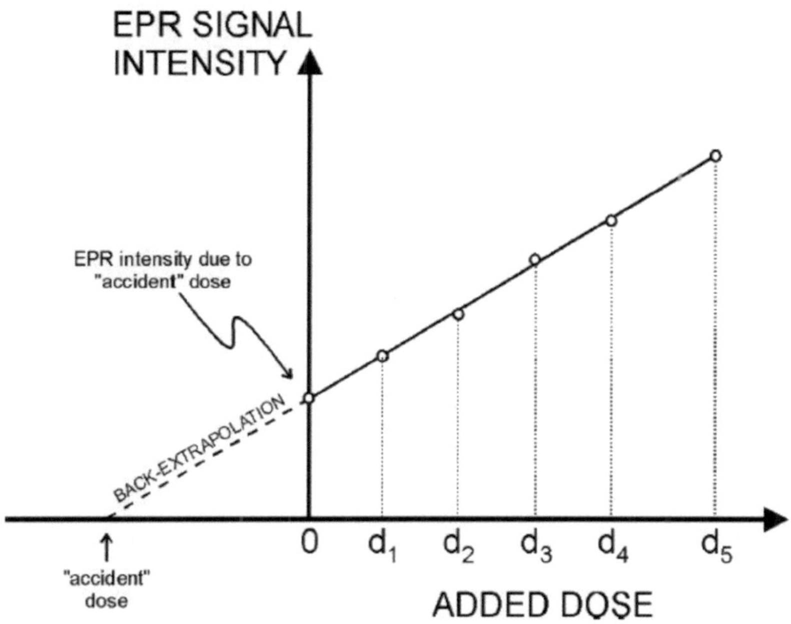

Fig. 8. The additive dose method for dose reconstruction is based on the re-irradiation (d_1 - d_5) of a tooth sample to obtain a sample-specific dose response curve, which is used to back extrapolate to the absorbed dose value (Desrosiers, M., and Schauer, D.A., 2001).

The dosimetric response of neutron irradiated human tooth enamel has been investigated (Khailov, Artem; et al., 2008). The neutron sensitivity (/Gy-100 mg) of human tooth enamel remained constant for various mean neutron energies ranging from 167 to 450 keV. Similarly, the EPR signal intensity remained independent of the neutron dose rate variation from 0.5 to 2.4 Gy/h (Rao F.H. Khan, et. al., 2004), other studies on the use of deciduous teeth have been performed (El-Faramawy, N.A., 2005, El-Faramawy, N., and Wieser, A., 2006).

5.4.2 Emergency dosimetry

There is growing awareness of the need for methodologies that can be used retrospectively to provide the dosimetry needed to carry out triage immediately after an event in which large numbers of people have potentially received clinically significant doses of ionizing radiation (Trompier François, et al., 2008b, Trompier François, et al., 2007a). Although some very promising approaches are being developed using biologically based parameters there also is recognition that such measurements have the potential to be confounded by other physiological and pathophysiological factors that are likely to be present in such event (Nicolalde, Roberto J; et al., 2008).

In contrast, the EPR measurements are based on physical changes in tissues whose magnitudes are not affected by the factors that can confound biologically based assessments. The EPR methods are based on the generation of stable free radicals, whose magnitude is proportional to the total dose of radiation received by the tissue, thereby allowing these tissues to be used as endogenous physical dosimeters (Nicolalde, Roberto J; et al., 2008).

5.4.2.1 Human subjects

5.4.2.1.1 In vivo EPR measurements of teeth

In vivo measurements of radiation-induced EPR signals in teeth is a safe technique (Ann Barry Flood, et al., 2007) and currently it utilizes a large permanent magnet (40 mT) and, in principle, this system could be deployed in the .field using a small vehicle (Dong, Ruhong; et al., 2008). While clones of this system (see Figure (9)) would be an effective component of large deployment teams, a smaller magnet system would facilitate wider distribution of this capability. The feasibility of such magnet systems has been demonstrated (Swartz et al., 2007). These are in a form that could be incorporated into a helmet-like structure that would fit over the head. An intraoral magnet is also being developed. It is anticipated that within several years, the technology will be advanced to a point where it may be possible to obtain sufficient sensitivity with lower frequencies and thus lower the requirements for the magnetic field (Benjamin B. et al., 2008). This would further decrease the size of the magnet that is needed.

A current laboratory-based system (Benjamin B.Williams, et al., 2007) can make measurements comfortably in human subjects with a 5-min acquisition time providing dose resolution of ±0.75 Gy (1 SD) and a threshold of not more than 2.0 Gy, with the result being immediately available. There are a number of areas in which improvements should be feasible within 1–2 years. Improvements that are in process include: increasing the sensitivity of the existing types of resonators and the number of teeth in which the measurement is made by changing the size and/or shape of the resonator (see Figure (10)), improving data analysis (Demidenko, E., et al., 2007), increasing microwave power, and reducing sources of noise. Dose resolution can be improved immediately by extending the time for the measurement, with the increase being proportional to the square root of the

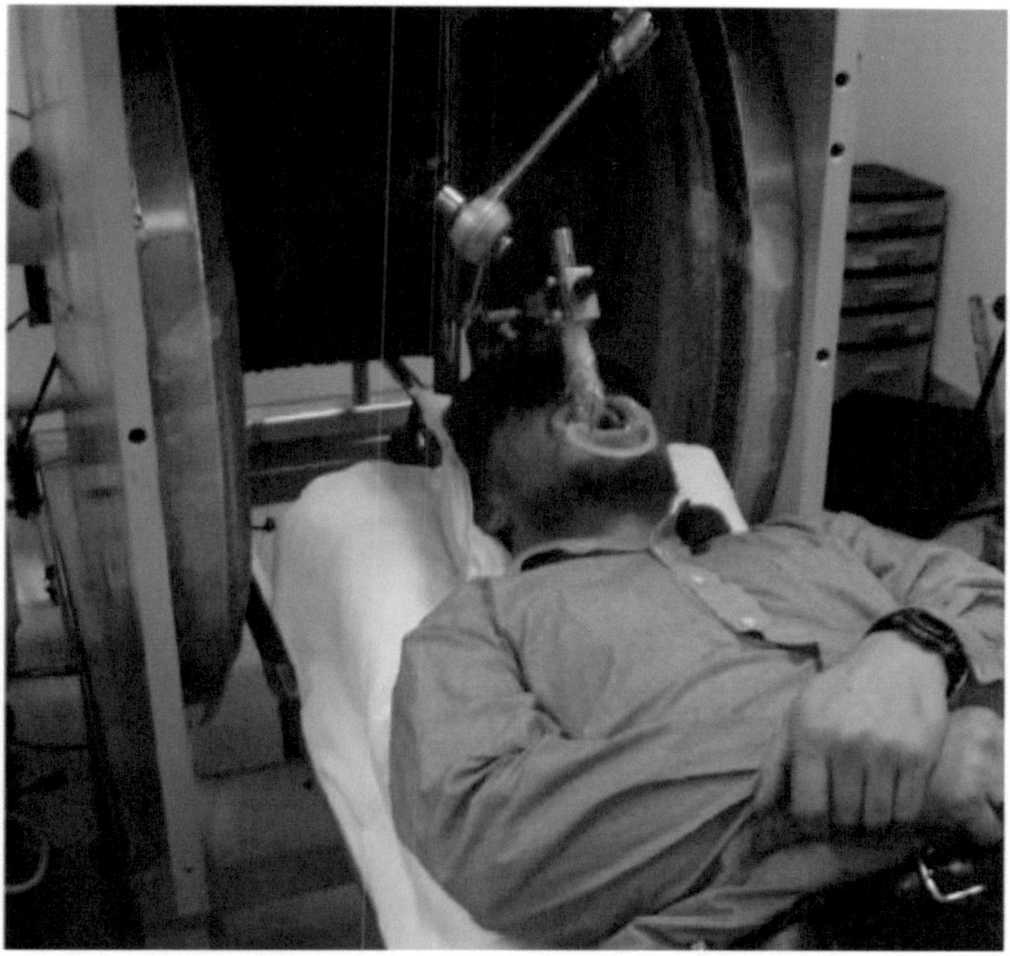

Fig. 9. Clinical spectrometer with a volunteer in it (Harold Swartz et al., 2007).

time of the measurements (i.e. increasing acquisition time from 5 to 20 min would increase the resolution by a factor of two) and by making the measurements in more than one tooth simultaneously. While the threshold, sensitivity, and accuracy can be improved further, there are some caveats that pertain to this method regardless of such improvements. The measured quantity is absorbed dose to teeth, not the critical organs of interest in radiation protection. This is not a problem if the exposure is homogeneous. In the event of an asymmetric exposure it may be feasible to utilize the Monte Carlo simulations of doses to human teeth from photon sources of eight standard irradiation geometries that have been performed and a set of dose conversion coefficients (DCCs) were calculated for 30 different tooth cells (Ulanovsky et al., 2005). DCCs were determined as ratios of tooth absorbed dose to air kerma for monoenergetic photon sources. To facilitate handling of the data set a software utility has been developed. The utility plots the DCC and computes conversion factors from enamel dose to air kerma and from enamel dose to organ dose for user-supplied discrete and continuous photon spectra.

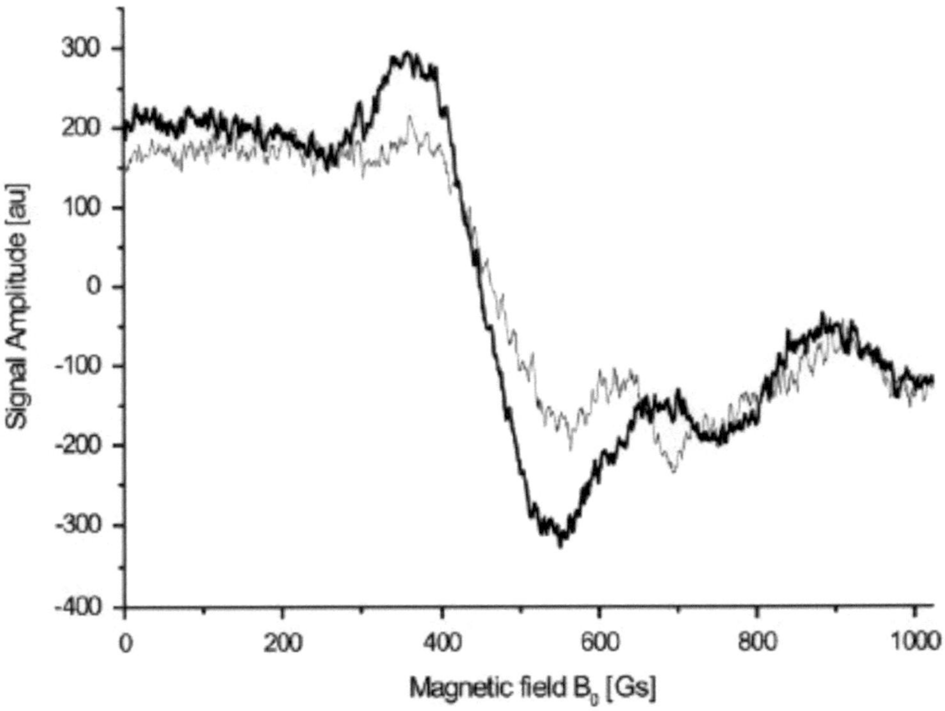

Fig. 10. Spectra from measuring two teeth (upper and lower) simultaneously (Harold Swartz et al., 2007).

The utility of EPR measurements for decision-making will depend on the homogeneity of the exposure and the type of radiation. The latter is noted because neutrons contribute very little to the EPR signal in teeth due to the low amount of hydrogen atoms in the enamel (Zdravkova et al., 2003; Trompier et al., 2004). If the dose has a major contribution from ingested or inhaled radionuclides, the dose delivered to the teeth may not closely reflect the dose to critical tissues (George A. Alexander et al., 2007).

5.4.2.1.2 Measurements in .fingernails (or toenails)

Although it was suggested as early as 1968 (Brady et al., 1968) that fingernails might be useful for after-the-fact dosimetry, only recently have the necessary studies been carried out to demonstrate convincingly that this approach has potential for use in the field for triage and perhaps even fairly precise determination of dose (Hongbin Li, et al., 2008). Preliminary results indicate that using simple cuttings from fingernails and X-band (9500 MHz) for the measurements, absorbed doses as of 1Gy with an uncertainty of ±0.50 Gy (1 SD) can be obtained with currently available techniques and instruments (Romanyukha et al., 2007; Trompier François et al., 2007b). If the use of fingernails for field dosimetry continues to develop, there should be no difficulty in constructing a field-deployable 9500MHz spectrometer for this purpose, which would be lightweight and automated for use by minimally trained individuals. The radiation-induced signals in fingernails are stable for at least several days (and much longer if the samples are collected within a few hours after the event and stored at low temperature). Because the measurements would be made in vitro, it should be possible to calibrate the radiation response of each sample by a simple procedure in which radiation is added to the sample. A potential advantage of measurements in fingernails, especially if combined with in vivo EPR dosimetry of teeth, include obtaining the measurement from a different location on the body (thereby providing a means to assess if there was an heterogeneous exposure).

Potential limitations to this approach may be overcome by simple modification of the collection process. For example, cutting of the fingernail can create a mechanically induced signal (MIS) that overlaps with the radiation-induced signal (RIS). However, the MIS decays rapidly and the decay is greatly accelerated by simple chemical treatment as shown in Figure (11), (Xiaoming He, et al., 2008, Dean Wilcox, et al., 2008). The influence of this MIS also can be removed by appropriate data processing because the shape is different from the RIS (Swarts Steven, et al., 2008). As is the case with any technique that requires removal of a sample from the subject, there is a potential for mislabeling the sample. This problem can be reduced by the development of automated procedures to rapidly remove any MIS and, if necessary, to calibrate the individual sample. Because only minimal manipulation of the sample is required and the measurement can be made within 5 min, it is feasible to determine the absorbed dose while the subject is still present. Finally, this method may not be applicable in children where nail volume is low (George A. Alexander, et al., 2007). Recently, doses in the range of 0.4 Gy could be detected using more sensitive EPR spectrometer with a resonator of high quality factor (Hirosuke Suzuki, 2008).

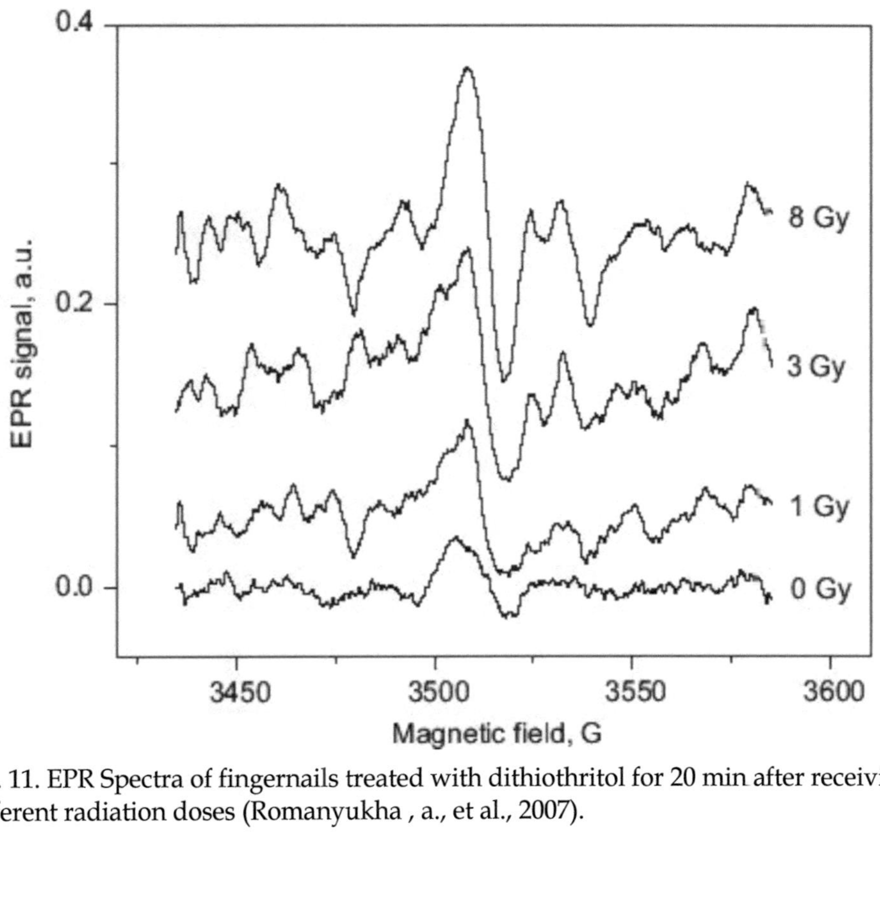

Fig. 11. EPR Spectra of fingernails treated with dithiothritol for 20 min after receiving different radiation doses (Romanyukha , a., et al., 2007).

5.4.2.1.3 Hai

Hair is composed mainly of alpha keratin same like fingernails, however its use as accidental dosimeter is limited because of the melanin pigments (Trivedi, A., and Grenstock, C., 1993). Some attempts were performed to study different hair types, with different pigments, but it was found that hair EPR signal exhibit some complexities that limit its use in dose evaluation. On the other hand, recently, melanin itself (in hair) was used to indicate the presence of radiation-induced radicals (Thomas Herrling et al., 2008).

5.4.2.1.4 Measurements in "biopsies" of teeth using 9500 MHz EPR

Many studies have demonstrated that retrospective measurements of dose by examination of isolated teeth with higher frequency EPR can provide very accurate estimates of dose at times ranging from immediately after the exposure to archeologically relevant times (Desrosiers and Schauer, 2001). The practical problem with this approach for acute dosimetry is the need to remove the tooth from the mouth. It now appears feasible, however, to obtain small samples from teeth rapidly and in a cosmetically acceptable manner. Small amounts can be used because of the increased sensitivity of higher frequency EPR and, there may be advantages in using frequencies even higher than 9500 MHz. Such a process could be very useful for triage and early assessment of dose to help in the determination of therapeutic intervention. Even if the technique of tooth biopsy does not fully meet the expectations, there may be situations where the value of the information that would be obtained would justify the removal of a tooth for in vitro measurement. The latter approach might be applicable in subjects for whom there are other indications of a potentially life-threatening dose and it is essential to verify the dose so that potentially risky therapies can be applied appropriately (George A. Alexander, et al., 2007).

5.4.2.1.5 Other tissues

Several reports in the literature have described the effects of radiation in workers who exposed their fingers to intense radioactive sources. The radiation injuries occurring after local exposure to a high dose (20 to 100 Gy) could lead to the need for amputation. It has been investigated the use of low-frequency EPR spectroscopy to evaluate non-invasively the absorbed dose. Low-frequency microwaves are indeed less absorbed by water and penetrate more deeply into living material (~10 mm in tissues using 1 GHz spectrometers). Preliminary results obtained with baboon and human fingers compared with human dry phalanxes placed inside a surface-coil resonator. The EPR signal increased linearly with the dose. The ratio of the slopes of the dry bone to whole finger linear regression lines was around 5. The detection limit achievable with the present spectrometer and resonator is around 60 Gy, which is well within the range of accidentally exposed fingers (Zdravkova, M. et Al., 2004).

5.4.2.2 Non-human subjects

This item may contain everything man uses: his food, clothes, tools, constructive materials, plastics (Trompier François; et al., 2008a) and so many things that can be found in the day life. These things around the man receive same (more or less) radiation dose in case of radiation accidents, hence evaluation of radiation doses using these objects may reflect to a close relation some details. Materials have been investigated by different techniques: sugar, wall bricks, roof tiles, plastics, watch glass, ruby present in watches, medicines carried by persons and shell button (Baffa, O. 1, et al., 2008). Sweeteners based on saccharine, cyclamate, stevia and aspartame may be used.

A preliminary study of EPR signals of watch glass carried out by Wu (Wu, K., et al., 1995) and Trompier François (Trompier François, et al., 2008c) had suggested that it was an appropriate dosimeter. A large number of watch glass and display windows of mobile phone with EPR techniques to study the variability of dosimetric properties among the different types of sample. Dose response, signal stability and effects of storage conditions was presented (Bassinet, Céline; et al., 2008).
Egg shell was used as accidental EPR dosimeter, as it contains Calcium carbonate $CaCO_3$ (Da Costa, Z.M., et al., 2007), dose response is linear from the range of 3 Gy to 1kGy, and no dose rate dependence was observed.

5.4.3 Environmental dosimetry

Concern regarding the possibility of criminal or terrorist use of nuclear materials has led to an interest in developing the capability to measure radiation dose in a variety of natural and man-made materials. One such novel EPR dosimeter is drywall, a common construction material composed largely of gypsum (calcium sulphate dihydrate). A radiosensitive EPR signal in drywall has been observed, and suitable dose measurement protocols have been developed (Thompson, Jeroen W. et al., 2008). In other study, investigation of radiation induced radicals in coral reefs revealed that microcrystalline aragonite radicals can be used for environmental dosimetry purposes (Sharaf, M.A. and Gamal M. Hassan. 2004). Some attempts to follow the changes in sealevel were performed via the $^{230}Th/^{234}U$–ESR combined analysis for corals (Bonnie A.B. et al., 2007).

6. Dating

ESR has increasingly contributed to interdisciplinary research, such as dating of geological and archaeological materials (Dobosz, B., and Krzyminiewski, R. et al., 2007). Geosciences benefit from the potential of ESR spectroscopy to date minerals (e.g. carbonate deposits, calcites, silica, phosphates, etc.) (Thompson, J. and Schwarcz, H.P. 2008), and fossils (e.g. shells, corals, bones, teeth) (Ikeya, 2002). Noticeable progress in dating has, for instance, recently been reported on ESR studies (Skinner, Anne R. 2008) of limestones from Cretaceous–Tertiary boundary on the extinction theory of the dinosaurs 65 million years ago (Griscom and Beltrán-Lopez, 2002), and on ESR dating of the most ancient human settlements in Europe by ESR spectroscopy of fossil herbivorous teeth, coupled with U-series measurements, to model the uptake of uranium (Falguéres et al., 2002). The very long lifetime of some signals allows ESR to provide information over the entire span of human evolution. A survey on the present state in geological dating using ESR is given by Skinner (2000) and Skinner et al. (2002) (Dieter F. Regulla, 2005).
Dating of megafauna can contribute to the better knowledge of megafauna presence in this region as well as to the events associated to the extinction of these species. In some cases ESR is an interesting possibility of dating, since no remains of compounds containing C-14 can be found (Oliveira, L., et al., 2008).
Also, dating of barnacles may reflect past sealevel and sea level changes (Skinner, Anne, et al., 2008). Knowing the dates of these changes can constrain the periods of regional hominid occupation and provide a better understanding of local and global environments. Also, dating of teeth may lead to the accurate determination of sedimentation rates, improving the age estimates for humids and other Paleolithic cultural deposits (Anne R. Skinner, et al., 2008).

7. Uncertainties

One of the problems in evaluation of uncertainty in EPR tooth dosimetry is that different factors that influence the total uncertainty cannot be separated. Another problem is that blank samples are absent resulting into a lack of calibration standards. Each tooth extracted for medical indications has its own age that starts from hydroxiapatite formation. Due to natural radiation exposure there are not any tooth enamel with zero dose. Therefore, each tooth accumulates some unknown dose doe to natural background (Shishkina E., et al., 2008).

For the reasons above, the accuracy of EPR dosimetry is not a trivial task. A new approach for estimation of the uncertainty using combination of physical and numerical (Monte Carlo simulation) experiments has been developed. The exp erimental basis of that work was the results of intercomparison of three EPR laboratories: GSF (Munich, Germany); ISS (Rome, Italy) and IMP (Ekaterinburg, Russia).

Results of the study show that EPR measurement uncertainty depends on both amplitude of signal and sample mass. Moreover, statistical modeling of EPR signal demonstrates the presence of not only a random error of signal evaluation but also a systematic bias of spectra processing that depend on amplitude (Shishkina E., et al., 2008).

For radiation therapy, a small uncertainty of the applied dose is required. The uncertainty budget for the alanine/ESR dosimetry system of the Physikalisch-Technische Bundesanstalt (PTB) was determined, which relies on the use of a reference sample. A method is also presented which allows a reduction of the influence of fading or other changes of the ESR amplitude of irradiated alanine probes with time. If certain conditions are met which are described in detail, a relative uncertainty of less than 0.5% can be reached for probes irradiated with ^{60}Co in the 5–25 Gy dose range, including the uncertainty of the primary standard (Krauss, A. 2006). First results for dose values between 2 Gy and 10 Gy are presented as well (Mathias Anton, 2006). More methods of evaluation of uncertainties of alanine dosimetry can be found elsewhere (Bartolotta, A., et al., 1993, Bergstrand, S., et al., 1998, Kojima, T., et al., 1999, Nagy, V., et al., 2002).

8. References

Anan M. Al-Karmi, M.A. Morsy. 2008. " EPR of gamma-irradiated polycrystalline alanine-in-glass dosimeter". Radiation Measurements 43, 1315 - 1318.

Ann Barry Flood, Shayan Bhattacharyya, Javier Nicolalde, R., Harold M. Swartz. 2007. "Implementing EPR dosimetry for life-threatening incidents: Factors beyond technical performance". Radiation Measurements 42, 1099 - 1109.

Baccaro, S., Brunella, V., Cecilia, A., Costa, L. 2003. "γ irradiation of poly(vinyl chloride) for medical applications". Nuclear Instruments and Methods in Physics Research B 208, 195–198

Baccaro, S., Cemmi, A., Colombi, C., Fiocca, M, Gambarini, G., Lietti, B., Rosi, G. 2004. "In-phantom dose mapping in neutron capture therapy by means of solid state detectors". Nuclear Instruments and Methods in Physics Research B 213, 666–669.

Baffa, O. 1, Oliveira, P.B.1 José, F.A.1 and Kinoshita, A. 2008. "An attempt to use sweeteners as a material for accident dosimetry". BioDos 2008, Dartmouth College, Hanover, NH., USA (7-11 September 2008).

Bartolotta, A., Fattibene, P., Onori, S., Pantaloni, M., Pettti, E. 1993. "Sources of uncertainity in therapy level alanine dosimetry". Appl. Radiat. Isot. 44 (1\2), 13-17.

Basly, J.P., Longy, I., Bernard, M., 1997. In.uence of radiation treatment on two antibacterial agents and four antiprotozoal agents: ESR dosimetry. Int. J. Pharm. 154, 109–113.

Bassinet, Céline; Trompier, François; Clairand, Isabelle. 2008. "Radiation accident dosimetry on glass by tl and epr spectrometry". BioDos 2008, Dartmouth College, Hanover, NH., USA (7-11 September 2008).

Benjamin B. Williams, Ruhong Dong, Maciej Kmiec, Greg Burke, Bruce Corliss, Eugene Demidenko, Oleg Grinberg, Javier Nicolalde, Jenna Pollack, Tim Raynolds, Ildar Salikhov, Artur Sucheta, Piotr Lesniewski, Harold Swartz. 2008. "In Vivo EPR Tooth Dosimetry". BioDos 2008, Dartmouth College, Hanover, NH., USA (7-11 September 2008).

Benjamin B.Williams, Artur Sucheta, Ruhong Dong,Yasuko Sakata, Akinori Iwasaki, Gregory Burke, Oleg Grinberg, Piotr Lesniewski, Maciej Kmiec, Harold M. Swartz. 2007. " Experimental procedures for sensitive and reproducible in situ EPR tooth dosimetry". Radiation Measurements 42, 1094 - 1098.

Bergstrand, S., Hole, E., Sagstuen, E. 1998. "A simple method for estimating dose uncertainty in ESR\ alanine dosimetry". Appl. Radiat. Isot. 49 (7), 845-854.

Bin Kang, Chang S, Dai Y, Chen D (2007b) Radiation synthesis and magnetic properties of novel $Co_{0.7}Fe_{0.3}$/chitosan compound nanoparticles for targeted drug carrier. Rad Phys Chem 76: 968–973

Boess, C., Bo¨ gl, K.W., 1996. In.uence of radiation treatment on pharmaceuticals: a review. Alkaloids, morphine derivatives and antibiotics. Drug Dev. Ind. Pharm. 22, 495–529.

Bogushevich, S., and Ugolev, I., 2000. “Inorganic EPR dosimeter for medical radiology”. Applied radiation and Isotopes. 52, 1217-1219.

Bonnie A.B. Blackwell, Steve J.T. Teng, Joyce A. Lundberg, Joel I.B. Blickstein, Anne R. Skinner. 2007. "Coupled $^{230}Th/^{234}U$-ESR analyses for corals: A new method to assess sealevel change". Radiation Measurements 42, 1250 - 1255.

Borgonove, A.F., Kinoshita, A., Chen, F., Nicolucci, P., Baffa, O. 2007. "Energy dependence of different materials in ESR dosimetry for clinical X-ray 10MV beam". Radiation Measurements 42, 1227 - 1232.

Borysheva, N., Ivannikov, A., Tikunov, D., Orlenko, S., Skvortsov, V., Stepanenko, V., Hoshi, M. 2007. "Taking into account absorbed doses in tooth enamel due to internal irradiation of human body by radioactive cesium isotopes at analysis EPR dosimetry data: Calculation by Monte-Carlo method". Radiation Measurements 42, 1190 - 1195.

Bradshaw, W., Cadena, E., Crawford, and Spetzler, H., 1962. “The use of alanine as a solid dosimeter”. Rad.Res., 17, 11-21.

Brady, J., Aarestad, N., Swartz, H., 1968. "In vivo dosimetry by electron spin resonance spectroscopy". Health Phys. 15, 43–47.

Breen, S., and Battista, J., 1999. “Feasibility of reading LiF thermoluminscent dosimeters by electron spin resonance”. Phys. Med. Boil., 44, 2063-2069 . UK.

Bruno Boizot, Guillaume Petite, Dominique Ghaleb, Georges Calas. 1998. " Radiation induced paramagnetic centres in nuclear glasses by EPR spectroscopy". Nuclear Instruments and Methods in Physics Research B 141, 580±584.

Bruno T. Rossi, Felipe Chen, Oswaldo Baffa. 2005. "A new 2 methylalanine-PVCESR dosimeter". Applied Radiation and Isotopes 62, 287–291

Callens, F. Vanhaelwyn, G., Mathys, P., Boesman, E. 1998. "EPR of carbonated derived radicals: Applications in dosimetry, dating and detection of irradiated food". Applied Magneic Resonance, 14, 235-254.

Cevc, P., Schara, M., Ravnik, N., 1972. "Electron paramagnetic resonance study of irradiated tooth enamel". Radiation Research. 51, 581.

Chen,F., Grae, C.F.O Baffa, O. 2007. "Response of L-alanine and 2-methylalanine minidosimeters for K-Band (24 GHz) EPR dosimetry". Nuclear Instruments and Methods in Physics Research B 264, 277–281

Ciesielski, B., and Wielopolski, L. 1994. "The effects of dose and radiation quality on the shape and power saturation of the EPR signal in alanine". Radiat. Res. 140, 105–111.

Costa, L. Brunella, V. Paganini, M., Baccaro, S., Cecilia, A. 2004. "Radical formation induced by γ radiation in poly(vinyl chloride) powder". Nuclear Instruments and Methods in Physics Research B 215 (2004) 471–478.

D'Oca M.C., Bartolotta A., Cammilleri C., Giuffrida S., Parlato A., Di Stefano V. 2008. "The additive dose method for dose estimation in irradiated oregano by thermoluminescence technique". Food Control. In Press.

Da Costa, Z.M., Pontuschka, W.M., Ludwig, V., Giehl, J.M., Da Costa, C.R., Duarte. E.L. 2007. "A study based on ESR, XRD and SEM of signal induced by gamma irradiation in eggshell". Radiation Measurements 42, 1233 - 1236.

Daniela Wagner, Mathias Anton, Hilke Vorwerk, Tammo Gsänger, Hans Christiansen, Bjoern Poppe, Clemens Friedrich Hess, Robert Michael Hermann. 2008. "In vivo alanine/electron spin resonance (ESR) dosimetry in radiotherapy of prostate cancer: A feasibility study". Radiotherapy and Oncology 88, 140–147.

Danilczuk, M., Gustafsson, H., Sastry, M.D., Lund, E., Lund A., 2008. "Ammonium dithionate—A new material for highly sensitive EPR dosimetry". Spectrochimica Acta Part A 69, 18–21.

Dean Wilcox, Xiaoming He, Jiang Gui, Andres Ruuge, Honbin Li, Benjamin Williams, Harold Swartz. 2008. " Dosimetry Based on EPR Spectral Analysis of Fingernail Clippings". BioDos 2008, Dartmouth College, Hanover, NH., USA (7-11 September 2008).

Deffner, U., and Regulla, F., 1980. "Influences of physical parameters on high-level amino acid dosimetry". Nuclear instruments and methods. 175, 134-135.

Demidenko, E., Williams, B.B., Sucheta, A., Dong, R., Swartz, H.M. 2007. "Radiation dose reconstruction from L-band in vivo EPR spectroscopy of intact teeth: Comparison of methods". Radiation Measurements 42, 1089 - 1093.

Desrosiers, M., 1989. "Gamma-irradiated seafood: identification and dosimeter by electron paramagnetic resonance spectroscopy" J. Agric. Food. Chem. 37, 96-100.

Desrosiers, M., Schauer, D.A., 2001. Electron paramagnetic resonance (EPR) biodosimetry. Nucl. Instrum. Meth. B 184 (1–2), 219–228.

Dobosz, B., and Krzyminiewski, R. 2007. "Linear transformation of EPR spectra as a method proposed for improving identification of paramagnetic species in ceramic". Applied Radiation and Isotopes 65, 392–396

Dong, Ruhong; Williams, Benjamin B.; Kmiec, Maciej; Demidenko, Eugene; Sucheta, Artur; Raynolds, Timothy; Nicolalde, Jean P.; Lesniewski, Piotr; Burke, Gregory; Swartz, Harold M. 2008. "Estimation of Radiation Doses of Teeth Using In Vivo EPR

Spectroscopy". BioDos 2008, Dartmouth College, Hanover, NH., USA (7-11 September 2008).

Eirik Malinen, Elin A. Hult, Eli O. Holea and Einar Sagstuen. 2003b. "Alanine Radicals, Part 4: Relative Amounts of Radical Species in Alanine Dosimeters after Exposure to 6–19 MeV Electrons and 10 kV–15 MV Photons" Radiation Research. 159, 149–153.

Eirik Malinen, Mojgan Z. Heydari, Einar Sagstuen and Eli O. Hole. 2003a. "Alanine Radicals, Part 3: Properties of the Components Contributing to the EPR Spectrum of X-Irradiated Alanine Dosimeters". RADIATION RESEARCH 159, 23–32.

El-Faramawy, N., 2008. "Investigation of some parameters influencing the sensitivity of human tooth enamel to gamma radiation using electron paramagnetic resonance". J. Radiat. Res. 49, 305-312.

El-Faramawy, N., and Wieser, A., 2006. 'The use of deciduous molars in EPR dose reconstruction". Radiat. Environ. Biophys. 44, 273–277.

El-Faramawy, N.A., 2005. "Comparison of γ and UV-light induced EPR spectra of enamel from deciduous molar teeth". Appl. Radiat. Isot. 62, 191–195.

Eva Stabell Bergstrand, Ken R Shortt, CarlKRoss and Eli Olaug Hole. 2003." An investigation of the photon energy dependence of the EPR alanine dosimetry system". Phys. Med. Biol. 48, 1753–1771.

Falguéres, C., Voinchet, P., Bahain, J.J., 2002. "ESR dating as a contributor to the chronology of the earliest humans in Europe". Adv. ESR Appl. 18, 67–76.

Fauconnet, A.L., Basly, J.P., Bernard, M., 1996. Gamma radiation induced effects on isoproterenol. Int. J. Pharm. 144, 123–125.

Gancheva, V. Yordanov, N.D., Callens, F., Vanhaelewyn, G., Raf., J. Bortolin, E., Onori, S., Malinen, E., Sagstuen, E., Fabisiak, S., Peimel-Stuglik, Z. 2008. " An international intercomparison on ''self-calibrated'' alanine EPR dosimeters". Radiation Physics and Chemistry 77, 357–364

George A. Alexander, Harold M. Swartz, Sally A. Amundson,William F. Blakely, Brooke Buddemeier, Bernard Gallez , Nicholas Dainiak, Ronald E. Goans, Robert B. Hayes, Patrick C. Lowry, Michael A. Noska, Paul Okunieff, Andrew L. Salner, David A. Schauer, Francois Trompier, KennethW. Turteltaub, Phillipe Voisin, Albert L.Wiley Jr., RuthWilkins. 2007. "BiodosEPR-2006 Meeting: Acute dosimetry consensus committee recommendations on biodosimetry applications in events involving uses of radiation by terrorists and radiation accidents". Radiation Measurements 42 (2007) 972 - 996.

Gibella, M., Crucq, A.S., Tilquin, B., 1993. De´ tection RPE de l'irradiation de me´dicaments. Journal de Chimie Physique et de Physico-Chimie Biologique 90, 1041–1053.

Gibella, M., Crucq, A.S., Tilquin, B., Stoker, P., Lesgards, G., Raffi, J., 2000. Electron spin resonance studies of some irradiated pharmaceuticals. Radiation Physics and chemistry, 58(1), 69-76.

Gordy. W., Ard, W., and Shields, H., 1955. "Microwave spectroscopy of biological substances. I: paramagnetic resonance in x-irradiated amino acids and proteins". Proc.Nat.Acad.Sci. USA.

Griscom, D.L., Beltra´ n-Lopez, V., 2002. "ESR spectra of limestones from the Cretaceous-Tertiary boundary: traces of a catastrophe". Int. Adv. ESR Appl. 18, 57–64.

Grišmanovs, V., Kumada, T., Tanifuji, T., and Nakagawa, T., 2000. "ESR spectroscopy of γ-irradiated Li_2TiO_3 ceramics". Radiation physics and chemistry. 58, 113-117.

Gunther, M.R., Kelman, D.J., Corbett, J.T., and Mason, R.P. (1995) Self-peroxidation of methmyoglobin results in formation of an oxygen-reactive tryptophan-centered radical. J. Biol. Chem. 270, 16075-16081.

Gustafsson, H., Lund, E., Olsson, S. 2008. "Lithium formate EPR dosimetry for verifications of planned dose distributions prior to intensity-modulated radiation therapy". Phys. Med. Biol. 53 4667-4682.

Hirosuke Suzuki, Katsuhisa Kato, Kenji Tamukai, Naoki Yoshida, Hiroaki Ohya, Kazunori Anzai, H. M. Swartz. 2008. "0.4 Gy and more sensitive Compact ESR system for irradiated nail". BioDos 2008, Dartmouth College, Hanover, NH., USA (7-11 September 2008).

Hongbin Li, Xiaoming He, Andres Ruuge, Jiang Gui, Ben Williams, Theodore E. MacVeagh, Dean E. Wilcox, Harold M. Swartz. 2008. "The Impact of Hydration and Microwave Power on EPR Signals in Fingernails/Toenails". BioDos 2008, Dartmouth College, Hanover, NH., USA (7-11 September 2008).

IAEA 2002 "Use of electron paramagnetic resonance dosimetry with tooth enamel for retrospective dose assessment" International Atomic Energy Authority, Vienna. IAEA-TECDOC-1331.

IAEA, (2004) 'Emerging applications of radiation in nanotechnology'. Proceedings of a consultants meeting held in Bologna, Italy, 22–25 March 2004, IAEA-TECDOC-1438

IAEA, 1978 "Food preservation by irradiation". International Atomic Energy Agency, Vienna, Austria, V. I and II.

ICGFI, 1999. "Facts about food irradiation" International Consultative Group on food irradiation.

ICRU, 1976. "Determination of absorbed dose in a patient irradiated by beams of x or gamma rays in radiotherapy procedures". Report 24, International Commission on Radiation Units and Measurements, Bethesda, MD.

Ikeya, M., 1993. "New applications of electron spin resonance-dating, dosimetry and microscopy". World scientific, Singapore.

Ikeya, M., 2002. "New prospects of ESR dosimetry and dating in 21th century". Adv. ESR Appl. 18, 3–7.

Jacobs, G.O., 1995. A review of the effects of gamma-radiation on pharmaceutical materials. J. Biomater. Appl. 10, 59–96.

Jacobs, G.P., 1983. Stability of cefazolin and other new cephalosporins following gamma irradiation. Int. J. Pharm. 17, 29–38.

Jaroslaw M. Wasikiewicz, Fumio Yoshii, Naotsugu Nagasawa, Radoslaw A. Wach, Hiroshi Mitomo. 'Degradation of chitosan and sodium alginate by gamma radiation, sonochemical and ultraviolet methods'. 2005. Radiation Physics and Chemistry 73 (2005) 287–295.

Jayakumara R, Prabaharana M, Reis RL, Mano JF (2005) Graft copolymerized chitosan—present status and applications. Carboh Polym 62: 142–158.

Jean-Michel Dolo, and Tristan Garcia. 2007. "Angular response of alanine samples: From powder to pellet". Radiation Measurements 42, 1201 - 1206.

Juárez-Calderón, J. Negrón-Mendoza, A., Ramos-Bernal, S. 2007. "Irradiation of ferrous ammonium sulfate for its use as high absorbed dose and low-temperature dosimeter". Radiation Physics and Chemistry 76, 1829–1832.

Kelman, D., J., DeGraz, J. A., and Mason, R.P. (1994) Reaction of myoglobin with hydrogen peroxide forms a peroxyl radical which oxidizes substrates. J. Biol. Chem. 269, 7458-7463.

Khailov, Artem; Ivannikov, Alexander; Skvortsov, Valeri; Stepanenko, Valeri; Tsyb, Anatoli; Hoshi, Masaharu. 2008. "The neutron dose conversion coefficients calculation forhuman tooth enamel in antropomorphic phantom". BioDose 2008, Dartmouth College, Hanover, NH., USA (7-11 September 2008).

Kizer, P.N., Morton, J.R., and Preston, K.F., 1991. "Electron paramagnetic resonance radiation dosimetry: possible inorganic alternatives to the EPR/alanine dosimeter". J. Chem. Soc. Faraday Trans. 87-19, 3147-3149.

Kojima, T., Tachibana, H., Haneda, N., Kawashima, I., Sharpe. 1999. "Uncertainity estimation in Co-60 gamma-ray dosimetry at JAERI involving a two-way dose range intercomparison study with NPL in the dose range 1-50 kGy". Radiation Physics and chemistry. 5, 619-626.

Kortov, V., Milman, I., Monakhov, A., and Sleasarev, A., 1993a. "Combined TSL-ESR MgO detectors for ionizing and UV radiations". Radiation protection dosimetry. 47, ¼, 273-276.

Kortov, V., Milman, I., Sleasarev, A., and Kijko, V., 1993b. "New BeO ceramics for TL ESR dosimetry". Radiation protection dosimetry. 47, ¼, 267-270.

Krauss, A. 2006. "The PTB water calorimeter for the absolute determination ofr absorbed dose to water in Co-60 radiation".Metrologia 43, 259.

Kurmaev EZ, Shin S, Watanabe M, Eguchi R, Ishiwata Y, Takeuchi TA, Moewes A, Ederer DL, Gao Y, Iwami M, Yanagihara M (2002) Probing oxygen and nitrogen bonding sites in chitosan by X-ray emission. J Electron Spectro Rel Phen 125: 133–138.

Lakshmi Prasuna, C.P., Chakradhar, R.P.S., Raoa,, J.L., Gopal, N.O 2008. "EPR as an analytical tool in assessing the mineral nutrients and irradiated food products-vegetables". Spectrochimica Acta Part A, in press.

Lanjanian, H., Ziaie, F., Modarresi, M., Nikzad, M., Shahvar, A., Durrani, S.A. 2008. "Atechnique to measure the absorbed dose in human tooth enamel using EPR method". Radiation Measurements 43, S648–S650.

Le Hai, Diep TB, Nagasawa N, Yoshii F, Kume T (2003) Radiation depolymerization of chitosan to prepare oligomers. Nucl Inst Meth Phys Res B 208: 466–470

Maghraby, A., 2003. "Characterization and calibration of some organic compounds for use in EPR dosimetry". Ph.D. thesis, Cairo University, Giza, Egypt.

Maghraby, A., 2007 "Identification of Irradiated Crab using EPR", radiation measurement,V. 42, 2, 220-224.

Maghraby, A., 2007. "A sensitive EPR dosimetry system based on sulfamic acid". Nuclear Instruments and Methods in Physics Research B 262, 46–50.

Maghraby, A., and Maha Anwar Ali. 2007. " Spectroscopic study of gamma irradiated bovine hemoglobin". Radiation Physics and Chemistry 76, 1600–1605.

Maghraby, A., Tarek, E., 2006. "A new EPR dosimeter based on sulfanilic acid". Radiation Measurements 41, 170– 176.

Maltar-Strmečkia, N. and Rakvin. B. 2005. "Thermal stability of radiation-induced free radicals in γ-irradiated L-alanine single crystals". Applied Radiation and Isotopes 63, 375–380.

Marrale, M., Braia, M., Gennaro, G., Triolo, A., Bartolotta, A. 2007. "Improvement of the LET sensitivity in ESR dosimetry for γ-photons and thermal neutrons through gadolinium addition". Radiation Measurements 42, 1217 - 1221.

Marta M, Patamia M, Lupi A, Antenucci M, et al (1996) Bovine hemoglobin cross-linked through the β-chains, functional and structural aspects The Journal of Biological Chemistry, 271(13), 7473–7478.

Mathias Anton 2008. "Postirradiation effects in alanine dosimeter probes of two different suppliers" Phys. Med. Biol. 53, 1241-1258.

Mathias Anton, 2005. "Development of a secondary standard for the absorbed dose to water based on the alanine EPR dosimetry system". Applied Radiation and Isotopes 62, 779–795.

Mathias Anton. 2006. "Uncertainties in alanine/ESR dosimetry at the Physikalisch-Technische Bundesanstalt". Phys. Med. Biol. 51 5419-5440.

McConnell, H., Heller, H., Cole, T., and Fessenden, R., 1960. J. Am. Chem. Soc., 82, 766.

Mehta, K., and Girzikowsky, R., 2000. "IAEA high-dose intercomparison in Co-60 field". Applied Radiation and Isotopes. 52, 1179-1184.

Miki T, Kai A and Ikeya M (1987) Electron spin resonance of blood stains and its application to the estimation of time after bleeding Forensic Science International, 35, 149-158.

Misra H and Fridovich I (1972) The generation of peroxide radical during the autoxidation of hemoglobin. J. Biol. Chem. 247, 6960-6962.

Miyagawa, I., and Gordy, W., 1959. J. Chem. Phys., 30, 159, 1960, ibids, 32, 255.

Miyagusku, L., Chen, F., Kuaye, A., Castilho, C.J.C., Baffa, O. 2007. "Irradiation dose control of chicken meat processing with alanine/ESR dosimetric system". Radiation Measurements 42, 1222 - 1226.

Miyazaki, T., Kaneko, T., Yoshimura, T., Crucq, A.S., Tilquin, B., 1994. Electron spin resonance study of radiosterilization of antibiotics: ceftazidime. J. Pharm. Sci. 83, 68–71.

Mojgan Z. Heydari, Eirik Malinen, Eli O. Hole, and Einar Sagstuen. 2002. "Alanine Radicals. 2. The Composite Polycrystalline Alanine EPR Spectrum Studied by ENDOR, Thermal Annealing, and Spectrum Simulations". J. Phys. Chem. A. 106, 8971-8977.

Morton, J., Ahlers, F., and Schneider, C., 1993. "ESR dosimetry with magnesium sulfate". Applied radiation and isotopes. 40, 851-857.

Nagy, V., Sholom, S., Chumak, V., Desroseirs, M. 2002. "Uncertainities in alanine dosimetry in the therapeutic dose range". Appl. Radiat. Isot. 56, 917-929.

Nagy,V., and Desrosiers, M., 1996. "Complex time dependence of the EPR signal of irradiated L-α-alanine". Applied radiation and isotopes. 47, 789-793.

Nam, J. and Regulla, D. 1989. "The significance of the international dose assurance service for radiation processing". Appl. Radiat. Isot. 40, 953–956.

Nette, H.P., Onori, S., Fattibene, Regulla, D., and Wieser, A., 1993. "Coordinated research efforts for establishing an international radiotherapy dose intercomparison service based on the alanine/ESR system". Appl. Rad. Isot. 44, 7.

Nicolalde, Roberto J; 1, Gougelet, Robert M; 2, Swartz, Harold M. 2008. "EPR Spectroscopy for the Process of Triaging Mass Casualties after a Catastrophic Nuclear Event: A Simulated Exercise". BioDos 2008, Dartmouth College, Hanover, NH., USA (7-11 September 2008).

Oliveira, L., Kinoshita, A., Lopes, R., Figueiredo, A., Baffa, O., 2008. "Dating of brazilian southern megafauna by ESR spectroscopy". BioDos 2008, Dartmouth College, Hanover, NH., USA (7-11 September 2008).

Onori, S., Bortolin, E., Lavalle, M., and Fuochi, P., 1998. "$CaSO_4$: Dy phosphor as a suitable material for EPR high dose assessment". Radiation physics and chemistry. 52, 1-6, 549-553.

Onori, S., Pantalon, M., Fattibene, P., Signoretti, E.C., Valvo, L., Santucci, M., 1996. ESR identi.cation of irradiated antibiotics: cephalosporins. Appl. Radiat. Isotop. 47, 1569–1572.

Parlato, A. Calderaro, E. Bartolotta, A._, D'Oca, M.C. Giuffrida, S.A. Brai,L. M. Tranchina, Agozzino, P. Avellone, G. Ferrugia, M. Di Noto, A.M. Caracappa. S. 2007. "Gas chromatographic/mass spectrometric and microbiological analyses on irradiated chicken". Radiation Physics and Chemistry 76 1463–1465.

Pavlenko, A Mironova-Ulmane, N Polakov D Riekstina, M. 2007. "Investigation of EPR signals on tooth enamel". J. Phys.: Conf. Ser. 93, 12047-12052.

Rachmilewitz E, Peisach J and Blumberg W (1971) Studies on the stability of oxyhaemoglobin A and its constituent chains and their derivatives J. Biol. Chem., 246, 3356-3366.

Ranjbar, A., Durrani, S., and Randle, K., 1999. "Electron spin resonance and thermoluminescence in powder of clear fused quartz: effect of grinding". Radiation measurements. 30, 73-81.

Rao F.H. Khan, Aslam, Rink, W.J., Boreham, D.R. 2004. "Electron paramagnetic resonance dose response studies for neutron irradiated human teeth". Nuclear Instruments and Methods in Physics Research B225, 528–534.

Regulla, D. F. 2005. "ESR spectrometry: a future-oriented tool for dosimetry and dating". Applied Radiation and Isotopes, 62, 117–127

Regulla, D., and Deffner, U. 1982. "Dosimetry by ESR spectroscopy of alanine". Appl. Radiat. Isot. 33, 1101–1114.

Reid, B.D., 1995. Gamma processing technology: an alternative technology for terminal sterilization of parenterals. PDA J. Pharm. Sci. Technol. 49, 83–89.

Romanyukha, A., Trompier, F., LeBlanc, B., Calas, C., Clairand,I., Mitchell, C., Smirniotopoulos, J.G., Swartz, H., 2007. "EPR dosimetry in chemically treated .fingernails". Radiat. Meas., 42, 1110-1113.

Romanyukha, A., Wieser, A., Regulla, D., 1996. "EPR dosimetry with different biological and synthetic carbonated materials". Radiation Protection Dosimetry, 65, 389-392.

Ruth M.D. Garciaa, Marc F. Desrosiers, John G. Attwood, David Steklenski, James Griggs, Andrea Ainsworth, Arthur Heiss, Paul Mellor, Deepak Patil, Jason Meiner. 2004. " Characterization of a new alanine .lm dosimeter: relative humidity and post-irradiation stability". Radiation Physics and Chemistry 71, 373–377.

Sagstuen, E., Eli O. Hole, Sølvi R. Haugedal, and William H. Nelson. 1997b. "Alanine Radicals: Structure Determination by EPR and ENDOR of Single Crystals X-Irradiated at 295 K". J. Phys. Chem. A 101, 9763–9772.

Sagstuen, E., Hole, E., Haugedal, S., Lund, A., Eid, O., and Erickson, R., 1997a. "EPR and ENDOR analysis of x-irradiated l-alanine and NaHC2O4. H2O. Simulation of microwave power dependence of satellite lines". Nukleonika 42, 353–372.

Schneider, C., 1994. "Electron spin resonance (ESR) spectroscopy applied to radiation dosimetry and other fields". Physikalisch-Technische Bundesanstalt (PTB-Bericht E-51), Braunschweig, Germany.

Schreiber, G., Wagner, U., Helle, N., Ammon, J., Buchholtz, H., Delinceé, H., Estendorfer, S., Von Grabowski, H., Kruspe, W., Mainczyk, K., Munz, H., Schleich, C., Vreden, N.,

Wiezorek., C., Bögl, K., 1993a. "Thermoluminescence analysis to detect irradiated fruit and vegetables-an intercomparaison study". Bericht des Instituts Für Sozialmedizin und Epidemiologie des Bundesgesundheitsamtes, Berlin, SozEp-Heft 3/1993.

Schreiber, G., Wagner, U., Leffke, N., Helle, N., Ammon, J., Buchholtz, H., Delinceé, H., Estendorfer, S., Fuchs, K., Von Grabowski, H., Kruspe, W., Mainczyk, K., Munz, H., Nootenboom, H., Schleich, C., Vreden, N., Wiezorek., C., Bögl, K., 1993b. "Thermoluminescence analysis to detect irradiated spices, herbs, and spice and herb mixtures - an intercomparison study". Bericht des Instituts Für Sozialmedizin und Epidemiologie des Bundesgesundheitsamtes, Berlin, SozEp-Heft 2/1993.

Schreiber, G., Ziegelmann, B., Quitzsch, G., Helle, N.,Bögl, K.,1993c. "Luminescence techniques to identify the treatment of foods by ionizing radiation". Food Structure, 12, 385.

Sharaf, M.A. and Gamal M. Hassan. 2004. " Radiation induced radical in barium sulphate for ESR dosimetry: a preliminary study". Nuclear Instruments and Methods in Physics Research B 225, 521–527.

Sharaf, M.A. and Gamal M. Hassan. 2004. "ESR dosimetric properties of modern coral reef". Nuclear Instruments and Methods in Physics Research B 217, 603–610.

Shin Toyoda, Eldana Tieliewuhan, Satoru Endo, Ken'ichi Tanaka, Kunio Shiraishi, Chuzo Miyazawa, Alexandre Ivannikov, Masaharu Hoshi, Kenzo Fujimoto, Makoto Akashi. 2008. "ESR dosimetry of enamel and dentin taken from victims of JCO accident". BioDos 2008, Dartmouth College, Hanover, NH., USA (7-11 September 2008).

Shishkina E., Ivanov D., Wieser A., Fattibene P. 2008. "Application of experimental and numerical methods for estimation of uncertainties in EPR tooth dosimetry". BioDos 2008, Dartmouth College, Hanover, NH., USA (7-11 September 2008).

Signoretti, E.C., Valvo, L., Fattibene, P., Onori, S., Pantalon, M., 1994. "Gamma radiation affects on cefuroxime and cefotaxime. Investigation on degradation and syn-anti izomerization. Drug Dev. Ind. Pharm. 20, 2493–2508.

Skinner, A.F., Blackwell, B.A.B., Chasteen, N.D., Brassard, P., 2002. "New clues to limits on ESR dating". Intern. Adv. ESR Appl. 18, 77–82.

Skinner, A.R., 2000. "ESR dating: is it still an 'experimental' technique?". Appl. Radiat. Isot. 52, 1311–1316.

Skinner, Anne R. 2008. "An overview of the relationship between ESR dating and other forms of dosimetry". BioDos 2008, Dartmouth College, Hanover, NH., USA (7-11 September 2008).

Skinner, Anne R. Blackwell, Bonnie A. B., Gong, Jane J. J., Blais-Stevens, Andrée. 2008. "A new dating proposal: electron spin resonance dating with pleistocene barnacles". BioDos 2008, Dartmouth College, Hanover, NH., USA (7-11 September 2008).

Ślawska-Waniewska A, Msiniewicz-Szablewska E, Nedelko N, Galazka-Friedman J and Friedman A, (2004) Magnetic studies of iron-entities in human tissues J. of magnetism and magnetic materials, 272-276, 2417-2419.

Sünnetcioğlu, M., Dadayli, D., Celik, S., Koksel, H., 1999. "Use of EPR spin probe technique for detection of irradiated wheat". Appl. Radiat. Isot. 50, 557-560.

Svistunenko, D A, Dunne J, Fryer M, Nicholls P, Reeder B, Wilson M, Bigott M, Cutruzzolà F and Cooper C (2002) Comparative study of tyrosine radicals in hemoglobin and myoglobins treated with hydrogen peroxide Biophysical Journal, 49, 2845-2855.

Svistunenko, D A. (2005) Reaction of haem containing proteins and enzymes with hydroperoxides: the radical view. Biochem. Biophys. Acta (bioenergetics). 1707, 127-155.

Svistunenko, D A. Chris E., Cooper (2004) A new method of identifying the site of tyrosyl radicals in proteins Biophysical journal, 87, 582-595.

Svistunenko, D A. Nathan A., Davies, Michael T. Wilson, Raz P. Stidwill, Mervyn Singer, and Chris E., Cooper (1997a) Free radical in blood: A measure of haemoglobin autoxidation *in vivo* J.Chem.Soc., Perkin trans. 2. 2539-2543.

Svistunenko, D A., Rakesh P., Sergey V., Voloshchenko, and Michael T. Wilson (1997b) The globin-based free radical of ferryl hemoglobin is detected in normal human blood. J. Biol. Chem. 272, 7114-7121.

Swarts, Steven; Black, Paul; and Bernhard, William. 2008. "Ex vivo analysis of irradiated finger nails: quantifying the radiation-induced signal in fingernails in the presence of an interfering signal from mechanically-induced radicals". BioDos 2008, Dartmouth College, Hanover, NH., USA (7-11 September 2008).

Swartz, H. Greg Burke, M. Coey, Eugene Demidenko, Ruhong Dong, Oleg Grinberg, James Hilton, Akinori Iwasaki, Piotr Lesniewski, Maciej Kmiec, Kai-Ming Lo, R. Javier Nicolalde, Andres Ruuge,Yasuko Sakata, Artur Sucheta, TadeuszWalczak, Benjamin B.Williams, Chad A. Mitchell, Alex Romanyukha, David A. Schauer. 2007. "In vivo EPR for dosimetry". Radiation Measurements 42, 1075 - 1084

Swartz, H.M., Burke, G., Coey, M., Demidenko, E., Dong, R., Grinberg, O., Hilton, J., Iwasaki, A., Lesniewski, P., Kmiec, M., Lo, K.-M., Nicolalde, R.J.. Ruuge, A., Sakata, Y., Sucheta, A., Walczak, T., Williams, B.B., Mitchell, C., Romanyukha, A., Schauer, D.A., 2007. "In vivo EPR for Dosimetry", Radiat. Meas., 42, .5. 23

Thomas Herrling, Katinka Jung , Jũrgen Fuchs. 2008. " The role of melanin as protector against free radicals in skin and its role as free radical indicator in hair". Spectrochimica Acta Part A 69, 1429–1435.

Thompson, J. and Schwarcz, H.P. 2008." Electron paramagnetic resonance dosimetry and dating potential of whewellite (calcium oxalate monohydrate)". Radiation Measurements 43, 1219 - 1225.

Thompson, Jeroen W.; Abu Atiya, Ibrahim; Boreham, Doug R. 2008. "Electron paramagnetic resonance (EPR) dosimetry of drywall". BioDos 2008, Dartmouth College, Hanover, NH., USA (7-11 September 2008).

Tilquin, B., Rollmann, B., 1996. Recherches a` conseiller pour I'application de la sterilization ionisante des me´ dicaments. Journal de Chimie Physique et de Physico-Chimie Biologique 93, 224–230.

Tor Arne Vestad, Eirik Malinen, Dag Rune Olsen, Eli Olaug Hole, Einar Sagstuen. 2004. "Electron paramagnetic resonance (EPR) dosimetry using lithium formate in radiotherapy: comparison with thermoluminescence (TL) dosimetry using lithium fluoride rods". Phys. Med. Biol. 49, 4701-4715.

Tristan Garcia and Jean-Michel Dolo. 2007. " Study of the in.uence of grain size on the ESR angular response in alanine radicals". Radiation Measurements 42, 1207 - 1212.

Trivedi, A., and Grenstock, C., 1993. "Use of sugars and hair for ESR emergency dosimetry". Applied Radiation and Isotopes. 44 (1-2), 85-90.

Trompier, F., Fattibene, P., Tikunov, D., Bartolotta, A., Carosi, A., Doca, M.C., 2004. EPR dosimetry in a mixed neutron and gamma radiation .field. Radiat. Prot. Dosim. 110, 437–442.

Trompier, F., Kornak, L., Calas, C., Romanyukha, A., LeBlanc, B., Clairand, I., Mitchell C.A., Swartz, H., 2007b. "Protocol for emergency EPR dosimetry in. fingernails". Radiat. Meas. 42, 1085-1088.

Trompier, F., Sadlo, J., Michalik, J., Stachowicz, W., Mazal, A. Clairand, I., Rostkowska, J., Bulski , W., Kulakowski, A., Sluszniak, J., Gozdz, S., Wojcik, A. 2007a. "EPR dosimetry for actual and suspected overexposures during radiotherapy treatments in Poland". Radiation Measurements 42, 1025 - 1028.

Trompier, François; Bassinet, Céline; Clairand, Isabelle. 2008a. "Radiation accident dosimetry on plastics by EPR spectrometry". BioDos 2008, Dartmouth College, Hanover, NH., USA (7-11 September 2008).

Trompier, François; Bassinet, Céline; Romanyukha, Alex; Clairand, Isabelle. 2008c. "Overview of physical and biophysical techniques for radiation accident dosimetry". BioDos 2008, Dartmouth College, Hanover, NH., USA (7-11 September 2008).

Ulanovsky, A., Wieser, A., Zankl, M., Jacob, P., 2005. "Photon dose conversion coefficients for human teeth in standard irradiation geometries". Health Phys. 9, 645–659.

Venkatesh B, Ramasamy S, Asokan R and Manoharan P (1997) Hemichromes in hemoglobin - An EPR study J. of Inorganic Biochemistry, 67, 121.

Veselka Ganchevaa, Einar Sagstuenb, Nicola D. Yordanov, 2006. "Study on the EPR/dosimetric properties of some substituted alanines". Radiation Physics and Chemistry 75, 329–335

Wajnberg E and Bemski G (1993) Electron spin resonance measurements of erythrocytes and hemoglobin stored at 77 K Naturwissenschaften, 80, 472-473.

WHO, 1988. "Food irradiation: a technique for preserving and improving safety of food" World Health Organization, Geneva, Switzerland.

WHO, 1994. "Safety and nutritional adequacy of irradiated food" World Health Organization, Geneva, Switzerland.

Wu, K., Sun, C.P., Shi, Y.M. 1995. "Dosimetric properties of watch glass: a potential practical ESR dosemeter for nuclear accidents". Radiat. Prot. Dosim., 59, 223-225.

Xiaoming He, Jiang Gui, Hongbin Li, Andres Ruuge, Dean Wilcox, Benjamin Williams, Harold Swartz. 2008. "Investigation of the Mechanically Induced EPR Signals in Fingernails/Toenails". BioDos 2008, Dartmouth College, Hanover, NH., USA (7-11 September 2008).

Yũrũs , S., Korkmaz, M., 2005. "Kinetics of radiation-induced radicals in gamma irradiated solid cefazolin sodium". Radiat. Eff. Defect. Solid. 160, 11–22.

Zavoisky, E., 1945. "Spin-magnetic resonance in paramagnetics". J. Phys., USSR, 9, 211.

Zdravkova, M Crokart, N Trompier, F Beghein, N Gallez, B Debuyst, R. 2004. "Non-invasive determination of the irradiation dose in fingers using low-frequency EPR". Phys. Med. Biol. 49, 2891-2898.

Zdravkova, M., Crokart, N., Trompier, F., Asselineau, B., Gallez, B., Gaillard- Lecanu, E., Debuyst, R., 2003. "Retrospective dosimetry after criticality accidents using low-frequency EPR: a study of whole human teeth irradiated in a mixed neutron and gamma-radiation .field" Radiat. Res. 160, 168–173.

Ziegelmann, B., Bögl, K., Schreiber, G., 1999. "TL and ESR signals of mollusk shells-correlations and suitability for the detection of irradiated foods". Radiation physics and chemistry, 54, 413-423.

31

Atmospheric Ionizing Radiation from Galactic and Solar Cosmic Rays

Christopher J. Mertens[1], Brian T. Kress[2], Michael Wiltberger[3], W. Kent Tobiska[4], Barbara Grajewski[5] and Xiaojing Xu[6]

[1]*NASA Langley Research Center, Hampton, Virginia*
[2]*Dartmouth College, Hanover, New Hampshire*
[3] *High Altitude Observatory, National Center for Atmospheric Research, Boulder, Colorado*
[4] *Space Environment Technologies, Pacific Palisades, California*
[5]*National Institute for Occupational Safety and Health, Cincinnati, Ohio*
[6]*Science Systems and Applications, Inc.*
USA

1. Introduction

An important atmospheric state variable, driven by space weather phenomena, is the ionizing radiation field. The two sources of atmospheric ionizing radiation are: (1) the ever-present, background galactic cosmic rays (GCR), with origins outside the solar system, and (2) the transient solar energetic particle (SEP) events (or solar cosmic rays), which are associated with eruptions on the Sun's surface lasting for several hours to days with widely varying intensity. Quantifying the levels of atmospheric ionizing radiation is of particular interest to the aviation industry since it is the primary source of human exposure to high-linear energy transfer (LET) radiation. High-LET radiation is effective at directly breaking DNA strands in biological tissue, or producing chemically active radicals in tissue that alter the cell function, both of which can lead to cancer or other adverse health effects (Wilson et al., 2003; 2005b). Studies of flight attendants have suggested adverse reproductive health outcomes (Aspholm et al., 1999; Lauria et al., 2006; Waters et al., 2000). The International Commission on Radiological Protection (ICRP) classify crews of commercial aircraft as radiation workers (ICRP, 1991). The US National Council on Radiation Protection and Measurements (NCRP) reported that among radiation workers monitored with recordable dose, the largest average effective dose in 2006 (3.07 mSv) was found in flight crew. In contrast, the average for the workers with the second largest effective dose, commercial nuclear power workers, was 1.87 mSv (NCRP, 2009). However, aircrew are the only occupational group exposed to unquantified and undocumented levels of radiation. Furthermore, the current guidelines for maximum public and prenatal exposure can be exceeded during a single solar storm event for commercial passengers on intercontinental or cross-polar routes, or by frequent use ($\sim$ 10-20 flights per year) of these high-latitude routes even during background conditions (AMS, 2007; Copeland et al., 2008; Dyer et al., 2009).

There is an important national need to understand and to predict the real-time radiation levels for the commercial aviation industry and the flying public, which has broad societal, public health, and economic benefits. NASA has met this need by developing the first-ever, real-time, global, physics-based, data-driven model for the prediction of biologically hazardous atmospheric radiation exposure. The model is called Nowcast of Atmospheric Ionizing Radiation for Aviation Safety (NAIRAS). In this chapter the underlying physics of the NAIRAS model is reviewed and some of the key results and applications of the model are presented. More specifically, the materials reviewed are the latest understanding in the physics and transport of atmospheric ionizing radiation from galactic and solar cosmic rays, the influence of space weather on the atmospheric ionizing radiation field, basic radiation dosimetry applied to atmospheric exposure, and the latest epidemiological understanding of radiation effects on pilots and aircrew.

2. Scientific and historical overview

2.1 Basics on cosmic rays and matter interactions

GCR consist of roughly 90% protons and 8% helium nuclei with the remainder being heavier nuclei and electrons (Gaisser, 1990). When these particles penetrate the magnetic fields of the solar system and the Earth and reach the Earth's atmosphere, they collide with air molecules and create cascades of secondary radiations of every kind (Reitz et al., 1993). The collisions are primarily due to Coulomb interactions of the GCR particle with orbital electrons of the air molecules, delivering small amounts of energy to the orbital electrons and leaving behind electron-ion pairs (Wilson et al., 1991). The ejected electrons usually have sufficient energy to undergo similar ionizing events. The cosmic ray ions lose a small fraction of their energy and must suffer many of these atomic collisions before slowing down. On rare occasions the cosmic ray ion will collide with the nucleus of an air molecule in which large energies are exchanged and the ion and nucleus are dramatically changed by the violence of the event. The remnant nucleus is highly disfigured and unstable, emitting further air nuclear constituents and decaying through the usual radioactivity channels (Wilson et al., 1991). One of the most important secondary particles created in GCR-air interactions is the neutron. Because of its charge neutrality, the neutron penetrates deep into the atmosphere, causing further ionization events along its path and contributing over half the atmospheric radiation exposure at typical commercial airline altitudes (Wilson et al., 2003). Furthermore, neutron exposures pose a relatively high health risk, since the massive low-energy ions resulting from neutron interactions always produce copious ions in the struck cell and repair is less efficient for these events (Wilson, 2000).

The intensity of the atmospheric radiations, composed of GCR primary and secondary particles, their energy distribution, and their effects on aircraft occupants vary with altitude, location in the geomagnetic field, and the time in the sun's magnetic activity (solar) cycle (Heinrich et al., 1999; Reitz et al., 1993; Wilson, 2000). The atmosphere provides material shielding, which depends on the overhead atmospheric depth. The geomagnetic field provides a different kind of shielding, by deflecting low-momentum charged particles back to space. Because of the orientation of the geomagnetic field, which is predominately dipolar in nature, the polar regions and high latitudes are susceptible to penetrating GCR (and SEP) particles. At each geographic location, the minimum momentum per unit charge (magnetic rigidity) a vertically incident particle can have and still reach a given location above the earth is called the vertical geomagnetic cutoff rigidity. The local flux of incident GCR at a given time varies widely with geomagnetic location and the solar modulation level. When solar activity is high, GCR flux is low, and vice versa. The dynamical balance between outward convective flux of solar wind and the inward diffusive flux of GCR is responsible for the anti-correlation between the incident GCR and the modulation level of solar cycle activity (Clem et al., 1996; Parker, 1965).

It is now generally understood that SEP events arise from coronal mass ejections (CME) from active regions on the solar surface (Kahler, 2001; Wilson et al., 2005b). The CME propagates through interplanetary space carrying along with it the local surface magnetic field frozen into the ejected mass. There is a transition (shock) region between the normal sectored magnetic structure of interplanetary space and the fields frozen into the ejected mass, where the interplanetary gas is accelerated forming the SEP. As the accelerated region passes an observation point, the flux intensity is observed to increase dramatically, and no upper limit in intensity is known within the shock region. The SEP energy spectrum obtained in the acceleration process is related to the plasma density and CME velocity. During a solar storm CME event, the number flux distribution incident at Earth's atmosphere is a combination of the GCR and SEP distributions. The SEP-air interaction mechanisms are the same as GCR-air interactions described above. The atmospheric radiations caused by a SEP also vary with altitude and geomagnetic field.

2.2 Commercial aircraft radiation exposure

GCR radiations that penetrate the atmosphere and reach the ground are low in intensity. However, the intensities are more than two orders of magnitude greater at commercial aircraft altitudes. At the higher altitudes of High Speed Civil Transport (HSCT), the GCR intensity is another two orders of magnitude higher (Wilson et al., 2003). When the possibility of high-altitude supersonic commercial aviation was first seriously proposed (The Supersonic Transport program proposed in 1961), Foelsche brought to light a number of concerns about associated atmospheric radiation exposure due to GCR and SEP, including the secondary radiations (Foelsche, 1961; Foelsche & Graul, 1962). Subsequently, Foelsche et al. (1974) conducted a detailed study of atmospheric ionizing radiation at high altitudes from 1965 to 1971 at the NASA Langley Research Center (LaRC). The study included a comprehensive flight program in addition to theoretical investigations. The measured data and theoretical calculations were integrated into a parametric Atmospheric Ionizing Radiation (AIR) model (Wilson et al., 1991). Prior to that study the role of atmospheric neutrons in radiation exposure was generally regarded as negligible (Upton et al., 1966). The LaRC studies revealed neutron radiation to be a major contributor to aircraft GCR exposure. Still the exposure levels were comfortably below allowable exposure limits for the block hours typical of airline crews of that time, except during a possible SEP event (less than 500 block hours were typical of the 1960's, although regulation allowed up to 1000 hours). Assessments of radiation exposure extending back to the 1930's for former Pan Am flight attendants were conducted by Waters et al. (2009) and Anderson et al. (2011).

There have been a number of significant changes since the original work of Foelsche (Wilson et al., 2003). A partial list of these changes, relevant to the development of the NAIRAS model, are: (1) the highly ionizing components of atmospheric radiations are found to be more biologically damaging than previously assumed and the associated relative biological effectiveness for fatal cancer has been increased (ICRP, 1991; ICRU, 1986); (2) recent animal and human studies indicate large relative biological effectiveness requiring protection for reproductive exposure (BEIR V, 1990; Chen et al., 2005; Fanton & Gold, 1991; Jiang et al., 1994; Ogilvy-Stuart & Shalet, 1993); (3) recent epidemiological studies (especially the data on solid tumors) and more recent atom-bomb survivor dosimetry have resulted in higher radiation risk coefficients for gamma rays (ICRP, 1991; NAS/NRC, 1980; UNSCEAR, 1988), resulting in lower proposed permissible limits (ICRP, 1991; NCRP, 1993); (4) subsequent to deregulation of the airline industry, flight crews are logging greatly increased hours (Barish, 1990; Bramlitt, 1985; Friedberg et al., 1989; Grajewski et al., 2011; Wilson & Townsend, 1988); and (5) airline crew members are now classified by some agencies as radiation workers (ICRP, 1991).

The last point (i.e., (5)) is particularly illuminating. Aircrews may even receive exposures above recently recommended allowable limits for radiation workers when flying the maximum allowable number of flight hours. Reviews (Hammer et al., 2009) and meta-analyses (Buja et al., 2005; 2006) of cancer studies among flight crew have found excesses of breast cancer and melanoma and possible excesses for other sites. Increasing risk trends with radiation exposure have not been consistently identified, likely due to the limitations of the exposure metrics used (Waters et al., 2000). There is further concern for reproductive health from exposure to high altitude and/or high-latitude flights, as the US National Institute for Occupational Safety and Health (NIOSH) continues to study adverse pregnancy outcomes among commercial flight attendants (Waters et al., 2000). US pilots have been exposed to increasing levels of ionizing radiation since the 1990s (Grajewski et al., 2011). Frequent-flyer business passengers are likely exposed to even higher doses than aircrew, since flight hours are not restricted for airline passengers. In addition, if a large SEP occurs during flight, both passengers and pregnant crew may greatly exceed allowable limits (Barish, 2004).

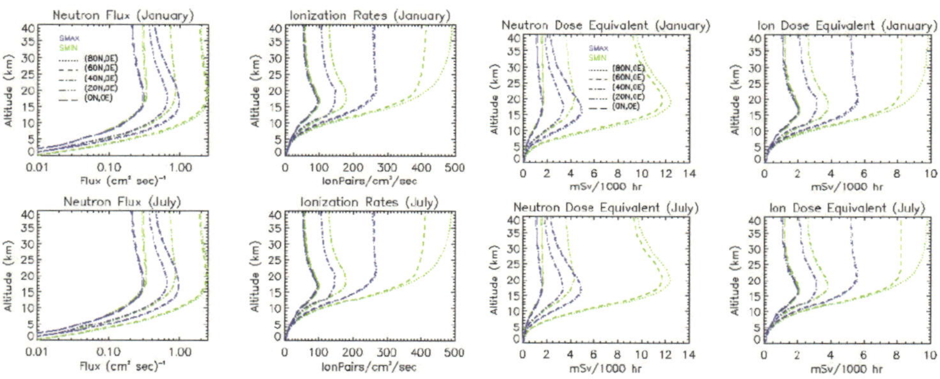

(a) AIR Neutron Flux and Ionization Rates (b) AIR Dose Equivalent Rates

Fig. 1. (a) Neutron flux and ionization rate profiles for various latitudes, solar activity levels, and seasons. The blue lines represent solar maximum conditions. The green lines represent solar minimum conditions. The top row corresponds to January atmospheric conditions while the bottom row corresponds to July. (b) Neutron and ion dose equivalent rates profiles for the same latitudes, solar activity levels, and seasons indicated in (a).

The original LaRC study (1965 to 1971) commissioned over 300 flights over most of the duration of solar cycle 20 on high-altitude aircraft and balloons to study both the background radiation levels over the solar cycle and to make measurements during SEP events. The LaRC flight package consisted of a 1-10 MeV neutron spectrometer, tissue equivalent ion chamber, and nuclear emulsion for nuclear reaction rates in tissue. Monte Carlo calculations (Lambiotte et al., 1971; Wilson et al., 1970) for incident GCR protons were used to extend the neutron spectrum to high energies. The measured data were combined with the theoretical calculations and integrated into the parametric AIR model, parameterized by neutron monitor count rate, vertical geomagnetic cutoff rigidity, and atmospheric depth. Solar cycle modulation of the GCR spectrum is parameterized by the ground-level neutron monitor count rates. Geomagnetic momentum shielding and overhead atmospheric shielding are parameterized by the vertical geomagnetic cutoff rigidity and atmospheric depth, respectively. The neutron flux (cm^{-2} sec^{-1}) component to the atmospheric radiations is converted to dose equivalent rate and total dose rate using 3.14 Sv cm^2 sec hr^{-1} and 0.5 Gy cm^2 sec hr^{-1}, respectively. The charged particle component of the atmospheric radiations is obtained from data taken by Neher (1961; 1967; 1971) and Neher & Anderson (1962) as compiled S. B. Curtis (Boeing 1969) and utilized by Wallance & Sondhaus (1978). The charge particle atmospheric ionization rates are directly converted to dose equivalent rate and total dose rate using measurement data from the tissue equivalent ion chamber. Nuclear stars in tissue are estimated from the nuclear emulsion measurement data after subtraction of the neutron-induced stars (Wilson et al., 1991). Recent updates to the parametric AIR model are described by Mertens et al. (2007a).

Figure 1a shows altitude profiles of neutron flux and ionization rates computed from the AIR model for summer and winter seasons at various latitudes in the northern hemisphere, for both solar maximum and solar minimum conditions. Figure 1b shows the corresponding profiles of dose equivalent rates. In this context, dose rate refers to the rate at which radiation energy is absorbed in human tissue per unit mass. The unit of dose is Gray (1 Gy = 6.24×10^{12} MeV kg^{-1}). Dose equivalent rate is the sum of the dose (Gy) from each radiation particle, which each particle dose weighted by a factor related to the potential to inflict biological damage (Mertens et al., 2007a). The unit of dose equivalent is Sievert (Sv). Dose equivalent is closely related to biological risk due to radiation exposure (Wilson, 2000). The dosimetric quantities are discussed in more detail in section 3.6.

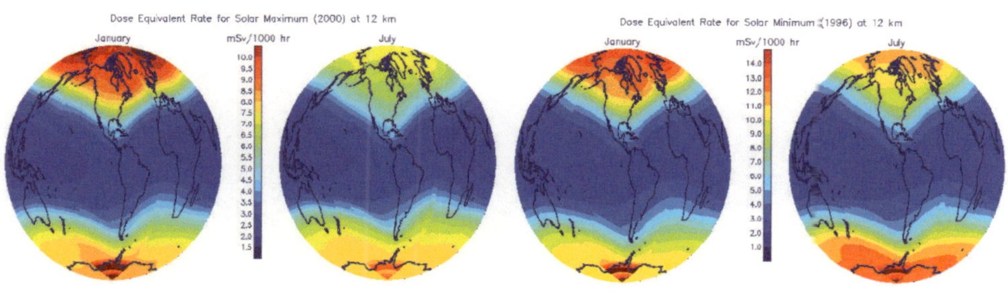

(a) AIR Dose Equivalent Rates: Solar Max (b) AIR Dose Equivalent Rates: Solar Min

Fig. 2. (a) Global distribution of dose equivalent rate (mSv/1000 hr) predicted by the parametric AIR model at 12 km for solar maximum conditions (year 2000) of cycle 23. (b) The same as in (a) but for solar minimum conditions (year 1996) of cycle 22.

The two most noticeable features in Figure 1 are: (1) the significant increase in flux, ionization rates, and dose equivalent rates at high-latitudes, and (2) the peak in these quantities occur near the typical cruising altitudes of commercial aircraft flying international routes (~ 10-12 km). The low altitude results are less reliable because of the limited altitude range of the balloon and flight measurements used to develop the AIR parameterizations. The first point is underscored in Figure 2, which shows global maps of the dose equivalent rates computed from the parametric AIR model for summer and winter northern hemisphere seasons at 12 km for both solar minimum and solar maximum conditions. The solar cycle variation in the dose equivalent rates is governed by modulation of the local interstellar spectrum of GCR particles by the heliospheric solar wind and interplanetary magnetic field (Mertens et al.,

2008; 2007a; Wilson et al., 2003; 1991). The latitudinal variation in the dose equivalent rates is determined by the low-momentum shielding provided by the Earth's magnetic field (Mertens et al., 2010a; 2009; 2008; 2007a; Wilson et al., 2003; 1991). The seasonal variations in the dose equivalent rates are determined by the seasonal variations in the overhead atmospheric mass at a fixed altitude (Mertens et al., 2008; 2007a; Wilson et al., 2003; 1991). These space weather and meteorological influences on atmospheric radiation exposure are discussed more fully in sections 3 and 4.

An important point to note is that the North Atlantic corridor region is one of the busiest in the world and comprises the most highly exposed routes in airline operations. This is clearly seen in Figure 2. High-latitude flights, polar flights, and flights over Canada are among the most highly exposed. The maximum GCR radiation exposure occurs during solar minimum conditions. The annual level of GCR exposure to pilots flying high-latitude routes and logging maximum flight hours (1000-hours) is sufficient to trigger individual monitoring and medical surveillance in the European Union (EU) states. Low latitude flights, on the other hand, are minimally exposed to cosmic ray radiation.

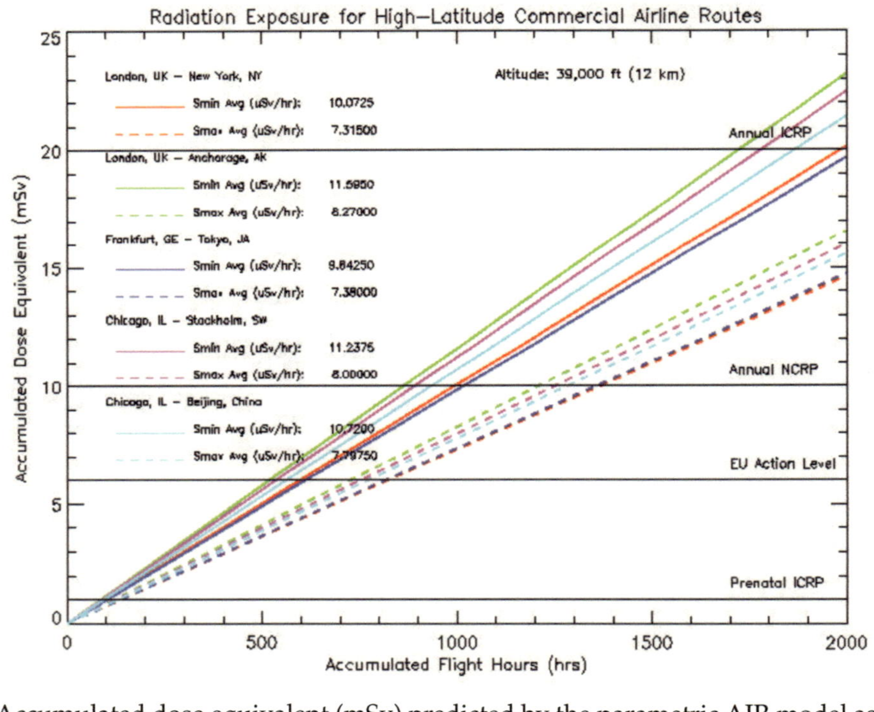

Fig. 3. Accumulated dose equivalent (mSv) predicted by the parametric AIR model as a function of accumulated flight time (hrs) at 12 km for representative high-latitude commercial routes. The solid lines denote solar minimum conditions (year 1996) while the dashed lines denote solar maximum (year 2000) conditions. The top two horizontal lines indicate the recommended ICRP and NCRP annual exposure limits for radiation workers, respectively. The bottom horizontal line shows the ICRP recommended prenatal and annual public exposure limits. The "EU Action Level" horizontal line is described in section 2.3

2.3 The NAIRAS decision support system

To place biological risk to high-latitude commercial aircrew and passengers from high-LET radiation exposure into context, consider the accumulated dose equivalent shown in Figure 3 for representative high-latitude commercial routes. The accumulated dose equivalent in Figure 3 is from GCR radiation exposure predicted by the parametric AIR model for both solar maximum and solar minimum conditions. The ICRP recommends 20 mSv as the annual occupational radiation worker limit (ICRP, 2008; 1991; Wilson, 2000; Wilson et al., 2003). The National Committee on Radiological Protection (NCRP) recommends 10 mSv as the annual occupational radiation worker limit (a conservative reconciliation of the NCRP 10*age occupational lifetime limit with its annual occupational limit of 50mSv/yr: see Wilson (2000); Wilson et al. (2003).). The recommended ICRP annual public limit and prenatal exposure limit is 1 mSv. These limits are shown as horizontal black lines in Figure 3.

The Council of the EU adopted Directive 96/20/EURATOM on 13 May 1996. Article 42 of the EU Directive imposes requirements relating to the assessment and limitation of aircrew cosmic ray radiation exposure (AMS, 2007). EU Member States were required to implement the Directive by 13 May 2000 through national legislation. Consistent with "best practice" radiation protection procedures, which is to keep all radiation exposures as low as reasonably achievable (i.e., the ALARA principle), the EU has also adopted an "action level" of 6 mSv/yr. The EU action level is also shown as a horizontal black line in Figure 3. For those likely to exceed 6 mSv/yr, individual record keeping and medical surveillance is required of the aircraft operators (Dyer & Lei, 2001). For exposures less than 6 mSv/yr, monitoring is only recommended and the actual implementation of these recommendation varies among the EU Member States (Meier et al., 2009). The EU Directive recommendation is that individual exposure is assessed by the aircraft operators if one is likely to exceed 1 mSv/yr, and that workers are educated on radiation health risks and work schedules are adjusted to ensure that the 6 mSv/yr level is not exceeded (EURADOS, 1996).

The results from the parametric AIR model in Figure 3 indicate that high-latitude flying commercial aircrew will trigger the "EU action level" for annual flight hours more than $\sim$ 500-600 hours during solar minimum and more than $\sim$ 800-900 hours during solar maximum conditions. The recommended ICRP annual public and prenatal limit is quite low. From Figure 3, the public/prenatal limit can be exceeded in 100 hours of flight time. For high-latitude and polar flights, the dose equivalent rate is on the order of $\sim$ 10 uSv/hr (Copeland et al., 2008; Dyer et al., 2009; Mertens et al., 2008; 2007a). An accumulated 100 hours of flight time can be accrued from 10-20 international flights with roughly 5-10 hours of flight time per flight. Thus, at an average high-latitude GCR exposure rate of 10 uSv/hr, the ICRP annual public and prenatal limit can be exceeded in 10-20 flights.

Recognizing the potential impact on present day passenger and crew exposures - due to the changes since the original work of Foelsche, as described in section 2.2, combined with the results shown in Figure 3 - further studies were initiated at LaRC. A new radiation measurement flight campaign was formulated and executed, consisting of a collaboration of fourteen institutions in five countries with a contribution of eighteen instruments to the flight program. New measurements and advances in theoretical modeling followed (Clem et al., 1996; Wilson et al., 2003), which culminated in the AIR workshop (Wilson et al., 2003).

Following the recent LaRC-sponsored AIR workshop, a number of recommendations for future work were put forth (Wilson et al., 2003). The recommendations relevant to the NAIRAS model development are: (1) utilize satellite input data to provide real-time mapping of GCR and SEP radiation levels to provide guidance in exposure avoidance; and (2) utilize state-of-the-art transport codes and nuclear databases to generate input data to the AIR model. The NAIRAS model addresses these two recommendations, but significantly goes beyond recommendation (2) by using physics-based, state-of-the-art transport code directly in simulating the atmospheric radiation exposure levels. The details of the NAIRAS model are given in section 3.

There are also economic consequences to the issue of aircraft radiation exposure. Over the last decade, airspace over Russian and China has opened up to commercial traffic, allowing for polar routes between North America and Asia (AMS, 2007). These cross-polar routes reduce flight time and operational cost; thus, the number of cross-polar commercial routes has increased exponentially. The typical cost savings from a cross-polar route from the US to China is between $35,000 and $45,000 per flight compared to the previous non-polar route (DOC, 2004). However, the polar region receives the largest quantity of radiation because the shielding provided by Earth's magnetic field rapidly approaches zero near the magnetic pole. On the other hand, the economic loss to an airline from rerouting a polar flight in response to an SEP warning, for example, can be a factor of three greater than the original cost-savings of flying the polar route if fuel stops and layovers are necessary. Thus, the cost to reroute a cross-polar route can be as much as $100,000 per flight (DOC, 2004). Consequently, an aircraft radiation prediction model must also be accurate to minimize radiation risks while simultaneously minimizing significant monetary loss to the commercial aviation industry.

The goal of the NAIRAS model is to provide a new decision support system for the National Oceanic and Atmospheric Administration (NOAA) Space Weather Prediction Center (SWPC) that currently does not exist, but is essential for providing the commercial aviation industry with data products that will enable airlines to achieve the right balance between minimizing flight cost while at the same time minimizing radiation risk. NOAA/SWPC is the main provider of real-time space weather nowcasts and forecasts, both nationally and internationally. However, there are no existing data or models, within NOAA/SWPC or outside NOAA/SWPC, that provide a comprehensive (i.e., comprehensive in terms of input observation data included and comprehensive in terms of the transport physics included in the real-time calculations), global, real-time assessment of the radiation fields that affect human health and safety. NAIRAS is the first-ever, global, real-time, data-driven model that predicts aircraft ionizing radiation exposure, including both GCR and SEP exposures. Thus, the NAIRAS model addresses an important national and international need with broad societal, public health and economic benefits. A detailed description of the NAIRAS model is given in the next section.

3. Description of NAIRAS model components

NAIRAS is an operational model for predicting aircraft radiation exposure from galactic and solar cosmic rays (Mertens et al., 2010a; 2009; 2008; 2007a). The NAIRAS prototype model development was funded by the NASA Applied Sciences / Aviation Weather Program. Real-time exposure rate graphical and tabular data products from the operational prototype are streaming live from the NAIRAS public web site at `http://sol.spacenvironment.net/~nairas/` (or, just google NAIRAS to locate web site). A subset of the NAIRAS real-time graphical products are available on the SpaceWx smartphone app for iPhone, IPad, and Android.

NAIRAS provides data-driven, global, real-time predictions of atmospheric ionizing radiation exposure rates on a geographic 1x1 degree latitude and longitude grid from the surface of the Earth to 100 km with a vertical resolution of 1 km. The real-time, global predictions are updated every hour. NAIRAS has adopted, as far as possible, the meteorological weather forecasting paradigm of combining physics-based forecast models with data assimilation techniques. Physics-based models are utilized within NAIRAS to transport cosmic rays through three distinct material media: the heliosphere, Earth's magnetosphere, and the neutral atmosphere. While the quantity of observations relevant to radiation exposure predictions is currently too sparse to apply data assimilation techniques per se, nevertheless, as much real-time measurement data as possible are utilized. The real-time measurement data are used to: (1) specify the ionizing radiation field at the boundaries of the aforementioned material media, and (2) characterize the internal properties of the aforementioned material media. The real-time measurements provide necessary observational constraints on the physics-based models that improve simulations of the transport and transmutations of cosmic ray radiation through the heliosphere, magnetosphere, and atmosphere.

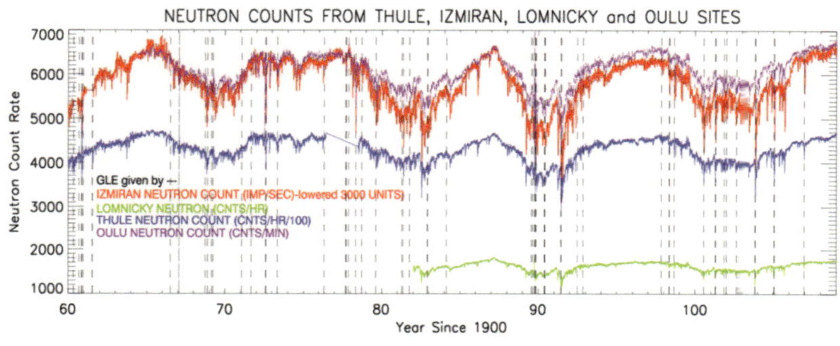

(a) Neutron Monitor Count Rates

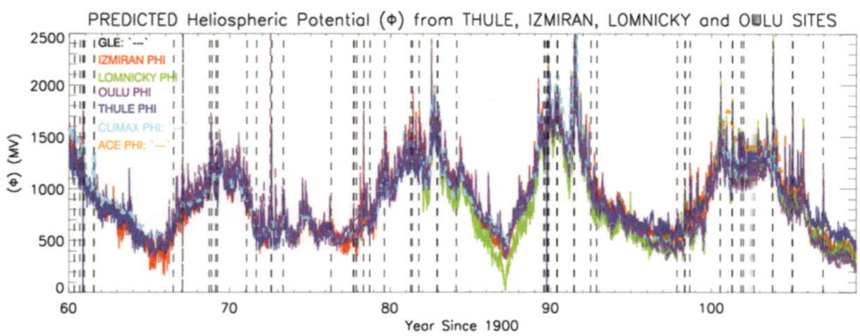

(b) Solar Modulation Parameter

Fig. 4. (a) Monthly-mean neutron monitor count rates from Thule, Izmiran (or Moscow), Lomnicky, and Oulu sites. The vertical dashed lines show ground level enhancement (GLE) events. The time period is 1960-2005. (b) Monthly-mean solar modulation parameter ($\Phi(t)$) predicted by the monthly-mean neutron monitor count rates shown in (a). The reference solar modulation parameters Φ_{CLIMAX} and Φ_{ACE} are also shown.

There are a number of models currently in use for calculating GCR radiation exposure at aircraft altitudes. The CARI-6 model utilizes a database of transport calculations generated by the deterministic LUIN code for a wide variety of geographic locations, altitudes, and solar activity levels (O'Brien et al., 1998; 2003). The EPCARD model is based on a similar approach, but uses the Monte Carlo FLUKA code for the transport calculations (Schraube et al., 1999). PC-AIRE is a semi-empirical model based on fits to measurement data (Lewis et al., 2002). Other aircraft radiation exposure models are described in the recent European Radiation Dosimetry Group report (Lindborg et al., 2004). Currently, the above models calculate SEP atmospheric radiation exposure post-storm on a case-by-case basis, although PC-AIRE incorporated low-earth orbit measurements to develop a simple extrapolation to SEP events (Lewis et al., 2002). Recently, Copeland et al. (2008) calculated adult and conceptus aircraft exposure rates for 170 SEP events for years 1986-2008 using the Monte Carlo MCNPX transport code.

The main differences that distinguish the NAIRAS model from the models discussed above are the following. Dynamical solar wind-magnetospheric interactions and the accompanying geomagnetic effects that govern the transport of cosmic rays through the magnetosphere are included in real-time in the NAIRAS radiation exposure calculations (Kress et al., 2004; 2010; Mertens et al., 2010a). Furthermore, the physics-based deterministic **H**igh Charge (**Z**) and **E**nergy **TR**a**N**sport code (HZETRN) is used in transporting cosmic rays through the atmosphere. The HZETRN transport calculations are continuously updated using real-time measurements of boundary condition specifications of the space radiation environment and of atmospheric density versus altitude (Mertens et al., 2008; 2007a). And finally, both GCR and SEP atmospheric radiation exposure predictions are included in real-time (Mertens et al., 2010a; 2009).

The remainder of this section contains a detailed description of the salient features of the NAIRAS physics-based modules that transport cosmic rays through the heliosphere, Earth's magnetosphere, and the neutral atmosphere. The input measurement data to the physics-based modules, used to specify boundary conditions of the material medium or to characterize the internal properties of material medium through which the cosmic rays are transported, are also described. The first material medium considered is the heliosphere. GCR are transported from outside the heliosphere to 1 AU using real-time measurements of ground-based neutron monitor count rates. This NAIRAS module is described in section 3.1. SEP are not transported per se but determined in the geospace environment in-situ using a combination of NOAA Geostationary Operational Environmental Satellite (GOES) and NASA Advanced Composition Explorer (ACE) ion flux measurements. This module is described in section 3.2. Both sources of cosmic rays, GCR and SEP, are transported through Earth's magnetosphere using a semi-physics-based geomagnetic shielding model. The geomagnetic shielding model utilizes real-time NASA/ACE solar wind and interplanetary magnetic field (IMF) measurements. This module is described in section 3.3. The cosmic rays that have passed through the heliosphere and magnetosphere are subsequently transported through the neutral atmosphere using the NASA LaRC HZETRN deterministic transport code. The internal properties of the atmosphere important to cosmic ray transport are provided in real-time by the global atmospheric mass density distribution obtained from the NOAA Global Forecasting System (GFS). These two NAIRAS modules are described in sections 3.4 and 3.5, respectively. Finally, the method of quantifying human exposure and the associated biological risk from the atmospheric ionizing radiation field is summarized in section 3.6.

3.1 Heliospheric GCR transport

GCR are transported through the heliosphere to the geospace environment using an expanded version of the Badhwar and O'Neill model (Badhwar & O'Neill, 1991; 1992; 1993; 1994; Badhwar & ONeill, 1996). The Badhwar and O'Neill model, which is simply referenced to as the GCR model, has been updated recently by O'Neill (2006) to use ground-based neutron monitor count rate measurements from the Climate neutron monitor site, in order to provide a measurement constraint on the simulated solar cycle modulation of the GCR spectrum at 1 AU. Comparisons between the GCR model and NASA/ACE measurements of the GCR spectra have shown that this step has enabled accurate predictions of GCR spectra in the geospace environment, at least on monthly to seasonal time scales (O'Neill, 2006). The NAIRAS team has extended the work of O'Neill (2006) by incorporating four high-latitude neutron monitor count rate measurements into the GCR model predictions at 1 AU. These additional high-latitude neutron monitor stations are Thule, Oulu, Izmiran (or Moscow), and Lomnicky. The reasons for utilizing these neutron monitor data are two-fold: (1) high-latitude locations are sensitive to the GCR spectral region most influenced by solar cycle variability, and (2) the data from these stations are available in real-time or near real-time.

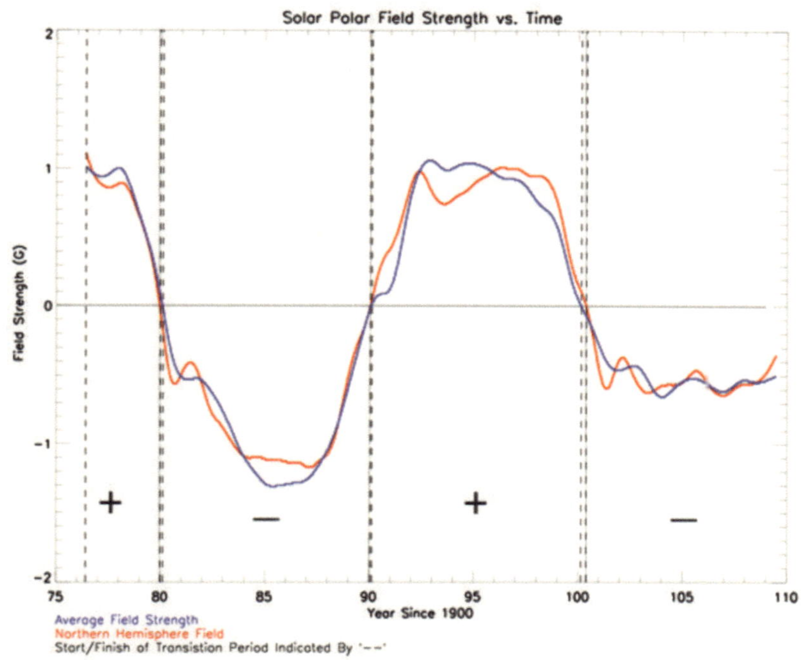

Fig. 5. Solar polar magnetic field data taken by measurements made at the Wilcox Solar Observatory. The blue line is the mean field strength of the northern and southern solar hemispheres. The red line is the northern hemisphere polar magnetic field strength. The '+' and '-' symbols between the vertical dashed lines indicate the time periods of solar positive and negative polarity, respectively. The transition region is the gap between the vertical dashed lines.

The GCR model propagates the local interstellar spectrum (LIS) of each element of the GCR composition to 1 AU by solving a steady-state, spherically symmetric Fokker-Planck transport equation, which accounts for diffusion, convection, and adiabatic deceleration of cosmic rays entering the heliosphere (Parker, 1965). The transport physics described above enables the temporal and spatial dependence of the GCR transport to be absorbed into a ratio of the diffusion coefficient to the bulk solar wind speed. The functional form of this ratio is given by

$$\tilde{k}(r,t) \equiv k(r,t)/V_{SW}(r,t) = (k_0/V_{SW})\beta R \left[1 + (r/r_0)^2\right] / \Phi(t) \tag{1}$$

where V_{SW} is the bulk solar wind speed (nominally set to 400 km/s for all time t), r is the distance from the sun in AU, t is time in years, k_0 and r_0 are constants, β is the particle's speed relative to the speed of light, R is the particle's magnetic rigidity in MV, and Φ is the so-called solar modulation parameter. Thus, the time-dependent behavior of the GCR spectral flux, due to the level of solar activity, is completely embedded in the solar modulation parameter. The solar modulation parameter is physically related to the energy that interstellar nuclei must have in order to overcome the heliospheric potential field, established by the large-scale structure of the IMF, and propagate through the heliosphere to the radius in question. The solar modulation parameter is determined by fitting the solution of the Fokker-Planck equation for a specified GCR nuclei to corresponding spectral flux measurements throughout the solar cycle, as described in the paragraph below (O'Neill, 2006).

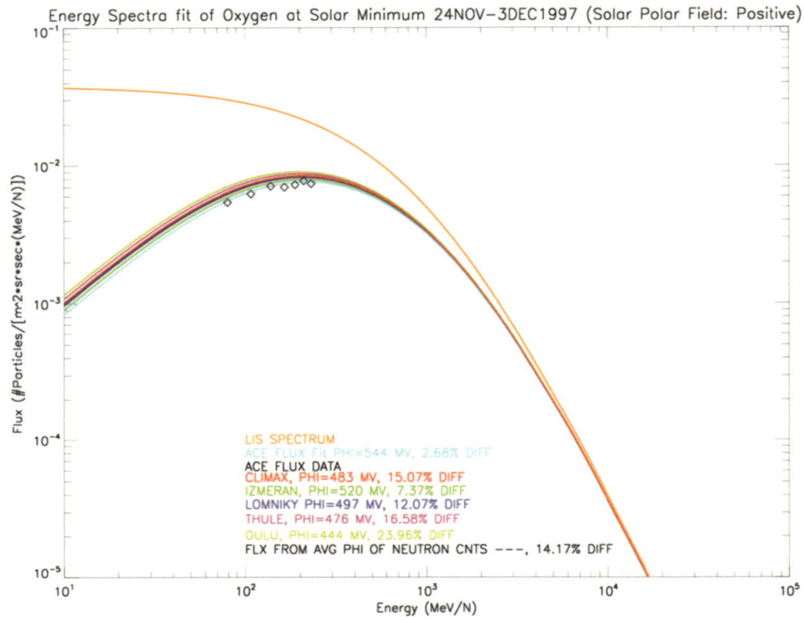

Fig. 6. NAIRAS GCR model spectral oxygen nuclei flux comparisons with ACE/CRIS measurements for solar minimum conditions. The GCR model flux was computed for neutron monitor count rates measured from the four high-latitude sites shown in Figure 4. Also shown are the oxygen spectral flux predicted by the reference Climax-based solar modulations parameter and the GCR model oxygen spectral flux computed using the average solar modulation parameter predicted by each neutron monitor site.

For a fixed parameterization of the LIS, the solar modulation parameter in (1) (i.e., $\Phi(t)$) was determined by fitting the solution of the steady-state Fokker-Planck equation for oxygen nuclei to measurements of the corresponding spectral flux. For energies below roughly 1 GeV (i.e. $\sim$ 50-500 MeV/nucleon), the measurement data were obtained from the Cosmic Ray Isotope Spectrometer (CRIS) instrument on the NASA/ACE satellite. For higher energies (1-35 GeV), the model was fit to data from the C2 instrument on the NASA High Energy Astrophysical Observatory (HEAO-3) satellite (Engelmann et al., 1990).

It is difficult to distinguish the GCR and solar components for protons and alpha spectra observed by CRIS. Fortunately, Lopate (2004) provided an extensive database of quiescent proton and alpha specta from IMP-8 measurements. Thus, the proton and alpha spectra in the

GCR model were fit to IMP-8 data. The high energy proton and alpha spectra were fit to the balloon-borne Isotope Matter-Antimatter Experiment (IMAX) measurements (Menn & et al., 2000).

Once the solar modulation parameter was derived based on the ACE/CRIS oxygen spectra, as described in the above paragraph, the LIS for the remaining elements (i.e., lithium (Z=3) through nickel (Z=28)) were similarly determined by fitting the solutions of the Fokker-Planck equation to the CRIS spectral flux measurements. A simple power law form of the differential LIS was assumed,

$$j_{\mathrm{LIS}}(E) = j_0 \beta^{\delta} (E + E_0)^{-\gamma} \tag{2}$$

where E is the particle kinetic energy per nucleon and E_0 is the rest mass energy per nucleon (938 MeV/n). The free parameters (γ,δ, and j_0) were determined from the fit of the GCR model to the CRIS measurements.

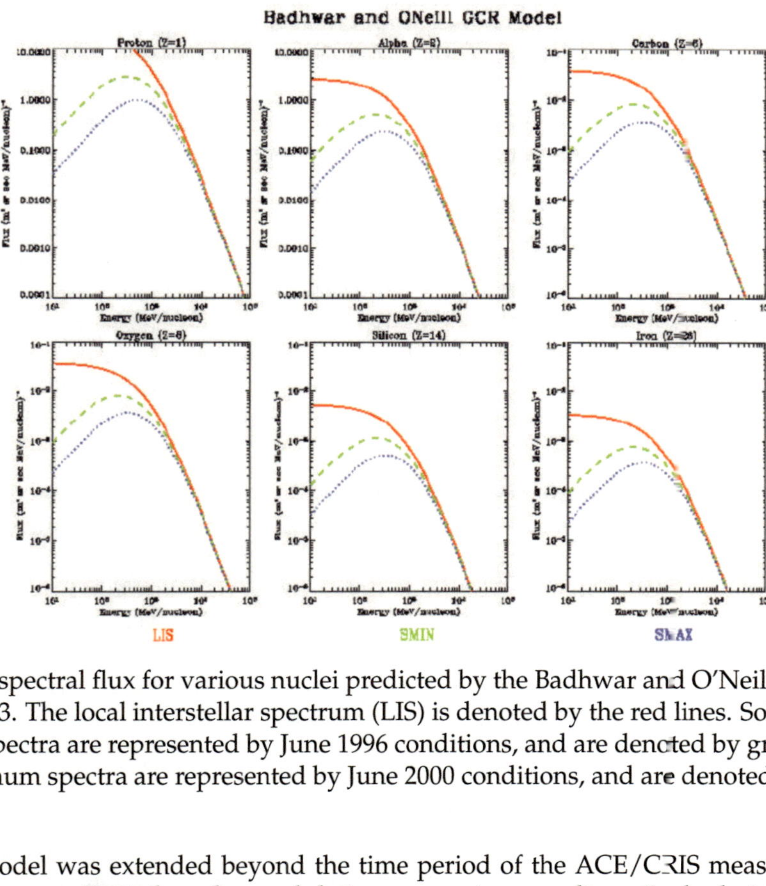

Fig. 7. GCR spectral flux for various nuclei predicted by the Badhwar and O'Neill model for solar cycle 23. The local interstellar spectrum (LIS) is denoted by the red lines. Solar minimum spectra are represented by June 1996 conditions, and are denoted by green lines. Solar maximum spectra are represented by June 2000 conditions, and are denoted by blue lines.

The GCR model was extended beyond the time period of the ACE/CRIS measurements in the following way. First, the solar modulation parameter was alternatively derived from the IMP-8 channel 7 (Z >8, high energy) measurements over three solar cycles from 1973 through 2001, and was calibrated against the solar modulation parameter derived from ACE/CRIS for the period of data overlap (1997.6 to 2001.8). GCR flux comparisons using both sets of solar modulation parameters correlated to within 98.9%. Next, linear fit coefficients were derived between the IMP-8 solar modulation parameter and Climax neutron monitor count rates from 1973-2001. The solar modulation parameter computed using the Climax neutron count rates correlated with the solar modulation parameter derived from IMP-8 data within 97%. Linear fits were derived for the three polarity states of the solar polar magnetic field: (1) positive solar cycle (outward field), (2) negative solar cycle (inward field), and (3) transition state (intermediate between positive and negative polarities with a high degree of modulation). The solar modulation parameter derived from Climax neutron count rates has been recently extended from 1958-2009. This extended Climax-based solar modulation parameter provides the reference solar modulation parameter from which to derive a real-time GCR model suitable for integration into the NAIRAS model.

Four neutron monitor sites - Thule, Oulu, Izmiran, and Lomnicky - were chosen to develop the NAIRAS GCR model. These high-latitude sites were chosen to maximize the solar cycle information content contained in the GCR spectrum and embedded in the ground-based neutron count rates. The neutrons detected on the ground are secondary particles produced by nuclear fragmentation reactions between the incoming GCR particles and the atmospheric constituents (Wilson et al., 1991). At high-latitudes the geomagnetic shielding of the incoming GCR particles is low. Thus, the information contained in the ground-level neutron counts on the low- and medium energy region of the GCR spectrum is high. This is highly desirable since this energy range of the GCR spectrum is most modulated by the solar wind and IMF, and thus closely related to the solar activity cycle (Mertens et al., 2008).

The real-time NAIRAS GCR model was developed by cross-correlating the Climax-based solar modulation parameter (denoted Φ_{CLIMAX}) with the neutron count rates measured at the four high-laitude sites mentioned above. Thus, linear fit coefficients were derived between Φ_{CLIMAX} and the neutron data at the four high-latitude sites. Monthly-mean neutron count rates at the four high-latitude sites are shown in Figure 4a. Figure 4b shows the solar modulation parameter predicted by the monthly-mean neutron count rates shown in Figure 4a. For comparison, the reference Climax-based solar modulation parameter (Φ_{CLIMAX}) is shown, along with the solar modulation parameter derived from NASA/ACE measurements (denoted Φ_{ACE}).

The heliospheric GCR diffusion coefficient depends on the large-scale structure of the IMF (Parker, 1965). As a result, the diffusion coefficient will depend on the polarity of the Sun's polar magnetic field (O'Neill, 2006). Consequently, to improve the accuracy of the neutron count rate fits to the reference Φ_{CLIMAX}, the data were sorted according to the polarity of the Sun's polar magnetic field, and three sets of fit coefficients were derived: (1) positive solar cycle (outward field), (2) negative solar cycle (inward field), and (3) transition state (intermediate between positive and negative polarities with a high degree of modulation), as described previously for the Φ_{CLIMAX} reference solar modulation parameter. Figure 5 shows the solar polar magnetic field data since 1978, which are obtained from measurements taken at the Wilcox Solar Observatory (WSO) located at Stanford University. The solar polar magnetic field data from WSO have been added to the NAIRAS input data stream.

Figure 6 shows a comparison between the NAIRAS GCR model spectral oxygen nuclei flux and measurements taken by ACE/CRIS for solar minimum conditions. Also shown is the oxygen spectral flux predicted by the reference solar modulation parameter (Φ_{CLIMAX}). Nominal NAIRAS operations uses the average solar modulation parameter determined from the available high-latitude neutron monitor count rate data. Figure 6 shows the nominal NAIRAS GCR oxygen spectral flux determined from all four high-latitude neutron sites (see Figure 4). In this case, the error in the nominal NAIRAS prediction of the incident GCR oxygen flux is comparable to the error of the predicted flux based on the reference solar modulation parameter ($\sim$ 15%).

A comprehensive set of error statics on the real-time NAIRAS GCR model has been compiled. Generally, the error in using the count rates from the high-latitude neutron monitor stations is comparable to the error in the Badhwar and O'Neill GCR model based on the reference solar modulation parameter. All errors are much larger for solar maximum conditions compared to solar minimum conditions. The main reason for this is that the ACE data are more suspect during solar maximum due to the combination of higher noise levels (GCR flux is minimum during solar maximum) and contamination by particles of solar origin.

Figure 7 shows the solar cycle variation in the GCR spectrum for several nuclei. The spectra were computed by the Badhwar and O'Neill model (Mertens et al., 2008; 2007a). The figure shows the LIS spectra and the spectra at solar maximum and solar minimum conditions. The solar cycle modulation of LIS as the GCR nuclei are transported through the heliosphere to 1 AU is clearly evident.

3.2 Geospace SEP fluence rate specification

The solar cosmic rays from SEP events are not transported from the Sun to the geospace environment using a model. Rather, in-situ satellite measurements of ion flux are used to constrain analytical representations of the SEP fluence rate spectrum. However, the analytical functions are guided by the state-of-the-art understanding of the origin, acceleration, and transport of energetic particles from the solar atmosphere to the geospace environment via the interplanetary medium.

The current understanding of SEP processes is that the energy spectrum is a result of injected particle seed populations that are stochastically accelerated in a turbulent magnetic field associated with a CME-driven interplanetary shock (Tylka & Lee, 2006). An analytical expression that represents the differential energy spectrum for this shock acceleration mechanism was given by Ellison & Ramaty (1985), where the spectrum has the form

$$\frac{d^2J}{dEd\Omega} = C_a\, E^{-\gamma_a} \exp\left(-E/E_0\right). \tag{3}$$

The differential energy spectrum on the left hand side of (3) has units of (cm^2-sr-hr-MeV/n)$^{-1}$, and the energy (E) has units of MeV/n (i.e., MeV/nucleon). The constant C_a is related to the injected seed population far upstream of the shock. The power-law energy dependence of the spectrum is due to shock acceleration of the seed population by random first-order Fermi acceleration (scattering) events in a turbulent magnetic field, with the power index (γ_a) related to the shock compression ratio. The exponential turnover in (3) represents high-energy limits to the the acceleration mechanism, such as escape from the shock region. Using the above analytical form, the three parameters (C_a, γ_a, and E_0) can be determined by fitting (3) to ion flux measurements.

Recently, Mewaldt et al. (2005) found that the Ellison-Ramaty spectral form failed to fit NOAA-GOES ion flux measurements at the highest energy channels during the Halloween 2003 SEP events. To circumvent this deficiency, Mewaldt et al. proposed using a double power-law spectrum. The low-energy spectrum is assumed to follow the Ellison-Ramaty form. The high-energy spectrum is assumed to have a power-law energy dependence with a different power index, such that

$$\frac{d^2J}{dEd\Omega} = C_b\, E^{-\gamma_b}. \tag{4}$$

The power-law expressions in (3)-(4) can be merged into one continuous spectrum by requiring that the differential energy spectra in (3)-(4) and their first derivatives are continuous at the merge energy. The result is given by the expression below:

$$\frac{d^2J}{dEd\Omega} = C\,E^{-\gamma_a} \exp\left(-E/E_0\right), \quad E \leq (\gamma_b - \gamma_a)E_0 \tag{5}$$

$$= C\,E^{-\gamma_b} \left\{ [(\gamma_b - \gamma_a)E_0]^{(\gamma_b - \gamma_a)} \exp\left(\gamma_b - \gamma_a\right) \right\}, \quad E > (\gamma_b - \gamma_a)E_0. \tag{6}$$

Physically, the double power-law spectrum in (5)-(6) represents SEP sources from two different injected seed populations. For example, the low-energy spectrum, with γ_a power index and the e-folding energy E_0, is likely associated with solar corona (solar wind) seed populations while the high-energy spectrum, with γ_b power index, is likely associated with flare suprathermal seed populations (Tylka et al., 2005).

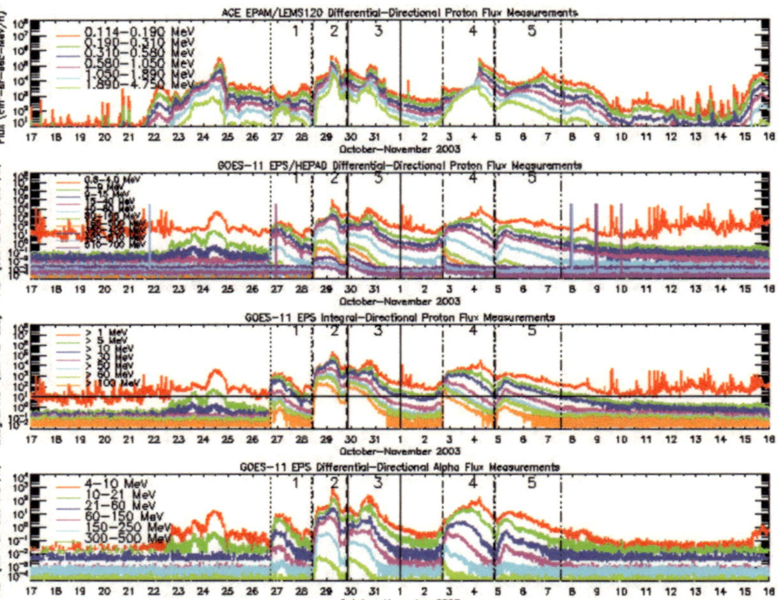

Fig. 8. Proton and alpha flux measurements used to derive the SEP fluence spectra. Row 1: ACE EPAM/LEMS120 differential-directional proton flux measurements. Row 2: GOES-11 EPS and HEPAD differential-directional proton flux measurements. Row 3: GOES-11 EPS integral-directional proton flux measurements. Row 4: GOES-11 EPS differential-directional alpha flux measurements. The different styled vertical lines bound the five SEP events during the Halloween 2003 solar-geomagnetic storm period, which are numbered in all panels. The horizontal line in Row 3 indicates the SEP threshold for the > 10 MeV integral proton flux channel.

Another widely used analytical representation of a SEP energy spectrum is the so-called Weibull distribution (Townsend et al., 2006; 2003). The Weibull distribution has been successful at fitting satellite ion flux measurements, and the differential energy spectrum is given by

$$\frac{d^2J}{dEd\Omega} = C\,k\,\alpha\,E^{\alpha-1} \exp\left(-kE^{\alpha}\right). \tag{7}$$

Notice that the Weibull distribution has an analytical form similar to the Ellison-Ramaty distribution in (3). The exponential energy dependence in the Weibull differential energy spectrum could be due to dissipation of the high-energy SEP ions by scattering from self-generated waves. However, further investigation is required to determine if this type of physical process is related to the Weibull distribution (Xapsos et al., 2000).

The NAIRAS model fits four analytical SEP spectral fluence rate functions to the satellite ion flux measurements. The analytical forms that are fit to the measurements are: (1) single power-law in (4), (2) Ellison-Ramaty in (3), (3) double power-law in (5)-(6), and (4) Weibull in (7). The free parameters for each analytical differential energy distribution are derived by a non-linear least-square fit to differential-directional ion flux measurements. The spectral fitting algorithm uses a Marquardt-Levenberg iteration technique (Brandt, 1999). The analytical form that yields the minimum chi-square residual in the fit to the ion flux measurements is the SEP spectral fluence rate distribution used in all subsequent model simulations.

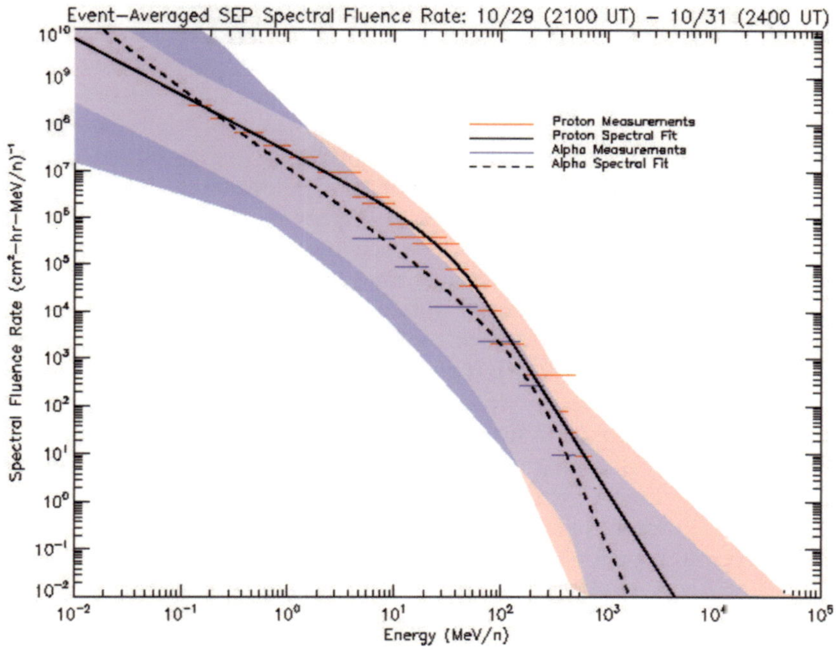

Fig. 9. Event-Averaged SEP spectral fluence rates for Halloween 2003 SEP event 3 [10/29/2003 (2100 UT) - 10/31/2003 (2400 UT)]. The shaded regions show the range of 1-hour averaged ion flux measurements and fitted spectra in the time interval of event 3. The peach shaded regions corresponds to proton flux measurements. The blue shaded regions corresponds to the alpha flux measurements.

NAIRAS utilizes available real-time measurements of proton and alpha differential-directional particle flux $(cm^2\text{-sr-sec-MeV/n})^{-1}$ for the SEP spectral fitting described above. SEP spectral fluence rates $(cm^2\text{-hr-MeV/n})$ incident on Earth's magnetosphere are obtained by time-averaging the particle flux measurements in 1-hr time bins and projecting the incident flux onto the vertical direction assuming an isotropic angular distribution for the solar ions. Low-energy proton data are obtained from the

Electron, Proton, and Alpha Monitor (EPAM) instrument onboard the NASA/ACE satellite (Gold & et al., 1998). EPAM is composed of five telescopes and the LEMS120 (Low-Energy Magnetic Spectrometer) detector in used in the SEP spectral ion fit, which measures ions at 120 degrees from the spacecraft axis. LEMS120 is the EPAM low-energy ion data available in real-time, for reasons described by Haggerty & Gold (2006). The other proton channels used in the SEP spectral fitting algorithm are obtained from NOAA/GOES Space Environment Monitor (SEM) measurements. The Energetic Particle Sensor (EPS) and the High Energy Proton and Alpha Detector (HEPAD) sensors on GOES/SEM measure differential-directional proton flux (Onsager & et al., 1996). Additional differential-directional proton flux measurement channels are generated by taking differences between the EPS integral proton flux channels. The channels used to derive SEP alpha spectral fluence rates are also obtained from EPS measurements. Five-minute averaged ACE and GOES data are used to derive the incident SEP spectral fluence rates.

Figure 8 shows the time variation of the proton and alpha flux measurements used to derive incident SEP spectral fluence rates for the Halloween 2003 storm period. The top panel displays the ACE low-energy proton flux measurements. The next two panels show the GOES-11 EPS and HEPAD proton flux spectra, and the integral proton flux measurements, respectively. By definition, a SEP event occurs when the >10 MeV integral proton flux exceeds 10 proton flux units (pfu $\equiv$ cm^{-2} sr^{-1} sec^{-1}) in three consecutive 5-minute periods (NOAA, 2009). The SEP event threshold is denoted by the horizontal line on the integral proton flux panel. There are a total of five SEP events during the Halloween 2003 storm period, which are denoted by the vertical lines in all panels in Figure 8. These events were associated with many simultaneous, complex phenomena such as solar flares, coronal mass ejections (CME), interplanetary shocks, and solar cosmic ray ground level events (GLE) (Gopalswamy et al., 2005). Different line styles are used to bound each of the five events, and the event number is shown between the vertical lines. Note that the onset of event 3 doesn't follow the conventional SEP threshold definition. It is clear from the integral proton flux that two events overlap: event 3 arrives before event 2 decreases below the SEP threshold level. However, there is an important distinguishing feature between the two events. That is, the beginning of our definition of event 3 is accompanied by a sudden increase in high-energy protons associated with the arriving SEP event, as noted by the sudden increase in the 510-700 MeV differential-directional proton flux measurements in Figure 8. Partitioning the simultaneous SEP events 2 and 3 into separate events is useful for the aircraft radiation exposure analysis discussed in section 4, since the high-energy portion of the differential-directional proton flux distribution penetrates deeper in the atmosphere.

In section 4, atmospheric ionizing radiation exposure during SEP event 3 [10/29 (2100 UT) - 10/31 (2400)] will be analyzed since the associated interaction between the arriving CME-driven interplanetary shock and Earth's magnetosphere caused the largest geomagnetic effects during the Halloween 2003 storm period, which is the focus of the case study in section 4.3. In order to isolate the geomagnetic effects, the event-averaged SEP spectral fluence rates is derived and shown in Figure 9. The horizontal lines in Figure 9 are the event-averaged differential ion fluence rate measurements. The width of the horizontal lines correspond to the energy width of the measurement channels. The black lines are the proton and alpha spectral fluence rates derived using the double power-law spectrum and fitting technique describe above. The shaded regions show the range of 1-hour averaged ion spectral fluence rates in the time interval of event 3. The peach colored region corresponds to the range of proton energy spectra and the blue colored region corresponds to the range of alpha energy spectra.

3.3 Magnetospheric cosmic ray transport

Lower energy cosmic rays are effectively attenuated by the geomagnetic field (internal field plus magnetospheric contributions) as these charges particles are transported through the magnetosphere and into the neutral atmosphere. The geomagnetic field provides a form of momentum shielding, or attenuation, by deflecting the lower-energy charged particles back out to space via the Lorentz force. This spectral filtering effect is quantified by a canonical variable, in the mathematical sense, called the geomagnetic cutoff rigidity. Once the cutoff rigidity is known, the minimum access energy to the neutral atmosphere is determined for each incident charged particle through the relativistic energy equation (Mertens et al., 2010a).

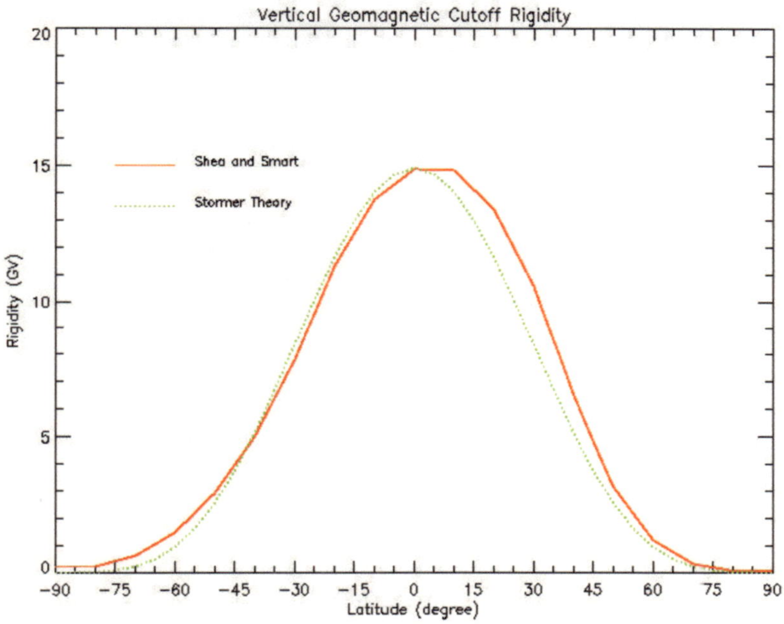

Fig. 10. Zonal-averaged vertical geomagnetic cutoff rigidity. The solid red line corresponds to the quiescent cutoff rigidities computed from particle trajectories and the IGRF model. The green dotted line is the analytical solution for the vertical cutoff rigidity using Störmer Theory.

The utility of the geomagnetic cutoff rigidity quantity is motivated by considering the motion of a charged particle in a magnetic field. The particle motion is determined by solving Newton's equation of motion for a charged particle subject to the Lorentz force. For a positively charged particle, the equation of motion is

$$\frac{d\mathbf{p}}{dt} = \frac{Ze}{c}\mathbf{v}\mathrm{x}\mathbf{B} \tag{8}$$

in cgs units. The bold-faced quantities are vectors and x designates the vector cross product. The charged particle momentum and velocity are **p** and **v**, respectively, and **B** is the magnetic field strength. The magnitude of the charge of an electron is denoted e and Z is the number of electron charge units. The equation of motion in (8) can be written, equivalently, as

$$\frac{R}{B}\frac{d\hat{\mathbf{v}}}{dt} = \hat{\mathbf{v}}\mathrm{x}\hat{\mathbf{B}} \tag{9}$$

where the ^ symbol denotes units vectors and

$$R \equiv \frac{pc}{Ze} \tag{10}$$

is defined as the rigidity. The canonical aspect of the rigidity is evident in the above equation. For a given magnetic field strength(B), charged particles with the same rigidity follow identical trajectories.

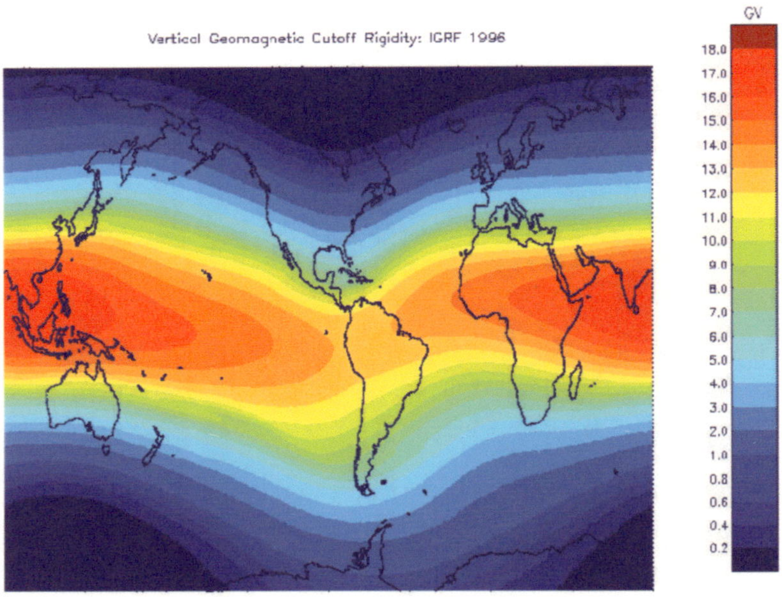

Fig. 11. Global grid of quiescent vertical geomagnetic cutoff rigidities (GV) calculated from charged particle trajectory simulations using the IGRF model for the 1996 epoch (solar cycle 23 minimum).

Motions of charged particles in a pure magnetic dipole field were examined by Störmer (1965). Because of the azimuthal symmetry in a pure dipole field, the azimuthal angular momentum is a conserved quantity. A main feature of Störmer theory is that regions of bounded and unbounded motion can be derived analytically from the integral of motion found from the conservation of azimuthal angular momentum (Störmer, 1965; VanAllen, 1968). It can be shown that the minimum rigidity that a vertically arriving particle must have in order to reach an altitude z above the Earth's surface is

$$R_{vc} = \frac{\mathcal{M}}{(R_e + z)^2} \cos^4 \lambda_m \approx 15 \cos^4 \lambda_m \,(\mathrm{GV}). \tag{11}$$

In the above equation, R_{vc} designates the vertical geomagnetic cutoff rigidity, $\mathcal{M}$ is the Earth's magnetic dipole moment, R_e is the average radius of the Earth, and λ_m denotes magnetic latitude. Therefore, vertically arriving charged particles with energies (E) less than the cutoff energy (E_{vc}) will be deflected by the Lorentz force and not reach altitude z. The cutoff energy for each charged particle of charge Z and mass number A is determined from the canonical cutoff rigidity through the relativistic energy equation, such that

$$E_{vc} = \left[\sqrt{R_{vc}^2 \left(Z/A \cdot \text{amu} \cdot c^2\right)^2 + 1} - 1\right] \cdot \text{amu} \cdot c^2, \tag{12}$$

where E is kinetic energy per nucleon (MeV/n), R_{vc} is vertical geomagnetic cutoff rigidity (MV), c is the speed of light in vacuum, and amu = 931.5 MeV/c^2 (atomic mass unit). Thus, the geomagnetic field has the effect of filtering out lower-energy charged particles as they are transported through the magnetosphere and into the neutral atmosphere.

The vertical cutoff rigidity at the Earth's surface derived from Störmer theory is plotted versus geographic latitude in Figure 10. The maximum cutoff rigidity is at the equator since a vertically arriving charged particle is perpendicular to the dipole magnetic field lines at the equator. The affect of the vector cross product in the Lorentz force in (8) is that charged particle motions perpendicular to magnetic field lines will experience the maximum deflection while particle motions parallel to the magnetic field will experience no deflecting force whatsoever. Figure 10 illustrates that a vertically arriving proton at the equator must have a kinetic energy of $\sim$ 15 GeV to arrive near the surface of the Earth. In the polar regions, vertically arriving charged particles are parallel to the magnetic field lines. Therefore, the cutoff rigidity is zero; particles of all energies can arrive at the Earth's surface in this case.

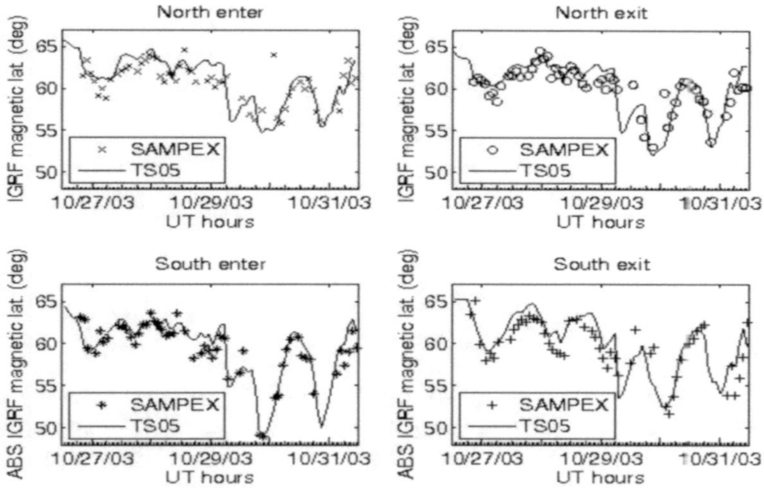

Fig. 12. The four panels show comparisons between cutoff latitudes determined in the TS05 geomagnetic field model and cutoff latitudes extracted from SAMPEX/PET energetic particle data during 28-31 October (Halloween) 2003 storms. Cutoff latitudes as SAMPEX enters and exists the north and south polar cap regions are shown separately.

The Earth's geomagnetic field is not a pure dipole field. On the contrary, the internal geomagnetic field is comprised of dipolar and non-dipolar contributions (Langlais & Mandea, 2000). The dipole moment is off-center and tilted with respect to the rotational axis. Furthermore, the geomagnetic field is distorted at large radial distances ($r \geq 4R_e$) by its interaction with the solar wind. A balance between the solar wind dynamic pressure and the magnetic field pressure, from the internal geomagnetic field, is established by inducing five magnetospheric current systems (Tsyganenko, 1989; 2002). These current systems generate their own magnetic fields which add vectorally to the internal geomagnetic field.

The complexities of the actual internal geomagnetic field, with dipolar and non-dipolar contributions, and the magnetospheric magnetic field contributions prohibit an analytical solution for the vertical geomagnetic cutoff rigidity. Numerical methods must be employed, as described below.

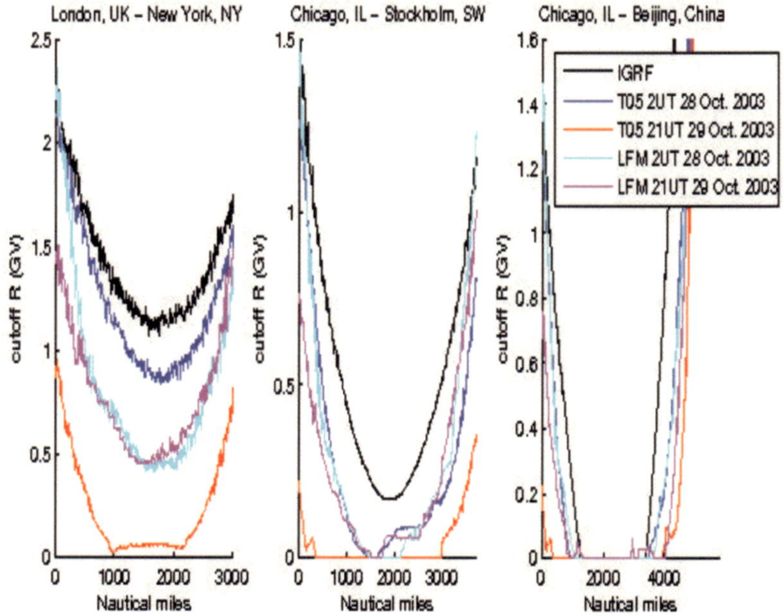

Fig. 13. Geomagnetic cutoff rigidities along three high-latitude commercial flight routes computed in three different magnetic field models: IGRF, TS05, and LFM MHD. The October 28 (0200 UT) calculations represent a magnetically quiet time, while the October 29 (2100 UT) calculations occur during a geomagnetic storm.

The internal geomagnetic field is specified by the International Geomagnetic Reference Field (IGRF) model (Langlais & Mandea, 2000). In the IGRF, the internal field is represented in terms of a magnetic potential function ($\Phi_M(r, \theta, \phi, t)$). Outside of the internal source region, i.e, for $r \geq R_e$, the magnetic potential function must be a solution of the Laplace equation. In spherical coordinates, this solution is expressed in the following form:

$$\Phi_M(r, \theta, \phi, t) = R_e \sum_{n=1}^{N} \sum_{m=0}^{n} \left(\frac{R_e}{r}\right)^{n+1} [g_n^m(t) \cos m\phi + h_n^m(t) \sin m\phi] \, P_n^m(\cos\theta). \quad (13)$$

In the above equation, r denotes the radial distance from the center of the Earth, θ and ϕ denote the geocentric colatitude and longitude at a given location. Schmidt-normalized associated Legendre functions of degree n and order m are denoted by $P_n^m(\cos\theta)$, and g_n^m and h_n^m denote the Gauss coefficients. The magnetic field components are given by the gradient of the potential function, such that

$$B_r = \frac{\partial \Phi_M}{\partial r}, \; B_\theta = \frac{1}{r}\frac{\partial \Phi_M}{\partial \theta}, \; B_\phi = \frac{1}{r \sin\theta}\frac{\partial \Phi_M}{\partial \phi}. \quad (14)$$

The Gauss coefficients are derived from a global set of magnetic field measurements, using the method of least-squares, and updated every five years (Langlais & Mandea, 2000). Secular variations in the Gauss coefficients are also derived from magnetic field measurements so that derivatives of the Gauss coefficients can be computed. In this way, the temporal dependence of the internal geomagnetic field is represented by

$$g_n^m(t) = g_n^m(T_0) + \dot{g}_n^m(t - T_0) \tag{15}$$

$$h_n^m(t) = h_n^m(T_0) + \dot{h}_n^m(t - T_0) \tag{16}$$

where $\dot{g}_n^m$ and $\dot{h}_n^m$ are first-order derivatives of the Gauss coefficients. The epoch of the IGRF model is denoted by T_0 and t is such that $T_0 \leq t \leq T_0 + 5$, where time is expressed in decimal years.

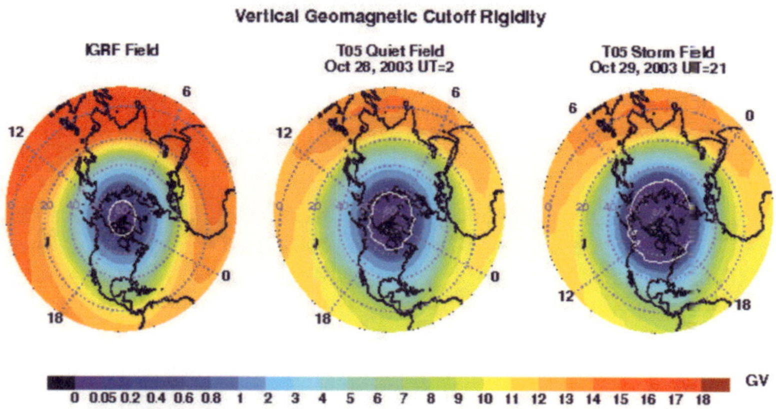

Fig. 14. Simulated vertical geomagnetic cutoff rigidity shown over the northern hemisphere in October 2003. The cutoff rigidities in the left column were calculated using the IGRF model. The cutoffs in the middle column were calculated using the TS05 model during a geomagnetically quiet period. Cutoff rigidities in the right column were calculated using the T05 model during the largest geomagnetically disturbed period of SEP event 3. Also shown are the magnetic latitude circles and the meridians at 0, 6, 12, and 18 magnetic local time.

The vertical cutoff rigidities in a realistic geomagnetic field are determined by numerical solutions of charged particle trajectories in the magnetic field using the techniques advanced by Smart & Shea (1994; 2005). Figure 11 shows the vertical cutoff rigidities at 20 km in the internal IGRF field. The longitudinal variations in the cutoff rigidity are due to a combination of geocentric offset and relative tilt of the magnetic dipole, with respect to the rotational axis, and the non-dipolar contributions to the internal geomagnetic field. A zonal-average of the IGRF cutoff rigidities in Figure 11 are compared to the analytical Störmer theory in Figure 10. The simple Störmer theory represents the latitudinal behavior of the vertical cutoff rigidity quite well. The displacement of the Störmer theory cutoffs in Figure 10 relative to the numerical solutions of the cutoffs in the IGRF field is due to the fact that the true dipole contribution to the internal geomagnetic field is off-centered and tilted with respect to the rotational axis.

NAIRAS real-time geomagnetic cutoff rigidities are computed from numerical solutions of charged particle trajectories in a dynamically varying geomagnetic field that includes both the internal magnetic field and the magnetospheric magnetic field contributions (Kress et al.,

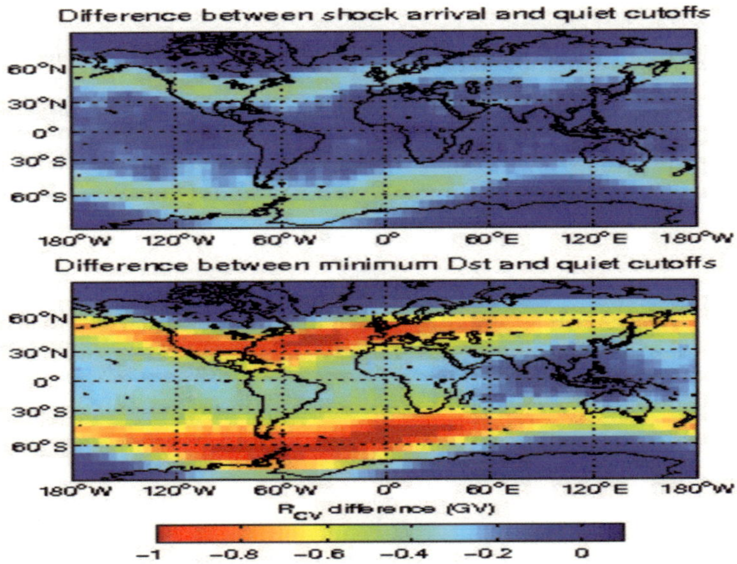

Fig. 15. Storm-quiet geomagnetic cutoff rigidities differences calculated using the TS05 magnetospheric magnetic field model. The magnetically quiet time cutoffs were calculated on October 28 (0200 UT), 2003. The storm cutoffs in the top panel were computed when the interplanetary shock arrived on October 29 (0612 UT), 2003. The storm cutoffs in the bottom panel were computed at the maximum build-up of the ring current on October 29 (2100 UT), 2003.

2010; Mertens et al., 2010a; 2009). The cutoff rigidity code was developed by the Center for Integrated Space Weather Modeling (CISM) at Dartmouth College. The CISM-Dartmouth geomagnetic cutoff model can be run using several different empirical, semi-physics-based, and physics-based models (Kress et al., 2004). In particular, the specification of the geomagnetic field due to Earth's internal field source is provided by the IGRF model [Langlais and Mandea, 2000], as discussed above. The real-time dynamical response of the magnetospheric magnetic field to solar wind conditions and IMF can be provided by the semi-physics-based TS05 model (Tsyganenko & Sitnov, 2005), or by the Lyon-Feder-Mobarry (LFM) global MHD (magnetohydrodynamic) simulation code (Fedder & Mobarry, 2004). Routines were developed and tested to couple the geomagnetic cutoff model with the different magnetic field models.

The LFM MHD code may be run as a stand alone model or coupled with other geospace models currently under development within CISM. For example, the LFM magnetospheric magnetic fields may be coupled with the Thermosphere-Ionosphere Nested Grid (TING) model (Wang et al., 2004) and/or with the Rice Convection Model (RCM) (Toffoletto et al., 2004), which models the ring current. The semi-physics-based TS05 model provides more accurate cutoff rigidities than the stand alone LFM MHD model, as determined by comparisons with satellite observations during a Halloween 2003 geomagnetic storm. Figure 12 shows comparisons between cutoff latitudes for $\sim$ 20 MeV protons computed using the TS05 model and measured by the Proton Electron Telescope (PET) instrument on the Solar Anomalous and Magnetospheric Particle Explorer (SAMPEX) satellite (Baker et al., 1993).

The agreement between SAMPEX/PET measurements and the TS05 model in Figure 12 are quite good. The reason the physics-based LFM MHD model doesn't calculate cutoffs as accurate as the semi-physics-based TS05 model is mainly due to the lack of a full kinetic description of the ring current in the MHD model, which typically causes the LFM fields to be too high. This is evident in Figure 13, which compares geomagnetic cutoff rigidities calculated along three representative high-latitude commercial flight routes from the IGRF, TS05, and LFM MHD magnetic field models. The small differences in cutoffs using the LFM MHD model between quiet and geomagnetic storm conditions is indicative of an inadequate modeling of the ring current build-up during the geomagnetic storm.

It is anticipated that the fully coupled LFM-RCM-TING model currently under development will significantly improve the simulations of cutoff rigidities compared to the stand alone LFM MHD model. Furthermore, the physics-based LFM-RCM-TING model will be able to incorporate short time-scale dynamics not included in semi-physics-based (empirical) magnetospheric magnetic field models. When the code development within CISM reaches sufficient maturity, the influence of short time-scale magnetospheric dynamics on the atmospheric ionizing radiation field using the fully coupled LFM-RCM-TING model will be assessed. For the present work, the simulated real-time geomagnetic cutoff rigidities are calculated using the TS05 model, and using the IGRF model for comparison.

Figure 14 shows the vertical cutoff rigidity over the northern hemisphere for three different models of the geomagnetic field during the Halloween 2003 storm period. The left column is cutoff rigidity computed using the IGRF field. Since total flight-path exposure at aviation altitudes do not change significantly ($< \sim 1\%$) for cutoffs less than 0.05 GV, the cutoffs are set to zero at geographic locations poleward of the 0.05 GV contour (see the bold-white 0.05 GV color contour in Figure 14). The middle column in Figure 14 shows the cutoff rigidities computed using the TS05 field under geomagnetically quiet conditions, October 28 (0200 UT), prior to the onset of the Halloween 2003 SEP event 3 (see section 3.2). One can see that even during magnetically quiet conditions, the cutoff rigidities predicted from the TS05 field are lower than predicted from the IGRF field, and the polar cap region (i.e., inside the bold-white 0.05 GV contour in Figure 14) is expanded to lower latitudes. A weaker field predicted by the TS05 model, compared to IGRF, is due in part to the diamagnetic effect of the magnetospheric ring current included in the TS05 model. Lower cutoff rigidities correspond to less momentum shielding and higher radiation exposure levels. The right column in Figure 14 shows the cutoff rigidities during peak geomagnetic storm conditions, October 29 (2100 UT), during SEP event 3. The cutoffs are lower at all latitudes compared to the two previous simulations, and the polar cap region has expanded to much lower latitudes than during the magnetically quiet period. These geomagnetic effects are discussed in more detail in section 4.3.

The difference in cutoffs between storm and quiet conditions is shown in Figure 15. The cutoffs were calculated using the TS05 model. The magnetically quiet period is the same as above, October 28 (0200 UT), 2003. The top panel shows the cutoff difference between the magnetically quiet time and the arrival of an interplanetary shock at the magnetosphere on October 29 (0612 UT), 2003. The bottom panel shows the storm-quiet cutoff difference when the Disturbed Storm Time (Dst) index is near its minimum on October 29 (2100 UT), 2003. The cutoffs are most suppressed at mid-latitudes during the night. The storm-quiet cutoff difference can be as much as ~ 1 GV, which has a significant effect on radiation exposure.

An important aspect of these model studies is our assessment of the impact of the changes in cutoff rigidity due to the magnetospheric field effect on atmospheric radiation exposure, and the identification of the need for accurate and computationally efficient geomagnetic cutoff rigidity models with solar wind-magnetospheric dynamical responses included. The $\sim$ 1 GV suppression in cutoff at mid-latitudes during a geomagnetic storm means that high-level SEP radiation exposure normally confined to the polar cap region will be extended to mid-latitudes. More details of these findings are included in section 4.3.

3.4 Atmospheric cosmic ray transport

The transport of cosmic rays through the neutral atmosphere is described by a coupled system of linear, steady-state Boltzmann transport equations, which can be derived on the basis of conservations principles (Wilson et al., 1991). The transport equation for the directional fluence $\phi_j(\mathbf{x}, \mathbf{\Omega}, E)$ of particle type j is given by (Mertens et al., 2008; 2007a)

$$\mathbf{\Omega} \bullet \nabla\Phi_j(\mathbf{x}, \mathbf{\Omega}, E) = \sum_k \int\int \sigma_{jk}(\mathbf{\Omega}, \mathbf{\Omega}', E, E')\Phi_k(\mathbf{x}, \mathbf{\Omega}', E')d\mathbf{\Omega}'dE' - \sigma_j(E)\Phi_j(\mathbf{x}, \mathbf{\Omega}, E) \quad (17)$$

where $\sigma_j(E)$ and $\sigma_{jk}(\mathbf{\Omega}, \mathbf{\Omega}', E, E')$ are the projectile-target macroscopic interaction cross sections. The $\sigma_{jk}(\mathbf{\Omega}, \mathbf{\Omega}', E, E')$ are double-differential particle production cross sections that represent all processes by which type k particles moving in direction $\mathbf{\Omega}'$ with energy E' produce a particle of type j moving in direction $\mathbf{\Omega}$ with energy E, including radioactive decay processes. The total interaction cross section $\sigma_j(E)$ for each incident particle type j is

$$\sigma_j(E) = \sigma_{j,at}(E) + \sigma_{j,el}(E) + \sigma_{j,r}(E), \quad (18)$$

where the first term refers to projectile collisions with atomic electrons of the target medium, the second term refers to elastic ion-nucleus scattering, and the third term contains all relevant nuclear reactions. The corresponding differential cross sections are similarly ordered.

Consider the transport of cosmic ray ions through the atmosphere. In this case, the second term in (18) represents elastic ion-nucleus Coulomb scattering between the incident ions and the atoms that comprise the neutral atmosphere. Figure 16 shows the characteristic elastic scattering length versus kinetic energy of various ions colliding with the neutral atmosphere (Mertens et al., 2008). Ion-nucleus scattering becomes important in the atmosphere only at low energies. For example, the length of the Earth's atmosphere in units of areal density is $\sim$ 1000 g/cm^2. Thus, Figure 16 shows that cosmic ray ions will not elastically scatter off an atmospheric nucleus before reaching the surface unless the ion kinetic energy is well below 1 MeV/amu. However, ions with kinetic energy less than 1 MeV/amu are stopped via ionization and/or atomic excitation energy loss processes at high altitudes before a scattering event can take place (see Figure 18). Multiple Coulomb scattering and coupling with ionization energy loss become important factors in the transport of ions within living tissue (Mertens et al., 2010b; 2007b), which are related to the degree of biological damage inflicted on sensitive components within the living cell. However, for cosmic ray transport through the atmosphere, the ion-nucleus scattering term in (18) can be neglected to a good approximation.

The principle mechanism for atomic interactions between the cosmic ray ions and the target medium is ionization and/or atomic excitation. This process is represented by the first term in (18). The result of this interaction is the transfer of energy from the projectile ions to the atomic electrons of the target medium via the Coulomb impulse force. Since the projectile ion mass is much greater than the electron mass, the ion travels essentially in a straight line as it looses energy through ionization of the target medium. The ionization and atomic excitation

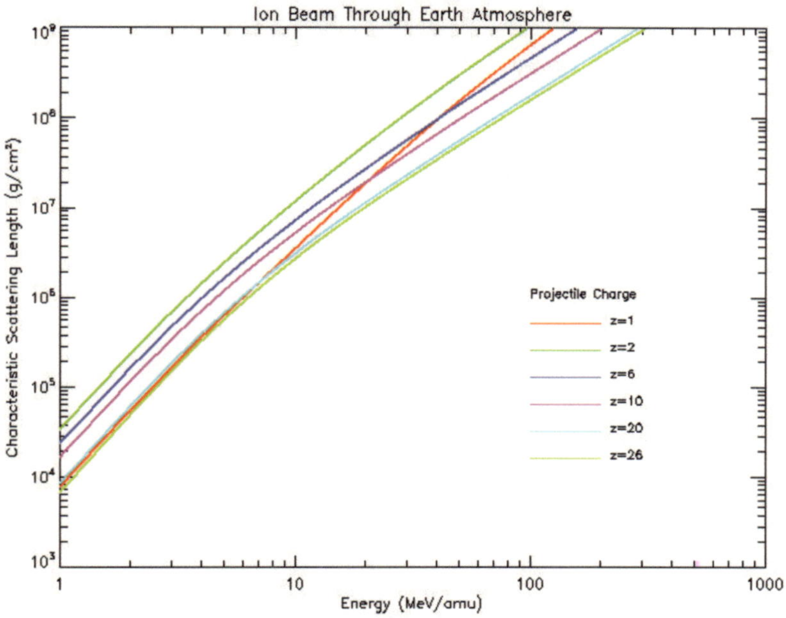

Fig. 16. Characteristic scattering length for ion beam transport through the Earth's atmosphere. The charge number (z) of the ion is specified in the legend.

energies, as well as the energies of ejected orbital electrons, are usually small in comparison to the incident ion kinetic energy. As a result, the ionization energy loss processes by which the projectile ions transfer energy to the target bound and/or ejected orbital electrons can be considered continuous. Because of this so-called continuous slowing down approximation (CSDA), the energy dE which is lost by the incident ion and transferred to the orbital electrons of the target medium by ionization and/or atomic excitation within an element of path dx is given by the stopping power, S, (Tai et al., 1997), i.e.,

$$S = -\frac{dE}{dx} = \frac{4\pi Z_P^2 Z_T e^4}{mv^2} N \left\{ B_0 - \frac{C(\beta)}{Z_T} + Z_P L_P(\beta) + Z_P^2 L_T(\beta) + \frac{1}{2}[G(M_P, \beta) - \delta(\beta)] \right\} \quad (19)$$

where

$$B_0 = \ln\left(\frac{2mc^2\beta^2}{I(1-\beta^2)}\right) - \beta^2. \quad (20)$$

In the above equations, Z_P and Z_T are the projectile ion charge and the number of electrons per target atom, respectively, v is the projectile velocity, c is the speed of light, $\beta = v/c$, N is the density of atoms in the target medium, and I is the mean ionization potential of the target medium. The electron charge and mass are denoted, respectively, by e and m.

The various terms in (19) have the following interpretation (Tai et al., 1997; Wilson et al., 1991). The B_0 term is the high-energy asymptotic limit of the stopping power assuming that the orbital electrons of the target atoms can be treated as essentially free electrons. This requires that the projectile's velocity be much greater than the orbital velocities of the bound atomic electrons, which is an inadequate approximation for inner shell electrons of heavy element target media. The $C(\beta)/Z_T$ term provides a correction for inner shell electrons. The $L_P(\beta)$ term arises from polarization of the target electrons by the incident ion, and is referred to as

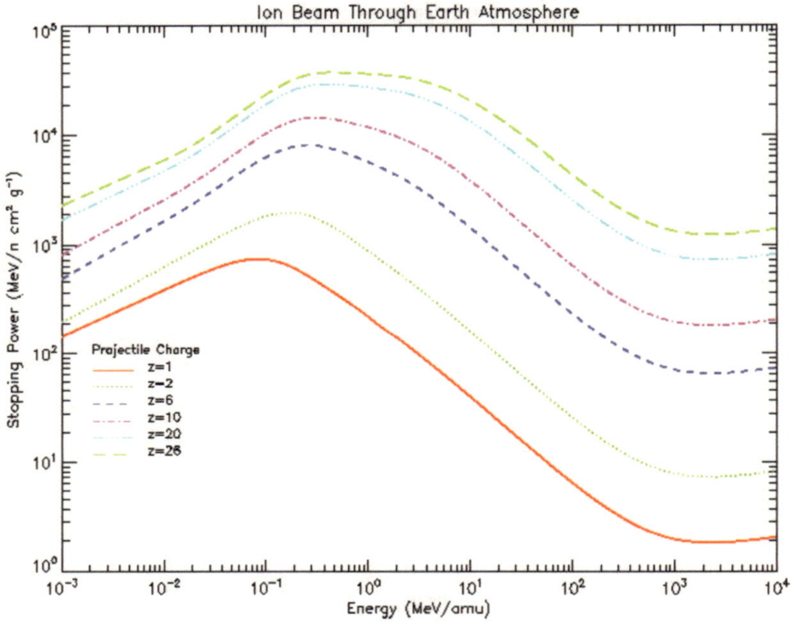

Fig. 17. Stopping power versus kinetic energy of incident ions on Earth's atmosphere. The charge number (z) of the ion is specified in the legend.

the Barkas effect. The $L_T(\beta)$ is the Bloch term which provides a correction to the assumption that the ejected orbital electrons in an ionization event can be represented as a plane wave for close collisions with the incident ion. The Mott term is denoted by $G(M_P, \beta)$, where M_P is the mass of the projectile ion, which includes a kinetic correction for the recoil of the target nucleus. The $\delta(\beta)$ term is a density correction that originates from the dielectric response of a solid target material to the electric field generated by the projectile ion. Finally, at low energy charge exchange processes begin to dominate, which leads to electron capture by the projectile ion and reduces the atomic excitation and/or ionization energy loss. This effect is included by introducing an effective charge for the projectile ion (Tai et al., 1997).

Figure 17 shows the stopping power for various cosmic ray ions incident on Earth's atmosphere. The stopping power decreases inversely with projectile energy between $\sim$ 100 keV/amu and 2 GeV/amu. The stopping power begins to increase with increasing projectile energy above 2 GeV/amu due to the relativistic corrections in the B_0 term in (19). The stopping power decreases for projectile energies less than $\sim$ 100 keV/amu due to electron capture by the projectile ion and the other correction terms in (19). The dependence of the stopping power on the projectile kinetic energy plays a major role in determining the spectral shape of the cosmic ray fluence rates in the atmosphere, as indicated by Figures 20 and 21.

The range of an ion is the mean path length traveled in the target medium before coming to rest after losing its initial kinetic energy through ionization and/or atomic excitation energy loss. In the CSDA, the range is defined by

$$R_j(E) = A_j \int_0^E \frac{dE'}{S_j(E')} \tag{21}$$

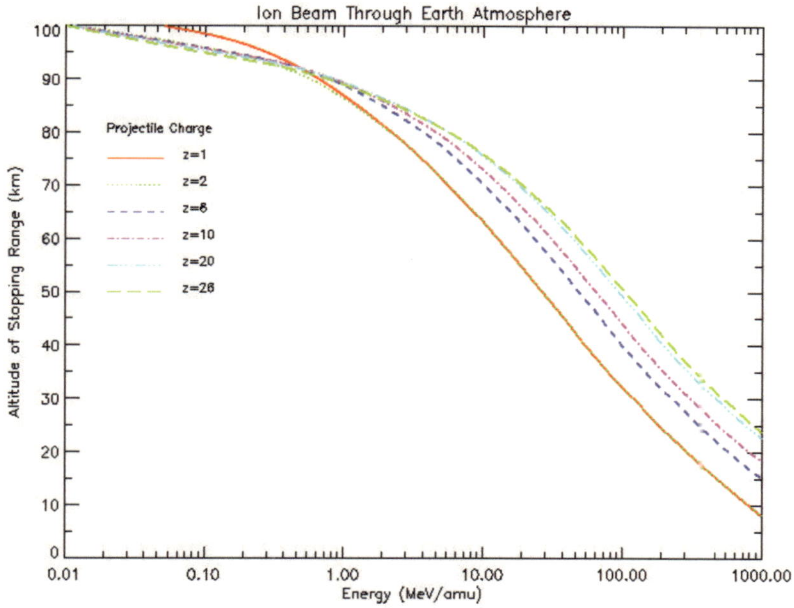

Fig. 18. Average range versus kinetic energy of an ion beam incident at the top of Earth's atmosphere. The charge number (z) of the ion is specified in the legend.

where A_j is the atomic mass number of ion particle type j. The above equation is referred to as the range-energy relation. Figure 18 shows the range of various cosmic ray ions incident on the neutral atmosphere as a function of ion kinetic energy. At 1 GeV/amu, only protons and alpha particles can reach the typical cruising altitudes of $\sim$ 10-12 km, or $\sim$ 200 g/cm^2, for commercial aircraft before coming to rest due to ionization energy loss. Below 1 GeV/amu, all cosmic ray ions loose their kinetic energy before reaching commercial aircraft cruising altitudes. For energies greater than 1 GeV/amu, the cosmic ray particle flux densities decrease with a power-law dependence on energy (see Figure 7). Although secondary charged particles can be produced by nuclear fragmentation reactions, which is represented by the third term in (18), the stopping range in Figure 18 explains why the heavy-ion fluence rates in Figure 20 are significantly less than the proton fluence rate. The same is true for the light-ions in Figure 21.

Two approximations have been made to the total ion-target interaction cross section in (18). First, elastic ion-nucleus scattering has been neglected for cosmic ray transport through the atmosphere. Second, the CSDA has been invoked in the representation of atomic ion-electron energy transfer collisions. As a consequence of these two approximations, the coupled Boltzmann transport equations in (17) can be expressed, alternatively, as

$$\overline{B}[\Phi_j(\mathbf{x},\mathbf{\Omega},E)] = \sum_k \int \int \sigma_{jk,r}(\mathbf{\Omega},\mathbf{\Omega}',E,E')\Phi_k(\mathbf{x},\mathbf{\Omega}',E')d\mathbf{\Omega}'dE' \qquad (22)$$

where

$$\overline{B}[\Phi_j(\mathbf{x},\mathbf{\Omega},E)] \equiv \left[\mathbf{\Omega}\bullet\nabla - \frac{1}{A_j}\frac{\partial}{\partial E}S_j(E) + \sigma_{j,r}(E)\right]\Phi_j(\mathbf{x},\mathbf{\Omega},E). \qquad (23)$$

The $\overline{B}[\Phi_j]$ in the above equations denote a differential operator acting on the directional fluence.

The differential operator in (22) can be inverted using the method of characteristics in order to transform the integro-differential equation into a Volterra-type integral equation (Wilson, 1977). As a result, the integral equation for cosmic ray transport is given by

$$\Phi_j(\mathbf{x},\mathbf{\Omega},E) = \frac{S_j(E_\gamma)P_j(E_\gamma)}{S_j(E)P_j(E)}\Phi_j(\mathbf{\Gamma}_{\Omega,\mathbf{x}},\mathbf{\Omega},E_\gamma)$$
$$+\sum_k \int_E^{E_\gamma} \frac{A_j P_j(E')}{S_j(E)P_j(E)}dE' \int_{E'}^{\infty} dE'' \int d\mathbf{\Omega}' \; \sigma_{jk,r}(\mathbf{\Omega},\mathbf{\Omega}',E',E'')$$
$$\text{x} \; \Phi_k[\mathbf{x}+(R_j(E)-R_j(E'))\mathbf{\Omega},\mathbf{\Omega}',E'']. \qquad (24)$$

In the above equation, $\mathbf{\Gamma}_{\Omega,\mathbf{x}}$ is a position vector of a point on the boundary surface and E_γ is given by

$$E_\gamma = R_j^{-1}[R_j + \mathbf{\Omega} \bullet (\mathbf{x} - \mathbf{\Gamma}_{\Omega,\mathbf{x}})] \qquad (25)$$

The R_j^{-1} operator in (25) is the inverse operation of obtaining the energy given the range using the range-energy relation in (21). The expression for the integral cosmic ray transport equation in (24) was made compact by introducing the total nuclear survival probability, which is defined by

$$P_j(E) \equiv \exp\left[-A_j \int_0^E \frac{\sigma_{j,r}(E')dE'}{S_j(E')}\right]. \qquad (26)$$

The first term in (24) describes the attenuation of the directional fluence specified at the boundary as a result of transport through the target medium. For atmospheric cosmic ray transport, an isotropic distribution is assumed for the directional fluence and the boundary specification is defined to be the cosmic ray particle fluence rates that have been transported through the heliosphere and magnetosphere and incident at the top of the neutral atmosphere. These incident cosmic ray ions are attenuated by ionization energy loss ($S(E)$) and nuclear absorption ($P(E)$), as indicated by the first term in (24). The second term in (24) describes the generation of type j particles from projectile-target nuclear fragmentation reactions by type k particles. The second term in (24) includes the production of type j particles from type k particles at all intervening positions between the boundary point and the position of observation, accounting for the attenuation by ionization energy loss and nuclear absorption in between the point of production of a type j particle and the observation point.

The representation of the relevant total nuclear absorption cross sections ($\sigma_{j,r}$) and nuclear fragmentation production cross sections ($\sigma_{jk,r}$) can not be expressed in a simple, compact form such as the stopping power in (19)-(20). Nevertheless, important insight into the influence of nuclear reactions on the atmospheric transport of cosmic rays can be gained by examining the probability of a nuclear reaction as a function of incident ion kinetic energy. The probability of a nuclear reaction is one minus the total nuclear survival probability in (26) (i.e., $1 - P_j(E)$), which is shown in Figure 19. For particles with kinetic energy below 100 MeV/amu, there is a small chance of a nuclear reaction. Recall from Figure 7 that the peak of the incident GCR spectrum is between $\sim$ 200-500 MeV/amu. At these energies, one out of every two particles will undergo some kind of nuclear reaction. For kinetic energies greater than 1 GeV/amu, nearly every particle will be subject to some type of nuclear reaction. Combining this discussion with the discussion of Figure 18, the only primary cosmic ray particles that can survive transport through the atmosphere and reach the cruising altitudes of typical commercial aircraft are protons with kinetic energy on the order of 1 GeV or greater. The

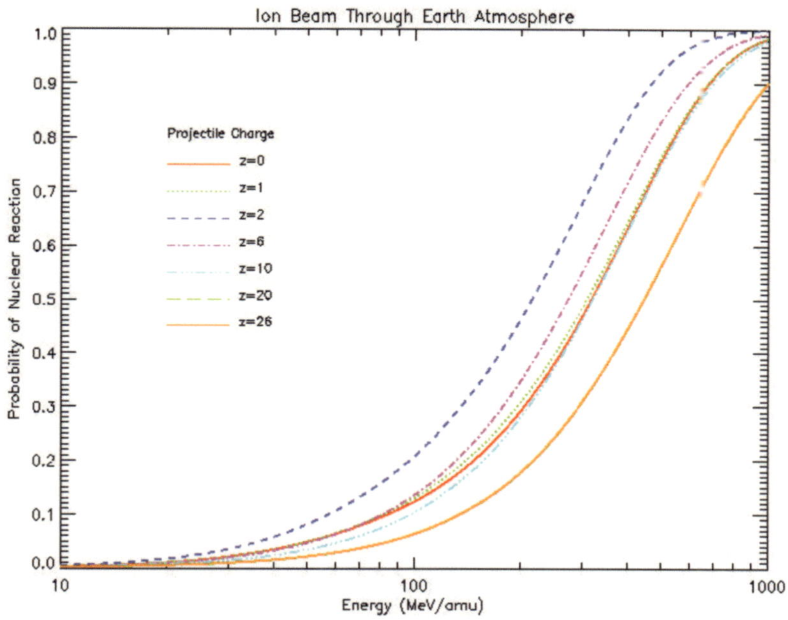

Fig. 19. Nuclear survival probability versus kinetic energy of an ion beam incident at the top of Earth's atmosphere. The charge number (z) of the ion is specified in the legend.

high-LET particles present at 10-12 km with energies less than 1 GeV/amu are secondary particles created at higher altitudes from nuclear fragmentation reactions most of which are neutrons.

The coupled cosmic ray integral transport equations in (24) are solved in the NAIRAS model using NASA LaRC's deterministic HZETRN code. Details of the early analytical and computation approaches to solving (24) are given by Wilson et al. (1995a; 1997; 1991; 2005a). The stopping power parameterization used in HZETRN is described by Tai et al. (1997). The nuclear cross sections for neutron and proton interactions are described extensively in Wilson et al. (1989). The model for calculating the heavy-ion nuclear fragmentation cross sections are described by Wilson et al. (1995b). HZETRN is used in a wide variety of radiation transport applications: e.g., the calculation of dosimetric quantities for assessing astronaut risk to space radiations on the International Space Station (ISS) and the Space Transportation System (STS) Shuttle, including realistic spacecraft and human geometry (Badavi et al., 2005; 2007a; Slaba et al., 2009; Wilson & et al., 2006). Extensive summaries of HZETRN laboratory and space-flight verification and validation are found in recent reports by Badavi et al. (2007a); Nealy et al. (2007); Wilson et al. (2005c;a).

The computation methods employed in HZETRN to solve the coupled cosmic ray integral transport equations in (24) are summarized below. The details are given in the references. The numerical procedures fall into two categories based on the fundamental physics of cosmic ray projectile-target nuclear interactions. The first category is heavy-ion transport. Ion beam experiments have shown that projectile fragments have an energy and direction very near to that of the incident heavy-ion projectile (Wilson, 1977; Wilson et al., 1995a). The observation of forward directed projectile fragments is the bases of the so-called straight-ahead approximation, where the integral over solid angle in (24) is neglected and

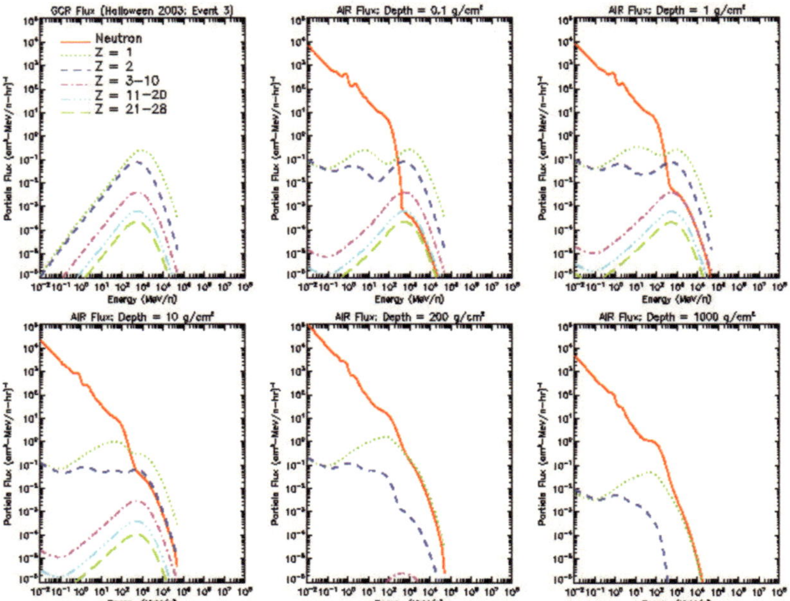

Fig. 20. Event-averaged GCR spectral fluence rates at zero vertical geomagnetic cutoff rigidity during the Halloween 2003 solar-geomagnetic storm [29 October 2003 (2100 UT) to 1 November (0000 UT)]. The panels show the fluence rates at different atmospheric depths. The typical cruising altitudes for commercial aircraft correspond to an atmospheric depth of roughly 200 g/cm^2. The fluence rates from different charge groups have been summed together to reduce the number of lines.

the transport is reduced to one-dimension along the direction of the incident heavy-ion beam. Moreover, the observation of equal velocity between the heavy-ion projectile and the projectile fragment suggests a delta function dependence in the projectile fragment production cross section, which effectively eliminates the integral over dE'' on the right-hand side of (24).

In addition to the approximations discussed in the previous paragraph, the target fragments are produced at low-energy and distributed nearly isotropically. At low energy, the target fragments do not travel far before coming to rest due to ionization energy loss. These observations justify a decoupling of the target and projectile fragments in the source term on the right-hand side of (24). The advantage of this decoupling is that the target fragments can be neglected in the heavy-ion transport procedure. The absence of the target fragments in the heavy-ion transport solution means that the summation over k type particles in (24) involves only projectiles with masses greater the the mass of the type j particle (i.e., $\sum_{k>j}$). These approximations enable a self-consistent solution of projectile fragment heavy-ion transport using backsubstitution and perturbation theory with rapid convergence. The contribution of the target fragments to the dosimetric quantities is included indirectly using the method described in section 3.6. Significant improvements in the accuracy and computational efficiency of HZETRN's heavy-ion numerical transport procedures have recently been made by Slaba et al. (2010c;d).

The numerical solution of the integral transport equation in (24) for light-particle projectiles does not permit the same approximations as the heavy-ion transport solution. Light-particles are defined in HZETRN as those particles with mass number $A \leq 4$ and charge number

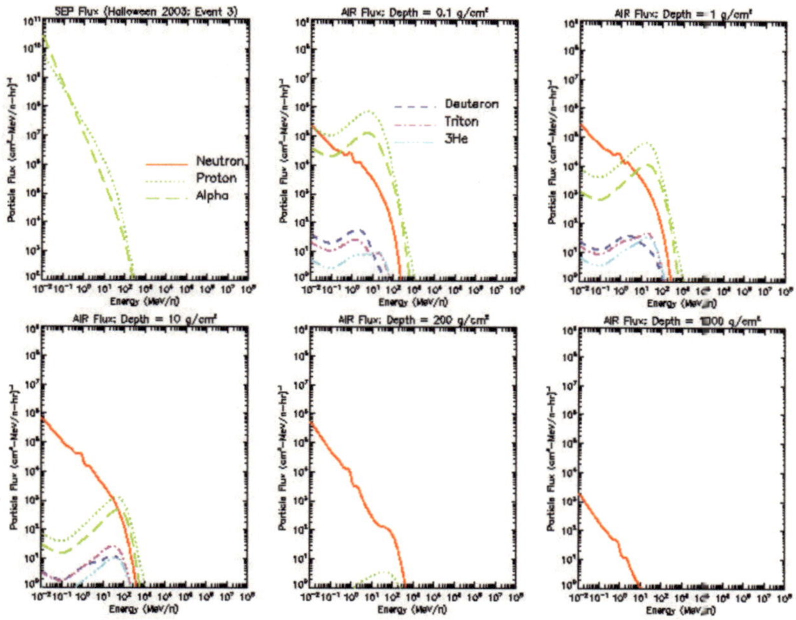

Fig. 21. Event-averaged SEP spectral fluence rates at zero vertical geomagnetic cutoff rigidity during the Halloween 2003 solar-geomagnetic storm [29 October 2003 (2100 UT) to 1 November (0000 UT)]. The panels show the fluence rates at different atmospheric depths. The typical cruising altitudes for commercial aircraft correspond to an atmospheric depth of roughly 200 g/cm^2.

$Z \leq 2$. Thus, there are six light-particles: neutrons and five charged particles (protons, deuterons, tritons, helium-3, and helium-4). The equal velocity relationship between projectile and projectile fragments is no longer valid for light-particle transport. This means that the energy integral over dE'' on the right-hand side of (24) must be explicitly evaluated, and both the projectile and target fragments are included in the numerical transport procedure. Moreover, the straight-ahead approximation can not be employed in light-particle transport. The integral over solid angle in (24) must be considered, which is especially important for low-energy neutron transport.

Important updates to the numerical solution of light-particle transport in HZETRN have been made recently by Slaba et al. (2010a;c;d;b). These updates have significantly improved accuracy and computation efficiency. The solution approach is to decompose the light-particle fluence into a straight-ahead component and an isotropic component. The transport solution for the light-particle straight-ahead component is described by Slaba et al. (2010c;d). The neutron fluence is further decomposed into semi-isotropic forward and semi-isotropic backward components. The numerical approach for solving the directionally coupled forward-backward neutron transport scheme for the semi-isotropic component is described by Slaba et al. (2010a;b), which also describes the solution of the charged particle isotropic component. The charged particle isotropic transport is approximated by assuming that the source term originates from nuclear fragmentation reactions between the target medium and the low-energy semi-isotropic neutrons, which turns out to be a good approximation (Slaba et al., 2010c;d).

In the NAIRAS model, there are 59 coupled transport equations in the HZETRN description of GCR transport through the atmosphere. This set includes transport equations for neutrons and GCR nuclear isotopes from protons through nickel (Z=28, A=58). Figure 20 shows the GCR spectral fluence at various atmospheric depths during the Halloween 2003 storm period. The top left panel shows the spectral fluence at zero vertical cutoff rigidity incident on the neutral atmosphere. The fluences in this panel are the predictions from NAIRAS GCR model (see section 3.1). There is no neutron fluence in this panel since neutrons are secondary particles created by projectile-target nuclear fragmentation reactions, as discussed above. The remaining panels show the GCR fluences (primaries + secondaries) at different depths within the atmosphere.

Since neutrons do not interact with the target medium via the Coulomb force, there is not an ionization threshold, or an atomic excitation threshold, or a nuclear Coulomb potential barrier to overcome. Thus, the neutrons are not brought to rest as the charged particles are. Neutrons continue to cascade down in energy through neutron-nucleus interactions and the low-energy neutron fluence continues to build, as is evident in Figure 20. The low-energy neutron fluence is quite large, even at a small atmospheric depth of 0.1 g/cm^2. The large low-energy neutron fluence at small atmospheric depth is dominated by backscattered neutrons generated at much larger depths, or at much lower altitudes in the atmosphere. At the atmospheric depth of typical cruising altitudes of commercial aircraft, the heavy-ions have largely disappeared, due to a combination of ionization energy loss and nuclear fragmentation reactions into lower energy, lighter particles. Refer to the discussion of Figures 18 and 19.

Solar cosmic rays consist mainly of protons and alpha particles. As a result, only the solution of the six light-particle coupled transport equations defined in HZETRN are required in the description of SEP transport through the atmosphere. Figure 21 shows the light-particle SEP spectral fluences at various atmospheric depths for the same time period during the Halloween 2003 storm as shown in the previous figure. The top left panel shows the spectral fluence at zero vertical cutoff rigidity incident on the neutral atmosphere. These SEP fluence rates were determined using the satellite in-situ ion flux measurements and the spectral fitting algorithm described in section 3.2. The remaining panels show the SEP fluences (primaries + secondaries) at different depths within the atmosphere. Similar to the GCR atmospheric transport properties, the large low-energy neutron fluence at small atmospheric depth for SEP events is due to the large backscattered neutron component, which also originates at larger penetrations depths. Furthermore, only nucleon (protons + neutrons) fluences remain at the typical cruising altitude of commercial aircraft, for the same reasons as previously described.

3.5 Meteorological data

This section describes the characterization of the internal properties of the atmosphere that are relevant to cosmic transport. The atmosphere itself provides shielding from incident charged particles. The shielding of the atmosphere at a given altitude depends on the overhead mass. Sub-daily global atmospheric depth is determined from pressure versus geopotential height and pressure versus temperature data derived from the National Center for Environmental Prediction (NCEP) / National Center for Atmospheric Research (NCAR) Reanalysis 1 project (Kalnay & et al., 1996). The NCAR/NCEP Reanalysis 1 project uses a state-of-the-art analysis/forecast system to perform data assimilation using past data from 1948 to the present. The data products are available 4x daily at 0, 6, 12, and 18 UT. The spatial coverage is 17 pressure levels in the vertical from approximately the surface (1000 hPa) to the

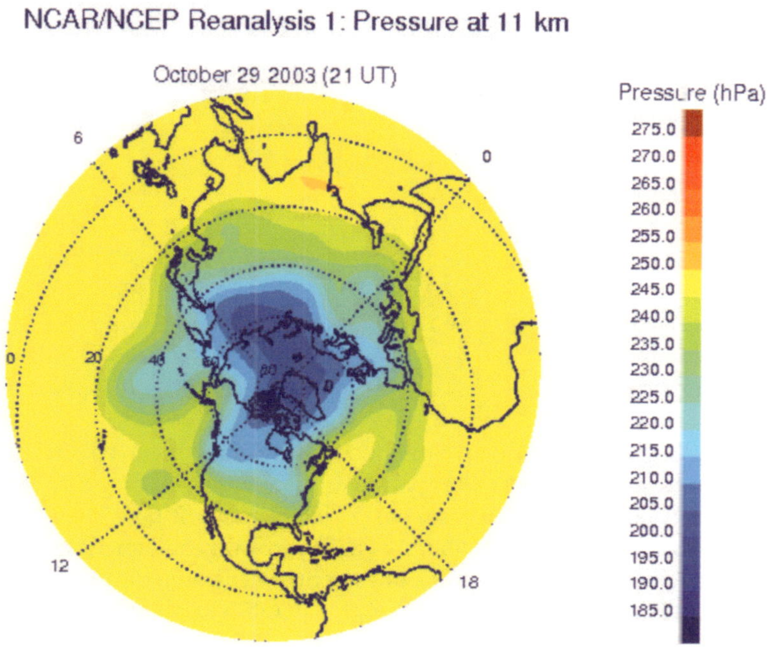

Fig. 22. NCAR/NCEP Reanalysis 1 pressure levels at 11 km corresponding to the date/time of the largest geomagnetically disturbed period of Halloween 2003 SEP event 3 (10/29/2003, 2100 UT) analyzed in section 4. Also shown are the magnetic latitude circles and the meridians at 0, 6, 12, and 18 magnetic local time.

middle stratosphere (10 hPa), while the horizontal grid is 2.5 x 2.5 degrees covering the entire globe.

NCAR/NCEP pressure versus geopotential height data is extended in altitude above 10 hPa using the Naval Research Laboratory Mass Spectrometer and Incoherent Scatter (NRLMSIS) model atmosphere (Picone et al., 2002). NCAR/NCEP and NRLMSIS temperatures are smoothly merged at 10 hPa at each horizontal grid point. NRLMSIS temperatures are produced at 2 km vertical spacing from the altitude of the NCEP/NCAR 10 hPa pressure surface to approximately 100 km. The pressure at these extended altitudes can be determined from the barometric law using the NRLMSIS temperature profile and the known NCAR/NCEP 10 hPa pressure level, which assumes the atmosphere is in hydrostatic equilibrium and obeys the ideal gas law. Finally, the altitudes and temperatures are linearly interpolated in log pressure to a fixed pressure grid from 1000 hPa to 0.001 hPa, with six pressure levels per decade. The result from this step is pressure versus altitude at each horizontal grid point from the surface to approximately 100 km.

Atmospheric depth (g/cm^2) at each altitude level and horizontal grid point is computed by vertically integrating the mass density from a given altitude to the top of the atmosphere. The mass density is determined by the ideal gas law using the pressure and temperature at each altitude level. The result from this step produces a 3-D gridded field of atmospheric depth. Atmospheric depth at any specified aircraft altitude is determined by linear interpolation along the vertical grid axis in log atmospheric depth. Figure 22 shows the atmospheric

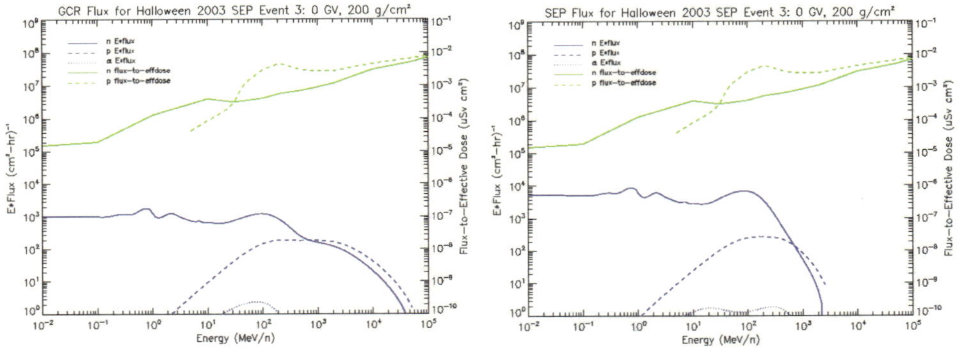

(a) GCR Component (b) SEP Component

Fig. 23. The green lines show the fluence-to-effective dose conversion coefficients. The blue lines show the product of the kinetic energy (MeV/amu) times the event-averaged spectral fluence rates for the Halloween 2003 storm period [29 October (2100 UT) to 1 November (0000 UT)]. The fluence rates were evaluated at zero vertical geomagnetic cutoff rigidity and at an atmospheric depth of 200 g/cm^2. The lines denote neutron quantities; dashed lines denote proton quantities; dotted lines denote alpha quantities. All quantities are shown for: (a) GCR component and (b) SEP component.

pressure over the northern hemisphere at 11 km on October 29, 2003 (2100 UT). This is the atmospheric data used in the exposure rate calculations in section 4.3.

3.6 Radiation dosimetry

The energy deposited in a target medium by the radiation field of particle j is the dose, which is given by

$$D_j(\mathbf{x}) = K \int_{\Omega} \int_0^{\infty} S_j(E)\Phi_j(\mathbf{x}, \mathbf{\Omega}, E) d\mathbf{\Omega} dE. \tag{27}$$

In the above equation, $S_j(E)$ is the target stopping power for particle j (Mev/g/cm^2) and K is a unit conversion factor (1.602x10^{-10}) to convert dose to units of Gray (1 Gy = J/kg). The target stopping power is given by (19)-(20), which is shown in Figure 17 for representative cosmic ray ions incident on Earth's atmosphere. Radiation health risk and the probability of biological damage depend not only on the absorbed dose, but also on the particle type and energy of the radiation causing the dose. This is taken into account by weighting the absorbed dose by a factor related to the quality of the radiation. The weighted absorbed dose has been given the name dose equivalent by the ICRP (ICRP, 1991). The unit of dose equivalent is the Sievert (Sv). Dose equivalent in tissue T from particle j ($H_{j,T}(\mathbf{x})$) is defined in terms of the tissue LET dependent quality factor Q, such that

$$H_{j,T}(\mathbf{x}) = \int_L Q(L) D_j(\mathbf{x}, L) dL, \tag{28}$$

where L is LET, which can be approximated by the stopping power in units of keV/um; $D_j(\mathbf{x}, L)$ is the spectral dose distribution from particle j in terms of LET, and $Q(L)$ is the tissue LET-dependent quality factor.

The relationship between the probability of biological damage and dose equivalent is found to also depend on the organ or tissue irradiated. A further dosimetric quantity, called the effective dose, is defined to include the relative contributions of each organ or tissue to the total biological detriment caused by radiation exposure. The effective dose ($E(\mathbf{x})$) is the sum of weighted dose equivalents in all the organs and tissues in the human body, such that

$$E(\mathbf{x}) = \sum_T \sum_j w_T H_{j,T}(\mathbf{x}). \tag{29}$$

The organ/tissue weighting factors are given in the ICRP 60 report (ICRP, 1991). A computationally efficient approach is to calculate the effective dose rates directly from the particle spectral fluence rates using pre-computed fluence-to-effective dose conversion coefficients. The NAIRAS model uses neutron and proton conversion coefficients tabulated by Ferrari et al. (1997a;b). The effective dose contributions from the other ions are obtained by scaling the proton fluence-to-effective dose conversion coefficients by Z_j^2/A_j, according to stopping power dependence on charge and mass in (19)-(20). All recommended ICRP radiation exposure limits are defined in terms of effective dose.

Figure 23 shows the proton and neutron fluence-to-effective dose conversion coefficients and the event-averaged GCR and SEP spectral fluence rates computed during the Halloween 2003 SEP event 3. The fluence rates were computed at zero vertical geomagnetic cutoff rigidity at an atmospheric depth of 200 g/cm^2, which is the depth corresponding to typical cruising altitudes of 10-12 km for commercial aircraft. Above 20 MeV, protons make a larger contribution to effective dose per unit fluence compared to neutrons. Below 20 MeV, the reverse is true. The fluence rates are shown in Figure 23 as a product of the fluence rates times the energy. This is a convenient representation on a log-log scale since the spectral integration with respect to log-energy, which is performed in order to obtain the effective dose rate, is proportional to energy multiplied by the fluence rate. In this representation, neutrons dominate below about 1 GeV. The peak in energy times the proton fluence rate is slightly larger for the SEP component as compared to the GCR component. At 100 MeV and below, the energy times the neutron fluence rate is nearly an order of magnitude greater for the SEP component compared to the GCR component.

The GCR and SEP normalized spectral and accumulated spectral effective dose rates are presented in Figure 24 for the event-averaged Halloween 2003 SEP event 3. Similar to the previous figure, the effective dose rate quantities were computed at zero vertical geomagnetic cutoff rigidity at an atmospheric depth of 200 g/cm^2. The spectral effective dose rates are normalized with respect to the peak in the spectrum. The peak in both the proton and neutron spectral effective dose rates occur between 100-200 MeV, which is true for both GCR and SEP contributions. Neutrons make the largest contributions to effective dose at energies below the peak in the spectrum. Protons make the largest contribution to effective dose at energies above the peak in the spectrum. The relative spectral contribution of protons and neutrons to effective dose holds for both the GCR and SEP components. Half of the total effective dose rate comes from spectral contributions at energies less than 100 MeV for the GCR component, and at energies less than roughly 20 MeV for the SEP component. Half of the neutron effective dose comes from spectral contributions at energies less than about 30 MeV for the GCR component, which is similar for the SEP component. Recall that the incident GCR and SEP proton fluence rates spectra are very different. For example, compare Figures 7 and 9. As a result, half of the proton effective dose rates comes from spectral contributions at energies less than 100 MeV for the SEP component, and at energies less than about 500 MeV for the GCR component.

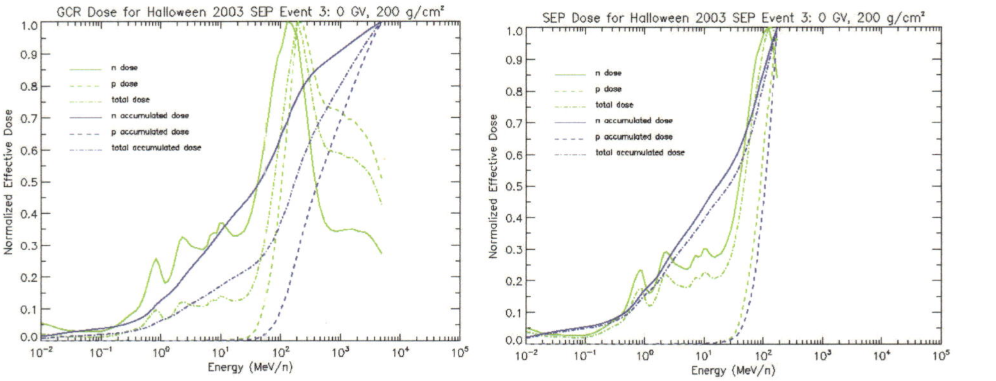

(a) GCR Component (b) SEP Component

Fig. 24. Normalized spectral effective dose rates evaluated during the Halloween 2003 storm period [29 October (2100 UT) to 1 November (0000 UT)] at zero vertical geomagnetic cutoff rigidity and at an atmospheric depth of 200 g/cm^2. The green lines show the spectral effective dose rate normalized to the peak in the spectrum. The blue lines are normalized to the spectrally integrated effective dose rate, and show the accumulated effective dose rate as the spectral effective dose rate is integrated over energy. The solid lines denote neutron quantities; dashed lines denote proton quantities; dash-dot lines denote total quantities.

4. Analysis of atmospheric ionizing radiation exposure

In this section NAIRAS predictions of aircraft radiation exposure are presented and analyzed for three distinct space weather phenomena. The suppression of GCR exposure due to a Forbush decrease is considered briefly in section 4.1. Maximum GCR exposure occurs during solar minimum, as discussed in section 2.1. Results during solar minimum are presented in section 4.2. The Halloween 2003 superstorm is used as a testbed for diagnosing the influence of geomagnetic storm effects on SEP atmospheric radiation exposure. Section 4.3 contains the analysis of the Halloween 2003 SEP exposure and geomagnetic storm effects. When possible, the NAIRAS predictions are compared to other models and/or onboard aircraft radiation measurements.

4.1 Case Study 1: GCR during forbush decrease

The Halloween 2003 superstorm was rich in the variety of simultaneous, and often competing, space weather influences (Gopalswamy et al., 2005). Some of the phenomena that occurred during this storm period that exert important influences on atmospheric ionizing radiation exposure are: SEP events, Forbush decreases, geomagnetic storms, cosmic ray anisotropy, and ground level enhancements (GLE). The topic of this section is aircraft radiation exposure during a Forbush decrease that occurred during the Halloween 2003 storm.

A Forbush decrease is a suppression of the GCR exposure due to the interaction of the solar wind with the incident GCR particles. At latitudes with cutoff rigidities greater than 1.0-1.2 GV, the SEP exposure rates are comparable to or less than the GCR exposure rates. Thus, at these latitudes, the total GCR+SEP dose rate can be less than the quiet-time dose rates prior to the SEP event during a Forbush decrease. Getley et al. (2005a;b) observed the apparent influence of a Forbush decrease on dosimetry measurements taken on Qantas Flight 107 from Los Angeles, California to New York, New York on October 29, 2003. These measurements

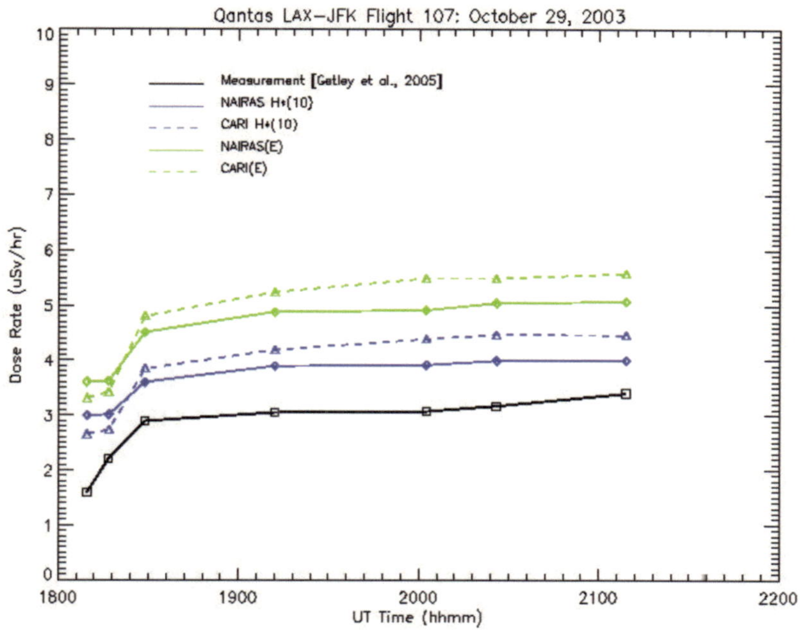

Fig. 25. Comparisons between NAIRAS, FAA/CARI, and TEPC measurements for Qantas Flight 107 from Los Angeles, California to New York, New York on October 29, 2003. The dosimetric quantities compared are effective dose rate (denoted E) and ambient dose equivalent rate (denoted H*(10)).

are well-suited for testing the ability of NAIRAS to model the reduction in the GCR exposure due to Forbush decreases.

Quantitative comparisons were made between NAIRAS GCR exposure rates and aircraft Tissue Equivalent Proportional Counter (TEPC) measurements taken by Getley et al. (2005a;b). The latitudes of the Qantas flight trajectory were too low to observe significant SEP radiation exposure. However, the GCR exposure was suppressed by a Forbush decrease. Figure 25 shows the comparisons between NAIRAS and the TEPC measurements along the Qantas flight trajectory. This figure shows comparisons of effective dose rate and ambient dose equivalent rate, which is a measurement-based proxy for the effective dose rate. Also shown in the figure are the dosimetric quantities computed from the FAA/CARI-6 model. The NAIRAS calculations are closer to the measurement data than the CARI model. The NAIRAS model can account for Forbush decrease effects by virtue of using real-time neutron monitor count rates in specifying the GCR fluence rates incident at the top of the neutral atmosphere, as discussed in section 3.1.

4.2 Case Study 2: GCR during solar minimum

In this section NAIRAS model predictions are compared with TEPC measurements during solar minimum conditions where the GCR exposure reaches its maximum during the solar cycle. TEPC measurements are taken on an equatorial flight and a high-latitude flight. The TEPC data were provided courtesy of Dr. Matthias Meier from The German Aerospace Corporation (DLR) in Cologne, Germany. The equatorial flight was from Dusseldorf,

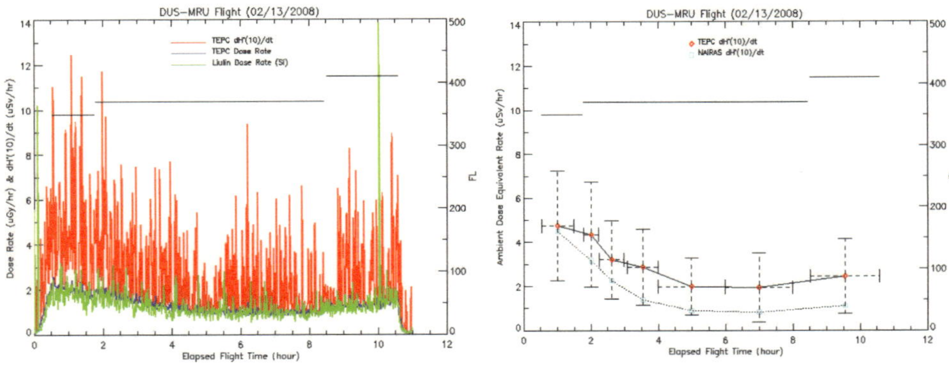

(a) 1-minute TEPC Data (b) TEPC/NAIRAS Comparison

Fig. 26. (a) TEPC and Liulin dosimetric measurements on a flight from Dusseldorf to Mauritius on February 13, 2008. The left ordinate axis indicate dose rates versus elapsed time of flight (abscissa). The TEPC data shown are: (blue line) absorbed dose rate (uGy/hr) and (red line) ambient dose equivalent rate (uSv/hr). The Liulin data are (green line) absorbed dose in silicon (uGy/hr). The right ordinate axis indicates the flight level versus elapsed flight time (abscissa). The flight levels are the horizontal black lines. (b) TEPC/NAIRAS comparisons of ambient dose equivalent rate for the same flight. The left ordinate axis indicates the ambient dose equivalent rate versus elapsed flight time (abscissa). The TEPC measurements are shown as red diamond symbols. The NAIRAS predictions are shown as light blue square symbols. The 1-minute TEPC data shown in the left panel have been averaged over roughly 1-hour periods. The horizontal error bars in the above figure correspond to the averaging interval. The vertical error bars represent one standard deviation in TEPC data.

Germany to the island nation of Mauritius on February 13, 2008, and the TEPC 1-minute data are shown in Figure 26a.

Initial comparison between NAIRAS and TEPC data are focused on the ambient dose equivalent, since this dosimetric quantity is a fairly reasonable proxy for effective dose - the quantity directly related to biological risk. Figure 26b shows the NAIRAS/TEPC ambient dose equivalent rate comparisons. The NAIRAS model predictions are within the statistical uncertainty of the TEPC measurements. Nevertheless, NAIRAS results are biased low with respect to the measurements. The largest differences are at the lowest latitudes near the equator. The equatorial low bias in the NAIRAS results may be due to an underprediction of electromagnetic cascade processes initiated by pion production. It is known that the generation and transport of these processes within the HZETRN component of NAIRAS needs improvement, and these processes are likely to be more important at low latitudes compared to higher latitudes. Improvements in HZETRN with modeling the transport and dosimetry of pion initiated electromagnetic cascade effects are currently underway.

The high-latitude dosimetric measurements were taken on a flight from Fairbanks, Alaska to Frankfurt, Germany on May 23, 2008. The 1-minute TEPC data are shown in Figure 27a. NAIRAS/TEPC ambient dose equivalent rate comparisons are shown in Figure 27b. The NAIRAS predictions are at the boundary of one standard deviation in the statistical uncertainty of the TEPC measurements. For the high-latitude comparison, the NAIRAS results are biased high with respect to the TEPC measurements. More comparisons between

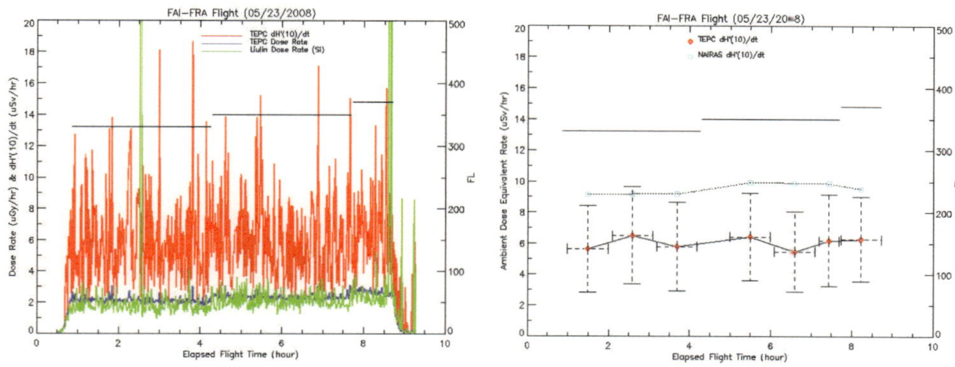

(a) 1-minute TEPC Data (b) TEPC/NAIRAS Comparisons

Fig. 27. (a) TEPC and Liulin dosimetric measurements on a flight from Fairbanks to Frankfurt on May 23, 2008. The left ordinate axis indicate dose rates versus elapsed time of flight (abscissa). The TEPC data shown are: (blue line) absorbed dose rate (uGy/hr) and (red line) ambient dose equivalent rate (uSv/hr). The Liulin data are (green line) absorbed dose in silicon (uGy/hr). The right ordinate axis indicates the flight level versus elapsed flight time (abscissa). The flight levels are the horizontal black lines. (b) NAIRAS/TEPC comparisons of ambient dose equivalent rate for the same flight. The left ordinate axis indicates the ambient dose equivalent rate versus elapsed flight time (abscissa). The TEPC measurements are shown as red diamond symbols. The NAIRAS predictions are shown as light blue square symbols. The 1-minute TEPC data shown in the left panel have been averaged over roughly 1-hour periods. The horizontal error bars in the above figure correspond to the averaging interval. The vertical error bars represent one standard deviation in TEPC data.

NAIRAS predictions and dosimetric measurements over a range of latitudes and solar modulation levels are required to fully characterize the uncertainty in the model predictions. Comprehensive flight campaigns are currently underway for extensive model verification and validation.

4.3 Case Study 3: A halloween 2003 SEP event

In this section NAIRAS model predictions are presented for SEP effective dose rates and accumulated effective dose along representative high-latitude commercial routes during the Halloween 2003 SEP event 3 [10/29 (2100 UT) - 10/31 (2400 UT)]. The incident SEP spectral fluence rates and meteorological data are fixed in time in the NAIRAS calculations presented in this section, which are given by the event-averaged spectral fluence rates and atmospheric depth-altitude data shown in Figures 9 and 22, respectively. On the other hand, the geomagnetic cutoff rigidity is allowed to vary in time along the flight trajectories, according to the magnetospheric magnetic field response to the real-time solar wind and IMF conditions (Kress et al., 2010). This enables the geomagnetic influence on SEP radiation exposure to be isolated. The Halloween 2003 superstorm is an ideal event to study geomagnetic effects since this event contained a major magnetic storm which was one of the largest of solar cycle 23. Geomagnetic effects on atmospheric ionizing radiation have not been sufficiently quantified in the past. A unique outcome of this analysis is that it was found that geomagnetic storm effects have a profound effect on SEP atmospheric radiation exposure, especially for flights

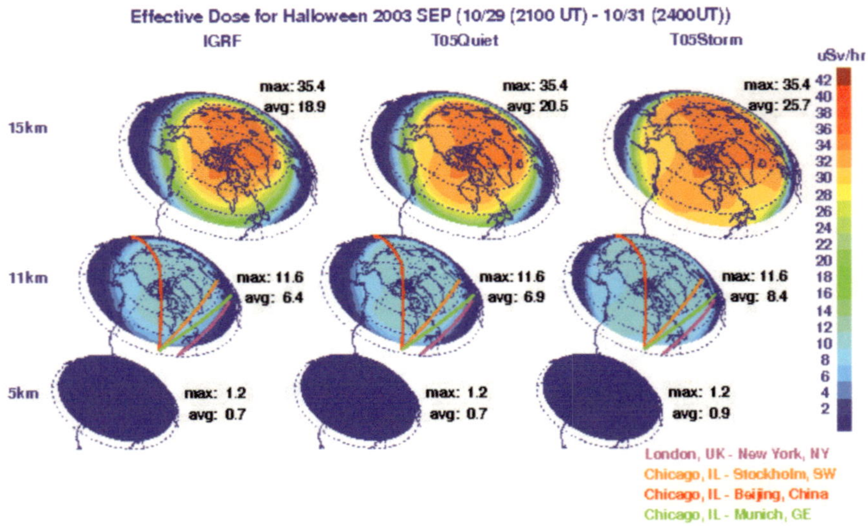

Fig. 28. Effective dose rates computed during Halloween 2003 SEP event 3. The three columns correspond to exposure rates calculated using the geomagnetic cutoff rigidities and magnetic field models shown in Figure 14. The three rows are exposure rates calculated at different altitudes. In each graph, the hemispheric average effective dose rate (uSv/hr) is indicated by the value next to "avg." The maximum exposure rate is indicated by the value next to "max". See text for definition of "avg" and "max."

along the North Atlantic corridor region connecting international flight from the east coast of the US with Europe. The details of this analysis are given in the section below.

4.3.1 Global SEP dose distribution

Global SEP atmospheric ionizing radiation exposure are obtained from a pre-computed database. The effective dose rates are calculated on a fixed 2-D grid in atmospheric depth and cutoff rigidity. The atmospheric depth grid extends from zero to 1300 g/cm^2, and the cutoff rigidity grid extends from zero to 19 GV. Both grids have non-uniform spacing with the highest number of grid points weighted toward low cutoff rigidities and tropospheric atmospheric depths. The real-time cutoff rigidities are computed on the same 2.5 x 2.5 horizontal grid as the NCEP/NCAR meteorological data. The pre-computed effective dose rates are interpolated to the real-time cutoff rigidity and atmospheric depth specified at each horizontal grid point.

Figure 28 shows global snapshots of atmospheric effective dose rates over the northern hemisphere polar region for the Halloween 2003 SEP event 3. The effective dose rates are shown at three altitudes and for three different magnetic field models used in the cutoff rigidity simulations. The left column shows exposure rates using the IGRF field. The middle column shows exposure rates computed for a geomagnetically quiet time prior to the onset of SEP event 3 using the TS05 field (October 28, 2003, 0002 UT). The right column shows the exposure rates using the TS05 field at the peak of the geomagnetic storm (October 29, 2003, 2100 UT) during SEP event 3. A typical cruising altitude for a commercial high-latitude flight is 11 km. Overlaid on the 11 km effective dose rate altitude surface are great circle routes for three representative high-latitude commercial flights: London, England (LHR) to New York,

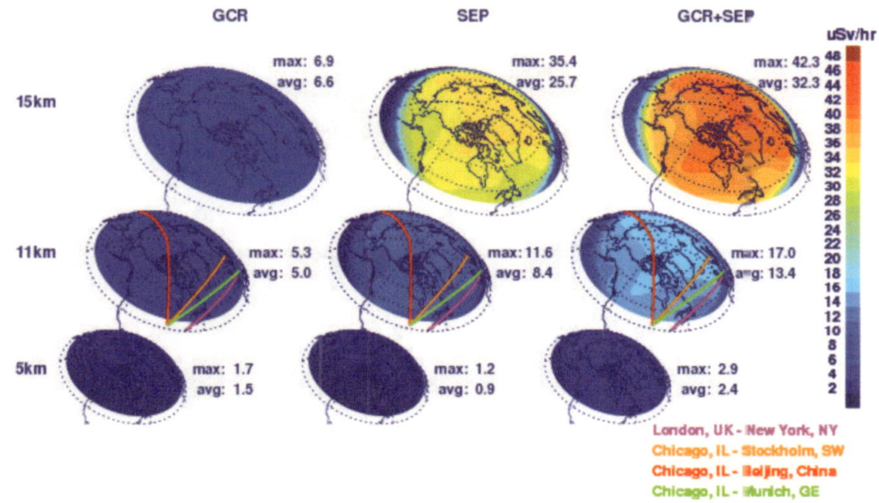

Fig. 29. Effective dose rates computed during Halloween 2003 SEP event 3. The first, middle, and last columns are the GCR, SEP, and total (GCR+SEP) effective dose rates, respectively. The three rows are exposure rates calculated at different altitudes. In each graph, the hemispheric average effective dose rate (uSv/hr) is indicated by the value next to "avg." The maximum exposure rate is indicated by the value next to "max". See text for definition of "avg" and "max."

New York (JFK) (5.75 hour flight time); Chicago, Illinois (ORD) to Stockholm, Sweden (ARN) (8.42 hour flight time), and a combination of two great circle routes from Chicago, Illinois (ORD) to Beijing, China (PEK) (13.5 hour flight time).

There are a number of striking features to be noted from Figure 28. First, the representation of the geomagnetic field has a significant influence on SEP atmospheric ionizing radiation exposure. Comparing the left and middle columns of Figure 28 shows that even during geomagnetically quiet periods, the magnetospheric magnetic field weakens the overall geomagnetic field with a concomitant increase in radiation levels. This is seen as a broadening of the open-closed magnetospheric boundary in the TS05 quiet field compared to the IGRF field. The cutoffs are zero in the region of open geomagnetic field lines. Thus, effective dose rates based on the IGRF field are underestimated even for magnetically quiet times. During strong geomagnetic storms, as shown in the third column of Figure 28, the area of open field lines are broadened further, bringing large exposure rates to much lower latitudes. Effective dose rates predicted using the IGRF model during a large geomagnetic storm can be significantly underestimated. The expansion of the polar region high exposure rates to lower latitudes, due to geomagnetic effects, is quantified by calculating hemispheric average effective dose rates from 40N to the pole. This is denoted by "avg" in Figure 28. At 11 km, there is roughly a 8% increase in the global-average effective dose rate using TS05 quiet-field compared to IGRF. During the geomagnetic storm, there is a $\sim$ 30% increase in the global-average effective dose rate using TS05 storm-field compared to IGRF.

A second important feature to note in Figure 28 is the strong altitude dependence due to atmospheric shielding. The exposure rates are very low at 5 km, independent of geomagnetic field model used. At 15 km, the exposure rates are significantly higher than at 11 km. Figure 28

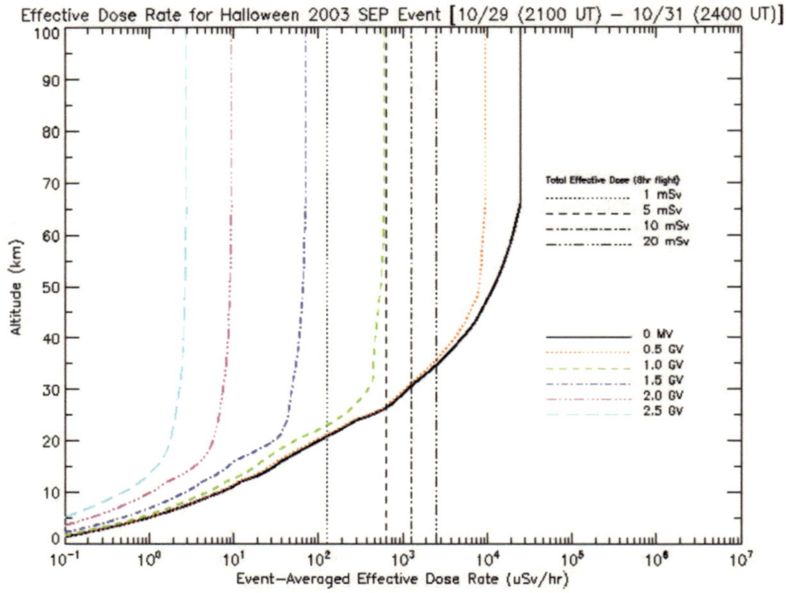

Fig. 30. Event-averaged SEP effective dose rates for Halloween SEP event 3 [(10/29/2003, 2100 UT) - 10/31/2003 (2400 UT)] as a function of altitude for various geomagnetic cutoff rigidities. Different vertical lines indicate constant exposure rates required to reach the corresponding total exposure levels indicated in the legend for a 8-hr flight.

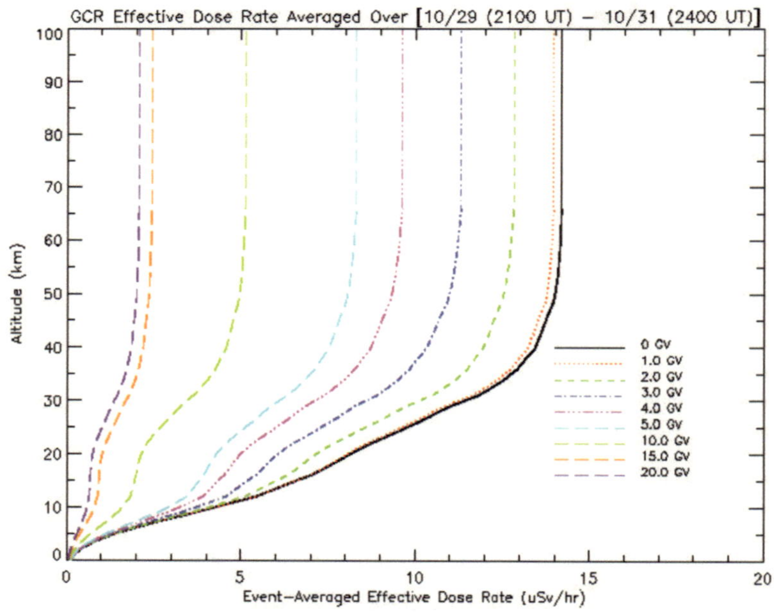

Fig. 31. Event-averaged GCR effective dose rates for Halloween SEP event 3 [(10/29/2003, 2100 UT) - 10/31/2003 (2400 UT)] as a function of altitude for various geomagnetic cutoff rigidities.

shows that the SEP effective dose rates increase (decrease) exponentially with increasing (decreasing) altitude. The SEP exposure rate altitude dependence is a fortunate feature for the aviation community, since radiation exposure can be significantly reduced by descending to lower altitudes. Private business jets will receive more radiation exposure than commercial aircraft if mitigation procedures are not taken, since business jet cruising altitudes are roughly 12-13 km. The altitude dependence of the SEP exposure rates is quantified in Figure 28 by showing the maximum effective dose rate at each altitude, which is the exposure rate at zero cutoff rigidity (i.e., in the polar region of open geomagnetic field lines) The maximum is denoted "max" in Figure 28. The exposure rate increases on average by 160% per km between 5 km and 11 km. Between 11 km and 15 km, the exposure rate increases on average by approximately 75% per km.

Figure 29 shows the event-averaged GCR and SEP effective dose rates separately for the Halloween 2003 SEP event 3. The total (GCR+SEP) effective dose rates are also shown. The SEP exposure rates were computed with the TS05 field at the peak of the geomagnetic storm (29 October 2003, 2100 UT) during SEP event 3, which are identical to the exposure rates from the third column of Figure 28. At 5 km, the GCR and SEP effective dose rates are comparable. The SEP exposure rates in the polar region at 11 km are roughly a factor of two greater than the corresponding GCR exposure rates. At 15 km over the polar region, the SEP exposure rates are greater than the GCR exposure rates by about a factor of five. Due to a substantially larger number of >100 MeV protons during the January 2005 storm event, the peak polar SEP effective dose rates are significantly greater during the January 2005 event compared to the Halloween 2003 storm period (Copeland et al., 2008).

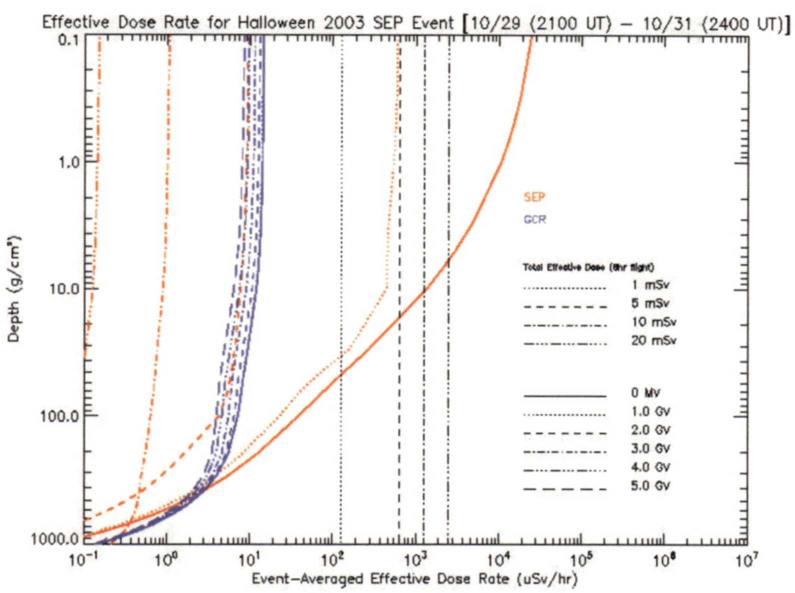

Fig. 32. Event-averaged GCR and SEP effective dose rates for Halloween SEP event 3 [(10/29/2003, 2100 UT) - 10/31/2003 (2400 UT)] as a function of altitude for various geomagnetic cutoff rigidities.

4.3.2 Dose on high-latitude flights

Before calculating radiation exposure along specified flight paths, it's constructive to examine a sample of effective dose rate profiles at different cutoff rigidities from our pre-computed database previously described for SEP event 3. Figure 30 shows SEP effective dose rates as a function of altitude for cutoff rigidities from zero to 2.5 GV. This figure clearly shows the exponential dependence of SEP exposure on both cutoff rigidity and altitude. The vertical lines indicate constant exposure rates necessary to receive a total exposure of 1, 5, 10, and 20 mSv on a 8 hour flight. A typical international, high-latitude flight is 8 hours. The rationale for choosing the total exposure identified with the vertical lines is as follows (Wilson et al., 2003): 20 mSv is the ICRP annual occupational radiation worker limit, 10 mSv is the NCRP annual occupational exposure limit, 5 mSv is the NCRP occasional public exposure limit, and 1 mSv is the ICRP annual public and prenatal exposure limit. The cutoff rigidities at high-latitudes are less than 1 GV. The typical commercial airline cruising altitudes correspond to an atmospheric depth between $\sim$ 200-300 g/cm^2. Consequently, one can see from Figure 30 that it's not possible for passengers on high-latitude commercial flights during the Halloween 2003 SEP events to approach or exceed the ICRP public and/or prenatal radiation exposure limit. However, the recommended ICRP annual public and prenatal exposure limits were exceeded during the January 2005 storm event (Copeland et al., 2008).

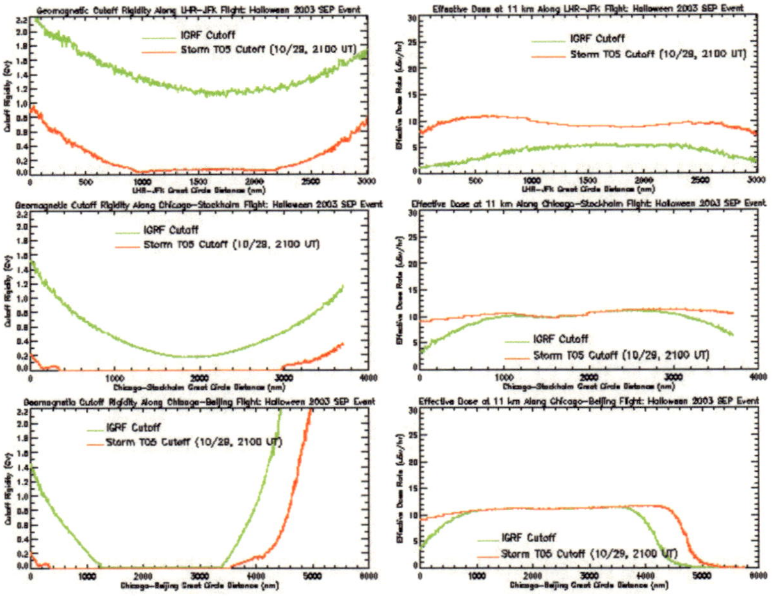

Fig. 33. Geomagnetic cutoff rigidities (left column) and effective dose rates (right column) calculated during Halloween 2003 SEP event 3 along three representative flight paths for a cruising altitude of 11 km. The green line represents cutoff rigidities and exposure rates calculated using the IGRF model. The red lines represent cutoffs and exposure rates computed using the TS05 model during the period of largest geomagnetic activity of event 3. The total flight times are the following: 5.75 hours from JFK-LHR, 8.42 hours for ORD-ARN, and 13.5 hours for ORD-PEK.

For comparison with the previous plot, Figure 31 shows the GCR effective dose rates as function of altitude for cutoff rigidities from zero to 20 GV. In contrast to the SEP exposure rates from Figure 30, the GCR exposure rates do not vary exponentially with respect to altitude or cutoff rigidity. The difference between the GCR and SEP exposure rate dependence on cutoff rigidity and altitude is elucidated further by placing both GCR and SEP effective dose rate profiles computed at the same cutoff rigidities on same figure, which is shown in Figure 32. The geomagnetic field is very effective at shielding SEP radiation. For example, at cutoff rigidities greater than 2 GV, which for North American corresponds to latitudes south of Canada and the upper regions of the United States, the SEP effective dose rates are already below the GCR exposure rates. The significance of geomagnetic storm effects is that the expansion of the polar cap region, or magnetosphere open-closed boundary, can bring the larger SEP exposure rates, normally confined within the polar cap region, to much lower latitude, even mid-latitudes for a large enough geomagnetic storm.

Figure 33 shows the cutoff rigidities and effective dose rates for the three representative high-latitude flights mentioned in the previous subsection, which were calculated along great circle routes. The left column is the cutoff rigidities along the flight paths and the right column is the corresponding effective dose rates along the flight paths. The cutoff rigidities include both latitude and time-dependent variations along the flight paths. The variations of the exposure rates along the flight paths include latitudinal variations in both atmospheric depth and cutoff rigidity. The temporal variations in cutoff rigidity also map into the variations of the exposure rates along the flight path. The top row shows results for the LHR-JFK flight, while the middle and bottom rows show results for the ORD-ARN and ORD-PEK flights, respectively. Each panel in Figure 33 shows cutoff rigidities and corresponding effective dose rates using the IGRF field (green lines) and the TS05 storm-field (red lines) in the cutoff calculations. The largest differences in flight-path cutoff rigidities between IGRF and TS05 storm-field models are for the LHR-JFK flight. The entire LHR-JFK flight path is near the magnetosphere open-closed boundary and is most sensitive to perturbations in cutoff rigidity due to geomagnetic effects. Consequently, the exposure rates along the LHR-JFK flight are most sensitive to geomagnetic effects. The ORD-PEK polar route is the least sensitive to geomagnetic suppression of the cutoff rigidity, since most of the flight path is across the polar cap region with open geomagnetic field lines. The influence of geomagnetic storm effects on the ORD-ARN flight is intermediate between a typical polar route and a flight along the north Atlantic corridor between the US and Europe.

The effective dose rates for the representative high-latitude flights in Figure 33 are within the range of exposure rates measured during other storm periods for similar flight paths. For example, Clucas et al. (2005) reported peak SEP dose rates in the 3.5-4.0 uSv/hr range for a LHR-JFK flight on July 14, 2000, a SEP event without a concomitant geomagnetic storm. From Figure 33, the peak dose rate for the LHR-JFK flight computed using the IGRF field is $\sim$ 4 uSv/hr. Clucas et al. also showed peak measured SEP dose rates on the order of 10-12 uSv/hr for a LHR-JFK flight during an April 2001 SEP event, which was accompanied by a geomagnetic storm. The average LHR-JFK effective dose rate in Figure 33 computed using the TS05 storm-field is 9.4 uSv/hr. Dyer et al. (2005) reported measured peak SEP dose rates on the order of 9.5 uSv/hr for a flight from Prague to New York during the April 2001 SEP event. Thus, our computed effective dose rates are in qualitative agreement with measured dose rates for similar flight paths during other storm periods.

The high sensitivity of SEP atmospheric dose rates to geomagnetic conditions near the open-closed magnetospheric boundary is also responsible for the high sensitivity of SEP dose rates to the exact flight path in the north Atlantic corridor region. For example, Dyer et al. (2007) found that the differences in peak dose rates between great circle and actual flight paths for LHR-JFK Concorde flights were a factor of five during the September 1989 SEP event and a factor of 2.5 during the October 1989 event, which was geomagnetically quiet. Furthermore, the difference in peak SEP dose rates between great circle and actual flight paths for the commercial flight from Prague to New York during the April 2001 event was a factor of two.

The actual flight paths for the Chicago to Munich flights reported by Beck et al. (2005) may have been equatorward of a great circle route, which could explain their low dose rates compared to our calculated LHR-JFK great circle route dose rates during the Halloween 2003 superstorm. During the Halloween 2003 SEP event 3, the Chicago-Munich flight measured a mean SEP dose rate of 3.6 uSv/hr and an accumulated dose of 0.032 mSv for the 8.75 hour flight. These results are comparable to our 5.75 hour JFK-LHR for geomagnetically quiet conditions presented in Figure 33. In other words, the Chicago-Munich flight didn't seem to experience the dose rate enhancement due to the large geomagnetic storm. By comparing the LHR-JFK dose rates in Figure 33 computed using the IGRF field and the TS05 storm-field with the dose rates computed for the ORD-ARN and ORD-PEK flights, it is clear that geomagnetic effects enable flights along the North Atlantic corridor, or near the magnetosphere open-closed boundary, to experience the same dose rates that are confined to the polar region under geomagnetically quiet conditions.

Another possible explanation for the relative difference in the Chicago-Munich measured dose rates reported by Beck et al. (2005) and our computed dose rates for the LHR-JFK flight is our use of the event-averaged incident SEP spectral fluence rate, which we employed for the purpose of isolating geomagnetic effects in our case study. It is clear from the shaded regions in Figure 9 that our event-averaged SEP spectra fluence rate is weighted more toward the peak ion flux measurements observed during SEP event 3. The measurements in Figure 8 show that the SEP ion flux rapidly decreased in time from the peak values present at the beginning of event 3. Thus, our use of an event-averaged incident SEP spectral fluence rate will tend to overestimate the accumulated effective dose over the flight-paths in Figure 33. However, employing this constraint on the incident SEP spectral fluence rate is necessary to unambiguously isolate the geomagnetic effects. Despite these caveats, the results discussed in this paper are within the current factor of two uncertainty in SEP atmospheric dose rates (Clucas et al., 2005).

5. Conclusions

The NAIRAS (Nowcast of Atmospheric Ionizing Radiation for Aviation Safety) prototype operational model is currently streaming live from the project public website (google NAIRAS). NAIRAS predicts biologically hazardous radiation exposure globally from the surface to 100 km in real-time. The NAIRAS model addresses an important national need with broad societal, public health, and economic benefits. Commercial aircrew are classified by European and international agencies as radiation workers, yet they are the only occupational group exposed to unquantified and undocumented levels of radiation. Furthermore, the current guidelines for maximum public and prenatal exposure can be easily exceeded during a single solar storm event for commercial passengers on intercontinental or polar routes, or by frequent use of these high-latitude routes even during background conditions. The NAIRAS model will provide a new decision support system that currently does not exist, but is essential for providing the commercial aviation industry with data products that will enable airlines to achieve the right balance between minimizing flight cost while at the same time minimizing radiation risk.

NAIRAS is a physics-based model that maximizes the use of real-time input data. GCR are transported from outside the heliosphere to 1 AU using real-time measurements of ground-based neutron monitor count rates. The SEP particle spectra are determined in-situ using a combination of NOAA/GOES and NASA/ACE ion flux measurements. Both sources of cosmic rays, galactic and solar, are transported through Earth's magnetosphere using a semi-physics-based geomagnetic shielding model. The geomagnetic shielding model utilizes real-time NASA/ACE solar wind and IMF measurements. The cosmic rays are transported from the magnetosphere through the neutral atmosphere using the NASA LaRC's HZETRN deterministic transport code. The real-time, global atmospheric mass density distribution is obtained from the NOAA Global Forecasting System. Global and flight path radiation exposure visualization and decision data products have been developed, which are available at the NAIRAS website.

Future research will focus on new science questions that emerged in the the development NAIRAS prototype operational model (Mertens et al., 2010c). The science questions identified by Mertens et al. (2010c) must be addressed in order to obtain a more reliable and robust operational model of atmospheric radiation exposure. Addressing these science questions require improvements in both space weather modeling and observations.

The Automated Radiation Measurements for Aviation Safety (ARMAS) is a new initiative to address the deficiencies in observations needed to improve the reliability and robustness of operational aircraft radiation exposure assessment. The ultimate goal of the ARMAS initiative is to integrate onboard radiation instruments into a global fleet of aircraft so that the radiation measurements can be downlinked in real-time and assimilated into the NAIRAS predictions of radiation exposure. A subsidiary goal of ARMAS is to provide a testbed to evaluate the accuracy and reliability of new generations of smaller, cheaper hardware technologies in measuring the atmospheric ionizing radiation field. The ARMAS initiative enables the NAIRAS model to adopt the successful meteorological paradigm for reliable and robust weather forecast, which is physics-based models combined with real-time data assimilation of meteorological fields. The NAIRAS/ARMAS approach is a space weather version of terrestrial weather forecasts. These efforts will occupy the NAIRAS team for the next decade or more.

Other research topics unrelated to biological risk from cosmic rays will also be addressed. Effort will also be directed toward predicting the risk that cosmic rays may pose to the operation of microelectronics instrumentation onboard aircraft, and the potential influence cosmic rays may have on the chemistry and climate of planetary atmospheres.

6. References

AMS (2007). Integrating space weather observations & forecasts into aviation operations, *Technical report*, American Meteorlogical Society Policy Program & SolarMetrics.

Anderson J. L., Waters, M. A, Hein, M. J., Schubauer-Berigan, M. K. & Pinkerton, L. E. (2011). Assessment of occupational cosmic radiation exposure of flight attendants using questionnaire data, *Aviat Space Environ Med*, 82: 1049–54.

Aspholm, R., Lindbohm, M. L., Paakkulainen, H., Taskinen, H., Nurminen, T. & Tiitinen, A. (1999). Spontaneous abortions among finnish flight attendants, *J. Occupational & Environmental Medicine* 41(6): 486–491.

Badavi, F. F., Nealy, J. E., de Angelis, G., Wilson, J. W., Clowdsley, M. S., Luetke, N. J., Cuncinotta, F. A., Weyland, M. D. & Semones, E. J. (2005). Radiation environment and shielding model validation for cev design, *Space 2005*, number AIAA 2005-6651, Am. Inst. of Aeronaut. and Astronaut., Long Beach, California.

Badavi, F. F., Tramaglina, J. K., Nealy, J. E., & Wilson, J. W. (2007a). Low earth orbit radiation environments and and shield model validation for iss, *Space 2007*, number AIAA 2007-6046, Am. Inst. of Aeronaut. and Astronaut., Long Beach, California.

Badhwar, G. D. & O'Neill, P. M. (1991). An improved model of galactic cosmic radiation for space exploration missions, *22nd International Cosmic Ray Conference*, number OG-5.2-13, pp. 643–646.

Badhwar, G. D. & O'Neill, P. M. (1992). An improved model of galactic cosmic radiation for space exploration missions, *Nuclear Tracks Radiat. Meas.*, 20: 403–410.

Badhwar, G. D. & O'Neill, P. M. (1993). Time lag of twenty-two year solar modulation, *23nd International Cosmic Ray Conference*, Vol. 3, pp. 535–539.

Badhwar, G. D. & O'Neill, P. M. (1994). Long term modulation of galactic cosmic radiation and its model for space exploration, *Adv. Space Res.* 14: 749–757.

Badhwar, G. D. & ONeill, P. M. (1996). Galactic cosmic radiation model and its applications, *Adv. Space Res.* 17: 7–17.

Baker, D. N., Mason, G. M., Figueroa, O., Colon, G., Watzin, J. G. & Aleman, R. M. (1993). An overview of the solar, anomalous, and magnetospheric particle explorer (sampex) mission, *IEEE Trans. Geosci. Remote Sen.* 31(3): 531–541.

Band, P. R., Spinelli, J. J., Ng, V. T. Y., Moody, J. & Gallagher, R. P. (1990). Mortality and cancer incidence in a cohort of commercial airline pilots, *Aviat. Space Environ. Med.* 61: 299–302.

Barish, R. J. (1990). Health physics concerns in commercial aviation, *Health Phys.* 59: 199–204.

Barish, R. J. (2004). In-flight radiation exposure during pregnancy, *Obstet. Gynecol.* 103: 1326–1330.

Beck, P., Latocha, M., Rollet, S. & Stehno, G. (2005). Tepc reference measurements at aircraft altitudes during a solar storm, *Adv. Space Res.* 36: 1627–1633.

BEIR V (1990). *National Researach Council. Health effects of exposure to low levels of ionizing radiation*, Washington, DC, National Academy Press.

Bramlitt, E. T. (1985). Commercial aviation crewmember radiation doses, *Health Phys.* 49: 945–948.

Brandt, S. (1999). *Data Analysis, Statistical and Computational Methods for Scientists and Engineers*, Springer-Verlag, New York.

Buja, A., Lange, J. H, Perissinotto, E., Rausa, G., Grigoletto, F., Canova, C. & Mastrangelo, G. (2005) . Cancer incidence among male military and civil pilots and flight attendants: an analysis on published data, *Toxicol Ind Health* 21: 273–82.

Buja, A., Mastrangelo, G., Perissinotto, E., Grigoletto, F., Frigo, A. C., Rausa, G., Marin, V., Canova. C. & Dominici, F. (2006). Cancer incidence among female flight attendants: a meta-analysis of the published data, *J Womens Health* 15: 98–105.

Chen, J., Lewis, B. J., Bennett, L. G. I., Green, A. R. & Tracy, B. L. (2005). Estimated neutron dose to embryo and foetus during commercial flight, *Radiat. Prot. Dosim.* 114(4): 475–480.

Clem, J., Clements, D. P., Esposito, J., Evenson, P., Huber, D., LHeureux, J., Meyer, P. & Constantin, C. (1996). Solar modulation of cosmic electrons, *Astrophys. J.* 464: 507.

Clucas, S. N., Dyer, C. S. & Lei, F. (2005). The radiation in the atmosphere during major solar particle events, *Adv. Space Res.* 36: 1657–1664.

Copeland, K., Sauer, H. H., Duke, F. E. & Friedberg, W. (2008). Cosmic radiation exposure on aircraft occupants on simulated high-latitude flights during solar proton events from 1 january 1986 through 1 january 2008, *Adv. Space Res.* 42: 1008–1029.

DOC (2004). Intense space weather storms october 19 - november 07, 2003, *Service assessment*, U.S. Department of Commerce, National Oceanic and Atmospheric Administration, National Weather Service, Silver Spring, Maryland.

Dyer, C., Hands, A., Lei, F., Truscott, P., Ryden, K. A., Morris, P., Getley, I., Bennett, L., Bennett, B. & Lewis, B. (2009). Advances in measuring and modeling the atmospheric radiation environment, *IEEE Trans. Nucl. Sci.* 56(8).

Dyer, C. & Lei, F. (2001). Monte carlo calculations of the influence on aircraft radiation environments of structures and solar particle events, *IEEE Trans. Nucl. Sci.* 48(6): 1987–1995.

Dyer, C., Lei, F., Hands, A., Clucas, S. & Jones, B. (2005). Measurements of the atmospheric radiation environment from cream and comparisons with models for quiet time and solar particle events, *IEEE Trans. Nucl. Sci.* 52(6): 2326–2331.

Dyer, C., Lei, F., Hands, A. & Truscott, P. (2007). Solar particle events in the qinetiq atmospheric radiation model, *IEEE Trans. Nucl. Sci.* 54(4): 1071–1075.

Ellison, D. C. & Ramaty, R. (1985). Shock acceleration of electrons and ions in solar flares, *Astrophys. J.* 298: 400–408.

Engelmann, J. J., Ferrando, P., Soutoul, A., Goret, P., Juliusson, E., Koch-Miramond, L., Lund, N., Masse, P., Peters, B., Petrou, N., & Rasmussen, I. L. (1990). Charge composition and energy spectra of cosmic-ray nuclei for elements from be to ni. results from heao-3-c2, *Astron. Astrophys.*, 233: 96–111.

European Radiation Dosimetry Group (EURADOS) (1996). Exposure of air crew to cosmic radiation. A report of EURADOS Working Group 11, *EURADOS Report 1996.01*. In: McAulay, I. R, Bartlett, D. T, Dietze, G. & et al. editors. European Commission Report Radiation Protection 85. Luxembourg: Office for Official Publications of the European Communities.

Fanton, J. W. & Golden, J. G. (1991). Radiation-induced endometriosis in Macaca mulatta, *Radiation Res*, 126: 141–146.

Fedder, J. G. L. J. A. & Mobarry, C. M. (2004). The lyon-fedder-mobarry (lfm) global mhd magnetospheric simulation code, *J. of Atmos. and Solar-Terrestrial Phys.* 66(15-16): 1333–1350.

Ferrari, A., Pelliccioni, M. & Pillon, M. (1997a). Fluence to effective dose conversion coefficients for neutrons up to 10 tev, *Radiat. Prot. Dos.* 71(3): 165–173.

Ferrari, A., Pelliccioni, M. & Pillon, M. (1997b). Fluence to effective dose and effective dose equivalent conversion coefficients for protons from 5 mev to 10 tev, *Radiat. Prot. Dos.* 71(2): 85–91.

Foelsche, T. (1961). Radiation exposure in supersonic transports, *Technical Report TN D-1383*, NASA.

Foelsche, T. & Graul, E. H. (1962). Radiation exposure in supersonic transports, *Atompraxis* 8: 365–380.

Foelsche, T., Mendell, R. B., Wilson, J. W. & Adams, R. R. (1974). Measured and calculated neutron spectra and dose equivalent rates at high altitudes Relevence to sst operations and space research, *Technical Report TN D-7715*, NASA.

Friedberg, W., Faulkner, D. N., Snyder, L., Jr., E. B. D. & OBrien, K. (1989). Galactic cosmic radiation exposure and associated health risk for air carrier crewmembers, *Aviat. Space Environ. Med.* 60: 1104–1108.

Gaisser, T. (1990). *Cosmic Rays and Particle Physics*, Cambridge University Press.

Getley, I. L., Duldig, M. L., Smart, D. F. & Shea, M. A. (2005a). Radiation dose along north america transcontinental flight paths during quiescent and disturbed geomagnetic conditions, *Space Weather* 3(S01004): doi:10.1029/2004SW000110.

Getley, I. L., Duldig, M. L., Smart, D. F. & Shea, M. A. (2005b). The applicability of model based aircraft radiation dose estimates, *Adv. Space Res.* 36: 1638–1644.

Gold, R. E. & et al. (1998). Electron, proton, and alpha monitor on the advanced composition and explorer satellite, *Space Sci. Rev.* 86: 541–562.

Gopalswamy, N., Yashiro, S., Liu, Y., Michalek, G., Vourlidas, A., Kaiser, M. L. & Howard, R. A. (2005). Coronal mass ejections and other extreme characteristics of the 2003 october - november solar eruptions, *J. Geophys. Res.* 110(A09S15): doi:10.1029/2004JA010958.

Grajewski, B., Waters, M. A., Yong, L. C., Tseng, C.-Y., Zivkovich, Z. & Cassinelli II, R. T. (2011). Airline pilot cosmic radiation and circadian disruption exposure assessment from logbooks and company records, *Ann. Occup. Hyg.* 55(5): 465–475.

Haggerty, D. K., E. C. R. G. C. H. & Gold, R. E. (2006). Quantitative comparison of ace/epam data from different detector heads: Implications for noaa rtse users, *Adv. Space Res.* 38: 995–1000.

Hammer, G. P, Blettner, M & Zeeb, H. (2009). Epidemiological studies of cancer in aircrew, *Radiat Prot Dosimetry* 136: 232–9.

Heinrich, W., Roesler, S. & Schraube, H. (1999). Physics of cosmic radiation fields, *Radiat. Prot. Dosim.* 86: 253–258.

ICRP (2008). *ICRP Publication 103: 2007 Recommendations of the International Commission on Radiological Protection*, ISBN 0-7020-3048-1, Elsevier.

ICRP (1991). *ICRP Publication 60: 1990 Recommendations of the International Commission on Radiological Protection*, Vol. 21(1-3), Pergamon Press.

ICRU (1986). *ICRU Report 40: The quality factor in radiation protection*, International Commission on Radiation Units and Measurements.

Jiang, T. N., Lord, B. I. & Hendry, J. H. (1994). Alpha particles are extremely damaging to developing hemopoiesis compared to gamma radiation, *Radiat. Res.* 137: 380–384.

Kahler, S. W. (2001). Origin and properties of solar energetic particles in space, *in* P. Song, H. J. Singer & G. L. Siscoe (eds), *Space Weather*, American Geophysical Union, Washington, DC.

Kalnay, E. & et al. (1996). The ncar/ncep 40-year reanalysis project, *Bull. Amer. Meteor. Soc.* 77: 437–470.

Kress, B. T., Hudson, M. K., Perry, K. L. & Slocum, P. L. (2004). Dynamic modeling of geomagnetic cutoff for the 23-24 november 2001 solar energetic particle event, *Geophys. Res. Lett.* 31(L04808): doi:10.1029/2003GL018599.

Kress, B. T., Mertens, C. J. & Wiltberger, M. (2010). Solar energetic particle cutoff variations during the 28-31 october 2003 geomagnetic storm, *Space Weather* 8(S05001): doi:10.1029/2009SW000488.

Lambiotte, J. J., Wilson, J. W. & Filipas, T. A. (1971). Proper-3c: A nucleon-pion transport code, *Technical Report TM X-2158*, NASA.

Langlais, B. & Mandea, M. (2000). An igrf candidate geomagnetic field model for epoch 2000 and a secular variation model for 2000-2005, *Earth Planets Space* 52: 1137–1148.

Lauria, L., Ballard, T. J., Caldora, M., Mazzanti, C. & Verdecchia, A. (2006). Reproductive disorders and pregnancy outcomes among female flight attedants, *Aviation, Space, and Envirnomental Medicine* 77(7): 533–559.

Lewis, B. J., Bennett, G. I., Green, A. R., McCall, M. J., Ellaschuk, B., Butler, A. & Pierre, M. (2002). Galactic and solar radiation exposure to aircrew during a solar cycle, *Radiat. Prot. Dosim.* 102(3): 207–227.

Lindborg, L., Bartlett, D. T., Beck, P., McAulay, I. R., Schnuer, K., Schraube, H. & (Eds.), F. S. (2004). Cosmic radiation exposure of aircraft crew: compilation of measured and calculated data. a report of eurados working group 5, European Radiation

Dosimetry Group, Lexembourg: Office for the Official Publications of the European Communities, European Communities.

Lopate, C. (2004). Private Communication.

Meier, M., Hubiak, M., Matthiä, D., Wirtz, M. & Reitz, G. (2009). Dosimetry at aviation altitudes, *Radiat. Prot. Dos.* 136(4): 251–255.

Menn, W. & et al. (2000). The absolute flux of protons and helium at the top of the atmosphere using imax, *Astrophys. J.* 533: 281–297.

Mertens, C. J., Kress, B. T., Wiltberger, M., Blattnig, S. R., Slaba, T. S., Solomon, S. C. & Engel, M. (2010a). Geomagnetic influence on aircraft radiation exposure during a solar energetic particle event in october 2003, *Space Weather* 8(S03006): doi:10.1029/2009SW000487.

Mertens, C. J., Moyers, M. F., Walker, S. A. & Tweed, J. (2010b). Proton lateral broadening distribution comparisons between grntrn, mcnpx, and laboratory beam measurements, *Adv. Space Res.* 45: 884–891.

Mertens, C. J., Tobiska, W. K., Bouwer, D., Kress, B. T., Solomon, S. C., Kunches, J., Grajewski, B., Gersey, B. & Atwell, W. (2010c). Nowcast of atmospheric ionizing radiation for aviation safety. Submitted as a white paper to The National Academies Decadal Strategy for Solar and Space Physics (Heliophysics) RFI.

Mertens, C. J., Tobiska, W. K., Bouwer, D., Kress, B. T., Wiltberger, M. J., Solomon, S. C. & Murray, J. J. (2009). Development of nowcast of atmospheric ionizing radiation for aviation safety (nairas) model, *1st AIAA Atmosheric and Space Environments Conference*, number AIAA 2009-3633, Am. Inst. of Aeronaut. and Astronaut., San Antonio, Texas.

Mertens, C. J., Wilson, J. W., Blattnig, S. R., Kress, B. T., Norbury, J. W., Wiltberger, M. J., Solomon, S. C., Tobiska, W. K. & Murray, J. J. (2008). Influence of space weather on aircraft ionizing radiation exposure, *46th Aerospace Sciences Meeting and Exhibit*, number AIAA 2008-0463, Am. Inst. of Aeronaut. and Astronaut., Reno, Nevada.

Mertens, C. J., Wilson, J. W., Blattnig, S. R., Solomon, S. C., Wiltberger, M. J , Kunches, J., Kress, B. T. & Murray, J. J. (2007a). Space weather nowcasting of atmospheric ionizing radiation for aviation safety, *45th Aerospace Sciences Meeting and Exhibit*, number AIAA 2007-1104, Am. Inst. of Aeronaut. and Astronaut., Reno, Nevada.

Mertens, C. J., Wilson, J. W., Walker, S. A. & Tweed, J. (2007b). Coupling of multiple coulomb scattering with energy loss and straggling in hzetrn, *Adv. Space Res.* 40: 1357–1367.

Mewaldt, R. A., Cohen, C. M. S., Labrador, A. W., Leske, R. A., Mason, G. M., Desai, M. I., Looper, M. D., Mazur, J. E., Selesnick, R. S., & Haggerty, D. K. (2005). Proton, helium, and electron spectra during the large solar particle events of october-november 2003, *J. Geophys. Res.* 110(A09S10): doi:10.1029/2005JA011038.

NAS/NRC (1980). *National Academy of Sciences/National Research Council: Health effects of exposures to low levels of ionizing radiation*, National Academy Press, Washington, DC. Committee on the Biological Effects of Ionizing Radiation, BEIR V.

NCRP (1993). *National Council on Radiation Protection and Measurements: Limitations of exposure to ionizing radiation*, Vol. 116, National Council on Radiation Protection and Measurements.

NCRP (2009). *National Council on Radiation Protection and Measurements: Ionizing Radiation Exposure of the Population of the United States*, NCRP Report No. 160, National Council on Radiation Protection and Measurements.

Nealy, J. E., Cucinotta, F. A., Wilson, J. W., Badavi, F. F., Dachev, T. P., Tomov, B. T., Walker, S. A., Angelis, G. D., Blattnig, S. R. & Atwell, W. (2007). Pre-engineering spaceflight validation of environmental models and the 2005 hzetrn simulations code, *Adv. Space Res.* 4: 1593–1610.

Neher, H. V. (1961). Cosmic-ray knee in 1958, *J. Geophys. Res.* 66: 4007–4012.
Neher, H. V. (1967). Cosmic-ray particles that changed from 1954 to 1958 to 1965, *J. Geophys. Res.* 72: 1527–1539.
Neher, H. V. (1971). Cosmic rays at high latitudes and altitudes covering four solar maxima, *J. Geophys. Res.* 76(7): 1637–1651.
Neher, H. V. & Anderson, H. R. (1962). Cosmic rays at balloon altitudes and the solar cycle, *J. Geophys. Res.* 67: 1309–1315.
NOAA (2009). National oceanic and atmospheric administration, space weather prediction center. Available from http://www.swpc.noaa.gov/ftpdir/indices/SPE.txt.
O'Brien, K., Friedberg, W., Smart, D. F. & Sauer, H. H. (1998). The atmospheric cosmic- and solar energetic particle radiation environment at aircraft altitudes, *Adv. Space Res.* 21: 1739–1748.
O'Brien, K., Smart, D. F., Shea, M. A., Felsberger, E., Schrewe, U., Friedberg, W. & Copeland, K. (2003). World-wide radiation doseage calculations for air crew memebers, *Adv. Space Res.* 31(4): 835–840.
Ogilvy-Stuart, A. L. & Shalet, S. M. (1991). Effect of radiation on the human reproductive system, *Environ Health Perspect Suppl*, 101(Suppl 2): 109–116.
O'Neill, P. M. (2006). Badhwar-oneill galactic cosmic ray model update based on advanced composition explorer (ace) energy spectra from 1997 to present, *Adv. Space Res.* 37: 1727–1733.
Onsager, T. G. & et al. (1996). Operational uses of the goes energetic particle detectors, in goes-8 and beyond, *in* E. R. Washwell (ed.), *SPIE Int. Soc. Opt. Eng.*, Vol. 2812, pp. 281–290.
Parker, E. N. (1965). The passage of energetic charged particles through interplanetary space, *Planet. Space Sci.* 13: 9–49.
Picone, J. M., Hedin, A. E., Drob, D. P. & Aikin, A. C. (2002). Nrlmsis-00 empirical model of the atmosphere: Statistical comparisons and scientific issues, *J. Geophys. Res.* 107(A12): 1468. doi:10/1029/2002JA009430.
Reitz, G., Schnuer, K. & Shaw, K. (1993). Editorial - workshop on radiation exposure of civil aircrew, *Radiat. Prot. Dosim.* 48: 3.
Schraube, H., Mares, V., Roesler, S. & Heinrich, W. (1999). Experimental verification and calculation of aviation route doses, *Radiat. Prot. Dosim.* 86(4): 309–315.
Slaba, T. C., Blattnig, S. R. & Aghara, S. K. (2010a). Coupled neutron transport for hzetrn, *Radiat. Meas.* 45: 173–182.
Slaba, T. C., Blattnig, S. R. & Badavi, F. F. (2010c). Faster and more accurate transport procedures for hzetrn, *J. Comput. Phys.* 229: 9397–9417.
Slaba, T. C., Blattnig, S. R. & Badavi, F. F. (2010d). Faster and more accurate transport procedures for hzetrn, *Technical Report TP-2010-216213*, NASA.
Slaba, T. C., Blattnig, S. R., Clowdsley, M. S., Walker, S. A. & Badavi, F. F. (2010b). An improved neutron transport algorithm for hzetrn, *Adv. Space Res.* 46: 800–810.
Slaba, T. C., Qualls, G. D., Clowdsley, M. S., Blattnig, S. R., Simonsen, L. C., Walker, S. W., & Singleterry, R. C. (2009). Analysis of mass averaged tissue doses in max, fax, and cam, and caf, *Technical Report TP-2009-215562*, NASA.
Smart, D. F. & Shea, M. A. (1994). Geomagnetic cutoffs: A review for space dosimetry calculations, *Adv. Space Res.* 14(10): 10,787–10,796.
Smart, D. F. & Shea, M. A. (2005). A review of geomagnetic cutoff rigidities for earth-orbiting spacecraft, *Adv. Space Res.* 36: 2012–2020.
Störmer, C. (1965). *The Polar Aurora*, Oxford at the Clarendon Press.

Tai, H., Bichsel, H., Wilson, J. W., Shinn, J. L., Cucinotta, F. A. & Badavi, F. F. (1997). Comparison of stopping power and range databases for radiation transport study, *Technical Report TP-3644*, NASA.

Toffoletto, F. R., Sazykin, S., Spiro, R. W., Wolf, R. A. & Lyon, J. G. (2004). Rcm meets lfm: initial results of one-way coupling, *J. of Atmos. and Solar-Terrestrial Phys.* 66(15-16): 1361–1370.

Townsend, L. W., Stephens, D. L., Hoff, J. L., Zapp, E. N., Moussa, H. M., Miller, T. M., Campbell, C. E. & Nichols, T. F. (2006). The carrington event: Possible doses to crews in space from a comparable event, *Adv. Space Res.* 38: 226–231.

Townsend, L. W., Zapp, E. N., Jr., D. L. S. & Hoff, J. L. (2003). Carrington flare of 1859 as a prototypical worst-case solar energetic particle event, *IEEE Trans. Nucl. Sci.* 50(6): 2307–2309.

Tsyganenko, N. A. (1989). Determination of magnetic current system parameters and development of experimental geomagneitc field models based on data from imp and heos satellite, *Planet Space Sci.* 37: 5–20.

Tsyganenko, N. A. (2002). A model of the near magnetosphere with dawn-dusk asymmetry: 1. mathematical structure, *J. Geophys. Res.* 107(A8): 1179. doi:10.1029/2001JA000219.

Tsyganenko, N. A. & Sitnov, N. I. (2005). Modeling the dynamics of the inner magnetosphere during strong geomagnetic storms, *J. Geophys Res.* 110: A03208. doi:10.1029/2004JA010798.

Tylka, A. J., Cohen, C. M. S., Dietrich, W. F., Lee, M. A., Maclennan, C. G., Mewaldt, R. A., Ng, C. K. & Reames, D. V. (2005). Shock geometry, seed populations, and the origin of variable elemental composition at high energies in large gradual solar particle events, *ApJ.* 625: 474–495.

Tylka, A. J. & Lee, M. A. (2006). Spectral and compositional characteristics of gradual and impuslive solar energetic particle events, in solar eruptions and energetic particles, *in* N. Gopalswamy, R. Mewaldt & J. Torsi (eds), *Solar Eruptions and Energetic Particles*, Vol. Geophysical Monograph 165, American Geophysical Union, Washington, DC.

UNSCEAR (1988). *UNSCEAR 1988 Report to the General Assembly: Sources, effects, and risks of ionizing radiation*, number E.88.IX.7, United Nations Scientific Committee on the Effects of Atomic Radiation, United Nations, New York.

Upton, A. C., Chase, H. B., Hekhuis, G. L., Mole, R. H., Newcombe, H. B., Robertson, J. S., Schaefer, H. J., Synder, W. S., Sondhaus, C. & Wallace, R. (1966). Radiobiological aspects of the supersonic transport, *Health Phys.* 12: 209–226.

VanAllen, J. A. (1968). *Physics of the Magnetosphere*, Vol. 10, Springer-Verlag, New York, New York, chapter Particle Description of the Magnetosphere.

Wallance, R. G. & Sondhaus, C. A. (1978). Cosmic ray exposure in subsonic air transport, *Aviation Space, and Environ. Med.* 74: 6494–6496.

Wang, W., Wiltberger, M., Burns, A. G., Solomon, S. C., Killeen, T. L., Maruyama, N. & Lyon, J. G. (2004). Initial results from the coupled magnetosphere-ionosphere-thermosphere model: thermosphere-ionosphere responses, *J. of Atmos. and Solar-Terrestrial Phys.* 66: 1425–1441.

Waters, M., Grajewski, B., Pinkerton, L. E., Hein, M. J. & Zivkovich, Z. (2009). Development of historical exposure estimates of cosmic radiation and circadian rhythm disruption for cohort studies of Pan- Am flight attendants, *Am J Ind Med* 52: 751–61.

Waters, M., Bloom, T. F. & Grajewski, B. (2000). The NIOSH/FAA working womens health study: Evaluation of the cosmic-radiation exposures of flight attendants, *Health Phys.* 79(5): 553–559.

Wilson, J. W. (1977). Analysis of the theory of high-energy ion transport, *Technical Report TN D-8381*, NASA.

Wilson, J. W. (2000). Overview of radiation environments and human exposures, *Heath Phys.* 79(5): 470–494.

Wilson, J. W., Badavi, F. F., Cucinotta, F. A., Shinn, J. L., Badhwar, G. D., Silberberg, R., Tsao, C. H., Townsend, L. W. & Tripathi, R. K. (1995a). Hzetrn: Description of a free-space ion and nucleon transport and shielding computer program, *Technical Report TP 3495*, NASA.

Wilson, J. W., Joes, I. W., Maiden, D. L. & Goldhagan, P. (eds) (2003). *Analysis, results, and lessons learned from the June 1997 ER-2 campaign*, NASA CP-2003-212155, NASA Langley Research Center.

Wilson, J. W., Lambiotte, J. J., Foelsche, T. & Filippas, T. A. (1970). Dose response functions in the atmosphere due to incident high-energy protons with applications to solar proton events, *Technical Report TN D-6010*, NASA.

Wilson, J. W., Mertens, C. J., Goldhagan, P., Friedberg, W., Angelis, G. D., Clem, J. M., Copeland, K. & Bidasaria, H. B. (2005b). Atmospheric ionizing radiation and human exposure, *Technical Report TP-2005-213935*, NASA.

Wilson, J. W., Miller, J. & Cucinotta, F. A. (eds) (1997). *Shielding strategies for human space exploration*, NASA Conference Publication 3360, NASA Johson Space Center.

Wilson, J. W. & Townsend, L. W. (1988). Radiation safety in commercial air traffic: A need for further study, *Health Phys.* 55: 1001–1003.

Wilson, J. W., Townsend, L. W., Nealy, J. E., Chung, S. Y., Hong, B. S., Buck, W. W., Lamkin, S. L., Ganapol, B. D., Khan, F. & Cucinotta, F. A. (1989). Bryntrn: A baryon ttansport model, *Technical report*, NASA.

Wilson, J. W., Townsend, L. W., Schimmerling, W., Khandelwal, G. S., Khan, F., Nealy, J. E., Cucinotta, F. A., Simonsen, L. C., Shinn, J. L., & Norbury, J. W. (1991). Transport methods and interactions for space radiation, *Technical Report RP-1257*, NASA.

Wilson, J. W., Tripathi, R. K., Cucinotta, F. A., Shinn, J. L., Badavi, F. F., Chun, S. Y., Norbury, J. W., Zeitlin, C. J., Heilbronn, L. & Miller, J. (1995b). Nucfrg2: An evalution of the semiempirical nuclear fragmentation database, *Technical Report TP-3533*, NASA.

Wilson, J. W., Tripathi, R. K., Mertens, C. J., Blattnig, S. R., Clowdsley, M. S., Cucinotta, F. A., Tweed, J., Heinbockel, J. H., Walker, S. A. & Nealy, J. E. (2005c). Verification and validation of high charge and energy (hze) transport codes and future development, *Technical Report TP-2005-213784*, NASA.

Wilson, J. W., Tweed, J., Walker, S. A., Cucinotta, F. A., Tripathi, R. K., Blattnig, S. & Mertens, C. J. (2005a). A benchmark for laboratory exposures with 1 a gev iron ions, *Adv. Space Res.* 35: 185–193.

Wilson, J. W. & et al. (2006). International space station: A testbed for experimental and computational dosimetry, *Adv. Space Res.* 37: 1656–1663.

Xapsos, M. A., Barth, J. L., Stassionopoulos, E. G., Messenger, S. R., Walters, R. J., Summers, G. P. & Burke, E. A. (2000). Chracterizing solar proton energy spectra for radiation effects applications, *IEEE Trans. Nucl. Sci.* 47(6): 2218–2223.

Part 6

Effects on Materials

32

Total Dose and Dose Rate Effects on Some Current Semiconducting Devices

Nicolas T. Fourches
Atomic Energy Commission, CEN Saclay, DSM/IRFU/SEDI, Gif sur Yvette, France

1. Introduction

This chapter will overview some of the aspects of total ionizing dose and dose rate effects on semiconducting devices especially those used in high energy and radiation physics. First, material aspects and interaction of Ionizing Radiation with Matter will be reviewed with emphasis on defect creation and carrier generation. Radiation induced defects are detrimental to device operation but electron-hole pair generation by impinging particles is the basis of all semiconductor detectors. In the second stage, problems related to transistor devices will be discussed with particular emphasis on gate oxide issues and on Silicon On Insulator technologies, Radiation-hardened by design. Although some of the phenomenological physics and chemistry related to ionizing radiation and its effects on Metal Oxide Semiconductors Structures were reviewed by Oldham (Oldham, 1989) and earlier authors (Ma & Dressendorfer, 1989), there is no finalized view of defect physics and chemistry in the device oxides, mainly because the related problems are very complex. However, most of the experiments regarding ionizing irradiation effects on electronics were made both at room temperature or above and at a relatively high dose rate (Messenger & Ash, 1986). For some present day practical applications in space or in high-energy physics low dose rates effects are important and are investigated to usefully complement single event effect studies. Effects at lower temperatures have been scarcely investigated (Saks et al., 1984) leading to a rather piecemeal knowledge of transport of photo-generated carriers and similarly for the activation energies of deep defects centers. As most important devices are silicon based, ionizing radiation effects studies of bulk silicon have been made in an early stage (Willardson, 1959; Sonder & Templeton, 1960; Cahn, 1959; and the Purdue Group). This is of special importance for silicon or other semiconductor detectors. Many good studies and reviews were made on semiconductors detectors mainly in the framework of Large Hadron Collider experiments (Wunstorf, 1997; Leroy, 2007). In the last part of this chapter, progress made on new pixel detectors such as Complementary Metal Oxide Semiconductor sensors will be reviewed because they constitute a very active field of research and development.

2. Interaction of Ionizing radiation with a semiconductor material

2.1 Energy deposition and electron-hole pair generation

The interaction of radiation and both non-ionizing and ionizing will be the starting point of this discussion. The main issue is related to the damage generated in the materials both by the non-ionizing and ionizing irradiation. These can be classified as permanent effects as they have a long lasting influence in a semiconductor (a few minutes to many years). Other effects should be regarded as transient effects, and have many consequences and applications. Ionization generates electron-hole pairs, which in a semiconductor are charge carriers and may be used for device operation, such as particle detectors. Generation of carriers can occur in a semiconductor with photons of energy of the order of the bandgap. At higher energies, electrons from inner shells are excited and result in a high number of photo-generated carriers. Similar processes occur with other charged particles, the total number of generated electron-hole pairs being given by a simple expression:

$$N=Ed/Eg \tag{1}$$

where Ed is the total energy deposited and Eg is the direct bandgap.
In practice, most of these pairs recombine, either directly or indirectly. An electric field applied to the device may dissociate these pairs and lead for a semiconductor detector to a current flow that can induce a signal according to the Ramo-Shockley theorem. Silicon, germanium, diamond counters, and other compound semiconductor detectors operate on this principle and have limited bulk sensitivity to Total Ionizing Dose (TID), but are most of the time very sensitive to Non-Ionizing-Effects.
The interaction of photons in the keV energy region with silicon dioxide results in the generation of electron-hole pairs that dissociate in the presence of the electric field applied to the gate of a MOS structure. In the case of charged particles with sufficient energy similar to protons, alphas, electrons, pions or so, the interaction with the oxide also generates electron-holes pairs according to ionization models originating in the Bethe-Bloch theory (Bethe, 1930,1934).
The Bethe-Bloch formula is valid at relatively large energies whereas in the lower range of energies, LSS (Lindhard,Scharff,Schiott, 1963) and Ziegler models should be taken into consideration, hence :

$$\frac{-dE}{dx} = Kz^2 \frac{Z}{A} \frac{1}{\beta^2} \left[\frac{1}{2} \ln \frac{2m_e c^2 \beta^2 \gamma^2 T_{max}}{I^2} - \beta^2 - \frac{\delta(\beta\gamma)}{2} \right] \tag{2}$$

dE/dx is the linear energy transfer, T_{max} is the maximum energy transmitted to a free electron in a collision, $\delta(\beta\gamma)$is the density effect correction to ionization energy loss, the mean excitation energy whose value are derived experimentally, related to oscillator strength, the other parameters being having their normal physical meaning. Ziegler published a review (Ziegler, 1988) of the detailed stopping laws derived from the original Bethe-Bloch formula. Except at very low energies where LSS models hold with a dominant contribution of nuclear collisions, most of the energy deposited results in ionization. This leads to applications in radiation detection but also to many detrimental effects in semiconducting devices exposed to harsh ionizing environments. These effects, together with the efforts devoted to overcome them, have been studied for many decades (since the early sixties), due to the advent of the satellite era. Thus, except for some recent applications in high-energy physics, most studies were motivated by space or military purposes. Therefore, whatever particle is responsible for the energy deposition, the results are expressed in terms of Total Ionizing Dose (TID) with units most often derived from the Gy(Si) (1J/Kg) or the rad(Si). A diagram summarizing the effects of radiation on materials is shown below. Both Ionizing radiation, charged particle effects and neutral particle (Non Ionizing) effects are taken into consideration (Fig. 1). The results are mainly lattice defects that have an impact on the properties of the material.

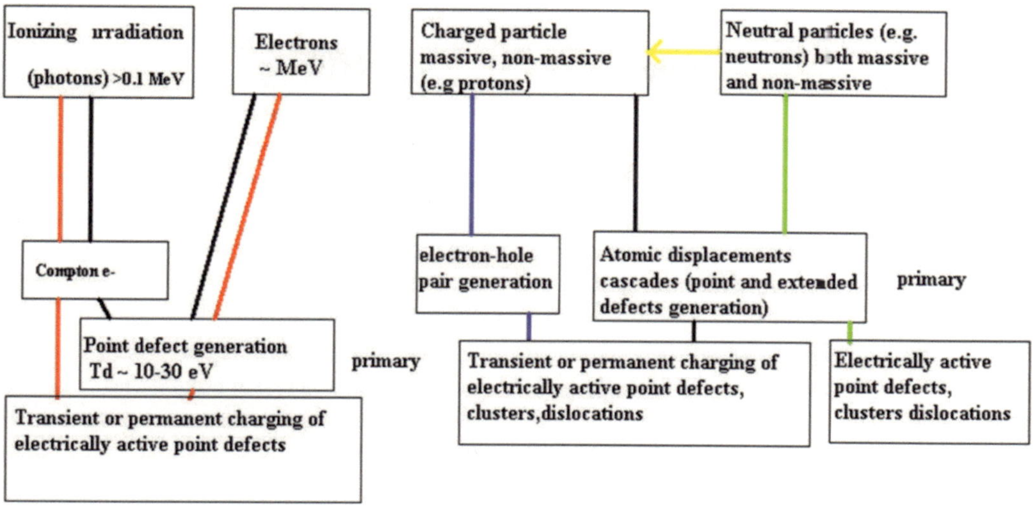

Fig. 1. Diagram summarizing the effects of ionizing radiation, charged particle and neutral particles on matter with no annealing considered.

2.2 Creation of defects by ionizing radiation

The point that should be discussed is whether or not ionizing irradiation directly creates point defects in the material. At high energy (~1 MeV) , the impinging particles will create energetic electrons or nuclear recoils that should result in atomic displacements (Fig. 2). This is valid either for uncharged or charged particles (neutron and protons, respectively). Defect creation has been reported even for neutrinos (Brüssler et al., 1989). Depending on the recoil energy single displacements or a displacement cascade will occur leading to the appearance of point or extended defects respectively

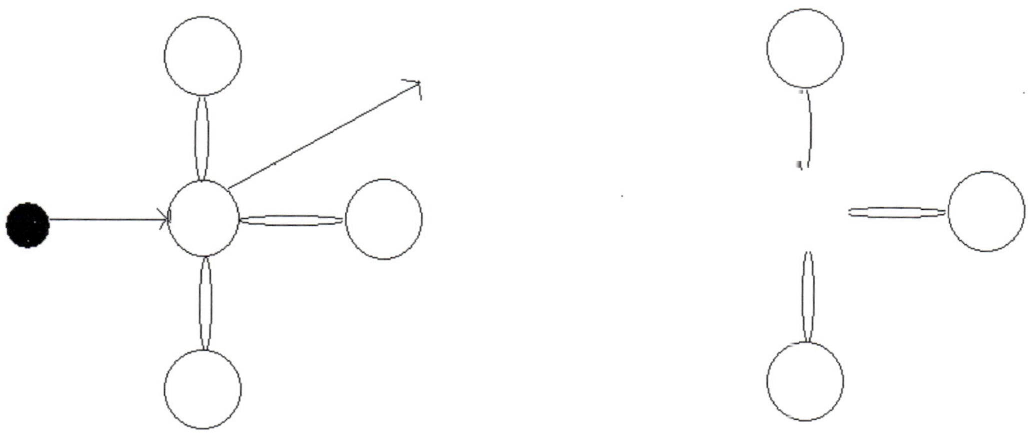

Fig. 2. Sketch of an interaction of a particle (charged or neutral, black on the schematic) with an atom in a plane lattice. A vacancy is created. Depending on the energy of the primary-knock-on the interstitial is ejected close to the vacancy and induces a Frenkel pair or result in a displacement cascade

Point defects such as vacancies or interstitials are not always stable in semiconductors such as germanium and silicon. Experiments with electron irradiation show a different picture when made either at room temperature (300 K) or at very low temperature (Mooney et al., 1983).The comparison with non-ionizing irradiation (neutron) shows a very close behavior (Fourches et al.,1991), because extended defect generation has a weak influence (Fourches,1995). Direct defect creation by ionizing particles such as electrons was widely studied because of fundamental considerations, but purely ionizing effects in semiconductors due to photons and defect creation due to photons has not been investigated so extensively (Sonder and Templeton,1960). For purely ionizing radiation such as gamma rays, this effect is clearly limited even in very sensitive HPGe detectors used for photons detection. In contrast, these materials are prone to degradation under non-ionizing irradiation, even when a very low fluence is considered (a few 10^9 /cm^2) (Fourches et al.,1991a, 1991b). These non-ionizing irradiation effects are often expressed in terms of non-ionizing energy loss (NIEL) in order to reduce the damage effects to a global parameter independent of the nature of the impinging particles. These are still the basis of many studies devoted to device and process development. Recent studies made in silicon for high energy developments or CMOS device improvements have resulted in some new data on the defect introduction rate of ^{60}Co photon irradiated silicon. According to Pintile et al. (Pintille et al. 2009), the introduction rate for room temperature induced interstitial related defect in silicon is of the order of 10^3 cm^{-3}/rad. This means that to obtain defect concentrations of the order of 10^{10} cm^{-3}, the TID should reach 10^7 rad (10 Mrad (Si)) which is a very important TID. This has two consequences. First, this means that the integrated photon flux required to degrade very defect sensitive detectors such as HPGe detectors is of the $10^{16}/10^{17}$ cm^{-2} which is a high value compared to other particle effect introduction rates (Fourches et al. 1991). Direct defect creation by photons in the MeV range should be considered as a negligible effect with respect to defect production by massive particles. Even in silica, widely used in MOS devices, the defects directly induced by irradiation are in negligible concentrations compared to precursors (E′ centers, positively charged particles) as long as ordinary oxides are considered. This is the reason oxygen vacancies in all the studies of ionizing radiation effects in MOS oxides focus on defect charging and charge carrier migration and annealing, properties which are mostly observed through electrical measurements.

2.3 Defect transformation and annealing: Device implication

Most devices using semiconductors operate at room temperature with the notable exception of cryogenic detectors such as HPGe. The concentration of defects in materials such as silicon, germanium and silicon dioxide depends significantly on the annealing processes of elementary defects created during irradiation. Most of the defects investigated have been point defects as the models used until now were simpler than for many atom defects such as dislocations and vacancy clusters (voids). Moreover, ionizing radiation studies have focused on electrical measurements either of macroscopic physical quantities or on the investigation of thermal relaxation behaviors, leading to the estimation of activation energies. The picture of the irradiated material is such that only a fraction of the defects introduced by irradiation remains in a stable form at room temperature. This has been established in a variety of semiconducting materials both elemental and compound. Most stable defects both in germanium and silicon are divacancy or vacancy related. Some doubt still exists in germanium because of the lack of symmetry sensitive experimental results such as uniaxial stress DLTS (Deep Level Transient Spectroscopy) or ESR (Electron Spin Resonance). Interstitial related defects were detected in later studies (Song et al. 1989) in silicon and interstitials related defects in germanium (Carvalho et al., 2007) were not clearly identified.

3. Devices questions and issues

3.1 Ionizing irradiation issues

3.1.1 Semiconductor detectors (new detector issues)

In the last twenty years, many groups undertook empirical investigations in order to determine the behavior of semiconductor detectors damaged by ionizing and charged particles. Most of these studies focused on detector characteristics directly related to operational performance. The huge leakage current increase observed after irradiation has a great impact on the overall system power management and furthermore the most important characteristic affected is the charge collection efficiency. The charge collection efficiency depends on the concentration of deep traps localized in the fiducial (i.e. the sensitive) volume of the detector (sensor). They act as recombination or trapping centers. In the full depletion mode often used in semiconductor or semi-insulating detectors the defects act as carrier generation centers. The carriers generated in the volume are responsible for the bulk leakage current of the detectors; surface leakage current related to ionization induced defects at the interface dominates in silicon detectors/sensors passivated by silicon dioxide. This is strong evidence for these effects in CMOS (Monolithic Active Pixel Sensors) sensors recently characterized (Fourches et al., 2009) which sensitivity to ionizing (gamma ray) radiation was compared to the sensitivity to neutron irradiation (Fig. 3). Clearly, the CMOS sensors are very sensitive to bulk damage, this being only dependent on the material characteristics and the bias scheme, these devices not being fully depleted. Ionizing irradiation effect (TID) can be strongly reduced by process and layout improvements (Ratti et al.2006, 2010).For other pixel detectors, in order to match future SuperLHC requirements, tolerance to TID and bulk damage can be enhanced by using structures that reduce the charge collection length (i.e; the average length taken by one carrier to be collected by the electrodes), using TSV (Through Silicon Via). Defect engineering has been fruitful in reducing the sensitivity to structural defects of many silicon detectors through impurity density modulation (ROSE Collaboration). Fewer results were published concerning high-purity germanium detectors for which early efforts were significant. Published results seems to show that annealing is not complete, although radiation induced point defects anneal out around 200°C (Mooney et al.,1983; Fourches et al.,1991) , clusters and other extended defects remain in the bulk of the crystal and only disappear at higher temperatures (Kuitunen et al.,2008). Other materials are used for semiconductor detectors, but have not attracted so much interest in terms of radiation tolerance. Diamond is the case in point but recent investigations have also been scarce (Velthuis, 2008). Their potential high radiation hardness is the result of higher atomic displacement energies, than for instance in silicon, and also the result of a higher band gap which leads in fewer electron-hole pairs generated. Applications are sought for LHC upgrades. III-V materials were also investigated (Bourgoin et al. 2001). In present times there is no definitive choice of either material or detector technology for future application in HEP (High Energy Physics), with respect to radiation hardness issues.

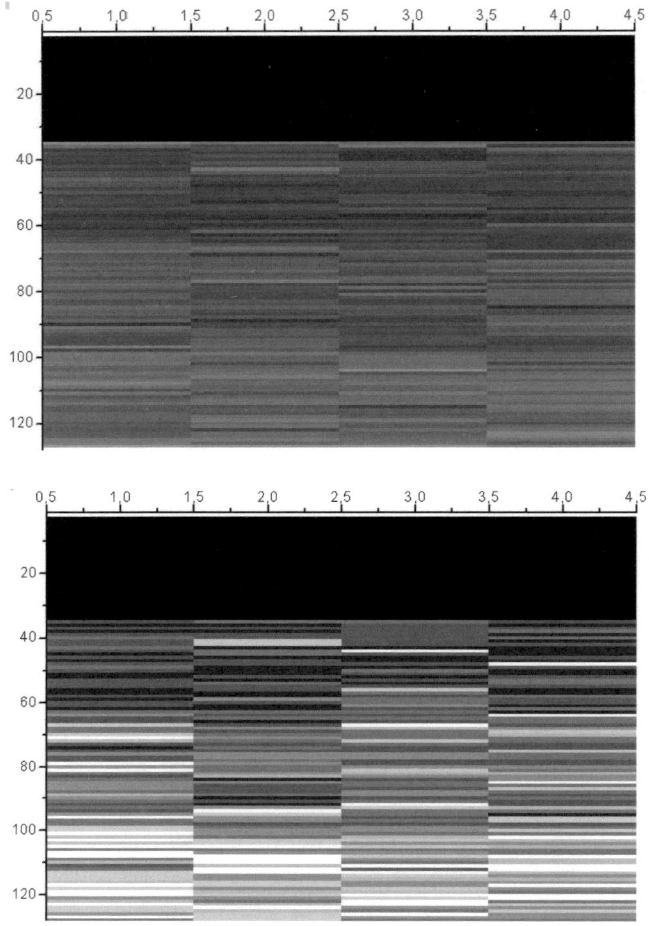

Fig. 3. comparison between ionizing (upper) and neutron (lower) irradiated pixels arrays. Note the uniform effects of ionizing (gamma) irradiation (after Fourches et al., 2008).

3.1.2 Microelectronic devices issues

The use of microelectronic devices, circuits, and processes in harsh ionizing irradiation environments is common for many decades now. Efforts were pursued for many years to obtain processes less sensitive to TID, Dose Rate Effects, and Single Event Effects (SEE) such as Single Event Upset, Latch-up, both destructive and non-destructive, and Single Event Gate Rupture (SEGR). A lot of review papers and books exist on that subject but this paragraph will emphasize only on key technological advancements and on the TID effects that are thought to be the dominant problem in accelerator-based experiments. SOS technologies (Silicon On Sapphire) and subsequent SOI (Silicon On Insulator) technologies were first developed to overcome SEE (Single Event Effects) with great success but some extra progress in purely bulk processes made high radiation hardness possible. Down-scaling the elementary device dimensions along with the Moore's law have decreased the sensitivity to TID provided special layout design precautions are taken. Most of the problems related to ionizing irradiation sensitivity are linked to MOS devices although bipolar and JFETs can be affected by TID (Fourches et al., 1998; Citterio et al. 1995; Dentan et al.1993,1996). Former MOS devices required gate special gate oxides, which were engineered in order to limit the total bulk positive charge induced by ionizing irradiation. Careful oxidation of the silicon has allowed a reduction of Pb (Pending Bonds) centers at the Si/SiO_2 interface and theretofore to a reduction of the interface charge. This results in a lower Threshold Voltage Shift (Fig. 4). It is also sometimes necessary to use closed shape MOS structures to limit leakage current when bulk processes are used.

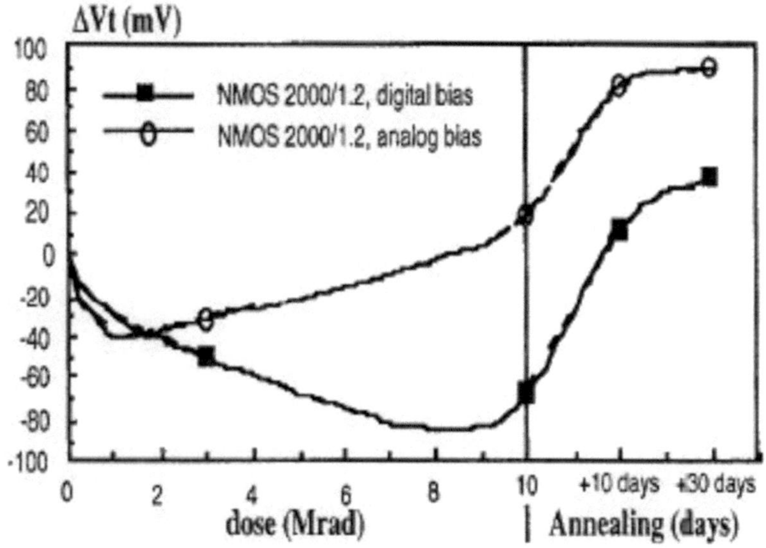

Fig. 4. Threshold voltage shift versus Total Ionizing Dose of a typical DMILL SOI NMOS device with different biasing schemes applied (out of M.Dentan et al.,1996). Digital bias corresponds to a nil source-drain voltage.

The R&D efforts made by many teams around the world have resulted in a very high radiation tolerance for n-channel SOI devices as well as p-channel ones. This has been established during the development of the DMILL technology.

3.2 MOS device hardening issues

The radiation hardness of MOS structures increases with the reduction of the oxide thickness, this being first observed in the 1970s and confirmed later (Saks et al., 1984). Simple models derived from classic carrier tunneling from a defect level and first used by Manzini and Modelli (Manzini & Modelli, 1983) in the reliability context show that the sensitivity to TID is negligible up to a few Mrads for oxide thicknesses less that 5 nm, which are common in modern sub-micron MOS low voltage processes.

Using the analytical expression published by the original Manzini and Modelli paper one can compute the threshold voltage shift of a MOS device versus oxide thickness for a given TID received assuming realistic e-h pair generation energies and total hole trapping with no electron trapping in the whole volume of the gate oxide. Fig. 5 summarizes the results. A commercially available computer code was used to obtain numerical results using a trapping depth of the order of 0.18 eV.

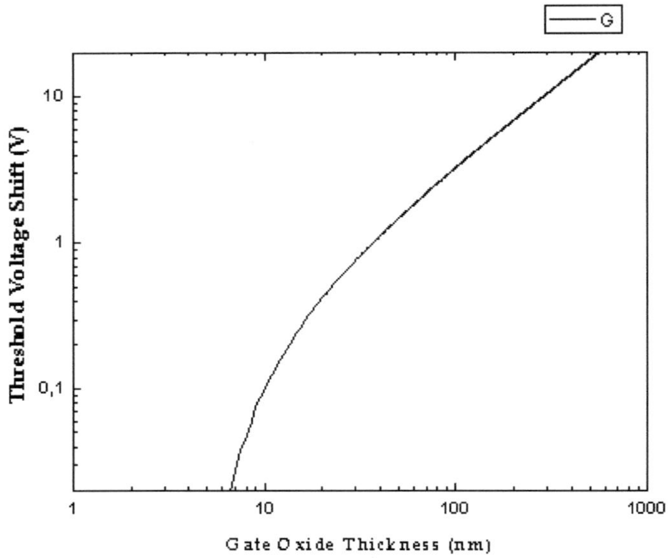

Fig. 5. Absolute threshold voltage shift versus gate Oxide Thickness (in nm) for a 100 krad Total Ionizing Dose received assuming the conditions introduced in the text.

3.3 Total dose and dose rate effects; Electronic devices

3.3.1 Ionizing effects at moderate cryogenic temperatures

Many applications require the operation of nano-microelectronic devices below room temperature. In the nineties, me and some of my colleagues embarked on the possibility of making a SOI process suitable for 90 K operation. n-channel and p-channel MOSFETs had

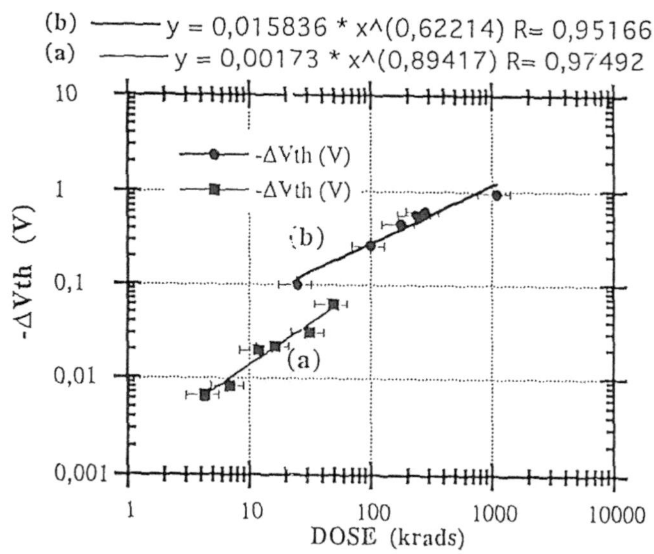

Fig. 6. Threshold voltage shift versus TID for different dose rates (a) low dose rate, (b) high dose rate. (b) high dose rate (50 krads/hour), (a) low dose rate 0.018 krads/hour both made at ~ 80 K on n-channel devices.(a + 5 V bias was maintained on the gate).

very good characteristics at these temperatures, making these devices usable for amplifier and detector signal readout applications (Fourches et al., 1997a,1997b). The questions arising were related to the TID radiation hardness at these temperatures. A thorough study was initiated on the effects of ionizing dose, dose rate effects, annealing and geometric effects. Fig. 6 and 7 demonstrate clearly the existence of favourable dose rate effects in the sense that a reduced dose rate results in a lower Vth shift at equivalent total ionizing dose.

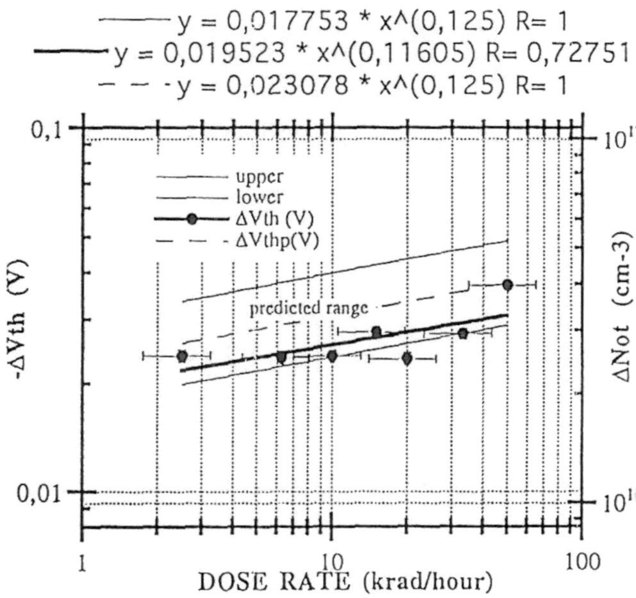

Fig. 7. Double scale plot showing the dose rate dependence of the threshold voltage shift in the range 1 krad/hour to 100 krad/hour. The net positive trapped oxide charge is indicated on the left scale. The shift follows a D'^{α} law. The law can be derived from CTRW (Continuous Time Random Walk) analytical approaches.

The TID response also strongly depends on the electric field E in the gate oxide. Fig. 8 clearly exhibits the difference in the slopes of the threshold voltage versus the Total Ionizing Dose (gamma rays) with electric Bias applied to the gate. At 0 V the dissociation of the electron-hole pairs is incomplete and a low charge generation yield results. When a moderately strong bias is applied (5V) the charge dissociation is almost complete and the slope is at its highest value. This effect is strongly reduced when the electric field is increased. Field assisted tunneling of trapped holes towards the silicon can explain the phenomenon in addition of field assisted hopping transport. The activation energy for dispersive transport is lowered when the electric field E is increased. ($E_a(E)=E_a-kE$). This results in a faster hole hopping frequency, between random sites.

Transport of holes within a temperature range extending from liquid nitrogen to room temperature, shows a temperature and field dependence, the same being true for the conduction of electrons in the oxide, these carriers being much more mobile than holes. The holes are trapped at low temperature, where they exhibit a very low mobility. They can be trapped either at E centers (oxygen vacancies) or self-trapped as polarons, which have a very low mobility particularly below room temperature. At higher temperatures, the mobility is enhanced and they can anneal out rapidly. Experiments, carried out in the 1990s devices on DMILL thick film SOI process showed the thermal anneal of trapped holes in the gate oxide (Fig. 9).

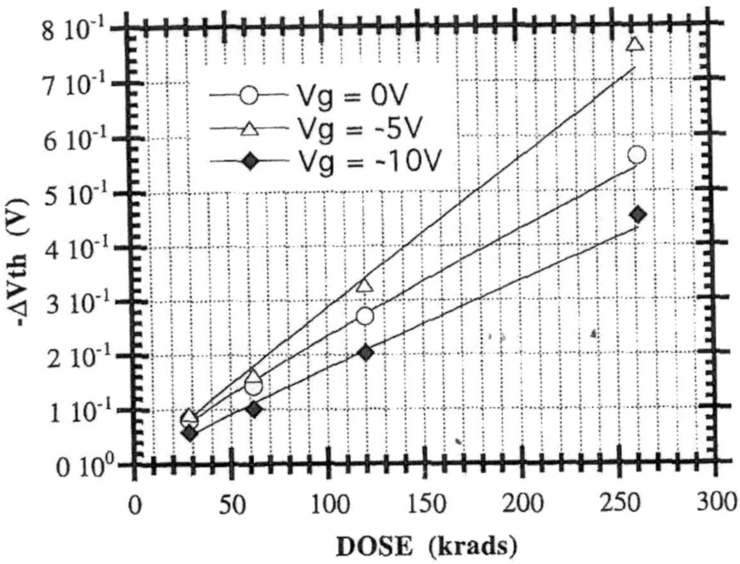

Fig. 8. Vth shift versus TID for p-channel devices with increasing voltage bias on the gate. The measurements were carried out at moderate cryogenic temperature. Note the importance of the shift with oxide thicknesses of 20 nm.

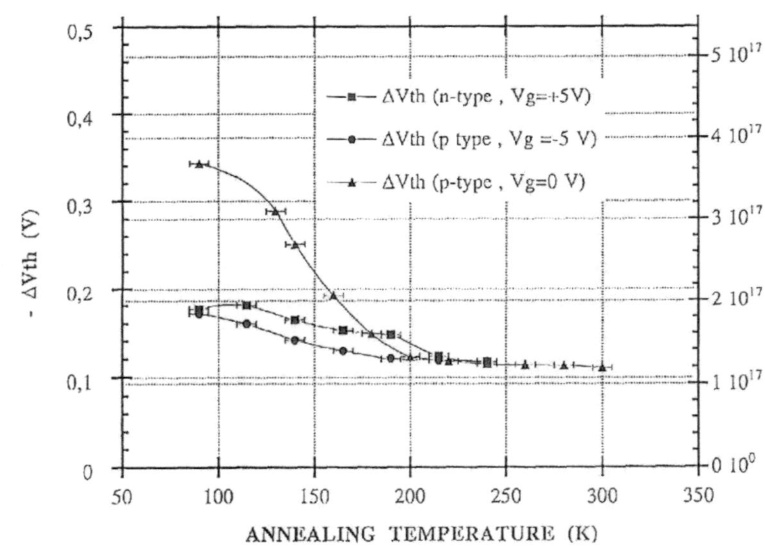

Fig. 9. Isochronal annealing of n-channel and p-channel transistors; (30 minutes for each temperature step); The Vth shift of the n and p channel devices with bias on was identical. The original Vth shift of the p- channel device with 0 V bias on was different due to a 300 krad TID compared to the 80 krad TID received by the former two devices.

The activation energy deduced from the annealing diagram is approximately 0.12 eV, this is close to the detrapping energy of the self-trapped holes and is very small compared with the silica bandgap. As this diagram shows most of the charge (at least 50 %) has annealed up to room temperature.

Early studies of hole transport in thick silica has confirmed the validity of the CTRW (Continuous Time Random Walk) theory (Sheer & Montroll, 1975) for hole-polaron transport in different conditions of field and temperature (McLean, Boesch, Jr. & Garrity in Ma & Dressendorfer,1989; McLean, Boesch, Oldham Ma & Dressendorfer,1989) . Analytical models were proposed a long time ago and it was reasonable to extend them to take into consideration the effects of dose rate, under continuous irradiation. Using signal theory and considering the physical processes as being causal a convolution of annealing effects and irradiation effects can be the basis of a simple analytical model. One obtains a dose and dose rate dependence of the threshold voltage shift of the form:

$$\Delta Vth \approx KoxD^{1-\alpha}D'^{\alpha} \tag{3}$$

Kox is a parameter that depends on the oxide characteristics, D the Total Ionizing Dose and D' the dose rate. See (Fourches, 1997b) for details.

A power law dependence on the dose and dose rate is obtained, with a value for derived from Fig. 7 of approximately 0.125, this value is lower than the values deduced from older experiments made on thicker oxides (Ma & Dressendorfer and corresponding chapter herein). Additionally it seems clear that the annealing process cannot be simply interpreted by a first order kinetic, as it would be the case for single deep traps. A saturation would occur at a very low TID level (Fourches, 2000), which is not observed in the measurements with an electric field applied. Further investigations show that saturation is clear at a much higher TID, for which threshold voltage shifts exceed 1 V. A numerical treatment was introduced (Fourches, 2000) taking into account the presence of deep hole and electron traps together with CTRW transport, which fits with the saturation observed on some devices exposed to very high TID. The existence of a strong annealing between 80-90 K and room temperature has consequences on the behavior of devices. Cryogenic operation of CMOS SOI electronics had been considered for Liquid Argon Calorimeters, but due to some of the shortcomings of the processes of the 1990's this could not be implemented.

3.3.2 Consequences on hardening issues at room temperature

These modeling issues have had some direct consequences on the hardening techniques to be used especially at low temperatures but also for the normal operation of MOS devices. First, it is clear that the Manzini-Modelli-Saks model is a key to these prediction issues. A good fit can be found between measurements results at low temperature and mathematical predictions leaving aside the annealing effects that result in the power law dependence. For instance a 20 nm oxide irradiated at high dose rate, leads to a Vth shift of 0.3 V at 100 krads according to our measurements. This is very close to the value deduced from Fig. 5. This experimental verification of the Manzini-Modelli model could only be made on modern hardened oxides, which exhibit a low active defect density, compared to early oxides, on which Saks et al. made the first experiments. On this basis on can deduce a roadmap for hardening issues to TID. Modern processes either bulk or SOI have thin gate oxides (at the expense of supply voltage range) and this improves radiation hardness. Leakage currents

due to interface and surface effects have been somehow reduced over the years by process improvements and layout techniques.
Other data obtained at 90 K show that the sensitivity to TID of the SOI process studied reduces when the dimensions of the device decrease. This is verified by plotting the prefactor Kox of expression (3) versus the gate length of each device. The exponent of the power law found less sensitive to the gate length. Fig. 10 and 11 show the results. The fact that downscaling the device dimensions has very favourable effects on tolerance to TID had a lot of implications for future processes both bulk and SOI. In retrospect it seems clear that today nanoscale or deep submicron technologies are intrisically TID hardened on the basis of some of the scaling laws discussed here and verified experimentally in the 1990's papers (Fourches, 1997) for Silicon On Insulator processes.

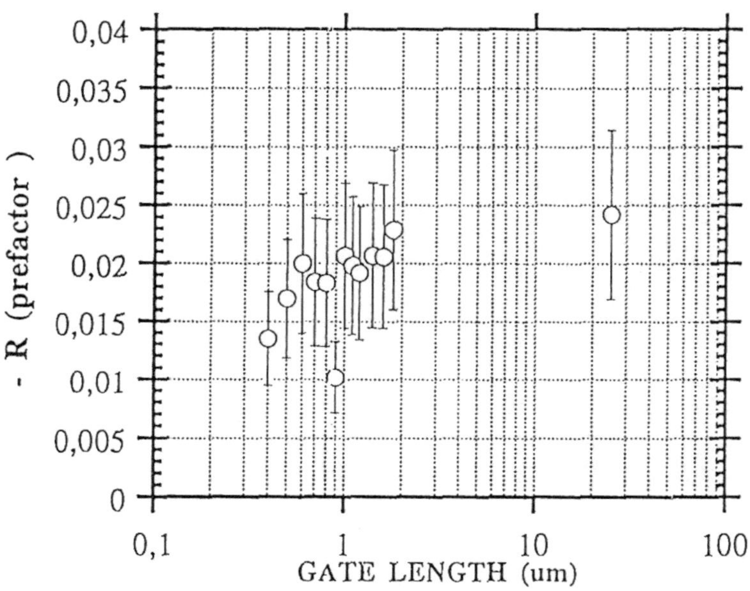

Fig. 10. Dependence of the pre-factor, with the gate length of a n-channel transistor, each transistor having received the same total ionizing dose for the same dose rates. It is clear that low gate length result in a reduced effect of radiation. The measurements were made at 90 K.

The interpretation of this effect at least at low temperature is relatively simple and was already introduced in (Fourches, 1997). Charge tunneling and simply migration is faster when the dimension of the device are lowered, resulting in an accelerated annealing of the positive charge. This should have favorable consequences even for room temperature operation of bulk device, because this simply concerns directly the properties of the gate oxide.
This is one of the reasons the TID response of MOS devices is a today an issue of now lower importance for most applications, except perhaps for the upgrades of the LHC, for Future Linear Colliders and for nuclear power plant applications. Most of the research focuses on Transient Effects, destructive or not, due to charged particles in available bulk processes. The SOI processes are immune to these transient effects by design, because of the device/device insulation systematically implemented.

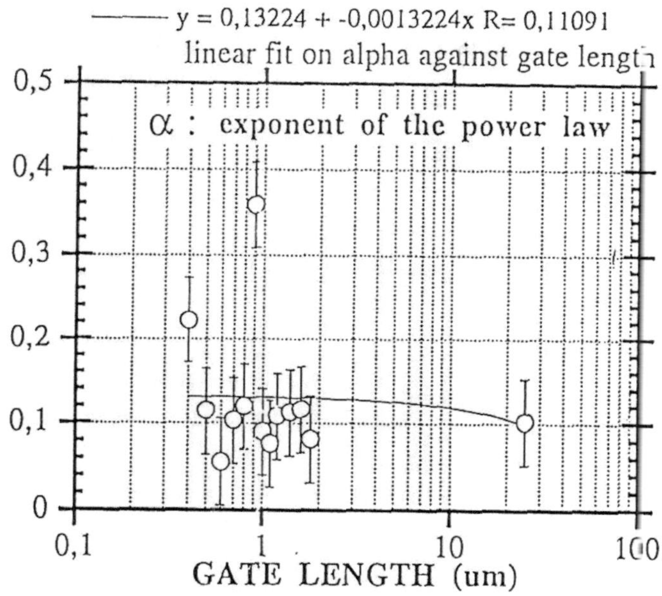

Fig. 11. Similar plot for the dependence of exponent with the power law. There is no clear exponent dependence on the gate length.

3.3.3 TID and dose rate effects on other devices

TID effects were reported and sustained efforts were made reduce them on bipolar transistors. DMILL technology was developed with a vertical npn bipolar transistor in order to obtain a radiation hard device, by a reduction of the passivation Si/SiO_2 interface. Despite this TID effects still appear with enhanced effects on npn devices due to the higher base current due to minority electron carriers, the oxide being positively charged. In addition to this dose rate effects were reported in 1991 on polysilicon bipolar transistors, and later enhanced low dose rate effects were confirmed (Pease, 2004). Studies on JFETs and MESFETs have also shown the existence of ionizing dose effects, extensive studies were not very developed (Citterio et al., 1995; Fourches, 1998), mainly because the main constraints with large and deep channel devices such as JFETs (Junction Field Effect Transistors) is their limited tolerance to displacement damage.

4. Irradiation: Detectors

4.1 Total dose and dose rate effects; Silicon detectors

In this paragraph emphasis will be on new semiconductor detectors such as CMOS sensors that are now good candidates for high radiation hardness, high spatial resolution inner pixel detectors for Vertexing purposes.

Sensitivity to Total Ionizing Dose for silicon detectors is clearly dependent on surface effects, the silicon surface being oxidized either naturally or for passivation purposes. Depending on the thickness of the oxide the sensitivity of the detector to leakage currents will be enhanced or limited. Recent studies have focused on CMOS Sensors (Fig.12) and CCDs. Sensitivity to bulk ionizing damage is limited and has not been thoroughly investigated except for charged particles (Hopkinson, 2000). The striking result is the limited Fixed

Pattern Noise increase due to ionizing irradiation, which is much less pronounced than it is when induced by displacement damage. Total Ionizing Dose induced displacements is often limited to the creation of Frenkel pairs and other point defects in the bulk that are less

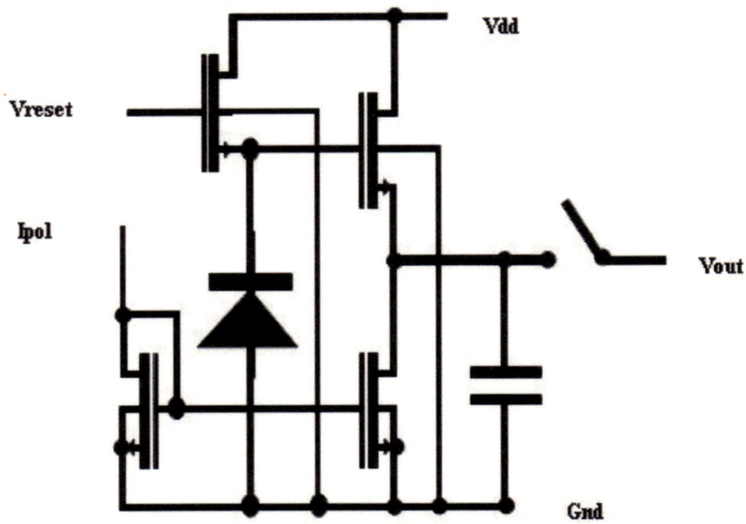

Fig. 12. Schematic structure of the CMOS sensor proposed for charged particle detection.

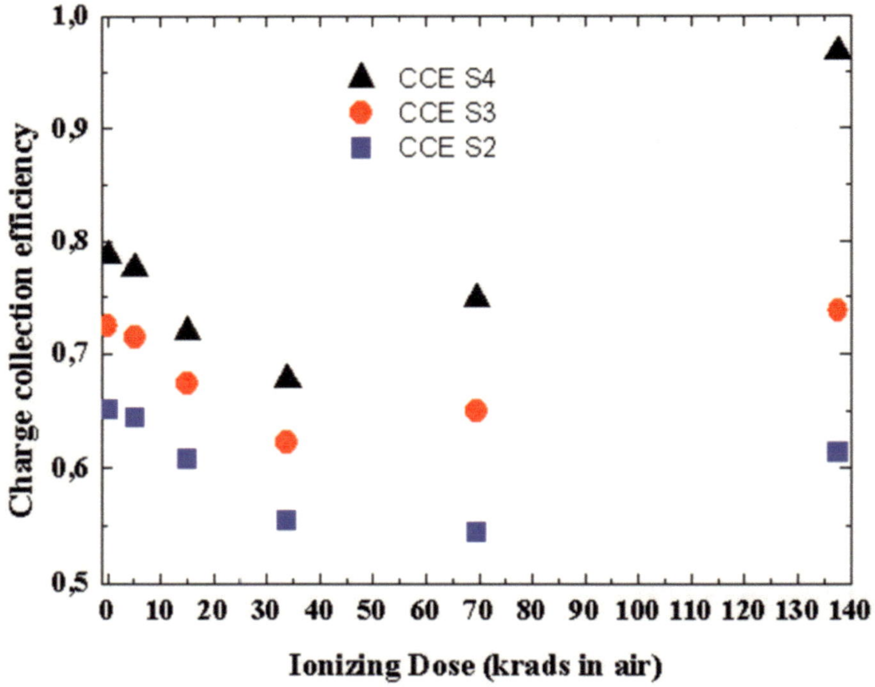

Fig. 13. CCE versus TID for different sensor diode sizes: (S4, S3, S2). The increase in the CCE along with the TID could find its origin in a high surface local electric field, which results in carrier multiplication.

effective in reducing the free drift length of the carriers than displacement cascades. For a detailed simulation study of deep defects effects on CMOS sensors see Fourches, 2009. This induces a reduction of the Charge Collection Efficiency (CCE) which characterizes the effectiveness of the sensor/detector to collect all the generated carriers. Ionizing Dose effects can reduce the charge collection efficiency through surface/interface carrier recombination. Fig. 13 seems to confirm this effect. The CMOS sensors degrade rapidly with TID. In this case the layout of the sensing diode was not optimized so high recombination and leakage current can occur even at small TID.

This Figure (Fig.13) shows that the degradation in charge collection efficiency are very fast with TID for these un-optimized pixel devices. This is due to the high surface leakage current that limits the current signal. CCE increase due to carrier multiplication has been recently reported and was the basis of possible radiation tolerant CMOS particle detectors (E. Villela et al., 2010).

The last figure (Fig.14) clearly shows that the spread of pedestals values is greater in neutron irradiated CMOS sensors than for ionizing irradiated ones. Moreover, the temporal noise increases very slowly with pedestal value, in contrast with the neutron irradiated pixels, for which low pedestal pixels correspond to low temporal noise pixels. These relatively new results obtained on very small pixels (a few tens of cubic microns) indicate that the temporal noise is very high after 137 krads and this originates in the shot noise induced by the interface leakage current. The pedestal value shift is probably due to the threshold voltage shift of the follower n-channel MOS transistors. This threshold voltage shift is much lower in the case of neutron-irradiated pixels as displacement damage has a limited impact on the threshold voltage of MOS transistors.

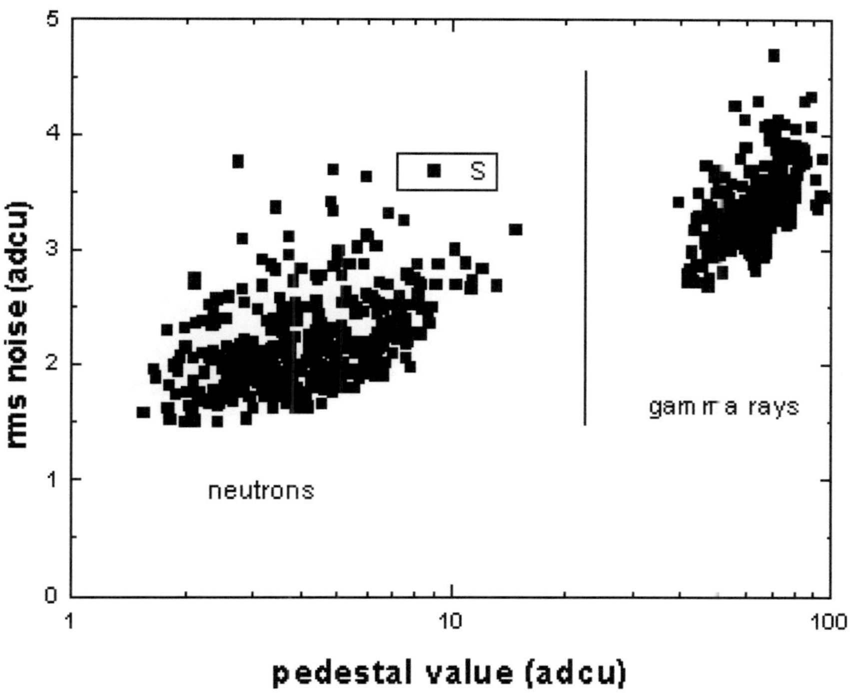

Fig. 14. Root Mean Square value of the temporal noise plotted versus the corresponding Pedestal (offset) values for pixels irradiated with neutrons (left) and pixels irradiated at a dose of 137 krads (Si)(right).

On other processes where radiation tolerance layout techniques were implemented, show a better response to ionizing irradiation. Up to now no long term dose rate effects were reported, nor studied (Fourches, 2008), on these novel CMOS detectors. But on the basis of other measurements the TID tolerance of CMOS sensors is higher than that of Charged Coupled Devices, making CMOS sensors more suitable for future ILC experiments. Radiation hardness up to several Mrads was reported on CMOS sensors using a Deep N-Well structure with charge amplifier readout (L. Ratti et al., 2010).
Other comments should follow concerning CMOS sensors and their counterparts CCDs and DEPFETs. In these devices transport is simply not due to the drift of charged carriers, carrier diffusion is also a very important component and has dramatic consequences on the overall migration length of carriers and subsequently on the signal build-up. Simulation with a software package may help obtaining predictive results but this was limited so far to displacement damage (Fourches, 2009).

4.2 Total dose and dose rate effects: Other detectors

Other materials have been proposed and used for particle detection and has been investigated, in terms of both ionizing irradiation-induced defects and displacement induced defects. Up to now, diamond has a good potential for charge particle detection and radiation tolerance, mainly because of the large amount of energy required to create a

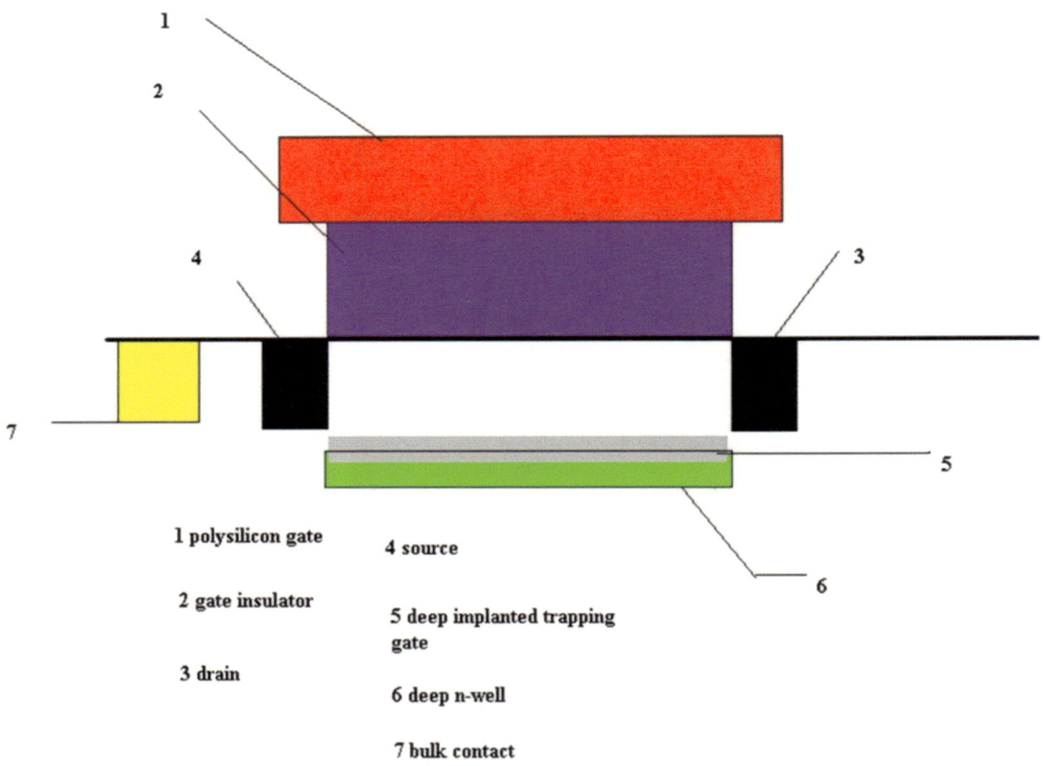

Fig. 15. Schematic structure of the deep trapping gate MOS detector structure.

vacancy-interstitial pair, and was proposed for LHC applications. Compound and alloy detectors have been not so successful because the density of residual defects remains high. This limits the CCE from the start, and affects the charge transport trough the fiducial zone. In spite of this, there have been recent studies of thick GaAs pixel detectors for medical applications (Bourgoin, 2001). Future proposals for high radiation hardness, both for ionizing radiation and displacement damage were recently presented (Fourches, 2010). The main idea is to reduce the fiducial value and to introduce a deep trapping gate below the channel of a MOS transistor which would act either as a detector (electron-hole pairs would be created along a charged particle track) or /and a volatile memory (Fig 15). As the trapped positive charge modulates the channel current of the MOS transistor the readout is simplified and can be implemented by a circuitry similar or to the ones used for CMOS sensors. As the deep trapping gates are made of a high density of structural defects, these result additionally in a high tolerance to displacement damage, confirmed by device simulation (Fourches, 2010).

5. Conclusion

Future developments about the TID effects are manifold. There is still some interest about the TID hardness at very high doses for high energy physics applications (SuperLHC). New nanoscale technologies, because of their reduced dimensions and the presumable existence of scaling laws will ease the development of new radiation hard ICs. On the detector side, similar advances are expected for reduced dimensions pixels, mainly for displacement damage but also for ionizing effects. Geiger mode Si detectors have been investigated by some researchers and are also promising (Ratti. L et al., 2006). Dose rate effects have been somehow investigated on silicon detectors for medical X ray imaging and much remains to be done to improve these sensors.

6. Acknowledgment

The author acknowledges the past contribution of the DMILL development team and the microelectronic group of the DAPNIA, together with the contributions of the LETI, DAM, IN2P3 and CERN. The author is grateful to the librarians for the help in bibliographic search. The author acknowledges the contributions of the CERI (Orléans), CERN, and the other members of the former R&D 29 collaboration. The author thanks R. Walker (EDELWEISS collaboration, Saclay) for proof checking part of the text.

7. References

Bethe H., Manchester, W. Heitler, Bristol, On The Stopping Power of Fast Particles and on the Creation of Positive Electrons, 83-112, Communicated by PAM Dirac , February 1934, *Proc. Roy. Soc.*.

Bethe H., Zur Theorie des Durchgangs Schneller Korpuskularstrahlen durch Materie, *Ann. Physik,* Vol. 5, p. 325 (1930) , Vol 397, (1930) p. 325-400

Bourgoin J.C. et al. Potential of thick GaAs epitaxial layers for pixel detectors, *Nuclear Instruments and Methods in Physics Research* A 458 (2001) 344-347.

Brüssler M., Metzner H., Sielemann R., Neutrino-Recoil Experiments in Germanium, *Mat. Sci. For. Trans. Tech. Publ. (Switzerland,)*Vols. 38-41, (1989) pp. 1205-1210

Cahn J., Irradiation Damage in Germanium and Silicon due to Electrons and Gamma Rays, *J. Appl. Phys.*, Vol. 30, N° 8, 1310-1316, August 1959

Carvalho A. et al. Self Interstitials in Germanium, *Phys. Rev. Lett.*, 99, (2007), 175502-1, 175502-4

Citterio M., Rescia S. & Radeka V., "Radiation Effects at Cryogenic Temperatures in Si-JFET, GaAs MESFET and MOSFET Devices", *IEEE Transactions On Nuclear Science*, Vol. 42, N° 6, December 1995 , p 2266-2270

Dentan, M. et al., Study of a CMOS-JFET Bipolar radiation hard analog technology suitable for high energy physics electronics, *IEEE Transactions On Nuclear Science*, Vol. 40, No. 6, 1993, 1555

Dentan M. et al , DMILL a mixed radiation hard BICMOS technology for high energy physics electronics, IEEE Transactions On Nuclear Science, Vol. 43, N° 3, June 1996, pp 1763-1767

Fourches N., et al. Design and Test of Elementary Digital Circuits Based on Monolithic SOI JFET's, *IEEE Transactions On Nuclear Science*, Vol. 45 , N° 1, February 1998, 41-49

Fourches N., Huck A. & Walter G., *IEEE Transactions on Nuclear Science* , Vol 38 , N° 6 , part 2, (1991) 1728

Fourches Nicolas T., "A Novel CMOS detector based on a deep trapping gate", Nuclear Science Symposium Record (Knoxville Tennessee), Oct. 30 2010-Nov. 6,2010, p655-658,http://dx.doi.org/10.1109/NSSMIC.2010.5873840

Fourches N. & Walter G., Bourgoin J.C., Neutron induced defects in high purity germanium, *J. Appl. Phys.* 69 (4) (1991) 2033

Fourches N., High defect density regions in neutron irradiated high purity germanium: characteristics and formation mechanisms , *J. Appl. Phys.* 77 (8) (1995) 3684

Fourches N., et al. Thick film SOI technology: characteristics of devices and performance of circuits for high energy physics at cryogenic temperatures; effects of ionizing radiation", *Nuclear Instruments and Methods in Physics Research*, A401 (1997) 229-237

Fourches N.T. et al. Radiation Induced Effects In A Monolithic Active Pixel Sensor: The Mimosa8 Chip, CEA/DSM/IRFU, CEN Saclay, 91191 GIF/YVETTE, France, (2008) http://arxiv.org/ftp/arxiv/papers/0805/0805.3934.pdf

Fourches N.T., Device Simulation of Monolithic Active Pixel Sensors: Radiation Damage Effects, *IEEE Transactions On Nuclear Science*, Vol. 56, Issue 6, (2009) 3743-3751

Fourches Nicolas T., "Charging in gate oxide under irradiation: A numerical approach" *Journal of Applied Physics* , Vol 88 , Number 9, (2000) 5410

Fourches Nicolas T., Charge buildup in metal oxide semiconductor structures at low temperature: influence of dose and dose rate, *Phys. Rev. B*, Vol 55, No 12, 1997, 7641

Hopkinson, G. R., Radiation effects in a CMOS active pixel sensor, *IEEE Transactions On Nuclear Science* ,Vol. 47, Issue 6, (2000), 2480-2484

Kuitunen K. et al., Divacancy clustering in neutron-irradiated and annealed *n*-type germanium, Phys. Rev. B 78, 033202-1, 033202-4 (2008)

Leroy, C & Rancoica Pier-Giorgio, Particle interaction and displacement damage in silicon devices operated in radiation environment, *Rep. Prog. Phys.* 70 (2007) 493-625

Lindhard J., Scharff M. and Schiott H. u, Mat. Fys. Medd. Dan. Vid. Selsk., 33, No. 14 (1963).

Ma T.P. & Dressendorfer P.V., Ionizing Radiation in MOS devices and circuits, (Wiley Interscience, New York, 1989)

Manzini S. & Modelli A., Tunneling discharge of Trapped Hole in Silicon Dioxide, Insulating Films On Semiconductors, J.F. Verweij and D.R.Wolters (editors), 1983, 112

Messenger George C. & Ash Milton S., "The Effects of Radiation on Electronic Systems", (Van Nostrand Reinhold, New York, 1986)

Mooney et al., Annealing of electron-induced defects in n-type germanium, *Phys.Rev. B28*, 6, (1983) 3272-3377

Oldham T.R, Ionizing Radiation Effects in MOS Oxides, (World Scientific, Singapore, 1999)

Pease R.L. et al., "Characterization of Enhanced Low Dose Rate Sensitivity (ELDRS), Effects Using Gated Lateral PNP Transistor Structure" , *IEEE Transactions On Nuclear Science*, Vol. 51, N°.6, December 2004, pp 3773-3780

Pintillie Iona et al. Radiation-induced point- and cluster-related defects with strong impact on damage properties of silicon detectors, *Nuclear Instruments and Methods in Physics* Research A 611 (2009) 52-68

Ratti L. et al., CMOS MAPS with fully integrated, hybrid-pixel-like analog front-end electronics, Stanford Linear Accelerator Center, Stanford, California, April 3, 2006

Ratti Lodovico et al., Front End Performance and Charge Collection Properties of Heavily Irradiated DNW MAPS, *IEEE Transactions On Nuclear Science*, Vol. 57, N°4, August 2010, p1781-1789

Saks N.S. Ancona M.G. & Modolo J.A et al., Radiation Effects On MOS capacitors with very thin oxides at 80 K, *IEEE Transactions On Nuclear Science*, Vol. NS-31, No. 6, 1984, 1249

Sher J.H. & Montroll E. W., Anomalous transit-time dispersion in amorphous solids, *Phys. Rev. B 12*, 2455-2477 (1975).

Sonder E. & L.C. Templeton L.C., Gamma irradiation of Silicon: I. Levels in n-Type Material containing oxygen, *J. Appl. Phys.*, Vol 31, No 7, 1960, 1279-1286

Song L. W., Zhan X. D., Benson B. W., & Watkins G. D., Bistable Defect in Silicon: The Interstitial-Carbon-Substitutional-Carbon Pair, *Physical, Review Lettters*, Vol 60, N° 5, February 1998, 1

Velthuis J.J. et al., "Radiation Hard diamond pixel detectors", *Nuclear Instruments and Methods in Physics Research A*, 591, (2008) 221-223

Vilella E. et al. "Readout Electronics for Low Dark Count Geiger Mode Avalanche Photodiodes Fabricated in Conventional HV-CMOS Technologies for Future Linear Colliders", Proceedings of TWEPP 2010, Aachen 20-24 September 2010

Willardson R. K. Transport Properties in Silicon & Gallium Arsenide, *J. Appl. Phys.*, Vol 30, N° 6, August 1959, 1158-1165

Wunstorf R. Radiation Hardness of Silicon Detectors: Current Status, *IEEE Transactions On Nuclear Science*, Vol. 44, N°3, June 1997, 806-814

Ziegler J.F., Stopping of Energetic Light Ions In Elemental Matter, *J. Appl. Phys., Appl.Phys. Rev*. Vol.85, N°3, 1February 1999, 1249-1272

Influence of Ionizing Radiation and Hot Carrier Injection on Metal-Oxide-Semiconductor Transistors

Momčilo Pejović[1], Predrag Osmokrović[2], Milica Pejović[1] and Koviljka Stanković[2]
[1]*University of Niš, Faculty of Electronic Engineering, Niš,*
[2]*University of Belgrade, Faculty of Electrical Engineering, Belgrade, Serbia*

1. Introduction

General characteristics of MOS components

Favourable characteristics of silicon dioxide (further referred to as oxide or SiO_2) as an almost irreplaceable dielectric in MOS (Metal-Oxide-Semiconductor) components have contributed to a considerable extent to the great success of the technology of manufacturing integrated circuits during the last decades. However, it has been proved that instabilities of electric charge in the gate oxide and at $Si-SiO_2$ interface, which inevitably occur due to the influence of ionizing radiation (γ and X-radiation, electrons, ions) (IR) and hot carrier injection (which includes Fowler-Nordheim high electric field Stress, avalanche hole injection, avalanche electron injection, and so on (HCI) during the operation of MOS transistors, lead to instabilities of their electric parameters and characteristics and present a serious problem to the reliability of MOS integrated circuits [1-4]. Studying these instabilities, particularly causes and mechanisms responsible for their occurrence, is very significant as it may help the manufacturer define appropriate technological parameters with the aim of increasing their reliability. Considering the rate of development and requirement that the components should be highly reliable, the investigation of their instabilities in normal working conditions is practically impossible. In order to determine, as quickly as possible, mechanisms that cause instabilities, accelerated reliability tests are used and they comprise of the application of strong electric fields (so-called electric stress) in static and impulse modes at room temperature and/or at elevated temperatures. In this context, reliability comprises probability that the component may fulfil its target function during a certain period and under certain conditions, i.e. that it may preserve the electric characteristics and the value of electric parameters under certain exploitation conditions during a certain period. Reliability is defined as a probabilistic category as no failure of the component can be predicted with certainty. Namely, if a group of components manufactured in almost the same conditions is subjected to stress, a smaller number of them will fail earlier (early failure), while a larger number of them will fail later. In order to remove from such a group of components those with the possibility of early failure, a selection procedure is applied to all the components. This procedure consists of a series of tests the most significant of which is the burn-in test. This test is carried out in two steps: short-term burning (10 – 30 hours) in normal and somewhat intense working conditions, in order to remove components liable to failure due to their manufacturing defects, and sufficiently long burning in tough working conditions in order to remove components with manifested functional failures. A significant fact that should be emphasized in relation to MOS components is that, although the high quality of components guarantees a certain level of their reliability, many defects may remain hidden and manifest during the application thus leading to the degradation of the components and their failure.

In recent times, the development of MOS integrated circuits shows a tendency of increasing the complexity and constantly decreasing dimensions, which comprises a need for increasingly thin gate oxides. In addition, discrete power MOS components have been also developed. At the beginning of their development, it was deemed that they would very soon replace bipolar transistors in power electronics. However, technological processes used in the manufacture of these transistors were more complex in relation to bipolar transistors so that a real progress in power electronics was made when VDMOS (Vertical-Double-Diffused MOS) transistors were developed because the manufacture of them was much simpler. Due to their favourable electric characteristics (great operative speed, low switch-on resistance, high switching voltage, etc.), the VDMOS transistors have found their application in the fields such as automotive industry [5]. Switching characteristics of VDMOS transistors at high frequencies (higher than 100 kHz [6]) have also proved very good owing to which these transistors are applied as switching components for high-frequency sources of supply used in medical electronics, engine control systems, switching sources of supply in telecommunication researches, etc.

In addition to its favourable properties, SiO_2 used in the manufacture of MOS components and integrated circuits has some shortcomings as well. Namely, instabilities of electric charge in the gate oxide and in interface traps, which are inevitable due to the presence of the IR or HCI processes, during the operation of MOS components and MOS integrated circuits, also lead to instabilities of their electric parameters and characteristics. Due to this, instabilities represent one of the most serious problems in relation to a reliable operation of MOS integrated circuits [2, 3] particularly when these components are used in radiation surroundings.

Researches have shown [3] that MOS components are very sensitive to different types of ionizing radiation and that their sensitivity significantly depends on the manufacturing technology. Additionally, some technological processes such as lithography with the use of a ray of electrons, X-rays, and processes in plasma, etc. included in the technological sequence for the production of integrated circuits with the aim of decreasing the dimensions of their components, may be significant sources of defects in the oxide [3].

The attention of today's researches on the impact of the IR and HCI processes on MOS components is directed in two ways. One way comprises the production of components with the highest possible resistance to these types of processes, while the other way leads towards the production of MOS transistors sensitive to IR in order to produce sensors and dosimeters of ionizing radiation [3]. It should be emphasized that the first operating principles of PMOS sensors and dosimeters of ionizing radiation were published as early as 1974 [7]. Unlike the influence of IR and HCI on the reliability of commercial MOS transistors which is the subject of many researches worldwide, PMOS transistors as IR dosimeters are investigated to a much lesser extent.

2. Origin and characteristics of defects and electrical charges in oxide and at $Si\text{-}SiO_2$ interface

Electrical charges in the gate oxide and at the $Si\text{-}SiO_2$ interface may be divided into four groups as follows [2, 8, 9]: *mobile ions, charges at the traps, fixed centres charges and charges at interface traps*. However, if a division of charges based on their influence on the current-voltage characteristics of MOS transistors is made, only charges in the gate oxide (including mobile ions, charges at the trap centres, and fixed charges) and charges at interface traps may be distinguished. The main classification of charges in the gate oxide and at interface traps of MOS transistors is shown schematically in Figure 1.

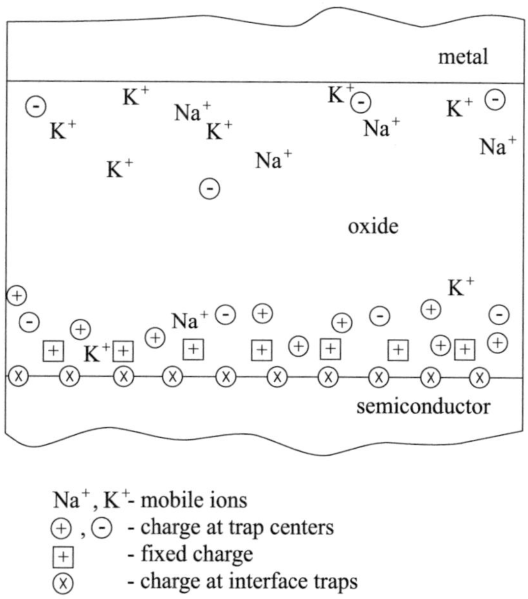

Fig. 1. Classification of charge at $Si-SiO_2$ interface and in the gate oxide of MOS transistor.

Mobile ions in the gate oxide are mainly positive ions of alkali metals such as Na^+ and K^+, as well as hydrogen ions. They may be incorporated into the oxide during the mere processing of the wafer, or they may reach the oxide, coming from the surrounding surfaces, during the processes of passivation and packing [9, 10]. The existence of these ions in the gate oxide is associated with the fact that during high temperature processes, they easily diffuse into the oxide, from contaminated surfaces. Mobile ions are not distributed uniformly in the oxide, but most of them are located in the semiconductor or metal whereby many more ions are located in the metal as the attractive force towards the metal is much higher than the attractive force towards the semiconductor.

The most important instability mechanism related to electric charges in the gate oxide is the migration of positive ions to the oxide-metal interface and oxide-semiconductor interface under the influence of the positive polarization of the gate and elevated temperature. A series of treatments are taken with the MOS technology in order to minimize these instabilities. Besides the compulsory annealing at high temperature and high level of cleanliness of the technological space and materials used, some modifications in the oxide manufacturing technology have been made in order to minimize instability effects.

Furthermore, a generally accepted procedure for minimizing mobile ions is the introduction of chlorine into the oxide, which is achieved by introducing a certain quantity of HCl or Cl_2 into the oxidation atmosphere or by passing oxygen through trichloroethylene.

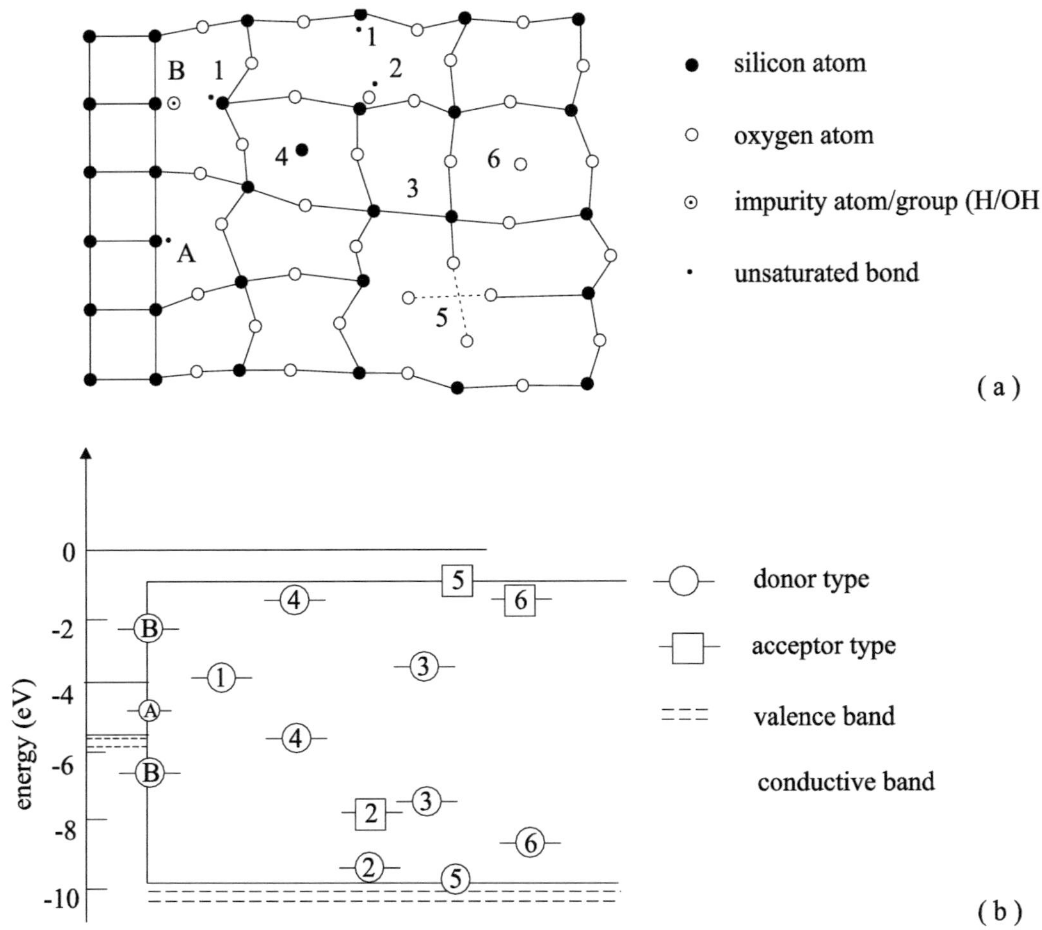

1-trivalent silicon $\equiv Si_o^\bullet$
2-non-bridging oxygen $\equiv Si_o - O^\bullet$
3-oxygen vacancy $V_o(\equiv Si_o^\bullet\,{}_o^\bullet Si \equiv)$
4-silicon intersticy Si_{oi}
5-silicon vacancy V_{Si_o}
6-oxygen intersticy O_i
A-trivalent silicon at interface $\equiv Si^\bullet$
B-silicon and impurity atom group $\equiv Si_s - H, \equiv Si_s - OH$

Fig. 2. Most common defects in oxide and at $Si - SiO_2$ interface: (a) two dimensional model of chemical bonds; (b) energy band model.

Charges at the trap centres are defects and admixed atoms in the oxide which have permissible energy levels in the forbidden zone of the oxide. They are mainly located near the Si – SiO_2 interface, while their number inside the oxide is negligible. Figure 2 shows the model of chemical bonding and diagram of energy zones that illustrate the most significant defects in the oxide and at the Si – SiO_2 interface [4, 11]. Although this is not adequately shown in Figure 2a, it should be taken into account that the Si – SiO_2 interface does not spread through only one atom layer, but rather through several of them. As Figure 2b shows, due to a great difference in the widths of the forbidden zones of silicon (1.12 eV) and oxide (9 eV), discontinuity of energy zones occurs at the interface, and charge carriers encounter a high energy barrier (around 3.2 eV for electrons and 4.7 eV for holes) which in normal conditions prevents any movement of electrons and holes from the semiconductor into the oxide.

The origin of the traps varies. They may be formed both on ≡Si-O- bonds and on defects in the oxide (oxygen vacancies, ≡Si-H and ≡Si-OH bonds, other admixed atoms, etc.). This means that there are different energy levels of the traps depending on the type and quality of the oxide. The traps may be neutral or charged either positively or negatively. During exploitation in certain conditions, they may become electrically charged by capturing electrons or holes, and neutralized by releasing the same, a consequence of which is the instability of the density of effective electric charge captured on them. Therefore, the defects are mainly donor-type defects meaning that they may be positively charged or neutral, so that total electric charge in the oxide is positive. It is important to note that holes may be captured more easily than electrons. Thus the ratio between the number of captured holes and the total number of holes injected into the oxide is approximately one, while, on the average, out of 10^5 injected electrons only less than one is captured. Nevertheless, instabilities of charges at the traps caused by the electron capture may be equal to instabilities caused by the capture of holes or even more pronounced. The reason for this lies in the fact that electrons may be more easily injected into the oxide, which increases the possibility of their capture.

The most significant mechanisms that lead to the occurrence of free carriers in the oxide may be classified into three groups: a) tunnelling of electrons and holes from the semiconductor or metal into the oxide, 2) injection of hot electrons and holes from the semiconductor into the oxide, and 3) generation of electron-hole pairs in the mere gate oxide which may be a consequence of various factors such as ionizing radiation effects, photo-generation or application of a strong electric field. To the traps in the oxide or to the valence or conductive zone of the oxide, holes from the valence zone of the semiconductor or electrons from the conductive zone of the semiconductor or metal may be tunnelled. The processes of tunnelling electrons or holes from the traps into the conductive or valence zone of the oxide, or directly into the semiconductor are possible, which by analogy leads to instability of the charge density at the traps.

There are four mechanisms related to the injection of hot carriers from the semiconductor into the gate oxide. These are: 1) substrate hot electron injection, 2) channel hot electron injection, 3) drain avalanche hot carrier injection (whereby the injection of holes is preferred when the normal component of the field is directed towards the oxide, while the injection of electrons is preferred when the field is directed towards the semiconductor), and 4) hot electron injection caused by secondary ionization by holes in the substrate. It should be noted that such changes in the density of charges at the traps are not the only form of instability that occur during the exploitation of MOS transistors. Namely, new traps may be formed which are neutral at the beginning, but which may become charged by capturing electrons and holes.

Fixed charge is located near the $Si-SiO_2$ interface at distances smaller than 8 nm and its density is constant regardless of changes in the threshold voltage of MOS components. This charge is always positive as it mainly consists of unsaturated bonds of silicon near the Si - SiO_2 interface. Fixed charge depends on the orientation of crystals and it is considerably higher with crystal orientation (111) than with crystal orientation (100). The quantity of fixed charge depends on the Technological conditions of oxidation (atmosphere and temperature). The oxide burn-in in nitrogen or argon at temperatures higher than 600°C decreases the density of fixed charge. This is one of the reasons for which the oxide burn-in has been introduced as a standard procedure in the manufacture of CMOS integrated circuits. It should be emphasized that instability of fixed charges is rarely manifested separately from changes in other charges in the oxide. Instabilities of this type are commonly followed by simultaneous changes in charges at interface traps and oxide traps. This is why in literature fixed charge is mentioned in a much broader sense, although this term also refers to charge in the traps particularly considering the instability of charge in them. Since the origin of fixed charge and traps is frequently the same, it may be assumed that they differ in instability mechanisms. Namely, instability of charge in the traps is a consequence of capturing and releasing free electrons and holes in the oxide, while the instability of fixed charge reflects in its increase as a result of appropriate chemical reactions. One of such reactions is the dissociation of the *Si – H* bond near the $Si - SiO_2$ interface. Under the influence of modest negative polarization of the gate and increased temperature, the *Si – H* group reacts with the *Si – O* group, whereby a interface trap is formed on the Si atom from the semiconductor, while released hydrogen and oxygen form an OH group which diffuses towards the interface. After the transfer of one electron into the semiconductor, the *Si* atom in the oxide remains positively charged and this represents the fixed charge. However, it is often impossible to determine a strict boundary between fixed charge and charge at the traps.

Charges at interface traps represent a significant group of electric charges in the oxide. Interface traps are located at the Si - SiO_2 interface the energy levels of which are in the forbidden zone of the semiconductor. These traps develop as a consequence of the semiconductor structure breaking because surface atoms remain with one unsaturated bond. This means that on clean areas of the semiconductor the number of interface traps is approximately equal to the number of surface atoms. However, thermal oxidation reduces the number of interface traps by several orders of magnitude as a great number of silicon atom bonds become saturated by bonding to oxygen atoms. From the aspect of the origin of interface traps, there is a complete analogy between them and fixed charge and thus with the trap centres. The difference is that fixed charge and the trap centres are in the oxide, while interface traps are at the Si - SiO_2 interface and they are easily accessible to free carriers from the semiconductor. The exchange of charge between interface traps and the semiconductor is relatively fast, so that the density of charge in interface traps at a given moment depends on the distribution of the interface traps energy levels and the position of the Fermi level in the semiconductor forbidden zone. Although the process of thermal oxidation considerably decreases the density of interface traps, an additional decrease is needed for a proper operation of MOS components. Today, a procedure commonly used is the oxide burn-in after metallization performed in nitrogen or in a mixture of nitrogen and hydrogen. In this manner, hydrogen is directly introduced in the oxide, or it is formed after the reaction of water and aluminium. This procedure decreases both the density of interface traps and density of fixed charge and therefore it is applied in almost all standard technologies. Another favourable circumstance is that oxides that increase in the presence of hydrogen chloride also contain, in addition to chlorine that neutralizes mobile ions, hydrogen that decreases the density of interface traps.

The mechanisms of instability of charges in interface traps are a consequence of the formation of new interface traps. Almost all of the described mechanisms may be monitored through the increase in the density of interface traps. The mechanism of fixed charge formation has already been described, whereby interface traps are formed simultaneously. The ion drift towards the Si - SiO_2 interface may lead to an increase in the density of interface traps. It has been proved that the processes of electric charge injection in the oxide most easily form new interface traps that may be achieved by the IR and HCI processes. It is important to point out that owing to the passage of electrons through the Si - SiO_2 interface new interface traps may be formed whereby electrons will not be captured by the traps, which means that the density of charge in them will not change.

As it has already been mentioned, interface traps exchange charge with the semiconductor very rapidly, while this is not the case with the trap centres. On the other hand, a considerable number of defects in the oxide are located in the close vicinity of the interface and they may, like interface traps and regardless of the fact that their microstructure is the same as that of defects deeper in the oxide, exchange charge with silicon. However, unlike real interface traps, these defects are separated from charges in silicon by an energy barrier so that the exchange of charge between them and silicon is carried out through carrier tunnelling whereby the duration of such a process depends to a great extent on the interface distance and the barrier height, and it may range from 1 μs to several years. The long period over which these defects exchange charge with silicon indicates that there is no strict boundary between the effects of charge in the oxide and the effects of interface traps on the electric parameters of MOS transistors. This is why in literature these defects are called boundary traps [12-15]. They may act both as the traps in the oxide and as interface traps. Such nature of the boundary traps may lead to problems in the interpretation of electric measuring results and to difficulties in separating the effects of interface traps and electric charge in the oxide since, depending on measuring conditions (values of the polarization voltage, signal frequency), the boundary traps may behave as interface traps or/and traps in the oxide. By combining appropriate measuring techniques and analysis it is possible to separate the effects of electric charge in the boundary traps [16, 17], which is particularly significant for the characterization of MOS components with a thin gate oxide, where all defects in the oxide are located near the interface and where these defects with their density values are comparable to or even exceed interface traps [14]. In case of thick oxides, as with power MOS transistors, the standard division into the traps in the oxide and interface traps is generally acceptable. In this case, the participation of boundary centres is that at the electric characterization of components will be registered as interface traps or the traps in the oxide, to a great extent depends on the measuring speed and the frequency of the signal that is used during the measurement.

3. Classification of the traps according to their effects on electric characteristics

According to the above, it may be concluded that all charges may be divided into two groups: charges in the oxide and charges at interface traps. Changes of charge densities in the oxide ΔN_{ot} and interface traps ΔN_{it} are directly responsible for changes of sub-threshold characteristics of MOS transistors during the IR and HCI processes. The participation of a certain type of defects in the values of ΔN_{ot} and ΔN_{it} primarily depends on their electric effects on the charge carriers in the channel of MOS transistors, as well as on their location in space. If the defects may capture carriers from the channel, due to which the sub-threshold characteristic slope decreases, it is deemed that they behave as interface traps and that value ΔN_{it} depends on them. If the defects act on carriers in the channel by attracting or rejecting them with Coulomb forces (depending on the mark of their electric charge), sub-threshold characteristics will be changed due to the change of ΔN_{ot}. It should be emphasized that the influence of defects developed in the IR and HCI processes on ΔN_{ot} and ΔN_{it} is determined primarily by their electric effects and after by their location. This indicates that some defects located at the *Si* – SiO_2 interface may behave as defects in the oxide (influence ΔN_{ot}), while some defects located in the oxide may behave as interface traps (influence ΔN_{it}) [3, 4, 18]. Although the influence of the location of defects on sub-threshold characteristics is smaller than their electric impact, it should not be neglected. For example, during recording sub-threshold characteristics (which is a fast process), only part of these defects may capture carriers from the channel.

The most common techniques for determining densities of the traps in the oxide and at the Si – SiO_2 interface are subthreshold-midgap (SMG) [19] and charge pumping (CP) techniques [20, 21] (a detailed description of these techniques is given in Section 6). Based on the position of the traps detected by these techniques, a new classification of the centres became necessary as with the use of the SMG technique both ΔN_{ot} (SMG) and ΔN_{it} (SMG) may be determined, while with the use of the CP technique only ΔN_{it} (CP) may be determined. The use of these two techniques is suitable as there is a great difference in effective frequencies (from several Hz for the SMG technique to 1 MHz for the CP technique). As a slower technique, SMG is used for determining two types of centres: fixed traps (FT) with a density of $\Delta N_{ft} \equiv \Delta N_{ot}$ (SMG) and switching traps (ST) with a density of $\Delta N_{st} \equiv \Delta N_{it}$ (SMG) [22, 23].

Fixed traps (FT) are traps in the oxide that do not exchange charges with Si substrate during recording sub-threshold characteristics by the SMG technique. They are commonly located deeper in the oxide, although they may also be near or at the Si – SiO_2 interface. A significant characteristic of FT is that they may be permanently annealed [22, 23].

Switching traps (ST) may be divided into faster switching traps (FST) and slower switching traps (SST). ST may exchange charges with Si substrate during the measurement. The speed of ST depends on their distance from the Si – SiO_2 interface [24]. This indicates that all defects developed in the IR or HCI processes, which are located near this interface and are capable of exchanging charges with the substrate, become part of ST. Furthermore, some ST may be located deeper in the oxide so that there is not sufficient time for them to exchange, during the the measurement, electric charges with carriers in the channel, meaning that ΔN_{ft} may contain a certain number of ST. FST which are located at the mere interface and which represent true interface traps (with density ΔN_{it}, i.e. meaning that ΔN_{it}(CP) $\equiv \Delta N_{fst}$). SST (with density ΔN_{sst}) located in the oxide, near the *Si* – SiO_2 interface are frequently called slow states (SS) [21], anomalous positive charge (APC) [25], switching oxide traps (SOT) [26] or border traps (BT) [24].

Finally, it should be pointed out that during the measurement by the SMG technique, ST exchange charges during the measurement meaning that in case of an n-channel MOS transistor (NMOS) these centres capture charges from the channel but releases them when the measurement is finished. It has been shown that ST may be temporarily annealed (compensation process) or permanently annealed (neutralization process), however not during the measurement but by exposing components to elevated temperatures.

3.1 Defects in SiO_2 and Si – SiO_2 interface caused by IR and HCI processes

3.1.1 Formation of electron-hole pairs in SiO_2

IR and HCI processes lead to the formation of similar defects in SiO_2 since the electrons play a crucial role in the energy transfer to the oxide in both cases, and the difference is in the electron energy. Namely, the secondary electrons released in the oxide by the gamma photons (IR), and the hot electrons (HCI), and, eventually, the secondary electrons released by the hot electrons play a main role in defect creations. However, it has been shown [22, 27-31] that a difference in electron energy has a significant influence on the created defect types. For instance, the highly energetic secondary electrons (the IR case) produce significantly less negatively charged FT than the low-energy hot electrons (the HCI case).

In the case of IR, the gamma photons interact with the electrons in the SiO_2 molecules mainly via Compton effect, releasing secondary electrons and holes, i.e. gamma photons break Si_0 – O and Si_0 – Si_0 covalent bonds in the oxide [3] (the index $_0$ is used to denote the oxide). The released electrons (so called "secondary electrons") which are highly energetic, may be recombined by holes at the place of production, or may escape recombination. The secondary electrons that escape recombination with holes travel some distance until they leave the oxide, losing their kinetic energy through the collisions with other secondary electrons or, what is more probable, with the bonded electrons in Si_0 – O and Si_0 – Si_0 covalent bonds in the oxide, releasing more secondary electrons (the latter bond represents an oxygen vacancy).

Each secondary electron, before it has left the oxide or been recombined by the hole, can break a lot of covalent bonds in the oxide producing a lot of new secondary highly energetic electrons, since its energy is usually much higher than an impact ionizing process energy (energy of 18 eV is necessary for the creation of one electron-hole pair [3], i.e. for electron ionization).

In the case of IR processes, gamma radiation breaks both covalent bonds in SiO_2 between oxygen atoms and weak $\equiv Si_0$ – H and $\equiv Si_0$ – OH bonds in the oxide. These processes develop in the following reactions [32, 33]

$$O_3 \equiv Si_o - O - Si_o \equiv \xrightarrow{h\nu} \equiv Si_o - O^\bullet + e^- + h^+, \tag{1}$$

$$\equiv Si_o - H \equiv \xrightarrow{h\nu} Si_o^\bullet + H^\bullet + e^- + h^+, \tag{2}$$

$$\equiv Si_o - OH \equiv \xrightarrow{h\nu} Si_o^\bullet + OH^\bullet + e^- + h^+. \tag{3}$$

As it can be noted, these reactions lead to the formation of electron-hole pairs (e^- - h^+). A part of the electrons formed by reactions (1)-(3) break covalent bonds in the oxide (reaction (1)). It is obvious that the secondary electrons play a more important role in bond breaking than highly energetic photons, as a consequence of the difference in their effective masses, i.e. in their effective cross section. The electrons leaving the place of production escape the oxide rapidly (for several picoseconds), but the holes remain in the oxide.

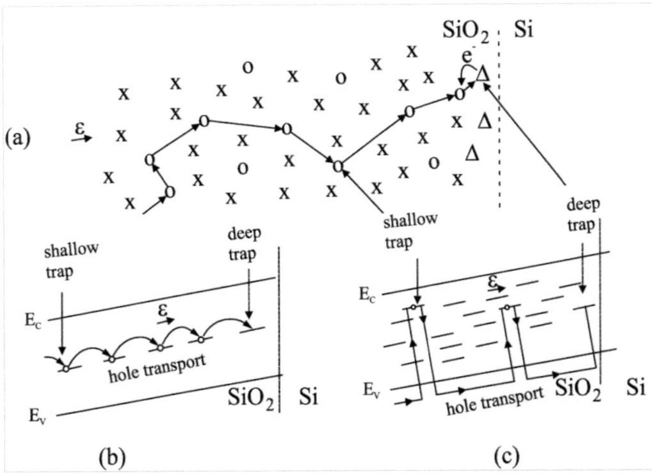

Fig. 3. Space diagram: (a) hole transport through the oxide bulk in the case of positive gate bias. "x" represents unbroken bonds and "o" represents broken bonds (trapped holes at shallow traps), respectively, and "Δ" represents the hole trap precursors near the interface (precursor of a deep trap). Energetic diagram: the hole transport (b) by tunneling between two localized traps and (c) by the oxide valence band.

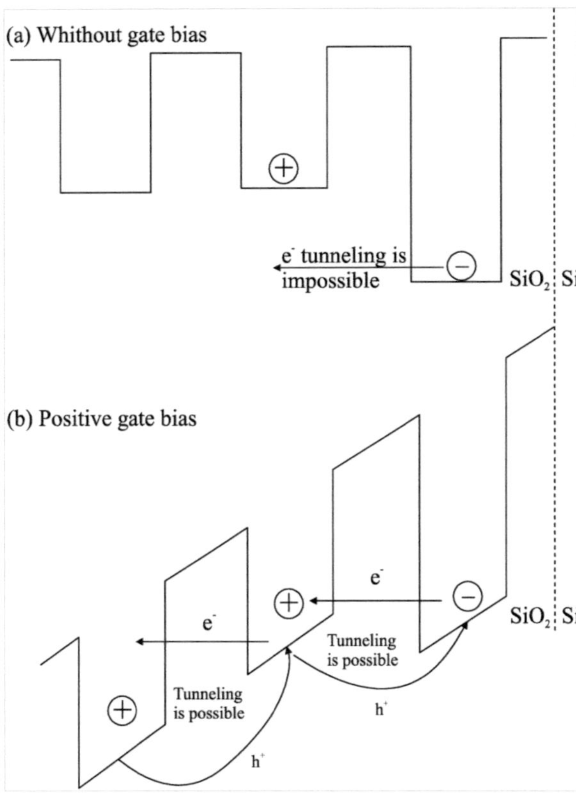

Fig. 4. The electron tunneling between adjacent centers: (a) shallow centre and (b) deep centre.

The holes released in the oxide bulk are usually only temporary, but not permanently captured at the place of production, since there are no energetically deeper centres in the oxide bulk. The holes move towards one of the interfaces (SiO_2 - Si or gate-SiO_2), depending on the oxide electric field direction, where they are trapped at energetically deeper trap hole centres (see Figures 3 and 4). Moreover, even in the zero gate voltage case, the electrical potential due to a work function difference between the gate and substrate is high enough for partial or complete holes moving towards the interface. In addition, some electrons could be trapped at the electron trapping centres of electron capture, but this is not very probable in the IR case [30].

In the case of HCI (e.g. Fowler-Nordheim tunnelling into the gate oxide), the hot electrons, having significantly lower energies than the secondary electrons created by IR, are injected either from the substrate or from the gate. Two cases should be considered: (1) thin oxide ($d_{ox} \leq 10$ nm) and (2) thick oxide ($d_{ox} > 10$ nm). In the former case, the hot electrons, tunnelling into the oxide conduction band, pass the oxide without collisions, since their travel distance is short and, consequently, the probability of the collision with the electrons in the covalent oxide bonds is low [34, 35]. The hot electrons reach the gate, if they are injected from the substrate, or reach the substrate, if they are injected from the gate, where they generate electron-hole pairs. The released (secondary) electrons are attracted by the gate or the substrate electrode, depending on the location of their release, while the generated holes are injected into the oxide.

In the case of thick oxide ($d_{ox} > 10$ nm), a hot electron passing through the oxide has to collide with the electrons bonded in the oxide, giving a certain amount of energy to them and performing the impact ionization [34, 36, 37]. The hot electrons, whose energies are usually higher than 9 eV representing the SiO_2 band gap, can break the Si_0 - O and Si_0 - Si_0 covalent bonds in the oxide by impact ionization and create electron-hole pairs.

The hot and secondary electrons after certain collisions with each other, and what are more probable, with bonded electrons, escape the oxide rapidly, but the holes transport towards one of the interfaces. The high electric field forces the holes to leave their production places, since there is no significant concentration of energetically deep trapping hole created in the oxide bulk, but these centres exist near the interfaces (see Figure 3). Besides, the hot and secondary electrons can produce more electron-hole pairs before they leave the oxide. In addition, some hot/secondary electrons could be trapped at electron centres creating negative charge, which is a very probable mechanism [31].

3.1.2 The defects created by electrons in the impact ionization process

The hot (HCI) and secondary electrons (IR and HCI) passing through the oxide bulk collide with bonded electrons in the most numerous bond: the non-strained silicon-oxygen bond, $\equiv Si_0 - O - Si \equiv$ (it could be also represented as $\equiv Si_0{\bullet\bullet}O{\bullet\bullet}Si_0 \equiv$, where •• represents two electrons making a covalent bond), by reaction [23]

$$\equiv \mathrm{Si_o - O - Si_o} \equiv \xrightarrow{e^-} \equiv \mathrm{Si_o - O^{\bullet}} + {}^{+}\mathrm{Si_o} \equiv +2e^-, \tag{4}$$

$$\mathrm{Si_o - O - Si_o} \equiv \xrightarrow{e^-} \equiv \mathrm{Si_o - O^{+}} + {}^{\bullet}\mathrm{Si_o} \equiv +2e^-, \tag{5}$$

where ≡, the indices $_0$ and $^\bullet$ denote three Si_0 - O bonds ($O_3 \equiv Si_0$ - O), silicon atom in the oxide and an electron, respectively.

Reaction (4) represents the most probable reaction in the oxide bulk, since these are the most numerous centres. The formed $\equiv Si_o - O^{+\bullet}Si_o \equiv$ complex is energetically very shallow, representing the temporary hole centre (the trapped holes can easily leave it [11]). The creation of $\equiv Si_o - O^{+\bullet}Si_o \equiv$ complex, given in reaction (5) is less probable (the $^{\bullet}$ denotes the unpaired electron, i.e. "uncoupled spin").
Nevertheless, the strained silicon-oxygen bond ($\equiv Si_0$ - O - $Si_0 \equiv$), mainly distributed near the interfaces, can also be easily broken by the passing hot/secondary electrons, usually creating the amphoteric non-bridging-oxygen (NBO) centre $\equiv Si_o - O^{\bullet}$, and positively charged E' centre, $\equiv Si_0^+$ [38] known as an E'_s centre [39] (the same process as in reaction (4)). A NBO centre is an amphoteric defect that could be more easily negatively charged than positively by trapping an electron:

$$\equiv Si_o - O^{\bullet} + e^- \rightarrow \equiv Si_o - O^- . \tag{6}$$

Obviously, the NBO centre, as an energetically deeper centre, is the main precursor of the negatively charged traps (defect) in the oxide bulk and interface regions.
A hot/secondary electron passing through the oxide can also collide with an electron in the strained oxygen vacancy bond $\equiv Si_o - Si_o \equiv$, i.e. $\equiv Si_o^{\bullet\bullet}Si_o \equiv$, which is a precursor of an E'_γ centre ($\equiv Si_o^{\bullet}$) [40], breaking this bond and knocking out an electron:

$$\equiv Si_o^{\bullet\bullet}Si_o \equiv \rightarrow \equiv Si_o^+ + {}^{\bullet}Si_o \equiv +2e^- \tag{7}$$

Moreover, this is a conventional structural model of the E'_γ centre: a hole trapped at an oxygen monovacancy ($\equiv Si_o^{+\bullet}Si_o \equiv$). The oxygen vacancy bonds are mainly distributed in the vicinity of interfaces.
The mentioned reactions could occur anywhere in the oxide: near the interfaces and in the oxide bulk. The trapped charge can be positive (oxide trapped holes) and negative (oxide trapped electrons) and the former is more important, since the holes trapping centres are numerous, including three types (E'_s, E'_γ and NOB centres), compared with one electron capture (NBO). The holes and electrons captured near the Si - SiO_2 interface have the greatest effect on MOSFET characteristics, since they have the strongest influence on the channel carriers.

3.1.3 Hole transport

The holes trapped at $\equiv Si_0^+$ centres form oxygen vacancies, and strained silicon-oxygen bonds are energetically deep and steady, at which the holes can remain for a longer time period, i.e. their filling with electrons is more difficult in relation to some shallowly captured holes. These centres exist near both interfaces, especially near the Si - SiO_2 interface. The holes created and trapped at the bulk defects (reaction (4)), representing energetically shallow centres, are forced to move towards one of the interface under the electric field, where they are trapped at deeper traps, since there are a lot of oxygen vacancies, as well as a lot of strained silicon-oxygen bonds near the interfaces, grouping all trapped positive charge there. The holes leave the energetically shallow centres in the oxide spontaneously and they are transported to the interface by a hopping process using either shallow centres in the oxide (Figure 3(b); the holes "hop" from one to another centre) or centres in the oxide valence band (Figure 3(c)) [41]. Figure 3 displays the holes transport in the space for the positive gate bias (a) and the energetic diagram for the possible mechanisms of this space process ((b) and (c)).

Figure 4 shows a possible hole (electron) tunnelling mechanism between adjacent centres: a shallow centre and deep centre. Figure 4(a) shows the case without gate bias, for the Si – SiO_2 interface. In this case, the holes, i.e. electrons tunnelling between these centres, are not possible. The holes are factious species representing the electron unoccupied places, and the hole movement is in fact the movement of bonded electrons, however the electron movement is the movement of free electrons.

When the MOS transistor is positively biased (Figure 4(b)), the bonded electron can tunnel from the deep centre to the shallow centre. It represents the hole tunnelling from shallow to deep centres, being permanently trapped at a deep centre. The electron, which is now in the shallow centre, can easily tunnel from this shallow centre to the next adjacent shallow centre, enabling the holes transport towards the interface. The pictures are similar for the negative gate bias [31].

3.1.4 The defects created by the holes

The holes can be created either in the oxide bulk (thick oxide) or in the gate/substrate and injected into the oxide bulk (thin oxide). Moving throughout the oxide, the holes can react with the hydrogen defect precursors ($\equiv Si_0$ – H and $\equiv Si_0$ – OH) and create the following defects [42].

$$\equiv Si_o - H + h^+ \rightarrow \equiv Si_o^\bullet + H^+, \tag{8}$$

$$\equiv Si_o - H + h^+ \rightarrow \equiv Si_o^+ + H^o, \tag{9}$$

$$\equiv Si_o - OH + h^+ \rightarrow \equiv Si_o - O^\bullet + H^+, \tag{10}$$

$$\equiv Si_o - OH + h^+ \rightarrow \equiv Si_o - O^+ + H^o, \tag{11}$$

The hot and secondary electrons may also interact, in a similar way, with these defect precursors. However, these precursors are not so important for the hot and secondary electrons, since their concentration is significantly lower than the concentration of the precursors representing the non-strained silicon-oxygen bonds. Moreover, it could not be expected that the holes break bonds in the non-strained silicon-oxygen bonds. Because of that, the precursors described in reactions (8-11) are only important for the holes that are transported through the oxide since the hydrogen is relatively weakly bounded and the holes can easily break these bonds. Reactions (8-11) are particularly important for the creation of the interface trap since they produce the hydrogen ion H^+ and hydrogen atom H°, which take part in defect creation at the Si – SiO_2 interface. The existence of these precursors in the oxide and, particularly, at the Si – SO_2 interface is very reasonable, since annealing in the hydrogen ambient is a standard step during the manufacture of numerous MOS devices.

When the holes reach the interface, they can break both the strained oxygen vacancy bonds $\equiv Si_0 - Si_0 \equiv$, forming $E_\gamma^{'}$ centres [38]

$$\equiv \mathrm{Si_o} - \mathrm{Si_o} \equiv + \mathrm{h}^+ \rightarrow \equiv \mathrm{Si_o^\bullet} + \equiv \mathrm{Si_o^+} \qquad (12)$$

and the strained silicon-oxygen bonds ≡ Si_0 – O – Si_0 ≡, creating the amphoteric NBO centres $\equiv \mathrm{Si_o} - \mathrm{O}^\bullet$ and $E_s^{'}$ ($\equiv \mathrm{Si_o^+}$) centres:

$$\equiv \mathrm{Si_o} - \mathrm{O} - \mathrm{Si_o} \equiv + \mathrm{h}^+ \rightarrow \mathrm{Si_o} - \mathrm{O}^\bullet \equiv \mathrm{Si_o^+} \qquad (13)$$

It could be assumed that the strained Si – O bonds, ≡ Si_0 – O – Si_0 ≡ and oxygen vacancy, ≡ Si_0 – Si_0 ≡, represent the main defect precursors in the oxide bulk and at the interfaces. However, these precursors exist mainly near the interface. It should be noted that $E_\gamma^{'}$, $E_s^{'}$ (reactions (12) and (13)) and NBO centre (reaction (13)) represent energetically deeper hole and electron trapping centres, respectively. The energetic levels of the defects created after the holes at $E_\gamma^{'}$ and $E_s^{'}$ centres and electrons at NBO centre, respectively, have been trapped, can be various. In the chemical sense same defects show different behaviours depending on the whole bond structure: the angles and distances between the surrounding atoms. It could be assumed that these defects represent fixed traps (FT) and slow switching traps (SST) [29, 30]. They can be positively and negatively charged, as well as neutral.

In the literature, there are many controversies about the oxide defects types induced by the IR and HCI processes and there are different names for the same or similar defects (see [23] for more details).

3.1.5 Gate oxide/substrate (SiO_2 – Si) interface

The defects at the Si — SiO_2 interface, known as the fast switching traps (FST), or true interface traps, represent an amphoteric defect $\mathrm{Si_3} \equiv \mathrm{Si_s^\bullet}$ a silicon atom $\equiv \mathrm{Si_s^\bullet}$ at the Si — SiO_2 interface back bonded to three silicon atoms from the substrate ≡ Si_3 and the FST are usually denoted as $\equiv \mathrm{Si_s^\bullet}$ or $\mathrm{Si}^\bullet$. They can be directly created by incident photons/hot electrons tunnelling from the substrate or the gate [43, 44], however this amount can be neglected. The direct creation of FST is only emphasized in the case of hot electrons for the positive gate bias applied to thin oxides ($d_{ox} < 10$ nm), where the electrons, tunnelling in the conduction band of oxide, pass the oxide without any collisions (the probability for collision process is small), accelerating themselves and reaching the interface with energy enough for an interface defect creation. Besides direct creation, the FST are mainly created by trapped holes (h^+ model) [45-48] and by hydrogen released in the oxide (hydrogen-released species model - *H*-model) [49-51].

The h^+ model proposes that a hole trapped near the interface create FST, suggesting that an electron-hole recombination mechanism is responsible [46]. Namely, when holes are trapped near the interface and electrons are subsequently injected from the substrate, recombination occurs. The interface trap may be formed from the energy released by this electron-hole recombination. It was supposed [23] that FST can be created by the reaction:

$$\equiv \mathrm{Si_o^+} + \mathrm{e}^+ + \equiv \mathrm{Si_s} - \mathrm{Si_s} \rightarrow \equiv \mathrm{Si_o^\bullet} + \equiv \mathrm{Si_o^\bullet} + \equiv \mathrm{Si_o^\bullet}\,. \qquad (14)$$

The main shortcoming of the h^+ model is its impossibility in explaining the delayed creation of FST [51]. However, this fact cannot completely disqualify this model, and we can suppose that part of interface traps are created by it.

The H model proposes that H^+ ions, released in the oxide by trapped holes (reactions (8) and (10)), drift towards the Si – SiO_2 interface under the positive electric field. When an H^+ ion arrives at the interface, it picks up an electron from the substrate, becoming a highly reactive hydrogen atom H° [52]:

$$H^+ + e^- \rightarrow H^\circ. \tag{15}$$

Also, according to the *H* model, the hydrogen atoms H° released in reactions (9) and (11) diffuse towards the Si – SiO_2 interface under the existing concentration gradient.
Highly reactive H° atoms react without an energy barrier at the interface producing FST throughout the following reactions [53-55]:
(i) The creation of interface trap $Si_s^\bullet$, when H° reacts with an interface trap precursor Si_s – H [52]

$$Si_s - H + H^o \rightarrow Si_o^\bullet + H_2. \tag{16}$$

or an interface trap precursor Si_s - OH:

$$Si_s - OH + H^o \rightarrow Si_o^\bullet + H_2O. \tag{17}$$

Many investigations have shown that the P_{b_0} and P_{b_1} centres which exist at the (100) interface represent interface traps [56-58]. The P_{b_0} defect has the following structure: $Si_s \equiv Si_{s_0}^\bullet$, but the structure of P_{b_1} defect, here denoted as $\equiv Si_{s_1}^\bullet$, is not known (the remaining three bonds of the $\equiv Si_{s_1}^\bullet$ defect can be bonded to various species [59]).
(ii) The passivation of the interface traps, when H° reacts with previously formed (fabrication process or IR and HCI processes induced) interface trap [53, 60]

$$Si_s^\bullet + H^o \rightarrow Si_s - H. \tag{18}$$

(iii) Dimerization of hydrogen, when H° reacts with another H° also existing near the interface [61, 62]

$$H^\circ + H^\circ \rightarrow H_2 \tag{19}$$

Reactions (16) and (17) are most probable at the start of the interface trap creation, since both have a large number of Si_s - H(OH) precursors, and the effective cross sections for reactions (16) and (17) are higher than for reactions (18) and (19) [52].

4. Annealing of MOS components after IR and HCI processes

For the electric characteristics of MOS components, changes of charge in the oxide and changes in interface traps are significant both during the IR and HCI processes and after the end of these processes. Some of the mechanisms occurring after the end of the IR and HCI processes remain active, and new mechanisms also develop and last for a longer period of time. The generation of charge in the oxide takes place as long as an IR or HCI process goes on. When these processes end, the density of charge in the oxide starts to decrease, while the generation of interface traps continues and this process may occur for a longer period of time [61, 63-73]. A process during which positive charge density in the oxide decreases with a simultaneous increase in interface traps density after an IR or HCI process, in which a large quantity of positive charge was incorporated in the oxide, is called annealing. This term is not fully adequate although it is frequently used, as only charge in the oxide is commonly "annealed" while the density of interface traps may increase. MOS components annealing may be performed spontaneously (in air atmosphere, at room temperature, without polarization), or at lowered or elevated temperatures, or in the presence of various intensity and direction electric field.

A number of experimental results [24, 74-80] have shown that the density of interface traps during annealing mainly does not change, while the density of trapped charge in the oxide decreases. Several models are proposed for the trapped charge annealing [75, 77]. It has been observed that during annealing part of trapped charge is permanently burnt-in (neutralized), while the other part may only be compensated, which leads to the effect of "inverse burn-in" (apparent increase in the density of trapped charge during annealing at gate negative polarization), which is attributed, by a group of authors, to the existence of the so-called Switching Oxide Trap (SOT) centres [77, 78]. Another group of authors attribute it to the existence of the so-called Anomalous Positive Charge (APC) centres [24, 79, 80]. Both groups explain inverse burn-in by the exchange of charge with substrate whereby such exchange, in case of SOT, is performed by electron tunnelling as the appropriate levels of these centres lay at the height of the substrate conductive zone [78], while APC are, similarly to interface traps, capable of performing direct exchange with the substrate as their energy levels are at the level of the forbidden zone [24, 79, 80]. Regardless of their different approaches, both groups have practically contributed to establishing the fact that SOT and APC are most probably the same centre, i.e. E' centre that, depending on the conditions of the experiment and measurement may exhibit various properties (burn-in, inverse burn-in or exchange of charges) and may be registered as a interface trap or the so-called boundary centre of capture.

A characteristic of these models of charge burn-in is that they are related to captured charge in the oxide and that they mainly do not exchange charge at interface traps, i.e. that they consider changes in the density of captured charge during annealing separately and independently of changes in the density of interface traps. This may be deemed the main disadvantage of this group of models, particularly because there are numerous experimental results that may be very accurately described by the above-described h^+ and H models. It is important to point out that according to h^+ and H models, the mechanisms of the formation of defects in the oxide and at the $Si - SiO_2$ interface take place not only during the IR or HCI processes, but after their cessation as well. This practically means that these two models are more comprehensive and more acceptable for explaining the annealing process than the charge burn-in model. It is important to emphasize here that most experimental results indicate the dominant role of hydrogen and that the H model in many cases gives a real picture of the processes in the oxide and at the $Si - SiO_2$ interface. Therefore, the processes of the captured charge density decrease and simultaneous increase in the density of interface traps, which take place within a shorter or longer time after the cessation of the IR or HCI processes, may be best described by the H model. However, this model may not describe the annealing processes in case in which the generation of interface traps begins and ends during mere tension so that their density do not increase during the annealing. In addition, none of the models mentioned so far may explain the phenomenon of a decrease in the density of interface traps during the annealing for a long period [81-83] and a latent increase in the density of interface traps [28,84]. Furthermore, none of the mentioned models may explain the phenomenon of an increase and then decrease in the density of interface traps during annealing for a longer period of time [85-88]. On the other hand, the latent generation of interface traps that occurs during annealing is manifested through rapid increase in their density after reaching apparent saturation, and simultaneous rapid decrease in the density of trapped charge, which is characteristic of MOS transistors the oxide of which contains a high concentration of oxygen and bounded hydrogen [87]. It is deemed that it is not the hydrogen originating from the mere oxide that is responsible for the latent increase in the density of interface traps, but rather the hydrogen which is, during the IR or HCI processes, released in the form of molecules in the adjusted layers (polysilicon gate, protective oxide). The hydrogen molecule diffuses towards the gate oxide and through it further towards the oxide-semiconductor interface in the vicinity of which it reacts with a positively charged centre of capture, according to the following reaction:

$$\equiv Si_0^{+} + H_2 \rightarrow \equiv Si_0 - H + H^{+}. \quad (20)$$

As it can be seen from the last reaction, the captured charge is neutralized and hydrogen ions are released which then diffuse towards the Si – SiO_2 interface and participate in the formation of a interface trap (reactions (16) and (17) [89]). The late increase in the density of interface traps is attributed to the long-lasting diffusion of hydrogen molecules which is particularly difficult through the polysilicon gate, as well as to a possibility that oxygen vacancies slow down the movement of hydrogen in the oxide [67, 89].
This explains the latent increase in the density of interface traps but cannot explain their decrease that occurs after the latent increase. Owing to this, based on experimental results of the annealing of irradiated power VDMOS transistors, the so-called Hydrogen-Water (H – W) model has been proposed, which can explain the behaviour of the densities of interface traps and captured charge during annealing including both the latent increase and the decrease in the density of interface traps [28, 85, 86]. According to this model, all free hydrogen (present in the oxide before the IR or HCI processes or released during these processes) is utilized for the so-called conventional generation of interface traps that ends with the cessation of the IR or HCI processes and/or in the initial stage of annealing immediately after the cessation of these processes. In the further stage of annealing, for initiating the latent generation of interface traps it is necessary and sufficient that a certain quantity (even a small one) of hydrogen ions appears at the interface. The origin of these ions will be discussed later. When these late ions reach the Si - SiO_2 interface under the influence of the electric field, they takeover electrons from the substrate (reaction 15) and form neutral atoms of hydrogen that may participate in one of the following processes:
(i) formation (depassivation) of interface traps (reaction (16)); (ii) pasivation of interface traps (reaction (18)); and (iii) dimerization (reaction (19)). Reaction (18) is more probable than the other two reactions because the concentration of $\equiv Si_s - H$ precursor is higher than the concentration of interface traps ($\equiv Si_s^{\bullet}$), as well as because the cross section of reactants is larger in this reaction in comparison to the other two, so that the density of interface traps begins to increase. In reactions (16) and (19), hydrogen molecules are released and therefore their concentration at the interface increases (concentration gradient occurs) due to which these molecules diffuse towards the inside of the oxide whereby on their way many of them react on with positively charged traps. The following reaction commonly occurs:

$$\equiv Si_0^+ + H_2 \rightarrow \equiv Si_0 - H + H^+. \quad (21)$$

Hydrogen ions formed in this reaction drift towards the interface, takeover electrons from the substrate and the processes (i) - (iii) are repeated. In this manner, for further formation of interface traps an additional source of H^+ ions is provided, and at the same time, burn-in of captured charge is performed, in compliance with results according to which the latent increase in the density of interface traps corresponds to the decrease in the density of captured charge.
Water molecules are deemed responsible for the decrease in the density of interface traps that occurs after the latent increase. Water molecules are mainly bounded (physically or chemically) in the thermal and/or protective CVD oxide. During annealing, they may be released (more expressed at higher temperatures) and after that they slowly diffuse towards the Si – $Si0_2$ interface where they finally react with interface traps according to the following reaction:

$$\equiv Si_s^\bullet + H_2O \rightarrow \equiv Si_s - OH + H^o . \quad (22)$$

As it can be seen from the last reaction, the final contribution of water molecules after long annealing is the passivation of interface traps. Namely, in reaction (22) one interface trap is passivated and a neutral hydrogen atom formed in the same reaction may react with (i) another neutral hydrogen atom (reaction (19)) which will not change the result of reaction (22), meaning that one interface trap remains passivated, (ii) with the $\equiv Si_s^\bullet$ defect (reaction (18)) with which one more interface trap is passivated, or (iii) with the $\equiv Si_s$ – H precursor (reaction (16), the only one with which the effect of reaction (22) is annulled). In this manner, the H – W model covers the latent increase and, afterwards, the decrease in the density of interface traps. Based on this model, it may be concluded that a small quantity of any type of hydrogen particles (ions, neutral atoms or molecules) may trigger the latent generation of interface traps. Although the authors of this model [28] mention a possibility that this generation is triggered by hydrogen ions captured in the traps in the oxide, or by hydrogen molecules formed in reaction (22) from the water molecule after its slow diffusion through the oxide, they still believe that for triggering the latent increase in the density of interface traps, hydrogen molecules originating from the protective CVD oxide and/or polysilicon gate are most probably responsible. The cause of their late arrival to the oxide-semiconductor interface may be the reduced speed of diffusion through polysilicon, even through the gate oxide (as a consequence of a decrease in the diffusion constant in the oxide owing to the IR processes), as well as a possibility that hydrogen molecules may have reduced speed by being captured at oxygen vacancies from which they are subsequently released.

5. Influence of charges in the oxide and interface traps on MOS transistors parameters

Charges in the oxide, as well as interface traps have a significant influence on the characteristics of MOS transistors. The transfer characteristic of MOS transistors in the saturation region may be described by the following expression [2, 4]:

$$I_D = \beta(V_G - V_T)^2 / 2 \, , \quad (23)$$

where β is an amplifying factor and may be expressed as

$$\beta = \frac{\mu W C_{ox}}{L} \ . \tag{24}$$

In expressions (23) and (24), I_D is drain current, V_G is voltage at the gate, V_T is threshold voltage, μ is electric mobility in the channel, $C_{ox} = \varepsilon_{ox}/d_{ox}$ is the capacitance of the gate oxide per area unit (ε_{ox} and d_{ox} represent dielectric constant and oxide thickness, respectively), while W and L are the width and effective length of the channel, respectively. Based on these expressions it may be noticed that charges in the oxide and in interface traps may influence the electric characteristics of MOS transistors only through the influence on threshold voltage and mobility of carriers in the channel. Positive charge in the oxide influences carriers in silicon by its electric field, attracting electrons towards the interface and rejecting holes from it. This leads to a change in threshold voltage which is decreased with NMOS, and increased (by absolute value) with PMOS transistors. At the same time, electrons approaching the interface with NMOS transistors should increase, while holes moving away with PMOS transistors should decrease the dispersal of carriers on the uneven areas of the interface, by which the mobility of carriers would be decreased with NMOS, and increased with PMOS transistors. However, it is deemed that charges in the oxide have a low influence on the carrier mobility and that their impact is not electrostatic but it is rather a consequence of the carrier capturing process [4, 90]. Namely, carrier capture in interface traps will not change the total amount of charge in the channel region, which means that no electric effect is produced. The influence of interface traps may be explained in the following way: interface traps capture a certain number of electrons or holes induced by a change in the surface potential caused by the gate voltage, due to which the formation of a channel requires higher gate voltage comprising an increase in the threshold voltage both with NMOS and PMOS transistors. At the same time, capturing carriers at the interface traps with both types of MOS transistors leads to a decrease in the number of conductive carriers in the channel (to a decrease in current) which is manifested in a decrease in the slope of transfer characteristics, a decrease in the amplifying factor and thus a decrease in the mobility of carriers μ. Namely, the greater the density of interface traps, the greater the number of captured carriers in relation to a total number of induced carriers, which is manifested in an apparently greater mobility degradation. Through such a conclusion, the real situation (decreased number of mobile carriers in the channel) is replaced with an apparent one (unchanged number of mobile carriers with reduced mobility), but the effect on the conductivity in the channel (i.e. on the transistor current) remains absolutely the same. Due to this, the use of a simpler model given in expressions (23) and (24), whereby the so-called effective (or apparent) mobility is used instead of the real one, is much more purposeful than complicated consideration of the real situation which would require the knowledge about the accurate function of the interface trap energy levels distribution in the forbidden zone of silicon. This claim is supported by the definition of the mobility of carriers according to which this mobility is not a physical value but only the coefficient of proportionality in the expression for the dependence of drift rate on electric field. Therefore, the value of mobility is not constant but it is rather necessary to adjust it in order to maintain the applicability of Ohm's law to the transport of current in the semiconductor [4]. Owing to this, the concept of effective mobility is common in modelling the effects of various factors on transport processes in the semiconductor, such as doping effects, effects of electric field,

high levels of injection, surface dispersal, crystal orientation, as well as effects of charge in the oxide and effects of interface traps [4, 91-93]. Further in this text, the term mobility will be used, but it will comprise the effective value.
The considerations given so far suggest that with NMOS transistors a decrease in mobility should be expected with an increase in the density of either interface traps or charge in the oxide. Considering the dominant influence of interface traps, a decrease in mobility in general should be also expected for PMOS transistors, even in the case in which the density of charge in the oxide increases somewhat more rapidly than the density of interface traps. In literature, however, there are examples of an increase in mobility, one of which was recorded in a study of PMOS transistors with a polysilicon gate [94]. In the same experiment, PMOS transistors with an aluminium gate were irradiated and an expected decrease in mobility was recorded, while an expected increase in threshold voltage was almost the same with both types of transistors. By additional analysis, a great increase in the density of charge in the oxide was determined, approximately equal with both types of samples while at the same time an increase in the density of interface traps was much greater with the samples with aluminium gates in relation to the samples with polysilicon gates. An increase in mobility with PMOS transistors with polysilicon gates was explained by their high resistance to the formation of interface traps, i.e. by a disproportionately great increase in the density of charge in the oxide in relation to the increase in the density of interface traps during the IR processes. This result is not in accordance with the conclusion that charge in the oxide has low influence on mobility. It may be deemed even controversial as similar results with PMOS transistors with polysilicon gates [95, 96] showed the opposite result, i.e. reduced mobility. However, regardless of the contradiction of these results, it may be concluded that charge in the oxide has significant influence on the mobility of carriers in the channel, although such influence is lower than the influence of interface traps. This is indicated by the results of researches with PMOS dosimetric transistors which show that a considerable increase in the density of charge in the oxide leads to an increase in mobility [97].
Many experimental results have shown that the influence of charge in the oxide and the influence of interface traps on the threshold voltage of MOS transistors may be modelled by the following expression [98]:

$$V_T = V_{T0} \mp \frac{qN_{ot}}{C_{ox}} + \frac{qN_{it}}{C_{ox}} \tag{25}$$

where V_{T0} is the threshold voltage of an ideal MOS transistor without any charge in the oxide and without interface traps, while the second and third terms include their influence, whereby the sigh "-" in front of the second term refers to an NMOS transistor, while the sign "+" refers to a PMOS transistor. Therefore, the influence of charge in the oxide (qN_{ot}) is included as voltage drop that creates an electric field on the oxide, i.e. qN_{ot}/C_{ox} . On the other hand, it is claimed that the influence of interface traps on carriers in the channel reflects in their capture, so a that part of carriers induced in the channel by the gate voltage remain immobile. If, for given gate voltage, the number of captured carriers from the channel per area unit is qN_{it}, then part of the gate voltage amounting to $qNit/C_{OX}$ is spent not on the formation of mobile carriers but rather on the formation of captured ones. By this amount the gate voltage required for the formation of a channel is increased (according to absolute value), which means that the influence of interface traps on threshold voltage is qN_{it}/C_{ox}.

In order to give consideration to the influence of charge in the oxide and of interface traps on mobility, several models may be used [99-101] which are based on the empirical expression for the dependence of effective mobility on charge in the oxide and the concentration of admixtures in the channel of NMOS transistors given in the paper [102]

$$\mu = \frac{\mu_0}{1+\alpha N_{ot}}, \tag{26}$$

where μ_0 = 3440 – 164 log N_A [cm²/(Vs)] is mobility without charge in the oxide (N_A is the concentration of acceptor impurities expressed in cm³) and $\alpha_{ot} = -1.04 \cdot 10^{-12} + 1.93 \log N_A$ [cm²]. Based on later researches [98] it has been shown that the above expression should be modified taking into account the contribution of interface traps as well. Therefore, mobility may be expressed in the following way:

$$\mu = \frac{\mu_0}{1+\alpha_{it}N_{it} \pm \alpha_{ot}N_{ot}}, \tag{27}$$

where coefficients α_{it} and α_{ot} depend on technology, and due to greater influence of charge on interface traps $\alpha_{it} > \alpha_{ot}$. The sign "+" in front of the third term in the denominator refers to NMOS transistors, while the sign "-" refers to PMOS transistors. This is in accordance with the observation above that charge in the oxide, being almost always positive, may lead to an increase in mobility with PMOS transistors (although this happens rarely owing to the dominant influence of charge on interface traps).

Due to the increase in interface traps, the leakage current of the inversely polarized connection increases as well. Leakage current consists of free electrons and holes generated in the depletion region of the inversely polarized junction, which are transferred through the field into the n-type or p-type semiconductor. The generation of free holes in the presence of interface traps takes place through the transfer of electrons from the valence band to the energy level of interface traps near the Fermi level, while the generation of free electrons takes place through the transfer of electrons from the energy level of interface traps near the Fermi level to the conductive band of the semiconductor. Therefore, electrons pass from the valence band into the conductive band through interface traps. This process leads to the occurrence of leakage current of the inversely polarized junction. The leakage current directly depends on the density of interface traps.

The low-frequency noise of MOS transistors also increases with the increase in the density of interface traps. This noise appears because interface traps capture carriers that move through the channel retaining them for a certain period of time and then emitting them back into the channel. This is a process of accidental nature and it modulates the channel current which may be particularly observed at low frequencies and high densities of interface traps.

The avalanche breakdown of the inversely polarized drain-substrate junction is a consequence of the presence of charge in the gate oxide originating from mobile ions, fixed charge, and charge in the traps. The sum of all these charges gives effective charge per area unit which is almost always positive. This charge transforms the drain-substrate p^+-n junction in PMOS transistors into the p^+-n^+ junction meaning that breakdown voltage is reduced. In NMOS transistors, the substrate is of p-type, and charge in the oxide depletes it

leading to a reduction of the field at the drain-substrate n^+-p junction and thus to an increase in avalanche breakdown voltage.
It is important to emphasize that it is impossible to give an appropriate typical quantitative picture of the described instabilities. There are differences not only in the values of changes in corresponding parameters, but also in the occurring instability mechanisms, and as it has been already mentioned these differences may lead to changes in an observed parameter even in opposite directions. The next important difference is in time dependencies of respective instabilities meaning that even if the same initial instability is manifested with this group of MOS transistors, the instabilities may significantly differ after some time. In addition, time dependences of respective parameters instabilities may have very complex forms as they are frequently defined by simultaneous activities of several mechanisms. Such differences in the behaviour of MOS transistor parameters are conditioned by differences in the manner of the gate oxide formation and by conditions of the component exploitation, and they may lead to qualitatively different forms of instabilities of the MOS transistor parameters.

6. Techniques of determining electric charge density in oxide and interface traps

Change of electric charge density in gate oxide and interface traps, which occur during IR and HCI processes, are studied through the application of certain techniques. An important role in that procedure is played by changes in electric parameters, primarily threshold voltage and amplification factor (considering that the influence of electric charge in oxide and interface traps on electric characteristics are transferred through them), which are determined on the basis of recorded I-V characteristics.
It is already known that a change in threshold voltage of MOS transistors due to IR and HCI processes can be expressed in the following way:

$$\Delta V_T = \Delta V_{ot} + \Delta V_{it}\,, \tag{28}$$

where ΔV_{ot} and ΔV_{it} are contributions to the change in threshold voltage due to the electric charge in oxide and in interface traps, respectively. By using the expression (25), the change in surface electric charge density in oxide ΔN_{ot} and the change in interface traps density ΔN_{it} can be determined as

$$\Delta N_{ot} = \pm\frac{C_{ox}}{q}\Delta V_{ot}\,,\ \Delta N_{it} = \frac{C_{ox}}{q}\Delta V_{it}\,, \tag{29}$$

where the signs "+" and "-" refer to p-channel and n-channel MOS transistors, respectively. ΔV_T (expression (28)) represents an experimentally determined value of threshold voltage $\Delta V_T = V_T - V_T(0)$ and the corresponding change in the electric charge densities in oxide $\Delta N_{ot} = N_{ot} - N_{ot}(0)$ and the interface traps $\Delta N_{it} = N_{it} - N_{it}(0)$ (expression (29)), where $V_T(0)$, $N_{ot}(0)$ and $N_{it}(0)$ are values of corresponding values prior to, and V_T, N_{ot} and N_{it} after IR or HCI processes.
There are several techniques used to determine electric charge density, and/or to separate the influence of electric charge density in gate oxide and interface traps, and each of these techniques has its own advantages and deficiencies [3,103]. Here we will describe the two

techniques most commonly used today. These are subthreshold midgap (SMG) technique which is used to determine electric charge density in gate oxide and interface traps, and charge pumping (CP) technique which is used only to determine change in interface traps density.

6.1 Subthreshold midgap technique

The subthreshold midgap (SMG) technique is based on the analysis of change in the subthreshold transfer characteristics of MOS transistors [19]. Theoretically, the subthreshold characteristic (dependence of drain current on the gate voltage for the given voltage on drain V_{DS}) can be described by the expression [2]:

$$I_D = \mu \frac{W}{L} \cdot \frac{N_A L_D kT}{\sqrt{2}} \left(\frac{n_i}{N_A} \right)^2 \exp\left(\frac{q\varphi_s}{kT} \right) \cdot \left[1 - \exp\left(-\frac{qV_{DS}}{kT} \right) \right] \cdot \sqrt{\frac{kT}{q\varphi_s}}, \tag{30}$$

where N_A is the concentration of impurities in the p-channel area, L_D is Debye length, φ_s is surface potential which represents the voltage function at the gate [2]. The last expression clearly expresses exponential dependence of drain current on the surface potential, and thereby on the gate voltage as well. The subthreshold characteristics of n-channel MOS transistor prior to and after being subjected to the IR process, as well as the necessary elements to explain the technique are given on the Figure 5 [19].

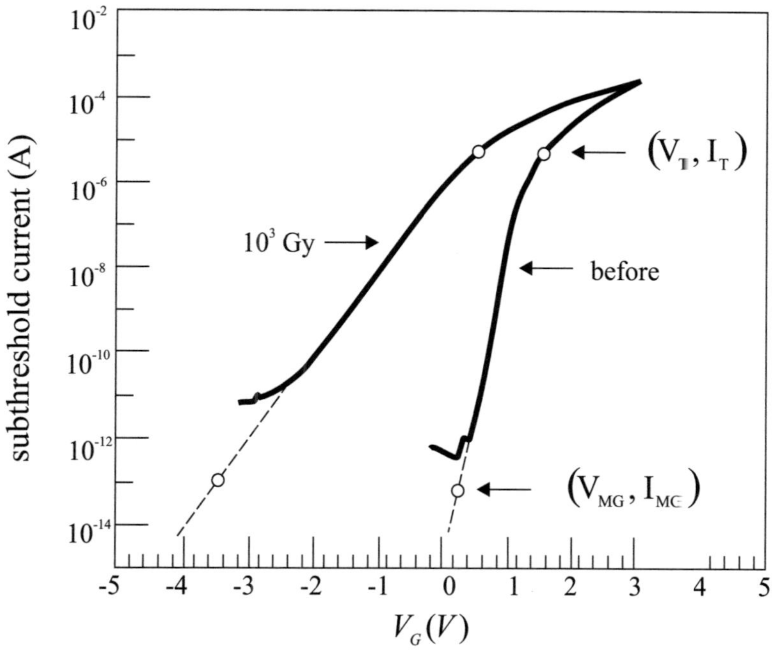

Fig. 5. Subthreshold current as a function of gate voltage, before irradiation and after gamma radiation dose of 10^3 Gy.

From the viewpoint of subthreshold characteristic, an increase in interface traps density, as already mentioned before, under the influence of IR or HCI processes, is manifested through the change in its slope. Namely, as an increase in gate voltage is followed by capture of carriers in interface traps, the "response" of surface potential will be decreased, and along with it, the drain current as well, to the changes in gate voltage, and therefore the voltage of subthreshold characteristic will be milder.

In case when Fermi level of semiconductors on the Si – SiO_2 interface is located in the middle of the forbidden zone, the total electric charge on interface traps, regardless of the disposition within the forbidden zone of the substrate, is equal to zero. This means that the shift on the voltage axis between two subthreshold characteristics in that case is solely the consequence of change in electric charge density in oxide. The gate voltage in which Fermi level on the *Si* – SiO_2 interface is located in the middle of the forbidden zone is marked with V_{MG} (midgap voltage), and is calculated as abscissa of the point (V_{MG}, I_{MG}) on the subthreshold characteristics. The above-mentioned point is determined on the basis of its ordinate, i.e. I_{MG} current (midgap current) which is calculated on the basis of the expression (30) by replacing appropriate values of surface potential $\varphi_s = \varphi_F$ and experimentally determined mobility μ. As the mobility of carriers depends on several parameters such as temperature, admixture concentrations, degree of component degradation (and thereby also the electric charge density in the oxide and interface traps), it must be determined experimentally for each component and for each radiation dose or duration of HCI process. It is determined from the slope of the curve $\sqrt{I_D} = f(V_G)$ [19], as the drain current of MOS transistors in saturation is given through the following expression:

$$I_D = \mu \frac{W}{2L} C_{ox} \left(V_{GS} - V_T\right)^2 . \qquad (31)$$

The I_{MG} current in V_{MG} voltage is very small (order of magnitude of picoampere or smaller, depending on the size of transistor), i.e. it is often significantly smaller than the lowest measured level of subthreshold current, determined by the leakage current. For this reason, the point (V_{MG}, I_{MG}) is located on the extension of the most linear part of subthreshold characteristics marked by the dashed line on Figure 5. Therefore, change in electric charge density in oxide, formed under the influence of IR or HCI process is generated as [19]

$$\Delta N_{ot} = \frac{C_{ox}}{q}\left(V_{MG}(0) - V_{MG}\right) \qquad (32)$$

where $V_{MG}(0)$ is the value of V_{MG} voltage prior to IR or HCI processes.

Other two important points on the Figure 5 are the points (V_T, I_T), which are located on subthreshold characteristics prior to and after the radiation. These points could be determined in a similar way as the points (V_{MG}, I_{MG}), and/or on the basis of the expression (30) and the definition of threshold voltage, and/or on the basis of the expression (30) and the definition of surface potential for the case of strong inversion. However, these points can be determined with highest reliability on the basis of knowledge about their abscissas, i.e. used generated values of threshold voltage, which avoid any ambiguity in the definition of threshold and/or start of strong inversion. The threshold voltage is most frequently determined as section of extrapolated dependence of the square root of drain current in MOS transistor in saturation from the threshold voltage at the gate (expression (31)), with V_G axis.

Presence of interface traps changes the slope of subthreshold characteristics, i.e. the difference between the voltage V_T and V_{MG} which represents the measure of "stretch-out" and is marked as V_{S0} (stretch-out voltage)

$$V_{S0} = V_T - V_{MG} \, , \tag{33}$$

Accordingly, an increase in electric charge density on interface traps which causes change in subthreshold characteristics inclination, i.e. V_{S0} voltage is now more simply determined on the basis of the expression [19]

$$\Delta N_{it} = \frac{C_{ox}}{q} (V_{S0} - V_{S0}(0)) \, , \tag{34}$$

where $V_{S0}(0)$ is the value of stretchout voltage prior to, and V_{S0} after IR or HCI process.

6.1.1 Charge-pumping technique

As opposed to the SMG technique, the charge-pumping (CP) technique does not give changes in charge densities in oxide and in interface traps, but is used solely to determine interface traps density while charge density in oxide can be subsequently determined on the basis of the expression (29) under the condition that the change in threshold voltage is known [20, 21, 104].

The charge-pumping effect can be explained on the basis of the diagram shown in Figure 6 [104]. The source and the drain of the transistor are short-circuited, and the p-n junction of source and drain with the substrate are inversely polarized with V_R voltage. In the absence of signal at the gate, under the influence of inverted polarization at the junction source-substrate and drain-substrate, the inverted saturation current of these connections will flow. When a train of rectangular pulses of sufficiently high amplitude is brought to gate (with pulse generator), a change of current direction in the substrate occurs. The intesity of that current is proportional with the pulse frequency, and "pumping" of the same amount of electric charge towards the substrate. As current cannot flow through oxide, the electric charge in the substrate come through p-n junctions of source and drain. In this way, in the case of n-channel MOS transistors, a channel is formed under the gate in positive pulse half-period, whereby electrons are captured on interface traps. During the negative half-period, when the channel area turns into the state of accumulation, mobile electrons from the channel are returned to the source and drain, and the captured electrons are recombined

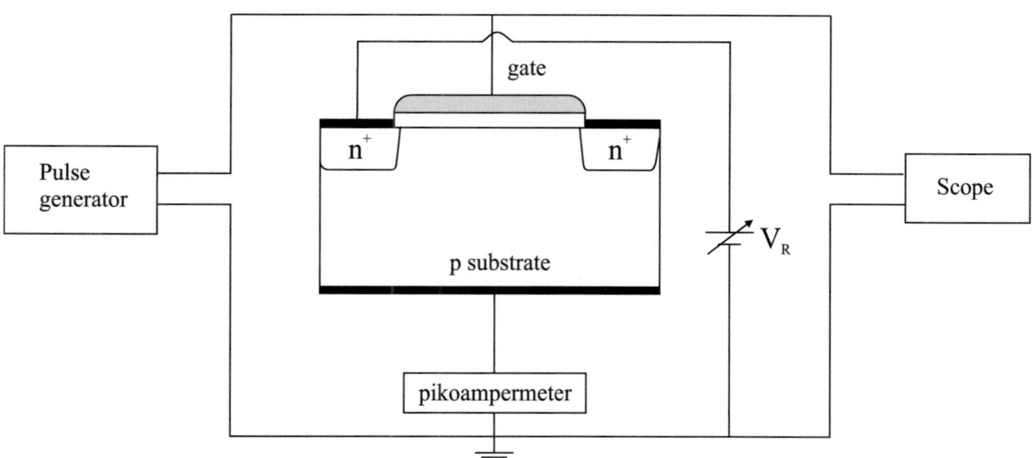

Fig. 6. Schematic diagram of charge pumping measurement.

with holes from the accumulated layer, thereby generating CP current which is proportional with the number of recombination centers, i.e. interface traps:

$$I_{CP} = q \cdot A_G \cdot f \cdot N_{it} \tag{35}$$

where A_G is gate surface, and f is pulse frequency. In order to avoid recombination with the channel electrons, it is necessary to ensure their return to the source and drain before overflow of cavities from the substrate occurs, which is accomplished by using reverse polarization of p-n junction or using a train of trapezoid pulses or triangular pulses with sufficient time for rise time t_r and fall time t_f pulse. However, part of the electrons whose capture is shallowest, are in the meantime thermally emitted into conductive band of the substrate, reducing the width of interface traps energy range measured by the CP technique, so that CP current is generated by interface traps within the range [21]

$$\Delta E = -2kT \ln\left(\upsilon_{th} n_i \sqrt{\sigma_n \sigma_p} \frac{|V_T - V_{FB}|}{|\Delta V_G|} \sqrt{t_r t_f} \right) \tag{36}$$

which is 0.5 eV from the middle of the forbidden band. In the expression (36) υ_{th} is thermal velocity, σ_n and σ_p are cross-section surfaces of carrier captures, n_i is self-concentration of carriers in the semiconductor, and ΔV_G is pulse height

Using of expression (35), interface traps density can be calculated on the basis of the measured maximum value of CP current. Maximum CP current is directly proportional to the pulse frequency, and a small-size transistor with usual state density needs a frequency of at least several kHz to enable the charge-pumping current level reach the order of magnitude of picoampere. Due to this, CP measuring is most often conducted with frequencies in the range between 100 kHz and around 1 MHz, whereby only real (FST) interface traps are registered (in some frequencies, CP is also contributed by part of SST which also captures electrons from the channel [105]).

As the CP technique requires a separate outlet for the substrate, it could be concluded that it is not applicable for power VDMOS transistors, in which the p-bulk is technologically connected to the source. However, thanks to the very structure of these transistors [106], the CP technique is applicable in a somewhat altered form; as shown in Figure 7 [25, 107]. It should be pointed out that VDMOS power transistor represents a parallel connection of a large number of cells (elementary transistor structures) with a large surface, which is especially suitable for the CP technique (higher level of current which is easier to measure).

The role of source and drain, as minority carrier in power VDMOS transistors (Figure 7) are taken over by p^- body areas of two adjacent cells, while the role of substrate (source of majority carriers) is played by n^- epitaxial layer between the cells which is directly connected to the n^+ area of the drain. In this case, the CP current is generated through recombination through interface traps which are located on the interface oxide – n^- substrate, and VDMOS structure acts as a PMOS transistor. In this way, the CP technique practically perform characterization of interface above the epitaxial layer, and not above the channel area, which must be taken into consideration when calculating interface traps density with use of the expression (35), whereby an adequate value of the active surface above the epitaxial layer should be taken as A_G. The oxide above the channel and the oxide above the epitaxial layer have the same characteristics and thickness, as they are made simultaneously in the production process, which also refers to the polysilicon gate. On the other hand, it is known that the surface interface traps do not depend on the type of semiconductor on which the oxide is formed. This means that, within the qualitative analysis, it can be considered that the conclusions attained on the basis of the CP measurements also refer to the interface above the channel surface.

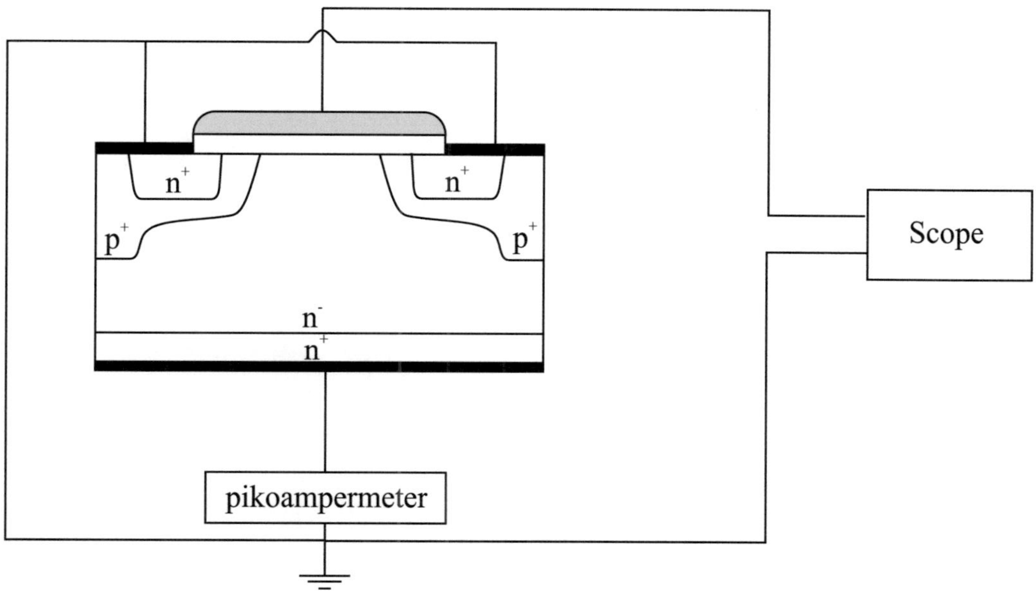

Fig. 7. Schematic diagram of charge pumping measurement for power VDMOS transistor.

7. Some results of IR and HCI processes and later annealing of MOS transistors

7.1 Behavior of CMOS transistors with Al gate from integrated circuits of the CD4007UB type in the course of IR process and subsequent annealing

The results of radiation with gamma rays, X-rays and electrons, as well as of subsequent annealing of MOS transistors with Al-gate from the integrated circuits CD4007UB are displayed in several papers [108-115], in which CMOS (NMOS and PMOS) transistors were used. This made it possible to directly evaluate the instability of electric parameters and characteristics of these integrated circuits on the basis of the data on instability of electric parameters and characteristics of these transistors.

Figures 8 and 9 display changes in the threshold voltage of CMS transistors during the radiation with gamma rays and later annealing on the temperature of 115°C for 0 V gate polarizations [111, 112]. As it can be seen, the threshold voltage in NMOS transistor (Figures 8) decreases down to the value of the absorbed dose of gamma radiation of 200 Gy, and then rises again, while in PMOS transistors (Figure 9], it continually rises. Change in threshold voltage also takes place during the annealing at elevated temperature, but in such a way as to approximate their values to the ones present prior to the radiation. It has been shown [112] that the annealing process can be accelerated by positive polarization on the gate.

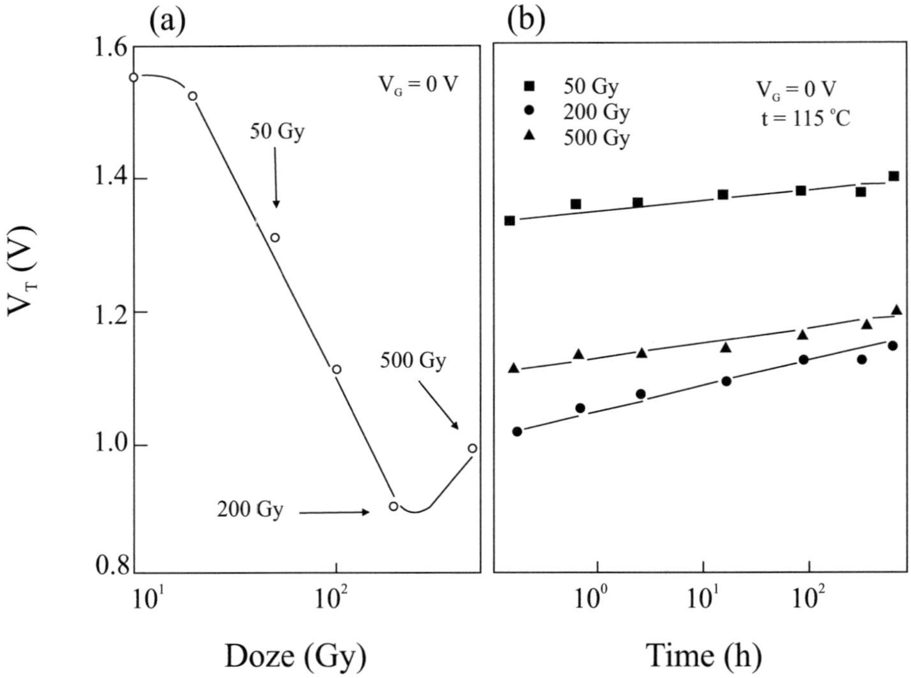

Fig. 8. Threshold voltage shift (ΔV_T) for NMOS transistor during gamma irradiation, V_G = 0 V (a) and thermal annealing, V_G = 0 V (b).

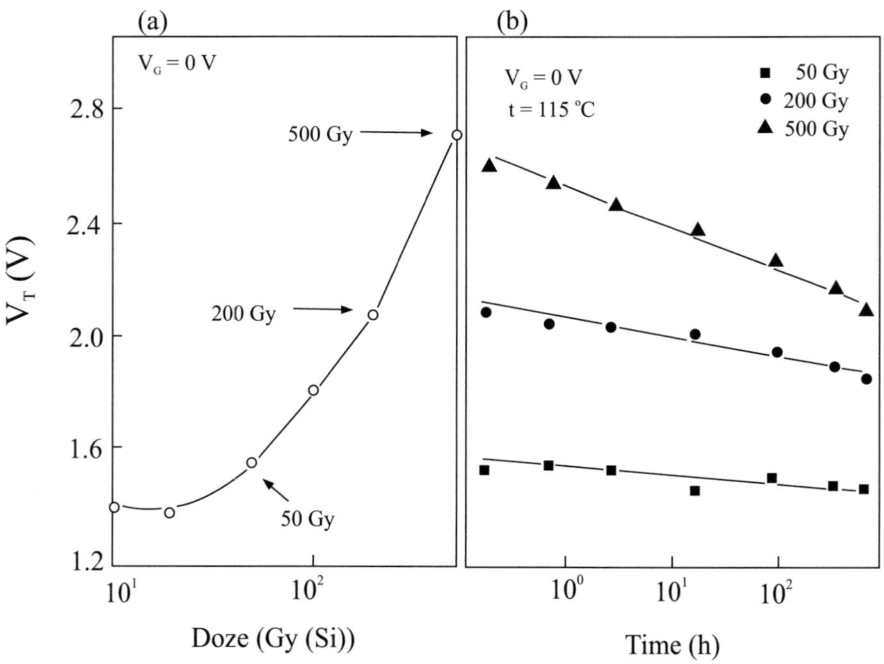

Fig. 9. Threshold voltage shift ΔV_T for PMOS transistor during gamma irradiation, V_G = 0 V (a) and thermal annealing, V_G = 0 V (b).

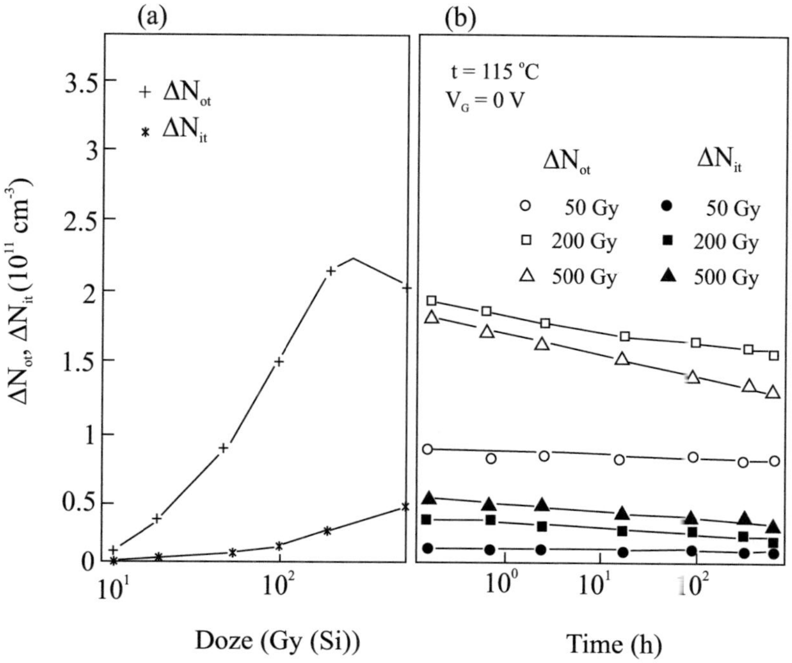

Fig. 10. The oxide trap (ΔN_{ot}) and interface traps (ΔN_{it}) density of NMOS transistor during gamma irradiation, V_G = 0 V (a) and thermal annealing, V_G = 0 V (b).

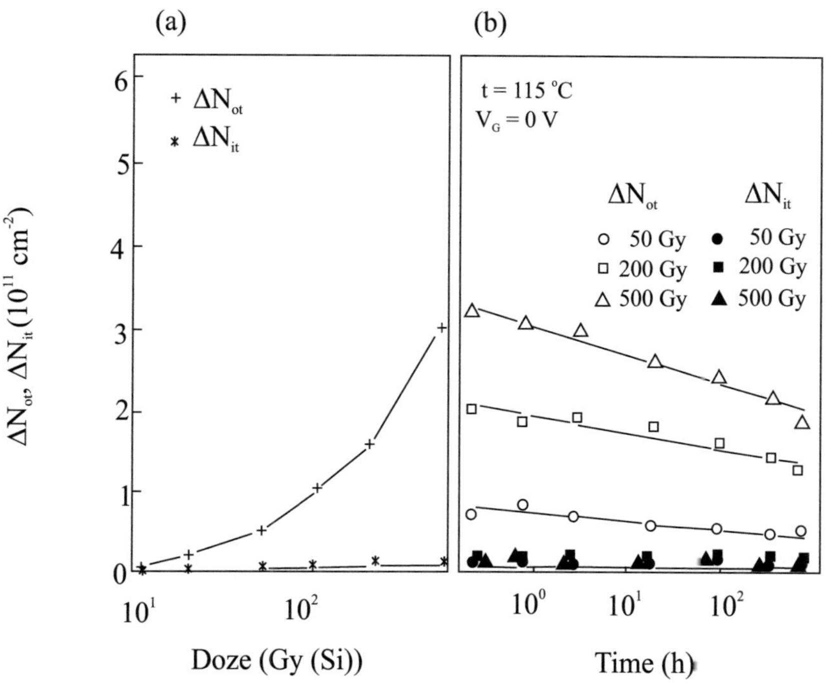

Fig. 11. The oxide trap (ΔN_{ot}) and interface traps (ΔN_{it}) density of PMOS transistor during gamma irradiation, V_G = 0 V (a) and thermal annealing, V_G = 0 V (b).

Figures 10 and 11 show changes in charge density in gate oxide and interface traps during radiation with gamma rays and later annealing in the temperature of 115°C for 0 V gate polarizations for the same transistors for which the change in threshold voltage (Figures 8 and 9) was monitored [111, 112]. The method described in the paper [90] was used to determine these densities. From the behavior of the curves in Figures 10 and 11, it was concluded that the annealing of radiated transistor is carried out through annealing of radiation defects described in detail in chapter 3. Namely, both the density of charge in the gate oxide and the density of interface traps show the tendency of decreasing during the burn in. It has been shown that the annealing of defects caused by gamma radiation depends not only on the conditions of transistor annealing, but on the conditions of their radiation as well, which can also be detected in Figures 10 and 11. Namely, the level of radiation exposure dictates the degree of annealing and the speed of the process of burn in the defects, so that the annealing is smaller in higher degrees of radiation exposure, considering the higher density of defects, but their annealing is more intense. This fact is supported by radiating a CMOS transistor with electrons and X-radiation whose energies amount to 10 MeV and later annealing with low-energy UV radiation [109].

In order to be able to consider the temperature annealing or annealing or annealing by low-energy UV radiation of gamma radiation degraded of electric characteristics of CMOS transistors as successful, it is necessary to ensure conditions in which the degraded electric parameters will become stable after annealing. It has been shown [110] that these conditions can be fulfilled by annealing on the temperature of 115° C and positive polarization at the 10 and 15 V gate. Check of characteristics stability was performed by continuing the annealing in an increased temperature and without polarization of the gate, and it was observed that the threshold voltage remains stable. Besides being definitive, the annealing process is accelerated by polarization at the 10 and 15 V gate. With lower voltage values in the continued temperature treatment of radiation-exposed transistors, the threshold voltage starts decreasing, which can most probably be attributed to the effect of the so-called "inverted" annealing.

7.2 Behavior of power VDMOS transistor during the IR process and the subsequent annealing

Electric parameters behavior of MOS transistors with polysilicon gate during gamma radiation was studied for many years [4]. For the commercial components can be said that they are mostly known. Figure 12 shows typical changes in threshold voltage during radiation of n-channel power VDMOS power transistors of the type EF1N10 at room temperature, and with gate voltages V_G = 0 and 10 V [116]. It can be seen that the threshold voltage decreases along with the increase in radiation dose and that the changes are more pronounced with higher values of gate voltage. It was also shown that the mobility is decreased during the radiation, and the changes are greater when the gate voltage is 10 V than when it is not applied.

Change in the oxide traps charge density density ΔN_{ot} and interface traps ΔN_{it} during the IR process for the same transistors in which the change in threshold voltage was monitored (Figure 12), determined by use of SMG technique, are displayed in Figures 13 and 14, respectively [116]. An increase in ΔN_{ot} and ΔN_{it} values during gamma radiation can be observed, and these changes are greater in the case V_G = 10 V. It has also been observed that the increase in the oxide trap density is substantially greater than the increase in interface traps density.

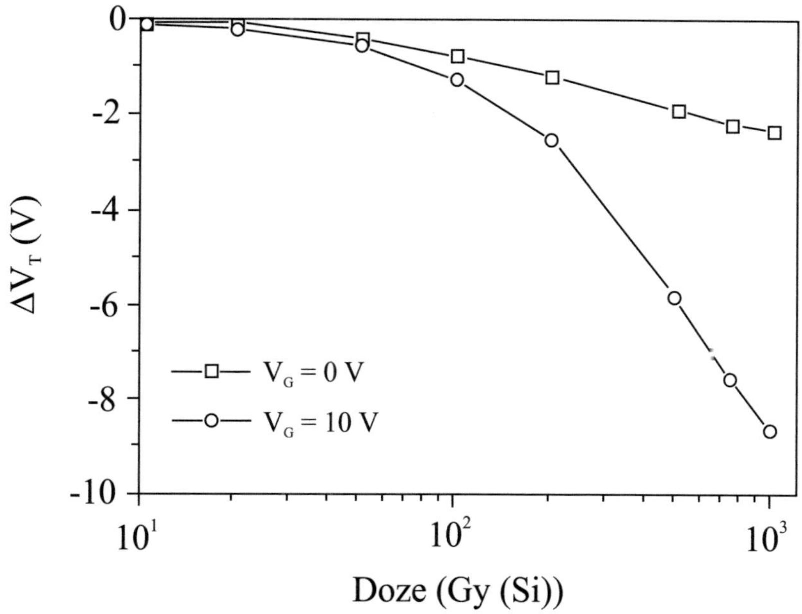

Fig. 12. Threshold voltage shift (ΔV_T) of n-channel power VDMOS transistor during gamma irradiation for V_G = 0 V and V_G = 10 V.

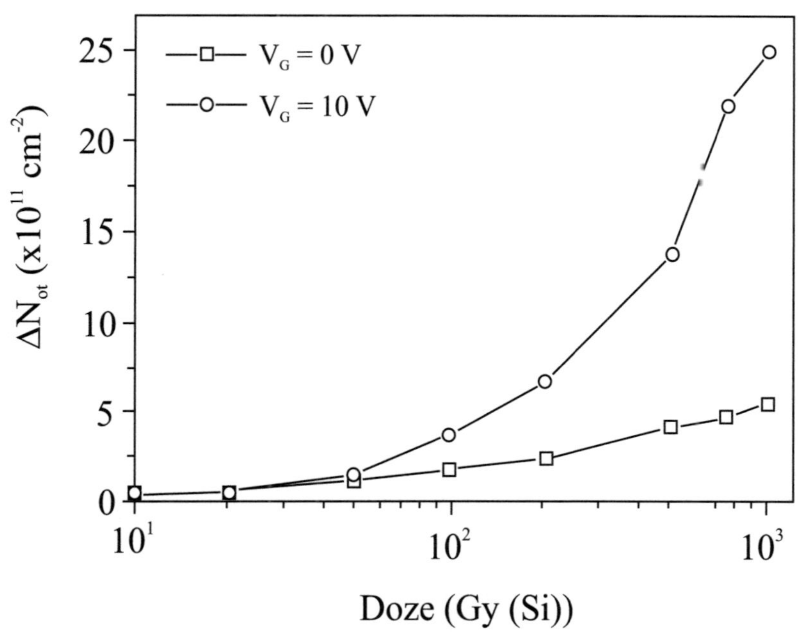

Fig. 13. The oxide traps charge density density (ΔN_{ot}) of n-channel power VDMOS transistor during gamma irradiation for V_G = 0 V and V_G = 10 V.

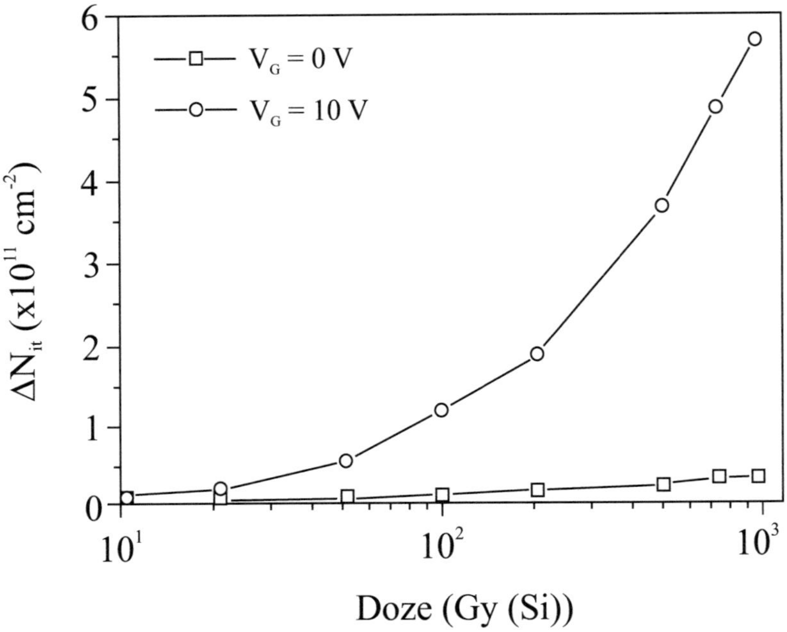

Fig. 14. The interface trap density (ΔN_{it}) of n-channel power VDMOS transistor during gamma irradiation for V_G = 0 V and V_G = 10 V.

Several papers [22, 27, 28, 85, 86, 88, 117-124] displayed the results of annealing the radiation-exposed power VDMOS transistors in room temperature and elevated temperatures. Figure 15 displays changes in threshold voltage of VDMOS power transistor of the type EFL1N10 when, during the gamma radiation, there was no voltage at the gate, and during the annealing it amounted to V_G = 10 V, while the annealing was conducted at room temperature, 55° C and 140° C. Figure 16 refers to the same annealing conditions, except that the voltage at the gate during the gamma radiation amounted to V_G = 10 V [27]. As it can be seen from these figures, a greater and faster change in threshold voltage ΔV_T occurs in case of higher temperature, and occurrence of super-recovery was detected only in transistors for which radiation was conducted when there was no polarization at the gate, and when their annealing was conducted on the temperature of 140° C.
The influence of temperature and voltage at the gate on the change in density of oxide traps charge ΔN_{ot} and interface traps ΔN_{it} for the same transistors for which a change in threshold voltage was monitored (Figures 15 and 16) are displayed in Figures 17 and 18 [27], and these changes are determined by using SMG technique. It can be seen that the density of interface traps after some annealing time starts increasing, while the time period prior to the start of increase can be very long. As mentioned before, this phenomenon of increase of ΔN_{it} value after some saturation is known in literature under the title latent increase (latent generation) of interface traps. The time interval prior to the occurrence of latent increase in interface traps density depends on the annealing temperature, it being shorter if the annealing temperature is higher. Also, a direct link between the latent increase in interface trap density and the latent decrease of density of the oxide trapped charge have been detected, as well as the fact that, after the latent increase in interface traps density, comes their decrease (passivization) during exposure to elevated temperature.

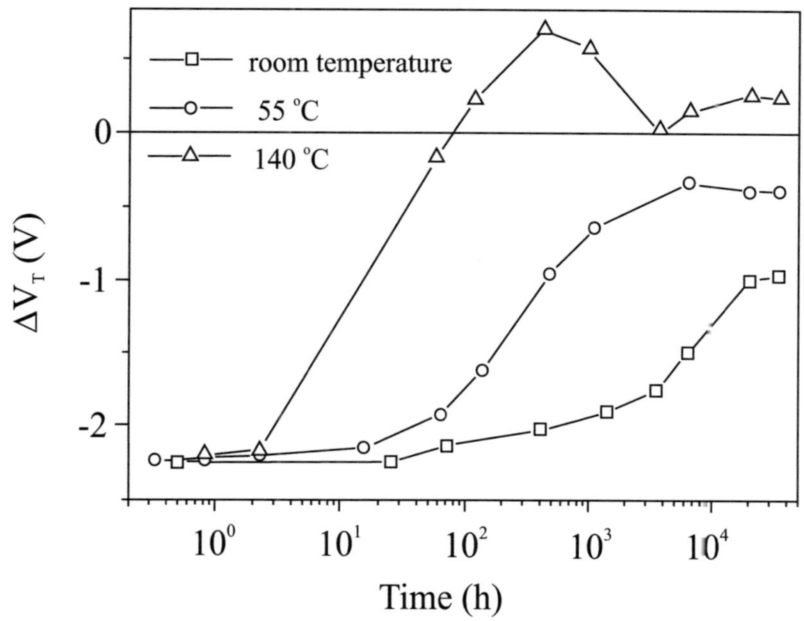

Fig. 15. Threshold voltage shift (ΔV_T) of n-channel power VDMOS transistor during annealing; V_G = 0 V during irradiation and V_G = 10 V during annealing

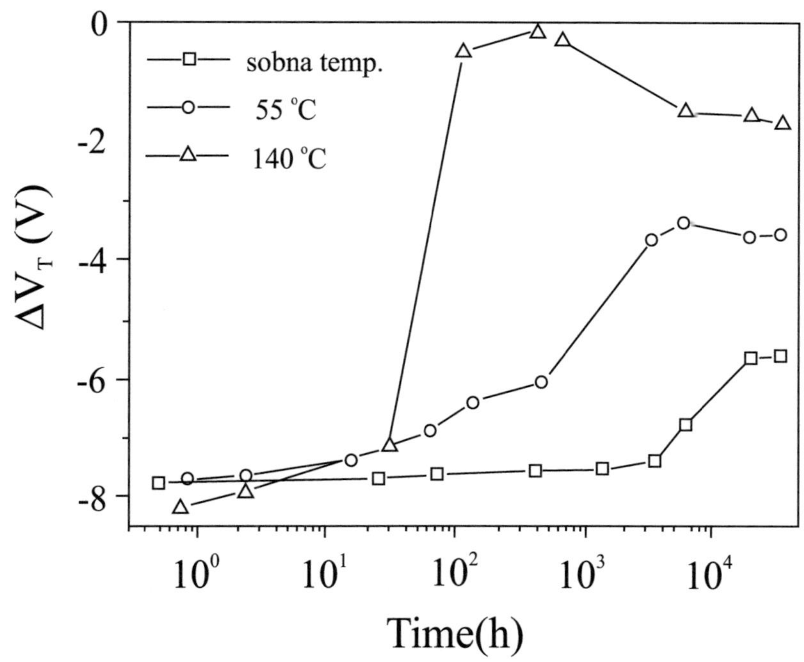

Fig. 16. Threshold voltage shift (ΔV_T) of n-channel power VDMOS transistor during annealing; V_G = 10 V during irradiation and annealing.

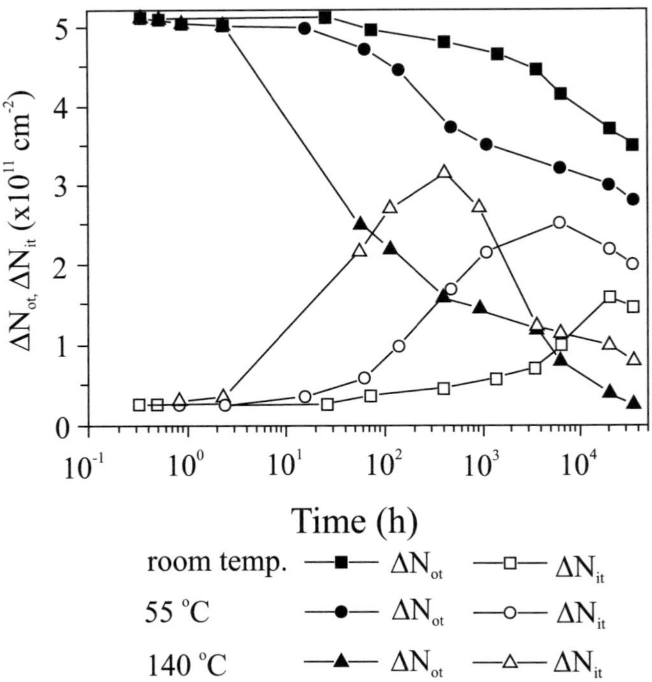

Fig. 17. The oxide traps charge density (ΔN_{ot}) and interface traps (ΔN_{it}) density during annealing for n-channel power VDMOS transistor irradiated for V_G = 0 V.

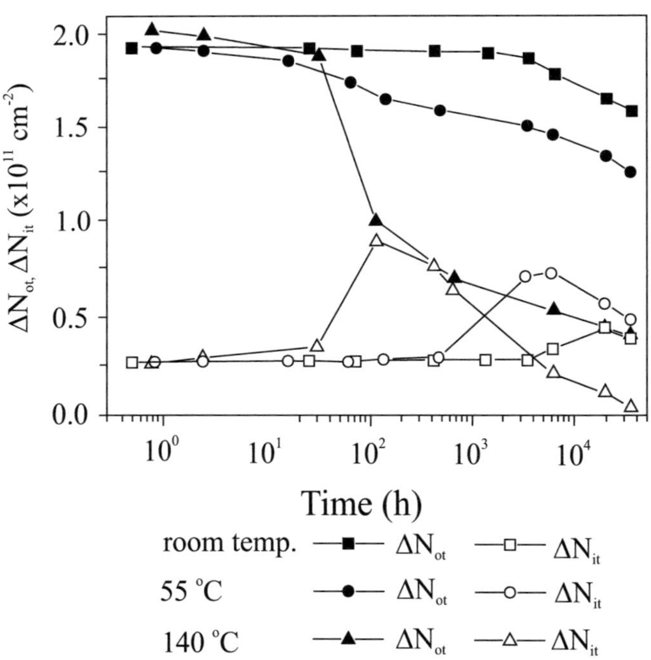

Fig. 18. The oxide traps charge density (ΔN_{ot}) and interface trap (ΔN_{it}) density during annealing for n-channel power VDMOS transistors irradiated at V_G = 10 V.

One of the important parameters which influence changes in density of oxide trapped charge ΔN_{ot} and interface traps ΔN_{it} during the annealing of power VDMOS transistors which have previously been exposed to gamma radiation is the value of voltage at the gate during the annealing. Experimental studies [119] have shown that the recovery of radiation-exposed transistors on the temperature of 140° C for the polarizations at the gate amounting to V_G = 0, 5 and 10 V, leads to latent increase in interface traps density and latent decrease in the density of oxide trapped charge. In the case V_G = 0, when only an electric field is present in the oxide due to the difference of work function of poly-Si gate and substrate, these changes are significantly smaller, when compared to the changes caused by the gate voltage of 5 and 10 V. The values of maximum densities of interface traps formed during the annealing, for V_G = 5 and 10 V have very few mutual differences. Also, there are certain differences in the time interval prior to the start of latent increase in ΔN_{it}.
Changes in densities of captured electric charge and interface traps shown in Figures 13 through 18 are very well described by H-W model [27, 28], whose more detailed description is given in chapter 4.

7.2.1 Isochronal annealing of power VDMOS transistors after the IR process

The isochronal annealing implies to the annealing of MOS transistors after IR or HCI processes with variable temperature. Figure 19 shows behavior of density of the oxide trapped charge and interface traps during an isochronal annealing of radiation-exposed power VDMOS transistors of the type EFP8N15 in the temperature range between 50° C and 290° C [121]. These densities were followed by using SMG technique. Duration of annealing at every temperature amounted to 5 min, while the temperature change amounted to 10° C, and the gate voltage during the annealing amounted to V_G = 10 V. As it can be seen, the

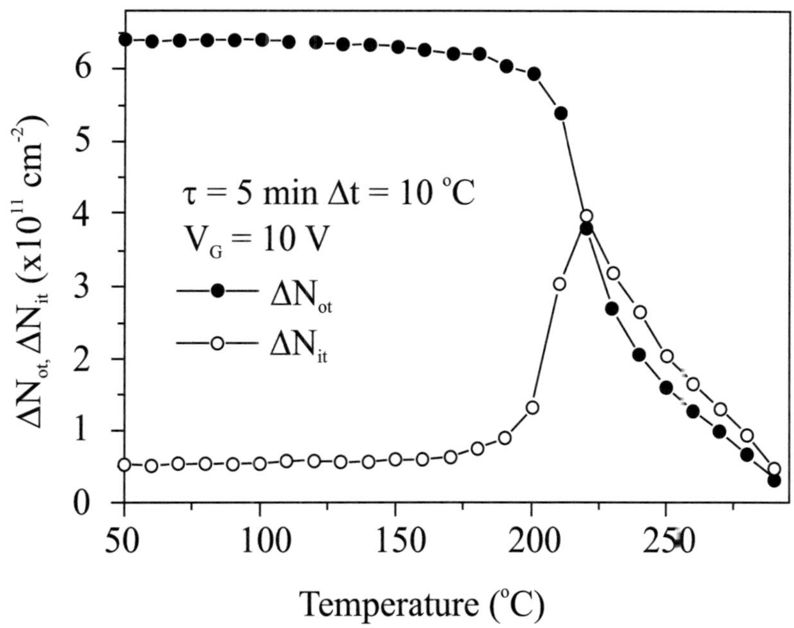

Fig. 19. The oxide traps charge density (ΔN_{ot}) and interface trap (ΔN_{it}) density for n-channel power VDMOS transistors during isochronal annealing determined using SMG technique.

values of ΔN_{ot} and ΔN_{it} are insignificantly changed up to the temperature of approximately 175° C, while at higher temperatures there occurs a rapid increase in interface traps densities, followed by decrease in the density of oxide trapped charge. The interface traps density starts decreasing from the temperature t=225° C. Figures 20 shows behavior of interface traps density with the temperature of isochronal annealing, generated by SMG and CP techniques. Similarities in behavior of these curves can be seen, but the values of ΔN_{it} generated by SMG technique are significantly higher, in accordance with the discussion given in chapter 6.

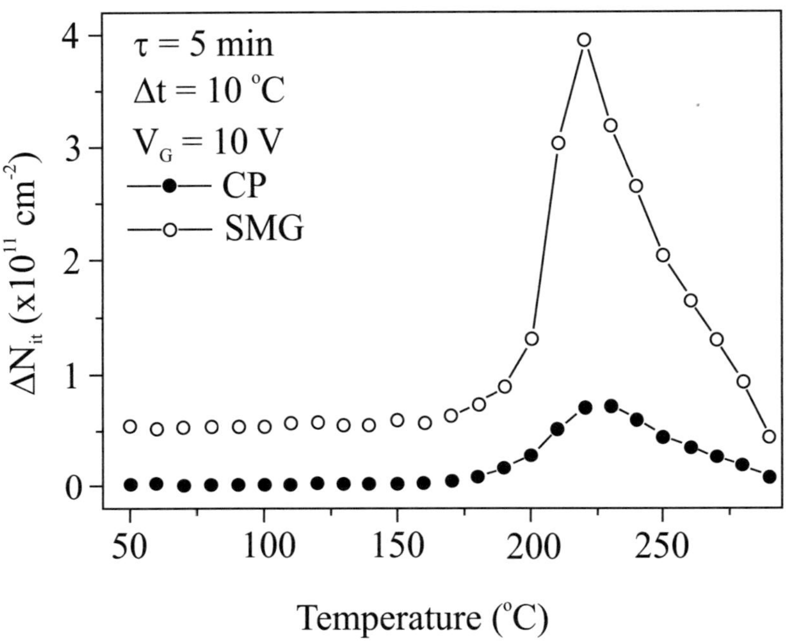

Fig. 20. The interface trap (ΔN_{it}) density for n-channel power VDMOS transistor during isochronal annealing determined using SMG and CP techniques.

7.2.2 Influence of voltage polarization at the gate during the annealing of power VDMOS transistors which had previously been subjected to gamma radiation

Figure 21 shows changes in the density of oxide trapped charge and interface traps during the annealing of power VDMOS transistor of the type EFP8N15 which had previously been subjected to the IR process. Annealing was conducted under alternating change in the voltage polarization at the gate and temperatures of 200° C and 250° C, while each phase, for one direction of electric field lasted for 1 h [116]. The densities of the oxide trapped charge and interface traps were determined by SMG technique. It can be seen from the figure that during the first phase for V_G = 10 V, there occurred a latent increase in interface traps density and latent decrease in the density of oxide trapped charge, for both temperatures. The first phase is similar to the results shown in Figures 17 and 18. During the second phase (V_G=-10 V), at the annealing temperature of 200° C, the value of ΔN_{it} continues to decrease, which could be expected on the basis of the H-W model, as the diffusion of water molecules is not dependent on the direction of electric field in oxide, while at the temperature of 250° C there is no change in this density, as almost all interface traps have already been passivated during the first phase. However, in the third phase (V_G=+10 V), the behavior of interface traps density is different from the behavior in the first phase. Namely, there is no latent increase in interface traps density, although the value of ΔN_{it} at first insignificantly raise at the very start of the annealing at the temperature of 250° C, and then continued to decrease, while at the temperature of 200° C, a mild increase was detected, and then slow decrease of this density (these changes are significantly smaller than during the first phase). On the basis of such behavior of interface traps and the application of H-W model, it was concluded that there were no hydrogen particles during the third phase (and especially no H_2 molecules) which could cause latent increase of ΔN_{it}.

The paper [116] also shows the results of ΔN_{it} generated through the application of CP technique for power VDMOS transistors of the type EFP8N15 annealing under the same conditions as shown in Figure 21. The behavior of this density is very similar to the behavior of the density generated through the SMG technique, but the values of ΔN_{it} generated with CP technique are significantly lower, which is in accordance with the sensitivity of the method discussed in chapter 6.

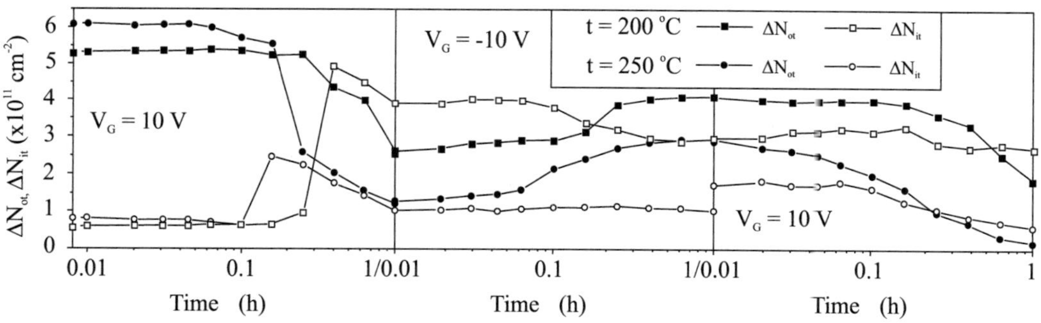

Fig. 21. The oxide traps charge density (ΔN_{ot}) and interface trap (ΔN_{it}) density during annealing with positive and negative polarization on the gate ($V_G = \pm 10$ V) for n-channel VDMOS transistors.

7.3 Behavior of power VDMOS transistor during the HCI process and subsequent annealing

Figure 22 shows the changes in threshold voltage of power VDMOS transistor EFL1N10 during the HCI process, with gate voltages of +80 V and -80 V [126]. It can be seen that the HCI process leads to significant changes in the threshold voltage ΔV_T. An initial decrease in the value of ΔV_T (during the first 40 min under positive, i.e. 20 min under negative gate voltage) and subsequent increase until the end of tension implies the occurrence of the "turn-around" effect. Similar behavior was also detected in power VDMOS transistor of the type IRF510 [29].

Figure 22 also shows the change in threshold voltage during the second HCI process. Namely, after the first HCI process, transistors were annealing at the temperature of 150° C in the duration of 3000 h, and then subjected to the HCI process again. It can be seen that the behavior is similar, while the changes in ΔV_T are more pronounced in the first HCI process, for both polarization signs at the gate.

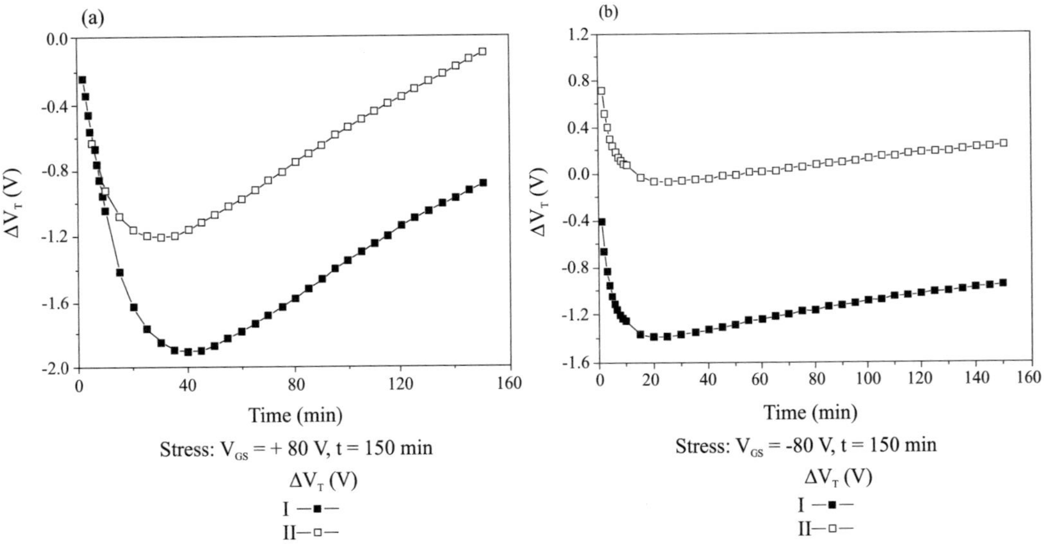

Fig. 22. Threshold voltage shift (ΔV_T) of n-channel power VDMOS transistors for positive ((a) $V_{GS} = 80$ V) and negative ((b) $V_{GS} = -80$ V) HCI during the first and second stress cycle.

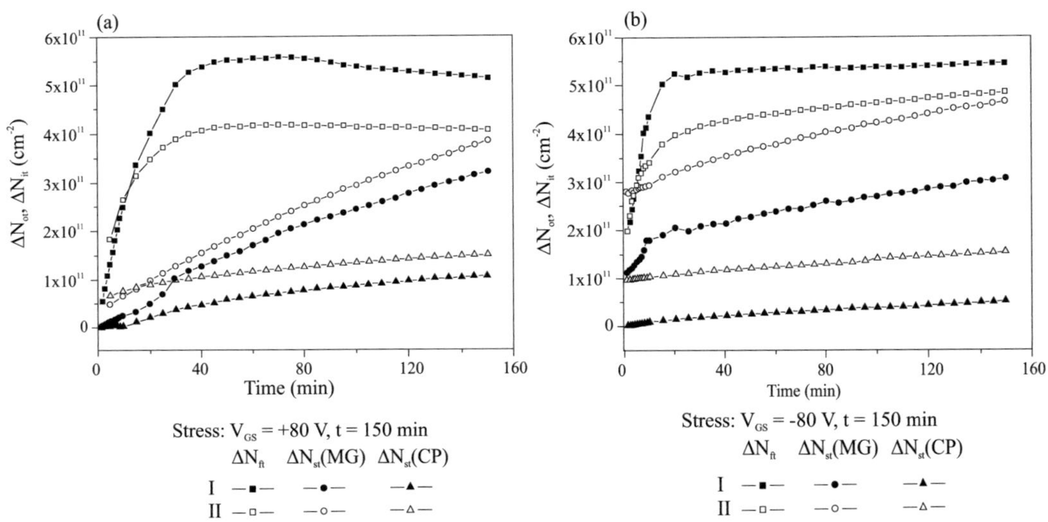

Fig. 23. The oxide traps charge density (ΔN_{ot}) and interface trap (ΔN_{it}) density of n-channel power VDMOS transistors for positive ((a) $V_{GS} = 80$ V) and negative ((b) $V_{GS} = -80$ V) HCI during the first and second stress cycle.

Figure 23 shows the changes in oxide trapped charge density in the gate oxide ΔN_{ot} determined by SMG technique and the changes in interface traps density ΔN_{it} (SMG) determined by SMG technique and ΔN_{it} (CP) determined by CP technique for the same transistors that were subjected to the HCI procedures during the monitoring of change in threshold voltage (Figure 22) [126]. From the figure it can be seen that the behaviors of ΔN_{it} (SMG) and ΔN_{it} (CP) are qualitatively the same, while there are significant quantitative differences between them. On the basis of the discussion given in section 3, the quantitative concurrences should not be expected, as the share of ST in the densities ΔN_{it} (SMG) and ΔN_{it} (CP) is different. Besides that, SMG and CP techniques in power VDMOS transistors register interface traps in two different areas of gate surface. SMG technique registers current in the p^- -area of the channel, while CP technique registers interface traps in the n^- -epi area of power VDMOS transistors.

Park et al [127] studied this problem in detail in a large number of similar samples and concluded that the initial values of threshold voltage in these two interfaces are significantly different. Another cause of quantitative difference in the values of ΔN_{it} (SMG) and ΔN_{it} (CP) is the fact that these two techniques record different areas of the silicon forbidden band. Namely, SMG technique records the defects in the area located at about 0.45 eV of the upper part of the silicon forbidden band, while CP technique records the defects at the same distance, but in the lower part of this area [127].

The initial increase ΔN_{ot} shown in Figure 23 is a consequence of the formation of positively charged FT at the early stage of HCI process [126]. ΔN_{it} achieves saturation at the later stage of this process, as the number of electrons captured by NBO centers increases. The increase in ΔN_{it} value for the entire duration of the HCI process is necessitated by the reaction of H^+ ions which are released in the oxide (H model) and the holes drifting towards the Si - SiO_2 interface and forming interface traps (h^+ model). Greater changes of ΔN_{it} (SMG) occur because this technique comprises all ST (ST=SST+FST), while ΔN_{it} (CP) comprises only FST.

Figures 24 shows the changes in threshold voltage of ΔV_T, while Figures 25 shows the changes in the density of oxide trapped charge ΔN_{ot} and the change in interface traps

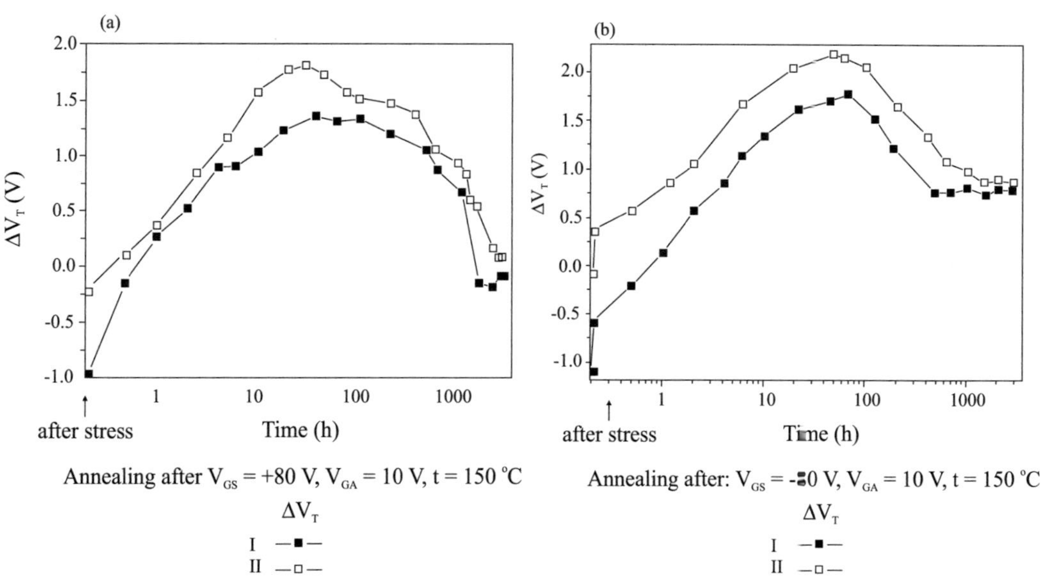

Fig. 24. Threshold voltage shift (ΔV_T) of n-channel power VDMOS transistor during the first and second thermal post HCI annealing cycle following (a) positive and (b) negative HCI.

densities determined by SMG technique (ΔN_{it} (SMG)) and CP technique (ΔN_{it} (CP)) during the annealing of power VDMOS transistor of the type EFL1N10 which had previously been subjected to the HCI process under gate voltages of +80 V and -80 V, annealing at the temperature of 150° C, and then subjected to HCI process again [126]. It can be seen that ΔV_T rises during the first 10 hours of recovery, then it decreases and achieves saturation in case of duration of 1.000 h of annealing, and that the behavior of ΔV_T is very similar in both polarization cases. The only difference is that ΔV_T achieves saturation sooner in the annealing preceded by HCI process with a negative gate polarization.

From Figure 25 it can be seen that there are no differences caused by the opposite sign of a gate polarization during the HCI process. Nor do the trends of the monitored changes differ during the first and the second recovery. As expected, the density of ΔN_{ot} decreases during annealing. ΔN_{it} (SMG) initially rises, while its decrease would occur during a longer annealing period, until the saturation is achieved. Very similar behavior also applies to the values of ΔN_{it} (CP), but this change is significantly smaller than the change ΔN_{it} (SMG) for the reasons discussed previously. The behavior of ΔN_{ot}, ΔN_{it} (SMG) and ΔN_{it} (CP) during annealing is very well described by the H-W model which details are given in chapter 5.

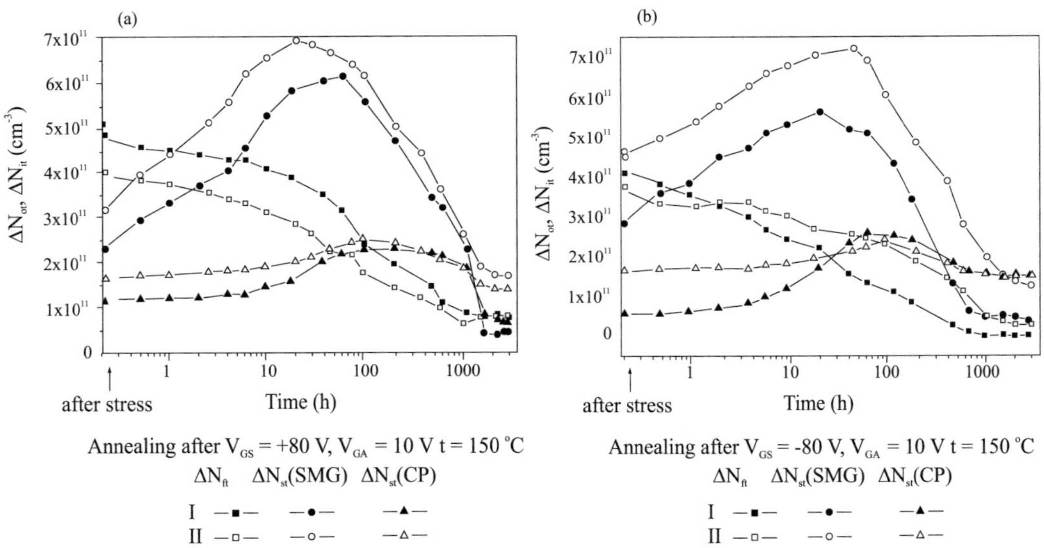

Fig. 25. The oxide traps charge density (ΔN_{ot}) and interface trap (ΔN_{it}) density of n-channel power VDMOS transistors during the first and second thermal post HCI annealing cycle following (a) positive and (b) negative HCI.

In the paper [22] shows comparative results of annealing of the n-channel power VDMOS transistor of the type EFL1N10, where a certain number of them is subjected to gamma radiation, and others to the HCI (Fowler-Nordheim high electric field stress) process. The changes in ΔN_{ot} , ΔN_{it} (SMG) and ΔN_{it} (CP) were monitored. It was noticed that, during annealing of transistors that had previously been subjected to gamma radiation there is a latent increase in interface trap density, both with SMG and CP technique. For transistors that had previously been subjected to the HCI process, a latent increase in interface trap density during the annealing occurs when only CP technique is used to determine their density. This shows that there is a difference in the nature of capture centers emerged during the IR and HCI processes. Finally, it should be pointed out that the behavior of ΔN_{it} (SMG) and ΔN_{it} (CP) after exposure to IR process given in the paper [127, 128], is identical to the behavior of ΔN_{it} (SMG) given in the paper [126] in case of the HCI process. The experiment was conducted with the samples that are structurally the same, but not identically processed as the samples in the work [126]. This comparison only illustrates the complicated nature of defect behavior occurring during the gamma radiation and the HCI process.

7.4 Application of PMOS transistor with Al-gate as a sensor and dosimeter of ionizing radiation

As early as 1974, there emerged an idea on the possibility of application of MOS transistors for the detection of the absorbed dose of ionizing radiation [7]. It was followed by the design and production of radiation-sensitive PMOS transistor with Al gate which is known under the title RADFET (radiation sensitive field effect transistors) which could be used as both sensor and dosimeter of ionizing radiation (gamma and X-radiation) [7]. These dosimeter transistors have so far found partial application in modern aircrafts [129, 130] (in which higher values of absorbed doses are measured) in medicine (in radiology, where they have the role of radiation sensors) [131, 132], in nuclear industry [132] and military [133, 134]. However, no PMOS dosimeter for small doses of radiation measuring, which could be used as a personal dosimeter has yet been realized.

The basic parameter for PMOS dosimetric transistor is the threshold voltage V_T, on basis of which the absorbed radiation dose is determined. The change in threshold voltage, as already said above, is a consequence of oxide trapped charge density ΔN_{ot} and interface traps ΔN_{it} caused by the IR process. In the case of PMOS transistors, the increase in these densities leads to the increase in V_T, as opposed to NMOS transistors, in which an increase of ΔN_{ot} reduces V_T, and increase of ΔN_{it} increases its value.

In general, the change in threshold voltage ΔV_T with an absorbed dose of radiation D is given by the following expression:

$$\Delta V_T = A \cdot D^n , \tag{37}$$

where A is the constant, and n is the degree of linearity. For $n = 1$, the constant A represents the sensitivity S of dosimetric transistor

$$S = \frac{\Delta V_T}{D} , \tag{38}$$

Figure 26 represents the single-point method, which is primarily used to determine the threshold voltage of PMOS dosimetric transistors [135]. It consists of the establishing constant current through the channel I_D (the value of 10 μA is usually taken) and measuring of V_0 voltage which corresponds with this current. It is also considered that the change in this voltage (ΔV_0) corresponds with the change in threshold voltage (ΔV_T).

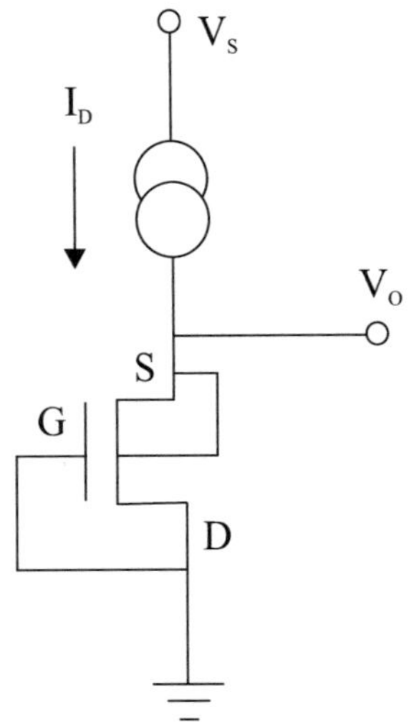

Fig. 26. The threshold voltage determination using single point method.

The second important parameter of PMOS dosimetric transistors is the recovery of threshold voltage after the radiation, which is known under the term fading f(t). It can be expressed in the following way:

$$f(t) = \frac{V_T(0) - V_T(t)}{V_T(0) - V_{T0}} = \frac{V_T(0) - V_T(t)}{\Delta V_T(0)}, \quad (39)$$

where V_{T0} is the threshold voltage prior to radiation, $V_T(0)$ immediately after irradiation, $V_T(t)$ after annealing time t and $\Delta V_T(0)$ is the threshold voltage shift immediately after irradiation of PMOS dosimetric transistor.
Fading can also be determined on the basis of the values ΔN_{ot} and ΔN_{it} [136, 137]

$$f(t) = \frac{\Delta N'_{ot}(t) + \Delta N'_{it}(t)}{\Delta N_{ot}(0) + \Delta N_{it}(0)}, \quad (40)$$

in which $\Delta N'_{ot}(t) = \Delta N_{ot}(0) - \Delta N_{ot}(t)$ ("the annealing part" of the oxide trapped charge) and $\Delta N'_{it}(t) = \Delta N_{it}(0) - \Delta N_{it}(t)$ ("the annealing part" of the interface traps), ΔN_{ot} (0) and ΔN_{it} (t) are the densities of oxide trapped charge and interface traps after the annealing for the time t, respectively, while ΔN_{it} (0) and ΔN_{it} (t) are adequate densities after the radiation, i.e. at the start of annealing.
The papers [136-143] show the results of exposure to gamma radiation and subsequent annealing of PMOS dosimetric transistors manufactured in the company Ei-Microelectronics, Nis, Serbia, while the papers [135, 143-145] show the results of radiation and annealing of PMOS dosimetric transistors manufactured in Tyndall National Institute, Cork, Ireland. Figure 27 shows change in threshold voltage with an increase in absorbed dose of gamma radiation for the transistors with oxide thickness of 1.23 μm [136]. It can be seen, that this dependence can be displayed in a coordinate system with log-log scale, with a straight line for all applied voltages at the gate. Also, changes in threshold voltage are greater in the case of positive voltages at the gate, although sensitivity grows with the increase of the absolute value V_G in both polarization modes (the smallest value applies to

the zero polarization). This means that the increase of ΔV_T takes place regardless of the direction of the electric field in oxide, but the size of these changes is dependent on it.

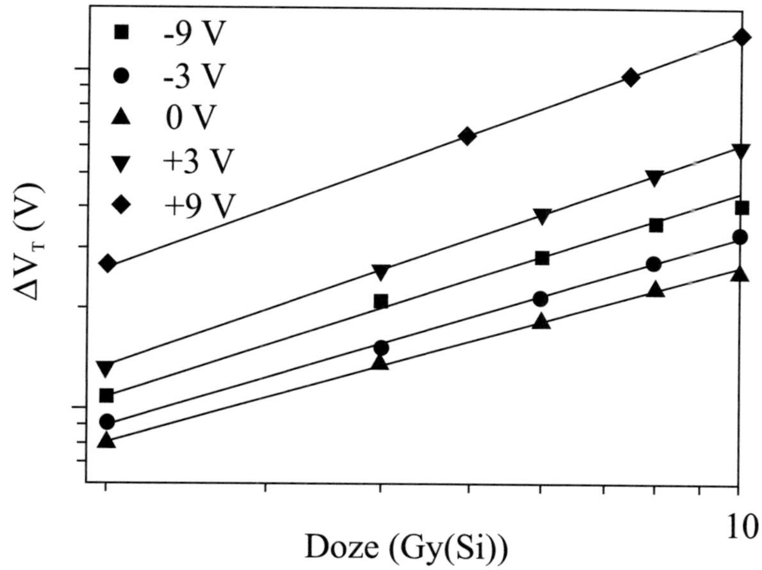

Fig. 27. Threshold voltage shift (ΔV_T) during irradiation of pMOS transistor, with 1.23 μm thick oxide.

Figure 28 shows change in threshold voltage as a function of absorbed dose of gamma radiation for different oxide thicknesses, for the gate voltage of V_G = 3 V [139]. It can be seen

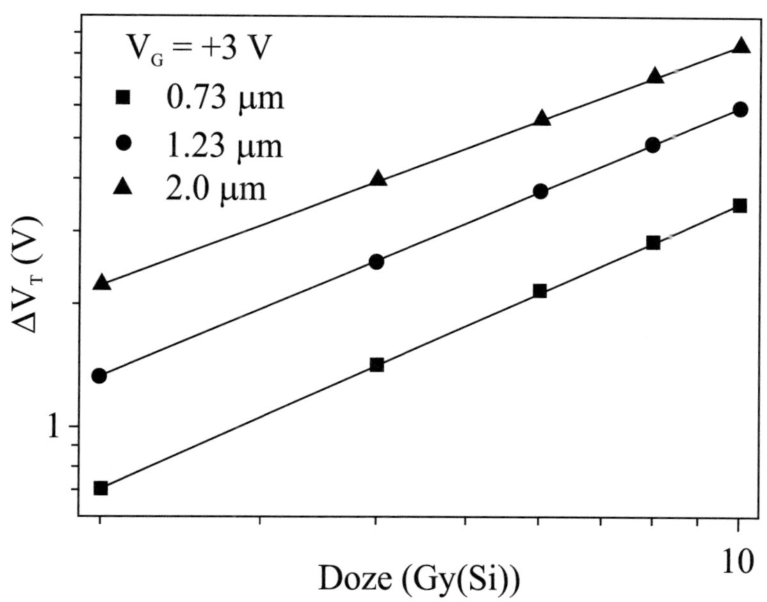

Fig. 28. Threshold voltage shift (ΔV_T) during irradiation of pMOS transistors with different oxide thickness and gate polarization of +3 V.

that the sensitivity grows along with the thickness of gate oxide, and as the tested PMOS dosimetric transistors have thermal oxide with the same thickness, these results show a significant role of CVD oxide in the change of threshold voltage. The gate voltage for these transistors was the same, i.e. the electric field in the thickest oxide was smallest, which affected their sensitivity.
The paper [135] shows changes in the threshold voltage of PMOS dosimetric transistors for the oxide thicknesses of 100 mm for the values of absorbed gamma radiation doses between 50 and 300 Gy, while the paper [143] stated this dependence for the dose range between 100 Gy and 500 Gy for gate voltages during radiation V_G=0 and 5 V (Figure 29). It was found that there is a linear dependence between the threshold voltage change and the absorbed radiation dose when the dose range amounted between 10 Gy and 500 Gy (Figure 29).

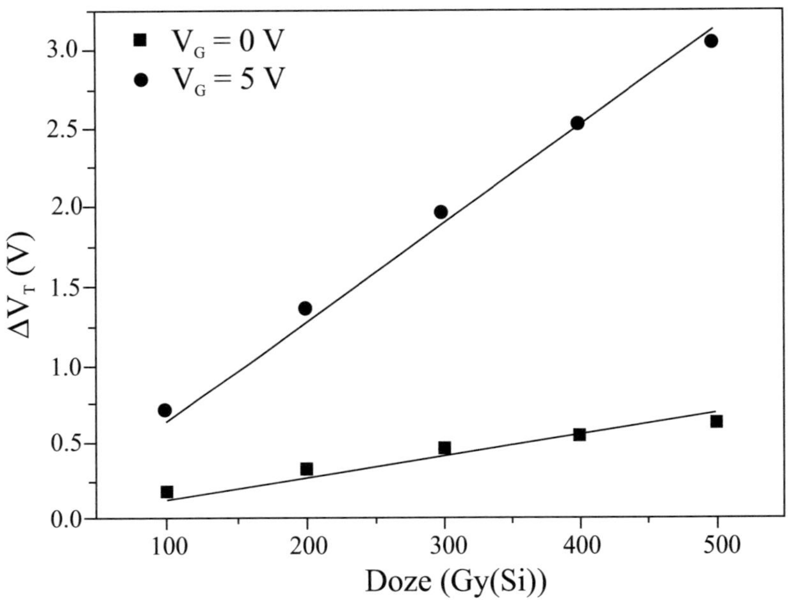

Fig. 29. Threshold voltage shift (ΔV_T) of pMOS transistors during irradiation for V_G = 0 V and V_G = 5 V.

Change of fading, as the other important parameter for PMOS dosimetric transistors, during annealing at room temperature is shown in figures 30 and 31 [116] for the transistors with oxide thickness of 1.23 and 2.0 µm. The radiation and recovery were conducted with the same voltage values at the gate. It can be seen that the transistors having oxide thickness of 2.0 µm show a more pronounced negative fading. On the basis of fading behavior of transistors with the same oxide thickness, but with different gate polarization, it is obvious that no concrete conclusion which would apply to all radiation-exposed transistors can be given. Lack of knowledge about the explicit form of dependence of fading behavior since the recovery time is not a deficiency of these PMOS transistors (which cannot be said for the sensitivity), as the practical application requires solely that it be smaller than some previously set value (usually, fading should be no less than ± 10% after three months of annealing at room temperature). On the basis of these results, it can be seen that for the oxide thickness of 2.0 µm, when the sensitivity is at its peak, the fading has a small value, and therefore in the case V_G=9 V, the sensitivity is S = 2 V/Gy for the absorbed dose D = 10 Gy(Si) and fading 5.9 % for annealing time of 3500 h

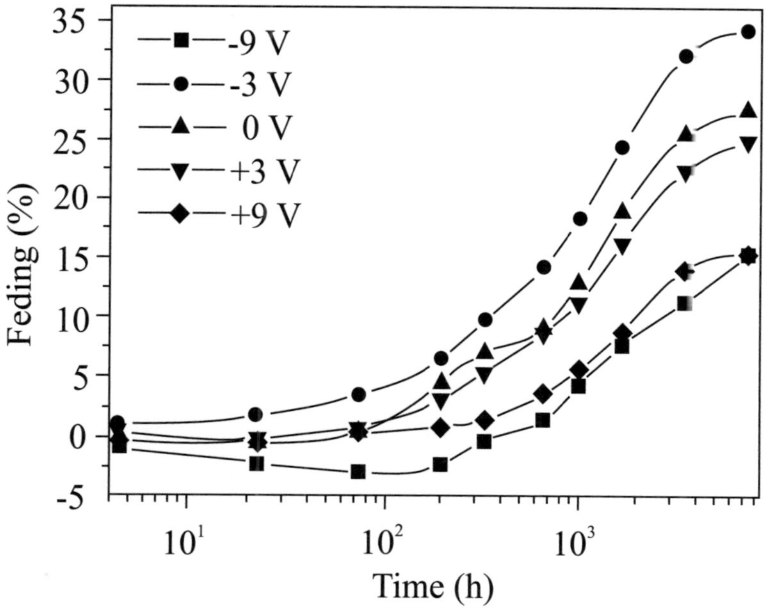

Fig. 30. Fading of irradiated pMOS transistors, with gate oxide thickness of 1.23 μm, during room temperature/bias annealing.

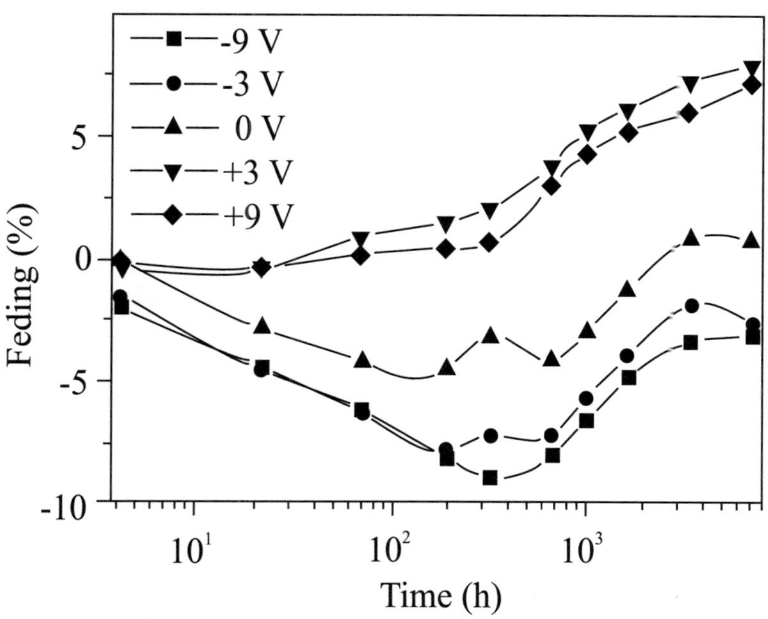

Fig. 31. Fading of irradiated pMOS transistors, with gate oxide thickness of 2.0 μm, during room temperature/bias annealing.

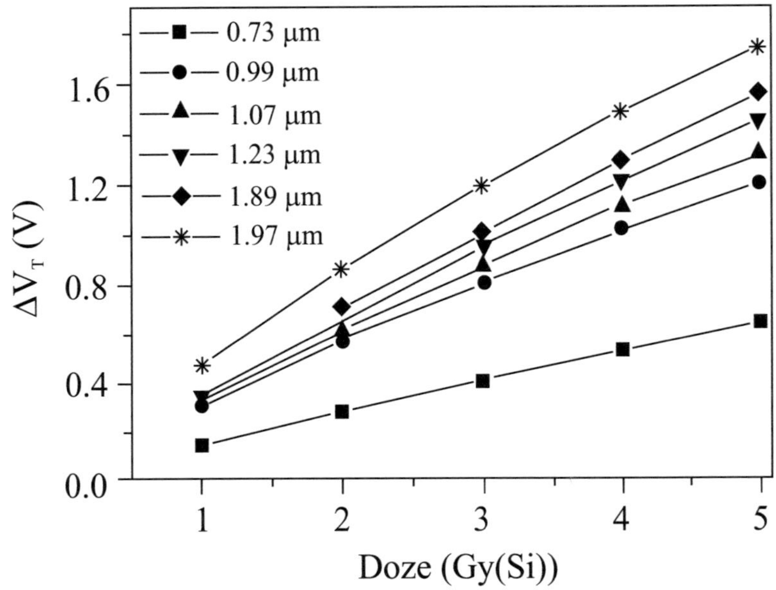

Fig. 32. Threshold voltage shifts (ΔV_T) during irradiation at dose rate $1.2 \cdot 10^{-3}$ Gy/s for different gate oxide thickness of pMOS transistors.

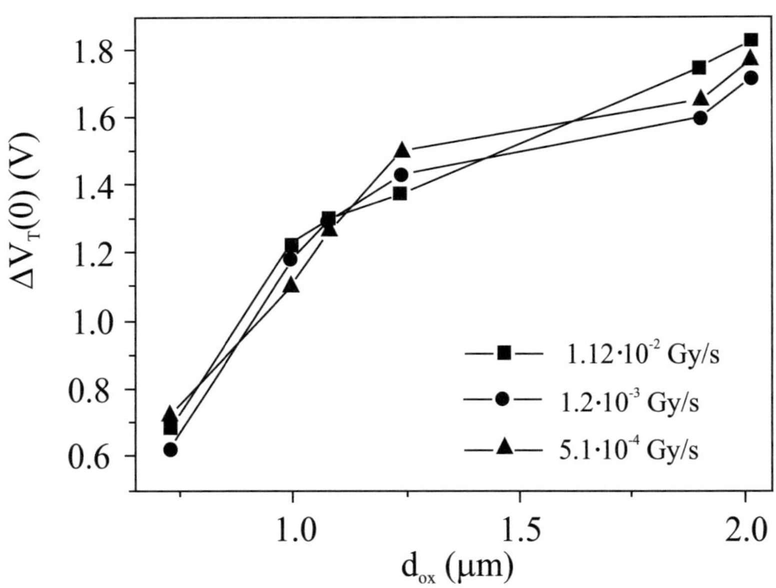

Fig. 33. Threshold voltage shifts (ΔV_T) for different pMOS transistors irradiated up to 5 Gy(Si).

Figure 32 shows changes in the threshold voltage from a dose of gamma radiation of PMOS dosimetric transistors with different gate oxide thicknesses (the thickness of thermal oxide was 0.3 μm, and the thickness of CVD oxide were different) during radiation with the dose

of 1.2 ·10-3 Gy [138]. It can be seen that the sensitivity grows along with the gate oxide thickness. This is clearly displayed in Figure 33 which shows changes in threshold voltage immediately after radiation, i.e. prior to annealing at room temperature for three dose speeds. On the basis of these figures, it can be concluded that the main influence on sensitivity is exerted by the total oxide thickness, and then the thickness of CVD oxide. Namely, the transistors with oxide thickness of 1.97 μm are more sensitive that a transistor with oxide thickness of 1.89 μm, although they have smaller thickness of CVD oxide (the same applies to the oxide thicknesses of 1.07 μm and 0.99 μm). A cause for such behavior can be the thickness and the type of thermal oxide.
As opposed to the results shown in Figures 32, in which the sensitivity of the tested PMOS transistors is given in the range between 1 and 5 Gy, Figures 34 shows their sensitivity in the dose range between 0.003 and 1 Gy for the case when there was no gate polarization during radiation [116].

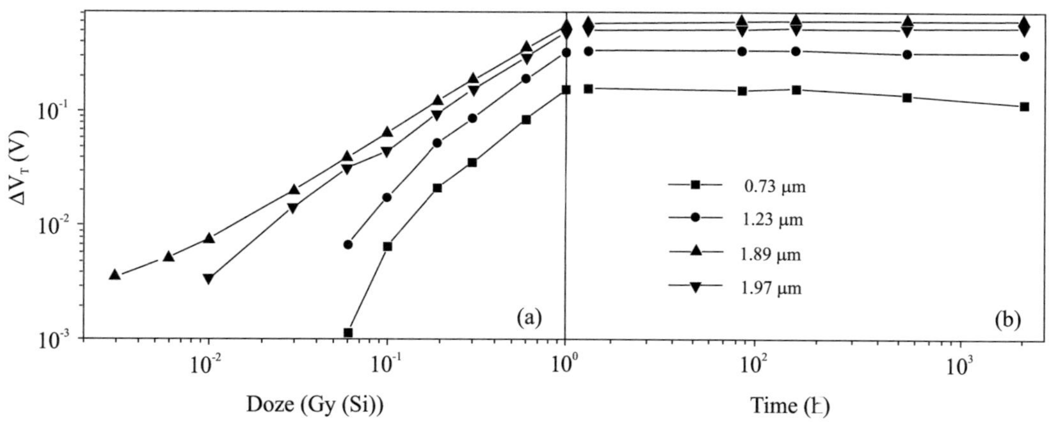

Fig. 34. Threshold voltage shifts (ΔV_T) of pMOS transistor during irradiation with dose rate of $7.51 \cdot 10^{-5}$ Gy/s (a) and annealing at room temperature (b), for different values of gate oxide thickness.

As it can be seen from the figure 34, the greatest sensitivity is displayed by the transistors with oxide thickness of 1.89 μm, which can register radiation doses in the order of magnitude cGy. As this transistor has the greatest thickness of CVD oxide, it can be assumed that this oxide would have a significant role in the area of small doses. The changes in threshold voltage during annealing of up to 2000 hours are insignificant for any oxide thickness (Figure 34b). Results with transistors of this oxide thickness show that their sensitivity increases with the increase of gate voltage during the gamma radiation, especially when the doses are higher than 0,01 Gy.

8. References

[1] J.R. Davis, Instabilities in MOS Devices, London: Gordon an Breach Science Publishers, 1981.
[2] S.M. Sze, Physics of Semiconductors Devices, New York: Wiley and Sons, 1981.

[3] T.P. Ma and P.V. Dressendorfer, Ionizin Radiation Effects in MOS Devices and Circuits, New York: Willey and Sons, 1989.
[4] S. Dimitrijev, Understanding Semiconductor Devices, New York: Oxford University Press, 2000.
[5] S. Gamerith and M. Polzl, Negative bias temperature stress on low voltage p-channel DMOS transistors and the role of nitrogen, Microelectron. Reliab., Vol. 42, pp. 1439-1443, 2002.
[6] B.J. Baliga, Switching lots of watts at high speed, IEEE Spectrum, Vol. 18, pp. 42-53, 1981.
[7] A.G. Holmes-Siedle, The space charge dosimeter- general principle of new method of radiation dosimetry, Nucl. Instrum. and Methods., Vol. NI-121, pp. 169-179, 1974.
[8] B.E. Deal, Standardized terminology for oxide charges associated with thermally oxidized silicon, IEEE Trans. Electron. Dev., Vol. ED-27, pp. 606-608, 1980.
[9] J.R. Davis, Instabilities in MOS Devices, London: Gordon and Breach Science Publishers, 1981.
[10] R.J. Kriegler, Ion instabilities in MOS structures, Proc. Int. Reliab. Phys. Symp., pp. 250-258, 1974.
[11] C.T. Sah, Origin of interface states and oxide charges generated by ionizing radiation, IEEE Trans. Nucl. Sci., Vol. NS-23, pp. 1563-1567, 1976.
[12] D.M. Fleetwood, Border traps in MOS devices, IEEE Trans. Nucl. Sci., Vol. NS-39, pp. 269-271, 1992.
[13] D.M. Fleetwood, P.S. Winokur, R.A. Reber Jr., T.L. Meisenheimer, J.R. Schwank, M.R. Shaneyfelt, L.C. Riewe, Effects of oxide traps, interface traps and border traps on metal-oxide-semiconductor devices, J. Appl. Phys., Vol. 73, pp. 5058-5074, 1993.
[14] D.M. Fleetwood, M.R. Shaneyfelt, W.L. Warren, J.R. Schwank, T.L. Meisenheimer, P.S. Winikur, Border traps: Issues for MOS radiation response and long-term reliability, Microelectron. Reliab., Vol. 35, pp. 403-408, 1995.
[15] D.M. Fleetwood and N.S. Saks, Oxide, interface and border traps in thermal, N_2O and N_2O -nitrogen oxides, J. Appl. Phys., Vol. 79, pp. 1583-1594, 1996.
[16] D.M. Fleetwood, P.S. Winokur, L.C. Riewe, R.A.. Reber Jr., Bulk oxide traps and border traps in metal-oxide-semiconducter capacitors, J. Appl. Phys., Vol. 84, pp. 6141-6148, 1996.
[17] D.M. Fleetwood, Revised model of thermally stimulated current in MOS capacitors, IEEE Trans. Nucl. Sci., Vol. NS-44, pp. 1826-1833, 1997.
[18] C.T. Wang, Hot carrier design considerations for MOS devices and circuits, New York: Van Nostrand Reinhold, 1992.
[19] J.P. McWhorter and P.S. Winokur, Simple technique for separating the effects of interface traps and trapped-oxide charge in metal-oxide semiconductor transistor, Appl. Phys. Lett., Vol. 48, pp. 133-135, 1986.
[20] M.A.B. Elliot, The use charge pumping currents to measure interface trap densities in MOS transistors, Solid-State Electron., Vol. 19, pp. 241-247, 1976.
[21] G. Groeseneken, H.E. Maes, N. Beltran and R.F. De Keersmaecker, A Reliable approach to charge-pumping measurements in MOS Transistors, IEEE Trans., Electron Dev., Vol. ED-31, pp. 42-53, 1984.
[22] G.S. Ristić, M.M. Pejović, A.B. Jakšić, Comparision between postirradiation annealing and post high electric field stress annealing od of n-channel power VDMOSFETs, Appl. Surface Science, Vol. 220, pp. 181-185, 2003.

[23] G.S. Ristić, Influence of ionizing radiation and hot carrier injection on metal oxide-semiconducter transistors, J. Phys. D: Appl. Phys., Vol. 41, 02300 (19 pp), 2008.
[24] L.P. Trombetta, F.J. Feigh, R.J. Zeto, Positive charge generation in metal-oxide-semiconductor capacitors, J. Appl. Phys., Vol. 69, pp. 2512-2521, 1991.
[25] O. Habaš, Z. Prijić, D. Pantić, N. Stojadinović, Charge-pumping characterization of SiO_2/Si interface virgin and irradiated power VDMOSFETs, IEEE Trans. Electr. Dev., Vol. ED-43, pp. 2197-2208, 1996.
[26] T. Uchino, M. Takashaki and T. Yoko, Model of oxygen-deficiency-related defects in SiO_2 glass, Phys. Rev. B, Vol. 62, pp. 2983-2986, 2000.
[27] M. Pejović and G. Ristić, Creation and passivation of interface traps in irradiated MOS transistors during annealing at different temperatures, Solid State Electr. Vol. 41, pp. 715-720, 1997.
[28] M. Pejović, G. Ristić and A. Jakšić, Formation and passivation of interface traps in irradiated n-channel power VDMOSFETs during thermal annealing, Appl. Surf. Sci., Vol. 108, pp. 141-148, 1997.
[29] G.S. Ristić, M.M. Pejović and A.B. Jakšić, Fowler-Nordheim high electric field stress of power VDMOSFETs, Solid-State Electron., Vol. 49, pp. 1140-1152, 2005.
[30] G.S. Ristić, M.M. Pejović and A.B. Jakšić, Defect behaviour in n-channel power VDMOSFETs during HEFS and thermal post-HEFS annealing, Appl. Surf. Sci., Vol. 252, pp. 3023-3032, 2006.
[31] G.S. Ristić, M.M. Pejović and A.B. Jakšić, Physico-chemical processes in metal-oxide-semiconductors with thick gate oxide during high electric field stress, J. Non. Cryst. Solids, Vol. 353, pp. 170-179, 2007.
[32] D.L. Griscom, Diffusion of radiolytic molecular hydrogen as a mechanism for the post-irradiation buildup of interface states in SiO_2-Si structures, J. Appl. Phys., Vol. 58, pp. 2524-2533, 1985.
[33] A.G. Revesz, Defect structure and irradiation behaviour of non-crystaline SiO_2, IEEE Trans. Nucl. Sci., Vol. NS-18, pp. 113-116, 1971.
[34] D.J. DiMaria, E. Cartier, D.A. Buchanan, Anode hole inection and trapping in silicon dioxide, J. Appl. Phys., Vol. 80, pp. 304-317, 1996.
[35] T.P. Chem, S.Li, S. Fung, C.D. Beling, K.F. Lo, Post-stress interface trap generation induced by oxide stress with FN injection, IEEE Trans. Electron Devices, Vol. ED-45, pp. 1972-1977, 1998.
[36] D.J. DiMaria, E. Cartier and D. Arnold, Impact ionization, trap creation, degradation and breakdown in silicon dioxide films on silicon, J. Appl. Phys., Vol. 73, pp. 3367-3384, 1993.
[37] D.J. DiMaria, Defect production, degradation and breakdown of silicon dioxide film, Solid-State Electron., Vol. 41, pp. 957-965, 1997.
[38] D.L. Griscom, Optical properties and structure of defects in silica glass, J. Ceram. Soc. Japan, Vol. 99, pp. 923-941, 1991.
[39] R. Helms and E.H. Poindexter, The silicon-silicon-dioxide system: its microstructure and imperfections, Rep. Progr. Phys., Vol. 57, pp. 791-852, 1994.
[40] R.A. Weeks, Paramagnetic resonance of lattice defects in irradiated quartz, J. Appl. Phys., Vol. 27, pp. 1376-1381, 1956.

[41] H.E. Boesch Jr., F.B. Mclean, J.M. McGarrity, G.A. Ausman Jr., Hole transport and charge relaxation in irradiated SiO_2 MOS capacitors, IEEE Trans. Nucl. Sci., NS-22, pp. 2163-2167, 1975.

[42] J.F. Conley, P.M. Lenahan, Electron spin resonance analysis of EP center interactions with H_2: Evidence for a localized EP center structure, IEEE Trans. Nucl. Sci., Vol. 42, pp. 1740-1743, 1995.

[43] D.A. Buchanan, A.D. Marwick, D.J. DiMaria, L. Dori, Hot-electron-induced hydrogen redistribution and defect generation in metal-oxide-semiconductor capacitors, J. Appl. Phys., Vol. 76, pp. 3595-3608, 1994.

[44] D.J. DiMaria, D.A. Buchanan, J.H Stathis, R.E. Stahlbush, Interface states induced by the presence of trapped holes near the silicon-silicondioxide interface, J. Appl. Phys., Vol. 7, pp. 2032-2040, 1995.

[45] S.K. Lai, Two carrier nature of interface-state generation in hole trapping and radiation damage, Appl. Phys. Lett., Vol. 39, pp. 58-60, 1981.

[46] S.K. Lai, Interface trap generation in silicon dioxide when electrons are captured by trapped hols, J. Appl. Phys., Vol. 54, pp. 2540-2546, 1983.

[47] S.T. Chang, J.K. Wu, S.A. Lyon, Amphoteric defects at Si-SiO_2 interface, Appl. Phys. Lett., Vol. 48, pp. 662-664, 1986.

[48] S.J. Wang, J.M. Sung, S.A. Lyon, Relationship between hole trapping and interface state generation in metal-oxide-silicon structure, Appl. Phys. Lett., Vol. 52, pp. 1431-1433, 1988.

[49] F.B. Mclean, A framework for understanding radiation-induced interface states in SiO_2 MOS structures, IEEE Trans. Nuclear Sci., Vol. NS-27, pp. 1651-1657, 1989.

[50] N.S. Saks, C.M. Dozier, D.B. Brown, Time dependence of interface trap formation in MOSFETs following pulsed irradiation, IEEE Trans. Nucl. Sci., Vol. NS-35, pp. 1168-1177, 1988.

[51] N.S. Saks, D.B. Brown, Interface trap formation via the two-stage H^+ process, IEEE Trans. Nucl. Sci., Vol. NS-36, pp. 1848-1857, 1989.

[52] D.L. Griscom, D.B. Brown, N.S. Saks, Nature of radiation-induced point defects in amorphous SiO_2 and their role in SiO_2-on-Si structures, The Physics and Chemistry of SiO_2 and the Si-SiO_2 interface, ed C.R. Helms and B.E. Deal (New York: Plenum) 1988.

[53] K.L. Brower , S.M. Myers, Chemical kinetics of hydrogen and (111) Si-SiO_2 interface defect, Appl. Phys. Lett., Vol. 57, pp. 162-164, 1990.

[54] J.H. Statis, E. Cartier, Atomic hydrogen reactions with P_b centers at the (100) Si-SiO_2 interface, Phys. Rev., Lett., Vol. 72, pp. 2745-2748, 1994.

[55] E.H. Poindexter, Chemical reactions of hydrogenous species in the Si-SiO_2 system, J. Non Cryst. Solids, Vol. 187, pp. 257-263, 1995.

[56] D. Wuillaume, D. Goguenheinm, J. B, Borgoin, Nature of defect generation by electric field stress at Si-SiO_2 interface, Appl. Phys. Lett., Vol. 58, pp. 490-492, 1991.

[57] E.H. Poindexter, P.J. Caplan, B.E. Deal, R.R. Razouk, Interface states and electron spin resonance centers in thermally oxidized (111) and (100) silicon wafers, J. Appl. Phys., Vol. 52, pp. 879-883, 1981.

[58] E.H. Poindexter, G.J. Gerardi, M.E. Ruecked, P.J. Caplan, N.M. Johnson, D.K. Biegelsen, Electronic traps and P_b centres at SiO_2/Si interface: Bend-gap energy distribution, J. Appl. Phys., Vol. 56, pp. 2844-2853, 1984.

[59] A. Stesmans, V.V. Afanasiev, Electron spin resonance features of interface defects in thermal (100) SiO_2/Si , J. Appl. Phys., Vol. 83, pp. 2449-2457, 1993.

[60] E. Cartie, J.H. Stathis, D.A. Buchanan, Passivation and depassivation of silicon dangling bonds at the SiO_2/Si inteface by atomic hydrogen, Appl. Phys. Lett., Vol. 63, pp. 1510-1512, 1993.

[61] D.L. Griscom, Diffusion of radiolytic molecular hydrogen as a mechanism for the post-irradiation buildup of interface states in SiO_2 -on-Si structures, J. Appl. Phys., Vol. 58, pp. 2524-2533, 1985.

[62] M.L. Reed, J.D. Plummer, SiO_2/Si interface trap production by low-temperature thermal processing, Appl. Phys. Lett., Vol. 51, pp. 514-516, 1987.

[63] N.S. Saks, S.M. Dozier, D.B. Brown, Time dependence of interface traps formation in MOSFETs following pulsed irradiation, IEEE Trans. Nucl. Sci., Vol. NS-35, pp. 1168-1177, 1988.

[64] D.B. Brown, N.S. Saks, Time dependence of radiation-induced interface trap formation in MOS devices as a function of oxide thickness and applied field, J. Appl. Phys., Vol. 70, pp. 3734-3747, 1991.

[65] R.E. Stahlbush, B.J. Mrstik, R.K. Lawrence, Post-irradiation behaviour of the interface state density and trapped positive sharge, IEEE Trans. Nucl. Sci., Vol. NS-37, pp. 1641-1649, 1990.

[66] B.J. Mrstik, R.W Rendul, Si-SiO_2 interface state generation during X-ray irradiation exposure to hydrogen ambient, IEEE Trans. Nucl. Sci., Vol. NS-38, pp. 1101-1110, 1991.

[67] R.E. Stalbush. A.H. Edwards, D.L. Griscom, B.J. Mrstik, Post-irradiation cracking of H_2 and formation of interface states in irradiated metal-oxide-semiconductor field-effect trnsistros, J. Appl. Phys., Vol. 73, pp. 658-667, 1993.

[68] N.S. Saks, R.B. Klein, R.E. Stahlbush, B.J. Mrstik, R.W. Rendell, Effects of post-stress hydrogen annealing of MOS oxides after Co-60 irradiation or Fowler-Nordheim injection, IEEE Trans. Nucl. Sci., Vol. NS-40, pp. 1341-1349, 1993.

[69] I.S. Al-Kofahi, J.F. Zhang, G. Groeseneken, Continuing degradation of the Si-SiO_2 interface after hot hole stress, J. Appl. Phys., Vol. 81, pp. 2686-2692, 1997.

[70] J.F. Zhang, I.S. Al-Kofahi, G. Groeseneken, Behaviour of hot stress Si-SiO_2 interface at elevated temperature, J. Appl. Phys., Vol. 83, pp. 843-850, 1998

[71] G. Van den Bosch, G. Groeseneken, H.E. Maes, R.B. Klein, N.S. Saks, Oxide and interface degradation resulting from substrate hot-hole injection in metal-oxide-semiconductor field-effect transistors at 295 and 77 K, J. Appl. Phys., Vol. 75, pp. 2073-2080, 1994.

[72] C.Z. Zhao, J.F. Zhang, G. Groesenecen, R. Degraeve, J.N. Ellis, C.D. Beech, Interface state generation after hole injection, J. Appl. Phys., Vol. 90, pp 328-336, 2001.

[73] J.F. Zhang, C.Z. Zhao, G. Groeseneken, R. Degraeve, Analysis of the kinetics for interface state generation following hole injection, J. Appl. Phys., Vol. 93, pp. 6107-6116, 2003.

[74] J.R. Schwank, P.S. Winikur, P.J. McWhorter, F.W. Sexton, P.V . Dressendorfer, D.C. Turpin, Physical mechanisms contributing to devices rebound, IEEE Trans. Nucl. Sci., Vol. NS-31, pp. 1434-1438, 1984.

[75] T.R. Oldham, A.J. Lelis, F.B. McLean, Spatial dependence of trapped holes determined from tunneling analysis and measured annealing, IEEE Trans. Nucl. Sci., Vol. NS-33, pp. 1203-1210, 1986.

[76] P.J. McWhorter, S.L. Miller, W.W. Miller, Modeling the annealing of radiation-induced trapped hols in a varying thermal environmental, IEEE Trans. Nucl. Sci., Vol. NS-37, pp. 1682-1689, 1990.

[77] A.L. Lelis, H.E. Boesch Jr., T.R. Oldham, F.B. Melean, Reversibility of trapped hole annealing, IEEE Trans. Nucl. Sci., Vol. NS-35, pp. 1186-1191, 1988.

[78] A.L. Lelis, T.R. Oldham, H.E. Boesch Jr., F.B. McLean, The nature of the trapped hole annealing process, IEEE Trans. Nucl. Sci., Vol. NS-36, pp. 1808-1815, 1989.

[79] R.K. Freitag, D. B. Brown, C.M. Dozier, Exsperimental evidence of two species of radiation induced trapped positive charge, IEEE Trans. Nucl. Sci., Vol. ED-40, pp. 1316-1322, 1993.

[80] R.K. Freitag, D. B. Brown, C.M. Dozier, Evidence for two types of radiation-induced trapped positive sharge, IEEE Trans. Nucl. Sci., Vol. NS-41, pp. 1828-1834, 1994.

[81] G. Singh, K.F. Galloway, T.J. Russel, Temperature-induced rebound in power MOSFETs, IEEE Trans. Nucl. Sci., Vol. NS-34, pp. 1366-1369, 1987.

[82] D.M. Fleetwood, Long-term annealing study of midgap interface-trap charge neutrality, Appl. Phys. Lett., Vol. 60, pp. 2883-2885, 1992.

[83] D.M. Fleetwood, M.R. Shaneyfelt, L.C. Riewe, P.S. Winokur, R.A. Reber, Jr., The role of border traps in MOS high-temperature postirradiation annealing response, IEEE Trans. Nucl. Sci., Vol. 40, pp. 1323-1334, 1993.

[84] J.R. Schwank, D.M. Fleetwood, M.R. Shaneyfelt, P.S. Winokur, C.L. Axness, L.C. Riewe, Latent interface-trap buildup and its implications for hardness assurance, IEEE Trans. Nucl. Sci., Vol. NS-39, pp. 1953-1963, 1992.

[85] G.S. Ristić, M.M. Pejović, A.B. Jakšić, Modeling of kinetics of creation and passivation of interface trap in metal-oxide-semicondictor transistors during postirradiation annealing, J. Appl. Phys., Vol. 83, pp. 2994-3000, 1998.

[86] G.S. Ristić, M.M. Pejović and A.B. Jakšić, Analysis of postirradiation annealing of n-channel power vertical-double-diffused metal-oxide semiconductor transistors, J. Appl. Phys., Vol. 87, pp. 3468-3477, 2000.

[87] D.M. Fleetwood, W.L. Waren, J.R. Schwank, P.S. Winokur, M.R. Shaneyfelt, L.C. Reewe, Effects of interface traps and border traps on MOS postirradiation annealing response, IEEE Trans. Nucl. Sci., Vol. NS-42, pp. 1698-1707, 1995.

[88] M. Pejović, A. Jakšić and G. Ristić, The behaviour of radiation-induced gate-oxide defects in MOSFETs during annealing at 140 °C, J. Non. Cryst. Solids, Vol. 240, pp. 182-192, 1998.

[89] J. Shwank, D.M. Fleetwood, M.R. Shaneyfelt, P.S. Winokur, Latent thermally activated interface trap generation in MOS devices, IEEE Electr. Devices Lett., Vol. ED-13, pp. 203-205, 1992.

[90] S. Dimitrijev, N. Stojadinivić, Analysis of CMOS transistors instabilities, Solid-State Electronics, Vol. 30, pp. 991-1003, 1989.

[91] B. J. Baliga, Modern Power Devices, New York: Johan Wiley and Sons, 1987.

[92] Y. Tsividis, Operation and modeling of the MOS transistors, New York: McGraw-Hill, 1987.

[93] N. Arora, MOSFET models for VLSI circuit simulation, Wien-New York: Springer Verlag, 1993.

[94] S. Golubović, S. Dimitrijev, D. Župac, M. Pejović, N. Stojadinović, Gamma radiation effects in CMOS transistors, 17th European Solid State Device Research Conf., ESSDERC 87, Bologna, Italy, pp. 725-728, 1987.

[95] D Župac, K.F. Galloway, R.D. Schrimpf, P. Augier, Effects of radiation-induced oxide-trapped charge on inversion layer hole mobility at 300 and 77 K, Appl. Phys. Lett., Vol. 60, pp. 3156-3158, 1992.

[96] D. Župac, K.F. Galloway, R.D. Schrimpf, P. Augier, Radiation induced mobility degradation in p-channel double-diffused metal-oxide-semiconducter power transistors at 300 and 77 K, J. Appl. Phys., Vol. 73, pp. 2910-2915, 1993.

[97] N. Stojadinović, M. Pejović, S. Golubović, G. Ristić, V. Davidović, S. Dimitrijev, Effect of radiation-induced oxide-trapped sharge on mobility in p-channel MOSFETs, Electron. Lett., Vol. 31, pp. 497-498, 1995.

[98] N. Stojadinović, S. Golubović, S. Djorić, S. Dimitrijev, Modeling radiation-induced mobility degradation in MOSFETs, Phys. Stat. Sol. (a), Vol. 169, pp. 63-66, 1998.

[99] K.F. Galloway, M. Gaitan, T.J. Russel, A simple model for separating interface and oxide charge effects on MOS device characteristics, IEEE Trans. Nucl. Sci., Vol. NS-31, pp. 1497-1501, 1984.

[100] F.C. Hsu and S. Tam, Relationship between MOSFET degradation and hot-electron-induced interface-state generation, IEEE Electr. Dev. Lett., Vol. EDL-5, pp. 50-52, 1984.

[101] A. Bellaaouar, G. Sarrabayrouse, P. Rossel, Influence of ionizing irradiation on the channel mobility of MOS trnsistors, Proc, IEE, Vol. 132, pp. 184-186, 1985.

[102] S.C. Sun, J.D. Plumer, Electron mobility in inversion and accumulation layers on thermally oxidized silicon surface, IEEE Trans. Electr. Dev., Vol. ED-27, pp. 1497-1507, 1980.

[103] D.K. Schroder, Semiconductor Material and Device Characterization, New-York, Wiley, 1990.

[104] J.S. Brugler, P.G.A. Jespers, Charge pumping in MOS devices, IEEE Trans. Electr. Dev., Vol. ED-16, pp. 297-302, 1969.

[105] R.E. Paulsen, R.R. Siergiej, M.L. French, M.H. White, Observation of near-interface oxide traps with the charge pumping technique, IEEE Electron Dev. Lett., Vol. 13, pp. 627-629, 1992.

[106] A. Jakšić, M. Pejović, G. Ristić, S. Raković, Latent interface-trap generation in commercial power VDMOSFETs, IEEE Trans. Nucl. Sci., Vol. NS-45, pp. 1365-1371, 1998.

[107] S.C. Witezak, K.F. Galloway, R.D. Schrimpf, J.R. Brews, G. Prevost, The determination of $Si\text{-}SiO_2$ interface trap density in irradiated four-terminal VDMOSFETs using charge pumping, IEEE Trans. Nucl. Sci., Vol. NS-43, pp. 2558-2564, 1996.

[108] S. Dimitrijev, S. Golubović, D: Župac, M. Pejović, N. Stojadinović, Analisis of gamma-radiation induced instability mechanisms in CMOS transistors, Solid-State Electr., Vol. 32, pp. 349-353, 1989.

[109] S. Golubović, M. Pejović, S. Dimitrijev, N. Stojadinović, UV- radiation annealing of the electron and X-irradiation demaged CMOS transistors, Phys. Stat. Sol. (a), Vol. 129, pp. 569-575, 1992.

[110] M. Pejović, G. Ristić, S. G. Golubović, A comparision between thermal annealing and UV-radiation annealing of γ-irradiation NMOS trnsistors, Phys. Stat. Sol. (a), Vol. 140, pp. K53-K57, 1993.

[111] M. Pejović, S. Golubović, G. Ristić, M. Odalović, Annealing of gamma-irradiated Al-gate NMOS transistors, Solid-State Electron., Vol. 37, pp. 215-216, 1994.

[112] M. Pejović, S. Golubović, G. Ristić, M. Odalović, Temperature and gate-bias effect on gamma-irradiated Al-gate metal-oxide-semiconductor transistors, Jpn. J. Appl. Phys., Vol. 33, pp. 986-990, 1994.

[113] S. Golubović, G. Ristić, M. Pejović, S. Dimitrijev, The role of interface traps in rebound mechanisms, Phys. Stat. Sol. (a), Vol. 143, pp. 333-339, 1994.

[114] M. Pejović, S. Golubović, G. Ristić, Temperature-induced rebound in Al-gate NMOS transistors, IEE Proc.-Circuits Devices Syst., Vol. 142, pp. 413-416, 1995.

[115] M. Pejović, A. Jaksić, G. Ristić, B. Baljošević, Processes in n-channel MOSFETs during postirradiation thermal annealing, Radiat. Phys. Chem., Vol. 49, pp. 521-525, 1997.

[116] G.S. Ristić, Radiation and postirradiation effects in power VDMOS transistors and PMOS dosimetric transistors, PhD, University of Niš, Faculty of Electronic Engineering, 1998.

[117] A. Jakšić, G. Ristić, M. Pejović, Analysis of the processes in power MOSFETs during γ-ray irradiation and subsquent thermal annealing, Phys. Stat. Sol. (a), Vol. 155, pp. 371-379, 1996.

[118] A. Jakšić, M. Pejović, G. Ristić and S. Raković, Latent interface-trap generation in commercial power VDMOSFETs, Abstract of 4th European Conf. RADECS 97, pp. A5-A6, 1997.

[119] G.S. Ristić, M.M. Pejović, A.B. Jakšić, Numerical simulation of creation-passivation kinetics of interface traps in irradiated n-channel power VDMOSFETs during thermal annealing with various gate bias, Microelectronics Engin., Vol. 40, pp. 51-60, 1998.

[120] G.S. Ristić, A.B. Jakšić, M.M. Pejović, Latent interface-trap builup in power VDMOSFETs: new experimental evidence and numerical simulation, European Conf. of Radiation and its Effects on Components and Systems, RADECS 99, Abbaye de Fontevraud, France, pp. H14-H17, 1999.

[121] A.B. Jakšić, M.M. Pejović, G.S. Ristić, Isothermal and isochronal annealing experiments on irradiated commercial power VDMOSFETs, IEEE Trans. Nucl. Sci., Vol. NS-47, pp. 659-666, 2000.

[122] A.B. Jakšić, G.S. Ristić, M.M. Pejović, New experimental evidence of latent interface-trap buildup in power VDMOSFETs, IEEE Trans. Nuclear Sci., Vol. NS-47, pp. 580-586, 2000.

[123] A. Jakšić, G. Ristić, M. Pejović, Rebound effect in power VDMOSFETs due to latent interface-trap generation, Electr. Lett., Vol. 31, pp. 1198-1199, 1995.

[124] A.B. Jakšić, M.M. Pejović, G.S. Ristić, Properties of latent interface-trap buildup in irradiated metal-oxide-semiconductor transistors determined by switching bias isothermal annealing experiments, Appl. Phys. Lett, Vol. 77, pp. 4220-4222, 2000.

[125] V. Danchenko, E.G. Stassinopouls, P.H. Fang, S.S. Brashears, Activation energies of thermal annealing of radiation-induced damage in n-and p-channels of CMOS integrated circuits, IEEE Trans. Nucl. Sci., Vol. NS-27, pp. 1658-1664, 1980.

[126] S.M. Aleksić, A. B. Jakšić, M.M. Pejović, Repeating of positive and negative high electruic stres and corresponding thermal post-stress annaling of the n-channel power VDMOSFETs, Solid State Electrn., Vol. 52, pp. 1197-1201, 2008.

[127] M.S. Park, I. Na, C.R. Wie, A comparision of ionizing radiation and high field stress effects in N-channel power vertical double-diffusen metal-oxide-semiconductor field-effect trnsistors, J. Appl. Phys., Vol. 97, 014503 (6pp) 2005.

[128] M.S. Park, C.R. Wie, Study of radiation effects in γ-ray irradiation power VDMOSFET by DCIV technique, IEEE Tran. Nucl. Sci., Vol. NS-48, pp. 2285-2293, 2001.

[129] A.K. Ward, N. Blower, L. Adams, J. Doutreleau, A. Holmes-Siedle, M. Pignol, J.J. Berneron, M. Mehlen, The meteosat-p2 radiation effects experiment, Proc. 40.th Congerss of the Inter. Astronautical Federation, Malaga, Spain, pp. 151-159, 1989.

[130] L. Adams, E.J. Daly, R. Harboe-Sorensen, A.G. Holms-Siedle, A.K. Ward, R.A. Bull, Measurements of SEU and total dose in geostationary orbit under normal and flore conditions, IEEE Trans. Nucl. Sci., Vol. NS-38, pp. 1686-1692, 1991.

[131] D.J. Glastone, L.M. Chin, A.G. Holes-Siedle, MOSFET radiation detectors used as patient radiation dose monitors during radiotherapy, 33rd Ann. Mtg. Am. Assoc. of Pyisics in Medicine, San Francisko, 1991.

[132] J.S. Leffler, S.R. Lindgrern, A.G. Holmes-Siedle, Applications of RADFET dosimeters to equipment radiation qualification and monitoring, Tran. of the American Nucl. Sosiety, Vol. 60, pp. 535-536, 1989.

[133] L.S. August, Desing criteria for high-dose MOS dosimeter for use in space, IEEE Trans. Nucl. Sci., Vol. NS-31, pp. 801-803, 1984.

[134] L.S. August, R.R. Curcle, Adventages of using a PMOS FET dosimeter in high-dose radiation effects testing, IEEE Trans. Nucl. Sci., Vol. NS-31, pp. 1113-1115, 1984.

[135] A. Jakšić, G. Ristić, M. Pejović, A. Mohammadzadeh, C. Sudre, W. Lane, Gamma-ray irradiation and post-irradiation reponse of high dose range RADFETs, IEEE Tran. Nucl. Sci., Vol. NS-49, pp. 1356-1363, 2002.

[136] G. Ristić, S. Golubović, M. Pejović, PMOS dosimeter with two-layer state oxide appeared at zero and negative bias, Electr. Lett., Vol. 30, pp. 295-296, 1994.

[137] G. Ristić, S. Golubović, M. Pejović, P-channel metal-oxide semiconductor dosimeter fading dependencies on gate bias and oxide thickness, Appl. Phys. Lett., Vol. 66, pp. 88-89, 1995.

[138] G. Ristić, A. Jakšić, M. Pejović, PMOS dosimetric transistors with two-layer oxide, Sensors and Actuators Vol. A63, pp. 123-134, 1997.

[139] G. Ristić, S. Golubović, M. Pejović, pMOS transistors for dosimetric application, Electr. Lett., Vol. 29, pp. 1644-1646, 1993.

[140] Z. Savić, B. Radjenović, M. Pejović, N. Stojadinović, The contribution of border traps to the threshold voltage shift in pMOS dosimetric transistors, IEEE Trans. Nucl. Sci., Vol. NS-42, pp. 1445-1454, 1995.

[141] G. Ristić, S. Golubović, M. Pejović, Sensitivity and fading of pMOS dosimeters with thick gate oxide, Sensors and Actuators, Vol. A51, pp. 153-158, 1996.

[142] G.S. Ristić, Thermal and UV annealing of irradiated pMOS dosimetic transistors, J. Phys. D: Appl. Phys., Vol. 42, 135101, (12pp), 2009.

[143] M.M. Pejović, M.M. Pejović and A.B. Jakšić, Radiation-sensitive field effect trasnsistors to gamma-ray irradiation, Nucl. Tecnol. and Radiat. Protection, Vol. 26, pp. 25-31, 2011.

[144] M.M. Pejović, S.M. Pejović, E.Č. Dolićanin, Đ. Lazarević, Gamma-ray irradiation and post-irradiation at room temperature response of pMOS dosimeters with thick gate oxides, Nucl. Tecnol. and Radiat. Protection, Vol. 26, 2011 (accepted to publication).

[145] M.M. Pejović, M.M. Pejović, A.B. Jakšić, Contribution of fixed traps to sensitivity of pMOS dosimeters during gamma ray irradiation and annealing at room temperature, Sensors and Actuators A: Physical, vol. 174, pp. 85-90, 2012.

34

New Developments in the Field of Radiochemical Ageing of Aromatic Polymers

Emmanuel Richaud, Ludmila Audouin,
Xavier Colin, Bruno Fayolle and Jacques Verdu
Arts et Metiers ParisTech, CNRS, PIMM,
France

1. Introduction

Polymers having an aromatic backbone polymers have a high mechanical strength and a high modulus. Their aromaticity increases their resistance for use in relatively severe conditions especially in aerospace and nuclear industry for which lifetime prediction is a key issue. For example, a challenge for nuclear plants is to extend lifetime from the initially planned 40 years duration to 50 or 60 years, which makes necessary to determine lifetime by a non-empirical method. Since polymers mechanical failure originates from chain scission or crosslinking of the backbone, the ideal method of lifetime prediction would first involve the elaboration of a kinetic model for chain scission and crosslinking. Then, the changes of molecular mass would be related to the changes of mechanical properties using the available laws of polymers physics. Lifetime would be then determined using a pertinent lifetime criterion. A noticeable difficulty comes here from the fact that oxidation, which plays a key major role in chain scission, is diffusion controlled and thus heterogeneously distributed in sample thickness. It is crucial, indeed, to determine experimentally and to predict this depth distribution of chain scission and crosslinking because it will play a key role on fracture properties.

This chapter will be henceforward devoted to the effect of aromaticity on radiostability, the effect on temperature on the chain scission/crosslinking competition, the diffusion limited oxidation (which will be illustrated by the effect of dose rate, atmosphere and sample thickness), then some concluding remarks on oxidative stability of aromatic polymers and the possible link with the absence of macromolecular mobility below T_g. We will start by some basics of radiochemistry which are necessary for the good understanding of this paper, and especially the quantitative treatment for crosslinking and chain scission.

2. Basics of radiochemical degradation

Let us first consider the reaction:

$$A + h\nu \rightarrow B$$

The radiochemical yield is defined as the number of B molecules that are generated per absorbed joule:

$$G' = \frac{n}{E'} = \frac{N}{N_{Av}} \cdot \frac{1}{E \times (1.6 \times 10^{-17})} = 10^{-7}.G \quad (1)$$

where:

- G′ is the radiochemical yield expressed in mol J^{-1},
- n is the number of moles of B which is formed by the radiochemical reaction,
- E′ is the amount of absorbed energy in J,
- E is the amount of absorbed energy in 100 eV,
- N is the number of B molecules,
- N_{Av} is the Avogadro's number,
- G is the yield in molecules per 100 eV absorbed.

Since the absorbed dose (denoted by δ and expressed in Gy) is defined as the amount of received energy (in J) per kilogram of polymer, the following equation can be derived:

$$\frac{n}{m} = 10^{-7}.G.\delta \quad (2)$$

So that:

$$r = \frac{dc}{dt} = 10^{-7}.G.I \quad (3)$$

r being the rate of a radiochemical event (in mol l^{-1} s^{-1}), I the dose rate in Gy s^{-1} and c the concentration of reacted or generated species (in mol kg^{-1} even if one often considers that it is the same than in mol l^{-1}).

2. Mathematical treatment for chain scission and crosslinking

Irradiation can provoke both chain scission and crosslinking, the relative proportion of these phenomena depending on many factors which will be presented in the following. The radical coupling may lead to trifunctionnal or tetrafunctionnal crosslinking noduli called respectively Y- and H-crosslinkings (Fig. 1):

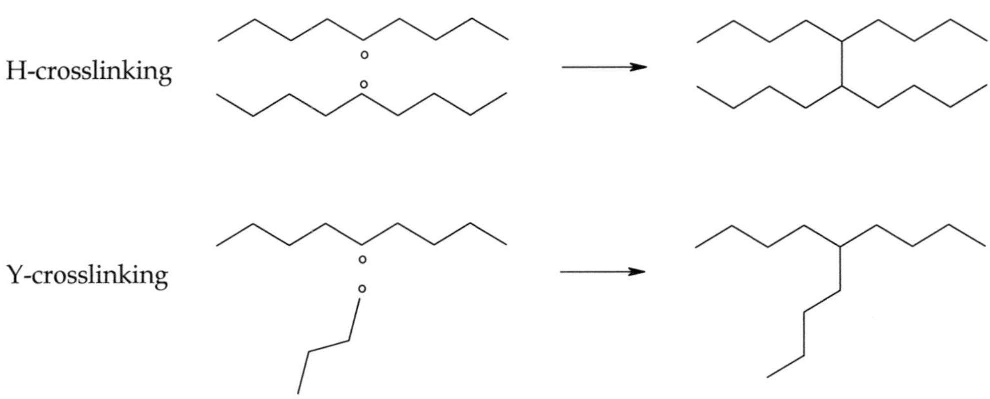

Fig. 1. Reaction of H- and Y- crosslinking.

The radiochemical yields for chain scissions G(s) and crosslinking G(x) in thick samples can be tentatively assessed using

- The Charlesby-Pinner's (Charlesby, 1960) from an analysis of the residual soluble fraction after beginning of sol-gel transition,
- Saito's equations (Saito, 1958) describing the average molar mass changes for soluble polymers before the sol-gel transition.

The corresponding mathematical treatment of these theories differs for Y- and H-crosslinking:

1. In the case of H- crosslinking mode:

$$\frac{1}{\overline{M}_N} - \frac{1}{\overline{M}_{N0}} = s - x = 10^{-7}.[G_H(s) - G_H(x)].\delta \tag{4}$$

$$\frac{1}{\overline{M}_W} - \frac{1}{\overline{M}_{W0}} = \frac{s}{2} - 2x = 10^{-7}.\left[\frac{G_H(s)}{2} - 2G_H(x)\right].\delta \tag{5}$$

$$w_S + w_S^{1/2} = \frac{G_H(s)}{2G_H(x)} + \frac{10^7}{\overline{M}_{W0}.G_H(x).\delta} \tag{6}$$

2. In the case of Y- crosslinking mode:

$$\frac{1}{\overline{M}_N} - \frac{1}{\overline{M}_{N0}} = s - x = 10^{-7}.[G_Y(s) - G_Y(x)].\delta \tag{7}$$

$$\frac{1}{\overline{M}_W} - \frac{1}{\overline{M}_{W0}} = \frac{s}{2} - x = 10^{-7}.\left[\frac{G_Y(s)}{2} - G_Y(x)\right].\delta \tag{8}$$

$$1 + 3w_S^{1/2} = \frac{2G_Y(s)}{G_Y(x)} + \frac{1.93 \times 10^7}{\overline{M}_{N0}.G_Y(x).\delta} \tag{9}$$

where:

- $w_S = 1 - w_I$ is the soluble fraction,
- $G_H(s)$, $G_H(x)$, $G_Y(s)$, and $G_Y(x)$ are the radiochemical yields expressed for 100 eV respectively for chain scissions and crosslinking for an H- and a Y- crosslinking mechanism,
- δ is the dose (Gy)
- M_{N0} and M_{W0} are respectively the initial number and weight average molar mass (kg mol^{-1}).

3. Effect of aromaticity on radiostability

The degradation of aromatic polymers was studied by many authors:

1. By monitoring mechanical properties (Sasuga et al., 1985, see Table 1):

These results give a first indication on the relative radiostability of aromatic polymers. Let us first mention that these lethal doses are considerably greater than those compiled by Wilski for polymer having an aliphatic backbone (Wilski, 1987). It can be attributed to the well-known stabilizing effect of aromatic groups which was first observed on methyl methacrylate-styrene copolymers (Alexander & Charlesby, 1954, Kellman et al., 1990,

Thominette et al., 1991). However, the results of these studies cannot be directly used in kinetic models for aromatic polymers.

Polymer	$\delta_{50\%}$	$\delta_{20\%}$
Kapton 500	10	90
Upilex	35	60
Ultem	1.5	4
A-Films	15	25
A-Paper	20	30
PEEK non-cryst	20	50
PEEK cryst	8	30
U-polymer	0.5	2
Udel-Polysulphone	0.75	1
PES	0.5	0.75
Noryl (modified PPO)	0.75	1.5

Table 1. Dose for reducing strain at break to 50% and 20% of initial value for various aromatic polymers submitted to 5.10^3 Gy s^{-1} in air (sample thickness ~ 0.1-0.2 mm) (Sasuga et al., 1985).

2. By radical appearance measured by Electron Spin Resonance spectroscopy (Heiland et al., 1996, see Fig. 2):

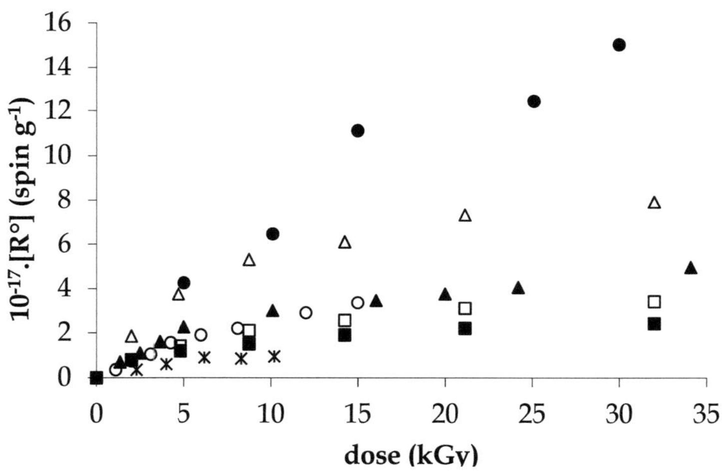

Fig. 2. Concentration in radicals versus dose measured in PPO (●), Kevlar (△), PSU (▲), PEPO (○), Ultem (□), Kapton (■), PEEK (∗) in aromatic polymers under 1 - 5 kGy h^{-1} dose rate at 77 K (Heiland et al., 1996).

Data presented in Fig. 2 permit the yield in radical build-up to be calculated according to the simplified scheme for radiochemical degradation in the absence of oxygen:

Initiation:	Polymer (PH) + hv → P°	r_i
Propagation:	P° + PH → PH + P°	k_P
β-scission:	P° → P° + s	k_R
Termination by coupling:	P° + P° → P–P + x	k_C
Termination by disproportionnation:	P° + P° → F + PH	k_D

Where F is a double bond.
The rate of radical formation is:

$$\frac{d[P°]}{dt} = r_i - 2k_t[P°]^2 \tag{10}$$

with:

$$k_t = k_C + k_D \tag{11}$$

Fig. 2 indicates the existence of an asymptote characterizing a steady-state at which the concentration in radicals is:

$$[P°]_\infty = \sqrt{\frac{r_i}{2k_t}} \tag{12}$$

Eq. 10 gives:

$$\frac{d[P°]}{\left(1+\frac{[P°]}{[P°]_\infty}\right)\cdot\left(1-\frac{[P°]}{[P°]_\infty}\right)} = \frac{r_i}{[P°]_\infty}\cdot dt \tag{13}$$

Which is integrated in:

$$\ln\left(\frac{1+\frac{[P°]}{[P°]_\infty}}{1-\frac{[P°]}{[P°]_\infty}}\right) = \frac{2r_i}{[P°]_\infty}\cdot t \tag{14}$$

At t ~ 0, [P°] ~ 0 so that:

$$\left(1+\frac{[P°]}{[P°]_\infty}\right)^2 = 1+\frac{2r_i}{[P°]_\infty}\cdot t \tag{15}$$

thus:

$$G(P°) = G_i = \left(\frac{d[P°]}{dt}\right)_{t\to 0} \tag{16}$$

The corresponding yields for radical build-up are given in Table 2.
Let us first mention that Table 1 results correspond to degradation under air, meanwhile Table 2 results correspond to degradation in inert environment, the possible influence of oxygen being discussed in the following. However, it can be checked that the lethal dose (or dose to reach an arbitrarily chosen threshold for a mechanical property) varies oppositely with the estimated G(P°) values.

Polymer	G(P°)
PPO	1.35
Kevlar®	1.25
PSU	0.75
Bis-A-PEPO	0.55
Ultem	0.50
Kapton®	0.40
PEEK	0.25

Table 2. Yields for radical build-up for aromatic polymers irradiated under 1 - 5 kGy h^{-1} dose rate at 77 K.

- By gas yield emission:

Concerning the yield for total gas emission, it can be checked that there is no great difference between gamma, proton and electron-beam irradiation (Hegazy et al., 1992b, Hill & Hopewell, 1996) and that these values are considerably lower than those measured for aliphatic polymers such as PE or PP (Schnabel, 1978). An example is worth to be mentioned (Schnabel, 1978): aliphatic polysulfones have a gas emission yield equal to 39 (poly butene-1-sulfone) or 71 (poly hexene-1-sulfone) meanwhile PES or PSU gas emission yields are lower than 1, illustrating here the protective effect by aromatic rings. However, the relative proportion of each emitted gas can vary due to some differences in temperature rising under irradiation. As it will be seen below, this difference due to irradiation nature has a lower influence on the radio induced degradation of polymer than temperature, nature of environment atmosphere and oxygen gradient in sample thickness.

I (kGy h^{-1})	Gaz	Kapton	Upilex-R	Upilex-S	PEEK-c	PEEK-a	PES	U-PS	U-Polymer
10	Total	24.0	22.0	91.0	39.0	54.0	46.0	150.0	480.0
	H_2	3.2	0.4	7.5	-	14.0	7.1	39.0	72.0
	N_2	5.1	9.7	14.0	6.4	-	-	-	-
	CO	5.4	2.4	1.4	12.0	6.0	16.0	19.0	220.0
	CO_2	8.1	4.8	15.0	4.3	24.0	19.0	25.0	180.0
	CH_4	1.0	0.1	0.3	0.2	0.3	0.3	16.0	12.0
	SO_2	-	-	-	-	-	12.0	13.0	-
6120	Total	25.0	17.0	18.0	31.0	39.0	69.0	210.0	460.0
	H_2	4.9	1.3	2.4	10.0	12.0	19.0	62.0	80.0
	N_2	0.2	0.1	2.8	-	-	-		-
	CO	3.4	2.2	2.4	5.1	5.5	8.3	32.0	220.0
	CO_2	10.0	3.9	8.8	9.5	16.0	16.0	16.0	120.0
	CH_4	0.8	-	0.3	0.2	0.2	0.4	13.0	32.0
	SO_2	-	-	-	-	-	23.0	43.0	-

Table 3. 103.G(gas) for irradiation at 30°C 100 µm films vacuum (Hegazy et al., 1992a)

3. Let us turn to the radiation-induced crosslinking observed in PEEK and PSU.

Fig. 3 depicts the changes in thermal behavior which are induced by radio-ageing for a thick PEEK sheath under air (Richaud et al., 2010a). As it will be seen later, the observed behavior is comparable to the one obtained for an irradiation under vacuum. Virgin and degraded samples were characterized by DSC heating-cooling-heating cycle.

- In the case of virgin sample, melting endotherms for the first and the second heating ramps are very comparable (peak temperature and enthalpy). The crystallization from molten state is characterized by a sharp exotherm at c.a. 300°C.
- In the case of 30.7 MGy irradiated sample, melting endotherm for the second heating ramp and crystallization exotherm are shifted to the lower temperatures with a lower transition enthalpy.

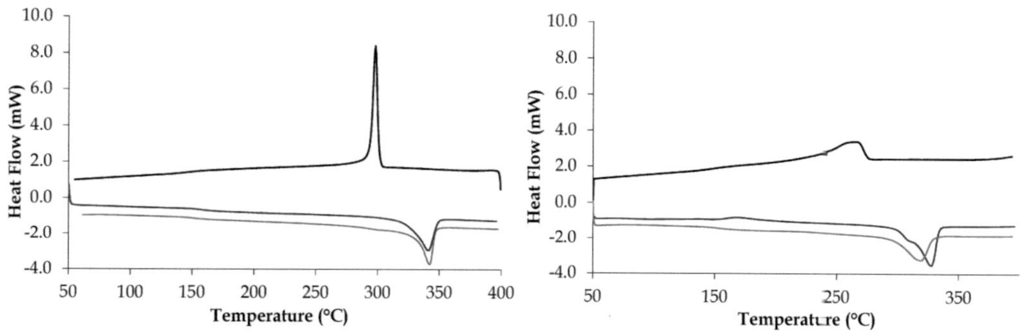

(a) (b)

Fig. 3. Heating-cooling-heating cycle by DSC for non-irradiated (a) and 30.7 MGy irradiated PEEK sample.

During irradiation of semi-crystalline polymers, chain scission and crosslinking occur only in amorphous phase, the crystalline phase undergoes only small changes so that melting characteristics remain almost unchanged at the first heating scan. After a first melting however, the sample is homogenized, and the changes of molar mass, branching ratio or crosslink density resulting from irradiation affect melting and crystallization. Irradiation effects become then observable by DSC. Here, the decrease of crystallization temperature and enthalpy is explained by the occurrence of radiation-induced crosslinking (Sasuga, 1991). This latter disfavors crystallization because it lowers the transport rate of chain segments from the melt to growing crystals. As a consequence, the degree of crystallinity and the lamellae thickness are decreased that explains the observed decrease of melting point and melting enthalpy.

Comparable data were obtained by Hegazy (Hegazy et al., 1992c). T_g values from this study are reported (Fig. 4):

It is well known that for a linear polymer, T_g is an increasing function of molar mass. According to Fox and Flory (Fox & Flory, 1950):

$$T_g = T_{g0} - \frac{k_{FF}}{\overline{M_N}} \tag{17}$$

Where:
- T_{g0} is the T_g of a virtual infinite polymer (K).
- M_N is the number average molar mass (kg mol^{-1}).
- k_{FF} is a constant characteristic of the chain chemical structure (K kg mol^{-1}).

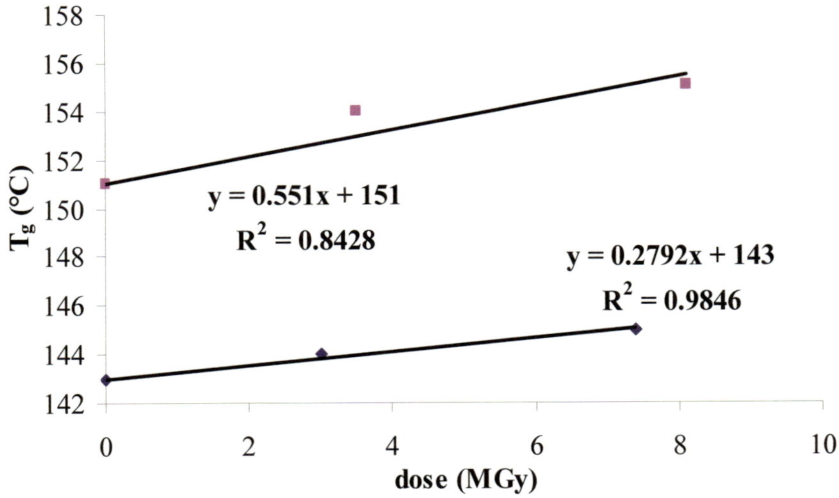

Fig. 4. T_g changes for PEEK amorphous (◆) and semi-crystalline (■) samples irradiated under vacuum.

For a network, T_g is an increasing function of crosslink density. According to Di Marzio (DiMarzio, 1964):

$$T_g = \frac{T_{gl}}{1 - k_{DM}.F.v_A} \tag{18}$$

Where:
- T_{gl} is the T_g of a virtual linear polymer containing all the structural units of polymer except crosslinks (K).
- F is a flex parameter linked to chain stiffness (kg mol^{-1}).
- v_A is the concentration in elastically active chains, which is calculated from crosslinks concentration (mol kg^{-1}).

Under vacuum, chain scission is generally negligible. Above gel point, it seems to us that T_g changes are given by combining Fox-Flory relationship and Saito's equation (assuming first that s << x):

$$\frac{dT_g}{d\delta} = \frac{dT_g}{dx} \cdot \frac{dx}{d\delta} = 10^{-7}.G(x).k_{FF} \tag{19}$$

One obtains:
- G(x) ~ 0.012 for amorphous PEEK,
- G(x) ~ 0.023 for semi-crystalline one.

It seemed to us interesting to compare radio-induced gelation for PEEK, PSU and PES (Fig. 5).

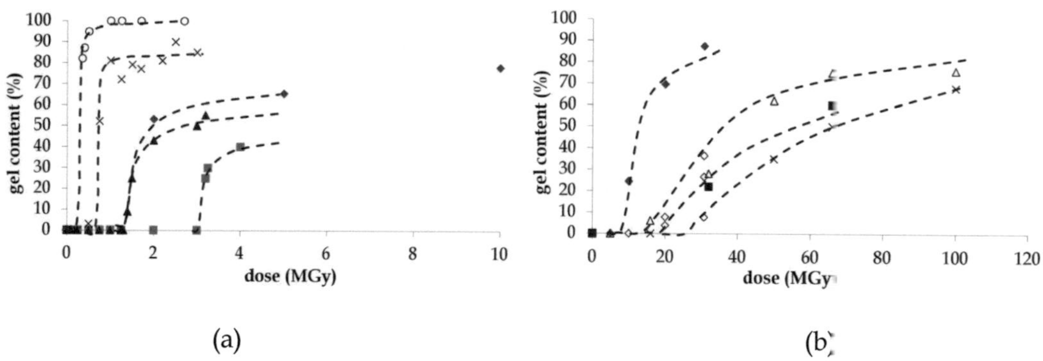

Fig. 5. Gel formation for PSU at 210°C (○), 180°C (×), 100°C (▲), 60°C (◆), room temperature (■) from Murakami & Kudo, 2007, Richaud et al., 2011), and PEEK (■: 24 kGy h^{-1}, ◆,△,×: 60 MGy h^{-1} for 3 different initial crystalline microstructures) samples irradiated at several temperatures (Vaughan & Stevens, 1995 , Richaud, 2010a).

From the above given mechanistic scheme where chain scission results from radical rearrangement and crosslinking results from radical coupling scheme, the following stationary rate expressions can be derived:

- Rate of chain scission:

$$\frac{ds}{dt} = \frac{k_R}{2} \cdot \left(\frac{r_i}{k_C + k_D} \right)^{1/2} \tag{20}$$

- Rate of crosslinking:

$$\frac{dx}{dt} = \frac{r_i}{2} \frac{k_C}{k_C + k_D} \tag{21}$$

Assuming that disproportionation is negligible compared to coupling for polymers under study, it comes:

$$\frac{dx}{dt} = \frac{r_i}{2} \tag{22}$$

So that : $G_i = 2G(x)$ in the absence of oxygen.

It seems that this relationship disagrees with experimental results in the case of aromatic polymers which will be discussed next.

As a conclusion of this short review of experimental results, it is clear that aromatic polymers belong the family of relatively radiostable polymers because primary radiochemical events, presumably radiolytic chain scissions of C-H bonds, have a radiochemical yield less than 0.5, i.e. 5 to 50 times lower than for aliphatic polymers. Aromatic groups are actually able to dissipate a great part of the absorbed energy into reversible processes (fluorescence, phosphorescence...). Data militate in favour of the following ranking:

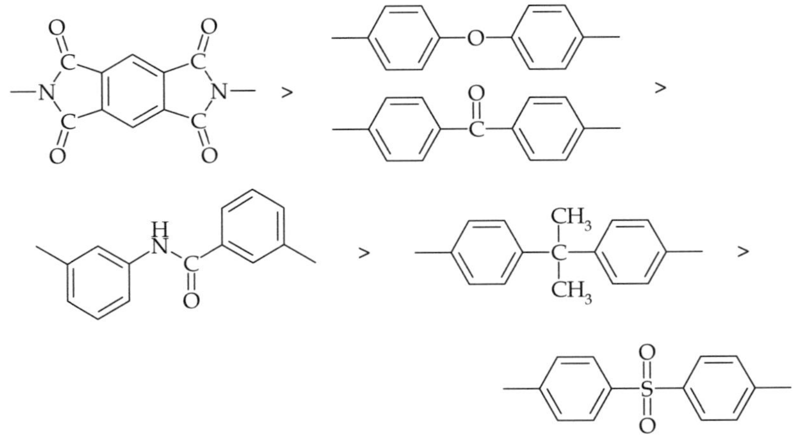

A comparison of the rate of gel content increase (Fig. 4) for samples irradiated at various temperatures and dose rates shows that gelation is more sudden in the case of PSU and more progressive in PEEK which means that radiation generated radicals would be more reactive for PSU than PEEK.

Data also show unambiguously:

- The protective effect of aromaticity: Kapton®, which is the most aromatic polymer, presents a very limited degradation, or even no significant degradation (Kang et al., 2008, Richaud et al, 2010b).
- The role of SO_2 moiety: experimental results for radiochemical yields values (Horie & Schnabel, 1984) shows that aliphatic polysulfones are among the less stable polymers with G(s) values on the order of 10. High yields of SO_2 emission are also observed indicating thus the probable existence of the weak (Li & Huang, 1999, Molnár et al., 2005) carbon-sulfone bond cleavage. It seems reasonable to suppose that such events occur also in PSU even though the yield is considerably lower owing to the well-known protective effect of aromatic nuclei illustrated for instance in studies on isobutylene-styrene.
- The role of isopopylidene group: gelation is undoubtedly due to the coupling of alkyl radicals. The difference in gelation rate would be explained by the nature of radicals: in PEEK, only aryl radicals can react meanwhile both aryl and primary methyl radical can react in PSU.

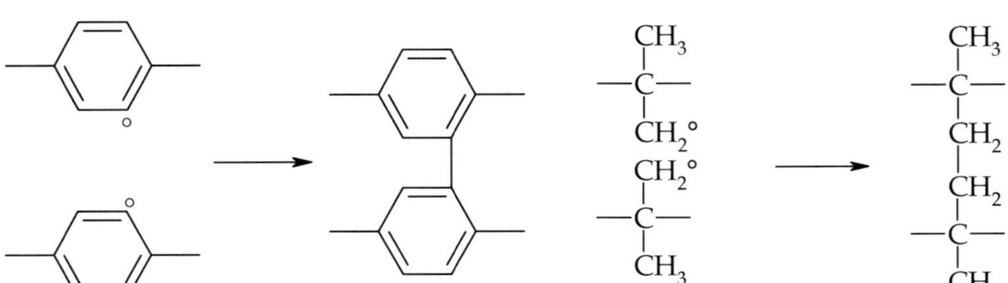

Theoretically, it seems that yields in gas emission G(gas) is half of yield radicals G(P°) as illustrated by proposal of degradation mechanisms below:

1. In isopropylidene containing materials:

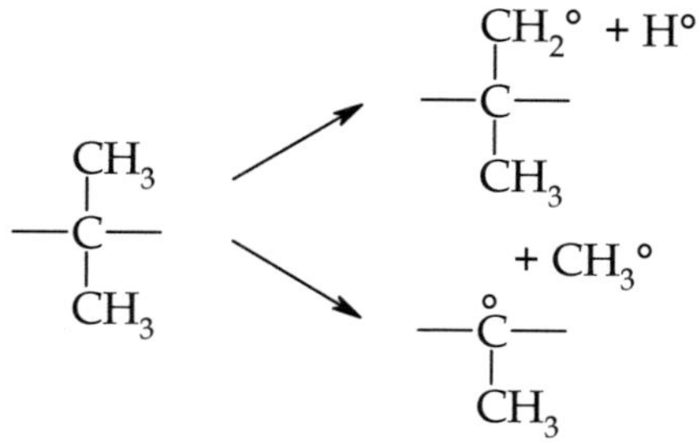

Then radicals react by coupling to give a gaseous molecule:

$$H^\circ + H^\circ \rightarrow H_2$$
$$CH_3^\circ + H^\circ \rightarrow CH_4$$
$$CH_3^\circ + CH_3^\circ \rightarrow C_2H_6$$

2. In sulfone containing polymer:

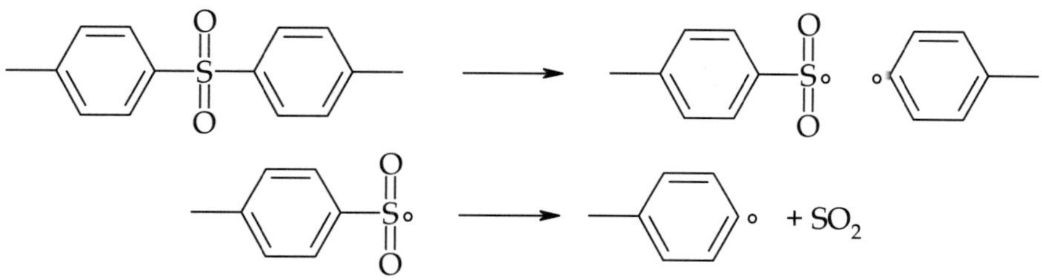

Analogous mechanisms could lead to carbon monoxide formation from PEEK.
Data for Kapton®, PEEK and PSU (Table 1 and Table 2) seem also to show that:

$$G(P^\circ) >> 2G(x) \text{ and } 2G(gas)$$

PEEK and PSU values gathered in this work are compared with those for polyethylene (Khelidj 2006), and those compiled for cyclohexane and benzene (Ferry, 2008) in Table 4:

thickness	T (K)	dose rate	atmosphere	$\delta_{50\%}$ (MGy)	δ_{gel} (MGy)	G (gas)
100 µm	298	0.1 kGy s^{-1}	He	> 0.3		
	423			> 0.15		
	463			0.1		
	503			0.1		
50 µm	298	1.7 Gy s^{-1}	vacuum		3	0.025
	373				1.25	0.07
	453				0.5	0.17
	483				< 0.25	2

Table 4. Yields for radicals, gas emission, crosslinking and double bond formation (corresponding to disproportionnation process). *: $G(H_2)$ is expected to be close to G(gas), **: G(F) is expected to be negligible in aromatic polymers.

These results call for the following comments:

1. The effect of aromaticity on radiostability is confirmed.
2. The ratio G(P°)/G(gas) is surprisingly lower than 1 for cyclohexane, is fairly close to 2 for benzene and PE, but increases for PSU and PEEK.

A possible explanation is the existence of very stable radicals that would be detected by ESR but would contribute neither to gas emission nor to crosslinking (see later) as for example:

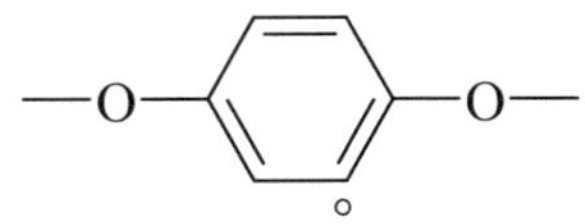

Let us also note that the absence of contribution to gas emission seems easily explainable because the probability for generating a short and volatile segments from such macroradicals is very low.

3. The ratio G(P°)/(G(x)+G(F)) is also clearly greater than 2 for PSU and particularly for PEEK meanwhile it is close to 1 in PE and in cyclohexane. The comparison with benzene suggests that this behaviour is due to the macromolecular structure of radicals preventing them to react by coupling to give crosslinking.

In fact, even if a negative concavity is observed in Fig. 1 and indicates the existence of a termination process even at 77 K, the relatively low value of G(x) indicates that segmental mobility would be reduced below T_g, which will be discussed in a first section dedicated to temperature effect and another one dealing with oxidizability.

4. Effect of temperature

Data in Table 5 confirm that lethal dose (here. the necessary dose to half the initial elongation at break) and gel dose vary oppositely with temperature meanwhile gas emission yield would increase.

thickness	T (K)	dose rate	atmosphere	δ_{50}% (MGy)	δ_{gel} (MGy)	G(gas)
100 µm	298	0.1 kGy s^{-1}	He	> 0.3	-	-
100 µm	423			> 0.15	-	-
100 µm	463			0.1	-	-
100 µm	503			0.1	-	-
50 µm	298	1.7 Gy s^{-1}	vacuum	-	3	0.025
50 µm	373			-	1.25	0.07
50 µm	453			-	0.5	0.17
50 µm	483			-	< 0.25	2

Table 5. Effect of temperature on lethal dose for PES (Sasuga & Hagiwara, 1987 and Murakami & Kudo, 2007).

As previously suggested, the only knowledge of changes in macroscopic (engineering) properties does not permit to ascribe the right cause of failure. Yields in chain scission and crosslinking permit to elucidate the degradation mechanism.

Let us first comment the T_g changes in PES (Li et al., 2006) presented in Fig. 6:

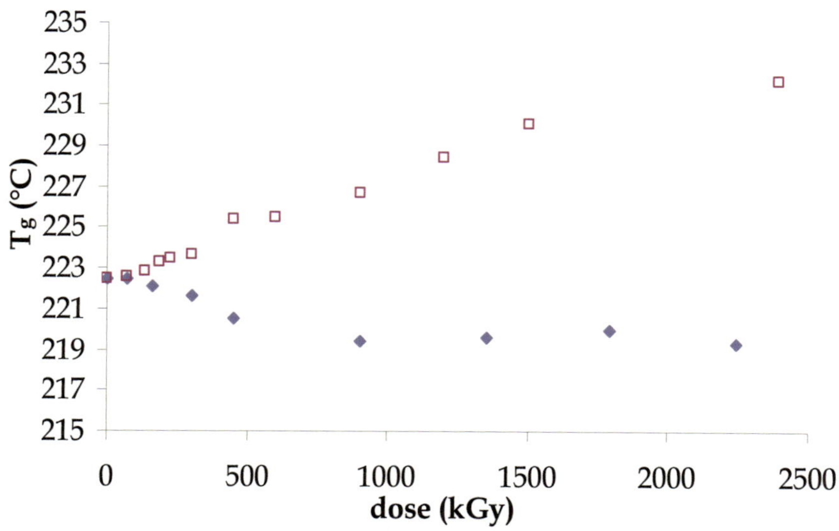

Fig. 6. Change in T_g for PES submitted to electron beam irradiation at room temperature (◆) or 230°C (□) (Li et al., 2006).

Before sol-gel transition, T_g changes are given by combining Fox-Flory law (Fox & Flory, 1950) with Saito's equation (Saito, 1958) irrespectively of H- or Y-crosslinking mode:

$$T_g - T_{g0} = -k_{FF}.\left(\frac{1}{\overline{M}_N} - \frac{1}{\overline{M}_{N0}}\right) = -10^{-7}.k_{FF}.[G(s) - G(x)].\delta \tag{23}$$

It turns into:

- for a pure crosslinking mechanism:

$$\frac{dT_g}{d\delta} = 10^{-7}.k_{FF}.G(x) \tag{24}$$

- for a pure chain scission mechanism:

$$\frac{dT_g}{d\delta} = -10^{-7}.k_{FF}.G(s) \tag{25}$$

Results presented in Fig. 5 indicate that crosslinking predominates above T_g, whereas chain scission predominates below T_g. Comparable exploitation can be done from other published results (Murakami & Kudo, 2007, Brown & O'Donnell, 1979):

All data converge towards the fact that G(x) increases with temperature. Despite some scattering, it also suggests that G(s)/G(x) decreases with temperature. According to Brown and O'Donnell (Brown & O'Donnell, 1979), it falls to 0 above T_g.

The results in Table 6 can be explained as follows: irradiation creates macroradicals. At low temperature, they react by an unimolecular process which generates a chain scission (possibly accompanied by gas emission). At high temperature, macromolecular mobility is sufficient to permit macroradicals to react by coupling to give a crosslinking. According to Zhen (Zhen, 2001) most polymers can only crosslink above their melting point.

dose rate	method	T (K)	$G_H(s)$	$G_Y(s)$	$G_H(x)$	$G_Y(x)$	G(s)/G(x)	thickness
8 Gy s^{-1}	Sol gel	308	0.03		0.04		0.75	3 mm
		353	0.05		0.05		1.00	3 mm
		398	0.3		0.2		1.50	3 mm
		493	0		0.67		0.00	3 mm
1.7 Gy s^{-1}	GPC	298	0.100	0.373	0.100	0.301	1.00	50 μm
		373	0.104	0.318	0.104	0.313	1.00	50 μm
		423	0.147	0.455	0.147	0.442	1.00	50 μm
		453	0.216	0.601	0.216	0.647	1.00	50 μm
1.7 Gy s^{-1}	Sol gel	298	0.067	0.065	0.104	0.147	0.64	50 μm
		373	0.170	0.138	0.245	0.312	0.69	50 μm
		453	0.075	0.268	0.170	0.575	0.44	50 μm

Table 6. Effect of temperature on radiochemical yields in chain scissions and crosslinking (envisaging the possibility of H- or Y- crosslinking mode) (Murakami & Kudo, 2007, Brown & O'Donnell, 1979).

Let us now discuss of the possibility of Y- and H-crosslinking.

- At 150°C, Hill (Hill et al., 1998) unambiguously showed that PSU crosslinks in Y-mode. A comparable analysis was performed by Li (Li et al., 2006). A proposal of mechanism is shown below:

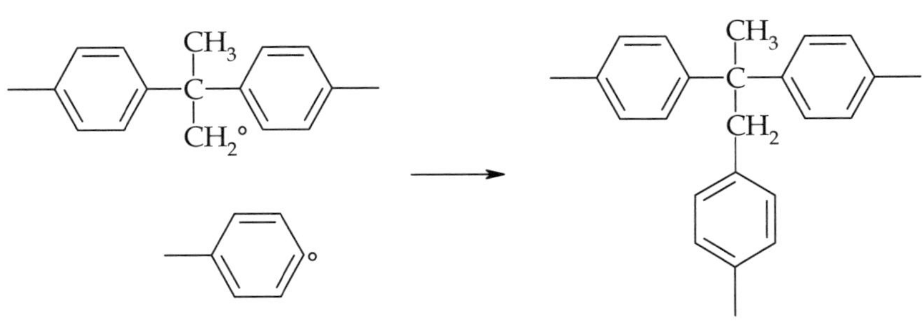

- Kudo (Murakami & Kudo, 2007) proposed a Y-mechanism also. However, by reexaminating his results using the above equations, his conclusion is questioned.
- Richaud (Richaud et al., 2011) proposed a H- crosslinking mechanism at 60°C, basing on the observation that if isopropopylidene groups are generated, the $\rightarrow C\text{–}CH_2^{\circ}$ group is considerably more reactive than aryl one and should react by coupling to give $\rightarrow C\text{-}CH_2\text{-}CH_2\text{-}C\leftarrow$ crosslinking bridges corresponding more to a H- crosslinking mechanism.

As a conclusion, it seems that elevating the temperature promotes the crosslinking. However, the nature of this latter and presumably the role of temperature remain unclear.

5. Effect of atmosphere and sample thickness

Results in Fig. 7 (Sasuga & Hagiwara, 1987) clearly show that oxygen accelerates the degradation of aromatic polymers.

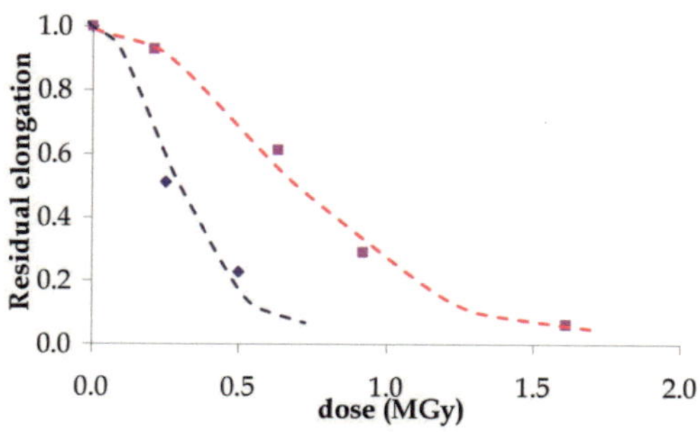

(a)

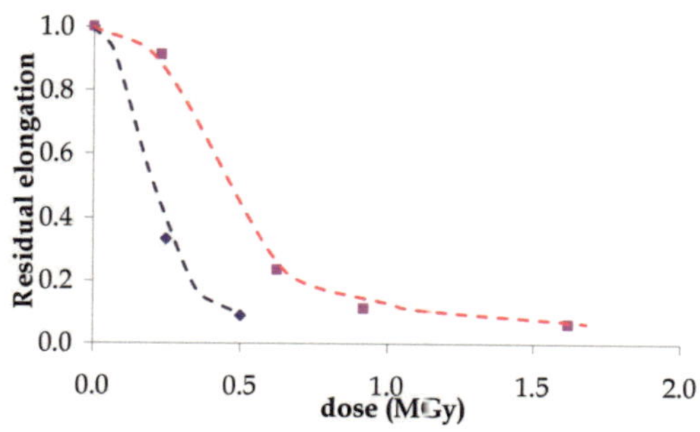

(b)

Fig. 7. Residual elongation as a function of dose for PSU (a) and PES (b) submitted to g-rays under air (■) or 0.7 MPa oxygen (◆).

The nature of the responsible process is suggested in a study on the degradation of PET films monitored by T_g measurements (Burillo et al., 2007): a strong T_g decrease is observed in air at the beginning and is certainly due to chain scission whereas a moderate T_g decrease is observed for irradiation under vacuum. As proposed by Sasuga (Sasuga, 1988), polymer radio-degraded under anaerobic conditions would undergo mainly crosslinking meanwhile they would undergo chain scissions when they are degraded under aerobic (oxidative) conditions.

The effect of atmosphere nature and the effect of thickness have the same origin linked to oxygen diffusion: thick samples present a diffusion limited oxidation (DLO), i.e. that their surface undergoes an oxidative degradation leading to chain scission meanwhile bulk undergoes an anaerobic radio-ageing generating radicals reacting only by coupling.

This effect was quantified by comparing the degradation of thin and thick samples by means of GPC (for molar mass assessment and subsequently chain scission and crosslinking yields assessment) and T_g (Richaud et al., 2011). T_g decreases for both 2 mm and 200 μm thick samples, but more significantly for thin films (Fig. 8) which is not surprising since T_g decreases with chain scissions and increases with crosslinking.

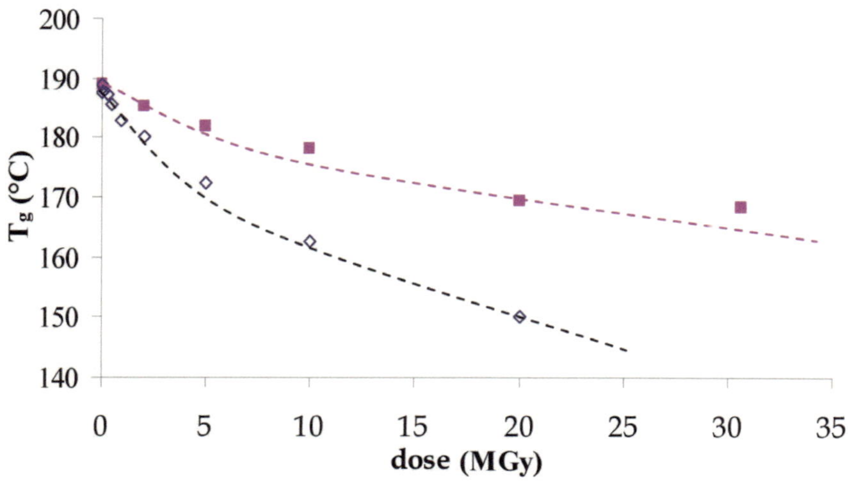

Fig. 8. T_g changes with dose for thin films (◆) or thick samples (■).

1. Case of thin films

Thin films undergo certainly a "pure" random chain scissions process. A graphical estimation (Fig. 5) gave for the thin samples (using k_{FF} ~ 220 K kg mol^{-1} for PSU):

$$\frac{dT_g}{d\delta} \sim \frac{T_g(0\ \text{MGy}) - T_g(5.0\ \text{MGy})}{5.0 \times 10^6} = 3.2\times10^{-6}\ \text{K Gy}^{-1}$$

from which one obtains G(s) ~ 0.15

2. Case of thick samples

Here, the simultaneous crosslinking compensates partially chain scissions consequences on T_g changes. For ideal networks having no dangling chains, the effect of crosslinking on T_g obeys the DiMarzio's law (DiMarzio,1964 – equation (18)) with:

- T_{gl} ~ 473 K
- F_{PSU} = 0.1105 kg mol^{-1}
- k_{DM} is the DiMarzio's constant: k_{DM} is close to 1
- ν is the elastically active chains concentration. For ideal tetrafunctional networks:

$$\nu = 2.x \tag{26}$$

x being the concentration of crosslinks events. Supposing that DiMarzio's equation can be applied for non-ideal networks, Eq. 23 turns into the Fox-Loshaek (Fox & Loshaek, 1955) equation for low conversion of the crosslinking process:

$$T_g = T_{gl} + k_{FL}.x \tag{27}$$

with the Fox-Loshaek constant k_{FL} being equal to:

$$k_{FL} = 2.T_{gl}.F.k_{DM} \tag{28}$$

so that: k_{FL} ~ 104 K mol kg^{-1}

Finally, the global changes of T_g would be given by:

$$\frac{dT_g}{d\delta} = \left(\frac{\partial T_g}{\partial s}\right).\frac{ds}{d\delta} + \left(\frac{\partial T_g}{\partial x}\right).\frac{dx}{d\delta} \tag{29}$$

so that:

$$\frac{dT_g}{d\delta} = 10^{-7}\left(k_{FL}.G(x) - k_{FF}.G(s)\right)$$

Using G(s) ~ 0.15, one can simulate the experimental T_g decrease:

$$\frac{dT_g}{d\delta} = 15.6\times10^{-7}\ \text{K Gy}^{-1}$$

with G(x) ~ 0.15.

In other words, effect of sample thickness is explained by the effect of oxygen diffusion (which will be developed in the next section): irradiation generates radicals. In the absence of oxygen (i.e. for the core a bulk sample), these later react together by coupling which gives a crosslinking phenomenon. When oxygen is present (which is the case for thin samples or the superficial layer of a bulk sample), it orientates the degradation towards an oxidative mechanism with the formation of peroxy radicals POO°, and POOH of which decomposition induces chain scissions.

6. Effect of dose rate

This effect is illustrated by some results by Tavlet (Tavlet, 1997) presented in Fig. 9:

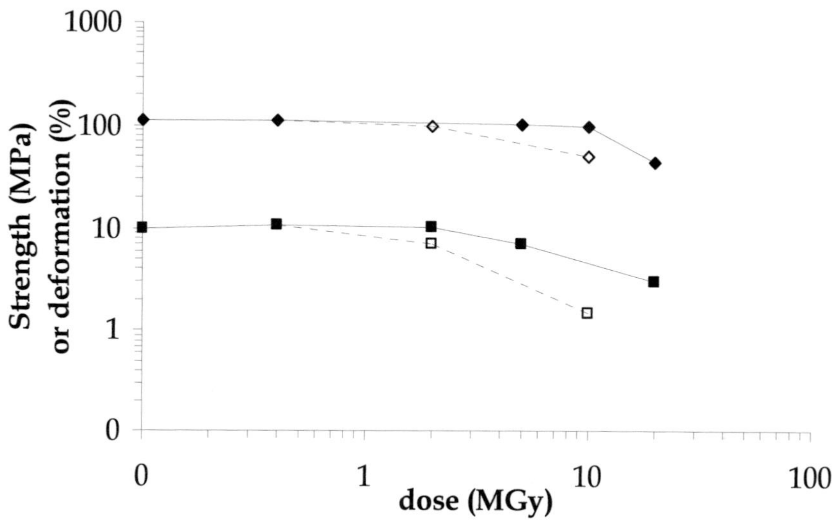

Fig. 9. Radiation degradation in an epoxy resin on strain (■,□), stress (◆.◇). Open symbols correspond to long term (low dose rate) ageing (Tavlet, 1997).

There are two effects to be distinguished:

- An increase in temperature involved by enhancing the dose rate,
- A physical and chemical effect linked to Diffusion Limited Oxidation, which will be described below.

From a simplified theory of diffusion controlled oxidation (Audouin et al, 1994), a rough estimation of oxygen consumption rate r_{OX} in films or superficial layers of thick samples can be done:

$$r_{OX} = \frac{D_{O2}.[O_2]_S}{TOL^2} \tag{30}$$

where:

- TOL is the thickness of oxidized layer (m),
- D_{O2} is the oxygen diffusion coefficient in polymer amorphous phase ($m^2\ s^{-1}$),
- $[O_2]_S$ is the equilibrium concentration under atmospheric pressure given by Henry's law:

$$[O_2] = s_{O2} \times P_{O2} \tag{31}$$

s_{O2} being the solubility coefficient of oxygen in polymer amorphous phase.

r_{OX} can be expressed from a simplified mechanistic scheme for radio-thermal oxidation:

$PH + h\nu \rightarrow P^\circ + ½ H_2$ r_i
$P^\circ + O_2 \rightarrow P^\circ$ k_2
$POO^\circ + PH \rightarrow POOH + P^\circ$ k_3
$POO^\circ + POO^\circ \rightarrow$ inact. prod. $+ O_2$ k_6

k_i being the rate constants (expressed in $l\ mol^{-1}\ s^{-1}$).
The oxidation rate for oxygen consumption can be expressed as:

$$r_{OX} = -\frac{d[O_2]}{dt} = -k_2.[P^\circ][O_2] + k_6.[POO^\circ]^2 \quad (32)$$

Using the classical steady-state assumption:

$$k_2.[P^\circ][O_2] = r_i + k_3.[POO^\circ].[PH] \quad (33)$$

$$r_i = 2k_6.[POO^\circ]^2 \quad (34)$$

it comes:

$$r_{OX} = \frac{r_i}{2} + k_3 \cdot \sqrt{\frac{r_i}{2k_6}} \cdot [PH] \quad (35)$$

knowing that: $r_{OX} = 10^{-7}.G(P^\circ).I$
I denoting the dose rate. Consequently, if:

- I is very high

$$\Rightarrow \quad r_{OX} \approx \frac{r_i}{2} \quad (36)$$

$$\Rightarrow \quad r_{OX} \propto I \quad (37)$$

$$\Rightarrow \quad TOL \propto I^{-1/2}$$

- I is moderate

$$\Rightarrow \quad r_{OX} \approx k_3 \cdot \sqrt{\frac{r_i}{2k_6}} \cdot [PH] \quad (38)$$

$$\Rightarrow \quad r_{OX} \propto I^{1/2} \quad (39)$$

$$\Rightarrow \quad TOL \propto I^{-1/4} \quad (40)$$

It is thus demonstrated that radiothermal oxidation is characterized by a skin-core structure: superficial layers are oxidized meanwhile bulk not. Results by Tavlet (Tavlet, 1997) are thus explained as follows: for irradiation performed under a relatively elevated dose rate, only a very thin polymer layer is degraded and time for drop of elongation at break is longer than for irradiation performed under a moderate dose rate. This reasoning is confirmed by the comparison of a bulk sample (4 mm) for which no dose rate effect is observed undoubtedly because the thickness of degraded superficial layer is negligible compared to the total thickness of bulk material. Examples of Oxidized layers measured for radiochemical degradation of aromatic polymers (Richaud et al., 2010a) are presented in Fig. 10 and confirm the existence of a Diffusion Limited Oxidation.

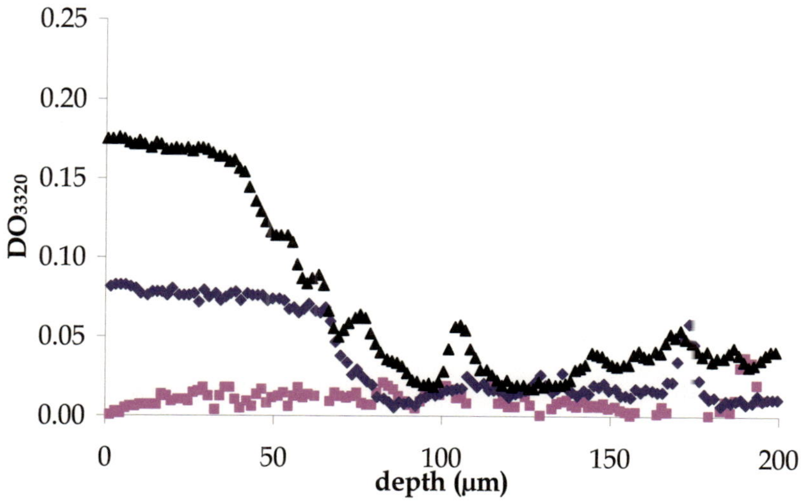

Fig. 10. Oxidation profiles in PEEK sheaths submitted to 10 MGy (■), 20 MGy (◆), 30.7 MGy (▲) γ-rays exposure (Richaud et al., 2010a).

7. On the radio-induced oxidizability of aromatic polymers

It seemed to us interesting to compare the k_3^2/k_6 ratio estimated from radiodegradation of aromatic polymers with polypropylene or polyethylene ones. Let us recall that k_3^2/k_6 describes the oxidizability of a polymer irrespectively of the oxidation mode (i.e. photo, thermo- or radio-induced). k_3^2/k_6 can be calculated from Eq. (32) using:

- $[PH] = \rho/M$, ρ being the polymer density and M the molar mass of repetitive unit,
- $r_i = 10^{-7}.G(P°).I$ from Heiland's results (Table 2),
- r_{OX} from TOL measured for example in FTIR mapping mode (Eq (30)).

One can use with a good approximation the values of oxygen permeation of BisPhenol A Polycarbonate values for in PEEK and PSU (Van Krevelen & Te Nijenhuis, 2009) at 60°C :

- $D_{O2} = 8.1\times10^{-8}$ m² s⁻¹
- $s_{O2} = 4\times10^{-8}$ mol l⁻¹ Pa⁻¹

For samples exposed under 1.5 bar oxygen, TOL were found close to:

- 100 μm in PSU (Richaud et al., 2011),
- 70 μm in PEEK (Fig. 10).

	TOL (m)	r_{OX} (mol l^{-1} s^{-1})	G_i (mol/100 eV)	r_i (mol l^{-1} s^{-1})	[PH] (mol l^{-1})	BDE (kJ mol^{-1})	k_3^2/k_6 (l mol^{-1} s^{-1})	k_3 (l mol^{-1} s^{-1})	k_6 (l mol^{-1} s^{-1})
PEEK	1.00E-04	5.2E-06	0.25	1.7E-07	12.9	460	1.8E-06	1.0E-04	5.6E-03
PSU	7.00E-05	1.1E-05	0.75	5.0E-07	5.6	410	1.3E-05	9.5E-03	6.8E+00

Table 7. Kinetic parameters of PEEK and PSU estimated from DLO theory.

so that one can calculate (Table 6):

- $k_3^2/k_6 = 1.3\times10^{-5}$ l mol⁻¹ s⁻¹ for PSU
- $k_3^2/k_6 = 1.8\times10^{-6}$ l mol⁻¹ s⁻¹ for PEEK

These orders of magnitude are surprisingly on comparable or higher than values reported in PP (Richaud et al., 2006) and PE (Khelidj, 2006). k_3 can be conveniently estimated from the paper by Korcek (Korcek et al., 1972) using the relations:

$\log k_3^{s\text{-}POO°} = 16.4 - 0.2\times BDE(C\text{-}H)$ at 30°C

$\log k_3^{t\text{-}POO°} = 15.4 - 0.2 \times BDE(C\text{-}H)$ at 30°C
$E_3^{s\text{-}POO°} = 0.55.(BDE(C\text{-}H) - 65)$
$E_3^{t\text{-}POO°} = 0.55.(BDE(C\text{-}H) - 62.5)$
s-POO° and t-POO° representing respectively secondary and tertiary peroxy radicals. Given the difference in Bond Dissociation Energies for C-H (denoted by BDE(C-H)) in benzene (Davico et al., 1995) or in $-CH_3$ (Korcek et al., 1972), it is clear that:

$$k_3(PEEK) < k_3(PSU) < k_3(PE) < k_3(PP)$$

Thus, it suggests that (Table 7):

$$k_6(PEEK, PSU) << k_6(PE, PP)$$

From a chemical point of view, the reactivity of aryl or primary peroxy radicals is not low enough to justify such values (Table 6). A possible explanation is that the termination between two POO° radicals is a process involving two rare species. The mobility of segments favouring the coupling could thus control the kinetics of the reaction. The low order of magnitude below T_g could be hence justified. Aromatic polymers could be interesting substrate for studying the Waite's theory:

$$-\frac{dC}{dt} = 4\pi N_A r_0 D\left(1+\frac{r_0}{\sqrt{\pi Dt}}\right)C^2 \quad (41)$$

where:

- r_0 is the cage radius, which corresponds to the maximal distance permitting the reaction: It would be on the order of $r_0 \sim$ 1-10 nm (Emanuel & Buchachenko, 1990),
- D is the diffusion coefficient of radical species. On the order of $D \sim 10^{-22} - 10^{-20}$ m^2 s^{-1} (Emanuel & Buchachenko, 1990),
- N_A is Avogadro's number.

According to this theory, kinetic constant depends on the diffusivity of a radical and in other words of macromolecular mobility.

8. Conclusions

A comparison of yields for gas emission, gel dose or critical dose for reaching an unacceptable level of mechanical properties shows that aromatic polymers radiochemical behaviour can be first explained by their aromatic character which is doubly favourable:

- Yield for radical formation is very low compared with aliphatic polymers, because aromatic groups are able to dissipate a great part of the absorbed energy into reversible processes (fluorescence, phosphorescence...).
- In the case of radio-oxidation, the propagation rate constant of hydrogen abstraction by peroxy radicals (k_3) is a decreasing function of the C-H bond dissociation energy, that takes a high value for aromatic hydrogen.

These positive effects are partially counterbalanced by the relatively low termination rate linked to the low segmental mobility in glassy state. It is noteworthy that P° + P° termination (which leads to crosslink when oxygen is lacking) is more efficient than expected because P° radicals can diffuse independently of segmental mobility by the valence migration process ($P_1° + P_2H \rightarrow P_1H + P_2°$).

This chapter also illustrates the effect of dose rate, thickness and oxygen concentration in polymers layers, which are linked by the theory of Diffusion Limited Oxidation: chain scissions predominate in aerobic degradation (i.e. thin sample, and low dose rate) and crosslinking in case of anaerobic degradation (thick sample, high dose rate). The effect of temperature can be considered as not partially understood.

9. References

Alexander A. & Charlesby A. (1954). Energy Transfer in Macromolecules Exposed to Ionizing Radiations. *Nature*, Vol.173, No.4404, (1954), pp. 578-579, ISSN 0028-0836

Audouin, L.; Langlois, V.; Verdu, J., de Bruijn, J.C.M. (1994). Role of oxygen diffusion in polymer ageing: kinetic and mechanical aspects. *Journal of Materials Science*, Vol.29, No.3, (January 1994), pp. 569-583, ISSN 0022-2461

Brown, J.R. & O'Donnell, J.H. (1979). Effect of gamma radiation on two aromatic polysulfones II. A comparison of irradiation at various temperatures in air-vacuum environments. *Journal of Applied Polymer Science*, Vol.23, No.9, (May 1979), pp. 2763-1775, ISSN 0021-8995

Burillo, G.; Tenorio, L.; Bucio, E.; Adem, E. & Lopez, G.P. (2007). Electron beam irradiation effects on poly(ethylene terephthalate). *Radiation Physics and Chemistry*, Vol.76, No.11-12, (November-December 2007), pp. 1728-1731, ISSN 0969-806X

Charlesby, A. (1960) *Atomic Radiation and Polymer*, Pergamon Press, Oxford, UK

Davico, G.E.; Bierbaum, V.M.; DePuy, C.H.; Ellison, G.B. & Squires, R.R. (1995). The C-H Bond Energy of Benzene. *Journal of American Chemical Society*, 1995, Vol.117, No.9, March 1995, pp. 2590–2599, ISSN 0002-7863

DiMarzio, E.A. (1964). On the second-order transition of a rubber. *Journal of Research of the National Bureau of Standards-A. Physics and Chemistry*, Vol .68A, No.6, 1964, pp 611-617, ISSN 0091-0635

Emanuel, N.M. & Buchachenko A.L. (1987). *Chemical Physics of Polymer Degradation and Stabilization (New Concepts in Polymer Science)*, VNU Science Press, ISBN 978-9067640923, Utrecht, The Netherlands

Ferry, M. (2008). *Comportement des verres cyclohexane/benzène et des copolymères éthylène/styrène sous rayonnements ionisants: Transferts d'énergie et d'espèces entre les groupements aliphatiques et aromatiques*, Thèse. Université de Caen

Fox, T.G. & Flory, P.J. (1950). Second-Order Transition Temperatures and Related Properties of Polystyrene. I. Influence of Molecular Weight. *Journal of Applied Physics*, Vol.21, No.6, (1950), pp. 581-591, ISSN 0021-8979

Fox, T.G. & Loshaek S. (1955). Influence of molecular weight and degree of crosslinking on the specific volume and glass temperature of polymers. *Journal of Polymer Science*, Vol.15, No.80 , (February 1955), pp. 371-390, ISSN 1099-0518

Hegazy, E.S.A.; Sasuga, T.; Nishii, M. & Seguchi, T. (1992a). Irradiation effects on aromatic polymers: 1. Gas evolution by gamma irradiation. *Polymer*, Vol.33, No.14, (1992), pp. 2897-2903, ISSN 0032-3861

Hegazy, E.SA.; Sasuga, T.; Nishii, M. & Seguchi, T. (1992b). Irradiation effects on aromatic polymers: 2. Gas evolution during electron-beam irradiation. *Polymer*, Vol.33, No.14, (1992), pp. 2904-2910, ISSN 0032-3861

Hegazy, E.S.A.; Sasuga, T. & Seguchi, T. (1992c). Irradiation effects on aromatic polymers: 3. Changes in thermal properties by gamma irradiation. *Polymer*, Vol.33, No.14, (1992), pp. 2911-2914, ISSN 0032-3861

Heiland, K.; Hill, D.J.T.; Hopewell, J.L.; Lewis, D.A., O'Donnell, J.H., Pomery, P.J. & Whittaker A.K. (1996). Measurement of Radical Yields To Assess Radiation Resistance in Engineering Thermoplastics, In: Polymer Durability, R.L. Clough, N.C. Billingham, K.T. Gillen, (Ed.), 637-649, American Chemical Society, ISBN 9780841231344, Washington DC, USA

Hill, D.J.T. & Hopewell, J.L. (1996). Effects of 3 MeV proton irradiation on the mechanical properties of polyimide films. *Radiation Physics and Chemistry*, Vol.48, No.5, (November 1996), pp. 533-537, ISSN 0969-806X

Hill, D.J.T.; Lewis, D.A.; O'Donnell, J.H. & Whittaker, A.K. (1998). The crosslinking mechanism in gamma irradiation of polyarylsulfone: evidence for Y-links. *Polymers for Advanced Technologies*, Vol.9, No.1, (January 1998), pp. 45–51, ISSN 1042-7147

Horie, K. & Wolfram Schnabel, W. (1984). On the kinetics of polymer degradation in solution: Part XI—Radiolysis of poly(olefin sulfones). *Polymer Degradation and Stability*, Vol.8, No.3, (1984), pp. 145-159, ISSN 0141-3910

Kang, P.H; Jeon, Y.K.; Jeun, J.P.; Shin, J.W. & Nho, Y.C. (2008). Effect of electron beam irradiation on polyimide film. *Journal of Industrial and Engineering Chemistry*, Vol.14, No.5, (September 2008), pp. 672-675, ISSN 1226-086X

Kellman, R.; Hill, D.T.J.; Hunter, D.S.; O'Donnell, J.H. & Pomery, P.J. (1991). Gamma Radiolysis of Styrene-co-Methyl Acrylate Copolymers. An Electron Spin Resonance Study, In: *Radiation Effects on Polymers*, R.L. Clough, S.W. Shalaby (Ed.), 119-134, American Chemical Society, ISBN 9780841221659, Washington DC, USA

Khelidj, N. (2006). *Vieillissement d'isolants de câbles en polyéthylène en ambiance nucléaire*, Thèse ENSAM de Paris

Korcek, S.; Chenier, J.H.B.; Howard, J.A. & Ingold, K.U. (1972). Absolute Rate Constants for Hydrocarbon Autoxidation. XXI. Activation Energies for Propagation and the Correlation of Propagation Rate Constants with Carbon–Hydrogen Bond Strengths. *Canadian Journal of Chemistry*, (1972), Vol.50, No.14, pp. 2285-2297, ISSN 0008-4042

Li, X.G. & Huang, M.R. (1999). Thermal degradation of bisphenol A polysulfone by high-resolution thermogravimetry. *Reactive and Functional Polymers*, Vol.42, No.1, (September 1999), pp. 59-64, ISSN 1381-5148

Li, J.; Oshima, A.; Miura, T. & Washio, M. (2006). Preparation of the crosslinked polyethersulfone films by high-temperature electron-beam irradiation. *Polymer Degradation and Stability*, Vol.91, No.12, (December 2006), pp. 2867-2873, ISSN 0141-3910

Molnár, G.; Botvay, A.; Pöppl, L.; Torkos, K.; Borossay, J.; Máthé, Á. & Török, T. (2005). Thermal degradation of chemically modified polysulfones. *Polymer Degradation and Stability*, Vol.89, No.3, (September 2005), pp. 410-417, ISSN 0141-3910

Murakami, K. & Kudo, H. (2007). Gamma-rays irradiation effects on polysulfone at high temperature. *Nuclear Instruments and Methods in Physics Research Section B: Beam*

Interactions with Materials and Atoms, Vol.265, No.1, (December 2007), pp. 125-129, ISSN 0168-583X

Richaud, E. ; Farcas, F. ; Bartoloméo, P. ; Fayolle, B. ; Audouin, L. & Verdu, J. (2006). Effect of oxygen pressure on the oxidation kinetics of unstabilised polypropylene. *Polymer Degradation and Stability*, Vol.91, No.2, (February 2006), pp. 398-405, ISSN 0141-3910

Richaud, E.; Ferreira, P.; Audouin, L.; Colin, X.; Verdu, J. & Monchy-Leroy, C. (2010a). Radiochemical ageing of poly(ether ether ketone). *European Polymer Journal*, Vol.46, No.4, (April 2010), pp. 731-743, ISSN 0014-3057

Richaud, E.; Audouin, L.; Colin, X.; Monchy-Leroy, C. & Verdu, J. 2010. Radiochemical Ageing of Aromatic Polymers PEEK, PSU and Kapton®. *AIP Conference Proceedings*, Vol.1255, (June 2010), pp. 10-12, ISSN 1551-7616

Richaud, E.; Colin, X.; Monchy-Leroy, C.; Audouin, L. & Verdu, J. (2011). Diffusion-controlled radiochemical oxidation of bisphenol A polysulfone. *Polymer International*, Vol.60, No.3, (March 2011), pp. 371–381, ISSN 0959-8103

Saito, O. (1958). On the Effect of High Energy Radiation to Polymers I. Cross-linking and Degradation. *Journal of the Physical Society of Japan*, Vol.13, No.2, (February 1958), pp. 198-206, ISSN 0031-9015

Sasuga, T.; Hayakawa, N.; Yoshida, K. & Hagiwara, M. (1985). Degradation in tensile properties of aromatic polymers by electron beam irradiation. *Polymer*, Vol.26, No.7, (July 1985), pp.1039-1045, ISSN 0032-3861

Sasuga, T. & Hagiwara, M. (1987). Radiation deterioration of several aromatic polymers under oxidative conditions. *Polymer*, Vol.28, No.11, (October 1987), pp. 1915-1921, ISSN 0032-3861

Sasuga, T. (1988). Oxidative irradiation effects on several aromatic polyimides. *Polymer*, Vol.29, No.9, (September 1988), pp. 1562-1568, ISSN 0032-3861

Sasuga, T. (1991). Electron irradiation effects on dynamic viscoelastic properties and crystallization behaviour of aromatic polyimides. *Polymer*, Vol.32, No.9, (1991), pp. 1539-1544, ISSN 0032-3861

Schnabel, W. (1978). Degradation by High Energy Irradiation, In: *Aspects of Degradation and Stabilization of Polymers*, H.H.G. Jellinek (Ed.), 149-190, Elsevier, ISBN13 978-0444415639, Oxford and New York

Tavlet, M. (1995). Aging of organic materials around high-energy particle accelerators. *Nuclear Instruments and Methods in Physics Research Section B: Beam Interactions with Materials and Atoms*, Vol.131, No.1-4, (August 1997), pp. 239-244, 0168-583X

Thominette, F.; Metzger, G.; Dalle, B. & Verdu, J. (1991). Radiochemical ageing of poly(vinyl chloride) plasticized by didecylphthalate. *European Polymer Journal*, Vol.27, No.1, (1991), pp. 55-59, ISSN 0014-3057

Van Krevelen, D.W. & Te Nijenhuis, K. (2009). *Properties of polymers, their correlation with chemical structure; their numerical estimation and prediction from additive group contributions, 4th Edition*, Elsevier, ISBN 978-0-08-054819-7, Amsterdam, The Netherlands

Vaughan, A.S. & Stevens, G.C. (1995). On crystallization, morphology and radiation effects in poly(ether ether ketone). *Polymer*, Vol.36, No.8, (1995), pp. 1531-1540, ISSN 0032-3861

Wilski, H. (1987). The radiation induced degradation of polymers. *Radiation Physics and Chemistry*, Vol.29, No.1, (1987), pp. 1-14, ISSN 0969-806X

Zhen, S.J. (2001). The effect of chain flexibility and chain mobility on radiation crosslinking of polymers. *Radiation Physics and Chemistry*, Vol.60, No.4-5, (2001), pp. 445-451, ISSN 0969-806X